Computational Electromagnetics

IEEE Press
445 Hoes Lane, PO Box 1331
Piscataway, NJ 08855-1331

1991 Editorial Board
Leonard Shaw, *Editor in Chief*
William C. Guyker, Editor, *Selected Reprint Series*

J. E. Brittain
S. H. Charap
R. C. Dorf
J. J. Farrell III
L. J. Greenstein
J. D. Irwin
W. K. Jenkins
S. Luryi
E. K. Miller
J. M. F. Moura
J. G. Nagle
J. D. Ryder
A. C. Schell
M. Simaan
M. I. Skolnik
G. S. Smith
Y. Sunahara
R. Welchel
J. W. Woods

Dudley R. Kay, *Executive Editor*
Carrie Briggs, *Administrative Assistant*
Karen G. Miller, *Assistant Editor*
Anne L. Reifsnyder, *Associate Editor*

IEEE Antennas and Propagation Society, *Sponsor*

AP-S Liaison for the IEEE PRESS

L. Wilson Pearson
Clemson University

Technical Reviewers for AP-S

John Fratamico
Science Applications International Corporation

Robert G. Olsen
Washington State University

Computational Electromagnetics

Frequency-Domain Method of Moments

Edited by

Edmund K. Miller
Los Alamos National Laboratory

Louis Medgyesi-Mitschang
McDonnell Douglas Corporation

Edward H. Newman
The Ohio State University

IEEE PRESS

A Selected Reprint Volume
IEEE Antennas and Propagation Society, *Sponsor*

The Institute of Electrical and Electronics Engineers, Inc., New York

345 East 47th Street, New York, NY 10017-2394

Printed in the United States of America

10 9 8 7 6 5 4 3 2 1

ISBN 0-87942-276-9
IEEE Order Number : PC0270-9

Library of Congress Cataloging-in-Publication Data

Computational electromagnetics : frequency-domain method of moments / edited by Edmund K. Miller, Louis N. Medgyesi-Mitschang, Edward H. Newman.
p. cm.
Includes bibliographical references and indexes.
ISBN 0-87942-276-9
1. Electromagnetism—Mathematical models. 2. Moments method (Statistics) 3. Integral equations—Numerical solutions.
I. Miller, E. K. (Edmund Kenneth) (date). II. Medgyesi-Mitschang, Louis N. (date). III. Newman, Edward H. (date).
IV. Title: Frequency-domain method of moments.
QC760.C65 1992
537—dc20 91-12159

The editors would like to dedicate this book to Professor Jack H. Richmond (1922–1990) for his pioneering work in the method of moments. After obtaining the Ph.D. in electrical engineering from The Ohio State University in 1955, Dr. Richmond remained as a professor in the Department of Electrical Engineering, and as a researcher with the ElectroScience Laboratory. He is probably best known for his development of the piecewise sinusoidal reaction formulation, and associated computer code, for the analysis of thin wires. However, he also made important contributions to the moment method analysis of dielectrics, rectangular plates, antennas over or under the earth, surface waves, etc. His contributions were recognized by his election as a Fellow of the IEEE in 1980 and his receipt of the IEEE Centennial Medal in 1984 and the George Sinclair Award in 1985. Professor Richmond will be sorely missed by the electromagnetics community, and in particular by those of us working in the method of moments.

Contents

PREFACE xi

INTRODUCTION 1

PART 1 THEORY 5

Toward Formulation of a General Diffraction Problem via an Integral Equation 7
A. W. Maue, *Zeitschrift für Physik* (1949)

Reaction Concept in Electromagnetic Theory 15
V. H. Rumsey, *The Physical Review* (1954)

Matrix Methods for Field Problems 24
R. F. Harrington, *Proceedings of the IEEE* (January 1967)

On Energy Conservation and the Method of Moments in Scattering Problems 38
N. Amitay and V. Galindo, *IEEE Transactions on Antennas and Propagation* (November 1969)

Origin and Development of the Method of Moments for Field Computation 43
R. F. Harrington, Chapter 1 in *Applications of the Method of Moments to Electromagnetic Fields,* The SCEEE Press, St. Cloud, FL (1980)

Effective Methods for Solving Integral and Integro-Differential Equations 48
D. R. Wilton and C. M. Butler, *Electromagnetics* (1981)

Potential Integrals for Uniform and Linear Source Distributions on Polygonal and Polyhedral Domains 60
D. R. Wilton, S. M. Rao, A. W. Glisson, D. H. Schaubert, O. M. Al-Bundak, and C. M. Butler, *IEEE Transactions on Antennas and Propagation* (March 1984)

Error Minimization and Convergence in Numerical Methods 66
D. G. Dudley, *Electromagnetics* (January 1985)

A Theorem on the Moment Methods 71
A. R. Djordjević and T. K. Sarkar, *IEEE Transactions on Antennas and Propagation* (March 1987)

PART 2 CONDUCTING BODIES 75

Scattering from Bodies of Revolution 77
M. G. Andreasen, *IEEE Transactions on Antennas and Propagation* (March 1965)

A Wire-Grid Model for Scattering by Conducting Bodies 86
J. H. Richmond, *IEEE Transactions on Antennas and Propagation* (November 1966)

Electromagnetic Modeling of Composite Wire and Surface Geometries 91
E. H. Newman and D. M. Pozar, *IEEE Transactions on Antennas and Propagation* (November 1978)

A Combined-Source Solution for Radiation and Scattering from a Perfectly Conducting Body 97
J. R. Mautz and R. F. Harrington, *IEEE Transactions on Antennas and Propagation* (July 1979)

Electromagnetic Scattering by Surfaces of Arbitrary Shape 107
S. M. Rao, D. R. Wilton, and A. W. Glisson, *IEEE Transactions on Antennas and Propagation* (May 1982)

Wire Grid Modeling of Surfaces 117
A. C. Ludwig, *IEEE Transactions on Antennas and Propagation* (September 1987)

Extended Integral Equation Formulation for Scattering Problems from a Cylindrical Scatterer 121
I. Toyoda, M. Matsuhara, and N. Kumagai, *IEEE Transactions on Antennas and Propagation* (November 1988)

PART 3 THIN WIRES **129**

On the Integral Equations of Thin Wire Antennas 131
K. K. Mei, *IEEE Transactions on Antennas and Propagation* (May 1965)

Theory of Conical Equiangular-Spiral Antennas: Part 1—Numerical Techniques 136
Y. S. Yeh and K. K. Mei, *IEEE Transactions on Antennas and Propagation* (September 1967)

Computer Techniques for Electromagnetic Scattering and Radiation Analyses 142
S. Gee, E. K. Miller, A. J. Poggio, E. S. Selden, and G. J. Burke, *IEEE International Electromagnetic Compatibility Symposium Record* (1971)

Accuracy-Modeling Guidelines for Integral-Equation Evaluation of Thin-Wire Scattering Structures 153
E. K. Miller, G. J. Burke, and E. S. Selden, *IEEE Transactions on Antennas and Propagation* (July 1971)

Radiation and Scattering by Thin-Wire Structures in the Complex Frequency Domain 156
J. H. Richmond, excerpted from *OSU Research Foundation Report RF 2902-10* (1973)

A Study of the Various Methods for Computing Electromagnetic Field Utilizing Thin Wire Integral Equations 170
T. K. Sarkar, *Radio Science* (1983)

Equivalent Line Current for Cylindrical Dipole Antennas and Its Asymptotic Behavior 180
R. E. Collin, *IEEE Transactions on Antennas and Propagation* (February 1984)

PART 4 PENETRABLE AND/OR LOSSY BODIES **185**

Scattering by a Dielectric Cylinder of Arbitrary Cross Section Shape 187
J. H. Richmond, *IEEE Transactions on Antennas and Propagation* (May 1965)

An Integral Equation Approach to Scattering from a Body of Finite Conductivity 195
K. M. Mitzner, *Radio Science* (1967)

Electromagnetic Fields Induced Inside Arbitrarily Shaped Biological Bodies 207
D. E. Livesay and K. M. Chen, *IEEE Transactions on Microwave Theory and Techniques* (December 1974)

Dielectric Coated Wire Antennas 215
J. H. Richmond and E. H. Newman, *Radio Science* (1976)

Electromagnetic Scattering from a Homogeneous Material Body of Revolution 223
J. R. Mautz and R. F. Harrington, *Arch. Elektron. Übertragungstech.* (1979)

Electromagnetic Scattering from Axially Inhomogeneous Bodies of Revolution 233
L. N. Medgyesi-Mitschang and J. M. Putnam, *IEEE Transactions on Antennas and Propagation* (August 1984)

A Tetrahedral Modeling Method for Electromagnetic Scattering by Arbitrarily Shaped Inhomogeneous Dielectric Bodies 243
D. H. Schaubert, D. R. Wilton, and A. W. Glisson, *IEEE Transactions on Antennas and Propagation* (January 1984)

Integral Equation Formulations for Imperfectly Conducting Scatterers 252
L. N. Medgyesi-Mitschang and J. M. Putnam, *IEEE Transactions on Antennas and Propagation* (February 1985)

Electromagnetic Scattering by Arbitrary Shaped Three-Dimensional Homogeneous Lossy Dielectric Objects 261
K. Umashankar, A. Taflove, and S. M. Rao, *IEEE Transactions on Antennas and Propagation* (June 1986)

PART 5 APERTURES 269

The Matrix Formulation of Scattering Problems 271
J. Van Bladel, *IEEE Transactions on Microwave Theory and Techniques* (March 1966)

Radiation from Apertures in Conducting Cylinders of Arbitrary Cross Section 277
R. F. Wallenburg and R. F. Harrington, *IEEE Transactions on Antennas and Propagation* (January 1969)

Electromagnetic Excitation of a Wire Through an Aperture-Perforated Conducting Screen 284
C. M. Butler and K. R. Umashankar, *IEEE Transactions on Antennas and Propagation* (July 1976)

Electromagnetic Penetration Through Apertures in Conducting Surfaces 291
C. M. Butler, Y. Rahmat-Samii, and R. Mittra, *IEEE Transactions on Antennas and Propagation* (January 1978)

Aperture Coupling in Bodies of Revolution 303
H. K. Schuman and D. E. Warren, *IEEE Transactions on Antennas and Propagation* (November 1978)

Electromagnetic Pulse Coupling Through an Aperture into a Two-Parallel-Plate Region 309
Y. Rahmat-Samii, *IEEE Transactions on Electromagnetic Compatibility* (August 1978)

PART 6 HYBRID MM/ALTERNATE GREEN'S FUNCTIONS 317

Radiation Patterns for Two Monopoles on a Perfectly Conducting Sphere 319
F. M. Tesche and A. R. Neureuther, *IEEE Transactions on Antennas and Propagation* (September 1970)

Computation of Radiation from Wire Antennas on Conducting Bodies 322
N. C. Albertsen, J. E. Hansen, and N. E. Jensen, *IEEE Transactions on Antennas and Propagation* (March 1974)

A Hybrid Technique for Combining Moment Methods with the Geometrical Theory of Diffraction 328
G. A. Thiele and T. H. Newhouse, *IEEE Transactions on Antennas and Propagation* (January 1975)

Wire Antennas in the Presence of a Dielectric/Ferrite Inhomogeneity 336
E. H. Newman and P. Tulyathan, *IEEE Transactions on Antennas and Propagation* (July 1978)

A Summary of Hybrid Solutions Involving Moment Methods and GTD 342
W. D. Burnside and P. H. Pathak, chapter in *Applications of the Method of Moments to Electromagnetic Fields,* The SCEEE Press, St. Cloud, FL (1980)

Hybrid Solutions for Scattering from Perfectly Conducting Bodies of Revolution 355
L. N. Medgyesi-Mitschang and D. S. Wang, *IEEE Transactions on Antennas and Propagation* (July 1983)

Modeling Antennas Near to and Penetrating a Lossy Interface 369
G. J. Burke and E. K. Miller, *IEEE Transactions on Antennas and Propagation* (October 1984)

General Integral Equation Formulation for Microstrip Antennas and Scatterers 379
J. R. Mosig and F. E. Gardiol, *IEE Proceedings* (1985)

Analysis of Planar Strip Geometries in a Substrate–Superstrate Configuration 388
D. R. Jackson and N. G. Alexópoulos, *IEEE Transactions on Antennas and Propagation* (December 1986)

A Note on Hybrid Finite Element Method for Solving Scattering Problems 397
J. M. Jin and V. V. Liepa, *IEEE Transactions on Antennas and Propagation* (October 1988)

Coupling of Finite Element and Moment Methods for Electromagnetic Scattering from Inhomogeneous Objects 401
X. Yuan, D. R. Lynch, and J. W. Strohbehn, *IEEE Transactions on Antennas and Propagation* (March 1990)

Validation of the Hybrid Quasi-Static/Full-Wave Method for Capacitively Loaded Thin-Wire Antennas 409
R. G. Olsen and P. D. Mannikko, *IEEE Transactions on Antennas and Propagation* (April 1990)

PART 7 NUMERICAL METHODS 417

Numerical Integration Methods 419
E. K. Miller and G. J. Burke, *IEEE Transactions on Antennas and Propagation* (September 1969)

Analysis of Various Numerical Techniques Applied to Thin-Wire Scatterers 422
C. M. Butler and D. R. Wilton, *IEEE Transactions on Antennas and Propagation* (July 1975)

Efficient Solution of Large Moments Problems: Theory and Small Problem Results 429
T. R. Ferguson, T. H. Lehman, and R. J. Balestri, *IEEE Transactions on Antennas and Propagation* (March 1976)

Survey of Numerical Methods for Solution of Large Systems of Linear Equations for Electromagnetic Field Problems 435
T. K. Sarkar, K. R. Siarkiewicz, and R. F. Stratton, *IEEE Transactions on Antennas and Propagation* (November 1981)

A Comparison of Spherical Wave Boundary Value Matching Versus Integral Equation Scattering Solutions for a Perfectly Conducting Body 445
A. C. Ludwig, *IEEE Transactions on Antennas and Propagation* (July 1986)

Generation of Wide-Band Data from the Method of Moments by Interpolating the Impedance Matrix 454
E. H. Newman, *IEEE Transactions on Antennas and Propagation* (December 1988)

Using Model-Based Parameter Estimation to Increase the Efficiency of Computing Electromagnetic Transfer Functions 459
G. J. Burke, E. K. Miller, S. Chakrabarti, and K. Demarest, *IEEE Transactions on Magnetics* (July 1989)

SELECTED BIBLIOGRAPHY 463

AUTHOR INDEX 499

SUBJECT INDEX 501

EDITORS' BIOGRAPHIES 507

Preface

THIS BOOK presents a selection of key articles in several different development and application areas designed to give the reader an integrated information source concerning moment-method (MM) modeling in computational electromagnetics (CEM). The articles were chosen to cover the analytical formulation and numerical implementation of various frequency-domain, integral-equation models, while also demonstrating their practical application. The book, divided into seven major sections, brings together in one volume a collection of papers whose sum total will provide a solid foundation in the theory, implementation, and application of MM techniques. It also includes an extensive bibliography of related material. The collection of papers in this book benefits the reader by providing in one place the analytical and numerical background necessary to come ''up to speed'' on a class of modeling tools that remains the most widely used among the kinds available to the computational electromagneticist. The increasing use of computational modeling for analysis and, more importantly, design mandates an appropriate familiarity among both working engineers and their managers with the background and capabilities of numerical modeling in electromagnetics.

A careful examination of more recent CEM literature shows that, as experimental results typically do not include detailed lists of the hardware used in their acquisition, numerical solutions similarly do not necessarily mention details of the MM models that were employed to obtain what are regarded as straightforward results. Because of this growing reliance on CEM, it is helpful to publish in one convenient resource volume a selection of articles that summarize the basic formulation, development, and representative applications of MM techniques for the more commonly encountered problems. The purpose of this book is to provide this resource. For practical reasons, this particular volume concentrates on frequency-domain, integral-equation models, an area that represents the bulk of CEM applications to date.

The book is organized into seven sections, the first dealing with theoretical aspects of CEM and the following six concentrating on various specific classes of problems as summarized here.

Section 1: Theory. It is the purpose of this section to provide a theoretical background for the moment method, including those papers that introduce or summarize various analytical issues involved in developing an integral equation, conditions necessary for model validity, the relationship of MM to other techniques, and the predominant characteristic of MM models as that of evaluating various kinds of field integrals.

Section 2: Conducting Bodies. The development of MM models naturally began with perfectly conducting two- and three-dimensional objects. Modeling of perfectly conducting 2D and 3D bodies in free space remains one of the most commonly encountered problems in CEM. This section includes formative papers of both types of problems, using wire grids as approximations to solid objects, hybrid objects involving wires attached to a surface, circumventing failure due to internal resonances by using combined-field treatments, and modeling of more general surface shapes using piecewise triangular models.

Section 3: Thin Wires. Wires represent a special class of conducting objects because of the simplification they offer and because so many actual problems do involve wires. Topics discussed in this section include models based on both the Hallén and electric field integral equations, the various kinds of basis and testing functions employed, a survey and comparison of some of these methods, and consideration of the limitations that arise from using the reduced kernel employed as the thin-wire approximation.

Section 4: Penetrable and/or Lossy Bodies. Modeling of penetrable objects increases problem complexity because of the need to deal with the fields of sources radiating in at least two different media. The section begins with an introduction of the volume-current approach, approximate treatments using the impedance boundary condition, and various approaches for homogeneous and inhomogeneous bodies.

Section 5: Apertures. Bodies with apertures that allow field penetration to the interior constitute one of the more demanding kinds of MM models. This section presents papers that address this basic problem from various perspectives, such as apertures in infinite planes in free space and separating different half spaces, apertures with wires behind, and apertures in bodies of revolution and planar cavities.

Section 6: Hybrid MM/Alternate Green's Functions. Hybrid techniques are those that involve use of different kinds of field propagators to model separate parts of an overall problem, such as combining the MM for a wire with GTD for scattering from nearby edges. Alternate Green's functions go beyond the one commonly used for an infinite medium by incorporating a special Green's function that satisfies field boundary conditions over special surfaces such as infinite plane, circular cylinder, or sphere. In either case, the goal is to reduce the number of unknowns needing solution in the MM model. Problems discussed here include a monopole attached to a sphere, use of the MM with GTD in various ways, and the problem of an infinite interface that arises when modeling antennas near the earth's surface.

Section 7: Numerical Methods. Various specialized numerical procedures can be important in making MM models more useful and computationally affordable. Some of the issues discussed in this section include the effects of various basis and testing functions, performing numerical integration more efficiently and accurately in computing the MM matrix, iteration as a way to solve an MM matrix more efficiently, the kinds of problems that arise in solving linear systems with increasing numbers of unknowns, and decreasing the number of frequency samples needed to define a transfer function.

As might be anticipated, the task of selecting the articles to be included in the book was not an easy one. The candidate papers for selection number at least in the hundreds, as computational work has expanded at what seems an exponential rate over the past 10 or 15 years. Because of this, the choice of what to include and what to omit was not an easy one. Although in some cases a particular selection might be considered an obvious one by most readers, in others there could have been several, or even many, possible choices to address a certain topic. Aside from scientific and application merit, two major factors were considered in article selection. These are the work's archival value and chronological precedence. The former concerns how long into the future a given article might be relevant and useful, while the latter relates to providing appropriate recognition to an original contribution. Another consideration is possible overlap with other references to CEM. Finally, where a choice needed to be made between the contributions of two or more different authors, everything else being equal, we tended to select the article of that individual who would otherwise not have been represented in the book. We have attempted to take all of these factors appropriately into account in selecting the final collection of articles included here. We hope that the papers that follow collectively represent a definitive exposition of moment-method, frequency-domain, integral-equation models in computational electromagnetics.

Special thanks are due to Louis Medgyesi-Mitschang for translating the article by Maue, ''Toward formulation of a general diffraction problem via an integral equation,'' which originally appeared in German. We also appreciate the support and encouragement of the AP-S Education Committee through whose auspices the book originated.

THE EDITORS

Introduction

COMPUTATIONAL methods have been important in science and engineering since before the development of Maxwell's equations. Of course, the advent of the digital computer has qualitatively changed the scope of problems for which numerical results can be obtained and has thereby transformed the tools available to the electromagnetics analyst, designer, and experimentalist. The first contributions reflecting this changing reality began appearing in the early to mid 1960s, due to authors such as Andreasen, Harrington, Oshiro, Mei, Richmond, Waterman, and Yee, one source of which is the August 1965 special issue of the PROCEEDINGS OF THE IEEE on "Radar Reflectivity." The number of articles presenting new techniques and results increased rapidly towards the end of that decade. A truly seminal publication was the book by Harrington, *Field Computation by Moment Methods* (Macmillan, 1968), which, together with his earlier book, *Time Harmonic Electromagnetic Fields* (McGraw-Hill, 1961), remain today two of the most frequently referenced sources in engineering electromagnetics.

The method of moments has become an important tool, and indeed may be properly claimed as forming the basis of all numerical tools in CEM. Although it has unfortunately become common practice to use the terms *moment method* and *integral-equation models* interchangeably, the MM can be more accurately described as a procedure by which both integral- and differential-equation-based models are developed. Whatever approach is followed, it is necessary, either implicitly or explicitly, to sample, discretize, and approximate the appropriate analytical differential or integral descriptions.

It is also recognized that CEM has been added to measurement and mathematical analysis as a means by which problems in electromagnetics are solved. While the term "method of moments" may be explicitly mentioned in the abstracts or index listings of relatively few of the papers published in such IEEE TRANSACTIONS as ANTENNAS AND PROPAGATION, ELECTROMAGNETIC COMPATIBILITY, MAGNETICS, and MICROWAVE THEORY AND TECHNIQUES, this only reflects the fact that its application is increasingly taken for granted. The same is true of the many other publications in which articles on CEM or wave-equation modeling can be found, such as:

A. E. U. (Archiv für Elektronik und Übertragungstechnik)
Applied Scientific Research
Canadian Journal of Physics
Electromagnetics
Electronics Letters
International Journal for Numerical Methods in Engineering
International Journal of Numerical Modelling: Electronic Networks, Devices, and Fields
Journal of the Acoustical Society of America
Journal of the Association for Computing Machinery
Journal of Applied Physics
Journal of the Applied Computational Electromagnetics Society
Journal of Computational Physics
Journal of Electromagnetic Waves and Applications
Journal of the Optical Society of America
Proceedings of the IEEE
Radio Science
The Institution of Electrical Engineers (various publications)
The SCEEE Press, St. Cloud, Florida
Transactions of IECE of Japan

DEVELOPING, USING, AND CHARACTERIZING COMPUTER MODELS

Computer models, or codes, might be viewed differently from both the developer's perspective and the user's perspective in terms of those issues that are most important to each of these communities. Developers deal primarily with the analytical and implementational steps of codes, while the users are more involved in computation and application. Both are concerned with validation, although users can always be counted on to apply codes in ways never envisioned by their developers, making validation an especially sensitive topic.

The Developer's Perspective

From the developer's perspective, the process of developing a numerical model involves a small number of basic steps whatever the details of a particular approach. These may basically be categorized as:

1. *Conceptualization.* It is at this step where physical principles, experimental observations, and so on are used to form hypotheses from which mathematical descriptions of the relevant phenomena can be devised.
2. *Formulation.* This step involves the evolution of the physical idea or mathematical description from its elementary form into one suitable for numerical or analytical evaluation. Various approximations may be utilized to make the subsequent analysis and/or computation easier.
3. *Numerical implementation.* At this stage the formulation is reduced to a form suitable for computation, leading to a computer code or algorithm. In the context of our present discussion, this step almost always involves developing a linear system of equations that are solved using matrix techniques.
4. *Computation/Application.* In the computation/application step, the limitations and "bugs" are discovered and

accuracy measures are established. Computation also involves approximation, but in a more ambiguous way than the previous steps, because a model is employed by the user to represent the reality of interest, and there is rarely a simple or obvious way to model most real problems. Furthermore, the numerical model itself is not solved exactly.

5. *Validation.* This step is probably the most crucial, as it establishes the degree to which the code can eventually be relied upon. It is an open-ended process since a code is usually applied to an expanding variety of problems. Several kinds of validation can be used, which include both internal (self-consistency) and external checks, with the latter including independent analytical, experimental, and numerical results.

Although all five of these steps are addressed to varying degrees in all the articles presented in this book, the first section on ''Theory'' concentrates more on the first three steps mentioned. The ''conceptualization'' step is certainly reflected in the paper by Maue, while the other selected papers in that section deal with various theoretical aspects of model formulation and development, as well as numerical implementation, computation, and validation.

The User's Perspective

From the user's perspective, selection and application of a specific model also involves another set of issues that need consideration. These include the general modeling approach that might be most appropriate, the kinds of applications intended for the model, and the problem characteristics the model is required to have. These issues can be succinctly summarized as follows.

1. *Approach.* Electromagnetic analysis can be conveniently organized according to how the fields are propagated in formulating and solving problems. The following appear to be the most important:
   a. A Green's function that leads to integral descriptions and equations, the approach to which this book is almost wholly addressed.
   b. The Maxwell curl equations that provide differential descriptions and equations, being described most often in the literature as ''finite-difference'' and ''finite-element'' models.
   c. Modal descriptions that involve analytical expansions in terms of classical functions, known as the ''T-matrix'' approach in acoustics, and more recently by the name ''generalized multipole technique'' in electromagnetics.
   d. Optical descriptions that involve rays, diffraction, refraction, and flux tubes, called by various names including ''geometrical optics,'' ''geometrical theory of diffraction (GTD),'' and high-frequency ''asymptotic techniques.''
2. *Application.* Electromagnetic applications can be subdivided into the following:
   a. Radiation, as in analyzing the fields of prescribed sources or excitations. This application, together with scattering, are the primary applications of MM models.
   b. Propagation, as in determining the fields at a distance. Although other solution techniques might be more common, integral equations can also be used for propagation, for example in surface-wave propagation over the ground.
   c. Scattering, as in assessing the effects of impedance discontinuities and perturbations. When MM impedance matrices are factored or otherwise solved independent of the right-hand side source term, then the model is equally applicable to scattering or radiation applications.
3. *Problem.* Finally, the problem can be characterized by factors such as:
   a. Dimensionality (1D, 2D, or 3D). CEM increasingly deals with 3D problems, although 2D modeling is still often used for purposes such as component design in the radar cross section (RCS) of large objects.
   b. Domain (time domain, frequency domain, spectral domain, etc.). At the time of this book's publication, most integral-equation modeling has been done in the frequency domain, while most differential-equation modeling is done using the time domain.
   c. Electrical properties of object or medium (dielectric, lossy, perfect conductor). Although perfectly conducting objects probably still receive the bulk of modeling attention, there is a rapidly increasing volume of work being done on lossy and/or penetrable objects.
   d. Object modeled (e.g., wire cylinder, body, waveguide, etc.). As a whole, wire objects, either as the geometry of actual interest or as a wire-grid approximation to a surface, probably remain the largest single class of applications.
   e. Object geometry (straight, polygonal, curved, arbitrary). With the growing availability of interactive computer-aided design (CAD) interfaces, the object geometries being modeled are becoming more complex. Nevertheless, most geometric descriptions remain piecewise or segmented approximations to general curvilinear objects.

Although these various attributes do not represent measures that are completely independent, they do seem to provide useful alternatives for cataloging the various kinds of modeling approaches used in CEM.

Characterizing a Computer Model

Among the various attributes that might be selected for characterizing and comparing computer models for electromagnetics or other applications, the following are most fundamental:

1. *Accuracy/Reliability.* Above all else, a modeling computation must possess acceptable, preferably known, and

better yet, "dialable" accuracy. This is an attribute to which all others, however desirable they might be, must be considered secondary, for invalid results have no value and can even be detrimental.

2. *Efficiency/Productivity.* Following accuracy as a desirable attribute is efficiency, both with respect to the *computer* resources needed for the modeling computation as well as with respect to the *human* resources needed to develop the input required to run the model and to interpret the results that the model produces. While it is obvious that a model that efficiently produces inaccurate results has no value, it is equally true that a model that produces acceptably accurate results but that requires computer and/or human resources incommensurate with the application has little more value.
3. *Utility/Applicability.* Finally, we must consider a model's utility in terms of the kinds of problems to which it can be applied. At this stage, general-purpose modeling codes and specialized single-problem codes become most differentiated. On the one hand, it is always easier to develop a model specialized for a particular problem that will be more accurate and/or efficient than a general-purpose code that can model that same problem. On the other hand, the more widely applicable a given modeling code becomes, and the more easily used it is, the greater its utility for the nonspecialist who does only infrequent modeling.

In some respects, the user of a computer code needing to do EM numerical modeling and simulation is in a position similar to that of an experimenter who needs various kinds of hardware to perform measurements. Both understand the needs of their particular problems in terms of the basic EM phenomena involved, but both also need to depend on complex tools that they did not design in order to accomplish their own particular goals. Just as the experimentalist doesn't need to know how to design a spectrum analyzer but only understand its functionality to use it productively, a code user shouldn't need to know *how* the code works but rather *what* it can do. In another context, a person needs to have only a well-defined but limited subset of knowledge of how an automobile works in order to be an effective driver. Many of the improvements being made by the automotive industry today are designed to improve automobile performance while requiring no further expertise on the part of the driver; i.e., the *how* part of the improvement is transparent to the user. In a fashion analogous to EM experimental hardware or automotive systems, user-oriented or user-friendly EM modeling software should also be designed to improve modeling performance without requiring users to be expert in the underlying modeling principles involved.

On a final note, it might be helpful to the reader to list a number of other books that the present volume complements. These include:

Harrington, R. F. *Field Computation by Moment Methods.* New York: Macmillan, 1968.

Mittra, R., ed. *Computer Techniques for Electromagnetics.* Elmsford, NY: Pergamon Press, 1973.

Mittra, R., ed. *Numerical and Asymptotic Techniques in Electromagnetics.* New York: Springer Verlag, 1975.

Uslenghi, P. L. E., ed. *Electromagnetic Scattering.* New York: Academic Press, 1980.

Strait, B. J., ed. *Applications of the Method of Moments to Electromagnetic Fields.* St. Cloud, FL: The SCEEE Press, 1980.

Stutzman, W. L. and G. A. Thiele. *Antenna Theory and Design.* New York: John Wiley & Sons, 1981.

Popovic, B. D., M. B. Dragovic, and A. R. Djordjevic. *Analysis and Synthesis of Wire Antennas.* Letchworth, Nertfordshire, England: Research Studies Press, 1982.

Skwirzyniski, J. K., ed. *Theoretical Methods for Determining the Interaction of Electromagnetic Waves with Structures.* Alphen aan den Rijn, The Netherlands and Rockville, MD: Sijthoff and Noordhoff, 1983.

Moore, J. and R. Pizer. *Moment Methods in Electromagnetics: Techniques and Applications.* New York: John Wiley & Sons, 1984.

Lo, Y. T. and S. W. Lee, eds. *Antenna Handbook.* New York: Van Nostrand Reinhold, 1987.

Hansen, R. C., ed. *Moment Methods in Antennas and Scattering.* Norwood, MA: Artech House, 1990.

Itoh, T., ed. *Numerical Techniques for Microwave and Millimeter-Wave Passive Structures.* New York: John Wiley & Sons, 1990.

Wang, J. H. H. *Generalized Moment Methods in Electromagnetics.* New York: John Wiley & Sons, 1991.

Part 1
Theory

THIS SECTION provides the theoretical background for the moment method (MM). Included here are introductory or summary discussions of the various analytical issues involved in developing an integral equation, consideration of the conditions necessary for model validity, the relationship of MM to other techniques, and the predominant characteristic of MM models as that of evaluating various kinds of field integrals.

The first paper, by Maue, ''Toward formulation of a general diffraction problem via an integral equation,'' can be regarded as a generalization of earlier work by Pocklington, Hallén, and others. It provides a departure point for most subsequent integral-equation development, to which most computational models, aside from the special case of wires, can trace their origin. The next paper, by Rumsey, ''Reaction concept in electromagnetic theory,'' is a classic paper that introduces the concept of reaction and shows how it can be used to derive an integral equation for MM solution of scattering by a dielectric body. The first paper by Harrington, ''Matrix methods for field problems,'' predates his widely referenced book published a year later, and introduces the general technique now known by engineering electromagneticists as the method of moments, illustrating its application to the charged plate and thin wire. The following article by Amitay and Galindo, ''On energy conservation and the method of moments in scattering problems,'' discusses a key issue in modeling, i.e., how to test the validity of a solution. In this case, an energy-balance approach is explored. Another paper by Harrington, ''Origin and development of the method of moments for field computation,'' provides a unifying development of the MM and shows how it relates to the world of variational techniques. A paper by Wilton and Butler, ''Effective methods for solving integral and integro-differential equations,'' presents an excellent tutorial introduction to the MM solution of integral and integro-differential equations. The examples chosen are simple, to permit ease of understanding and computation, and yet contain many of the features found in electromagnetics, such as the logarithmically singular kernel. The paper, ''Potential integrals for uniform and linear source distributions on polygonal and polyhedral domains,'' by Wilton *et al.*, evaluates typical line, surface, and volume integrals that are common to the evaluation of many MM impedance matrices, the modeling step that most differentiates among the wide variety of integral-equation–based models. The important issue of error characterization in MM modeling is discussed by Dudley in ''Error minimization and convergence in numerical methods.'' This paper discusses in a very general way the meaning of and conditions for convergence of MM and related methods. This theory section concludes with ''A theorem on the moment methods,'' by Djordjevic and Sarkar, who make the important point that everything we do in numerical methods is essentially point matching, even though one may start from a quite different philosophy.

Toward Formulation of a General Diffraction Problem via an Integral Equation

by A.W. Maue

With Two Figures.

(Received 25 April 1949)

The problem of diffraction of scalar and electromagnetic waves by bodies is formulated with the aid of an integral equation, which is valid for a covering on the surface of the body. In the electromagnetic case with the body assumed to be perfectly conducting, this covering is the surface current. The special cases of edges and the diffraction by finite surface patches will be discussed. As an example, the diffraction by a cylinder will be examined.

Let a wave illuminate a closed body. The question to be asked is what diffraction phenomena arise. This general question can be delineated more specifically 1) by the body shape and 2) by the boundary conditions valid on the surface.

In the traditional approach to this problem, one introduces special coordinates conforming to the shape of the body and which contain the body surface as a coordinate surface. The solution of the wave equation leads to special functions which are only known and tabulated for the simplest body shapes.

The general problem under consideration is a boundary value problem in three-dimensional space (wave-equation + boundary condition + radiation condition at infinity). The goal of the subsequent derivations is to reduce the three-dimensional problem to a two-dimensional one in which a surface function, instead of a volume function, becomes the unknown. This unknown function is defined on the surface of the body and satisfies an integral equation. This approach makes sense, particularly for complicated body shapes, since then the form of the body, not the simple wave equation, becomes the actual problem. In view of the generality of the diffraction problem under consideration, it appears desirable to examine these relationships from a standpoint other than the customary one, even if the question of utility for practical calculations is left open at this time.

The appearance of an integral equation in place of a differential equation for the wave equation represents a complication. Balancing this are two simplifications: 1.) The number of independent variables is reduced by one, and the introduction of a special spatial coordinate grid is unnecessary. 2.) The only condition on the desired (unknown) function is that it satisfy the integral equation. Thus at the outset, the problem revolves around the determination of a single function, in contrast to the original volumetric formulation where all solutions to a differential equation must be found from which a particular one is subsequently selected through the boundary conditions.

The use of an integral equation for formulation of a boundary value problem is a well-known general mathematical method. Magnus[1] and Fock[2] applied it recently to the diffraction problem.

In the following, the emphasis is placed on a parallel development of scalar and vector (electromagnetic) diffraction problems. The starting point is the Kirchhoff integral formulation and its generalization for electromagnetic waves. These expressions are a mathematical statement of Huygen's principle. They are mostly used for the approximate treatment of diffraction problems (Kirchhoff's Diffraction Theory[3]) but are themselves valid exactly. In introducing these expressions and the general properties of the associated integrals, we omit proofs and refer instead to the work of Franz[4] who also treated scalar and electromagnetic waves in a parallel manner. The two characteristics of the integral given by Franz, are supplemented by a third property which can be inferred from the special singular character of the Green's function by direct calculation.

1. The Scalar Diffraction Problem

Following Kirchhoff, the solution u of the wave equation

$$\Delta u + k^2 u = 0 \tag{1}$$

in a spatial region can be obtained when u and $\partial u/\partial n$ are known on the boundary of the region. This is realized with the aid of an integral over the boundary. If P is the observation point, Q is the source point and n is the inward normal, then

$$u(P) = \int \left\{ u(Q) \frac{\partial G(P,Q)}{\partial n_Q} - G(P,Q) \frac{\partial u(Q)}{\partial n_Q} \right\} dQ. \tag{2}$$

holds. $G(P,Q)$ is the Green's function of the unbounded (free) space where it does not matter which of the expressions

$$G(P,Q) = \frac{\cos kr}{4\pi r}, \quad \frac{e^{\pm ikr}}{4\pi r} \qquad (r = r_{PQ}) \tag{3}$$

is used.

For the problem posed above, we apply (2) for the desired solution u and for the exterior of the diffracting body. If we exclude the radiation source with a small sphere, the boundary of the domain under consideration has three parts: the surface of the body, the surface of the small sphere, and a surface at infinity. Since u satisfies the radiation condition, the integral over the last surface vanishes when we specifically choose

$$G = \frac{e^{ikr}}{4\pi r} \tag{4}$$

for a time variation of $e^{-j\omega t}$ ($\omega = kc$, $c =$ phase velocity of the waves). The integral over the small spherical surface yields the incident wave u_0. If we designate the still unknown boundary values of u and $\partial u/\partial n$ on the diffracting body with f and g, one obtains

$$u(P) = u_0(P) + \int \left\{ \frac{\partial G(P,Q)}{\partial n_Q} f(Q) - G(P,Q)\, g(Q) \right\} dQ, \tag{5}$$

where the integration spans the body surface and n points outward from the body. Thus the sought-after spatial (volumetric) function u is reduced to two surface quantities f and g, which remain to be determined.

The type of integral that arises in (5) is

$$v(P) = \int \left\{ \frac{\partial G(P,Q)}{\partial n_Q} f(Q) - G(P,Q)\, g(Q) \right\} dQ \tag{6}$$

Reprinted with permission from *Zeitschrift für Physik*, vol. 126, pp. 601–618, 1949. Translation rights from German into English with the permission of Springer-Verlag.

In general, this integral over a closed surface for arbitrary functions f and g and either expression in (3) for G has the following properties:
1. v satisfies the wave equation (1) on both sides of the surface.
2. On the surface the boundary conditions are satisfied

$$\left\{\begin{aligned} & v_1 - v_2 = f \\ & \left(\frac{\partial v}{\partial n}\right)_1 - \left(\frac{\partial v}{\partial n}\right)_2 = g, \end{aligned}\right. \tag{7}$$

where the subscript 1 corresponds to the side of the surface to which the normal n points.
3. On the surface itself v and $\partial v/\partial n$ are equal to the average of the boundary values of both surfaces:

$$\left\{\begin{aligned} & v = \tfrac{1}{2}(v_1 + v_2) \\ & \frac{\partial v}{\partial n} = \frac{1}{2}\left(\left(\frac{\partial v}{\partial n}\right)_1 + \left(\frac{\partial v}{\partial n}\right)_2\right). \end{aligned}\right. \tag{8}$$

Of these assertions, the third requires further explanation. The first equation in (8) has direct meaning. Since G and $\partial G/\partial n$ go to infinity only as $1/r$ when P and Q approach the surface, the integral (6) converges even when P lies on the surface. In contrast the expression

$$\left(\frac{\partial v}{\partial n}\right)_P = \int\left\{\frac{\partial^2 G(P,Q)}{\partial n_Q\,\partial n_P} f(Q) - \frac{\partial G(P,Q)}{\partial n_P} g(Q)\right\} dQ \tag{8a}$$

obtained by differentiating the integrand in (6) with respect to n_P, diverges for points P on the surface. Nevertheless, the expression allows itself to be recast through partial integration, so that a specific value can be ascribed on the surface which is given by the second equation in (8). Through two partial integrations, the divergence can be completely eliminated. After one partial integration, the integral is conditionally convergent when P lies on the surface. To obtain the proper value of the integral, one must excise P with a small circle centered at P, whose radius is allowed to go to zero. In our context (see Section 4) it is useful to confine ourselves to a single partial integration. Thus, one has to convert differentiation with respect to P in the first term to one with respect to Q:

$$\frac{\partial^2 G}{\partial n_Q\,\partial n_P} = -(\mathfrak{n}_Q \operatorname{grad}_Q (\mathfrak{n}_P \operatorname{grad}_Q G)).$$

Using the Nabla notation, one obtains through use of the vector identity

$$([\mathfrak{A}\,\mathfrak{B}], [\mathfrak{C}\,\mathfrak{D}]) = (\mathfrak{A}\,\mathfrak{C})(\mathfrak{B}\,\mathfrak{D}) - (\mathfrak{A}\,\mathfrak{D})(\mathfrak{B}\,\mathfrak{C})$$

the representation

$$\frac{\partial^2 G}{\partial n_Q\,\partial n_P} = -(\mathfrak{n}_Q \nabla_Q)(\mathfrak{n}_P \nabla_Q) G = ([\mathfrak{n}_Q \nabla_Q], [\mathfrak{n}_P \nabla_Q]) G - (\mathfrak{n}_Q \mathfrak{n}_P)(\nabla_Q \nabla_Q) G.$$

For the second term on the right, one uses the wave equation (1) valid for G. One integrates by parts the part of the integral associated with the first term by applying the ∇_Q on the left with a change of sign to f instead of to the terms to its right in the last equation. One obtains

$$\begin{aligned} \left(\frac{\partial v}{\partial n}\right)_P = & -\int([\mathfrak{n}_P \operatorname{grad}_Q G(P,Q)], [\mathfrak{n}_Q \operatorname{grad} f(Q)])\, dQ + \\ & + k^2 \int G(P,Q)(\mathfrak{n}_P \mathfrak{n}_Q) f(Q)\, dQ - \\ & + \int \frac{\partial G(P,Q)}{\partial n_P} g(Q)\, dQ. \end{aligned} \tag{9}$$

We apply the conclusions obtained for the integral of type (6) to the representation (5) of the solution of the boundary value problem. First we see that (1) is automatically satisfied. Further, it is required that on the body u and $\partial u/\partial n$ actually take on the outer boundary values f and g, or what is the equivalent as a result of (7) that both take on the inner boundary value 0. Since (5) satisfies the wave equation in the interior of the body, it is sufficient to require that the boundary value for either u or $\partial u/\partial n$ be 0. It follows immediately, excluding the special case that k is an eigenvalue of the interior of the body, that u vanishes identically in the interior of the body and that therefore the other quantity u or $\partial u/\partial n$ automatically takes on the inner boundary value of 0. Finally due to (8), we can instead require that on the body surface itself $u = 1/2\ f$ or that $\partial u/\partial n = 1/2\ g$. In view of (5), (6), and (9), this leads to the two equivalent integral equations

$$\frac{1}{2} f(Q) = u_0(P) + \int\left\{\frac{\partial G(P,Q)}{\partial n_Q} f(Q) - G(P,Q) g(Q)\right\} dQ \tag{10}$$

and

$$\begin{aligned} \frac{1}{2} g(P) = \left(\frac{\partial u_0}{\partial n}\right)_P - & \int([\mathfrak{n}_P \operatorname{grad}_Q G(P,Q)], \\ & [\mathfrak{n}_Q \operatorname{grad} f(Q)])\, dQ + \\ & + k^2 \int G(P,Q)(\mathfrak{n}_P \mathfrak{n}_Q) f(Q)\, dQ - \\ & - \int \frac{\partial G(P,Q)}{\partial n_P} g(Q)\, dQ, \end{aligned} \tag{11}$$

associated with the surface of the body. Each of these equations, together with one of the foregoing boundary conditions associated with the body, uniquely determines the surface quantities f and g, since all the requirements mandated of f and g are satisfied and the boundary value problem is assumed to be uniquely solvable.

Expressing the boundary condition in the form

$$a\,u + b\,\frac{\partial u}{\partial n} = 0, \tag{12}$$

there arises besides (10) and (11), a second stipulation for f and g

$$a f + b g = 0. \tag{13}$$

In the most general case, a and b are complex functions on the body surface. They are to be chosen such that the boundary value problem has a unique solution. Thus independent scattering from the body without outside illumination is not possible. If specifically a and b are real, there is neither transmission nor scattering at the body, but only pure reflection.

Of the two equations (10) and (11), generally (10) is the simpler one to use. Equation (10) is a pure integral equation of the second kind while (11) contains derivatives of f. Only for the boundary condition $u = 0$ is (11) preferable. Namely in this case, (10) degenerates to an integral equation of the first kind and the derivatives of the unknown vanish in (11). Making the choice between (10) and (11) in this manner, one avoids hereby the previously introduced conditionally convergent integrals.

Finally, let us also examine the two simplest special cases of the boundary condition. The formulation of the diffraction problem obtained for the boundary condition $u = 0$ is

$$f(P) = 0, \quad \frac{1}{2} g(P) + \int \frac{\partial G(P,Q)}{\partial n_P} g(Q)\, dQ = \left(\frac{\partial u_0}{\partial n}\right)_P \tag{14}$$

and for the boundary condition $\partial u/\partial n = 0$

$$\frac{1}{2} f(P) - \int \frac{\partial G(P, Q)}{\partial n_Q} f(Q)\, dQ = u_0(P), \quad g(P) = 0. \tag{15}$$

In the following discussions of the scalar diffraction problem, we restrict ourselves to these simple cases.

All the derived expressions are valid also where the diffracting body is infinite in extent. If the radiation source is at a finite distance, then G and f vanish at infinity as $1/r$, $\partial G/\partial n$ and g vanish at least as $1/r$ which together with the oscillations of the exponential in G is sufficient for the convergence of the surface integrals. If the radiation source is at infinity, one can use a convergence factor which one allows subsequently to approach unity.

2. The Electromagnetic Diffraction Problem

In vacuum, Maxwell's equations for the electromagnetic field $\mathfrak{E}$, $\mathfrak{H}$ can be written as

$$\operatorname{rot} \mathfrak{E} = i k \mathfrak{H}, \quad \operatorname{rot} \mathfrak{H} = -i k \mathfrak{E}. \tag{16}$$

The field inside a spatial region can be specified when the tangential components of $\mathfrak{E}$ and $\mathfrak{H}$ on the boundary surface are known. We deduce the relationships corresponding to the Kirchhoff integral formulation (2) from Eqs. (17) and (18) of Franz (Ref. 4, p. 602) which we specialize for a homogeneous medium and rewrite in our notation as:

$$\left\{\begin{aligned} \mathfrak{E} &= \int [\operatorname{grad}_Q G\, [\mathfrak{E}\, \mathfrak{n}_Q]]\, dQ - \\ &\quad - \frac{1}{ik} \operatorname{rot} \int [\operatorname{grad}_Q G\, [\mathfrak{H}\, \mathfrak{n}_Q]]\, dQ, \\ \mathfrak{H} &= \int [\operatorname{grad}_Q G\, [\mathfrak{H}\, \mathfrak{n}_Q]]\, dQ + \\ &\quad + \frac{1}{ik} \operatorname{rot} \int [\operatorname{grad}_Q G\, [\mathfrak{E}\, \mathfrak{n}_Q]]\, dQ. \end{aligned}\right. \tag{17}$$

G is again given by (3), n is the inward normal. To simplify the expressions, the explicit dependence of the individual quantities on the source point P and the integration point Q is omitted. This will befollowed when there is no chance of misunderstanding. G is always dependent on P and Q, the other quantities are dependent on Q when they are under the integral, and on P when they are outside the integral. Correspondingly, differentiation of these quantities under the integral are with respect to Q, and outside the integral with respect to P. In the differentiation of G and the normal vector n, the subscript P or Q is attached.

We assume that the diffracting body is perfectly conducting so that on the body surface

$$[\mathfrak{E}\, \mathfrak{n}] = 0 \tag{18}$$

For the boundary value of the tangential components of $\mathfrak{H}$, with n as the outward normal on the body the following holds

$$[\mathfrak{H}\, \mathfrak{n}] = -\mathfrak{J}. \tag{18a}$$

$\mathfrak{J}$ is the electric surface current and has the role of the unknown surface layer. The incident wave is $\mathfrak{E}_0$, $\mathfrak{H}_0$. Applying (17) to the diffraction problem then yields

$$\left\{\begin{aligned} \mathfrak{E} &= \mathfrak{E}_0 + \frac{1}{ik} \operatorname{rot} \int [\operatorname{grad}_Q G, \mathfrak{J}]\, dQ, \\ \mathfrak{H} &= \mathfrak{H}_0 - \int [\operatorname{grad}_Q G, \mathfrak{J}]\, dQ, \end{aligned}\right. \tag{19}$$

where

$$\mathfrak{J} \perp \mathfrak{n} \tag{20}$$

and G is again given by the expression in (4).

As in (5) so also in (19), the form of the expression ensures that the differential equations required for the exterior region, that is Eq. (16), are now satisfied. Moreover, (19) also satisfies these equations in the interior of the body. Furthermore, for the tangential components of the integral expression arising in (19)

$$\mathfrak{A} = \int [\operatorname{grad}_Q G, \mathfrak{J}]\, dQ \tag{21}$$

and for its curl, the following boundary conditions hold on the surface,

$$[\mathfrak{A}\, \mathfrak{n}]_1 - [\mathfrak{A}\, \mathfrak{n}]_2 = \mathfrak{J}, \quad [\operatorname{rot} \mathfrak{A}, \mathfrak{n}]_1 - [\operatorname{rot} \mathfrak{A}, \mathfrak{n}]_2 = 0. \tag{22}$$

provided only that $\mathfrak{J}$ satisfies (20) but is arbitrary otherwise. The subscript 1 denotes the side of the surface to which n is pointing.

For the curl to have meaning on the surface itself, it must be recast. Changing the differentiation with respect to P to one with respect to Q and using a vector identity yields

$$\begin{aligned} \operatorname{rot}_P [\operatorname{grad}_Q G, \mathfrak{J}] &= G \left[\overset{\leftarrow}{V}_P \left[\overset{\leftarrow}{V}_Q \mathfrak{J}\right]\right] = -G \left[\overset{\leftarrow}{V}_Q \left[\overset{\leftarrow}{V}_Q \mathfrak{J}\right]\right] \\ &= -G \overset{\leftarrow}{V}_Q \left(\overset{\leftarrow}{V}_Q \mathfrak{J}\right) + G \left(\overset{\leftarrow}{V}_Q \overset{\leftarrow}{V}_Q\right) \mathfrak{J}. \end{aligned}$$

Here the arrows denote that ∇ operates on G and not on F. In the second term on the right (1) can be used for G. In the integral one performs a partial integration on the first term and applies the right-most ∇ with a change of sign to $\mathfrak{J}$ instead of to the function on the left. One obtains hereby

$$\operatorname{rot} \mathfrak{A} = \int \operatorname{div} \mathfrak{J} \cdot \operatorname{grad}_Q G\, dQ - k^2 \int G \cdot \mathfrak{J}\, dQ. \tag{23}$$

The integrals (21) and (23) like (9), are only conditionally convergent on the surface and are to be understood as limit values in the same sense as (9). Singularities arise in the normal components of (21) and in the tangential components of (23). The values on the surface are again the averages of the boundary values on either side. Specifically for the tangential components on the surface, in the second equation with reference to (22), it holds that

$$\left\{\begin{aligned} [\mathfrak{A}\, \mathfrak{n}] &= \tfrac{1}{2} ([\mathfrak{A}\, \mathfrak{n}]_1 + [\mathfrak{A}\, \mathfrak{n}]_2) \\ [\operatorname{rot} \mathfrak{A}, \mathfrak{n}] &= [\operatorname{rot} \mathfrak{A}, \mathfrak{n}]_1 = [\operatorname{rot} \mathfrak{A}, \mathfrak{n}]_2. \end{aligned}\right. \tag{24}$$

As a consequence of (18) and (18a) it is required that the quantities $[\mathfrak{E} n]$ and $[\mathfrak{H} n]$ deduced from (19), take the limiting boundary values of 0 and $-\mathfrak{J}$ on the body. From (19), (21), and (22) the corresponding inner boundary values for both quantities are zero. Corresponding exactly to the conditions in the scalar diffraction problem, it is again sufficient to require that the inner boundary value of one of the two quantities vanish because (19) satisfies (16) inside the body, which then guarantees that the field (19) identically vanishes inside the body. It is assumed here as in Section 1 that k is not an eigenvalue of the interior region of the body. With reference to (24), the values associated with $[\mathfrak{E}n]$ and $[\mathfrak{H}n]$ on the surface are 0 and $-1/2\,\mathfrak{F}$. Following (19), (21) and (23), two integral equations on the body surface are obtained. Written explicitly they are,

$$\left\{\begin{aligned} & ik\int G(P,Q)\,[\mathfrak{F}(Q),\mathfrak{n}_P]\,dQ + \\ & \quad + \frac{1}{ik}\int [\mathrm{grad}_Q\, G(P,Q),\mathfrak{n}_P]\,\mathrm{div}\,\mathfrak{F}(Q)\,dQ = \\ & -[\mathfrak{E}_0(P),\mathfrak{n}_P], \end{aligned}\right. \tag{25}$$

$$\begin{aligned} & \tfrac{1}{2}\mathfrak{F}(P) - \int [[\mathrm{grad}_Q\, G(P,Q),\mathfrak{F}(Q)],\mathfrak{n}_P]\,dQ = \\ & -[\mathfrak{H}_0(P),\mathfrak{n}_P]. \end{aligned} \tag{26}$$

Since we have already taken into consideration the boundary condition (18), each of these equations constitutes a complete formulation of the boundary value problem. Equation (26) is the simpler equation. It is a pure integral equation of the second kind and contains no conditionally convergent integrals.

3. Diffracting Objects with Edges

Until now it was tacitly assumed that the normal direction is uniquely determined at every point on the surface, that is, the surface is everywhere finitely curved. Now it will be assumed that there are edges.

The kernel of the integral equations (14), (15), and (26) for a smooth surface has a singularity only at $Q=P$ and there it behaves as l/r_{PQ}. The inhomogeneous terms $\partial u_0/\partial n$, u_0 and $-[\mathfrak{H}_0 n]$ are continuous on the surface of the body. The unique solvability of the integral equations by the continuous functions g, f, and $\mathfrak{F}$, arose from a derivation of the equations, based on two assumptions, namely, that the posed boundary value problem is uniquely solvable and that k is not an eigenvalue of the interior of the body. In the case of edges, the kernel becomes singular there and likewise the quantities $\partial u_0/\partial n$ and $-[\mathfrak{H}_0 n]$ since there is a jump in the normal direction on the edge. Additional discontinuities of the kernel itself need not have special impact on the solution, since an integral composed of such a discontinuous kernel is nevertheless in general continuous. Thus, for example, the solution f from (15) is continuous in the presence of edges. But precisely because of this insensitivity of the integral to the singularities of the kernel, the Eqs. (14) and (16) with a singular inhomogeneous term can be solved only with functions which become infinite on the edge.

One goes from a smooth body to one with edges by taking the limit of a curvature that is finite everywhere to one that is piecewise infinite. To handle such a transition rigorously from the standpoint of an integral equation would be difficult and unnecessarily complicated and is therefore avoided. Rather, the simple methods of planar potential theory will be applied to extend the integral equation formulation for a diffraction problem to bodies with edges.

Two questions need to be answered: first, one must elucidate whether the solutions f, g, and $\mathfrak{F}$ become singular on the edge and what is the nature of these singularities. Second, the possibility of the existence of line integrals along the edge, in addition to the surface integrals in the integral equation, must be discussed.

The solution u of the wave equation (1) in the neighborhood of a point on the edge will be examined. The extent of the domain under consideration is small compared to the radius of curvature of the edge contour and that of the proximal surfaces and it is also small compared to the wavelength. In this small region, u can be taken as the solution of the two-dimensional potential equation on a plane perpendicular to the edge (the x-y plane); that is, it can be considered specifically independent of the z coordinate in the direction of the edge. Then using the planar polar coordinates (r,ϕ), u can be constructed from the functions

$$\begin{array}{llll} r^{\pm\nu}\cos\nu\varphi, & r^{\pm\nu}\sin\nu\varphi, & \text{for} & \nu>0 \\ 1, & \lg r & \text{for} & \nu=0. \end{array} \tag{27}$$

The admissible values of ν are specified from the boundary conditions on the body (see Fig. 1). Exactly one-half of the expansion coefficients required for the construction of the solution u for a diffraction problem, are always determined in principle by the conditions at a large distance from the diffracting object (illuminating source + radiation condition) and the other half by the conditions on the object. Hence one may eliminate all the functions in (27) with negative exponents in r (including the function $lg\ r$) by setting the corresponding numerical coefficients to zero, so that u also remains finite on the edge. Thereby f remains also finite and thus the case of the boundary condition $\partial u/\partial n=0$ is disposed of. For the boundary condition $u=0$ we need to set

$$g=\frac{\partial u}{\partial n}=-\frac{1}{r}\frac{\partial u}{\partial|\varphi|}, \tag{28}$$

which goes to infinity as $r^{\nu-1}$ for values of $\nu<1$ for the remaining terms in (27). The only function to which this applies, and which satisfies the boundary condition $u=0$, is

$$r^{\nu}\cdot\cos\nu\varphi \quad \text{with} \quad \nu=\frac{\pi}{\pi+2\alpha}, \tag{29}$$

where α is defined in Fig. 1. We summarize the derived results for the behavior of f and g on the edge as

$$f\ \text{finite}, \qquad g\lesssim s^{-\frac{2\alpha}{\pi+2\alpha}}, \tag{30}$$

where s denotes the distance from the edge. (30) also holds for the case of a body with reentrant edges; then α is negative.

It should also be noted, that (30) does not play the role of a limiting condition on the edge. Boundary or limiting conditions are unnecessary for integral equations. The purpose of (30) is rather to limit the class of functions that are allowed as solutions. Thus not the continuity of f and $g\bullet s^{\frac{2\alpha}{\pi+2\alpha}}$ is required, but only their finiteness. The continuity of the solution follows automatically as in the case of smooth bodies where f and g become continuous even when one only requires piecewise continuity of them.

The question regarding a possible added term in the integral equation associated with the edge remains to be clarified. For f such a term is to be excluded at once since f remains finite. For g the presence of such an added term would mean that in an infinitesimally small neighborhood of the edge, the integral

$$\int g(Q)\,dQ \tag{31}$$

is finite. To arrive at a more rigorous formulation, some preliminaries are necessary.

In the x-y plane, the cartesian coordinates are conformally mapped to a curvilinear coordinate system ξ, η with the line element

$$ds^2=e^2(d\xi^2+d\eta^2) \tag{32}$$

The surface of the body is given by the (bent) η-axis where $\xi=0$ holds (see Fig. 1). Additionally on the edge itself, $\eta=0$. If one considers an arbitrary curve $\xi=\xi_0$ instead of $\xi=0$ as the surface of considers an arbitrary curve $\xi=\xi_0$ instead of $\xi=0$ as the surface of the body, then the lowest order solution in ξ, η for the potential equation

$$\frac{\partial^2 u}{\partial \xi^2}+\frac{\partial^2 u}{\partial \eta^2}=0 \qquad 33)$$

satisfying the edge condition $u=0$ is simply

$$u=\xi-\xi_0. \qquad 34)$$

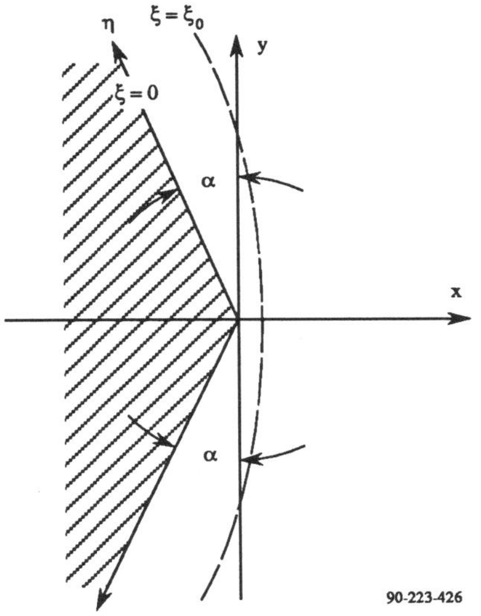

Fig. 1. Cross-sectional cut perpendicular to the edge.

This solution agrees with (29) for $\xi_0=0$, that is, with the admissible solutions (27) with lowest order in r.

To answer the question of the edge term, one first has to extend (31) for $\xi_0\neq 0$ over a finite domain $-\eta_0<\eta<\eta_0$. Then one passes from a rounded edge to a knife edge ($\xi_0\to 0$), finally shrinking the finite domain of integration ($\eta_0\to 0$). From (32) and (34), one obtains

$$g=\frac{\partial u}{\partial n}=\frac{1}{e}\frac{\partial u}{\partial \xi}=\frac{1}{e} \qquad 35)$$

and furthermore for $\xi=\xi_0$ using (32)

$$ds=e\cdot d\eta. \qquad 36)$$

From this one obtains the desired integral (31) in the notation of the planar problem independent of ξ_0 as

$$\int_{-\eta_0}^{\eta_0} g\,ds=2\eta_0, \qquad 37)$$

which vanishes for $\eta_0\to 0$. Hence, the edge does not produce an additional term for g either. Since the $\xi-\eta$ grid is infinitely dense on the edge, the scaling magnitude e vanishes here. It is therefore of prime importance in the above result, that e in (35) and (36) behave reciprocally. Should second instead of first derivatives appear in (35) this would no longer be the case.

As in the scalar case, the electromagnetic one can be treated as a planar problem and reduced to such. To this end, one introduces a unit vector z perpendicular to the x-y plane which together with the normal vector n and $[zn]$ form orthogonal coordinate axes on the body surface.

Decomposition of the surface current $\mathfrak{J}$ and the tangential components of $\mathfrak{H}_0$ into edge-normal and edge-parallel terms yields

$$\begin{cases}\mathfrak{H}_0=H^0_\perp\,[\mathfrak{z}\,\mathfrak{n}]+H^0_\parallel\,\mathfrak{z}+H^0_n\,\mathfrak{n},\\ \mathfrak{J}=F_\perp\,[\mathfrak{z}\,\mathfrak{n}]+F_\parallel\,\mathfrak{z}.\end{cases} \qquad 38)$$

Through this decomposition (26) splits into two equations where deleting the term

$$\int F_\perp \frac{\partial G}{\partial z_Q}[\mathfrak{n}_Q\,\mathfrak{n}_P]\,dQ \qquad 39)$$

gives

$$\frac{1}{2}F_\parallel+\int\frac{\partial G}{\partial n_P}F_\parallel\,dQ=H^0_\perp, \qquad 40)$$

$$\frac{1}{2}F_\perp-\int\frac{\partial G}{\partial n_Q}F_\perp\,dQ=-H^0_\parallel. \qquad 41)$$

The expression (39) contains the term $\partial G/\partial z_0$. Through integration by parts one can shift the differentiation with respect to z to $F_\perp$. Since in the spirit of the planar approximation $F_\perp$ is taken as constant in the direction parallel to the edge, (39) is consequently eliminated. A remainder term resulting from the partial integration does not arise since the edge measured along z does not extend to infinity but encompasses the body. Thus z represents a periodic variable.

Equations (40) and (41) have the same form as (14) and (15) and result from them when one replaces g, f, $\partial u_0/\partial n$, u_0 by $F_\parallel$, $F_\perp$, $H^0_\perp$, $-H^0_\parallel$, respectively. As $\partial u_0\partial n$ so does $H^0_\perp$ have a jump discontinuity on the edge; while $\mathfrak{H}_0$ itself is unchanging, n has a jump discontinuity. In contrast, $H^0_\parallel$ like u_0 is continuous since Z has a constant direction. Analogous to (30) one has

$$F_\perp \text{ finite}, \qquad F_\parallel \lesssim s^{-\frac{2\alpha}{\pi+2\alpha}}. \qquad 42)$$

With reference to (38), it is evident that the surface current along the edge may become infinite, while the current perpendicular to the edge remains finite.

In connection with the calculations of Meixner[1] for the electromagnetic diffraction on a circular disk, it is interesting to consider the field in the neighborhood exterior to the edge, which corresponds to the unbounded component $F_\parallel$. From (18a) and the second equation of (38), $F_\parallel$ is the tangential magnetic field along the bent η-axis in Fig. 1. The normal component of the magnetic field vanishes on the body as a consequence of (18) and the first equation of (16). Thus collecting the complex cartesian components of $\mathfrak{H}$ (see Fig. 1) yields

$$H_x-iH_y=-iF_\parallel e^{\mp i\alpha}, \qquad 43)$$

where the $\mp$ signs in the exponent correspond to the positive or negative η-axis, respectively. The solution of the potential equation with (43) as a boundary condition, is

$$H_x-iH_y=-iC_1(x+iy)^{-\frac{2\alpha}{\pi+2\alpha}}. \qquad 44)$$

where C_1 denotes the proportionality factor omitted in (42) and specifically assumed here to be real. The energy density of the field is proportional to the magnitude

$$H_x^2+H_y^2=C_1^2\cdot r^{-\frac{4\alpha}{\pi+2\alpha}} \qquad (45)$$

Since α can be at most $\pi/2$, the energy density is integrable over the plane and thus satisfies the Meixner condition required of a solution of the diffraction problem. This requirement of finite field energy density allows Meixner to choose from a large number of solutions of Maxwell's equations which are singular on the edge those that are physically meaningful and sufficient. Through (45), the field behavior is more precisely specified.

A totally similar situation holds for the electric field $\mathfrak{E}$, it corresponds to the solution $F_\perp$ of (41), which is finite and odd at the edge and which behaves like $\pm\, s^{\frac{\pi}{\pi+2\alpha}}$. From the associated magnetic field H_z one computes from (16) the infinite electric field on the edge,

$$E_x - i E_y = C_2 (x + i y)^{-\frac{2\alpha}{\pi + 2\alpha}} \qquad 46)$$

for real C_2.

4. The Diffraction from Edged Planar Surfaces

Edged planar surfaces constitute a special case of a body with edges. In this special case $\alpha = \pi/2$ so that on the edge the following hold

$$\left\{\begin{array}{ll} f \ \text{finite}, & g \lesssim s^{-1/2}, \\ F_\perp \ \text{finite}, & F_\parallel \lesssim s^{-1/2}. \end{array}\right. \qquad 47)$$

In Eq. (15), which will be considered next, the integral spans both sides of the planar surface, where we distinguish between the upper surface (+) and the lower surface (−). Since the integration (source) point Q covers the same surface twice, $\partial G/\partial n_Q$ takes on the same value on either surface except for sign. It is therefore useful to introduce the notation f_+ and f_- for the values of f on the upper and lower surfaces and integrating only once over the surface using the difference $f_+ - f_-$. If we specify in this integration the normal direction as pointing out of the upper surface, then the integral in (15) becomes

$$\int \frac{\partial G}{\partial n_Q} (f_+ - f_-)\, d Q . \qquad 48)$$

One point remains to be clarified here. Consider that a surface was formed by bringing together the upper and lower surfaces of a body of finite thickness d (see Fig. 2). Let the field point P lie on the upper surface. Then in integrating over this surface, P is viewed as a surface point; the integration over the lower surface is different. Here the limiting value of the integral is used which is obtained as P approaches the lower surface from above. If one considers the field point P to be identical with the surface point, then one has to correct the term associated with the lower surface in (48) with the difference between its value on the upper boundary and that on the surface. Then following from (6), (7), and (8), one adds the term $-1/2\, f_-$ to (48). Then using the shorthand notation,

$$f_+ + f_- = \psi, \qquad f_+ - f_- = \varphi \qquad 49)$$

one obtains from (15), the integral equation for the boundary condition $\partial u/\partial n = 0$ as

$$\frac{1}{2}\psi - \int \frac{\partial G}{\partial n_Q} \varphi\, d Q = u_0 . \qquad 50)$$

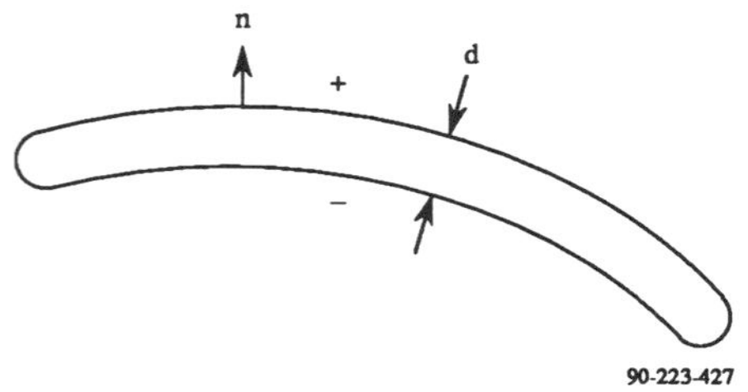

Fig. 2. Limiting process for an edged surface.

If one specifies the field point before passing to the limit on the lower surface, instead of on the upper one, one obtains the same equation since (50) is symmetric relative to both sides. To obtain a second equation, one can also consider the term linear in d as one passes to the limit $d \to 0$ (see Fig. 2). It is simpler to use Eq. (11) with $g = 0$ which leads to the same result.

Proceeding similarly as in the derivation of (50) and using the fact that due to $g = 0$ in (9), (7) and (8), the integrals associated with the integration surfaces do not have jump discontinuities, one obtains as the second equation

$$\int ([\mathfrak{n}_P \operatorname{grad}_Q G], [\mathfrak{n}_Q \operatorname{grad} \varphi])\, d Q - k^2 \int G (\mathfrak{n}_P \mathfrak{n}_Q) \varphi\, d Q = \frac{\partial u_0}{\partial n_P} . \qquad 51)$$

For a wave u in the exterior region, (5) yields

$$u = u_0 + \int \frac{\partial G}{\partial n_Q} \varphi\, d Q . \qquad 52)$$

Since ψ is not present in either (51) or (52), its evaluation is entirely unnecessary and Eq. (50) is redundant.

The situation is totally comparable for the boundary condition $u = 0$ and for the electromagnetic case. Of the two unknown surface layers and the two integral equations, only a specific one is needed. Hence in a particular instance only the layer corresponding to will be introduced and only the integral equation corresponding to (51) need be considered. Using the abbreviation

$$g_+ + g_- = \gamma \qquad 53)$$

equation (10) with $f = 0$ yields

$$\int G \gamma\, d Q = u_0 . \qquad 54)$$

In the derivation one uses the fact that from (6), (7), and (8) the integral is continuous on the integration surface. From (5) the spatial wave is

$$u = u_0 - \int G \gamma\, d Q . \qquad 55)$$

In (54) the field point lies on the surface; in (55) it is in space.

The integral equation for the electromagnetic case arises from (25) by noting the continuity of the integrals on the surface which follows from (23) and (24). Using the abbreviation

$$\mathfrak{J} = \mathfrak{J}_+ + \mathfrak{J}_- \qquad 56)$$

one obtains an equation valid on the surface as

$$i k \int G [\mathfrak{J}\, \mathfrak{n}_P]\, d Q + \frac{1}{i k} \int [\operatorname{grad}_Q G, \mathfrak{n}_P] \cdot \operatorname{div} \mathfrak{J}\, d Q = - [\mathfrak{E}_0\, \mathfrak{n}_P] . \qquad 57)$$

From (19) the magnetic field in space becomes

$$\mathfrak{H} = \mathfrak{H}_0 - \int [\operatorname{grad}_Q G, \mathfrak{J}]\, d Q . \qquad 58)$$

Here it is physically obvious that the field is determined only by the total current $\mathfrak{J}$ and that the difference $\mathfrak{J}_+ - \mathfrak{J}_-$ plays no role.

The behavior of the newly introduced surface layers on the edge contour follows from (47):

$$\begin{array}{ll} \varphi \ \text{finite}, & \gamma \lesssim s^{-1/2}, \\ J_\perp \ \text{finite}, & J_\parallel \lesssim s^{-1/2}. \end{array} \qquad 59)$$

For the convergence of the second integral in (57) on the boundary, it is essential that $J_\perp$ remain finite; since $J_\perp \sim s^{-1/2}$ would lead to $\operatorname{div} \mathfrak{J} \sim s^{-3/2}$.

While in the foregoing sections the option always existed to use the simpler of the two available integral equations, such a choice no longer exists for the case of diffraction from finite surfaces. One is generally forced to use Eqs. (51), (54), and (57). These equations are inconvenient for three reasons.

First, conditionally convergent integrals arise in (51) and (57) of the type discussed in Section 1 whose values must be determined by passing to a limit. The approach, described in Section 1 where the integrals were subjected to a two-fold rather than a single partial integration before P can be admitted to be a point on the surface, would generate a new divergence at the boundary while totally eliminating the one at $P=Q$. Second, (51) and (57) are not pure integral equations since derivatives of the unknowns are present. Third, all three integral equations are of the first kind.

In the special case of a flat finite surface, (51) simplifies to

$$\int (\mathrm{grad}_Q\, G, \mathrm{grad}\, \varphi)\, dQ - k^2 \int G \varphi\, dQ = \frac{\partial u_0}{\partial n_P}, \qquad 60)$$

while (54) and (57) remain unchanged.

5. The Diffraction by a Cylinder as an Example

Finally, the foregoing general relationships will be illustrated through a familar example. For this we choose the electromagnetic diffraction by a perfectly conducting cylinder of infinite length, subject to perpendicular illumination.

The cylinder radius is a; its axis is along the z-axis of the cartesian coordinate system x,y,z. The incident radiation is along the x-direction. Depending on the polarization, the magnetic field of the incident radiation is

$$\mathfrak{H}_{0y} = A\, e^{ikx} \qquad \text{or} \qquad \mathfrak{H}_{0z} = B\, e^{ikx} \qquad 61)$$

with arbitrary amplitude A or B. The integral equation (26) decomposes into two equations associated with the two cases of polarization. Both cases can be interpreted as examples of two scalar diffraction problems with the boundary conditions $u=0$ and $\partial u/\partial n=0$. One obtains these equations in the same way as Eqs. (40) and (41). The integral corresponding to (39) vanishes since $\mathfrak{J}$ and likewise $\mathfrak{H}_0$ are independent of z and $\partial G/\partial z_0$ is odd with respect to $z_Q - z_P$. Using the cylindrical coordinates r, ϕ, z and denoting the field point P with upper case letters and the integration point Q with lower case, the equations are:

$$\left\{\begin{aligned} &\frac{1}{2} F_z(\Phi) + a \int \left(\frac{\partial G(r_{PQ})}{\partial R}\right)_{R=r=a} \cdot F_z(\varphi)\, d\varphi\, dz = \mathfrak{H}_{0\varphi}(a,\Phi), \\ &\frac{1}{2} F_\varphi(\Phi) - a \int \left(\frac{\partial G(r_{PQ})}{\partial r}\right)_{R=r=a} \cdot F_\varphi(\varphi)\, d\varphi\, dz = -\,\mathfrak{H}_{0z}(a,\Phi). \end{aligned}\right. \qquad 62)$$

Carrying out the integration over z, the Green's function (4) in three dimensional space reduces to the one for a plane,

$$\begin{aligned} &\bar{G}(P,Q) = \frac{i}{4} H_0^{(1)}(k\, r_{PQ}), \\ &r_{PQ} = \sqrt{R^2 + r^2 - 2Rr\cos(\varphi - \Phi)} \end{aligned} \qquad 63)$$

$H_0^{(1)}$ is the Hankel function of the first kind and order zero. Its factor in (63) is evaluated so that at $P=Q$, $\bar{G}$ behaves like $\frac{1}{2\pi}\, lg\, \frac{\text{const.}}{r_{PQ}}$. Due to the symmetry of r_{PQ} in R and r,

$$\frac{\partial \bar{G}}{\partial R} = \frac{\partial \bar{G}}{\partial r} = \frac{1}{2}\left(\frac{\partial \bar{G}}{\partial R} + \frac{\partial \bar{G}}{\partial r}\right) = \frac{1}{2}\frac{\partial}{\partial a}\left\{\bar{G}_{R=r=a}\right\}. \qquad 64)$$

for $R=r=a$. On the cylindrical surface one sets

$$r_{PQ} = 2a\left|\sin\frac{\chi}{2}\right| \qquad \text{with} \qquad \chi = \varphi - \Phi \qquad 65)$$

and referring to (61) and (63)-(65), (62) becomes

$$\left.\begin{aligned} &\frac{1}{2} F_z(\Phi) + \frac{ia}{8}\int_0^{2\pi} \frac{\partial}{\partial a} H_0^{(1)}\left(2ka\sin\frac{\chi}{2}\right) \cdot F_z(\Phi+\chi)\, d\chi = A\, e^{ika\cos\Phi} \\ &\frac{1}{2} F_\varphi(\Phi) - \frac{ia}{8}\int_0^{2\pi} \frac{\partial}{\partial a} H_0^{(1)}\left(2ka\sin\frac{\chi}{2}\right) \cdot F_\varphi(\Phi+\chi)\, d\chi = -B\, e^{ika\cos\Phi}. \end{aligned}\right. \qquad 66)$$

Expanding the right hand sides and the unknown functions in a Fourier series:

$$\left\{\begin{aligned} &e^{ika\cos\Phi} = \sum_{n=-\infty}^{\infty} i^n J_n(ka)\, e^{in\Phi}, \\ &e^{ika\cos\Phi}\cos\Phi = \sum_{n=-\infty}^{\infty} i^{n-1} J_n'(ka)\, e^{in\Phi}, \\ &F_z(\Phi) = \sum_{n=-\infty}^{\infty} a_n e^{in\Phi}, \qquad F_\varphi(\Phi) = \sum_{n=-\infty}^{\infty} b_n e^{in\Phi}, \end{aligned}\right. \qquad 67)$$

and using the abbreviation

$$C_n = \frac{ia}{4}\int_0^{2\pi} \frac{\partial}{\partial a} H_0^{(1)}\left(2ka\sin\frac{\chi}{2}\right) e^{in\chi}\, d\chi \qquad 68)$$

one obtains by equating coefficients

$$\left\{\begin{aligned} &\tfrac{1}{2}(1 + C_n)\, a_n = A\, i^{n-1} J_n'(ka), \\ &\tfrac{1}{2}(1 - C_n)\, b_n = -B\, i^n J_n(ka). \end{aligned}\right. \qquad 69)$$

J_n is the n-th order Bessel function. Here and subsequently, primes denote differentiation.

One evaluates (68) most easily with the aid of the addition theorem

$$\begin{aligned} &H_0^{(1)}\left(\sqrt{R^2 + r^2 - 2Rr\cos\chi}\right) = \\ &\sum_{n=-\infty}^{\infty} J_n(kr)\, H_n^{(1)}(kR)\, e^{in\chi} \\ &\text{für} \quad R \geq r, \quad 0 < \chi < 2\pi, \end{aligned} \qquad 70)$$

that for $R=r=a$ becomes

$$H_0^{(1)}\left(2ka\sin\frac{\chi}{2}\right) = \sum_{n=-\infty}^{\infty} J_n(ka)\, H_n^{(1)}(ka)\, e^{in\chi} \qquad 71)$$

Using the fact that

$$J_{-n} = (-1)^n \cdot J_n, \qquad H_{-n}^{(1)} = (-1)^n \cdot H_n^{(1)} \qquad 72)$$

one obtains

$$C_n = \frac{i\pi ka}{2}\left[J_n'(ka)\, H_n^{(1)}(ka) + J_n(ka)\, H_n^{(1)\prime}(ka)\right]. \qquad 73)$$

The identity

$$J_n'(ka)\, H_n^{(1)}(ka) - J_n(ka)\, H_n^{(1)\prime}(ka) = \frac{2}{i\pi ka} \qquad 74)$$

for the Wronskian further yields

$$\begin{cases} 1 + C_n = i\pi k a J_n'(ka) H_n^{(1)}(ka), \\ 1 - C_n = -i\pi k a J_n(ka) H_n^{(1)\prime}(ka), \end{cases} \tag{75}$$

Using this, the solutions for the surface currents follow from (69) and (67) as

$$\begin{cases} F_z(\Phi) = -\dfrac{2A}{\pi k a} \displaystyle\sum_{n=-\infty}^{\infty} \dfrac{i^n e^{in\Phi}}{H_n^{(1)}(ka)} \\ F_\varphi(\Phi) = -\dfrac{2iB}{\pi k a} \displaystyle\sum_{n=-\infty}^{\infty} \dfrac{i^n e^{in\Phi}}{H_n^{(1)\prime}(ka)} \end{cases} \tag{76}$$

From (76) one can compute the magnetic field of the diffracted wave $\mathfrak{H}^b = \mathfrak{H} - \mathfrak{H}_0$ with the use of the second equation in (19), cast in a convenient form as

$$\begin{aligned} \mathfrak{H}_r^b(R,\Phi) &= \frac{a}{R} \int \frac{\partial \bar{G}(R,\Phi; a,\varphi)}{\partial \Phi} F_z(\varphi)\, d\varphi \\ \mathfrak{H}_\varphi^b(R,\Phi) &= -a \int \frac{\partial \bar{G}(R,\Phi; a,\varphi)}{\partial R} F_z(\varphi)\, d\varphi \\ \mathfrak{H}_z^b(R,\Phi) &= -a \int \frac{\partial \bar{G}(R,\Phi; a,\varphi)}{\partial a} F_\varphi(\varphi)\, d\varphi \end{aligned} \tag{77}$$

In the first equation of (77), $\mathrm{grad}_Q\, G = -\mathrm{grad}_P\, G$ is used. Substituting $\bar{G}$ from (63), the integrals can be evaluated with the aid of (70) yielding the expressions

$$\begin{cases} \mathfrak{H}_r^b(R,\Phi) = \dfrac{A}{kR} \displaystyle\sum_{n=-\infty}^{\infty} n\, i^n \dfrac{J_n(ka)}{H_n^{(1)}(ka)} H_n^{(1)}(kR)\, e^{in\Phi} \\ \mathfrak{H}_\varphi^b(R,\Phi) = iA \displaystyle\sum_{n=-\infty}^{\infty} i^n \dfrac{J_n(ka)}{H_n^{(1)}(ka)} H_n^{(1)\prime}(kR)\, e^{in\Phi} \\ \mathfrak{H}_z^b(R,\Phi) = -B \displaystyle\sum_{n=-\infty}^{\infty} i^n \dfrac{J_n'(ka)}{H_n^{(1)\prime}(ka)} H_n^{(1)}(kR)\, e^{in\Phi} \end{cases} \tag{78}$$

which are in agreement with the results obtained from the customary methods of calculation.

References

1. W. Magnus, "Regarding the Questions of Uniqueness of a Boundary Value Problem of $\Delta u + k^2 u = 0$, J. German Mathematical Publishers (Jder, dtsch. Math.-Ver.) **52**, 177 (1943).
2. V. Fock, J. Physics **10** (1946) – Bull. Acad. Sci. USSR, S. Phys. **10**, 171 (1946).
3. A systematic approximation approach, which constitutes a first approximation (or correction) to the Kirchhoff diffraction theory, is given by W. Franz, Z. Physik. **125**, 563 (1949).
4. W. Franz, Z. Naturforsch. **3a**, 500 (1948).
5. J. Meixner, Z. Naturforsch. **3a**, 506 (1948).

(Ed. note: Translated by LNMM)

Reaction Concept in Electromagnetic Theory

V. H. Rumsey
Ohio State University, Columbus, Ohio

(Received September 21, 1953; revised manuscript received March 8, 1954)

A physical observable called the reaction is defined to simplify the formulation of boundary value problems in electromagnetic theory. To illustrate its value it is used to obtain formulas for scattering coefficients, transmission coefficients, and aperture impedances. An approximate solution to problems of this type is obtained by replacing the correct source (of the scattered field for example) with an approximate source which is adjusted so that its reaction with certain "test" sources is correct. This insures that the approximate source "looks" the same as the correct source according to the physical tests which are inherent in the problem. The formulas so obtained have a stationary character (for the cases considered) and thus the results could also be obtained from a variational approach. However the physical approach has two important advantages. It is general whereas the variational technique has to be worked out for each problem. It is conceptually simple and leads directly to results which might not be uncovered by the variational approach because of the complexity of the mathematical formulation. The problem of scattering by a dielectric body is used to illustrate this latter point.

INTRODUCTION

The classical analysis of electromagnetic waves is based on the theory of fields which satisfy Maxwell's equations. It is interesting to examine this concept from the point of view of an experimenter whose objective is to use the theory to correlate his measurements. Suppose that we attempt to measure the field radiated by some source of electromagnetic energy by observing the signal received at the terminals of an antenna placed at the point of observation. By moving the antenna around we can obtain a considerable amount of information about the given field, but it is very difficult to relate this information to the classical field parameters, e.g., the electric field. Indeed from a literal point of view the postulate of electric field might be questioned on the grounds that any experiment designed to measure the electric field at a point must necessarily consist of measuring the effect of the field over a small but finite region, and therefore the postulate is incompatible with the process of performing an observation. This suggests that it is desirable to introduce into the theory a fundamental observable which represents measurements which can actually be performed.

DEFINITIONS AND PROPERTIES OF REACTION

We introduce a quantity, called the "reaction," which is defined as follows. Let the source of a monochromatic electromagnetic field consist of the volume distributions of electric and magnetic current $d\mathbf{J}$ and $d\mathbf{K}$ (i.e., the electric dipole moment contained in a volume V is equal to the vector $\iiint_V d\mathbf{J}$). (The word source is used in the sense that everywhere in the given region there is no field when the source is absent. Thus currents which may be induced in various parts of the region are not counted as sources because they vanish when the true source is turned off.) Let $\mathbf{E}(a)$ and $\mathbf{H}(a)$ represent the electric and magnetic fields generated by the source distributions $d\mathbf{J}(a)$ and $d\mathbf{K}(a)$, which we call the source a for simplicity. Similarly for some other source b, which generates a field at the same frequency and in the same environment. Define a complex number [the usual $\exp(\pm i\omega t)$ time convention is employed], denoted by $\langle a,b\rangle$ as follows:

$$\langle a,b\rangle=\iiint_V[\mathbf{E}(b)\cdot d\mathbf{J}(a)-\mathbf{H}(b)\cdot d\mathbf{K}(a)], \qquad (1)$$

where the volume V contains the source a.

The reciprocity theorem[1] states that

$$\langle a,b\rangle=\langle b,a\rangle, \qquad (2)$$

provided that all media are isotropic and that a and b can be contained in a finite volume.

The scalar $\langle a,b\rangle$ is a measure of the reaction (or coupling) between the sources a and b. We think of "reaction" as a physical observable like mass, length, charge, etc.: Eq. (1) is to be understood as a formula for the measure of reaction. For example, in electrostatic theory, let the source a consist of the volume distribution of charge $dq(a)$: similarly for the source b. Then we could define the static reaction (a,b) by the relation

$$(a,b)=\iiint\mathbf{E}(b)dq(a), \qquad (3)$$

which is analogous to Eq. (1). In this case the physical observable which (a,b) represents is the resultant force exerted by a on b. In the oscillating case let the source a consist of a unit current generator connected to the terminals of some antenna. Then it follows from Eq. (1) that $\langle a,b\rangle$ is equal to the open circuit voltage generated at the antenna terminals by the source b. Again for static fields the electric field is equal to (u,a) the reaction between the given source a and an infinitesimal unit charge u placed at the point of observation.

[1] S. A. Schelkunoff, *Electromagnetic Waves* (D. Van Nostrand Publishing Company, New York, 1943), p. 477. The proof given applies here with a minor extension.

Reprinted with permission from *The Physical Review*, vol. 94, no. 6, pp. 1483–1491, 1954.

For oscillating fields the component of electric field in a particular direction is equal to $\langle u,a\rangle$, the reaction between the given source a and an infinitesimal electric dipole u, of unit moment, placed parallel to this direction at the point of observation.

The complex number $\langle a,b\rangle$ consists of the sum of products of the form $V(a)I(b)$ where $V(a)$ represents the voltage generated by a across the terminals of a current generator of strength $I(b)$. If

$$V(a)=|V(a)|\exp(i\alpha),\quad I(b)=|I(b)|\exp(i\beta),$$

then the corresponding explicit time functions which these complex numbers represent are

$$V(a,t)=|V(a)|\cos(\omega t+\alpha),\quad I(b,t)=|I(b)|\cos(\omega t+\beta),$$

and the complex number,

$$V(a)I(b)=|V(a)I(b)|\exp[i(\alpha+\beta)],$$

represents the explicit time function,

$$|V(a)I(b)|\cos(2\omega t+\alpha+\beta)$$
$$=|V(a)I(b)|[2\cos(\omega t+\alpha)\cos(\omega t+\beta)-\cos(\alpha-\beta)].$$

The first term in the square brackets is the instantaneous rate at which b works against a, and the second term is the average rate per cycle at which b works against a. Thus, $\langle a,b\rangle$ represents the instantaneous minus the average rate at which b expends energy on a. Note that if $\langle a,b\rangle=0$, then no energy is transferred from a to b at any time.

Observe that

$$\langle a,(b+c)\rangle=\langle a,b\rangle+\langle a,c\rangle, \tag{4}$$

where a, b, and c represent any three sources radiating at the same frequency in the same region. Note also that if A represents any scalar and Aa represents the source a increased in strength by the factor A, then

$$\langle Aa,b\rangle=A\langle a,b\rangle. \tag{5}$$

Before proceeding any further with the formal development let us now discuss how the concept of reaction can be used, in order to see what further developments are needed.

APPLICATIONS OF THE REACTION CONCEPT

Consider the problem of scattering by a perfectly conducting body, of surface S, which is irradiated by the source g (see Fig. 1). Let $\mathbf{J}(c)$ represent the surface distribution of electric current which is induced on the scatterer by g. The scattered field, $\mathbf{E}(c)$, is defined as the field that would be generated by $\mathbf{J}(c)$ (acting as a source) if the scatterer were absent (the total field is then the superposition of the incident and scattered fields). We postulate that if $\mathbf{J}(c)$ were known the scattered field $\mathbf{E}(c)$ could be calculated. It is known that the tangential component of $\mathbf{E}(c)$ on S is equal to minus the tangential component of the incident electric field, because the scatterer is a perfect conductor. Thus, although $\mathbf{J}(c)$ is unknown it is possible to calculate the reaction between the source $\mathbf{J}(c)$ and some known current distribution, $\mathbf{J}(a)$, on the surface S, for

$$\langle a,c\rangle=\iint_S \mathbf{J}(a)\cdot\mathbf{E}(c)dS \tag{6}$$

and the tangential component of $\mathbf{E}(c)$ on S is known.

FIG. 1. A hypothetical surface S in the presence of a source g.

Suppose that we wish to calculate the "echo," i.e., the signal at g due to the scattered field. If we think of g as a unit current generator connected to the terminals of some antenna, the open circuit voltage at these terminals generated by the scattered field is equal to $\langle g,c\rangle$. Thus the problem is to calculate $\langle g,c\rangle$. Let $\mathbf{J}(a)$ represent an assumed distribution of electric current on S which we propose to adjust so that it approximates $\mathbf{J}(c)$. We would like to adjust $\mathbf{J}(a)$ so that

$$\langle g,a\rangle=\langle g,c\rangle, \tag{7}$$

for then the echo obtained by substituting $\mathbf{J}(a)$ for $\mathbf{J}(c)$ would be correct. Obviously this is too much to expect: indeed we find that Eq. (7) cannot be enforced because we cannot calculate $\langle g,c\rangle$. We can interpret Eq. (7) as the condition that the approximate source a (i.e., $\mathbf{J}(a)$) should "look" the same as the correct source c (i.e., $\mathbf{J}(c)$) to the source g, in the sense that a and c produce the same signal at g. Thus we can think of g as a "test" source which is used to test for any difference between a and c. This suggests that we regard Eq. (7) as a special case of the more general restriction

$$\langle x,a\rangle=\langle x,c\rangle, \tag{8}$$

which expresses the condition that a and c should "look" the same to an arbitrary test source x. The problem is now a matter of trying to enforce Eq. (8) for every "available" test source, i.e., every x for which $\langle x,a\rangle$ and $\langle x,c\rangle$ can be calculated. The only sources in the problem are g, c and a. We have seen that g is not "available" because $\langle g,c\rangle$ cannot be calculated and we find that c is not available because $\langle c,c\rangle$ cannot be calculated. Thus a is the only "available" test source. We therefore adjust the approximation a to satisfy the condition

$$\langle a,a\rangle=\langle a,c\rangle \tag{9}$$

and use the value of a so obtained in place of the correct source c.

To carry out the calculation of the echo we assume some current distribution $\mathbf{J}(a)$ whose level can be adjusted, i.e., let

$$\mathbf{J}(a)=U\mathbf{J}(u),\quad\text{or}\quad a=Uu, \tag{10}$$

where $\mathbf{J}(u)$ is fixed and U is an adjustable constant. Substituting for a in Eq. (9) gives

$$U=\langle u,c\rangle/\langle u,u\rangle. \tag{11}$$

Now the echo $\sim -\langle c,g\rangle$, where

$$\begin{aligned}-\langle c,g\rangle &= \int\int_S \mathbf{J}(c)\cdot\mathbf{E}(c)dS \\ &= \langle c,c\rangle \simeq \langle a,a\rangle = U^2\langle u,u\rangle = \langle u,c\rangle^2/\langle u,u\rangle \\ &= \left[\int\int_S \mathbf{J}(u)\cdot\mathbf{E}(c)dS\right]^2 \Big/ \\ &\qquad \int\int_S \mathbf{J}(u)\cdot\mathbf{E}(u)dS. \end{aligned} \tag{12}$$

In Eq. (12) $\mathbf{J}(u)$ represents the assumed current distribution, $\mathbf{E}(u)$ represents the electric field generated by $\mathbf{J}(u)$ (in the absence of the scatterer), and $\mathbf{E}(c)$ represents the tangential component at S of the incident electric field.

The approximation can be improved by starting from an assumed distribution $\mathbf{J}(a)$ which contains a number of adjustable constants. Thus, let

$$a=Ll+Mm+\cdots, \tag{13}$$

where L, M, $\cdots$ represent adjustable constants and, l, m, $\cdots$ represent fixed source distributions which are assumed. The problem is to find the linear combination of l, m, $\cdots$, denoted by a, which best approximates the correct source $\mathbf{J}(c)$, denoted by c. Here we can enforce the conditions:

$$\begin{aligned}\langle a,l\rangle &= \langle c,l\rangle, \\ \langle a,m\rangle &= \langle c,m\rangle, \\ \cdot\quad &\quad \cdot \\ \cdot\quad &\quad \cdot \\ \cdot\quad &\quad \cdot \end{aligned} \tag{14}$$

which ensure that a and c "look" the same to l, m, $\cdots$. If the assumed set of source distributions l, m, $\cdots$ constitute a complete orthogonal set, the equations (14) represent the condition that a and c are identical in every respect. Substituting for a from (13) in (14) gives the following equations for the constants:

$$\begin{aligned}L\langle l,l\rangle+M\langle m,l\rangle+\cdots &= \langle c,l\rangle, \\ L\langle l,m\rangle+M\langle m,m\rangle+\cdots &= \langle c,m\rangle, \\ \cdot \qquad\quad \cdot \qquad\quad &\quad \cdot \\ \cdot \qquad\quad \cdot \qquad\quad &\quad \cdot \\ \cdot \qquad\quad \cdot \qquad\quad &\quad \cdot \end{aligned} \tag{15}$$

In terms of a matrix notation, the constants L, M, $\cdots$ are given explicitly by

$$\begin{bmatrix} L \\ M \\ \cdot \\ \cdot \\ \cdot \end{bmatrix} = \begin{bmatrix} \langle l,l\rangle & \langle l,m\rangle & \cdots \\ \langle m,l\rangle & \langle m,m\rangle & \cdots \\ \cdot & \cdot & \\ \cdot & \cdot & \\ \cdot & \cdot & \end{bmatrix}^{-1} \begin{bmatrix} \langle l,c\rangle \\ \langle m,c\rangle \\ \cdot \\ \cdot \\ \cdot \end{bmatrix}. \tag{16}$$

By substituting for L, M, $\cdots$ in (13), the approximate value of the echo is given by (see Eq. (12)):

$$\langle a,a\rangle=\langle a,c\rangle=(\langle l,c\rangle\langle m,c\rangle\cdots) \times \begin{bmatrix} \langle l,l\rangle & \langle l,m\rangle & \cdots \\ \langle m,l\rangle & \langle m,m\rangle & \cdots \\ \cdot & \cdot & \\ \cdot & \cdot & \\ \cdot & \cdot & \end{bmatrix}^{-1} \begin{bmatrix} \langle l,c\rangle \\ \langle m,c\rangle \\ \cdot \\ \cdot \\ \cdot \end{bmatrix}. \tag{17}$$

These results can also be obtained by means of the variational technique. For example, let δa represent a slight change of the source distribution represented by a. If both a and $a+\delta a$ satisfy Eq. (9), then $2\langle a,\delta a\rangle = \langle a,\delta a\rangle+\langle \delta a,a\rangle=\langle \delta a,c\rangle$ to the first order. If δa represents a slight change about the correct distribution c, we can substitute c for a in this equation and thus obtain $\langle \delta a,c\rangle=0$. Hence, the expressions $\langle a,a\rangle$ and $\langle a,c\rangle$ are stationary for variations of a about c, if a satisfies Eq. (9), i.e., if $a=u\langle u,c\rangle/\langle u,u\rangle$ [see Eqs. (10) and (11)]. Thus the expression $x=\langle a,c\rangle=\langle a,a\rangle=\langle u,c\rangle^2/\langle u,u\rangle$ is stationary for variations of the assumed distribution $\mathbf{J}(u)$ about the correct distribution $\mathbf{J}(c)$, and Eqs. (14) and (15) can be obtained by setting $\partial x/\partial L=0$, $\partial x/\partial M=0$, etc. However, the fact that an expression is stationary for variations of an assumed distribution about the correct distribution does not justify the assumption that it will yield the "best" approximation when the assumed distribution is completely arbitrary. The reaction approach does show that the approximation is the best in a physical sense, i.e., in the sense that the approximate source "looks" the same as the correct source to any source in the problem which can be used for such an observation. More precisely, the reaction between the approximate source and every available test source is correct, it being understood that a test source is "available" if its reaction with the correct source can be calculated. Note that this comparison of the variational method with the reaction method is, so far, based on the specific problem of scattering by a perfect conductor. There are other problems (some of which we consider later) where a number of different approximations can be obtained from the variational method, and there is no way of deciding which approximation is best. In the reaction approach such problems yield an excessive number of test sources, i.e., the number of independent test sources exceeds the number of adjustments which can be made in the approximation, so that it is not possible to make the reaction with every test source correct. In this case it is necessary to decide what selection of the available test sources is most likely to yield the best approximation. From the physical point of view the answer is clearly that selection which most nearly represents the actual physical observation which we are trying to approximate. The physical approach has the added advantages of being general (whereas the stationary formulation has to be established for each specific problem), and of providing a simple understanding of the type of approximation which is being used. In short, the fundamental advantage of the reaction method is its conceptual simplicity which leads directly to results which might not be uncovered by the variational approach because of the complexity of the mathematical formulation. These points are illustrated by the examples which follow.

TRANSMISSION CALCULATIONS

The problem of scattering by a perfect conductor can be used to illustrate the formulation of transmission problems. We are now interested in the signal received at an arbitrary point whereas in the echo problem the point of reception was at the given source g. As before the total field is represented as the sum of contributions from g and c (the induced electric current distribution on S) radiating as if the scatterer were absent. The contribution from g can be calculated easily since g is given. We therefore put up a source h at the point of observation (see Fig. 2) and the problem is to calculate the reaction $\langle c,h\rangle$ evaluated in the absence of the scatterer. Let the notation be as follows: c generates the same field as g inside of S (as before), d generates the same field as h inside of S, a is the approximation for c (as before), b is the approximation for d. The sources a, b, c, and d are all distributions of electric current on S, and all sources radiate as if the scatterer were absent. We therefore adjust the approximations a and b to satisfy the conditions:

$$\langle a,x\rangle=\langle c,x\rangle, \tag{18}$$

and

$$\langle b,y\rangle=\langle d,y\rangle, \tag{19}$$

where x and y represent any test sources inside of S. The available test sources are represented by $x=a$, $x=b$, $y=a$ and $y=b$. Since Eq. (18) cannot be satisfied for both values of x, we have to decide which value is likely to give the better approximation: similar remarks apply to Eq. (19). Now the quantity which we are trying to approximate is $\langle c,h\rangle$ where

$$\langle c,h\rangle=\int\!\!\int_S \mathbf{J}(c)\cdot\mathbf{E}(h)dS$$
$$=\int\!\!\int_S \mathbf{J}(c)\cdot\mathbf{E}(d)dS=\langle c,d\rangle. \tag{20}$$

To approximate $\langle c,d\rangle$ we replace c by its approximation a, or d by b, or both. Thus there are three possible approximations which are represented by $\langle a,d\rangle$, $\langle c,b\rangle$ and $\langle a,b\rangle$. We see that these are all the same if we

FIG. 2. Two sources g and h in the presence of a conductor.

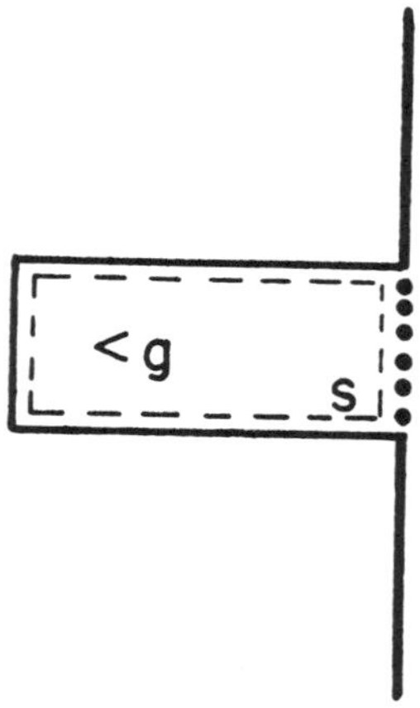

FIG. 3. A source g radiating through a waveguide.

choose $x=b$ and $y=a$ which is therefore the best choice.

An explicit formula for $\langle c,d\rangle$ is obtained by setting

$$a=Uu \quad\text{and}\quad b=Vv, \tag{21}$$

where U and V are adjustable constants. Proceeding as before [see Eq. (10)], we obtain

$$\langle c,d\rangle\simeq\langle c,v\rangle\langle d,u\rangle/\langle u,v\rangle, \tag{22}$$

where u and v represent assumed electric current distributions. In terms of integrals over S, Eq. (22) becomes

$$\langle c,d\rangle\simeq\left[\int\!\!\int_S \mathbf{E}(g)\cdot\mathbf{J}(v)dS\right]\left[\int\!\!\int_S \mathbf{E}(h)\cdot\mathbf{J}(u)dS\right]\Big/ \int\!\!\int_S \mathbf{E}(u)\cdot\mathbf{J}(v)dS, \tag{23}$$

where $\mathbf{J}(u)$ and $\mathbf{J}(v)$ are the current distributions which we assume to approximate the current distributions induced on the scatterer by g and h, respectively.

IMPEDANCE CALCULATIONS

Suppose that the given source g radiates into space through a length of uniform wave guide as shown in Fig. 3. This represents a type of problem which has been treated successfully by means of the variational approach.[2]

We postulate that it is impractical to calculate the field which is radiated through the aperture (the aperture is shown as the line of dots in Fig. 3), but that it is practical to calculate the field (in the waveguide) that would be obtained if the aperture were covered with a conducting plate.

Let $\mathbf{J}(c)$ represent the electric current distribution that would be induced by g on a conducting plate covering the aperture. Let $\mathbf{E}(g,1)$ represent the field generated by g with the plate in position, $\mathbf{E}(g,2)$ represent the field generated by g with the plate removed, and $\mathbf{E}(c,2)$ the field that would be generated

[2] *Waveguide Handbook*, edited by N. Marcuvitz (McGraw-Hill Book Company, Inc., New York, 1951).

by $\mathbf{J}(c)$ (acting as a source) with the plate removed. Then,

$$\mathbf{E}(g,1)=\mathbf{E}(g,2)+\mathbf{E}(c,2), \tag{24}$$

for points in the wave-guide region (inside of the dashed curve S shown in Fig. 3). If g consists of a unit current generator connected to a pair of terminals, then it follows from (24) that

$$Z_1=V_1=V_2+\langle c,g\rangle=Z_2+\langle c,g\rangle, \tag{25}$$

where V_1 and V_2 are the voltages at these terminals with and without the plate respectively, Z_1 and Z_2 are the corresponding impedances, and $\langle c,g\rangle$ is evaluated with the plate removed. The problem is to calculate an approximation for Z_2, assuming that Z_1 is known. For example, suppose that the wave-guide structure consists of a biconical horn as illustrated in Fig. 4. The source g consists of a unit current generator connected to the input. Then

$$Z_1=-i\mu^{\frac{1}{2}}\epsilon^{-\frac{1}{2}}\pi^{-1}\log(\cot\alpha/2)\tan\omega l\mu^{\frac{1}{2}}\epsilon^{-\frac{1}{2}},$$
$$\mathbf{J}(c)=[2\pi l\cos\omega l\mu^{\frac{1}{2}}\epsilon^{\frac{1}{2}}\sin\theta]^{-1}\boldsymbol{\theta},$$

where l and θ are given in Fig. 4, ω represents the frequency, μ and ϵ represent the permeability and inductive capacity of the medium inside of S, and $\boldsymbol{\theta}$ represents a unit vector in the spherical coordinate system shown in Fig. 4.

The specific example which we have chosen to illustrate impedance calculations is representative of the general problem in which g is inside of as illustrated in Fig. 5. It is essentially the same as the scattering problem in which g is outside of S. Here, as in the scattering problem, $-\mathbf{J}(c)$ generates the same field as g on the source-free side of S. Here, $\langle g,c\rangle$ cannot be calculated because tangential $\mathbf{E}(g)$ at S is unknown, whereas $\langle g,c\rangle$ cannot be calculated in the scattering problem because $\mathbf{J}(c)$ is unknown. In this problem we therefore assume some approximation for $\mathbf{E}(c)$, the tangential component of electric field in the aperture (or over S in the more general terminology). We postulate that it is possible to calculate $\mathbf{E}(a)$, the field which fits this assumed distribution of tangential $\mathbf{E}(c)$ on both sides of S. By definition the tangential component of $\mathbf{E}(a)$ at S is continuous but the tangential component of $\mathbf{H}(a)$ obviously is not continuous. In short, the field $\mathbf{E}(a)$ is generated by a certain distribution of electric currents $\mathbf{J}(a)$ on S which can be calculated from the assumed distribution of tangential electric field. If this assumed distribution were correct, then $\mathbf{J}(a)$ would turn out to be identical to $\mathbf{J}(c)$. We can see now that the impedance problem is formally identical to the scattering problem, with the modification that the approximate source a is now required to look the same as the correct source c to a test source x which is outside of S. Thus Eq. (9) applies again:

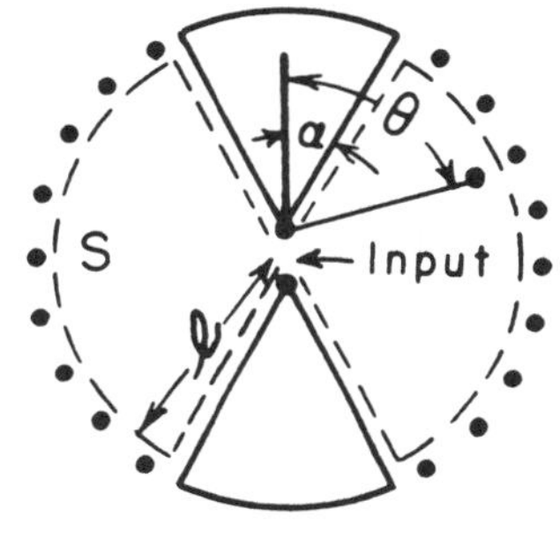

FIG. 4. A biconical horn.

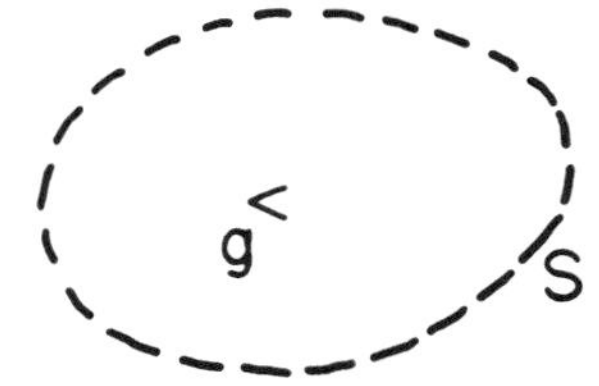

FIG. 5. A source g inside of a closed surface S.

$$\langle a,a\rangle=\langle a,c\rangle. \tag{9}$$

Instead of Eq. (10), we now put

$$\mathbf{E}(a)=U\mathbf{E}(u), \tag{10a}$$

where $\mathbf{E}(u)$ represents the assumed distribution of tangential $\mathbf{E}(c)$ over S and U is an adjustable constant. Equation (10a) is equivalent to Eq. (10) because it implies that

$$a=Uu, \tag{10}$$

as before. The approximation for $\langle c,g\rangle$ given by (12) also applies except that we express $\langle u,c\rangle$ in the form $\iint_S \mathbf{E}(u)\cdot\mathbf{J}(c)dS$. Thus, the approximation for the impedance obtained from Eq. (25) (replacing c by a) is

$$Z_2=Z_1-\langle u,c\rangle^2/\langle u,u\rangle. \tag{26}$$

The problem of transmission through the aperture can be treated by placing a test source h at the point of observation outside of S and proceeding as in Eqs. (18)–(23).

SCATTERING BY A DIELECTRIC

The problem of scattering by a dielectric brings out some interesting features which have not been encountered up to this point. Let the given source g radiate in the presence of a dielectric scatterer of surface S, as in Fig. 1. The scattered field is again defined as the difference between the fields generated by g with and without the scatterer. The problem is to find a source, a, which generates a field outside of S which is approximately equal to the scattered field (and such that it is possible to calculate the field generated by a).

Let $\mathbf{E}(g,1)$ represent the field generated by g in the presence of the scatterer, and

$$\mathbf{J}(c)=\mathbf{n}\times\mathbf{H}(g,1),\quad \mathbf{K}(c)=\mathbf{E}(g,1)\times\mathbf{n}, \tag{27}$$

where $\mathbf{n}$ represents a unit vector normal to S pointing inwards. The surface distributions of electric and magnetic current $\mathbf{J}(c)$ and $\mathbf{K}(c)$, are the equivalent currents of Schelkunoff.[3] Let the combination of $\mathbf{J}(c)$ and $\mathbf{K}(c)$

[3] S. A. Schelkunoff, Phys. Rev. 56, 308 (1939).

be represented by the symbol c. Then the combination of g and $-c$ generates a field which is zero everywhere inside of S. Therefore we can remove the scatterer without affecting the field generated by $(g-c)$. It follows that the scattered field is equal to the field which c generates outside of S, in the absence of the scatterer. Thus the problem is to find an approximation for c. Observe that c generates the same field as g inside of S, regardless of the medium inside of S. Thus we require that the approximation for c, represented by a, should look the same as g to a test source x inside of S, and for the purpose of simplifying calculations we can perform this test in "free space" (see Fig. 6).

At this point it is desirable to extend the notation [introduced in Eq. (24)] in which a number represents the environment, as follows. Let $\langle p,1,q\rangle$, $\langle p,2,q\rangle$, and $\langle p,3,q\rangle$ represent the reaction between sources p and q in the presence of the scatterer, in free-space, and in an infinite homogeneous dielectric medium, respectively. Thus the test illustrated by Fig. 6 is represented by the equation

$$\langle g,2,x\rangle=\langle a,2,x\rangle. \tag{28}$$

Observe that c generates zero field outside of S, in the presence of the scatterer. The region outside of S can therefore be filled with the same material as the inside of S without affecting the result. We therefore require a to satisfy the requirement,

$$\langle a,3,y\rangle=0, \tag{29}$$

where y represents any test source outside of S (see Fig. 7).

The coarsest approximation is obtained by representing a in the form

$$a=Ll+Mm, \tag{30}$$

where l and m represent the assumed distributions of electric and magnetic current on S respectively, and L and M are adjustable constants. Then the possible choices of x and y are represented by $x=l$, $x=m$, $x=a$, $y=l$, $y=m$, and $y=a$ (only four of which are independent). Observe that if we set $x=l$ in the expression $\langle a,2,x\rangle$, we must interpret this as the limit obtained as l approaches a from the inside of S as illustrated in Fig. 8. Similarly, the substitute for y must be taken just outside of S. An expression of the form $\langle a,2,a\rangle$ is evaluated by imagining two sources a slightly displaced from each other in the manner of Fig. 8. Considering the symmetry of Eqs. (28) and (29), the obvious

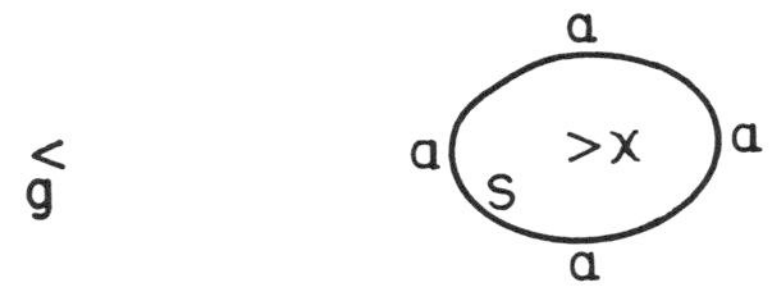

FIG. 6. A test source x set up to detect the difference between sources a and g.

choice appears to be $x=a$ and $y=a$. A further point in support of this choice is the fact that the echo$=\langle g,2,c\rangle$ $=\langle c,2,c\rangle$ is approximated by $\langle c,2,a\rangle$ or $\langle a,2,a\rangle$ and the choice $x=a$ makes these two results the same. The choice $y=a$ is also supported by the fact that if a were correct (i.e., if $y=c$), its radiation would be confined to the region occupied by the source under test (which is also a), thus producing, in a sense, the maximum irradiation of the source under test.

Substituting $x=a$ in Eq. (28) and $y=a$ in Eq. (29) gives the following equations for the constants L and M:

$$L\langle g,2,l\rangle+M\langle g,2,m\rangle=L^2\langle l,2,l\rangle+LM\langle (l),2,m\rangle +LM\langle l,2,(m)\rangle+M^2\langle m,2,m\rangle, \tag{31}$$

$$0=L^2\langle l,3,l\rangle+LM\langle (l),3,m\rangle +LM\langle l,3,(m)\rangle+M^2\langle m,3,m\rangle, \tag{32}$$

where the notation (l) in $\langle (l),2,m\rangle$ indicates that l is just inside of m (see Fig. 8), e.g.,

$$\langle (l),2,m\rangle=\int\int_S \mathbf{J}(l)\cdot\mathbf{E}(m,2,\text{internal})dS, \tag{33}$$

where

$$\mathbf{n}\times[\mathbf{E}(m,2,\text{internal})-\mathbf{E}(m,2,\text{external})]=\mathbf{K}(m) \quad [\text{see Eq. (27)}]. \tag{34}$$

The approximation for the echo is then given by substituting a for c in $\langle g,2,c\rangle$.

It can be shown that the formula for the echo obtained in this way is stationary for variations of the assumed distribution of electric current $\mathbf{J}(l)$ and magnetic current $\mathbf{K}(m)$ about the correct distribution. The interesting point is that a number of stationary formulas for the echo can be derived, all of which are based on assumed distributions $\mathbf{J}(l)$ and $\mathbf{K}(m)$ and involve the same "free space" calculations, but differ in the way that the calculations are combined. Here is a case where the physical approach through the reaction concept leads to a result which probably would not have been uncovered by the variational technique, although, once the result has been established, it is possible to see how it could have been obtained by means of a variational approach.

Another approach is to set down the equations for continuity of tangential $\mathbf{E}$ and $\mathbf{H}$ which are:

$$\begin{aligned}\mathbf{n}\times\mathbf{E}(g,2)-\mathbf{n}\times\mathbf{E}(c,2,\text{external}) &=\mathbf{n}\times\mathbf{E}(c,3,\text{internal}),\\ \mathbf{n}\times\mathbf{H}(g,2)-\mathbf{n}\times\mathbf{H}(c,2,\text{external}) &=\mathbf{n}\times\mathbf{H}(c,3,\text{internal}).\end{aligned} \tag{35}$$

We multiply these equations by tangential $\mathbf{H}$ and $\mathbf{E}$, respectively, and integrate over S. This is represented by the equation

$$\langle g,2,z\rangle-\langle (a),2,z\rangle=\langle a,3,(z)\rangle, \tag{36}$$

where z represents a test source distributed over S and the correct source c has been replaced by the approximate source a. Here it is necessary to use every available test source, i.e., $z=l$ and $z=m$ in order to evaluate L and M. We then obtain two equations which are essentially the same as those used by Crowley[4] for a certain class of impedance calculations. The formula for the echo obtained from Eq. (36) is stationary but it is different from that given by Eqs. (28) and (29). An important point of difference is that the approximation given by Eq. (36) does not give the same value for the two possible approximations for the echo, $\langle g,2,a\rangle$ and $\langle a,2,a\rangle$.

There is a simple connection between Eqs. (28), (29), and (36). It is obtained from the relation

$$\langle(a),2,b\rangle-\langle a,2,(b)\rangle=\langle(a),3,b\rangle-\langle a,3,(b)\rangle, \qquad (37)$$

which we shall establish [see Eq. (47)]. In Eq. (37), a and b represent any two sources, and 2 and 3 represent any two environments. Thus, Eq. (36) can be written in the form

$$\langle g,2,z\rangle=\langle a,2,(z)\rangle+\langle(a),3,z\rangle. \qquad (38)$$

Bearing in mind that in Eqs. (28) and (29) x is inside of S and y is outside of S, we see that Eq. (36) is satisfied if Eq. (28) is satisfied for $x=z$ and Eq. (29)

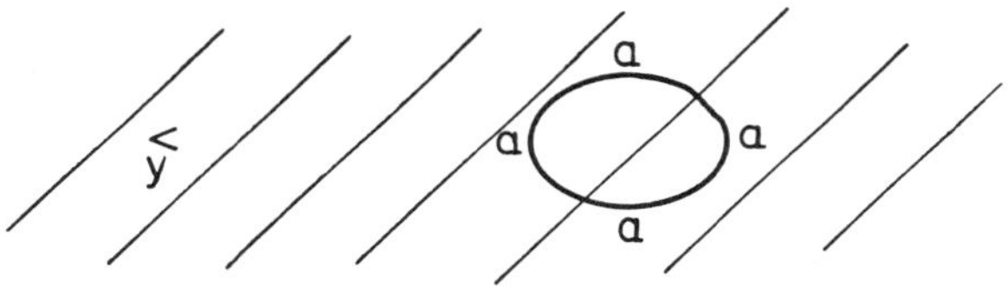

FIG. 7. A test source y set up to detect the external field from a.

for $y=z$. In this sense, Eqs. (28) and (29) embody the information contained in Eq. (36). We obtain different formulas for the echo because we choose to enforce Eq. (36) for two different values of z, $z=l$ and $z=m$, whereas we enforce Eqs. (28) and (29) for $x=a$ and $y=a$. To put it another way, if $x=y=z=a$ then the same value of a is obtained from any two of the three equations (28), (29), and (36). It is interesting to note that the value of $\langle g,2,a\rangle$ (the echo) obtained from (36) is stationary provided Eq. (36) is enforced for the single value of z, $z=a$. On the other hand, if (28) and (29) are enforced for $x=a$ and $y=a$, then all three quantities $\langle g,2,a\rangle$, $\langle a,2,a\rangle$ and $\langle a,3,a\rangle$ are stationary.

To give a specific comparison between these two approaches, the echo from an infinite plane dielectric slab for a plane wave at normal incidence, has been computed from Eqs. (28) and (29) with $x=a$ and $y=a$, and from Eq. (36) with $z=l$ and $z=m$. We note at the

FIG. 8. The source l just inside of the source a.

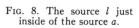

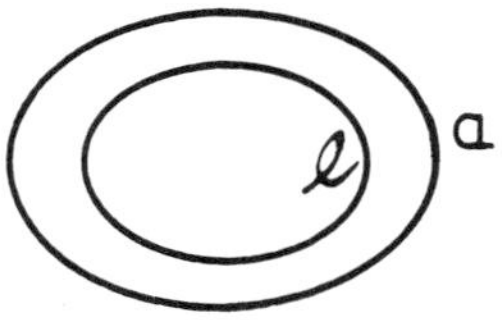

outset that the results of this comparison must be regarded as suggestive rather than conclusive for not only is the example specific, as opposed to general, but the assumed distribution is also specific. Note also that the assumed distribution, which is usually a function of position, in this case consists of four discrete values corresponding to the front and the back faces of the slab. Thus, if our approximation contains four independent adjustable constants, any set of tests which determines them uniquely is bound to yield the correct solution.

The two methods were compared on the basis of the type of approximation represented by Eq. (30), which contains two adjustable constants. A crude approximation is obtained by assuming equal and opposite electric and magnetic currents on the front and back faces, e.g., the distributions l and m can be represented by the pair of values $(+1, -1)$. For thin slabs and low dielectric constants both methods give an echo which is four times the correct value, and for low dielectric constants, the maximum echo, as a function of slab thickness, is about twice the correct value. The situation here is that the differences between the two methods are overshadowed by the crudeness of the assumed distribution, which emphasizes the overriding importance of starting from an assumed distribution which is nearly correct. A better approximation is obtained by assuming that l and m are represented by the distribution $(+1, -\exp it)$ where t represents the electrical thickness, i.e., $t=\omega\mu^{\frac{1}{2}}\epsilon^{\frac{1}{2}}d$, where $d=$ thickness of slab and μ and ϵ are the constants of the dielectric. This is the type of assumption used in physical optics and neglects reflection from the back face. For an incident field of one volt per meter the following formulas are obtained. The correct solution is

$$|\langle g,2,c\rangle|=\frac{2K|1-\exp 2it|}{\eta_0|1-K^2\exp 2it|}. \qquad (39)$$

The approximation given by Eqs. (28) and (29) is

$$|\langle g,2,a\rangle|=\frac{2K|1-\exp(it/1+K)|^2}{\eta_0|1-2\exp(2it/1+K)+\exp 2it|}. \qquad (40)$$

The approximation given by Eq. (36) is

$$|\langle g,2,a\rangle| =\frac{2K|1-\exp it/1+K|^2|1-\exp 2it|}{\eta_0|1-2\exp(2it/1+K)+\exp(4it/1+K)-K^2\exp(4it)+2K^2\exp[2it(2+K)/1+K]-K^2\exp(4it/1+K)|}. \qquad (41)$$

[4] T. H. Crowley, "Variational Impedance Calculations," Antenna Laboratory Report 478–5, Ohio State University (unpublished).

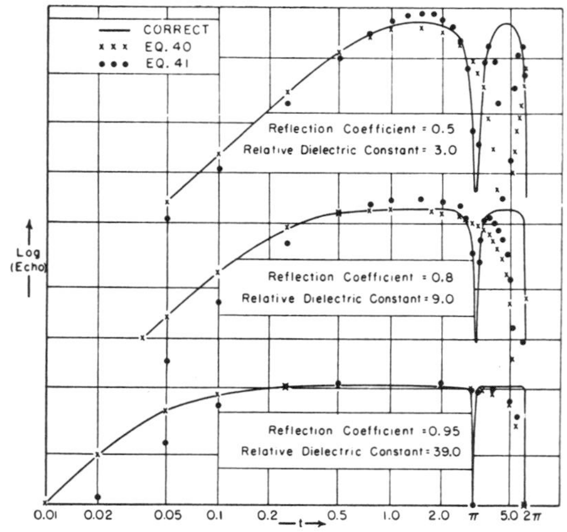

FIG. 9. The echo from a dielectric slab *versus* thickness.

In Eqs. (39), (40), and (41), K represents the reflection coefficient for $t=\infty$; i.e.,

$$K=(\epsilon_r-1)/(\epsilon_r+1), \tag{42}$$

where ϵ_r = relative dielectric constant of the dielectric.

These results are plotted in Fig. 9. It can be seen that the error in Eq. (40) is insignificant for $t\leqslant 1$ but becomes progressively worse for larger values of t until there is practically no correlation with the correct result. Equation (41) is in error for small values of t by the factor $(K+1)$, but, unlike Eq. (40), it is correct in the vicinity of $t=\pi$, although otherwise it shows practically no correlation with the correct result for large values of t. The behavior at $t=\pi$ is interesting because the assumed distribution is correct at this value of t, and therefore both (40) and (41) might be expected to reduce to the correct result at $t=\pi$. The failure of Eq. (40) at $t=\pi$ is due to the fact that the external field generated by l or m in a homogeneous dielectric is zero. Thus, while Eq. (29) is certainly satisfied it does not yield any information about the constants L and M, i.e., for $t=\pi$ the test represented by Eq. (29) turns out to be trivial for the particular assumed distribution under consideration. This suggests that instead we should put the test source y inside of S, but then we find that in order to arrive at an enforceable condition we are brought back to Eq. (36).

FURTHER PROPERTIES OF THE REACTION

The foregoing example of scattering by a dielectric shows the need for a more elaborate formulation to handle cases where Eq. (1) is inadequate. When a and b [see Eq. (1)] consist of surface distributions over the same surface S the field at a due to b is discontinuous and consequently the integral in Eq. (1) is not defined. In this case there are two possible values for the reaction which are obtained by imagining that a and b are slightly separated. If a is imagined to be just outside of b, we have

$$\langle a,(b)\rangle=\int\!\!\int_S [\mathbf{J}(a)\cdot\mathbf{E}(b,\text{external}) - \mathbf{K}(a)\cdot\mathbf{H}(b,\text{external})]dS; \tag{43}$$

and for b just outside of a, we have

$$\langle(a),b\rangle=\int\!\!\int_S [\mathbf{J}(a)\cdot\mathbf{E}(b,\text{internal}) - \mathbf{K}(a)\cdot\mathbf{H}(b,\text{internal})]dS. \tag{44}$$

It can be shown[5] that

$$\langle a,(b)\rangle=\langle(b),a\rangle, \tag{45}$$

under the same conditions as are required for the usual reciprocity theorem [see Eq. (2)]. It follows that

$$\langle(a),b\rangle-\langle a,(b)\rangle = \int\!\!\int_S [\mathbf{J}(a)\times\mathbf{K}(b)-\mathbf{J}(b)\times\mathbf{K}(a)]\cdot\mathbf{n}dS. \tag{46}$$

Thus

$$\langle(a),1,b\rangle-\langle a,1,(b)\rangle=\langle(a),2,b\rangle-\langle a,2,(b)\rangle, \tag{47}$$

where 1 and 2 represent any two environments.

The formula (1) is also inadequate if a or b cannot be contained in a finite volume. For example the reaction between infinite traveling wave line sources can be defined as follows. Let the source distribution be represented by $\mathbf{J}(x,y)\exp\gamma z$ and $\mathbf{K}(x,y)\exp\gamma z$, where the constant γ is such as to represent waves which are attenuated as they travel along the z axis, and the medium is independent of z but is otherwise heterogeneous. Then

$$\langle a,b\rangle=\int\!\!\int_\Sigma [\mathbf{J}(a,x,y)\sigma\mathbf{E}(b,x,y) + \mathbf{K}(a,x,y)\sigma\mathbf{H}(b,x,y)]dxdy, \tag{48}$$

where σ represents the matrix

$$\sigma=\begin{bmatrix}1&0&0\\0&1&0\\0&0&-1\end{bmatrix},$$

and the surface Σ contains all sources.[6] This formula differs from Eq. (1) but can be applied in the same way: it represents the same physical observable. Modifications similar to Eqs. (43) and (44) apply if a and b are at the same place.

Returning to the case where a and b are at different places, let S represent any surface which separates a and b (e.g., all parts of a are inside of S and all parts of b are outside of S). Then

$$\langle a,b\rangle=\int\!\!\int_S [\mathbf{H}(a)\times\mathbf{E}(b)-\mathbf{H}(b)\times\mathbf{E}(a)]\cdot\mathbf{n}dS, \tag{49}$$

[5] T. H. Crowley, J. Appl. Phys. **25**, 119 (1954).
[6] V. H. Rumsey, J. Appl. Phys. **24**, 1358 (1953).

where $\mathbf{n}$ points away from a towards b. This illustrates another aspect of the reaction concept, for Eq. (49) states that the flux of the vector in brackets is conserved (provided S separates a and b), and $\langle a,b\rangle$ represents the conserved property. This principle of conservation is self-evident from a physical point of view, for Eq. (42) can be interpreted as stating that the reaction between a and b is the same as the reaction between a and the equivalent sources $\mathbf{n}\times\mathbf{E}(b)=\mathbf{K}(b)$ and $\mathbf{H}(b)\times\mathbf{n}=\mathbf{J}(b)$ which generate the same field as b at a. It is apparent that the reaction between a and b is the same as the reaction between a and any source which produces the same field as b at a.

The reaction concept can be extended to anisotropic media by using a more general form of the reciprocity theorem (which was brought to the author's attention by M. H. Cohen). Let $[\mathbf{E}(a),\mathbf{H}(a)]$ represent the field generated by the source distribution $[d\mathbf{J}(a),d\mathbf{K}(a)]$ as before. Let $[\mathcal{E}(a),\mathcal{H}(a)]$ be the field generated by the same source when all media are replaced by the corresponding media whose macroscopic constants μ (permeability), ϵ (dielectric constant), and σ (conductivity), are the transposes of the original constants. [If $\tilde{\epsilon}$ represents the transpose of ϵ, and $\mathbf{A}$ and $\mathbf{B}$ represent any two vectors, then the scalar product $\mathbf{A}\cdot(\epsilon\mathbf{B})\equiv\mathbf{B}\cdot(\tilde{\epsilon}\mathbf{A})$]. Then the formula for the reaction is

$$\langle a,b\rangle=\int\int\int[\mathbf{E}(b)\cdot d\mathbf{J}(a)-\mathbf{H}(b)\cdot d\mathbf{K}(a)], \quad (50)$$

subject to the same conditions as Eq. (1) except that the media need not be isotropic.

ACKNOWLEDGMENT

It is a pleasure to acknowledge many helpful discussions with T. H. Crowley, and the partial support of a contract between the Signal Corps and the Ohio State University Research Foundation.

Matrix Methods for Field Problems

ROGER F. HARRINGTON, SENIOR MEMBER, IEEE

Abstract—**A unified treatment of matrix methods useful for field problems is given. The basic mathematical concept is the method of moments, by which the functional equations of field theory are reduced to matrix equations. Several examples of engineering interest are included to illustrate the procedure. The problem of radiation and scattering by wire objects of arbitrary shape is treated in detail, and illustrative computations are given for linear wires. The wire object is represented by an admittance matrix, and excitation of the object by a voltage matrix. The current on the wire object is given by the product of the admittance matrix with the voltage matrix. Computation of a field quantity corresponds to multiplication of the current matrix by a measurement matrix. These concepts can be generalized to apply to objects of arbitrary geometry and arbitrary material.**

I. Introduction

THE USE of high-speed digital computers not only allows one to make more computations than ever before, it makes practicable methods too repetitious for hand computation. In the past much effort was expended to analytically manipulate solutions into a form which minimized the computational effort. It is now often more convenient to use computer time to reduce the analytical effort. Almost any linear problem of analysis can be solved to some degree of approximation, depending upon the ingenuity and effort expended. In other words, the methods are known, but much work remains to be done on the details.

It is the purpose of this paper to give a brief discussion of a general procedure for solving linear field problems, and to apply it to some examples of engineering interest. The procedure is called a *matrix method* because it reduces the original functional equation to a matrix equation. The name *method of moments* has been given to the mathematical procedure for obtaining the matrix equations. Sometimes the procedure is called an approximation technique, but this is a misnomer when the solution converges in the limit. It is only the computational time for a given accuracy which differs from other solutions, as, for example, an infinite power series. Of course, the method can also be used for truly approximate solutions, that is, ones which do not converge in the limit.

The mathematical concepts are conveniently discussed in the language of linear spaces and operators. However, an attempt has been made to minimize the use of this language, so that readers unfamiliar with it may better follow the discussion. Those concepts which are used are defined as they are introduced. Detailed expositions of linear spaces and operators may be found in many textbooks [1]–[3].

In this paper, only equations of the inhomogeneous type

$$L(f) = g \tag{1}$$

will be considered. Here L is a *linear operator*, g is the *excitation* or *source* (known function), and f is the *field* or *response* (unknown to be determined). The problem is said to be *deterministic* if the solution is unique, that is, if only one f is associated with each g. The problem of *analysis* involves determining f when L and g are given. The problem of *synthesis* involves determining L when f and g are specified. This paper deals only with analysis.

The method of moments gives a general procedure for treating field problems, but the details of solution vary widely with the particular problem. The examples of this paper have been chosen not only because they illustrate these details, but also because they are problems of engineering interest. It is hoped that these examples will allow the reader to solve similar problems, and also will suggest extensions and modifications suitable for other types of problems. While the examples are all taken from electromagnetic theory, the procedures apply to field problems of all kinds.

II. Formulation of Problems

Given a deterministic problem of the form (1), it is desired to identify the operator L, its domain (the functions f on which it operates), and its range (the functions g resulting from the operation). Furthermore, one usually needs an *inner product* $\langle f, g\rangle$, which is a scalar defined to satisfy[1]

$$\langle f, g\rangle = \langle g, f\rangle \tag{2}$$

$$\langle \alpha f + \beta g, h\rangle = \alpha\langle f, h\rangle + \beta\langle g, h\rangle \tag{3}$$

$$\langle f^*, f\rangle \begin{cases} > 0, & \text{if } f \neq 0 \\ = 0, & \text{if } f = 0 \end{cases} \tag{4}$$

where α and β are scalars, and $*$ denotes complex conjugate. The *norm* of a function is denoted $\|f\|$ and defined by

$$\|f\| = \sqrt{\langle f, f^*\rangle}. \tag{5}$$

It corresponds to the Euclidean vector concept of length. The *metric* d of two functions is

$$d(f, g) = \|f - g\| \tag{6}$$

Manuscript received September 15, 1966. This invited paper is one of a series planned on topics of general interest.—*The Editor*. This work was supported partly by Contract AF 30(602)-3724 from the Rome Air Development Center, Griffiss Air Force Base, N. Y., and partly by the National Science Foundation under Grant GK-704.

The author is with the Department of Electrical Engineering, Syracuse University, Syracuse, N. Y.

[1] The usual definition of inner product in Hilbert space corresponds to $\langle f^*, g\rangle$ in our notation. For this paper it is more convenient to show the conjugate operation explicitly wherever it occurs, and to define the adjoint operator without conjugation.

Reprinted from *Proc. IEEE*, vol. 55, no. 2, pp. 136–149, February 1967.

and corresponds to the Euclidean vector concept of distance between two points. It is important for discussing the convergence of solutions.

Properties of the solution of (1) depend on properties of the operator L. The *adjoint operator* L^a and its domain are defined by

$$\langle Lf, g\rangle = \langle f, L^a g\rangle \tag{7}$$

for all f in the domain of L. An operator is *self adjoint* if $L^a = L$ and the domain of L^a is that of L. An operator is *real* if Lf is real whenever f is real. An operator is *positive definite* if

$$\langle f^*, Lf\rangle > 0 \tag{8}$$

for all $f \neq 0$ in its domain. It is *positive semidefinite* if $>$ is replaced by $\geq$ in (8), *negative definite* if $>$ is replaced by $<$ in (8), etc. Other properties of operators will be identified as they are needed.

If the solution to $L(f)=g$ exists and is unique for all g, then the *inverse operator* L^{-1} exists such that

$$f = L^{-1}(g). \tag{9}$$

If g is known, then (9) represents the solution to the original problem. However, (9) is itself an inhomogeneous equation for g if f is known, and its solution is $L(f)=g$. Hence, L and L^{-1} form a pair of operators, each of which is the inverse of the other.

Facility in formulating problems using the concepts of linear spaces comes only with practice, which will be provided by the examples in later sections. For the present, a simple abstract example will be considered, so that mathematical concepts may be illustrated without bringing physical concepts into the picture.

Example: Given $g(x)$, find $f(x)$ in the interval $0 \leq x \leq 1$ satisfying

$$-\frac{d^2 f}{dx^2} = g(x) \tag{10}$$

and

$$f(0) = f(1) = 0. \tag{11}$$

This is a boundary value problem for which

$$L = -\frac{d^2}{dx^2}. \tag{12}$$

The range of L is the space of all functions g in the interval $0 \leq x \leq 1$ which are being considered. The domain of L is the space of those functions f in the interval $0 \leq x \leq 1$, satisfying the boundary conditions (11), and having second derivatives in the range of L. The solution to (10) is not unique unless appropriate boundary conditions are included. In other words, both the differential operator and its domain are required to define the operator.

A suitable inner product for this problem is

$$\langle f, g\rangle = \int_0^1 f(x)g(x)\,dx. \tag{13}$$

It is easily shown that (13) satisfies the postulates (2) to (4), as required. Note that the definition (13) is not unique. For example,

$$\int_0^1 w(x)f(x)g(x)\,dx \tag{14}$$

where $w(x) > 0$ is an arbitrary weighting function, is also an acceptable inner product. However, the adjoint operator depends on the inner product, and it can often be chosen to make the operator self adjoint.

To find the adjoint of a differential operator, form the left-hand side of (7), and integrate by parts to obtain the right-hand side. For the present problem

$$\begin{aligned}\langle Lf, g\rangle &= \int_0^1 \left(-\frac{d^2 f}{dx^2}\right) g\,dx \\ &= \int_0^1 \frac{df}{dx}\frac{dg}{dx}\,dx - \left[\frac{df}{dx}g\right]_0^1 \\ &= \int_0^1 f\left(-\frac{d^2 g}{dx^2}\right)dx + \left[f\frac{dg}{dx} - g\frac{df}{dx}\right]_0^1.\end{aligned} \tag{15}$$

The last terms are boundary terms, and the domain of L^a may be chosen so that these vanish. The first boundary terms vanish by (11), and the second vanish if

$$g(0) = g(1) = 0. \tag{16}$$

It is then evident that the adjoint operator to (12) for the inner product (13) is

$$L^a = L = -\frac{d^2}{dx^2}. \tag{17}$$

Since $L^a = L$ and the domain of L^a is the same as that of L, the operator is self adjoint.

It is also evident that L is a real operator, since Lf is real when f is real. That L is a positive definite operator is shown from (8) as follows:

$$\begin{aligned}\langle f^*, Lf\rangle &= \int_0^1 f^*\left(-\frac{d^2 f}{dx^2}\right)dx \\ &= \int_0^1 \frac{df^*}{dx}\frac{df}{dx}\,dx - \left[f^*\frac{df}{dx}\right]_0^1 \\ &= \int_0^1 \left|\frac{df}{dx}\right|^2 dx.\end{aligned} \tag{18}$$

Note that L is a positive definite operator even if f is complex.

The inverse operator to L can be obtained by standard Green's function techniques.[2] It is

$$L^{-1}(g) = \int_0^1 G(x, x')g(x')\,dx' \tag{19}$$

where G is the Green's function

$$G(x, x') = \begin{cases} x(1 - x'), & x < x' \\ (1 - x)x', & x > x' \end{cases}. \tag{20}$$

[2] See, for example, Friedman [2], ch. 3.

One can verify that (19) is the inverse operator by forming $f=L^{-1}(g)$, differentiating twice, and obtaining (10). Note that no boundary conditions are needed on the domain of L^{-1}, which is characteristic of most integral operators. That L^{-1} is self adjoint follows from the proof that L is self adjoint, since

$$\langle Lf_1, f_2\rangle = \langle g_1, L^{-1}g_2\rangle. \tag{21}$$

Of course, the self-adjointness of L^{-1} can also be proved directly. It similarly follows that L^{-1} is positive definite whenever L is positive definite, and vice versa.

III. Method of Moments

A general procedure for solving linear equations is the *method of moments* [4]. Consider the deterministic equation

$$L(f) = g \tag{22}$$

where L is a linear operator, g is known, and f is to be determined. Let f be expanded in a series of functions $f_1, f_2, f_3, \cdots$ in the domain of L, as

$$f = \sum_n \alpha_n f_n \tag{23}$$

where the α_n are constants. The f_n are called *expansion functions* or *basis functions*. For exact solutions, (23) is usually an infinite summation and the f_n form a complete set of basis functions. For approximate solutions, (23) is usually a finite summation. Substituting (23) into (22), and using the linearity of L, one has

$$\sum_n \alpha_n L(f_n) = g. \tag{24}$$

It is assumed that a suitable inner product $\langle f, g\rangle$ has been determined for the problem. Now define a set of *weighting functions*, or *testing functions*, $w_1, w_2, w_3, \cdots$ in the range of L, and take the inner product of (24) with each w_m. The result is

$$\sum_n \alpha_n \langle w_m, Lf_n\rangle = \langle w_m, g\rangle \tag{25}$$

$m=1, 2, 3, \cdots$. This set of equations can be written in matrix form as

$$[l_{mn}][\alpha_n] = [g_m] \tag{26}$$

where

$$[l_{mn}] = \begin{bmatrix} \langle w_1, Lf_1\rangle & \langle w_1, Lf_2\rangle & \cdots \\ \langle w_2, Lf_1\rangle & \langle w_2, Lf_2\rangle & \cdots \\ \cdots & \cdots & \cdots \end{bmatrix} \tag{27}$$

$$[\alpha_n] = \begin{bmatrix} \alpha_1 \\ \alpha_2 \\ \vdots \end{bmatrix} \qquad [g_m] = \begin{bmatrix} \langle w_1, g\rangle \\ \langle w_2, g\rangle \\ \vdots \end{bmatrix}. \tag{28}$$

If the matrix $[l]$ is nonsingular its inverse $[l^{-1}]$ exists. The α_n are then given by

$$[\alpha_n] = [l_{nm}^{-1}][g_m] \tag{29}$$

and the solution for f is given by (23). For concise expression of this result, define the matrix of functions

$$[\tilde{f}] = [f_1 \quad f_2 \quad f_3 \quad \cdots] \tag{30}$$

and write

$$f = [\tilde{f}_n][\alpha_n] = [\tilde{f}_n][l_{nm}^{-1}][g_m]. \tag{31}$$

This solution may be exact or approximate, depending upon the choice of the f_n and w_n. The particular choice $w_n=f_n$ is known as *Galerkin's method* [5], [6].

If the matrix $[l]$ is of infinite order, it can be inverted only in special cases, for example, if it is diagonal. The classical eigenfunction method leads to a diagonal matrix, and can be thought of as a special case of the method of moments. If the sets f_n and w_n are finite, the matrix is of finite order, and can be inverted by known computational algorithms.

One of the main tasks in any particular problem is the choice of the f_n and w_n. The f_n should be linearly independent and chosen so that some superposition (23) can approximate f reasonably well. The w_n should also be linearly independent and chosen so that the products $\langle w_n, g\rangle$ depend on relatively independent properties of g. Some additional factors which affect the choice of f_n and w_n are a) the accuracy of solution desired, b) the ease of evaluation of the matrix elements, c) the size of the matrix that can be inverted, and d) the realization of a well-conditioned matrix $[l]$.

Example: Consider again the problem stated by (10) and (11). For a power-series solution, choose

$$f_n = x^{n+1} - x \tag{32}$$

$n=1, 2, 3, \cdots, N$, so that the series (23) is

$$f = \sum_{n=1}^{N} \alpha_n(x^{n+1} - x). \tag{33}$$

Note that the term $-x$ is needed in (34), else the f_n will not be in the domain of L, that is, the boundary conditions will not be satisfied. For testing functions, choose

$$w_n = f_n = x^{n+1} - x \tag{34}$$

in which case the method is that of Galerkin. In Section V it is shown that the w_n should be in the domain of the adjoint operator. Since L is self adjoint for this problem, the w_n should be in the domain of L, as are those of (34). Evaluation of the matrix (27) for the inner product (13) and L given by (12) is straightforward. The resultant elements are

$$l_{mn} = \langle w_m, Lf_n\rangle = \frac{mn}{m+n+1}. \tag{35}$$

A knowledge of the matrix elements (35) is fully equivalent to the original differential equation. Hence, a matrix formulation for the problem has been obtained. For any particular excitation g, the matrix excitation $[g_m]$ has elements given by

$$g_m = \langle w_m, g\rangle = \int_0^1 (x^{m+1} - x)g(x)\,dx \tag{36}$$

and a solution to the boundary value problem is given by (31). This solution is a power series, exact if f can be expressed as a power series. In general, it is an infinite power-series solution, in which case a finite number of terms gives an approximate solution. The nature of the approximation is discussed in Section V.

IV. Special Techniques

As long as the operator equation is simple, application of the method of moments gives solutions in a straightforward manner. However, most field problems of engineering interest are not so simple. The physical problem may be represented by many different operator equations, and a suitable one must be chosen. Even then the form of L may be very complicated. There are an infinite number of sets of expansion functions f_n and testing functions w_n that may be chosen. Finally, there are mathematical approximations that can be made in the evaluation of the matrix elements of l_{mn} and g_m. In this section a number of special techniques, helpful for overcoming some of these difficulties, will be discussed in general terms. Some of these concepts will be used in the electromagnetic field problems considered later.

Point-Matching: The integration involved in evaluating the $l_{mn}=\langle w_m, Lf_n\rangle$ of (27) is often difficult to perform in problems of practical interest. A simple way to obtain approximate solutions is to require that (24) be satisfied at discrete points in the region of interest. This procedure is called a *point-matching method.* In terms of the method of moments, it is equivalent to using Dirac delta functions as testing functions.

Subsectional Bases: Another approximation useful for practical problems is the *method of subsections.* This involves the use of basis functions f_n each of which exists only over subsections of the domain of f. Then each α_n of the expansion (23) affects the approximation of f only over a subsection of the region of interest. This procedure often simplifies the evaluation and/or the form of the matrix $[l]$. Sometimes it is convenient to use the point-matching method of the preceding section in conjunction with the subsection method.

Extended Operators: As noted earlier, an operator is defined by an operation (for example, $L=-d^2/dx^2$) plus a domain (space of functions to which the operation may be applied). We can *extend the domain* of an operator by redefining the operation to apply to new functions (not in the original domain) so long as this extended operation does not change the original operation in its domain. If the original operator is self adjoint, it is desirable to make the extended operator also self adjoint. By this procedure we can use a wider class of functions for solution by the method of moments. This becomes particularly important in multivariable problems (fields in multidimensional space) where it is not always easy to find simple functions in the domain of the original operator.

Approximate Operators: In complex problems it is sometimes convenient to approximate the operator to obtain solutions. For differential operators, the finite difference approximation has been widely used [7]. For integral operators, an approximate operator can be obtained by approximating the kernel of the integral operator [5]. Any method whereby a functional equation is reduced to a matrix equation can be interpreted in terms of the method of moments. Hence, for any matrix solution using approximation of the operator there will be a corresponding moment solution using approximation of the function.

Perturbation Solutions: Sometimes the problem under consideration is only slightly different (perturbed) from a problem which can be solved exactly (the unperturbed problem). A first-order solution to the perturbed problem can then be obtained by using the solution to the unperturbed problem as a basis for the method of moments. This procedure is called a *perturbation method.* Higher-order perturbation solutions can be obtained by using the unperturbed solution plus correction terms in the method of moments. Sometimes this is done as successive approximations by including one correction term at a time, but for machine computations it is usually easier to include all correction terms at once.

V. Variational Interpretation

It is known that Galerkin's method ($w_n=f_n$) is equivalent to the Rayleigh-Ritz variational method [5], [6]. The method of moments is also equivalent to the variational method, the proof being essentially the same as that for Galerkin's method. The application of these techniques to electromagnetic field problems is known as the reaction concept [8], [9].

An interpretation of the method of moments in terms of linear spaces will first be given. Let $\mathcal{S}\,(Lf)$ denote the range of L, $\mathcal{S}\,(Lf_n)$ denote the space spanned by the Lf_n, and $\mathcal{S}\,(w_n)$ denote the space spanned by the w_n. The method of moments (25) then equates the projection of Lf onto $\mathcal{S}\,(w_n)$ to the projection of the approximate Lf onto $\mathcal{S}\,(w_n)$. In other words, both the approximate Lf and the exact Lf have equal components in $\mathcal{S}\,(w_n)$. The difference between the approximate Lf and the exact Lf is the error, which is orthogonal to $\mathcal{S}\,(w_n)$. Because of this orthogonality, a first-order change in the projection produces only a second-order change in the error. In Galerkin's method, $\mathcal{S}\,(w_n)=\mathcal{S}\,(f_n)$, and the distance from the approximate Lf to the exact Lf is minimized. In general, the method of moments does not minimize the distance from the approximate f to the exact f, although it may in some special cases.

The variational approach to the same problem is as follows. Given an operator equation $Lf=g$, it is desired to determine a functional of f (number depending on f)

$$\rho(f)=\langle f, h\rangle \tag{37}$$

where h is a given function. If h is a continuous function, then $\rho(f)$ is a *continuous linear functional.* Now let L^a be the adjoint operator to L and define an adjoint function f^a (adjoint field) by

$$L^a f^a = h. \tag{38}$$

By the calculus of variations, it can then be shown that [6]

$$\rho = \frac{\langle f, h\rangle\langle f^a, g\rangle}{\langle Lf, f^a\rangle} \tag{39}$$

is a variational formula for ρ with stationary point (37) when f is the solution of $Lf=g$ and f^a the solution to (38). For an approximate evaluation of ρ, let

$$f = \sum_n \alpha_n f_n \qquad f^a = \sum_m \beta_m w_m. \tag{40}$$

Substitute these into (39), and apply the Rayleigh-Ritz conditions $\partial\rho/\partial\alpha_i = \partial\rho/\partial\beta_i = 0$ for all i. The result is that the necessary and sufficient conditions for ρ to be a stationary point are (25), [6]. Hence, the method of moments is identical to the Rayleigh-Ritz variational method. Sometimes the method of moments is called a *direct method*, in contrast to variational approaches which are often rather circuitous.

The above variational interpretation can be used to give additional insight in how to choose the testing functions. It is evident from (38) and (40) that the w_n should be chosen so that some linear combination of them can closely represent the adjoint field f^a. When we calculate f itself by the method of moments, h of (37) is a Dirac delta function, ρ of (37) is no longer a continuous linear functional, and f^a of (38) is a Green's function. This implies that some combination of the w_n must be able to approximate the Green's function. Since a Green's function is usually poorly behaved, one should expect computation of a field by the method of moments to converge less slowly than computation of a continuous linear functional. This is found to be the case.

VI. Electrostatics

This section is a general discussion of electrostatic problems according to the operational formulation. The static electric intensity $\mathbf{E}$ is conveniently found from an electrostatic potential ϕ according to

$$\mathbf{E} = -\nabla\phi \tag{41}$$

where ∇ is the gradient operator. In a region of constant permittivity ε and volume change density ρ, the electrostatic potential satisfies the *Poisson equation*

$$-\varepsilon\nabla^2\phi = \rho \tag{42}$$

where ∇^2 is the Laplacian operator. For unique solutions, boundary conditions on ϕ are needed. In other words, the domain of the operator must be specified.

For now, consider fields from charges in unbounded space, in which case

$$r\phi \to \text{constant as } r \to \infty \tag{43}$$

for every ρ of finite extent, where r is the distance from the coordinate origin. The differential operator formulation is therefore

$$L\phi = \rho \tag{44}$$

where

$$L = -\varepsilon\nabla^2 \tag{45}$$

and the domain of L is those functions ϕ whose Laplacian exists and which have $r\phi$ bounded at infinity according to (43). The well-known solution to this problem is

$$\phi(x, y, z) = \iiint \frac{\rho(x', y', z')}{4\pi\varepsilon R}\, dx'dy'dz' \tag{46}$$

where $R = \sqrt{(x-x')^2 + (y-y')^2 + (z-z')^2}$ is the distance from a source point (x', y', z') to a field point (x, y, z). Hence, the inverse operator to L is

$$L^{-1} = \iiint dx'dy'dz' \frac{1}{4\pi\varepsilon R}. \tag{47}$$

It is important to keep in mind that (47) is inverse to (45) only for the boundary conditions (43). If the boundary conditions are changed, L^{-1} changes. Also, the designation of (45) as L and (47) as L^{-1} is arbitrary, and the notation could be reversed if desired.

A suitable inner product for electrostatic problems is

$$\langle\phi, \psi\rangle = \iiint \phi(x, y, z)\psi(x, y, z)\, dxdydz \tag{48}$$

where the integration is over all space. That (48) satisfies the required postulates (2), (3), and (4) is easily verified. It will now be shown that L is self adjoint for this inner product. From the left-hand side of (7)

$$\langle L\phi, \psi\rangle = \iiint (-\varepsilon\nabla^2\phi)\psi\, d\tau \tag{49}$$

where $d\tau = dxdydz$. Green's identity is

$$\iiint_V (\psi\nabla^2\phi - \phi\nabla^2\psi)\, d\tau = \oiint_S \left(\psi\frac{\partial\phi}{\partial n} - \phi\frac{\partial\psi}{\partial n}\right) ds \tag{50}$$

where S is the surface bounding the volume V and n is the outward direction normal to S. Let S be a sphere of radius r, so that in the limit $r\to\infty$ the volume V includes all space. For ϕ and ψ satisfying boundary conditions (43), $\psi\to C_1/r$, and $\partial\phi/\partial n\to C_2/r^2$ as $r\to\infty$. Hence, $\psi\partial\phi/\partial n\to C/r^3$ as $r\to\infty$, and similarly for $\phi\partial\psi/\partial n$. Since $ds = r^2 \sin\theta\, d\theta d\phi$ increases only as r^2, the right-hand side of (50) vanishes as $r\to\infty$. Equation (50) then reduces to

$$\iiint \psi\nabla^2\phi\, d\tau = \iiint \phi\nabla^2\psi\, d\tau \tag{51}$$

from which it is evident that the adjoint operator L^a is

$$L^a = L = -\varepsilon\nabla^2. \tag{52}$$

Since the domain of L^a is that of L, the operator L is self adjoint. The concept of self adjointness in this case is related to the physical concept of reciprocity.

It is evident from (45) and (47) that L and L^{-1} are real operators. It will now be shown that they are also positive definite, that is, they satisfy (8). As discussed in Section II, this need be shown only for L or L^{-1}. For L, form

$$\langle \phi^*, L\phi \rangle = \iiint \phi^*(-\varepsilon\nabla^2\phi)\, d\tau \tag{53}$$

and use the vector identity $\phi\nabla^2\phi = \nabla\cdot(\phi\nabla\phi) - \nabla\phi\cdot\nabla\phi$ plus the divergence theorem. The result is

$$\langle \phi^*, L\phi \rangle = \iiint_V \varepsilon\nabla\phi^*\cdot\nabla\phi\, d\tau - \oiint_S \varepsilon\phi^*\nabla\phi\cdot ds \tag{54}$$

where S bounds V. Again take S a sphere of radius r. For ϕ satisfying (43), the last term of (54) vanishes as $r\to\infty$ for the same reasons as in (50). Then

$$\langle \phi^*, L\phi \rangle = \iiint \varepsilon|\nabla\phi|^2\, d\tau \tag{55}$$

and, for ε real and $\varepsilon > 0$, L is positive definite. In this case positive definiteness of L is related to the concept of electrostatic energy.

VII. Charged Conducting Plate

Consider a square conducting plate 2a meters on a side and lying on the $z=0$ plane with center at the origin, as shown in Fig. 1. Let $\sigma(x, y)$ represent the surface charge density on the plate, assumed to have zero thickness. The electrostatic potential at any point in space is

$$\phi(x, y, z) = \int_{-a}^{a} dx' \int_{-a}^{a} dy' \frac{\sigma(x', y')}{4\pi\varepsilon R} \tag{56}$$

where $R = \sqrt{(x-x')^2 + (y-y')^2 + z^2}$. The boundary condition is $\phi = V$ (constant) on the plate. The integral equation for the problem is therefore

$$V = \int_{-a}^{a} dx' \int_{-a}^{a} dy' \frac{\sigma(x', y')}{4\pi\varepsilon\sqrt{(x-x')^2 + (y-y')^2}} \tag{57}$$

$|x| < a$, $|y| < a$. The unknown to be determined is the charge density $\sigma(x, y)$. A parameter of interest is the capacitance of the plate

$$C = \frac{q}{V} = \frac{1}{V}\int_{-a}^{a} dx \int_{-a}^{a} dy\, \sigma(x, y) \tag{58}$$

which is continuous linear functional of σ.

A straightforward development of a subsection and point-matching solution [10] will first be given, and later it will be interpreted in terms of more general concepts. Consider the plate divided into N square subsections, as shown in Fig. 1. Define functions

$$f_n = \begin{cases} 1 & \text{on } \Delta s_n \\ 0 & \text{on all other } \Delta s_m \end{cases} \tag{59}$$

and let the charge density be represented by

$$\sigma(x, y) \approx \sum_{n=1}^{N} \alpha_n f_n. \tag{60}$$

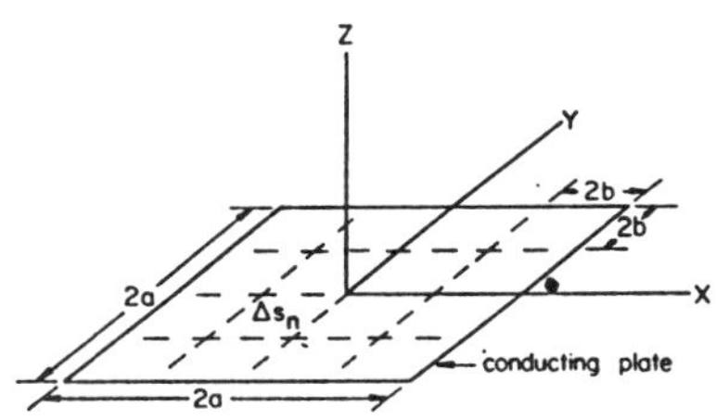

Fig. 1. A square conducting plate.

Substituting (60) into (57) and satisfying the resultant equation at the midpoint (x_m, y_m) of each Δs_m, one obtains the set of equations

$$V = \sum_{n=1}^{N} l_{mn}\alpha_n, \qquad m = 1, 2, \cdots, N \tag{61}$$

where

$$l_{mn} = \int_{\Delta x_n} dx' \int_{\Delta y_n} dy' \frac{1}{4\pi\varepsilon\sqrt{(x_m - x')^2 + (y_m - y')^2}}. \tag{62}$$

Note that l_{mn} is the potential at the center of Δs_m due to a uniform charge density of unit amplitude over Δs_n. A solution to the set (61) gives the α_m, in terms of which the charge density is approximated by (60). The corresponding capacitance of the plate, approximating (58), is

$$C \approx \frac{1}{V}\sum_{n=1}^{N} \alpha_n \Delta s_n = \sum_{mn} l_{nm}^{-1} \Delta s_n. \tag{63}$$

This result can be interpreted as stating that the capacitance of an object is the sum of the capacitances of all its subsections plus the mutual capacitances between every pair of subsections.

To translate the above results into the language of linear spaces and the method of moments, let

$$f(x, y) = \sigma(x, y) \tag{64}$$

$$g(x, y) = V, \quad |x| < a, \quad |y| < a \tag{65}$$

$$L(f) = \int_{-a}^{a} dx' \int_{-a}^{a} dy' \frac{f(x', y')}{4\pi\varepsilon\sqrt{(x-x')^2 + (y-y')^2}}. \tag{66}$$

Then $L(f) = g$ is equivalent to (57). A suitable inner product, satisfying (2) to (4), for which L is self adjoint, is

$$\langle f, g \rangle = \int_{-a}^{a} dx \int_{-a}^{a} dy\, f(x, y)\, g(x, y). \tag{67}$$

To apply the method of moments, use the function (59) as a subsectional basis, and define testing functions as

$$w_m = \delta(x - x_m)\delta(y - y_m) \tag{68}$$

which is the two-dimensional Dirac delta function. Now the elements of the $[l]$ matrix (27) are those of (62) and the $[g]$ matrix of (28) is

$$[g_m] = \begin{bmatrix} V \\ V \\ \vdots \\ V \end{bmatrix}. \tag{69}$$

The matrix equation (26) is, of course, identical to the set of equations (61). In terms of the inner product (67), the capacitance (58) can be written as

$$C = \frac{\langle \sigma, \phi \rangle}{V^2} \tag{70}$$

since $\phi = V$ on the plate. Equation (70) is the conventional stationary formula for the capacitance of a conducting body [11].

For numerical results, the l_{mn} of (62) must be evaluated. Let $2b = 2a/\sqrt{N}$ denote the side length of each Δs_n. The potential at the center of Δs_n due to unit charge density over its own surface is

$$\begin{aligned} l_{nn} &= \int_{-b}^{b} dx \int_{-b}^{b} dy \frac{1}{4\pi\varepsilon\sqrt{x^2+y^2}} \\ &= \frac{2b}{\pi\varepsilon} \ln(1+\sqrt{2}) = \frac{2b}{\pi\varepsilon}(0.8814). \end{aligned} \tag{71}$$

This derivation uses Dwight 200.01 and 731.2 [12]. The potential at the center of Δs_m due to unit charge over Δs_n can be similarly evaluated, but the formula is complicated. For most purposes it is sufficiently accurate to treat the charge on Δs_n as if it were a point charge, and use

$$l_{mn} \approx \frac{\Delta s_n}{4\pi\varepsilon R_{mn}} = \frac{b^2}{\pi\varepsilon\sqrt{(x_m-x_n)^2+(y_m-y_n)^2}} \qquad m \neq n. \tag{72}$$

This approximation is 3.8 percent in error for adjacent subsections, and has less error for nonadjacent ones. Table I shows capacitance, calculated by (63) using the α's obtained from the solution of (61), for various numbers of subareas. The second column of Table I uses the approximation (72), the third column uses an exact evaluation of the l_{mn}. A good estimate of the true capacitance is 40 picofarads. Figure 2 shows a plot of the approximate charge density along the subareas nearest the center line of the plate, for the case $N = 100$ subareas. Note that σ exhibits the well-known square root singularity at the edges of the plate.

TABLE I

CAPACITANCE OF A SQUARE PLATE (PICOFARADS PER METER)

No. of subareas	$C/2a$ approx. l_{mn}	$C/2a$ exact l_{mn}
1	31.5	31.5
9	37.3	36.8
16	38.2	37.7
36	39.2	38.7
100		39.5

VIII. ELECTROMAGNETIC FIELDS

The operator formulation of electromagnetic fields is analogous to that of electrostatic fields, but considerably more complicated. For the time-harmonic case, $e^{j\omega t}$ variation, the Maxwell equations are[3]

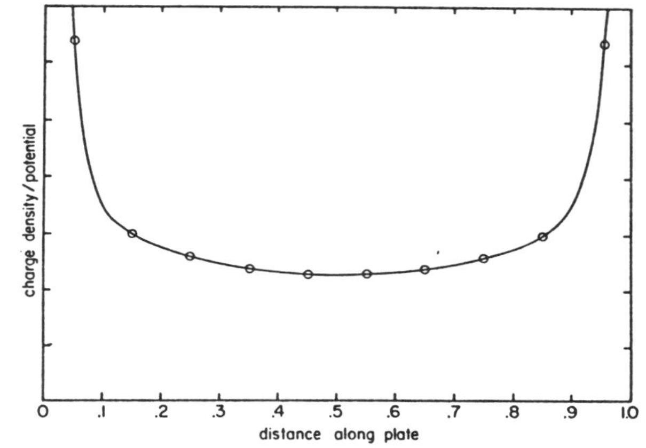

Fig. 2. Approximate charge density on subareas closest to the centerline of a square plate.

$$\begin{aligned} \nabla \times E &= -j\omega\mu H \\ \nabla \times H &= j\omega\varepsilon E + J \end{aligned} \tag{73}$$

where E is the electric field, H the magnetic field, and J the electric current density. Equations (73) can be combined into a single equation for E as

$$\frac{-1}{j\omega} \nabla \times (\mu^{-1} \nabla \times E) - j\omega\varepsilon E = J. \tag{74}$$

This is of the form

$$L(E) = J \tag{75}$$

where the operator L is evident from (74). For a specific case, let the permittivity and permeability be that of free space, that is, $\varepsilon = \varepsilon_0$ and $\mu = \mu_0$. The domain of L must be restricted by suitable differentiability conditions on E, and boundary conditions on E must be given. To be specific, let these boundary conditions be the radiation condition, that is, the field must represent outward traveling waves at infinity.

The inverse operator is the well-known potential integral solution to (74), which is

$$E = L^{-1}(J) = -j\omega A - \nabla\Phi \tag{76}$$

where

$$A(r) = \mu \iiint J(r') \frac{e^{-jk|r-r'|}}{4\pi|r-r'|} d\tau' \tag{77}$$

$$\Phi(r) = \frac{1}{\varepsilon} \iiint \rho(r') \frac{e^{-jk|r-r'|}}{4\pi|r-r'|} d\tau' \tag{78}$$

$$\rho = \frac{-1}{j\omega} \nabla \cdot J. \tag{79}$$

These equations can be combined into a single equation

$$E = L^{-1}(J) = \iiint \Gamma(r, r') \cdot J(r') d\tau' \tag{80}$$

where Γ is the dyadic Green's function. However, the derivation of (80) involves an interchange of integration and differentiation which restricts the domain of L^{-1} more than

[3] Only the case of electric sources is considered in this paper. The more general case of electric and magnetic sources is treated by the reaction concept [8], [9].

necessary [13]. It is often better to consider (76) to (78) as the basic equations, with (80) as symbolic of them.

A suitable inner product for electromagnetic field problems is

$$\langle E, J\rangle = \iiint E \cdot J \, d\tau \tag{81}$$

which is the quantity called *reaction*. Note that (81) satisfies postulates (2), (3), and (4). The concept of reciprocity is a statement of the self-adjointness of L^{-1}, that is,

$$\langle L^{-1}J_1, J_2\rangle = \langle J_1, L^{-1}J_2\rangle. \tag{82}$$

The operator L is also self adjoint, since (82) can be written as

$$\langle E_1, LE_2\rangle = \langle LE_1, E_2\rangle. \tag{83}$$

Other properties of L can be determined as the need arises.

IX. Wires of Arbitrary Shape

An important engineering problem is the electromagnetic behavior of thin wire objects. A general analysis of such objects according to the method of moments is presented in this section. The impressed field is considered arbitrary, and hence both the antenna and scatterer problems are included in the solution. The distinction between antennas and scatterers is primarily that of the location of the source. If the source is at the object it is viewed as an antenna; if the source is distant from the object it is viewed as a scatterer.

So that the development of the solution may be easily followed, it is given with few references to the general theory. Basically, it involves a) an approximation of the exact equation for conducting bodies by an approximate equation valid for thin wires, b) replacement of the derivatives by finite difference approximations, yielding an approximate operator, c) use of pulse functions for expansion functions, to give a step approximation to the current and charge, and d) the use of point-matching for testing.

A particularly descriptive exposition of the solution can be made in terms of network parameters. To effect a solution, the wire is considered as N short segments connected together. The end points of each segment define a pair of terminals in space. These N pairs of terminals can be thought of as forming an N port network, and the wire object is obtained by short-circuiting all ports of the network. One can determine the impedance matrix for the N port network by applying a current source to each port in turn, and calculating the open circuit voltages at all ports. This procedure involves only current elements in empty space. The admittance matrix is the inverse of the impedance matrix. Once the admittance matrix is known, the port currents (current distribution on the wire) are found for any particular voltage excitation (applied field) by matrix multiplication.

An integral equation for the charge density σ_s and current J_s on a conducting body S in a known impressed field E^i is obtained as follows. The scattered field E^s, produced by σ_s and J_s, is expressed in terms of retarded potential integrals, and the boundary condition $n \times (E^i + E^s) = 0$ on S is applied. This is summarized by

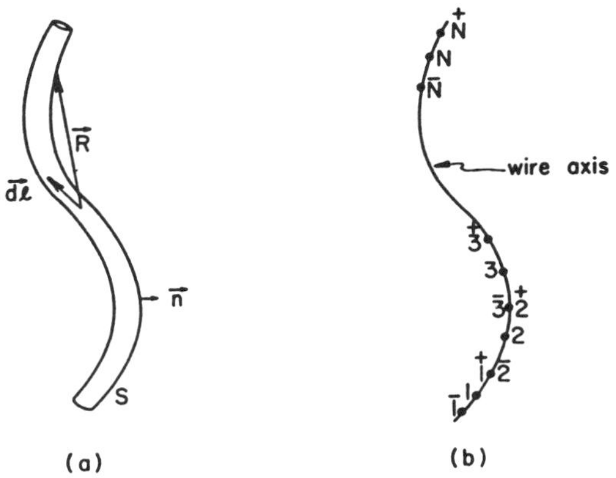

Fig. 3. (a) A wire scatterer. (b) The wire axis divided into N segments.

$$E^s = -j\omega A - \nabla\phi \tag{84}$$

$$A = \mu \oiint_S J_s \frac{e^{-jkR}}{4\pi R}\, dS \tag{85}$$

$$\phi = \frac{1}{\varepsilon} \oiint s\sigma_s \frac{e^{-jkR}}{4\pi R}\, dS \tag{86}$$

$$\sigma_s = \frac{-1}{j\omega} \nabla_s \cdot J_s \tag{87}$$

$$n \times E^s = -n \times E^i \quad \text{on } S. \tag{88}$$

Figure 3(a) represents an arbitrary thin-wire scatterer, for which the following approximations are made. a) The current is assumed to flow only in the direction of the wire axis. b) The current and charge densities are approximated by filaments of current I and charge σ on the wire axis. c) The boundary condition (88) is applied only to the axial component of E at the wire surface. To this approximation, (84) to (88) become

$$-E_l^i = -j\omega A_l - \frac{\partial \phi}{\partial l} \quad \text{on } S \tag{89}$$

$$A = \mu \int_{\text{axis}} I(l) \frac{e^{-jkR}}{4\pi R}\, dl \tag{90}$$

$$\phi = \frac{1}{\varepsilon} \int_{\text{axis}} \sigma(l) \frac{e^{-jkR}}{4\pi R}\, dl \tag{91}$$

$$\sigma = \frac{-1}{j\omega} \frac{dI}{dl} \tag{92}$$

where l is the length variable along the wire axis, and R is measured from a source point on the axis to a field point on the wire surface.

A solution to the above equations is obtained as follows. Integrals are approximated by the sum of intergrals over N small segments, obtained by treating I and q as constant over each segment. Derivatives are approximated by finite differences over the same intervals used for integration. Figure 3(b) illustrates the division of the wire axis into N segments, and defines the notation. If a wire terminates, the boundary condition $I=0$ is taken into account by starting the first segment 1/2 interval in from the end of the wire. This is suggested in Fig. 3(b) by the extra 1/2 interval shown at each end. The nth segment is identified by its starting point $\bar{n}$, its midpoint n, and its termination $\dot{n}$. An increment Δl_n denotes that between $\bar{n}$ and $\dot{n}$, $\Delta l_{\bar{n}}$ and $\Delta l_{\dot{n}}$ denote increments shifted 1/2 segment minus or plus along l. The desired approximations for (89) to (92) are then

$$-E_l^i(m) \approx -j\omega A_l(m) - \frac{\phi(\dot{m}) - \phi(\bar{m})}{\Delta l_m} \tag{93}$$

$$A(m) = \mu \sum_n I(n) \int_{\Delta l_n} \frac{e^{-jkR}}{4\pi R}\, dl \tag{94}$$

$$\phi(\dot{m}) \approx \frac{1}{\varepsilon} \sum_n \sigma(\dot{n}) \int_{\Delta l_{\dot{n}}} \frac{e^{-jkR}}{4\pi R}\, dl \tag{95}$$

$$\sigma(\dot{n}) \approx \frac{-1}{j\omega}\left[\frac{I(n+1) - I(n)}{\Delta l_{\dot{n}}}\right] \tag{96}$$

with equations similar to (95) and (96) for $\phi(\bar{m})$ and $\sigma(\bar{n})$.

The σ's are given in terms of the I's by (96), and hence (93) can be written in terms of the $I(n)$ only. One can view the N equations represented by (93) as the equations for an N port network with terminal pairs $(\dot{n}, \bar{n})$. The voltages applied to each port are approximately $E^i \cdot \Delta l_n$. Hence, by defining

$$[I] = \begin{bmatrix} I(1) \\ I(2) \\ \vdots \\ I(N) \end{bmatrix} \qquad [V] = \begin{bmatrix} E^i(1)\cdot l_1 \\ E^i(2)\cdot l_2 \\ \vdots \\ E^i(N)\cdot l_N \end{bmatrix} \tag{97}$$

one can rewrite (93) in matrix form as

$$[V] = [Z][I]. \tag{98}$$

This corresponds to the method of moment representation (26), with $[Z]$ corresponding to $[l]$, $[V]$ to $[g]$, and $[I]$ to $[\alpha]$. The elements of the matrix $[Z]$ can be obtained by substituting (94) through (96) into (93) and rearranging into the form of (98). Alternatively, one can apply (93) through (96) to two isolated elements and obtain the impedance elements directly. This latter procedure will be used because it is somewhat easier to follow.

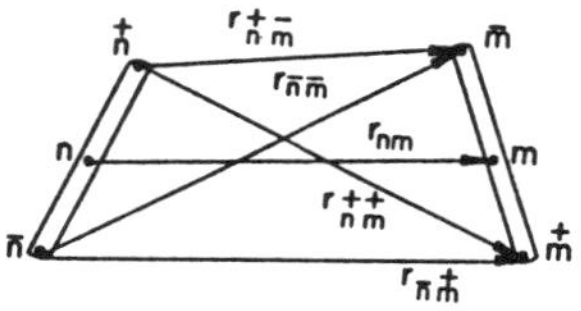

Fig. 4. Two segments of a wire scatterer.

Consider two representative elements of the wire scatterer, as shown in Fig. 4. The integrals in (94) and (95) are of the same form, and are denoted by

$$\psi(n, m) = \frac{1}{\Delta l_n} \int_{\Delta l_n} \frac{e^{-jkR_{mn}}}{4\pi R_{mn}}\, dl_n. \tag{99}$$

Symbols + and − are used over m and n when appropriate. Evaluation of the ψ in general is considered in the Appendix. Let element n of Fig. 4 consist of a current filament $I(n)$, and two charge filaments of net charge

$$q(\dot{n}) = \frac{1}{j\omega} I(n) \qquad q(\bar{n}) = \frac{-1}{j\omega} I(n) \tag{100}$$

where $q = \sigma \Delta l$. The vector potential at m due to $I(n)$ is, by (94),

$$A = \mu I(n) \Delta l_n \psi(n, m). \tag{101}$$

The scalar potentials at $\dot{m}$ and $\bar{m}$ due to the charges (100) are, by (95)

$$\begin{aligned} \phi(\dot{m}) &= \frac{1}{j\omega\varepsilon}[I(n)\psi(\dot{n}, \dot{m}) - I(n)\psi(\bar{n}, \dot{m})] \\ \phi(\bar{m}) &= \frac{1}{j\omega\varepsilon}[I(n)\psi(\dot{n}, \bar{m}) - I(n)\psi(\bar{n}, \bar{m})]. \end{aligned} \tag{102}$$

Substituting from (101) and (102) into (93), and forming $Z_{mn} = E^i(m) \cdot \Delta l_m / I(n)$, one obtains

$$Z_{mn} = j\omega\mu \Delta l_n \cdot \Delta l_m \psi(n, m) + \frac{1}{j\omega\varepsilon}[\psi(\dot{n}, \dot{m}) - \psi(\bar{n}, \dot{m}) - \psi(\dot{n}, \bar{m}) + \psi(\bar{n}, \bar{m})]. \tag{103}$$

This result applies for self impedances $(m = n)$ as well as for mutual impedances. When the two current elements are widely separated, a simpler formula based on the radiation field from a current element can be used.

The wire object is completely characterized by its impedance matrix, subject, of course, to the approximations involved. The object is defined by $2N$ points on the wire axis, plus the wire radius. The impedance elements are calculated by (103), and the voltage matrix is determined by the impressed field, according to (97). The current at N points on the scatterer is then given by the current matrix, obtained from the inversion of (98) as

$$[I] = [Y][V] \qquad [Y] = [Z]^{-1}. \tag{104}$$

Once the current distribution is known, parameters of interest such as field patterns, input impedances, echo areas, etc., can be calculated by numerically evaluating the appropriate formulas.

X. Wire Antennas

A wire antenna is obtained when the wire is excited by a voltage source at one or more points along its length. Hence, for an antenna excited in the nth interval, the applied voltage matrix (97) is

$$[V^s] = \begin{bmatrix} 0 \\ \vdots \\ V_n \\ \vdots \\ 0 \end{bmatrix} \tag{105}$$

i.e., all elements zero except the nth, which is equal to the source voltage. The current distribution is given by (104), which for the $[V]$ of (105) becomes

$$[I] = V_n \begin{bmatrix} Y_{1n} \\ Y_{2n} \\ \vdots \\ Y_{Nn} \end{bmatrix}. \tag{106}$$

Hence, the nth column of the admittance matrix is the current distribution for a unit voltage source applied to the nth interval. Inversion of the impedance matrix therefore gives simultaneously the current distributions when the antenna is excited in any arbitrary interval along its length. The diagonal elements Y_{nn} of the admittance matrix are the input admittances of the wire object fed in the nth interval, and the Y_{mn} are the transfer admittances between a port in the mth interval and one in the nth interval.

The radiation pattern of a wire antenna is obtained by treating the antenna as an array of N current elements $I(n)\Delta l_n$. By standard formulas, the far-zone vector potential is given by

$$A = \frac{\mu e^{-jkr_0}}{4\pi r_0} \sum_n I(n)\Delta l_n e^{jkr_n \cos \xi_n} \tag{107}$$

where r_0 and r_n are the radius vectors to the distant field point and to the source points, respectively, and ξ_n are the angles between r_0 and r_n. The far-zone field components are

$$E_\theta = -j\omega A_\theta \qquad E_\phi = -j\omega A_\phi \tag{108}$$

where θ and ϕ are the conventional spherical coordinate angles.

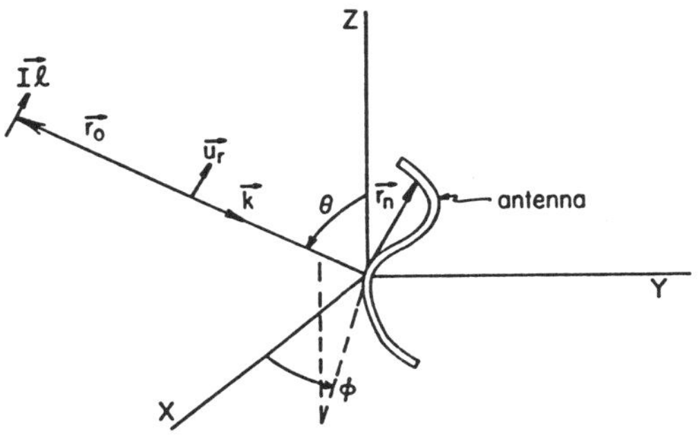

Fig. 5. A wire antenna and distant dipole.

An alternative derivation of the radiation pattern can be obtained by reciprocity. Figure 5 represents a distant current element Il_r (subscripts r denote "receiver"), adjusted to produce the unit plane wave

$$E^r = u_r e^{-jk_r \cdot r_n} \tag{109}$$

in the vicinity of the antenna. Here u_r is a unit vector specifying the polarization of the wave, k_r is a wave number vector pointing in the direction of travel of the wave, and r_n is the radius vector to a point n on the antenna. By reciprocity,

$$E_r = \frac{1}{Il} \int_{\text{antenna}} E^r \cdot I \, dl \tag{110}$$

where E_r is the u_r component of E from the antenna, and I is the current on the antenna. The constant $1/Il$ is that needed to produce a plane wave of unit amplitude at the origin, which is

$$\frac{1}{Il} = \frac{\omega\mu e^{-jkr_0}}{j4\pi r_0}. \tag{111}$$

A numerical approximation to (110) is obtained by defining a voltage matrix

$$[V^r] = \begin{bmatrix} E^r(1)\cdot\Delta l_1 \\ E^r(2)\cdot\Delta l_2 \\ \vdots \\ E^r(N)\cdot\Delta l_N \end{bmatrix} \tag{112}$$

where E^r is given by (109), and expressing (110) as the matrix product

$$E_r = \frac{\omega\mu e^{-jkr_0}}{j4\pi r_0}[\tilde{V}^r][I] = \frac{\omega\mu e^{-jkr_0}}{j4\pi r_0}[\tilde{V}^r][Y][V^s] \tag{113}$$

where $[\tilde{V}]$ denotes the transpose of $[V]$. Note that $[V^r]$ is the same matrix for plane-wave excitation of the wire. Equation (113) remains valid for an arbitrary excitation $[V^s]$; it is not restricted to the single source excitation (105).

The power gain pattern for the u_r component of the radiation field is given by

$$g(\theta, \phi) = \frac{4\pi r_0^2}{\eta} \frac{|E_r(\theta, \phi)|^2}{P_{\text{in}}} \tag{114}$$

where $\eta = \sqrt{\mu/\varepsilon}$ is the intrinsic impedance of space, and P_{in} is the power input to the antenna (* denotes conjugate)

$$P_{\text{in}} = \text{Re}\,\{[\tilde{V}^s][I^*]\} = \text{Re}\,\{[\tilde{V}^s][Y^*][V^{s*}]\}. \tag{115}$$

For the special case of a single source, (105), P_{in} becomes simply Re $(|V_n|^2 Y_{nn})$. Using (113) and (115) in (114), one has

$$g(\theta, \phi) = \frac{\eta k^2}{4\pi} \frac{|[\tilde{V}^r(\theta, \phi)][Y][V^s]|^2}{\text{Re}\,\{[\tilde{V}^s][Y^*][V^{s*}]\}} \tag{116}$$

where $[V^r(\theta, \phi)]$ is given by (112) for various angles of incidence θ, ϕ. Equation (116) gives the gain pattern for only a single polarization of the radiation field. If the total power gain pattern is desired, the g's for two orthogonal polarizations may be added together.

Computations for linear wire antennas have been made using the formulas of this section, and good results obtained. For far-field quantities, such as radiation patterns, as few as 10 segments per wavelength give accurate results. (Radiation patterns are continuous linear functionals, that is, they depend on the weighted integral of the antenna current.) For the current itself, convergence was slower. A typical result for a half-wave antenna was about four percent change in going from 20 to 40 segments, less for other lengths. Faster convergence can be obtained by going from a step approximation to a piecewise-linear approximation to the current. This modification was used for most of the computations, of which Fig. 6 is typical. It shows the input admittance to a center-fed linear antenna with length-to-diameter ratio 74.2 ($\Omega = 2 \log L/a = 10$) using 32 segments.[4]

[4] Because of the extra 1/2 interval at each wire end, this corresponds to an $N = 31$ solution.

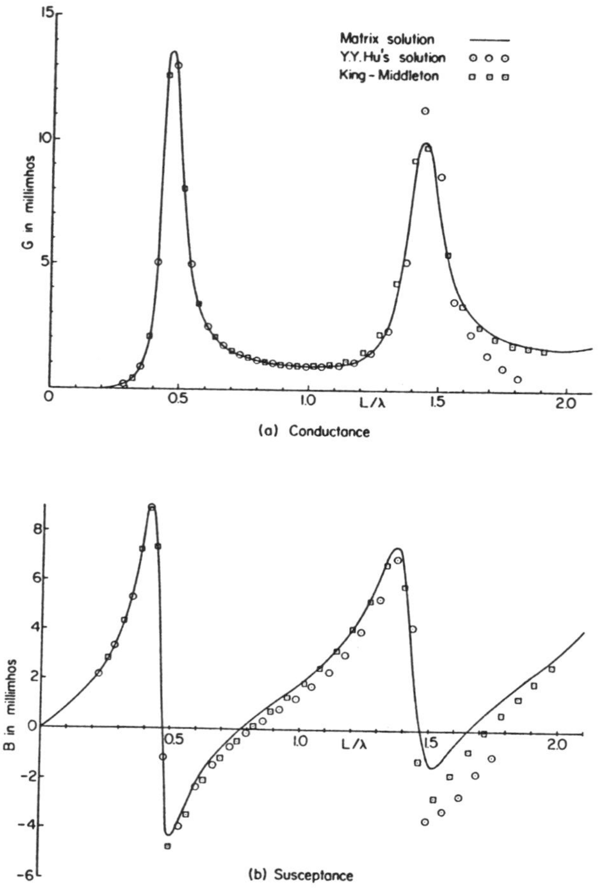

Fig. 6. Input admittance for a center-fed linear antenna of length L and diameter $L/72.4$.

For the points tested, it was almost identical to the 64 segment solution using (103). It is compared to the second-order variational solution of Y. Y. Hu [14], and to the second iteration of Hallén's equation by King and Middleton [15]. The conductances are in close agreement except for Hu's solution $L > 1.3\lambda$, in which case her trial functions are inadequate. The input susceptances are in poorer agreement, which is to be expected because each solution treats the gap differently. The matrix solution of this paper treats it as if it were one segment in length. Hu's solution contains no trial function which can support a singularity in current at the gap, hence gives a low gap capacitance. The King-Middleton method is an iterative procedure, and hence B depends on the number of iterations. Many more computations, as well as a description of the piecewise-linear modification for the current, can be found in the original report [16].

XI. Wire Scatterers

Consider now the field scattered by a wire object in a plane wave incident field. Figure 7 represents a scatterer and two distant current elements, Il_t at the transmitting point, $\boldsymbol{r}_t$, and Il_r at the receiving point $\boldsymbol{r}_r$. The Il_t is adjusted to produce a unit plane wave at the scatterer

$$E^t = \boldsymbol{u}_t e^{-j\boldsymbol{k}_t \cdot \boldsymbol{r}_n} \tag{117}$$

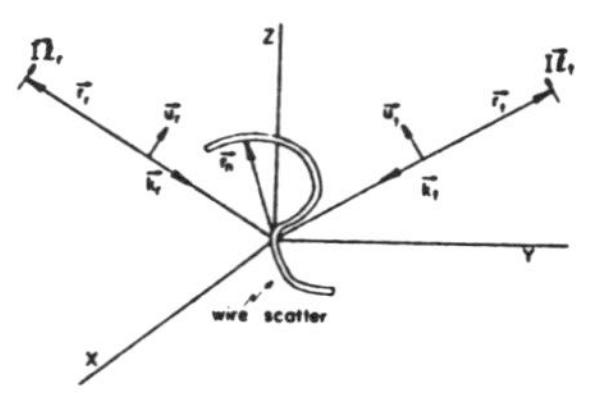

Fig. 7. Definitions for plane-wave scattering.

where the notation is analogous to that of (109). The voltage excitation matrix (97) is then

$$[V^t] = \begin{bmatrix} E^t(1)\cdot \Delta l_1 \\ E^t(2)\cdot \Delta l_2 \\ \vdots \\ E^t(N)\cdot \Delta l_N \end{bmatrix} \tag{118}$$

and the current $[I]$ is given by (104) with $[V]=[V^t]$. The field produced by $[I]$ can then be found by conventional techniques.

The distant scattered field can also be evaluated by reciprocity, the same in the antenna case. A dipole Il_r at the receiving point is adjusted to produce the unit plane wave (109) at the scatterer. The scattered field is then given by (113) with $[V^s]$ replaced by $[V^t]$, that is,

$$E_r = \frac{\omega\mu e^{-jkr_r}}{j4\pi r_r}[\tilde{V}^r][Y][V^t]. \tag{119}$$

A parameter of interest is the bistatic scattering cross section σ, defined as that area for which the incident wave contains sufficient power to produce the field E_r by omnidirectional radiation. In equation form, this is

$$\begin{aligned}\sigma &= 4\pi r_r^2 |E_r|^2 \\ &= \frac{\eta^2 k^2}{4\pi}|[\tilde{V}^r][Y][V^t]|^2.\end{aligned} \tag{120}$$

For the monostatic cross section, set $[V^r]=[V^t]$ in (120). The cross section depends on the polarization of the incident wave and of the receiver. A better description of the scatterer can be made in terms of a scattering matrix.

Another parameter of interest is the total scattering cross section σ_t, defined as the ratio of the total scattered power to the power density of the incident wave. The total power radiated by $[I]$ is given by (115) for any excitation; therefore the scattered power is given by (115) with $[V^s]$ replaced by $[V^t]$. The incident power density is $1/\eta$, hence

$$\sigma_t = \eta \operatorname{Re}[\tilde{V}^t][Y^*][V^{t*}]. \tag{121}$$

Note that σ_t is dependent on the polarization of the incident wave.

Computations for linear wire scatterers have been made using the same $[Y]$ matrix as for antennas. Again far-field quantities, such as echo areas, converged rapidly, with good results obtained with as few as 10 segments per wavelength. Computation of the current converged less rapidly than far-field quantities, but more rapidly than did computation of the current on antennas. This is because the impressed field

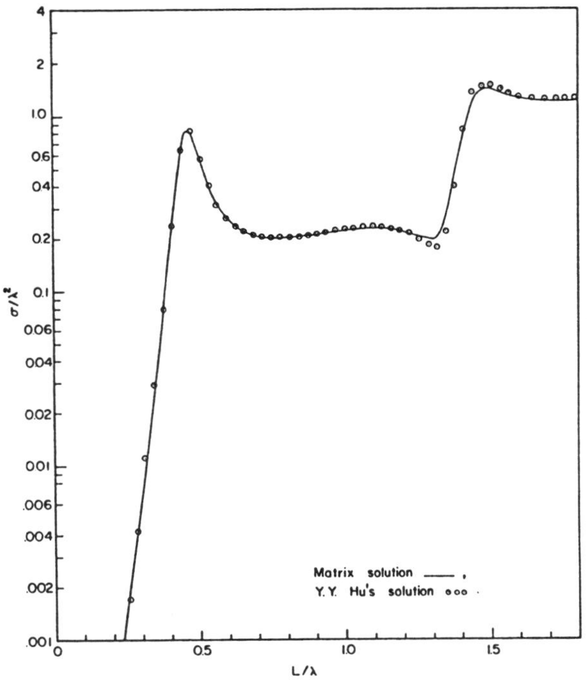

Fig. 8. Echo area of a wire scatterer of length L and diameter $L/72.4$, broadside incidence.

E^i for plane-wave scattering is a well-behaved function, compared with the impulsive impressed field of the antenna problem. Figure 8 shows the echo area for the same wire as was used for an antenna in Fig. 6. Hence, length-to-diameter ratio is 74.2, and a 32 segment piecewise-linear current approximation was used. It is compared to Hu's second-order variational solution [14]. Again good agreement is obtained in the range $L<1.3\lambda$, for which Hu's trial functions are adequate, and a slight discrepancy shows up for $L>1.3\lambda$. Additional computations for linear wire scatterers are given in the original report [16].

XII. Discussion

The method of reducing a functional equation to a matrix equation, and then inverting the matrix for a solution, is particularly well suited to machine computation. Furthermore, the inverted matrix is a representation of the system for arbitrary excitation, hence all responses are solved for at once. As demonstrated by the treatment of wire objects of arbitrary shape, one can also obtain solutions for classes of systems.

In electromagnetic theory, the interpretation of the solution in terms of generalized network parameters is quite general, and applies to bodies of arbitrary shape and arbitrary material. This generalization has been discussed in another paper [17]. The network representation is also useful for the treatment of loaded bodies, both with lumped loads [18] and with continuous loading. Examples of continuously loaded bodies are dielectric coated conductors, magnetic coated conductors, and imperfect conductors.

The solution for wires of arbitrary shape, Section IX, is a first-order solution to the appropriate integrodifferential equation. Higher-order solutions can be obtained by using better-behaved expansion and/or testing functions, and by taking into account the curvature of the wire within the elementary segments. For a general solution, it appears to be more convenient to use a numerical procedure than an analytical procedure. This numerical procedure can be implemented by further subdividing each wire segment, and summing the contributions from the finer subdivisions to obtain the elements of $[Z]$.

As the order of solution is increased, much of the complication comes from the treatment of singularities. The derivative of the current (i.e., charge) is discontinuous at wire ends and at any voltage source along the wire. In the first-order solution this problem has not been accurately treated, and computations appear to justify that this procedure is permissible. For example, at the end of a wire the solution (103) treats the charge as an equivalent line segment extending 1/2 interval beyond the current. The actual charge is singular (or almost so), and could be treated by a special subroutine. While this modification is simple, a similar modification for voltage sources along the wire is not practicable for a general program. This is because the impedance matrix would then depend on the location of the source instead of being a characteristic of the wire object alone. On the basis of experience, it appears that a first-order solution with no special treatment of singularities is adequate for most engineering purposes. This is particularly true for far-zone quantities, such as radiation patterns and echo areas, which are relatively insensitive to small errors in the current distribution.

A number of other electromagnetic field problems have been treated in the literature by procedures basically the same as the method of moments with point matching. Some of these problems are scattering by conducting cylinders [19], [20], scattering by dielectric cylinders [21], [22], and scattering by bodies of revolution [23]. Also available in the literature is an alternative treatment of linear wire scatterers, using sinusoidal expansion functions [24], and an alternative treatment of wire antennas of arbitrary shape, using an equation of the Hallén type [25].

Appendix—Evaluation of ψ

An accurate evaluation of the scalar ψ function of (99) is desired. Let the coordinate origin be located at the point n, and the path of integration lie along the z axis. Then

$$\psi(m, n) = \frac{1}{8\pi\alpha}\int_{-\alpha}^{\alpha} \frac{e^{-jkR_{mn}}}{R_{mn}}\,dz' \tag{122}$$

where

$$2\alpha = \Delta l_n \tag{123}$$

$$R_{mn} = \begin{cases} \sqrt{\rho^2 + (z-z')^2} & m \neq n \\ \sqrt{a^2 + (z')^2} & m = n \end{cases} \tag{124}$$

and a = wire radius. The geometry for these formulas is given in Fig. 9.

One approximation to the ψ's can be obtained by expanding the exponential in a Maclaurin series, giving

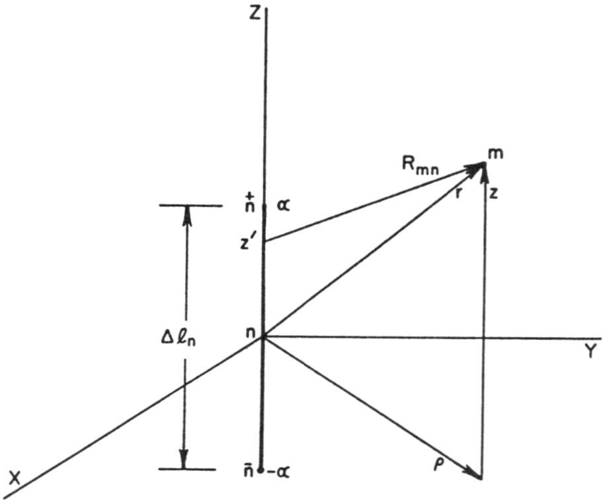

Fig. 9. Geometry for evaluating $\psi(m, n)$.

$$\psi = \frac{1}{8\pi\alpha}\int_{-\alpha}^{\alpha}\left(\frac{1}{R_{mn}} - jk - \frac{k^2}{2}R_{mn} + \cdots\right)dz'. \quad (125)$$

The first term is identical with the static potential of a filament of charge. The second term is independent of R_{mn}. Hence, a two-term approximation of (122) is

$$\psi(m, n) \approx \frac{1}{8\pi\alpha}\log\left[\frac{z + \alpha + \sqrt{\rho^2 + (z + \alpha)^2}}{z - \alpha + \sqrt{\rho^2 + (z - \alpha)^2}}\right] - \frac{jk}{4\pi}. \quad (126)$$

If $r = \sqrt{\rho^2 + z^2}$ is large and $\alpha \ll \lambda$, then

$$\psi(m, n) \approx \frac{e^{-jkr}}{4\pi r}. \quad (127)$$

For a first-order solution, one can take (126) as applying for small r, say $r \leq 2\alpha$, and (127) for large r, say $r > 2\alpha$.

For higher-order approximations, more rapid convergence can be obtained by taking a phase term e^{-jkr} out of the integrand. Then

$$\begin{aligned}\psi &= \frac{e^{-jkr}}{8\pi\alpha}\int_{-\alpha}^{\alpha}\frac{e^{-jk(R_{mn}-r)}}{R_{mn}}\,dz' \\ &= \frac{e^{-jkr}}{8\pi\alpha}\int_{-\alpha}^{\alpha}\left(\frac{1}{R_{mn}} - \frac{jk(R_{mn} - r)}{R_{mn}} - \frac{k^2(R_{mn} - r)^2}{2R_{mn}} + \cdots\right)dz'. \quad (128)\end{aligned}$$

Term by term integration gives

$$\psi(m, n) = \frac{e^{-jkr}}{8\pi\alpha}\left[I_1 - jk(I_2 - rI_1) - \frac{k^2}{2}(I_3 - 2rI_2 + r^2I_1) + j\frac{k^3}{6}(I_4 - 3rI_3 + 3r^2I_2 - r^3I_1) + \cdots\right] \quad (129)$$

where

$$I_1 = \log\left[\frac{z + \alpha + \sqrt{\rho^2 + (z + \alpha)^2}}{z - \alpha + \sqrt{\rho^2 + (z - \alpha)^2}}\right] \quad (130)$$

$$I_2 = 2\alpha \quad (131)$$

$$I_3 = \frac{\alpha + z}{2}\sqrt{\rho^2 + (\alpha + z)^2} + \frac{\alpha - z}{2}\sqrt{\rho^2 + (z - \alpha)^2} + \frac{\rho^2}{2}I_1 \quad (132)$$

$$I_4 = 2\alpha\rho^2 + \frac{2\alpha^3 + 6\alpha z^2}{3}. \quad (133)$$

An expansion of the type (129) is theoretically valid for all r, but it fails numerically for large r because it involves subtractions of almost equal numbers. For $\rho < a$, one should set $\rho = a$ in the expansion.

An expression suitable for large r is obtained by expanding (122) in a Maclaurin series in z' as

$$\psi = \frac{1}{8\pi\alpha}\int_{-\alpha}^{\alpha}\left[f(0) + f'(0)z' + \frac{1}{2!}f''(0)(z')^2 + \cdots\right]dz' \quad (134)$$

where

$$f(z') = \frac{e^{-jk\sqrt{\rho^2 + (z - z')^2}}}{\sqrt{\rho^2 + (z - z')^2}}.$$

When a five-term expansion of (134) is integrated term by term, there results

$$\psi = \frac{e^{-jkr}}{4\pi r}\left[A_0 + jk\alpha A_1 + (k\alpha)^2A_2 + j(k\alpha)^3A_3 + (k\alpha)^4A_4\right] \quad (135)$$

where

$$\begin{aligned}A_0 &= 1 + \frac{1}{6}\left(\frac{\alpha}{r}\right)^2\left[-1 + 3\left(\frac{z}{r}\right)^2\right] + \frac{1}{40}\left(\frac{\alpha}{r}\right)^4\left[3 - 30\left(\frac{z}{r}\right)^2 + 35\left(\frac{z}{r}\right)^4\right] \\ A_1 &= \frac{1}{6}\left(\frac{\alpha}{r}\right)\left[-1 + 3\left(\frac{z}{r}\right)^2\right] + \frac{1}{40}\left(\frac{\alpha}{r}\right)^3\left[3 - 30\left(\frac{z}{r}\right)^2 + 35\left(\frac{z}{r}\right)^4\right] \\ A_2 &= -\frac{1}{6}\left(\frac{z}{r}\right)^2 - \frac{1}{40}\left(\frac{\alpha}{r}\right)^2\left[1 - 12\left(\frac{z}{r}\right)^2 + 15\left(\frac{z}{r}\right)^4\right] \\ A_3 &= \frac{1}{60}\left(\frac{\alpha}{r}\right)\left[3\left(\frac{z}{r}\right)^2 - 5\left(\frac{z}{r}\right)^4\right] \\ A_4 &= \frac{1}{120}\left(\frac{z}{r}\right)^4.\end{aligned} \quad (136)$$

For accuracy of better than one percent, one can use (129) for $r < 10\alpha$ and (135) for $r \geq 10\alpha$.

An alternative derivation of the type of (135) can be obtained as follows. For $r > z'$, one has the expansion

$$\frac{e^{-jkR_{mn}}}{-jkR_{mn}} = \sum_{n=0}^{\infty}(2n + 1)j_n(kz')h_n^{(2)}(kr)P_n\left(\frac{z}{r}\right) \quad (137)$$

where j_n are the spherical Bessel functions of the first kind, $h_n^{(2)}(kr)$ are the spherical Hankel functions of the second kind, and $P_n(z/r)$ are the Legendre polynomials. If (137) is substituted into (122) and integrated term by term, there results

$$\psi(m, n) = \frac{1}{4\pi j} \sum_{n=0}^{\infty} b_n h_n^{(2)}(kr) P_n\left(\frac{z}{r}\right) \tag{138}$$

where

$$b_n = \frac{2n+1}{2\alpha} \int_{-k\alpha}^{k\alpha} j_n(x)\, dx. \tag{139}$$

Equation (138) can be rearranged into the form of (135), although the recurrence formulas for $h_n^{(2)}$ and P_n make computation directly from (138) almost as easy.

Acknowledgment

The computer programming for the numerical results was done by J. Mautz.

References

[1] B. Z. Vulikh, *Introduction to Functional Analysis for Scientists and Technologists*, I. N. Sneddon, trans. Oxford: Pergamon, 1963.
[2] B. Friedman, *Principles and Techniques of Applied Mathematics*. New York: Wiley, 1956.
[3] J. W. Dettman, *Mathematical Methods in Physics and Engineering*. New York: McGraw-Hill, 1962.
[4] L. Kantorovich and G. Akilov, *Functional Analysis in Normed Spaces*, D. E. Brown, trans. Oxford: Pergamon, 1964, pp. 586–587.
[5] L. Kantorovich and V. Krylov, *Approximate Methods of Higher Analysis*, C. D. Benster, trans. New York: Wiley, 1964, ch. 4.
[6] D. S. Jones, "A critique of the variational method in scattering antennas," *IRE Trans. on Antennas and Propagation*, vol. AP-4, pp. 297–301, July 1956.
[7] Forsythe and Wasov, *Finite Difference Methods for Partial Differential Equations*. New York: Wiley, 1960.
[8] V. H. Rumsey, "The reaction concept in electromagnetic theory," *Phys. Rev.*, ser. 2, vol. 94, pp. 1483–1491, June 15, 1954.
[9] R. F. Harrington, *Time-Harmonic Electromagnetic Fields*. New York: McGraw-Hill, 1961, pp. 340–345.
[10] D. K. Reitan and T. J. Higgins, "Accurate determination of the capacitance of a thin rectangular plate," *Trans. AIEE (Communication and Electronics)*, pt. I, vol. 75, pp. 761–766, 1956 (Jan. 1957 section).
[11] J. Van Bladel, *Electromagnetic Fields*. New York: McGraw-Hill, 1964, p. 96.
[12] H. B. Dwight, *Tables of Integrals and Other Mathematical Data*. New York: Macmillan, 1947.
[13] J. Van Bladel, "Some remarks on Green's dyadic for infinite space," *IRE Trans. on Antennas and Propagation*, vol. AP-9, pp. 563–566, November 1961.
[14] Yueh-Ying Hu, "Back-scattering cross section of a center-loaded cylindrical antenna," *IRE Trans. on Antennas and Propagation*, vol. AP-6, pp. 140–148, January 1958.
[15] R. W. P. King, *The Theory of Linear Antennas*. Cambridge, Mass.: Harvard University Press, 1956, p. 172.
[16] R. F. Harrington et al., "Matrix methods for solving field problems," Rome Air Development Center, Griffiss AFB, N. Y., final rept. under Contract AF 30(602)-3724, March 1966.
[17] R. F. Harrington, "Generalized network parameters in field theory," *Proc. Symposium on Generalized Networks*, MRIS series, vol. 16. Brooklyn, N. Y.: Polytechnic Press, 1966.
[18] ——, "Theory of loaded scatterers," *Proc. IEE (London)*. vol. 111, pp. 617–623, April 1964.
[19] K. K. Mei and J. G. Van Bladel, "Scattering by perfectly-conducting rectangular cylinders," *IEEE Trans. on Antennas and Propagation*, vol. AP-11, pp. 185–192, March 1963.
[20] M. G. Andreasen, "Scattering from parallel metallic cylinders with arbitrary cross sections," *IEEE Trans. on Antennas and Propagation*, vol. AP-12, pp. 746–754, November 1964.
[21] J. H. Richmond, "Scattering by a dielectric cylinder of arbitrary cross section shape," *IEEE Trans. on Antennas and Propagation*, vol. AP-13, pp. 334–341, May 1965.
[22] ——, "TE-wave scattering by a dielectric cylinder of arbitrary cross-section shape," *IEEE Trans. on Antennas and Propagation*, vol. AP-14, pp. 460–464, July 1966.
[23] M. G. Andreasen, "Scattering from bodies of revolution," *IEEE Trans. on Antennas and Propagation*, vol. AP-13, pp. 303–310, March 1965.
[24] J. H. Richmond, "Digital computer solutions of the rigorous equations for scattering problems," *Proc. IEEE*, vol. 53, pp. 796–804, August 1965.
[25] K. K. Mei, "On the integral equations of thin wire antennas," *IEEE Trans. on Antennas and Propagation*, vol. AP-13, pp. 374–378, May 1965.

On Energy Conservation and the Method of Moments in Scattering Problems

NOACH AMITAY, MEMBER, IEEE, AND VICTOR GALINDO, MEMBER, IEEE

Abstract—Electromagnetic scattering problems, including waveguide discontinuity, phased array, and scattering (exterior type) problems, are frequently described by integral equations that can be solved by the Ritz–Galerkin or generalized method of moments. Under appropriate conditions, it has been shown that reciprocity and variational properties are, in fact, preserved in the approximate solutions. It is shown here that in the Ritz-Galerkin method, energy is also conserved under certain conditions, even in those scattering problems where reciprocity does not exist. Hence energy conservation cannot serve as a check for accuracy of a numerical solution obtained by the Ritz method or other related methods.

I. Introduction

ELECTROMAGNETIC (EM) scattering problems, which at one time were insoluble by direct analysis, can often be solved numerically by either the Ritz–Galerkin method or, more generally, the method of moments [1]–[8]. These methods have become increasingly useful for complex problems when coupled with the use of high-speed and large-storage computers. At one time, the connection between the Ritz and Galerkin methods of solution to the integral equation formulation of these problems was not well understood, while reciprocity was considered a good check of the accuracy of the solution. However, Jones [9], Kantorovich and Krylov [10], and others have pointed out that the classical Ritz and Galerkin methods are equivalent formulations, and that the approximate solution obtained by this method satisfies the reciprocity theorem.

In this paper we show that under certain conditions, the approximate Ritz–Galerkin solutions for a broad class of EM scattering problems not only satisfy reciprocity, but also satisfy the law of conservation of energy.[1] It is shown further that even if reciprocity is not satisfied, energy is still conserved in the approximate solutions. Hence, conservation of energy can only be used to check computational round-off errors but is not at all a measure of the accuracy of a Ritz–Galerkin solution.

These results will be illustrated for a broad class of EM scattering problems including waveguide discontinuity [7], infinite phased array [1]–[3], and exterior type [4], [5] scattering problems. In many of these cases, the operators involved are of complex–symmetric or symmetric form.

Manuscript received January 28, 1969; revised April 22, 1969. This work was supported by the U. S. Army Materiel Command under Contract DAHC-60-69-C-0008.

The authors are with Bell Telephone Laboratories, Inc., Whippany, N. J. 07981.

[1] Actually, the unitary condition is satisfied by the approximate solutions when an equivalent lossless scattering matrix can be defined.

II. Waveguide Discontinuity and Infinite Phased-Array Problems

The integral equation for these problems can generally be written in operator notation [13] as

$$y = Lx \tag{1}$$

where L denotes an integral operator, x the unknown scattered field, and y the known incident field. For both waveguide discontinuity and infinite phased-array problems, (1) may appear in functional form [3], [11], [12] as

$$2\hat{z} \times H_i(r) = \iint_{\text{aperture}} \Big[\sum_{n=1}^{\infty} Y_n \varphi_n(r) \varphi_n(r') + \sum_{-\infty}^{\infty} Y_m' \psi_m(r) \psi_m^*(r') \Big] \cdot E(r')\, dr'. \tag{2}$$

The quantities H_i and E are, respectively, the incident tangential magnetic and the total tangential electric fields in the discontinuity aperture, as shown in Fig. 1. The $\{\varphi_n(r)\}$ are orthonormal waveguide modal functions, and the $\{\psi_m(r)\}$ are either another set of orthonormal waveguide modal functions or, in the case of an infinite phased array, a set of Floquet type modes.[2] For waveguides, the $\{\psi_m\}$ are (or can be set) pure real, whereas for the phased-array case they are complex. Y_n and Y_m', the modal admittances, are either pure real (propagating modes) or pure imaginary (evanescent modes). Consequently, for waveguide discontinuities, L in (2) is complex–symmetric, i.e., $L(r,r') = L(r',r)$ where L is a complex operator. For infinite phased arrays, L in (2) can readily be put in complex-symmetric form provided that the $\{\varphi_n\}$ possess point symmetry [3], i.e., $\varphi_n(r) = \pm\varphi_n(-r)$, over a symmetric aperture.

The first step in solving (1) by either a Ritz or general moments method is to approximate the unknown x by a finite set of linearly independent expansion functions $\{\eta_i: i = 1,\cdots,I\}$ so that

$$x \approx \sum_{i=1} a_i \eta_i \tag{3}$$

with the unknown coefficients a_i to be determined. This, substituted in (1), gives

$$y \approx \sum_{i=1}^{I} a_i L\eta_i. \tag{4}$$

[2] The Floquet modes are complex and orthonormal in the sense $(\psi_m^*, \psi_n) = \delta_{mn}$, where δ_{mn} is the Kronecker delta.

Reprinted from *IEEE Trans. Antennas Propagat.*, vol. AP-17, no. 6, pp. 747–751, November 1969.

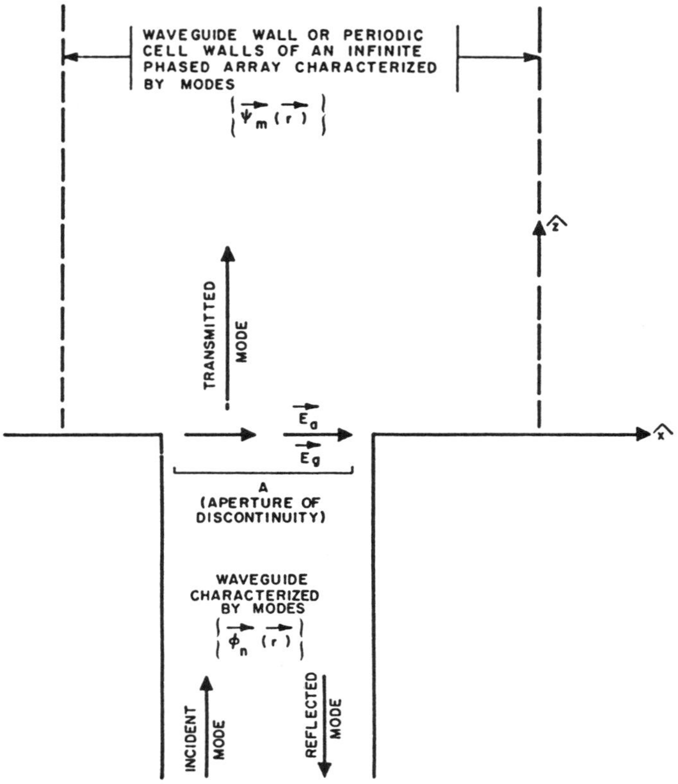

Fig. 1. Waveguide discontinuity and infinite phased-array scattering by a discontinuity at the aperture A.

In the Ritz–Galerkin method, I moments of (4) with respect to the $\{\eta_i\}$ are taken.

With an appropriate moment or scalar product defined, for example, as [13]

$$(\omega,z) \equiv \iint_{\text{aperture}} \omega(r) \cdot z(r)\, dr \tag{5}$$

(4) becomes[3]

$$(\eta_j^*,y) = \sum_{i=1}^{I} a_i(\eta_j^*,L\eta_i), \quad j = 1,\cdots,I. \tag{6}$$

Thus the integral operator L is approximated by a finite square matrix (Ith order degenerate operator) with elements L_{ji} equal to $(\eta_j^*,L\eta_i)$. If L is a symmetric operator, i.e., $L = \tilde{L}$ which is the necessary condition for reciprocity, the matrix elements $L_{ji} \neq L_{ij}$. However, the use of the Ritz–Galerkin method, as previously defined, leads to an approximate matrix wherein the elements $L_{ji} = L_{ij}$, and therefore the approximate solutions satisfy reciprocity [9].

Let us now consider what is meant by *energy conservation* with respect to (2), the discontinuity and phased-array scattering equation. If we assume, for convenience, that a single mode φ_1 is incident from the waveguides and that only one mode propagates therein, and also that only one mode propagates in the other region,[4] then we may expand the tangential aperture fields (Fig. 1):

$$\boldsymbol{E} \equiv \boldsymbol{E}_g = (1 + R)\varphi_1 + \sum_{n=2}^{\infty} b_n\varphi_n \tag{7}$$

and

$$-\hat{z} \times \boldsymbol{H}_g = (1 - R)Y_1\varphi_1 - \sum_{n=2}^{\infty} b_nY_n\varphi_n \tag{8}$$

and

$$\boldsymbol{E} = \boldsymbol{E}_a = \sum_{-\infty}^{\infty} A_m\psi_m \tag{9}$$

$$-\hat{z} \times \boldsymbol{H}_a = \sum_{-\infty}^{\infty} A_mY_m'\psi_m \tag{10}$$

where R is the reflection coefficient and the subscripts g and a pertain to the two regions in Fig. 1. The integral equation (2) is of course found by setting $\boldsymbol{E}_g = \boldsymbol{E}_a$ and $\hat{z} \times \boldsymbol{H}_g = \hat{z} \times \boldsymbol{H}_a$ (while utilizing the orthonormality properties of the $\{\varphi_n\}$ and $\{\psi_m\}$). If this is done, then it follows that the z directed Poynting's vector at the aperture integrates to

$$\iint_A \hat{z} \cdot \boldsymbol{E}_g \times \boldsymbol{H}_g^*\, da = (1 - |R|^2)Y_1 + (R - R^*)Y_1 - \sum_{n=2}^{\infty} |b_n|^2Y_n^* \tag{11}$$

and is equal to

$$\iint_A \hat{z} \cdot \boldsymbol{E}_a \times \boldsymbol{H}_a^*\, da = |A_0|^2Y_0' + \sum_{-\infty}^{\infty}{}' |A_m|^2Y_m'^* \tag{12}$$

where in the primed summation the zero mode is excluded. The real parts of (11) and (12) represent conservation of *propagating* energy

$$1 - |R|^2 = |A_0|^2(Y_0'/Y_1) \tag{13}$$

since Y_1 and Y_0' are real, and the imaginary parts of (11) and (12) represent a "conservation" of time averaged stored energy

$$(R - R^*)Y_1 - \sum_{n=2}^{\infty} |b_n|^2Y_n^* = \sum_{-\infty}^{\infty}{}' |A_m|^2Y_m'^*. \tag{14}$$

Suppose, now, that we apply the method of moments to the integral equation (2), as outlined in (3) through (6). The resulting matrix equation is then

$$\begin{pmatrix} \cdot \\ \cdot \\ \cdot \\ y_j \\ \cdot \\ \cdot \\ \cdot \end{pmatrix} = \begin{bmatrix} \\ \\ \\ L_{ji} \\ \\ \\ \\ \end{bmatrix} \begin{pmatrix} \cdot \\ \cdot \\ \cdot \\ a_i \\ \cdot \\ \cdot \\ \cdot \end{pmatrix} \tag{15}$$

[3] Note that if $\{\eta_j\}$ are real, then (6) follows the Ritz–Galerkin method as defined earlier. When the $\{\eta_j\}$ are complex, the moments are taken with respect to the $\{\eta_j\}$ for reasons that will become clear later.

[4] The procedure and results apply equally for an arbitrary number of propagating modes in each region.

where

$$y_j = (2\hat{z} \times H_i, \eta_j^*) = 2Y_1(\varphi_1, \eta_j^*) \tag{16}$$

and

$$L_{ji} = \sum_{n=1}^{N} Y_n(\eta_j^*, \varphi_n)(\varphi_n, \eta_i) + \sum_{m=-M_1}^{M_2} Y_m'(\eta_j^*, \psi_m)(\psi_m^*, \eta_i). \tag{17}$$

Note that the sums in the approximating matrix element L_{ji} have been truncated and are finite in (17).[5] The approximations used to derive (15) are

$$E \approx E_g = (1 + R')\varphi_1 + \sum_{n=2}^{N} b_n'\varphi_n = \sum_{i=1}^{I} a_i\eta_i$$

$$E \approx E_a = \sum_{-M_1}^{M_2} A_m'\psi_m = \sum_{i=1}^{I} a_i\eta_i \tag{18}$$

so that

$$b_n' \equiv \sum_{i=1}^{I} a_i(\varphi_n, \eta_i)$$

$$(1 + R') \equiv \sum_{i=1}^{I} a_i(\varphi_1, \eta_i)$$

$$A_m' \equiv \sum_{i=1}^{I} a_i(\psi_m^*, \eta_i) \tag{19}$$

where R', b_n', and A_m' are the approximate solutions corresponding to the true R, b_n, and A_m in (7) through (10). The magnetic fields H_g and H_a follow from Maxwell's curl equations as

$$-\hat{z} \times H_g = (1 - R')Y_1\varphi_1 - \sum_{n=2}^{N} b_n'Y_n\varphi_n$$

and

$$-\hat{z} \times H_a = \sum_{-M_1}^{M_2} A_m'Y_m'\psi_m. \tag{20}$$

In order to determine whether the approximate fields in (18), (19), and (20) satisfy the conservation of energy

$$\iint_A \hat{z} \cdot E_g \times H_g^* \, da = \iint_A \hat{z} \cdot E_a \times H_a^* \, da \tag{21}$$

we substitute these fields in (21) and obtain

$$2 \sum_{i=1}^{I} Y_1 a_i(\eta_i, \varphi_1) = \sum_{i=1}^{I} \sum_{j=1}^{I} a_i a_j^* \cdot \Big[\sum_{n=1}^{N} Y_n^*(\eta_i, \varphi_n)(\varphi_n, \eta_j^*) + \sum_{-M_1}^{M_2} Y_m'^*(\eta_i, \psi_m^*)(\psi_m, \eta_j^*) \Big]. \tag{22}$$

[5] This is equivalent to approximating the operator L in (2) by the same truncation.

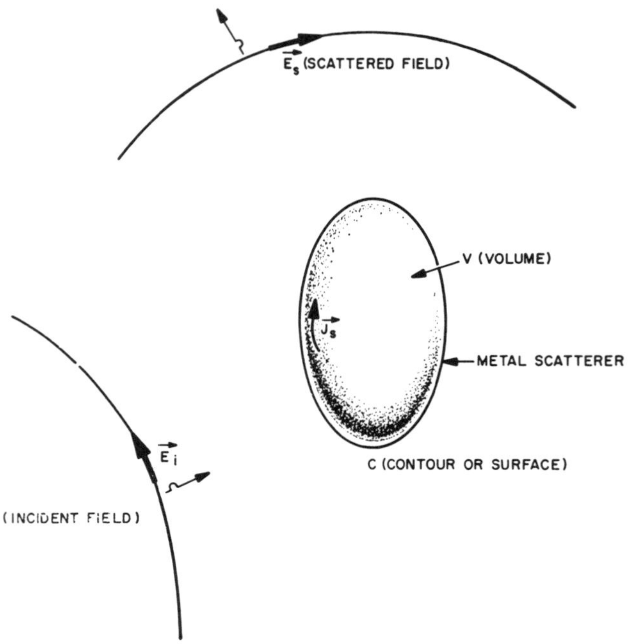

Fig. 2. Exterior type scattering problem geometry.

Equation (22) is recognized as a quadratic form derived from (15),

$$\{\cdots a_i \cdots\} \begin{pmatrix} \cdot \\ \cdot \\ \cdot \\ y_i \\ \cdot \\ \cdot \\ \cdot \end{pmatrix}^* = \{\cdots a_i \cdots\} \Big[L_{ij} \Big]^* \begin{pmatrix} \cdot \\ \cdot \\ \cdot \\ a_j \\ \cdot \\ \cdot \\ \cdot \end{pmatrix}^* . \tag{23}$$

The choice of $L_{ij} = (\eta_i^*, L\eta_j)$ leads to conservation of energy when $\{\eta_j\}$ are complex. As noted earlier, reciprocity is not satisfied, whereas it would be if $L_{ij} = (\eta_i, L\eta_j)$. If we consider the original integral equation represented approximately in the subspace $\{\eta_i: i = 1, \cdots, I\}$ as

$$2\hat{z} \times H_i^* = L^*E^* \tag{24}$$

then (23) is of the following quadratic form:

$$2E \cdot \hat{z} \times H_i^* = E \cdot L^*E^*. \tag{25}$$

Clearly then, (21) is an equality, and energy is conserved by the approximate solutions. In most practical cases, real $\{\eta_j\}$ are used so that the approximate solutions obtained by the Ritz–Galerkin method inherently satisfy reciprocity *and* conservation of energy.

III. Exterior Type Scattering Problems

The results derived for the scattering operators of discontinuity and infinite phased-array problems apply, for the most part, to exterior type [4], [5], [14] scattering problems. One difference, however, is the form of the energy conservation equation. A brief discussion of the exterior scattering problem may therefore prove useful. We will confine the discussion to metallic scatterers, the extension to dielectric scatterers being straightforward.

Consider an arbitrary incident electric field E_i upon a metallic scatterer with finite surface C and volume V as shown in Fig. 2.

The scattered electric field E_s is related to a tangential surface current distribution J_s on the surface C by

$$-E_s(r) = \iint_C G_0(r,r') \cdot J_s(r')\, ds \tag{26}$$

where $G_0(r,r')$ is the free-space dyadic Green's function. Taking r on the surface C, we then have the integral equation for J_s:

$$E_i = L_s J_s \quad \text{(in operator form)}$$

or

$$E_i(r_s) = \iint_C G_0(r_s,r') \cdot J_s(r')\, ds \quad \text{(in functional form)} \tag{27}$$

where E_i and J_s are tangential to the surface C.

An expression for the conservation of energy may now be given by integrating the outward Poynting's vector over the surface C. The real part of this integral must vanish for any lossless scatterer (metallic, dielectric, etc.), since no net power is generated or absorbed in volume V. Hence

$$P_r = \text{Re} \iint_C E \times H^* \cdot \hat{n}\, ds = 0 \tag{28}$$

where $\hat{n}$ is an outward normal from C. Equation (28) implies conservation of energy. With the total electric and magnetic fields given, respectively, as $E = E_i + E_s$ and $H = H_i + H_s$, we find that

$$P_r = \text{Re} \iint_C (E_i \times H_s^* + E_s \times H_i^* + E_s \times H_s^*) \cdot \hat{n}\, ds = 0 \tag{29}$$

since $\text{Re} \iint_C E_i \times H_i^* \cdot \hat{n}\, ds \equiv 0$. For the metallic scatterer, (29) can be reduced to

$$P_r = -\text{Re} \iint_C (E_i + E_s) \cdot J_s^*\, ds = 0 \tag{30}$$

where (30) clearly vanishes for exact solutions of J_s and E_s since $E_i = -E_s$ on C.

Suppose we now solve the integral equation (27) by the method of moments. We assume an approximate current J_{s_a} given by

$$J_{s_a} = \sum_{n=1}^{N} A_n \eta_n \tag{31}$$

where $\{\eta_n\}$ form an orthonormal set of functions over the surface C such that $(\eta_m^*,\eta_n) = \delta_{mn}$. By substituting (31) in (27) and taking moments, we obtain a matrix equation for the A_n:

$$B_m = (\eta_m^*, E_i) = \sum_{n=1}^{N} A_n(\eta_m^*, L_s\eta_n). \tag{32}$$

An approximate scattered field on the surface, E_{s_a}, is derivable from the approximate current found in (32) by using (26)

$$-E_{s_a} = \sum_{n=1}^{N} A_n L_s \eta_n. \tag{33}$$

In this case, of course, $\hat{n} \times (E_{s_a} + E_i)$ does not vanish on C. To check the conservation of energy, we find

$$P_r = -\text{Re} \iint_C (E_i + E_{s_a}) \cdot J_{s_a}^*\, ds$$

$$= -\text{Re} \iint_C \left(\sum_{m=1}^{\infty} B_m \eta_m - \sum_{m=1}^{N} A_m L_s \eta_m \right) \cdot \sum_{n=1}^{N} A_n^* \eta_n^*\, ds$$

so that

$$P_r = \text{Re} \sum_{n=1}^{N} A_n^* \left[B_n - \sum_{m=1}^{N} A_m (\eta_n^*, L_s \eta_m) \right] = 0 \tag{34}$$

by virtue of (32). Thus, energy is conserved in this method of moments solution of exterior type scattering problems.

Note that (34) in a matrix form is

$$P_r = \text{Re} \left[J_{s_a}^* \cdot E_i - J_{s_a}^* \cdot L_s J_{s_a} \right] = 0 \tag{35}$$

in the subspace spanned by the set $\{\eta_m : m = 1, \cdots, N\}$.

IV. Conclusion

We have shown the conditions under which energy is conserved by the Ritz–Galerkin and related method of moments solutions to the integral equations for scattering from waveguide discontinuities, phased arrays, and exterior type scatterers. In these cases, conservation of energy[6] can be used only as a check for computational round-off errors but not as a check for the accuracy of the approximate solution.

In most practical cases where real expansion functions are used, the Ritz–Galerkin method clearly *preserves the symmetry* of a scattering operator, whenever it exists. Consequently, the stationary (or variational) property and the reciprocity property of approximate solutions derived in this manner are preserved as well as conservation of energy. Even when the scattering operator is not symmetric, such as for media containing plasma or ferrite materials or certain nonsymmetric phased arrays, it is shown that energy is nevertheless conserved in the approximate solutions. Furthermore, even in the case of *lossy* scatterers, energy is still conserved.

The conservation theory for the approximate solutions obtained here can be shown to apply to the scattering equations of quantum mechanics and other physical problems also.

Acknowledgment

The authors are grateful to Dr. C. P. Bates for his constructive criticism of this paper.

[6] Or any of the unitary conditions of an equivalent lossless scattering matrix.

References

[1] V. Galindo and C. P. Wu, "Numerical solutions for an infinite phased array of rectangular waveguides with thick walls," *IEEE Trans. Antennas and Propagation*, vol. AP-14, pp. 149–158, March 1966.

[2] B. L. Diamond, "Resonance phenomena in waveguide arrays," *Proc. 1967 IEEE Internatl. Symp. on Antennas and Propagation* (Ann Arbor, Mich.), pp. 110–115.

[3] V. Galindo and C. P. Wu, "Integral equations and variational expressions for arbitrary scanning of regular infinite arrays," *IEEE Trans. Antennas and Propagation* (Communications), vol. AP-14, pp. 392–394, May 1966.

[4] J. H. Richmond, "Scattering by a dielectric cylinder of arbitrary cross section shape," *IEEE Trans. Antennas and Propagation*, vol. AP-13, pp. 334–341, May 1965.

[5] K. K. Mei and J. G. Van Bladel, "Scattering by perfectly conducting rectangular cylinders," *IEEE Trans. Antennas and Propagation*, vol. AP-11, pp. 185–192, March 1963.

[6] R. F. Harrington, *Field Computation by Moment Methods*. New York: Macmillan, 1968.

[7] N. Amitay and V. Galindo, "Application of a new method for approximate solutions and error estimates to waveguide discontinuity and phased array problems," *Radio Sci.*, vol. 3, pp. 830–843, August 1968.

[8] W. J. Cole, E. R. Nagelberg, and C. M. Nagel, "Iterative solution of waveguide discontinuity problems," *Bell Sys. Tech. J.*, vol. 46, pp. 649–672, March 1967.

[9] D. S. Jones, "A critique of the variational method in scattering problems," *IRE Trans. Antennas and Propagation*, vol. AP-4, pp. 297–301, July 1956.

[10] L. V. Kantorovich and V. I. Krylov, *Approximate Methods of Higher Analysis*. New York: Interscience, 1964.

[11] C. P. Wu and V. Galindo, "Surface-wave effects on dielectric sheathed phased arrays of rectangular waveguides," *Bell Sys. Tech. J.*, vol. 47, pp. 117–142, January 1968.

[12] V. Galindo and C. P. Wu, "Dielectric loaded and covered rectangular waveguide phased arrays," *Bell Sys. Tech. J.*, vol. 47, pp. 93–116, January 1968.

[13] B. Friedman, *Principles and Techniques of Applied Mathematics*. New York: Wiley, 1956.

[14] K. M. Mitzner, "An integral equation approach to scattering from a body of finite conductivity," *Radio Sci.*, vol. 2, pp. 1459–1470, December 1967.

Origin and Development of the Method of Moments for Field Computation

ROGER F. HARRINGTON

ELECTRICAL AND COMPUTER ENGINEERING DEPARTMENT, SYRACUSE UNIVERSITY, SYRACUSE, NY 13210

Abstract—**A short history is given of the development of mathematical methods related to the method of moments and used for electromagnetic field computation. A brief description of the general theory is given, emphasizing various viewpoints which lead to different names given to the method. Its relationships to the Rayleigh–Ritz variational method and to the perturbation method are also shown and discussed.**

I. History

THIS section will, of course, be a personal view of the origin and development of the method of moments for field computation. The basic idea of taking a linear functional equation and representing it by a linear matrix equation is relatively old. Galerkin, a Russian mechanical engineer, developed his method around 1915, before it had a firm mathematical basis. Quantum mechanics, developed in the 1920s, used many of the ideas of linear spaces and their extension to Hilbert spaces. However, before the advent of the high-speed computer, these methods were not popular because of the tedious computation required for their use. They were often thought of as last-resort numerical methods, to be used only if everything else failed. In truth, however, they are no more numerical than other so-called analytical methods, at least if used properly. They merely emphasize a different aspect of mathematics—that of linear spaces and orthogonal projections.

Perhaps the first large-scale effort to solve practical electromagnetic engineering problems was undertaken during World War II at the M.I.T. Radiation Laboratory. Schwinger and others applied the variational method to microwave problems [1], and Marcuvitz organized the solutions to many of these problems into the popular *Waveguide Handbook* [2]. Rumsey formalized some of the concepts into a more compact notation in his "reaction concept" [3].[1] Researchers, including myself [4], used these variational methods for a number of practical problems.

In the mid-1960s, several researchers started solving the electromagnetic field equations by numerical methods. Mei and Van Bladel used a subsectional and point-matching method to compute the scattering from rectangular cylinders [5]. Others were developing similar methods that were not, until later, made public [6], [7]. The accuracy obtained from these numerical solutions was impressive, but, being brought up on variational solutions, I thought that even greater accuracy could be obtained by the latter method. I had been taught that "a parameter could be calculated correct to the second order if a function correct to the first order is used in a stationary formula." Hence, if I used a numerical solution for the current in a stationary formula for scattered field, I should get an order of magnitude higher accuracy for it than obtainable from a numerical solution. Or so I thought. I tried it for the simple case of scattering from a cylinder, and to my surprise I got exactly the same answer as obtained from the numerical solution. Something was wrong in my reasoning.

Also during the early 1960s, I was teaching a course on the use of linear spaces for applied mathematics, using Friedman's book [8]. When it became available in the English translation, I also studied the Russian book by Kantorovich and Krylov [9]. It became apparent to me that Galerkin's method was formally equivalent to the Rayleigh–Ritz variational method, and also to Rumsey's reaction concept. But the numerical methods being used by researchers in electromagnetic theory were not really Galerkin's method. They used the apparently cruder methods, such as subsectional expansion and point matching. Were these also variational methods?

The answer was yes, at least in concept. There was no good reason why one had to choose expansion and testing functions the same, as was done in both Galerkin's method and in the Rayleigh–Ritz variational method. It was easier to prove mathematical theorems when they were the same, but it made solutions more difficult to calculate. One was really free to choose expansion and testing functions separately for computational convenience, and still claim that the solution was stationary in form. Next came the question as to what to call the general method. Certainly others had used it in the past, and I didn't want to introduce new jargon. After a search of the literature, I decided that the exposition most closely analogous to what I was using was that given by Kantorovich and Akilov [10]. They called it the "method of moments," and hence that is the name I chose.

Now that I understood fairly well the theory of the method of moments, and was learning the power of it, I started preparing some class notes for use in a course on electromagnetic theory at the Rome Air Development Center. Some of the students there shared my enthusiasm for the method, and initiated a contract to formulate the theory as it applied to electromagnetic field problems. The result was a rather extensive report, "Matrix Methods for Solving Field Problems" [11]. An extract of this work appeared as the paper "Matrix Methods for Field Problems" [12].[2] An expansion of this work was organized into the monograph *Field Computation by Moment Methods* [13]. Since then, the general method

[1] See pp. 15–23 of this book for a reprint of this paper—The Editors.

[2] See pp. 24–37 of this book for a reprint of this paper.

The paper, "Origin and development of the method of moments for field computation" by R. F. Harrington—Chapter 1 in *Applications of the Method of Moments to Electromagnetic Fields*, B. J. Strait, ed. appears with 18 other papers on Moment Methods in a 1980 SCEEE Press book entitled, *Applications of the Method of Moments to Electromagnetic Fields*; the book is still in print and is available from the SCEEE Press, 1101 Massachusetts Avenue, St. Cloud, FL 34769.

of moments has come into widespread use for practical electromagnetic engineering problems.

II. The Method of Moments

Most solutions of functional equations can be interpreted in terms of projections onto subspaces of functional spaces. For computation, these subspaces must necessarily be finite dimensional. For theoretical work, they may be infinite dimensional. The general concept of solution of equations by projection onto subspaces has a number of different names. Some of the more common ones are the *method of moments,* the *method of weighted residuals,* the *method of projections,* and the *Petrov–Galerkin method.* We shall show in the next section it is also formally equivalent to the Rayleigh–Ritz variational method.

Consider the deterministic equation

$$Lf = g \tag{1}$$

where L is a linear operator, g is a known function, and f is an unknown function to be determined. Let f be represented by a set of functions $\{f_1, f_2, f_3, \cdots\}$ in the domain of L as the linear combination

$$f = \sum_j \alpha_j f_j \tag{2}$$

where the α_j are scalars to be determined. The f_j are called *expansion functions* or *basis functions.* For approximate solutions (2) is a finite summation; for exact solutions it is usually an infinite summation. Substituting (2) into (3), and using the linearity of L, we have

$$\sum_j \alpha_j Lf_j = g \tag{3}$$

where the equality is usually approximate. Now define a set of *testing functions* or *weighting functions* $\{w_1, w_2, w_3, \cdots\}$ in the range of L. Take the inner product (usually an integration) of (3) with each w_i, and use the linearity of the inner product to obtain

$$\sum_j \alpha_j \langle w_i, Lf_j\rangle = \langle w_i, g\rangle \tag{4}$$

$i = 1, 2, 3, \cdots$. This set of equations can be written in matrix form as

$$[l]\vec{\alpha} = \vec{g} \tag{5}$$

where $[l]$ is the matrix

$$[l] = [\langle w_i, Lf_j\rangle] \tag{6}$$

and $\vec{\alpha}$ and $\vec{g}$ are column vectors

$$\vec{\alpha} = [\alpha_j] \tag{7}$$

$$\vec{g} = [\langle w_i, g\rangle]. \tag{8}$$

If $[l]$ is nonsingular, its inverse exists, and $\vec{\alpha}$ is given by

$$\vec{\alpha} = [l]^{-1}\vec{g}. \tag{9}$$

The solution for f is then given by (2). For concise notation, define the row vector of functions

$$\tilde{f} = [f_j]. \tag{10}$$

Write (2) as $f = \tilde{f}\vec{\alpha}$ and substitute from (9). The result is

$$f = \tilde{f}[l]^{-1}\vec{g}. \tag{11}$$

This solution may be either approximate or exact depending upon the choice of expansion functions and testing functions.

The method of moments may also be applied to eigenvalue equations. Such applications are discussed in detail in Chapter 7 of [13].

The name *method of moments* derives from the original terminology that $\int x^n f(x)\, dx$ is the nth moment of f. When x^n is replaced by an arbitrary w_n, we continue to call the integral a moment of f. The name *method of weighted residuals* derives from the following interpretation. If (3) represents an approximate equality, then the difference between the exact and approximate Lf's is

$$g - \sum_j \alpha_j Lf_j = r \tag{12}$$

which is called the *residual* r. The inner products $\langle w_i, r\rangle$ are called the *weighted residuals*. Now (4) is obtained by setting all weighted residuals equal to zero.

The name *method of projections* comes from the concept of orthogonal projections onto subspaces in the theory of linear spaces. The expansion functions f_j generate a subspace in which we approximate f, and the testing functions w_i generate a subspace into which we project Lf. The method of moments then sets the residual equal to the null vector in the testing subspace. The inner products with each w_i in (4) are proportional to the w_i components of (3). Hence, we can alternatively say that each component of the residual is zero in the testing subspace. Still a third way of stating it is that the residual is orthogonal to every w_i.

The name *Petrov–Galerkin's method* comes from the fact that Petrov extended Galerkin's method, described in the next section, to the case of operators not self-adjoint. If the domain of the adjoint operator is different from the domain of the original operator, then one must choose the testing functions different from the expansion functions. This leads to the general method of moments.

III. Specializations of the Method of Moments

As mentioned at the beginning of the preceding section, most solutions can be interpreted in terms of the method of moments. The eigenfunction method is the special case for which the expansion functions are taken to be eigenfunctions of L, and the testing functions eigenfunctions of the adjoint operator L^*. Of course, it is usually difficult to find these eigenfunctions, but once they are known the solution of $Lf = g$ is simple. In this case the matrix $[l]$ of (5) is diagonal, with elements equal to the eigenvalues λ_i when the eigenfunctions are normalized or biorthogonalized. The inverse $[l]^{-1}$ then has diagonal elements λ_i^{-1}, and the moment solution (11) reduce to the eigenfunction solution.

When the domains of L and L^* are the same, we can choose $w_i = f_i$, and the specialization is known as *Galerkin's method*. When L is self-adjoint, this has the advantage of making $[l]$ a symmetric matrix. Since the treatment of symmetric matrices is easier than nonsymmetric matrices, particularly for eigenvalue problems, this can be a theoretical advantage. However, for computations the evaluation of the elements of $[l]$ may be difficult when Galerkin's method is used, and this often outweighs the advantage of keeping $[l]$ symmetric.

The simplest specialization for computation is the *point matching* or *collocation method*. This basically involves satisfying the approximate representation (3) at discrete points in the region of interest. In terms of the method of moments, this is formally equivalent to choosing the testing functions to be Dirac delta functions. The integrations represented by the inner products now become trivial, which is the major advantage of this method. The solution is sometimes sensitive to the points at which the equation is matched, which is the major disadvantage of the method. However, if a judicious choice of points of match is taken, the method often gives answers of good accuracy.

Another possibility is that of minimizing the length or norm of the residual, given by (12). If the usual inner product is used, the procedure is called a *least squares method*. It is evident from (12) that minimization of r is equivalent to finding the shortest distance from g to the subspace generated by the Lf_j, $j = 1, 2, 3, \cdots$. Hence, by the projection theorem, the least norm is obtained by taking the $w_i = Lf_i$ in the method of moments. This same conclusion can be obtained by the methods of calculus as follows:

$$\begin{aligned} \|r\|^2 &= \langle g - \Sigma\alpha_i Lf_i,\, g - \Sigma\alpha_j Lf_j\rangle \\ &= \langle g, g\rangle - 2\Sigma\alpha_i\langle Lf_i, g\rangle + \Sigma\Sigma\alpha_i\alpha_j\langle Lf_i, Lf_j\rangle. \end{aligned} \tag{13}$$

The minimum is obtained by setting the derivatives of $\|r\|^2$ with respect to each α_k equal to zero. This gives

$$\frac{\partial\|r\|^2}{\partial\alpha_k} = 0 = -2\langle Lf_k, g\rangle + 2\Sigma\alpha_i\langle Lf_k, Lf_i\rangle \tag{14}$$

$k = 1, 2, 3, \cdots$. These equations are identical to (4) if $w_i = Lf_i$, and the statement of equivalence is proved.

IV. The Rayleigh–Ritz Variational Method

Given $Lf = g$, we often want only a functional of f (number dependent on f) instead of f itself. It is then possible to construct a *variational formula* for the functional, which has the property that first-order errors in f give rise to second-order errors in the functional. This we demonstrate by means of the *calculus of variations*, the essential concepts of which are summarized in the next paragraph.

Given a function f, and a second function f_1, only slightly different from f, then the *variation of* f is defined as

$$\delta f = f - f_1. \tag{15}$$

Just as f is analogous to a vector with infinitely many components, δf is analogous to the differential of a function of infinitely many variables. It is readily shown that variations satisfy the same rules as differentials in ordinary calculus. For example,

$$\delta(fg) = f\,\delta g + g\,\delta f \tag{16}$$

and so on. The *fundamental theorem of the calculus of variations* states that if

$$\langle\delta f, g\rangle = 0 \tag{17}$$

for all possible δf in the inner product space, then $g = 0$.

Now suppose we are given $Lf = g$, and we wish to calculate the continuous linear functional of f

$$\rho = \langle h, f\rangle \tag{18}$$

where h is a known continuous function. For an approximation, let $f = \alpha\phi$, where ϕ is a *trial function* (guess at f) and α is a scalar. We then use the method of moments and test this guess with a testing function w, setting

$$\alpha\langle w, L\phi\rangle = \langle w, g\rangle. \tag{19}$$

This gives us a value of α, whence our approximation to ρ, denoted J, becomes

$$J = \frac{\langle w, g\rangle\langle h, \phi\rangle}{\langle w, L\phi\rangle}. \tag{20}$$

Our next problem is to determine how to pick w. We want (20) to be insensitive to variations in both ϕ and w, hence variations of J with respect to ϕ and w must independently vanish. Keeping w constant and varying ϕ, we have

$$\delta J = \frac{1}{\langle w, L\phi\rangle^2}\big(\langle w, L\phi\rangle\langle w, g\rangle\langle h, \delta\phi\rangle - \langle w, L\delta\phi\rangle\langle w, g\rangle\langle h, \phi\rangle\big) \tag{21}$$

which can be simplified to

$$\delta J = J\left(\frac{\langle h, \delta\phi\rangle}{\langle h, \phi\rangle} - \frac{\langle L^*w, \delta\phi\rangle}{\langle L^*w, \phi\rangle}\right) \tag{22}$$

where L^* is the adjoint operator to L. It is apparent that $\delta J = 0$ if

$$L^*w = h \tag{23}$$

which is the *adjoint equation*. Keeping ϕ constant and varying w in (20), we have by analogous steps $\delta J = 0$ if

$$L\phi = g \tag{24}$$

which is the original equation. Hence, (20) is called a *variational formula* for ρ, with *stationary point* (18) when ϕ is the solution to (24) and w is the solution to the adjoint equation (23). Note that ρ can also be written as

$$\rho = \langle w, g\rangle \tag{25}$$

since $\langle h, f\rangle = \langle L^* w, f\rangle = \langle w, Lf\rangle = \langle w, g\rangle$. Hence, we can think of ρ as a functional either of f or of w.

Suppose now we wish to use the stationary property of (20) to obtain approximations to (18). Trial functions ϕ and w correct to the first order lead to second-order approximations of ρ. If we wish more accurate results, the *Rayleigh–Ritz procedure* can be used as follows. Let trial functions ϕ and w be represented by linear combinations

$$\phi = \sum_j \alpha_j f_j$$
$$w = \sum_i \beta_i w_i \tag{26}$$

and substitute into (20). After simplification, we have

$$J = \frac{\left(\sum \beta_i \langle w_i, g\rangle\right)\left(\sum \alpha_j \langle h, f_j\rangle\right)}{\sum\sum \beta_i \alpha_j \langle w_i, Lf_j\rangle}. \tag{27}$$

Now the stationary point is found by setting

$$\frac{\partial J}{\partial \alpha_k} = \frac{\partial J}{\partial \beta_k} = 0 \tag{28}$$

for $k = 1, 2, 3, \cdots$. This gives a set of equations for determining α_j and β_i, and hence the approximations to ρ

$$\rho_f = \sum_j \alpha_j \langle h, f_j\rangle \qquad \text{from (18)}$$
$$\rho_w = \sum_i \beta_i \langle w_i, g\rangle \qquad \text{from (25)}$$
$$\rho_J = \frac{\rho_f \rho_w}{\sum\sum \beta_i \alpha_j \langle w_i, Lf_j\rangle} \qquad \text{from (27)}. \tag{29}$$

Now let $J = N/D$, and evaluate the derivatives of (27) with respect to α_k

$$\frac{\partial J}{\partial \alpha_k} = \frac{1}{D^2}\left[D\langle h, f_k\rangle \sum \beta_i \langle w_i, g\rangle - N \sum \beta_i \langle w_i, Lf_k\rangle\right]$$
$$= \frac{1}{D}\left[\rho_w \langle h, f_k\rangle - \rho_J \sum \beta_i \langle w_i, Lf_k\rangle\right] \tag{30}$$

and, similarly, the derivatives with respect to β_k

$$\frac{\partial J}{\partial \beta_k} = \frac{1}{D}\left[\rho_f \langle w_k, g\rangle - \rho_J \sum \alpha_j \langle w_k, Lf_j\rangle\right]. \tag{31}$$

Equations (28) are satisfied if we set $\rho_f = \rho_w = \rho_J$ in (30) and (31), whence

$$\sum \beta_i \langle f_k, L^* w_i\rangle = \langle f_k, h\rangle$$
$$\sum \alpha_j \langle w_k, Lf_j\rangle = \langle w_k, g\rangle \tag{32}$$

$k = 1, 2, 3, \cdots$. Comparing the second of these with (4), we see that it is the moment solutions for $Lf = g$. Similarly, the first of (32) is the moment solution for the adjoint equation $L^* w = h$. Hence, if we take the f_j as a basis for f and the w_i as a basis for w, the *Rayleigh–Ritz variational method is formally identical to the method of moments.*

Finally, it may be noted that (20) is only one of several functionals having stationary value ρ. For example, another one is

$$J = \langle h, \phi\rangle + \langle w, g\rangle - \langle w, L\phi\rangle. \tag{33}$$

It is left for the reader to show that $\partial J = 0$ when ϕ satisfies (24) and w satisfies (23), and that the stationary point of J is ρ of (18).

V. The Perturbation Method

If the problem under consideration is only slightly different (perturbed) from one which can be solved exactly (the unperturbed problem), then an approximate solution to the perturbed problem can be obtained by using the unperturbed solution as an expansion function in the method of moments. Such a procedure is called a *perturbation method*. More accurate solutions can be obtained by using the unperturbed solution plus correction terms in the method of moments.

Suppose we have the deterministic problem $Lf = g$, and an unperturbed problem

$$L_o f_o = g \tag{34}$$

for which the solution f_o is known. Let $M = L - L_o$ be the difference operator, and express the original problem as

$$Lf = (L_o + M)f = g. \tag{35}$$

For a first-order perturbation solution, let $f = \alpha f_o$ and apply the method of moments. This gives, for an arbitrary testing function w,

$$\alpha(\langle w, L_o f_o\rangle + \langle w, Mf_o\rangle) = \langle w, g\rangle.$$

Now, by (34), $\langle w, Lf_o\rangle = \langle w, g\rangle$, and hence the above equation reduces to

$$\alpha = 1 - \frac{\langle w, Mf_o\rangle}{\langle w, g\rangle + \langle w, Mf_o\rangle}. \tag{36}$$

If the perturbation is small, the second term in the denominator of (36) will be small compared to the first term, and $f = \alpha f_o$ becomes

$$f \approx \left(1 - \frac{\langle w, Mf_o\rangle}{\langle w, g\rangle}\right) f_o. \tag{37}$$

This is the general first-order perturbation solution. For higher-order solutions, choose $f_1 = f_o$ in the method of moments, and $f_2, f_3, \cdots$ serve as correction terms. The advantage of a perturbation solution over other moment solutions lies in the faster convergence of the perturbation solution.

It is often desirable to make the perturbation solution also a variational one. This is accomplished as follows. If L is self-adjoint, choose $w = f_o$ in (37), otherwise choose $w = f_o^*$, the solution to the adjoint equation. Then, according to the concepts of Section IV, the solution (37) is also variational.

VI. Conclusion

The preceding sections give a short history of the method of moments as applied to electromagnetic field problems, and a brief description of some of the underlying theory. Since most methods of solution can be interpreted in terms of the method of moments, it is not possible to say just when it started. The numerical methods and the variational methods can be traced back to Maxwell's time. The viewpoint of using projections from an infinite dimensional function space onto a finite dimensional subspace was put on a firm mathematical basis by Hilbert, and used extensively in quantum mechanics in the 1920s. It was the development of high-speed computers since World War II that has made the method so powerful for solving practical engineering problems.

Perhaps something should be said about the use of the word ''formal'' when discussing the equivalence of the method of moments to variational methods. When I manipulate formulas, I assume that the required parameters exist. When I discuss functionals, I assume that they are continuous, and so on. If integrals do not exist, or if functionals are not continuous, then the derivations are no longer valid. Engineers and physicists are notorious for using mathematics in situations where it has not yet been proven to apply, and even where it does not apply in a rigorous sense. For example, we use variational formulas for the field at a point, or the current at a terminal, even though they are not continuous functionals of the field or current. Such usage sometimes leads to computational trouble, and sometimes does not. When trouble is encountered, one should return to fundamentals to seek the cause. Usually, it can be found in the violation of some explicit or implicit assumption made in the mathematical derivation of the formulas.

References

[1] D. S. Saxon, ''Notes on Lectures by Julian Schwinger—Discontinuities in Waveguides,'' Massachusetts Institute of Technology, 1945.

[2] N. Marcuvitz, *Waveguide Handbook,* Rad. Lab. Series Vol. 10, McGraw-Hill Book Co., N.Y., 1951.

[3] V. H. Rumsey, ''The Reaction Concept in Electromagnetic Theory,'' *Phys. Rev.,* Series 2, vol. 94, pp. 1483-1491, June 1954.

[4] R. F. Harrington, *Time-Harmonic Electromagnetic Fields,* McGraw-Hill Book Co., N.Y., 1961.

[5] K. K. Mei and J. Van Bladel, ''Scattering by Perfectly Conducting Rectangular Cylinders,'' *IEEE Trans.,* vol. AP-11, pp. 185-192, March 1963.

[6] M. G. Andreasen, ''Scattering from Parallel Metallic Cylinders with Arbitrary Cross Sections,'' *IEEE Trans.,* vol. AP-12, pp. 746-754, November 1964.

[7] F. K. Oshiro, ''Source Distribution Techniques for the Solution of General Electromagnetic Scattering Problems,'' *Proc. First GISAT Symp.,* vol. 1, Mitre Corp., pp. 83-107, 1965.

[8] B. Friedman, *Principles and Techniques of Applied Mathematics,* John Wiley and Sons, N.Y., 1956.

[9] L. V. Kantorovich and V. I. Krylov, *Approximate Methods of Higher Analysis,* Translated by C. D. Benster, John Wiley and Sons, N.Y., 1964.

[10] L. V. Kantorovich and G. P. Akilov, *Functional Analysis in Normed Spaces,* Translated by D. E. Brown, Pergamon Press, Oxford, pp. 586-587, 1964.

[11] R. F. Harrington *et al.,* ''Matrix Methods for Solving Field Problems,'' Final Report, Contract AF 30(602)-3724, Rome Air Development Center, March 1966.

[12] R. F. Harrington, ''Matrix Methods for Field Problems,'' *Proc. IEEE,* vol. 55, pp. 136-149, February 1967.

[13] R. F. Harrington, *Field Computation by Moment Methods,* the Macmillan Co., N.Y., 1968.

Effective Methods for Solving Integral and Integro-Differential Equations

DONALD R. WILTON and CHALMERS M. BUTLER

DEPARTMENT OF ELECTRICAL ENGINEERING, UNIVERSITY OF MISSISSIPPI, UNIVERSITY, MS 38677

Abstract—**Effective numerical methods for solving simple integral and integro-differential equations with logarithmically singular kernels are presented tutorially in this paper. Fundamental features and interrelationships of the methods are discussed. Procedures for applying the techniques to practical equations of electromagnetics are suggested.**

1. Introduction

The development of numerical methods for solving integral equations in electromagnetics has been the subject of intensive research for more than fifteen years. During these years, careful analysis has paved the way for the development of efficient and effective numerical methods and, of equal importance, has provided a solid foundation for a thorough understanding of the techniques. Progress in both understanding and procedures has been so great that equations which could be solved only by researchers ten years ago now can be solved handily by undergraduate engineering students.

In this paper are described simple but highly effective methods for solving integral and integro-differential equations of a class having principal properties usually found in equations encountered in electromagnetics. Features of the methods are discussed and their interrelationships are delineated. The presentation is tutorial and should be readily accessible to an applications-oriented audience, but is is hoped that the reader more deeply steeped in numerical techniques would benefit from it also.

To illustrate numerical techniques, the authors outline sample methods for solving simple one-dimensional equations whose kernels exhibit logarithmic singularities. These equations are referred to as *generic* equations and are selected for consideration here for two reasons. First, practically important one-dimensional electric field integral equations and integro-differential equations of electromagnetics possess logarithmically singular kernels and, since the nature of the kernel dictates major properties of an integral equation, the salient features of the simpler generic equations replicate those of the electric field equations. Second, all but one of the generic equations have known exact solutions which provide a basis for ascertaining the accuracy of numerical solutions. Exactly how techniques for solving the generic equations can be applied to practical equations is discussed.

2. Simple Solution of Integral Equation

As an introductory example, a simple but effective numerical method is developed for the elementary integral equation

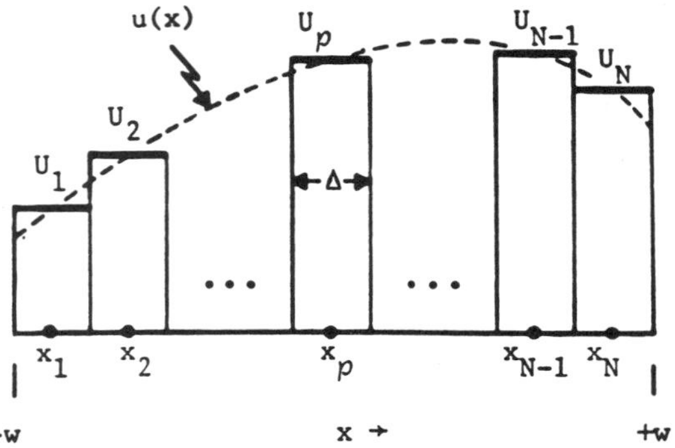

Fig. 1. Weighted pulse-function representation of $u(x)$: $u(x_n) \approx U_n \Pi_n(x)$.

$$-\frac{1}{2\pi}\int_{-w}^{w} u(x')\ln|x-x'|\,dx' = f(x), \qquad x \in (-w, w) \tag{1}$$

in which $u(x)$ is the unknown, $f(x)$ is the forcing function, and $-\dfrac{1}{2\pi}\ln|x-x'|$ is the kernel. In the solution procedure u is approximated by a linear combination of known, linearly independent basis functions. In this example, pulses [1] are employed which means that $u(x)$ is represented in $(-w, w)$ by a piecewise-constant approximation as

$$u(x) \approx \sum_{n=1}^{N} U_n \Pi_n(x) \tag{2}$$

in which the U_n's are unknown constants and the pulse functions Π_n's are defined by

$$\Pi_n(x) = \begin{cases} 1, & x \in (x_n - \Delta/2,\ x_n + \Delta/2) \\ 0, & \text{otherwise} \end{cases}. \tag{3}$$

The approximation is illustrated in Fig. 1. The interval $(-w, w)$ is divided into N segments of equal length $\Delta = 2w/N$ with their centers at

$$x_n = -w + \Delta\left(n + \tfrac{1}{2}\right), \quad n = 1, 2, \cdots, N. \tag{4}$$

Subject to the above approximation of $u(x)$, (1) becomes

$$\sum_{n=1}^{N} U_n s_n(x) \approx f(x), \quad x \in (-w, w) \tag{5}$$

Reprinted with permission from *Electromagnetics*, vol. 1, pp. 289–308, 1981.

where

$$s_n(x) = -\frac{1}{2\pi}\int_{x_n-\Delta/2}^{x_n+\Delta/2} \ln|x - x'|\,dx' \tag{6}$$

which can be integrated readily. Eq. (5) is enforced exactly at N points x_m, $m = 1, 2, \cdots, N$, in the interval $(-w, w)$, as illustrated in Fig. 1, to obtain the following set of N equations in N unknowns $\{U_n\}$:

$$\sum_{n=1}^{N} U_n S_{mn} = F_m, \quad m = 1, 2, \cdots, N. \tag{7}$$

In (7), $F_m = f(x_m)$ and $S_{mn} = s_n(x_m)$ or

$$S_{mn} = \frac{\Delta}{2\pi}\Bigg\{1 - \ln\Delta - \tfrac{1}{2}\ln\left|(m-n)^2 - \tfrac{1}{4}\right| - (m-n)\ln\left(\frac{|m-n+\frac{1}{2}|}{|m-n-\frac{1}{2}|}\right)\Bigg\}. \tag{8}$$

In matrix notation (7) is

$$[S_{mn}][U_n] = [F_m]. \tag{9}$$

The solution of (7) or (9) yields $\{U_n\}$, knowledge of which enables one to approximate $u(x)$ by means of (2).

Approximate solutions of (1) with $f = 1$ and $f = x$, determined by the numerical method outlined above, are presented graphically in Fig. 2 for various values of N together with the exact solutions (Table 1) for comparison. For $N = 5$ the actual pulse approximation is plotted but only values of U_n are indicated at the pulse centers for $N = 9$ and $N = 20$. A few points are omitted to avoid obscuration. One should note the improvement in the solution as N is increased and should observe that the numerical results exhibit approximately the edge condition near $x = \pm w$.

From (8) one sees that S_{mn} depends on the difference $(m - n)$ rather than on m and n individually which means that $[S_{mn}]$ is a *Toeplitz* matrix and that its elements depend on a single index $(m - n)$: $S_{mn} = S_{m-n}$. Since $[S_{mn}]$ also is observed to be *symmetric*,

$$S_{mn} = S_{m-n} = S_{n-m} = S_{|m-n|}, \tag{10}$$

from which we conclude that there are only N distinct elements in $[S_{mn}]$. All elements can be found from knowledge of those of the first (or last) row or of the first (or last) column, an observation that can be used to reduce the time and labor needed to compute the matrix elements. If we choose to compute the elements of the first row, then all others can be determined from

$$S_{mn} = S_{1,|m-n|+1}. \tag{11}$$

The properties above are direct results of (i) dependence of the kernel upon $|x - x'|$, (ii) even-function nature of $\Pi_n(x)$ about its center, (iii) equal-width pulses, and (iv) location of match points at pulse centers. A third interesting property of $[S_{mn}]$ is that it is *diagonally strong*. That is, the magnitude of the main diagonal elements is greater than that of any of the off-diagonal elements ($|S_{nn}| > |S_{mn}|$, $m \neq n$). Usually computing the inverse of such a matrix or solving the associated set of linear equations is a very stable process.

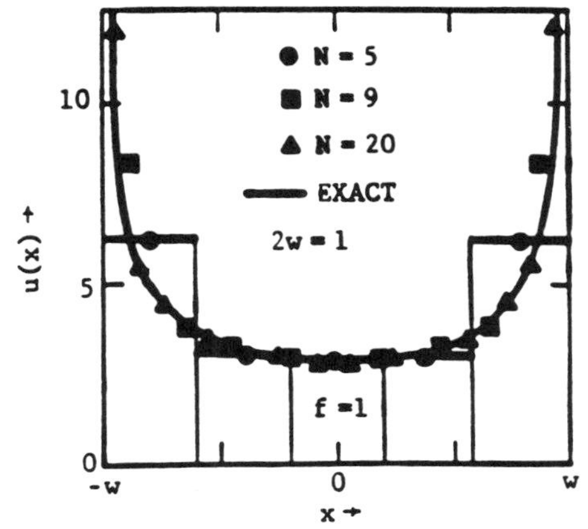

Fig. 2(a). Solution of Eq. (1) with $f(x) = 1$ for different N (pulse expansion/point matching).

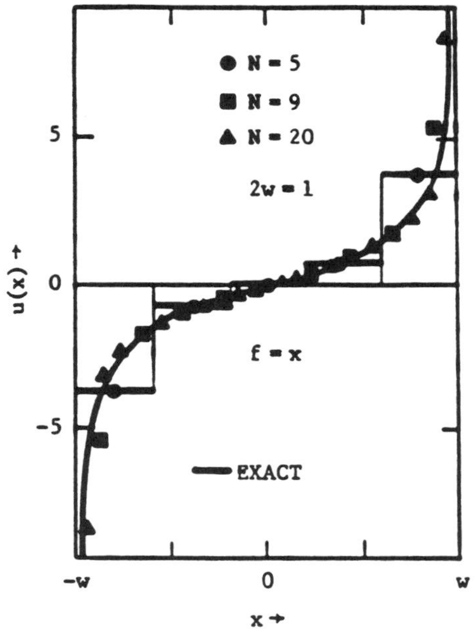

Fig. 2(b). Solution of Eq. (1) with $f(x) = x$ for different N (pulse expansion/point matching).

3. Solution of Integral Equation—Numerically Computed Matrix Elements

In the numerical solution of (1) the matrix elements $[S_{mn}]$ are determined from (8), but, in solving practical integral equations, integrals which are encountered can rarely be evaluated analytically. In such instances numerical integration methods are necessary but they must be employed with great care, especially when the integral is improper. An

TABLE I
INTEGRAL EQUATION SOLUTION

Case i	Unknown $u_i(x)$	Forcing Function $f_i(x)$
0	$\frac{2}{w\ln(2/w)}\frac{1}{\sqrt{1-(x/w)^2}}$	1
1	$\frac{2}{w}(x/w)\frac{1}{\sqrt{1-(x/w)^2}}$	(x/w)
2	$\frac{2}{w}\left[2(x/w)^2+\frac{1}{2\ln\left(\frac{2}{w}\right)}-1\right]\frac{1}{\sqrt{1-(x/w)^2}}$	$(x/w)^2$
3	$\frac{3}{w}[2(x/w)^3-(x/w)]\frac{1}{\sqrt{1-(x/w)^2}}$	$(x/w)^3$

Integral Equation:

$$-\frac{1}{2\pi}\int_{-w}^{w} u_i(x')\ln|x-x'|\,dx' = f_i(x),\ x\in(-w,w);\quad w\neq 2$$

integral equation with such a kernel is

$$\frac{1}{4\pi}\int_{-h}^{h} u(z')\,k(z-z')\,dz' = f(z),\quad z\in(-h,h) \tag{12}$$

in which the kernel is

$$k(z-z') = \frac{1}{2\pi}\int_{-\pi}^{\pi}\frac{1}{R}\,d\phi' \tag{13}$$

with

$$R = \left[(z-z')^2 + 4a^2\sin^2\frac{\phi'}{2}\right]^{1/2}. \tag{14}$$

With $f(z)$ a constant, (12) is essentially the equation for the charge density on a conducting tube of length $2h$ and radius $2a$ which is charged to a constant potential.

Below, we outline a procedure for solving (12) by the pulse expansion/point-matching technique. Eqs. (1) and (12) are of the same form but have different kernels so the sole difference in the solution procedures lies in the computation of the elements of the coefficient matrices. In the present case the matrix is denoted $[S_{mn}]$ and its elements are

$$S_{mn} = \frac{1}{4\pi}\int_{z_n-\Delta/2}^{z_n+\Delta/2} k(z_m-z')\,dz'$$
$$= \frac{1}{4\pi}\int_{-\Delta/2}^{\Delta/2} k(z_m-z_n-z')\,dz' \tag{15}$$

where $\Delta = 2h/N$ is the length of the N segments into which the interval $(-h, h)$ is partitioned and where z_n and z_m designate pulse centers and match points, respectively:

$$z_p = -h + \Delta(p-\tfrac{1}{2}),\quad p = 1, 2, \cdots, N. \tag{16}$$

A procedure to solve (12) must incorporate a method to perform numerically the double integration of (15), which, unless done with care, can be quite laborious. The difficulty in evaluating (15) numerically is due to the singular behavior of $k(z-z')$ at $z' = z$. To circumvent these problems we observe that

$$k(z-z') \xrightarrow[\left(\frac{z-z'}{2a}\right)\to 0]{} -\frac{1}{\pi a}\ln\left(\frac{|z-z'|}{8a}\right) \tag{17}$$

and we replace $k(z-z')$ by

$$k(z-z') = \left\{k(z-z') + \frac{1}{\pi a}\ln\left(\frac{|z-z'|}{8a}\right)\right\} - \frac{1}{\pi a}\ln\left(\frac{|z-z'|}{8a}\right) \tag{18}$$

which is formed by subtracting (17) from and adding it to $k(z-z')$. Clearly the right side of (18) is equal to $k(z-z')$ and it possesses two significant properties which we use to advantage. First, the term in the braces taken compositely is very slowly varying and therefore can be integrated numerically with ease. Second, the remaining term, which contains the troublesome singularity, is analytically integrable. Hence, with (18) in (15) we arrive at

$$\mathscr{S}_{mn} = \frac{\Delta}{4\pi^2 a}\ln(8a) + \frac{1}{2\pi a}S_{mn} + \frac{1}{4\pi}\int_{-\Delta/2}^{\Delta/2}\left\{k(\Delta[m-n]-z') + \frac{1}{\pi a}\ln(|\Delta[m-n]-z'|/8a)\right\}dz' \tag{19}$$

where $\mathscr{S}_{mn}$ is given in (8). Even though the integrand in (19) is very smooth, $k(\Delta[m-n]-z')$ still is an integral and hence evaluation of (19) requires in part double numerical integration. Fortunately, this difficulty can be overcome also by converting (19) to

$$\mathscr{S}_{mn} = \frac{1}{4\pi^2 a}[\Delta \ln(8a) + 2\pi S_{mn}] + \frac{1}{4\pi^2 a}\int_{-\Delta/2}^{\Delta/2}\{\beta_{mn}\mathscr{K}(\beta_{mn}) + \ln(|\Delta[m-n]-z'|/8a)\}\,dz' \quad (20)$$

in which $\mathscr{K}$ is the *complete elliptic integral of the first kind* [2] with modulus

$$\beta_{mn} = \left\{1 + \left(\frac{\Delta[m-n]-z'}{2a}\right)^2\right\}^{-1/2}. \quad (21)$$

The advantage of (20) over (19) is that there exist accurate approximations for $\mathscr{K}(\beta)$ [3] of simple form which are easy to integrate. The elements $\mathscr{S}_{mn}$ can be determined from (20) for all combinations of indices and for any radius a. The numerical integration method employed in the computation must be one which does not require evaluation of the integrand at $z'=0$ (identically), since for $m=n$ both the elliptic integral and the logarithm function in the braces are unbounded as $z' \to 0$. Even though (20) may be employed for all values of m and n, some increase in efficiency can be realized by employing the approximation

$$\mathscr{S}_{mn} \approx \frac{1}{4\pi}\ln\left(\frac{|m-n|+\frac{1}{2}}{|m-n|-\frac{1}{2}}\right), \quad \Delta|m-n| \gg 2a \quad (22)$$

whenever $\Delta|m-n| \gg 2a$.

Observe that $[\mathscr{S}_{mn}]$ is symmetric and Toeplitz,

$$\mathscr{S}_{mn} = \mathscr{S}_{|m-n|}, \quad (23)$$

and hence all its elements can be determined from those of the first (or last) row or column as in (11). Also $[\mathscr{S}_{mn}]$ is diagonally strong.

With a technique for computing $[\mathscr{S}_{mn}]$ now in hand, the procedure for solving (12) is identical to that for solving (1)—in fact, the same computer program may be employed but with a subroutine for computing S_{mn} replaced by one for computing $\mathscr{S}_{mn}$. Sample solutions of (12) are displayed in Fig. 3. Notice that $u(z)$ exhibits the known edge behavior proportional to $[1-|z/h|]^{-1/2}$ at interior points near $+h$ and $-h$.

4. Solution of First Order Integro-Differential Equation

In this section techniques for solving an integro-differential equation are presented. The simple equation

$$-\frac{1}{2\pi}\frac{d}{dx}\int_{-w}^{w} u(x')\ln|x-x'|\,dx' = f(x), \quad x \in (-w, w) \quad (24)$$

is to be solved subject to the constraint

$$\int_{-w}^{w} u(x)\,dx = 0. \quad (25)$$

by integration, (24) can be converted to

$$-\frac{1}{2\pi}\int_{-w}^{w} u(x')\ln|x-x'|\,dx' = \mathscr{f}(x) + C, \quad x \in (-w, w) \quad (26)$$

where C is an unknown constant of integration and where

$$\mathscr{f}(x) = \int_0^x f(x')\,dx'. \quad (27)$$

Eq. (24) reminds one of a differential equation with the needed constraining condition (25) while the pair (26) and (25) can be viewed as two simultaneous equations with unknowns u and C. Apart from C, (26) is the familiar equation (1) but the presence of the differential operator in (24) is a feature which warrants careful investigation.

The role of the constant C in (26) can be clarified by a simple example. Let $f=1$, then $\mathscr{f}=x$ and the right side of (26) becomes $C+x$. From Table 1, it is clear that $u(x) = Cu_0(x) + wu_1(x)$. To satisfy (25) C must be zero, so the solution is $u(x) = wu_1(x)$.

Because (26) is so similar to (1), a method for solving it is considered before (25) is addressed. By expanding u in pulses (2) and enforcing (26) at match points x_m at pulse centers, one obtains

$$\sum_{n=1}^{N} U_n S_{mn} - C = \mathscr{f}(x_m), \quad m = 1, 2, \cdots, N \quad (28)$$

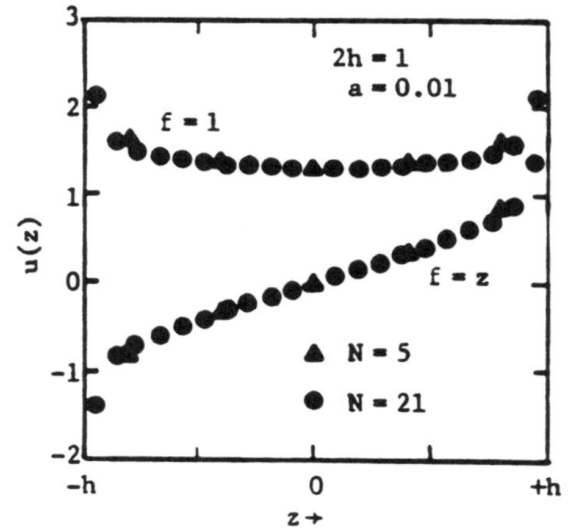

Fig. 3. Solution of Eq. (12) for different forcing functions.

where S_{mn} are the elements (8). Also, with u approximated by (2), the constraining equation (25) becomes

$$\Delta \sum_{n=1}^{N} U_n = 0. \tag{29}$$

Eqs. (28) and (29) can be written as a matrix equation

$$\begin{bmatrix} S_{11} & S_{12} & S_{13} & \cdots & S_{1N} & -1 \\ S_{21} & S_{22} & S_{23} & \cdots & S_{2N} & -1 \\ \vdots & \vdots & \vdots & & \vdots & \vdots \\ S_{N1} & S_{N2} & S_{N3} & \cdots & S_{NN} & -1 \\ \Delta & \Delta & \Delta & & \Delta & 0 \end{bmatrix} \cdot \begin{bmatrix} U_1 \\ U_2 \\ \vdots \\ U_N \\ C \end{bmatrix} = \begin{bmatrix} \mathscr{f}(x_1) \\ \mathscr{f}(x_2) \\ \vdots \\ \mathscr{f}(x_N) \\ 0 \end{bmatrix} \tag{30}$$

in which the role of C as an unknown is clear.

We now outline a method of solving (24) and (25) directly. With u expanded in pulses (24) can be expressed as

$$\sum_{n=1}^{N} U_n \frac{d}{dx} \mathscr{s}_n(x) \approx f(x), \quad x \in (-w, w). \tag{31}$$

Since there are N unknowns $\{U_n\}$, $N-1$ equations must be obtained from the numerical processing of (31) to be solved simultaneously with (29). To obtain $N-1$ equations from (31), one might be tempted to place match points at $x_n \pm \Delta/2$ midway between pulse centers, but $\left\{\frac{d}{dx} \mathscr{s}_n(x)\right\}_{x=x_n \pm \Delta/2}$ is unbounded. Hence, the simple point-matching scheme applied to (1) cannot be applied to (24), if we insist that u be represented by pulses. Perhaps it has occurred to the reader that there are other deficiencies in the point-matching method. A simple example is the case of an equation whose forcing function happens to pass through zero at several of the selected match points. One remedy for this difficulty is to replace enforcement of an equation at a match point by enforcement of its average at several nearby points. That is, the mth equation of (7) would be replaced by one which equates the average of the left member of (5) to that of the right member, with both averages determined from values computed at points near x_m. Contributions from more and more points in a fixed interval about x_m an be included in the averaging process until, in the limit, one is in effect equating the integrals over the interval of the two sides of (5). Below we apply this procedure of equating averages to Eq. (31) and we discover that it effectively replaces the differential operator by a difference operator.

$N-1$ equations are to be obtained from the above mentioned enforcements of equality of averages of the two sides of (31). To this end, we multiply both sides of (31) by $\Pi_{m+1/2}(x)$, which is a pulse of width Δ and centered at

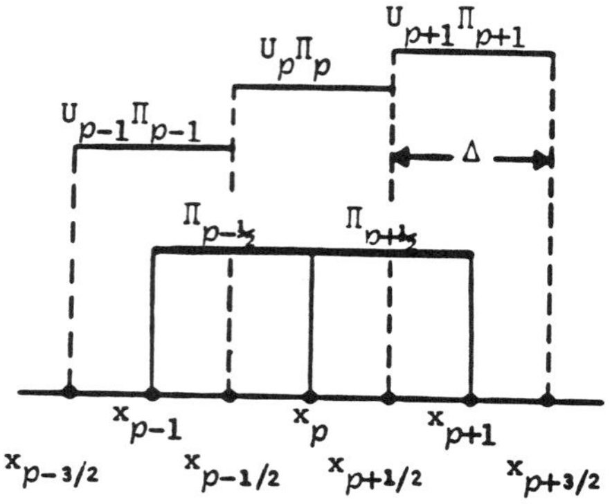

Fig. 4. Illustration of the weighted expansion pulses $U_p\Pi_p$ and the pulses $\Pi_{p\pm 1/2}$ used in the enforcement procedure.

$x_{m+1/2}$ midway between centers of expansion pulses, and integrate the resulting product over $(-w, w)$ to obtain

$$\Delta \sum_{n=1}^{N} U_n S'_{mn} = F_m, \quad m = 1, 2, \cdots, N-1 \tag{32}$$

where

$$F_m = \int_{x_m}^{x_{m+1}} f(x)\, dx \tag{33}$$

and

$$S'_{mn} = \int_{x_m}^{x_{m+1}} \frac{d}{dx} \mathscr{s}_n(x)\, dx = \frac{1}{\Delta}[\mathscr{s}_n(x_{m+1}) - \mathscr{s}_n(x_m)]$$
$$= \frac{1}{\Delta}[S_{m+1,n} - S_{mn}]. \tag{34}$$

The so-called testing pulses $\Pi_{m+1/2}$ and expansion pulses Π_n are illustrated in Fig. 4. Often $f(x)$ is sufficiently smooth over (x_m, x_{m+1}) to permit the approximation $F_m \approx \Delta f(x_{m+1/2})$ in which case (32) becomes

$$\sum_{n=1}^{N} U_n S'_{mn} \approx f(x_{m+1/2}), \quad m = 1, 2, \cdots, N-1. \tag{35}$$

Eq. (29) and the $N-1$ equations (32) can be solved for the N coefficients U_n or, when f is sufficiently smooth, one solves (29) and (35). In matrix form (35) is

$$\begin{bmatrix} S'_{11} & S'_{12} & \cdots & S'_{1N} \\ S'_{21} & S'_{22} & \cdots & S'_{2N} \\ \vdots & \vdots & & \vdots \\ S'_{N-1,1} & S'_{N-1,2} & \cdots & S'_{N-1,N} \\ \Delta & \Delta & \cdots & \Delta \end{bmatrix} \cdot \begin{bmatrix} U_1 \\ U_2 \\ \vdots \\ U_N \end{bmatrix} \approx \begin{bmatrix} f(x_{3/2}) \\ f(x_{5/2}) \\ \vdots \\ f(x_{N-1/2}) \\ 0 \end{bmatrix}. \tag{36}$$

For larger and larger N, taken to achieve a converged solution, the approximation $F_m \approx \Delta f(x_{m+1/2})$ becomes better and better in general whenever f is well behaved, thereby justifying ultimate utilization of (35) or (36).

Even though $\Delta S'_{mn}$ results directly from integrating $\Pi_{m'+1/2}(x)\frac{d}{dx}t_n(x)$ over $(-w, w)$ without approximation, S'_{mn} can be interpreted as $\frac{d}{dx}t_n(x)$ at $x_{m+1/2}$ with $\frac{d}{dx}$ replaced by the finite-difference operator. This interpretation is clear from (34). Hence, (35) can be viewed as (31) with the derivative replaced by the difference and enforced at the points $x_{m+1/2}$. Or, as suggested by (34), (32) can be obtained directly by subtracting the mth equation from the $(m+1)$th equation of (28).

5. Solution of a Second Order Integro-Differential Equation

In this section we consider techniques for solving the integro-differential equation

$$-\frac{1}{2\pi}\frac{d^2}{dx^2}\int_{-w}^{w} u(x')\ln|x-x'|\,dx' = f(x), \qquad x\in(-w, w) \tag{37}$$

where the unknown $u(x)$ must satisfy the boundary conditions

$$u(w) = u(-w) = 0. \tag{38}$$

Our interest in (37) stems from the fact that it has features analogous to those frequently found in integral equations of electromagnetics. For the sake of definiteness, however, (37) may be viewed as a low-frequency approximation of the integro-differential equation for the current induced on a conducting strip of width $2w$ excited by TE illumination [4]. The boundary condition (38) corresponds to the requirement that the current $u(x)$ must vanish at the strip edges.

Paralleling the integration of (24) to obtain (26), we eliminate the derivatives in (37) by regarding the equation as a differential equation with the integral as the unknown and solve the equation to obtain

$$-\frac{1}{2\pi}\int_{-w}^{w} u(x')\ln|x-x'|\,dx' = A + Bx + \int_{-w}^{x} f(x')(x-x')\,dx', \qquad x\epsilon(-w, w). \tag{39}$$

If A and B were known, (39) would have the same form as (1) and hence $u(x)$ could be found by any procedure used to solve the latter. Since A and B are unknown, however, $u(x)$ can only be determined in terms of A and B, and the additional information necessary to fix these constants must come from the boundary conditions (38). This may be illustrated by solving (39) explicitly for $f(x) = a + bx$. In this case, (39) becomes

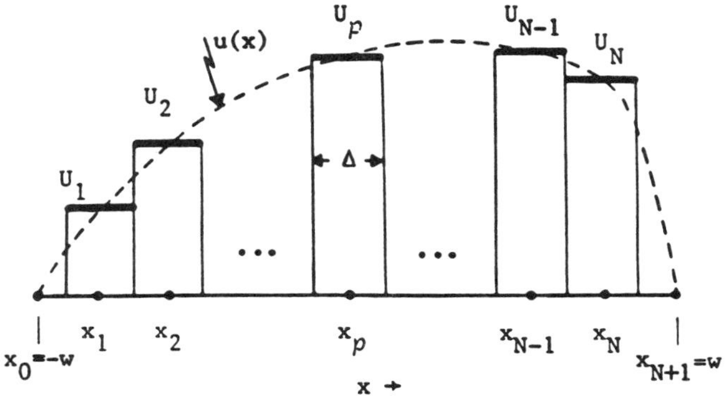

Fig. 5. Piecewise-constant representation of $u(x)$ with $u(\pm w) = 0$.

$$-\frac{1}{2\pi}\int_{-w}^{w} u(x')\ln|x-x'|\,dx' = A' + B'x + a\frac{x^2}{2} + b\frac{x^3}{6} \tag{40}$$

where A' and B' are new constants which absorb contributions of the particular integral to the homogeneous solution of (37). Using Table 1, one may obtain the solution of (40) by superposition of the responses due to individual driving terms on the right-hand side:

$$u(x) = \frac{1}{\sqrt{w^2-x^2}}(C + Dx + 2ax^2 + bx^3). \tag{41}$$

Since (40) has the same kernel as (1) and (26) considered earlier, it is not surprising that the solution *appears* to be singular at $x = \pm w$. However, the boundary conditions (38) require that $u(x)$ vanish at $x = \pm w$ and this is possible only if the quantity in parentheses in (41) vanishes at these points. Enforcing this condition yields

$$u(x) = \sqrt{w^2-x^2}\,(-2a - xb), \tag{42}$$

in which it is seen that $u(x)$ vanishes as $\sqrt{w-|x|}$ as x approaches the endpoints $\pm w$. The *derivative* of $u(x)$, however, is singular at and varies as $(w-|x|)^{-1/2}$ near the endpoints.

5.1 Solution of Integral Equation

In this section we develop a numerical approach for solving (39) subject to (38). The similarity of (39) to (1) and (26) suggests that we again employ a pulse function representation of $u(x)$ and use the point-matching procedure to enforce (39). In contrast to the solution procedures for (1) and (26), however, $u(x)$ is now required to satisfy (38) as well. This is accomplished by dividing the interval $(-w, w)$ into $N+1$ segments of equal length $\Delta = 2w/(N+1)$ with segment endpoints located at (Fig. 5)

$$x_n = -w + n\Delta, \quad n = 0,\cdots,N+1, \tag{43}$$

and by representing $u(x)$ in the piecewise-constant approximation

$$u(x) \approx \sum_{n=1}^{N} U_n \Pi_n(x) \tag{44}$$

with $\Pi_n(x)$ defined as in (3). Note that the points x_n in (43) and Fig. 5 are located such that no pulse exists in the intervals of length $\Delta/2$ about the endpoints $x = \pm w$. By this construction the constraint (38) is automatically satisfied by the piecewise constant representation (44).

If (44) is substituted into (39) and the resulting equation is point-matched at $x = x_0, \cdots, x_{N+1}$, one obtains $(N + 2)$ linear equations in the $(N + 2)$ unknowns $U_1, \cdots, U_N$, A, and B:

$$\sum_{n=1}^{N} U_n S_{mn} - A - Bx_m = \mathscr{F}_m, \quad m = 0, 1, \cdots, N+1, \tag{45a}$$

or in matrix form

$$\begin{bmatrix} S_{01} & S_{02} & \cdots & S_{0N} & -1 & -x_0 \\ S_{11} & S_{12} & \cdots & S_{1N} & -1 & -x_1 \\ \vdots & \vdots & & \vdots & \vdots & \vdots \\ S_{N1} & S_{N2} & \cdots & S_{NN} & -1 & -x_N \\ S_{N+1,1} & S_{N+1,2} & \cdots & S_{N+1,N} & -1 & -x_{N+1} \end{bmatrix} \cdot \begin{bmatrix} U_1 \\ U_2 \\ \vdots \\ U_N \\ A \\ B \end{bmatrix} = \begin{bmatrix} \mathscr{F}_0 \\ \mathscr{F}_1 \\ \vdots \\ \mathscr{F}_N \\ \mathscr{F}_{N+1} \end{bmatrix} \tag{45b}$$

in which

$$\mathscr{F}_m = \int_{-w}^{x_m} f(x')(x_m - x')\, dx' \approx \Delta \sum_{p=0}^{m-1} f(x_p)(x_m - x_p). \tag{46}$$

The elements S_{mn} are defined as in (8), and the approximation in (46) may be used if $f(x)$ is slowly varying over each subinterval. Fig. 6 shows numerically computed solutions of (45) for various values of N with $f(x) = 1$ and x. For comparison the exact solution obtained from (42) is also shown.

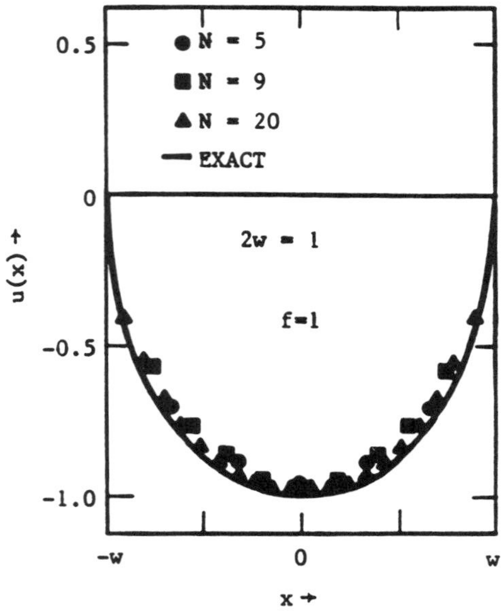

Fig. 6(a). Solution of Eq. (45) with $f(x) = 1$ for different N (pulse expansion/point matching).

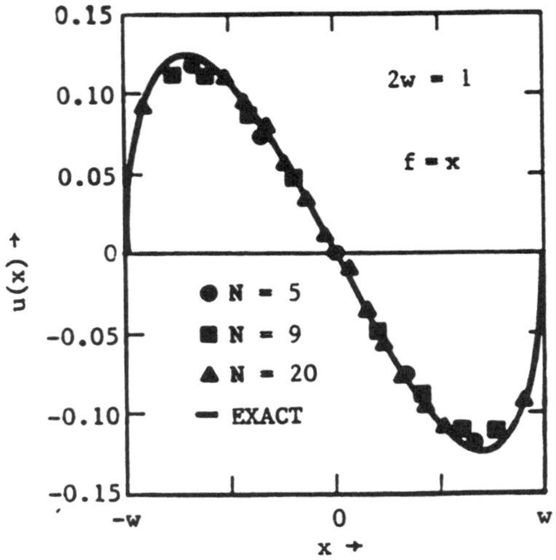

Fig. 6(b). Solution of Eq. (45) with $f(x) = x$ for different N (pulse expansion/point matching).

5.2 *Solution of Integro-Differential Equation*

In this section we examine several procedures for directly solving the integro-differential equation (37) subject to (38). An advantage of solving (37) directly is that the superfluous constants A and B in (39) are not needed. In practical cases involving coupled equations, a large number of such constants may appear, and avoiding them may substantially save computer storage and computation time. Furthermore, in problems involving junctions of several structures methods for the direct solution of the integro-differential equations are much simpler than those for the corresponding integral equations. That the direct approach imposes little or no penalty in numerical efficiency is demonstrated by revealing approximate equivalences among the numerically processed integral equation and the several integro-differential equations at various stages of the numerical developments.

Triangle Expansion and Pulse Testing Functions. The direct solution procedure developed to solve (24) suggests a method for solving (37) if we put the latter equation into a form similar to that of the former. To this end, one of the derivatives in (37) is brought under the integral sign and onto the kernel. Since

$$\frac{d}{dx}\ln|x-x'| = \frac{1}{x-x'} = -\frac{d}{dx'}\ln|x-x'|, \quad (47)$$

the resulting integral is convergent only if interpreted in a Cauchy principal value sense. (Both derivatives in (37) cannot be brought onto the kernel since the resulting integral would not converge.) The last form in (47) permits the resulting integral to be integrated by parts which, with (38), yields

$$-\frac{1}{2\pi}\frac{d}{dx}\int_{-w}^{w}\frac{d}{dx'}u(x')\ln|x-x'|\,dx' = f(x), \qquad x\in(-w,w). \quad (48)$$

Note that the unbounded derivative of $u(x)$ at $x = \pm w$ does not allow us to repeat this procedure with the remaining derivative outside the integral. Eq. (48) now has the same form as (24) but with the derivative $u'(x)$ rather than $u(x)$ itself playing the role of the unknown. Indeed, as (41) illustrates, $u'(x)$ in the present problem has the same singular edge behavior as did $u(x)$ in (24). Thus, paralleling the solution of (24), we represent $u(x)$ in (48) in such a way that its *derivative* is piecewise constant and test the equation with pulse functions. But if $u'(x)$ is approximated as piecewise constant, then $u(x)$ itself must be approximated as piecewise linear. This may be conveniently done through use of the *triangle functions*

$$\Lambda_n(x) = \begin{cases} 1 - \dfrac{|x-x_n|}{\Delta}, & x\in(x_{n-1}, x_{n+1}) \\ 0, & \text{otherwise} \end{cases} \quad (49)$$

which are of unit height and defined on intervals of width 2Δ. With these basis functions $u(x)$ may be approximated as

$$u(x) \approx \sum_{n=1}^{N} U_n\Lambda_n(x). \quad (50)$$

The piecewise-linear approximation of $u(x)$ by triangles and the resulting piecewise-constant approximation of $u'(x)$ are illustrated in Fig. 7. Note that the boundary conditions $u(\pm w) = 0$ are automatically satisfied in this representation. It is easily demonstrated that the derivative of $\Lambda_n(x)$ is

$$\Lambda_n'(x) = -\frac{1}{\Delta}\left[\Pi_{n+1/2}(x) - \Pi_{n-1/2}(x)\right] \quad (51)$$

and hence from (50)

$$u'(x) \approx -\frac{1}{\Delta}\sum_{n=1}^{N} U_n\left[\Pi_{n+1/2}(x) - \Pi_{n-1/2}(x)\right] = \sum_{n=0}^{N}\frac{U_{n+1}-U_n}{\Delta}\Pi_{n+1/2}(x) \quad (52)$$

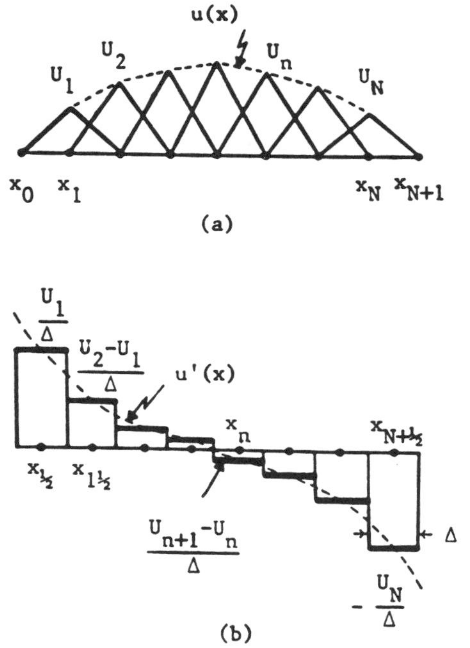

Fig. 7. The use of triangle basis functions yields (a) a piecewise linear approximation of $u(x)$ and (b) a piecewise constant approximation of $u'(x)$.

where $\Pi_{n+1/2}(x)$ is a unit pulse of width Δ centered at $x_{n+1/2}$, midway between the points x_n and x_{n+1}, and $U_0 = U_{N+1} \equiv 0$. If (52) is substituted into (48) and the resulting equation is integrated against $\Pi_m(x)$, i.e., tested with $\Pi_m(x)$, for $m = 1,\cdots,N$, property (10) permits the resultant equation to be manipulated into the form

$$\Delta\sum_{n=1}^{N} U_n S''_{mn} = \int_{-w}^{w} f(x)\Pi_m(x)\,dx \approx \Delta f(x_m), \qquad m = 1,2,\cdots,N \quad (53)$$

where the elements S''_{mn} are defined as

$$S''_{mn} = \frac{1}{\Delta^2}\left[S_{m+1,n} - 2S_{mn} + S_{m-1,n}\right] \equiv \frac{\Delta^2 S_{mn}}{\Delta x^2}. \quad (54)$$

The notation $\Delta^2/\Delta x^2$ is introduced to denote the operation of forming a second central finite difference, which approximates a second derivative in terms of finite differences. The approximation of the right-hand side in (53) improves with increasing N and may be made if $f(x)$ is sufficiently smooth. Eq. (53) constitutes a linear system of N equations in N unknowns which can be solved for the unknown coefficients $\{U_n\}$.

Eqs. (45) and (53) are two different linear systems of equations from which approximate solutions of (37) and (38) may be determined. Remarkably, however the system (53) with the approximated right-hand side is derivable from (45) if its right-hand side is approximated as in (46). Hence the two systems with approximated right-hand sides are equivalent and must have identical solutions. To establish this equivalence we apply the operator $\Delta(\Delta^2/\Delta x^2)$ to (45) with approximated right-hand side (46). That is, we multiply the mth equation of (45) by $-2/\Delta$ and add to it $1/\Delta$ times the $(m+1)$th and $(m-1)$th equations for $m = 1, 2, \cdots, N$. The left-hand side of the resultant equation is identical to the left-hand side of (53) whereas the right-hand side is identical to the approximate form of the right-hand side of (53) since

$$\sum_{p=0}^{m} f(x_p)(x_{m+1}-x_p) - 2\sum_{p=0}^{m-1} f(x_p)(x_m - x_p)$$
$$+\sum_{p=0}^{m-2} f(x_p)(x_{m-1}-x_p)$$
$$= f(x_{m-1})(x_{m+1}-x_{m-1}) + f(x_m)(x_{m+1}-x_m)$$
$$-2f(x_{m-1})(x_m - x_{m-1})$$
$$= \Delta f(x_m). \tag{55}$$

Since the system (53) can be obtained from a linear combination of equations (45), the two systems must have identical solutions. Hence Fig. 6 also serves to illustrate the solutions of (53) for $f(x) = 1$ and x.

Before proceeding to develop other approaches for solving (37), we pause to reflect on the procedures developed to solve both (24) and (37). We chose pulse functions to represent the unknown in (24) but found that the resultant integral was not differentiable at the desired match points. Hence we smoothed the derivative by averaging the equation, i.e., testing it with pulse functions, in each subinterval containing the derivative singularity. In (37), however, the presence of the additional derivative dictated the choice of the smoother triangle functions to represent the unknown, thus sufficiently smoothing the integral that the pulse testing approach could again be used. It seems plausible, however, that pulse functions could have been used to represent the unknown if smoother testing functions—for example, triangle functions—had been chosen. After all, testing involves multiplication of the equation by a testing function followed by an integration of the product, and integration by parts should permit us to transfer some of the burden of differentiability to the testing function. This idea is examined further in the next section.

Pulse Expansion and Triangle Testing. In this section we outline a numerical procedure for solving (37) and (38) using pulse functions to represent $u(x)$ and triangle functions to test the equation. The effect of testing a second derivative with a triangle function may be seen by evaluating the following integral involving the second derivative of an arbitrary function $h(x)$ defined on $(-w, w)$ and tested with $\Lambda_m(x)$:

$$\int_{-w}^{w} \frac{d^2h}{dx^2}\Lambda_m\,dx$$
$$= \Delta\left[\frac{h(x_{m+1}) - 2h(x_m) + h(x_{m-1})}{\Delta^2}\right]$$
$$= \Delta\frac{\Delta^2 h_m}{\Delta x^2}. \tag{56}$$

This result is easily obtained by one integration by parts over the various subintervals defining $\Lambda_m(x)$ in (49). Now let $u(x)$ be represented by pulse functions as in (44), substitute the representation into (37), and test the resulting equation with $\Lambda_m(x)$ for $m = 1, 2, \cdots, N$. Use of (56) allows the left-hand side to be written as

$$\Delta\sum_{n=1}^{N} U_n \frac{\Delta^2}{\Delta x^2}{\scriptstyle\mathcal{A}}_n(x_m) = \Delta\sum_{n=1}^{N} U_n\frac{\Delta^2 S_{mn}}{\Delta x_2}$$
$$= \Delta\sum_{n=1}^{N} U_n S''_{mn} \tag{57}$$

whereas the right-hand side is

$$\int_{-w}^{w} f(x)\Lambda_m(x)\,dx \approx \Delta f(x_m). \tag{58}$$

We have already seen that application of the finite difference operator $\Delta(\Delta^2/\Delta x^2)$ to the left-hand side of (45a) yields (57). Application of the operator to the right-hand side (with exact $\mathscr{F}_m$) yields

$$\frac{1}{\Delta}\int_{-w}^{x_{m-1}} f(x')(x_{m-1}-x')\,dx'$$
$$-\frac{2}{\Delta}\int_{-w}^{x_m} f(x')(x_m - x')\,dx'$$
$$+\frac{1}{\Delta}\int_{-w}^{x_{m+1}} f(x')(x_{m+1}-x')\,dx'$$
$$= \frac{1}{\Delta}\int_{x_{m-1}}^{x_m} f(x')(x - x_{m-1})\,dx'$$
$$+\frac{1}{\Delta}\int_{x_m}^{x_{m+1}} f(x')(x_{m+1}-x')\,dx'$$
$$= \int_{-w}^{w} f(x')\Lambda_m(x')\,dx' \approx \Delta f(x_m) \tag{59}$$

which is identical to (58). Since (57) and (58) can be obtained (without approximation) from linear combinations of equations (45a), the solutions to the two systems must be identical. Further, with the approximations on the right-hand side, the solution of these two systems is equivalent to that of (53).

We have seen in this section that smooth testing functions can be chosen to compensate for the choice of less well-behaved expansion functions. On the other hand, if we had

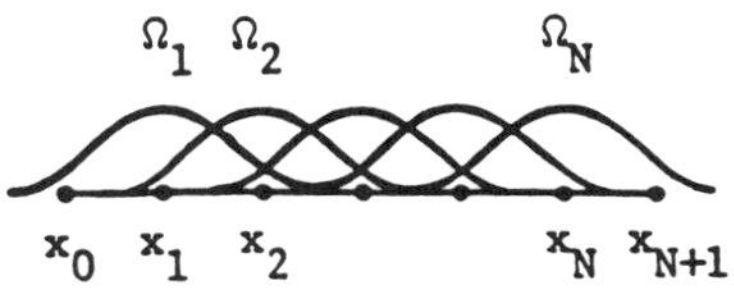

Fig. 8. The spline functions $\Omega_n(x)$.

chosen a smoother expansion function to begin with, perhaps we could have eliminated testing with a pulse function and used the simpler point-matching procedure instead. This possibility is explored next.

Spline Function Expansion and Point-Matching. In this section we investigate the use of a set of expansion functions which are smoother than triangles. These functions, called *spline functions*, are continuous and have a continuous first derivative. They are defined as

$$\Omega_n(x) = \begin{cases} \frac{1}{2}\left(\frac{x - x_{n-3/2}}{\Delta}\right)^2, & x \in (x_{n-3/2}, x_{n-1/2}) \\ \frac{3}{4} - \left(\frac{x - x_n}{\Delta}\right)^2, & x \in (x_{n-1/2}, x_{n+1/2}) \\ \frac{1}{2}\left(\frac{x - x_{n+3/2}}{\Delta}\right)^2, & x \in (x_{n+1/2}, x_{n+3/2}) \\ 0, & \text{otherwise} \end{cases} \tag{60}$$

and are shown in Fig. 8. From the figure it is seen that $\Omega_1(x)$ and $\Omega_N(x)$ extend past the ends of the interval $(-w, w)$. This deficiency can be remedied by defining special basis functions for the end intervals, but we are not presently concerned with such details. We shall merely expand $u(x)$ as

$$u(x) = \sum_{n=1}^{N} U_n \Omega_n(x) \tag{61}$$

and extend the domain of integration in (37) to the interval $(-w - \Delta/2, w + \Delta/2)$. The smoothness of the representation (61) renders the integral in (37) sufficiently smooth that it can be differentiated twice throughout the solution interval. Hence we substitute (61) into (37), point-match the resulting equation at x_m, $m = 1, \cdots, N$, and multiply both sides by Δ to obtain a linear system of equations for U_n. The matrix elements are easily computed by bringing the derivatives inside the integral and integrating by parts, yielding

$$-\frac{\Delta}{2\pi}\frac{d^2}{dx^2}\int_{w-\Delta/2}^{w+\Delta/2} \Omega_n(x') \ln|x - x'|\, dx' \,\Big|_{x=x_m}$$

$$= -\frac{\Delta}{2\pi}\int_{w-\Delta/2}^{w+\Delta/2} \frac{d^2}{dx'^2}\Omega_n(x') \ln|x_m - x'|\, dx'$$

$$= \Delta S''_{mn}. \tag{62}$$

The right-hand side of the mth equation is just $\Delta f(x_m)$. Again we obtain a system of equations identical to the approximate form of (53). However, though the values $\{U_n\}$ are the same as those found by the other procedures with the approximated right-hand sides, the values of $u(x_m)$ as computed from (61) are now $u(x_m) \approx (U_{m-1} - 6U_m + U_{m+1})/8$ whereas (44) and (50) yielded $u(x_m) \approx U_m$. This implies that the present scheme yields a somewhat smoother solution than the previous solutions because each point $u(x_m)$ in the present scheme consists of a weighted average of three adjacent values of the previous solutions.

We observe that the point-matching procedure applied to an integro-differential equation may be interpreted as testing the equation with delta functions $\Delta\delta(x - x_m)$, $m = 1, 2, \cdots, N$, which, by their sifting property, merely select the points x_m at which equality is enforced. In the next section we carry to extreme the idea that smoothness in the expansion functions can be traded for smoothness in the testing functions by choosing a delta-function representation of the unknown combined with spline testing functions.

Delta-Function Expansion and Spline Testing. If the spline functions are chosen as testing functions, it seems reasonable to expect that the resulting smoothing of the left-hand side of (37) may allow us to use expansion functions less well-behaved than the pulse functions. Accordingly, let us attempt to approximate $u(x)$ by delta-functions as

$$u(x) \approx \sum_{n=1}^{N} U_n \Delta\delta(x - x_n), \tag{63}$$

where $U_n\Delta$ is the total moment of $u(x)$ in the interval $(x_{n-1/2}, x_{n+1/2})$. Next (63) is substituted into (37) and the resulting equation is tested with $\Omega_m(x)$ (again the domain of the testing interval is extended to $(w - \Delta/2, w + \Delta'/2)$). The resulting left-hand side is

$$\sum_{n=1}^{N} U_n \int_{-w-\Delta/2}^{w+\Delta/2} \frac{d^2}{dx^2} \cdot\left[-\frac{1}{2\pi}\int_{-w}^{w} \Delta\delta(x' - x_n) \ln|x - x'|\, dx'\right] \cdot \Omega_m(x)\, dx = \Delta \sum_{n=1}^{N} U_n S''_{mn} \tag{64}$$

and the right-hand side can be approximated as

$$\int_{-w}^{w} f(x)\Omega_m(x)\, dx \approx f(x_m)\int_{-w-\Delta/2}^{w+\Delta/2} \Omega_m(x)\, dx = \Delta f(x_m). \tag{65}$$

Hence the system of equations so obtained is again equivalent to (53).

In summary, we have seen in these last four subsections

TABLE II
EXPANSION/TESTING COMBINATIONS YIELDING IDENTICAL SOLUTIONS OF EQ. (37)

Expansion functions	Testing functions
$\Delta\delta(x-x_n)$	$\Omega_m(x)$
$\Pi_n(x)$	$\Lambda_m(x)$
$\Lambda_n(x)$	$\Pi_m(x)$
$\Omega_n(x)$	$\Delta\delta(x-x_m)$

that with the approximations

$$\begin{aligned}\int_{-w}^{w} f(x)\Delta\delta(x-x_m)\,dx &\approx \int_{-w}^{w} f(x)\Pi_m(x)\,dx\\ &\approx \int_{-w}^{w} f(x)\Lambda_m(x)\,dx\\ &\approx \int_{-w-\Delta/2}^{w+\Delta/2} f(x)\Omega_m(x)\,dx\\ &\approx \Delta f(x_m) \qquad (66)\end{aligned}$$

involving the right-hand side of (37), each combination of expansion and testing functions selected yielded an identical system of equations with identical solution coefficients $\{U_m\}$. The various combinations selected are summarized in Table 2, where one notes that as one decreases the smoothness of the expansion functions, one must correspondingly increase the smoothness of the testing functions in order to obtain a comparable solution. While other considerations may dictate the actual choice of basis and testing functions in more advanced problems, with minor modifications the general observations of this section remain valid in these cases also.

6. Observations

In this final section we briefly relate some of the properties of the generic equations of previous sections to corresponding properties of electromagnetic equations. To this end, consider the electric field integral equation for scattering by a perfectly conducting surface S illuminated by a time-harmonic incident field $\bar{E}^{\text{inc}}$:

$$\frac{1}{j\omega\mu\epsilon}\left[(k^2+\nabla\nabla\cdot)\bar{A}\right]_{\tan} = -\bar{E}^{\text{inc}}_{\tan}(\bar{r}),\ \bar{r}\in S \qquad (67)$$

where

$$\bar{A}(\bar{r}) = \frac{\mu}{4\pi}\int_S \bar{J}(\bar{r}')\frac{e^{-jkR}}{R}\,dS' \qquad (68)$$

and

$$R = |\bar{r}-\bar{r}'|. \qquad (69)$$

Eqs. (67) and (68), together with the boundary condition

$$\bar{J}\cdot\hat{n} = 0 \qquad (70)$$

at each point on the boundary of S with $\hat{n}$ a unit vector tangent to S and normal to the boundary, constitute an integro-differential equation for the unknown induced current $\bar{J}$ on S. Although (67) is a more complicated equation in that it is a vector equation which applies on a two dimensional surface, it has a number of features common to (37). The most important of these features are the following:

1) Both (37) and (67) have singular kernels, the former of logarithmic type, the latter of the type $1/R$. When the problem has symmetry such that $\bar{J}$ is independent of one dimension, $1/R$ can be integrated in that dimension to obtain a one-dimensional integral equation with a logarithmically singular kernel closely corresponding to (37). With or without symmetry, the singular nature of the kernel determines the behavior, i.e., the edge conditions, of the current in the neighborhood of any boundaries and/or discontinuities of S.
2) In both (37) and (67), second derivative operators act on the integral. Thus, as in (37), (67) requires a certain degree of smoothness in representing $\bar{J}$ before the derivative operator may be applied to $\bar{A}$, or a testing procedure (or finite differences) must be used to effect the smoothing. As was found in solving (37), smoothness in the expansion functions may be traded for smoothness in the testing functions.

If (67) is specialized to a thin tubular cylinder parallel to the z-axis and of length $2h$ and radius a, the thin wire assumptions lead to [5]

$$\frac{1}{j\omega\epsilon}\left(\frac{d^2}{dz^2}+k^2\right)\int_{-h}^{h} I(z')K(z-z')\,dx' = -E_z^{\text{inc}}(z),\quad z\in(-h,h), \qquad (71)$$

where the total wire current $I(z)$ satisfies the boundary conditions

$$I(\pm h) = 0 \qquad (72)$$

and where

$$\begin{aligned}K(z-z') &= \frac{1}{2\pi}\int_{-\pi}^{\pi}\frac{e^{-jkR}}{R}\,d\phi\\ &= \mathcal{k}(z-z') + K_1(z-z') \qquad (73)\end{aligned}$$

with

$$R = \sqrt{(z-z')^2 + 4a^2\sin^2\frac{\phi}{2}}. \qquad (74)$$

Note in (73) that the term $\mathcal{k}(z-z')$, defined in (13), contains a logarithmic singularity. The term $K_1(z-z')$ is non-singular (in fact, vanishes at $z=z'$) and may be viewed as a perturbation on $\mathcal{k}(z-z')$. Matrix elements involving integration over the singular point in K may be computed as in (18)–(20) with an additional contribution from the integral of

$$K_1(z - z') = \frac{1}{2\pi} \int_{-\pi}^{\pi} \left(\frac{e^{-jkR} - 1}{R} \right) d\phi. \qquad (75)$$

The resulting double integral is easily integrated numerically.

Our analysis of the previous section suggests that any of the expansion and testing function combinations of that section as summarized in Table 2 may also be used to solve (71). Note that the additional integral arising from testing the integral associated with the k^2 term in (71) could be eliminated by using one of the approximations (66) since the integral in (71) (but not its derivative) is relatively smooth.

With minor modifications, observations similar to the above may be made for all electric field integral equation formulations in scattering and radiation, including, for example, scattering and radiation by strips [6], wires [7], and bodies of revolution [6]. Integral and integro-differential equations which are mathematically equivalent to (67) also arise in analyses of apertures [8, 9]. Equations in investigations of scattering by dielectric bodies also involve the integro-differential operator of (67) and, hence, similar techniques are applicable [6].

7. Acknowledgment

The authors gratefully acknowledge the expert assistance cheerfully provided by Ms. Tina Begley in the preparation of the manuscript.

References

[1] Harrington, R. F. (1968), *Field Computations by Moment Methods*, 229 pp., Macmillan, New York.

[2] Dwight, H. B. (1961), *Tables of Integrals and Other Mathematical Data*, p. 180, Macmillan, New York.

[3] Abramowitz, M., and I. A. Stegun (1964), *Handbook of Mathematical Functions*, p. 591, U.S. Government Printing Office, Washington, D.C.

[4] Butler, C. M., and D. R. Wilton (1980). General analysis of narrow strips and slots, *IEEE Trans. Antennas Propagat.*, AP-28(1), 42–48.

[5] Butler, C. M., and D. R. Wilton (1975), Analysis of various numerical techniques applied to thin-wire scatterers, *IEEE Trans. Antennas Propagat.*, AP-23(4), 534–540.

[6] Glisson, A. W., and D. R. Wilton (1980), Simple and efficient numerical methods for problems of electromagnetic radiation and scattering from surfaces, *IEEE Trans. Antennas Propagat.*, AP-28(5), 593–603.

[7] Wilton, D. R., and C. M. Butler (1976), Efficient numerical techniques for solving Pocklington's equation and their relationships to other methods, *IEEE Trans. Antennas Propagat.*, AP-24(1), 83–86.

[8] Butler, C. M., Y. Rahmat-Samii, and R. Mittra (1978), Electromagnetic penetration through apertures in conducting surfaces, *IEEE Trans. Antennas Propagat.*, AP-26(1), 82–93.

[9] Butler, C. M., and K. R. Umashankar (1976), Electromagnetic penetration through an aperture in infinite, planar screen separating two half spaces of different electromagnetic properties, *Radio Sci.*, 11(7), 611–619.

Potential Integrals for Uniform and Linear Source Distributions on Polygonal and Polyhedral Domains

DONALD R. WILTON, SENIOR MEMBER, IEEE, S. M. RAO, ALLEN W. GLISSON, MEMBER, IEEE, DANIEL H. SCHAUBERT, SENIOR MEMBER, IEEE, O. M. AL-BUNDAK AND CHALMERS M. BUTLER, FELLOW, IEEE

Abstract—**Formulas for the potentials due to uniform and linearly varying source distributions defined on simply shaped domains are systematically developed and presented. Domains considered are infinite planar strips, infinite cylinders of polygonal cross sections, planar surfaces with polygonal boundaries, and volumetric regions with polyhedral boundaries. The expressions obtained are compact in form and their application in the numerical solution of electromagnetics problems by the method of moments is illustrated.**

INTRODUCTION

In the numerical solution of problems in electromagnetics, one is frequently faced with the evaluation of static potential integrals associated with source distributions defined on elementary source regions such as line segments, polygons, and polyhedrons. In two dimensions, for example, potential integrals due to constant or linearly varying source density distributions confined to cylinders whose cross-sections are line segments or polygons are often required. In three dimensions, potentials due to sources confined to planar polygonal or polyhedral regions are required. Potential integrals for such distributions have found use, for example, in the numerical solution of the following problems:

1) static or quasi-static electric and magnetic problems formulated as integral equations [1], [2], [3];
2) the evaluation of "self-term" contributions to the moment matrix in time-harmonic electromagnetic radiation and scattering problems [4], [5];
3) time domain-problems in which time-retardation effects over subdomains are neglected [6].

In this communication, analytical expressions are obtained for all of the most commonly encountered of these potential integrals. In a few instances, the formulas given have appeared previously in the literature. Nevertheless, we include them for completeness together with the observation that the derivations given and the expressions obtained are generally much more concise than those appearing elsewhere in the literature. These features permit the expressions to be easily checked and translated into computer programs. The integrals are also expressed in terms of readily identifiable geometrical quantities, which facilitates programming the expressions and evaluating them in certain limiting cases.

Manuscript received July 13, 1983.

D. R. Wilton was with the Department of Electrical Engineering, University of Mississippi, University, MS. He is now with the Department of Electrical Engineering, University of Houston, Houston, TX 77004.

S. M. Rao was with the Department of Electrical Engineering, University of Mississippi University, MS. He is now with the Rochester Institute of Technology, Rochester, NY 14623.

A. W. Glisson is with the Department of Electrical Engineering, University of Mississippi, University, MS 38677.

D. H. Schaubert was with the Division of Electronic Products, Bureau of Radiological Health, Rockville, MD. He is now with the Department of Electrical and Computer Engineering, University of Massachusetts, Amherst, MA 01003.

O. M. Al-Bundak was with the Department of Electrical Engineering, University of Mississippi, University, MS.

C. M. Butler was with the Department of Electrical Engineering, University of Mississippi, University, MS. He is now with the Department of Electrical Engineering, University of Houston, Houston, TX 77004.

EVALUATION OF POTENTIAL INTEGRALS

Surface Sources Distributed on an Infinite Strip

Let sources be distributed on an infinite strip such as that shown in Fig. 1(a). The source density is assumed to be invariant along the direction of the strip axis. Projection of the strip onto a plane $\mathcal{P}$ with unit normal $\hat{\mathbf{n}}$ parallel to the strip axis defines a line segment C, a generator of the strip, depicted in greater detail in Fig. 2. The strip potential is evaluated as a superposition of that due to a distribution of elemental line sources. Each line source is assumed to contribute a potential proportional to $\ln P$, where $P = |\boldsymbol{\rho} - \boldsymbol{\rho}'|$ and $\boldsymbol{\rho}$ is the projection of $\mathbf{r}$ onto $\mathcal{P}$ while $\boldsymbol{\rho}'$ is the similar projection of $\mathbf{r}'$. Position vector $\mathbf{r}$ is the vector from the origin to the observation point while $\mathbf{r}'$ is that to a source point on the strip. The perpendicular distance from the point located by $\boldsymbol{\rho}$ to the line segment C or its extension is designated P^0, as shown in Fig. 2. C is parameterized by the arc length variable l' measured from the plane which is perpendicular to the extension of C and which passes through the point located by $\mathbf{r}$. P^0 and l' represent, in effect, a pair of rectangular coordinates in $\mathcal{P}$ locating points on C. In terms of l', the endpoints of C are located at l^+ and l^-.

Distances measured in $\mathcal{P}$ from $\boldsymbol{\rho}$ to given endpoints $\boldsymbol{\rho}^+$ and $\boldsymbol{\rho}^-$ of C are denoted P^+ and P^-, respectively. The quantities P^0, $\hat{\mathbf{P}}^0$, $P^{\pm}$, and $l^{\pm}$ of Fig. 2 are readily calculated in terms of $\boldsymbol{\rho}$, $\boldsymbol{\rho}^+$, and $\boldsymbol{\rho}^-$ by the following sequence of computations:

$$\hat{\mathbf{l}} = \frac{\boldsymbol{\rho}^+ - \boldsymbol{\rho}^-}{|\boldsymbol{\rho}^+ - \boldsymbol{\rho}^-|}, \ \hat{\mathbf{u}} = \hat{\mathbf{l}} \times \hat{\mathbf{n}}, \ l^{\pm} = (\boldsymbol{\rho}^{\pm} - \boldsymbol{\rho}) \cdot \hat{\mathbf{l}},$$

$$P^0 = |(\boldsymbol{\rho}^{\pm} - \boldsymbol{\rho}) \cdot \hat{\mathbf{u}}|, \quad P^{\pm} = |\boldsymbol{\rho}^{\pm} - \boldsymbol{\rho}| = \sqrt{(P^0)^2 + (l^{\pm})^2},$$

$$\hat{\mathbf{P}}^0 = \frac{(\boldsymbol{\rho}^{\pm} - \boldsymbol{\rho}) - l^{\pm}\hat{\mathbf{l}}}{P^0}.$$

Note that $\hat{\mathbf{u}} = \pm\hat{\mathbf{P}}^0$, the sign depending on which end of C corresponds to the vector $\boldsymbol{\rho}^+$. Other quantities appearing in Fig. 2 are used in subsequent sections.

An integral proportional to the potential of a uniform source distribution on the strip is now easily evaluated in terms of the quantities defined:

$$\int_C \ln P \, dl' = \int_{l^-}^{l^+} \ln \sqrt{(P^0)^2 + (l')^2} \, dl'$$
$$= l^+ \ln P^+ - l^- \ln P^- + P^0 \left(\tan^{-1} \frac{l^+}{P^0} - \tan^{-1} \frac{l^-}{P^0} \right) - (l^+ - l^-). \tag{1}$$

This and the following integral, though trivial to evaluate, are

Reprinted from *IEEE Trans. Antennas Propagat.*, vol. AP-32, no. 3, pp. 276–281, March 1984.

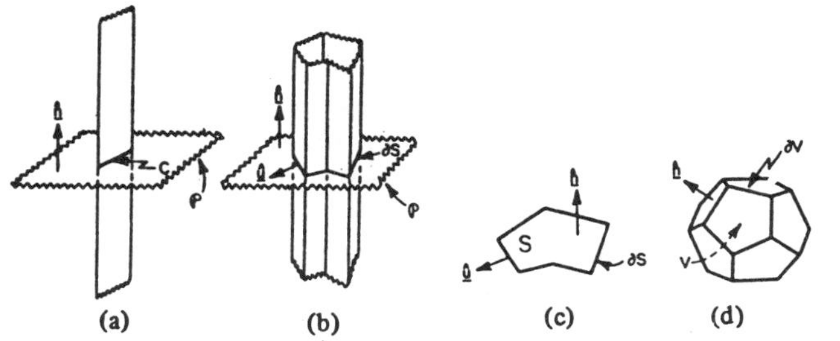

Fig. 1. Four examples. (a) An infinite strip with line-segment cross section C. (b) An infinite cylinder with polygonal cross section S. (c) A planar surface S with polygonal boundary ∂S. (d) A volumetric region V with polyhedral boundary ∂V.

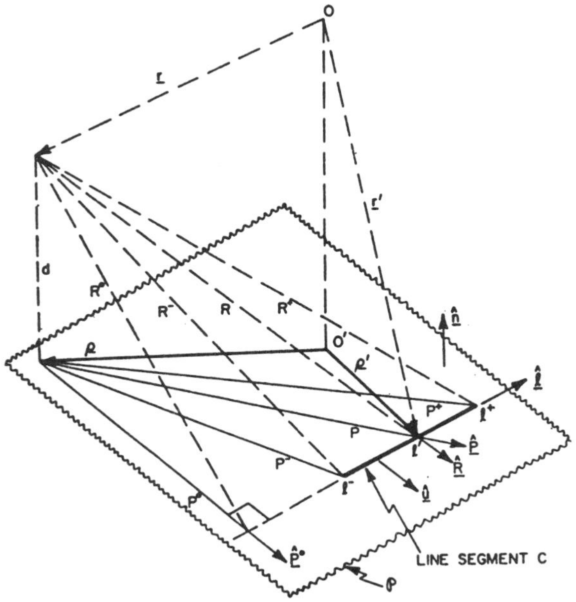

Fig. 2. Geometrical quantities associated with the line segment C lying in the plane P. The observation point for the potential is located by the position vector $\mathbf{r}$ with respect to the coordinate origin 0.

included for completeness and to facilitate introduction of the geometrical quantities used in the computation of potential integrals in subsequent sections.

To treat a source distribution which varies linearly on C, we consider the vector-valued integral

$$\int_C \hat{\mathbf{l}} l' \ln P \, dl' = \hat{\mathbf{l}} \int_{l^-}^{l^+} l' \ln \sqrt{(P^0)^2 + (l')^2} \, dl'$$

$$= \frac{\hat{\mathbf{l}}}{2} \{(P^+)^2 \ln P^+ - (P^-)^2 \ln P^-$$

$$- \tfrac{1}{2}[(l^+)^2 - (l^-)^2]\} \qquad (2)$$

where $\hat{\mathbf{l}}$ is the unit vector tangent to C in the direction of increasing l'. The integral in (2) is given a vector character merely to emphasize its correspondence to (4), (6), and (8) which follow. Obviously, the potential of a more general linear variation of the source distribution can easily be synthesized as an appropriate linear combination of (1) and (2).

As an application of (1), consider the numerical evaluation of the integral

$$\int_{l^-}^{l^+} H_0^{(2)}(kP) \, dl'$$

arising when pulse functions are used as basis functions in a solution by the method of moments of two-dimensional time-harmonic electromagnetic scattering problems [7]. $H_0^{(2)}$, the two-dimensional Green's function, is the Hankel function of zero order and second kind; $k = \omega\sqrt{\mu\epsilon}$ is the wavenumber associated with the assumed $\exp(j\omega t)$ time variation. If ρ is on the source strip, the integrand of the above integral is singular at $\rho = \rho'$ as evidenced by the small argument expansion of $H_0^{(2)}$:

$$H_0^{(2)}(kP) \xrightarrow[P \to 0]{} 1 - j\frac{2}{\pi}\left(\ln \frac{kP}{2} + \gamma\right)$$

where $\gamma = 0.5772 \ldots$ is Euler's constant. When P vanishes within the integration domain, the integral is difficult to evaluate numerically due to the presence of the logarithmic singularity. Subtraction of this singular term from the integrand leaves a bounded integrand, however, and hence the integral may be cast into a form which can be numerically evaluated by rewriting it as

$$\int_{l^-}^{l^+} H_0^{(2)}(kP) \, dl' = \int_{l^-}^{l^+} \left[H_0^{(2)}(kP) + j\frac{2}{\pi} \ln P \right] dl'$$

$$- j\frac{2}{\pi} \int_{l^-}^{l^+} \ln P \, dl'$$

where the logarithmically singular contribution has been subtracted from and added to the original integral. The integrand of the first integral on the right of the equality is nonsingular and can be numerically integrated readily; that of the second integral is singular but the integral is merely (1) and hence is analytically integrable.

Volume Sources Distributed Within a Polygonal Cylinder

We consider next potential integrals associated with volumetric source distributions confined to the interior of an infinite cylinder having a polygonal cross section S (cf., Fig. 1(b)). Source densities are assumed to be invariant along the cylinder axis, which is oriented parallel to $\hat{\mathbf{n}}$, a unit vector normal to S. The boundary of the polygonal cross section S is denoted ∂S and has a right-hand orientation with respect to $\hat{\mathbf{n}}$. It comprises a series of line segments, the ith one of which is denoted by $\partial_i S$.

Evaluation of the potential integral is accomplished by employing an appropriate Gauss integral theorem (divergence theorem, etc.) to first transform an integration over S to one over the boundary ∂S of S. Application of such theorems requires that the integrand be continuously differentiable on the domain of integration S. This is not the case, however, when the endpoint of the vector ρ falls in S or on ∂S. In this case, a region S_ϵ, defined as the intersection of S and a small disk of radius ϵ centered at ρ, is isolated for separate treatment. Integrals over S_ϵ or its boundary, ∂S_ϵ, may be evaluated by expression them in terms of a local polar coordinate system with its origin centered at the point ρ. Such integrals occasionally involve the angular extent $\alpha(\rho)$ of the circular arc portion of ∂S_ϵ lying within S. Thus, if the point located by ρ falls outside S, S_ϵ is empty and $\alpha(\rho) = 0$; if it falls inside S, $\alpha(\rho) = 2\pi$ (Fig. 3(a)); if it falls on ∂S but not at

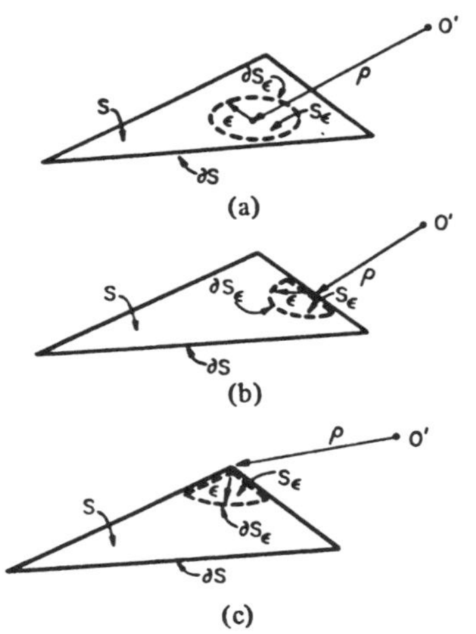

Fig. 3. Region S_ϵ corresponding to a point ρ. (a) Point inside a triangle S. (b) Point on an edge of S. (c) Point at a vertex of S.

a corner, $\alpha(\rho) = \pi$ (Fig. 3(b)); if it falls at a vertex of S, $\alpha(\rho)$ becomes the angle between the two edges of S meeting at the vertex (Fig. 3(c)). In the following it is convenient always to assume that the point located by ρ lies within or on the boundary of S; whenever this is not the case, the correct derivation is recovered by assuming $\epsilon = 0$ and $\alpha(\rho) = 0$.

In the evaluation of each of the integrals of this and the following sections, the strategy is to apply Gauss' theorems a sufficient number of times to reduce the integrals on the boundary edges of the original integration domain. In the present case, these are the edges $\partial_i S$ of the polygon. Consequently, the geometry defined in Fig. 2 is also applicable to the present problem if the line segment C there is taken to be $\partial_i S$, the ith edge of the planar polygon S with normal $\hat{\mathbf{n}}$. In the following, we also append a subscript i to any quantity defined in Fig. 2 when the quantity is to be associated with the ith edge.

With these preliminaries in hand, the potential integral for a uniform source distribution may be evaluated as follows:

$$\int_S \ln P \, dS'$$

$$= \lim_{\epsilon \to 0} \frac{1}{2} \int_{S-S_\epsilon} \nabla_s' \cdot \left[\left(P \ln P - \frac{P}{2} \right) \hat{\mathbf{P}} \right] dS'$$

$$+ \lim_{\epsilon \to 0} \int_{S_\epsilon} \ln P \, dS' = \lim_{\epsilon \to 0} \frac{1}{2} \int_{\partial(S-S_\epsilon)} \left(P \ln P - \frac{P}{2} \right)$$

$$\hat{\mathbf{P}} \cdot \hat{\mathbf{u}} \, dl' + \lim_{\epsilon \to 0} \alpha(\rho) \frac{\epsilon^2}{2} \left(\ln \epsilon - \tfrac{1}{2} \right)$$

$$= \frac{1}{2} \sum_i \int_{\partial_i S} \left(P \ln P - \frac{P}{2} \right) \hat{\mathbf{P}} \cdot \hat{\mathbf{u}}_i \, dl'$$

$$= \frac{1}{2} \sum_i \mathbf{P}_i^0 \cdot \hat{\mathbf{u}}_i \int_{\partial_i S} \left(\ln P - \tfrac{1}{2} \right) dl'$$

$$= \frac{1}{2} \sum_i \mathbf{P}_i^0 \cdot \hat{\mathbf{u}}_i \Bigg[l_i^+ \ln P_i^+ - l_i^- \ln P_i^-$$

$$+ P_i^0 \left(\tan^{-1} \frac{l_i^+}{P_i^0} - \tan^{-1} \frac{l_i^-}{P_i^0} \right) - \frac{3}{2} (l_i^+ - l_i^-) \Bigg] \quad (3)$$

where $\mathbf{P}_i^0 = P_i^0 \hat{\mathbf{P}}_i^0$, $\hat{\mathbf{P}}$ is the vector $(\rho' - \rho)/P$, and $\hat{\mathbf{u}}$ is the outward normal vector on ∂S lying in the plane of S (cf., Fig. 2). We have also used the fact that $\hat{\mathbf{u}}_i \cdot \hat{\mathbf{P}} P = \mathbf{P}_i^0 \cdot \hat{\mathbf{u}}_i$ is constant on $\partial_i S$. The summation is over all the edges of S. The surface divergence operator, $\nabla_s' \cdot (\cdots)$, involves differentiations with respect to source (primed) coordinates only. Note that the limit of the integral over the circular arc portion of $\partial(S - S_\epsilon)$ vanishes since the integrand remains bounded whereas the domain of integration vanishes as $\epsilon \to 0$. A limiting argument convinces one that the contribution to the sum in (3) from any edge is zero if ρ lies on the edge or on its extension. The steps used to evaluate (3) are common to the remaining integrals of this and subsequent sections and hence are summarized here as follows.

1) The integral over S is partitioned into two integrals over $S - S_\epsilon$ and S_ϵ, respectively. It is convenient to view each integral as a limit with $\epsilon \to 0$, although the value of the sum is, of course, independent of ϵ.

2) The integrand of the integral on $S - S_\epsilon$ is written as the differential (involving the surface del operator ∇_s) of some quantity, and a Gauss integral theorem (in this case, the surface divergence theorem [8]) is used to transform the integral into one over the boundary $\partial(S - S_\epsilon)$ of $S - S_\epsilon$. The orientation of the integration path is assumed to be right-handed with respect to $\hat{\mathbf{n}}$.

3) The limit of the integral over S_ϵ is evaluated; this limit vanishes by inspection when the integrand is bounded and the domain of integration vanishes. When the integrand is unbounded, the integral is evaluated explicitly by the procedure discussed at the beginning of this section and the limit of the integral as $\epsilon \to 0$ is then determined.

4) The limit of the integral over $\partial(S - S_\epsilon)$ is evaluated; occasionally the contribution from the portion of ∂S_ϵ in S vanishes since the integrand remains bounded while the domain of integration vanishes. When this is not the case, the limit of the integral is explicitly evaluated in polar coordinates.

5) The remaining integral over ∂S is decomposed into a sum of line integrals over the subboundaries (edges) $\partial_i S$.

6) The line integral over $\partial_i S$ is evaluated in terms of the geometrical quantities defined in Fig. 2.

For a source density varying linearly over a polygonal cylinder, two independent directions for the linear variation can be considered. Both cases can be treated simultaneously, however, by considering the *vector-valued* integral

$$\int_S (\rho' - \rho) \ln P \, dS'$$

$$= \lim_{\epsilon \to 0} \frac{1}{2} \int_{S-S_\epsilon} \nabla_s' \left(P^2 \ln P - \frac{P^2}{2} \right) dS'$$

$$+ \lim_{\epsilon \to 0} \int_{S_\epsilon} \hat{\mathbf{P}} P \ln P \, dS'$$

$$= \lim_{\epsilon \to 0} \frac{1}{2} \int_{\partial(S-S_\epsilon)} \left(P^2 \ln P - \frac{P^2}{2} \right) \hat{\mathbf{u}} \, dl'$$

$$= \frac{1}{2} \sum_i \hat{\mathbf{u}}_i \int_{\partial_i S} \left(P^2 \ln P - \frac{P^2}{2} \right) dl'$$

$$= \frac{1}{2} \sum_i \hat{\mathbf{u}}_i \Bigg\{ [(P_i^0)^2 + \tfrac{1}{3}(l_i^+)^2] l_i^+ \ln P_i^+ - [(P_i^0)^2 + \tfrac{1}{3}(l_i^-)^2] l_i^- \ln P_i^- + \tfrac{2}{3}(P_i^0)^3 \left(\tan^{-1} \frac{l_i^+}{P_i^0} - \tan^{-1} \frac{l_i^-}{P_i^0} \right) - \tfrac{5}{18}[(l_i^+)^3 - (l_i^-)^3] - \tfrac{7}{6}(P_i^0)^2 (l_i^+ - l_i^-) \Bigg\} \tag{4}$$

which is also evaluated via steps 1)-6) and with the aid of [9, eq. (623.2)]. In this case, the appropriate Gauss integral theorem transforms the surface integral of the (surface) gradient of a scalar to a vector line integral around the surface boundary [8]. If ρ is on the ith edge or its extension, $P_i^0 = 0$, $P_i^\pm = |l_i^\pm|$, and the corresponding arctangent terms make no contribution to the sum in (4).

If the actual source distribution varies linearly with distance from some arbitrary point ρ_a along the direction of the unit vector $\hat{\mathbf{a}}$ in the plane of S, then the distribution is proportional to $\hat{\mathbf{a}} \cdot (\rho' - \rho_a)$. The potential hence is proportional to

$$\int_S \hat{\mathbf{a}} \cdot (\rho' - \rho_a) \ln P \, dS'$$

$$= \hat{\mathbf{a}} \cdot \int_S (\rho' - \rho) \ln P \, dS' + \hat{\mathbf{a}} \cdot (\rho - \rho_a) \int_S \ln P \, dS'$$

which is seen to be a linear combination of (3) and (4).

Surface Sources Distributed on Polygons

If surface sources are distributed on a planar polygon S, (cf., Fig. 1(c)), the potential of an elemental source on S is no longer of logarithmic form. Instead, the potential observed at a point $\mathbf{r}$ due to an elemental source on S ar $\mathbf{r}'$ is proportional to $1/R = 1/|\mathbf{r} - \mathbf{r}'|$. To facilitate evaluation of the potential integrals, the distances $R_i^0 = \sqrt{(P_i^0)^2 + d^2}$ and $R_i^\pm = \sqrt{(P_i^\pm)^2 + d^2}$, the latter associated with the endpoints of $\partial_i S$ (the ith edge of S), are introduced (Fig. 2). The distance d is the height of the observation point above the plane of S, measured positively in the direction of $\hat{\mathbf{n}}$, and may be calculated by

$$d = \hat{\mathbf{n}} \cdot (\mathbf{r} - \mathbf{r}_i^\pm)$$

where $\mathbf{r}_i^+ (\mathbf{r}_i^-)$ is a given position vector to the upper (lower) endpoint of $\partial_i S$. The vectors $\hat{\mathbf{l}}_i$ and $\rho_i^\pm$ are now defined in terms of the line segment endpoints as

$$\hat{\mathbf{l}}_i = \frac{\mathbf{r}_i^+ - \mathbf{r}_i^-}{|\mathbf{r}_i^+ - \mathbf{r}_i^-|}$$

and

$$\rho_i^\pm = \mathbf{r}_i^\pm - \hat{\mathbf{n}}(\hat{\mathbf{n}} \cdot \mathbf{r}_i^\pm).$$

The potential due to a uniform source distribution on S is now found to be proportional to

$$\int_S \frac{dS'}{R}$$

$$= \lim_{\epsilon \to 0} \int_{S - S_\epsilon} \nabla_s' \cdot \left(\frac{R}{P} \hat{\mathbf{P}} \right) dS' + \lim_{\epsilon \to 0} \int_{S_\epsilon} \frac{dS'}{R}$$

$$= \lim_{\epsilon \to 0} \int_{\partial(S - S_\epsilon)} \frac{R}{P} \hat{\mathbf{P}} \cdot \hat{\mathbf{u}} \, dl' + \lim_{\epsilon \to 0} \alpha(\rho)(\sqrt{\epsilon^2 + d^2} - |d|)$$

$$= -\alpha(\rho)|d| + \sum_i \int_{\partial_i S} \frac{R}{P} \hat{\mathbf{P}} \cdot \hat{\mathbf{u}}_i \, dl'$$

$$= -\alpha(\rho)|d| + \sum_i \mathbf{P}_i^0 \cdot \hat{\mathbf{u}}_i \int_{\partial_i S} \left(\frac{1}{R} + \frac{d^2}{P^2 R} \right) dl'$$

$$= -\alpha(\rho)|d| + \sum_i \hat{\mathbf{P}}_i^0 \cdot \hat{\mathbf{u}}_i \left[P_i^0 \ln \frac{R_i^+ + l_i^+}{R_i^- + l_i^-} + |d| \left(\tan^{-1} \frac{|d| l_i^+}{P_i^0 R_i^+} - \tan^{-1} \frac{|d| l_i^-}{P_i^0 R_i^-} \right) \right]$$

$$= \sum_i \hat{\mathbf{P}}_i^0 \cdot \hat{\mathbf{u}}_i \left[P_i^0 \ln \frac{R_i^+ + l_i^+}{R_i^- + l_i^-} - |d| \times \left(\tan^{-1} \frac{P_i^0 l_i^+}{(R_i^0)^2 + |d| R_i^+} - \tan^{-1} \frac{P_i^0 l_i^-}{(R_i^0)^2 + |d| R_i^-} \right) \right]. \tag{5}$$

The last integral above may be evaluated with the aid of [9, eqs. (200.01), (387.)] Note the appearance of the "residue" contribution involving the angle $\alpha(\rho)$ of the portion of ∂S_ϵ in S. This term is combined with the sum by expressing it as a sum of the angles between adjacent vertices around the polygon,

$$\alpha(\rho) = \sum_i \hat{\mathbf{P}}_i^0 \cdot \hat{\mathbf{u}}_i \left(\tan^{-1} \frac{l_i^+}{P_i^0} - \tan^{-1} \frac{l_i^-}{P_i^0} \right),$$

and employing the identity

$$\tan^{-1} \frac{l_i^\pm}{P_i^0} - \tan^{-1} \frac{|d| l_i^\pm}{P_i^0 R_i^\pm} = \tan^{-1} \frac{P_i^0 l_i^\pm}{(R_i^0)^2 + |d| R_i^\pm}$$

where each arctangent function is evaluated on its principal branch. If $\mathbf{r}$ is on an edge or its extension, it is easily shown that the contribution to the sum in (5) from that edge vanishes.

Integral (5) has been evaluated for triangular patches by several authors [1], [2], [3], [6], [10], but the formula presented here has advantages in terms of accuracy, conciseness, and convenience for numerical work. By contrast, the formula presented in [1], for example, is found to be inaccurate for moderate-to-large separation distances between observation and source points because it contains many terms which must approximately cancel one another. The present form is also symmetric in the vertex indices, which not only is intuitively satisfying, but also has the practical consequence of rendering it suitable for programming using loop operations when the vertex information is stored in column vector form. This latter feature stands in contrast to the formulas given in [1], [2], [3], [10], which would be difficult to extend to arbitrary polygonal regions.

The corresponding integral for linearly varying source distributions is again evaluated in vector form:

$$\int_S \frac{\rho' - \rho}{R}\, dS'$$

$$= \lim_{\epsilon\to 0} \int_{S-S_\epsilon} \nabla_s' R\, dS' + \lim_{\epsilon\to 0} \int_{S_\epsilon} \frac{\rho'-\rho}{R}\, dS'$$

$$= \lim_{\epsilon\to 0} \int_{\partial(S-S_\epsilon)} R\hat{\mathbf{u}}\, dl' = \lim_{\epsilon\to 0} \int_{S_\epsilon} \frac{P}{R}\, \mathbf{P}\, dS'$$

$$= \sum_i \hat{\mathbf{u}}_i \int_{\partial_i S} R\, dl'$$

$$= \frac{1}{2} \sum_i \hat{\mathbf{u}}_i \left[(R_i^0)^2 \ln \frac{R_i^+ + l_i^+}{R_i^- + l_i^-} + l_i^+ R_i^+ - l_i^- R_i^- \right] \tag{6}$$

where the last integral is evaluated by [9, eq. (230.01)]. Comments similar to those in the previous paragraph apply in comparing (6) to the formula given in [2] for the potential due to a linear source distribution on a triangle.

In the solution of electromagnetic scattering problems by the method of moments with triangular patch modeling [4], the above integrals are used in the numerical evaluation of moment matrix elements. For example, one may evaluate the time-harmonic scalar potential due to a uniform charge density distributed on a triangular patch T by the same subtraction-and-addition of the singularity approach used in the two-dimensional case. The scalar potential is proportional to

$$\int_T \frac{e^{-jkR}}{R}\, dS' = \int_T \frac{e^{-jkR} - 1}{R}\, dS' + \int_T \frac{dS'}{R}.$$

The first integral on the right has a bounded integrand for every observation point and hence can be integrated numerically; the second integral is a special case of (5) and hence can be analytically evaluated.

An application of (6) is also found in [4]. There vector-valued basis functions are introduced which are proportional to the vector $\rho' - \rho_m$, where ρ_m is the projection onto the plane of triangle T of the position vector $\mathbf{r}$ to the mth vertex of T. Vector potentials due to these basis functions are then proportional to

$$\int_T (\rho' - \rho_m) \frac{e^{-jkR}}{R}\, dS'$$

$$= \int_T (\rho' - \rho) \frac{e^{-jkR}}{R}\, dS' + (\rho - \rho_m) \int_T \frac{e^{-jkR}}{R}\, dS'$$

$$= \int_T (\rho' - \rho) \frac{e^{-jkR} - 1}{R}\, dS' + (\rho - \rho_m) \int_T \frac{e^{-jkR} - 1}{R}\, dS'$$

$$+ \int_T \frac{\rho' - \rho}{R}\, dS' + (\rho - \rho_m) \int_T \frac{dS'}{R}.$$

The last two integrals are merely (5) and (6) with S specialized to T, and the two preceding ones always have bounded integrands which can be numerically integrated. The discussion following (4) with regard to synthesizing the potential of a source distribution with arbitrary linear variation in a polygon S also applies to (6).

It is also noted that the integrals (5) and (6) are used directly in time-domain formulations of electromagnetic scattering problems, as illustrated in [6].

Volume Sources Distributed on a Polyhedron

Consider next a distribution of sources within an arbitrary polyhedral region V (cf., Fig. 1(d)). The boundary of V is denoted by ∂V and has an outward unit normal $\mathbf{n}$. ∂V comprises a number of faces, the jth one of which is designated $\partial_j V$. The jth face, in turn, is a polygon having a boundary $\partial\partial_j V$ which comprises a number of edges, the ith one of which is designated $\partial_i\partial_j V$. The strategy employed to evaluate potentials of source distributions over V is merely an extension of that of the previous sections: Gauss integral theorems are first used to transform integrals over V into integrals over $\partial V = \Sigma_j \partial_j V$. Then integrals over $\partial_j V$ are transformed into integrals over the polyhedron edges, $\partial_i\partial_j V$. When the observation point $\mathbf{r}$ is in V or on ∂V, however, before the appropriate Gauss theorem may be applied it is necessary to exclude for separate treatment a region V_δ, the intersection of V and a sphere of radius δ centered at $\mathbf{r}$ (cf., Fig. 4). The boundary of this region is designated ∂V_δ, and the solid angle of the spherical sector of ∂V_δ contained in V is designated $\Omega(\mathbf{r})$.

With these considerations, the potential of a uniform source distributed in a polyhedron V is found to be proportional to

$$\int_V \frac{dV'}{R}$$

$$= \lim_{\delta\to 0} \frac{1}{2} \int_{V-V_\delta} \nabla' \cdot \hat{\mathbf{R}}\, dV' + \lim_{\delta\to 0} \int_{V_\delta} \frac{dV'}{R}$$

$$= \lim_{\delta\to 0} \frac{1}{2} \int_{\partial(V-V_\delta)} \hat{\mathbf{R}} \cdot \hat{\mathbf{n}}\, dS' + \lim_{\delta\to 0} \frac{\delta^2}{2} \Omega(\mathbf{r})$$

$$= \frac{1}{2} \sum_j \int_{\partial_j V} \hat{\mathbf{R}} \cdot \hat{\mathbf{n}}_j\, dS' = -\frac{1}{2} \sum_j d_j \int_{\partial_j V} \frac{dS'}{R}$$

$$= \frac{1}{2} \sum_j d_j \left\{ \sum_i \hat{\mathbf{P}}_{ij}^0 \cdot \hat{\mathbf{u}}_{ij} \times \left[|d_j| \left(\tan^{-1} \frac{P_{ij}^0 l_{ij}^+}{(R_{ij}^0)^2 + |d_j| R_{ij}^+} - \tan^{-1} \frac{P_{ij}^0 l_{ij}^-}{(R_{ij}^0)^2 + |d_j| R_{ij}^-} \right) - P_{ij}^0 \ln \frac{R_{ij}^+ + l_{ij}^+}{R_{ij}^- + l_{ij}^-} \right] \right\} \tag{7}$$

where $\hat{\mathbf{R}}$ is the unit vector from $\mathbf{r}$ to $\mathbf{r}'$. The double subscript ij denotes a quantity associated with the ith edge of the jth face of V, while a quantity associated with the face only carries the single subscript j. Note that the last equality in (7) follows from (5). The logarithmic terms associated with the two faces attached to a given edge can be combined in (7) to halve the number of logarithm evaluations.

The case in which V is a rectangular parallelepiped has also been considered by MacMillan [11] and Waldvogel [12]. The above result can easily be reduced to that of MacMillan when $\mathbf{r}$ is exterior to V, but is apparently in disagreement with his result for $\mathbf{r}$ interior to V. It appears that his formula is not intended to apply to the case in which the observation point falls in the source region. Waldvogel has also obtained an expression for the potential when V is an arbitrary polyhedron [13]. Equation (7) is equivalent to his expression, but is obtained in a much simpler fashion. Okon has derived formulas for the potentials of uniformly charged tetrahedra and parallelepipeds [14]. His formulas, however, are much too lengthy to compare with (7). In contrast to the formula given in (7), we point out that the derivations of the formulas in [11]-[14] consume an average of seven pages of text, while the final formulas for the potentials in [11] and [14] average more than five pages of text.

The corresponding integral for a linearly varying source distribution is evaluated in vector form as

$$\int_V \frac{\mathbf{r}' - \mathbf{r}}{R}\, dV'$$

$$= \lim_{\delta\to 0} \int_{V - V_\delta} \nabla' R\, dV' + \lim_{\delta\to 0} \int_{V_\delta} \hat{\mathbf{R}}\, dV'$$

$$= \lim_{\delta\to 0} \int_{\partial(V - V_\delta)} R\hat{\mathbf{n}}\, dS' = \sum_j \hat{\mathbf{n}}_j \int_{\partial_j V} R\, dS'$$

$$= \sum_j \hat{\mathbf{n}}_j \left[\lim_{\epsilon\to 0} \int_{\partial_j V - S_{\epsilon j}} \nabla'_s \cdot \left(\frac{R^3}{3P_j} \hat{\mathbf{P}}_j \right) dS' \right.$$

$$\left. + \lim_{\epsilon\to 0} \int_{S_{\epsilon j}} R\, dS' \right]$$

$$= \frac{1}{3} \sum_j \hat{\mathbf{n}}_j \left[\lim_{\epsilon\to 0} \int_{\partial(\partial_j V - S_{\epsilon j})} \frac{R^3}{P_j} \hat{\mathbf{P}}_j \cdot \hat{\mathbf{u}}_j\, dl' \right]$$

$$= \frac{1}{3} \sum_j \hat{\mathbf{n}}_j \left[-\alpha_j(\mathbf{r}) |d_j|^3 + \sum_i \int_{\partial_i\partial_j V} \frac{R^3}{P_j} \hat{\mathbf{P}}_j \cdot \hat{\mathbf{u}}_{ij}\, dl' \right]$$

$$= \frac{1}{3} \sum_j \hat{\mathbf{n}}_j \left[-\alpha_j(\mathbf{r}) |d_j|^3 \right.$$

$$\left. + \sum_i \mathbf{P}^0_{ij}\, \hat{\mathbf{u}}_{ij} \int_{\partial_i\partial_j V} \left(R + \frac{d_j^2}{R} + \frac{d_j^4}{P_j^2 R} \right) dl' \right]$$

$$= \frac{1}{3} \sum_j \hat{\mathbf{n}}_j \left\{ \sum_i \hat{\mathbf{P}}^0_{ij} \cdot \hat{\mathbf{u}}_{ij} \left[\frac{P^0_{ij}[(R^0_{ij})^2 + 2d_j^2]}{2} \right.\right.$$

$$\times \ln \frac{R^+_{ij} + l^+_{ij}}{R^-_{ij} + l^-_{ij}} + \frac{P^0_{ij}}{2} (l^+_{ij} R^+_{ij} - l^-_{ij} R^-_{ij}) - |d_j|^3$$

$$\left.\left. \times \left(\tan^{-1} \frac{P^0_{ij} l^+_{ij}}{(R^0_{ij})^2 + |d_j| R^+_{ij}} - \tan^{-1} \frac{P^0_{ij} l^-_{ij}}{(R^0_{ij})^2 + |d_j| R^-_{ij}} \right) \right] \right\}. \tag{8}$$

Note that the last integral in (8) has terms of the same form as terms of (5) and (6).

As an application, integrals (7) and (8) are used in the evaluation of matrix elements in the moment matrix derived in [5]. There current and charge sources are distributed in tetrahedral volumes and the time-harmonic vector and scalar potentials they produce are required. The procedure for performing the numerical integrations is simply the generalization to a three-dimensional source region of that described following (6).

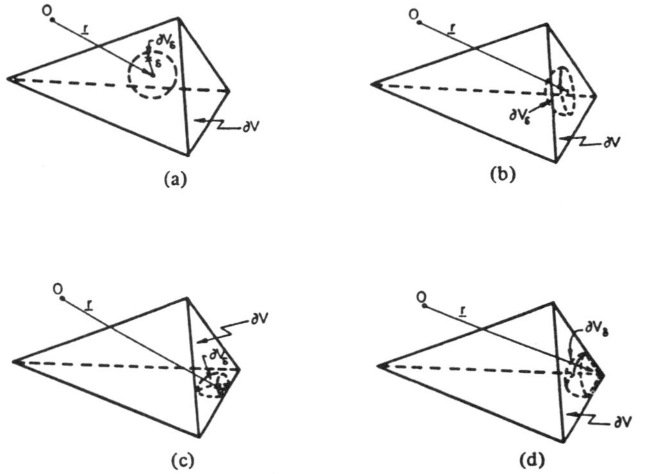

Fig. 4. Region V_δ corresponding to a point $\mathbf{r}$. (a) Point inside a tetrahedron V. (b) Point on a face of V. (c) Point on an edge of V. (d) Point at a vertex of V.

SUMMARY

Formulas for the potentials due to uniform and linearly varying source distributions defined on domains of simple shape are systematically derived and presented. Domains considered are infinite planar strips, infinite cylinders of polygonal cross sections, planar surfaces with polygonal boundaries, and volumetric regions bounded by polyhedrons. Of particular note is the compactness of the formulas obtained. Applications of the formulas to the numerical solution of electromagnetics problems by the method of moments are given.

REFERENCES

[1] S. M. Rao, A. W. Glisson, D. R. Wilton, and B. S. Vidula, "A simple numerical solution procedure for statics problems involving arbitrary-shaped surfaces," *IEEE Trans. Antennas Propagat.*, vol. AP-27, no. 5, pp. 604–608, Sept. 1979.

[2] E. E. Okon and R. F. Harrington, "The polarizabilities of electrically small apertures of arbitrary shape," *IEEE Trans. Electromagn. Compat.*, vol. EMC-23, no. 4, pp. 359–366, Nov. 1981.

[3] ——, "A method of computing the capacitance of flat discs of arbitrary shape," *Electromagn.*, vol. 1, no. 2, pp. 229–241, Apr.-June 1981.

[4] S. M. Rao, D. R. Wilton, and A. W. Glisson, "Electromagnetic scattering by surfaces of arbitrary shape," *IEEE Trans. Antennas Propagat.*, vol. AP-30, no. 3, pp. 409–418, May 1982.

[5] D. Schaubert, D. R. Wilton, and A. W. Glisson, "A tetrahedral modeling method for electromagnetic scatering by arbitrarily shaped inhomogeneous dielectric bodies," *IEEE Trans. Antennas Propagat*, vol. AP-32, no. 1, pp. 77–85, Jan. 1984.

[6] S. M. Rao, "Electromagnetic scattering and radiation of arbitrarily-shaped surfaces by triangular patch modeling," Ph.D. dissertation, Univ. Mississippi, Aug. 1980.

[7] R. F. Harrington, *Field Computation by Moment Methods.* New York: Macmillan, 1968.

[8] J. Van Bladel, *Electromagnetic Fields.* New York: McGraw-Hill, 1964, pp. 502–504.

[9] H. B. Dwight, *Tables of Integrals and Other Mathematical Data.* New York: Macmillan, 1961.

[10] A. B. Birtles, B. J. Mayo, and A. W. Bennett, "Computer technique for solving 3-dimensional electron-optics and capacitance problems," *Proc. Inst. Elec. Eng.*, vol. 120, no. 2, pp. 213–220, Feb. 1973.

[11] W. D. MacMillan, *The Theory of the Potential.* New York: Dover, 1958.

[12] J. Waldvogel, "The Newtonian potential of a homogeneous cube," *J. Appl. Math. Phys.*, (ZAMP), vol. 27, pp. 867–871, 1976.

[13] ——, "The Newtonian potential of homogeneous polyhedra," *J. Appl. Math. Phys.* (ZAMP), vol. 30, pp. 388–398, 1979.

[14] E. E. Okon, "The potential due to uniform distributions in some polygons and polyhedra," in preparation.

Error Minimization and Convergence in Numerical Methods

D. G. DUDLEY

ELECTROMAGNETICS LABORATORY, DEPARTMENT OF ELECTRICAL AND COMPUTER ENGINEERING, UNIVERSITY OF ARIZONA, TUCSON, ARIZONA 85721, USA

Abstract—**Operators that occur in the solution to boundary value problems in electromagnetic theory are reviewed in Hilbert space. Various types of convergence are discussed. The Method of Moments is reviewed and the special cases of Galerkin's Method, the Rayleigh–Ritz Method, and the Method of Least Squares are included. Error minimization and convergence in the various cases are emphasized. A classic electromagnetic example is discussed and operator characteristics in the quasistatic limit are noted.**

1. Introduction

In the study of linear electromagnetic boundary value problems, one is faced with the inversion of a linear operator equation $Lu = f$, where L is a differential, integral, or integrodifferential operator. If the inversion is possible and if the result is unique, u is called the *solution* to the problem. In many important situations, however, the geometric complexity is such that an analytic solution is intractable. The researcher in such event must therefore consider the use of methods to obtain a *numerical approximation* to the solution. This paper is concerned with the issue of convergence of numerical approximation and the relationship, if any, between approximation and solution.

Numerical techniques can be applied directly to the differential equations comprising the electromagnetic model; alternatively, they can be applied to integral or integro-differential equations, derivable from the differential forms by use of a Green's theorem. Unfortunately, except in quasistatic limits, the operator seldom has mathematical properties that allow firm statements concerning properties of the approximation. In addition, in the numerical process, the operator L is often represented by a matrix A, where the mathematical relationship between the domain of the exact operator L and the domain of its approximation A is often unclear. The result is that, even if properties of the inversion of A can be established, it is difficult to relate them to the inversion of L.

A further complication occurs when the size of the matrix A is large enough that direct inversion methods become intractable. In this case, iterative methods [1] are attractive, but they can add further uncertainty in attempts to relate numerical approximation to solution.

The paper begins with a review of some basic concepts of operators in Hilbert space, including various types of convergence. Next, the Method of Moments is introduced in a general way with Galerkin's Method, the Rayleigh–Ritz Method, and the Method of Least Squares emerging as special cases. Emphasis is placed on what can and cannot be said regarding convergence of the Method of Moments and its specializations. An example of application of the basic ideas is then given in a classic electromagnetic example: Diffraction by a slit in an infinite screen. Included is the important special case of the quasistatic limit. The paper ends with some comments about numerical convergence.

2. Operators in Hilbert Space

The purposes of this section are to review some principles of operators in Hilbert space and to establish notation. Only those topics relevant to the discussion herein are mentioned; the reader is referred to the references [2], [3], [4] for completeness.

This paper is based upon concepts in a general separable Hilbert space $\mathscr{H}$ and two specific Hilbert spaces as follows: $\mathscr{L}_2(a, b)$, the space of absolutely square integrable complex-valued functions on the interval (a, b); and $\mathscr{E}_n$, Euclidean space of dimension n, where the n-components are complex numbers. The following are possible *inner products* for these two spaces. For $f(x), g(x) \in \mathscr{L}_2(a, b)$, and $w(x)$ a suitable *weight function*,

$$\langle f, g\rangle = \int_a^b f(x)\bar{g}(x)w(x)\,dx \qquad (1)$$

For $x, y \in \mathscr{E}_n$ and $x = [\xi_1, \xi_2, \ldots, \xi_n]$, $y = [\eta_1, \eta_2, \ldots, \eta_n]$,

$$\langle x, y\rangle = \sum_{k=1}^{n} \xi_k \bar{\eta}_k \qquad (2)$$

Although inner product conventions vary in the literature, for $x, y \in \mathscr{H}$ the following inner product property [2] is consistent with the definitions in (1) and (2):

$$\langle x, \alpha y\rangle = \bar{\alpha}\langle x, y\rangle \qquad (3)$$

For $x \in \mathscr{H}$, the following is the *norm* $\|x\|$ induced by the inner product:

$$\|x\| = \sqrt{\langle x, x\rangle} \qquad (4)$$

For L a linear operator in $\mathscr{H}$, let $\mathscr{D}_L$ and $\mathscr{R}_L$ be the domain and range of L, respectively. Then, for $u \in \mathscr{D}_L$ and $f \in \mathscr{R}_L$, this paper is concerned specifically with

$$Lu = f \qquad (5)$$

Let the *adjoint operator* L^* be defined by

$$\langle Lu, v\rangle = \langle u, L^*v\rangle \qquad (6)$$

for $u \in \mathscr{D}_L$, $v \in \mathscr{D}_L$. The operator L is *symmetric* if for any $u, v \in \mathscr{D}_L$

$$\langle Lu, v\rangle = \langle u, Lv\rangle \qquad (7)$$

Reprinted with permission from *Electromagnetics*, vol. 5, pp. 89–97, 1985. © Hemisphere Publishing Corp.

Such a definition may not include all the elements in $\mathscr{D}_{L^*}$. If, however, in addition to (7), $\mathscr{D}_L = \mathscr{D}_{L^*}$, the operator is *selfadjoint*.

An important collection of operators for which there are established convergence criteria are non-negative, positive, and positive-definite operators. The reader is cautioned that there is little uniformity of notation concerning these operators in the literature. For the purposes herein, an operator L is *non-negative* if $\langle Lx, x\rangle \geq 0$, for all $x \in \mathscr{D}_L$. An operator is *positive* if $\langle Lx, x\rangle > 0$, for all $x \neq 0$ in $\mathscr{D}_L$. It is easy to show that a positive operator is symmetric. An operator is *positive-definite* if $\langle Lx, x\rangle \geq c^2 \| x\|^2$, for $c > 0$ and $x \in \mathscr{D}_L$.

A special inner product and norm [3], associated with positive and positive-definite operators, are useful in relating convergence criteria. Define the *energy inner product* with respect to the operator L by

$$[x, y] = \langle Lx, y\rangle \tag{8}$$

With this inner product definition, $\mathscr{D}_L$ becomes a Hilbert space $\mathscr{H}_L$ that is dense in $\mathscr{H}$. The associated *energy norm* in $\mathscr{H}_L$ is given by

$$| x | = \sqrt{\langle Lx, x\rangle} \tag{9}$$

For positive-definite operators, there is an easily proved important relationship between norms, as follows:

$$\| x\| \leq \frac{1}{c} | x | \tag{10}$$

Among the many forms of convergence criteria, there are several types that are particularly useful in numerical methods in electromagnetics. For a sequence $\{u_n\} \subset \mathscr{H}$, u_n *converges to u* is written

$$u_n \to u \tag{11}$$

and means that

$$\lim_{n\to\infty} \| u_n - u\| = 0 \tag{12}$$

The statement u_n *converges in energy to u* is written

$$u_n \overset{e}{\to} u \tag{13}$$

and means that

$$\lim_{n\to\infty} | u_n - u | = 0 \tag{14}$$

The statement u_n *converges weakly to u* is written

$$u_n \overset{w}{\to} u \tag{15}$$

and means that for every $g \in \mathscr{H}$

$$\lim_{n\to\infty} |\langle u_n - u, g\rangle | = 0 \tag{16}$$

It is straightforward to show the following relationships among the types of convergence:

A. If $\| Lu_n\|$ is bounded, convergence implies convergence in energy.
B. Convergence implies weak convergence.
C. Convergence in energy implies $Lu_n \overset{w}{\to} f$. The weak convergence is for those g, defined by (16), in $\mathscr{H}_L$. If, however, $\| Lu_n\|$ is bounded, then $Lu_n \overset{w}{\to} f$ in $\mathscr{H}$.
D. If L is positive-definite, convergence in energy implies convergence.

Properties A, B and the first half of C follow from the Cauchy-Schwarz-Bunjakowski inequality. The second half of C is based on the Hilbert space $\mathscr{H}_L$ being dense in $\mathscr{H}$. (See [3], page 24-25). Property D is a direct result of (10).

3. The Method of Moments

The purpose of this section is to introduce the Method of Moments in a general way and develop various special cases. Emphasis is on convergence and error minimization. Included is a short discussion of some methods for inverting matrices.

An approximate solution to $Lu = f$ is given by the following procedure. For L an operator in $\mathscr{H}$, consider

$$Lu - f = 0 \tag{17}$$

where $u \in \mathscr{D}_L$, $f \in \mathscr{R}_L$. Define the linearly independent sets $\{\phi_k\}_{k=1}^n \subset \mathscr{D}_L$ and $\{w_k\}_{k=1}^n \subset \mathscr{R}_L$, where ϕ_k and w_k are called *expansion* functions and *weighting* functions, respectively. Define a sequence of approximants to u by

$$u_n = \sum_{k=1}^{n} \alpha_k \phi_k \quad n = 1, 2, \cdots \tag{18}$$

A matrix equation is formed in (17) by the condition that, upon replacement of u by u_n, the left side shall be orthogonal to the sequence $\{w_k\}$. If L is linear, the result is the matrix equation of the *Method of Moments* [5], [6], viz:

$$\sum_{k=1}^{n} \alpha_k\langle L\phi_k, w_m\rangle = \langle f, w_m\rangle \quad m = 1, 2, \ldots, n \tag{19}$$

Note that the *exact* operator equation (17) in a Hilbert space $\mathscr{H}$ has been transformed into an *approximate* operator equation on Hilbert space $\mathscr{E}_n$, viz:

$$Ax = b \tag{20}$$

where, in usual matrix form,

$$x = (\alpha_1 \quad \alpha_2 \quad \cdots \quad \alpha_n)^T \tag{21}$$

$$b = (\langle f, w_1\rangle \quad \langle f, w_2\rangle \quad \cdots \quad \langle f, w_n\rangle)^T \tag{22}$$

$$A = [a_{mk}] \tag{23}$$

where T denotes *transpose* and a_{mk} are the individual matrix elements, given by

$$a_{mk} = \langle L\phi_k, w_m\rangle \tag{24}$$

A principal question is the convergence of (18) when the sequence $\{\alpha_k\}$ is determined by solution of the matrix equation (20).

In the special case where the expansion functions are identical to the weighting functions, the result is *Galerkin's Method* [3], viz:

$$\sum_{k=1}^{n} \alpha_k\langle L\phi_k, \phi_m\rangle = \langle f, \phi_m\rangle \quad m = 1, 2, \ldots, n \tag{25}$$

If nothing more is known about the linear operator L, nothing in general can be said concerning the convergence of the approximants u_n to the solution u. If, however, L is positive (see [3], [5], [6] for examples with Laplace's equation) and the sequence $\{\phi_k\}$ is complete in $\mathscr{H}_L$, Galerkin's Method becomes the *Rayleigh–Ritz Method*. In the usual derivation of Rayleigh–Ritz, accomplished by minimization of a quadratic functional, it is proved [3] that the approximants u_n in (38) converge in energy, viz:

$$\lim_{n\to\infty} |u_n - u| = 0 \tag{26}$$

By Property C, Section 2, the result in (26) implies that $Lu_n \overset{w}{\to} f$. Unfortunately, nothing can be said about the nearness of u_n to u. If, however, L is positive-definite (see [3] for examples with Laplace's equation), by Property D, Section 2, the approximants converge, viz:

$$\lim_{n\to\infty} \|u_n - u\| = 0 \tag{27}$$

It is a matter of concern that many of the interesting and practical problems in electromagnetics involve operators that are neither positive-definite nor positive. Indeed, many of them are complex.

For the more general operators often encountered in electromagnetics, a positive operator can be produced by the following procedure. Since L always has an adjoint L^*, multiplication of both sides of (5) by L^* produces

$$L^*Lu = L^*f \tag{28}$$

for any $f \in \mathscr{D}_{L^*}$. Provided that $Lu = 0$ has none but the trivial solution, it is easy to show that the operator L^*L is positive. Indeed, $\langle L^*Lu, u\rangle = \|Lu\|^2 > 0$, unless $Lu = 0$. But, $Lu = 0$ implies $u = 0$.

The Method of Moments applied to L^*L gives

$$\sum_{k=1}^{n} \alpha_k \langle L^*L\phi_k, w_m\rangle = \langle L^*f, w_m\rangle \quad m = 1, 2, \ldots, n \tag{29}$$

The Galerkin specialization follows immediately, viz:

$$\sum_{k=1}^{n} \alpha_k \langle L^*L\phi_k, \phi_m\rangle = \langle L^*f, \phi_m\rangle \quad m = 1, 2, \ldots, n \tag{30}$$

Since L^*L is positive, if the sequence $\{\phi_k\}$ is complete in $\mathscr{D}_{L^*L}$, equation (30) is the Rayleigh–Ritz method and convergence in energy $u_n \overset{e}{\to} u$ is assured, viz:

$$\lim_{n\to\infty} |u_n - u| = 0 \tag{31}$$

where the energy norm is with respect to the operator L^*L. By properties of the adjoint, (30) can also be written

$$\sum_{k=1}^{n} \alpha_k \langle L\phi_k, L\phi_m\rangle = \langle f, L\phi_m\rangle \quad m = 1, 2, \ldots, n \tag{32}$$

which is the result in the *Method of Least Squares*, more usually derived [7] by minimization of

$$\|Lu_n - f\|^2$$

It is easy to show that (31) implies that

$$\lim_{n\to\infty} \|Lu_n - f\| = 0 \tag{33}$$

so that $Lu_n \to f$. Unless the operator L^*L is positive-definite, nothing can be said concerning the convergence of u_n to u.

In all of the above forms of the Method of Moments, one is faced with the inversion of the matrix equation approximating $Lu = f$, namely,

$$Ax = b \tag{34}$$

Although it is beyond the scope of this paper to discuss this problem in detail, a few remarks are in order. If the size of the matrix is small enough (less than approximately 100 × 100 on minicomputers and 250 × 250 on large machines), there are accurate, established algorithms for performing the inversion. The preferred method according to [1] when **A** is dense and unstructured, as is the case in many electromagnetic problems, is Gaussian elimination with either partial or complete pivoting, possibly supplemented by iterative improvement. Such algorithms are readily available, for example in the widely distributed IMSL Library [11]. (In IMSL, Routine LEQT1C contains partial pivoting; LEQ2C adds iterative improvement.) Reference [1] contains a complete description of the method and an error analysis.

Because of the size of the matrices often encountered in the Method of Moments, iterative methods are often required in their inversion. Among the collection of iterative techniques, the class of conjugate direction methods [1], [12], [13], in particular the conjugate gradient method, has recently received considerable attention in the electromagnetic literature. A few cautions are in order. First, a principal feature of the conjugate gradient method is that the matrix A must be positive. Just as in the Method of Least Squares in the previous section, this requirement can be forced by multiplication of (20) by the adjoint operator, viz:

$$A^*Ax = A^*b \tag{35}$$

This operation has the undesirable effect of squaring the condition number of A. Second, it is well known that, because of roundoff errors, the finite termination aspect associated with the method does not occur in practice. The method is therefore a true iterative method [14] with potential problems in rate of convergence for matrices with high condition numbers. Golub and Van Loan [1] suggest a preconditioning strategy for such cases. Third, in practice it is extremely difficult to compare accuracies and rates of convergence under the condition of finite precision arithmetic [15] and comparisons without the finite precision assumption are at worst irrelevant or at best misleading.

Finally, it is crucial to note that, regardless of the various convergence criteria associated with iterative or direct methods, there is no direct relation between algorithm convergence properties associated with the inversion of $Ax = b$ and convergence properties of the original problem $Lu = f$, except in the special case where L itself is positive or positive-definite.

4. Example—Slit Diffraction

The purpose of this section is to give an example of an electromagnetic problem and indicate the difficulties with approximation that are typical of numerical methods in electromagnetics. The problem is the classic one of diffraction of a plane wave by a slit of infinite axial length and of width a, contained in a perfectly conducting screen. Consider the case of TM_z polarization (H_y, E_x, E_z), where the screen is in the x-y plane and the slit width is x-directed. It is well known [8] that all field components can be obtained by solving an integral equation on $\mathscr{L}_2(-d, d)$ for the x-directed electric field u in the aperture, viz:

$$Lu = f \tag{36}$$

$$f = 2\eta e^{iv\cos\phi} \tag{37}$$

$$L = \int_{-d}^{d} (\cdot) H_0^{(2)}(|v - q|)\, dq \tag{38}$$

where η is the intrinsic impedance of free space, ϕ is the angle in the x-z plane relative to the x-axis and

$$d = \frac{ka}{2} \tag{39}$$

$$v = kx \tag{40}$$

where k is the wave number $2\pi/\lambda$. Numerical methods have often been employed to produce an approximation to the solution to this integral equation. Since the kernel is only weakly singular (logarithmic), a simple application of the Method of Moments yields such an approximation. Since, however, the operator L is non-selfadjoint, legitimate questions arise concerning the accuracy of the approximation. Because the theory of non-selfadjoint operators is not well understood, it is not possible, based on any known mathematically sound convergence criteria, to compare the approximation obtained in the Method of Moments to the exact solution. This difficulty is typical of convergence problems in numerical electromagnetics.

This unfortunate situation has a resolution in the quasistatic limit. For ka small enough [8], the integral equation in (36) becomes

$$L_Q \hat{u} = f \tag{41}$$

where

$$L_Q = \int_{-p}^{p} (\cdot) k(t, s)\, d\mu(s) \tag{42}$$

$$u(s) = \hat{u}(s) w(s) \tag{43}$$

$$f_Q(t) = \frac{i\gamma\eta\pi}{2} \exp\left(i\frac{2t}{\gamma}\cos\phi\right) \tag{44}$$

$$k(t, s) = \log(|t - s|) \tag{45}$$

$$w(s) = \frac{1}{\sqrt{(p^2 - s^2)}} \tag{46}$$

$$s = \frac{\gamma q}{2} \tag{47}$$

$$t = \frac{\gamma v}{2} \tag{48}$$

$$p = \frac{\gamma ka}{4} \tag{49}$$

$$\log \gamma = 0.57721566\ldots \ (\textit{Euler's constant}) \tag{50}$$

and where the measure $d\mu$ is given in terms of the weight function w by

$$d\mu(s) = w(s)\, ds \tag{51}$$

It is straightforward to show that, for $k(t, s)$ given by (45),

$$\int_{-p}^{p}\int_{-p}^{p} |k(t, s)|^2\, d\mu(s)\, d\mu(t) < \infty \tag{52}$$

(The proof will be deferred until later in the paper.) Therefore, $k(t, s)$ is a *Hilbert–Schmidt* kernel [9] with respect to the weight function w. Such a kernel generates a *compact* (completely continuous) integral operator L_Q. In addition, since $k(t, s) = \bar{k}(s, t)$, the kernel is symmetric. The operator L_Q therefore has the following properties [10]:

A. The eigenvalues $\{\lambda_k\}$ of L_Q are real.
B. The eigenfunctions $\{\phi_k\}$ of L_Q corresponding to different eigenvalues are orthogonal.
C. The number of eigenvalues is either finite or, if infinite, the only limit point is $\lambda = 0$.

In the case where the number of eigenvalues is infinite, the approximation u_n to the aperture field given by

$$u_n = \sum_{k=1}^{n} \frac{\langle f, \phi_k \rangle}{\lambda_k} \phi_k \tag{53}$$

converges to u, viz:

$$\lim_{n\to\infty} \|u_n - u\| = 0 \tag{54}$$

provided

$$\sum_{k=1}^{\infty} \frac{|\langle f, \phi_k \rangle|^2}{\lambda_k^2} < \infty \tag{55}$$

where inner products are, with respect to the weight function w, defined in (46). The result in (53) yields an approximation with known convergence properties provided that the eigenfunctions can be found. The restriction given by (55) is a well-known result for first-kind integral equations [9], [10]. Although it could be a serious limitation in general, it causes no difficulties for the variety of forcing functions normally encountered in diffraction problems.

Fortunately this example is a member of a small class of integral equations in electromagnetics where the eigenfunctions are known. Indeed [16],

$$L_Q T_k(s/p) = \lambda_k T_k(s/p) \tag{56}$$

where

$$\lambda_k = \begin{cases} -\pi \log(2/p) & k = 0 \\ -\pi/k & k \neq 0 \end{cases} \tag{57}$$

and where T_k are the Chebyshev polynomials, orthogonal with respect to the weight function $w(s)$. Note that the limit point of the eigenvalues is zero, as predicted by Property C above. When used in conjunction with the operator L_Q, the eigenfunctions provide the "natural" expansion of the aperture field in the sense that convergence of the approximation (53) to the exact solution is assured, provided (55) is satisfied.

Finally, the Hilbert–Schmidt property (52) of the kernel $k(t, s)$ is easy to show *a posteriori*. A Chebyshev polynomial expansion of the kernel in the variable s, followed by evaluation of coefficients in the usual manner yields

$$\log(|t-s|) = \sum_{k=0}^{\infty} \epsilon_k \lambda_k T_k(t/p) T_k(s/p) \quad (58)$$

where ϵ_k is *Neumann's number*. Substitution of (58) into (52) gives

$$\int_{-p}^{p}\int_{-p}^{p} (\log|t-s|)^2 \, d\mu(s)\, d\mu(t) = \sum_{n=0}^{\infty} \lambda_n^2 < \infty \quad (59)$$

The first equality is expected, since the integral over the square of the kernel of a compact selfadjoint integral operator produces the sum of the squares of the eigenvalues [10]. The inequality follows since the sum of the squares of the eigenvalues, given by (57), yields a closed form result.

6. Discussion

This paper has been concerned with solution and approximation. Classes of operators have been discussed where there are firm mathematical properties providing the necessary convergence links. When the operators are more general and there exist no available mathematical convergence criteria, difficulties arise in the determination of the relationship between the numerical approximation and the solution. In these cases, however, there are recognized procedures used by numerical electromagnetic researchers resulting in forms of convergence by computer experiment. These procedures can be divided into two categories:

A. Self checks
B. Comparative checks

In self checks, the computer user chooses a specific value of n and computes u_n. The parameter n is then increased until successive increases no longer bring about a significant change in the result. One criterium is the value of $\|u_n\|/n$ as a function of increasing n. In comparative checks, the computer user compares the numerical result with solutions or limiting values that have been obtained by other methods, preferably analytical.

These "numerical convergence" methods are far from mathematically satisfying. There are, however, many important electromagnetic problems beyond the range of operators with "nice" mathematical properties. In the absence of a more complete theory of non-selfadjoint operators, numerical convergence will continue to be often the only convergence available. In this event, the numerical researcher must recognize certain limitations in the result. Indeed, it is misleading to refer to the result as *solution* when in fact it is *numerical approximation* with no firm mathematical estimate of nearness to solution.

Acknowledgments

This paper is an extension of remarks by the author, made while introducing the special session on Error Minimization and Convergence in Numerical Methods at the North American Radio Science Meeting, Vancouver, in June 1985. The author would like to thank A. K. Gautesen whose informal review of the paper led to several improvements in its content. A portion of the work was accomplished while the author was a Summer Visitor at Lawrence Livermore National Laboratory. The paper was computer typeset by the author using the algorithm $\mathrm{T_EX}$, with consulting assistance by M. E. Poggio.

References

[1]. Golub, G. H. and C. F. Van Loan, *Matrix Computations*, The Johns Hopkins University Press, Baltimore, Chapt. 10, 1983.

[2] Helmberg, G., *Introduction to Spectral Theory in Hilbert Space*, North-Holland, Amsterdam, 1969.

[3] Mikhlin, S. G., *The Problem of the Minimum of a Quadratic Functional*, Holden-Day, San Francisco, 1965.

[4] Akhiezer, N. I. and I. M. Glazman, *Theory of Linear Operators in Hilbert Space*, vol. 1, Frederick Unger, New York, 1961.

[5] Harrington, R. F., *Field Computation by Moment Methods*, Macmillan, New York, 1968.

[6] Harrington, R. F., "Matrix methods for field problems," *Proc. IEEE*, vol. 55, pp. 136–149, 1967.

[7] Stakgold, I., *Boundary Value Problems of Mathematical Physics*, vol. 2, Macmillan, Sect. 8.10, 1967.

[8] Butler, C. M. and D. R. Wilton, "General analysis of narrow strips and slots," *IEEE Trans. Antennas Propagat.*, vol. AP-28, pp. 42–48, 1980.

[9] Stakgold, I., *Boundary Value Problems of Mathematical Physics*, vol. 1, Macmillan, Chapt. 3, 1967.

[10] Hochstadt, H., *Integral Equations*, Wiley-Interscience, New York, Chapt. 3, 1973.

[11] *User's Manual, IMSL Library*, vol. 2, IMSL Customer Relations, NBC Building, 7500 Bellaire Blvd., Houston, TX 77036, Routines LEQT1C, LEQ2C, 1984.

[12] Gill, P. E., W. Murray, and M. H. Wright, *Practical Optimization*, Academic Press, London, Sect. 4.8.3, 1981.

[13] Luenberger, D. G., *Optimization by Vector Space Methods*, John Wiley and Sons, New York, pp. 290–297, 1969.

[14] Reid, J. K., "On the method of conjugate gradients for the solution of large sparse systems of linear equations," in *Large Sparse Sets of Linear Equations*, J. K. Reid (editor), Academic Press, New York, pp. 231–254, 1971.

[15] Wozniakowski, H., "Roundoff error analysis of a new class of conjugate gradient algorithms," *Linear Algebra and its Applic.*, vol. 29, pp. 507–529, 1980.

[16] Butler, C. M., "General solutions of the narrow strip (and slot) integral equations," *IEEE Trans. Antennas Propagat.*, vol. AP-33, pp. 1085–1090, 1985.

A Theorem on the Moment Methods

ANTONIJE R. DJORDJEVIĆ AND TAPAN K. SARKAR,
SENIOR MEMBER, IEEE

Abstract—**The inner product involved in the moment methods is usually an integral, which is evaluated numerically by summing the integrand at certain discrete points. In connection with this inner product, a theorem is proved, which states that the overall number of points involved in the integration must not be smaller than the number of unknowns involved in the moment method. If these two numbers are equal, a point-matching solution is obtained, irrespective of whether one has started with Galerkin's method or the least squares method. If the number of points involved in the integration is larger than the number of the unknowns, a weighted point-matching solution is obtained.**

I. Introduction

The moment methods have been widely used for solving linear operator equations in many electromagnetic problems [1]. However, sometimes certain simple facts are overlooked, which might obscure the true nature of the final solution obtained by the method used, or even lead to erroneous results or conclusions.

For example, one can think, one is using the least squares technique or the Galerkin method, but the resulting solution can be identical to a point-matching solution because of a careless evaluation of the inner product. Or one might obtain unstable results due to the same negligence.

The aim of this communication is to present a simple and almost obvious theorem on the moment methods, which might be helpful in avoiding the above problems.

II. Formulation of the Theorem

Let us consider the operator equation

$$Lf = g, \tag{1}$$

where L is a linear operator (which includes certain boundary conditions), $g(X)$ is a known function (excitation), $f = f(Y)$ is an unknown function (the solution to be found), and X and Y are points in multidimensional spaces.

Equation (1) can be solved by applying the moment methods [1]. As the first step of this approach, we have to approximate the unknown function f by a finite sum:

$$f(Y) = \sum_{i=1}^{n} a_i f_i(Y), \tag{2}$$

where a_i are unknown coefficients to be determined, $f_i(Y)$ are known expansion functions which form a suitable basis, and in the limit when $n \to \infty$ must be able to represent the true solution, $f(Y)$ [2], [3]. In addition, the expansion in (2) has to satisfy certain boundary conditions as required by the original equation (1). Now we substitute (2) into (1), to obtain

$$\sum_{i=1}^{n} a_i L f_i = g. \tag{3}$$

The second step in the moment methods is to compute the coefficients $\{a_i\}$ so that the approximate equation (3) is satisfied in some sense. To that purpose we have to define an inner product of two arbitrary functions $u(X)$ and $v(X)$, $\langle u, v\rangle$, belonging to the range of the operator L. Next, we take the inner products of (3) with some suitable weighting functions $w_j(X)$, $j = 1, \cdots, n$, which form a functional basis, belonging to the range of the operator L. Thus we obtain a set of linear equations in $\{a_i\}$:

$$\sum_{i=1}^{n} a_i \langle w_j, Lf_i\rangle = \langle w_j, g\rangle, \qquad j = 1, \cdots, n, \tag{4}$$

which can be solved for $\{a_i\}$ by using either direct methods (e.g., the Gaussian elimination, or the LU transform), or by using iterative methods (e.g., the conjugate-gradient method), which are suitable for very large systems of equations.

The inner product $\langle u, v\rangle$ is usually an integral of the product of functions $u(X)$ and $v(X)$. In very few cases the inner product in (4) can be evaluated analytically, and in most practical problems it is evaluated numerically. This involves only samples of the integrand at certain points. In other words, the numerical integration formulas used to evaluate the inner product can be written in the general form as

$$\int_D p(X)\, dD = \sum_{k=1}^{m} b_k p(X_k), \tag{5}$$

where D is the domain over which the integration is performed, X is a point in that domain, b_k are weighting coefficients, and X_k are points at which the samples of the function $p(X)$ are evaluated. The function $p(X)$, in our case, equals $u(X)v(X)$.

If the same integration formula is applied to both inner products of (4), we have

$$\sum_{i=1}^{n} a_i \sum_{k=1}^{m} w_j(X_k) b_k L f_{i|X=X_k} = \sum_{k=1}^{m} w_j(X_k) b_k g(X_k), \qquad j = 1, \cdots, n. \tag{6}$$

Let us introduce the following matrices:

$$[F] = [Lf_{i|X=X_k}]_{n\times m}, \tag{7}$$

$$[W] = [w_j(X_k)]_{n\times m}, \tag{8}$$

$$[B] = \text{diag}\,(b_1, \cdots, b_m), \tag{9}$$

$$[G] = [g(X_k)]_{m\times 1}, \tag{10}$$

$$[A] = [a_i]_{n\times 1}. \tag{11}$$

The system (6) can now be written in a compact form as

$$[W][B][F]^t[A] = [W][B][G], \tag{12}$$

where the superscript t denotes the transpose. Let us also denote

$$[V] = [W][B]. \tag{13}$$

Manuscript received October 30, 1985; revised March 4, 1986. This work was supported in part by the Office of Naval Research under contract N00014-79-C-0598.

A. R. Djordjević is with the Department of Electrical Engineering, University of Belgrade, P.O. Box 816, 11001 Belgrade, Yugoslavia.

T. K. Sarkar is with the Department of Electrical and Computer Engineering, Syracuse University, Syracuse, NY 13210.

IEEE Log Number 8612490.

Reprinted from *IEEE Trans. Antennas Propagat.*, vol. AP-35, no. 3, pp. 353–355, March 1987.

Now we have the following equation instead of (12):

$$[V][F]'[A]=[V][G]. \qquad (14)$$

The matrix $[V]$ can be considered as a weighting matrix, which multiplies the system of linear equations

$$[F]'[A]=[G]. \qquad (15)$$

If $m > n$, the system (15) is, generally, overdetermined. Note that equations (15) are, essentially, point-matching equations, which are obtained by postulating that the approximate equation (3) is satisfied at points $X = X_k$, $k = 1, \cdots, m$. The purpose of multiplying the system (15) by $[V]$ is to obtain a system of n equations in n unknowns. The solution to (14) can be regarded as a weighted point-matching solution.

Note that it is not necessary that the same integration formula be used for each of equations (4). If different formulas are used, then, instead of (13), the elements of the matrix $[V]$ are evaluated as

$$v_{jk}=w_j(X_{jk})b_{jk}, j=1, \cdots, n, k=1, \cdots, m, \qquad (16)$$

where X_{jk} and b_{jk}, $k = 1, \cdots, m$, are the coefficients of the integration formula used for the jth equation.

Let us consider (14). In order that this equation have a unique solution for $[A]$, the matrix $[V][F]'$ has to be regular. According to the Binet–Cauchy theorem [4], the matrix $[V][F]'$ will be regular if the condition

$$m \geq n \qquad (17)$$

is fulfilled, and if rank $[V]$ = rank $[F] = n$. If the condition (17) is violated, a unique solution does not exist (although a minimum-norm solution can be found, which might be useful in certain cases [5]).

Thus we have proved the following theorem.

If the integrals representing the inner product in a moment method solution to a linear operator equation are evaluated numerically, the overall number of points involved in the integration must not be smaller than the number of the unknown coefficients.

IV. Discussion

As the first consequence of the above theorem, let us consider the special case when $m = n$. If the inverse matrix $[V]^{-1}$ exists, both sides of (14) can be multiplied by $[V]^{-1}$, and a $n \times n$ system of linear equations is obtained. This system is, essentially, a system of point-matching equations, and, hence, the solution to (14) is identical to the point-matching solution. An example where such a situation can occur, is the following procedure. Let us adopt the expansion in (2) to be a piecewise-constant approximation (usually referred to as a pulse approximation). In that case $f_i(Y)$ is zero everywhere except over a small domain of Y, where it is constant (usually, equal to unity). If the Galerkin method is used, the weighting functions are $w_i = f_i$. The weighting functions being nonzero only over a small domain, the inner products in (4) are sometimes evaluated by using the midpoint rule, i.e., by using only one integration point per inner product. Obviously, the final result is identical to a point-matching solution, for which the matching points coincide with the integration points in the above Galerkin procedure. A similar result can be obtained if the least squares technique is used. In this case we have $w_i = (Lf_i)^*$, where the asterisk denotes the conjugate-complex value. If the pulse approximation is adopted, we have to evaluate the integrals representing the inner products over the whole domain of X where the original equation (1) is to hold, unlike the Galerkin procedure, where the integration has to be performed only over the domain where f_i is nonzero. However, the overall number of the integration points must be larger than n, unless we wish to obtain a point-matching solution (for $m = n$). Of course, this result is valid whatever weighting functions are utilized (triangular, piecewise-sinusoidal, etc.), as long as the inner products are evaluated by using numerical quadrature formulas.

The need for taking only a few integration points can arise not only in order to increase the speed of the computations, but also because of certain problems associated with the kind of the approximation adopted for the solution, which are not always clearly recognized. For example, in solving a wire-antenna problem, a piecewise-constant approximation of the current distribution can be used. (A good survey of the methods for the analysis of wire antennas is given in [6].) If the exact kernel is taken, then Lf_i has a nonintegrable singularity at the edge of the domain where f_i is nonzero. This precludes the use of both the Galerkin and the least squares technique, because the resulting inner products diverge! Yet, if the integration in evaluation of an inner product is confined to the interior of the domain where f_i is nonzero, acceptable results might be obtained, although the accuracy of the results can be much worse than with a point-matching solution (in addition to requiring a much longer CPU time), and the final result, obviously, is *not* a Galerkin (or least squares) solution (because such a solution does not exist). A singularity, though square integrable, also occurs if a piecewise-linear approximation (i.e., triangular approximation) is adopted, which can have an adverse effect especially to a least squares solution.

Another important fact which follows from (16) is that the weighting coefficients of the integration formula multiply the values of the weighting functions w_i. In other words, the weighting functions are to a certain extent modified by the formulas for the numerical integration. The only exception is if the repeated midpoint rule is used on equally sized subdomains, in which case all the coefficients b_k are equal.

If the repeated midpoint rule is used, then the matrix $[V]$ for the Galerkin and for the least squares solution has special, simple forms. Thus, for the Galerkin solution we have $[V] = \Delta D[f_i(Y_j)]$, while for the least squares solution $[V] = \Delta D[F]^*$, where ΔD is the size of subdomains over which the functions f_i are nonzero. It is worth noting that the least squares solution is now equivalent to solving the overdetermined system (15) in the least squares sense. It is well known that with such a procedure the resulting system matrix $[F]^*[F]'$ is positive definite and therefore the system (14) can be solved by iterative methods. However, the condition number of that matrix might become very large [7].

Finally, an analogous theorem can be formulated in connection with the moment method solution of an integral equation of the general form

$$\int_D f(Y)g(X, Y)\,dD=h(X)+\Lambda f(X), \qquad (18)$$

where Λ is a parameter. Namely, if the unknown function f is approximated according to (2), then the overall number of points involved in the numerical integration must not be smaller than n.

IV. Conclusion

The inner product involved in the moment methods is usually an integral, which is evaluated numerically. Certain precautions have to be taken in evaluating the inner product in order to obtain a valid solution. In connection with this a theorem is proved, which states that the overall number of points involved in the integration must not be smaller than the number of unknowns involved in the moment method. If these two numbers are equal, a point-matching solution is obtained, rather than the desired moment method solution (e.g., a Galerkin or a least squares solution). If the number of points involved in the integration is larger than the number of the unknowns, a weighted point-matching solution is obtained. This conclusion remains valid whatever weighting functions are utilized, as long as the inner product is evaluated by using numerical quadrature formulas.

References

[1] R. F. Harrington, *Field Computation by Moment Methods.* New York: Macmillan, 1968.

[2] T. K. Sarkar, "A note on the choice of weighting functions in the method of moments," *IEEE Trans. Antennas Propagat.*, vol. AP-33, pp. 436–441, Apr. 1985.

[3] T. K. Sarkar, A. R. Djordjević, and E. Arvas, "On the choice of expansion and weighting functions in the method of moments," *IEEE Trans. Antennas Propagat.*, vol. AP-33, pp. 988–996, Sept. 1985.

[4] D. S. Mitrinović and D. Ž. Djoković, *Polynomials and Matrices.* Belgrade, Yugoslavia: Naučna knjiga, 1962, p. 262 (in Serbo-Croatian).

[5] N. Dorny, *A Vector Space Approach to Models and Optimization.* New York: Krieger, 1980.

[6] E. K. Miller and F. J. Deadrick, in *Numerical and Asymptotic Techniques in Electromagnetics*, R. Mittra, Ed. Berlin: Springer-Verlag, 1975, ch. 4.

[7] G. Golub and C. F. van Loan, *Matrix Computations.* Baltimore, MD: Johns Hopkins Univ. Press, 1983.

Part 2
Conducting Bodies

DEVELOPMENT of MM models naturally began with perfectly conducting 2D and 3D objects. This section includes formative papers of both types of problems, using wire grids as approximations to solid objects, hybrid objects involving wires attached to a surface (circumventing failure due to internal resonances by using combined-field treatments), and modeling of more general surface shapes using piecewise triangular models.

The first paper is ''Scattering from bodies of revolution,'' by Andreasen, in which a mixed-domain expansion (entire domain in ϕ and subdomain in the generating curve of the body) for the surface currents is used in the MM solution of an electric field integral equation (EFIE). This paper outlines the basic analytical/numerical issues that are relevant in the computer implementation of MM techniques, and is the point of departure for further development by subsequent investigators. The next paper, by Richmond, ''A wire-grid model for scattering by conducting bodies,'' provides the analog using a wire representation for a solid surface. The currents are constrained to be on the wire itself, satisfying Kirchhoff's node law. The concepts in the previous two papers are merged in ''Electromagnetic modeling of composite wire and surface geometries,'' by Newman and Pozar. A solution is developed for open or closed surfaces with wire attachments. The sinusoidal reaction formulation is extended to include junction terms. An important consideration in numerical solutions of integral equations, describing the radiation or scattering from closed bodies, is the uniqueness of the solution at the internal resonances of the corresponding cavity. Mautz and Harrington, in ''A combined-source solution for radiation and scattering from a perfectly conducting body,'' show how an exterior Dirichlet problem can be formulated providing uniqueness at all frequencies. This formulation uses a combination of electric and magnetic currents to produce either the outside radiated fields or the scattered fields on a body. It is shown that the resulting combined-source operator is the adjoint of the combined-field operator. This approach is particularly efficacious for radiation (embedded antenna) problems. Rao, Wilton, and Glisson, in ''Electromagnetic scattering by surfaces of arbitrary shape,'' formulate the EFIE for a triangular patch representation spanning the body. The patches are planar. The subdomain expansion on the patches are free of line or point charges at the subdomain boundaries. An alternative to patch modeling is the representation of a 3D surface with wire grids. Ludwig, in ''Wire grid modeling of surfaces,'' discusses the issue of wire diameter and spacing to achieve reliable modeling results. The final paper in this section, ''Extended integral equation formulation for scattering problems from a cylindrical scatterer,'' by Toyoda *et al.*, develops the boundary element method to yield unique solutions for a class of surface integral equations.

Scattering from Bodies of Revolution

MOGENS G. ANDREASEN, MEMBER, IEEE

***Abstract*—The problem of scattering of a plane electromagnetic wave from an arbitrary metallic body of revolution is solved by a theoretical method for arbitrary incidence and polarization. The method permits numerical computations by high-speed digital computers, and examples are given. The incident wave is expanded in cylindrical modes, and an integral equation is solved for the induced current distribution of each mode. The scattering cross section, including the back-scattering or radar cross section, is found by summation of the mode scattered fields. The method is limited to a maximum perimeter length of twenty wavelengths.**

While the cases discussed in the paper pertain to perfectly conducting bodies, other surface boundary conditions, an arbitrary surface impedance or coatings by lossy dielectrics, can also be treated with equal precision.

I. Introduction

THIS WORK is concerned with a theoretical evaluation of the electromagnetic scattering from an arbitrary metallic body of revolution. This is a problem which has been given considerable attention over the past twenty years because of its significance for numerous applications in radar techniques in general. More specifically, the problem has become important in recent years for tracking and discriminating between space vehicles and objects.

The general scattering problem is a very complicated mathematical boundary-value problem which so far has resisted exact analytical treatment except in such special cases as the sphere and the spheroid. In these cases the surface of the scatterer coincides with a coordinate surface of an orthogonal coordinate system in which the vector wave equation can be solved by the method of separation of variables. The general scattering problem for an arbitrary metallic body of revolution is solved here by an integral equation method which is, in principle, an exact method. The method takes its starting point in an expansion of the incident wave in a set of orthogonal cylindrical TE and TM modes propagating along the symmetry axis of the body considered. Each of these incident modes will induce a current distribution on the surface of the body. Due to the mode orthogonality, the total induced current distribution is represented by the sum of the induced mode current distributions. Each mode current distribution satisfies two coupled integral equations. These integral equations are extremely complicated for an analytical treatment. However, they can be reduced to a set of simultaneous linear complex equations which can be solved by a digital computer. The accuracy thus obtained depends upon the number of equations used, but can, in principle, be increased as much as desired by increasing the number of equations sufficiently.

The only limitation of the numerical integral equation method for solving the general scattering problem is the speed of presently available computers. At the present state-of-the-art, the general problem cannot be solved economically if it is necessary to solve more than about one hundred linear equations. Since the mode current density does always have two orthogonal components to be determined, this means that each mode current distribution can be sampled a maximum of about fifty times. Most often, the sampling points along the perimeter of the cross section of the body can be spaced as much as one sixth of a wavelength apart without greatly impairing the accuracy of the solution. As a consequence, the method permits treating bodies with a maximum cross-sectional perimeter of about twenty wavelengths.

As indicated above, the general scattering problem for a metallic body of revolution is solved by expanding the incident wave in cylindrical modes and solving the scattering problem for each mode independently. This approach may seem to be an unnecessary complication of the problem. The scattering problem could have been formulated without expanding the incident wave, assuming the current distribution to be sampled in a sufficiently large number of points on the surface of the body. However, with a maximum average distance between neighboring sampling points of about one sixth of a wavelength and a maximum of fifty sampling points, the maximum linear dimension of a body that can be treated by this method is less than one wavelength. By expanding the incident wave in cylindrical modes, each of which induces a current distribution with known azimuthal symmetry properties, the scattering problem can be solved for a much larger body as described above.

Moreover, even if the digital computer would permit the solution of the scattering problem without expanding the incident wave, it would still be of a great advantage economically to perform the expansion. The reason for this is that the computer time needed to solve a set of simultaneous linear equations is proportional to the cube of the number of equations. Accordingly, it is more economical to solve several integral equations with a moderate number of sampling points, as is done when the expansion is carried out, than it is to solve a single integral equation with a large number of sampling points, as is necessary when no expansion is used.

Manuscript received May 21, 1964; revised August 11, 1964, and August 31, 1964. The work reported here was carried out at TRG-West, Menlo Park, Calif., and was supported in part by the U. S. Air Force under Contract No. AF 19(628)-500 on Subcontract No. 294 from Lincoln Laboratory, Massachusetts Institute of Technology, Lexington, Mass.

The author is with TRG-West, a division of TRG, Inc., Menlo Park, Calif.

Reprinted from *IEEE Trans. Antennas Propagat.*, vol. AP-13, no. 2, pp. 303–310, March 1965.

II. The Integral Equations

The integral equations for scattering by a rotationally symmetric body will be derived from the vector potential of the current distribution. Assuming the surface current distribution on the scatterer surface to be $\overline{K}$, the vector potential of the scattered field is[1]

$$\overline{A}^s(\bar{r}') = \frac{\mu}{4\pi}\int_A \overline{K}(\bar{r})\,\frac{e^{-jkR}}{R}\,da \tag{1}$$

where the integration is extended over the surface of the body. R is the distance between the source point with radius vector $\bar{r}$ and the observation point with radius vector $\bar{r}'$, both points being on the surface of the body (see Fig. 1). $k=2\pi/\lambda$ is the free-space propagation constant. The time factor $\exp(j\omega t)$ is suppressed.

The scalar potential is defined by

$$\psi^s(\bar{r}') = \frac{1}{4\pi\epsilon}\int_A \eta(\bar{r})\,\frac{e^{-jkR}}{R}\,da \tag{2}$$

where $\eta(\bar{r})$ is the surface charge density on the scatterer. The continuity equation requires the following relation between $\overline{K}$ and η to be satisfied

$$\eta(\bar{r}) = -\frac{\nabla_t\cdot\overline{K}(\bar{r})}{j\omega} \tag{3}$$

where ∇_t is the differential operator tangential to the scatterer surface.

The scattered electric field can now be expressed by

$$\overline{E}^s(\bar{r}') = -j\omega\overline{A}^s(\bar{r}') - \nabla'\psi^s(\bar{r}') \tag{4}$$

Let the electric field of the incident wave be $\overline{E}^i$. The boundary condition at the surface of the scatterer is then

$$\hat{t}_n'\cdot[j\omega\overline{A}^s(\bar{r}') + \nabla_t'\psi^s(\bar{r}')] = \hat{t}_n'\cdot\overline{E}^i(\bar{r}'), \qquad n=1,2 \tag{5}$$

where $(\hat{t}_1, \hat{t}_2)$ are unit vectors of a local orthogonal coordinate system on the surface of the scatterer. The integral equations (5) can be reduced as follows by introducing the expressions (1) and (2) for the potentials

$$\hat{t}_n'\cdot\overline{E}^i(\bar{r}') = \frac{j\omega\mu}{4\pi}\,\hat{t}_n'\cdot\left[\int_A \overline{K}(\bar{r})\,\frac{e^{-jkR}}{R}\,da + \frac{1}{k^2}\nabla_t'\int \frac{e^{-jkR}}{R}\,\nabla_t\cdot\overline{K}(\bar{r})da\right] \qquad n=1,2 \tag{6}$$

To obtain a further reduction of this expression for a body of revolution, let us introduce the local coordinate system (t, ϕ) (see Fig. 1). We assume that the incident plane wave has been expanded in a set of azimuthal modes and consider the scattering of the mth mode. The electric field components along $\hat{t}$ and $\hat{\phi}$ of the mth mode of the incident wave can be expressed as a sum of an even mode and an odd mode as follows:

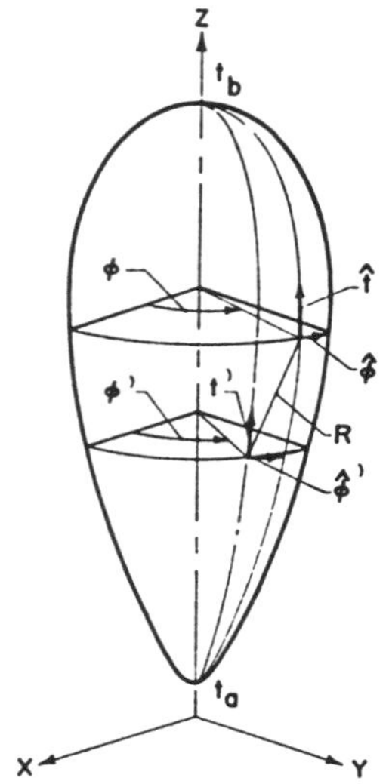

Fig. 1. Coordinates used on scatterer surface.

$$E_t = E_1(t)\,{\cos \atop \sin}\,(m\phi) \tag{7}$$

$$E_\phi = E_2(t)\,{\sin \atop -\cos}\,(m\phi) \tag{8}$$

The current distribution on the scatterer generated by this incident field is of the form

$$\overline{K} = K_1(t)\,{\cos \atop \sin}\,(m\phi)\hat{t} + K_2(t)\,{\sin \atop -\cos}\,(m\phi)\hat{\phi} \tag{9}$$

To obtain the reduction of (6), introduce

$$\nabla_t = \hat{t}\frac{\delta}{\delta t} + \frac{\hat{\phi}}{\rho}\frac{\delta}{\delta\phi} \tag{10}$$

and (9) and (10) in (6). Also, introducing the angle v between $\hat{t}$ and $\hat{z}$, and the following expressions for the unit vectors

$$\left.\begin{aligned} \hat{t} &= \sin v\cos\phi\hat{x} + \sin v\sin\phi\hat{y} + \cos v\hat{z} \\ \hat{t}' &= \sin v'\cos\phi'\hat{x} + \sin v'\sin\phi'\hat{y} + \cos v'\hat{z} \\ \hat{\phi} &= -\sin\phi\hat{x} + \cos\phi\hat{y} \\ \hat{\phi}' &= -\sin\phi'\hat{x} + \cos\phi'\hat{y} \end{aligned}\right\} \tag{11}$$

we find after tedious calculations the following pair of integral equations to be satisfied by the current components $K_1(t)$ and $K_2(t)$:

$$\begin{aligned} -j\frac{4\pi}{\zeta}E_1(t') &= \int_{kt_a}^{kt_b}\{[\sin v\sin v'(G_{m-1}+G_{m+1}) + 2\cos v\cos v'G_m] \\ &\quad\cdot K_1(t) - \sin v'(G_{m-1}-G_{m+1})K_2(t)\}k\rho d(kt) \\ &\quad + 2\frac{d}{d(kt')}\int_{kt_a}^{kt_b}\left[\frac{dK_1(t)}{d(kt)} + \frac{\sin v}{k\rho}K_1(t) \right. \\ &\quad \left. + \frac{m}{k\rho}K_2(t)\right]G_mk\rho d(kt) \end{aligned} \tag{12}$$

[1] Stratton, J. A., *Electromagnetic Theory*, New York: McGraw-Hill, 1941.

and

$$-j\frac{4\pi}{\zeta}E_2(t') = \int_{kt_a}^{kt_b}[-\sin v(G_{m-1}-G_{m+1})K_1(t)+(G_{m-1}+G_{m+1})K_2(t)] \cdot k\rho d(kt) - 2\frac{m}{k\rho'}\int_{kt_a}^{kt_b}\left[\frac{dK_1(t)}{d(kt)}+\frac{\sin v}{k\rho}K_1(t) + \frac{m}{k\rho}K_2(t)\right]G_m k\rho d(kt) \quad (13)$$

where $\zeta=\sqrt{\mu/\epsilon}$ is the intrinsic impedance of free space. In these equations we have introduced a Green's function defined by

$$G_m = \int_0^{\pi}\frac{e^{-jkR}}{kR}\cos(m\phi)d\phi \quad (14)$$

This Green's function can be interpreted physically as the scalar field radiated by a scalar ring source with harmonically varying source strength along the ring. The integral in (14) cannot be reduced analytically except in special cases. However, for $t\neq t'$, the integrand is a regular function and the integral can be evaluated by numerical integration. For $t=t'$, the Green's function has an integrable singularity. The integration along the t axis [see (12) and (13)] therefore must be done analytically in the immediate vicinity of $t=t'$. The details of this evaluation[2] will not be dealt with here.

The reduction of the integral equations [(12) and (13)] to a set of simultaneous linear equations is trivial. The integrals in (12) and (13) can be approximated by a weighted sum of sampled values of the integrand. That is, by a weighted sum of N-sampled values K_{1n} and K_{2n} of the current components $K_1(t)$ and $K_2(t)$, as follows:

$$\sum_{n=1}^{N}[A_{pn}K_{1n}+B_{pn}K_{2n}] = -j\frac{4\pi}{\zeta}E_1(t_p') \quad (15)$$

$$\sum_{n=1}^{N}[C_{pn}K_{1n}+D_{pn}K_{2n}] = -j\frac{4\pi}{\zeta}E_2(t_p') \quad (16)$$

Requiring these relations to hold in the N-sampling points along the t axis, $p=1, 2, \cdots, N$, equations (15) and (16) represent a system of $2N$-simultaneous linear equations from which the N values of $K_1(t)$ and the N values of $K_2(t)$ can be determined. The choice of the weighting factors A_{pn}, B_{pn}, C_{pn} and D_{pn} is, of course, very important for the accuracy of the solution. For $p\neq n$, the weighting factors can be found using the trapezoidal or the Simpson rule of integration. For $p=n$, the weighting factors have to be evaluated analytically because of the singularity at $t=t'$.

[2] Andreasen, M. G., Scattering from rotationally symmetric metallic bodies, Final Rept, Subcontract 294, AF 19(628)-500, TRG-West, Menlo Park, Calif. Apr 1964.

III. Expansion of the Incident Wave

In the derivation of the integral equations [(12) and (13)] we have tacitly assumed that the plane wave incident upon the metallic body of revolution can be expanded in a set of azimuthal modes. According to (7) and (8), the tangential electrical field component of the mth incident mode at the surface of the metallic body was assumed to be of the form

$$\overline{E}_{is}{}^m = \hat{t}E_1(t)\begin{matrix}\cos\\ \sin\end{matrix}(m\phi) + \hat{\phi}E_2(t)\begin{matrix}\sin\\ -\cos\end{matrix}(m\phi) \quad (17)$$

where the upper function symbol applies for parallel polarization and the lower symbol applies for perpendicular polarization. We shall now show that the plane incident wave can, in fact, be expanded in a discrete set of cylindrical modes, each mode having a different azimuthal mode number, while all modes propagating with the same velocity along the symmetry axis of the body.

The incident plane wave is assumed to propagate in the direction $\theta=\pi-\theta_i$, $\phi=0$, and to be polarized in a direction making an angle θ_p with the unit vector $-\hat{\theta}$ (see Fig. 2). Let the amplitude of the magnetic field of the incident wave be unity. The amplitude of the electric field is then $E_0=\zeta=120\pi$. To simplify the expansion of the plane wave, let us separate the wave into a wave polarized parallel to the x-z plane of incidence, and a wave polarized perpendicular to this plane, as follows:

$$\overline{E}_i = -E_i^{\parallel}\hat{\theta}+E_i^{\perp}\hat{y} \quad (18)$$

where

$$E_i^{\parallel} = \zeta\cos\theta_p(\cos\theta_i\hat{x}+\sin\theta_i\hat{z})e^{-jk(z\cos\theta_i - x\sin\theta_i)} \quad (19)$$

and

$$E_i^{\perp} = \zeta\sin\theta_p\hat{y}e^{-jk(z\cos\theta_i - x\sin\theta_i)} \quad (20)$$

The parallel-polarized wave is a TM wave (no magnetic field component along the z axis) and is expandable in cylindrical TM modes. The perpendicular-polarized wave is expandable in cylindrical TE modes (no electric field component along the z axis). The last exponential in (19) and (20) can be expanded in a Fourier series[1]

$$e^{jkx\sin\theta_i} = e^{jk\rho\cos\phi\sin\theta_i} = e^{-jkz\cos\theta_i}\sum_{m=0}^{\infty}\epsilon_m j^m J_m(k\rho\sin\theta_i)\cos(m\phi) \quad (21)$$

where J_m is the Bessel function or order m, and ϵ_m is Neumann's number

$$\epsilon_m = \begin{cases}1 & \text{for } m=0\\ 2 & \text{for } m>0\end{cases} \quad (22)$$

Defining the tangential electric field at the surface of the body by a sum of mode fields of the form (17), we find for the parallel-polarized wave

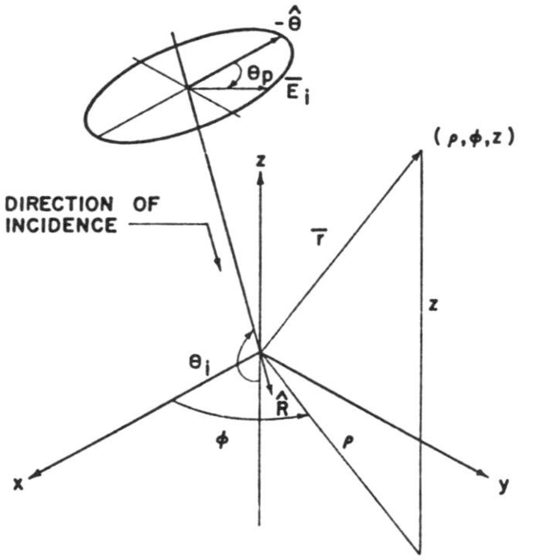

Fig. 2. Coordinates for the incident wave.

$$E_1(l) = -\cos\theta_p e^{-jkz\cos\theta_i}\epsilon_m j^{m+1} \cdot [j\sin\theta_i\cos vJ_m(k\rho\sin\theta_i) \quad (23)$$
$$+ \cos\theta_i \sin vJ_m'(k\rho\sin\theta_i)] \quad (24)$$

$$E_2(l) = \cos\theta_p\cos\theta_i e^{-jkz\cos\theta_i}\epsilon_m j^{m+1} m \frac{J_m(k\rho\sin\theta_i)}{k\rho\sin\theta_i} \quad (25)$$

where (ρ, ϕ, z) is a point on the surface of the body of revolution.

For the perpendicular-polarized wave we find similarly

$$E_1(l) = -\sin\theta_p\sin v e^{-jkz\cos\theta_i}\epsilon_m j^{m+1} m \frac{J_m(k\rho\sin\theta_i)}{k\rho\sin\theta_i} \quad (26)$$

$$E_2(l) = \sin\theta_p e^{-jkz\cos\theta_i}\epsilon_m j^{m+1} J_m'(k\rho\sin\theta_i) \quad (27)$$

This completes the derivation of the expansion formula for the incident plane wave. Because of the rotational symmetry of the body considered, the cylindrical modes are uncoupled. That is, a particular incident mode induces no mode current with a mode number different from the mode number of the incident mode. This is the reason why the scattering problem can be solved mode for mode, with subsequent application of the superposition principle.

While the mode expansion is an infinite series, the series can be truncated in practice. Due to the asymptotic behavior of the Bessel function for small arguments, the series can be truncated approximately for

$$m = k\rho_{max}\sin\theta_i + 6 \quad (28)$$

where ρ_{max} is the maximum radial dimension of the body. This formula holds only for $k\rho_{max}\sin\theta_i \gtrsim 3$. For small values at $k\rho_{max}\sin\theta_i$ m approaches 1 for $\theta_i = 0$. In the computer program developed, m is actually determined by searching a table of the Bessel function.

IV. The Scattering Cross Section

So far, we have been concerned only with the calculation of the current distribution induced on the surface of the metallic body. The determination of the current distribution is only a convenient intermediate step for finding the scattering cross section of the body, but it also provides valuable physical insight in the scattering phenomena. Once the current distribution has been found, it is a relatively simple matter to determine the scattering cross section of any direction in space by integration of each mode current distribution and summation of the scattered mode fields. The details of this calculation will be dealt with only briefly.

The scattering cross section in a particular direction (θ_0, ϕ_0) can be defined by

$$\sigma_s(\theta_0, \phi_0) = 4\pi R_0^2 \left| \frac{E_s(R_0, \theta_0, \phi_0)}{E_0} \right|^2 \quad (29)$$

where E_s is the scattered electric field in the direction (θ_0, ϕ_0) at a large distance R_0 from the metallic body, and E_0 is the electric field intensity of the incident wave. The scattered field is a sum of mode fields, each mode field being radiated by a mode current distribution. The relation between the scattered mode field and the mode current is the well-known integral expression[3]

$$\frac{1}{\zeta}\bar{E}_s(\bar{r}') = -\frac{j}{4\pi}\int_A \bar{K}(\bar{r})\cdot[\epsilon - \hat{R}\hat{R}]\frac{e^{-jkR}}{kR}k^2 da \quad (30)$$

where $\bar{K}$ is the current density, ϵ the unit dyad, $\hat{R}$ the unit vector pointing in the direction (θ_0, ϕ_0), and R the distance from the point $\bar{r}$ on the metallic body to the observation point $\bar{r}'$. This expression can be reduced considerably by introducing the current distribution (9) and separating the ϕ dependence. After tedious calculations, the scattering cross section in the direction (θ_0, ϕ_0) is then found to be expressed by

$$\frac{\sigma_s(\theta_0, \phi_0)}{\lambda^2} = \frac{1}{16\pi}\left[\left|\sum_{m=0}^{\infty} j^m \begin{matrix}\cos\\ \sin\end{matrix}(m\phi_0)T_{1m}(\theta_0)\right|^2 + \left|\sum_{m=0}^{\infty} j^m \begin{matrix}\sin\\ -\cos\end{matrix}(m\phi_0)T_{2m}(\theta_0)\right|^2\right] \quad (31)$$

where

$$T_{1m}(\theta_0) = \int_{kl_a}^{kl_b}\{K_{1m}(l)[\cos\theta_0\sin v(J_{m+1}(w) - J_{m-1}(w)) + j2\sin\theta_0\cos vJ_m(w)] + K_{2m}(l)\cos\theta_0(J_{m+1}(w) + J_{m-1}(w))\}k\rho \cdot e^{jkz\cos\theta_0}d(kl) \quad (32)$$

[3] Silver, S., *Microwave Antenna Theory and Design*, MIT Radiation Lab Ser No. 12, New York: McGraw-Hill, 1949.

and

$$T_{2m}(\theta_0) = \int_{kt_a}^{kt_b} \{ K_{1m}(t) \sin v [J_{m+1}(w) + J_{m-1}(w)] + K_{2m}(t)[J_{m+1}(w) - J_{m-1}(w)] \} k\rho \cdot e^{jkz \cos \theta_0} d(kt) \qquad (33)$$

where we have introduced the abbreviation

$$w = k\rho \sin \theta_0 \qquad (34)$$

The integrals in (32) and (33) are most easily evaluated numerically by a digital computer.

V. Numerical Results

The integral equations [(12) and (13)] have been programmed for a digital computer. The resulting computer program will solve the general scattering problem for virtually any cross section of a metallic body of revolution. The validity of the program has been checked by comparing the results for the induced current distribution on a sphere with exact analytical results which can be derived for this special case by the method of separation of variables. These results are shown in Fig. 3 for a 1.0-wavelength diameter sphere. The two components K_1 and K_2 of the current density [see (19)] are plotted as functions of a normalized surface coordinate s which has the value 0 on the front of the sphere and the value 1 on the back of the sphere. The agreement between the results obtained by the two methods is seen to be quite good.

Figure 4 shows the cross section of a cylindrical rod with hemispherical caps at the ends. The rod is 0.4-wavelength diameter and is 1.5 wavelengths long. The current distribution induced on this rod by a plane wave incident along the axis of symmetry of the rod was computed by the scattering program. The numerical results are shown in Fig. 5. The circumferential current component K_2 is seen to be very small on the cylindrical part of the rod, as expected, while the current component K_1 along the contour of the rod is seen to oscillate about the value 2. The oscillation of the current component K_1 represents a standing wave that is due to reflection from the end of the rod and has a period of one-half wavelength, as expected.

The cross section of another body used to test the scattering program is shown in Fig. 6. The body consists of a cone capped by a sphere in each end. The body is 2 wavelengths long and has a maximum diameter of 1.2 wavelengths. The current distribution induced on this body was computed by the scattering program both when the plane wave is incident upon the fat end of the conesphere (direction of incidence 1 in Fig. 6) and when it is incident from the opposite side (direction of incidence 2 in Fig. 6). The numerical results are shown in Figs. 7 and 8, respectively. It is interesting to note that the shadow effect is much less pronounced when the plane wave is incident upon the fat end of the conesphere than when it is incident from the opposite end. This is expected and can be explained qualitatively by the dependence of the decay exponent of the creeping waves in the shadow zone upon the princiapl radii of curvature of the surface of the body. The circumferential current component K_2 is seen to be relatively small at the small end of the conesphere as expected. The current distributions also exhibit standing-wave effects due to reflection from the far end of the conesphere.

Figure 9 shows the cross section of a circular disk with rounded edge. The disk is 1.1 wavelengths in diameter and is 0.1 wavelength thick. The current distribution induced on this body was computed by the scattering program and is shown in Fig. 10. The scattering results for the disk are particularly interesting because they exhibit physical phenomena not present on the sphere, the rod or the conesphere treated above. One notices that the circumferential current component K_2 has a sharp peak at the edge of the disk. This peak was, of course, expected. In the limit of zero thickness of the disk, the circumferential current component will have an integrable singularity at the edge. The circumferential current component K_2 also is seen, as expected, to be attenuated much faster than the radial component K_1 in the shadow zone.

Figures 3, 5, 7, 8 and 10 show only the magnitude of the current components. The phase distributions were, of course, also computed but have not been plotted.

The determination of the induced current distribution is the principal and most difficult task in any scattering problem. Once this current distribution has been found, it is a relatively simple matter to determine the scattering cross section for any direction in space by integration of each mode current distribution and summation of the scattered mode fields according to the formulas given in Section IV.

The scattering patterns for the three bodies considered previously, the rod, the conesphere and the disk, are plotted in Figs. 11–14. These figures show the scattering cross section in the plane of incidence ($\phi = 0°$) and in the plane perpendicular to the plane of incidence ($\phi = 90°$) vs. the angle of observation θ (0° for the forward direction, 180° for the backward direction).

The scattering pattern for the cylindrical rod is shown in Fig. 11. The pattern exhibits several lobes due to interference between the forward and backward traveling waves on the rod. It may be interesting to compare the value of the scattering cross section in the forward direction with the value that can be predicted optically. Optically, for the forward scattered field, the rod is equivalent to an aperture with the same area as the rod cross section. The scattering cross section in the forward direction can then be expressed by

$$\sigma = GA \qquad (35)$$

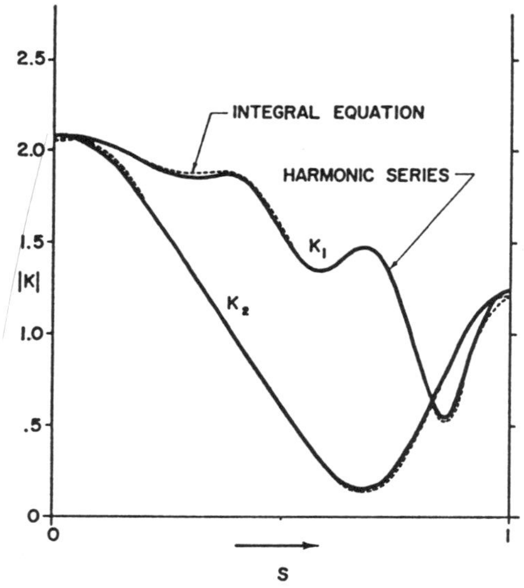

Fig. 3. Current distribution on sphere 1.0λ diameter

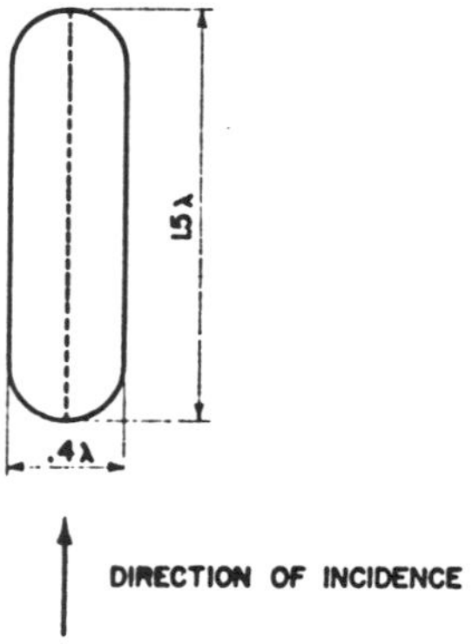

Fig. 4. Cylindrical rod with rounded ends.

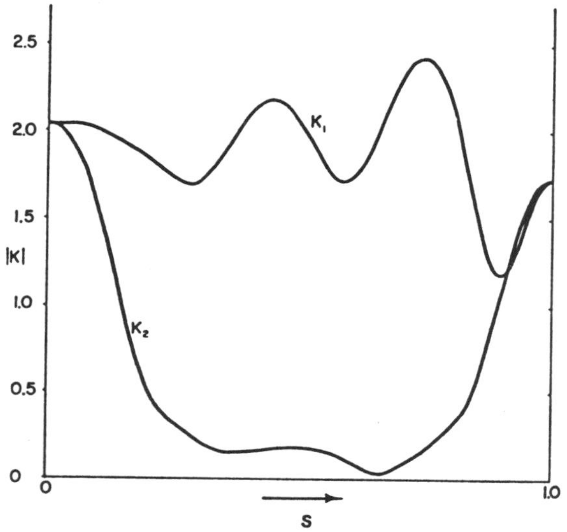

Fig. 5. Current distribution induced on cylindrical rod shown in Fig. 4.

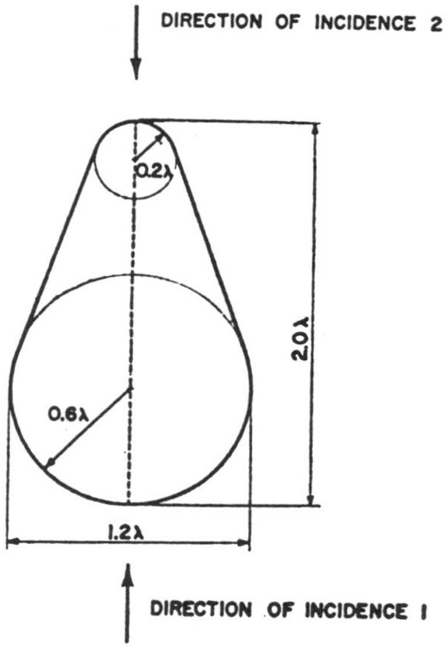

Fig. 6. Cone with spherical caps.

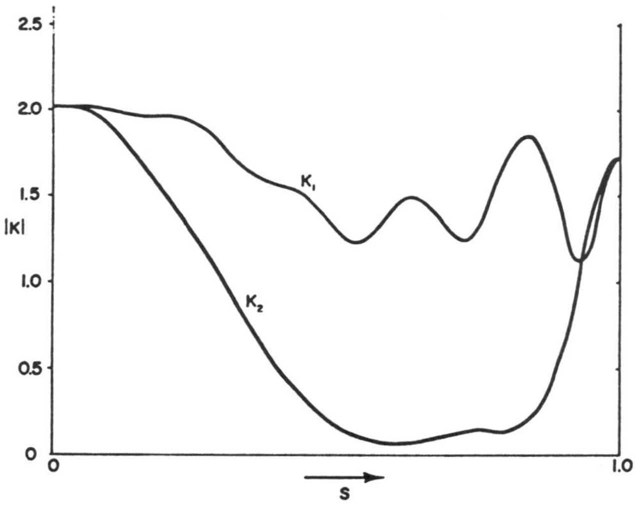

Fig. 7. Current distribution induced on cone-sphere for direction of incidence 1 in Fig. 6.

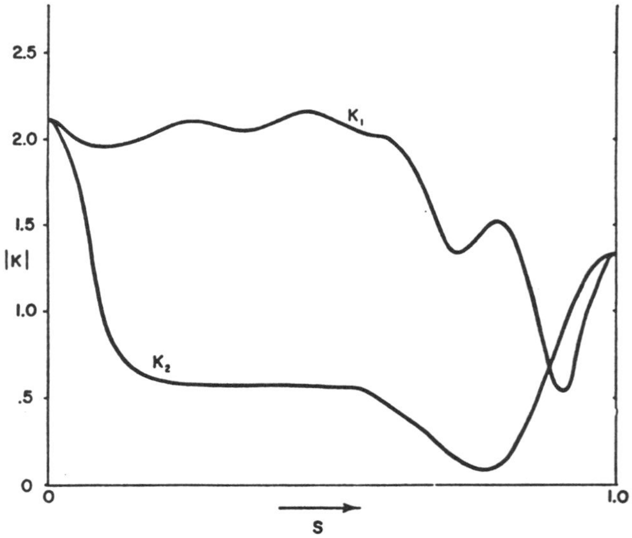

Fig. 8. Current distribution induced on cone-sphere for direction of incidence 2 in Fig. 6.

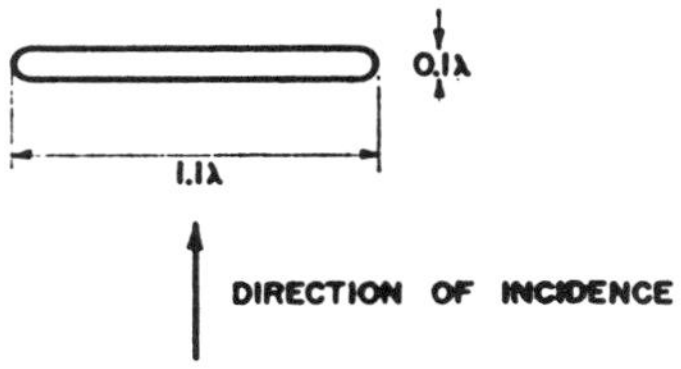

Fig. 9. Circular disk with rounded edge.

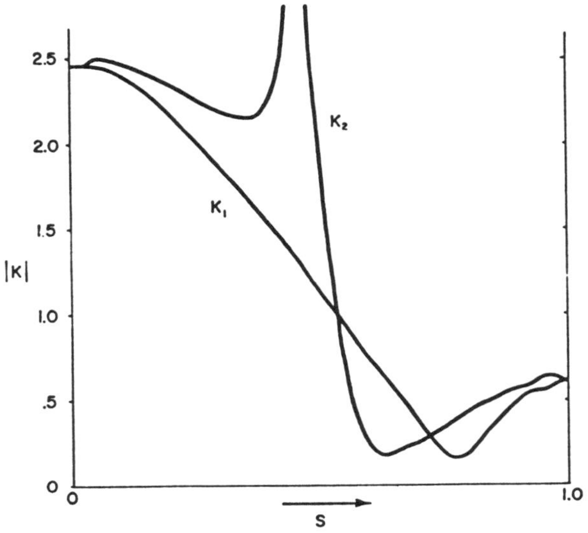

Fig. 10. Current distribution induced on circular disk shown in Fig. 9.

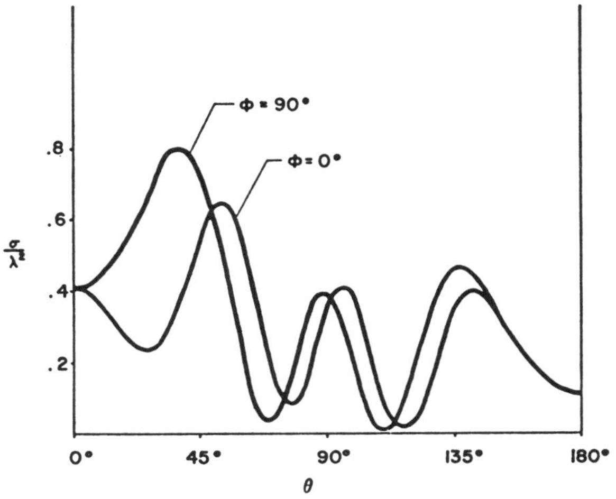

Fig. 11. Scattering pattern for cylindrical rod shown in Fig. 4.

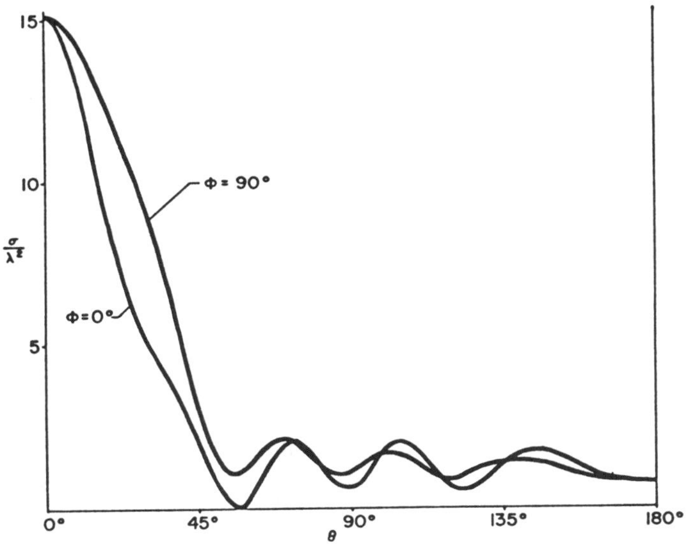

Fig. 12. Scattering pattern for cone-sphere for direction of incidence 1 in Fig. 6.

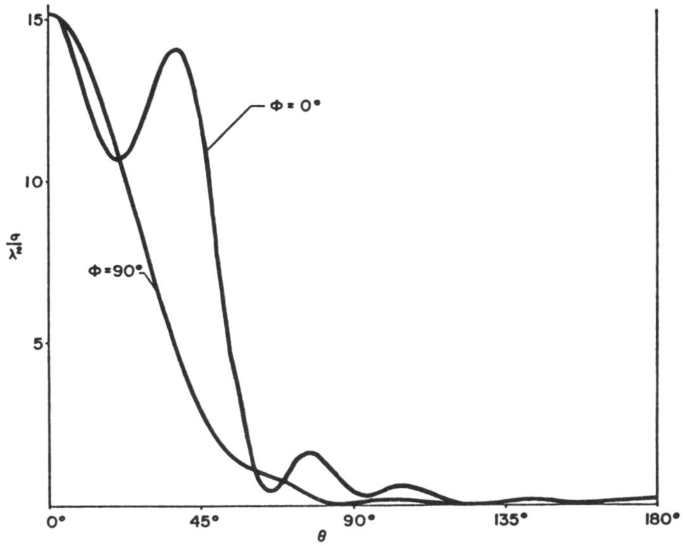

Fig. 13. Scattering pattern for cone-sphere for direction of incidence 2 in Fig. 6.

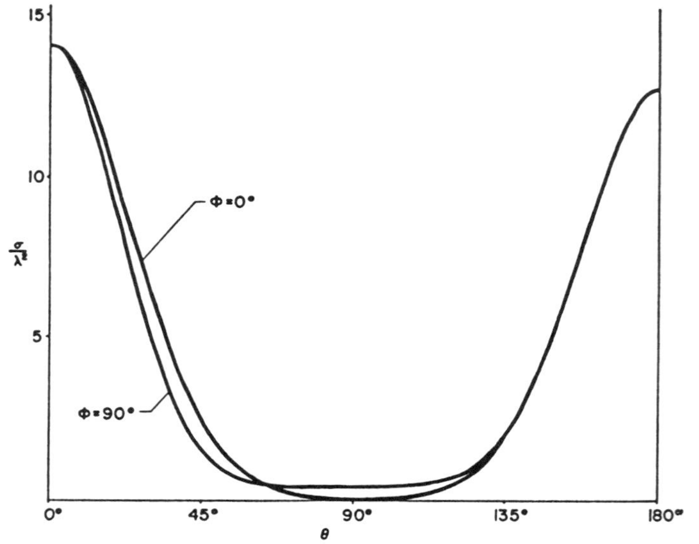

Fig. 14. Scattering pattern for circular disk shown in Fig. 9.

where A is the aperture area and G is the power gain of the aperture

$$G = \frac{4\pi}{\lambda^2} A \tag{36}$$

Accordingly, we find $\sigma/\lambda^2 = 0.12$ which agrees very poorly with the value 0.41 read from Fig. 11. The reason for this discrepancy is that the rod is only 0.4λ diameter and that scattering from the far end of the rod is, therefore, not at all negligible.

The scattering patterns for the conesphere are shown in Figs. 12 and 13. As expected, they exhibit a strong main lobe due to the shadowing. It is interesting to note that for direction of incidence 2, the pattern has a strong sidelobe for $\theta = 37.5°$. This lobe is the result of reflection from the conical part of the body. Optically, the sidelobe should appear for $\theta = 37°$. According to the reciprocity theorem, the forward scattering cross section must be the same for directions of incidence 1 and 2. Although the induced current distributions for the two cases are entirely different (see Figs. 7 and 8), the computed values of the forward scattering cross section were found to agree to 0.4 per cent, or 0.02 dB. The optical scattering cross section in the forward direction according to (35) and (36) is $\sigma/\lambda^2 = 16.1$ which agrees reasonably well with the value 15.2 read from Figs. 12 and 13.

The scattering patterns for the disk are shown in Fig. 14. As expected, the patterns exhibit a strong lobe both in the forward direction and in the backward direction. The optical scattering cross section in the forward direction is $\sigma/\lambda^2 = 11.4$ according to (35) and (36). The actual value read from Fig. 14 is 14.1. The discrepancy is principally due to the strong edge effect of the disk.

The numerical results shown so far are for axial incidence only. To illustrate the use of the computer program for nonaxial incidence, we have computed the back-scattering cross section vs. the angle of incidence for a prolate spheroid 2.833λ long and 1.417λ overall diameter. The results are shown in Fig. 15. It is interesting to note that for perpendicular polarization the back-scattering cross section exhibits oscillations. These oscillations can be ascribed to interference between the specularly reflected wave and creeping waves and show up much more strongly for perpendicular polarization than for parallel polarization because the decay exponent for the creeping waves is much smaller for perpendicular polarization than it is for parallel polarization.

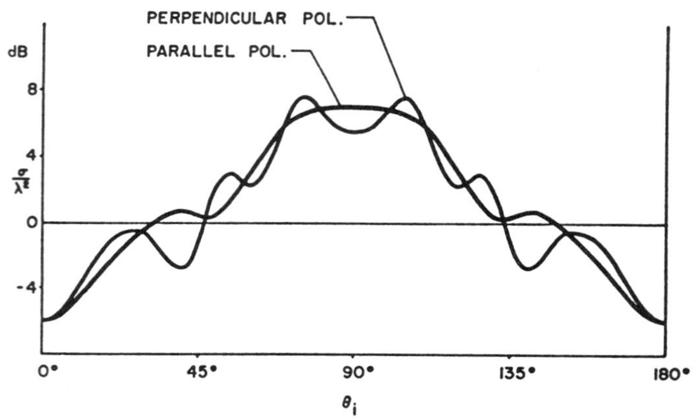

Fig. 15. Back-scattering cross section of prolate spheroid vs. angle of incidence.

The bistatic scattering cross section for the spheroid was computed for several directions. The cross sections for any two directions were found to agree by better than 0.15 dB.

VI. Conclusion

The integral equation method presented here permits the exact evaluation of the scattering of a plane electromagnetic wave by a metallic body of revolution, for any direction of incidence, any polarization and any longitudinal cross section of the body. An analytical solution of this general problem does not exist and there is little hope that such solution can be found. The solution obtained here has been made possible by extensive use of a fast digital computer.

Although the results discussed specifically in this paper pertain to scattering from perfectly conducting bodies, the methods described can be applied with relative ease, and equal effectiveness, to bodies having other boundary conditions, such as coating of lossy dielectric materials.

Acknowledgment

The author wishes to acknowledge many helpful discussions with his colleagues at TRG.

Correction to "Scattering from Bodies of Revolution"

Recent work on the evaluation of scattering from electrically large bodies of revolution has revealed an error in the computer program used to generate the numerical results in the above paper [1]. The error found would affect only azimuth mode numbers $m \geq 5$ and, therefore, will change only the curves in Fig. 15 which shows the back-scattering cross section of a 2-to-1 spheroid. The correct curves replacing Fig. 15 are shown here in Figs. 1 and 2.

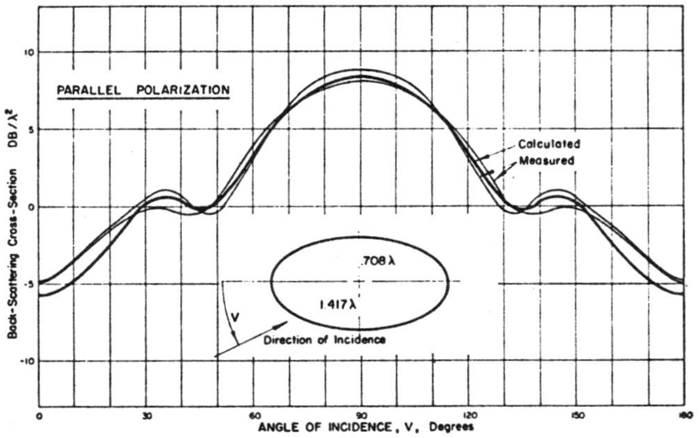

Fig. 1. Back-scattering cross section of 2 to 1 spheroid for parallel polarization.

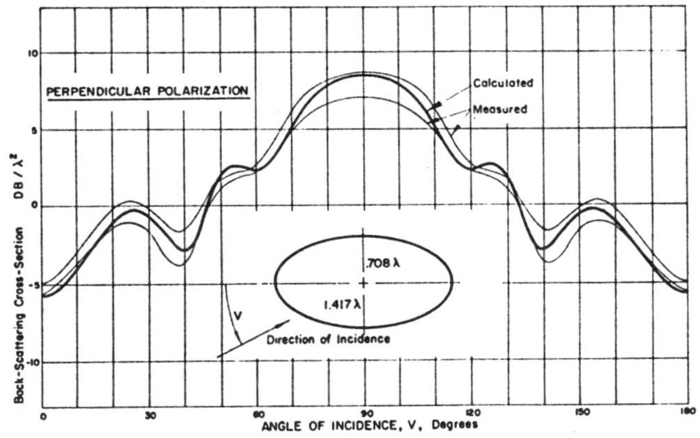

Fig. 2. Back-scattering cross section of 2 to 1 spheroid for perpendicular polarization.

Since the calculated results on the spheroid were first published, we have had an opportunity to check the calculations against experimental measurements. Figures 1 and 2 show two measured curves, one of which actually corresponds to the angle of incidence region from 180 degrees to 360 degrees. The agreement between the calculated results and the measured results is seen to be quite good. The measurements were made at Lockheed Missiles and Space Company, Sunnyvale, Calif. I would like to thank Dr. D. Levine, Lockheed, for the permission to include these measured results with the corrected theoretical results.

The back-scattering cross section for nose-on incidence on the spheroid has been compared also with results obtained by P. C. Waterman [2] by a different theoretical method [3]. The results are found to differ by 0.1 dB only. The nose-on back-scattering cross section has also been calculated by E. M. Kennaugh and D. L. Moffatt [4]. Their method is an approximate method, however, and their result for the above spheroid differs by 2 dB from our result.

Manuscript received May 19, 1966.

M. G. Andreasen
TRG Control Data Corp.
Menlo Park, Calif.

References

[1] M. G. Andreasen, "Scattering from bodies of revolution," *IEEE Trans. on Antennas and Propagation*, vol. AP-13, pp. 303–310, March 1965.
[2] P. C. Waterman, MITRE Corp., Boston, Mass., and M. G. Andreasen, TRG, Menlo Park, Calif., private communication.
[3] P. C. Waterman, "Matrix formulation of electromagnetic scattering," *Proc. IEEE*, vol. 53, pp. 805–812, August 1965.
[4] E. M. Kennaugh and D. L. Moffatt, "Axial echo area of the prolate spheroid," *Proc. IEEE (Correspondence)*, vol. 52, pp. 1252–1253, October 1964.

A Wire-Grid Model for Scattering by Conducting Bodies

JACK H. RICHMOND, SENIOR MEMBER, IEEE

Abstract—**A point-matching solution is developed for scattering by conducting bodies of arbitrary shape. Numerical results are included for the backscatter echo area of circular and square wire loops, circular and square plates, spheres, and hemispheres. The results show good agreement with experimental data. An efficient calculation procedure is achieved by using a wire-grid model instead of a continuous conducting surface. A system of linear equations is generated by enforcing the boundary conditions at the center of each wire segment of the grid, and a digital computer is employed to solve these equations for the currents on the segments. Then it is straightforward to calculate the distant scattered field and the echo area.**

I. Introduction

CONSIDERABLE progress has been achieved in recent months in using the high-speed digital computer to obtain accurate solutions for the scattered fields of various conducting bodies. Contributions in this area have been made by Andreasen [1] for bodies of revolution, by Angelakos and Baghdasarian [2] for circular wire loops, by Richmond [3] for straight wires, and by Waterman [4] for bodies of arbitrary shape.

This paper describes a practical technique for calculating the scattered fields of wire loops, conducting plates, and bodies of arbitrary shape. An efficient calculation procedure is achieved by using a wire-grid model to approximate the continuous conducting surface. A system of linear equations is generated by forcing the tangential electric field intensity to vanish at the center of each wire segment of the grid. These equations are solved with the aid of a digital computer to determine the currents on the wires. Then it is relatively simple to calculate the distant scattered field and the echo area.

The results obtained in this manner are presented in the form of backscatter echo-area curves for circular and square wire loops, circular and square plates, spheres, and hemispheres. The results show good agreement with experimental data.

In the wire-grid model, the elementary scatterer is a short slender wire with a uniform current density on its surface. The scattered field of such an element is considered next.

II. The Field of a Wire Segment with Uniform Surface Current Density

Consider a perfectly conducting wire segment of length "s" and radius "a" to have a harmonic electric current induced uniformly over its surface. Let the axis of the wire coincide with the z axis and let the center of the wire be the coordinate origin as shown in Fig. 1. We assume a time dependence given by $e^{j\omega t}$. If the current density is

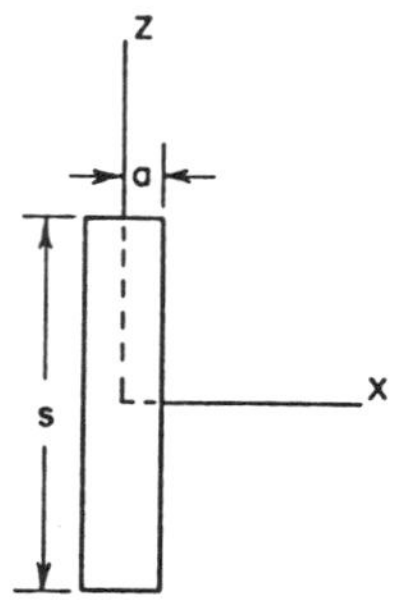

Fig. 1. A wire segment of length s and radius a has uniform current density on its surface.

$$J = \hat{z}J \tag{1}$$

where J is a constant and $\hat{z}$ denotes a unit vector parallel with the z axis, the current in amperes is given by

$$I = 2\pi aJ. \tag{2}$$

In general, the field of this source is expressed as a surface integral over the surface of the wire. In our application, however, the general expressions simplify as follows.

$$E_\rho(\rho, \phi, z) = \frac{\rho I\sqrt{\mu/\epsilon}}{4\pi jk}(1 + jkr)\left.\frac{e^{-jkr}}{r^3}\right|_{r_1}^{r_2} \tag{3}$$

$$E_\phi(\rho, \phi, z) = 0 \tag{4}$$

$$E_z(\rho, \phi, z) = \frac{I\sqrt{\mu/\epsilon}}{4\pi jk} \cdot \int_{-s/2}^{s/2} [2r^2(1 + jkr) - (\rho^2 + a^2)(3 + 3jkr - k^2r^2)]\frac{e^{-jkr}}{r^5}dt \tag{5}$$

where ρ, ϕ and z are the cylindrical coordinates of the observation point,

$$k = \omega\sqrt{\mu\epsilon} \tag{6}$$

$$r = \sqrt{\rho^2 + a^2 + (z - t)^2} \tag{7}$$

$$r_1 = \sqrt{\rho^2 + a^2 + (z + s/2)^2} \tag{8}$$

and

$$r_2 = \sqrt{\rho^2 + a^2 + (z - s/2)^2}. \tag{9}$$

These expressions for the field are accurate if $\rho = 0$ or if

Manuscript received May 3, 1966; revised July 18, 1966. This work was supported in part by the Air Force Systems Command, United States Air Force, under Contract AF 19(628)-4883 with The Ohio State University Research Foundation.

The author is with the Antenna Laboratory, Department of Electrical Engineering, The Ohio State University, Columbus, Ohio.

Reprinted from *IEEE Trans. Antennas Propagat.*, vol. AP-14, no. 6, pp. 782–786, November 1966.

r_1^2 and r_2^2 are large in comparison with the quantity $a\rho$.

Now that suitable expressions are available for the scattered field of a wire segment, we are in a position to describe the point-matching solution for scattering by a wire-grid body.

III. The Point-Matching Solution for Wire-Grid Bodies

The scattered field of a wire-grid body can be generated by the electric surface currents which are induced on the wire segments, radiating in free space. If each wire segment is very short in comparison with the wavelength, it is reasonable to use the approximation that the current density is uniformly distributed over the surface of each segment. In this case, the scattered field of each segment is given by (3) through (9).

In a rigorous solution, the tangential electric field intensity vanishes everywhere on the surface of each perfectly conducting wire segment. If the wire radius is small, it is found that accurate results can be obtained by forcing the tangential electric field to vanish at just one point at the center of each segment.

In order to enforce the boundary conditions, we must develop expressions for the tangential component of the electric field intensity radiated by segment j (with unit current) when the observation point is at the center of segment i. This quantity is denoted by S_{ij} and is called a "scattering coefficient." Suitable expressions for the S_{ij} are derived in the Appendix.

Using a wire-grid model for a conducting body of arbitrary shape, we generate a system of linear equations by setting the tangential electric field intensity equal to zero at the center of each wire segment of the grid. The number of equations N will be equal to the total number of segments, and the ith equation is given by

$$\sum_{j=1}^{N} S_{ij} I_j = -E_i^i \tag{10}$$

where I_j is the current induced on segment j, and E_i^i represents the tangential component of the incident electric field intensity at the center of segment i. Thus, (10) forces the scattered field to cancel the incident field at the center of segment i.

After this system of equations has been solved, the distant scattered field can be calculated by means of the equations given in Section IV.

IV. The Distant Scattered Field

If the observation point (with spherical coordinates r, θ, and ϕ) is at a great distance from the scattering body, the vector potential of the field from segment i is given by

$$\mathbf{A}_i = \hat{\imath} A_i \tag{11}$$

$$A_i = \frac{\mu s_i I_i}{4\pi r} e^{-jkr} e^{jk(x_i \sin\theta\cos\phi + y_i \sin\theta\sin\phi + z_i\cos\theta)} \tag{12}$$

where $\hat{\imath}$ is a unit vector parallel with segment i, s_i is the length of segment i, I_i denotes the current on segment i, and (x_i, y_i, z_i) are the coordinates of the center of segment i. The scattered electric field intensity of the wire-grid body is

$$E_\theta^s = -j\omega \sum_{i=1}^{N} (\cos\theta\cos\phi\cos\alpha_i\cos\beta_i + \cos\theta\sin\phi\cos\alpha_i\sin\beta_i + \sin\theta\sin\alpha_i) A_i \tag{13}$$

$$E_\phi^s = j\omega \sum_{i=1}^{N} (\sin\phi\cos\beta_i - \cos\phi\sin\beta_i) A_i \cos\alpha_i \tag{14}$$

where the angles α_i and β_i specify the orientation of segment i as defined in the Appendix.

Next we give a brief description of a computer program based on this point-matching technique.

V. A Digital-Computer Program

Based on the equations in the Appendix and in Sections II to IV, a digital computer program has been developed for scattering by a wire-grid body of arbitrary shape. The input data consist of a list of the parameters s_i, α_i, β_i, x_i, y_i, and z_i for the various wire segments in the body, and the angles θ and ϕ which specify the direction of propagation of the incident plane wave. Finally, the polarization of the incident wave is specified.

This computer program is useful for calculating the scattering properties of wire loops (open, closed, plane, or nonplanar loops), conducting plates of various shapes, and open or closed conducting surfaces such as the hemisphere and the spheroid. In its present form, the program will handle a maximum of 100 wire segments when an IBM 7094 computer is employed. Thus, solutions can be obtained for bodies with a maximum surface area of approximately one square wavelength. When the conducting body is a figure of revolution, the maximum size is greatly increased.

In the next sections we present some of the echo-area data calculated with the aid of this computer program.

VI. Scattering by Wire Loops

Figures 2 and 3 show the backscatter echo area of circular loops for the broadside aspect and for oblique incidence, respectively. The circular loop was approximated by a regular polygon with 32 sides. In Fig. 2 it may be noted that the calculations show satisfactory agreement with the experimental data of Kouyoumjian [5]. The data in Fig. 3 also show satisfactory agreement with the few experimental measurements that are available. That is, the calculations for ϕ polarization show that the echo area for the broadside aspect $(\theta=0)$ is 16.25 dB above that for the edge aspect $(\theta=90°)$, in close agreement with the 16.3-dB figure reported by Kouyoumjian [5].

Figure 4 shows the echo area of square loops for the broadside aspect, calculated with the computer program described above. The square loop was divided into 40 wire segments of equal length. No experimental data are available for comparison, but the results in Fig. 4 appear to be reasonable in every respect. From a comparison of Figs. 2 and 4 it is found that square and circular loops of equal area have nearly the same backscatter echo area for the broadside aspect.

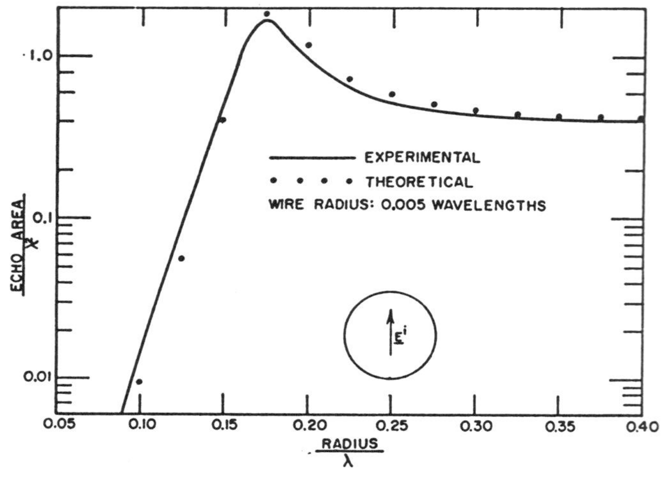

Fig. 2. Backscatter echo area of circular wire loops at the broadside aspect.

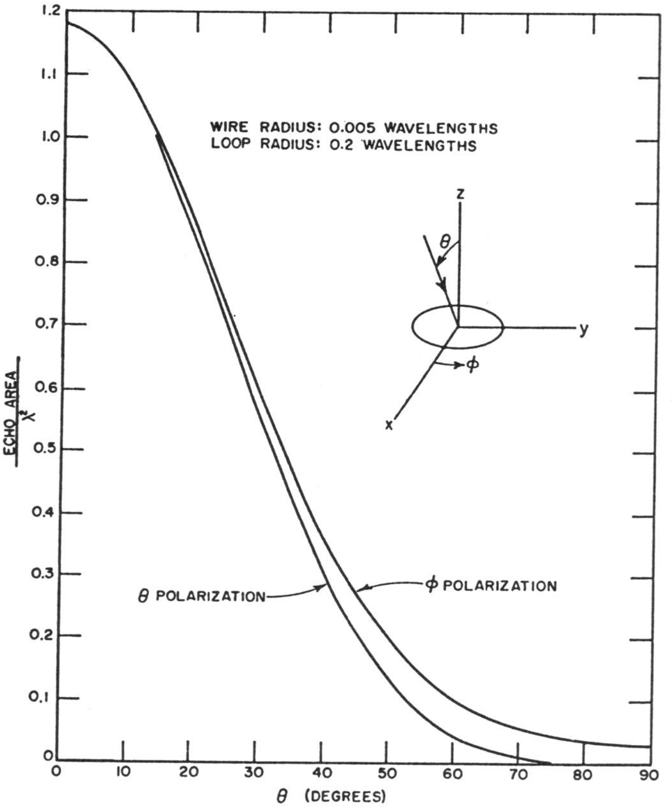

Fig. 3. Calculated backscatter echo area of circular wire loop vs. angle of incidence θ. The axis of propagation of the incident wave is in the xz plane and the loop lies in the xy plane.

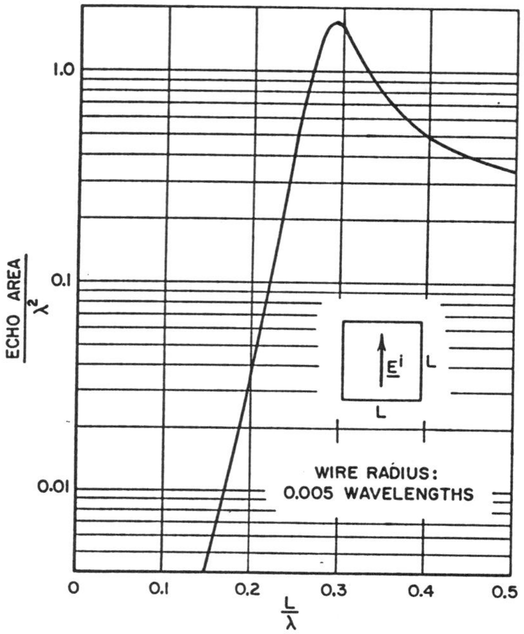

Fig. 4. Calculated backscatter echo area of square wire loops at the broadside aspect.

VII. Scattering by Conducting Plates

Figures 5 and 6 show the backscatter echo area of circular and square plates for the broadside aspect, calculated by the point-matching technique. It may be noted that the calculations show reasonably good agreement with the experimental data reported by Kouyoumjian [6].

As shown in Fig. 5, the circular plate was modeled with a system of regular polygons (with 20 sides) and radial wires. The wire radius was taken to be 0.005λ. In each case, the radii of the polygons were taken to be 0.05λ, 0.10λ, 0.15λ, etc. The radius of the equivalent circular plate is taken to be the same as the radius of the largest polygon in the wire-grid model. Thus, a circular plate of radius 0.15λ is modeled with a wire grid consisting of three polygons and two rings of radial wire segments. This wire grid has a hole in the center of radius 0.05λ. Evidently a hole of this size has little effect on the backscattering properties of a circular plate or the corresponding wire-grid model if the outer radius is 0.15λ or more.

As shown in Fig. 6, the square plate was modeled with a wire grid consisting of 8 horizontal wires and 8 vertical wires. The wire diameter was taken to be 1/50 of the width L.

VIII. Scattering by Bodies of Revolution

Figures 7 and 8 show the calculated backscatter echo area of perfectly conducting spheres and hemispheres. It may be noted in Fig. 7 that the point-matching solution shows satisfactory agreement with the rigorous solution for the sphere. The calculated phase angle of the backscattered field agrees with the rigorous solution for the sphere with an error of less than 4°.

The calculations for the sphere are based on a wire-grid model consisting of an array of regular polygons (with 20 sides) and conical rings of straight wire segments as shown in Fig. 7. The number of polygons is $70R/\lambda$ where R denotes the radius of the sphere. The wire radius is 0.005λ. The calculations were performed with an IBM 7094 computer. It took 0.4 minute to obtain the solution for a sphere of radius 0.2λ and 2.9 minutes for a radius of 0.4λ.

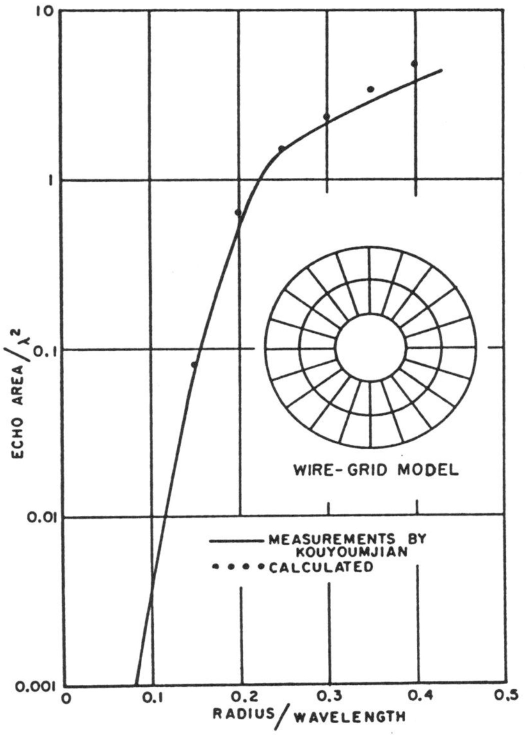

Fig. 5. Backscatter echo area of thin circular conducting plates for the broadside aspect.

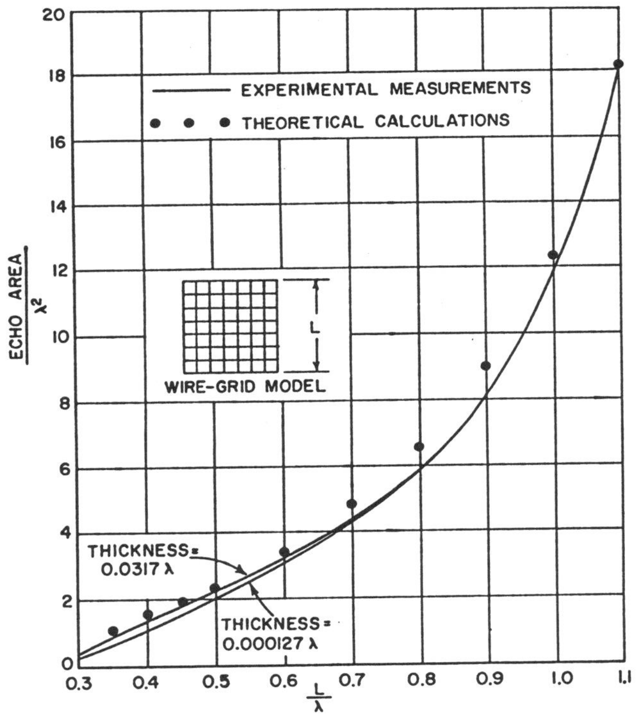

Fig. 6. Echo area of perfectly conducting square plates at the broadside aspect.

The results shown in Fig. 8 agree with measured data reported by Blore and Musal [7] for a plane wave incident on the convex surface of a conducting hemisphere. No published data appear to be available for concave hemispheres with radii less than 0.4λ. The calculation

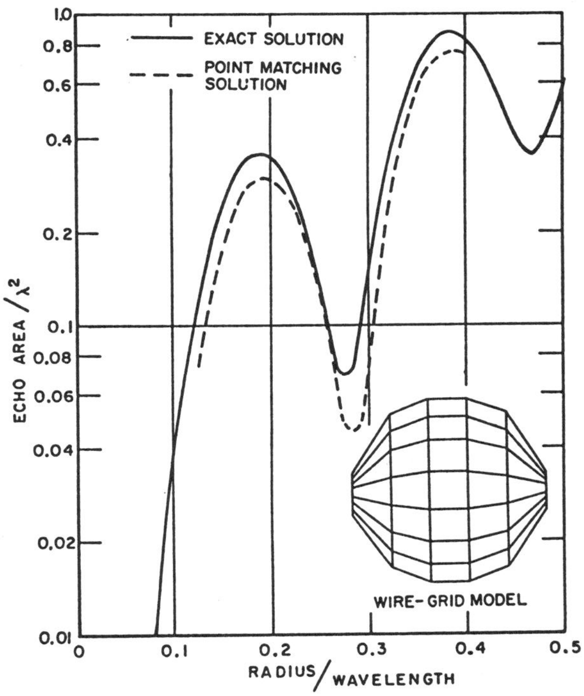

Fig. 7. Backscatter echo area of perfectly conducting spheres.

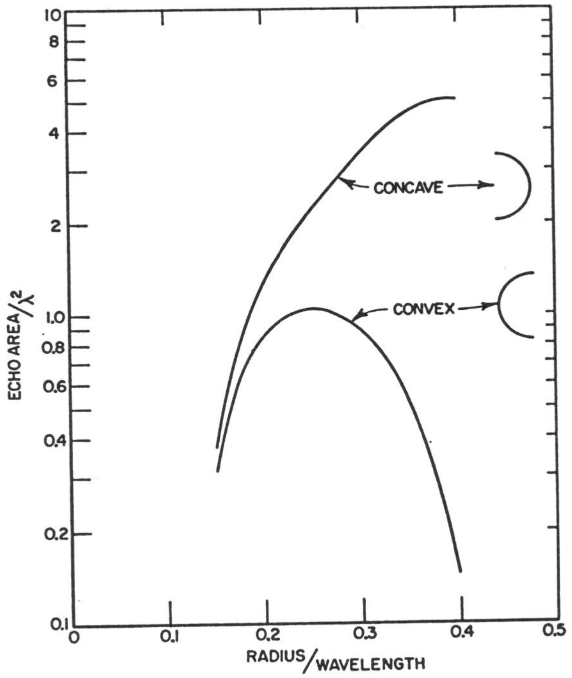

Fig. 8. Calculated backscatter echo area of perfectly conducting hemispheres.

time was 6.2 minutes for each of the curves in Fig. 8.

The solutions for the spheres, hemispheres, and circular plates were obtained with the aid of a single computer program which will handle an arbitrary body of revolution for on-axis incidence. A considerable saving in computation time and cost is achieved in this case by noting that the current varies as cos ϕ on each polygon and sin ϕ on each conical ring [1]. (The angle ϕ is measured from the x axis toward the y axis, and the incident wave is assumed to propagate in the z direction and to be polarized in the y direction. Each polygon lies in a plane which is parallel with the xy plane.)

Appendix

The Scattering Coefficients

Equations (3) through (9) are useful for calculating the scattering coefficients S_{ij}. Before using these equations, however, we must calculate the cylindrical coordinates of the center of segment i in a system having its origin at the center of segment j and its z axis parallel with the axis of segment j.

Let the points at the centers of segments i and j have rectangular coordinates (x_i, y_i, z_i) and (x_j, y_j, z_j). The distance between these points is

$$r_{ij} = \sqrt{x_{ij}^2 + y_{ij}^2 + z_{ij}^2} \tag{15}$$

where

$$x_{ij} = x_i - x_j, \quad y_{ij} = y_i - y_j, \quad \text{and} \quad z_{ij} = z_i - z_j. \tag{16}$$

The radial vector extending from point j to point i is

$$\mathbf{r}_{ij} = \hat{x}x_{ij} + \hat{y}y_{ij} + \hat{z}z_{ij}. \tag{17}$$

Let $\hat{\imath}$ and $\hat{\jmath}$ denote unit vectors parallel with the axes of segments i and j, respectively. The unit vector $\hat{\imath}$ can be specified by its angle α_i with respect to the xy plane and the angle β_i between the x axis and the projection of $\hat{\imath}$ on the xy plane. Thus,

$$\hat{\imath} = \hat{x} \cos \alpha_i \cos \beta_i + \hat{y} \cos \alpha_i \sin \beta_i - \hat{z} \sin \alpha_i. \tag{18}$$

The unit vector $\hat{\jmath}$ is given by a similar expression in terms of α_j and β_j.

At this point it is convenient to define a rotated cylindrical coordinate system (ρ', ϕ', z') with an origin at (x_j, y_j, z_j) and with the z' axis coinciding with the axis of segment j. The pertinent coordinates of the center of segment i in this system are given by

$$z' = x_{ij} \cos \alpha_j \cos \beta_j + y_{ij} \cos \alpha_j \sin \beta_j - z_{ij} \sin \alpha_j \tag{19}$$

$$\rho' = \sqrt{r_{ij}^2 - z'^2}. \tag{20}$$

The field of segment j at an observation point at the center of segment i can be written as

$$\mathbf{E}_{ij} = \hat{\rho}' E_\rho' + \hat{\jmath} E_z' \tag{21}$$

where

$$\hat{\rho}' = (\mathbf{r}_{ij} - z'\hat{\jmath})/\rho' \tag{22}$$

and E_ρ' and E_z' are given by (3) through (9) with ρ and $\hat{z}$ replaced by ρ' and z'.

Finally, the scattering coefficient S_{ij} is the component of $\mathbf{E}_{ij}$ which is tangential to segment i:

$$S_{ij} = \hat{\imath} \cdot \mathbf{E}_{ij}. \tag{23}$$

From (18), (21), and (23),

$$\begin{aligned} S_{ij} = (E_z' - z'E_\rho'/\rho')&(\cos \alpha_i \cos \beta_i \cos \alpha_j \cos \beta_j \\ &+ \cos \alpha_i \sin \beta_i \cos \alpha_j \sin \beta_j + \sin \alpha_i \sin \alpha_j) \\ &+ E_\rho'(x_{ij} \cos \alpha_i \cos \beta_i + y_{ij} \cos \alpha_i \sin \beta_i \\ &- z_{ij} \sin \alpha_i)/\rho'. \end{aligned} \tag{24}$$

References

[1] M. G. Andreasen, "Scattering from bodies of revolution," *IEEE Trans. on Antennas and Propagation*, vol. AP-13, pp. 303–310, March 1965.

[2] A. Baghdasarian and D. J. Angelakos, "Scattering from conducting loops and solution of circular loop antennas by numerical methods," *Proc. IEEE*, vol. 53, pp. 818–822, August 1965.

[3] J. H. Richmond, "Digital computer solutions of the rigorous equations for scattering problems," *Proc. IEEE*, vol. 53, pp. 796–804, August 1965.

[4] P. C. Waterman, "Matrix formulation of electromagnetic scattering," *Proc. IEEE*, vol. 53, pp. 805–812, August 1965.

[5] R. G. Kouyoumjian, "Back-scattering from a circular loop," The Ohio State University Engineering Experiment Station, Columbus, Bull. 162, November 1956.

[6] ——, "The calculation of the echo areas of perfectly conducting objects by the variational method," Ph.D. dissertation, The Ohio State University, Columbus, pp. 63–73, November 1953.

[7] W. E. Blore and H. M. Musal, "The radar cross section of metal hemispheres, spherical segments, and partially capped spheres," *IEEE Trans. on Antennas and Propagation (Communications)*, vol. AP-13, pp. 478–479, May 1965.

Electromagnetic Modeling of Composite Wire and Surface Geometries

E. H. NEWMAN, MEMBER, IEEE, AND D. M. POZAR, STUDENT MEMBER, IEEE

***Abstract*—A moment method solution to the problem of radiation or scattering from geometries consisting of open or closed surfaces, wires, and wire/surface junctions is presented. The method is based on the sinusoidal reaction formulation. Several examples of input impedance calculations illustrate the versatility, accuracy, and computational efficiency of the method.**

I. INTRODUCTION

THE METHOD of moments is perhaps the most widely used tool for the electromagnetic modeling of bodies which are not large in terms of a wavelength. The advantages of moment methods are accuracy, versatility, and the ability to compute near- as well as far-zone parameters.

The most widely used forms of the method of moments are the thin-wire computer programs [1]-[3]. These programs are ideal for modeling most wire antennas. By forming a wire-grid model they also can be used to model solid surfaces. Unfortunately, the size of the surface which can be modeled practically is severely limited since it requires many wires to accurately model a solid surface. Further, even a very fine wire mesh may not yield accurate near-zone parameters, such as current distribution or impedance.

More recently, surface-patch models have been developed for modeling solid surfaces. The advantage of the surface patch solution is that fewer unknowns are required per square wavelength of surface area. Wang *et al.* [4] have developed a model for wires and plates based on the sinusoidal reaction formulation. Albertsen *et al.* [5] used pulse expansion modes to model wires and plates, including the case where a wire touches a plate. Burke and Poggio have incorporated these results into a user-oriented computer code [3]. Parhami *et al.* have treated the problem of a wire surface junction using the finite difference technique [6].

The purpose of this paper is to present a technique for modeling wires and surfaces, including the case where the wires contact the surfaces, which is sufficiently accurate to compute impedance. The technique is based on the sinusoidal reaction formulation and is applicable to open as well as closed surfaces. Continuity of current is enforced on the wires, on the surfaces and at the wire/surface junctions. The accuracy and versatility of the model is demonstrated by several numerical examples. Attachment points are restricted to be somewhat removed from an edge.

II. THEORY

The Reaction Method

Consider a geometry of arbitrarily shaped scatterers in a homogeneous medium as shown in Fig. 1. S is the surface enclosing the scatterers, and $\hat{n}$ is the unit outward normal to S. The scatterers may consist of solid surfaces and thin wires, and the wires may contact the surfaces. For now, the surfaces will be considered closed. The sources $(\mathbf{J}_i, \mathbf{M}_i)$ generate the fields $(\mathbf{E}, \mathbf{H})$ in the presence of the scatterers. The $e^{j\omega t}$ time dependence is suppressed. From the surface-equivalence theorem the field interior to the surface S will vanish without changing the exterior fields if the surface current densities

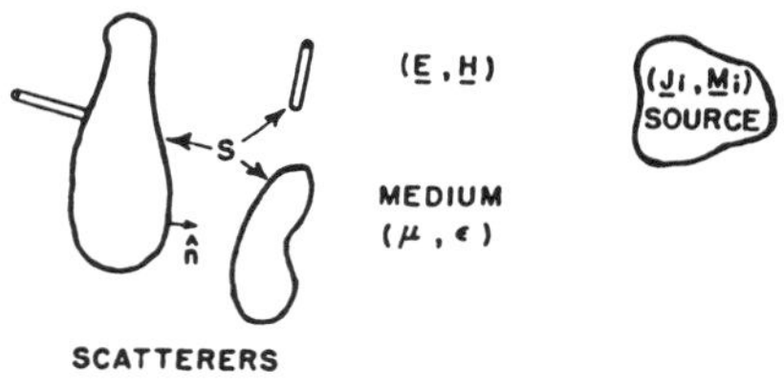

Fig. 1. General problem geometry.

$$\mathbf{J}_s = \hat{n} \times \mathbf{H}, \tag{1}$$

$$\mathbf{M}_s = \mathbf{E} \times \hat{n} \tag{2}$$

are introduced on the surface S. The scattering obstacles may then be replaced by the ambient medium without altering the field anywhere. The scattered field, radiated by $(\mathbf{J}_s, \mathbf{M}_s)$ in the ambient medium, is defined as

$$\mathbf{E}_s = \mathbf{E} - \mathbf{E}_i \tag{3}$$

$$\mathbf{H}_s = \mathbf{H} - \mathbf{H}_i. \tag{4}$$

An electric test source, which radiates the fields $(\mathbf{E}_T, \mathbf{H}_T)$ in the ambient medium, is now placed in the interior region of S. Noting that this test source has zero reaction with the sources $(\mathbf{J}_s, \mathbf{M}_s)$ and $(\mathbf{J}_i, \mathbf{M}_i)$, the reaction integral equation [4] is obtained:

$$\iint_S (\mathbf{J}_s \cdot \mathbf{E}_T - \mathbf{M}_s \cdot \mathbf{H}_T)\, ds + \iiint_V (\mathbf{J}_i \cdot \mathbf{E}_T - \mathbf{M}_i \cdot \mathbf{H}_T)\, dv = 0 \tag{5}$$

where the volume integral is over the source volume. The reaction equation (5) is used to determine the unknown surface currents $(\mathbf{J}_s, \mathbf{M}_s)$. For conductors of finite conductivity the impedance boundary condition

$$\mathbf{M}_s = Z_s \mathbf{J}_s \times \hat{n} \tag{6}$$

is used where Z_s may be a function of position. For simplicity, only perfect conductors are considered here; thus $Z_s = \mathbf{M}_s = 0$.

In deriving (5) it was assumed that all surfaces were closed so that Schelkunoff's equivalence theorem could be used. However, it can be shown that (5) applies equally well for open surfaces. An open surface, such as a fictitious plate of

Manuscript received October 14, 1977; revised February 23, 1978. This work was supported in part by the U.S. Army Research Office and in part by the Ohio State University Research Foundation, under Grant DAAG29-76-G-0331.

The authors are with the ElectroScience Laboratory, Department of Electrical Engineering, the Ohio State University, Columbus, OH 43212.

Reprinted from *IEEE Trans. Antennas Propagat.*, vol. AP-26, no. 6, pp. 784–788, November 1978.

zero thickness, can be considered to be a limiting case of a real plate of finite thicknes as the plate thickness goes to zero. In general, different currents exist on the top and bottom surfaces of the real plate. As the plate thickness goes to zero, the fields radiated by the top and bottom currents become identical to the fields radiated by a single surface current located on the plate center. This single surface current is the vector sum of the top and bottom surface currents and is the current which must be determined to treat open surfaces. If electric test sources are used in (5), then $\mathbf{J}_s$ will be the vector sum of the current on the top and bottom surfaces [7].

The integral equation (5) is solved by the moment method. The unknown current $\mathbf{J}_s$ is expanded in a set of N basis (expansion) functions,

$$\mathbf{J}_s = \sum_{n=1}^{N} I_n \mathbf{J}_n, \tag{7}$$

and (5) is enforced for N electric test sources placed in S. Thus (5) reduces to the set of simultaneous linear equations

$$\sum_{n=1}^{N} I_n Z_{mn} = V_m; \qquad m = 1, 2 \cdots N \tag{8}$$

where

$$Z_{mn} = -\iint_n \mathbf{J}_n \cdot \mathbf{E}_m \, ds \tag{9}$$

$$V_m = \iiint_V (\mathbf{J}_i \cdot \mathbf{E}_m - \mathbf{M}_i \cdot \mathbf{H}_m) \, dv \tag{10}$$

where $(\mathbf{E}_m, \mathbf{H}_m)$ are the fields of the mth test source radiating in the medium (μ, ϵ) and the integration in (9) is over the surface of the nth expansion mode.

The expansion and test modes used here are identical. Thus the method is a Galerkin method, and a symmetric impedance matrix $[Z]$ results. The form of the expansion/test functions will now be defined. It is this form which ultimately determines the accuracy and efficiency of the solution as well as the types of geometries which may be modeled.

Expansion/Testing Functions

Three basic types of modes are used: wire dipole modes, surface dipole modes, and a special attachment mode whenever a wire connects to a surface. With this choice of functions geometries consisting of flat surfaces, thin wires, and wire-surface connections may be modeled. A piecewise flat approximation can be made to model singly curved surfaces. Note that all of the modes will involve sinusoids with the free-space wavenumber. This allows all but one of the integrations required to find the $\mathbf{E}_m$ to be done in closed form, thus avoiding the very difficult $1/r^3$ singularity or second derivative associated with finding the extreme near-zone fields of an electric current source, as is required to evaluate self or overlapping impedance elements Z_{mn}. The one integration not available in closed form is associated with the disk component of the attachment mode, as discussed below.

1) Thin-Wire Mode: The wire mode used is identical to that used by Richmond [1]. It is a piecewise-sinusoidal V-dipole consisting of two sinusoidal monopoles. A V-dipole with a 180° internal angle lying on the z-axis is shown in Fig. 2(a).

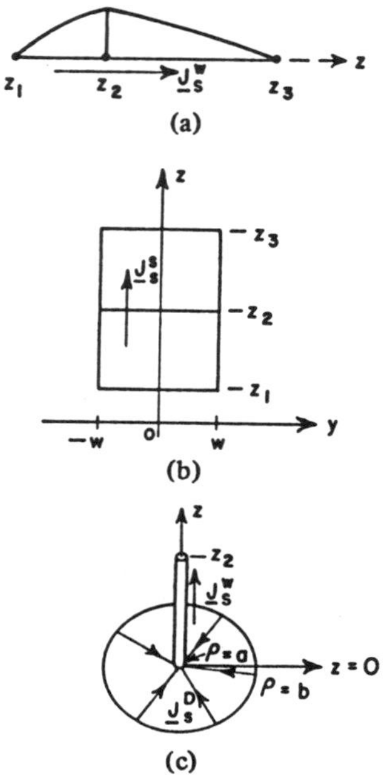

Fig. 2. Expansion and test modes. (a) Thin-wire V-dipole. (b) Sinusoidal surface V-dipole. (c) Attachment dipole.

Using the thin-wire approximation the current on this dipole is given by

$$\mathbf{J}_s^w = \frac{\hat{z}}{2\pi a}\left[P_1 \frac{\sin k(z - z_1)}{\sin k(z_2 - z_1)} + P_2 \frac{\sin k(z_3 - z)}{\sin k(z_3 - z_2)}\right] \tag{11}$$

where P_1 and P_2 represent pulse functions with unit value when $z_1 < z < z_2$ and $z_2 < z < z_3$, respectively, and are zero elsewhere. Also, a is the wire radius and $k = 2\pi/\lambda$. This choice of mode has the advantages that the near-zone fields and the Z_{mn} are known in closed form [8]. These modes are placed in an overlapping array on the wire ensuring continuity of current on the wire.

2) Surface-Patch Mode: The surface-patch mode is a surface V-dipole consisting of two sinusoidal surface monopoles. A surface V-dipole with interior angle of 180° is shown in Fig. 2(b). The current on this dipole is given by

$$\mathbf{J}_s^s = \hat{z}\,\frac{kP_1 \sin k(z - z_1) \cos ky}{2 \sin k(z_2 - z_1) \sin kw} + \hat{z}\,\frac{kP_2 \sin k(z_3 - z) \cos ky}{2 \sin k(z_3 - z_2) \sin kw} \tag{12}$$

where P_1 and P_2 represent unit pulse functions as before. This mode is similar to that used by Wang *et al.* [4] except that the sinusoidal variation is with the speed of light and a cosine variation is used transverse to $\mathbf{J}_s^s$ instead of the constant variation used by Wang *et al.* Because of this the fields due to the current of (12) may be found in closed form [9]. Two orthogonal and overlapping arrays of the surface-patch modes are placed on the surface, allowing a two-dimensional vector surface current density. Results presented in the next section indicate that reasonable accuracy can be obtained with $2w$ or $z_2 - z_2$ or $z_3 - z_2$ as large as 0.25 λ.

3) Attachment Mode: When a wire is attached to a surface a special attachment mode is introduced. The purpose of the attachment mode is twofold: first is to establish continuity of current at the wire/surface junction and second is to insure that in the immediate vicinity of the attachment the surface current density has the proper $\hat{\rho}$ polarization and $1/\rho$ dependence. This mode consists of two parts: a wire monopole and a disk monopole, as shown in Fig. 2(c). The wire monopole current density is similar to the thin-wire monopole mode:

$$\mathbf{J}_s^w = \frac{1}{2\pi a}\,\frac{\sin k(z_2 - z)}{\sin kz_2}\,\hat{z}. \tag{13}$$

Thus

$$\mathbf{J}_s^D = \frac{-\sin k(b-\rho)}{2\pi\rho\,\sin k(b-a)}\,\hat{\rho}, \qquad a \leqslant \rho \leqslant b \tag{14}$$

where a, b are the inner and outer radii of the annulus. Note that the total current on the disk at $\rho = a$ is equal to the total current on the wire segment at $z = 0$ insuring continuity of current at the attachment. Also observe that the $\sin k(b - \rho)$ function in the numerator of (14) forces the disk current to be zero at the disk edge, $\rho = b$. It is this property which allows the disk to be placed on the surface and still maintain continuity of current on the surface. The $\sin k(b - \rho)$ function was chosen (rather than, say, $b - \rho$ or $\sin \alpha(b - \rho)$) since this permits the fields of the disk to be obtained with only one numerical integration.

The attachment mode is applied by placing it directly on the surface-patch modes wherever the wire meets the surface. Attachment points are not restricted to be in the center of or at the corners of the surface patch modes. However, a more detailed treatment of the attachment mode, including the edge singularity, would be required for attachment points less than about 0.1 λ from an edge. The inner radius a of the annulus corresponds to the wire radius, and results presented below indicate that the outer radius b has little effect on the final result if it is chosen to be between 0.1 λ to 0.25 λ (see Fig. 3).

With these expansion functions, current continuity in the direction of $\mathbf{J}_s$ is always maintained, although the current may not be continuous in directions orthogonal to the direction of current flow. Thus $\nabla \cdot \mathbf{J}_s = -j\omega\rho_s$ is always finite, and no line charges appear on the surface.

For antennas, feeds may be placed at the endpoints of any wire segment, including the attachment segment. For example, a monopole on a ground plane may be modeled by inserting a generator at the base of the monopole between the attachment mode wire segment and disk. In the work presented here the delta-gap feed was used, although the magnetic frill model also could be used.

The scatterer or antenna geometry is thus modeled by N_w wire modes, N_s surface-patch modes, and N_A attachment modes. Then, $N = N_w + N_s + N_A$, and (8) is solved for the unknown current samples I_n by standard Gauss–Jordan elimination.

III. NUMERICAL EXAMPLES

Several numerical examples will now be presented which illustrate the accuracy and versatility of the solution presented in Section II. In each case, the number of modes used is summarized in Table I. Note that N_s is the total number of surface-patch dipoles. Thus there are $N_s/2$ dipoles in each of the two orthogonal directions on the surface. The data presented are input impedance or admittance, since this tends to be a sensitive indicator of overall accuracy.

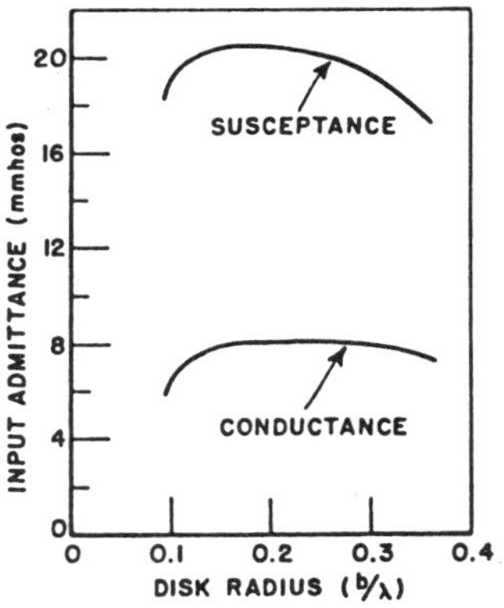

Fig. 3. Input admittance for geometry of Fig. 4, with $N = 14$, versus disk radius. $f = 160$ MHz.

TABLE I
PARAMETERS OF CALCULATIONS

Figure/Curve	N_W	N_S	N_A	N (Total Unknowns)
3/ — — —	2	40	1	43
3/ ———	1	12	1	14
3/ ----	121	-	-	121
5/• • •	2	12	1	15
6/• • •	4	47	1	52
7/ ———	4	12	1	17
8/ ———	4	12	2	18
9/ ———	7	77	3	87

In the experimental models the wires were brass rods, 0.0008 m in radius, with a conductivity of 15 × 10^6 mho/m. The finite conductivity of the wires was taken into account in the manner used by Richmond [10]. The metal plates were made of brass or aluminum, and their finite conductivity was not taken into account. In all cases the driven wire element was driven with a coaxial feed with outer radius 0.0047 m. All measurements were made by the authors.

The calculated results were all made using the same general purpose computer code. No advantage of inherent symmetries in the geometries was made. The resulting symmetries in the expansion mode currents offered a useful check on the calculations.

Fig. 4 presents the input admittance for a monopole on a square finite ground plane versus frequency in the vicinity of the first resonance. The calculation was done with $N = 14$ and with 43 unknowns and is compared with measured data and with a wire-grid model which used 121 unknowns. This example shows the accuracy and rapid convergence of the patch solution as well as a savings in storage over the wire-grid model. The run time for the $N = 14$ case was about 25 min on a Datacraft 6024/3, which is equivalent to about 2–3 min on an IBM 370-165.[1] In comparison the wire-grid model took about 40 min on the Datacraft 6024/3.

Fig. 3 shows the effect of the attachment disk radius b on the input admittance of the same geometry as Fig. 4. This graph shows that the disk radius is not critical and usually can be selected between 0.1 λ → 0.25 λ.

The impedance versus height of a monopole on a 1.4 λ square ground plane is shown in Fig. 5. These calculations are compared with the input impedance of a monopole of the same height on an infinite ground plane computed by image

[1] Recent improvements in the technique and computer code have reduced the run time by about a factor of 6, while increasing accuracy.

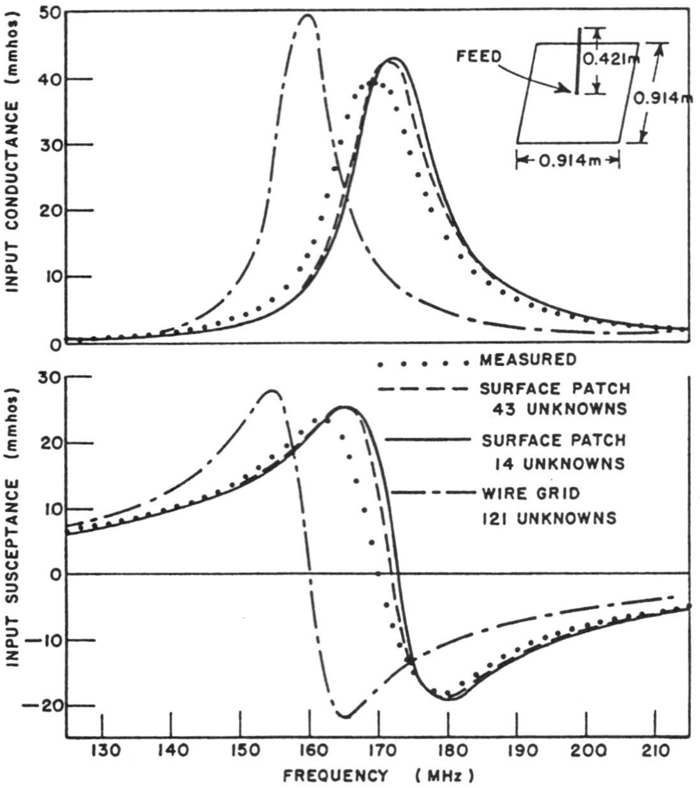

Fig. 4. Input admittance versus frequency for monopole (h = 0.421 m, a = 0.0008 m) on square ground plane (9.914 × 0.914 m), compared with measured data and wire-grid modeling. b = 0.2 λ.

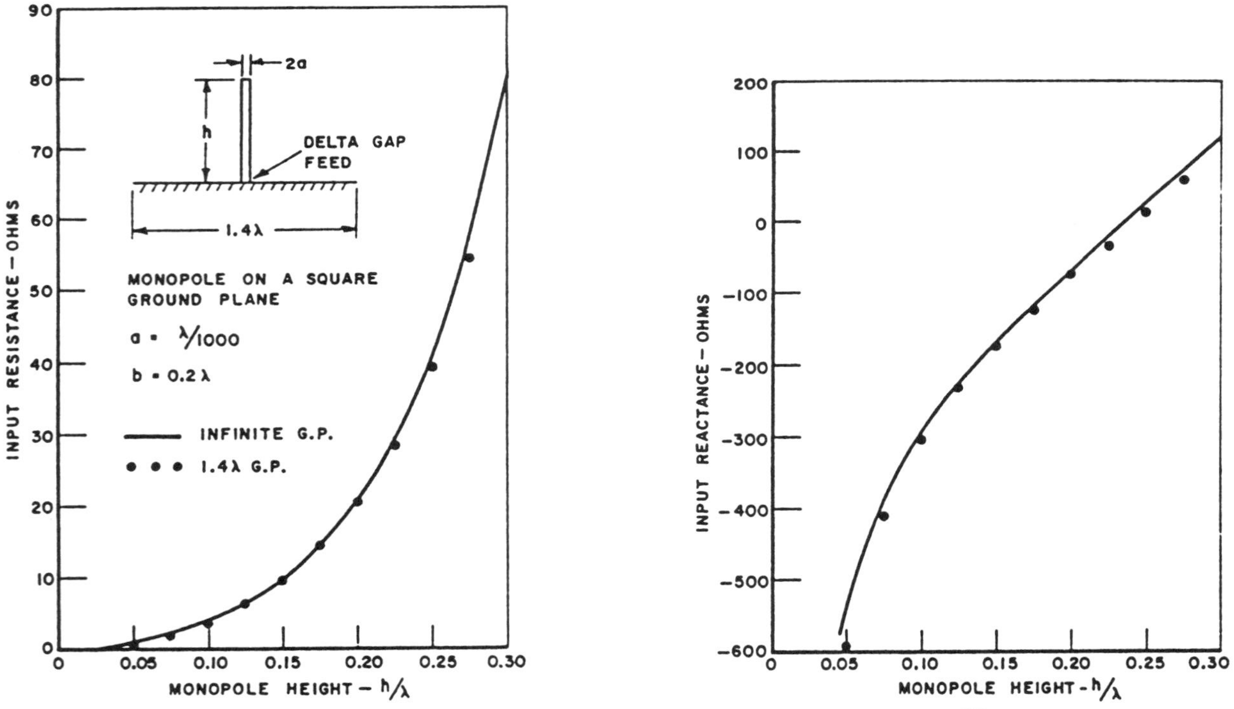

(a) (b)

Fig. 5. (a) Input resistance versus monopole height for monopole centered on 1.4 λ square ground plane compared with same monopole on infinite ground plane as computed by image theory. (b) Input reactance versus monopole height for monopole centered on 1.4 λ square gound plane compared with same monopole on infinite ground plane, as computed by image theory.

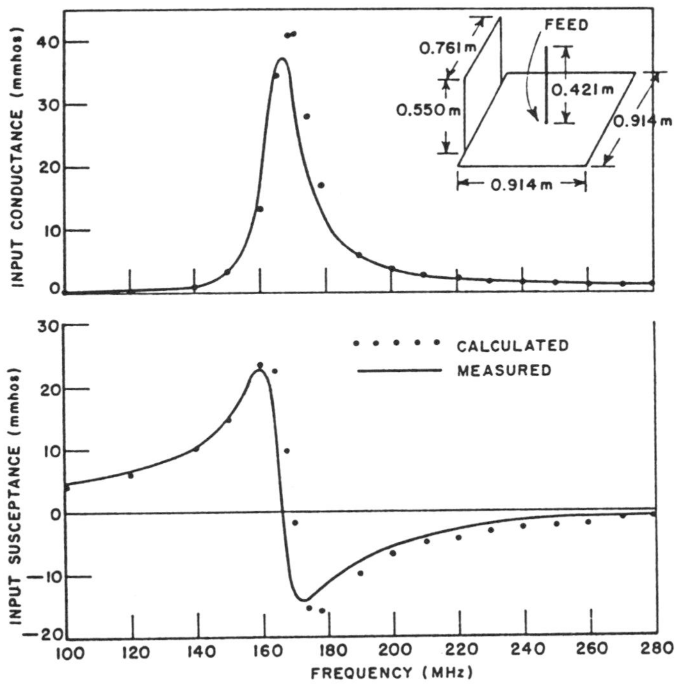

Fig. 6. Input admittance versus frequency for geometry of Fig. 4 with attached reflected plate compared with measured values.

theory. Since the 1.4 λ ground plane is fairly large, good agreement is obtained.

Fig. 6 shows the input admittance for a geometry identical to that of Fig. 3, except that a reflecting plate is attached to one side of the ground plane. Note that the width of the reflecting plate differs from that of the ground plane and that the theory and measurements are in close agreement. Overlap or hinge surface-patch dipoles are used to insure continuity of current at the plate to plate junction.

Fig. 7 shows the input admittance of the geometry of Fig. 3 with the addition of a parasitic wire element 1.5 cm above the ground plane and parallel to the fed monopole. The admittance is shown as a function of the separation d of the two wires. Fig. 8 shows a similar geometry except that the parasitic element now touches the ground plane.

Fig. 9 shows the measured and calculated input impedance of a T-bar fed, rectangular cavity-backed slot antenna opening into a finite ground plane.

IV. CONCLUSION

A moment method solution to the problem of radiation or scattering from geometries consisting of wires, open or closed surfaces, and wire/surface junctions has been presented. The technique is a Galerkin solution based on the sinusoidal reaction formulation. The expansion and test modes were chosen to obtain continuity of current on the composite structure, the proper $\hat{\rho}/\rho$ behavior of the surface current density at attachment points, and facilitate the numerical evaluation of the impedance matrix. Numerical examples presented illustrate the accuracy, versatility, stability, and computational efficiency of the method. Future work will center on further improving the speed and computational efficiency of the method and adding the presence of dielectrics.

ACKNOWLEDGMENT

The authors wish to acknowledge the assistance of J. H. Richmond for helpful discussions and computer subroutines.

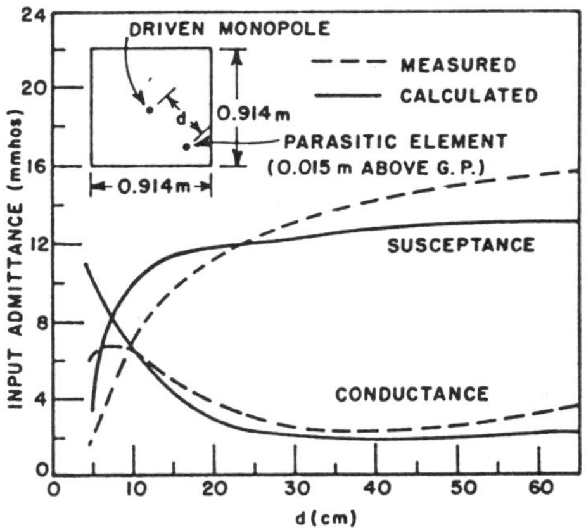

Fig. 7. Input admittance versus separation of driven monopole and parasitic wire element 0.015 m above the 0.0914 × 0.914 m ground plane. Driven monopole height = 0.421 m. Parasitic wire length = 0.842 m. f = 150 MHz.

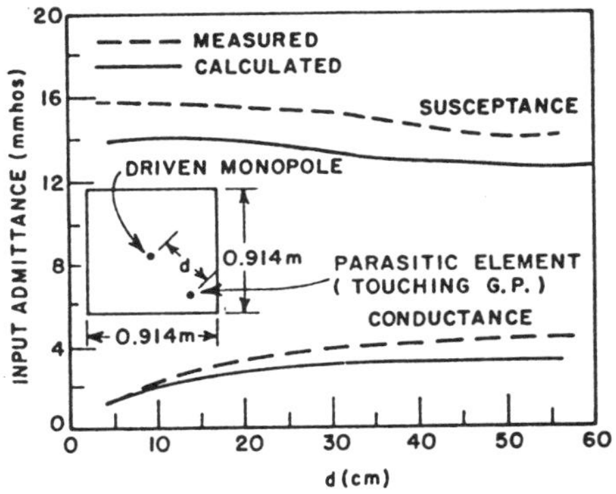

Fig. 8. Input admittance versus separation of driven monopole and parasitic wire element touching 0.914 × 0.914 m ground plane. Driven monopole height = 0.421 m. Parasitic wire height = 0.842 m. f = 150 MHz.

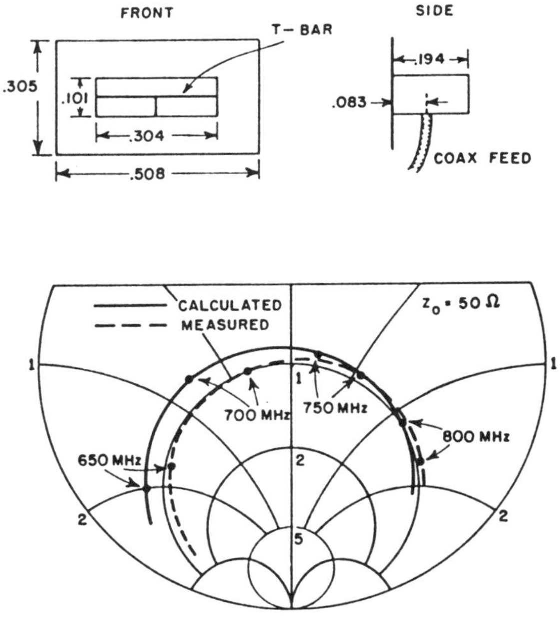

Fig. 9. Geometry and input impedance of T-bar fed slot antenna. Note: dimensions in meters.

REFERENCES

[1] J. H. Richmond, "A wire-grid model for scattering by conducting bodies," *IEEE Trans. Antennas Propagat.*, vol. AP-14, pp. 782–786, Nov. 1966.

[2] H. H. Chao and B. S. Strait, "Computer programs for radiation and scattering by arbitrary configurations of bent wires," Scientific Rep. 7 AFCRL-70-0374, Syracuse Univ., Syracuse, NY, Sept. 1970.

[3] G. J. Burke and A. J. Poggio, "Numerical electromagnetic code—Method of moments," Naval Ocean Systems Center tech. document 116, AFWL-TR-76-320, July 1977.

[4] N. N. Wang, J. H. Richmond, and M. C. Gilreath, "Sinusoidal reaction formulation for radiation and scattering from conducting surfaces," *IEEE Trans. Antennas and Propagat.*, vol. AP-23, pp. 376–382, May 1975.

[5] N. C. Albertsen, J. E. Hansen, and N. E. Jensen, "Computation of radiation from wire antennas on conducting bodies," *IEEE Trans. Antennas and Propagat.*, vol. AP-22, pp. 200–206, Mar. 1974.

[6] P. Parhami, Y. Rahmat-Samii, and R. Mittra, "Investigation of antennas on a finite ground plane," *AP-S Int. Symp.*, Univ. of Massachusetts Amherst, Oct. 11–15, 1976, pp. 511–514.

[7] W. A. Davis and R. Mittra, "A new approach to the thin scatterer problem using the hybrid equations," *IEEE Trans. Antennas and Propagat.*, vol. AP-25, May 1977.

[8] J. H. Richmond and N. H. Geary, "Mutual impedance of nonplanar-skew sinusoidal dipoles," *IEEE Trans. Antennas and Propagat.*, vol. AP-23, May 1975.

[9] J. H. Richmond, D. M. Pozar, and E. H. Newman, "Rigorous near-zone field expressions for rectangular sinusoidal monopole," *IEEE Trans. Antennas and Propagat.*, vol. AP-26, p. 509, May 1978.

[10] J. H. Richmond, "Radiation and scattering by thin-wire structures in the complex frequency domain," Rep. 2902-10, Ohio State Univ. ElectroScience Laboratory, Dep. of Electrical Engineering, prepared under Grant NGL 35-008-138 for National Aeronautics and Space Administration, July 1973.

Correction to "Electromagnetic Modeling of Composite Wire and Surface Geometries"

E. H. NEWMAN, MEMBER, IEEE, AND D. M. POZAR, STUDENT MEMBER, IEEE

In the above paper[1] Table I is incorrect, and the corrected version is shown below.

TABLE I
PARAMETERS OF CALCULATIONS

FIGURE/CURVE	N_W	N_S	N_A	N(TOTAL UNKNOWNS)
3/ ———	2	12	1	15
4/ — — —	2	40	1	43
4/ ———	1	12	1	14
4/ —·—	121	-	-	121
5/ •••••	2	84	1	87
6/ •••••	2	21	1	24
7/ ———	4	12	1	17
8/ ———	4	12	2	18
9/ ———	7	77	3	87

Manuscript received March 21, 1979.

[1] E. H. Newman and D. M. Pozar, *IEEE Trans. Antennas Propagat.*, vol. AP-26, no. 6, pp. 784–789, Nov. 1978.

A Combined-Source Solution for Radiation and Scattering from a Perfectly Conducting Body

JOSEPH R. MAUTZ, SENIOR, MEMBER, IEEE AND ROGER F. HARRINGTON, FELLOW, IEEE

Abstract—A combined-source solution is developed for electromagnetic radiation and scattering from a perfectly conducting body. In this solution a combination of electric and magnetic currents, called the combined source, is placed on the surface S of the conducting body. The combined-source operator equation is obtained from the E-field boundary-value equation. It is shown that the solution to this operator equation is unique at all frequencies. The combined-field operator equation also has a unique solution, but it is not directly applicable to the aperture radiation problem. The H-field and E-field operator equations fail to give unique solutions at frequencies corresponding to the resonant frequencies of a cavity formed by a hollow conductor of the same shape. The combined-source operator equation is solved by the method of moments. The solution, valid for a three-dimensional closed surface S, is then applied to a surface of revolution. Examples of numerical computations are given for a sphere, a cone-sphere, and a finite circular cylinder.

I. INTRODUCTION

THE H-FIELD integral equation [1]–[3] and the E-field integral equation [3]–[5] have been used extensively to calculate the electromagnetic radiation and scattering from perfectly conducting bodies. Both equations fail to give solutions at frequencies corresponding to the resonant frequencies of a cavity formed by a hollow conductor of the same shape [6]. One way to overcome this difficulty for the scattering problem is to use a combined-field solution [3], [6]. This method is not convenient for aperture problems. An alternative method, called the combined-source solution, more convenient for aperture problems, is derived in this paper.

A combined-source formulation for the solution of the three-dimensional exterior Dirichlet problem for the Helmholtz equation is suggested in [7]. Numerical solutions to the two-dimensional version of this combined-source formulation have appeared in [8] and [9]. The solution presented in [8] is for the scalar Helmholtz equation. Although the solution presented in [9] is specifically addressed to the two-dimensional electromagnetic scattering problem, this solution is still a solution to the scalar Helmholtz equation because the two-dimensional vector wave equation separates into two scalar Helmholtz equations.

We develop here a general combined-source formulation valid for three-dimensional electromagnetic radiation and scattering from a closed surface S. This formulation uses a combination of electric and magnetic currents to produce either the radiated field outside S or the scattered field outside S, depending on whether the electromagnetic problem is one of radiation or scattering. In the radiation problem the tangential electric field on S is specified. For the scattering problem the surface S is assumed to be perfectly conducting so that the tangential component of the scattered electric field neutralizes that of the incident electric field on S.

In Section II it is shown that, like the combined-field solution [6], the combined-source solution is unique at all frequencies, even the internal resonances of the surface S. It is also shown in Section II that the combined-source operator is the adjoint of the combined-field operator. The Appendix shows the relationship of the combined-source operator equation for the two-dimensional scattering problem to the hybrid potential equation which appears in [9].

II. COMBINED-SOURCE OPERATOR EQUATION

The E-field operator equation for the electric current $\mathbf{J}^c$ induced on the surface S of a perfectly conducting body by an incident electric field $\mathbf{E}^i$ is

$$-\mathbf{E}_{\tan}(\mathbf{J}^c) = \mathbf{E}_{\tan}^{i} \qquad \text{on } S, \tag{1}$$

where $\mathbf{E}(\mathbf{J}^c)$ is the electric field due to $\mathbf{J}^c$ and the subscript tan denotes the tangential component on S. Unfortunately, (1) fails if the frequency is such that S is the surface of a resonant cavity, because then an electric current on S which produces no tangential electric field on S can exist. To circumvent this difficulty we simulate the field outside S due to $\mathbf{J}^c$ by the field radiated by an electric current $\mathbf{J}$ and a magnetic current $\mathbf{M}$ placed simultaneously on S. Then (1) is replaced by

$$-\mathbf{E}_{\tan}(\mathbf{J}, \mathbf{M}) = \mathbf{E}_{\tan}^{i} \qquad \text{just outside } S, \tag{2}$$

where $\mathbf{E}(\mathbf{J}, \mathbf{M})$ is the electric field due to the electric current $\mathbf{J}$ and the magnetic current $\mathbf{M}$.

A relationship between $\mathbf{J}$ and $\mathbf{M}$ on S is required to complete the formulation (2). This relationship should ensure that $\mathbf{E}(\mathbf{J}, \mathbf{M})$ is nontrivial on S. If $\mathbf{J}$ and $\mathbf{M}$ were constant electric and magnetic currents on an infinite plane and if $\mathbf{J} \cdot \mathbf{J}^*$ and $\mathbf{M} \cdot \mathbf{M}^*$ were specified, then the relation

$$\mathbf{M} = \alpha \mathbf{n} \times \mathbf{J}, \tag{3}$$

where $\mathbf{n}$ is the unit normal vector to the plane and α is a nonnegative real constant, would maximize $\mathbf{E}_{\tan}(\mathbf{J}, \mathbf{M}) \cdot \mathbf{E}_{\tan}^*(\mathbf{J}, \mathbf{M})$ on the positive normal side of the plane. For the relationship between $\mathbf{J}$ and $\mathbf{M}$ on S we choose (3) with $\mathbf{n}$ equal to the outward unit normal vector on S and α equal to a positive real constant to be determined later.

It will now be shown that (3) ensures that $\mathbf{E}_{\tan}(\mathbf{J}, \mathbf{M})$ is nontrivial just outside S whenever $(\mathbf{J}, \mathbf{M})$ is nontrivial on S. This is done by showing that the assumption that there is a nontrivial $(\mathbf{J}, \mathbf{M})$ for which

$$\mathbf{E}_{\tan}(\mathbf{J}, \mathbf{M}) = 0 \qquad \text{just outside } S \tag{4}$$

Manuscript received June 2, 1978; revised January 25, 1979. This work was supported by the National Science Foundation, Washington, D.C., under Grant ENG-76-04588.

The authors are with the Department of Electrical and Computer Engineering, Syracuse University, Syracuse, NY 13210.

Reprinted from *IEEE Trans. Antennas Propagat.*, vol. AP-27, no. 4, pp. 445–454, July 1979.

leads to a contradiction. If (4) is true, then (J, M) radiates an electromagnetic field (E, H) which satisfies

$$\left.\begin{aligned} \mathbf{n} \times \mathbf{H} &= -\mathbf{J} \\ \mathbf{E} \times \mathbf{n} &= -\mathbf{M} \end{aligned}\right\} \quad \text{just inside } S. \tag{5}$$

The first of equations (5) follows from the fact that, since there are no external resonances, (4) implies that n × H is zero just outside S. The complex power flow P associated with (E, H) into the region interior to S is given by

$$P = -\iint_S \mathbf{E} \times \mathbf{H}^* \cdot \mathbf{n}\, dS. \tag{6}$$

Substitution of (5) into (6) gives

$$P = -\iint_S \mathbf{M} \cdot \mathbf{n} \times \mathbf{J}^*\, ds. \tag{7}$$

Because there are no sources inside S,

$$\operatorname{Re}(P) \geqslant 0. \tag{8}$$

In view of (3), (7) contradicts (8).

As a first step in the determination of α, let

$$\mathbf{E}_{\tan}(\mathbf{J}, \mathbf{M}) = \mathbf{E}_{\tan}(\mathbf{J}, \mathbf{O}) + \mathbf{E}_{\tan}(\mathbf{O}, \mathbf{M}), \tag{9}$$

where the two terms on the right-hand side of (9) are the contributions to $\mathbf{E}_{\tan}(\mathbf{J}, \mathbf{M})$ due to J and M, respectively. If the frequencies at which S is the perfectly conducting surface of a resonant cavity are called resonant frequencies, then at each of these resonant frequencies there is a nontrivial J and a nontrivial M such that

$$\mathbf{E}_{\tan}(\mathbf{J}, \mathbf{O}) = 0 \qquad \text{just outside } S \tag{10}$$

$$\mathbf{E}_{\tan}(\mathbf{O}, \mathbf{M}) = 0 \qquad \text{just outside } S. \tag{11}$$

However, if J and M are related by (3), then there is no frequency at which (10) and (11) hold simultaneously. Hence $\mathbf{E}_{\tan}(\mathbf{J}, \mathbf{O})$ and $\mathbf{E}_{\tan}(\mathbf{O}, \mathbf{M})$ of (9) compliment each other in that when one is zero, the other fills in for it. This mutual "filling in" will be most effective if the general order of magnitude of $\mathbf{E}_{\tan}(\mathbf{J}, \mathbf{O})$ is the same as that of $\mathbf{E}_{\tan}(\mathbf{O}, \mathbf{M})$. If the radiation of an infinite plane sheet of constant current is indicative of these orders of magnitude, then the ratio of the magnitude of the operator $\mathbf{E}_{\tan}(\mathbf{J}, \mathbf{O})$ to that of the operator $\mathbf{E}_{\tan}(\mathbf{O}, \mathbf{M})$ is roughly the characteristic impedance η of the medium. Hence the choice

$$\alpha = \eta \tag{12}$$

will make the general order of magnitude of the quantity $\mathbf{E}_{\tan}(\mathbf{J}, \mathbf{O})$ the same as that of the quantity $\mathbf{E}_{\tan}(\mathbf{O}, \mathbf{M})$.

With (12), (3) becomes

$$\mathbf{M} = \eta \mathbf{n} \times \mathbf{J}. \tag{13}$$

The substitution of (13) into (2) and the division of (2) by η yields

$$-\frac{1}{\eta} \mathbf{E}_{\tan}(\mathbf{J}, \eta \mathbf{n} \times \mathbf{J}) = \frac{1}{\eta} \mathbf{E}_{\tan}{}^{i} \qquad \text{just outside } S. \tag{14}$$

This is the operator equation for the combined-source formulation for scattering from a perfectly conducting body.

For aperture radiation a tangential electric field, called the aperture electric field, is specified over a portion of the closed surface S. This portion of S is called the aperture or apertures. For aperture radiation the combined-source current (J, M) is called upon to produce the aperture electric field just outside the aperture or apertures and zero tangential electric field elsewhere just outside S. Recall that in the scattering problem the tangential electric field of (J, M) is supposed to cancel the tangential component of the incident electric field just outside S. Hence the operator equation for the combined-source formulation for external radiation from an aperture cut out of a closed conducting surface is obtained by replacing $\mathbf{E}_{\tan}{}^{i}$ in (14) by the negative of the aperture electric field.

It will be shown next that the combined-source operator L^{cs} on the left-hand side of (14) defined by

$$L^{cs}(\mathbf{J}) = -\frac{1}{\eta} \mathbf{E}_{\tan}(\mathbf{J}, \eta \mathbf{n} \times \mathbf{J}) \tag{15}$$

is the adjoint of the combined-field operator L^{cf} on the left-hand side of [6, eq. (18)] with $\alpha = 1$ therein. This combined-field operator L^{cf} is defined by

$$L^{cf}(\mathbf{J}) = -\mathbf{n} \times \mathbf{H}(\mathbf{J}) - \frac{1}{\eta} \mathbf{E}_{\tan}(\mathbf{J}). \tag{16}$$

The electric field in (15) is evaluated just outside its magnetic current source $\eta \mathbf{n} \times \mathbf{J}$, whereas the magnetic field H(J) in (16) is evaluated just inside its electric current source J. An operator $L^{\dagger}$ is said to be the adjoint of another operator L if

$$\iint_S \mathbf{J}_1 \cdot L^{\dagger}(\mathbf{J}_2)\, ds = \iint_S \mathbf{J}_2 \cdot L(\mathbf{J}_1)\, ds \tag{17}$$

for every $\mathbf{J}_1$ and $\mathbf{J}_2$. The proof that L^{cs} is the adjoint of L^{cf} hinges on reciprocity. From reciprocity,

$$\iint_S \mathbf{J}_1 \cdot \left(-\frac{1}{\eta} \mathbf{E}(\mathbf{J}_2, \eta \mathbf{n} \times \mathbf{J}_2)\right) dS = -\frac{1}{\eta} \iint_S \mathbf{J}_2 \cdot \mathbf{E}(\mathbf{J}_1)\, ds + \iint_S (\mathbf{n} \times \mathbf{J}_2) \cdot \mathbf{H}(\mathbf{J}_1)\, dS, \tag{18}$$

where $\mathbf{J}_1$ is an electric current just outside S, $\mathbf{J}_2$ is an electric current on S, and $\eta \mathbf{n} \times \mathbf{J}_2$ is a magnetic current on S. Equation (18) can be rewritten as

$$\iint_S \mathbf{J}_1 \cdot \left(-\frac{1}{\eta} \mathbf{E}(\mathbf{J}_2, \eta \mathbf{n} \times \mathbf{J}_2)\right) dS = \iint_S \mathbf{J}_2 \cdot \left(-\mathbf{n} \times \mathbf{H}(\mathbf{J}_1) - \frac{1}{\eta} \mathbf{E}(\mathbf{J}_1)\right) dS. \tag{19}$$

It is now evident from (19) and (15)–(17) that the combined-source operator L^{cs} is the adjoint of the combined-field operator L^{cf}.

III. METHOD OF MOMENTS SOLUTION

In view of (15) the operator equation (14) for the combined-source formulation can be written as

$$L^{cs}(\mathbf{J}) = \frac{1}{\eta}\mathbf{E}_{\tan}{}^{i} \qquad \text{on } S. \tag{20}$$

The method of moments solution to (20) is obtained by first letting

$$\mathbf{J} = \sum_{j=1}^{N} I_j \mathbf{J}_j, \tag{21}$$

where $\mathbf{J}_j$ is a known vector expansion function on S and I_j is an unknown coefficient. Next, the dot product of (20) with a known vector testing function $\mathbf{W}_i$ on S is integrated over S. The resulting equation is

$$\sum_{j=1}^{N} I_j \iint_S \mathbf{W}_i \cdot L^{cs}(\mathbf{J}_j)\, dS = \frac{1}{\eta}\iint_S \mathbf{W}_i \cdot \mathbf{E}^i\, dS. \tag{22}$$

As i runs from 1 to N, (22) generates the matrix equation

$$[Z^{cs}]\vec{I} = \vec{V}, \tag{23}$$

where $\vec{I}$ and $\vec{V}$ are $N \times 1$ column vectors and $[Z^{cs}]$ is an $N \times N$ square matrix. The jth element of $\vec{I}$ is I_j, the ith element of $\vec{V}$ is given by

$$V_i = \frac{1}{\eta}\iint_S \mathbf{W}_i \cdot \mathbf{E}^i\, dS, \tag{24}$$

and the ijth element of $[Z^{cs}]$ is given by

$$Z_{ij}{}^{cs} = \iint_S \mathbf{W}_i \cdot L^{cs}(\mathbf{J}_j)\, dS. \tag{25}$$

The solution $\vec{I}$ to (23) determines the combined-source electric current $\mathbf{J}$ according to (21).

The combined-source moment matrix $[Z^{cs}]$ will now be expressed in terms of the combined-field moment matrix $[Z^{cf}]$. The ijth element of the combined-field moment matrix $[Z^{cf}]$ is given by

$$Z_{ij}{}^{cf} = \iint_S \mathbf{W}_i \cdot L^{cf}(\mathbf{J}_j)\, dS, \tag{26}$$

where, as in (25), $\mathbf{J}_j$ is the jth expansion function and $\mathbf{W}_i$ is the ith testing function. Because the combined-source operator L^{cs} is the adjoint of the combined-field operator L^{cf}, (26) becomes

$$Z_{ij}{}^{cf} = \iint_S \mathbf{J}_j \cdot L^{cs}(\mathbf{W}_i)\, dS. \tag{27}$$

Comparison of (25) and (27) shows that the combined-source matrix $[Z^{cs}]$ is the transpose of the combined-field matrix which would result if the set of expansion functions and the set of testing functions were interchanged.

IV. ELECTRIC CURRENT

Unfortunately, the electric current $\mathbf{J}$ given by the combined-source moment solution (21) has no physical significance. An electric current $\mathbf{J}^c$ which does have physical significance is defined as follows for both the radiation and scattering problems.

For the scattering problem,

$$\mathbf{J}^c = \mathbf{n} \times \mathbf{H}^i + \mathbf{n} \times \mathbf{H}(\mathbf{J}, \mathbf{M}), \tag{28}$$

where $\mathbf{H}^i$ is the incident magnetic field and $\mathbf{H}(\mathbf{J}, \mathbf{M})$ is the magnetic field just outside S due to the combined source $(\mathbf{J}, \mathbf{M})$ on S. Note that $\mathbf{J}^c$ as defined by (28) is the electric conduction current induced on the scattering surface S of the perfectly conducting body.

For the radiation problem two different electric conduction currents can be defined. The first of these is $\mathbf{n} \times (\mathbf{H}^+ - \mathbf{H}^-)$ where $\mathbf{H}^+$ is the magnetic field $\mathbf{H}(\mathbf{J}, \mathbf{M})$ just outside S, and $\mathbf{H}^-$ is the magnetic field just inside S as obtained from the solution to the interior radiation problem. This electric current radiates the true field outside S. The sheet of magnetic current $\mathbf{E} \times \mathbf{n}$ placed just outside the perfectly conducting closed surface S will also radiate the true field outside S provided that $\mathbf{E}$ is the specified aperture electric field. The second electric conduction current is the electric current induced on S by this sheet of magnetic current. For the radiation problem $\mathbf{J}^c$ is defined to be this second electric conduction current. This $\mathbf{J}^c$ is $\mathbf{n} \times \mathbf{H}$ where $\mathbf{H}$ is the magnetic field just outside S. Hence for the radiation problem

$$\mathbf{J}^c = \mathbf{n} \times \mathbf{H}(\mathbf{J}, \mathbf{M}), \tag{29}$$

where $\mathbf{H}(\mathbf{J}, \mathbf{M})$ is the magnetic field just outside S due to the combined source $(\mathbf{J}, \mathbf{M})$ on S.

The quantity $\mathbf{n} \times \mathbf{H}(\mathbf{J}, \mathbf{M})$ appearing in (28) and (29) will be evaluated just outside S in the following manner. Let

$$\mathbf{n} \times \mathbf{H}(\mathbf{J}, \mathbf{M}) = \sum_{j=1}^{N} C_j \mathbf{J}_j, \tag{30}$$

where the coefficients C_j remain to be determined. The integral of the dot product of (30) with $\mathbf{W}_i$ over S is

$$\iint_S \mathbf{W}_i \cdot \mathbf{n} \times \mathbf{H}(\mathbf{J}, \mathbf{M})\, dS = \sum_{j=1}^{N} C_j \iint_S \mathbf{W}_i \cdot \mathbf{J}_j\, dS. \tag{31}$$

From duality

$$\mathbf{H}(\mathbf{J}, \mathbf{M}) = \frac{1}{\eta}\mathbf{E}\left(\frac{1}{\eta}\mathbf{M}, -\eta\mathbf{J}\right). \tag{32}$$

Substitution of (13) into (32) gives

$$\mathbf{H}(\mathbf{J}, \mathbf{M}) = \frac{1}{\eta}\mathbf{E}(\mathbf{n} \times \mathbf{J}, \eta\mathbf{n} \times (\mathbf{n} \times \mathbf{J})). \tag{33}$$

Comparison of (15) and (33) shows that

$$H_{tan}(J, M) = -L^{cs}(n \times J). \tag{34}$$

Substitution of (34) into (31) gives

$$\iint_S (n \times W_i) \cdot L^{cs}(n \times J)\, dS = \sum_{j=1}^{N} C_j \iint_S W_i \cdot J_j\, dS. \tag{35}$$

In view of (21), (35) generates the matrix equation

$$[Z^{cs'}]\vec{I} = [D]\vec{C} \tag{36}$$

as i goes from 1 to N. Here $[Z^{cs'}]$ is the combined-source moment matrix which would result if the set of expansion functions J_j were replaced by $n \times J_j$ and the set of testing functions W_i were replaced by $n \times W_i$. More explicitly, the ijth element of $[Z^{cs'}]$ is given by

$$Z_{ij}^{cs'} = \iint_S (n \times W_i) \cdot L^{cs}(n \times J_j)\, dS. \tag{37}$$

In (36), $[D]$ is the $N \times N$ square matrix whose ijth element is given by

$$D_{ij} = \iint_S W_i \cdot J_j\, dS. \tag{38}$$

Also, $\vec{I}$ is the column vector which satisfies (23) and $\vec{C}$ is the $N \times 1$ column vector of the unknown coefficients C_j. The jth element of $\vec{C}$ is C_j.

The matrix equation (36) determines $\vec{C}$. The desired $n \times H(J, M)$ is given by (30) where C_j is the jth element of the solution vector $\vec{C}$ to the matrix equation (36).

V. FIELD MEASUREMENT

The electric field $E(J, M)$ outside S is obtained from

$$E(J, M) \cdot Il_r = \iint_S E(Il_r) \cdot J\, dS - \iint_S H(Il_r) \cdot M\, dS. \tag{39}$$

This equation is the statement of reciprocity between the combined source (J, M) and a receiving electric current element Il_r placed at the point r_r at which the l_r component of $E(J, M)$ is to be evaluated. Here $E(Il_r)$ is the electric field due to Il_r and $H(Il_r)$ is the magnetic field due to Il_r.

Substitution of (21) and (13) into (39) gives

$$E(J, M) \cdot Il_r = \eta[\tilde{R} + \tilde{R}']\vec{I}, \tag{40}$$

where $\tilde{R}$ and $\tilde{R}'$ are the transposes of the $N \times 1$ column vectors $\vec{R}$ and $\vec{R}'$, respectively. The jth elements of $\vec{R}$ and $\vec{R}'$ are given by

$$R_j = \frac{1}{\eta} \iint_S E(Il_r) \cdot J_j\, dS \tag{41}$$

$$R_j' = \iint_S (n \times H(Il_r)) \cdot J_j\, dS. \tag{42}$$

From (23)

$$\vec{I} = [Z^{cs}]^{-1}\vec{V} \tag{43}$$

such that (40) becomes

$$E(J, M) \cdot Il_r = \eta[\tilde{R} + \tilde{R}'][Z^{cs}]^{-1}\vec{V}, \tag{44}$$

where the ith element of $\vec{V}$ is given by (24). Equation (44) is the general field measurement result valid for either arbitrary incident field excitation or for arbitrary aperture field excitation.

If the excitation is the incident electric field $E(Il_t)$ of a transmitting electric current dipole Il_t placed at the point r_t then, according to (24),

$$V_i = \frac{1}{\eta} \iint_S E(Il_t) \cdot W_i\, dS. \tag{45}$$

In this case (44) is the field measurement result for dipole scattering.

The statement of reciprocity between the transmitting electric current element Il_t and the receiving electric current element Il_r for radiation in the presence of the conducting surface S is

$$E(J, M) \cdot Il_r = E(J, M)_r \cdot Il_t. \tag{46}$$

Here (J, M) is the correct combined-source current when the incident electric field is $E(Il_t)$, and $(J, M)_r$ is the correct combined-source current when the incident electric field is $E(Il_r)$. If (44) were exact as it stands and also as it would be if Il_t and Il_r were interchanged, then (46) would imply that

$$[\tilde{R} + \tilde{R}'][Z^{cs}]^{-1}\vec{V} = [\tilde{R} + \tilde{R}']_t[Z^{cs}]^{-1}[\vec{V}]_r. \tag{47}$$

The subscript t on the right-hand side of (47) denotes that Il_r is replaced by Il_t in (41) and (42). The subscript r in (47) denotes that Il_t is replaced by Il_r in (45). In general, (47) is only approximately true. Hence the combined-source moment solution for dipole scattering does not always satisfy reciprocity exactly.

With regard to (46) it will be shown that

$$E(J, M) \cdot Il_r = E(J)_r \cdot Il_t, \tag{48}$$

where the left-hand side of (48) is the field measurement result (44) and the right-hand side of (48) is the corresponding field measurement result obtained from the combined-field moment solution with the transmitting and receiving current elements Il_t and Il_r interchanged and with the sets of expansion and testing functions interchanged. The combined-field operator equation for the electric current $(\mathbf{J})_r$ is

$$L^{cf}(\mathbf{J})_r = \frac{1}{\eta}\mathbf{E}_{tan}(I\mathbf{l}_r) + \mathbf{n} \times \mathbf{H}(I\mathbf{l}_r). \tag{49}$$

Letting

$$(\mathbf{J})_r = \sum_{j=1}^{N} I_j \mathbf{W}_j \tag{50}$$

and testing (49) with $\mathbf{J}_i$, $i = 1, 2, \cdots, N$, we obtain

$$[\tilde{Z}^{cs}]\vec{I} = \vec{R} + \vec{R}', \tag{51}$$

where $\vec{I}$ is the column vector of the unknown coefficients I_j, and $\vec{R}$ and $\vec{R}'$ are given by (41) and (42). The fact that the combined-source moment matrix $[Z^{cs}]$ is the transpose of the combined-field moment matrix which would result if the sets of expansion and testing functions were interchanged was used to obtain (51). From reciprocity

$$\mathbf{E}(\mathbf{J})_r \cdot I\mathbf{l}_t = \iint_S \mathbf{E}(I\mathbf{l}_t) \cdot (\mathbf{J})_r \, dS. \tag{52}$$

Solution of (51) for I and the substitution of (50) into (52) gives

$$\mathbf{E}(\mathbf{J})_r \cdot I\mathbf{l}_t = \eta \tilde{V}[\tilde{Z}^{cs}]^{-1}[\vec{R} + \vec{R}'], \tag{53}$$

where $\vec{V}$ is given by (45). The desired result (48) is a consequence of the fact that the right-hand sides of (53) and (44) are equal. Thus the dipole scattering obtained from the combined-source moment solution is equal to the dipole scattering obtained from the combined field moment solution with the transmitting and receiving current elements interchanged and with the sets of expansion and testing functions interchanged.

It is apparent from the development (49)–(53) of the combined-field moment solution that, for dipole scattering, the combined-source excitation vector $\vec{V}$ is equal to the combined-field measurement vector with the set of expansion functions replaced by the set of testing functions and the receiving current element replaced by the transmitting current element. Similarly, the combined-source measurement vector $[\vec{R} + \vec{R}']$ is equal to the combined-field excitation vector with the set of testing functions replaced by the set of expansion functions and with the transmitting current element replaced by the receiving current element.

The choice

$$I\mathbf{l}_t = \frac{1}{G_t}\mathbf{u}^t, \tag{54}$$

where $\mathbf{u}^t$ is a unit vector in the $\mathbf{l}_t$ direction and

$$G_t = \frac{-je^{-jkr_t}}{4\pi r_t}, \tag{55}$$

makes the excitation vector $\vec{V}$ essentially independent of r_t for large r_t. Similarly, the measurement vector in (44) will be essentially independent of r_r for large r_r if

$$I\mathbf{l}_r = \frac{1}{G_r}\mathbf{u}^r, \tag{56}$$

where $\mathbf{u}^r$ is a unit vector in the $\mathbf{l}_r$ direction and

$$G_r = \frac{-je^{-jkr_r}}{4\pi r_r}. \tag{57}$$

For large r_t the transmitter field of (54) approaches the plane wave given by

$$\mathbf{E}(I\mathbf{l}_t) = k\eta \mathbf{u}^t e^{-j\mathbf{k}_t \cdot \mathbf{r}} \tag{58}$$

$$\mathbf{H}(I\mathbf{l}_t) = (\mathbf{k}_t \times \mathbf{u}^t)e^{-j\mathbf{k}_t \cdot \mathbf{r}}, \tag{59}$$

where k is the propagation constant, $\mathbf{k}_t$ is the propagation vector, and $\mathbf{r}$ is the radius vector from the origin in the vicinity of S. Similarly, for large r_r the receiver field of (56) approaches

$$\mathbf{E}(I\mathbf{l}_r) = k\eta \mathbf{u}^r e^{-j\mathbf{k}_r \cdot \mathbf{r}} \tag{60}$$

$$\mathbf{H}(I\mathbf{l}_r) = (\mathbf{k}_r \times \mathbf{u}^r)e^{-j\mathbf{k}_r \cdot \mathbf{r}}, \tag{61}$$

where $\mathbf{k}_r$ is the propagation vector. Substitution of (56) into (44) gives

$$\mathbf{E}(\mathbf{J}, \mathbf{M}) \cdot \mathbf{u}^r = \eta G_r[\tilde{R} + \tilde{R}'][Z^{cs}]^{-1}\vec{V}. \tag{62}$$

For incident plane wave excitation the scattering cross section σ for a $\mathbf{u}^r$ polarized receiver is defined by

$$\sigma = 4\pi r_r^2 \frac{|\mathbf{E}^s \cdot \mathbf{u}^r|^2}{|\mathbf{E}^i|^2}, \tag{63}$$

where $\mathbf{E}^i$ is the incident electric field and $\mathbf{E}^s$ is the scattered electric field a large distance r_r from the origin. In view of (57) substitution of (62) for $\mathbf{E}^s \cdot \mathbf{u}^r$ and (58) for $\mathbf{E}^i$ in (63) gives

$$\frac{\sigma}{\lambda^2} = \frac{1}{16\pi^3}\left|[\tilde{R} + \tilde{R}'][Z^{cs}]^{-1}\vec{V}\right|^2, \tag{64}$$

where λ is the wavelength $2\pi/k$. Here $[Z^{cs}]$ is the combined-source moment matrix whose elements are given by (25). The elements of $\vec{V}$, $\vec{R}$, and $\vec{R}'$ are given by the substitution of (58), (60), and (61) into (24), (41), and (42), respectively. Hence

$$V_i = k \iint_S \mathbf{W}_i \cdot \mathbf{u}^t e^{-j\mathbf{k}_t \cdot \mathbf{r}} \, dS \tag{65}$$

$$R_j = k \iint_S \mathbf{J}_j \cdot \mathbf{u}^r e^{-j\mathbf{k}_r \cdot \mathbf{r}} \, dS \tag{66}$$

$$R_j' = -\iint_S (\mathbf{n} \times \mathbf{J}_j) \cdot (\mathbf{k}_r \times \mathbf{u}^r)e^{-j\mathbf{k}_r \cdot \mathbf{r}} \, dS. \tag{67}$$

For aperture excitation the gain G seen by a $\mathbf{u}^r$ polarized receiver is defined by

$$G = \frac{4\pi r_r^2 \mid \mathbf{E} \cdot \mathbf{u}^r \mid^2}{\eta P}, \tag{68}$$

where $\mathbf{E}$ is the radiated electric field a large distance r_r from the origin and P is the time-average power radiated. In view of (57) substitution of (62) for $\mathbf{E} \cdot \mathbf{u}^r$ in (68) gives

$$G = \frac{\eta}{4\pi P} \left| [\tilde{R} + \tilde{R}'][Z^{cs}]^{-1} \vec{V} \right|^2. \tag{69}$$

Again, $[Z^{cs}]$ is the combined-source moment matrix whose elements are given by (25). The elements of $\vec{R}$ and $\vec{R}'$ are still given by (66) and (67). However, the elements of $\vec{V}$ are now the result of the substitution of the negative of the aperture electric field for $\mathbf{E}^i$ in (24). Thus

$$V_i = \frac{-1}{\eta} \iint_S \mathbf{W}_i \cdot \mathbf{E}^a \, dS, \tag{70}$$

where $\mathbf{E}^a$ is the aperture electric field. The time-average power P associated with the combined source $(\mathbf{J}, \mathbf{M})$ can be calculated from either the induced electromagnetic force (EMF) method, the integration of the far-field power pattern, or the integration of the Poynting vector just outside S. Details of each of these methods can be found in [10].

VI. EXAMPLES

In a research report [11] the formulas for the combined-source electric current $\mathbf{J}$, the induced electric current $\mathbf{J}^c$, the scattering cross section σ, the gain G, and the various complex powers associated with $\mathbf{J}$ are specialized to a body of revolution. Computer programs to calculate these quantities are also described and listed in the report [11]. Some computational results obtained with these programs are given in this section.

Fig. 1 shows the body of revolution and an orthogonal coordinate system (t, ϕ) with unit vectors $\mathbf{u}_t$ and $\mathbf{u}_\phi$ on S. Fig. 2 shows the transmitter bearing $(\theta_t, 0)$ and the receiver bearing (θ_r, ϕ_r) for arbitrary plane wave scattering. In Fig. 2, $\mathbf{k}_t$ and $\mathbf{k}_r$ are the propagation vectors of plane waves coming from the transmitter and receiver locations.

Fig. 3 shows the electric current $\mathbf{J}^c$ induced by a plane wave axially incident on a conducting sphere for which $ka = 2.75$, where a is the radius of the sphere. The first resonance of the spherical cavity is at $ka = 2.744$. If the incident magnetic field is in the y direction, the induced electric current $\mathbf{J}^c$ defined by (28) is of the form

$$\mathbf{J}^c = \mathbf{u}_\theta J_\theta \cos\phi + \mathbf{u}_\phi J_\phi \sin\phi, \tag{71}$$

where neither J_θ nor J_ϕ depend on ϕ. In (71) $\mathbf{u}_\theta$ and $\mathbf{u}_\phi$ are unit vectors in the θ and ϕ directions, respectively. Fig. 3(a) shows the combined-source solution for J_θ/H_y versus θ. Fig. 3(b) shows the combined-source solution for J_ϕ/H_y versus θ. Here H_y is the y component of the incident magnetic field at the center of the sphere. In Fig. 3 the label theta on the horizontal axis stands for θ and $\theta = 0$ is the forward scattering direction. In Fig. 3 the symbols X and + denote the magnitude and the phase, respectively, of the combined-source solution. The solid curves represent the "exact" Mie series solution.

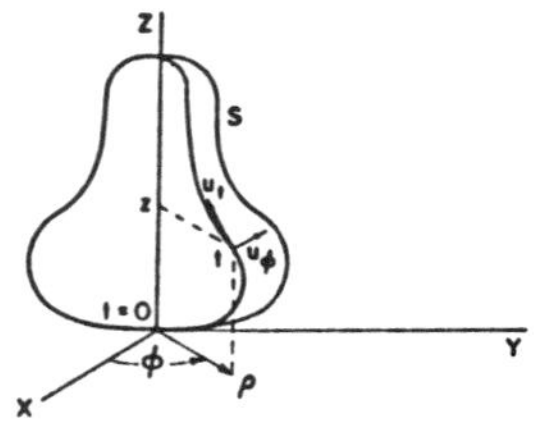

Fig. 1. Body of revolution and coordinate system.

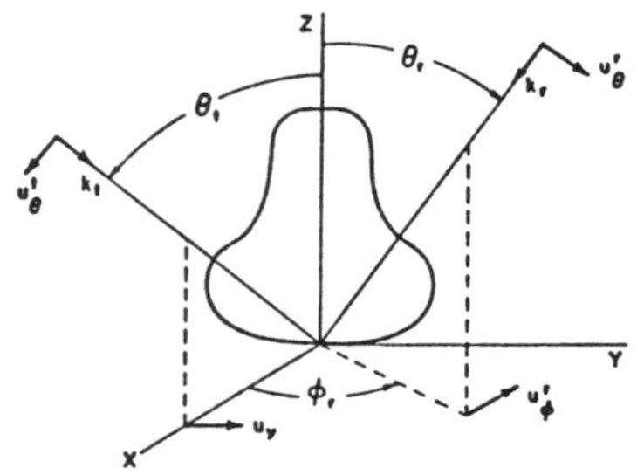

Fig. 2. Plane wave scattering by a conducting body of revolution.

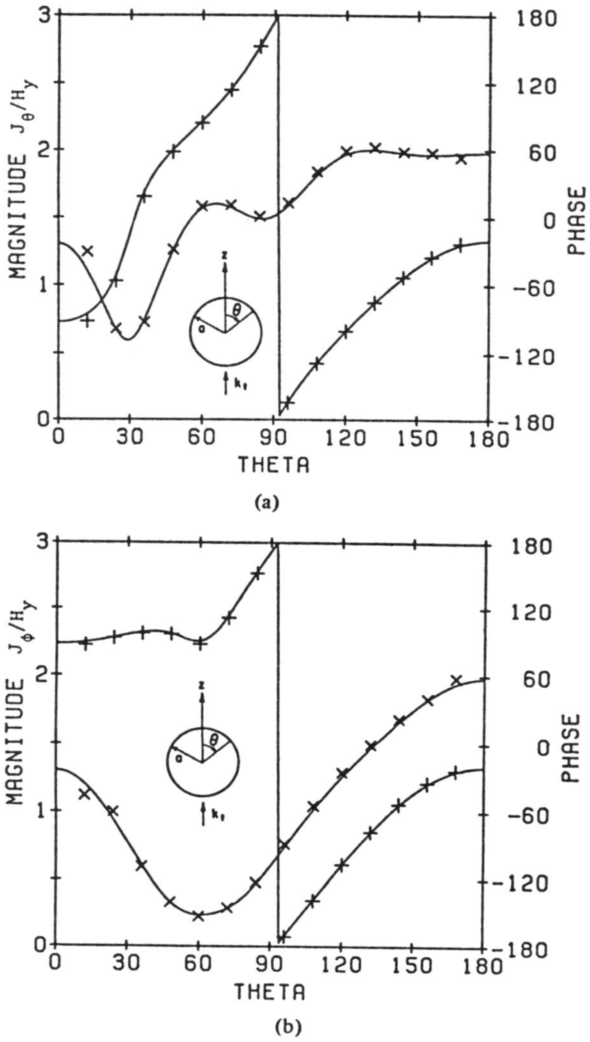

Fig. 3. Electric current J_θ/H_y and J_ϕ/H_y induced on a conducting sphere, $ka = 2.75$, by an axially incident plane wave. Solid curves represent exact solution. Symbols x and + denote magnitude and phase, respectively, of combined-source solution.

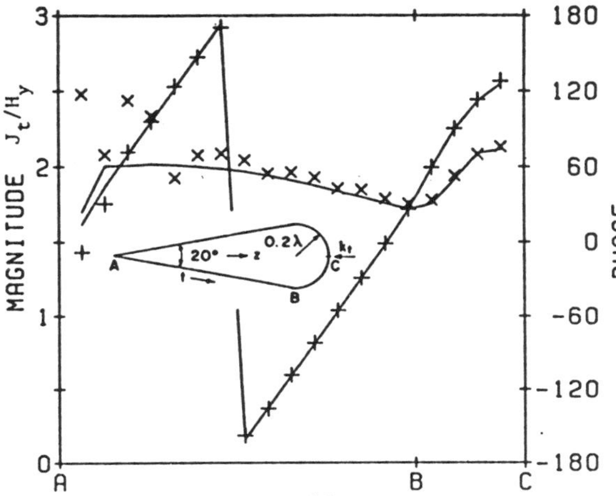

(a)

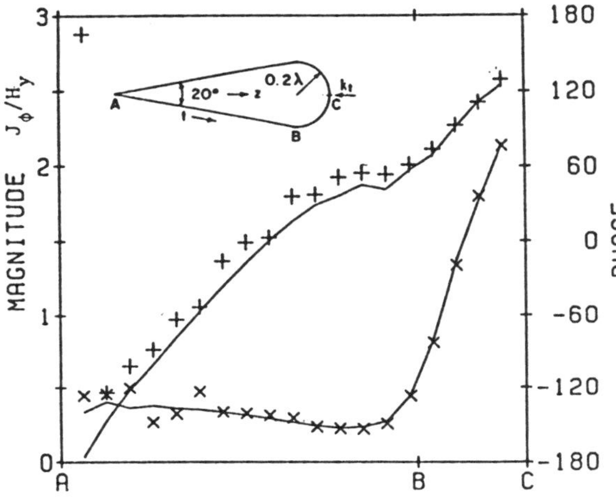

(b)

Fig. 4. Electric current J_t/H_y and J_ϕ/H_y induced on a conducting cone-sphere, cone angle = 20°, sphere radius = 0.2λ, by a plane wave axially incident on the sphere. Solid curves show combined-source solution. Symbols x and + denote magnitude and phase, respectively, of E-field solution.

The combined-source solution gives currents which differ only slightly from those computed by the combined-field solution [6]. In [6, fig. 5(a)], the electric current error Δ of the combined-field solution for the conducting sphere is linear from $ka = 2.70$ to $ka = 2.80$. Similarly, the electric current error Δ of the combined-source solution for the conducting sphere is also linear from $ka = 2.70$ to $ka = 2.80$. The values of Δ at $ka = 2.70$ and $ka = 2.80$ are given in Table I. Although not quite as accurate as the combined-field solution for ka between 2.70 and 2.80, the combined-source result for $\mathbf{J}^c$ is likewise unaffected by the first resonance at $ka = 2.744$. Both the combined-source and the combined-field results for J_θ and J_ϕ were obtained by covering the generating curve with 14 equally spaced triangular functions. For integrations in t each triangular function was sampled at four points. A 20-point Gaussian quadrature formula was used for integrations with respect to ϕ between the limits $\phi = 0°$ and $\phi = 180°$.

Fig. 4 shows the electric current $\mathbf{J}^c$ induced on a conducting cone-sphere by a plane wave axially incident on the sphere end. As shown in the inserts of Fig. 4, the cone angle of this cone-sphere is 20° and the sphere radius is 0.2 wavelength. The cone and sphere are joined such that the tangent to the

TABLE I
ELECTRIC CURRENT ERROR Δ

ka	Δ combined field	Δ combined source
2.70	0.0245	0.0347
2.80	0.0263	0.0358

generating curve is continuous at the cone-to-sphere junction. If the incident magnetic field is in the y direction, the induced electric current $\mathbf{J}^c$ defined by (28) is of the form

$$\mathbf{J}^c = \mathbf{u}_t J_t \cos\phi + \mathbf{u}_\phi J_\phi \sin\phi. \tag{72}$$

Both J_t/H_y and J_ϕ/H_y are plotted versus t where, as shown in the inserts, t is the arc length along the generating curve. Here H_y is the y component of the incident magnetic field at the tip of the cone. The solid curves in Fig. 4 represent the combined-source solution, and the symbols X and + denote the magnitude and the phase, respectively, of the E-field solution as tabulated in [12, p. 66]. Because J_ϕ was smoothed as explained in [12, p. 33] this E-field solution for $|J_\phi|$ differs from that in [13, fig. 13]. An H-field solution for $|J_\phi|$ appears in [3, p. 218].

The combined-source results for J_t and J_ϕ in Fig. 4 were obtained by covering the generating curve with 19 equally spaced triangular functions. For integrations in t each triangular function was sampled at four points. A 20-point Gaussian quadrature formula was used for the integrations with respect to ϕ.

The induced electric current and gain for a thin rotationally symmetric aperture at the equator of a conducting sphere of size $ka = 2.75$ driven by a minus θ directed electric field are shown in Fig. 5(a) and 5(b). In Fig. 5(a), the symbols X and + denote the real and imaginary parts of $2\pi a\eta J_\theta{}^c/V$ versus θ as obtained from the combined-source solution. Here $J_\theta{}^c$ is the θ component of $\mathbf{J}^c$ defined by (29), and V is the voltage across the aperture. At $\theta = 90°$, $2\pi a\eta J_\theta{}^c/V$ reduces to the input admittance normalized by dividing by the admittance $1/\eta$ of free space. In Fig. 5(b) the symbol X denotes the gain pattern for the aperture problem of Fig. 5(a) as obtained from the combined-source solution. The solid curves in Fig. 5(a) and 5(b) represent the exact Mie series solution. For comparison the E-field solution for the same sphere and same aperture as in Fig. 5(a) and 5(b) is shown in Fig. 5(c) and 5(d). In Fig. 5(c) and 5(d) the symbols X denote the E-field solution and the solid curves denote the Mie series solution. The E-field solution in Fig. 5(c) and 5(d) is poor because $ka = 2.75$ is very close to the first resonance which occurs at $ka = 2.744$.

The moment solutions of Fig. 5 use 18 expansion functions equally spaced on the generating curve such that the ninth and tenth expansion functions straddle the aperture. Both the impulsive aperture field at the equator and the aperture field

$$\mathbf{E}^a = \frac{-V}{a\Delta\theta}\mathbf{u}_\theta, \qquad 90° - \frac{\Delta\theta}{2} \leqslant \theta \leqslant 90° + \frac{\Delta\theta}{2}, \tag{73}$$

where $0 \leqslant \Delta\theta \leqslant 180°/19$, give exactly the same moment solution. For convenience the aperture field (73) with $\Delta\theta = 180°/19$ was used to calculate the Mie series solution.

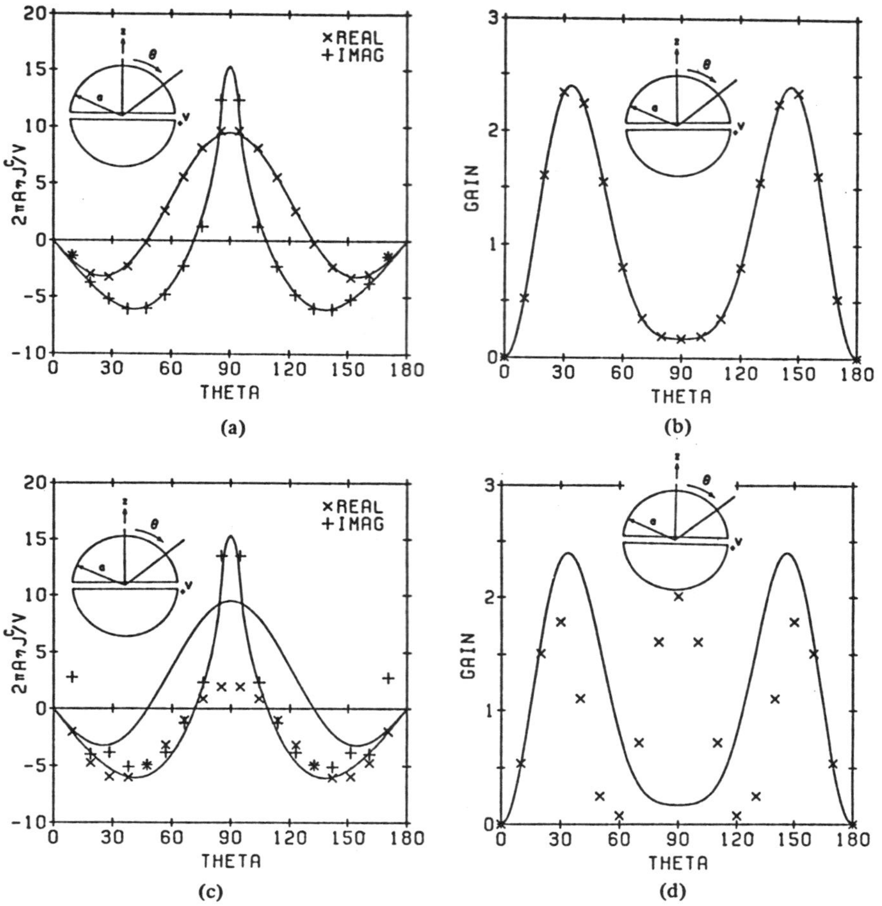

Fig. 5. Electric current $J_\theta{}^c$ and gain for a rotationally symmetric aperture at the equator of a conducting sphere, ka = 2.75. Solid curves represent exact solution. Symbols x and + denote combined-source solution in (a) and (b), and E-field solution in (c) and (d). First resonance of spherical cavity is ka = 2.744.

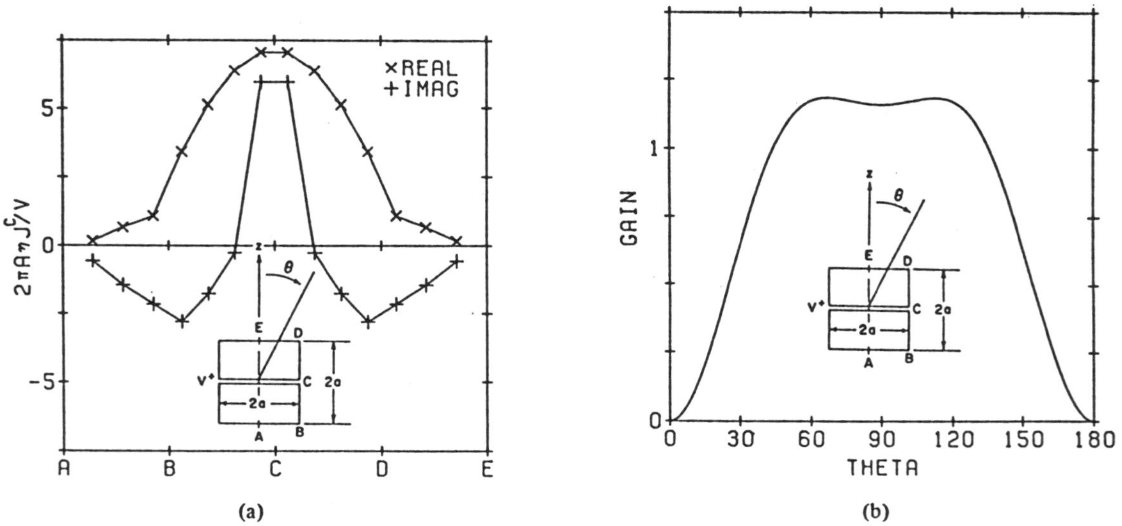

Fig. 6. Electric current $J_t{}^c$ and gain for a rotationally symmetric aperture at the center of a conducting cylinder of length $2a$ and radius $a = \lambda/4$, combined-source solution.

Fig. 6 shows the combined-source solution for the induced electric current $\mathbf{J}^c$ defined by (29) and for the gain of an axially symmetric aperture at the center of a conducting cylinder whose length is twice its radius a. Here $a = \lambda/4$. In Fig. 6(a) the symbols $\times$ and $+$ denote the real and imaginary parts of $2\pi a \eta J_t^c/V$ versus t, where J_t^c is the t component of $\mathbf{J}^c$ and V is the voltage across the aperture. These symbols have been connected by straight lines in Fig. 6(a) to improve readability. Fig. 6(b) shows the gain pattern for the aperture problem of Fig. 6(a) as obtained from the combined-source solution. There are 14 triangular expansion functions on the generating curve of the finite cylinder in Fig. 6. These triangular functions are not equally spaced. The generating curve consists of 7 equal-length intervals on each end cap of the cylinder and 16 equal-length intervals on the lateral side. The domain of each expansion function consists of four of these intervals. For the moment solutions of Figs. 5 and 6 the integrations with respect to t were approximated by sampling each triangular function at four points, and a 20-point Gaussian quadrature formula was used for the integrations with respect to ϕ. The approximate (moment solution) gain patterns in Figs. 5 and 6 were calculated from (69) with P obtained from the numerical integration of the far-field power pattern.

VII. DISCUSSION

A combined-source solution has been developed for electromagnetic radiation and scattering problems involving only the region external to a closed surface S. As such, this solution is not affected by resonances of the region internal to S. Because the combined source contains a magnetic current which implies a discontinuity of the tangential electric field, the combined-source formulation does not apply to scattering from "zero-thickness" bodies, like plates or washers, nor to aperture radiation from a conducting surface which was not closed before the aperture was cut out of it. The combined-source formulation could be adapted to electromagnetic problems involving only the region interior to a closed conducting surface, but then the solution would deteriorate near the internal resonances.

The combined-source formulation is closely related to the combined-field formulation [6]. For instance, the combined-source operator is the adjoint of the combined-field operator. From this it follows that the combined-source moment matrix is the transpose of the combined-field matrix which would result if the set of expansion functions and the set of testing functions were interchanged. For dipole scattering the combined-source excitation vector is the combined-field measurement vector which would result if the set of expansion functions were replaced by the set of testing functions and if the receiving current element were replaced by the transmitting current element. The combined-source measurement vector is the combined-field excitation vector which would result if the set of testing functions were replaced by the set of expansion functions and if the transmitting current element were replaced by the receiving current element. It follows from the above three statements that the combined-source solution for dipole scattering is the combined-field solution which would result if the sets of expansion and testing functions were interchanged and if the transmitting and receiving current elements were interchanged.

One disadvantage of the combined-source solution in comparison with the combined-field solution is that the combined source $(\mathbf{J}, \mathbf{M})$ has no physical significance. The physically significant electric currents defined by (28) and (29) require calculation of $\mathbf{n} \times \mathbf{H}(\mathbf{J}, \mathbf{M})$ from $(\mathbf{J}, \mathbf{M})$. In view of (34) this calculation is especially simple for the body of revolution expansion functions because the operation $(\mathbf{n} \times)$ is closed with respect to those expansion functions. However, the approximation (30) is still present. This could explain why the electric current error for the combined-source solution is slightly larger than that for the combined-field solution in Table I.

One advantage of the combined-source solution over the combined-field solution is that the combined-source formulation is directly applicable to the aperture radiation problem in which the aperture electric field is specified. One must recast the aperture radiation problem into an equivalent scattering problem in order to apply the combined-field solution. In this equivalent scattering problem the surface S is perfectly conducting everywhere and the incident field is the field radiated by the magnetic current $\mathbf{E}^a \times \mathbf{n}$ in free space, where $\mathbf{E}^a$ is the specified electric field in the aperture.

APPENDIX

COMBINED-SOURCE OPERATOR EQUATION FOR THE TWO-DIMENSIONAL SCATTERING PROBLEM

For the two-dimensional case the surface S is the cylindrical surface generated by moving a line parallel to the z axis around a closed curve C in the xy plane. Let $\mathbf{u}_z$ be a unit vector in the z direction and let $\mathbf{u}_c$ be the unit tangent vector on C. As observed from the positive z axis inside the cylinder, C is oriented counterclockwise.

For the transverse magnetic (TM) case in which

$$\mathbf{E}_{\tan}{}^i = \mathbf{u}_z E_z{}^i, \tag{A1}$$

it is sufficient to let

$$\mathbf{J} = \mathbf{u}_z u, \tag{A2}$$

where u is an unknown function on C. In view of (A1) and (A2) the combined-source operator equation (14) has only a z component which is

$$\frac{1}{\eta}\,\mathbf{u}_z \cdot \nabla \times \mathbf{F} + jkA_z = \frac{1}{\eta} E_z{}^i, \tag{A3}$$

where

$$\mathbf{F} = -\eta \int_C G(\rho, \rho')u(\rho')\mathbf{u}_c(\rho')\, dC', \tag{A4}$$

$$A_z = \int_C G(\rho, \rho')u(\rho')\, dC', \tag{A5}$$

and

$$G(\rho, \rho') = \frac{1}{4j} H_0{}^{(2)}(k\,|\,\rho - \rho'\,|). \tag{A6}$$

Here k is the propagation constant, ρ is the radius vector in the xy plane, and $H_0{}^{(2)}$ is the Hankel function of the second kind. After some manipulation (A3) becomes

$$\frac{u(\rho)}{2} + \int_C \left(\frac{\partial}{\partial n'} G(\rho, \rho') + jkG(\rho, \rho') \right) u(\rho')\, dC' = \frac{1}{\eta} E_z{}^i \tag{A7}$$

for ρ on C. Here $\partial/\partial n'$ is the component of the gradient in the direction of the exterior normal vector at ρ'. The above equation is similar to [9, eq. 10].

For the transverse electric (TE) case in which $\mathbf{E}_{\tan}{}^i$ has only a $\mathbf{u}_c$ component, the relation

$$\mathbf{E}_{\tan} \cdot \mathbf{u}_c = \frac{j\eta}{k} \frac{\partial H_z}{\partial n} \tag{A8}$$

between the electric and magnetic fields $\mathbf{E}_{\tan}$ and $\mathbf{u}_z H_z$ valid for both the incident field and the field due to the combined source is used to write the $\mathbf{u}_c$ component of (14) as

$$\frac{-j}{k} \frac{\partial}{\partial n} H_z(\mathbf{J}, \eta \mathbf{n} \times \mathbf{J}) = \frac{j}{k} \frac{\partial}{\partial n} H_z{}^i \qquad \text{just outside } S, \tag{A9}$$

where $\partial/\partial n$ is the component of the gradient in the direction of the exterior normal at ρ. In (A9) H_z is the z component of the magnetic field due to the electric current $\mathbf{J}$ and the magnetic current $\eta \mathbf{n} \times \mathbf{J}$. Also, $H_z{}^i$ is the z compoinent of the incident magnetic field. If

$$\mathbf{J} = \mathbf{u}_c u, \tag{A10}$$

where u is an unknown function on C, then (A9) becomes

$$\frac{-jku(\rho)}{2} + jk \int_C \frac{\partial}{\partial n} G(\rho, \rho') u(\rho')\, dC' + FP \int_C \frac{\partial^2}{\partial n \partial n'} G(\rho, \rho') u(\rho')\, dC' = \frac{\partial}{\partial n} H_z{}^i \tag{A11}$$

for ρ on C. Here FP is defined by [9, eq. (12)]. This FP is necessary to eliminate the electric field due to the electric charge which is implied when a small portion of the electric current centered about the field point is removed in order to calculate the second integral on the left-hand side of (A11). Equation (A11) is similar to [9, eq. (11)]. It is apparent from (A11) that the factor k is missing from the first two terms on the left-hand side of [9, eq. (11)].

REFERENCES

[1] F. K. Oshiro and K. M. Mitzner, "Digital computer solution of three-dimensional scattering problems," in *Symposium Dig., 1967 IEEE Int. Antennas Propagat. Symp.*, Oct. 1967, pp. 257-263.

[2] P. L. E. Uslenghi, "Computation of surface currents on bodies of revolution," *Alta Freq.*, vol. 39, no. 8, pp. 1-12, Aug. 1970.

[3] A. J. Poggio and E. K. Miller, "Integral equation solutions of three-dimensional scattering problems," in *Computer Techniques for Electromagnetics*, R. Mittra, Ed. Oxford: Pergamon Press, 1973, ch. 4.

[4] M. G. Andreasen "Scattering from bodies of revolution," *IEEE Trans. Antennas Propagat.*, vol. AP-13, no. 2, pp. 303-310, Mar. 1965.

[5] R. F. Harrington, *Field Computation by Moment Methods.* New York: Macmillan, 1968.

[6] J. R. Mautz and R. F. Harrington, "*H*-field, *E*-field, and combined-field solutions for conducting bodies of revolution," *A.E.U.* (Germany), vol. 32, no. 4, pp. 157-164, Apr. 1978.

[7] H. Brakhage and P. Werner, "Über das Dirichletsche Aussenraumproblem für die Helmholtzsche Schwingungsgleichung," *Archiv d. Math.*, vol. 16, pp. 325-329, 1965.

[8] D. Greenspan and P. Werner, "A numerical method for the exterior Dirichlet problem for the reduced wave equation," *Arch. Ration. Mech. Anal.*, vol. 23, pp. 288-316, 1966.

[9] J. Bolomey and W. Tabbara, "Numerical aspects on coupling between complementary boundary value problems," *IEEE Trans. Antennas Propagat.*, vol. AP-21, pp. 356-363, May 1973.

[10] J. R. Mautz and R. F. Harrington, "A combined-source solution for radiation and scattering from a perfectly conducting body," Report TR-78-3, Dept. of Electrical and Computer Engineering, Syracuse University, Syracuse, NY 13210, April 1978.

[11] —"Application of the combined-source solution to a conducting body of revolution," Dep. Elec. Comp. Eng., Syracuse Univ., Syracuse, NY, Rep. TR-78-6, June 1978.

[12] R. F. Harrington and J. R. Mautz, "Radiation and scattering from bodies of revolution," Rep. AFCRL-69-0305, D.D.C. No. AD 895 670, Contract No. F19628-67-C-0233 with Air Force Cambridge Res. Lab., July 1969; Also, Carl Baum, Ed., Air Force Weapons Lab., Interaction Note 188.

[13] J. R. Mautz and R. F. Harrington, "Radiation and scattering from bodies of revolution," *Appl. Sci. Res.*, vol. 20, pp. 405-435, June 1969.

Electromagnetic Scattering by Surfaces of Arbitrary Shape

SADASIVA M. RAO, DONALD R. WILTON, SENIOR MEMBER, IEEE, AND ALLEN W. GLISSON, MEMBER, IEEE

Abstract—**The electric field integral equation (EFIE) is used with the moment method to develop a simple and efficient numerical procedure for treating problems of scattering by arbitrarily shaped objects. For numerical purposes, the objects are modeled using planar triangular surfaces patches. Because the EFIE formulation is used, the procedure is applicable to both open and closed surfaces. Crucial to the numerical formulation is the development of a set of special subdomain-type basis functions which are defined on pairs of adjacent triangular patches and yield a current representation free of line or point charges at subdomain boundaries. The method is applied to the scattering problems of a plane wave illuminated flat square plate, bent square plate, circular disk, and sphere. Excellent correspondence between the surface current computed via the present method and that obtained via earlier approaches or exact formulations is demonstrated in each case.**

I. INTRODUCTION

ENGINEERS AND researchers in electromagnetics have been quick to take advantage of the expanding capabilities of digital computers over the past two decades by developing effective numerical techniques applicable to a wide variety of practical electromagnetic radiation and scattering problems. As new computer developments dramatically increase computational capabilities, however, it becomes less cost effective to develop highly efficient but specialized codes for treating certain classes of geometries than to use less efficient but existing general purpose codes that can handle a wide variety of problems. For these reasons there has been a growing interest in the use and development of computer codes for treating scattering by arbitrarily shaped conducting bodies.

To date, the most notable approaches for treating such problems have used integral equation formulations in conjunction with the method of moments. The body surface in these approaches is generally modeled either as a wire mesh—the so-called wire-grid model—or as a surface partitioned into smooth or piecewise-smooth patches—the so-called surface patch model.

The wire-grid modeling approach has been remarkably successful in treating many problems, particularly in those requiring the prediction of far-field quantities such as radiation patterns and radar cross sections [1]. Not only is the connectivity of a wire-grid model easily specified for computer input, but the approach also has the advantage that all numerically computed integrals in the moment matrix are one dimensional. However, the approach is not well suited for calculating near-field and surface quantities such as surface current and input impedance. Some of the problems encountered include the presence of fictitious loop currents in the solution, ill-conditioned moment matrices and incorrect currents at the cavity resonant frequencies of the scatterer [2], and difficulties in interpreting computed wire currents and relating them to equivalent surface currents. The accuracy of wire-grid modeling has also been questioned on theoretical grounds [3]. Most of these difficulties can be either wholly or partially overcome by surface patch approaches, however, which account for much of the recent activity in this area.

Several approaches to surface patch modeling have been reported in the literature. Knepp and Goldhirsh [4] partitioned a conducting surface into nonplanar quadrilateral patches and employed the magnetic field integral equation (MFIE) to solve the electromagnetic scattering problem. Albertsen *et al.* [5] solved for the current and computed radiation patterns for satellite structures with attached wire antennas, booms, and solar panels. They employed a hybrid formulation in which the MFIE, with planar quadrilateral surface patches, was used to model the satellite, and the electric field integral equation (EFIE) was used to treat the wire antennas. Their approach also forms the basis for the arbitrary surface treatment of the widely used numerical electromagnetic code (NEC) developed at the Lawrence Livermore Laboratory [6]. Wang *et al.* [7] used an EFIE formulation and modeled relatively complex surfaces by means of planar rectangular patches. Newman and Pozar [8] extended the use of the well-known piecewise-sinusoidal basis functions of thin-wire theory to the treatment of surfaces in their EFIE formulation for surfaces with attached wires. Sankar and Tong [9] employed planar triangular patches to model a square plate and pointed out the possibility of extending their approach to arbitrary bodies. Their formulation, based on a variational formula for the current made stationary with respect to a set of trial functions, is equivalent to a Galerkin solution of the EFIE. Wang [10], [11] employed planar triangular patches in conjunction with the MFIE, but used basis functions containing the phase variation of the incident field in each patch, which unfortunately yield a moment matrix dependent on the incident field. Jeng and Wexler [12] suggested using the MFIE and nonplanar triangles to model arbitrary surfaces, while Singh and Adams [13] proposed the use of planar quadrilateral patches and sinusoidal basis functions in conjunction with the EFIE for the same purpose.

In arbitrary surface modeling the EFIE has the advantage of being applicable to both open and closed bodies, whereas the MFIE applies only to closed surfaces. On the other hand, for arbitrarily shaped objects the EFIE is considerably more difficult to apply than the MFIE. In fact, of the above authors, only Wang *et al.* and Newman and Pozar have actually applied the EFIE to nonplanar structures—and the use of rectangular patches limits their approaches to structures with curvature in one dimension only.

Manuscript received May 28, 1980; revised August 6, 1981. This work was supported by the Rome Air Development Center, Griffiss AFB, NY, under Contract No. F30602-78-C-0148.

S. M. Rao was with Syracuse University, Syracuse, NY, on leave from the Department of Electrical Engineering, University of Mississippi, University, MS 38677. He is now with the Department of Electrical Engineering, Rochester Institute of Technology, Rochester, NY 14623.

D. R. Wilton was with Syracuse University, Syracuse, NY, on leave from the Department of Electrical Engineering, University of Mississippi, University, MS 38677.

A. W. Glisson is with the Department of Electrical Engineering, University of Mississippi, University, MS 38677.

Reprinted from *IEEE Trans. Antennas Propagat.*, vol. AP-30, no. 3, pp. 409–418, May 1982.

The difficulites with the EFIE stem primarily from the presence of derivatives appearing in conjunction with a singular kernel in the integral equation. For example, if the vector basis functions which represent the surface current are not constructed so that their normal components are continuous across surface edges, then the continuity equation demands the presence of line or point charges at such edges. These fictitious charges, when present, usually cause anomalies or inconsistencies in the solution. Although the approaches of Wang *et al.* and Newman and Pozar are free of these difficulties, their use of rectangular patches, with their consequent limitation to surfaces with curvature in one dimension only, is too restrictive for many applications.

For modeling arbitrarily shaped surfaces, planar triangular patch models, an example of which is shown in Fig. 1, are particularly appropriate. Some of the advantages of triangular patch surface modeling have been noted by Sankar and Tong [9], as well as by Wang [10], and are similar to those of wire-grid modeling. For example, triangular patches are capable of accurately conforming to any geometrical surface or boundary, the patch scheme is easily specified for computer input, and a varying patch density can be used according to the resolution required in the surface geometry or current. Although planar quadrilateral (nonrectangular) patches share some of these features, it is difficult to construct basis functions defined on them which are free of line charges. Furthermore, the vertices of planar quadrilaterals cannot be independently specified—a restriction that is a severe inconvenience to the modeler.

In this paper, a numerical solution of the problem of scattering by either open or closed arbitrarily shaped conducting bodies is presented. The approach combines the advantages of triangular patch modeling and the EFIE formulation, and results in an algorithm which is both simple and efficient. Crucial to the approach is the development of special basis functions defined on triangular patches which are free of fictitious line or point charges and which are analogous to the so-called "rooftop" functions used in rectangular patch models [14].

In the following section the EFIE formulation is presented, the special set of basis functions is developed, and the method of moments [15] is applied to obtain a linear system of equations for the surface current. In Section III, numerical results are presented for scattering by a flat square plate, a bent square plate, a circular disk, and a sphere. Section IV summarizes the contents of the paper.

II. ELECTRIC FIELD FORMULATION

In this section an integral equation for the surface current induced on a conducting scatterer is derived from boundary conditions on the electric field. To solve the integral equation by the method of moments, a set of expansion functions and a testing procedure are developed and used to derive the elements of the moment matrix. Finally, the numerical computation of the moment matrix elements is discussed.

Electric Field Integral Equation

Let S denote the surface of an open or closed perfectly conducting scatterer with unit normal $\hat{\mathbf{n}}$. An electric field $\mathbf{E}^i$, defined to be the field due to an impressed source in the absence of the scatterer, is incident on and induces surface currents $\mathbf{J}$ on S. If S is open, we regard $\mathbf{J}$ as the vector sum of the surface currents on opposite sides of S; hence the normal component of $\mathbf{J}$ must vanish on boundaries of S. The scattered electric field

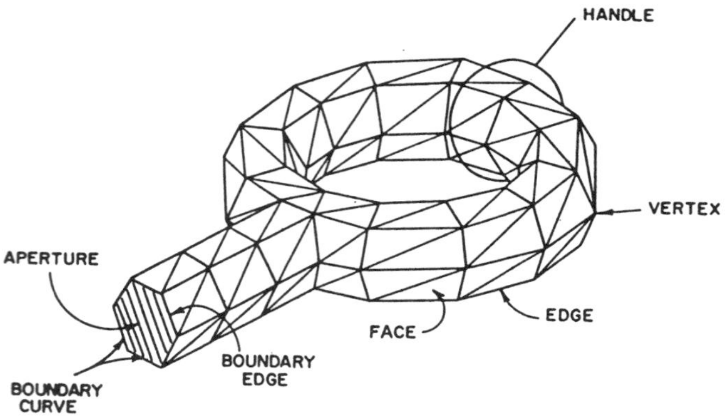

Fig. 1. Arbitrary surface modeled by triangular patches.

$\mathbf{E}^s$ can be computed from the surface current by

$$\mathbf{E}^s = -j\omega\mathbf{A} - \nabla\Phi \tag{1}$$

with the magnetic vector potential defined as

$$\mathbf{A}(\mathbf{r}) = \frac{\mu}{4\pi}\int_S \mathbf{J}\,\frac{e^{-jkR}}{R}\,dS' \tag{2}$$

and the scalar potential as

$$\Phi(\mathbf{r}) = \frac{1}{4\pi\epsilon}\int_S \sigma\,\frac{e^{-jkR}}{R}\,dS'. \tag{3}$$

A harmonic time dependence $\exp(j\omega t)$ is assumed and suppressed, and $k = \omega\sqrt{\mu\epsilon} = 2\pi/\lambda$, where λ is the wavelength. The permeability and permittivity of the surrounding medium are μ and ϵ, respectively, and $R = |\mathbf{r} - \mathbf{r}'|$ is the distance between an arbitrarily located observation point $\mathbf{r}$ and a source point $\mathbf{r}'$ on S. Both $\mathbf{r}$ and $\mathbf{r}'$ are defined with respect to a global coordinate origin O. The surface charge density σ is related to the surface divergence of $\mathbf{J}$ through the equation of continuity,

$$\nabla_s \cdot \mathbf{J} = -j\omega\sigma. \tag{4}$$

We derive an integrodifferential equation for $\mathbf{J}$ by enforcing the boundary condition $\hat{\mathbf{n}} \times (\mathbf{E}^i + \mathbf{E}^s) = \mathbf{0}$ on S, obtaining

$$-\mathbf{E}_{\tan}{}^i = (-j\omega\mathbf{A} - \nabla\Phi)_{\tan}, \quad \mathbf{r} \text{ on } S. \tag{5}$$

Equation (5), with (2)–(4), constitutes the so-called electric field integral equation. One notes that the presence of derivatives on the current in (4) and on the scalar potential in (5) suggests that care should be taken in selecting the expansion functions and testing procedure in the method of moments.

Development of Basis Functions

In this section we discuss a set of basis functions introduced by Glisson [16] which is suitable for use with the EFIE and triangular patch modeling. We assume that a suitable triangulation, defined in terms of an appropriate set of faces, edges, vertices, and boundary edges, as illustrated in Fig. 1, has been found to approximate S. (More detailed considerations concerning the mathematical representation and topological properties of triangular patch models may be found in [17].)

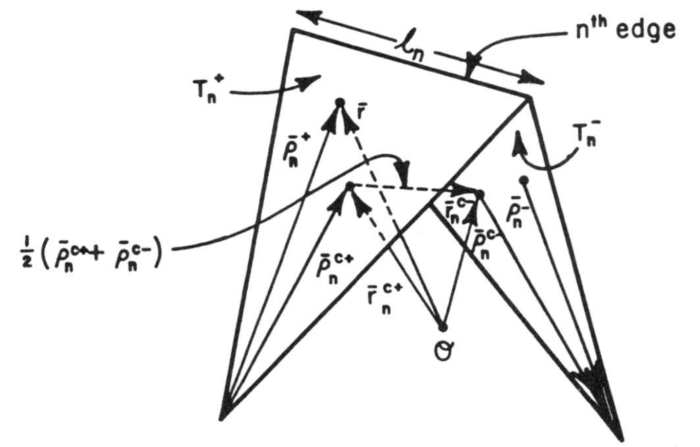

Fig. 2. Triangle pair and geometrical parameters associated with interior edge.

It is convenient to start our development by noting that each basis function is to be associated with an *interior edge* (i.e., nonboundary edge (cf Fig. 1)) of the patch model and is to vanish everywhere on S except in the two triangles attached to that edge. Fig. 2 shows two such triangles, T_n^+ and T_n^-, corresponding to the nth edge of a triangulated surface modeling a scatterer. Points in T_n^+ may be designated either by the position vector $\mathbf{r}$ defined with respect to $\mathcal{O}$, or by the position vector ρ_n^+ defined with respect to the free vertex of T_n^+. Similar remarks apply to the position vector ρ_n^- except that it is directed toward the free vertex of T_n^-. The plus or minus designation of the triangles is determined by the choice of a positive current reference direction for the nth edge, the reference for which is assumed to be from T_n^+ to T_n^-.[1] We define the vector basis function associated with the nth edge as

$$\mathbf{f}_n(\mathbf{r}) = \begin{cases} \dfrac{l_n}{2A_n^+}\rho_n^+, & \mathbf{r} \text{ in } T_n^+ \\ \dfrac{l_n}{2A_n^-}\rho_n^-, & \mathbf{r} \text{ in } T_n^- \\ 0, & \text{otherwise,} \end{cases} \tag{6}$$

where l_n is the length of the edge and $A_n^\pm$ is the area of triangle $T_n^\pm$. (Note that we use the convention, followed throughout the paper, that subscripts refer to edges while superscripts refer to faces.) The basis function $\mathbf{f}_n$ is used to approximately represent the surface current, and we list and discuss below some properties which make it uniquely suited to this role.

1) The current has no component normal to the boundary (which excludes the common edge) of the surface formed by the triangle pair T_n^+ and T_n^-, and hence no line charges exist along this boundary.
2) The component of current normal to the nth edge is constant and continuous across the edge as may be seen with the aid of Fig. 3, which shows that the normal component of $\rho_n^\pm$ along edge n is just the height of triangle $T_n^\pm$ with edge n as the base and the height expressed as $(2A_n^\pm)/l_n$. This latter factor normalizes $\mathbf{f}_n$ in (6) such that its flux density normal to edge n is unity, ensuring continuity of current normal to the edge. This result, together with 1), implies that *all* edges of T_n^+ and T_n^- are free of line charges.

[1] For orientable surfaces, the current reference direction may be obtained from the connection matrix used to describe the triangulation scheme. This matrix merely lists the vertices linked by each edge, the order of appearance of the vertices for each edge effectively assigning an orientation to the edge. The direction of the cross product of this edge orientation vector with the surface normal in each adjacent triangle may be taken as the positive current reference direction in that triangle for the basis function associated with the edge.

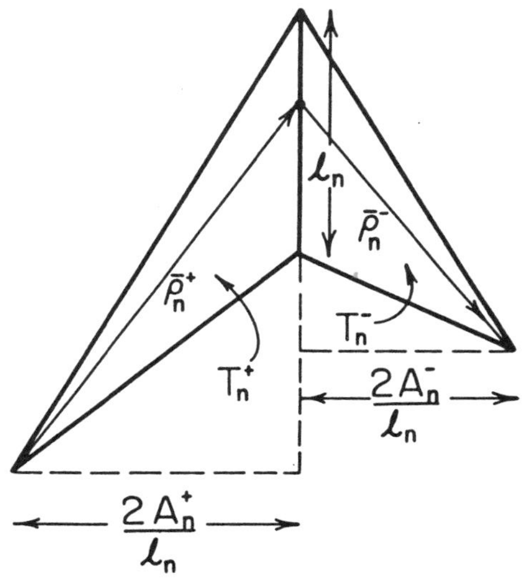

Fig. 3. Geometry for construction of component of basis function normal to edge.

3) The surface divergence of $\mathbf{f}_n$, which is proportional to the surface charge density associated with the basis element, is

$$\nabla_s \cdot \mathbf{f}_n = \begin{cases} \dfrac{l_n}{A_n^+}, & \mathbf{r} \text{ in } T_n^+ \\ -\dfrac{l_n}{A_n^-}, & \mathbf{r} \text{ in } T_n^- \\ 0, & \text{otherwise,} \end{cases} \tag{7}$$

since the surface divergence in $T_n^\pm$ is $(\pm 1/\rho_n^\pm)\partial(\rho_n^\pm f_n)/\partial\rho_n^\pm$. The charge density is thus constant in each triangle, the total charge associated with the triangle pair T_n^+ and T_n^- is zero, and the basis functions for the charge evidently have the form of *pulse doublets* [14].
4) The *moment* of $\mathbf{f}_n$ is given by $(A_n^+ + A_n^-)\mathbf{f}_n^{\text{avg}}$ where

$$\begin{aligned}(A_n^+ + A_n^-)\mathbf{f}_n^{\text{avg}} &\equiv \int_{T_n^+ + T_n^-} \mathbf{f}_n \, dS \\ &= \frac{l_n}{2}(\rho_n^{c+} + \rho_n^{c-}) \\ &= l_n(\mathbf{r}_n^{c+} - \mathbf{r}_n^{c-})\end{aligned} \tag{8}$$

and $\rho_n^{c\pm}$ is the vector between the free vertex and the centroid of $T_n^\pm$ with ρ_n^{c-} directed toward and ρ_n^{c+} directed away from the vertex, as shown in Fig. 2, and $\mathbf{r}_n^{c\pm}$ is the vector from $\mathcal{O}$ to the centroid of $T_n^\pm$. Equation (8) may be most easily derived by expressing the integral in terms of area coordinates, to be discussed below.

The current on S may be approximated in terms of the $\mathbf{f}_n$ as

$$\mathbf{J} \cong \sum_{n=1}^{N} I_n \mathbf{f}_n(\mathbf{r}) \tag{9}$$

where N is the number of interior (nonboundary) edges. Since a basis function is associated with each nonboundary edge of the triangulated structure, up to three basis functions may have nonzero values within each triangular face. But at a given edge only the basis function associated with that edge has a current component *normal to the edge* since, according to 1), all other basis currents in adjacent faces are parallel to the edge. Furthermore, since the normal component of $\mathbf{f}_n$ at the nth edge is unity, *each coefficient I_n in (9) may be interpreted as the normal component of current density flowing past the nth edge.* Also, we see that the basis functions are independent in each triangle since the current normal to the nth edge, I_n in (9), is an independent quantity. At surface boundary edges, the sum of the normal components of current on opposite sides of the surface cancel because of current continuity. Therefore we neither define nor include in (9) contributions from basis functions associated with such edges.

Because of the considerable variation in the direction of the flow lines of $\mathbf{f}_n$ within a triangle, it is not at first obvious that a linear superposition of basis functions is capable of representing, say, a constant current flowing in an arbitrary direction within a triangle. That this is possible, however, can be seen with the aid of Fig. 4, which shows a triangle T^q with edges arbitrarily labeled 1, 2, and 3 (in effect, we employ here a "local indexing scheme," in contrast to the "global indexing scheme" used earlier). With the vectors ρ_1, ρ_2, and ρ_3 as shown, the basis functions in T^q are $\mathbf{f}_i = (l_i/2A^q)\rho_i$, $i = 1, 2, 3$, where A^q is the triangle area and where, for simplicity, the current reference directions are assumed to be out of the triangle for each edge. It is apparent from the figure and the definition of $\mathbf{f}_i$ that the linear combinations $l_2\mathbf{f}_1 - l_1\mathbf{f}_2$ and $l_3\mathbf{f}_1 - l_1\mathbf{f}_3$ are *constant* vectors for every point $\mathbf{r}$ in T^q and are parallel to sides 3 and 2, respectively. Since the two composite forms are linearly independent (i.e., nonparallel), a constant vector of arbitrary magnitude and direction within T^q may be synthesized by an appropriate linear combination of the two forms, as asserted.

Testing Procedure

The next step in the method of moments is to select a testing procedure. We choose as testing functions the expansion functions $\mathbf{f}_n$ developed in the previous section. With a symmetric product defined as

$$\langle \mathbf{f}, \mathbf{g} \rangle \equiv \int_S \mathbf{f} \cdot \mathbf{g}\, dS, \tag{10}$$

(5) is tested with $\mathbf{f}_m$, yielding

$$\langle \mathbf{E}^i, \mathbf{f}_m \rangle = j\omega \langle \mathbf{A}, \mathbf{f}_m \rangle + \langle \nabla\Phi, \mathbf{f}_m \rangle. \tag{11}$$

If one makes use of a surface vector calculus identity [18] and the properties of $\mathbf{f}_m$ at the edges of S, the last term in (11) can be rewritten as

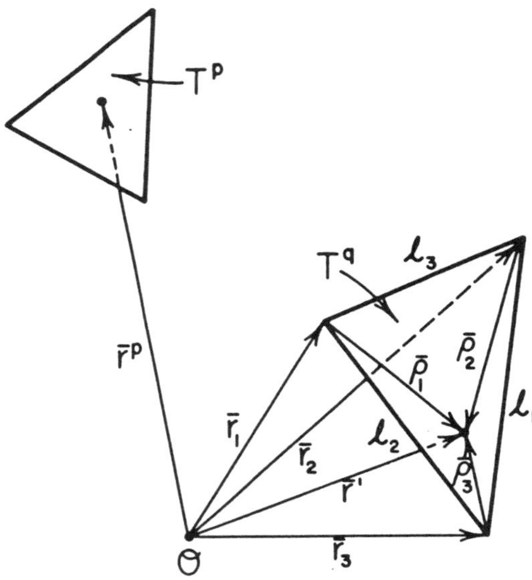

Fig. 4. Local coordinates and edges for source triangle T^q with observation point in triangle T^p.

$$\langle \nabla\Phi, \mathbf{f}_m \rangle = -\int_S \Phi \nabla_s \cdot \mathbf{f}_m\, dS. \tag{12}$$

With (7), the integral in (12) may now be written and approximated as follows:

$$\begin{aligned}\int_S \Phi \nabla_s \cdot \mathbf{f}_m\, dS &= l_m \left(\frac{1}{A_m^+} \int_{T_m^+} \Phi\, dS - \frac{1}{A_m^-} \int_{T_m^-} \Phi\, dS \right) \\ &\cong l_m [\Phi(\mathbf{r}_m^{c+}) - \Phi(\mathbf{r}_m^{c-})].\end{aligned} \tag{13}$$

In (13) the average of Φ over each triangle is approximated by the value of Φ at the triangle centroid. With similar approximations, the vector potential and incident field terms in (11) may be written as

$$\begin{aligned}\left\langle \begin{Bmatrix} \mathbf{E}^i \\ \mathbf{A} \end{Bmatrix}, \mathbf{f}_m \right\rangle &= l_m \left[\frac{1}{2A_m^+} \int_{T_m^+} \begin{Bmatrix} \mathbf{E}^i \\ \mathbf{A} \end{Bmatrix} \cdot \rho_m^+\, dS + \frac{1}{2A_m^-} \int_{T_m^-} \begin{Bmatrix} \mathbf{E}^i \\ \mathbf{A} \end{Bmatrix} \cdot \rho_m^-\, dS \right] \\ &\cong \frac{l_m}{2} \left[\begin{Bmatrix} \mathbf{E}^i(\mathbf{r}_m^{c+}) \\ \mathbf{A}(\mathbf{r}_m^{c+}) \end{Bmatrix} \cdot \rho_m^{c+} + \begin{Bmatrix} \mathbf{E}^i(\mathbf{r}_m^{c-}) \\ \mathbf{A}(\mathbf{r}_m^{c-}) \end{Bmatrix} \cdot \rho_m^{c-} \right],\end{aligned} \tag{14}$$

where the integral over each triangle is eliminated by approximating $\mathbf{E}^i$ (or $\mathbf{A}$) in each triangle by its value at the triangle centroid and carrying out integrations similar to those used to obtain (8). With (12)-(14), (11) now becomes

$$j\omega l_m \left[\mathbf{A}(\mathbf{r}_m^{c+}) \cdot \frac{\boldsymbol{\rho}_m^{c+}}{2} + \mathbf{A}(\mathbf{r}_m^{c-}) \cdot \frac{\boldsymbol{\rho}_m^{c-}}{2} \right] + l_m [\Phi(\mathbf{r}_m^{c-}) - \Phi(\mathbf{r}_m^{c+})] = l_m \left[\mathbf{E}^i(\mathbf{r}_m^{c+}) \cdot \frac{\boldsymbol{\rho}_m^{c+}}{2} + \mathbf{E}^i(\mathbf{r}_m^{c-}) \cdot \frac{\boldsymbol{\rho}_m^{c-}}{2} \right], \quad (15)$$

which is the equation enforced at each triangle edge, $m = 1, 2, \cdots, N$.

We remark that another interpretation of the testing procedure arriving at (15) is also possible. One may equate line integrals of the form $\int_{C_m} \mathbf{F} \cdot d\mathbf{r}$, where $\mathbf{F}$ represents the right and left sides of (5), and C_m is the piecewise linear path from the point $\mathbf{r}_m^{c+}$ to the midpoint of edge m and thence to $\mathbf{r}_m^{c-}$. $\mathbf{E}^i$ and $\mathbf{A}$ can be approximated along each portion of the path by their respective values at the triangle centroids. The resulting equality, apart from the factor l_m, is (15). Under either interpretation, the testing procedure reduces the differentiability requirement on Φ in (5) by integrating $\nabla\Phi$ first, the procedure having been constructed with this goal in mind. The purpose of approximations (13) and (14) is to eliminate surface integrals of the potential quantities, allowing a double surface integral to be approximated by a quantity involving a single surface integral in the numerical computation of the moment matrix elements. These approximations are justified by observing that the potentials are locally smooth within each subdomain, as follows from their integral definitions and the locally smooth nature of the source representation in terms of the basis functions [14].[2]

Matrix Equation Derivation

Substitution of the current expansion (9) into (15) yields an $N \times N$ system of linear equations which may be written in matrix form as

$$ZI = V \quad (16)$$

where $Z = [Z_{mn}]$ is an $N \times N$ matrix and $I = [I_n]$ and $V = [V_m]$ are column vectors of length N. Elements of Z and V are given by

$$Z_{mn} = l_m \left[j\omega \left(\mathbf{A}_{mn}^{+} \cdot \frac{\boldsymbol{\rho}_m^{c+}}{2} + \mathbf{A}_{mn}^{-} \cdot \frac{\boldsymbol{\rho}_m^{c-}}{2} \right) + \Phi_{mn}^{-} - \Phi_{mn}^{+} \right] \quad (17)$$

[2] Note that if the approximations (13) and (14) had not been made, the procedure leading to (15) would have been identical to Galerkin's method since the basis and testing functions chosen are identical [15]. For the EFIE, the matrix would then satisfy the symmetry property $Z_{mn} = Z_{nm}$, but this desirable property is lost due to the approximations made. Since Z_{mn} and Z_{nm} are different approximations to the same quantity, however, then their average also approximates the quantity, and one tempting possibility is to average Z with its transpose, thus restoring the symmetry property to the moment matrix. This approach, however, has not to date been tested.

$$V_m = l_m \left(\mathbf{E}_m^{+} \cdot \frac{\boldsymbol{\rho}_m^{c+}}{2} + \mathbf{E}_m^{-} \cdot \frac{\boldsymbol{\rho}_m^{c-}}{2} \right) \quad (18)$$

where

$$\mathbf{A}_{mn}^{\pm} = \frac{\mu}{4\pi} \int_S \mathbf{f}_n(\mathbf{r}') \frac{e^{-jkR_m^{\pm}}}{R_m^{\pm}} \, dS', \quad (19)$$

$$\Phi_{mn}^{\pm} = -\frac{1}{4\pi j\omega\epsilon} \int_S \nabla_s' \cdot \mathbf{f}_n(\mathbf{r}') \frac{e^{-jkR_m^{\pm}}}{R_m^{\pm}} \, dS', \quad (20)$$

$$R_m^{\pm} = |\mathbf{r}_m^{c\pm} - \mathbf{r}'|$$

and

$$\mathbf{E}_m^{\pm} = \mathbf{E}^i(\mathbf{r}_m^{c\pm}). \quad (21)$$

For plane wave incidence, we set

$$\mathbf{E}^i(\mathbf{r}) = (E_\theta \hat{\theta}_0 + E_\phi \hat{\phi}_0) e^{j\mathbf{k}\cdot\mathbf{r}} \quad (22)$$

where the propagation vector $\mathbf{k}$ is

$$\mathbf{k} = k(\sin\theta_0 \cos\phi_0 \hat{\mathbf{x}} + \sin\theta_0 \sin\phi_0 \hat{\mathbf{y}} + \cos\theta_0 \hat{\mathbf{z}}) \quad (23)$$

and $(\hat{\theta}_0, \hat{\phi}_0)$ defines the angle of arrival of the plane wave in terms of the usual spherical coordinate convention. Unit vectors θ_0 and ϕ_0 are *constant* vectors which coincide with the usual spherical coordinate unit vectors only at points on the line from O in the direction of $\mathbf{k}$.

Once the elements of the moment matrix and the forcing vector V are determined, one may solve the resulting system of linear equations (16) for the unknown column vector I. The elements of Z in (16) may be evaluated by naively computing Z_{mn} directly by (17) (with aid of (19) and (20)) for each index combination m and n. However, as shown in the next section, this procedure is extremely inefficient since the integrals required for each combination of m and n are generally required for a number of other combinations as well.

Efficient Numerical Evaluation of Matrix Elements

Evaluation of each matrix element Z_{mn} associated with edges m and n involves integrations over triangles $T_n^{\pm}$ with observation points located at the centroids of triangles $T_m^{\pm}$. One is easily convinced that some of the same integrals required for an element Z_{mn} are also needed to compute an element Z_{rs}, $r \neq m$, $s \neq n$, if edge r happens to be an edge of T_m^{+} or T_m^{-} while edge s is an edge of T_n^{+} or T_n^{-}. Indeed, if one focuses attention on a single pair of *faces* rather than on a pair of *edges*, one observes that the integrals evaluated for a source face with scalar and vector potentials observed at the centroid of another face are involved in all the elements Z_{mn} having edges n as (nonboundary) edges of the source triangle and edges m as the (nonboundary) edges of the observation triangle. Thus, the total number of matrix elements requiring evaluation of the same potential integrals can be as large as nine. Clearly then, it is far more efficient to compute the required potential integrals by face-pair combinations, rather than directly compute single elements of Z by edge-pair combinations. For each face-pair combination, the potential integrals may be multiplied by the appropriate coefficients (cf (17)) and their contributions accumulated in the appropriate elements of Z as they are computed.

In accordance with the above discussion, consider the evaluation of the vector and scalar potential integrals for a given source and observation face combination. Fig. 4 illustrates such a face pair with an observation point in face p and with source currents residing in face q. Each of the three basis functions which may exist simultaneously in T^q is proportional to one of the vectors ρ_1, ρ_2, or ρ_3 defined in the figure. Each vector ρ_i, $i = 1, 2, 3$, is shown directed away from its associated vertex in the figure, but would be directed toward the vertex if the current reference direction for the associated edge were into the triangle. Consequently,

$$\rho_i = \pm(\mathbf{r}' - \mathbf{r}_i), \qquad i = 1, 2, 3, \tag{24}$$

where the positive sign is used if the positive current reference direction is out of T^q and the negative sign is used otherwise. We wish to evaluate the magnetic vector potential,

$$\mathbf{A}_i^{pq} = \frac{\mu}{4\pi} \int_{T^q} \left(\frac{l_i}{2A^q}\right) \rho_i \frac{e^{-jkR^p}}{R^p}\, dS', \tag{25}$$

and the electric scalar potential,

$$\Phi_i^{pq} = -\frac{1}{4\pi j\omega\epsilon} \int_{T^q} \left(\frac{l_i}{A^q}\right) \frac{e^{-jkR^p}}{R^p}\, dS', \tag{26}$$

associated with the ith basis function on face q observed at the centroid of face p. In (25) and (26),

$$R^p = |\mathbf{r}^{cp} - \mathbf{r}'| \tag{27}$$

where $\mathbf{r}^{cp}$ is the position vector of the centroid of face p.

Integrals (25) and (26) are most conveniently evaluated by transforming from the global coordinate system to a local system of coordinates defined within T^q. To define these coordinates, note that the vectors ρ_i in Fig. 4 divide T^q into three subtriangles of areas A_1, A_2, and A_3, with l_1, l_2, and l_3, respectively, as one of their sides. The areas are not independent, however, since they must satisfy $A_1 + A_2 + A_3 = A^q$. We now introduce the so-called *normalized area coordinates* [19]

$$\xi = \frac{A_1}{A^q}, \quad \eta = \frac{A_2}{A^q}, \quad \zeta = \frac{A_3}{A^q}, \tag{28}$$

which, because of the area constraint, must satisfy

$$\xi + \eta + \zeta = 1. \tag{29}$$

Note that all three coordinates vary between zero and unity in T^q and that at the triangle corners $\mathbf{r}_1$, $\mathbf{r}_2$, and $\mathbf{r}_3$, the triplet (ξ, η, ζ) takes on the values (1, 0, 0), (0, 1, 0), and (0, 0, 1), respectively. The transformation from Cartesian to normalized area coordinates may be written in vector form as

$$\mathbf{r}' = \xi\mathbf{r}_1 + \eta\mathbf{r}_2 + \zeta\mathbf{r}_3, \tag{30}$$

where ξ, η, and ζ are subject to the constraint (29). It can easily be shown that surface integrals over T^q transform as follows:

$$\int_{T^q} g(\mathbf{r})\, dS = 2A^q \int_0^1 \int_0^{1-\eta} g[\xi\mathbf{r}_1 + \eta\mathbf{r}_2 + (1 - \xi - \eta)\mathbf{r}_3]\, d\xi\, d\eta. \tag{31}$$

With (24), (27), (30), and (31), (25) and (26) may now be written as

$$\mathbf{A}_i^{pq} = \pm\frac{\mu l_i}{4\pi}(\mathbf{r}_1 I_\xi^{pq} + \mathbf{r}_2 I_\eta^{pq} + \mathbf{r}_3 I_\zeta^{pq} - \mathbf{r}_i I^{pq}) \tag{32}$$

and

$$\Phi_i^{pq} = \mp\frac{l_i}{j2\pi\omega\epsilon} I^{pq}, \tag{33}$$

where

$$I^{pq} = \int_0^1 \int_0^{1-\eta} \frac{e^{-jkR^p}}{R^p}\, d\xi\, d\eta, \tag{34a}$$

$$I_\xi^{pq} = \int_0^1 \int_0^{1-\eta} \xi \frac{e^{-jkR^p}}{R^p}\, d\xi\, d\eta, \tag{34b}$$

$$I_\eta^{pq} = \int_0^1 \int_0^{1-\eta} \eta \frac{e^{-jkR^p}}{R^p}\, d\xi\, d\eta, \tag{34c}$$

$$I_\zeta^{pq} = I^{pq} - I_\xi^{pq} - I_\eta^{pq}. \tag{34d}$$

Thus only three independent integrals, (34a)–(34c), must be numerically evaluated for each combination of face pairs p and q. The three integrals, in turn, contribute to up to nine elements of Z in (16). For a closed object with N edges the number of independent integrals computed is $4N^2/3$. By contrast, the edge-by-edge approach would require the evaluation of $12N^2$ integrals or nine times as many. Numerical evaluation of the integrals (34a)–(34c) may be accomplished by using numerical quadrature techniques specially developed for triangular domains [20]. However, for the terms in which $p = q$ the integrands are singular, and for these cases the singular portion of each integrand must be removed and integrated analytically [17].

III. NUMERICAL RESULTS

In this section numerical results are presented for surface current distributions induced on selected scatterers under plane wave illumination. Although one radar cross-section example is given, emphasis in the examples is on the calculation of current distributions, not only because of their practical value in problems such as electromagnetic compatibility and nuclear electromagnetic pulse (EMP) penetration, but also because we believe that the calculation of accurate surface currents is a much more stringent test of a numerical approach than is ultimate calculation of far-field quantities. The geometries considered are a conducting square plate, a bent plate, a circular disk, and a sphere. The plate and disk problems involve open surfaces and therefore are a test of the EFIE approach when edges are present. The disk also serves as an example of a structure with a curved boundary, while the sphere exemplifies both a closed surface and a doubly curved surface. Both problems are examples of surfaces not amenable to rectangular patch modeling. A summary of the computational resources required to calculate results for several of the examples is given at the end of the section.

Flat Plate

Figs. 5 and 6 show the dominant component current distributions along the two principal cuts on a square plate illuminated by a normally incident plane wave. For comparison, the solution of Glisson [16], obtained using rectangular patches, is also given. The number of patches listed in the figures refers to the number of *charge* patches in the earlier solution of Glisson and to the number of triangles (also equal to the number of charge patches) in the present solution. Note that these quantities play similar roles in the two approaches. No comparison of the convergence rate of the two approaches should be inferred from the figures since both solutions are already well converged for the number of unknowns used. Note also that the density of data points appearing in the figures for the triangular patch solution is not truly indicative of the linear density of the subdomains. This is because, in effect, we show data points only for every other edge, i.e., only for those edges where the current reference direction vector is parallel to the current component we wish to observe.

Fig. 5 shows the current induced on a plate 0.15 λ on each side. At this low frequency the current distribution is largely determined by the edge conditions and hence this case provides a good test of the technique's capacity for handling surface edges. We note the absence of any anomalies in the computed distribution near the plate edges. The elimination of such anomalies is attributed to the use of basis functions in which the expansion coefficients are not associated with current components parallel to plate edges and to a testing procedure in which potentials are not evaluated at edges [14].

Fig. 6 shows corresponding results for a 1.0 λ square plate. From the figure, one observes that the edge behavior of the current distribution is confined to a relatively smaller region near the edges than for the 0.15 λ plate and that the current on the interior portion of the plate has begun to exhibit the physical optics-plus standing wave distribution characteristic of the higher frequencies. Also shown for comparison are the corresponding results reported in [21].[3] Note that because of the placement of subdomains, the component of current normal to the plate edge in the solution of [21] vanishes prematurely in the subdomain nearest the edge.

Fig. 7 compares the calculated radar cross section (RCS) with the thin plate measurements of Kouyoumjian [22] and the computations of Rahmat-Samii and Mittra [23] for a square plate. In the latter case, we suspect that the premature vanishing of the current near the plate edge causes the underestimation in the RCS observed in the figure for lower frequencies. Kouyoumjian has also computed the RCS from a variational formula and we find no discernable difference between his computations and our results for square plates whose sides are smaller than 0.4 λ, the range for which his formula should be most accurate. Also shown in Fig. 7 is a plot of the

[3] The data reported in [21] is actually for the electric field in a square aperture in a ground plane but has been converted to an equivalent electric current on a square conducting plate via the duality of the plate and aperture problems.

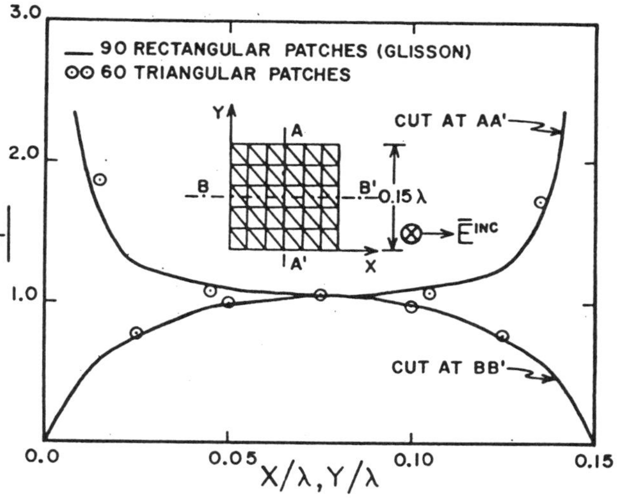

Fig. 5. Distribution of dominant component of current on 0.15 λ square flat plate.

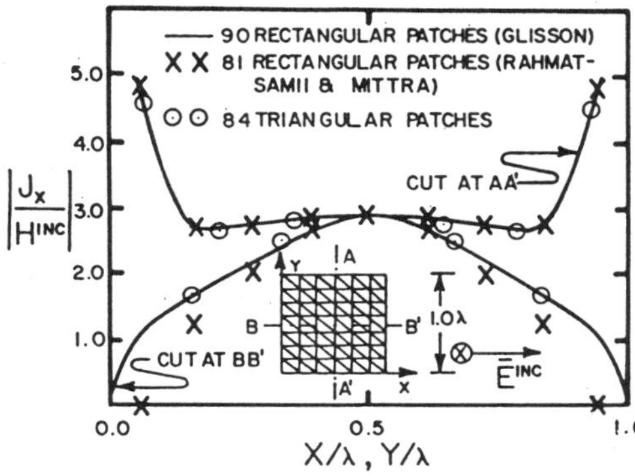

Fig. 6. Distribution of dominant component of current on 1.0 λ square flat plate.

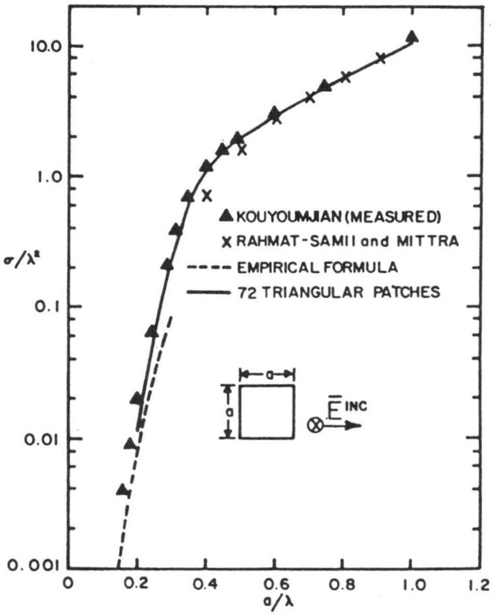

Fig. 7. Monostatic radar cross section versus frequency for square plate, normal incidence. Measurements are for plate of thickness 0.000127 λ.

RCS for a square plate given by an empirical formula derived from measurements [24].

While there is generally very good agreement between the various results reported in Figs. 6 and 7, we have not been able to obtain a favorable comparison with the corresponding current distribution and RCS calculations of Wang *et al.* [7]. There are several reasons to suspect their calculations. First, they employ a set of basis functions which would generally cause spurious oscillations in the current distribution along the direction of current flow, and which would hence be expected to adversely affect the solution for the current distribution and, ultimately, the RCS. Secondly, their model ignores the presence of cross-polarized components of surface current on the plate. Finally, in the RCS case, the excellent agreement which they obtain between computed and measured radar cross section is unfortunately based on a comparison to the thick plate rather than the thin plate measurements of Kouyoumjian [22] for which their theory more appropriately applies and which we have shown in Fig. 7.

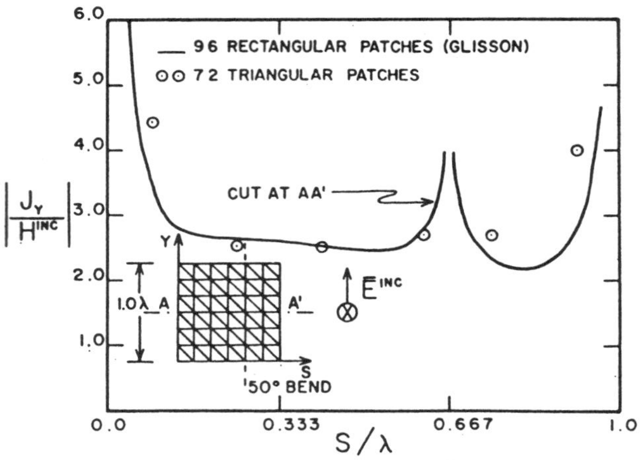

Fig. 8. Distribution of dominant component of current on 1.0 λ bent square plate.

Bent Plate

Fig. 8 shows the dominant component current distribution along a cut through the symmetry plane of a bent square plate. The bend is parallel to and located a distance of one-third the plate width from one edge, and a plane wave with the electric field polarized parallel to the bend is incident normal to the larger section of the bent plate. The smaller plate section is bent through an angle of 50° toward the shadow side of the plate. Other frequencies, polarizations, and angles of incidence have been examined, and the resulting current distributions show good correspondence with those of Glisson [16].

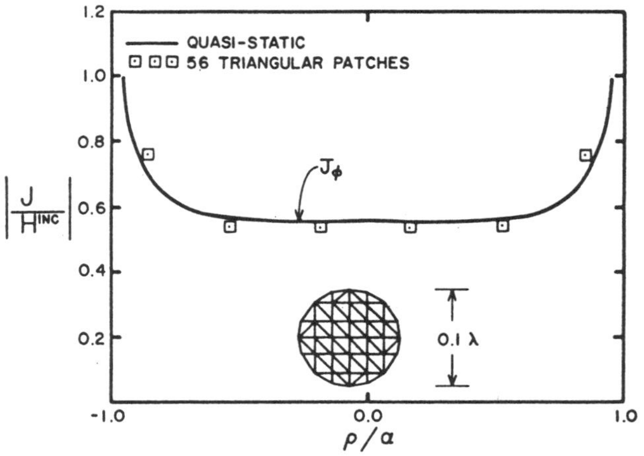

Fig. 9. Distribution of current on 0.05 λ radius circular disk.

Circular Disk

Fig. 9 shows the computed current distribution on a circular disk illuminated by a normally incident plane wave. The component J_ϕ is shown along a diameter cut oriented perpendicular to the incident electric field vector. Also shown for comparison is the quasi-static solution valid at low frequencies [25].

Sphere

Fig. 10 shows the computed current distribution along the principal cuts on a 0.2 λ radius conducting sphere. The cases of axial and equatorial incidence are both considered in order to observe the influence of the triangulation scheme on the solution. Also shown for comparison is the exact eigenfunction solution. Results for both illuminations are in very good agreement with the exact solution.

Although we have not attempted to solve the sphere problem at frequencies near its cavity resonances, we expect the usual difficulties associated with the EFIE formulation at the internal resonance frequencies of closed bodies to arise. To alleviate the problem, one might make use of the present treatment of the EFIE operator in a formulation which combines it with the MFIE operator in such a way as to eliminate the singularity which is present in both operators alone [26].

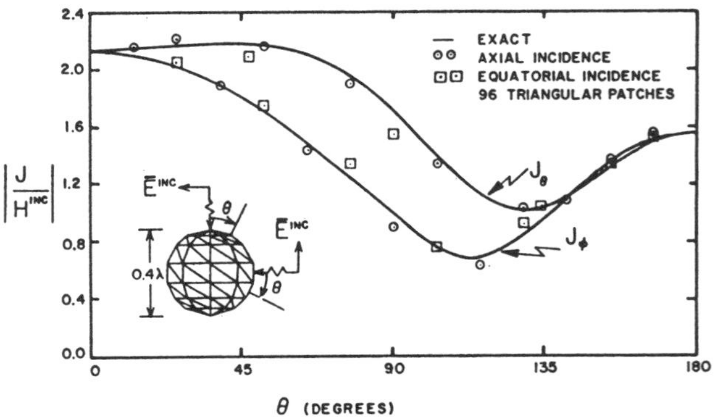

Fig. 10. Distribution of current components on 0.2 λ radius conducting sphere.

Computational Aspects

Table I summarizes data on the number of triangular patches, number of unknowns, and the computation time required on a Univac 1100/83 computer to generate the results of Figs. 5, 6, 8, and 10. The timing data for Fig. 6 were not re-

TABLE I
SUMMARY OF COMPUTATION DATA FOR SELECTED EXAMPLE PROBLEMS

Case (Figure Number)	Number of Patches N_f	Number of Boundary Edges N_b	Number of Unknowns N†	Matrix Fill Time (s)	Matrix Inversion Time (s)	Total CPU Time (s)
5	60	22	79	65	20	92
6	84	26	113	127*	57*	195*
8	72	24	96	94	35	139
10	96	0	144	163	118	294

* Estimated from convergence study.
† $N = (3N_f - N_b)/2$.

corded during computation but are estimated from the convergence study discussed below. The present version of the computer code has not been completely optimized with respect to computation speed, and it should be possible to increase this speed by at least a factor of two or three by 1) decreasing the order of integration used to calculate the potential integrals when R^p in (34) is large compared to source patch size, and 2) replacing the matrix inversion procedure, used primarily for diagnostic purposes, by a linear equation solving procedure.

A convergence study of the current distribution on the 1.0 λ square plate of Fig. 6 was carried out using 18, 32, 50, and 72 patches corresponding to 21, 40, 65, and 96 unknowns, respectively. It was found that, except for the nearly singular component of current nearest the plate edge, the computed currents in the sequence of computations differed from the results of Glisson (cf Fig. 6) by no more than 10 percent.

IV. SUMMARY

In this paper, the electric field integral equation (EFIE) is used with the method of moments to develop a simple and efficient numerical procedure for treating problems of scattering by arbitrarily shaped objects. For numerical purposes the objects are modeled by planar triangular surface patch models. Because the EFIE formulation is used, the procedure is applicable to both open and closed bodies. Crucial to the formulation is the development of a set of special subdomain basis functions which are defined on pairs of adjacent triangular patches and yield a current representation free of line or point charges at subdomain boundaries.

The approach is applied to the scattering problems of plane wave illumination of a flat square plate, a bent square plate, a circular disk, and a sphere. Comparisons of surface current density with previous computations or exact formulations show good correspondence in each case.

A listing of the computer code is available from the authors.

ACKNOWLEDGMENT

The authors are grateful to Mr. Robert Baxter and Dr. Deanne Pecora of the Lockheed Missiles and Space Company, Inc., for providing us with the results of their timing and convergence study summarized in Table I.

REFERENCES

[1] J. H. Richmond, "A wire-grid model for scattering by conducting bodies," *IEEE Trans. Antennas Propagat.*, vol. AP-14, no. 6, pp. 782–786, Nov. 1966.

[2] E. K. Miller and F. J. Deadrick, "Some computational aspects of thin-wire modeling," in *Numerical and Asymptotic Techniques in Electromagnetics*, R. Mittra, Ed. New York: Springer-Verlag, 1975, ch. 4.

[3] K. S. H. Lee, L. Marin, and J. P. Castillo, "Limitations of wire-grid modeling of a closed surface," *IEEE Trans. Electromagn. Compat.*, vol. EMC-18, no. 3, pp. 123–129, Aug. 1976.

[4] D. L. Knepp and J. Goldhirsh, "Numerical analysis of electromagnetic radiation properties of smooth conducting bodies of arbitrary shape," *IEEE Trans. Antennas Propagat.*, vol. AP-20, no. 3, pp. 383–388, May 1972.

[5] N. C. Albertsen, J. E. Hansen, and N. E. Jensen, "Computation of radiation from wire antennas on conducting bodies," *IEEE Trans. Antennas Propagat.*, vol. AP-22, no. 2, pp. 200–206, Mar. 1974.

[6] G. J. Burke and A. J. Poggio, "Numerical Electromagnetic Code (NEC) - method of moments," Naval Ocean Systems Center, San Diego, CA, Tech. Document 116, July 1977.

[7] N. N. Wang, J. H. Richmond, and M. C. Gilreath, "Sinusoidal reaction formulation for radiation and scattering from conducting surfaces," *IEEE Trans. Antennas Propagat.*, vol. AP-23, no. 3, pp. 376–382, May 1975.

[8] E. H. Newman and D. M. Pozar, "Electromagnetic modeling of composite wire and surface geometries," *IEEE Trans. Antennas Propagat.*, vol. AP-26, no. 6, pp. 784–789, Nov. 1978.

[9] A. Sankar and T. C. Tong, "Current computation on complex structures by finite element method," *Electron. Lett.*, vol. 11, no. 20, pp. 481–482, Oct. 1975.

[10] J. J. H. Wang, "Numerical analysis of three-dimensional arbitrarily-shaped conducting scatterers by trilateral surface cell modelling," *Radio Sci.*, vol. 13, no. 6, pp. 947–952, Nov.–Dec. 1978.

[11] J. J. H. Wang and C. Papanicolopulos, "Surface patch modeling of scatterers of arbitrary shapes," *Antennas Propagat. Soc. Int. Symp. Digest*, Univ. Washington, Seattle, WA, June 1979, pp. 159–162.

[12] G. Jeng and A. Wexler, "Finite element, boundary integral equation analysis of scattering problems," *URSI Symp. on Electromagnetic Wave Theory*, Stanford Univ., Stanford, CA, June 1977, pp. 179–181.

[13] J. Singh and A. T. Adams, "A non rectangular patch model for scattering from surfaces," *IEEE Trans. Antennas Propagat.*, vol. AP-27, no. 4, pp. 531–535, July 1979.

[14] A. W. Glisson and D. R. Wilton, "Simple and efficient numerical methods for problems of electromagnetic radiation and scattering from surfaces," *IEEE Trans. Antennas Propagat.*, vol. AP-28, no. 5, pp. 593–603, Sept. 1980.

[15] R. F. Harrington, *Field Computation by Moment Methods*. New York: Macmillan, 1968.

[16] A. W. Glisson, "On the development of numerical techniques for treating arbitrarily-shaped surfaces," Ph.D. dissertation, Univ. Mississippi, 1978.

[17] D. R. Wilton, S. S. M. Rao, and A. W. Glisson, "Electromagnetic scattering by arbitrary surfaces," Rome Air Development Center, Griffiss AFB, NY, Tech. Rep. RADC-TR-79-325, Mar. 1980.

[18] J. Van Bladel, *Electromagnetic Fields*. New York: McGraw-Hill, 1964, p. 502.

[19] O. C. Zienkiewicz, *The Finite Element Method in Engineering Science*. New York: McGraw-Hill, 1971.

[20] P. C. Hammer, O. P. Marlowe, and A. H. Stroud, "Numerical integration over simplexes and cones," *Math. Tables Aids Comp.*, vol. 10, pp. 130–137, 1956.

[21] C. M. Butler, Y. Rahmat-Samii, and R. Mittra, "Electromagnetic penetration through apertures in conducting surfaces," *IEEE Trans. Antennas Propagat.*, vol. AP-26, no. 1, pp. 82–93, Jan. 1978.

[22] R. G. Kouyoumjian, "The calculation of the echo areas of perfectly conducting objects by the variational method," Ph.D. dissertation, The Ohio State Univ., 1953.

[23] Y. Rahmat-Samii and R. Mittra, "Integral equation solution and RCS computation of a thin rectangular plate," *IEEE Trans. Antennas Propagat.*, vol. AP-22, no. 4, pp. 608–610, July 1974.

[24] C. T. Ruck, Ed., *Radar Cross Section Handbook*. New York: Plenum, 1970, p. 523.

[25] J. J. Bowman, T. B. A. Senior, and P. L. E. Uslenghi, *Electromagnetic and Acoustic Scattering by Simple Shapes*. Amsterdam: North-Holland, 1969, pp. 576–577.

[26] J. R. Mautz and R. F. Harrington, "*H*-field, *E*-field, and combined field solutions for conducting bodies of revolution," *AEÜ*, vol. 32, no. 4, pp. 157–164, Apr. 1978.

Wire Grid Modeling of Surfaces

ARTHUR C. LUDWIG

Abstract—**When a surface is numerically modeled with a wire grid, results are sensitive to the wire diameter. It is shown that the best accuracy is obtained when the wire satisfies the "same surface area" rule of thumb, for the canonical problem of scattering (or radiation) from an infinite circular cylinder. It is important to note that wires that are too thick are just as bad as wires that are too thin. It is also shown that the boundary value match between wires is not a reliable check on the validity of far-field results. Finally, data are given on the effect of wire spacing. Results are obtained from exact solutions of both the true problem and the wire grid model, thus isolating the effects of wire grid modeling per se.**

I. Introduction

A common application of the method of moments (MM) integral equation technique involves modeling a surface with a wire grid [1]. Two perplexing problems arose when this approach was tried, using the Livermore Numerical Electromagnetics Code (NEC) [2].

1) Some results are quite sensitive to the grid wire diameter; although a rule of thumb exists for selecting the diameter, a convincing justification for the choice appears to be lacking.
2) Although the $\bar{E}$-field at the center of a grid wire could be forced to match the correct boundary value, the field *between* wires was sometimes off by a factor of two or more; did this imply an error in the other results?

This paper presents the results of an exact canonical problem which sheds light on these two issues. In addition, data are presented on the effect of the spacing between grid wires.

Users of NEC are aware of a rule of thumb where the surface area of the wires parallel to one linear polarization is made equal to the surface area of the solid surface being modeled [3]. For a square wire grid, suitable for any polarization, the total surface area of the wires is then twice the area of the solid surface. This will be called the "same surface area" rule of thumb, even though it can be argued that "twice surface area" is a better description. The published literature on the wire size question appears to be relatively sparse. One reference which explicitly deals with the wire size question simply suggests using thick wires which just touch at the minimum grid spacing [4]. An excellent paper by Lee *et al.* provides equations which may be used to derive wire size, but does not explicitly use them for this purpose [5]. Lee *et al.* show that the natural frequencies of a solid sphere and wire grid sphere are in general different; they can be made equal if the wire size satisfies the "same surface area" rule of thumb. However, in the same paper, equations are given which reveal that equalizing the matrix elements for the inductive field contribution requires a diameter 8 percent greater than the rule of thumb; alternatively, an equivalent circuit representation suggests a diameter of 8 percent or 82 percent less than the rule of thumb to equalize the inductive or capacitive values, respectively. Therefore, the results of Lee *et al.* are somewhat ambiguous.

Moore and Pizer [6] draw a similar conclusion from the work of Lee *et al.*; they then state that empirical results suggest that the same surface area rule of thumb is a *minimum,* and up to five times the minimum wire size may be necessary to get good agreement with experiment. The results given here show that this is not necessarily so; a wire that is too thick creates just as much error as a wire that is too thin.

Wire grids have also been used in lieu of solid surfaces in practical antenna designs, and published design criteria suggest a wire diameter about 16 percent smaller than the same surface area rule of thumb [7]. In this context, it should be noted that the numerical wire grid is not a perfect model of a real wire grid, since the boundary condition is typically matched numerically at the center of the wire rather than the surface of the wire, and even for a small diameter there can be a large circumferential variation of current on a real wire. For example, a plane wave normally incident on a wire dipole only 0.075 wavelengths in diameter creates a current with nearly a 2:1 variation in current magnitude around the circumference [8].

Although this paper provides support for the same surface area rule of thumb, it has also been shown recently that when this involves using different wire sizes on the same body, a better result is obtained using constant wire size even though the rule is then violated [8]. Clearly the question is complex, and while this paper is intended to add new information, it certainly does not provide the final answer to the wire size question.

The second issue, the boundary value match between wires, seems to have drawn even less attention in the literature. The results given here do provide a definite answer to this question: a solution providing a poor match between wires may still provide excellent accuracy for fields even a short distance away from the surface. Therefore, a poor boundary value match does not necessarily imply an error in other results.

II. The Canonical Problem

An infinite circular cylinder is selected as the geometry of the canonical problem. A "true" electric current on the cylinder, and the electromagnetic fields inside and outside the

Manuscript received October 10, 1986; revised March 10, 1987.
The author is with the General Research Corporation, 5383 Hollister Avenue, P. O. Box 6770, Santa Barbara, CA 93160.
IEEE Log Number 8715706.

Reprinted from *IEEE Trans. Antennas Propagat.*, vol. AP-35, no. 9, pp. 1045–1048, September 1987.

cylinder are defined. The true fields are solutions of Maxwell's equations, and exactly satisfy the boundary conditions on the surface of the cylinder: 1) continuity of the tangential $\bar{E}$ field; and 2) discontinuity of the tangential $\bar{H}$ field associated with the electric current.

The true problem is then modeled using a wire grid, and the basic wire MM assumptions:

1) wire currents flow only in the axial direction, and are uniform around the circumference.
2) the $\bar{E}$ field at the center of each wire is forced to be identically equal to the true $\bar{E}$ field.

However, instead of actually obtaining the solution with a MM technique, the problem is constructed such that an exact solution is easily found. Therefore, any resulting errors are due solely to the grid model itself, including the two basic assumptions given above, as distinct from other potential MM error sources such as the choice of basis functions, matrix inversion, etc.

For the true problem as given above, the incident field has been ignored, and the only concern is the currents on the cylinder and the fields produced by these currents. By postulating different incident fields, the true problem will represent several possible physical problems. For example, to represent a perfect electrical conductor scattering problem, one would postulate an incident field with a value on the cylindrical surface that is equal in magnitude and 180° out of phase with the $\bar{E}$-field produced by the current. (Note that due to the circular symmetry assumed below, the incident field must be a cylindrical wave, but the same basic concept could be adapted for an incident plane wave.) For an imperfect conductor, the postulated incident field need only be multiplied by a complex constant to obtain the desired relationship between the fields and currents on the cylinder surface. To represent an antenna problem, the incident field is zero and the assumed currents are postulated as the source. In all cases, the incident field is identical for the true problem and wire grid model, and it is the difference in the fields produced by the currents which creates error; therefore, we can ignore the incident field.

The exact solution for the true problem, which is defined to be a uniform surface current, will be presented first. Then, the true current will be modeled with a wire grid, and an exact solution for the grid model is obtained, based on the two assumptions above. The two exact solutions are then compared to determine differences between the true current and the wire grid model.

A. The True Problem

The true problem is defined to be a uniform surface current on an infinite circular cylinder of radius a

$$\bar{K} = K_a \hat{i}_z. \tag{1}$$

The exact solution for the fields is given for $r > a$ by

$$\bar{E}_1 = E_a H_0^{(2)}(kr)\hat{i}_z$$

$$\bar{H}_1 = -E_a \frac{j}{Z} H_0^{(2)\prime}(kr)\hat{i}_\theta \tag{2}$$

and for $r < a$ by

$$\bar{E}_2 = E_a \frac{H_0^{(2)}(ka)}{J_0(ka)} J_0(kr)\hat{i}_z$$

$$\bar{H}_2 = -E_a \frac{j}{Z} \frac{H_0^{(2)}(ka)}{J_0(ka)} J_0'(kr)\hat{i}_\theta \tag{3}$$

where the fields are related to the current by

$$K_a = \frac{E_a}{Z}\left[Y_1(ka) - \frac{Y_0(ka)}{J_0(ka)} J_1(ka)\right]. \tag{4}$$

J, Y, and H are the standard Bessel functions [9]; k and Z are the propagation constant and free-space impedance, respectively; coordinates are shown in Fig. 1.

B. The Wire Grid Model Solution

The behavior of the true surface current will now be modeled by n wires of radius b, as illustrated in Fig. 1. The current on each of the n wires is assumed to be uniform around the circumference; therefore, each wire has an exact field solution given by (1)–(4) with "a" replaced by "b." The new scale factor E_b is determined by forcing the total $\bar{E}$-field at the center of each wire to be exactly the same as the true $\bar{E}$-field given by (2) or (3) with $r = a$. Due to symmetry, E_b must be the same for each of the n wires. Therefore, it is a simple matter to sum up the field for the n wires at the center of any one wire, equate it to the true field, and solve for E_b. Once E_b is known, it is also simple to compute the exact current and/or field at any point in space for the n grid wires, and to compare with the exact true results.

III. Discussion and Results

Obviously, the wire grid will not correctly model every feature of the true problem. Two features which are typically important are 1) the far-field radiation, and 2) the total current. These are very closely related, as would be expected. However, it is also important to be able to validate results, and two validation techniques which have been used successfully with a completely different approach are to 1) check the boundary condition match, and 2) check the field inside the body [8].

Originally, it was speculated that if the wire grid was accurately modeling a surface, then the boundary condition must be approximately satisfied over the entire surface—between the wires as well as the wire centers. It turns out that this is false. (Even with the benefit of hindsight, the reasons for this are not obvious.)

In contrast, the field inside the body is an excellent indicator of accuracy—at least for the selected canonical problem. In fact, the error in the field at the center of the cylinder is the same as the error in the far-field radiation to four or five decimal places for the cases evaluated. The surprising aspect of this is that the far field is more closely related to the field at the center of the cylinder than it is to the total current. Again, it is not obvious that this should be true.

The selected problem has one critical characteristic; at values of ka which are roots of $J_0(ka)$, the surface current

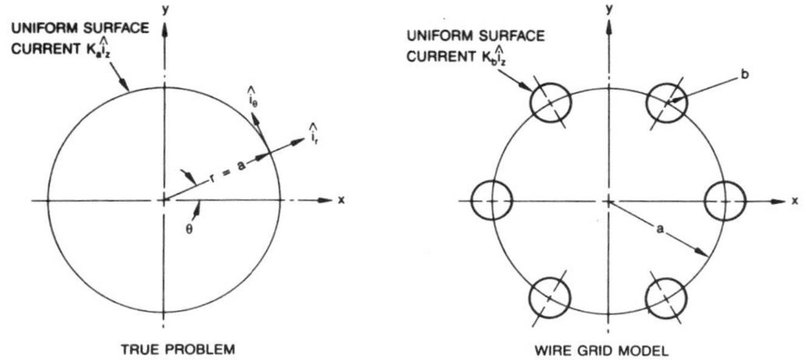

Fig. 1. The canonical problem.

does not radiate at all, and the solution represents a resonant field totally contained inside a cylindrical cavity. The error is strongly dependent on how close ka is to such a resonant value. Data will be presented for representative cases.

A. Error Versus Grid Wire Size

The difference between the true results and wire grid model results as a function of wire size is shown in Fig. 2. The selected value of $ka = 10$ is nearly between two roots of $J_0(ka)$ so this represents a strongly radiating case. The value of $n = 50$ corresponds to a wire spacing of 0.2 wavelengths, which as shown later, is a reasonable value. The error is shown for the total current—i.e., the current density $\bar{K}$ times the total surface circumference—and for the field at $r = 0$ and $r \to \infty$. As mentioned above, the error in the field at $r = 0$ and $r \to \infty$ is virtually identical. The error is near zero when $nb = a$, i.e., when the total surface area of the wire grid is the same as the surface area of the cylinder being modeled. Note that the results are in error for wires which are either too large or too small, as stated in the introduction.

The total current is in error by 0.08 dB, or about 1 percent, at this point, reaching its minimum error at a slightly smaller wire size. The phase error is due totally to a phase shift in the wire grid fields at the wire center, and is the same for all three parameters. The error is less than 0.2 degrees at $nb = a$. These data are fairly typical for values of ka well away from roots of J_0, although the details change. For example, with $ka = 1.2$ and $n = 6$, the magnitudes of the errors are very similar, but the sign of the amplitude error is reversed—the far-field error is 0.35 dB at $nb/a = 0.5$ and -0.28 dB at $nb/a = 1.5$. For a large diameter ($ka = 51$ and $n = 255$), the error is smaller by about a factor of two, but otherwise similar.

Also shown in Fig. 2 is the error in the field at $r = a$ exactly between two wires—the boundary value check. It is seen that this error has little correlation with the errors of primary interest. The situation is even worse when results are compared for various values of ka. At $ka = 1.2$, the error reverses sign—meaning the field is larger between wires than at a wire center—and at $ka = 51$, the error is very small and well correlated with the other errors. In other words, it is all over the map, and is worthless as an indicator. This is an unfortunate peculiarity of wire grid modeling that the fields near the surface are not modeled accurately. For other techniques where the near field is accurate, the boundary value check is an important and valuable method for validating the solution [8].

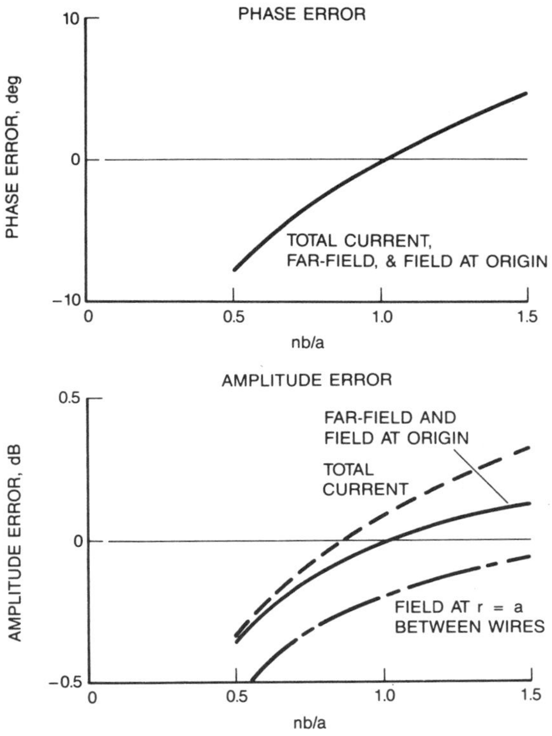

Fig. 2. Effect of wire grid size, $ka = 10$, $n = 50$.

A value $ka = 9$ is close to a root (8.65372) and the ratio of the radiated field to the total current is down 8.7 dB compared to $ka = 10$. The error for this case is shown in Fig. 3 and it is seen that the magnitude of the amplitude error has grown dramatically compared to Fig. 2. (The phase error is nearly unchanged.) However, the error is still small for $nb/a = 1.0$. Again, the field between wires is off by a large margin.

The accuracy of the near-field fields was also checked at two more points inside the cylinder, and two points outside the cylinder. The points inside were at a radius $r = a/2$, and at angles θ corresponding to a wire center, and to a point between wires. The points outside were at the same angles but at a radius $r = 2a$. The error in the wire grid field was about the same at the new points as at $r = 0$, and highly correlated with the far-field error. Therefore, the near field is quite accurately modeled at all of the points tested except for the point on the surface itself.

B. Error Versus Grid Spacing

Error in the far-field radiation is shown as a function of grid spacing in Fig. 4. The grid wire size satisfies the $nb = a$ criterion. Again, the error is dependent on how close ka is to a root of J_0. The three values $ka = 1.2$, 10, and 51 are all nearly between roots and the error is similar for the three cases. One value, $ka = 9$, which is near a root, is also shown and the relative error is much larger. The normalized grid spacing variable n/ka used in Fig. 4 corresponds to the number of grid wires per wavelength along the circumference. For the three strongly radiating cases, a surprisingly low value of three points power wavelength provides good results. Five points per wavelength also provides good results for the weakly radiating case $ka = 9$.

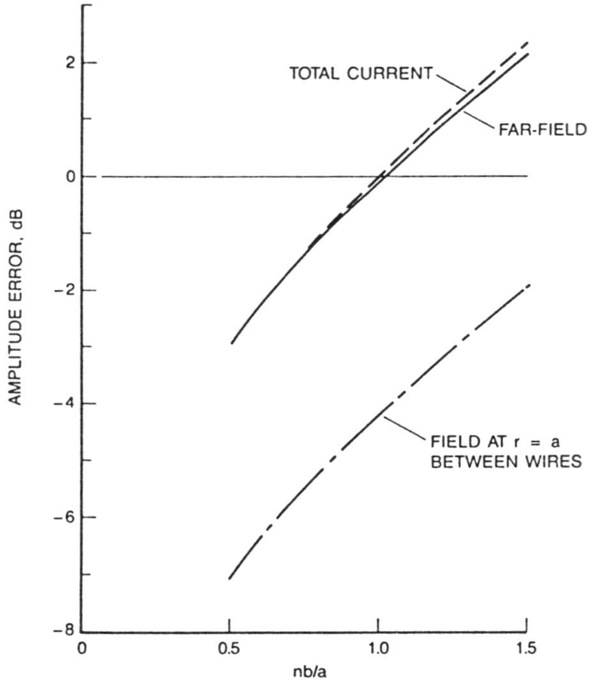

Fig. 3. Effect of wire grid size, $ka = 9$, $n = 45$.

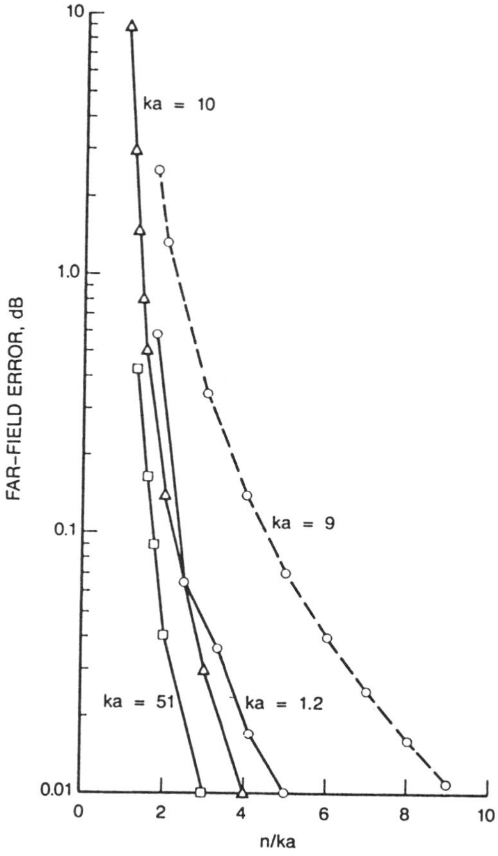

Fig. 4. Effect of grid spacing.

Another effect of the grid spacing is that more grid wires decreases the sensitivity to wire size. For $ka = 9$, and a wire size of half the rule of thumb value, the error in the far-field is 2.9 dB using five wires per wavelength. Using 10 wires per wavelength reduces the error to 1.6 dB. The field between wires also appears to be much closer to the true value when more wires are used.

Poggio and Miller [10] have found empirically that 10 to 20 wires per wavelength are necessary for good accuracy [10]. The larger number of wires may be due to differences in the problems (e.g., the presence of edges) or related to the sensitivity effect found here.

IV. Conclusion

The virtue of the canonical problem considered here is that it allows an exact solution for both the true problem and wire grid model, and this isolates the effect of wire grid modeling per se. The main drawback is that it is a special case, and perhaps the results will not hold for other problems. The same basic approach could be applied to other canonical problems, such as a plane wave incident on a cylinder, where exact solutions can still be obtained, albeit with somewhat more difficulty.

With the above caveat in mind, the results certainly enhance confidence in the "same surface area" wire size rule of thumb. In addition, the results indicate that five wires per wavelength generally provide accurate results, although more wires are better, and decrease error sensitivity to the wire size. Finally, near fields are modeled accurately at points tested inside and outside the cylinder, but accuracy is poor on the cylinder between the grid wires.

Acknowledgment

Several discussions with Dr. E. K. Miller of the University of Kansas have contributed to the evolution of this work. Dr. G. Burke of the Lawrence Livermore Laboratory was very helpful in providing information on prior work on this topic.

References

[1] J. H. Richmond, "A wire grid model for scattering by conducting bodies," *IEEE Trans. Antennas Propagat.,* vol. AP-14, no. 6, pp. 782–786, Nov. 1966.

[2] G. J. Burke and A. J. Poggio, "Numerical electromagnetics code (NEC)—Method of moments," Naval Ocean Syst. Center, San Diego, CA, NOSC Tech. Document 116, Jan. 1981.

[3] R. A. Pearlman and R. B. Young, "Numerical electromagnetic code applied to RCS predictions," presented at Low Observables Symp., Orlando, FL, June 1986.

[4] D. H. Forgan, *Computation of the Performance of HF Aerials Mounted in Aircraft,* Royal Aircraft Establishment Tech. Rep. 74077, August 1974.

[5] K. S. H. Lee, Marin, and J. P. Castillo, "Limitations of wire-grid modeling of a closed surface," *IEEE Trans. Electromag. Compat.,* vol. EMC-18, no. 3, pp. 123–129, Aug. 1976.

[6] J. Moore and R. Pizer, *Moment Methods in Electromagnetics.* Research Studies Press, 1984, secs. 1.1.2 and 6.4.

[7] Y. T. Lo, "TV receiving antennas," in *The Engineering Handbook,* H. Jasik, Ed. New York: McGraw-Hill, 1961, ch. 24.

[8] A. C. Ludwig, "A comparison of spherical wave boundary value matching versus integral equation scattering solutions for a perfectly conducting body," *IEEE Trans. Antennas Propagat.,* vol. AP-34, no. 7, pp. 857–865, July 1986.

[9] M. Abramowitz and Z. Stegun, Eds., "Handbook of mathematical functions," Nat. Bur. Stand., U.S. Government Printing Office, Washington, DC, June 1964.

[10] A. J. Poggio and E. K. Miller, "Integral equation solutions of three-dimensional scattering problems," in *Computational Techniques for Electromagnetics,* R. Mittra, Ed. Elmsford, NY: Pergamon, 1973, ch. 4.

Extended Integral Equation Formulation for Scattering Problems from a Cylindrical Scatterer

ICHIHIKO TOYODA, MASANORI MATSUHARA, AND NOBUAKI KUMAGAI, FELLOW, IEEE

Abstract—**A new extended integral equation is developed for electromagnetic scattering from a perfectly conducting cylinder and a dielectric cylinder. The conventional surface integral equations cannot yield unique solutions when the wavenumber of the electromagnetic wave is equal to an eigenwavenumber of the system. Several methods to overcome this difficulty have been presented, but each method includes some drawbacks. We propose a numerical method in which the boundary element method is applied to the extended integral equations with the observation points lying on a closed surface inside the scatterer. It is shown that the extended integral equations have unique solutions for any given wavenumber. As examples, plane wave scattering from a perfectly conducting elliptic cylinder, a dielectric elliptic cylinder, and a dielectric rectangular cylinder is numerically analyzed.**

I. Introduction

THE INTEGRAL equation formulation is a practical numerical approach that has been widely used for many electromagnetic scattering problems. According to the choice of the observation points, the integral equations can be classified into two types: surface integral equations [1]–[4] and extended integral equations [5]–[7]. The surface integral equations with the observation points lying on the surface of the scatterer are suited for a simple numerical approach, such as the boundary element method. However, the surface integral equations have a difficulty. They cannot yield unique solutions when the wavenumber of the electromagnetic wave is equal to an eigenwavenumber of the system. To make surface integral equations yield unique solutions for all wavenumbers, two different approaches are available: one is to use the combined surface integral equations [3], [4], [8], [9] and another is to add some interior conditions to the surface integral equations [10]–[12]. However, both approaches require much calculation. In addition, the combined surface integral equation method requires the manipulation of terms containing the second derivative of the Green's function.

On the other hand, the extended integral equations, where the observation points are inside the scatterer, always have unique solutions for any value of the given wavenumber. However, the extended integral equations have only been analyzed by the T matrix method so far, and there is no simple numerical approach, such as the boundary element method, that can be applied to the extended integral equations. Additionally, the T matrix method yields an ill-conditioned matrix when the ratio of the largest dimension of the scatterer to the smallest dimension becomes a large value [13]. If we put the observation points in the scatterer uniformly, we can apply the boundary element method to the extended integral equations. However, in this case the matrix equations are still ill conditioned.

Manuscript received September 9, 1987; revised March 4, 1988.

The authors are with the Department of Communication Engineering, Faculty of Engineering, Osaka University, Yamada Oka, Suita, Osaka 565, Japan.

IEEE Log Number 8823434.

In this paper, we propose a numerical method in which the boundary element method is applied to the extended integral equations with the observation points lying on a closed surface near the boundary of the scatterer. In this case the matrix equations are not ill conditioned for a flat scatterer, but the extended integral equations do not yield unique solutions for specific wavenumbers. However, this difficulty can be avoided easily by shifting the closed surface on which we put the observation points in the scatterer. This method is theoretically concise and is suited for the boundary element method, which is a general numerical approach for solving integral equations.

In numerical examples, plane wave scattering from a perfectly conducting elliptic cylinder, a dielectric elliptic cylinder, and a dielectric rectangular cylinder has been numerically analyzed, and satisfactory results have been obtained.

II. Integral Equations

Consider a scatterer uniform along the z axis as shown in Fig. 1. The surface of the scatterer is Γ, the exterior region is Ω_e, and the interior region is Ω_i in the cross section of the scatterer. We assume that the electromagnetic field is either an E wave (TM) or an H wave (TE). It is known that $E_x = E_y = H_z = 0$ in the case of the E wave and $H_x = H_y = E_z = 0$ in the case of the H wave. Let $f = E_z$ in the case of the E wave and $f = H_z$ in the case of the H wave. For simplicity, only the scattering from a dielectric cylinder will be discussed. Then f in the exterior region Ω_e, namely, $f = f_e$, satisfies the following Helmholtz equation

$$\nabla^2 f_e + k_e^2 f_e = -g_e, \qquad \text{in } \Omega_e \tag{1}$$

where k_e is the wavenumber of the wave in Ω_e and g_e denotes the source in Ω_e. $f = f_i$ in the interior region Ω_i satisfies

$$\nabla^2 f_i + k_i^2 f_i = 0, \qquad \text{in } \Omega_i \tag{2}$$

where k_i is the wavenumber of the wave in Ω_i.

Reprinted from *IEEE Trans. Antennas Propagat.*, vol. 36, no. 11, pp. 1580–1586, November 1988.

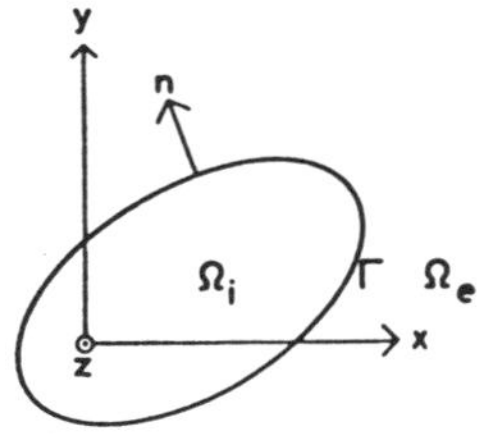

Fig. 1. Cylindrical scatterer and coordinate system.

The boundary conditions along the boundary Γ require

$$f_e = f_i$$
$$\frac{1}{\mu_e}\frac{\partial f_e}{\partial n}\left(=\frac{1}{\mu_e}\mathbf{n}\cdot\nabla f_e\right)=\frac{1}{\mu_i}\frac{\partial f_i}{\partial n}, \qquad \text{on } \Gamma \tag{3}$$

in the case of E wave scattering, and

$$f_e = f_i$$
$$\frac{1}{\epsilon_e}\frac{\partial f_e}{\partial n}=\frac{1}{\epsilon_i}\frac{\partial f_i}{\partial n}, \qquad \text{on } \Gamma \tag{4}$$

in the case of H wave scattering. Here $\mathbf{n}$ denotes a unit outward normal vector to the boundary Γ, and μ_e, μ_i, ϵ_e, and ϵ_i are the permeability and permittivity of the medium in Ω_e and Ω_i, respectively.

Now we introduce the two-dimensional Green's function in free space to be written in the form:

$$G_k(\mathbf{p},\mathbf{q}) = -\frac{j}{4}H_0^{(2)}(k|\mathbf{p}-\mathbf{q}|) \tag{5}$$

where $\mathbf{p}$ is the position vector of the observation point and $\mathbf{q}$ is the position vector of the source point. $H_0^{(2)}(\cdot)$ is the Hankel function of the second kind of zeroth order.

Substitution of (1) and (5) into Green's theorem leads to the following integral expressions. In the case of $\mathbf{p} \in \Omega_e$,

$$\oint_\Gamma \left\{G_{k_e}(\mathbf{p},\mathbf{q})\frac{\partial f_e(\mathbf{q})}{\partial n_q}-\frac{\partial G_{k_e}(\mathbf{p},\mathbf{q})}{\partial n_q}f_e(\mathbf{q})\right\}dq = f_e^{\text{inc}}(\mathbf{p}) - f_e(\mathbf{p}) \tag{6}$$

and in the case of $\mathbf{p} \in \Omega_i$,

$$\oint_\Gamma \left\{G_{k_e}(\mathbf{p},\mathbf{q})\frac{\partial f_e(\mathbf{q})}{\partial n_q}-\frac{\partial G_{k_e}(\mathbf{p},\mathbf{q})}{\partial n_q}f_e(\mathbf{q})\right\}dq = f_e^{\text{inc}}(\mathbf{p}) \tag{7}$$

where

$$f_e^{\text{inc}}(\mathbf{p}) = \int_{\Omega_e} G_{k_e}(\mathbf{p},\mathbf{q})g_e(\mathbf{q})\,dQ. \tag{8}$$

Substitution of (2) and (5) into Green's theorem leads to the following integral expressions. In the case of $\mathbf{p} \in \Omega_i$,

$$\oint_\Gamma \left\{G_{k_i}(\mathbf{p},\mathbf{q})\frac{\partial f_i(\mathbf{q})}{\partial n_q}-\frac{\partial G_{k_i}(\mathbf{p},\mathbf{q})}{\partial n_q}f_i(\mathbf{q})\right\}dq = f_i(\mathbf{p}) \tag{9}$$

and in the case of $\mathbf{p} \in \Omega_e$,

$$\oint_\Gamma \left\{G_{k_i}(\mathbf{p},\mathbf{q})\frac{\partial f_i(\mathbf{q})}{\partial n_q}-\frac{\partial G_{k_i}(\mathbf{p},\mathbf{q})}{\partial n_q}f_i(\mathbf{q})\right\}dq = 0. \tag{10}$$

Here, $\oint_\Gamma dq$ denotes the line integration with respect to $\mathbf{q}$ along the boundary Γ, and $\int_{\Omega_e} dQ$ denotes the surface integration with respect to $\mathbf{q}$ over the region Ω_e. $\partial/\partial n_q$ denotes the normal derivative with respect to $\mathbf{q}$ on the boundary Γ. $f_e^{\text{inc}}(\mathbf{p})$ is the incident field which is produced at point $\mathbf{p}$ by the source g_e, and $f_e^{\text{inc}}(\mathbf{p})$ is a known quantity in the scattering problems.

Equations (6) and (9) mean that the field at any point in Ω_e and Ω_i can be determined by the field f and $\partial f/\partial n$ on the boundary Γ. Application of the boundary conditions (3) or (4) to the integral expressions in (7) and (10) leads to the extended integral equations for determining f and $\partial f/\partial n$ on the boundary Γ. Substitution of the boundary conditions (3) or (4) into (7) and (10) leads to

$$\oint_\Gamma \left\{G_{k_e}(\mathbf{p},\mathbf{q})\frac{\partial f_e(\mathbf{q})}{\partial n_q}-\frac{\partial G_{k_e}(\mathbf{p},\mathbf{q})}{\partial n_q}f_e(\mathbf{q})\right\}dq = f_e^{\text{inc}}(\mathbf{p}), \qquad \mathbf{p} \in \Omega_i \tag{11a}$$

$$\oint_\Gamma \left\{G_{k_i}(\mathbf{p},\mathbf{q})\,\gamma\,\frac{\partial f_e(\mathbf{q})}{\partial n_q}-\frac{\partial G_{k_i}(\mathbf{p},\mathbf{q})}{\partial n_q}f_e(\mathbf{q})\right\}dq = 0, \qquad \mathbf{p} \in \Omega_e. \tag{11b}$$

Here, γ is μ_i/μ_e in the case of the E wave and ϵ_i/ϵ_e in the case of the H wave. Thus f_e and $\partial f_e/\partial n$ on the boundary Γ can be determined by solving the extended integral equation in (11). After f and $\partial f/\partial n$ on the boundary Γ are determined, substitution of f and $\partial f/\partial n$ into (6) and (9) determines the electromagnetic field f in Ω_e and Ω_i.

If enforced at all observation points $\mathbf{p}$ in the interior region Ω_i, the extended integral equation in (11) always has a unique solution. However, in practice, (11) has to be applied to a finite number of observation points $\mathbf{p}$. If the extended integral equation in (11) is applied to a finite number of points $\mathbf{p}$ uniformly arranged in Ω_i, a unique solution can always be obtained for any given wavenumber [14]. However, in this case the extended integral equation becomes numerically unstable because most of the points $\mathbf{p}$ are distant from the boundary Γ and the kernels of the integral equation become smooth functions of $\mathbf{q}$. Since the kernels should be singular functions of $\mathbf{q}$ or nearly singular functions of $\mathbf{q}$ for numerical stability, all of the points $\mathbf{p}$ should be near the boundary Γ.

In this paper we propose the extended integral equations whose observation points are on closed lines Γ_i and Γ_e near the boundary Γ in Ω_i and Ω_e. The closed lines Γ_i and Γ_e are shown in Fig. 2.

The extended integral equation proposed here is

$$\oint_\Gamma \left\{G_{k_e}(\mathbf{p},\mathbf{q})\frac{\partial f_e(\mathbf{q})}{\partial n_q}-\frac{\partial G_{k_e}(\mathbf{p},\mathbf{q})}{\partial n_q}f_e(\mathbf{q})\right\}dq = f_e^{\text{inc}}(\mathbf{p}), \qquad \mathbf{p} \in \Gamma_i \tag{12a}$$

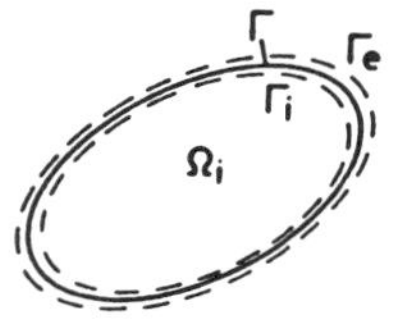

Fig. 2. Closed lines Γ_i and Γ_e.

$$\oint_\Gamma \left\{ G_{k_i}(\mathbf{p}, \mathbf{q})\, \gamma \frac{\partial f_e(\mathbf{q})}{\partial n_q} - \frac{\partial G_{k_i}(\mathbf{p}, \mathbf{q})}{\partial n_q} f_e(\mathbf{q}) \right\} dq = 0, \qquad \mathbf{p} \in \Gamma_e. \quad (12b)$$

There are some specific wavenumbers for which the extended integral equation in (12) does not have a unique solution. However, as will be discussed in next chapter, this difficulty can be easily avoided by shifting the closed line Γ_i in the interior of the scatterer.

Taking the limit as the closed lines Γ_i and Γ_e approach the boundary Γ, the extended integral equation in (12) becomes well-known surface integral equation:

$$⨍_\Gamma \left\{ G_{k_e}(\mathbf{p}, \mathbf{q}) \frac{\partial f_e(\mathbf{q})}{\partial n_q} - \frac{G_{k_e}(\mathbf{p}, \mathbf{q})}{\partial n_q} f_e(\mathbf{q}) \right\} dq + \frac{1}{2} f_e(\mathbf{p}) = f_e^{\text{inc}}(\mathbf{p}), \qquad \mathbf{p} \in \Gamma \quad (13a)$$

$$⨍_\Gamma \left\{ G_{k_i}(\mathbf{p}, \mathbf{q})\, \gamma \frac{\partial f_e(\mathbf{q})}{\partial n_q} - \frac{G_{k_i}(\mathbf{p}, \mathbf{q})}{\partial n_q} f_e(\mathbf{q}) \right\} dq - \frac{1}{2} f_e(\mathbf{p}) = 0, \qquad \mathbf{p} \in \Gamma. \quad (13b)$$

Here $⨍_\Gamma\, dq$ denotes the principal value of the line integral with respect to $\mathbf{q}$ along the boundary Γ. Though the surface integral equation in (13) is obtained by direct integration of the scattering problems (1)–(4) which always have unique solutions, there are always specific wavenumbers for which (13) does not have a unique solution even if we apply (13) to all observation points $\mathbf{p}$ on the boundary Γ.

III. Uniqueness of Solutions

The integral equations in (12) and (13) were derived by direct integration of the differential equations, and unknown quantities are the physical quantities f_e and $\partial f_e/\partial n$ on the boundary Γ. Therefore, solutions always exist. However, there is no guarantee that the solutions are always unique.

Consider the interior Dirichlet problem

$$\nabla^2 u + k_e^2 u = 0, \qquad \text{in } \Omega_i \quad (14a)$$

$$u = 0, \qquad \text{on } \Gamma_i. \quad (14b)$$

Assume that $u(\mathbf{p})$ is given by

$$u(\mathbf{p}) = \oint_\Gamma \left\{ G_{k_e}(\mathbf{p}, \mathbf{q}) \xi(\mathbf{q}) - \frac{\partial G_{k_e}(\mathbf{p}, \mathbf{q})}{\partial n_q} \chi(\mathbf{q}) \right\} dq, \qquad \mathbf{p} \in \Omega_i \quad (15)$$

with the single-layer distribution $\xi(\mathbf{q})$ and the double-layer distribution $\chi(\mathbf{q})$ on the boundary Γ. Here $\xi(\mathbf{q})$ and $\chi(\mathbf{q})$ satisfy the following equation:

$$\oint_\Gamma \left\{ G_{k_i}(\mathbf{p}, \mathbf{q}) \gamma \xi(\mathbf{q}) - \frac{\partial G_{k_i}(\mathbf{p}, \mathbf{q})}{\partial n_q} \chi(\mathbf{q}) \right\} dq = 0, \qquad \mathbf{p} \in \Gamma_e. \quad (16a)$$

It is easy to confirm that (15) satisfies (14a) by direct substitution. Besides, the condition that (15) satisfies (14b) implies that

$$\oint_\Gamma \left\{ G_{k_e}(\mathbf{p}, \mathbf{q}) \xi(\mathbf{q}) - \frac{\partial G_{k_e}(\mathbf{p}, \mathbf{q})}{\partial n_q} \chi(\mathbf{q}) \right\} dq = 0, \qquad \mathbf{p} \in \Gamma_i. \quad (16b)$$

Equation (14) has a unique solution $u = 0$ for most wavenumbers. Hence (16) also has a unique solution $\xi = \chi = 0$ for most wavenumbers. However, (14) and (16) do not have unique solutions for specific wavenumbers, i.e., the eigenwavenumbers of the interior Dirichlet problem related to $u = 0$ on Γ_i. Equation (16) is the homogeneous equation associated with the inhomogeneous integral equation in (12). Hence the extended integral equation in (12) also does not have a unique solution for the eigenwavenumbers of the interior Dirichlet problem related to $u = 0$ on Γ_i. This difficulty can be easily avoided by shifting the closed line Γ_i in the interior of the scatterer and thus shifting the eigenwavenumber. At this time, the position of the closed line Γ_e is unrelated to this difficulty, and the position of Γ_e is always arbitrary.

The surface integral equation in (13) is not as good as (12) because (13) does not have a unique solution for the eigenwavenumbers of the interior Dirichlet problem related to $u = 0$ on Γ. By allowing Γ_i to move with the wavenumber, one can, as indicated in the previous paragraph, always make the solution to (12) unique.

IV. Numerical Examples

A. Plane Wave Scattering from an Elliptic Cylinder

In this section the boundary element method solutions to the extended integral equations in (12), (29), and (30) are developed for plane wave scattering from an elliptic cylinder as shown in Fig. 3. The incident plane wave is $f_e^{\text{inc}} = \exp(-jk_e x)$ which travels in the $+x$ direction. First, we set N nodes $\{\mathbf{q}_n, n = 1, 2, \cdots, N\}$ uniformly along the boundary Γ, and let the unknown functions $(1/k_e)(\partial f_e/\partial n)$ and f_e on the boundary Γ be expanded in a series of N basis functions $\{S_n(\mathbf{q}), n = 1, 2, \cdots, N\}$:

$$\frac{1}{k_e} \frac{\partial f_e(\mathbf{q})}{\partial n_q} = \sum_{n=1}^{N} \phi_n S_n(\mathbf{q})$$

$$f_e(\mathbf{q}) = \sum_{n=1}^{N} \psi_n S_n(\mathbf{q}), \qquad \mathbf{q} \in \Gamma \quad (17)$$

where $\{\phi_n$ and $\psi_n, n = 1, 2, \cdots, N\}$ are unknown constants which can be determined by substituting (17) into (12), (29),

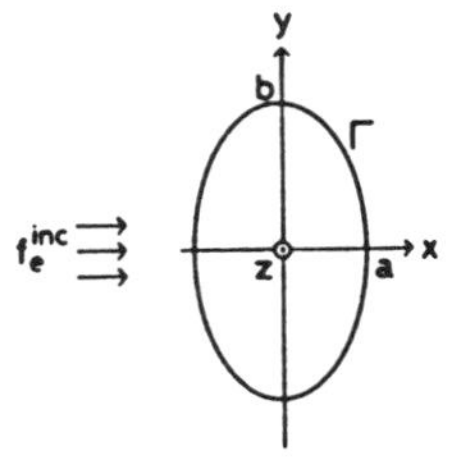

Fig. 3. Elliptic cylindrical scatterer.

and (30). Locally defined basis functions are desirable; thus we use B spline functions of third order [15]. The second derivatives of the unknown functions $(1/k_e)(\partial f_e/\partial n)$ and f_e on the boundary Γ which are expressed in (17) with B spline functions of third-order $S_n(\mathbf{q})$ are continuous since all of the nodes $\mathbf{q}_n$ are simple nodes.

Next, we uniformly set N observation points $\{\mathbf{p}_{in}$ and $\mathbf{p}_{en}$, $n = 1, 2, \cdots, N\}$ on the closed lines Γ_i and Γ_e, respectively. The closed line Γ_i lies inside the scatterer, and is concentric with the boundary Γ. The closed line Γ_e lies outside the scatterer and is concentric with the boundary Γ. The ratio of Γ_i to Γ is α_i, and the ratio of Γ_e to Γ is α_e. The observation points $\mathbf{P}_{in}$ and $\mathbf{p}_{en}$ are

$$\mathbf{p}_{in} = \alpha_i \mathbf{q}_n$$

$$\mathbf{p}_{en} = \alpha_e \mathbf{q}_n, \qquad n = 1, 2, \cdots, N \tag{18}$$

where $0 < \alpha_i < 1$ and $\alpha_e > 1$.

Then the extended integral equations in (12), (29), and (30) can be written in the form of the matrix equations in (19)–(21), respectively:

$$F \begin{bmatrix} \Phi \\ \Psi \end{bmatrix} \left(= \begin{bmatrix} A & B \\ C & D \end{bmatrix} \right) = \begin{bmatrix} E \\ 0 \end{bmatrix} \tag{19}$$

$$A\Phi = E \tag{20}$$

$$B\Psi = E. \tag{21}$$

Here A, B, C, and D in (19)–(21) are the $N \times N$ matrices whose elements are called $a_{nn'}$, $b_{nn'}$, $c_{nn'}$ and $d_{nn'}$ respectively. Φ, Ψ, and E are $N \times 1$ column vectors whose elements are ϕ_n, ψ_n, and e_n. Here,

$$a_{nn'} = \oint_\Gamma k_e G_{k_e}(\mathbf{p}_{in}, \mathbf{q}) S_{n'}(\mathbf{q})\, dq$$

$$b_{nn'} = -\oint_\Gamma \frac{\partial G_{k_e}(\mathbf{p}_{in}, \mathbf{q})}{\partial n_q} S_{n'}(\mathbf{q})\, dq$$

$$c_{nn'} = \oint_\Gamma \gamma k_e G_{k_i}(\mathbf{p}_{en}, \mathbf{q}) S_{n'}(\mathbf{q})\, dq$$

$$d_{nn'} = -\oint_\Gamma \frac{\partial G_{k_i}(\mathbf{p}_{en}, \mathbf{q})}{\partial n_q} S_{n'}(\mathbf{q})\, dq$$

$$e_n = f_e^{\text{inc}}(\mathbf{p}_{in}). \tag{22}$$

Φ and Ψ can be obtained by solving (19)–(21), and substitution of Φ and Ψ into (17) leads to the unknown functions $(1/k_e)(\partial f_e/\partial n)$ and f_e on the boundary Γ.

The relation between wavenumber k_e and condition number $\kappa(F)$ is shown in Fig. 4 with the parameter α_i indicating the position of the closed line Γ_i in the scatterer. Here,

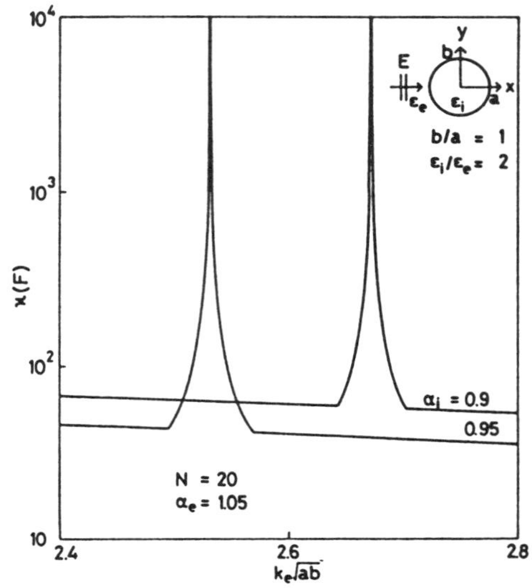

Fig. 4. Relation between wavenumber and condition number.

$$\kappa(F) = \|F\| \cdot \|F^{-1}\|. \tag{23}$$

In this paper, the Euclidean norm is used. The condition number is the index of the stability of matrices. When the matrix equations become singular and do not have unique solutions, the condition number becomes infinitely large. In the case of $\alpha_i = 0.95$, the condition number becomes large near the eigenwavenumber $k_e\sqrt{ab} = 2.5314(=2.4048/0.95)$ of the interior Dirichlet problem related to $u = 0$ on Γ_i, and stable unique solutions cannot be obtained. However, we can obtain stable solutions near this eigenwavenumber by changing α_i to, for example, 0.9. In the case of $\alpha_i = 0.9$, the solutions become unstable near the eigenwavenumber $k_e\sqrt{ab} = 2.6720(=2.4048/0.9)$. Consequently, we carry out the numerical computation by making the condition number as an index and change the position of Γ_i a little for wavenumbers where the condition number becomes large. Then we can obtain stable solutions for any wavenumber.

The scattered wave is given by

$$f_e^{\text{sca}}(\mathbf{p}) = -\oint_\Gamma \left\{ G_{k_e}(\mathbf{p}, \mathbf{q}) \frac{\partial f_e(\mathbf{q})}{\partial n_q} - \frac{\partial G_{k_e}(\mathbf{p}, \mathbf{q})}{\partial n_q} f_e(\mathbf{q}) \right\} dq, \qquad \mathbf{p} \in \Omega_e \tag{24}$$

from (6), where $(1/k_e)(\partial f_e/\partial n)$ and f_e are the unknown functions on the boundary. These functions can be obtained by solving the matrix equations (19)–(21). If we move $\mathbf{p}$ to a remote point on the negative x axis, i.e., $\mathbf{p} = -\mathbf{i}_x r$, $r \to \infty$, in (24), the radar cross section σ can be written as follows:

$$\sigma = 2\pi r |f_e^{\text{sca}}|^2 = \frac{k_e}{4} \left| \oint_\Gamma \left\{ j\mathbf{i}_x \cdot \mathbf{n} f_e + \frac{1}{k_e} \frac{\partial f_e}{\partial n} \right\} \cdot \exp(-jk_e \mathbf{i}_x \cdot \mathbf{q})\, dq \right|^2. \tag{25}$$

Table I shows radar cross section $\sigma/\pi\sqrt{ab}$ obtained from the

TABLE I
RADAR CROSS SECTION $\sigma/\pi\sqrt{ab}$

	$N = 50$ (percent)	$N = 20$ (percent)	Exact
E wave scattering from a perfectly conducting cylinder	1.046058176 ($<10^{-8}$)	1.046059094 (8.78×10^{-5})	1.046058176
H wave scattering from a perfectly conducting cylinder	0.986628624 (1.15×10^{-4})	0.986369131 (2.64×10^{-2})	0.986629760
E wave scattering from a dielectric cylinder	0.538973053 (3.20×10^{-4})	0.538637870 (6.25×10^{-2})	0.538974776
H wave scattering from a dielectric cylinder	0.090594321 (1.01×10^{-4})	0.090559997 (3.80×10^{-2})	0.090594412

$k_e\sqrt{ab} = 3$, $b/a = 1$, $\epsilon_i/\epsilon_e = 2$, $\alpha_i = 0.95$, $\alpha_e = 1.05$.

present method and the exact solution in the case of a circular cylinder. The value of the radar cross section computed by the present method is shown in the upper row, and the corresponding error is shown in the lower row. The computed results approach the exact solution as N increases. The error is less than 0.001 percent when $N = 50$, thus very accurate solutions are obtained. In practice, the matrices are $(N/2) \times (N/2)$ for a perfectly conducting cylinder, and $N \times N$ for a dielectric cylinder because of the symmetry of the system. Figs. 5(a)–5(d) show radar cross sections $\sigma/\pi\sqrt{ab}$ as functions of the wavenumber $k_e\sqrt{ab}$.

B. Plane Wave Scattering from a Dielectric Rectangular Cylinder

In this section the boundary element method solutions to the extended integral equation (12) are developed for plane wave scattering from the dielectric rectangular cylinder shown in Fig. 6. The incident plane wave is $f_e^{\text{inc}} = \exp(-jk_e x)$ which travels the $+x$ direction. It is expected that $(1/k_e)(\partial f_e/\partial n)$ is discontinuous and f_e is continuous but that the first derivative of f_e is discontinuous at the corners of the rectangular cylinder. In the case of using B spline functions, continuity of the function depends on the multiplicity of the nodes. First of all, we set N nodes $\{\mathbf{q}_n, n = 1, 2, \cdots, N\}$ uniformly on the boundary Γ and at the corners. We let the nodes at the corners be four-fold nodes for $(1/k_e)(\partial f_e/\partial n)$ and three-fold nodes for f_e. Then we let the unknown functions $(1/k_e)(\partial f_e/\partial n)$ and f_e on the boundary Γ be expanded in a series of $N + 12$ third-order B-spline functions $\{S_n^{(1)}(\mathbf{q}), n = 1, 2, \cdots, N + 12\}$ and $N + 8$ third-order B-spline function $\{S_n^{(2)}(\mathbf{q}), n = 1, 2, \cdots, N + 8\}$ as follows:

$$\frac{1}{k_e}\frac{\partial f_e(\mathbf{q})}{\partial n_q} = \sum_{n=1}^{N+12} \phi_n S_n^{(1)}(\mathbf{q}),$$

$$f_e(\mathbf{q}) = \sum_{n=1}^{N+8} \psi_n S_n^{(2)}(\mathbf{q}), \qquad \mathbf{q} \in \Gamma. \tag{26}$$

In this case, the function of $(1/k_e)(\partial f_e/\partial n)$ and the first derivatives of f_e are discontinuous at the corners.

Then we put $N + 12$ observation points $\{\mathbf{p}_{in}$ and $\mathbf{p}_{en}, n = 1, 2, \cdots, N + 12\}$ on the closed lines Γ_i and Γ_e, respectively. The closed line Γ_i lies inside the scatterer and is concentric with the boundary Γ. The closed line Γ_e lies outside the scatterer, and is concentric with the boundary Γ. The ratio of Γ_i to Γ is α_i and the ratio of Γ_e to Γ is α_e. The observation points $\mathbf{p}_{in}$ and $\mathbf{p}_{en}$ are

$$\mathbf{p}_{in} = \alpha_i \mathbf{q}_n$$

$$\mathbf{p}_{en} = \alpha_e \mathbf{q}_n \tag{27}$$

except at the corners, where $0 < \alpha_i < 1$ are $\alpha_e > 1$. We set four observation points $\mathbf{p}_{in}$ and $\mathbf{p}_{en}$ near each corner.

Then the extended integral equation (12) becomes (19). In this case, the coefficient matrix of (19) is not a square matrix. Therefore, (19) multiplied by F^* which is the conjugate trasnpose matrix of F is (28) where F^*F is a square matrix.

$$F^*F\begin{bmatrix}\Phi\\ \Psi\end{bmatrix} = F^*\begin{bmatrix}E\\ 0\end{bmatrix}. \tag{28}$$

Φ and Ψ are obtained by solving (28), and substitution of Φ and Ψ into (26) leads to the unknown functions $(1/k_e)(\partial f_e/\partial n)$ and f_e on the boundary Γ. The solution of (28) is the least-squares solution of (19).

Fig. 7 shows the field distributions of $(1/k_e)(\partial f_e/\partial n)$ and f_e on the boundary Γ, in the case of E-wave scattering from the dielectric rectangular cylinder. The function of $(1/k_e)(\partial f_e/\partial n)$ and the first derivative of f_e are discontinuous at the corners T_2 and T_3. Therefore, it is obviously effective to use the B-spline functions.

V. CONCLUSION

In this paper, a method of overcoming the various difficulties of integral equation formulations for electromagnetic scattering problems is described. The conventional surface integral equations cannot yield unique solutions when the wavenumber of the electromagnetic wave is equal to the eigenwavenumbers of the system. Previous investigators presented several methods of overcoming this difficulty, but each method has some drawbacks. For example, matrix equations become ill conditioned or need much calculation, etc. In this paper we apply the extended integral equations to observation points on closed surfaces inside and outside the scatterer but near its boundary and show that unique solutions are obtained for any wavenumber. As examples, plane wave

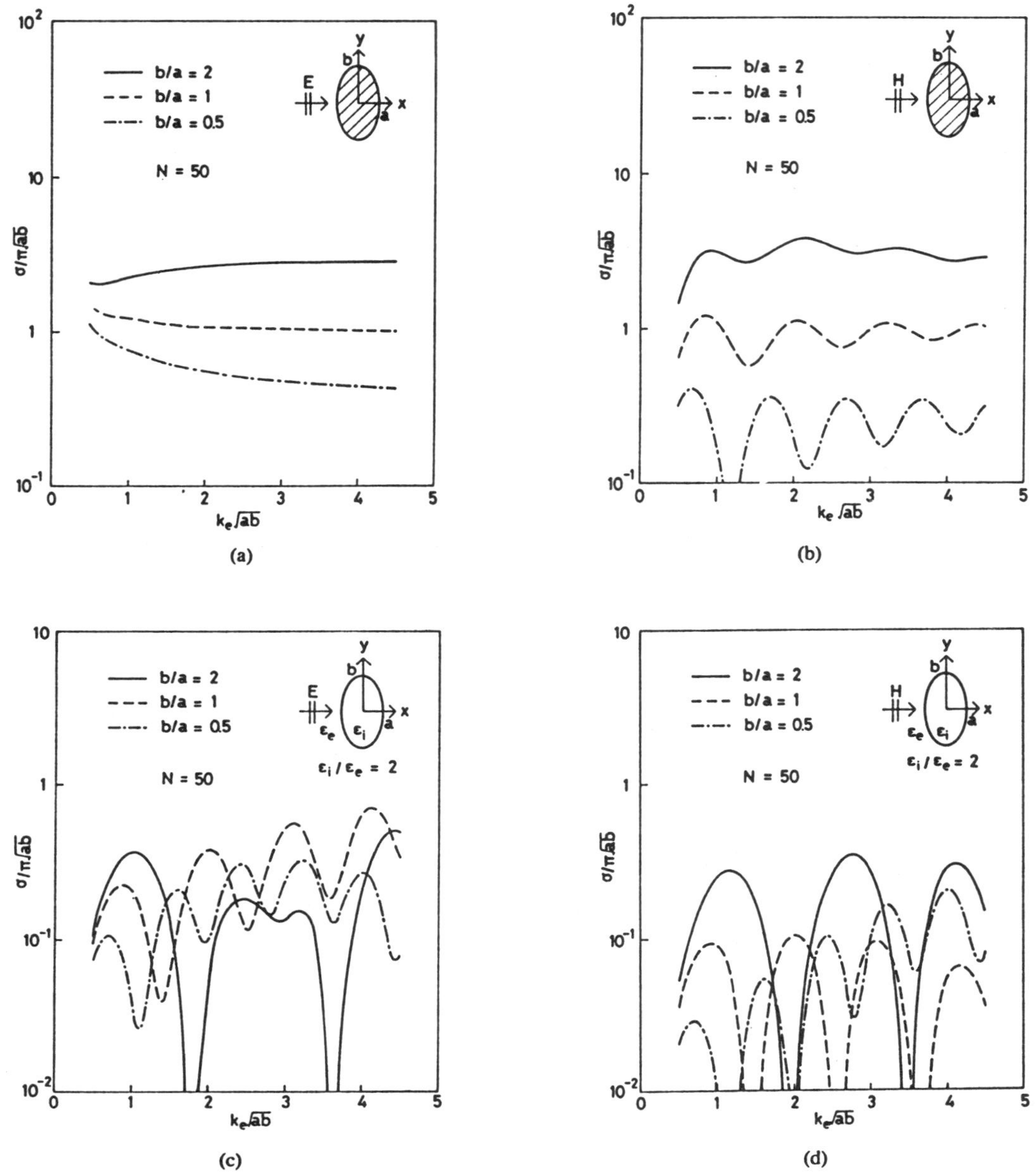

Fig. 5. Radar cross section $\sigma/\pi\sqrt{ab}$. (a) E wave scattering from a perfectly conducting cylinder. (b) H wave scattering from a perfectly conducting cylinder. (c) E wave scattering from a dielectric cylinder. (d) H wave scattering from a dielectric cylinder.

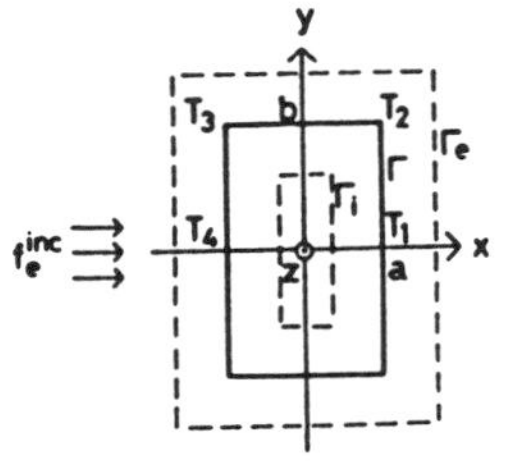

Fig. 6. Dielectric rectangular cylindrical scatterer.

scattering from a perfectly conducting elliptic cylinder, a dielectric elliptic cylinder, and a dielectric rectangular cylinder is numerically analyzed by the boundary element method whose basis functions are third-order B-spline functions. We show that the present method can easily overcome the difficulties of the integral equation formulation. Besides, we show that a very accurate solution can be obtained and that the B-spline functions are appropriate basis functions for expanding a function with a discontinuity.

Appendix

The extended integral equations for a perfectly conducting scatterer are

$$\oint_{\Gamma} G_{k_e}(\mathbf{p}, \mathbf{q}) \frac{\partial f_e(\mathbf{q})}{\partial n_q} dq = f_e^{\text{inc}}(\mathbf{p}), \qquad \mathbf{p} \in \Gamma_i \tag{29}$$

in the case of E-wave scattering and

$$-\oint_{\Gamma} \frac{\partial G_{k_e}(\mathbf{p}, \mathbf{q})}{\partial n_q} f_e(\mathbf{q})\, dq = f_e^{\text{inc}}(\mathbf{p}), \qquad \mathbf{p} \in \Gamma_i \tag{30}$$

in the case of H-wave scattering.

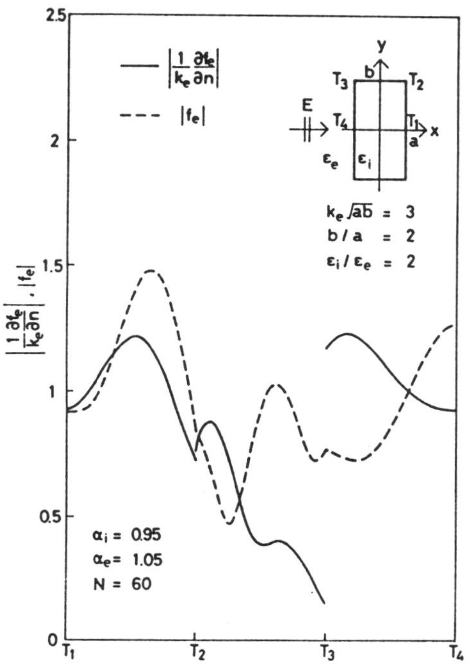

Fig. 7. Field distributions f_e and f_e/n on the boundary.

References

[1] K. K. Mei and J. G. Van Bladel, "Scattering by perfectly-conducting rectangular cylinders," *IEEE Trans. Antennas Propagat.*, vol. AP-11, pp. 185–192, Mar. 1963.

[2] M. G. Andreasen, "Scattering from parallel metallic cylinders with arbitrary cross sections," *IEEE Trans. Antennas Propagat.*, vol. AP-12, pp. 746–754, Nov. 1964.

[3] A. J. Poggio and E. K. Miller, "Integral equation solutions of three-dimensional scattering problems," in *Computer Techniques for Electromagnetics*, R. Mittra, Ed. Oxford, UK: Pergamon, 1973, ch. 4.

[4] J. R. Mautz and R. F. Harrington, "*H*-field, *E*-field, and combined-field solutions for conducting bodies of revolution," *AEÜ*, vol. 32, pp. 157–164, 1978.

[5] P. C. Waterman, "Matrix formulation of electromagnetic scattering," *Proc. IEEE*, vol. 53, pp. 805–812, Aug. 1965.

[6] N. Okamoto, "Matrix formulation of scattering by a homogeneous gyrotropic cylinder," *IEEE Trans. Antennas Propagat.*, vol. AP-18, pp. 642–649, Sept. 1970.

[7] K. A. Al-Badwaihy and J. L. Yen, "Extended boundary condition integral equations for perfectly conducting and dielectric bodies: Formulation and uniqueness," *IEEE Trans. Antennas Propagat.*, vol. AP-23, pp. 546–551, July 1975.

[8] C. Müller, *Foundations of the Mathematical Theory of Electromagnetic Waves.* Berlin: Springer, 1969.

[9] J. R. Mautz and R. F. Harrington, "Electromagnetic scattering from a homogeneous material body of revolution," *AEÜ*, vol. 33, pp. 71–80, 1979.

[10] H. A. Schenck, "Improved integral formulation for acoustic radiation problems," *J. Acoust. Soc. Amer.*, vol. 44, pp. 41–58, Jan. 1968.

[11] R. Mittra and C. A. Klein, "Stability and convergence of moment method solutions," in *Numerical and Asymptotic Techniques in Electromagnetics*, R. Mittra, Ed. New York: Springer, 1975, ch. 5.

[12] N. Morita, "Resonant solutions involved in the integral equation approach to scattering from conducting and dielectric cylinders," *IEEE Trans. Antennas Propagat.*, vol. AP-27, pp. 869–871, Nov. 1979.

[13] D. J. N. Wall, "Methods of overcoming numerical instabilities associated with the *T*-matrix method," in *Acoustic, Electromagnetic and Elastic Wave Scattering—Focus on the T-Matrix Approach*, V. K. Varadan and V. V. Varadan, Eds. New York: Pergamon, 1980, pp. 269–286.

[14] I. Toyoda, M. Matsuhara, and N. Kumagai, "Analysis of scalar wave scattering problems by means of integral equation formulation," Tech. Rep. Inst. Elec. Eng. Japan, EMT-86-83, Oct. 1986 (in Japanese).

[15] C. de Boor, *A Practical Guide to Splines.* New York: Springer-Verlag, 1978.

Part 3
Thin Wires

WIRES represent a special class of conducting objects because of the simplification they offer, and because so many actual problems do involve wires. Topics discussed in this section include models based on both the Hallén and electric field integral equations, the various kinds of basis and testing functions employed, a survey and comparison of some of these methods, and consideration of the limitations that arise from using the reduced kernel employed as the thin-wire approximation.

The section begins with the widely referenced paper by Mei, "On the integral equations of thin wire antennas." This paper describes one of the first MM treatments of a thin-wire formulation, in this case based on the Hallén integral equation. Next is the paper by Yeh and Mei, continuing development of the Hallén integral equation, "Theory of conical equiangular-spiral antennas: Part 1—Numerical techniques." This paper introduced the three-term, subdomain current expansion (constant, sine, and cosine) now employed in the widely used numerical electromagnetics code (NEC), and one of the first generalized wire treatments. The paper by Gee *et al.*, "Computer techniques for electromagnetic scattering and radiation analyses," presents the theoretical background and a variety of applications for the original version of NEC, which is based on the electric field integral equation (EFIE) with a three-term current expansion and point matching. It includes near-field computations, lumped and distributed impedance loading, wire-grid models, and interface effects. The problem of assessing required sampling density for subdomain wire modeling is addressed by Miller *et al.* in "Accuracy-modeling guidelines for integral-equation evaluation of thin-wire scattering structures." Presented in this paper is the numerical convergence achieved for several kinds of wire objects modeled as scatterers, using randomly sampled scattered fields. The contribution by Richmond, "Radiation and scattering by thin-wire structures in the complex frequency domain," is excerpted from OSU Research Foundation report RF 2909-10, 1973, and has not appeared in the open literature. It describes the piecewise sinusoidal reaction formulation for radiation and scattering by thin wires, including the reaction integral equation, the piecewise sinusoidal expansion/testing functions, and the MM matrix elements. Simple modifications for finite wire conductivity, dielectric shells, and lumped loads are also included. Sarkar presents a useful comparison of alternatives available for wire problems in "A study of the various methods for computing electromagnetic field utilizing thin wire integral equations." Collin points out in "Equivalent line current for cylindrical dipole antennas and its asymptotic behavior," that the common practice of filament testing, i.e., enforcing the integral equation on the wire centerline, results in an integral equation with no exact solution. The result is an MM solution that exhibits relative convergence, so that, as the number of current expansion functions increases, the MM solution initially converges to a value close to the exact result, and then diverges.

On the Integral Equations of Thin Wire Antennas

K. K. MEI, MEMBER, IEEE

***Abstract*—The feasibility of direct numerical calculations of antenna integral equations is investigated. It is shown that integral equation of Hallen's type is the most adequate for such applications. The extension of Hallen's integral equation to describe thin wire antennas of arbitrary geometry is accomplished, and results are presented for dipole, circular loops, and equiangular spiral antennas.**

INTRODUCTION

During the past seven years, the advancement of antenna design has been characterized by an exhaustive utilization of antenna geometry. Broadband antennas are notable examples. In the study of antenna theory, a knowledge of the current distribution is of fundamental importance. Such data may be obtained either by measurement or by solving the antenna integral equation. Integral equations are difficult to solve even for the simplest case of a dipole antenna. However, as a result of the development in modern high speed computers, the range of application of the integral equation method has been greatly enlarged. The purpose of this paper is to present an investigation of the feasibility of direct numerical calculations of antenna integral equations. To simplify the discussion, the trapezoidal rule of integration is assumed throughout, although it is realized that in a practical calculation better integration schmes, such as quadratic rule, etc., may need to be used. Typical results of calculations are presented.

NUMERICAL SOLUTIONS OF DIPOLE ANTENNAS

It is well known that the axial component of the electric field produced by the current on a cylindrical dipole antenna [1] is given by

$$\frac{d}{dz}\int_L \oint \frac{dJ(z')}{dz'} G(z, c; z'c')dc'dz' + k^2 \int_L \oint J(z')G(z, c; z', c')dc'dz' = j\omega\epsilon E(z) \quad (1)$$

where the symbol $J(z')$ represents the surface current density, $\oint dc'$ represents the integration around the periphery of the cylinder, and $G(z, c; z', c')$ is the free space Green's function,

$$G(z, c; z', c') = \frac{e^{-jk|\bar{r}-\bar{r}'|}}{4\pi|\bar{r} - \bar{r}'|}$$

Manuscript received August 11, 1964; revised November 30, 1964. This manuscript originally appeared as Internal Technical Memorandum M-79 at the Electronics Research Lab., University of California. The research reported here was made possible through support received from the National Science Foundation under Grant GP-2203.

The author is with the Dept. of Electrical Engineering, Electronics Research Lab., University of California, Berkeley, Calif.

For simplicity we shall omit the integration $\oint dc'$ in the discussion that follows, i.e., the symbol $\int_L dz'$ will represent the surface integral over the cylinder.

When the electric field on the surface of the antenna is considered, (1) reduces to

$$\frac{d}{dz}\int_L J'(z')G(z, z')dz' + k^2 \int_L J(z')G(z, z')dz' = -j\omega\epsilon E_z{}^i(z) \quad (2)$$

where $E_z{}^i(z)$ is the electric field produced by the generator. Equation (2) is an integrodifferential equation for the current, which may be solved numerically by a combination of the difference equation method and the numerical integration method. The disadvantage of such an approach is that difference equations are generally unstable and critical to the errors in the approximation. An alternative approach is to transform (2) into a pure integral equation. Equation (2) may be readily transformed into such an equation of one of several familiar forms. The one used by Pocklington [2] is

$$\int_L J(z')\left[\frac{\partial^2}{dz^2} G(z, z') + k^2G(z, z')\right]dz' = -j\omega\epsilon E_z{}^i(z) \quad (3)$$

integrating both sides of (3), say from 0 to z, gives

$$\int_L J(z')\left[\frac{\partial}{\partial z} G(z, z') + k^2 \int_0^z G(\xi, z')d\xi\right]dz' = -j\omega\epsilon \int_0^z E_z{}^i(\xi)d\xi + A \quad (4)$$

The integral equation used by Hallen [1] is,

$$\int_L J(z')G(z, z')dz' = B\cos kz - \frac{jV}{2Z_0}\sin k|z| \quad (5)$$

In these integral equations, the constants of integration A and B are to be determined by the condition that the current vanishes at both ends of the antenna; V and Z_0 are, respectively, the voltage applied and the intrinsic impedance of free space.

The numerical solution of an integral equation may be effected by approximating the integration with a finite sum at n different points. The resulting algebraic equations will have the following form [3], [4]:

$$\begin{aligned} K_{11}J(z_1) + K_{12}J(z_2) + \cdots + K_{1n}J(z_n) &= F(z_1) \\ K_{21}J(z_1) + K_{22}J(z_2) + \cdots + K_{2n}J(z_n) &= F(z_2) \\ \cdots\cdots\cdots\cdots\cdots\cdots \\ K_{n1}J(z_1) + K_{n2}J(z_2) + \cdots + K_{nn}J(z_n) &= F(z_n) \end{aligned} \quad (6)$$

Reprinted from *IEEE Trans. Antennas Propagat.*, vol. AP-13, no. 3, pp. 374-378, May 1965.

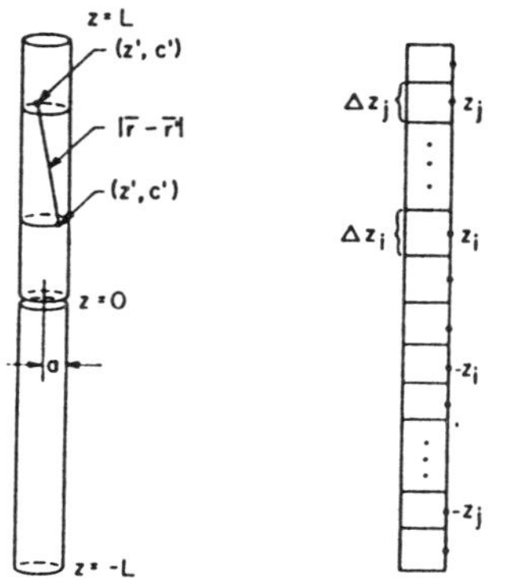

Fig. 1. Relevant of a dipole antenna and its subdivisions

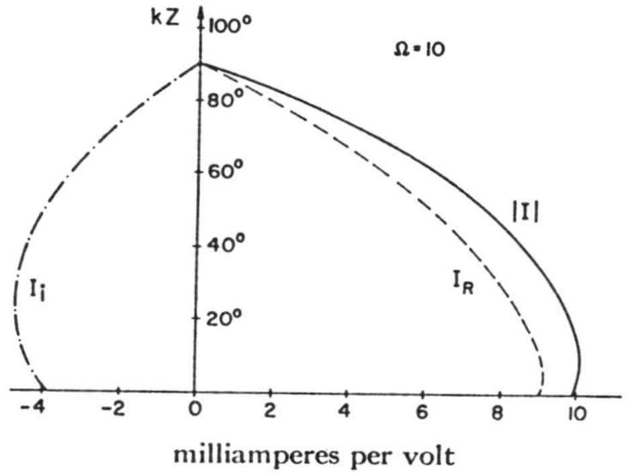

Fig. 2. Current distribution $I=I_R+jI_i$ on a dipole antenna of parameters $\Omega=2\log 2L/a=10$, $kL=\pi/2$

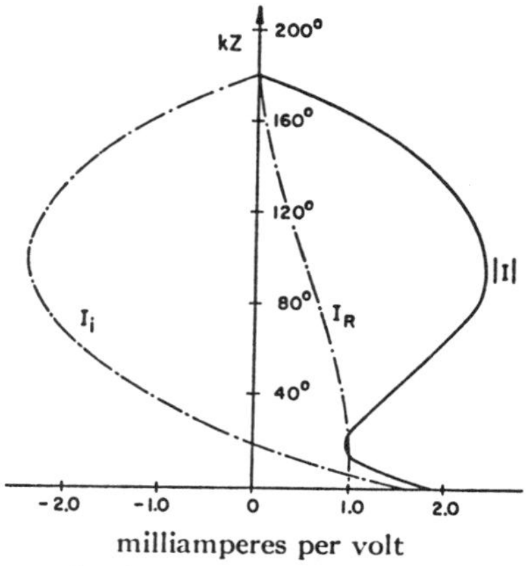

Fig 3. Current distribution on a dipole antenna of parameters $\Omega=2\log 2L/a=10$, $kL=\pi$.

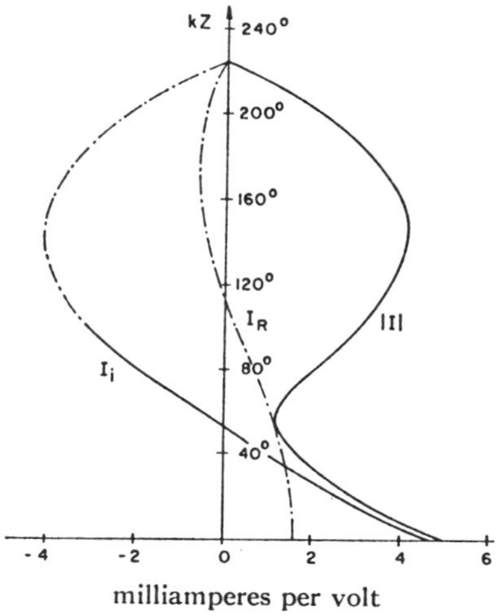

Fig. 4. Current distribution $I=I_R+jI_i$ on a dipole antenna of parameters $\Omega=2\log 2L/\theta=10$, $kL=5\pi/4$.

The matrix elements K_{ij} and $F(z_i)$ for (3), (4), and (5) are given, respectively, as

$$\left.\begin{aligned} K_{ij} &= \int_{\Delta z_j}\left(\frac{\partial^2}{\partial z^2}+k^2\right)G(z_i,z')dz' \\ F(z_i) &= -j\omega\epsilon E_z{}^i(z_i) \end{aligned}\right\} \qquad (7)$$

$$\left.\begin{aligned} K_{ij} &= \int_{\Delta z_j}\left[\frac{\partial}{\partial z}G(z_i,z')+k^2\int_0^{z_i}G(\xi,z')d\xi\right]dz' \\ F(Z_i) &= -j\omega\epsilon\int_0^{z_i}E_z{}^i(\xi)d\xi + A \end{aligned}\right\} \qquad (8)$$

$$\left.\begin{aligned} K_{ij} &= \int_{\Delta z_j}G(z_i,z')dz' \\ F(z_i) &= B\cos kz_i - \frac{jV}{2Z_0}\sin k\,|\,z_i\,| \end{aligned}\right\} \qquad (9)$$

where Δz_j's, the subdivisions of the antenna, as shown in Fig. 1 are sufficiently small so that the current in each may be considered constant.

We notice that the integral in (7) does not converge at $i=j$. Whether the often used approximation for a thin antenna of radius a,

$$\oint_c G(z,c;z',c')dc' \approx 2\pi a\left\{\frac{e^{-jk[(z-z')^2+a^2]^{1/2}}}{4\pi[(z-z')^2+a^2]^{1/2}}\right\} \qquad (10)$$

can be applied in the divergent integral of (7) is open to question [5]–[7]. An inspection of (6), (7), and (10) indicates that such approximation will not lead to the correct solution. This is so because, if approximations (7) and (10) are used in the limit of small radius a (6) approaches a diagonal matrix. That is to say, for a very thin antenna, the solution of (6) would then give $J(z)\partial E_z{}^i(z)$, which is not compatible with the well founded knowledge of antenna current distributions.

The improper integrals in (8) and (9) at $i=j$ may be integrated by using Cauchy's principal value. In these cases, we may also use the approximation (10). Actual computations based on such an approximation indeed give correct results. This possibly accounts for the fact that approximations (7) and (10) have been successfully used in variational form [8], [9], since the variational formulation introduces an additional integration, which in effect suppresses the divergent nature of the integral.

Of particular importance in the inversion of a large matrix is the problem of round-off errors accumulated through large number of arithmetic operations. In general, the round-off errors depend on the orientations of the hyperplanes represented by each row of the matrix, in the n-dimensional vector space. Qualitatively speaking, the round-off errors will be small if the hyperplanes are essentially perpendicular to one another, and the reverse is true if two or more of them are almost parallel [10]. Inspection of (8) indicates that for small radius a the coefficient K_{ij} will be small for $i<j$, and large for $i \geq j$. Hence, in the limit of a very thin antenna, the matrix elements described by (8) approach those of a triangular matrix. For the same situation, however, the matrix elements described by (9) approach those of a diagonal matrix, which is certainly superior to a triangular one in view of the previous consideration on computational errors. We shall, therefore, use integral (5) as the basis of our calculations.

A few typical results of calculation on dipole antennas are shown in Figs. 2–4. It is of interest to note that calculations based on the model of a slice generator excitation [1], and those based on the model of a magnetic loop current excitation [11] have no noticeable differences in their results.

Arbitrary Thin Wire Antennas

The extension of (3) and (4) to describe a general curved wire antenna is immediate, provided a curved cylindrical coordinate system is used. Figure 5 describes such a coordinate system, where s is the arc length measured from the feed gap, and $\hat{s}$ is the unit tangent vector at s. If the radius a of the wire is sufficiently small so that the current density may be considered to be uniform around the periphery of the wire, the corresponding integral (3) and (4) for a curved wire antenna are, respectively, [5]

$$\int_L J(s')\left[\frac{\partial^2}{\partial s \partial s'} G(s, s') - k^2 G(s, s')\hat{s}\cdot\hat{s}'\right] ds' = j\omega\epsilon E_s{}^i(s) \quad (11)$$

and

$$\int_L J(s')\left[\frac{\partial}{\partial s'} G(s, s') - k^2 \int_0^s G(\xi, s')\hat{\xi}\cdot\hat{s}' d\xi ds'\right] = j\omega\epsilon \int_0^s E_s{}^i(\xi) d\xi + A \quad (12)$$

The extension of (5) to describe a general curved wire antenna is not so apparent. The complication arises in that the kernel of the closed-cycle type is essential in the conventional way of deriving integral (5). Such a kernel has the special property

$$\frac{\partial}{\partial s} K(s, s') = -\frac{\partial}{\partial s'} K(s, s') \quad (13)$$

The structures which give rise to kernels of this type are limited to straight wires, circular arcs, and helical wires [9]. In the following we shall attempt to generalize (5) so as to include wire antennas of arbitrary geometry.

In accord with the assumptions of a thin wire antenna, the tangential component of the vector potential and scalar potential on the antenna are given, respectively, as

$$A_s(s) = \int_L J(s')G(s, s')\hat{s}\cdot\hat{s}' ds' \quad (14)$$

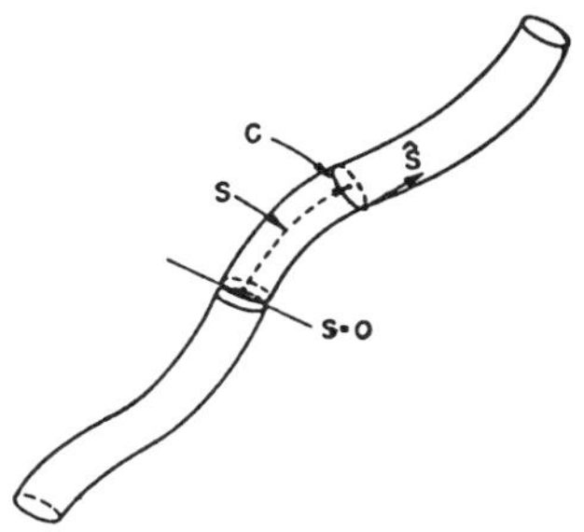

Fig. 5. A curved cylindrical coordinate system.

and

$$\phi(s) = \frac{-1}{j\omega\epsilon}\int_L \frac{dJ(s')}{ds'} G(s, s') ds' \quad (15)$$

We define a scalar function $\Phi(s)$ as

$$\Phi(s) = -j\omega\epsilon \int_0^s \phi(\xi) d\xi = \int_0^s \int_L \frac{dJ(s')}{ds'} G(\xi, s') d\xi \quad (16)$$

Integrating (16) by parts and considering $J(s)$ to vanish at both ends, we obtain

$$\Phi(s) = -\int_0^s \int_L J(s') \frac{\partial G(\xi, s')}{\partial s'} d\xi \quad (17)$$

For the s component of the electric field on the antenna to vanish, it is required that

$$E_s(s) + E_s{}^i(s) = 0 \quad (18)$$

where $E_s{}^i(s)$ is the s component of the incident electric field when the antenna is receiving, or it is the impressed field of the source if the antenna is transmitting.

From the well-known equation

$$E_s(s) = -\nabla_s \phi - j\omega\mu A_s$$

we have

$$k^2 A_s(s) - j\omega\epsilon \frac{d\phi(s)}{ds} = -j\omega\epsilon E_s{}^i(s) \quad (19)$$

or

$$\frac{d^2\Phi(s)}{ds^2} = -k^2 A_s(s) - j\omega\epsilon E_s{}^i(s) \quad (20)$$

Adding $k^2\Phi(s)$ to both sides of (20), we obtain

$$\frac{d^2\Phi(s)}{ds^2} + k^2\Phi(s) = k^2(\Phi(s) - A_s(s)) - j\omega\epsilon E_s{}^i(s) \quad (21)$$

The solution of (21) is

$$\Phi(s) = C \cos ks + D \sin k|s| + \int_0^s k(\Phi(\xi) - A_\xi(\xi)) \sin k(s - \xi) - \frac{j}{Z_0}\int_0^s E_\xi{}^i(\xi) \sin k(s - \xi) d\xi \quad (22)$$

Since $\Phi(0)=0$, we see that the constant C must vanish. Now consider the integration

$$F(s) = k\int_0^s \Phi(\xi)\sin k(s-\xi)d\xi$$
$$= -k\int_0^s\int_0^\xi\int_L J(s') \cdot\frac{\partial G(\eta, s')}{\partial s'}ds'd\eta\sin k(s-\xi)d\xi \qquad (23)$$

After changing the order of integration in (23), we obtain

$$F(s) = -k\int_L\int_0^s\int_\eta^s J(s')\frac{\partial G(\eta, s')}{ds'}\sin k(s-\xi)d\xi d\eta ds'$$
$$= -\int_L\int_0^s J(s')\frac{\partial G(\eta, s')}{ds'}[1-\cos k(s-\eta)]d\eta ds'$$
$$= \Phi(s) + \int_L\int_0^s J(s')\frac{\partial G(\eta, s')}{ds'}\cos k(s-\eta)d\eta ds' \qquad (24)$$

Next we consider the integration

$$H(s) = k\int_0^s A_\xi(\xi)\sin k(s-\xi)d\xi$$
$$= k\int_0^s\int_L J(s')G(\xi, s')\hat{\xi}\cdot\hat{s}'\sin k(s-\xi)d\xi ds' \qquad (25)$$

Integration by parts gives

$$H(s) = \int_L J(s')G(\xi, s')\hat{\xi}\cdot\hat{s}'\cos k(s-\xi)\Big|_{\xi=0}^{\xi=s}ds'$$
$$-\int_0^s\int_L\left[\frac{\partial G(\xi, s')}{d\xi}\hat{\xi}\cdot\hat{s}' + G(\xi, s')\frac{\partial(\hat{\xi}\cdot\hat{s}')}{d\xi}\right]J(s') \cdot\cos k(s-\xi)d\xi$$
$$= \int_L J(s')G(s, s')\hat{s}\cdot\hat{s}'ds'$$
$$-\int_L J(s')G(0, s')\hat{o}\cdot\hat{s}'\cos ks ds'$$
$$-\int_0^s\int_L\left[\frac{\partial G(\xi, s')}{\partial\xi}\hat{\xi}\cdot\hat{s}' + G(\xi, s')\frac{\partial(\hat{\xi}\cdot\hat{s}')}{\partial\xi}\right]J(s') \cdot\cos k(s-\xi)d\xi \qquad (26)$$

Substituting (24) and (26) into (2), we obtain the integral equation for the current,

$$\int_L J(s')\pi(s, s')ds' = D\sin k|s|$$
$$+\int_L J(s')G(0, s')\hat{o}\cdot\hat{s}'\cos ks ds'$$
$$-\frac{j}{Z_0}\int_0^s E_\xi{}^i(\xi)\sin k(s-\xi)d\xi \qquad (27)$$

where

$$\pi(s, s') = G(s, s')\hat{s}\cdot\hat{s}' - \int_0^s\left[\frac{\partial G(\xi, s')}{\partial\xi}\hat{\xi}\cdot\hat{s}' + \frac{\partial G(\xi, s')}{\partial s'} + G(\xi, s')\frac{\partial(\hat{\xi}\cdot\hat{s}')}{\partial\xi}\right] \cdot\cos k(s-\xi)d\xi \qquad (28)$$

The term $D\sin k|s|$ represents the effect of a slice generator which is redundant when the integral of $E_\xi{}^i$ is present. Indeed, if $E_\xi{}^i(\xi) = V/2\delta(\xi)$, where $\delta(\xi)$ is the Dirac delta function, we have

$$\frac{-j}{Z_0}\int_0^s E_\xi{}^i(\xi)\sin k(s-\xi)d\xi = \frac{-jV}{2Z_0}\sin k|s| \qquad (29)$$

which is consistent with (5).

To show that (27) reduces to (5) for a dipole antenna, we assume the source to be a slice generator, and notice that in this particular case

$$\frac{\partial G(\xi, s')}{\partial\xi} = -\frac{\partial G(\xi, s')}{\partial s'}$$

and

$$\hat{\xi}\cdot\hat{s} = 1$$

Hence, (27) becomes

$$\int_L J(z')G(z, z')dz'$$
$$= \int_L J(z')G(0, z')dz'\cos kz - \frac{jV}{2Z_0}\sin k|z| \qquad (30)$$

Comparing (30) with (5), we have

$$B = \int_L J(z')G(0, z')dz'$$

which may be shown to be correct by considering (5) at $z=0$. Consequently, the term $\int_L J(x')G(0, z')dz'$ should be replaced by a constant, which has to be determined by the condition of the current at the ends of the antenna, otherwise the solution of the integral equation will not be unique. Therefore, the integral equation describing an arbitrary thin wire antenna is

$$\int_L J(s')\pi(s, s')ds'$$
$$= C'\cos ks - \frac{j}{Z_0}\int_0^s E_\xi{}^i(\xi)\sin k(s-\xi)d\xi \qquad (31)$$

The specialization of (31) to a circular loop antenna also agrees with that derived by Adachi [12].

A further check of the integral equation may be effected as following. We differentiate (31) twice with respect to s, and make use of the differential relation,

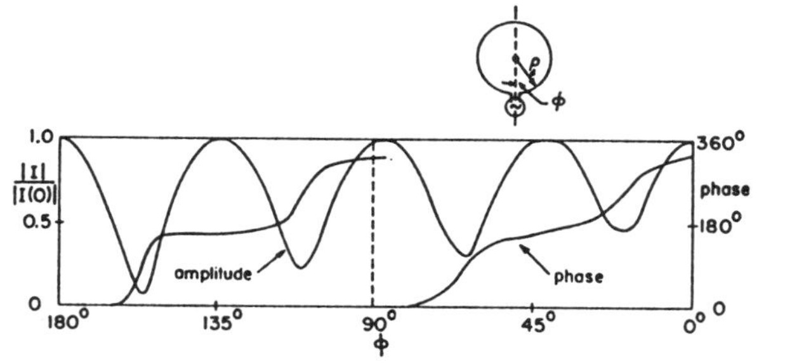

Fig. 6. Amplitude and phase of the current on a circular loop antenna of the parameter $\Omega = 2 \log 8/a\rho = 15$ (ρ = radius of the loop, a = radius of wire) $k\rho = 4.0$.

$$\frac{d}{dx}\int_0^{f(x)} g(x, x')dx'$$

$$= \int_0^{f(x)} \frac{\partial}{\partial x} g(x, x')dx' + g(x, x')\,|_{x'=x} f'(x) \qquad (32)$$

and obtain

$$k^2 \int_L \int_0^s J(s') \left[\frac{\partial G(\xi, s')}{\partial \xi}\, \hat{\xi}\cdot\hat{s}' + \frac{\partial G(\xi, s')}{\partial s'} + G(\xi, s') \frac{\partial(\hat{\xi}\cdot\hat{s}')}{\partial \xi} \right]$$

$$\cdot \cos k(s - \xi) d\xi ds' - \frac{\partial}{\partial s}\int_L J(s') \frac{\partial G(s, s')}{\partial s'} ds'$$

$$= -k^2 C' \cos ks + j\frac{k^2}{Z_0}\int_0^s E_\xi{}^i(\xi)$$

$$\cdot \sin k(s - \xi) d\xi - j\omega\epsilon E_\xi{}^i(s) \qquad (33)$$

Multiplying (31) by k^2 and adding the result to (33), results in

$$k^2 \int_L J(s') G(s, s') \hat{s}\cdot\hat{s}' ds'$$

$$- \frac{\partial}{\partial s}\int_L J(s') \frac{\partial G(s, s')}{\partial s'} ds = -j\omega\epsilon E_s{}^i(s) \qquad (34)$$

which is essentially (19). Therefore, the integral (31) is shown to be the correct one.

Applications

Equation (31) has been applied to circular loop antennas [13] and equiangular spiral antennas. The representative results are shown in Figs. 6 and 7.

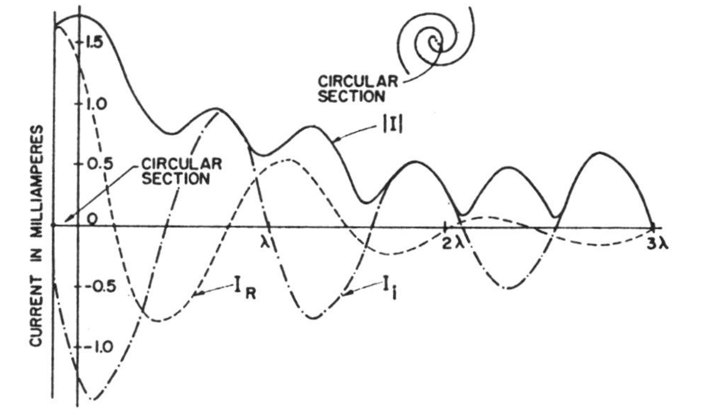

Fig. 7. Current distribution of a 3λ-arm equiangular spiral antenna or $r = ce^{a\phi}$, with $a = 0.2$, $c = 0.05\lambda$, radius of wire 0.025λ.

Acknowledgment

The author wishes to take this opportunity to thank his former professor, Dr. Jean G. Van Bladel, for his patient teaching and guidance. He is also grateful to Prof. V. H. Rumsey for encouragement and to Prof. D. J. Angelakos for valuable comments. The assistance of Y. H. Yeh and S. H. Lee in computer programming is acknowledged.

References

[1] King, R., *The Theory of Linear Antennas.* Cambridge, Mass.: Harvard University Press, 1956.
[2] Schelkunoff, S. A., *Advanced Antenna Theory.* New York: Wiley, 1952, p 132.
[3] Hildebrand, F. B., *Method of Applied Mathematics.* New York: Prentice-Hall, 1954, pp 448–451.
[4] Mei, K. K., and J. G. Van Bladel, Scattering by perfectly-conducting rectangular cylinders, *IEEE Trans. on Antennas and Propagation,* vol AP-11, Mar 1963, pp 185–192.
[5] Aharoni, J., *Antennae—An Introduction to Their Theory.* Oxford, England: Clarendon Press, 1946, pp 133–135.
[6] Van Bladel, J. G., Some remarks on Green's dyadic for infinite space, *IRE Trans. on Antennas and Propagation,* vol AP-9, Nov 1961, pp 563–566.
[7] Bouwkamp, C. J., Diffraction theory, *Rept. Progr. in Phys.,* vol 17, 1954, pp 35–100.
[8] Tai, C. T., A new interpretation of the integral equation formulation of cylindrical antennas, *IRE Trans. on Antennas and Propagation,* vol. AP-3, Jul 1955, pp 125–127.
[9] Tang, C. H., Input impedance of arc antennas and short helical radiators, *IEEE Trans. on Antennas and Propagation,* vol AP-12, Jan 1964, pp 2–9.
[10] Crandell, S. H., *Engineering Analysis.* New York: McGraw-Hill, 1956, sec 1.3, pp 15–18.
[11] Albert, G. E., and J. L. Synge, The general problem of antenna radiation and the fundamental integral equation with application to an antenna of revolution, pt 1, *Quart. Appl. Math.,* vol 6, Jul 1948, pp 117–132.
[12] Adachi, S., and Y. Mushiake, Theoretical formulation of circular loop antennas by integral equation method, *Sci. Repts. Res. Insts. Tohoku Univ. Ser. A,* B-(Elect. Comm.), vol 9, no 1, pp 9–18.
[13] Baghdasarian, A., and D. J. Angelakos, Scattering and radiation from conducting loops, to be published as an Electronics Research Lab. Rept 65-1, University of California, Berkeley.

Theory of Conical Equiangular-Spiral Antennas Part I—Numerical Technique

YU SHUAN YEH, MEMBER, IEEE, AND KENNETH K. MEI, MEMBER, IEEE

Abstract—**The integral equation method is applied to find the rigorous solutions of the current distributions on conical, equiangular-spiral antennas of arbitrary spiral parameter and cone angle. With a transcendental interpolation function, antennas up to 10λ in armlength can be calculated. Comparisons of calculated and experimental results are presented, indicating excellent agreement. The computer programming resulting from this investigation thus replaces painstaking procedures of design, experimentation, and optimization of equiangular-spiral antennas by a few minutes of computer calculations.**

Introduction

EQUIANGULAR-spiral antennas have many applications, yet they have defied theoretical analysis for many years. The first successful theoretical investigation of a planar, equiangular-spiral antenna was conducted by Cheo *et al.* [1], who used a spiral anisotropic sheet as the mathematical model for the antenna. With some improvements on Cheo's analysis, Bernard and Ishimaru [2] and Laxpati and Mittra [3] were able to solve the antenna problem including dipole and slot excitations. The analysis of conical, equiangular-spiral antennas was first attempted by Bickel [4] with the same model as Cheo's, and later by Mittra and Klock [5] with the WKB method which considers the conical spiral as a perturbation to a helix. While all these analyses provide valuable qualitative information on equiangular-spiral antennas, the design data they can provide are considerably limited.

In this paper we analyze the equiangular-spiral antenna by means of the integral equation technique previously reported by Mei [6]. The integral equation method is applicable to antennas of general shape, but the conventional numerical method for solving the integral equation is limited to antennas of about 4λ in armlength using the IBM 7094 computer, which is short of that of a practical spiral antenna. In Part I of this paper, a numerical technique is presented which enables us to use the integral equation technique to calculate conical, equiangular-spiral antennas up to 10λ in armlength. The technique is then used to calculate conical, equiangular-spiral antennas of arbitrary cone angle and spiral parameter.

The numerical method of analysis has many advantages over experimental investigations in that it gives accurate and inexpensive detailed results of current. Excellent experimental results on the impedance and radiation pattern of conical equiangular antennas were reported by Dyson [7], [8]. Our numerical results agree remarkably well with Dyson's measurements. The theoretical calculations not only reduce the labor and cost of obtaining design data for the spiral antennas, but also take over the job of testing and readjusting for optimization in the design process. The numerical method thus replaces painstaking procedures of fabrication and measurement by a few minutes of computer calculations. The computer program resulting from this investigation is available upon request.[1]

In Part II of this paper, the general behavior of the current distribution will be discussed, which leads to a method of approximation for the input impedances.

Manuscript received February 10, 1967; revised April 24, 1967. The research reported here was wholly supported by the National Science Foundation under Grant GK-676.

Y. S. Yeh is with the Division of Engineering and Applied Physics, Harvard University, Cambridge, Mass. He was formerly with the Electronics Research Laboratory, University of California, Berkeley, Calif.

K. K. Mei is with the Department of Electrical Engineering and Electronics Research Laboratory, University of California, Berkeley, Calif.

Geometry of the Antenna

The geometry of an equiangular-spiral antenna is shown in Fig. 1. Arms 1 and 2 of the spiral are described by $r=r_0e^{a\phi}$ and $r=r_0e^{a(\phi-\pi)}$, respectively. For convenience, length s on arm 1 will be measured positive and that on arm 2 will be negative. The spiral geometry terminates at positions p and p', and a straight wire is inserted between these two points. The generator is located at the middle of p-p'.

The geometric parameters that may affect the performance of the antennas are θ_0=cone angle, a=spiral constant, r_0=starting radius of the antenna, and δ=angular width of the arm. Parameter δ is defined in the following.

In the enlarged view of an equiangular-spiral antenna (Fig. 2) curve 1–1′ is the center line of the spiral wire and is defined by $r=r_0e^{a\phi}$. The wire radius R is proportional to r and is given by $R=cr$. The radial line 0–1 intersects the wire circumference at radii r' and r'' which are given approximately by

$$r' = r(1 - c/\sin \alpha) \qquad (1)$$

$$r'' = r(1 + c/\sin \alpha), \qquad (2)$$

where $\sin \alpha=(1+a^2/\sin^2 \theta_0)^{-1/2}$.

We define δ from the relation

$$r'' = r'e^{a\delta}; \qquad (3)$$

thus

$$\delta = \frac{1}{a} \log \left(\frac{\sin \alpha + c}{\sin \alpha - c}\right). \qquad (4)$$

[1] Requests should be addressed to K. K. Mei, Electronics Research Laboratory, University of California, Berkeley, Calif. 94720

Reprinted from *IEEE Trans. Antennas Propagat.*, vol. AP-15, no. 5, pp. 634–639, September 1967.

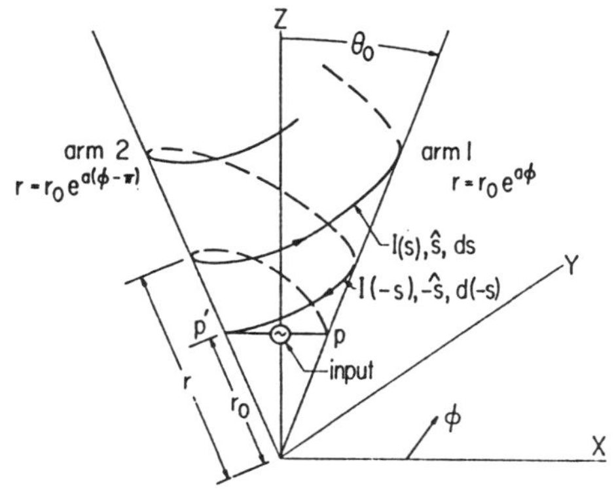

Fig. 1. Conical spiral antenna with associated parameters.

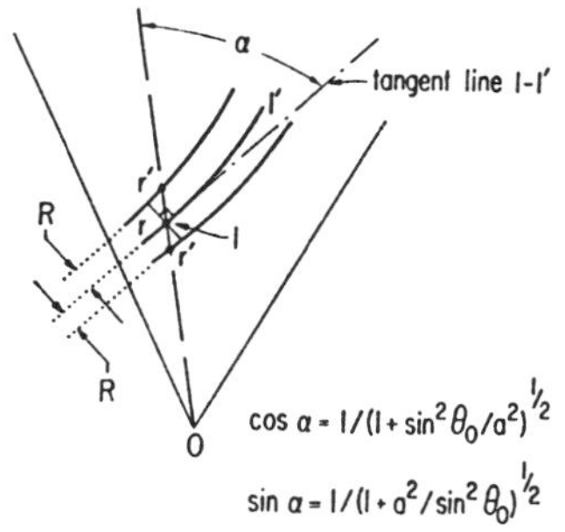

Fig. 2. Enlarged view of an antenna arm.

Since the quotient $c/\sin\alpha$ is always much smaller than 0.05, δ can then be simplified to

$$\delta \simeq 2c/a \sin \alpha. \tag{5}$$

Once δ is known, the radius of the wire at any point on the antenna can be calculated by

$$R = cr = \frac{\delta a \sin \alpha}{2} r. \tag{6}$$

Integral Equations

The integral equation of a thin wire antenna of arbitrary geometry is [6]

$$\int_L I(s')\pi(s, s')ds' = B \cos ks - \frac{jV_0}{2Z_0} \sin ks, \tag{7}$$

where

$$\pi(s, s') = G(s, s')\hat{s}\cdot\hat{s}' - \int_0^s \left\{\frac{\partial G(t, s')}{\partial t} \hat{l}\cdot\hat{s}' + \frac{\partial G(t, s')}{\partial s'} + G(t, s') \frac{d\hat{l}}{dt}\cdot\hat{s}'\right\} \cos k(s - t)dt \tag{8}$$

$$G(s, s') = \frac{e^{-jk|\bar{r}-\bar{r}'|}}{4\pi|\bar{r} - \bar{r}'|}, \tag{9}$$

in which $\hat{s}$, $\hat{s}'$, and $\hat{l}$ are all unit vectors along the wire of the antenna.

Because of the symmetry of the spiral antenna, the currents on the two arms are equal in magnitude but opposite in sense. Because of our definition of s, the current on arm 2 is considered positive in the direction toward the generator. Thus we have $I(s) = I(-s)$, and the integral equation may be written as

$$\int_0^L I(s')F(s, s')ds' = B \cos ks - \frac{jV_0}{2Z_0} \sin ks, \tag{10}$$

where

$$F(s, s') = \pi(s, s') + \pi(s, -s'). \tag{11}$$

The unit vector $\hat{l}$ has step discontinuities at points p and p'. Thus $\pi(s, s')$ should be expressed as

$$\begin{aligned}\pi(s, s') &= G(s, s')\hat{s}\cdot\hat{s}' \\ &- G(s_p, s')(\hat{s}_p^+ - \hat{s}_p^-)\cdot\hat{s}' \cos k(s - s_p)H(s - s_p) \\ &- P\int_0^s \left[\frac{\partial G(t, s')}{\partial t} \hat{l}\cdot\hat{s}' + \frac{\partial G(t, s')}{\partial s'} + G(t, s') \frac{d\hat{l}}{dt}\cdot\hat{s}'\right] \\ &\cdot \cos k(s - t)dt,\end{aligned} \tag{12}$$

where s_p is the length s at p, P denotes the principal value of the integration, and $H(s-s_p)$ is the Heaviside step function.

Basic Geometric Factors

The geometrical relations required to evaluate the integral equation are presented in the following with reference to Fig. 1. The length along the conical spiral curve $r = r_0e^{a\phi}$ is most conveniently defined as starting from the tip of the cone. Thus,

$$s = Qr = Qr_0e^{a\phi}, \tag{13}$$

where Q is known as the slowness factor with

$$Q = (1 + \sin^2 \theta_0/a^2)^{1/2}. \tag{14}$$

The length at p is $s_p = Qr_0$. The length at the input of the antenna is therefore $s_{in} = s_p - r_0 \sin \theta_0$ instead of $s = 0$. Consequently, the lower limit of the integrations in (7) and (8) must be changed from 0 to s_{in}.

In the evaluation of the kernel of the integral equation, the values of $\bar{r}$, $\hat{s}$, $d\hat{s}/ds$ must be known. For the spiral geometry they are

$$\bar{r} = r \sin \theta_0 \cos \phi\hat{x} + r \sin \theta_0 \sin \phi\hat{y} + r \cos \theta_0\hat{z} \tag{15}$$

$$\hat{s} = \frac{1}{Q}\left[\sin \theta_0(\cos \phi - \sin \phi/a) \cdot\hat{x} + \sin \phi_0(\sin \phi + \cos \phi/a)\hat{y} + \cos \theta_0\hat{z}\right] \tag{16}$$

$$\frac{d\hat{s}}{ds} = \frac{1}{raQ^2}\left[- \sin \theta_0(\sin \phi + \cos \phi/a) \cdot\hat{x} + \sin \phi_0(\cos \phi - \sin \phi/a)\hat{y}\right]. \tag{17}$$

The above formulas are true for $s \geq 0$ (or s on arm 1). For $s < 0$, we only have to realize that arm 2 is essentially arm 1 rotated by 180° with respect to the Z axis. Thus,

$$(\widehat{-s}) = -\left[s_x(-\hat{x}) + s_y(-\hat{y}) + s_z(\hat{z})\right]$$
$$= \frac{1}{Q}\Big[\sin\theta_0(\cos\phi - \sin\phi/a) \cdot \hat{x} + \sin\theta_0(\sin\phi - \cos\phi/a)\hat{y} - \cos\theta_0\hat{z}\Big]. \qquad (18)$$

It should be noted that $(\widehat{-s}) \neq -\hat{s}$.

Numerical Analysis

The numerical methods of solving integral equations generally consist of reducing the integral equation to a finite set of algebraic equations. Let

$$\int_a^b I(s')\pi(s, s')ds' = f(s) \qquad (19)$$

be the integral equation to be considered. We assume that $I(s')$ can be approximated by a set of N linearly independent functions $\phi_j(s')$ such that

$$I(s') \simeq \sum_{j=1}^{N} A_j\phi_j(s'), \qquad (20)$$

where $A_j(j=1, \cdots, N)$ are constants.

Substituting (20) into (19), we obtain

$$\sum_{j=1}^{N} A_j \int_a^b \phi_j(s')\pi(s, s')ds' \simeq f(s). \qquad (21)$$

We have thus reduced the problem to that of finding the N constants A_j, which can be done in many different ways. The easiest method of determining A_j with a digital computer is by collection, which consists of enforcing (21) at N different points in (a, b). Therefore, we obtain N linear equations

$$\sum_{j=1}^{N} A_j \int_a^b \phi_j(s')\pi(s_i, s')ds' = f(s_i), \quad i = 1, 2, \cdots, N \qquad (22)$$

which may be solved for A_j by standard methods. The choice of ϕ_j depends on many factors such as convenience in computation, stability of the resulting matrix, speed of convergence, etc. A convenient set of ϕ_j is the function $U_j(s)$ which is defined as

$$U_j(s) = 1 \quad \text{for } s \text{ in } \Delta s_j,$$
$$= 0 \quad \text{otherwise.} \qquad (23)$$

The function $U_j(s)$ is shown in Fig. 3. Mei has shown that the stability of the matrix in (22), using U_j, does not deteriorate with increasing N if $\pi(s, s')$ is singular at $s=s'$ [9]. However, the above method has the disadvantage of slow convergence. A better integration scheme would be to replace the flat top of U_j by a curve using various interpolation techniques as shown in Fig. 4. For example, using quadratic curves, we obtain a set of functions $Q_j(s)$ defined as

$$Q_j(s) = A_j + B_j(s - s_j) + C_j(s - s_j)^2 \quad \text{for } s \text{ in } \Delta s_j,$$
$$= 0 \quad \text{otherwise,} \qquad (24)$$

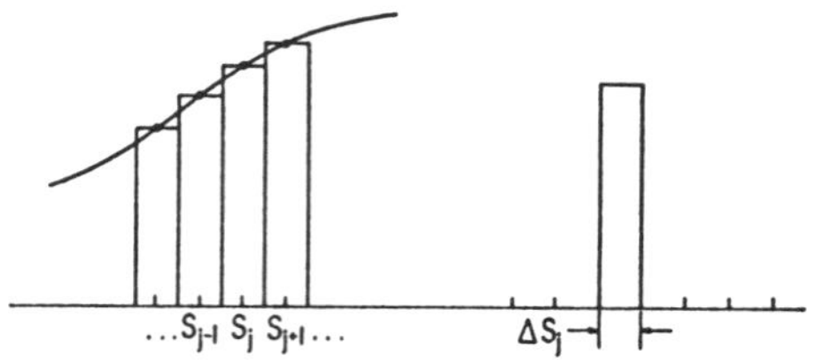

Fig. 3. Trapezoidal approximation of a function and the function $U_j(s)$.

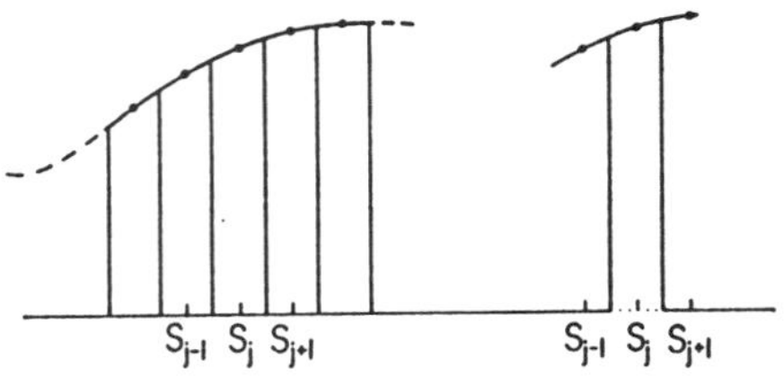

Fig. 4. Quadratic approximation of a function and the function $Q_j(s)$.

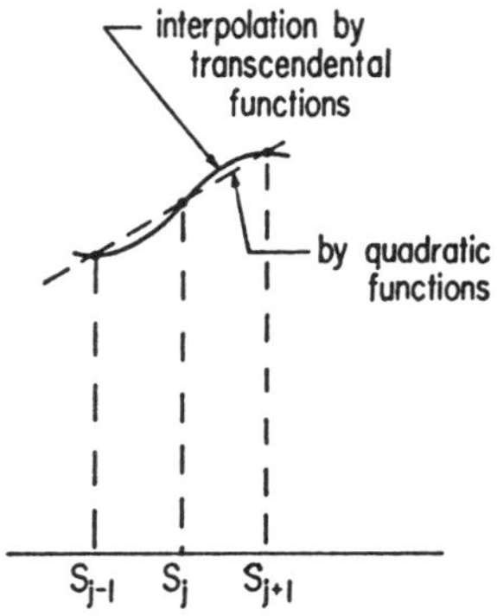

Fig. 5. An illustration of the difference between transcendental and quadratic interpolations.

where the constants A_j, B_j, and C_j may be solved in terms of $I(s_{j-1})$, $I(s_j)$, and $I(s_{j+1})$. The rate of convergence depends primarily on the local similarity between the interpolating curve and the solution of the integral equation. From this point of view, we conclude that the quadratic curves are inadequate for the type of integral equations we are concerned with, in that the currents on the antennas are oscillatory. The respresentation of a curve near a point of inflection by a quadratic curve is illustrated in Fig. 5. We see that it is indeed rather poor.

Owing to the fact that the current on the antenna propagates along the wire with approximately the velocity of light, we propose a transcendental interpolation function, which results in the set of functions $T_j(s)$, defined as

$$T_j(s) = A_j + B_j \sin k(s - s_j) + C_j \cos k(s - s_j) \quad \text{for } s \text{ in } \Delta s_j,$$
$$= 0 \quad \text{otherwise} \qquad (25)$$

where the constants A_j, B_j, and C_j are such that $T_j(s_j) = I(s_j)$ and the continuation of $T_j(s)$ outside Δs_j gives the values of $I(s)$ at s_{j-1} and s_{j+1}. Thus,

$$I_{j-1} = A_j - B_j \sin kd_{j-1} + C_j \cos kd_{j-1}$$
$$I_j = A_j + C_j \qquad (26)$$
$$I_{j+1} = A_j + B_j \sin kd_{j+1} + C_j \cos kd_{j+1}$$

where $d_{j-1}=s_j-s_{j-1}$ and $d_{j+1}=s_{j+1}-s_j$. Solving for A_j, B_j, and C_j, and substituting in (25), we obtain

$$I(s) = X(s)I_{j-1} + Y(s)I_j + Z(s)I_{j+1} \quad \text{for } s \text{ in } \Delta s_j \quad (27)$$

where

$$X(s) = \frac{1}{H}[- \sin kd_{j+1} + (1 - \cos kd_{j+1}) \sin k(s - s_j) + \sin kd_{j+1} \cos k(s - s_j)]$$

$$Y(s) = \frac{1}{H}[\sin k(d_{j-1} + d_{j+1}) + (\cos kd_{j+1} - \cos kd_{j-1}) \sin k(s - s_j) - (\sin kd_{j+1} + \sin kd_{j-1}) \cos k(s - s_j)] \quad (28)$$

$$Z(s) = \frac{1}{H}[- \sin kd_{j-1} + (\cos kd_{j-1} - 1) \sin k(s - s_j) + \sin kd_{j-1} \cos k(s - s_j)], \quad (29)$$

in which $H=\sin k(d_j+d_{j-1})-\sin kd_j-\sin kd_{j-1}$.

With the above approximation, the integral equation (10) should be written as

$$\sum_{j=1}^{n} \int_{\Delta s_j} I(s')F(s_i, s')ds' = B \cos ks_i - j\frac{V_0}{2z_0} \sin ks_i$$

$$i = 1, 2, 3, \cdots, n. \quad (30)$$

The integration on the left side may be carried out in any of the known numerical schemes. In this investigation we have used Lobatto's four-point formula [10], which is known to minimize the mean square error when the end points are fixed.

Convergence

It must be emphasized that the sinusoidal interpolation by no means assumes that the current is sinusoidal. The sinusoidal curves are merely convenient functions for interpolating the current between sampling points. A comparison is made of the mean square error incurred by the parabola and sinusoidal interpolations when applied to a function of the form $e^{-xa} \sin (bx+c)$. Calculations with $a=0$, $0.4/\lambda$, $0.8/\lambda$, $1.2/\lambda$, and $b=0.8k_0$, $0.9k_0$, k_0, $1.1k_0$, $1,2k_0$ for different values of c over a range of $x=0.4\lambda$ indicate a decisive advantage to using the sinusoidal interpolation.

Actual calculation shows that the sinusoidal interpolation converges much faster than the trapezoidal method. For example, to calculate the current distribution on a quarter-wavelength dipole, 20 subdivisions are needed for the trapezoidal method. The sinusoidal interpolation achieves the same accuracy with only six subdivisions.

In the calculation of the current distribution on the spiral antenna, one must notice that the beginning turns have a circumference much less than a wavelength. To allow for the rapid variation of the kernel in this range, the segment length Δs_j is taken according to geometrical degrees. For example, the length Δs_j may be taken as 75° in electrical or geometrical length, whichever is the shorter.

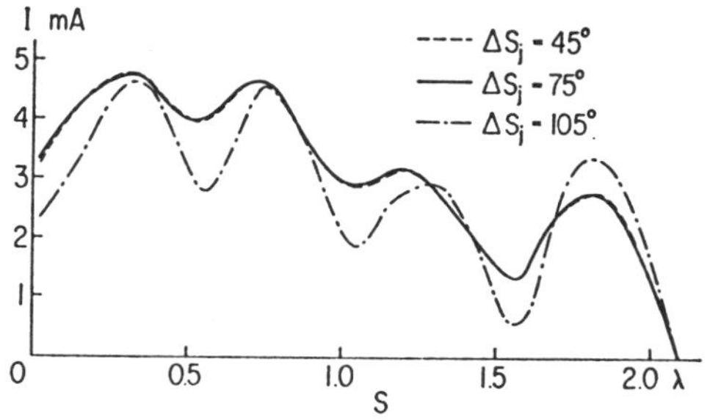

Fig. 6. Convergence test for a conical spiral with $\theta_0=10°$, $a=0.053$, $\delta=41.2°$, and $r_0=0.316\lambda$.

Tests of convergence of one antenna using a different segment length are shown in Fig. 6. It can be seen that $\Delta s_j=75°$ is sufficient to bring about convergent results. Under this approximation, antennas up to 10 wavelengths can be investigated.

End Condition

An additional equation is obtained by applying the boundary condition at the end of the antenna, i.e., $I(L)=0$. Denote L by s_{n+1}; then (27) gives

$$I(L) = 0 = X(L)I_{n-2} + Y(L)I_{n-1} + Z(L)I_n. \quad (31)$$

Equations (30) and (31) form an $(n+1)\times(n+1)$ matrix equation that can be solved by matrix inversion.

Singularities of the Kernel

As mentioned previously, the integration over each segment is performed with Lobatto's integration formula. In view of the rapid change of the Green's function when s' goes through s_i, the integration must be treated more carefully when s' approaches s_i. The proper treatment is to use the surface integration. However, it has been shown for cylindrical wires [11] that when the surface current is replaced by a line current along the center of the antenna arm and the integration is carried out by Simpson's rule, the surface integration

$$\int_{\Delta s_j} \int_c J(s')G(s_j, s', c, c')dc'ds'$$

differs from the approximate integration

$$\int_{\Delta s_j} I(s')G(s_j, s')ds'$$

by less than 1 percent if Δs_j is bigger than $2.5R$. The step length in this calculation is of the order of R.

It is also noticed from (8) that besides the single integration mentioned above there is a double integration of the form

$$\int_0^L I(s') \int_0^s \frac{\partial G(t, s')}{\partial t} \hat{t}\cdot\hat{s}' + \frac{\partial G(t, s')}{\partial s'} + G(t, s') \frac{d\hat{t}}{dt}\cdot\hat{s}' \cdot \cos k(s - t)dtds'. \quad (32)$$

This might cause trouble when dt and ds' are in the same segment. However, it can be shown that when $t=s'+\epsilon$, and $\epsilon \to 0$,

$$\frac{\partial G(t, s')}{\partial t}\hat{t}\cdot\hat{s}' + \frac{\partial G(t, s')}{\partial s'} \to \frac{\epsilon^3}{D^3} \to \frac{\epsilon^3}{\sqrt{\epsilon^2 + R^2}^3} \tag{33}$$

and

$$G(t, s')\frac{d\hat{t}}{dt}\cdot\hat{s}' \to \frac{\epsilon}{D} \to \frac{\epsilon}{\sqrt{\epsilon^2 + R^2}}. \tag{34}$$

Equations (33) and (34) indicate that the kernel of the double integration is well behaved and there is no sharp peak in the integrand.

Results

A. Antennas of Constant Wire Radius

A typical current distribution and phase progression along a constant wire radius antenna is shown in Fig. 7. The length s is measured in terms of wavelength from the tip of the cone, hence the antenna need not start at $s=0$.

The magnitude of the current remains constant for approximately half a wavelength and then starts to decrease. The antenna can therefore be separated into input and attenuation regions. The phase velocity can be approximated by free-space wave velocity up to the beginning of the attenuation region. It is noted that this phase progression characteristic is not influenced by a change of wire thickness and therefore depends only on θ_0 and a.

To provide a better understanding of the decay rate, the current distribution is plotted on a logarithmic scale (Fig. 8). The slope of the current curve is therefore a measure of the local attenuation rate along the antenna. Another current distribution of the same antenna with a different wire radius is also included as a point of comparison.

Of the two antennas considered, the thicker antenna has a lower input impedance but a higher attenuation rate. In both cases the attenuation rate decreases with distance from the input. This is due to the fact that when the wave goes away from the input, the angular armwidth decreases and hence there is less attenuation.

Since the angular armwidth influences the attenuation rate and is not frequency-independent for the constant wire radius antenna, such an antenna cannot be expected to have frequency-independent performance, although frequency-independent operation within a limited bandwidth is possible with some uniform wire structures.

The radiation pattern of the antenna with R_0 (radius of the wire) $=0.0058\lambda$ is shown in Fig. 9. The radiation pattern is essentially backfire with a small forward lobe which is due to the reflection from the end. Dyson's measured pattern is also presented and is shown to be in excellent agreement with the theoretical result. The radiation pattern of the $R_0=0.0096\lambda$ antenna is only slightly narrower but has a much smaller forward lobe because of its faster attenuation.

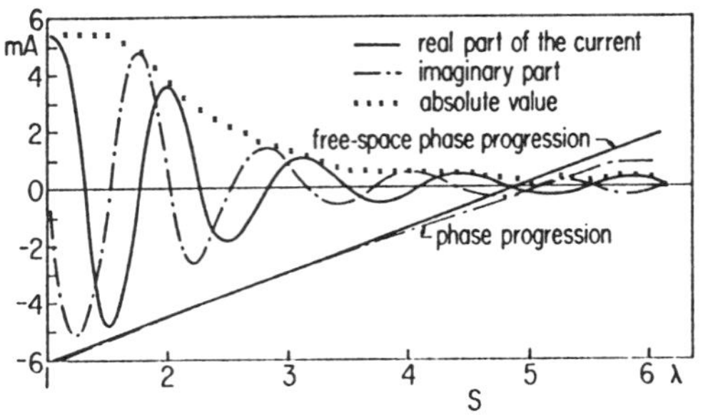

Fig. 7. Current distribution for a constant wire radius conical spiral antenna with $\theta=10°$, $a=0.053$, $r_0=0.316\lambda$, and $R_0=0.0096\lambda$.

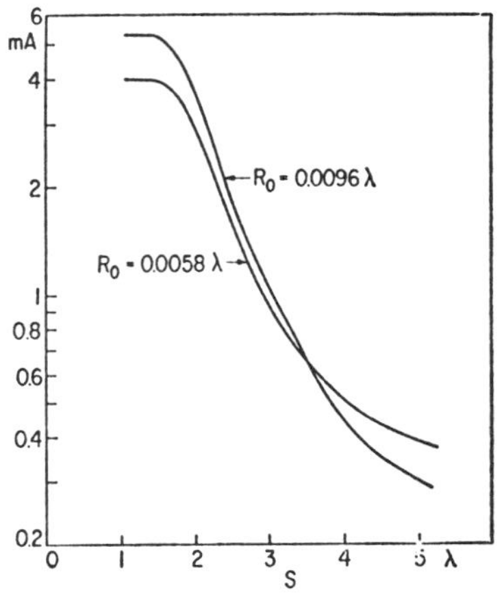

Fig. 8. Current distribution on a logarithmic scale for two constant wire radius conical spiral antennas having the same geometric parameters given in Fig. 7 but different wire radius R_0.

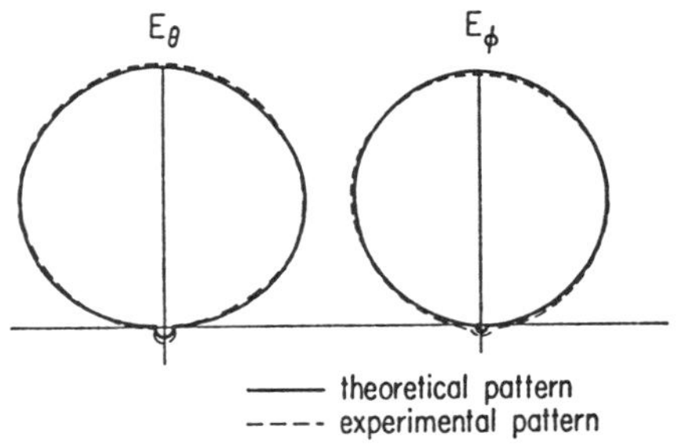

Fig. 9. Comparison of the theoretical and experimental radiation pattern of a constant wire radius conical spiral antenna with $\theta=10°$, $a=0.053$, $r_0=0.316\lambda$, and $R_0=0.0056\lambda$. (The experimental curve is taken from Fig. 7 in Dyson [7] at $f=600$ MHz.)

B. Antennas of Expanding Wire Radius

If the wire radii of the antenna expand linearly with the radial dimensions of the antenna, the structure is then invariant, in terms of wavelength, under a change of frequency. Consequently, one may expect that the shape of the current distribution is also invariant. For example, an antenna is specified by $\theta_0=10°$, $a=0.062$, $\delta=42.2°$, and $r_0=0.2\lambda$ under $f=f_1$. If the frequency is changed to $f=(\frac{1}{2})f_1$, the only parameter changed is r_0 which is now given by $r_0=0.1\lambda$. These two current distributions are presented in

Fig. 10. The invariance of the current distribution is obvious and a strong indication that investigation of the spiral antenna at one frequency is sufficient for the prediction at other frequencies. We notice that in those antennas the current decays exponentially with constant rate.

Conclusion

The numerical technique of solving the antenna integral equation for the current distribution on conical, equiangular-spiral antennas has been presented. By introducing a new interpolation technique we have extended the capability of the integral equation method to analyze antennas up to 10λ in armlength with the IBM 7094 computer. The theoretical and experimental results agree remarkably well, which confirms the accuracy of the calculations. It is of interest to note that it takes only 2.5 minutes to analyze a 5λ conical, equiangular-spiral antenna with the IBM 7094 computer. The computer program resulting from this investigation is certain to result in great savings of labor in the design of equiangular-spiral antennas. This computer program is also available on request. The discussion on the general behavior of the current on conical log-spiral antenna will be presented in Part II.

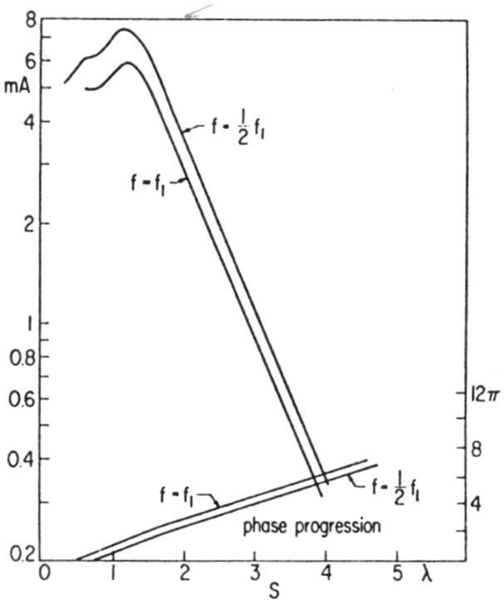

Fig. 10. Current distribution on a logarithmic scale for a conical spiral antenna with $\theta_0 = 10°$ and $\delta = 42.2°$ under two different frequencies.

References

[1] R. R. S. Cheo, W. J. Welch, and V. H. Rumsey, "A solution to the frequency independent antenna problem," *IRE Trans. Antennas and Propagation*, vol. AP-9, pp. 527–534, November 1961.

[2] G. D. Bernard and A. Ishimaru, "A class of equiangular spiral antenna excited by a vertical dipole," Dept. of Elec. Engrg., University of Washington, Seattle, Tech. Rept. 92, 1964.

[3] S. R. Laxpati and R. Mittra, "A study of the equiangular spiral antenna," Dept. of Elec. Engrg., University of Illinois, Urbana, Tech. Rept. AFAL-TR-65-330, February 1966.

[4] S. H. Bickel, "Application of the Lebedev transform to the conical spiral antenna problem," Ph.D. dissertation, Dept. of Elec. Engrg., University of California, Berkeley, July 1964.

[5] R. Mittra and P. W. Klock, "A theoretical study of the conical spiral antenna," Dept. of Elec. Engrg., University of Illinois, Urbana, Tech. Rept. AFAL-TR-66-114, May 1966.

[6] K. K. Mei, "On the integral equations of thin wire antennas," *IEEE Trans. Antennas and Propagation*, vol. AP-13, pp. 374–378, May 1965.

[7] J. D. Dyson, "The unidirectional equiangular spiral antenna," *IRE Trans. Antennas and Propagation*, vol. AP-7, pp. 329–334, October 1959.

[8] ——, "The characteristics and design of the conical log-spiral antenna," *IEEE Trans. Antennas and Propagation*, vol. AP-13, pp. 488–499, July 1965.

[9] K. K. Mei, "Scattering of radio waves by rectangular cylinders," Ph.D. dissertation, Dept. of Elec. Engrg., University of Wisconsin, Madison, July 1962.

[10] M. Abramovitz and I. A. Stegun, Eds., *Handbook of Mathematical Functions*. NBS Appl. Math. Ser. 55, March 1965.

[11] R. W. P. King, *Theory of Linear Antennas*. Cambridge, Mass.: Harvard University Press, 1956, p. 16.

COMPUTER TECHNIQUES FOR ELECTROMAGNETIC SCATTERING AND RADIATION ANALYSES

Sherman Gee, Edmund K. Miller*, Andrew J. Poggio**, Edward S. Selden, and Gerald J. Burke
MBAssociates, P. O. Box 196, San Ramon, California 94583
*Lawrence Radiation Laboratory, Livermore, CA **Cornel Aeronautical Laboratory, Buffalo, NY

ABSTRACT

Computer techniques for solving electromagnetic radiation and scattering problems in the resonance region employing the thin-wire electric-field integral equation are discussed and demonstrated. The versatility and wide applicability of this approach for electromagnetic computer modeling are emphasized, and numerous sample results are presented to illustrate the method's accuracy and utility.

INTRODUCTION

Tremendous strides have been made in recent years in adapting the rapidly growing computer technology toward solving electromagnetic problems of increased scope and complexity. Computer-aided solutions are now possible for many problems of practical interest which have resisted reduction to classically sought closed-form solution.

This paper is devoted to providing some insight into the power and value of computer techniques, and the diverse nature and scope of the electromagnetic problems toward which these methods are applicable. Since the computer revolution embraces nearly the entire electromagnetics spectrum, a subject much too broad to be examined in one paper, we shall restrict our attention to scattering and radiation analysis in the resonance region where analytical solutions are generally unavailable. Out attention will further be focussed on an area which has profoundly benefited from the digital computer, the integral-equation formulation. Finally, our discussion will be devoted to one method or approach -- the thin-wire version of the electric-field integral equation.

This rather parochial viewpoint is taken so that the limited detail associated with broad coverage may be replaced by an indepth presentation which will convey the power and flexibility of such computer-oriented methods. At the same time, the limitations of this kind of analysis can be more fully stated so as not to mislead the reader as to its capabilities. This presentation then is not so much a review paper as it is a concise picture of the important role of computer modeling in electromagnetics.

THEORETICAL FOUNDATION

Since our primary concern here is the computer implementation of a numerical procedure and its typical applications rather than the accompanying theoretical preliminaries, we shall just outline the derivation of the integral equation used in the numerical modeling procedure. The formulation requires an integral relationship between the to-be-determined induced sources associated with an object and their resulting (secondary) fields. By evaluating the secondary field over a region where the total field behavior is known via boundary relations or continuity conditions, we are able to relate the unknown sources to the driving (primary) field which causes them. This leads to an integral equation for the unknown source distribution in terms of the specified primary field.

The starting point of our analysis in the thin-wire version of the electric-field integral equation given by

$$\hat{s}(\underline{r})\cdot\underline{E}^{p}(\underline{r}) = \frac{i\omega\mu}{4\pi}\int_{C(\underline{r})} I(s')\left[\hat{s}(\underline{r})\cdot\hat{s}(\underline{r}') - \frac{1}{k^2}\frac{\partial^2}{\partial s\partial s'}\right]g(\underline{r},\underline{r}')ds' \tag{1}$$

where $\hat{s}(\underline{r})$ is a unit tangent vector at $\underline{r}$, $\underline{E}^p$ is the primary field, I is the current, $C(\underline{r})$ is the structure geometry and is here used to imply an integration over all structures wires, $k = \omega\sqrt{\mu_0\varepsilon_0}$ is the wave number with an assumed $\exp(i\omega t)$ time variation, and s' is a length variable along $C(\underline{r})$. The function $g(\underline{r},\underline{r}')$ is the free-space Green's function $[\exp(-ik|\underline{r} - \underline{r}'|)/|\underline{r} - \underline{r}'|]$ specialized to the thin-wire geometry; i.e., the observation point $\underline{r}$ is assumed to be on the wire surface and the source point $\underline{r}'$ on the wire axis (see Figure 1). In this way, the singularity in $g(\underline{r},\underline{r}')$ is never encountered.

This particular thin-wire integral equation has been chosen for the analysis of wire structures because: 1) it is easily applicable to general geometries [or $C(\underline{r})$], 2) it maintains accuracy for small wire

Reprinted from *IEEE Internat. Electromagn. Compat. Symp. Rec.*, pp. 122–131, 1971.

radius/wavelength values, and 3) the required current integration is analytically possible for certain types of current variation. Note the $\underline{E}^p$ is arbitrary and may be due to an incident plane wave (in which case the structure's scattering characteristics are obtained) or due to localized excitation (from which the radiation properties of the structure as an antenna are derived).

An approximate solution for I(s') in (1) can be obtained by reducing the integral equation to an Nth order system of linear equations in which the N unknowns are sampled values of the structure currents. The N equations are generated by enforcing the integral equation at N points (wire segments) on the structure (N values of $\underline{r}$). The coefficients in these equations are interpretable as mutual impedances and are dependent on the structure geometry (see Appendix A). While there are many methods for accomplishing this reduction of the integral equation to a linear system (representable in matrix form), they differ only in detail and the computational effort required to obtain the matrix elements. Common to all such methods are the representation of the current in terms of its sampled values; a matching of the integral equation over the structure in some prescribed fashion; the numerical calculation of the N^2 "mutual impedance" coefficients; and subsequent solution of the linear system via inversion, factorization or iteration. A method for the expansion of the current in terms of sampled values is presented in Appendix B.

It may be appreciated that the process of generating a linear system to replace the integral equation is essentially one of evaluating the tangential electric field at point $\underline{r}$ on the structure due to the current on a segment around $\underline{r}'$, with $\underline{r}$ and $\underline{r}'$ scanned over the structure geometry C($\underline{r}$). The successful numerical modeling of a structure requires its representation by a collection of straight wire segments which is electromagnetically indistinguishable, within some specified accuracy criteria, from the original structure. Thus not only structures consisting entirely of straight wires, but curved wire structures, and solid surface objects as well, or combinations of these, are amenable to modeling in this fashion.

If the linear system representing the integral equation is written in matrix form as [Z]I] = E] and the admittance matrix is found as $[Y] = [Z]^{-1}$, it is a simple matter to evaluate the current distribution on the structure for any primary field through the matrix operation I] = [Y]E]. All other electromagnetic field quantities such as near and far fields, polarization, impedance, etc. can then be found through other far-less demanding operations. These processes are entirely consistent with Maxwell's equations and except for the integral-equation solution method do not involve any a priori, and possibly, restrictive assumptions.

The procedure outlined above is of course not unique; every scattering and antenna problem involves a similar series of operations for its solution. Nor are the special features of the wire modeling approach new; the topic has been rather extensively documented (Richmond, 1966; Tanner and Andreasen, 1967). The main contribution of this presentation is to outline the method's computer implementation and to provide a convincing demonstration of its scope of applicability as evidenced by a variety of practical problems for which the technique has been profitably employed. The development of a dependable computer method is a process which requires an awareness of potential numerical limitations and continued willingness to subject the computed results to experimental comparison as the scope of application is expanded. The numerical results included here are a representative cross-section of test cases of practical interest. The extensive validation to which the numerical procedure has been subjected allows its confident use as a computer-experimental tool.

COMPUTER IMPLEMENTATION

Development of a computer program based on the integral-equation method is relatively straightforward. The difficulties in efficiently using the approach come from the many options open to the user pertaining to the desired output data (near or far fields, polarizations; bistatic or monostatic patterns; etc.) and the preparation of input data for the specific geometry of interest.

Prior to the generation of input data a suitable wire model to represent the structure must be developed. Modeling guidelines have evolved in the course of using the technique; a discussion of the numerical convergence with decreasing segment length for wire scatterers is presented elsewhere (Miller, et al., 1971). The wire structure geometry is then defined for the computer model in terms of the segment endpoints in cartesian coordinates, wire radius, and the load impedance if any, for each of the N structure segments, which are numbered 1, . . . , N for identification. Segment electrical-connection information, required for the current-interpolation procedure (see Appendix B), is supplied by two N-integer connection arrays, one of which pertains to the positive end of each of the segments relative to its current reference direction, and the other to the negative end. Positive integers, for example, are used for two-segment junctions and denote the actual segment number to which the given segment is connected. Negative integers are used to denote connection of the segment to two or more other segments, identifiable by the same negative integer.

For structures such as the wire-grid helicopter (se Figure 9), the data is generally derived segment-by-segment from scale drawings of the structure. This is an obvious area where future improvements could increase the effectiveness of computer modeling. An interactive graphics approach whereby the structure geometry data is automatically developed by the computer is possibly the most likely method of overcoming this problem. Many structures however, consist largely of straight and/or curved wires or other systematically varying shapes which can be described by parametric equations so that subroutines can be used to generate these portions of the structure geometry. As a check on the geometry data for a new structure, a computer-drawn plot of the data can be generated and/or the data can be checked by a subroutine which verifies the compatibility of the connection and coordinate data with the program conventions (e.g. to verify that the electrical connections are physically correct). It is thus possible to eliminate virtually all errors in the geometry generator before proceeding with the actual calculation.

Having thus prescribed the structure being modeled, the subsequent steps in its numerical solution are the calculation of the impedance matrix, the matrix solution, the computation of the current, and finally the desired-field evaluation. All of these operations (besides the initial integrations required for the impedance-matrix evaluation) involve matrix manipulations. The storage and computer time required for them can be substantially reduced by structure symmetries which lead to a reduction in: 1) impedance-matrix calculation time ~ N^2/s; 2) linear-system solution time ~ N^3/s^2; and 3) current and field evaluation ~ N/s, where s is the number of symmetric sections which comprise the structure.

It is possible to effect additional solution efficiencies by specializing a computer program to a certain type or class of structures or by judiciously employing physically acceptable approximations in the solution procedure. A choice is frequently required between reducing solution cost for specific problems and developing a program of broad applicability. Generally speaking, it is desirable to have available a general purpose program which is adaptable with a minimum of change to particular problem applications.

The modeling capability of the integral equation (1) can be extended to problems other than conducting structures in free space by suitably modifying the Green's function for the problem of interest. The simplest extension is to a lossy medium, accomplished by using a complex wavenumber in the integral equation. A more complicated problem is that of an antenna near an interface of two semi-infinite media of different electrical properties, in which case a Green's function which involves the Sommerfeld integral can be used (Sommerfeld, 1964). Furthermore, the time-dependent equivalent of (1), obtainable by Fourier transformation, can be used for transient studies. Examples from all of these extensions are shown below.

In concluding this section, we wish to emphasize that numerical methods both complement and supplement experimental measurement and classical theoretical analysis. It is not true to claim that the demonstration of a computer technique's validity for a limited range of problems proves it capable of solving any problem to which the technique might be applied. There are as yet many areas of uncertainty associated with computer modeling, and many rather simple problems for which satisfactory programs have not yet been developed. Techniques like that discussed above, while even now satisfyingly versatile, require continued improvement and extension. Modeling capabilities of this kind, if judiciously employed, do provide the initial phase of what eventually will become true computer experimentation.

SCOPE OF APPLICABILITY

The numerical solution of the integral equation allows in-depth analyses of a wide variety of electromagnetic problems. Radiation and scattering problems including mutual coupling can be treated with almost equal facility. A highly complex structure such as a helicopter (c.f. next section) has been successfully modeled employing the wire-grid equivalence described earlier. The use of wire-grid modeling provides a great flexibility in the type of structures which can be analyzed since conducting surfaces and structural details such as corners and wire appendages can all be effectively represented using wire segments. An indication of the wide scope of applicability of the technique is provided in Figure 2, where representative structures are listed with the required computation times and costs (on a CDC-6600 computer and a usage charge of $1000/hour).

Both antenna radiation and scattering calculations are shown in the figure. The first structure shown is a zigzag dipole with a total wire length of 0.7λ. The input impedance at six different frequencies was determined for the computation time and cost indicated. The second structure is a fore-shortened log-periodic antenna where the four longest dipole elements are inductively loaded, demonstrating the modeling of impedance loading. The third configuration represents an unbalanced conical spiral antenna with wide tape arms, one of which is terminated by a reactive load. The wide tape arms are modeled here using two thin wires which conform to the edges of the tape. Additional cross wires can be placed on each arm to account for the transverse currents if they prove significant. The fourth structure is the wire-grid equivalent of a parabolic reflector; the indicated computation time was required to calculate a single antenna pattern using a dipole feed. Distortions of localized regions of the reflecting surface and its effect on the electromagnetic characteristics can be easily analyzed by appropriately modifying the geometric data. The fifth structure is a wire-grid representation of a slotted conducting plate for which the backscatter RCS at two frequencies, polarizations, and aspect angles were calculated. This suggest that the numerical approach is applicable to problems concerning electromagnetic coupling through surface apertures since the surface currents and aperture field distributions can be made available. The final structure is a "squirrel cage" which is a broadband scatterer for penetration-aid applications. The speed and accuracy with which the numerical computations are performed allow large-scale parametric studies to be implemented which are highly suited to the iteration approach required in decoy design where the RCS of the scatterer is tailored to best match some desired (vehicle) RCS.

It is hoped that these examples provide an indication of the versatility of the numerical approach as applied to structures of various shapes and sizes. The next section presents additional examples in greater detail so that one can obtain an appreciation of the different types of solutions which are realizable.

TYPICAL RESULTS

The usefulness of the integral-equation technique considered above is best illustrated with sample results. Several types of applications are discussed in the following paragraphs.

The determination of possible corona discharge in the vicinity of an antenna requires a knowledge of the antenna near fields. Figure 3 is a plot of the normalized radial electric field of a loop antenna in the plane of the loop at an angle of 90 degrees from the source. Comparison with the published results of Fante, et al. (1969) is shown. The near fields for a slightly more complicated configuration is shown in Figure 4. Two 9.3 inch diameter loops are each loaded with 16 capacitive elements uniformly distributed along the conductor such that the loop resonates at 170 MHz. The upper loop is excited, and the lower loop, which is separated by 11 inches, acts as a parasitic element. The near fields (1 watt input power) are computed at a radial distance (from the z axis) equal to one inch plus the loop outer diameter. The fields are evaluated as a function of azimuthal angle φ from the z axis in the plane of the top loop (0), in the mid-plane (Δ), and in the plane of the bottom loop (x). The fields are respectively normalized to the three values (volts/meter) at the top of the figure. With this type of information one can predict where breakdown is most likely to occur and what, if any, techniques can be used to reduce this possibility.

Antenna input impedance can be calculated quire readily once the current on the antenna has been determined. For instance, Figure 5 shows the effect of the transmission line on the input admittance as a function of antenna electrical length and a comparison with results obtained using an ideal voltage source on the antenna. A complicated structure such as Loran-C antenna presents a more challenging problem. Computations have been performed to evaluate the input impedance of that antenna excited near the base of the 625-foot mast. The 24 radial arms of catenary shape are joined at the top of the antenna. Figure 6 shows a comparison between the computed impedance and experimental data (supplied by the U. S. Coast Guard). Data such as this is useful in the design stage, and allows the performance of numerical "experiments" to determine the characteristics of various antenna configurations.

The calculation of radiation and scattering patterns is straightforward once the admittance matrix, Y, for a given structure, has been obtained. Upon determining the current induced by the primary source (via multiplication of Y times the primary-field vector), the secondary fields are obtained as an integration (analytic for the far field) of the structure currents. A bistatic pattern requires only one current evaluation, whereas the monostatic variation necessitates finding the current for each viewing angle desired. Figure 7 shows the backscatter RCS of a straight wire with bow-tie termination. Only data for $0 \leq \Theta \leq 90$ degrees is shown because of the structure symmetry. The predictions are very accurate even the region of the deep null.

Figure 8 illustrates the backscatter RCS as a function of aspect angle for the wire teepee shown, where a is the wire radius. The numerical data are plotted and compared with experimental data obtained by

Micronetics, Inc. Two sets of experimental data are shown in Figure 8, representing clockwise and counterclockwise rotation of the angle Θ.

Airborne platforms are generally complicated structures and accurate modeling, particularly in the resonance region, is required to accurately predict their electromagnetic characteristics. Wire-grid models sufficiently represent solid surfaces when the open regions are small in terms of wavelengths (Richmond, 1966). Figure 9 shows numerical results obtained from a wire-grid model of an OH-6A helicopter compared with experimental data obtained by Collins Radio.

The interaction of an antenna and its environment can be quite pronounced when the environment is in the form of a ground plane. Although it is quite simple to predict antenna characteristics in the presence of a perfectly conducting ground plane, this simplicity is not realized when ground losses must be taken into account. The classical technique for handling the lossy-ground problem is contained in the Sommerfeld integral (Sommerfeld, 1964) which become part of the integral-equation kernel. However, for antennas with any degree of complexity, the integrals can not be analytically evaluated and numerical integration is difficult and inefficient. An alternate, approximate technique, based on the Fresnel reflection coefficients, has been developed and has been found to yield reliable results (Burke, et al., 1970). Figure 10 compares results using both methods as applied to a two-element parasitic array of vertical antennas. The outstanding feature of the results presented is that the approximate technique reduced computation time by about two orders of magnitude compared to the time required employing the Sommerfeld formulation, with no significant loss of accuracy.

The transient (time domain) analysis of antennas and scatterers is another area of interest where the digital computer provides a unique capability. Using a Gaussian pulse for the exciting field, the time-dependent currents induced on a structure can be evaluated by solving an integral equation which is the Fourier transform of Equation (1), or which is developed directly from the time-dependent form of Maxwell's equations. Techniques for solving this type of equation are well known (Bennett and Weeks, 1968; Sayre, 1969; Poggio and Miller, 1970). It is a simple matter to evaluate the impulse response in the frequency domain from this information. In effect, the transfer function for the electromagnetic device is determined which can then be used to evaluate characteristics for other exciting waveforms. The advantage of the time-domain analysis is that the frequency-domain characteristics over a broad bandwidth can be determined with one relatively simple calculation.

Figure 11a shows the feedpoint current variation as a function of time for a dipole antenna excited with a Gaussian pulse at the center of the antenna. When the Fourier transform of this current is divided by the Fourier transform of the Gaussian-pulse input voltage, the frequency-domain input admittance is obtained as shown in Figure 11b. The results are compared with independent results as provided by King and Middleton (King, 1956). The bandwidth limitation and accuracy of the results are affected by time-sample size, space-sample size, the length of time over which the current is evaluated, and the subsequent accuracy with which this sampled data is transformed.

An example of a scattering application is shown in Figure 12. The problem of determining the time-domain back-scattered fields (or equivalently, the frequency-domain RCS) from a crown band for axial incidence is presented in this figure. The time-dependent response to a Gaussian pulse is presented in a) while the Fourier transform is used to determine the frequency-domain response in b) over a finite bandwidth. Again, a comparison with independently computed results shows good agreement.

CONCLUSIONS

The results shown above serve to illustrate the potential usefulness of the thin-wire, electric-field, integral equation for the numerical analysis of a variety of scattering and antenna problems. This approach, when combined with an efficient numerical method for its solution together with the speed and size of current digital computers, provides an economical and reliable alternative to, and complement for, the experimental study of a wide range of practical problems. These problems can include both thin-wire structures and wire-grid models of solid-surface objects. Particular advantages are offered by the numerical method to determine such properties as average RCS, current distributions, input impedance, near-field behavior, etc., which may be difficult and expensive to measure experimentally.

APPENDIX A: NUMERICAL PROCEDURE

An outline of the numerical solution procedure for the induced current via reduction of the integral equation to a linear system is presented here.

The current integration in Equation (1) extends over the entire structure and produces the tangential electric field at any point on the surface. In mathematical terms, the current or source points lie in the domain, and the field or observation points in the range, of the integral operator. An intuitive approach to solving the current is provided by: (1) approximating the current in terms of unknown sampled values and specified functional variation on the structure; and (2) enforcing the resulting integrals to match the integral equation in a pointwise sense (assuming a collocation solution) over the range of the integral operator. This procedure generates a set of linear equations for the unknown sampled current values and demonstrates at its simplest, the essence of the method of moments. Our discussion will be limited here to sub-sectional collocation, a version of the moment method, which is discussed in more general terms by Harrington (1969) and Kantorovich and Krylov (1964).

Let the actual structure current by approximated by

$$I(s') \approx \sum_{n=1}^{N} a_n f_n(s') \qquad \text{(A1)}$$

where the a_n are constants to be determined and the f_n are the basis or trial functions. The f_n may be defined over the entire domain of the integral, or over a sequence of sub-domains. It is the latter approach which is used in the sinusoidal current expansion employed for the foregoing calculations. Upon substitution of (A1) into the integral equation (1), and requiring exact equality between the right- and left-hand sides of the equation at N points, p_m, m = 1, . . . , N over the wire structure, we obtain

$$\frac{i\omega\mu_o}{4\pi}\sum_{n=1}^{N} a_n \int_C f_n(s')\left[\hat{s}(p_m)\cdot\hat{s}' + \frac{1}{k^2}\frac{\partial^2}{\partial s^2}\right] g(p_m,s')ds' = \hat{s}(p_m)\cdot\underline{E}^p(p_m);\ m = 1,\ldots,N \qquad \text{(A2)}$$

which gives N equations for the N quantities a_n. This process of point-fitting the integral equation involves the use of delta functions for the weighting or testing functions employed in the approximation of the rigorous integral equation.

The linear system (A2) can be rewritten in matrix form as

$$\sum_{n=1}^{N} Z_{mn} a_n = E_m;\ m = 1,\ldots,N$$

where Z and E are given by

$$Z_{mn} = \frac{i\omega\mu_o}{4\pi}\int_C f_n(s'))\left[\hat{s}(p_m)\cdot\hat{s}' + \frac{1}{k^2}\frac{\partial^2}{\partial s^2}\right] g(p_m,s')ds',$$

and

$$E_m = \hat{s}(p_m)\cdot\underline{E}^p(p_m).$$

Note that Z has the dimensions of impedance; hence its characterization as an impedance matrix. It represents the tangential electric field at point p_m on the structure due to the current term a_n. A solution for the a_n simply follows as

$$a_n = \sum_{m=1}^{N} Y_{nm} E_m;\ n = 1,\ldots,N$$

where $Y = Z^{-1}$ is known as the admittance matrix. Since Y is derivable without dependence upon E, it truly approximates the electromagnetic response of the structure for which it has been obtained for any exciting primary field $\underline{E}^p$.

APPENDIX B: CURRENT INTERPOLATION

Let the current on segment j be expressed as

$$I_j(s') = A_j + B_j \sin k(s'-s_j) + C_j \cos k(s'-s_j) \qquad \text{(B1)}$$

with s_j the midpoint coordinate (Yeh and Mei, 1967). Also, let segment j be connected to segments j-1 and j+1 at its minus and plus reference ends respectively with the reference directions on all three segments the same. Evaluation of I_j at s_j, s_{j-1}, and s_{j+1} results in

$$A_j + C_j = I_j,$$

$$A_j - B_j s_{j-1} + C_j c_{j-1} = I_{j-1}, \qquad \text{(B2)}$$

$$A_j + B_j s_{j+1} + C_j c_{j+1} = I_{j+1}$$

where d_j is the length of the jth segment and

$$s_{j\pm1,j} = \sin[k(d_{j\pm1} + d_j)/2],\ c_{j\pm1,j} = \cos[k(d_{j\pm1} + d_j)/2].$$

Solution for A_j, B_j, and C_j in terms of I_{j-1}, I_j, and I_{j+1} provides an equation of the form

$$I_j(s') = X_j(s')I_{j-1} + Y_j(s')I_j + Z_j(s')I_{j+1} \qquad \text{(B3)}$$

where X_j, Y_j, Z_j contain the coefficients A_j, B_j, C_j. The system of equations which results from the collocation solution to the integral equation is thus seen to involve as unknowns the N current samples at the center of the N segments into which the structure is divided.

The extension of the interpolation procedure to multiple junctions is straightforward. Consider the case where segment j is connected to m segments numbered j+1, . . . , j+m at its plus end and the single segment j-1 at its minus end. Then only the equation representing (B1) evaluated at s_{j+1} is modified and becomes

$$A_j + \frac{1}{m}\sum_{i=j+1}^{j+m}[B_j s_{i,j} + C_j c_{i,j}] = \frac{1}{m}\sum_{i=j+1}^{j+m} I_i \qquad \text{(B4)}$$

which comes from interpolating I_j to the midpoints of the m connected wires where it is equated to the average midpoint value. A solution for the A_j, B_j, and C_j in terms of the midpoint currents I_{j-1}, I_{j+1}, . . . , I_{j+m} follows as before. A multiple junction at the minus end of the segment is similarly treated.

Equation (B3) is substituted into the integral equation and, after summing over the N segments and collecting terms in I_j, gives

$$\frac{i\omega\mu_o}{4\pi}\sum_{j=1}^{N} I_j\left[\int_{d_{j+1}} X_{j+1}K_{j+1}ds' + \int_{d_j} Y_j K_j ds' + \int_{d_{j-1}} Z_{j-1}K_{j-1}ds'\right] = \hat{s}(p_m)\cdot\underline{E}^p(p_m) \qquad \text{(B5)}$$

with

$$K_j = \left[\hat{s}(p_m)\cdot\hat{s}_j' + \frac{1}{k^2}\frac{\partial^2}{\partial s^2}\right] g(p_m,s').$$

The bracketed term in (B5) is interpretable as an impedance. It relates the current I_j, interpolated into neighboring segments, to the field at observation point p_m, m = 1, . . . , N.

ACKNOWLEDGMENT

The authors gratefully acknowledge the contributions of Drs. G. M. Pjerrou, A. R. Neureuther and Mr. H. L. Bosserman who were instrumental in the original development of the thin-wire computer program.

REFERENCES

Bennett, C. L. Jr., and W. L. Weeks, Technical Report TR-EE68-11, Purdue University, Lafayette, Indiana (1968).

Burke, G. J., S. Gee, E. K. Miller, A. J. Poggio, and E. S. Selden, Proceedings 20th Annual Symposium on USAF Antenna R&D, Allerton Park, Monticello, Illinois (1970).

Fante, R. L., J. J. Otazo, and J. T. Mayhan, Radio Science, 4, 697 (1969).

Harrington, R. F., Proc. IEEE, 55, 136 (1967).

Kantorovich, L. V. and V. F. Krylov, "Approximate Methods of Higher Analysis," Interscience, New York (1964).

King, R. W. P., "The Theory of Linear Antennas," 149, Harvard University Press, Cambridge, Massachusetts (1956).

Miller, E. K. and B. J. Maxum, Final Report No. ECOM-0456-1, USAECOM, Ft. Monmouth, N.J. (1970).

E. K. Miller, G. J. Burke, and E. S. Seldon, IEEE Trans. on Ant. Prop., AP-19, 534 (1971).

Poggio, A. J. and E. K. Miller, Tech Memo MB-TM-70/20, MBAssociates, San Ramon, California (1970).

Richmond, J. H., IEEE Trans. Ant. Prop., AP-14, 782 (1966).

Sayre, E. P., Technical Report TR-69-4, Syracuse University (1969).

Sommerfeld, A., "Partial Differential Equations in Physics", Academic Press, New York (1964).

Tanner, R. L. and M. G. Andreason, IEEE Spectrum, 4, 53 (1967).

Yeh, Y. S. and K. K. Mei, IEEE Trans. Ant. Prop., AP-15, 634 (1967).

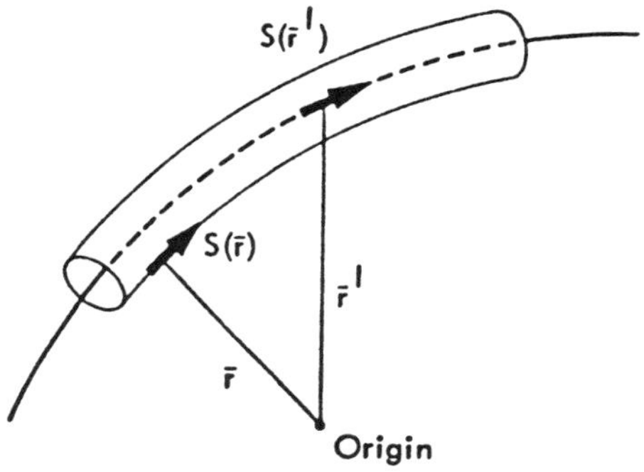

FIGURE 1.
GEOMETRY FOR THIN WIRE ELECTRIC
FIELD INTEGRAL EQUATION

	No. Freqs (F) Impedance Z	No. Patterns [P]	No. Angles (k)		Backscatter RCS $\sigma_n(\pi)$ n Polarization	
Structure		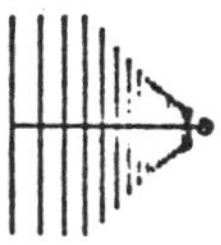		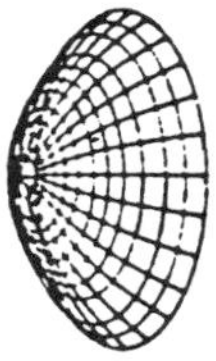	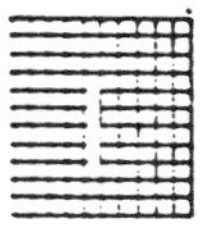	
Output Data	Z 6 (F) 0.7 λ	2[P(θ, Φ)] 19.4 λ	4[P(θΦ)] 3.8 λ	1[P(θ, Φ)] 113.6 λ	$\sigma_2(\pi)$ 2(F,k) 12 λ	$\sigma_2(\pi)$ 23(k) 4.1 λ
CDC 6600 Time	39 Sec.	70 Sec.	68 Sec.	147 Sec.	77 Sec.	62 Sec.
Computation Cost	$9	$17	$16	$35	$18	$15

FIGURE 2. DIVERSITY OF STRUCTURES AND COST COMPARISONS

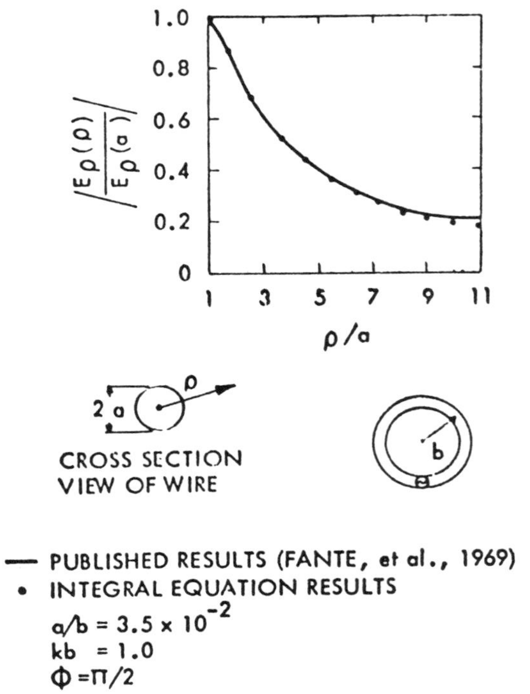

FIGURE 3.
LOOP ANTENNA NEAR FIELDS

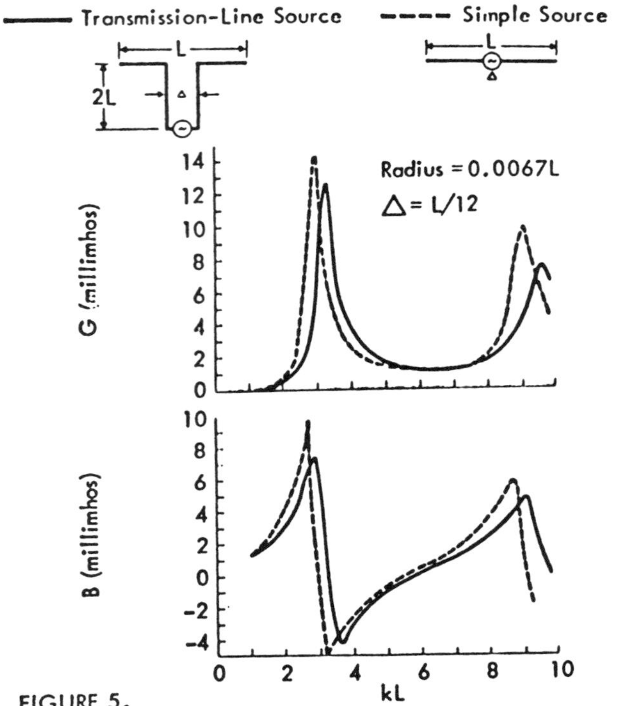

FIGURE 5.
ADMITTANCE VERSUS kL INCLUDING TRANSMISSION LINE EFFECT (G IS THE CONDUCTANCE AND B THE SUSCEPTANCE)

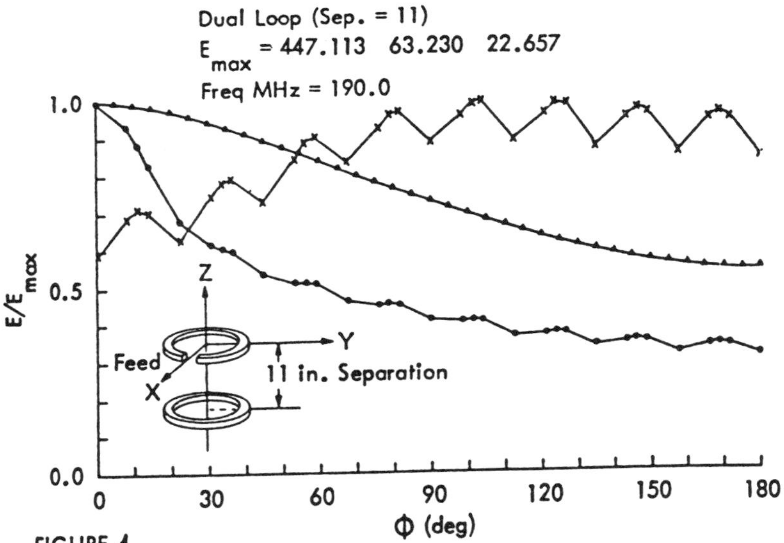

FIGURE 4.
NEAR FIELD VARIATION IN THE VICINITY OF A CAPACITIVELY-LOADED DUAL-LOOP ANTENNA

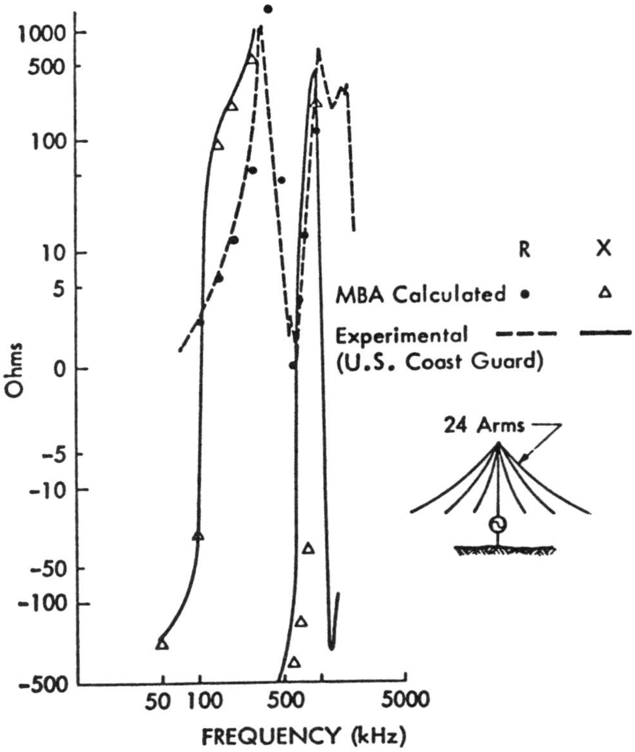

FIGURE 6.
INPUT IMPEDANCE OF LORAN C ANTENNA OVER PERFECTLY CONDUCTING GROUND

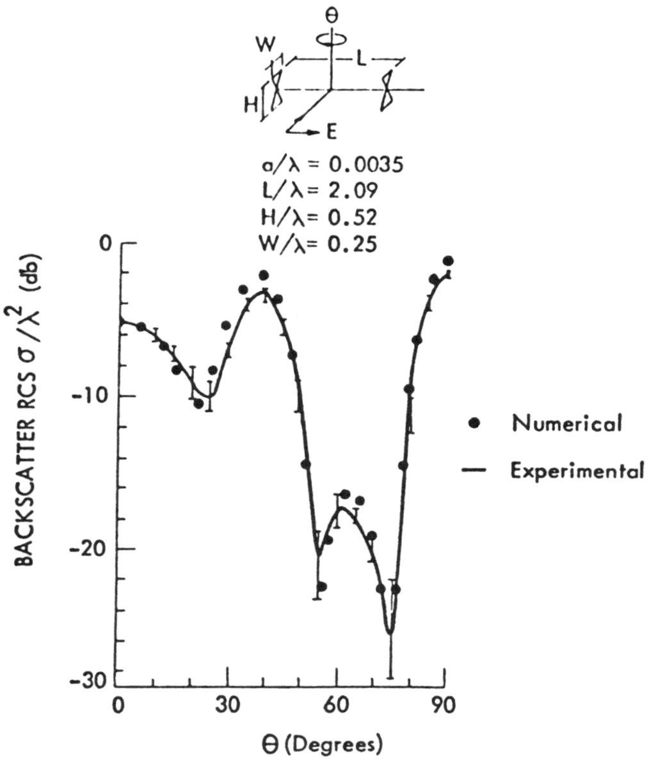

FIGURE 7.
RCS FOR STRAIGHT WIRE WITH BOW-TIE TERMINATIONS

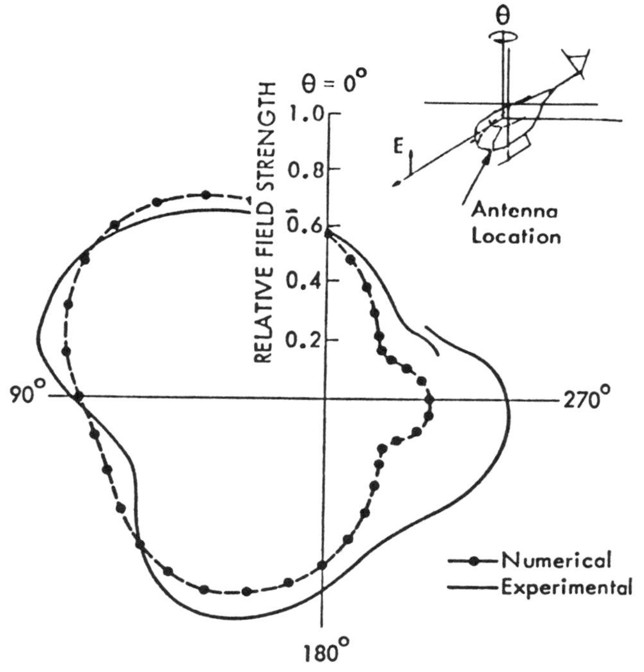

FIGURE 9.
RADIATION PATTERN FOR TOWEL BAR HOMING ANTENNA ON OH-6A HELICOPTER

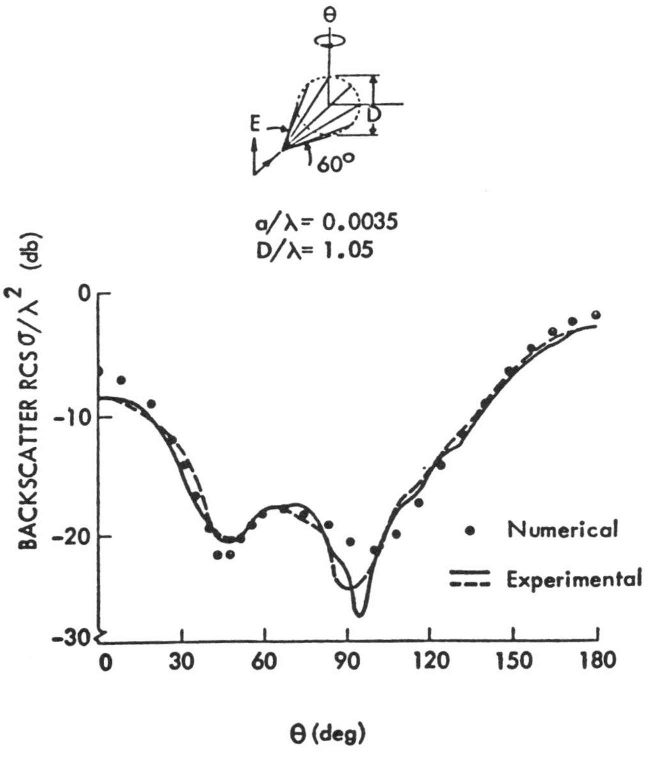

FIGURE 8.
RCS FOR WIRE TEE-PEE

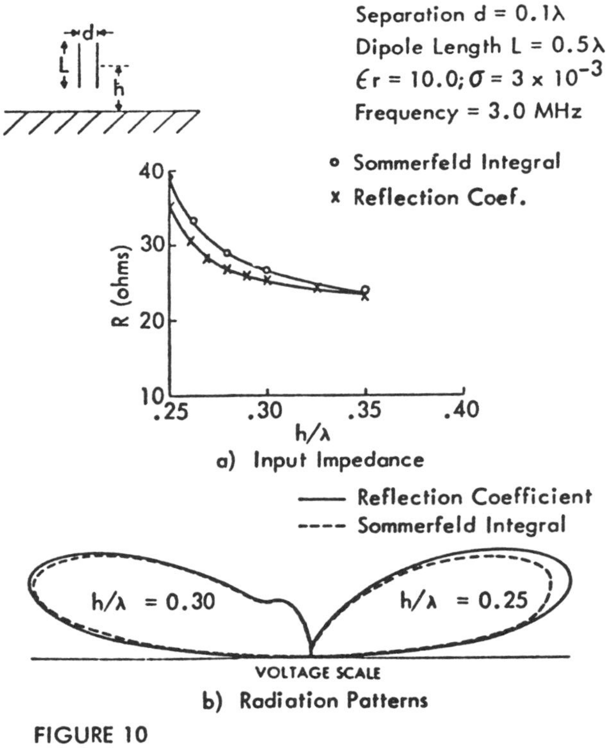

FIGURE 10
INPUT IMPEDANCE AND RADIATION PATTERNS FOR AN ARRAY OF TWO HALF-WAVE DIPOLES OVER LOSSY GROUND

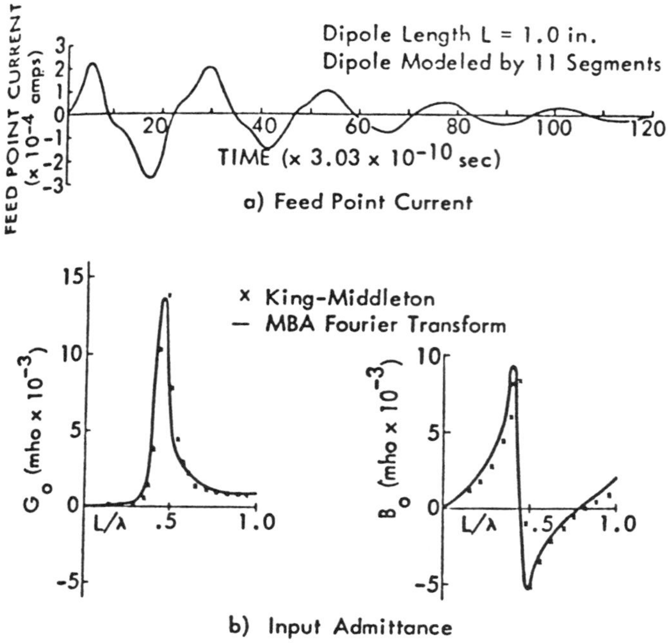

FIGURE 11.
DIPOLE ANTENNA EXCITED BY A GAUSSIAN PULSE

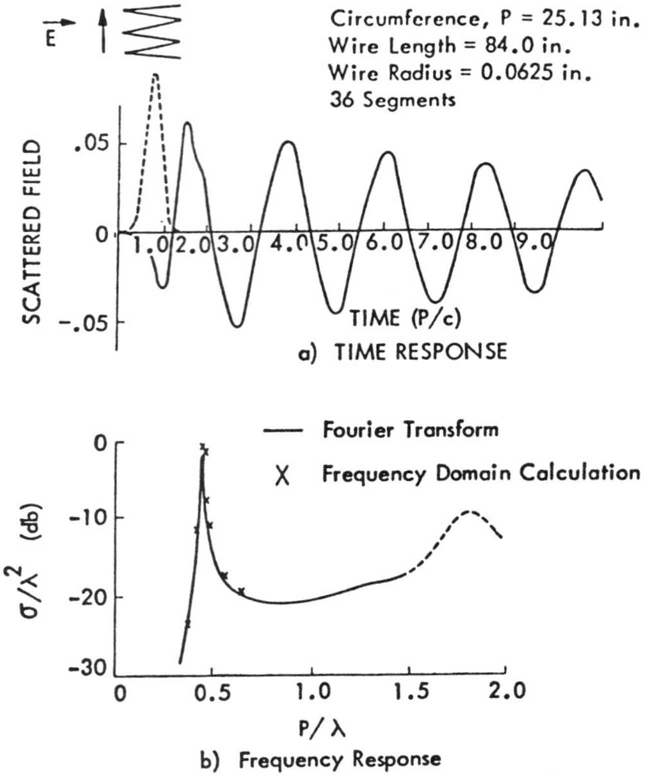

FIGURE 12.
SCATTERING OF A GAUSSIAN PULSE BY A CROWN BAND

Accuracy–Modeling Guidelines for Integral-Equation Evaluation of Thin-Wire Scattering Structures

Abstract—**A numerical study to determine accuracy and modeling criteria for the integral-equation analysis of thin-wire scatterers is described. Results obtained using a collocation solution method with sinusoidal current interpolation show that 6–18 current samples per wavelength are sufficient to produce radar cross section results with absolute numerical convergence accuracies on the order of 10 percent or less (~0.4 dB) depending upon structure complexity.**

I. Introduction

With the widespread availability of the digital computer, the application of numerical solution techniques to the evaluation of scattering and radiation problems has now become routine. However, disregarding the limitations inherent in the numerical modeling of physical problems may unknowingly result in the generation of invalid data. Therefore, in order to avoid such pitfalls, it is prudent to develop guidelines for the numerical approach which can provide some assurance that the calculations obtained meet certain minimum accuracy requirements. It is the purpose of this communication to present such guidelines for the analysis of thin-wire structures via the thin-wire approximation to the electric field integral equation (EFIE).

II. Solution Technique

The formulation and numerical solution of the thin-wire EFIE and its application to both thin-wire structures and wire-grid models of solid-surface structures has been outlined by Richmond [4]. A presentation of typical results obtained from this approach is given by Miller *et al.* [2], where the validity of the technique is demonstrated for a wide variety of structures, both for radiation and scattering problems, by comparison with experimental measurement. Therefore, we restrict our discussion here to only those aspects of the numerical method necessary to proper interpretation of the significance of the results which are given.

The thin-wire EFIE relates the current induced on a thin-wire structure to a specified (source) tangential electric field distribution over the structure. On applying a version of the method of moments [1], collocation for example, a linear system of equations is generated in which the N unknowns are a sampled sequence of current values on the structure, and the forcing function is similarly a sampled sequence of the source field. The $N \times N$ coefficient matrix relating the currents to the fields may be viewed as an impedance matrix for the structure, which is thus characterized in the manner of an N-port network. Inversion of the impedance matrix yields the corresponding admittance matrix which allows computation of the structure current distribution for any source field.

The solution accuracy of the current distribution, scattered fields, driving point impedance, etc., is a function, for a given solution method, of the segmentation scheme or sampling sequence used to derive the impedance matrix, and of the total number N of samples used. Since the calculation time required to obtain the impedance matrix varies as N^2 for a general structure, and as N^3 for the subsequent computation of the admittance matrix, it is clear that N should be minimized consistent with the desired solution accuracy. It is not possible to establish *a priori* what value of N will yield acceptably accurate results for a given structure. Therefore, a numerical experiment which determines accuracy dependence upon N for various classes of structures may be useful. Such a study was reported by Neureuther *et al.* [3] for linear antennas and scatterers; a report on results for a variety of structure geometries is presented here.

Manuscript received August 10, 1970; revised November 27, 1970.

III. Numerical Results

The particular solution technique employed for the EFIE in this study is that of collocation, using sinusoidal interpolation for the current (see Yeh and Mei [5]) and delta-function weights. Since the accuracy of the overall numerical procedure used has already been established for structures of the kind considered here (see Miller *et al.* [2]), we will restrict the data presented to numerical results only.

Accuracy data was obtained for the 10 structures illustrated in Fig. 1. In order to reduce the sensitivity of the data to a particular incident-wave configuration, an average backscatter radar cross section was computed as a function of aspect angle and polarization. The number of segments used to model a given structure was systematically varied, where for all structures except the spiral, the segments were nearly all (within 5 percent) of equal length. Because the segmentation scheme used for the spiral is based on equal subtended angles for each segment to obtain a polygonal model, the segment lengths varied by a maximum factor of 6.5. Note that since the curved geometries (circular ring, helix, spiral, and squirrel-cage) are modeled with straight wire segments, the model changes with segmentation, i.e., a circle can be modeled as a square, pentagon, hexagon, etc. In these cases the total wire perimeter length P was held constant as the segmentation was varied.

The average cross section σ_{ref} obtained using the maximum number of segments N_{ref} was taken as the reference value. The relative error Δ is then defined by

$$\Delta \equiv \frac{|\sigma_N - \sigma_{ref}|}{\sigma_{ref}}$$

where σ_N is the average cross section obtained using N segments to model the structure. For the data to be presented here, $N_{ref} > 1.25\, N$ so that the values obtained for Δ for the largest values of N shown should still be meaningful.

For the convenience of those who wish to use these results as a check against their own programs, the cross section values for the specified values of angle of incidence and polarization are included in Table I. These data pertain to the case where $N = N_{ref}$.

A plot of Δ as a function of the number of segments used per wavelength N_λ (i.e., $N_\lambda = N/P$) for the 10 structures investigated is shown in Fig. 2. The Δ curves for the various structures are seen to exhibit similar trends with increasing N_λ, approaching straight lines of approximately equal slope. A transition region toward more steeply sloping lines for smaller values of N_λ is also apparent.

Generally speaking, the data may be separated into three relatively distinct groups or classes in terms of the relative error-segmentation dependence. The most accurate results are obtained for the simplest scatterer geometries—the straight wire, helix, and circular ring—where $\Delta \lesssim 10^{-2}$, for $N_\lambda > 25$. Intermediate accuracy is obtained for the slightly more complex geometries such as the V dipole, trifin, etc., where $\Delta \sim 10^{-1}$, for $N_\lambda > 20$. Finally, a third class of scatterers which fits neither of these is suggested by the results obtained for the squirrel cage and conical spiral, where, for $\Delta \sim 10^{-1}$, $N_\lambda \sim 15$ and 50, respectively. For both of these structures, however, the trend of Δ with increasing N_λ appears to tend toward the intermediate structure results.

The straight portion of the curves in Fig. 2 can be approximated by $\Delta \sim A \exp(-BN_\lambda)$ for large enough N_λ, where A and B are constants. For most of the structures, B is relatively independent of geometry, having a value of approximately 0.031, while A, on the other hand, is dependent upon the structure class and to some extent upon the particular geometry within that class. The constant A is largest for the conical spiral and smallest for the circular ring.

It appears that the spiral results differ most significantly from the others because the constant subtended angle segmentation used for the expanding spiral results in a monotonic increase in segment length along the spiral. Thus the accuracy is limited

Reprinted from *IEEE Trans. Antennas Propagat.*, vol. AP-19, no. 4, pp. 534–536, July 1971.

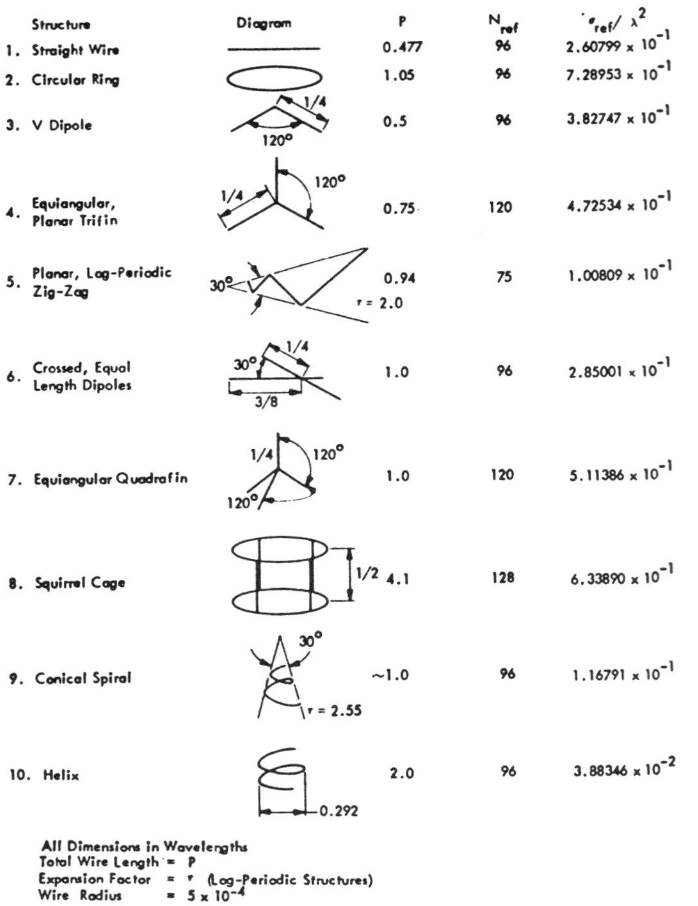

Fig. 1. Scatterer geometries.

TABLE I

Structure	Incident	E Field Alignment	σ/λ^2
Straight wire	normal to wire	wire axis	0.75865
Circular ring	ring axis	ring diagonal	1.6961
	ring diagonal	ring plane	0.24962
V dipole	bisector in dipole plane (on vertex)	dipole plane	0.53768
Trifin	one arm	structure plane	0.54058
Zig-zag	axis in zig-zag plane (on small end)	zig-zag plane	0.11666
Crossed dipoles	normal to structure plane	evenly divided dipole	0.26762
Quadrafin	one arm	one arm	0.56824
Squirrel cage	axis	plane of two opposite ribs	1.3889
Conical spiral	axis	normal to plane containing wire ends	0.04628
Helix	axis	plane containing ends	0.01733

primarily by the longest segments. If an N_λ value is used for the spiral based on the longest segment length used, rather than on the average segment length as shown on Fig. 2 (the actual N_λ value would be decreased to a new value $N_\lambda' \approx 0.6\, N_\lambda$), then the spiral curve would fall in the intermediate class results. Note that the use of equal-length segments to model the spiral may be somewhat impractical for accurate modeling of the small-radius tip while at the same time using a reasonable number of segments near

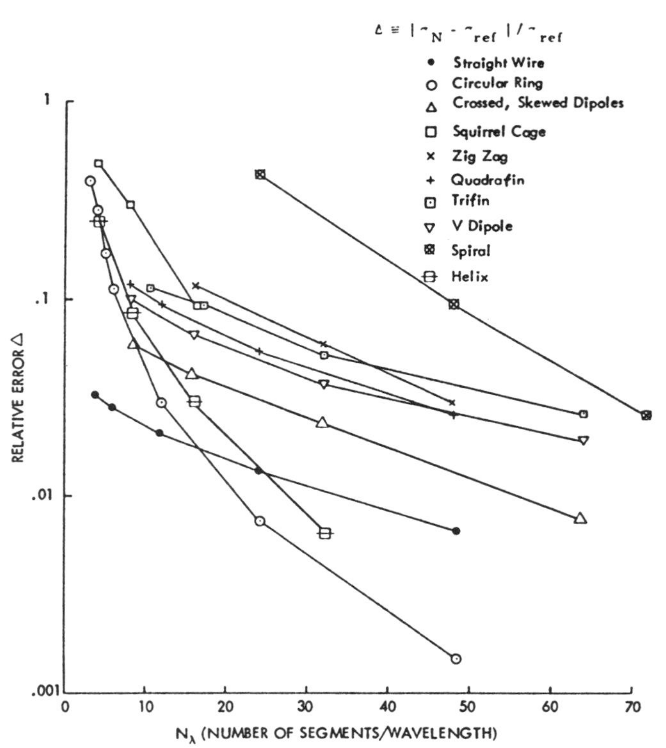

Fig. 2. Relative error as function of number of segments per wavelength.

the large end. Possibly a compromise which uses equal-angular segments to a certain point along the spiral, and then equal-length segments beyond, as suggested by Yeh and Mei [5], would be advantageous. Experience gained in the analysis of structures similar to those included here, and others as well, indicates that maximum solution efficiency and accuracy is obtained using nearly equal length segments.

A comment is appropriate concerning the implications of the preceding scattering calculation results for the analysis of the same structures excited as antennas. The numerical method, it has been noted, leads to an Nth-order admittance matrix for an N-segment structure. This matrix allows calculation of the induced structure current for any specified distribution of tangential electric field, and is thus suited to treating the structure as either an antenna or a scatterer.

It is clear then that the numerical convergence of the admittance matrix as a function of segmentation is also source independent. However, the cross section calculation effectively averages over the source and current distributions on the structure, and may thus demonstrate a convergence trend which differs from that of individual elements in the admittance matrix. As a matter of fact, the well-known logarithmic susceptance dependence upon the feed-gap size exhibited by a linear antenna may also result from using the collocation solution method and a one-segment source [i.e., susceptance α log (N_λ)]. Thus, a convergence study pertaining to antenna characteristics, input admittance, for example, may be somewhat inconclusive since the segmentation dependence is due to both the source model and numerical convergence.

IV. Conclusions

Because of the wide variation which can be encountered in practical problems, it is impossible to establish with absolute certainty the modeling or segmentation requirements to realize a specified numerical accuracy when treating an arbitrary structure. However, on the basis of studies such as the present one, it is possible to broadly identify the necessary conditions for obtaining the accuracy range which can be reasonably expected of the numerical results. Solution economy appears to be maximized, broadly speaking, by using a segmentation where practical, which results in all segments being of nearly equal lengths. The numerical accuracy which is then realizable for a typical thin-wire structure can be expected to be in the range 10^{-1}–10^{-2} using 6–~18 segments per wavelength, depending upon structure complexity.

E. K. Miller
G. J. Burke
E. S. Selden
MBAssociates
San Ramon, Calif. 94583

References

[1] R. W. Harrington, *Field Computation by Moment Methods.* New York: Macmillan, 1968.
[2] E. K. Miller, G. J. Burke, B. J. Maxum, G. M. Pjerrou, E. S. Selden, and A. R. Neureuther, "On the thin-wire electric field integral equation," Submitted for publication to *IEEE Trans. Antennas Propagat.*
[3] A. R. Neureuther, B. D. Fuller, G. D. Hakke, G. Hohmann, *et al.*, "A comparison of numerical methods for thin wire antennas," presented at the 1968 Fall URSI Meeting, Berkeley, Calif.
[4] J. H. Richmond, "A wire-grid model for scattering by conducting bodies," *IEEE Trans. Antennas Propagat.*, vol. AP-14, Nov. 1966, pp. 782–788.
[5] Y. S. Yeh and K. K. Mei, "Theory of conical equiangular spiral antennas," *IEEE Trans. Antennas Propagat.*, vol. AP-15, Sept. 1967, pp. 634–639.

Radiation and Scattering by Thin-Wire Structures in the Complex Frequency Domain *

JACK H. RICHMOND, Fellow, IEEE

October 23, 1990

Abstract

Piecewise-sinusoidal expansion functions and Galerkin's method are employed to formulate a solution for an arbitrary thin-wire configuration in a homogeneous conducting medium. The analysis is performed in the real or complex frequency domain. In antenna problems, the solution determines the current distribution, impedance, radiation efficiency, gain and far-field pattern. In scattering problems, the solution determines the absorption cross section, scattering cross section, and the polarization scattering matrix. This report presents the electromagnetic theory for thin wires.

1 Introduction

In 1932 Carter [1] used the induced emf formulation and a sinusoidal current distribution to derive expressions for the mutual impedance between half-wave dipoles. H.E. King [2] extended these results in 1957 to parallel dipoles in echelon with arbitrary wire lengths. Baker and LaGrone [3] employed numerical integration for skew dipoles.

These two-segment sinusoidal solutions have been useful for coupled dipoles with lengths up to about 0.5λ. In this range, the current distribution on each thin-wire antenna is in most cases not greatly disturbed by the other. More sophisticated techniques are required, however, for longer antennas and for more complex wire configurations. One approach is to reduce the integral equation to a system of simultaneous linear equations. The unknown constants in these equations are usually samples of the current function $I(\ell)$ or the coefficients in a modal expansion for the current distribution.

In 1967 Yeh and Mei [4] employed such techniques to analyze the conical-spiral antenna. Programs for arbitrary thin-wire configurations were developed by Tanner and Andreasen [5], Miller and Morton [6], Chao and Strait [7], and Richmond [8,9,10,11,12]. In References [4], [5] and [6] the current distribution on each wire segment has the form $I = A + B\cos k\ell + C\sin k\ell$. Reference [7] uses a piecewise-linear expansion. References [8] through [12] use a piecewise-sinusoidal expansion with $I = A\cos k\ell + B\sin k\ell$. Others who have employed the piecewise-sinusoidal expansion include Butler [13] and Imbriale and Ingerson [14].

This paper presents the electromagnetic theory for thin-wire antennas and scatterers. The ambient medium is considered to be isotropic, linear and homogeneous. The analysis is performed in the frequency domain, and the generator or incident wave may have a real or complex frequency. The solution satisfies Kirchhoff's current law on the wire structure, and has favorable properties of convergence and computational efficiency. The computer programs will be presented in a future report.

*This paper has been abstracted by E.H. Newman from the Ohio State University, ElectroScience Lab report 2902-10, prepared under Grant No. NGL 36-008-138 with the National Aeronautics and Space Administration, Langley Research Center, July 1973.

The author is with the ElectroScience Lab, The Ohio State University Department of Electrical Engineering.

With no significant loss of generality, the wire structure is considered to be a generalized polygon assembled from straight wire segments. The formulation and the program have been tested extensively in radiation and scattering problems with various dipoles, loops, arrays and wire-grid models of plates, spheres, cones, aircraft and ships. Although the air-earth or air-water interface is not considered, the theory and program are useful in many situations involving buried or submerged antennas and scatterers.

A piecewise-sinusoidal expansion is used for the current distribution. The matrix equation $ZI = V$ is generated by enforcing reaction tests with a set of sinusoidal dipoles located in the interior region of the wire. Since the test dipoles have the same current distribution as the expansion modes, this may be regarded as an application of Galerkin's method [15]. However, the physical ideas of Rumsey's reaction concept [16] were more inspirational in this development than the mathematical ideas of the moment method.

On each thin-wire structure, we define a set of terminals or current-sampling points. Terminals are defined at each corner or bending point, at each junction where several straight wires intersect, and at the wire endpoints. For accuracy, no segment should have a length much greater than $\lambda/4$. Thus, a long segment may be subdivided by defining additional sampling points.

With several terminals defined in this manner, the wire structure is a multiport system. The elements in the open-circuit impedance matrix are calculated by numerical integration when appropriate, or by closed-form expressions in terms of exponential integrals. The impedance matrix is inverted to obtain the short-circuit admittance matrix.

The sinusoidal reaction formulation was developed earlier for wire structures in free space. The generalization to wires in a conducting medium with complex frequency is based on electromagnetic similitude and analytic continuation.

The next section presents the reaction integral equation for thin wires. The remaining text defines the sinusoidal expansion and testing functions and develops the theory for wire structures with lumped loading and finite conductivity. Numerical results are displayed for the echo area, radiation efficiency and impedance of a straight wire in a conducting medium.

2 The Reaction Integral Equation

Let S denote the closed surface of the wire structure, and let V denote the interior volumetric region. In the presence of the wire, an external source $(\mathbf{J}_i, \mathbf{M}_i)$ generates the field $(\mathbf{E}, \mathbf{H})$. When radiating in the homogeneous medium (μ, ϵ) without the wire, this source generates the incident field $(\mathbf{E}_i, \mathbf{H}_i)$. The scattered field is defined as follows:

$$\mathbf{E}_s = \mathbf{E} - \mathbf{E}_i \tag{1}$$

$$\mathbf{H}_s = \mathbf{H} - \mathbf{H}_i \tag{2}$$

These fields are considered to be time-harmonic with the same frequency. The time dependence $e^{j\omega t}$ or e^{st} is suppressed.

From the surface-equivalence theorem of Schelkunof [17], the interior field will vanish without disturbing the exterior field $(\mathbf{E}, \mathbf{H})$ if we introduce the following surface-current densities

$$\mathbf{J}_s = \hat{n} \times \mathbf{H} \tag{3}$$

$$\mathbf{M}_s = \mathbf{E} \times \hat{n} \tag{4}$$

on the surface S. (The unit vector $\hat{n}$ is directed outward on S.) In this situation, we may replace the wire structure with homogeneous medium (μ, ϵ) without disturbing the field anywhere. When $\mathbf{J}_s$ and $\mathbf{M}_s$ radiate in the homogeneous medium, they generate the field $(\mathbf{E}_s, \mathbf{H}_s)$ in the exterior and $(-\mathbf{E}_i, -\mathbf{H}_i)$ in the interior region.

Now let us place a test source (or probe) in the interior region V and consider its reaction with the other sources. If the test source has electric current density $\mathbf{J}_m$ and magnetic current density $\mathbf{M}_m$,

$$\int\int (\mathbf{J}_m \cdot \mathbf{E}_s - \mathbf{M}_m \cdot \mathbf{H}_s)\, ds = -\int\int (\mathbf{J}_m \cdot \mathbf{E}_i - \mathbf{M}_m \cdot \mathbf{H}_i)\, ds. \tag{5}$$

In Equation (5) $(\mathbf{E}_s, \mathbf{H}_s)$ denotes the field generated by $(\mathbf{J}_s, \mathbf{M}_s)$, and the integrals extend over the surface of the test source. Equation (5) is one form of the reaction integral equation (RIE). If we enforce Equation (5) with a set of delta-function electric test sources, the RIE reduces to the well-known electric field integral equation (EFIE). If we enforce Equation (5) with a set of delta-function magnetic test sources, the RIE reduces to the well-known magnetic field integral equation (MFIE). Thus, the RIE is more general than the EFIE or the MFIE. In other words, Equation (5) states that the interior test source has zero reaction with the other sources.

From Equation (5) and the reciprocity theorem, we obtain another form of the reaction integral equation:

$$\oint_S (\mathbf{J}_s \cdot \mathbf{E}^m - \mathbf{M}_s \cdot \mathbf{H}^m)\, ds + \int\int\int (\mathbf{J}_i \cdot \mathbf{E}^m - \mathbf{M}_i \cdot \mathbf{H}^m)\, dv = 0 \tag{6}$$

where $(\mathbf{E}^m, \mathbf{H}^m)$ is the field of the test source radiating in the homogeneous medium. This reaction integral equation was developed by Rumsey [16] in 1954. For thin-wire problems we shall employ Equation (6) with electric test sources.

In the wire structure, let each segment have a circular cylindrical surface. At each point on the composite cylindrical surface of the wire, it is convenient to define a right-handed orthogonal coordinate system with unit vectors $(\hat{n}, \hat{\phi}, \hat{\ell})$ where $\hat{n}$ is the outward normal vector, $\hat{\ell}$ is directed along the wire axis and

$$\hat{\phi} = \hat{\ell} \times \hat{n}. \tag{7}$$

Thus $(\hat{n}, \hat{\phi}, \hat{\ell})$ correspond directly with the unit vectors $(\hat{\rho}, \hat{\phi}, \hat{z})$ usually employed in the circular-cylindrical coordinate system.

To simplify the integral equation, we assume the wire radius "a" is much smaller than the wavelength λ, and the wire length is much greater than the radius. Furthermore, we shall neglect the integrations over the flat end surfaces of the wire, neglect the circumferential component J_ϕ of the surface-current density, and consider the axial component J_ℓ to be independent of ϕ. (For thick wires, a more detailed treatment is essential for the ϕ-dependent current modes and the integrations over the junction regions and the open ends of the wire. A more elaborate formulation may also be required if one wire passes within a few diameters of another, or if a wire is bent to form a small acute angle.) In view of these approximations the current density on the wire structure is related to the current as follows:

$$\mathbf{J}_s(\ell) = \frac{\hat{\ell} I(\ell)}{2\pi a} = \frac{\mathbf{I}(\ell)}{2\pi a} \tag{8}$$

where ℓ is a metric coordinate measuring position along the wire axis, and $I(\ell)$ is the total current (conduction plus displacement).

On a perfectly conducting wire, the magnetic current density $\mathbf{M}_s$ vanishes. If the wire has finite conductivity, we take

$$\mathbf{E} = Z_s \mathbf{J}_s \tag{9}$$

for the tangential electric field on S, where Z_s is the surface impedance for exterior excitation. From Equations (4), (7), (8) and (9),

$$\mathbf{M}_s = Z_s \mathbf{J}_s \times \hat{n} = \frac{\hat{\phi} Z_s I(\ell)}{2\pi a}. \tag{10}$$

By virtue of Equations (8) and (10), Equation (6) reduces to

$$-\int_0^L I(\ell)\left(E_\ell^m - Z_s H_\phi^m\right) d\ell = V_m \tag{11}$$

where L denotes the overall wire length and

$$V_m = \iiint (\mathbf{J}_i \cdot \mathbf{E}^m - \mathbf{M}_i \cdot \mathbf{H}^m)\, dv \tag{12}$$

$$E_\ell^m = \frac{1}{2\pi}\int_0^{2\pi} \hat{\ell}\cdot \mathbf{E}^m d\phi \tag{13}$$

$$H_\phi^m = \frac{1}{2\pi}\int_0^{2\pi} \hat{\phi}\cdot \mathbf{H}^m d\phi. \tag{14}$$

The sinusoidal reaction formulation for thin wires is based on the integral Equation (11). In this equation the known quantities are $\mathbf{E}^m$, $\mathbf{H}^m$, V_m and Z_s. The current distribution $I(\ell)$ is regarded as an unknown function. To permit a solution for the current distribution, the following sections define suitable test sources and expansion modes.

3 The Sinusoidal Test Sources

For a test source we choose a filamentary electric dipole with a sinusoidal current distribution. This is not a wire dipole, but merely an electric line source in the homogeneous medium. The sinusoidal dipole is probably the only finite line source with simple closed-form expressions for the near-zone fields. (See the Appendices.) Furthermore, the mutual impedance between two sinusoidal dipoles is available in terms of exponential integrals, and the piecewise-sinusoidal function is evidently close to the natural current distribution on a perfectly conducting thin wire. These factors governed the choice of test sources.

A typical test source is a V dipole with unequal arm lengths and terminals at the vertex. The current is zero at the endpoints and rises sinusoidally to a maximum at the terminals. The terminal current is one ampere, and the current distribution has a slope discontinuity at the terminals.

For the linear test dipole illustrated in Figure 1, the current distribution is $\mathbf{I}(z) = \mathbf{F}(z)$ where

$$\mathbf{F}(z) = \frac{\hat{z}P_1 \sinh\gamma(z-z_1)}{\sinh\gamma d_1} + \frac{\hat{z}P_2 \sinh\gamma(z_3-z)}{\sinh\gamma d_2}. \tag{15}$$

$P_1(z)$ is a pulse function with unit value for $z_1 < z < z_2$ and zero value elsewhere. The pulse function P_2 has unit value for $z_2 < z < z_3$ and vanishes elsewhere. The segment lengths are $d_1 = z_2 - z_1$ and $d_2 = z_3 - z_2$. The current distribution on a V test dipole is

$$\mathbf{F}(\ell) = \frac{\hat{\ell}_1 P_1 \sinh\gamma(\ell-\ell_1)}{\sinh\gamma d_1} + \frac{\hat{\ell}_2 P_2 \sinh\gamma(\ell_3-\ell)}{\sinh\gamma d_2}. \tag{16}$$

In Equations (15) and (16), γ denotes the complex propagation constant of the homogeneous exterior medium:

$$\gamma = s\sqrt{\mu\epsilon}. \tag{17}$$

It is only with this value for γ that the sinusoidal test sources have the advantages mentioned earlier.

The test dipole is located in the interior region of the wire structure. To simplify the integrations in Equations (13) and (14), we place the test dipole on the wire axis.

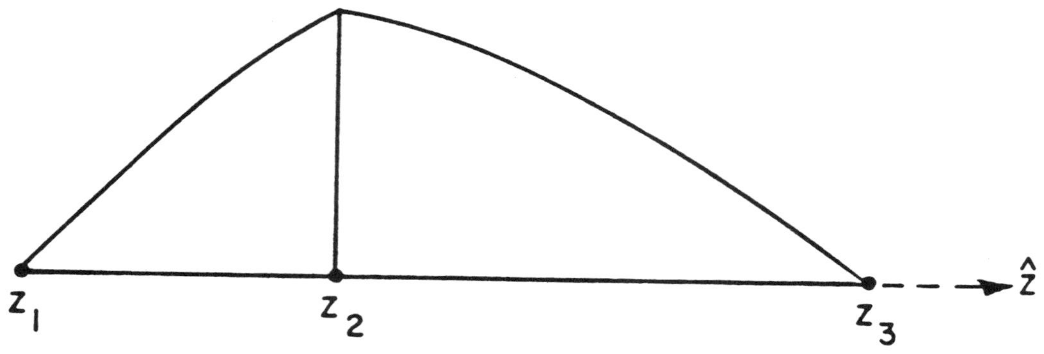

Figure 1: A linear test dipole and its sinusoidal current distribution. The endpoints are at z_1 and z_3 with terminals at z_2.

A typical problem requires not just one but several test dipoles located at different positions along the wire axis to form an overlapping array. When test dipole m radiates in the homogeneous medium, it generates the field $(\mathbf{E}^m, \mathbf{H}^m)$. Each test dipole has the same frequency as the true source. Using N test dipoles, Equation (11) is enforced for each one. Thus, Equation (11) represents a system of N simultaneous integral equations with $m = 1, 2, \cdots N$. In other words, Equation (11) requires each test dipole in the array to have the correct reaction with the true source.

4 The Sinusoidal Expansion Functions

The current distribution on the wire structure is expanded in a finite series as follows:

$$\mathbf{I}(\ell) = \sum_{n=1}^{N} I_n \mathbf{F}_n(\ell) \tag{18}$$

where the normalized expansion functions $\mathbf{F}_n(\ell)$ are the same as the test-dipole current distributions in Equation (16). Since each expansion function extends over just a two-segment portion of the wire structure, these functions are subsectional bases. Since N is finite, Equation (18) may be considered either as an expansion or an approximation, depending on the context. In Equation (18), the coefficients I_n are complex constants which represent samples of the current function $I(\ell)$. If the wire segments are short in comparison with the wavelength, the sinusoidal bases resemble the triangular bases of the piecewise-linear model.

Figure 2 illustrates a current distribution $I(\ell)$, its two-mode approximation $I'(\ell)$ and the normalized expansion functions $F_1(\ell)$ and $F_2(\ell)$. It may be noted that $I(\ell)$ is a smooth function except at generators, lumped loads and wire corners. The piecewise-sinusoidal expansion has slope discontinuities at these appropriate locations and also at each intermediate sampling point. With favorable circumstances, the calculated samples I_n will be accurate and the corresponding piecewise-sinusoidal current distribution $I'(\ell)$ will be satisfactory for far-field calculations. For near-zone field analysis, however, one may abandon the sinusoidal interpolation and model the current distribution $I(\ell)$ with a smooth function fitting the calculated samples. In this process, one should not smooth out the slope discontinuities at the generators, lumped loads or wire corners.

By inserting Equation (18) into Equation (11), we obtain the following system of simultaneous linear algebraic equations:

$$\sum_{n=1}^{N} I_n Z_{mn} = V_m \qquad \text{where } m = 1, 2, \cdots N \tag{19}$$

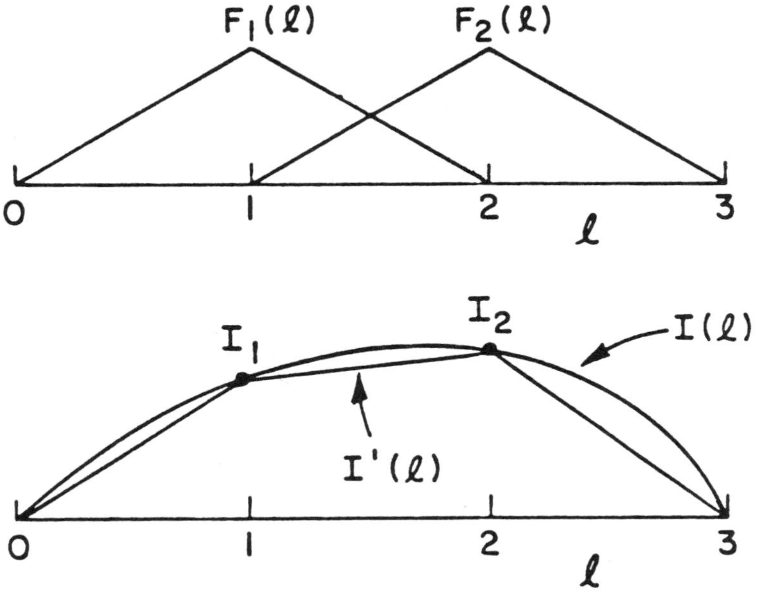

Figure 2: The expansion functions $F_1(\ell)$ and $F_2(\ell)$, the current distribution $I(\ell)$ and the two-mode approximation $I'(\ell)$.

$$Z_{mn} = -\int_n F_n(\ell)\left(E_\ell^m - Z_s H_\phi^m\right) d\ell. \tag{20}$$

In Equation (20), the integral extends over the two segments in the range of the expansion mode F_n. Equation (19) can be expressed in matrix form as $ZI = V$ where Z denotes the square impedance matrix, I is the current column and V is the voltage column.

5 The Impedance Matrix

The elements in the open-circuit impedance matrix are denoted Z_{mn}. By convention, the first and second subscripts indicate the row and column, respectively. Thus, Z_{mn} denotes the mutual impedance between test dipole m and expansion mode n.

The expansion modes form an array of overlapping tubular dipoles located on the wire surface. Each tubular dipole has a sinusoidal distribution of electric surface-current density and an associated magnetic surface-current density. If the test dipoles had the same shapes and positions as the expansion dipoles, the reciprocity theorem could be invoked to demonstrate the symmetry of the impedance matrix. Since the filamentary test dipoles differ from the tubular expansion dipoles, our impedance matrix is not precisely symmetric.

In practice, we regain symmetry by taking a short-cut in calculating the elements in the impedance matrix. From Equations (13), (14) and (20), Z_{mn} is expressed as an integration over the composite circular-cylindrical surface of the tubular expansion dipole n. The short-cut is accomplished by approximating the surface integral with a line integral. Thus, to reduce computational costs, we approximate Z_{mn} by the mutual impedance between two filamentary V dipoles, one located on the wire axis and the other on the wire surface. For a straight wire, the mutual impedance is independent of the circumferential position of the dipole located on the surface. For a bent wire, however, the mutual impedance is sometimes quite sensitive to the circumferential position. Via numerical experiments, we have found a suitable position such that the circumferential integrations in Equations (13) and (14) are adequately approximated from a single sample of the integrand.

With the approximation mentioned above, the impedances Z_{mn} are given by classical induced emf theory [1,2,3]. The line integral in Equation (20) is usually performed with numerical integration. When dipoles m and n are close together, however, a closed-form expression (in terms of exponential integrals) is employed for Z_{mn}.

6 Lumped Loads

Suppose lumped loads are inserted in the wire structure at the current sampling points. These linear impedances may be active or passive. The voltage drop across a lumped load has the same effect as inserting an equivalent voltage generator at that point, so a detailed analysis of a lumped load will involve a magnetic frill, ring or tube as in Section 8.

In the simplest model, the delta gap, the lumped loads introduce a new term in the mutual impedance element Z_{mn} so that Equation (19) is replaced with the following

$$\sum_{n=1}^{N} I_n \bar{Z}_{mn} = V_m. \tag{21}$$

where

$$\bar{Z}_{mn} = Z_{mn} \pm Z_{mn}^{\ell\ell} \tag{22}$$

On the right-hand side of Equation (22), the first term Z_{mn} denotes the mutual impedance via electromagnetic coupling as discussed in Section 5. The new term $Z_{mn}^{\ell\ell}$ denotes the lumped load impedance threaded by both dipole mode currents I_m and I_n at their terminals. The plus sign is appropriate in Equation (22) if the current expansion functions $\mathbf{F_m}$ and $\mathbf{F_n}$ are parallel through this load, and the minus sign is correct if they are antiparallel at this point. Thus, the effect of lumped loading is accounted for simply by modifying the elements in the square impedance matrix. Of course, $Z_{mn}^{\ell\ell}$ vanishes unless $m = n$ or dipole modes m and n have their terminals at the same end of a shared segment.

7 Wires with Finite Conductivity

The surface impedance Z_s will vanish unless the wire has finite conductivity. From Equation (20),

$$Z_{mn} = Z'_{mn} - \int_n F_n E_\ell^m \, d\ell \tag{23}$$

where

$$Z'_{mn} = Z_s \int_n F_n H_\phi^m \, d\ell. \tag{24}$$

Although longitudinal inhomogeneities in the surface impedance and wire radius offer no difficulties, it is assumed in Equation (24) and hereafter that these parameters are independent of ℓ. The integral in

Equation (24) extends over the two wire segments in the range of expansion mode F_n. From Equation (14) and Ampere's law, a suitable approximation is

$$H_\phi^m = \frac{F_m(\ell)}{2\pi a} \tag{25}$$

$$Z'_{mn} = \frac{Z_s}{2\pi a}\int_{m,n} F_n(\ell)F_m(\ell)d\ell \tag{26}$$

where region (m, n) is the wire surface shared by dipoles m and n. This region covers two intersecting segments if m and n are equal. If m and n differ, the shared region covers at most one wire segment. For a diagonal element, m and n are equal and Equations (16) and (26) yield

$$Z'_{mm} = \frac{Z_s}{8\pi\gamma a}\left[\frac{\sinh(2\gamma d_1) - 2\gamma d_1}{\sinh^2 \gamma d_1} + \frac{\sinh(2\gamma d_2) - 2\gamma d_2}{\sinh^2 \gamma d_2}\right] \tag{27}$$

where d_1 and d_2 are the lengths of the two wire segments occupied by mode m. For an off-diagonal element, a suitable approximation is $Z'_{mn} = 0$ if modes m and n do not share a segment. If they share one segment and have terminals at the same end of this segment (length d),

$$Z'_{mn} = \frac{[\sinh(2\gamma d) - 2\gamma d]Z_s}{8\pi\gamma a \sinh^2 \gamma d}. \tag{28}$$

If modes m and n share one segment and have terminals at opposite ends of this segment,

$$Z'_{mn} = \frac{(\gamma d \cosh \gamma d - \sinh \gamma d)Z_s}{4\pi\gamma a \sinh^2 \gamma d}. \tag{29}$$

If $\mathbf{F}_m$ and $\mathbf{F}_n$ are antiparallel on the shared segment, a minus sign must be inserted on the right-hand side of Equations (28) or (29).

If Equation (26) is employed, it is obvious that the square matrix Z'_{mn} will be symmetric. From Equation (23) and the reciprocity theorem, the matrix Z_{mn} (and $\bar{Z}_{mn}$ with lumped loading) will also be symmetric if the tubular expansion dipoles are approximated by filamentary dipoles located on the wire surface. This symmetry alleviates computational expenses and storage requirements.

8 The Excitation Voltages

From Equation (12) and reciprocity, the excitation voltages are given by

$$V_m = \int_m \mathbf{F}_m \cdot \mathbf{E}_i d\ell \tag{30}$$

where $\mathbf{E}_i$ denotes the incident field generated by $(\mathbf{J}_i, \mathbf{M}_i)$ radiating in the homogeneous medium. The integration extends over both arms or segments of test dipole m.

If the incident field is generated by a distant source with spherical coordinates (r_o, θ_o, ϕ_o),

$$\mathbf{E}_i = \mathbf{E}_o \exp(\gamma \mathbf{r} \cdot \hat{r}_o) \tag{31}$$

where $\mathbf{E}_o$ is a vector constant, $\hat{r}_o$ is a unit vector from the coordinate origin to the distant source, and $\mathbf{r}$ is the radial vector from the origin to the observation point. From Equations (30) and (31), the excitation voltages induced by an incident plane wave are

$$V_m = \int_m \mathbf{F}_m \cdot \mathbf{E}_o \exp(\gamma \mathbf{r} \cdot \hat{r}_o) d\ell. \tag{32}$$

Now consider the field $\mathbf{E}^m$ generated by test dipole m when radiating in the homogeneous medium. Using the vector potential, we find the field at the distant point (r_o, θ_o, ϕ_o) to be

$$\mathbf{E}^m = -\frac{s\mu e^{-\gamma r_o}}{4\pi r_o} \int_m \mathbf{F}_m \exp(\gamma \mathbf{r} \cdot \hat{r}_o) d\ell \tag{33}$$

where the radial component is to be suppressed. From Equations (32) and (33),

$$V_m = -\frac{4\pi r_o}{s\mu} e^{\gamma r_o} \mathbf{E}_o \cdot \mathbf{E}^m. \tag{34}$$

Equations (32) and (34) are useful in plane-wave scattering problems. If the source $(\mathbf{J}_i, \mathbf{M}_i)$ is near the wire structure, we have a near-zone scattering problem and employ Equation (30).

A wire structure is usually called a scatterer if the source $(\mathbf{J}_i, \mathbf{M}_i)$ is located some distance away, and an antenna if the source is at the wire surface. In electromagnetic theory, however, there is no fundamental distinction between the scattering problem and the antenna problem. The antenna problem is merely an extreme example for near-zone scattering.

If an antenna is fed with a parallel-wire transmission line, the transmission line is properly considered to be part of the radiating system. In addition to the TEM mode, higher-order modes will exist on the transmission line. Thus, moment methods are employed to determine the current distribution on the transmission line as well as on the antenna. If a wire antenna is fed through a coaxial cable, the source may be modeled as a magnetic surface-current density $\mathbf{M}_i$ on the aperture surface of the coaxial feed. The shape of the source $\mathbf{M}_i$ is determined by the details of the terminal region. Thus, a voltage generator may be modeled as a magnetic disk or a magnetic tube as indicated by Otto [18] in 1968.

Consider a wire antenna driven by a voltage generator v_i located at one of the current sampling points ℓ_i. The generator voltage v_i is considered positive if it tends to force a current in the direction of the expansion mode $\mathbf{F}_i(\ell)$. From Equations (12) and (30), the excitation voltages are

$$V_m = \int_m \mathbf{F}_m \cdot \mathbf{E}_i d\ell = -\int\int \mathbf{M}_i \cdot \mathbf{H}^m ds \tag{35}$$

where the line integral extends over test dipole m and the surface integral extends over the magnetic source $\mathbf{M}_i$. If the magnetic source is approximated by a loop encircling the wire and $\mathbf{M}_i$ is uniform around the loop, then

$$\mathbf{M}_i = -\hat{\phi} v_i. \tag{36}$$

If the loop has small radius b, Equation (35) reduces to

$$V_m = b v_i \oint_i \hat{\phi} \cdot \mathbf{H}^m d\phi. \tag{37}$$

If displacement currents are neglected, Equation (37) and Ampere's law yield

$$V_m = v_i F_m(\ell_i). \tag{38}$$

From Equation (38), all the excitation voltages V_m vanish except one: $V_i = v_i$. Although this simple result is often adequate, the accuracy and convergence of the solution may be improved by modeling the source as a magnetic disk or tube (instead of a loop) and using Equation (35). The approximate result in Equation (38) may be regarded as the delta-gap model.

9 Radiation Efficiency and Echo Area

In this section the frequency is considered real. That is, $s = j\omega$ and ω is real. Furthermore, let **E** and **H** denote the rms field intensities. Both the wire and the surrounding homogeneous medium may have finite conductivities.

When the wire structure is excited as an antenna, the time-average power input is the sum of the powers delivered at the various ports:

$$P_i = \text{Real} \sum_{n=1}^{N} v_n I_n^*. \tag{39}$$

The time-average power dissipated in the wire antenna is

$$P_d = \text{Real} \oint_S (\mathbf{E} \times \mathbf{H}^*) \cdot \mathbf{ds} = \frac{R_s}{2\pi a} \int_0^L \mathbf{I} \cdot \mathbf{I}^* dl \tag{40}$$

where S denotes the closed surface of the wire, $\mathbf{ds}$ is directed inward on S, and R_s is the surface resistance. The last form in Equation (40) is a convenient approximation based on Equations (3), (8) and (9).

Suppose the voltage generators are modeled with tubular magnetic current sources with radii slightly larger than that of the wire. The wire need not have any gaps at the terminals. In calculating the power dissipated in the wire via Equation (40), we integrate over the wire surface S and consider the magnetic sources $\mathbf{M}_i$ to be outside S. The time average power radiated from the antenna to the exterior region is

$$P_r = \text{Real} \oint_{S'} (\mathbf{E} \times \mathbf{H}^*) \cdot \mathbf{ds} \tag{41}$$

where surface S' lies just outside surface S so the magnetic sources $\mathbf{M}_i$ are in the interior region of S'. From Equations (39), (40) and (41) and Poynting's theorem,

$$P_i = P_d + P_r. \tag{42}$$

If the ambient medium has finite conductivity, the "radiated power" P_r is actually dissipated in the exterior region. The radiation efficiency may be defined as the ratio of the power radiated to the power input:

$$E_r = P_r / P_i. \tag{43}$$

If Equation (43) is employed, the antenna will have perfect efficiency unless it is constructed of dissipative media.

An alternative definition has been proposed by Tsao [19] as follows. It is reasonable to consider a certain portion of the exterior-region dissipation to be a propagation phenomenon rather than intrinsically an antenna problem. Therefore, let the time-average power radiated by the antenna be defined as follows:

$$P_R = e^{2\alpha r} \text{Real} \oint (\mathbf{E} \times \mathbf{H}^*) \cdot \mathbf{ds} \tag{44}$$

where $\gamma = \alpha + j\beta$ and the integration covers a spherical surface with radius r centered at the antenna. With this definition, the radiated power is independent of the range r in the far-zone region. Now the radiation efficiency is defined by

$$E_R = P_R / P_i. \tag{45}$$

The definition in Equation 45 penalizes the antenna not only for power dissipated in the antenna structure, but also for excess near-zone losses. Thus, even a perfectly conducting antenna may have imperfect efficiency.

For a target in a homogeneous conducting medium, it is convenient to define the radar cross section (or echo area) as follows:

$$\sigma = \lim_{r \to \infty} 4\pi r^2 e^{2\alpha r} S_s / S_i \quad (46)$$

where S_s and S_i denote the time-average power densities in the scattered and incident fields, respectively. When echo area data are presented for a specific target, it is necessary to specify the location of the point in space where S_i is evaluated and from which the range r is measured. Equation 46 reduces to the standard definition for a target in a lossless medium as the attenuation constant α tends to zero. Without the factor $\exp(2\alpha r)$, the echo area would vanish for every finite target in a conducting medium.

10 Numerical Results

In comparison with antennas in free space, relatively little data are available for structures in conducting medium. Therefore, it is not necessary or desirable to choose a complicated configuration to illustrate trends. Figure 3 illustrates the backscatter echo area of a perfectly conducting straight wire for broadside incidence with Equation 46. For defining the range r and the incident power density S_i, the coordinate origin is located at the center of the wire. At the highest frequency (100 Mhz), the wire length is approximately equal to the wavelength in the ambient medium. As the conductivity of the medium increases, the echo area decreases at first and then increases. It seems reasonable that the highly resonant properties of the thin straight wire should decrease and finally disappear. All the numerical results presented in this section were calculated with the sinusoidal reaction technique. For these calculations, the wire was divided into a number J of segments where $J = 6$ or

$$J = 3.2|\gamma h| \quad (47)$$

(whichever is larger) where h denotes the half-length of the wire.

Figure 4 illustrates the radiation efficiency of a perfectly conducting center-fed linear dipole. The radiated power was calculated via Equation 44 by integration of the power density over a far-zone sphere. Again the center of the wire was selected as the coordinate origin for measuring the range r. The power input was obtained from the terminal current and voltage. The low efficiency observed in the low-frequency range is attributed to the excess near-zone losses of the short uninsulated dipole. As the conductivity of the medium increases, the efficiency decreases at first and then increases. This is not surprising in view of the discussion following Equation 45

Figure 5 shows the resistance and reactance of the center-fed linear dipole mentioned above.

11 Summary

Rumsey's reaction integral equation is discussed, and it is pointed out that it is more general than the electric field integral equation or the magnetic field integral equation. The sinusoidal reaction formulation is presented for an arbitary thin-wire structure in a conducting medium. The wire structure may have finite conductivity and lumped loading. The analysis is performed in the real or complex frequency domain, and it covers both the antenna and the scattering situations. A fundamental distinction is indicated between the moment voltages and impedances and the multiport voltages and impedances.

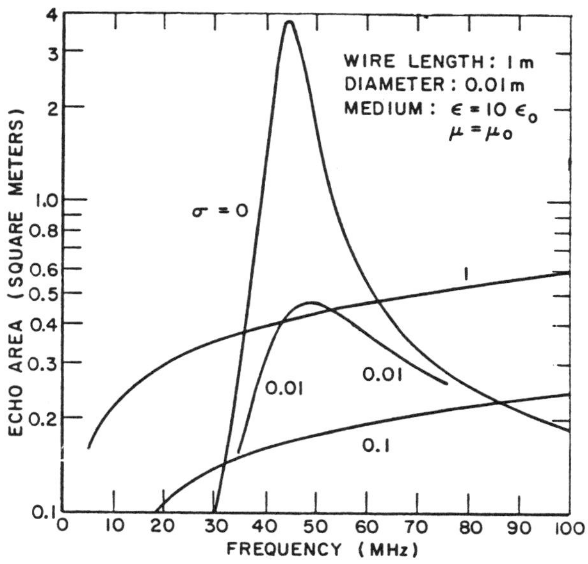

Figure 3: Broadside backscatter for uninsulated perfectly-conducting straight wire in homogeneous medium with conductivity σ in mhos/m.

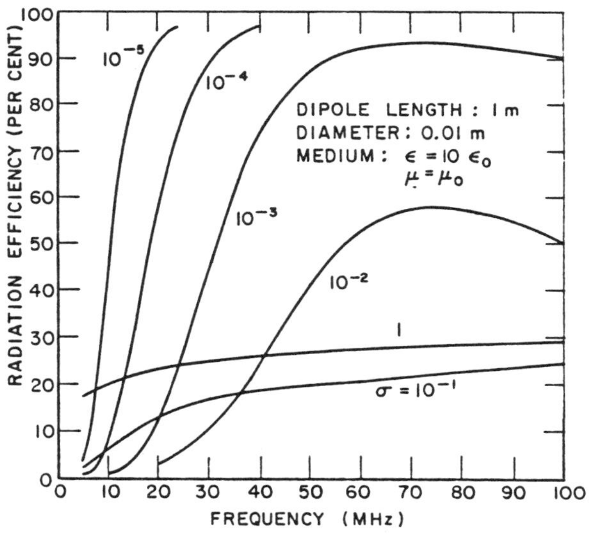

Figure 4: Radiation efficiency of uninsulated perfectly-conducting center-fed linear dipole in a homogeneous medium with conductivity σ in mhos/m.

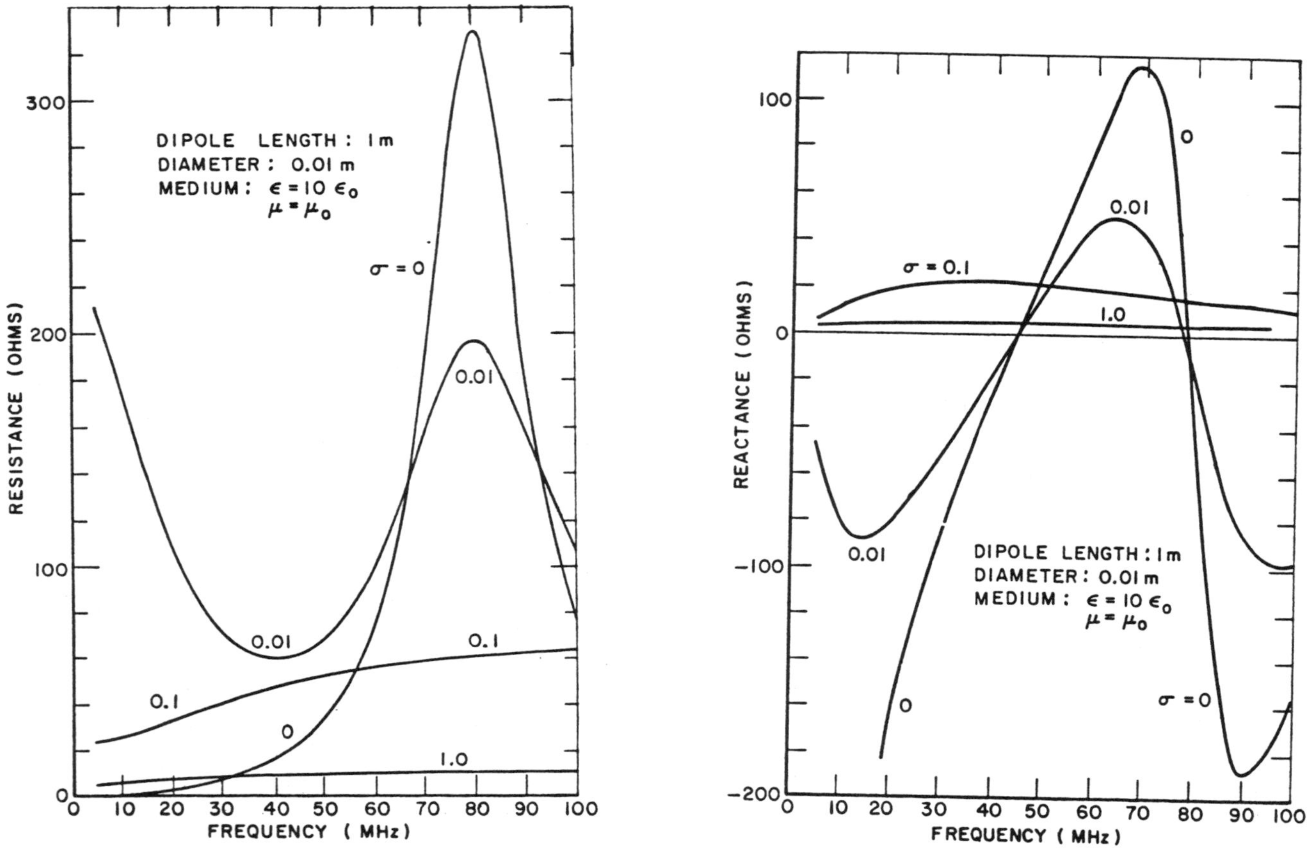

Figure 5: Resistance and Reactance of uninsulated perfectly-conducting center-fed linear dipole in a homogeneous medium with conductivity σ in mhos/m.

References

[1] Carter, P.S., "Circuit Relations in Radiating Systems and Applications to Antenna Problems," *IRE Proc.*, Vol. 20, (June 1932), pp. 1004–1041.

[2] King, H.E., "Mutual Impedance of Unequal Length Antennas in Echelon," *IRE Trans.*, Vol. AP-5, (July 1957), pp. 306–313.

[3] Baker, H.C. and A.H. LaGrone, "Digital Computation of the Mutual Impedance Between Thin Dipoles," *IRE Trans.*, Vol. AP-10, (March 1962), pp. 172–178.

[4] Yeh, Y.S. and K.K. Mei, "Theory of Conical Equiangular-Spiral Antenna, Part I - Numerical Technique," *IEEE Trans.*, Vol. AP-15, (September 1967), pp. 634–639.

[5] Tanner, R.L. and M.G. Andreasen, "Numerical Solution of Electromagnetic Problems," *IEEE Spectrum*, (September 1967), pp. 53–61.

[6] Miller, E.K. and J.B. Morton, "The RCS of a Metal Plate with a Resonant Slot," *IEEE Trans.*, Vol. AP-18, (March 1970), pp. 290–292.

[7] Chao, H.H. and B.J. Strait, "Radiation and Scattering by Configurations of Bent Wires with Junctions," *IEEE Trans.*, Vol. AP-19, (September 1971), pp. 701–702.

[8] Richmond, J.H., "Theoretical Study of V Antenna Characteristics for the ATS-E Radio Astronomy Experiment," Report 2619-1, 13 February 1969, The Ohio State University ElectroScience Laboratory, Department of Electrical Engineering; prepared under Contract NAS5-11543 for National Aeronautics and Space Administration, Goddard Space Flight Center, Greenbelt, Maryland. (N69-19062)

[9] Richmond, J.H., "Computer Analysis of Three-Dimensional Wire Antennas," Report 2708-4, 22 December 1969, The Ohio State University ElectroScience Laboratory, Department of Electrical Engineering; prepared under Contract DAAD05-69-C-0031 for Department of the Army, Ballistic Research Laboratory, Aberdeen Proving Ground, Maryland.

[10] Richmond, J.H. and N.H. Geary, "Mutual Impedance Between Co-Planar-Skew Dipoles," *IEEE Trans.*, Vol. AP-18, (May 1970), pp. 414-416.

[11] Richmond, J.H., "Coupled Linear Antennas with Skew Orientation," *IEEE Trans.*, Vol. AP-18, (September 1970), pp. 694–696.

[12] Richmond, J.H., "Admittance Matrix of Coupled V Antennas," *IEEE Trans.*, Vol. AP-18, (November 1970), pp. 820-821.

[13] Butler, C.M., "Currents Induced on a Pair of Skew Crossed Wires," *IEEE Trans.*, Vol. AP-20, (November 1972), pp. 731-736.

[14] Imbriale, W.A. and P.G. Ingerson, "On Numerical Convergence of Moment Solutions of Moderately Thick Wire Antennas Using Sinusoidal Basis Functions," *IEEE Trans.*, Vol. AP-21, (May 1973), pp. 363-366.

[15] Kantorovich, L. and V. Krylov, Approximate Methods of Higher Analysis," New York: Wiley, (1964), Chapter 4.

[16] Rumsey, V.H., "Reaction Concept in Electromagnetic Theory," *Physical Review*, Vol. 94, (June 15, 1954), pp. 1483–1491.

[17] Schelkunoff, S.A., "On Diffraction and Radiation of Electromagnetic Waves," *Physical Review*, Vol. 56, (August 15, 1939).

[18] Otto, D.V., "Fourier Transform Method in Cylindrical Antenna Theory," *Radio Science*, (New Series), Vol. 3, No. 11, (November 1968), pp. 1050–1057.

[19] Tsao, C.K.H., "Radiation Resistance of Antennas in Lossy Media," *IEEE Trans.*, Vol. AP-19, (May 1971), pp. 443–444.

A study of the various methods for computing electromagnetic field utilizing thin wire integral equations

Tapan K. Sarkar

Department of Electrical Engineering, Rochester Institute of Technology, Rochester, New York 14623

(Received May 6, 1982; revised September 3, 1982; accepted September 3, 1982.)

In this paper we analyze the numerical aspects of the various methods that have been utilized to analyze thin wire antennas. First, we derive the properties of the operators for Pocklington's and Hallen's integral equation. On the basis of these properties, we discuss the various iterative methods used to find current distribution on thin wire structures. An attempt has been made to resolve the question of numerical stability associated with various entire domain and subdomain expansion functions in Galerkin's method. It has been shown that the sequence of solutions generated by the iterative methods monotonically approaches the exact solution provided the excitations chosen for these problems are in the range of the operator. Such a statement may not hold for Galerkin's methods if the inverse operator is unbounded. Moreover, if the excitation function is not in the range of the operator, then the sequence of solutions forms an asymptotic series. Examples have been presented to illustrate this point.

1. INTRODUCTION

Over the past few years several methods have been developed by many researchers [*Gray*, 1966; *Harrington*, 1968; *King and Middleton*, 1946, 1948; *Schelkunoff*, 1952; *Siegel and Labus*, 1934; *Wu*, 1969] to analyze scattering and radiation from thin wire structures. In this presentation we investigate the properties of the integro-differential equations that arise for the various techniques developed so far. The motivation for this work is to study the causes of the numerical instabilities that sometimes arise in the solution of the current distribution on thin wire structures. The numerical instabilities may be either due to an incorrect application of the numerical techniques or due to the operator equation being actually ill posed. In a recent note, *Jones* [1981] has claimed that the Hallen's integral equation is a well-posed problem and that the sources of numerical instabilities lie with the particular numerical technique used to solve an operator equation. As we shall presently demonstrate, the proof presented by Jones is not complete. Jones did not consider all the aspects of a well-posed problem. For a problem to be well posed, three conditions have to be met by the operator equations. According to *Stakgold* [1979, p. 58], the three conditions are defined as follows:

> When dealing with boundary value problems we shall still be faced with these three questions:
> (1) Is there at least one solution (existence)?
> (2) Is there at most one solution (uniqueness)?
> (3) Does the solution depend continuously on the data?
> If the answer to this trio of questions is affirmative the problems is said to be well-posed (otherwise ill-posed). Until recently it was sound dogma to require that every real physical problem be well posed. However it is now understood that ill-posed problems occur frequently in practice but that their physical interpretation and mathematical solution are somewhat more delicate.

Jones, in his paper, addresses only the first two questions. In his note he did not check whether the inverse operator in Hallen's integral equation is bounded or not! It is based on the third statement of Stakgold that the Hallen's integral is ill posed. We prove later that the operator involved in Hallen's integral equation is compact, and hence its inverse is unbounded. Therefore, by definition, Hallen's integral equation is an ill-posed problem. *Tikhonov and Dimitriyev* [1968] were the first to recognize that Hallen's integral equation is ill posed and developed a 'self-regularization' procedure to solve that integral equation. In summary, if any numerical instability is observed for Hallen's integral equation, one cannot put the blame entirely on the numerical procedure utilized to obtain a solution.

Two well-known problems which are not well posed are the Dirichlet problem for the wave equa-

Paper number 2S1379.
0048-6604/83/0102-1379$08.00

T. K. Sarkar, *Radio Science*, vol. 18, pp. 29–38, 1983, copyright by the American Geophysical Union.

tion and the Cauchy problem for the Laplace equation. In fact, for the latter problem Hadamard showed that a global smooth solution cannot exist unless the initial data satisfy certain compatibility conditions [*Bloom*, 1981].

In this paper, we investigate the properties of Pocklington's integro-differential operator and Hallen's integral operator. We also discuss the advantages of an iterative method and the direct method (Galerkin's method) of solving the two operator equations and the numerical stabilities of the various order of the solutions given by the two techniques. Finally, we investigate the convergence properties of the approximate solutions when the excitation is not in the range of the operator.

2. PROPERTY OF THE POCKLINGTON E FIELD OPERATOR

The Pocklington integral equation for the current on the surface of an antenna can be written by equating the total tangential electric field on the conductor surface to zero, i.e.,

$$E^i_{\tan} + E^s_{\tan} = 0 \tag{1}$$

where the subscript represents the tangential component of the electric field and the superscripts i and s stand for the incident and scattered fields, respectively. By assuming a time variation of the form exp ($j\omega t$), equation (1) can be rewritten for the tubular antenna of length L and radius a as [*Jones*, 1979]

$$k^2 \int_{-L/2}^{+L/2} dz' I(z') G(z\ z') + \frac{\partial^2}{\partial z^2} \int_{-L/2}^{+L/2} dz' I(z') G(z, z') = j\omega 4\pi\varepsilon E^i_{\tan}(z) \qquad -\frac{L}{2} \le z \le +\frac{L}{2} \tag{2}$$

where

$$G(z, z') = \text{Green's function} = \frac{1}{2\pi} \int_0^{2\pi} \frac{\exp(-jkR)}{R}\, d\phi \tag{3}$$

$$R = \left[(z - z')^2 + 4a^2 \sin^2 \frac{\phi}{2} \right]^{1/2} \tag{4}$$

and $k = (2\pi)/\lambda$. In the terms of an operator, equation (1) can be written as

$$PI = k^2 P_1 I + P_2 I = V \tag{5}$$

where P_1 and P_2 represent the operators in the first and second parts of the integral in (2), respectively. First, we would like to investigate the properties of the operator P in (5), since the method of solution for $I(z')$ in (2) is dependent on whether P is bounded or unbounded.

If a constant C (independent of $I(z')$) exists such that the following inequality is always satisfied,

$$\|P\| = \max \frac{\|PI\|}{\|I\|} = \max_{\|I\|=1} \|PI\| \le C \tag{6}$$

then the operator P is said to be bounded [*Stakgold*, 1979, p. 296]. If such a constant C exists, which is the maximum of all possible $\|PI\|$ with the constraint $\|I\| = 1$, then we say that the operator P is bounded with respect to the norm $\|\cdot\|$. The two norms that we shall be dealing with are the $\mathscr{L}^2$ norm and the Chebyshev norm. The $\mathscr{L}^2$ norm is defined as

$$\|I\|_{\mathscr{L}^2} = \left[\int_{-L/2}^{+L/2} |I(z)|^2\, dx \right]^{1/2} \tag{7}$$

and the Chebyshev norm is defined as

$$\|I\|_T = \max_{-L/2 \leqslant z \leqslant +L/2} |I(z)| \tag{8}$$

When no subscripts are used, then it could be either of the two norms. If we are using the $\mathscr{L}^2$ norm, then we are restricting the domain of the operator P to elements which are in $\mathscr{L}^2$ (or square integrable). This does not imply that $I(z')$ cannot be infinite within the range $-L/2 \le z' \le +L/2$. However, only those type of singularities are permitted in $I(z')$ which are square integrable. Any function which is not square integrable is excluded from the domain of P (as they are not in $\mathscr{L}^2$). On the other hand, if we use the Chebyshev norm, then the function has to be bounded. Under the Chebyshev norm any unbounded function cannot be in the domain of the operator. Thus the function log z is in $\mathscr{L}^2$ as it is square integrable (see definition (7)) but not in the domain of functions satisfying the Chebyshev norm. Physically, then, convergence of a sequence of functions under the $\mathscr{L}^2$ norm yields least squares convergence, whereas convergence under the Chebyshev norm yields pointwise convergence.

Examination of the Green's function reveals that the kernel has a singularity. The singularity can be observed by rewriting the kernel as [*Schelkunoff*, 1952, p. 141]

$$G(z, z') = \frac{1}{2\pi} \int_0^{2\pi} \frac{1}{R}\, d\phi - \frac{1}{2\pi} \int_0^{2\pi} \frac{1 - e^{-jkR}}{R}\, d\phi \tag{9}$$

The first term in equation (9) can be transformed to a complete elliptic integral of the first kind. Thus we find

$$G(z, z') = \frac{p}{\pi a} F\left(\frac{\pi}{2}, p\right) - \frac{1 - e^{-jk|z-z'|}}{|z - z'|}$$
$$+ \text{ terms of the order of } k^2a^2 \tag{10}$$

where

$$p = \frac{2a}{[4a^2 + (z - z')^2]^{1/2}} \tag{11}$$

As $z \to z'$, the Green's function behaves as

$$G(z, z') \to \frac{1}{\pi a} \log \frac{4[(z - z')^2 + 4a^2]^{1/2}}{|z - z'|} + G_2 \tag{12}$$

where G_2 contains terms which are bounded and hence square integrable. In the immediate vicinity of $z \approx z'$,

$$G(z, z') \to -\frac{1}{\pi a} \log |z - z'| + G_3 \tag{13}$$

where G_3 contains terms which are bounded. So the singularity of the kernel is manifested through the log function in (13). Since log functions are square integrable, we find

$$\| P_1 \|_{\mathscr{L}^2} \leq \frac{1}{\pi a} \left| \int_{-L/2}^{+L/2} dz \int_{-L/2}^{+L/2} dz' \cdot \{\log |z - z'| + G_3\}^2 \right|^{1/2} = C < \infty \tag{14}$$

where C is a constant. Hence P_1 is bounded under the $\mathscr{L}^2$ norm. Therefore P_1 is a Hilbert Schmidt operator as it has a square integrable kernel [*Stakgold*, 1979, p. 352]. It can also be shown that a Hilbert Schmidt operator is a compact operator [*Stakgold*, 1979, p. 353]. Under the Chebyshev norm,

$$\| P \|_T \leq \max_{-L/2 \leq z \leq L/2} \frac{1}{\pi a} \left| \int_{-L/2}^{+L/2} dz' \log |z - z'| \right| + \left| \int_{-L/2}^{+L/2} dz' G_3 \right|$$
$$\leq \max_{-L/2 \leq z \leq L/2} \frac{1}{\pi a} \left\{ \left(z + \frac{L}{2}\right) \log \left(z + \frac{L}{2}\right) + \left(\frac{L}{2} - z\right) \log \left(\frac{L}{2} - z\right) - L \right\} + \text{const}$$
$$\leq M \text{ (const)} \quad \forall -\frac{L}{2} \leq z \leq \frac{L}{2} \tag{15}$$

Hence the operator P_1 is also bounded under the Chebyshev norm.

Next, consider the second integral in (2). We have

$$P_2 I \approx \frac{\partial}{\partial z} \int_{-L/2}^{+L/2} dz' \frac{\partial I(z')}{\partial z'} \cdot \left\{ -\frac{1}{\pi a} \log |z - z'| + G_3 \right\}$$
$$= \bar{\int}_{-L/2}^{L/2} dz' \cdot \frac{\partial I(z')}{\partial z'} \cdot \frac{\partial}{\partial z} \cdot \left\{ -\frac{1}{\pi a} \log |z - z'| + G_3 \right\} \tag{16}$$

where the bar over the second integral represents a principal value. It is clear that the operator P_2 in (16) is unbounded under the Chebyshev norm because $(\partial I)/(\partial z')$ is unbounded as $z \to \pm L/2$. Thus there exists no constant C for all $-L/2 \leq z \leq +L/2$ (because the charge $(\partial I)/(\partial z') \to \infty$ at the edges):

$$\| P_2 \|_T \leq C < \infty$$

Also, the operator P_2 is unbounded under $\mathscr{L}^2$ norm as $(\partial I)/(\partial z')$ and $\partial/(\partial z)$ $\{\log |z - z'|\}$ are not square integrable.

However, if the antenna has no edges (and the effect of end caps is neglected), then $(\partial I)/(\partial z')$ is everywhere bounded and square integrable. Even then we show that P_2 is unbounded.

Sneddon has shown through theorem 8 [*Sneddon*, 1962, p. 234] that if $f(z)$ is square integrable over $-L/2 \leq z \leq +L/2$ and zero everywhere else, then the formula

$$\bar{f}_H(z) = -\frac{1}{\pi} \frac{d}{dz} \int_{-\infty}^{+\infty} f(t) \log \frac{|t - z|}{|t|} dt \tag{17}$$

defines almost everywhere a function $\bar{f}_H(z)$ which is also square integrable, and

$$\| f(z) \|_{\mathscr{L}^2} = \| \bar{f}_H(z) \|_{\mathscr{L}^2}$$

Hence

$$\| P_2 \| \simeq \frac{1}{\pi a} \left\| \frac{\partial I}{\partial z} \right\|$$

Even though $(\partial I)/(\partial z)$ is always bounded, the ratio $\| \partial I/\partial z \| / \| I \|$ is not bounded, and hence $P = P_1 + P_2$ is an unbounded operator.

Therefore even if the antenna has no sharp edges, the operator P is unbounded under the $\mathscr{L}^2$ norm. Hence the Pocklington E field operator, $P = P_1 + P_2$, is unbounded both under the $\mathscr{L}^2$ and the Chebyshev norm. Since it is numerically difficult to solve an unbounded operator equation, perhaps that is why Hallen considered the potential equation. The operator for Hallen's integral equation is a bounded operator and hence easy to solve numerically. This we show next.

3. PROPERTY OF THE HALLEN OPERATOR

Hallen transformed Pocklington's equation as given by (2) into the following integral equation

$$\int_{-L/2}^{+L/2} dz'\, I(z')\, G(z, z') = D \cos kz + F \sin kz + \frac{jk}{\omega} \int_{-L/2}^{z} E^i_{\tan}(z') \sin k(z - z')\, dz' \tag{18}$$

where D and F are obtained from the boundary conditions (i.e., $I(\pm L/2) = 0$). We define the Hallen operator as (from (2))

$$HI = \int_{-L/2}^{+L/2} dz'\, I(z')\, G(z, z') = P_1 I \tag{19}$$

Hence the operator H is bounded under the $\mathscr{L}^2$ and the Chebyshev norm. Also, H is a compact operator under the $\mathscr{L}^2$ norm.

It is important to note however, that the unknown $I(z')$ in (18) is hidden in D and F. To illustrate this further, if we consider a delta gap excitation for the antenna, then (18) becomes [*Wu*, 1969, p. 325]

$$\int_{-L/2}^{+L/2} dz'\, I(z')\, G(z, z') = A \sin k\,|z| + D \cos kz$$

where A is known and D is unknown. Observe at $z = 0$

$$D = \int_{-L/2}^{+L/2} dz'\, I(z')\, G(0, z')$$

If the operator H is bounded, then D will be finite. If one wishes, then perhaps one can transfer D to the left-hand side of the equation and thus form an additional part of the operator H. But since D is a part of the operator H, whatever bound holds for H also holds for D.

Next we estimate a bound for $\| H \|$ both under the $\mathscr{L}^2$ and Chebyshev norm. We observe

$$\| H \|_T \leq \max_{-L/2 \leq z \leq +L/2} \frac{1}{2\pi} \int_{-L/2}^{+L/2} dz' \int_0^{2\pi} d\phi\, \frac{\exp(-jkR)}{R}$$

$$\lesssim \max_{-L/2 \leq z \leq +L/2} \frac{1}{2\pi} \int_{-L/2}^{+L/2} dz' \int_0^{2\pi} d\phi\, \frac{1}{R} \approx \frac{L}{a} \tag{20}$$

Schelkunoff has obtained a similar estimate for (16) but utilizing the reduced kernel. By utilizing the reduced kernel [*Schelkunoff*, 1952, p. 144] one obtains

$$\| H \|_T \lesssim \max\left[2 \log \frac{L}{a}, \log \frac{2L}{a}\right] + \text{terms of the order of } a^2$$

$$\lesssim 2 \log \frac{L}{a} \qquad \frac{L}{a} \gg 1 \tag{21}$$

Under the $\mathscr{L}^2$ norm we obtain

$$\| H \|_{\mathscr{L}^2} \leq \left[\int_{-L/2}^{+L/2} dz \int_{-L/2}^{+L/2} dz' \cdot \left\{\int_0^{2\pi} d\phi\, \frac{\exp(-jkR)}{2\pi R}\right\}^2\right]^{1/2}$$

$$\lesssim \left[\int_{-L/2}^{+L/2} dz \int_{-L/2}^{+L/2} dz' \left\{\int_0^{2\pi} \frac{d\phi}{2\pi R}\right\}^2\right]^{1/2} \approx \frac{L}{a} \tag{22}$$

An estimate similar to (17) can be obtained for the reduced kernel.

In our analysis, we do not consider the reduced kernel because *Schelkunoff* [1952, p. 149] has shown that an integral equation with the reduced kernel mathematically has no solutions (i.e., the excitation is not in the range of the operator). However, if one solves such problems numerically, one does indeed get a numerical solution. In section 5 we discuss the convergence properties of the numerical solutions in which the excitation is not in the range of the operator.

As the Hallen integral operator H is a bounded operator, unlike the Pocklington E field operator (which is unbounded), it may be computationally much easier to solve Hallen's integral equation than Pocklington's equation.

4. SOLUTION OF HALLEN'S INTEGRAL EQUATION

4.1. *By Iterative Methods*

It is well known that if H is a compact invertible operator (under $\mathscr{L}^2$ norm) on an infinite dimensional space, then its inverse is often unbounded [*Stakgold*, 1979, p. 353]. Hence the problem of the solution of (18) in the $\mathscr{L}^2$ norm is ill posed. If a problem is ill posed under the $\mathscr{L}^2$ norm, then it is definitely ill posed under the Chebyshev norm. However, it can be regularized in the following way. We take (18) and cast it in the form

$$HI = Q \tag{23}$$

$$I_n = [U - \tau H] I_{n-1} + \tau Q \tag{24}$$

with a starting guess of $I_0 = Q$; U is the identity operator. The sequence I_n generated by (24) converges to a solution I which satisfies $HI = Q$ for all Q in the range of H [*Jones*, 1979, p. 196]. The sequence generated by (24) always converges to I provided

$$\| [U - \tau H] \| < 1$$

or

$$|\tau| \cdot \| H \| = \| \tau H - U + U \| \leq \| U - \tau H \| + \| U \| < 2$$

or

$$\frac{1}{|\tau|} > \frac{\| H \|}{2} \tag{25}$$

In (25), $\| H \|$ could be either the $\mathscr{L}^2$ or the Chebyshev norm depending on the type of convergence desired. For all values of $1/|\tau| > \| H \|/2$ and Q in the range of the operator H, the iterative process defined by (24) will always converge monotonically to a solution $I(z')$, if it exists. This has been shown in theorem 1 (in the appendix). By the terms of theorem 1, the iterative process will converge for any starting value I_0 if

$$\frac{1}{|\tau|_{\mathscr{L}^2}} > \frac{\| H \|_{\mathscr{L}^2}}{2} \quad \text{or} \quad \frac{1}{|\tau|_{\mathscr{L}^2}} \approx \frac{L}{a} \tag{26}$$

when convergence is desired in the $\mathscr{L}^2$ norm. For the convergence in the Chebyshev norm, it is required that

$$\frac{1}{|\tau|_T} > \frac{\| H \|}{2} \quad \text{or} \quad \frac{1}{|\tau|_T} \approx \frac{L}{a} \text{ for the exact kernel} \tag{27a}$$

For the reduced kernel, however, we have

$$\frac{1}{|\tau|_T} > \log \frac{L}{a} \quad \text{or} \quad \frac{1}{|\tau|_T} \approx 2 \log \frac{L}{a} \tag{27b}$$

Hallen in his classic iterative scheme chose the value of τ as given by (27*b*) [*Schelkunoff*, 1952; *Wu*, 1969]. A detailed description of how D is solved for at each iteration is given [*Wu*, 1969, p. 326]. Observe that if $1/\tau > \| H \|/2$ or $\| U - \tau H \| < 1$, then the iterations defined by (8.149) and (8.150) of Wu would always converge for any starting I_0.

Other researchers have chosen different values of τ. For example, *Gray* [1966] chose

$$\frac{1}{\tau} = \text{Re}\left[2 \log \frac{L}{a} - 2\gamma - 2 \log \frac{kL}{2} - j\pi + 2Ei\left(+\frac{jkL}{2}\right)\right] \tag{28}$$

where γ is Euler's constant and Ei is the exponential integral. *King and Middleton* [1946, 1948] decided to make

$$\frac{1}{\tau} = \int_{-L/2}^{L/2} G\left(\frac{L}{2} - \frac{\lambda}{4}, z'\right) \sin k\left(\frac{L}{2} - z'\right) dz'$$
$$\approx \frac{2 \log L}{a} \text{ for the exact kernel} \tag{29}$$

whereas *Siegel and Labus* [1934] chose

$$\frac{1}{\tau} = 2 \log \frac{L}{a} - \text{Cin}(kL) - 1 - \frac{\sin kL}{kL} \tag{30}$$

where Cin is the special form of the cosine integral. Finally, *Schelkunoff* [1952], after a careful analysis, decided

$$\frac{1}{\tau} = 2 \log \frac{L}{a} - \text{Cin}(kL) - 1 - \frac{\sin kL}{kL} - j\,\text{Si}(kL) + j\frac{1 - \cos kL}{kL} \tag{31}$$

In general, it really does not make any difference whatsoever, what value of τ one chooses, one is guaranteed to have pointwise convergence or convergence in the mean, depending on whether one chooses τ according to (27) or (26). This of course assumes that a solution to the problem exist, i.e., Q is in the range of H.

In summary, the iterative method converts $HI = Q$, a Fredholm equation of the first kind, to $I_n = BI_{n-1} + \tau Q$. a Fredholm equation of the second kind. The advantage of the equation of the second kind is that not only is $\| [U - B] \|$ bounded, but also its inverse $\| [U - B] \|$ is bounded, provided unity is not an eigenvalue of B and $\| B \|$ is bounded. Mathematically, one has regularized the problem by the introduction of the parameter τ. With this regularization scheme the convergence of the sequence I_n is monotonic. The method converges as long as $\tau < a/L$.

Finally, we conclude by noting that as the iterative process continues the unknown D and F in (18) are determined as outlined by *Wu* [1969].

4.2 *By Galerkin's Method*

The next generation of the methods were developed primarily by *Harrington* [1968] under the generic name of 'moment methods.' This very popular versatile method has been excellently documented by *Harrington* [1968]. In Galerkin's method, the unknown function I is expressed as

$$I_N(z) = \sum_{i=1}^{N} \alpha_i \Psi_i(z) \tag{32}$$

where $\Psi_i(z)$ are known functions which may extend from $-L/2 \leq z \leq +L/2$ or could span only a partial portion of the domain of z, i.e., $-L/2 < \sigma_1 \leq z \leq \sigma_2 < +L/2$. In the former case, Ψ_i's become entire domain functions, whereas in the latter Ψ_i's are called subdomain basis functions. We solve for $I_n(z)$ by solving for the unknowns α_i in (32). We also convert the infinite dimensional problem $HI = Q$ to a finite dimensional problem by replacing I with I_N, i.e., we solve the following equation $\sum_{i=1}^{N} \alpha_i H\Psi_i = Q$ in the finite dimensional space spanned by the basis functions Ψ_i, $i = 1, 2, \cdots, N$. We next find a unique solution in finite dimensional space by weighting the residual $\sum_{i=1}^{N} \alpha_i H\Psi_i - Q$ to zero in the following way

$$\sum_{i=1}^{N} \alpha_i \langle H\Psi_i, \Psi_j \rangle = \langle Q, \Psi_j \rangle \qquad j = 1, 2, \cdots, N \tag{33}$$

In a matrix form

$$[G]\,[\alpha] = [V] \tag{34}$$

where $[G] = [\langle H\Psi_i, \Psi_j \rangle]$ and

$$[V] = [\langle Q, \Psi_j \rangle]$$

and the inner product is defined as

$$\langle \phi_i, \phi_j \rangle = \int_{-L/2}^{+L/2} dz\, \phi_i(z)\, \phi_j(z)$$

The unknown α's in (34) are obtained as

$$[\alpha] = [G]^{-1}[V] \tag{35}$$

The next question that normally arises is whether the sequence I_N defined in (32) approaches any limit I as $N \to \infty$. And secondly, whether I satisfies the equation $HI = Q$. We cannot talk about convergence in the Chebyshev metric (as defined in (8)) because a Chebyshev norm cannot be derived from an inner product [*Stakgold*, 1979, p. 272]. In other words, in an inner product space we cannot define a Chebyshev norm. Hence we shall be talking about only the $\mathscr{L}^2$ norm for Galerkin's method. So we shall be discussing convergence in the mean. Galerkin's method guarantees the weak convergence of the residuals (from (33)), i.e.,

$$\lim_{N \to \infty} \langle HI_N - Q, \Psi_j \rangle \to 0 \qquad j = 1, 2, \cdots, N \tag{36}$$

However, if H is a bounded operator (i.e., $\| H \|_{\mathscr{L}^2} \leq$ const $< \infty$, then (36) implies strong convergence of the residuals, i.e.,

$$\lim_{N \to \infty} \| HI_N - Q \|_{\mathscr{L}^2} \to 0 \tag{37}$$

This has been proved by *Mikhlin* [1964]. Physically, (37) implies that as $N \to \infty$, the total potential on the surface of the conductor for Hallen's method converges to zero in a least squares fashion.

Unfortunately, in Galerkin's method the convergence of the residuals to zero in (37) does not imply the convergence of I_N to a solution I of $HI = Q$. The convergence of $I_N \to I$ in the domain of H is possible if and only if $\| H^{-1} \|_{\mathscr{L}^2}$ is bounded, as

$$\| I_N - I \|_{\mathscr{L}^2} \leq \| H^{-1} \|_{\mathscr{L}^2} \cdot \| HI_N - Q \|_{\mathscr{L}^2} \tag{38}$$

So if $\| H^{-1} \|_{\mathscr{L}^2}$ is unbounded, even though the residuals go to zero, the sequence of solutions I_N may not converge to I. This is in contrast to the iterative methods, where monotonic convergence to I is guaranteed if τ and Q are chosen as prescribed.

Since $\| H^{-1} \|_{\mathscr{L}^2}$ is unbounded in this case, the application of Galerkin's method to $HI = Q$ may not guarantee that $\| I_N - I \|_{\mathscr{L}^2} \to 0$ as $N \to \infty$. In other words, there is no quantitative way to describe the convergence of $I_N \to I$ as various expansion functions are chosen for Ψ_i. Hence we address the question: For a fixed order of approximation N, how should one choose a set of expansion functions Ψ_i such that the round off and the truncation error in the numerical computation of α in (32) is a minimum?

Suppose the Gram matrix E is generated by the basis functions $[E_{ij} = \langle \Psi_i, \Psi_j \rangle]$; then we show in the appendix (theorem 2) that

$$\text{cond}\,[G] \leq \text{cond}\,[\hat{H}].\ \text{cond}\,[E] \tag{39}$$

i.e., the condition number of the Galerkin matrix G in (34) is bounded by the condition number of the operator $\hat{H}$ in the finite N dimensional space and the Gram matrix E. Equation (39) is valid only in the finite N dimensional space spanned by Ψ_i. It is important to note that even though H may not have any eigenvalues in an infinite dimensional space, it has at least an eigenvalue on a finite dimensional space [*Stakgold*, 1969, p. 332]. If the homogeneous equation $HI = 0$ has only the trivial solution $I = 0$ and $\| H \|_{\mathscr{L}^2}$ is bounded, then cond $[\hat{H}] < \infty$ and the inequality in (39) has meaning because the right-hand side of (39) can never be infinity.

So (39) directly implies the following:

1. Use of an orthonormal set of basis functions Ψ_i for the current implies

$$\text{cond}\,[G] \leq \text{cond}\,[\hat{H}] \tag{40}$$

i.e., the problem would not be worse conditioned than the original problem. Note that cond $[E] = 1$

for subdomain basis functions like pulses or entire domain orthonormal basis functions like $2^{1/2}/\pi \sin(m\pi z)$ for $m = 1, 2, \cdots, N$. Equation (40) also implies that the solution of $HI = Q$ by Galerkin's method in a finite dimensional space may be a better conditioned problem than the original problem posed in the finite dimensional space N.

Also, from (40) there is no way to tell whether the Galerkin matrix G associated with the entire domain basis functions would be more ill conditioned than the Galerkin matrix associated with the pulse functions.

2. Use of subdomain basis functions like triangles or piecewise sinusoids may deteriorate the condition number of the Galerkin matrix $[G]$ from that of the original problem. This is because cond $[E] > 1$ for these cases.

For the case when Ψ_i's are chosen as piecewise triangles, then E is a tridiagonal matrix of the form

$$\begin{bmatrix} P & Q & 0 & \cdot \\ Q & P & Q & \cdot \\ 0 & Q & P & \cdot \end{bmatrix}$$

where $P = 2\Delta z/3$ and $Q = \Delta z/6$ and $\Delta z = L/(N + 1)$. Since the jth eigenvalue of a tridiagonal matrix is given by [*Jones*, 1979, p. 70]

$$\lambda_j = P + 2Q \cos\left(\frac{j\pi}{N + 1}\right)$$

we have

$$\text{cond } [E]_{\text{triangles}} \le \frac{|P| + 2|Q|}{|P| - 2|Q|} = 3 \tag{41}$$

Hence for all values of N the Galerkin matrix due to piecewise triangle expansion functions may have a condition number which at most can be 3 times that of the original problem, i.e.,

$$\text{cond } [G] \le 3 \text{ cond } [H]$$

For the piecewise sinusoids, however,

$$P = \frac{2k\,\Delta z - \sin 2k\,\Delta z}{2k \sin^2 k\,\Delta z}$$
$$Q = \frac{\sin k\,\Delta z - k\Delta z \cos k\,\Delta z}{2k \sin^2 k\,\Delta z} \tag{42}$$

In this case cond $[E]$ is bounded by

$$\text{cond } [E]_{\text{sinusoids}} \le \frac{|2k\,\Delta z - \sin 2k\,\Delta z| + 2|\sin k\,\Delta z - k\,\Delta z \cos k\,\Delta z|}{|2k\,\Delta z - \sin 2k\,\Delta z| - 2|\sin k\,\Delta z - k\,\Delta z \cos k\,\Delta z|} \tag{43}$$

In the limit $\Delta z \to 0$

$$\text{cond } [E]_{\text{sinusoids}} \le 3 \tag{44}$$

Thus (44) implies that as the dimension of the problem becomes large, the Galerkin matrix due to piecewise sinusoids is no less numerically ill conditioned than the matrix produced by piecewise triangles. It may be quite possible that for a particular value of N the Galerkin matrix due to piecewise sinusoidal functions may be better conditioned than that of the piecewise triangles or even that of the pulse functions.

In the above analysis an attempt has been made to provide a worst case theoretical bound for the condition number of the various matrices of interest.

It is important to stress that the problem we have addressed here is not which set of basis functions would provide the best approximation for the current, but which type of expansion functions would give rise to a well-conditioned Galerkin matrix G which will be easy to invert numerically. This is because truncation and round off error associated with the solution of (34) is directly related to cond $[G]$.

5. IS A SOLUTION POSSIBLE IF THE EXCITATION IS NOT IN THE RANGE OF THE OPERATOR?

We discuss the question of existence of a solution for the current on the antenna structure when we try to excite it with a source which is not in the range of the operator H. Clearly, if the excitation is not in the range of the operator, then mathematically a solution does not exist. But numerically one could always find a solution to the integral equation. This numerical solution has some very interesting properties as outlined in theorems 3 and 4 (in the appendix). If we try numerically to solve an integral equation $HI = Q$ with Q not in range of H, then the sequence of solutions I_N diverges even though the residuals $HI_N - Q$ associated with $HI = Q$ may approach zero monotonically. This has been proved in theorem 3 of the appendix. In theorem 4, we develop further properties of the solution I_N. There we prove that the sequence I_N indeed forms an asymptotic series. The asymptotic series has the property that it converges at first and then as more and more terms are included in the series, the series actually diverges. Even though the theorems 3 and 4 have been proved for the iterative methods, they are also valid for Galerkin's method. We now present some examples to illustrate when Q is in the range of operator and when it is not.

As an example, consider the radiation problem where an antenna of length L and radius a is excited by a delta gap at the center. If we consider Pocklington's equation, then clearly the delta function excitation is not in the range of the operator. Since the delta function is not square integrable, it is not in the range of the operator under both the $\mathscr{L}^2$ norm and the Chebyshev norm. Hence if one attempts to solve Pocklington's equation for a delta function excitation, then according to theorems 3 and 4 the sequence of solutions diverges and $\lim_{N\to\infty} \| I_N \| \to \infty$. This is indeed true, because the admittance approaches infinity as the capacitance of a delta gap is infinity.

Hallen's integral equation for a delta function excitation is given by [*Wu*, 1969, p. 321]

$$\int_{-L/2}^{+L/2} dz'\, I(z')\, G(z, z') = A \sin k|z| + D \cos kz \tag{45}$$

where A is a known constant and D is unknown. It is seen that the right-hand side of (45) has a discontinuous derivative with respect to z, whereas the left-hand side has a continuous derivative with respect to z. Hence the delta function excitation is not in the range of the operator. Perhaps this is the reason why the solution yielded by the iterative methods of Hallen [*Schelkunoff*, 1952] and *King and Middleton* [1946, 1948] seemed to diverge as the solution progressed. Whether an aribtrary excitation is in the range of the operator is difficult to verify both theoretically and numerically. When the excitation is not in the range of the operator, we obtain a solution which diverges in an asymptotic sense, i.e., the solution seems to converge at first and then diverges. However, this postulate may be difficult to verify numerically for certain problems. As an example, consider the partial sum of the series

$$S_N = 1 + \frac{1}{2} + \frac{1}{3} + \frac{1}{4} + \cdots + \frac{1}{N}$$

The partial sum S_N diverges as $N \to \infty$. This is because if we look at the following M terms of the series we find

$$\frac{1}{M+1} + \frac{1}{M+2} + \frac{1}{M+3} + \cdots + \frac{1}{2M}$$
$$> \frac{1}{2M} > \frac{1}{2M} + \frac{1}{2M} + \cdots + \frac{1}{2M} = \frac{1}{2}$$

Hence

$$S_\infty > 1 + \frac{1}{2} + \frac{1}{2} + \frac{1}{2} + \cdots = \infty$$

However, if we program the series on the computer and ask the computer to give us a result when the addition of the $N + 1$ term does not change the partial sum by 10^{-10} (say), we would get a convergent result!

In conclusion, we must try to learn theoretically as much about the problem as possible. Numerical methods may be applied as a last resort as it may be the only way to obtain a solution easily. The convergence of the numerically computed results is determined to a large extent by the theoretical analysis of the problem rather than apparent convergences in numerical computations.

6. CONCLUSIONS

In summary, we have brought out the following features.

1. The thin wire E field Pocklington integral operator is unbounded, whereas the Hallen E field operator is bounded.
2. The inverse operator for Hallen's equation is unbounded.
3. A discussion of the various iterative methods showing how the Fredhold equation of the first kind has been converted to a Fredholm equation of the second kind is presented.
4. The conditions under which the iterative methods converge both for the $\mathscr{L}^2$ norm and the Chebyshev norm have been presented.
5. The monotonic rate of convergence of the sequence of solutions associated with iterative methods has been established for certain values of τ and for $Q \in R(H)$.
6. The numerical stability in the solution of the matrix equations for Galerkin's method for various expansion functions is examined.
7. The sequence of solutions I_N forms an asymptotic series for both the iterative and Galerkin's method when the excitation is not in the range of the operator.

APPENDIX

Theorem 1. For all $Q \in$ range of H, the sequence I_n generated by the recursion

$$I_{n+1} = [U - \tau H]\, I_n - \tau Q \triangleq BI_n + Q' \tag{46}$$

where U is the identity matrix (and $\| \tau H \| < 2$) with the initial guess $I_0 = Q'$ converges to I_e (the exact solution, if it exists) in the norm, i.e., $\lim_{n\to\infty} \| I_n - I_e \| \to 0$ and the convergence is strictly monotone increasing, i.e., $I_k \uparrow I_e$.

Proof. The iterative process (46) converges as long as the norm of B is less than 1. It is clear that if $|\tau| \cdot \| H \| < 2$, then $\| B \| = \| U - \tau H \| < 1$. Now we have

$$I_e - I_{n+1} = I_e - BI_n - Q' = B[I_e - I_n] \tag{47}$$

By taking the norm of both sides and simplifying,

$$\| I_e - I_{n+1} \| \leq \| B \| \cdot \| I_e - I_n \| \leq \{\| B \|\}^{n+1} \cdot \| I_e - I_0 \| \tag{48}$$

Since $\| B \| < 1$, as $n \to \infty$ we have

$$\lim_{n\to\infty} \| I_e - I_{n+1} \| = 0 \tag{49}$$

and thus I_{n+1} converges to the exact solution.

That $I_k \uparrow I_e$ is seen easily as

$$\varepsilon_{n+1} \triangleq I_e - I_{n+1} \tag{50}$$

and

$$\varepsilon_n \triangleq I_e - I_n \tag{51}$$

are related by

$$\varepsilon_{n+1} = B\varepsilon_n \tag{52}$$

and so

$$\| \varepsilon_{n+1} \| \leq \| B \| \cdot \| \varepsilon_n \| \leq \| \varepsilon_n \| \tag{53}$$

and with equality if and only if $\varepsilon_n = 0$. It follows that if n_0 is the smallest integer for which $\| \varepsilon_{n+1} \| = \| \varepsilon_n \|$, then $\varepsilon_n = 0$ for $n \geq n_0$, and $\| \varepsilon_{n+1} \| < \| \varepsilon_n \|$ for $n < n_0$, i.e., $I_n \uparrow I_e$, and theorem 1 is proved.

Theorem 2. Consider the operator equation $HI = Q$ in a finite dimensional space N. Let the unknown I be expanded in terms of the normalized basis functions ψ_i. Define cond $[\hat{H}] = (\lambda_{\max}[\hat{H}])/(\lambda_{\min}[\hat{H}])$ in the given N dimensional space.

Let cond $[G]$ and cond $[E]$ be the condition numbers of the Galerkin matrix $[G_{ij} = \langle \hat{H}\psi_i, \psi_j \rangle]$ and of the Gram matrix $[E_{ij} = \langle \psi_i, \psi_j \rangle]$, respectively. Then according to *Richter* [1978],

$$\text{cond } [G] \leq \text{cond } [\hat{H}]. \text{ cond } [E] \tag{54}$$

Proof. Let $I = \sum_{i=1}^{N} \alpha_i \psi_i$; then from *Stakgold* [1979, p. 341]

$$|\langle \hat{H}I, I \rangle| = \left| \sum_{i=1}^{N} \sum_{j=1}^{N} \alpha_i \bar{\alpha}_j \langle \hat{H}\psi_i, \psi_j \rangle \right| \leq \| \hat{H} \| \cdot \left\| \sum_{i=1}^{N} \alpha_i \psi_i \right\|^2$$

$$= \| \hat{H} \| \cdot \langle EI, I \rangle \leq \| \hat{H} \| \cdot \lambda_{\max}[E] \cdot \| \alpha \|^2 \tag{55}$$

Since $\langle I, I \rangle = \| \alpha \|^2$ we have

$$\frac{|\langle \hat{H}I, I \rangle|}{\langle I, I \rangle} \leq \| \hat{H} \| \cdot \lambda_{\max}[E] \tag{56}$$

from which it follows that

$$\lambda_{\max}[G] \leq \| \hat{H} \| \cdot \lambda_{\max}[E] \tag{57}$$

Also, since

$$|\langle \hat{H}I, I \rangle| \geq \frac{\left\| \sum_{i=1}^{N} \alpha_i \psi_i \right\|}{\| \hat{H}^{-1} \|}$$

$$= \frac{\langle EI, I \rangle}{\| \hat{H}^{-1} \|} \geq \frac{\lambda_{\min}[E] \cdot \| \alpha \|^2}{\| \hat{H}^{-1} \|} \tag{58}$$

so

$$\frac{|\langle \hat{H}I, I \rangle|}{\langle I, I \rangle} \geq \frac{\lambda_{\min}[E]}{\| \hat{H}^{-1} \|} \tag{59}$$

from which it follows that

$$\lambda_{\min}[G] \geq \frac{\lambda_{\min}[E]}{\| \hat{H}^{-1} \|} \tag{60}$$

Hence we have

$$\frac{\lambda_{\max}[G]}{\lambda_{\min}[G]} \leq \| \hat{H} \| \cdot \| \hat{H}^{-1} \| \cdot \frac{\lambda_{\max}[E]}{\lambda_{\min}[E]}$$

$$\text{cond } [G] \leq \text{cond } [\hat{H}] \cdot \text{cond } [E] \tag{61}$$

Theorem 3. If $Q \notin R(H)$, then the sequence of approximations I_n generated by

$$I_{n+1} = BI_n + Q' \tag{62}$$

with the initialization $I_0 = Q'$ yields the following relationships

$$\text{(i)} \quad \lim_{n\to\infty} \| R_{n+1} - R_n \| = 0 \qquad R_n = HI_n - Q \tag{63}$$

and

$$\text{(ii)} \quad \lim_{n\to\infty} \| I_n \| = \infty \tag{64}$$

Proof. (i) We have

$$R_{n+1} - R_n = H[I_{n+1} - I_n) \tag{65}$$

Since

$$I_0 = Q'$$

$$I_1 = BI_0 + Q' = (B + U)Q'$$

$$\vdots$$

$$I_n = [B^n + B^{n-1} + \cdots + B + U]Q'$$

then

$$I_{n+1} - I_n = B^{n+1}Q$$

Since the operator H is bounded [i.e., $\| H \| < M$], we have $\| R_{n+1} - \| R_n \| = HB^{n+1}Q \leq M \| B^{n+1}Q' \| \leq M \cdot \| B^{n+1} \| \cdot \| Q' \|$. Hence $\lim_{n\to\infty} \| R_{n+1} - R_n \| = 0$ as $\lim_{n\to\infty} \{\| B^n \|\} \to 0$.

(ii) If $\lim_{n\to\infty} \| I_n \| = \infty$ does not hold, then we have $\lim_{n\to\infty} \| I_n \| < \infty$ (i.e., a bounded sequence). Thus there is a subsequence I'_n which is bounded in norm. Now if we put the operator equation in a Hilbert space setting (now we can only talk about the $\mathscr{L}^2$ norm), and since a Hilbert space is weakly compact [*Mikhlin*, 1964] one can always extract from I'_n another sequence I''_n which converges weakly to some element I of the Hilbert space, i.e., $I''_n \xrightarrow{\omega} I$. Also, we have from (i) $\lim_{n\to\infty} HI_n \xrightarrow{S} Q$ (strong convergence in norm). However, as H is a bounded operator, we have

$$\lim_{n\to\infty} HI''_n \xrightarrow{\omega} HI \tag{66}$$

and also

$$\lim_{n\to\infty} HI_n \xrightarrow{S} Q \tag{67}$$

Since the weak and strong limits of a sequence must coincide, $HI''_n = Q$. This means $Q \in R(H)$, a contradiction.

Theorem 4. The sequence I_n as derived in (ii) of theorem 3 indeed forms an asymptotic series (i.e., for obtaining a meaningful solution the series has to be truncated after a finite number of terms, otherwise the results may be worse).

Proof. To demonstrate the source of divergence in I_n, we assume

$$HI_0 = Q_0 \qquad \text{with} \quad I_1 = Q_0 + \Delta Q$$

then

$$I_n = I_0 - [B]^n\, I_0 + \theta_{n-1} \qquad n \geq 1 \tag{68}$$

where

$$\theta_{n-1} = \sum_{i=0}^{n-1} [B]^i\, \Delta Q \tag{69}$$

If $Q_0 + \Delta Q \notin R(H)$, then by theorem 3, $\lim_{n\to\infty} \| \theta_{n-1} \| = \infty$, since $\lim_{n\to\infty} [B]^n I_0 = 0$. Observe that this holds irrespective of the size of $|\Delta Q|$. Now the error in the iterates is obtained as

$$\varepsilon_n = I_0 - I_n = [B]^n\, I_0 - \theta_{n-1} \tag{70}$$

Note that the norm of the first term is monotonically decreasing, and thus it is evident that the algorithm should be terminated after a certain optimum number of steps. Unfortunately, the exact number of iterations depends on the particular Q under consideration and the growth rate of $\| \theta_{n-1} \|$ versus the decay rate of $\| [B]^n\, I_0 \|$.

Acknowledgments. This work has been supported by the Office of Naval Research under contract N00014-79-C-0598.

REFERENCES

Bloom, F. (1981), Ill-posed problems for integro-differential equations in mechanics and electromagnetic theory, in *SIAM Studies in Applied Mathematics*, vol. 3, Society for Industrial and Applied Mathematics, Philadelphia, Pa.

Gray, M. C. (1966), A modification of Hallen's solution of the antenna problem, *J. Appl. Phys.*, *15*, 61–65.

Harrington, R. F. (1968), *Field Computation by Moment Methods*, Macmillan, New York.

Jones, D. S. (1979), *Methods in Electromagnetic Wave Propagation*, Clarendon, Oxford.

Jones, D. S. (1981), Note on the integral equation for a straight wire antenna, *IEE Proc.*, *128*(2), 114–116.

King, R. W. P., and D. Middleton (1946), The cylindrical antenna: Current and impedance, *Q. Appl. Math.*, *3*, 302–305.

King, R. W. P., and D. Middleton (1948), Corrections, *Q. Appl. Math.*, *6*, 192.

Mikhlin, S. G. (1964), *The Minimum of a Quadratic Functional*, Holden-Day, San Francisco, Calif.

Richter, G. R. (1978), Numerical solution of integral equations of the first kind with nonsmooth kernels, *SIAM J. Numer. Anal.*, *15*(3), 511–522.

Schelkunoff, S. A. (1952), *Advanced Antenna Theory*, John Wiley, New York.

Siegel, L., and J. Labus (1934), Scheinwiderstand von Antennen, *Hochfrequenztech. Electroakust.*, *43*, 166–172.

Sneddon, I. N. (1962), *The Use of Integral Transforms*, McGraw-Hill, New York.

Stakgold, I. (1979), *Green's Functions and Boundary Value Problems*, John Wiley, New York.

Tikhonov, A. N., and V. I. Dimitriyev (1968), A method for calculating the current distribution in a system of linear dipoles and of the radiation pattern in such a system (in Russian), *Univ. Vychisl. Trentr. Sobrink Rabot*, *10*, 3–8. (English translation, FSTC-HT-23-261-70, AD 709614, Defense Documentation Center, Cameron Station, Alexandria, Va.)

Wu, T. T. (1969), Introduction to linear antennas, in *Antenna Theory*, Part 1, edited by R. Collin, McGraw-Hill, New York.

Equivalent Line Current for Cylindrical Dipole Antennas and its Asymptotic Behavior

R. E. COLLIN, FELLOW, IEEE

***Abstract*—The concept of an equivalent line source for representing the current on a cylindrical dipole antenna is introduced. It is shown that the Fourier series coefficients of this equivalent line source are related to those of the actual antenna current by exponentially growing factors that grow rapidly for the higher order harmonics. This is used to explain why Hallén's integral equation using an approximate kernel does not have an exact solution. The asymptotic behavior of the Fourier coefficients is established. An explanation of why approximate solutions to the approximate integral equation often provide good results for the current and input impedance is also given.**

INTRODUCTION

Fig. 1 illustrates a cylindrical dipole antenna of length $2l_0$ and radius a. A uniform concentrated electric field is impressed over a small region at the center. When the wall thickness is negligible, Hallén's integral equation for the total current density $J(z')$ on the antenna is

$$\int_0^{2\pi}\int_{-l_0}^{l_0} G(z,z',\phi,\phi')J(z')\,dz' = C\cos k_0 z - \frac{jY_0}{2}\int_{-l_0}^{l_0} E_i(z')\sin k_0|z-z'|\,dz'. \tag{1}$$

In this equation E_i is the z directed incident field, $Y_0 = (\epsilon_0/\mu_0)^{1/2}$ and C is a constant to be determined so that the edge condition at $z = \pm l_0$ is satisfied, i.e., such that $J(\pm l_0) = 0$. For a simplified mathematical model we will assume that E_i is constant on the surface of the antenna and equal to $V/2b$ over a band of length $2b$ at the center. In this case (1) becomes

$$\int_0^{2\pi}\int_{-l_0}^{l_0} G(z,z',\phi,\phi')J(z')\,dz' = C\cos k_0 z - \frac{jY_0V}{2k_0b}\begin{cases}1-\cos k_0b\cos k_0z, & -b\le z\le b\\ \sin k_0b\sin k_0|z|, & b\le|z|\le l_0.\end{cases} \tag{2}$$

In (1) and (2) G is the free space Green's function $e^{-jkR}/4\pi R$ where

$$R = [(z-z')^2 + 2a^2 - 2a^2\cos(\phi-\phi')]^{1/2}.$$

In the case of a vanishingly small bandlength $2b$ the impressed electric field $E_i = V/2b$ is replaced by a delta function $V\delta(z)$, and the last term in (2) becomes $(-j/2)Y_0V\sin k_0|z|$. The model used here is that used by many authors. It does not correspond to an impressed field across a finite gap as often stated. In the latter case the edge condition would require the input current to be zero and thus yield an infinite impedance.

The integral equation (1) is a standard scattering formulation. The scattered magnetic field $H_{s\phi}$ satisfies the relation

$$H_{s\phi}(a^+) - H_{s\phi}(a^-) = J \tag{3a}$$

where a^+ and a^- represent r as the surface $r = a$ is approached from the exterior or interior. On the interior surface the current

Manuscript received May 20, 1983; revised September 27, 1983.

The author is with the Electrical Engineering and Applied Physics Department, Case Institute of Technology, Cleveland, OH 44106.

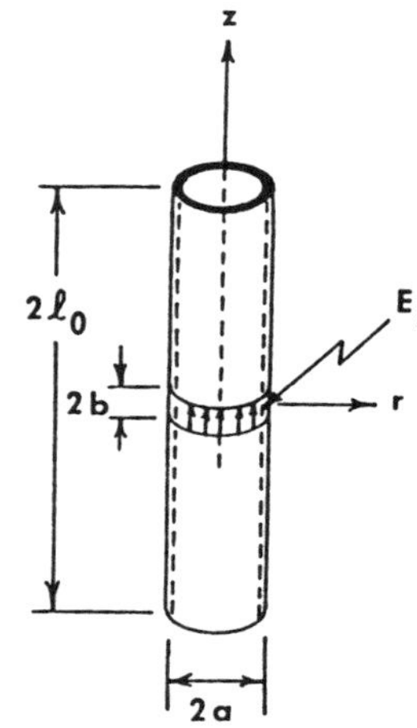

Fig. 1. The tubular cylindrical dipole antenna.

density is given by

$$-\mathbf{a}_r \times (\mathbf{H}_s + \mathbf{H}_i) = \mathbf{J}^- \tag{3b}$$

while on the exterior surface

$$\mathbf{a}_r \times (\mathbf{H}_s + \mathbf{H}_i) = \mathbf{J}^+ \tag{3c}$$

where $\mathbf{H}_i$ is the incident magnetic field. Near $z = 0$ the interior current is essentially zero so $J(0) \approx J^+(0)$ when $a \ll \lambda_0$.

Wu and King use a formulation in which the vector potential function A_z is the total potential, both incident and scattered [1], [2]. In this formulation there is an extra source term on the right side of (1) and (2). In their model the total electric field equals a constant over a vanishing small band on the outside. This field can be maintained by a small concentrated band of azimuthally directed magnetic current on the outside. This magnetic current loop is equivalent to a disk of uniform z directed current filaments from which the contribution to the total vector potential can be found. The disk of axially directed current elements has a double layer of charge associated with it and accounts for the discontinuity in the scalar potential in the Wu-King theory. In the Wu-King theory the incident field produced by the magnetic current loop could be calculated and then the scattering formulation could be applied.

In order to simplify the integral equation the distance parameter

$$R = [(z-z')^2 + 2a^2 - 2a^2\cos(\phi-\phi')]^{1/2}$$

in the Green's function $G = e^{-jk_0R}/4\pi R$ is often replaced by $R_0 = [(z-z')^2 + a^2]^{1/2}$. This is usually done on the basis that the current $2\pi aJ(z')$ can be assumed to be concentrated on the z axis as a line source $I_A(z') = 2\pi aJ(z')$. Elliott makes this approximation on the basis of a demonstration that the error made in using the approximate kernel instead of the exact kernel is small [3]. However, the proof is based on the invalid assumption that the current $J(z')$ is slowly varying, an assumption which is incorrect at the input $z' = 0$ and the two ends where $J(z')$ can have a very rapid variation. Taylor and Wilton use an extended boundary condition argument to support the formulation [4]. This extended boundary condition does not yield a unique solution in general if the scattered field is made to cancel the incident field along a single line only.

Reprinted from *IEEE Trans. Antennas Propagat.*, vol. AP-32, no. 2, pp. 200–204, February 1984.

It is well-known and has been documented in the literature that the approximate integral equation does not have a solution [2], [5], [6]. It is a curious happening that in spite of this approximate solutions, analytic or numerical, of this nonvalid integral equation often gives results for the current distribution and input impedance that agrees very well with experimental data. Another interesting aspect is the well-known fact that for a delta function applied field the input current has a logarithmic singularity [2]. Nevertheless, approximate solutions are still apparently acceptable. From the point of view of numerical work the difference between $\sin k_0 |z|$ and the functions on the right side in (1) and (2) is very small and as long as the numerical work is limited to determining the current in an approximate manner only the logarithmic singularity does not show up. For thicker antennas when l_0/a is around 10, numerical computations often show rapidly rising current values near the input. The impedance calculated under these circumstances must be viewed with some caution as to its validity.

The main thrust of this communication is to view the use of the approximate integral equation from a different perspective, namely, rather than assuming that the actual antenna current $2\pi aJ(z')$ is concentrated on the z axis the question is asked whether or not there exists some equivalent line current that will produce the same field for $r \geqslant a$. If such a line current exists, its relationship to the total antenna current $2\pi aJ(z')$ needs to be established.

In the next section we will show that an equivalent line current must have Fourier coefficients that grow exponentially for the higher harmonic terms. Thus, this equivalent line current and along with it the approximate integral equation are useful only for those cases when the current distribution can be adequately approximated by a finite Fourier series. For a thin antenna as many as 100 harmonic terms or more may be used but for a thick antenna the number of harmonic terms that can be used is much less.

From a physical point of view it is quite clear that the field from a line current on the z axis can not have the singular behavior that is required at the edge where $r = a$, $z = \pm l_0$. This was pointed out by Schelkunoff many years ago [5].

EQUIVALENT LINE CURRENT

Let the solution to the integral equation (2) be $J(z)$. The Fourier transform of this current is $\hat{J}(w)$ where

$$\hat{J}(w) = \int_{-l_0}^{l_0} e^{jwz} J(z)\, dz. \tag{4}$$

We can solve for the vector potential $A_z(r, z)$ in the Fourier transform domain and relate $\hat{J}(w)$ to $\hat{A}_z(a, w)$. The result is

$$2\pi a\hat{J}(w) = \frac{4j\hat{A}_z(a, w)}{\mu_0 J_0(a\sqrt{k_0^2 - w^2}) H_0^2(a\sqrt{k_0^2 - w^2})}. \tag{5}$$

The same procedure may be applied to relate the Fourier transform $\hat{I}(w)$ of the equivalent line source to the same vector potential and gives

$$\hat{I}(w) = \frac{4j\hat{A}_z(a, w)}{\mu_0 H_0^2(a\sqrt{k_0^2 - w^2})}. \tag{6}$$

In the above J_0 is a Bessel function and H_0^2 is the Hankel function. A comparison of (5) and (6) shows that the equivalent line source that would produce the same field for $r \geqslant a$ has a Fourier transform given by

$$\hat{I}(w) = J_0(a\sqrt{k_0^2 - w^2})2\pi a\hat{J}(w). \tag{7}$$

For $k_0 a$ small and $w^2 \leqslant k_0^2$ the spectral content of the equivalent line source and the total antenna current is essentially equal. However, for $w^2 \gg k_0^2$ and such that $|w|a$ is of order unity or greater we have the asymptotic relationship, for w real,

$$\hat{I}(w) \sim \frac{e^{a|w|}}{\sqrt{|w|a2\pi}} 2\pi a\hat{J}(w). \tag{8}$$

Thus the spectral content of $\hat{I}(w)$ grows exponentially for the high spatial frequencies. This is the reason why the approximate integral equation does not have a solution. Essentially this same result was obtained by Wu [2]. The growth in the spectral density is physically caused by the fact that line currents that vary rapidly with z produce fields that decay rapidly in an exponential manner in the radial direction. Hence the high frequency spectral components of $I(z)$ must grow exponentially if these components are required to make a significant contribution to the field at $r = a$.

The current $J(z)$ can be expanded in a Fourier cosine series of the form

$$J(z) = \sum_{n=1,3\cdots}^{\infty} J_n \cos\frac{n\pi z}{2l_0}.$$

The equivalent line current has a similar expansion

$$I(z) = \sum_{n=1,3\cdots}^{\infty} I_n \cos\frac{n\pi z}{2l_0}.$$

The sampling theorem shows that the Fourier coefficients are given by the sample values of the Fourier transform, thus

$$l_0 J_n = \hat{J}(n\pi/2l_0)$$

$$l_0 I_n = \hat{I}(n\pi/2l_0).$$

It follows that

$$I_n = 2\pi aJ_n J_0(a\sqrt{k_0^2 - (n\pi/2l_0)^2}) \tag{9}$$

and that for large n

$$I_n \sim \frac{e^{n\pi a/2l_0}}{\sqrt{2\pi^2 na/2l_0}} 2\pi aJ_n \tag{10}$$

and hence the Fourier coefficients of the equivalent line source grow exponentially for large n if the J_n decrease algebraically only.

For a thin antenna, say $l_0/a = 100$, the factor multiplying $2\pi aJ_n$ in (10) equals 2 for n around 130. Thus the exponential growth only begins when the high order harmonics beyond $n = 130$ are considered. For a thick antenna with $l_0/a = 10$ the exponential growth takes over when n is small, around 13 for this specific example.

However, if the approximate integral equation is used for a thick antenna and a relatively large number of the I_n have been found, the corresponding Fourier coefficients for the actual antenna current may be obtained from (9) and this technique would allow a more realistic evaluation of the antenna current and input impedance to be made in terms of the J_n coefficients.

The discussion above shows in a fairly clear manner why the

approximate integral equation gives good results for thin antennas when analytical and numerical solutions are limited to finding only a relatively small number of the Fourier coefficients of the current distribution. For these the $2\pi a J_n$ are very nearly equal to the I_n.

Unfortunately, there is no criteria available as to where to stop the numerical solution such that the approximation yields valid results. Plots of input resistance and reactance as a function of the number of expansion functions used in the method of moments tend to show stable values for these parameters after some finite number of expansion functions have been employed. If the approximation is continually refined by increasing the number of expansion functions beyond this point the solution will diverge. If the applied field extends over a finite band of length $2b$ the antenna input current remains finite. However, solutions to the approximate integral equation must ultimately diverge if carried out to a sufficiently high order even if the applied field extends over a finite band.

The asymptotic behavior of the Fourier coefficients are dependent on the nature of the applied field and the edge conditions at the antenna ends and can be readily established as shown in the next section.

ASYMPTOTIC BEHAVIOR OF THE FOURIER COEFFICIENTS

Let the scattered electric field $E_z(a, z)$ be represented by $\psi(z)$ for z in the intervals $|z| \geqslant l_0$. For z close to $\pm l_0$ and assuming infinitly thin walls $\psi(z)$ behaves like $(z^2 - l_0^2)^{-1/2}$ as required by the edge condition. The Fourier transform $\hat{\psi}(w)$ is thus asymptotic to $w^{-1/2}$ for large w. In the interval $|z| < l_0$ the scattered electric field $E_z(a, z)$ equals the negative of the applied field $E_i(z)$. In the Fourier transform domain we have

$$j\omega\epsilon_0\mu_0 \hat{E}_z(a, w) = (k_0^2 - w^2)\hat{A}_z(a, w)$$
$$= -j\omega\epsilon_0\mu_0[\hat{E}_i(w) - \hat{\psi}(w)]. \quad (11)$$

For $|w|a$ large we can use the asymptotic expressions for the Bessel functions in (5) which along with (11) shows that the asymptotic value of $\hat{J}(w)$ is given by

$$\hat{J}(w) \sim 2jk_0Y_0 \frac{[\hat{E}_i(w) - \hat{\psi}(w)]}{w}. \quad (12)$$

For a delta function applied field $\hat{E}_i(w) = V$. Hence the asymptotic value of the Fourier coefficients for the tubular cylindrical antenna for this case is

$$J_n \sim \frac{j4k_0Y_0V}{n\pi} - \frac{j2k_0Y_0K}{l_0(n\pi/2l_0)^{3/2}} \quad (13)$$

where K is a suitable constant. The dominant coefficients give [7]

$$2\pi aJ(z) \sim j8k_0aY_0V \sum_{n=1,3,\cdots}^{\infty} \frac{\cos n\pi z/2l_0}{n}$$
$$= -j4k_0aY_0V \ln(\tan \pi z/4l_0) \quad (14)$$

which verifies that the input current has a logarithmic singularity [2].

When the applied field exists over a band of length $2b$ we have

$$\hat{E}_i(w) = V\frac{\sin wb}{wb} \quad (15)$$

and the corresponding J_n arising from this are asymptotic to $1/n^2$. In this case the input current remains finite and the Fourier coefficients are dominated by the second term in (13). However, the Fourier series for the equivalent line current will still diverge so the approximate integral equation will not have an exact solution.

The function $(\sin wb)/wb$ leads to a Fourier series which has coefficients proportional to n^{-2} for large n. This asymptotic behavior does not take over before wb is of order π which corresponds to $n > 2l_0/b$. If $b = a$ this occurs at about the same point that the exponential growth of the Fourier coefficients for the equivalent line current takes effect. It follows that the Fourier series coefficients for the case of a field applied over a length $2b$ are the same as for a delta function applied field up to harmonics or order $2l_0/b$, which for a narrow band can be as large as 100 or more.

SOLID CYLINDRICAL ANTENNA

For a solid cylindrical antenna with flat end caps the radial end cap currents also contribute to the z component of the scattered electric field. Let E_{z1} be due to J and E_{z2} be due to the radial currents. We then have $E_z = E_{z1} + E_{z2} = -E_i$ for $-l_0 < z < l_0$ and equals ψ otherwise. In the Fourier transform domain $(k_0^2 - w^2)\hat{A}_z(a, w) = j\omega\epsilon_0\mu_0\hat{E}_{z1}(a, w) = j\omega\epsilon_0\mu_0[-\hat{E}_i(w) + \hat{\psi}(w) - \hat{E}_{z2}(w)]$. In this case ψ behaves like $(z^2 - l_0^2)^{-1/3}$ and $\hat{\psi}$ is asymptotic to $w^{-2/3}$. In the Appendix it is shown that $\hat{E}_{z2}(w)$ has a component with the same asymptotic behavior. Thus, the second Fourier series coefficients given in (13) will have $(n\pi/2l_0)^{5/3}$ in place of $(n\pi/2l_0)^{3/2}$ in the denominator. Hence for the solid antenna the Fourier series converges somewhat faster. The radial currents must be included in order to obtain the correct edge behavior. In addition to currents satisfying the edge condition given above a current that is finite at the edge and flows from the cylindrical surface on to the end caps may be present. This current produces a contribution to E_{z2} that behaves like $(z - l_0)^{-1}$ as z approaches l_0. This singularity is cancelled by an equal and opposite one present in the field E_{z1} produced by the constant axial current at $z = \pm l_0$ as shown in the Appendix.

DISCUSSION AND CONCLUSION

The concept of an equivalent line current for a cylindrical antenna was introduced and its relationship to the total current on the antenna established. It was found that the Fourier coefficients for the equivalent line current are related to those for the antenna current by exponentially growing factors. For thin antennas the exponential growth does not take over until very high order harmonics are considered, n greater than 100 or more. It also was shown how the Fourier coefficients for the antenna current can be obtained from those for the equivalent line current. The asymptotic behavior of the Fourier coefficient was also established. The logarithmic singularity in the input current for a delta function applied field was verified from this asymptotic behavior.

The reason why approximate solutions of the approximate integral equation for a thin antenna with an applied delta function field are good is that the right side of (1) does not change very much for different applied fields as long as this field is highly concentrated near the center. Thus the Fourier series coefficients in the expansion of the current are very nearly the same for a fairly large number of the lowest order harmonics.

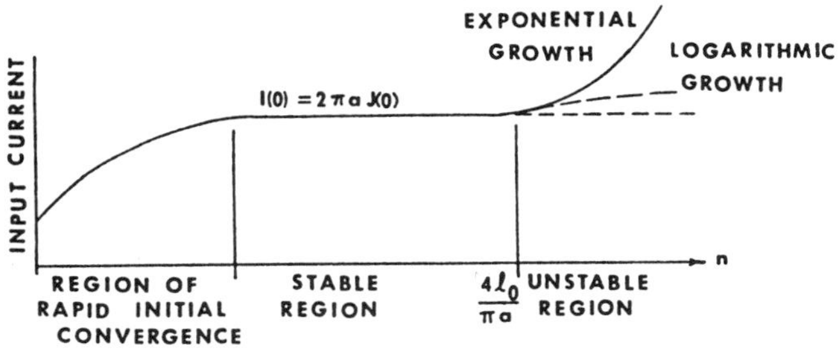

Fig. 2. Qualitative behavior of equivalent line current and antenna current at the input as a function of the number of harmonics used in its expansion.

It is only when very high order coefficients are determined that the influence of the exact behavior of the applied field manifests itself. For these coefficients the solution is no longer convergent. In the case of thin antennas, the Fourier series expansion for the equivalent line current converges rapidly enough initially that stable values for the input impedance can be found before the ultimate exponential growth of the higher order coefficients takes effect.

Fig. 2 shows in a qualitative way the expected behavior of the equivalent line current at $z = 0$ and the actual antenna current at $z = 0$. Initially a stage of relatively rapid convergence as a function of the number of harmonics used in the expansion occurs. This is followed by a stage where the current remains essentially constant and represents the converged result quite accurately. The final stage occurs when n is of order $4l_0/\pi a$ at which point the coefficients in the expansion of $I(z)$ grow exponentially. The Fourier series for $I(z)$ has the characteristics of an asymptotic series. The antenna current will remain stable as long as the applied field is not a delta function. If the applied field is a delta function the antenna current will eventually show a logarithmic growth. For a thin antenna the stable region is quite wide but for a thick antenna the stable region is narrow or may not even exist. In the latter case the Fourier coefficients for the antenna current can be found from those of the equivalent line source by the relation (9). The point at which the logarithmic growth begins to show up is around the same harmonic number at which the exponential growth in the Fourier coefficients for the equivalent line source takes effect.

APPENDIX

Consider a circular disk of radius a on which a radial current $J_r(r)\delta(z - l_0)$ exists. In the Fourier transform domain this current produces a radial vector potential function given by

$$\hat{A}_r(r, w) = -j \frac{\mu_0 \pi}{2} e^{jwl_0} \int_0^a J_r(r') J_1(r'\sqrt{k_0^2 - w^2})\, dr' \times H_1^2(r\sqrt{k_0^2 - w^2}), \quad r \geqslant a.$$

A reasonable approximation for the current density J_r that will satisfy the edge condition that it vanishes like the 2/3 power of the distance from the edge $r = a$ is

$$J_r(r) = I_0 \left(\frac{r}{a}\right)^2 [1 - (r/a)^2]^{2/3}$$

where I_0 is a constant.

With this current distribution the integration can be performed to give [8]

$$\hat{A}_r(r, w) = -j \frac{\mu_0 \pi}{2} e^{jwl_0} H_1^2(r\sqrt{k_0^2 - w^2}) \times aI_0 2^{2/3} \Gamma(5/3) J_{8/3}(a\sqrt{k_0^2 - w^2}) \cdot (a\sqrt{k_0^2 - w^2})^{-5/3}.$$

The z component of electric field contributed by $\hat{A}_r$ is given by

$$j\omega\epsilon_0\mu_0 \hat{E}_{z2}(r, w) = -jw \frac{1}{r} \frac{\partial}{\partial r} r\hat{A}_r(r, w).$$

From this expression it is readily found that $\hat{E}_{z2}(a, w)$ is asymptotic to $w^{-2/3}$. Since $E_{z2}(a, z)$ from J_r must be cancelled by $E_{z1}(a, z)$ arising from the z directed current $J(z)$ on the surface $r = a$ near the end $z = l_0$ it follows that $E_{z1}(a, z)$ from $J(z)$ must have the same asymptotic behavior. Hence for the solid cylindrical antenna the edge condition requires the Fourier coefficients to behave like $n^{-5/3}$ for large n.

A current distribution that remains finite at $r = a$ is $J_r = -(I_0/2\pi a^2)r$. Such a current distribution produces a radial vector potential function given by

$$\hat{A}_r(r, w) = \frac{j\mu_0}{4} I_0 e^{jwl_0} \frac{J_2(a\sqrt{k_0^2 - w^2}) H_1^2(r\sqrt{k_0^2 - w^2})}{\sqrt{k_0^2 - w^2}}.$$

From this potential function an axial electric field E_{z2} is produced that has an asymptotic behavior at $r = a$, as z approaches l_0, that may be derived as follows: $E_{z2}(a, z)$ is given by

$$E_{z2}(a, z) = \frac{I_0 Z_0}{k_0} \frac{\partial}{\partial z} \int_{-\infty}^{\infty} \left[\frac{1}{r} \frac{\partial}{\partial r} rH_1^2(r\sqrt{k_0^2 - w^2})\right]_a \times J_2(a\sqrt{k_0^2 - w^2}) e^{-jw(z - l_0)}\, dw.$$

First let $\lambda = w(z - l_0)$, $dw = d\lambda/(z - l_0)$ and let $z - l_0$ approach zero. λ then becomes large which allows the Bessel functions to be replaced by their asymptotic values for large arguments. If we then return to the variable w it is found that

$$E_{z2}(a, z) \sim \frac{-I_0 Z_0}{8\pi^2 k_0 a} \frac{\partial}{\partial z} \left[\int_{-\infty}^{\infty} \frac{e^{-jw(z - l_0)}}{\sqrt{k_0^2 - w^2}}\, dw + \frac{j}{a} \int_{-\infty}^{\infty} \frac{e^{-jw(z - l_0)}}{k_0^2 - w^2}\, dw\right].$$

The first integral equals $2jK_0(jk_0|z - l_0|)$ where K_0 is the modified Bessel function. The second integral can be evaluated by residue theory. In this way it is found that as z approches l_0

$$E_{z2} \sim \frac{jI_0 Z_0}{4\pi^2 k_0 (z - l_0) a} + \frac{jI_0 Z_0}{8\pi k_0 a^2} e^{-jk_0|z - l_0|} sg(z - l_0)$$

where sgu equals 1 or -1 according to whether u is greater or less than zero. The singularity in E_{z2} is too great to be allowed. However, if the axial current density on the cylindrical surface $r = a$ also has a constant term equal to $I_0/2\pi a$ it is readily shown that this current produces an axial electric field E_{z1} that behaves asymptotically like

$$E_{z1} \sim \frac{-jI_0 Z_0}{4\pi^2 k_0 (z - l_0) a}$$

when z approaches l_0 and hence cancels that coming from the radial current that remains finite at $r = a$. Note that the electric field near the end of a thin antenna would take on the characteristics of that from a point charge unless the current vanished at the end when a approaches zero.

ACKNOWLEDGMENT

This investigation was carried out while the author was on sabbatical leave at The Ohio State University. The courtesy and support provided by Professor H. C. Ko is greatly appreciated.

REFERENCES

[1] T. T. Wu and R. W. P. King, "The thick tabular transmitting antenna," *Radio Sci.*, vol. 2, pp. 1061–1065, 1967.

[2] T. T. Wu, "Introduction to linear antennas," in *Antenna Theory*, part I, R. E. Collin and F. J. Zucker, Eds. New York: McGraw-Hill, New York, 1969, ch. 8.

[3] R. S. Elliott, *Antenna Theory and Design*. Englewood Cliffs, NJ: Prentice Hall, 1981, ch. 7.

[4] C. D. Taylor and D. R. Wilton, "The extended boundary condition solution of the dipole antenna of revolution," *IEEE Trans. Antennas Propagat.*, vol. AP-20, pp. 772–776, Nov. 1972.

[5] S. A. Schelkunoff, *Advanced Antenna Theory*. New York: Wiley, 1952, ch. 5.

[6] R. H. Duncan and F. A. Hinchey, "Cylindrical antenna theory," *NBS J. Res.*, vol. 64D, no. 5, pp. 569–584, Sept.–Oct. 1960.

[7] R. E. Collin, *Field Theory of Guided Waves*. New York: McGraw-Hill, 1960, p. 580.

[8] I. S. Gradshteyn and I. M. Ryzhik, *Table of Integrals, Series, and Products*. New York: Academic, 1965, p. 688.

Part 4
Penetrable and/or Lossy Bodies

MODELING of penetrable bodies increases problem complexity because of the need to deal with the fields of sources radiating in at least two different media. The section begins with an introduction of the volume-current approach, approximate treatments using the impedance boundary condition, and various approaches for homogeneous and inhomogeneous bodies.

The fundamental concepts of MM solutions for penetrable bodies are developed in Richmond's early paper, ''Scattering by a dielectric cylinder of arbitrary cross section shape.'' The adaptation of Rytov and Leontovich's classis analysis of the impedance boundary condition (IBC) to integral equation formulation is the subject of Mitzner's paper on ''An integral equation approach to scattering from a body of finite conductivity.'' The effect of IBC on curved surfaces is also examined. Livesay and Chen apply numerical methods to a class of medically oriented problems in ''Electromagnetic fields induced inside arbitrarily shaped biological bodies.'' Richmond and Newman develop the reaction formulation in ''Dielectric coated wire antennas.'' The insulating shell is modeled by equivalent volume polarization currents. Both the shell and the wire are assumed electrically thin. The testing and expansion bases are filamentary sinusoidal functions, resulting in significant analysis simplifications. Mautz and Harrington, in ''Electromagnetic scattering from a homogeneous material body of revolution,'' explore the efficacy of various combinations of surface integral equations that arise for penetrable bodies. This is an especially well-written resource paper in this area. The MM formulation for penetrable bodies with rotational symmetry, where the penetrable region is discretely varying along the body, is described in ''Electromagnetic scattering from axially inhomogeneous bodies of revolution,'' by Medgyesi-Mitschang and Putnam. The paper provides the numerical implementation of the junction condition between regions, a numerical analog of the Meixner condition. ''A tetrahedral modeling method for electromagnetic scattering by arbitrarily shaped inhomogeneous dielectric bodies,'' by Schaubert *et al.*, solves a 3D integral equation for penetrable bodies with tetrahedral volume elements. The dielectric region must be piecewise homogeneous. Approximate Galerkin testing is used. The elements are chosen to satisfy the correct jump condition at the interfaces between different dielectric media. In ''Integral equation formulations for imperfectly conducting scatterers,'' by Medgyesi-Mitschang and Putnam, a Galerkin solution is developed for the Leontovich, resistive sheet, and magnetically conducting sheet boundary conditions. Special properties for the Galerkin operators are derived. Finally, a combined (electric and magnetic) field-surface integral formulation is developed for dielectric bodies by Umashankar *et al.* in ''Electromagnetic scattering by arbitrary shaped three-dimensional homogeneous lossy dielectric objects.'' A triangular patch is used to model the arbitrary surface of the body. An extensive discussion is given of the attributes and requirements of the patch basis functions.

Scattering by a Dielectric Cylinder of Arbitrary Cross Section Shape

JACK H. RICHMOND, SENIOR MEMBER, IEEE

Abstract—**The theory and equations are developed for the scattering pattern of a dielectric cylinder of arbitrary cross section shape. The harmonic incident wave is assumed to have its electric vector parallel with the axis of the cylinder, and the field intensities are assumed to be independent of distance along the axis. Solutions are readily obtained for inhomogeneous cylinders when the permittivity is independent of distance along the cylinder axis.**

Although other investigators have approximated the field within the dielectric body by the incident field, we treat the total field as an unknown function which is determined by solving a system of linear equations.

In the case of the dielectric cylindrical shell of circular cross section, this technique yields results which agree accurately with the exact classical solution. Scattering patterns are also presented in graphical form for a dielectric shell of semicircular cross section, a thin homogeneous plane dielectric sheet of finite width, and an inhomogeneous plane sheet. The effects of surface-wave excitation and mutual interaction among the various portions of the dielectric shell are included automatically in this solution.

I. Introduction

ALTHOUGH RIGOROUS solutions are available for scattering by homogeneous dielectric cylinders of circular and elliptical cross section, only approximate solutions exist for homogeneous or inhomogeneous cylinders of other shapes with finite cross section dimensions. A ray-optics technique is commonly employed by radome designers. Each ray passing through the dielectric body is assumed to undergo the same reflection and phase delay that is experienced by a plane wave passing through an infinitely wide plane sheet of the same thickness and with the same angle of incidence. This ray-optics method often provides reasonably accurate results for slightly curved dielectric shells, but it is inadequate for rapidly curving shells and the edge region of a truncated shell.

The ray-optics solution has been refined and extended considerably by Kouyoumjian, Peters and Thomas. [1] Their technique has proven quite successful with circular dielectric cylinders, spheres, and a few other shapes. However, the method becomes somewhat complicated in the general case and it does not always provide accurate results for small or irregular dielectric bodies.

Rhodes [2] has developed an iteration technique for dielectric scattering problems. The first-order solution is obtained by approximating the total field in the dielectric body by the incident field, and then calculating the scattered field by considering the equivalent volume currents in the dielectric region to radiate in unbounded free space. Useful results are obtained for thin dielectric shells having a dielectric constant near unity. By a similar technique, Andreasen [3] has obtained data for thin spherical shells; Stickler [4] has calculated the scattering patterns of dielectric cylinders of rectangular cross section and Philipson [5] has made an analysis of the scattering properties of thin dielectric rings.

Using a variational formulation, Cohen [6] has obtained accurate solutions for the circular dielectric cylinder. This approach becomes rather complicated, however, and the calculations become lengthy when the dielectric body has an arbitrary shape and dielectric constant.

The technique developed in this paper is based on the integral equation for the field of a harmonic source in the presence of a dielectric cylinder of arbitrary cross section shape. The dielectric cylinder is divided into square cells which are small enough so that the electric field intensity is nearly uniform in each cell. The total electric field intensity within each cell is initially considered to be an unknown quantity. A system of linear equations is obtained by enforcing at the center of each cell the condition that the total field must equal the sum of the incident and scattered fields. This system of equations is solved with the aid of a digital computer to evaluate the electric field intensity in each cell. It is then a rather simple procedure to calculate the scattered field at any other point in space. The technique has the following advantages.

1) The solution approaches the exact solution if a sufficiently large number of cells is employed.

2) Solutions are obtained for a dielectric shell of arbitrary shape as quickly and systematically as for a circular shell.

3) Simply by inserting the appropriate equations for the incident field, one obtains the solution for any two-dimensional source (such as a line source, any array of line sources, or a plane-wave source) in the presence of a dielectric cylinder.

4) Solutions are readily obtained for dielectric shells of tapered thickness and inhomogeneous dielectric shells.

5) The effects of surface-wave excitation and interaction among the various parts of the dielectric cylinder are included automatically in the solution.

Manuscript received December 9, 1963; revised November 9, 1964. This work was supported in part by the Research and Technology Division, Air Force Systems Command, Wright-Patterson Air Force Base, Ohio, under Contract AF 33(616)-7614 with The Ohio State University Research Foundation.

The author is with the Dept. of Electrical Engineering, Antenna Lab., The Ohio State University, Columbus, Ohio.

Reprinted from *IEEE Trans. Antennas Propagat.*, vol. AP-13, no. 3, pp. 334–341, May 1965.

6) Accurate solutions are obtained for dielectric cylinders having cross section dimensions up to several wavelengths. The ray-tracing or geometrical-optics methods often fail when the cross section dimensions are on the order of one wavelength or less.

7) Once a solution has been obtained for any particular source location, a relatively simple calculation will provide the new solution corresponding to a rotated or translated source.

The following sections develop the theory and the equations involved in this technique and present numerical results for homogeneous and inhomogeneous dielectric cylinders of various cross section shapes.

II. The Basic Theory

Consider a harmonic wave incident in free space on a dielectric cylinder of arbitrary cross section as suggested in Fig. 1. The time-factor $e^{j\omega t}$ is understood. It is assumed that the incident electric field intensity $\mathbf{E}^i$ has only a z component and it is not a function of z, where the z axis is taken to be parallel with the axis of the cylinder. That is,

$$\mathbf{E}^i = \hat{z}E^i(x, y) \tag{1}$$

The dielectric cylinder is assumed to have the same permeability as free-space ($\mu = \mu_0$). The dielectric material is assumed to be linear and isotropic, but it may be inhomogeneous with respect to the transverse coordinates as follows:

$$\epsilon = \epsilon(x, y) \tag{2}$$

where ϵ represents the complex permittivity.

Let $\mathbf{E}$ represent the total field; that is, the field set up by the source in the presence of the dielectric cylinder. The "scattered field" is defined to be the difference between the total and the incident fields. Thus,

$$\mathbf{E} = \mathbf{E}^i + \mathbf{E}^s \tag{3}$$

Under the assumed conditions, the total and the scattered electric field intensities will have only z components.

The scattered field $\mathbf{E}^s$ may be generated by an equivalent electric current J radiating in unbounded free space, where

$$J = j\omega(\epsilon - \epsilon_0)\mathbf{E} \tag{4}$$

with ω representing the angular frequency $2\pi f$. This equivalent current density is often called the "polarizarion current."

The field of an electric current filament dI parallel with the z axis in free space is given by

$$d\mathbf{E}^s = -\hat{z}(\omega\mu/4)H_0^{(2)}(k\rho)dI \tag{5}$$

where $H_0^{(2)}(k\rho)$ is the Hankel function of order zero, ρ is the distance from the current filament to the observation point and $k = \omega\sqrt{\mu_0\epsilon_0} = 2\pi/\lambda$. The free-space wavelength is denoted by λ. The increment of electric current which generates the scattered field is given by

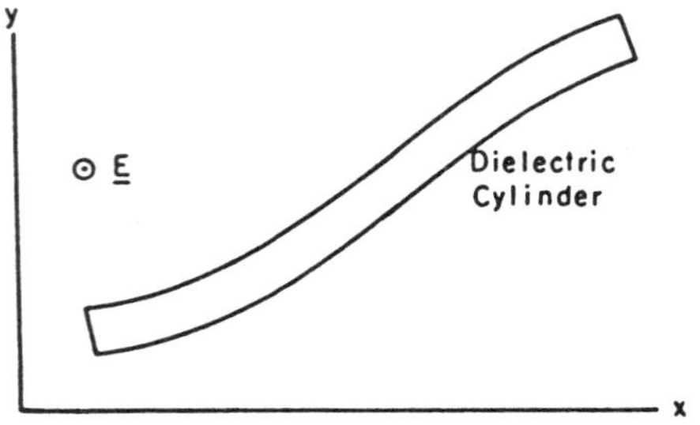

Fig. 1. Cross section of a dielectric cylinder showing the coordinate system.

$$dI = J\,dS = j\omega(\epsilon - \epsilon_0)E\,dS \tag{6}$$

where dS is the increment of surface area on the cross section of the dielectric cylinder. From (5) and (6), the scattered field is given by

$$E^s(x, y) = -(jk^2/4)\iint(\epsilon_r - 1)E(x', y')H_0^{(2)}(k\rho)dx'dy' \tag{7}$$

where (x, y) and (x', y') are the coordinates of the observation point and the source point, respectively, ϵ_r is the complex relative dielectric constant ($\epsilon_r = \epsilon/\epsilon_0$) and

$$\rho = \sqrt{(x - x')^2 + (y - y')^2} \tag{8}$$

The integration in (7) is to be performed over the cross section of the dielectric cylinder. In the inhomogeneous case the relative dielectric constant is considered to be a function of the source coordinates

$$\epsilon_r = \epsilon_r(x', y')$$

Equation (7) is valid for the scattered field at any point inside or outside the dielectric region. The integral equation for the total field E is obtained from (3) and (7),

$$E(x, y) + (jk^2/4)\iint(\epsilon_r - 1)E(x', y')H_0^{(2)}(k\rho)dx'dy' = E^i(x, y) \tag{9}$$

Let us divide the cross section of the dielectric cylinder into cells sufficiently small so that the dielectric constant and the electric field intensity are essentially constant over each cell. The division into cells is indicated in Fig. 2. If (9) is enforced at the center of cell m, the following expression is obtained:

$$E_m + (jk^2/4)\sum_{n=1}^{N}(\epsilon_n - 1)E_n \cdot \iint_{\text{cell } n} H_0^{(2)}(k\rho)dx'dy' = E_m^i \tag{10}$$

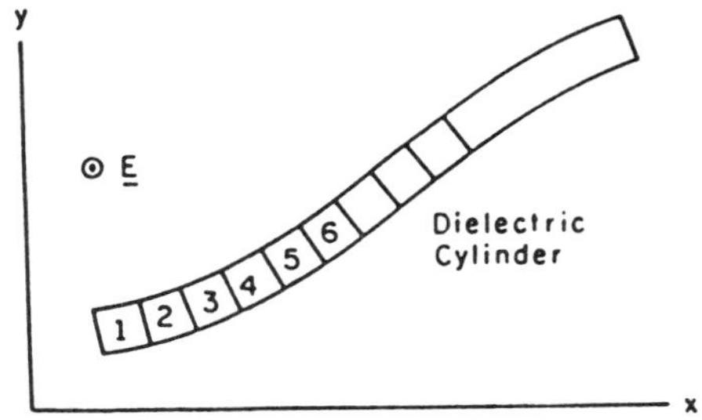

Fig. 2. The cross section of the dielectric cylinder is divided into small cells.

where ϵ_n and E_n represent the complex relative dielectric constant and the electric field intensity at the center of cell n and

$$\rho = \sqrt{(x' - x_m)^2 + (y' - y_m)^2} \tag{11}$$

By taking $m = 1, 2, 3, \cdots, N$, (10) yields a system of N linear equations, where N represents the total number of cells. These can be solved to determine the total electric field intensity at the center of each cell ($E_1, E_2, E_3, \cdots, E_N$). Having thus determined the total field $E(x, y)$ in the dielectric region, it is then possible to calculate the scattered field of the dielectric cylinder at any point in space by means of (7). The details of the solution are described in the following section.

III. Surface Integrals of the Hankel Function

The surface integrals in (10) can be evaluated by numerical integration formulas such as the trapezoidal rule and Simpson's rule. These methods have been employed successfully, but it was found that the calculations are quite lengthy, and care must be exercised in integrating through the singularity that exists when the observation point is at the center of cell n. The region of integration (over cell n) is square or rectangular in the simplest case, and a closed-form solution for this integral is not known. A simple solution is, however, available for the integral of the zero-order Hankel function over a circular region. It is given by

$$\begin{aligned}(jk^2/4) \underset{\text{cell } n}{\int_0^{2\pi} \int_0^a} H_0^{(2)}(k\rho)\rho' d\rho' d\phi' \\ &= (j/2)[\pi k a H_1^{(2)}(ka) - 2j] \qquad \text{if } m = n \\ &= (j\pi k a/2) J_1(ka) H_0^{(2)}(k\rho_{mn}) \qquad \text{if } m \neq n \end{aligned} \tag{12}$$

where ρ is given by (11) and ρ' and ϕ' are polar coordinates based on a coordinate origin at the center of cell n. The first solution given in (12) applies if the observation point is at the center of the circular region (i.e., if $m = n$). The second solution applies if the observation point is at a distance ρ_{mn} from the center of the circular region, where ρ_{mn} is greater than the radius a of the circular region. Numerical calculations have shown that little error is incurred in approximating square cells with circular cells of the same cross section area to take advantage of the simple expressions given in (12). The distance between the centers of cells m and n is given by

$$\rho_{mn} = \sqrt{(x_m = x_n)^2 + (y_m - y_n)^2} \tag{13}$$

Now the system of linear equations represented by (10) can be written in the following form:

$$\sum_{n=1}^{N} C_{mn} E_n = E_m{}^i \qquad \text{with } m = 1, 2, \cdots, N \tag{14}$$

If a_n represents the radius of the equivalent circular cell which has the same cross section area as cell n, the coefficients C_{mn} are given by

$$C_{mn} = 1 + (\epsilon_m - 1)(j/2)[\pi k a_m H_1^{(2)}(k a_m) - 2j] \qquad \text{if } n = m \tag{15}$$

$$C_{mn} = (j\pi k a_n/2)(\epsilon_n - 1) J_1(k a_n) H_0^{(2)}(k\rho_{mn}) \qquad \text{if } n \neq m \tag{16}$$

IV. Formulating the Scattered Field

Once the system of linear equations (14) has been solved, the scattered field of the dielectric cylinder can be calculated at any point in space by means of (7). To simplify the surface integral in (7), it is convenient again to divide the cylinder into small cells whose cross section shape is approximately square. From (7) and (12), the scattered field at any point outside the dielectric body is given by

$$\begin{aligned} E^s(x, y) \\ &= -j(\pi k/2) \sum_{n=1}^{N} (\epsilon_n - 1) E_n a_n J_1(k a_n) H_0^{(2)}(k\rho_n) \end{aligned} \tag{17}$$

where ϵ_n represents the average dielectric constant over cell n, a_n is the radius of the circle having the same area as cell n and ρ_n is the distance from the observation point to the center of cell n,

$$\rho_n = \sqrt{(x - x_n)^2 + (y - y_n)^2} \tag{18}$$

The distant scattering pattern of the dielectric cylinder is obtained by employing the asymptotic form for the Hankel function of large argument and taking

$$\rho_n = \rho_0 - x_n \cos\phi - y_n \sin\phi \tag{19}$$

where ρ_0 and ϕ are the polar coordinates of the distant observation point. The distant scattered field is given by

$$\begin{aligned} E^s(\rho_0, \phi) &= -j(\pi k/2)\sqrt{2j/\pi k \rho_0} \\ &\quad \cdot e^{-jk\rho_0} \sum_{n=1}^{N} (\epsilon_n - 1) E_n a_n J_1(k a_n) \\ &\quad \cdot e^{jk(x_n \cos\phi + y_n \sin\phi)} \end{aligned} \tag{20}$$

The plane-wave scattering properties of any cylindrical body of infinite length are conveniently described in terms of the echo width [7] which is denoted by $W(\phi)$ and is defined as follows:

$$W(\phi) = \operatorname*{limit}_{\rho_0 \to \infty} 2\pi\rho_0 \left| \frac{E^s(\rho_0, \phi)}{E^i} \right|^2 \tag{21}$$

From (20) and (21), the echo width of a dielectric cylinder of arbitrary cross section shape is given as follows:

$$W(\phi) = \frac{\pi^2 k}{|E^i|^2} \left| \sum_{n=1}^{N} (\epsilon_n - 1) E_n a_n J_1(k a_n) e^{jk(x_n \cos\phi + y_n \sin\phi)} \right|^2 \tag{22}$$

V. Numerical Results for the Circular Cylindrical Shell

Using (14), (15), (16), (20) and (22), numerical solutions were obtained for a circular dielectric cylindrical shell. Figure 3 shows the electric field distribution in the dielectric shell set up by an incident plane wave, and Fig. 4 shows the distant scattering pattern. Figure 4 also shows the exact classical solution which involves an infinite series of cylindrical mode functions. It may be noted that the integral-equation technique yields results which show excellent agreement with the exact solution. In view of the nature of the integral-equation formulation, it is believed that highly accurate results can also be obtained for dielectric cylinders of other cross section shapes where an exact solution is not available.

In Fig. 3 it may be noted that the electric field intensity in the dielectric region varies from 0.8 to 1.58 in magnitude. The incident plane wave has an electric field intensity of 1.0. Thus, poor accuracy is to be expected if the electric field intensity in the dielectric region is approximated by the incident electric field intensity.

In addition to the plane-wave solution previously described, the integral-equation technique readily yields solutions when the incident wave is the field of a line source or any array of line sources of infinite length. The line sources are assumed to be parallel with the axis of the dielectric cylinder. Figure 5 shows the scattering pattern of a circular dielectric cylindrical shell when the incident field is generated by a nearby line source having an omnidirectional pattern. In Fig. 5 the echo width is referred to the incident power density at the center of the dielectric cylindrical shell. Similar calculations show that the scattering pattern approaches the plane-wave solution shown in Fig. 4 when the line source is at a great distance from the dielectric shell. This provides one check on the accuracy and validity of the integral-equation technique as applied to the dielectric cylinder in the presence of a line source.

VI. Numerical Results for Semicircular Cylindrical Shell

Figure 6 shows the plane-wave scattering pattern of a semicircular cylindrical shell, computed with the integral-equation technique developed in Sections II and III. This particular semicircular dielectric shell has much smaller echo width than the corresponding circular shell illustrated in Fig. 4. However, it is possible to obtain a considerable increase in the backscatter echo width of a semicircular shell (compared with the circular shell of the same radii) by proper choice of the inner and outer radii and the wavelength. No exact solution is available for comparison with the results shown in Fig. 6, and the geometric-optics solution is not likely to be accurate for a shell of such small cross section dimensions.

It should be pointed out that the solution for the semicircular shell (or any other cross section shape) is as simple, straightforward and systematic as that for the circular cylindrical shell when the integral-equation technique is employed. This represents a distinct advantage over the boundary-value solution, the geometric-optics solutions and the variational solutions.

VII. Numerical Results for Plane Homogeneous Dielectric Slabs

Figure 7 shows the calculated scattering patterns of a plane dielectric slab of finite thickness and width for the case where a plane wave is incident normally on the slab. Results are shown both for lossless and dissipative homogeneous slabs.

For comparison, the physical optics solutions are also shown. The physical optics solution is based on the approximation that the electric field intensity within the dielectric body is the same as that in a dielectric slab of infinite width. Thus, the edge effects and surface-wave phenomena are usually neglected in the physical-optics approximation in contrast with the integral-equation solution. Although accurate agreement is not to be expected between the two methods, reasonably good agreement is observed in Fig. 7.

It should be mentioned that slightly dissipative dielectric cylinders are handled as easily as lossless dielectric bodies with the integral-equation technique. In problems involving perfectly conducting or highly conducting scatterers, a somewhat different formulation has been employed with good success. Some modifications must be made in the techniques developed here to study the scattering properties of highly dissipative dielectric bodies, since in this case it is not reasonable to assume the electric field intensity is nearly uniform over each cell of the type employed here.

Figure 8 shows the scattering patterns of the same homogeneous dielectric slab for grazing incidence. Re-

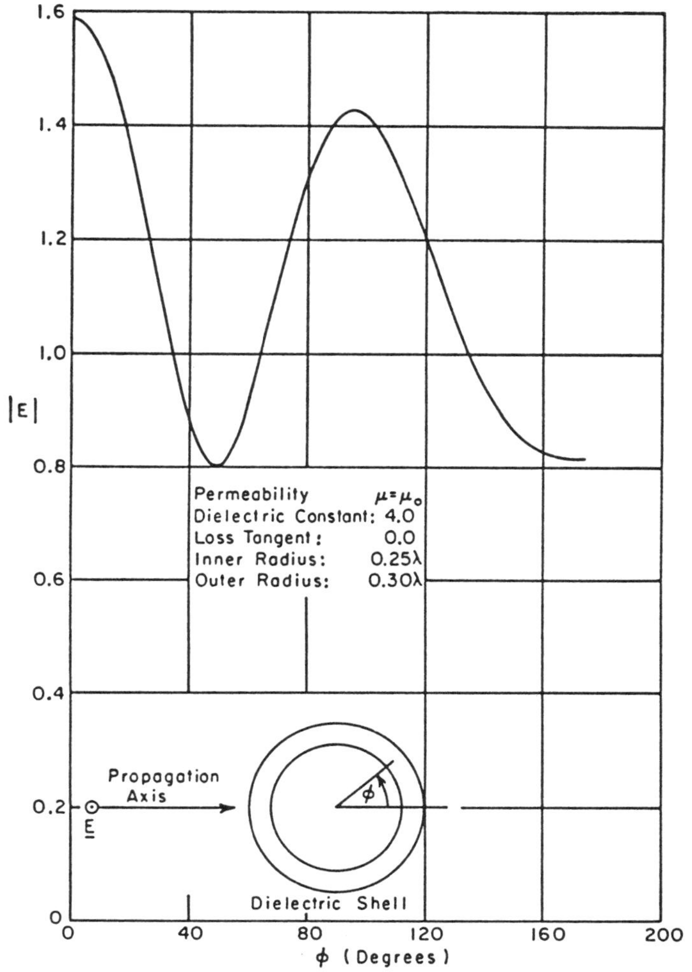

Fig. 3. Electric field distribution in circular dielectric cylindrical shell with plane-wave incident, calculated with the integral-equation technique.

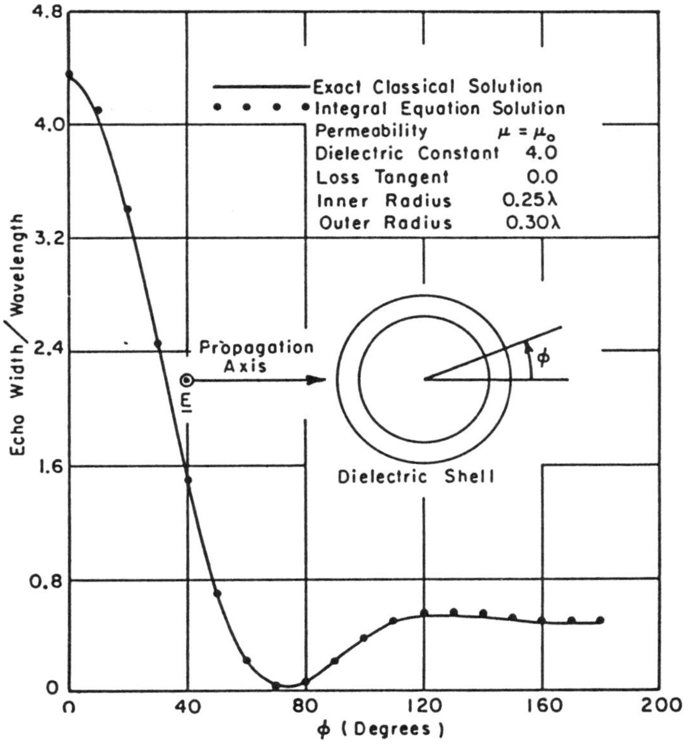

Fig. 4. Distant scattering pattern of circular dielectric cylindrical shell with plane-wave incident.

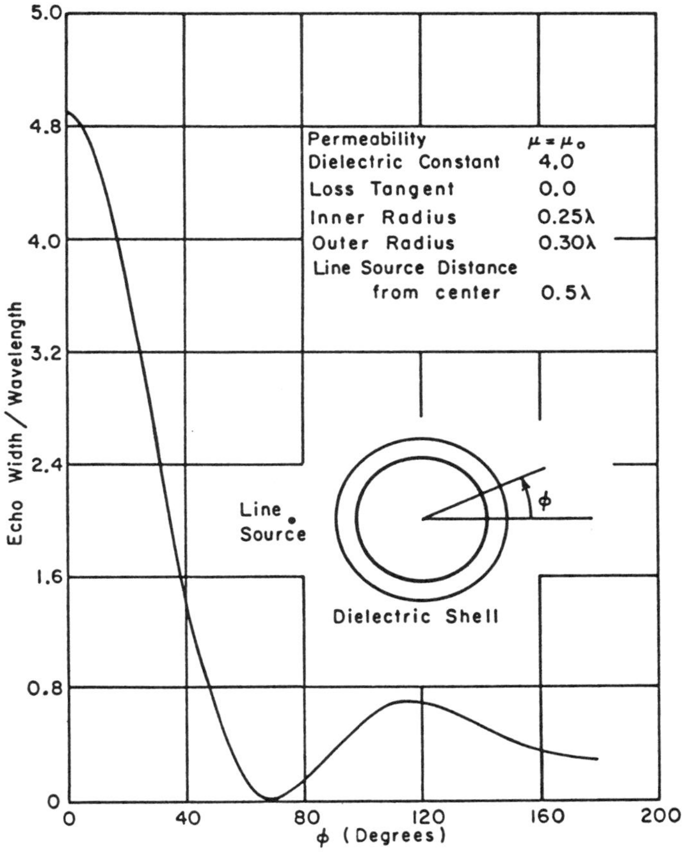

Fig. 5. Scattering pattern of a circular dielectric cylindrical shell in the presence of a nearby parallel line source, calculated with the integral-equation technique.

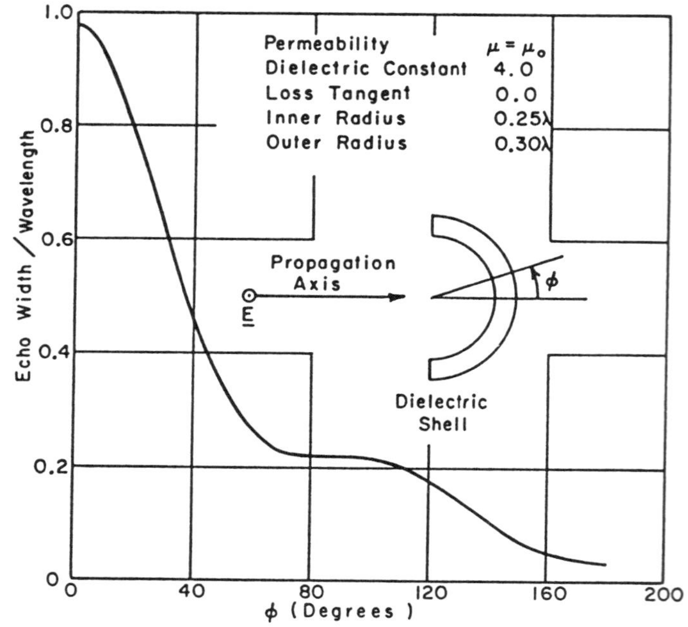

Fig. 6. Scattering pattern of a semicircular dielectric cylindrical shell with plane-wave incident, calculated with the integral-equation technique.

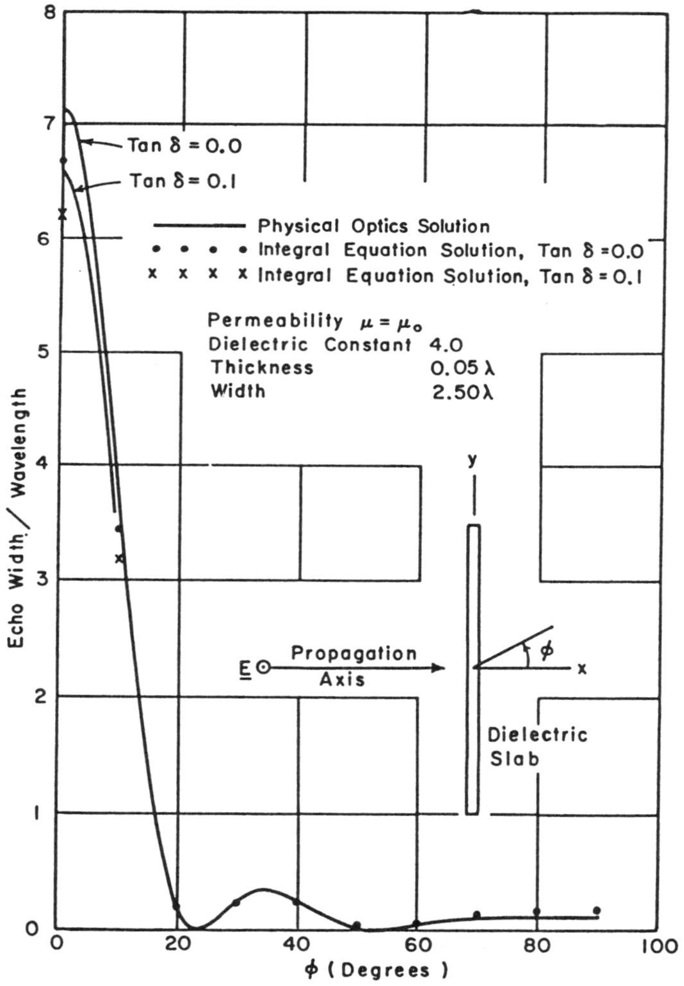

Fig. 7. Calculated scattering patterns of a homogeneous plane dielectric slab of finite thickness and width with a plane wave having normal incidence.

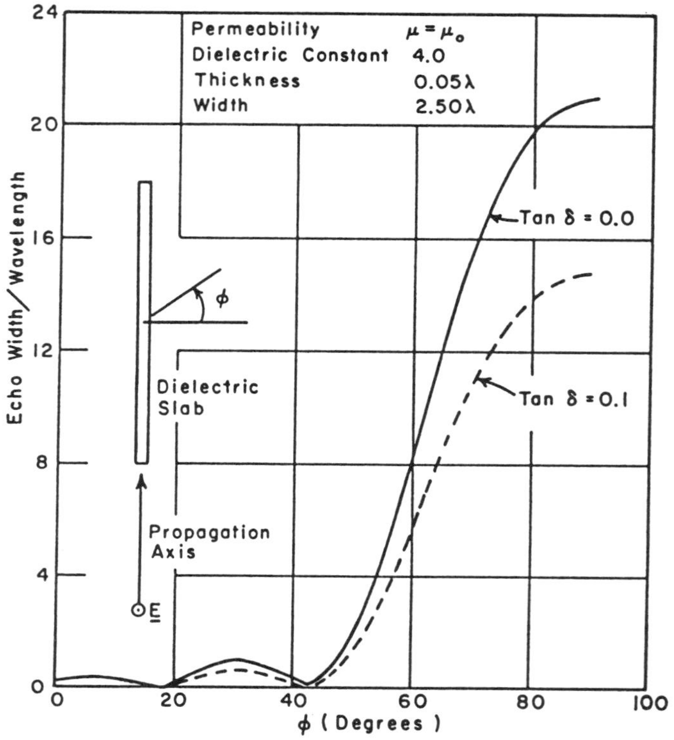

Fig. 8. Calculated scattering patterns of a homogeneous plane dielectric slab with a plane wave at grazing incidence.

sults are shown both for the lossless and dissipative cases. The physical-optics solution is not shown for this case simply because the optical solutions do not yield useful results for grazing incidence unless considerable effort is made to include the effects of surface-wave excitation in the dielectric slab. It may be noted in Fig. 8 that the forward scattering intensity is much greater for grazing incidence than for normal incidence, and that the results are much more sensitive to the loss tangent ($\tan \delta$) for grazing incidence.

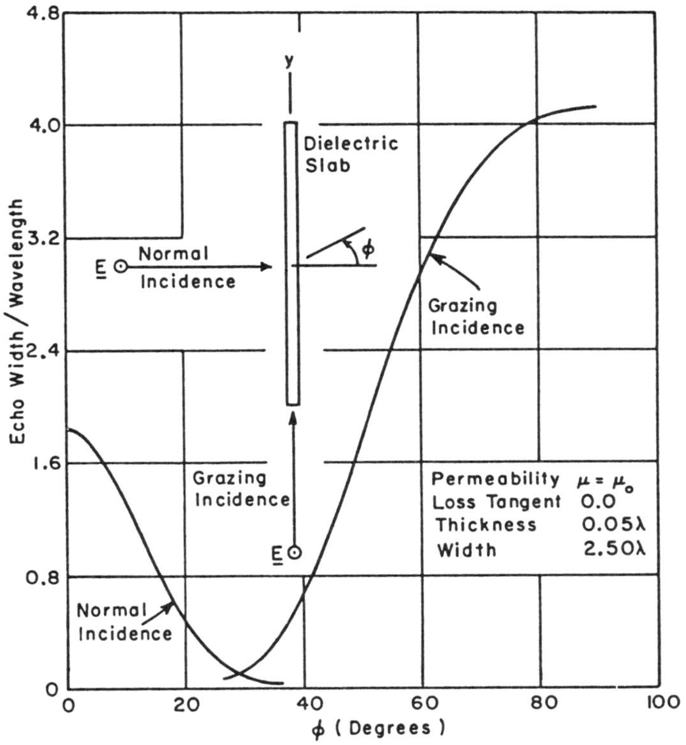

Fig. 9. Calculated scattering patterns of an inhomogeneous plane dielectric slab with a plane wave at normal and grazing incidence. Dielectric constant varies linearly from 4.0 at center to 1.0 at edges of slab.

VIII. Numerical Results for Plane Inhomogeneous Dielectric Slabs

Figure 9 shows the scattering patterns for a lossless inhomogeneous dielectric slab for normal incidence and grazing incidence. The dielectric constant is assumed to vary linearly from 4.0 at the center of the slab to 1.0 at the edges, being a function of the y coordinate only. A physical-optics solution for this problem would be somewhat involved. With the integral-equation technique the solution is straightforward and systematic.

IX. Conclusions

A technique is developed for calculating the scattered fields of a dielectric cylinder or cylindrical shell of arbitrary cross section shape. The solution is based on the integral equation for the problem, and it involves the solution of a system of linear equations. The first step in the solution is to determine the electric field distribution within the dielectric body. This field distribution is often of considerable interest in itself, since it promotes an understanding of the phenomena which are important in the scattering process. For example, any surface waves which may be excited in the dielectric body are revealed in the standing waves in the field distribution, and the edge effects are evident in the field perturbations near the edges of a dielectric shell.

The solution is best accomplished with the aid of a digital computer. This integral-equation solution is more systematic, more general and more accurate than the optical methods or the variational solutions. For an inhomogeneous dissipative dielectric shell of any cross section shape, one merely inserts in the equations the coordinates which define the shape of the shell, and the appropriate dielectric constant and loss tangent associated with each point in the dielectric body. The incident wave may be a plane wave or any combination of plane waves, or it may be the field generated by a parallel line source or any array of parallel line sources. One merely inserts the value of the incident field associated with each point in the dielectric body. The effects of surface waves and edge diffraction are included automatically in the solution.

Examples are included to show the scattering patterns of circular and semicircular dielectric cylindrical shells and plane dielectric slabs of finite thickness and width. Results are illustrated for inhomogeneous and dissipative dielectric slabs as well as their simpler counterparts.

The technique can be extended to apply to incident waves having arbitrary polarization, to dielectric bodies having a permeability differing from that of free space, and even to three-dimensional scattering problems. There will, however, be a corresponding increase in the complexity and the computational costs.

This integral-equation technique is applicable primarily to problems in which the cross section area of the dielectric cylinder is not too large. The dielectric cross section is divided into cells which are square or nearly so. For accurate results, the edge dimensions of each cell should not exceed $0.2/\sqrt{\epsilon_r}$ wavelengths. If, for example, the dielectric constant is 4.0, a solution can be obtained at reasonable cost for a dielectric shell of thickness up to 0.1 wavelength and width up to 10 wavelengths. This will involve the solution of 100 linear equations, which is not an unreasonable number. By employing interpolation formulas, it is possible to handle a shell of thickness 0.1 wavelength and width up to 40 wavelengths by solving a set of 100 linear equations. The computation time is determined by the cross section area of the dielectric cylinder. It is believed that a cross section area up to four square wavelengths can be handled at reasonable cost and without difficulty by this technique with the aid of a computer such as the IBM 7094.

References

[1] Kouyoumjian, R. G., L. Peters, Jr., and D. T. Thomas, A modified geometrical optics method for scattering by dielectric bodies, *IEEE Trans. on Antennas and Propagation*, vol AP-11, Nov 1963, pp 690–703

[2] Rhodes, D. R., On the theory of scattering by dielectric bodies, Rept. 475-1, Antenna Lab., The Ohio State University, Columbus, Jul 1953.

[3] Andreasen, M. G., Back-scattering cross section of a thin, dielectric, spherical shell, *IRE Trans. on Antennas and Propagation*, vol AP-5, Jul 1957, pp 267–270.

[4] Stickler, D. C., Electromagnetic diffraction by dielectric strips, *IRE Trans. on Antennas and Propagation*, vol AP-6, Jan 1958, pp 148–151.

[5] Philipson, L. L., An analytical study of scattering by thin dielectric rings, *ibid.*, pp 3–8.

[6] Cohen, M. H., Application of the reaction concept to scattering problems, *IRE Trans. on Antennas and Propagation*, vol AP-3, Oct 1955, pp 193–199.

[7] Harrington, R. F., Time-Harmonic Electromagnetic Fields. New York: McGraw-Hill, 1961, pp 358–359.

An Integral Equation Approach to Scattering From a Body of Finite Conductivity[1]

K. M. Mitzner

Northrop Norair, Hawthorne, Calif. 90250, U.S.A.

(Received January 31, 1967; revised July 19, 1967)

The problem of scattering from a homogeneous body is formulated in terms of two coupled integral equations relating the effective electric and magnetic surface currents $\mathbf{K}_e$ and $\mathbf{K}_m$. The formulation chosen, in which each equation involves the constitutive parameters of only one medium, is especially suited to the case of a high conductivity scatterer. From the equation for the conducting medium, one can derive, under increasingly restrictive assumptions, first an explicit expression for $\mathbf{K}_m$ in terms of $\mathbf{K}_e$, then a curvature-dependent boundary condition relating the two currents, and finally the usual Leontovich boundary condition. Numerical results for scattering from circular cylinders of small radius show the advantage of the curvature-dependent condition over the Leontovich condition.

1. Introduction

The fundamental work on scattering from a body of large but finite conductivity is contained in papers by Rytov (1940) and Leontovich (1948). Starting from an asymptotic series representation of the field, they derive the impedance boundary condition

$$\mathbf{K}_m = Z_c \mathbf{n} \times \mathbf{K}_e, \tag{1.1}$$

which is frequently referred to as the Leontovich boundary condition. Here the tangent fields at the surface S of the conductor have been expressed in terms of the effective electric and magnetic surface currents

$$\mathbf{K}_e = -\mathbf{n} \times \mathbf{H} \quad \text{and} \quad \mathbf{K}_m = \mathbf{n} \times \mathbf{E}. \tag{1.2}$$

The unit vector $\mathbf{n}$ is the normal *into* the conductor, and time dependence $e^{-i\omega t}$ is assumed.

The surface impedance Z_c is just the wave impedance in the conductor. This can be expressed in terms of the constitutive parameters as

$$Z_c = (\mu^*/\epsilon^*)^{1/2}, \text{ with } \epsilon^* = \epsilon + i\sigma/\omega,\ \mu^* = \mu + i\tau/\omega. \tag{1.3}$$

In the body of this paper, we shall consider only the case of a nonferrous metallic conductor for which ϵ and τ are negligible and for which $|Z_c|$ is less than the free-space impedance Z_0. Then the most significant parameter characterizing propagation in the conductor is the skin depth

$$\delta = 1/(\tfrac{1}{2}\,\omega\mu\sigma)^{1/2}, \tag{1.4}$$

in terms of which the wave impedance and the wave number are

$$Z_c = \tfrac{1}{2}\,(1-i)\,\omega\mu\delta, \qquad k_c = (1+i)/\delta. \tag{1.5}$$

The extension of our results to more general media is considered briefly in section 6.

[1] This work was supported in part by the Air Force Avionics Laboratory, Wright-Patterson Air Force Base, under contract AF33(615)-3797.

K. M. Mitzner, *Radio Science*, vol. 2, pp. 1459–1470, 1967, published by the National Bureau of Standards in cooperation with the U.S. National Committee of the International Scientific Radio Union (URSI).

One of the requirements for validity of the Leontovich condition is that the radii of curvature of the body be large compared to δ. A modification to treat smaller radii of curvature is, however, implicit in Rytov (1940) and is given explicitly (but with an error of a factor of 2) by Leontovich (1948). The corrected form of this boundary condition is

$$K_{mu} = -(1-p)\, Z_c K_{ev}, \qquad K_{mv} = (1+p)\, Z_c K_{eu} \tag{1.6}$$

for a homogeneous conductor. Here

$$p = \frac{1}{2}\, i(\kappa_v - \kappa_u)/k_c = \frac{1}{4}\,(1+i)\,\delta\,(\kappa_v - \kappa_u), \tag{1.7}$$

u and v are principal curvature coordinates so oriented that

$$\mathbf{e}_u \times \mathbf{e}_v = \mathbf{n}, \tag{1.8}$$

and κ_u and κ_v are the principal curvatures (Van Bladel, 1964), defined as positive when $\mathbf{n}$ points toward the center of curvature. Since $\mathbf{n}$ points into the conductor, the κ are positive where the conducting body is convex.

Higher approximations can be obtained formally by the asymptotic approach, but it is not clear that the asymptotic series is accurate enough for these to be meaningful. In any case, they involve surface derivatives of the $\mathbf{K}$ and are thus not well suited to computation. Furthermore, the asymptotic approach cannot be used to study the transition from the approximate boundary condition to the exact formulation.

In this paper we shall adopt an integral equation approach which does not have these limitations. The first step then is to choose an appropriate integral formulation of scattering from an interface. This choice is not trival, for there are various mathematically equivalent formulations.

For the problem at hand, which is characterized by considerable attenuation in one medium (and by $|Z_c|/Z_0 < 1$), the appropriate formulation consists of the two coupled integral equations

$$\tfrac{1}{2}\mathbf{K}_m + L_c\mathbf{K}_m - Z_c M_c \mathbf{K}_e = 0, \tag{1.9}$$

$$\tfrac{1}{2}\mathbf{K}_e - L_0\mathbf{K}_e - (1/Z_0) M_0 \mathbf{K}_m = \mathbf{K}_{e0} = -\mathbf{n}\times\mathbf{H}^{\text{inc}}. \tag{1.10}$$

Here $\mathbf{H}^{\text{inc}}$ is the incident field. The operators L and M are defined by

$$L\mathbf{K} = \int_{\bar{S}} dS'\,\mathbf{n}\times(\nabla' G \times \mathbf{K}'), \tag{1.11}$$

$$M\mathbf{K} = (1/ik)\int_{\bar{S}} dS'\,\mathbf{n}\times[k^2 G\mathbf{K}' - (\kappa_u' + \kappa_v')(\nabla' G)(\mathbf{n}' - \mathbf{n})\cdot\mathbf{K} + (\mathbf{K}' - \mathbf{K})\cdot\nabla_S'\nabla' G], \tag{1.12}$$

with G the Green's function $\qquad G = e^{ikR}/4\pi R, \quad R = |\mathbf{r} - \mathbf{r}'|, \qquad (1.13)$

and with the subscript on the operator indicating whether k_c or the free-space k_0 is used. The observation point $\mathbf{r}$ is on S. Primes denote functions of the integration variable $\mathbf{r}'$; $\int_{\bar{S}}$ indicates the integral over the punctured surface as defined in the appendix; ∇_S' is the tangential gradient with respect to $\mathbf{r}'$.

In section 2, we shall prove that (1.9) and (1.10) hold at any point of S around which the curvature is continuous, provided that there are no sources in the conductor and no surface current sources on S.

The integral eq (1.9) is used in section 3 to obtain three approximate expressions for $\mathbf{K}_m$ in terms of $\mathbf{K}_e$. The simplest is the Leontovich condition of (1.1), which we show to be valid to within errors of $O(\delta^2 k_0^2)$, $O(\delta^2/h^2)$, and $O(\delta\kappa)$, where h is the distance to the nearest significant source and κ is the larger of $|\kappa_u|$ and $|\kappa_v|$. The last error term is reduced to $O(\delta^2\kappa^2)$ in the curvature-dependent boundary condition

$$(1+p)K_{mu} = -Z_c K_{ev}, \qquad (1-p)K_{mv} = Z_c K_{eu}; \tag{1.14}$$

this is essentially equivalent to (1.6), the two equations differing only by an $O(\delta^2\kappa^2)$ term. The most accurate approximation, including second-order corrections in δk_0, δ/h, and $\delta\kappa$, is given by

$$(1+p)K_{mu}\mathbf{e}_u + (1-p)K_{mv}\mathbf{e}_v = 2Z_c M_c \mathbf{K}_e; \tag{1.15}$$

here the components of $\mathbf{K}_m$ are still given explicitly in terms of $\mathbf{K}_e$, but in a manner more complicated than an impedance boundary condition. Since (1.15) involves a surface integral, it clearly cannot be obtained by the essentially local approach of Leontovich and Rytov.

Substituting any of the approximate boundary conditions into the integral equation (1.10), we find that the resulting equation for $\mathbf{K}_e$ differs from the equation for a perfect conductor (on which $Z_c = 0$) by an $O(Z_c/Z_0)$ term. Thus, if $|Z_c| \ll Z_0$, we can assume that $\mathbf{K}_e$ is the same as on a perfect conductor. This simplification has indeed been used in most work with impedance boundary conditions.

For larger Z_c, the determination of $\mathbf{K}_e$ becomes more difficult. One approach, direct numerical solution of the integral equation (1.10), is of special interest. Indeed, it was the motivating factor behind the present study. We discuss this approach in section 4.

In section 5 we solve the problem of scattering from a circular cylinder three ways—exactly, using the Leontovich condition, and using the curvature-dependent condition. The results show that the curvature-dependent condition is, in general, more accurate than the Leontovich condition for cylinders of small radius.

In section 6 we discuss the modifications in the theory when (1.5) does not hold and when Z_c/Z_0 is large.

2. Development of the Integral Formulation

The main purpose of this section is to prove (1.9) and (1.10) under the assumptions already cited. However, as a necessary and useful preliminary, we present the integral expressions relating the field in a homogeneous region V to the tangential fields on the boundary S of the region. These equations are

$$\mathbf{E} - L^*(\hat{\mathbf{n}} \times \mathbf{E}) + ZM^*(\hat{\mathbf{n}} \times \mathbf{H}) = \mathbf{E}^{\text{inc}}, \tag{2.1}$$

$$\mathbf{H} - L^*(\hat{\mathbf{n}} \times \mathbf{H}) - (1/Z)M^*(\hat{\mathbf{n}} \times \mathbf{E}) = \mathbf{H}^{\text{inc}}, \tag{2.2}$$

with

$$L^*\mathbf{F} = \int_S dS'\nabla'G \times \mathbf{F}', \qquad M^*\mathbf{F} = -(1/ik)\int_S dS'(k^2GI - \nabla\nabla'G)\cdot\mathbf{F}'. \tag{2.3}$$

Here $\hat{\mathbf{n}}$ is the unit normal out of V and I is the unit dyadic. Only the incident field due to sources in V is included in $\mathbf{E}^{\text{inc}}$ and $\mathbf{H}^{\text{inc}}$; thus, the incident fields are zero when V is the conducting body.

Equations (2.1) and (2.2) are valid at at any interior point of V but specifically are not valid on S itself.

The foregoing equations are readily derived from eq (17) of the well-known paper of Maue (1949); it is only necessary to introduce the source terms and then use the identities

$$\nabla^2 G = -k^2 G \text{ for } \mathbf{r} \neq \mathbf{r}', \qquad \nabla G = -\nabla' G. \tag{2.4}$$

Equations (2.1) and (2.2) are doubly important. First, they can be used to find the field anywhere in space once the surface currents $\mathbf{K}_e$ and $\mathbf{K}_m$ are known. Secondly, they are the starting point for the derivation of (1.9) and (1.10).

Now let us derive the integral eq (1.10). The first step is to note the identity

$$\int_S dS' \, [(\mathbf{K} \cdot \nabla_S') \, (\mathbf{n}^* \times \nabla' G) + (\mathbf{n}' \cdot \mathbf{K}) (\kappa_u' + \kappa_v') \, \mathbf{n}^* \times \nabla' G] = 0, \tag{2.5}$$

where $\mathbf{K}$ and $\mathbf{n}^*$ are vectors which are independent of $\mathbf{r}'$. Like (2.1) and (2.2), (2.5) is valid at any interior point of V but not on S itself. The identity is easily derived from eq (43) of Van Bladel (1964): we set into that equation

$$f = \mathbf{e} \cdot (\mathbf{n}^* \times \nabla' G), \tag{2.6}$$

with $\mathbf{e}$ an arbitrary constant vector; then we take the dot product with $\mathbf{K}$; after straightforward algebraic manipulations to move $\mathbf{K}$ under the integral sign and $\mathbf{e}$ outside the integral sign, we obtain an equation which is the dot product of $\mathbf{e}$ with (2.5); but since $\mathbf{e}$ is arbitrary, this result can only be true if (2.5) holds, and the proof is thus completed.

To continue now the derivation of (1.10), we write (2.2) for the free-space region V_0 and cross-multiply by $\mathbf{n}(\mathbf{r}_s)$, where $\mathbf{r}_s$ is on S. Then we add to the left-hand side the identically zero integral of (2.5) with

$$\mathbf{K} = -(1/ik_0 Z_0) \, \mathbf{K}_m(\mathbf{r}_s), \qquad \mathbf{n}^* = \mathbf{n}(\mathbf{r}_s). \tag{2.7}$$

Rearranging the integral terms and taking the limit as $\mathbf{r} \to \mathbf{r}_s$, we *formally* obtain (1.10).

To show that the formal limiting process is valid, we must demonstrate the existence of the integrals defining L and M in (1.11) and (1.12). The most important part of this demonstration is for the $(\mathbf{K}' - \mathbf{K})$ term in (1.12). Using (A.13) and (A.7), with (A.7) for $n = 3$ playing the critical role, we see that the integral in question exists provided the second derivatives of $\mathbf{K}$ (in this case $\mathbf{K}_m$) exist. This is assured if the curvature of S and the second derivatives of $\mathbf{K}_{e0}$ are continuous near $\mathbf{r}$. In practical situations, continuity of the curvature and the absence of surface current sources are sufficient to assure that $\mathbf{K}_{e0}$ has continuous second derivatives. Thus the integral exists under the assumptions cited in section 1.

Furthermore, by straightforward application of the theorems and techniques given by Kellogg (1953) for the treatment of surface distributions, we can show that the other integrals in (1.11) and (1.12) exist under the same assumptions. Thus the derivation of (1.10) is completed.

The integral eq (1.9) can be derived by a completely analogous process starting with (2.1) for the conducting region V_c. Since we are assuming there are no sources in V_c, (1.9) has no source term.

The crucial step in the derivation of the integral equations is the use of the integral (2.5) to cancel the nonintegrable singularity which would otherwise occur. This is an extension to the case of surface integral equations of a standard technique for one-dimensional integral equations which is discussed by Kantorovich and Krylov (1958). From a computational point of view, this method of eliminating the singularity is clearly superior to that of Maue (1949), which leads to an integro-differential equation.

3. The Approximate Expressions for $\mathbf{K}_m$

We shall now derive from (1.9) the three approximate expressions for $\mathbf{K}_m$ given in (1.1), (1.14), and (1.15). The derivations are based on the fact that

$$|G_c| = e^{-R/\delta}. \tag{3.1}$$

Thus we can approximate the operators L_c and M_c by operators L_c^t and M_c^t in which the integration is carried out over a subdomain S_t of S with dimensions of order δ. The approximation errors are

$$\epsilon_L^t = (L_c - L_c^t)\mathbf{K}_m, \quad \epsilon_M^t = (M_c - M_c^t)\mathbf{K}_e. \tag{3.2}$$

We choose a disc-like subdomain of the type described in the appendix. We further require that S_t be smooth enough to be represented in the form (A.10), but it should be noted that some of our results can be derived with looser assumptions on the smoothness.

Using various equations given in the appendix, we can now write the expanded form

$$L_c^t\mathbf{K}_m = -\int_{\bar{D}} dD'(\mathbf{n}\cdot\mathbf{n}')^{-1}(e^{(i-1)R/\delta}/4\pi\rho^2)b_1(\rho/R)^3\{(\mathbf{n}\cdot\mathbf{s}^*)\mathbf{K}_m - [(\mathbf{n}-\mathbf{n}')\cdot\mathbf{K}_m]\mathbf{s}^*$$
$$+(\mathbf{n}\cdot\mathbf{s}^*)(\mathbf{K}_m' - \mathbf{K}_m) - [(\mathbf{n}-\mathbf{n}')\cdot(\mathbf{K}_m' - \mathbf{K}_m)]\mathbf{s}^*\}. \tag{3.3}$$

The approximation of this expression is straightforward but rather tedious, and thus we shall give only a verbal sketch here. We begin by making the further substitutions of (A.15) through (A.19). We also note from (A.11) and (A.20) that $\boldsymbol{\alpha}'$ is cubic, $\boldsymbol{\alpha}_u'$ and $\boldsymbol{\alpha}_v'$ quadratic in $\sin\theta$ and $\cos\theta$. The various parts of the integral in (3.3) are now estimated using (A.6), (A.7), and (A.9). All δ^2 terms disappear because the integrands have odd periodicity in θ.

If we now combine (3.2) and (3.3), we can write the result of the approximation process as

$$L_c\mathbf{K}_m = \mathbf{J}_m(L) + \boldsymbol{\epsilon}_m + \boldsymbol{\epsilon}_m^*. \tag{3.4}$$

Here

$$\boldsymbol{\epsilon}_m = \boldsymbol{\epsilon}_L^t - \mathbf{J}_m(L-D), \tag{3.5}$$

$$\boldsymbol{\epsilon}_m^* = O(\delta^3\lambda\kappa\mathbf{K}_m) + O[\delta^3\lambda(\mathbf{K}_m^+ - \mathbf{K}_m^-)/\rho] + O[\delta^3\kappa(\mathbf{K}_m^+ - 2\mathbf{K}_m + \mathbf{K}_m^-)/\rho^2] + O(\delta^3|\nabla_s\alpha|\mathbf{K}_m); \tag{3.6}$$

$\mathbf{K}^+$ and $\mathbf{K}^-$ are given by (A.8);

$$\lambda = O(\kappa^2) + O(\alpha); \tag{3.7}$$

and

$$\mathbf{J}_m(P) = \int_{\bar{P}} dD'(e^{(i-1)\rho/\delta}/4\pi\rho)[1-(i-1)\rho/\delta]\{\tfrac{1}{2}(\kappa_u\cos^2\theta + \kappa_v\sin^2\theta)\mathbf{K}_m$$
$$-[\kappa_u\cos^2\theta(\mathbf{K}_m\cdot\mathbf{e}_u)\mathbf{e}_u + \kappa_v\sin^2\theta(\mathbf{K}_m\cdot\mathbf{e}_v)\mathbf{e}_v] \tag{3.8}$$

for a disc-shaped subdomain P of the tangent plane L. For $P = L$, we have

$$\mathbf{J}_m(L) = \tfrac{1}{2}p[(\mathbf{K}_m\cdot\mathbf{e}_u)\mathbf{e}_u - (\mathbf{K}_m\cdot\mathbf{e}_v)\mathbf{e}_v]. \tag{3.9}$$

The estimation of $M_c\mathbf{K}_e$ is carried out in a similar manner, with (A.13) now also coming into play. All terms involving δ disappear because of odd periodicity. We thus find

$$M_e\mathbf{K}_e = \mathbf{J}_e(L) + \epsilon_e + \epsilon_e^*, \tag{3.10}$$

where the difinition of ϵ_e is analogous to (3.5), and

$$\epsilon_e^* = O\ (\delta^2\kappa^2\mathbf{K}_e) + O\ [\delta^2(\mathbf{K}_e^+ - 2\mathbf{K}_e + \mathbf{K}_e^-)/\rho^2], \tag{3.11}$$

$$\mathbf{J}_e(P) = (1-i)\int_P dD'\,(e^{(i-1)\rho/\delta}/4\pi\rho^2)\ (\rho/\delta)\ \mathbf{n}\times\mathbf{K}_e, \tag{3.12}$$

$$\mathbf{J}_e(L) = \tfrac{1}{2}\ \mathbf{n}\times\mathbf{K}_e. \tag{3.13}$$

To proceed further, we must argue that it is legitimate in most practical problems to express the order terms in the more directly meaningful forms

$$O[\delta(\mathbf{K}^+ - \mathbf{K}^-)/\rho] = O(\mathbf{K})\ [O(\delta k_0) + O(\delta/h) + O(\delta\kappa)],$$

$$O\ [\delta^2(\mathbf{K}^+ - 2\mathbf{K} + \mathbf{K}^-)/\rho^2] = O(\mathbf{K})\ [O(\delta^2 k_0^2) + O(\delta^2/h^2) + O(\delta^2\lambda)]. \tag{3.14}$$

Consider first the case $$k_0 \gg 1/h, \qquad k_0 \gg \kappa, \qquad k_0^2 \gg \lambda, \tag{3.15}$$

for which the boundary looks essentially flat and the incident wave is essentially plane; here the order terms involving differences can be estimated by taking the derivatives of the physical optics field, and we get the δk_0 error terms by considering the limiting case of grazing incidence. When $1/h$ is the dominant quantity, the rate of change of the field depends on the curvature of the incident wave, and this leads to the δ/h terms. When κ dominates, the rate of change can be estimated from quasi-statics, and this leads to the $\delta\kappa$ and $\delta^2\lambda$ terms.

When none of the quantities is dominant, the situation is more complicated. In the absence of a thorough mathematical study, we must fall back on experience, which tells us there is no general tendency for the tangential derivatives of the field to be exceptionally large relative to the field itself in the resonance region. Thus we would expect (3.14) to be applicable in the resonance region. Some data confirming this expectation are presented in section 5.

To further simplify matters, we note that the Leontovich boundary condition is valid at least as an order relationship for all problems of interest here. Thus $\mathbf{K}_e$ can be replaced by $\mathbf{K}_m/Z_c$ in the error terms.

Now we can rapidly confirm the various approximations for $\mathbf{K}_m$. Setting (3.4) into (1.9) gives

$$\tfrac{1}{2}\mathbf{K}_m + \mathbf{J}_m\ (L) - Z_cM_c\mathbf{K}_e + (\epsilon_m + \epsilon_m^*) = 0, \tag{3.16}$$

which is just the explicit expression (1.15) with the error terms included. Transforming ϵ_m^* by means of (3.14), we verify that (1.15) is indeed correct to second order in δk_0, δ/h and $\delta\kappa$.

Further substituting (3.10) into (3.16) and applying (3.13) and (3.14) to the error terms, we next obtain the curvature dependent boundary condition (1.14), with errors indeed of second order in δk_0, δ/h, and $\delta\kappa$. Finally, the Leontovich condition (1.1) is obtained by omitting the correction term p.

We are assuming here that t can be made large enough so that ϵ_m and ϵ_e are at worst of the order of the other error terms. This will not always be the case, and thus the present theory must be modified when treating such bodies as a thin shell and such features as the borders of a narrow slit. If there is a question as to whether the ϵ are small enough in a particular problem, they can be estimated after the fact from the calculated values of $\mathbf{K}_e$ and $\mathbf{K}_m$.

4. Integral Equation Approach to the Calculation of $\mathbf{K}_e$

Expressing $\mathbf{K}_m$ in terms of $\mathbf{K}_e$ is only the first step in solving the scattering problem. We then must find $\mathbf{K}_e$. In the next section we solve a simple problem by separation of variables. Here we discuss the much more general integral equation approach.

The basic equation in this approach is (1.10). We first eliminate $\mathbf{K}_m$ using the appropriate approximate expression. Then we solve the resulting equation numerically with the aid of a digital computer.

A computer program for solving (1.10) is presently under development. The groundwork has been laid in recent work on scattering from a perfect conductor (Andreasen, 1964, 1965; Oshiro and Su, 1965; Oshiro and Cross, 1966; Oshiro, Torres, and Heath, 1966). In extending this work to (1.10), it is important to express the operator M in a form convenient for numerical computation. The most suitable form is given in (1.12).

In eliminating $\mathbf{K}_m$, there is no necessity to use the same approximation everywhere on the body. Equation (1.1) can be used in regions of low curvature, (1.14) for somewhat higher curvature, and (1.15) for even higher curvature. Indeed there may be regions where the full integral equation (1.9) is needed to express $\mathbf{K}_m$ in terms of $\mathbf{K}_e$.

It should be noted, however, that (1.9) and (1.10) constitute a desirable integral formulation of the interface problem only when there is high attenuation in one medium. For an interface between two low-loss dielectrics, the formulation of Müller (1957), in which the parameters of both media appear in both equations, is usually preferable.

5. Example—The Field on a Circular Cylinder

Leontovich (1948) introduced the curvature correction as a means of estimating the error in the Leontovich boundary condition. To the best of this author's knowledge, it has never actually been used in computation. Thus there is no quantitative information as to the effect of the curvature corrections. In view of this, let us investigate here the problem of scattering from an infinite circular cylinder, comparing the exact values of the field on the surface with the values obtained using the curvature-dependent and Leontovich boundary conditions.

The cylinder problem is especially simple because the principal curvatures are just

$$\kappa_\theta = 1/a, \qquad \kappa_z = 0, \tag{5.1}$$

where a is the radius. Then, from (1.14), we get the curvature-dependent boundary condition

$$E_\theta = -[Z_c/(1+\tfrac{1}{2}\, i/q)]H_z, \qquad E_z = [Z_c/(1-\tfrac{1}{2}\, i/q)]H_\theta, \tag{5.2}$$

with

$$q = (1+i)a/\delta = k_c a. \tag{5.3}$$

Deleting the (i/q) term gives the Leontovich condition.

The cylinder problem is readily solved by standard techniques, and we shall omit the details. For a plane wave normally incident at $\theta = 0$ with $\mathbf{E}$ parallel to the cylinder axis, we have

$$\mathbf{E}/E_0 = \mathbf{e}_z\left[e^{-i\tau\cos\theta} + \sum_{n=-\infty}^{\infty} a_n H_n^{(1)}(\tau)e^{in(\theta-\pi/2)}\right], \tag{5.4}$$

with

$$a_n = -[J_n(\tau) - c_n J_n'(\tau)]/[H_n^{(1)}(\tau) - c_n H_n^{(1)\prime}(\tau)], \qquad \tau = k_0 a, \tag{5.5}$$

and the prime denoting differentiation with respect to the argument. The exact solution is obtained by setting

$$c_n = (Z_c/Z_0)J_n(q)/J'_n(q), \tag{5.6}$$

the curvature dependent approximation by setting

$$c_n = i(Z_c/Z_0)/(1 - \tfrac{1}{2}\, i/q), \tag{5.7}$$

and the Leontovich approximation by deleting the $O(1/q)$ correction.

Similarly, for a wave with **H** parallel to the axis, $(\mathbf{H}/H_0)$ is given by (5.4) and (5.5), but now the c_n are

$$c_n = (Z_0/Z_c)J_n(q)/J'_n(q) \quad \text{[Exact Solution]}, \tag{5.8}$$

$$c_n = i(Z_0/Z_c)(1 + \tfrac{1}{2}\, i/q) \quad \text{[Curvature-Dependent Approximation]}, \tag{5.9}$$

and again the Leontovich approximation is obtained by deleting the $O(1/q)$ correction.

The data given here are for the case $\mu = \mu_0$, so that

$$Z_c/Z_0 = \tfrac{1}{2}(1 - i)k_0\delta. \tag{5.10}$$

Figure 1 shows a typical result for a cylinder of small radius, $k_0a = 0.1$, with parallel **E** polarization. Here, even though δ/a takes the fairly large value 0.5, the curvature-dependent boundary condition is accurate to 5 percent in magnitude, 4° in phase. The Leontovich condition is markedly less accurate in both magnitude and phase. If we hold k_0a constant and decrease δ/a, then both approximations improve, but the results for the curvature-dependent condition improve more rapidly. Thus for $\delta/a = 0.3$, the maximum error in the curvature-dependent condition is about one percent and 0.3°, but the *minimum* error in the Leontovich condition is still about 6 percent and 4°.

In figures 2 and 3, we have kept $k_0a = 0.1$ and increased δ/a to 0.75 and 1.25 respectively. Here we are indeed pushing the limits of the theory. Even so, the approximate boundary conditions give fair accuracy, especially in the illuminated region. Again the curvature-dependent boundary condition gives a better overall approximation.

In figure 4, $\delta/a = 0.5$ as in figure 1, but $k_0\delta$ has been increased from 0.05 to 0.15. Once more, the curvature-dependent condition gives the better approximation. For both approximate boundary conditions, the maximum magnitude error is much greater than in the $k_0\delta = 0.05$ case. That is, even though $k_0\delta \ll \delta/a$ in both cases, the increase in $k_0\delta$ strongly affects the magnitude error. Additional data indicate that this is a general effect, that, at least for the cylinder problem, the magnitude error is more sensitive to $k_0\delta$ than to $\kappa\delta$.

In figure 5, we consider parallel **H** polarization. For $k_0a = 1$ and $\delta/a = 0.3$, both approximate boundary conditions give tangential **H** to within 2 percent and 2°, an error too small to plot. The curve for $\delta/a = 0$ (perfect conductor) is included to show that **H** is indeed shifted significantly. These results confirm the validity of the impedance boundary conditions in the resonance region.

6. Extension to General Attenuating Media

In deriving the approximations (1.1), (1.14), and (1.15), we assumed that ϵ and τ are negligible, so that (1.4) and (1.5) are valid. We further assumed $|Z_c| < Z_0$.

Let us remove first the restriction on ϵ and τ. Then the three approximate expressions are formally valid provided p, Z_c, and k_c are expressed in terms of ϵ^* and μ^* rather than of δ. In most cases, the error terms are the same except that δ is replaced by $1/k_I$, where k_I is the imaginary part of k_c. However, when the real part k_R is considerably larger than k_I, we find that $O(\delta^2\lambda)$ must

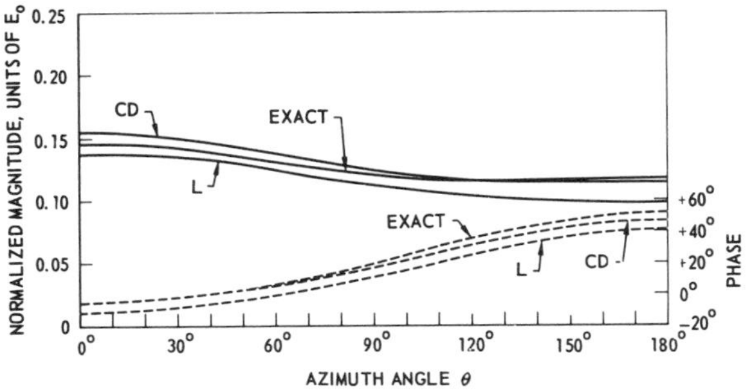

FIGURE 1. *Magnitude and phase of* E_z *on the surface of a cylinder for* $k_0a = 0.1$, $\delta/a = 0.5$ ($k_0\delta = 0.05$).

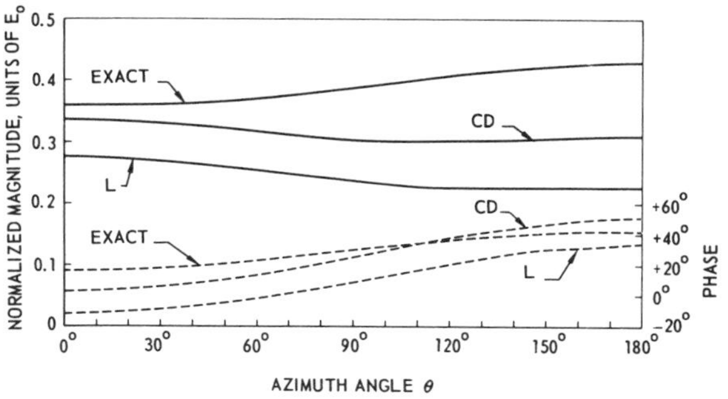

FIGURE 3. *Magnitude and phase of* E_z *on the surface of a cylinder for* $k_0a = 0.1$, $\delta/a = 1.25$ ($k_0\delta = 0.125$).

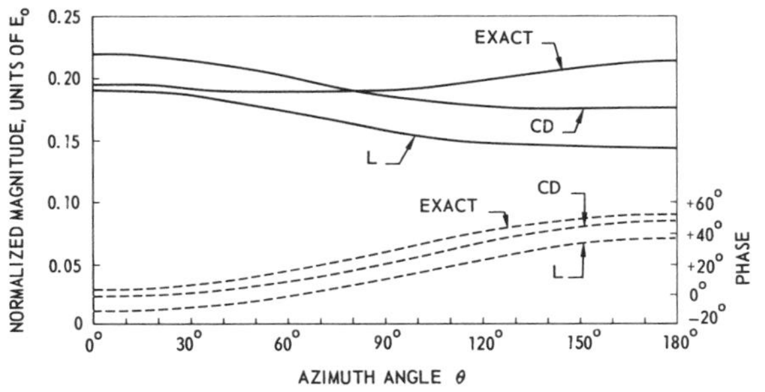

FIGURE 2. *Magnitude and phase of* E_z *on the surface of a cylinder for* $k_0a = 0.1$, $\delta/a = 0.75$ ($k_0\delta = 0.075$).

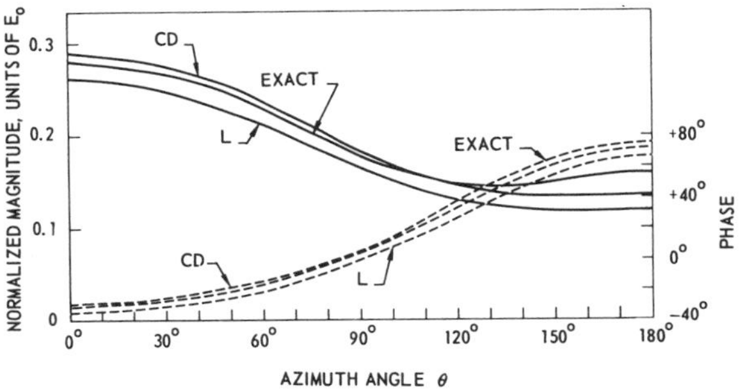

FIGURE 4. *Magnitude and phase of* E_z *on the surface of a cylinder for* $k_0a = 0.3$, $\delta/a = 0.5$ ($k_0\delta = 0.15$).

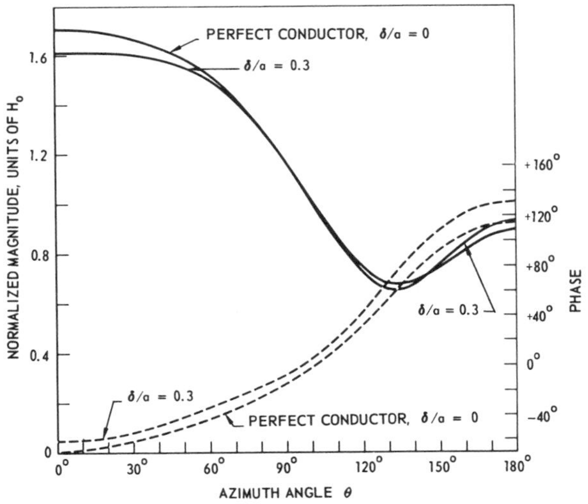

FIGURE 5. *Magnitude and phase of* H_z *on the surface of a cylinder for* $k_0a = 1.0$, $\delta/a = k_0\delta = 0, 0.3$.

be changed to $O\,[|k_c|\lambda/k_l^3]$; this more stringent condition is necessary so that an approximation analogous to (A.18) will hold.

Now let us consider the case $|Z_c| > Z_0$. Here it is best to replace (1.9) and (1.10) by the mathematically equivalent dual formulation

$$\tfrac{1}{2}\mathbf{K}_e + L_c\mathbf{K}_e + (1/Z_c)M_c\mathbf{K}_m = 0, \tag{6.1}$$

$$\tfrac{1}{2}\mathbf{K}_m - L_0\mathbf{K}_m + Z_0M_0\mathbf{K}_e = \mathbf{K}_{mo}. \tag{6.2}$$

The reason is that we want to minimize the coupling of M_0 into the second equation of the formulation, and the coupling is $O(Z_c/Z_0)$ for (1.10) but $O\,(Z_0/Z_c)$ for (6.2).

From (6.1) we can derive the Leontovich condition and a curvature-dependent condition which differs from (1.14) only by a second order term. The analog to (1.15) is

$$(1+p)\;K_{eu}\mathbf{e}_u + (1-p)\;K_{ev}\mathbf{e}_v = -\,(2/Z_c)\;M_c\mathbf{K}_m. \tag{6.3}$$

7. Concluding Remarks

Our main purpose here has been to present an integral equation formulation of scattering from a good conductor (or other highly attenuating medium) and to derive from it approximate relationships between $\mathbf{K}_m$ and $\mathbf{K}_e$ at the surface.

The integral equation formulation is given by (1.9) and (1.10) for small Z_c and by the dual equations (6.1) and (6.2) for large Z_c. The operators appearing in these equations are defined in (1.11) and (1.12). Equation (1.12), which appears to be a new result, is especially important, since it is much better suited for numerical work than is the commonly-encountered integro-differential operator of Maue (1949).

The dual explicit expressions (1.15) and (6.3) also appear to be new results. It should be emphasized that these expressions are not just of theoretical interest, but can be used directly in numerical solution techniques.

The boundary conditions (1.1) and (1.14) were derived under assumptions essentially equivalent to those of Leontovich and Rytov. The most significant point here is the introduction of a new analytical technique for obtaining approximate boundary conditions from integral equations. Other uses of this technique include determination of boundary conditions which describe reflection and transmission by a conducting shell (Mitzner, 1967).

8. Appendix

We shall here consider the geometry of S in the vicinity of a surface point $\mathbf{r}$ at which the normal is continuous. To this end, we introduce a tangent-normal cartesian coordinate system (x, y, z) with center at $\mathbf{r}$, at which point

$$(\mathbf{e}_x, \mathbf{e}_y, \mathbf{e}_z) = (\mathbf{e}_u, \mathbf{e}_v, \mathbf{n}). \tag{A.1}$$

We also introduce an associated cylindrical coordinate system (ρ, θ, z). We then define S_t as the subdomain of S contained within the cylinder

$$-t \leqslant z \leqslant t,\; \rho \leqslant t, \tag{A.2}$$

with t chosen small enough so that S_t is a simply-connected surface whose z-coordinate is a single-valued function $f(\rho, \theta)$.

Let L be the tangent plane at $\mathbf{r}$, and let D be the disc in L with center $\mathbf{r}$ and radius t. Then D is the projection of S_t on L, and we can express an integral over S_t as an integral over D, provided we take account of the element scaling factor

$$dD' = (\mathbf{n}\cdot\mathbf{n}')\,dS'. \tag{A.3}$$

Now let g be a function satisfying the odd periodicity condition

$$g(\rho,\theta-\pi) = -g(\rho,\theta), \tag{A.4}$$

and let h be a function satisfying the even periodicity condition

$$h(\rho,\theta-\pi) = h(\rho,\theta). \tag{A.5}$$

Then we have the integral formulas

$$\int_{\bar{D}} dD'\mathbf{K}'g/\rho^n = \lim_{\Delta\to 0}\int_\Delta^t d\rho \int_0^\pi d\theta(\mathbf{K}^+ - \mathbf{K}^-)g/\rho^{n-1}, \tag{A.6}$$

$$\int_{\bar{D}} dD'(\mathbf{K}' - \mathbf{K})h/\rho^n = \lim_{\Delta\to 0}\int_\Delta^t d\rho \int_0^\pi d\theta(\mathbf{K}^+ - 2\mathbf{K} + \mathbf{K}^-)h/\rho^{n-1}. \tag{A.7}$$

Here $$\mathbf{K}^+ = \mathbf{K}[\rho,\theta,f(\rho,\theta)], \qquad \mathbf{K}^- = \mathbf{K}[\rho,\theta-\pi,f(\rho,\theta-\pi)]. \tag{A.8}$$

The symbol $\int_{\bar{D}}$ indicates the integral over the punctured surface, that is, the limit as $\Delta\to 0$ of the integral over that part of D outside a sphere of radius Δ centered at $\mathbf{r}$.

Note specifically that if $\mathbf{K}$ and h are bounded on D and if the second derivatives of $\mathbf{K}$ exist in a neighborhood of $\mathbf{r}$, then the integral in (A.7) exists for $n=3$.

Another formula which is very useful for studying integrals on D and $\bar{D}$ is

$$\int_0^t d\rho\ \rho^n\ \exp\{(i-1)\rho/\delta\} = O(\delta^{n+1}). \tag{A.9}$$

Before proceeding further, we require that the surface be smooth enough at $\mathbf{r}$ and throughout S_t so that it can be represented in the form

$$z = f(\rho,\theta) = \tfrac{1}{2}\rho^2(\kappa_u\cos^2\theta + \kappa_v\sin^2\theta) + \rho^3\alpha', \tag{A.10}$$

where the κ are the principal curvatures at $\mathbf{r}$,

$$\alpha' = \frac{1}{6}(\alpha'_{30}\cos^3\theta + 3\alpha'_{21}\cos^2\theta\sin\theta + 3\alpha'_{12}\cos\theta\sin^2\theta + \alpha'_{03}\sin^3\theta), \tag{A.11}$$

and the α_{ij}'s are slowly varying functions of ρ and θ.

We then readily find that, on S_t,

$$\nabla' G_c = (1/4\pi R^2)e^{(i-1)R/\delta}(\rho/R)b_1\mathbf{s}^* \tag{A.12}$$

and $$\mathbf{n}\times(\mathbf{C}\cdot\nabla'_S)\nabla' G_c = (1/4\pi R^3)e^{(i-1)R/\delta}[b_1\mathbf{C}^*\times\mathbf{n} - b_2(\mathbf{C}^*\cdot\mathbf{s}^*)\mathbf{s}^*\times\mathbf{n}], \tag{A.13}$$

where
$$\mathbf{C}^* = \mathbf{n}' \times (\mathbf{C} \times \mathbf{n}') \tag{A.14}$$

is the component of $\mathbf{C}$ normal to $\mathbf{n}'$,

$$b_1 = 1 - (i-1)R/\delta, \qquad b_2 = [3 - 3(i-1)R/\delta - 2i(R/\delta)^2](\rho/R)^2, \tag{A.15}$$

and
$$s^* = (\mathbf{r} - \mathbf{r}')/\rho = -\{(\cos\theta \mathbf{e}_x + \sin\theta \mathbf{e}_y) + [\tfrac{1}{2}\rho(\kappa_u \cos^2\theta + \kappa_v \sin^2\theta) + \rho^2\alpha']\mathbf{n}\}. \tag{A.16}$$

Furthermore,
$$\rho/R = 1 + \beta_1\rho^2, \tag{A.17}$$

$$e^{(i-1)R/\delta} = e^{(i-1)\rho/\delta}[1 + (\rho/\delta)\beta_2\rho^2], \tag{A.18}$$

and
$$(\mathbf{n}\cdot\mathbf{n}')^{-1} = 1 + \beta_3\rho^2, \qquad \mathbf{n} \times \mathbf{n}' = (\mathbf{n} \times \vec{\gamma}_1)\rho,$$

$$\mathbf{n} - \mathbf{n}' = \rho[(\kappa_u \cos\theta + \tfrac{1}{2}\rho\alpha'_u)\mathbf{e}_x + (\kappa_v \sin\theta + \tfrac{1}{2}\rho\alpha'_v)\mathbf{e}_y] + \rho^3\vec{\gamma}_2 + \beta_4\rho^2\mathbf{n}. \tag{A.19}$$

Here
$$\alpha'_u = \alpha'_{30}\cos^2\theta + 2\alpha'_{21}\cos\theta\sin\theta + \alpha'_{12}\sin^2\theta,$$

$$\alpha'_v = \alpha'_{21}\cos^2\theta + 2\alpha'_{12}\cos\theta\sin\theta + \alpha'_{03}\sin^2\theta, \tag{A.20}$$

where the fact that these two expressions are quadratic in $\sin\theta$ and $\cos\theta$ is more important than the exact forms, and

$$\beta_i = O(\kappa^2), \qquad \vec{\gamma}_1 = O(\kappa), \qquad \vec{\gamma}_2 = O(\kappa^3) + O(|\nabla_S\alpha|). \tag{A.21}$$

9. References

Andreasen, M. G. (1964), Scattering from parallel metallic cylinders with arbitrary cross sections, IEEE Trans. Ant. Prop. **AP-12**, No. 6, 746–754.

Andreasen, M. G. (1965), Scattering from bodies of revolution, IEEE Trans. Ant. Prop. **AP-13**, No. 2, 303–310.

Kantorovich, L. V., and V. I. Krylov (1958), Approximate Methods of Higher Analysis, trans. by C. D. Benster, 101–102 (Interscience Publishers, Inc., New York, N.Y.).

Kellogg, O.D. (1953), Foundations of Potential Theory, ch. 6 (Dover Publications, Inc., New York, N.Y.).

Leontovich, M. A. (1948), Approximate boundary conditions for the electromagnetic field on the surface of a good conductor, Investigations on Radiowave Propagation, Pt. II, 5–12 (Printing House of the Academy of Sciences, Moscow). Trans. V. A. Fock, Diffraction, refraction, and reflection of radio waves, Appendix, Air Force Cambridge Res. Center TN-57-102 (1957).

Maue, A.-W. (Aug. 1949), Zur Formulierung eines allgemeinen Beugungsproblems durch eine Integralgleichung, Z. Physik, **126**, 601–618. (Translation available from this author.)

Mitzner, K. M. (1967), Effective boundary conditions describing reflection and transmission by a conducting shell of arbitrary shape, paper presented at Spring URSI meeting, Ottawa, Canada.

Müller, C. (1957), Grundprobleme der Mathematischen Theorie Elektromagnetischer Schwingungen, sec. 23 (Springer-Verlag, Berlin).

Oshiro, F. K., and R. G. Cross (1966), A source distribution technique for solution of two-dimensional scattering problems, Northrop Norair Rept. NOR 66–74.

Oshiro, F. K., and C. W. Su (1965), A source distribution technique for the solution of general electromagnetic scattering problems, Northrop Norair Rept. NOR 65–271.

Oshiro, F. K., F. P. Torres, and H. C. Heath (1966), Numerical procedures for calculating radar cross section of arbitrarily shaped three-dimensional geometries, Air Force Avionics Lab. Tech. Rept. AFAL-TR-66-162, Vol. I.

Rytov, S. M. (1940), Computation of the skin effect by the perturbation method, Zhur. Eksp. i Teoret. Fiz., **10**, No. 2, 180–189. (Translation available from this author.)

Van Bladel, J. (1964), Electromagnetic Fields, App. 2 (McGraw-Hill Book Co., Inc., New York, N.Y.).

(Paper 2–12–312)

Electromagnetic Fields Induced Inside Arbitrarily Shaped Biological Bodies

DONALD E. LIVESAY, MEMBER, IEEE, AND KUN-MU CHEN, SENIOR MEMBER, IEEE

Abstract—**A theoretical method has been developed to determine the electromagnetic field induced inside heterogeneous biological bodies of irregular shapes. A tensor integral equation for the electric field inside the body was derived and solved numerically for various biological models.**

I. INTRODUCTION

During the past few years, researchers have been engaged in a controversy over the possible hazards of nonionizing electromagnetic radiation. Humans and animals have exhibited a variety of physiological reactions to electromagnetic radiation, but investigators have been largely unable to determine whether these effects were produced thermally or nonthermally. The problem arises mainly because it is difficult to determine the electromagnetic field intensity inside an arbitrary configuration of tissues. Without knowing the field intensity, it is virtually impossible to determine the heat generation and the temperature inside the tissue structure; hence, it is difficult to judge whether the observed effect is thermal or nonthermal.

When the human body or a biological system is illuminated by an electromagnetic wave, an electromagnetic field is induced inside the body and an electromagnetic wave is scattered externally by the body. Since the human body or a biological system is an irregularly shaped heterogeneous conducting medium with frequency-dependent permittivity and conductivity, the distribution of the internal electromagnetic field and the scattered electromagnetic wave will depend on the body's physiological parameters and geometry, as well as the frequency and polarization of the incident wave.

The mathematical complexity of the problem has led researchers in this area to investigate simple models of tissue structures. Some commonly used models are the plane slab [3], [4], and the dielectric cylinder [5].

Although analyzing simple models does increase our understanding of microwave absorption and scattering by biological tissues, the results have limited applicability to arbitrary physiological systems. In our study, we have developed a general technique for calculating the electric field induced inside an arbitrary biological body by an incident electromagnetic wave. After deriving a tensor integral equation for the internal electric field, we proceeded to solve the equation numerically by the method of moments [7].

We note that Richmond [8], [9] has carried out a similar moment method solution to a two-dimensional integral equation for infinite dielectric cylinders with arbitrary cross sections. An integral-equation approach to scattering by dielectric rings has been examined by Van Doeren [10].

Manuscript received May 2, 1974; revised August 26, 1974. This work was supported in part by the National Science Foundation under Grant ENG 74-12603.

The authors are with the Department of Electrical Engineering and Systems Science, Michigan State University, East Lansing, Mich. 48824.

II. DERIVATION OF INTEGRAL EQUATION

Consider a finite body of arbitrary shape, with permittivity $\epsilon(r)$ and conductivity $\sigma(r)$, illuminated in free space by a plane electromagnetic wave as shown in Fig. 1. The induced current in the body gives rise to a scattered field E^s, which may be accounted for by replacing the body with an equivalent free-space current density J_{eq}, given by

$$J_{eq}(r) = [\sigma(r) + j\omega(\epsilon(r) - \epsilon_0)]E(r) = \tau(r)E(r). \quad (1)$$

The first term of (1) is the conduction current and the second term represents the polarization current. ϵ_0 is the free-space permittivity and $E(r)$ is the total electric field inside the body.

The scattered field inside the body may be expressed in terms of J_{eq} by using the free-space tensor Green's function $\mathsf{G}(r,r')$. However, when the field point is inside the body, E^s must be evaluated with special care because of singularity and uniqueness problems. According to Van Bladel's paper [6], the scattered field E^s at an arbitrary point inside the body can be expressed as

$$E^s(r) = \int_V J_{eq}(r') \cdot \left[\text{PV}\mathsf{G}(r,r') - \frac{\mathsf{I}\delta(r - r')}{3j\omega\epsilon_0} \right] dV'$$

$$= \text{PV} \int_V J_{eq}(r') \cdot \mathsf{G}(r,r')\, dV' - \frac{J_{eq}(r)}{3j\omega\epsilon_0} \quad (2)$$

where

$$\mathsf{G}(r,r') = -j\omega\mu_0 \left[\mathsf{I} + \frac{\nabla\nabla}{k_0^2} \right] \psi(r,r') \quad (3)$$

$$\psi(r,r') = \frac{\exp(-jk_0 |r - r'|)}{4\pi |r - r'|} \quad (4)$$

$$k_0 = \omega(\mu_0\epsilon_0)^{1/2}$$

μ_0 is the permeability of free space, and the PV symbol

Reprinted from *IEEE Trans. Microwave Theory Tech.*, vol. MTT-22, no. 12, pp. 1273–1280, December 1974.

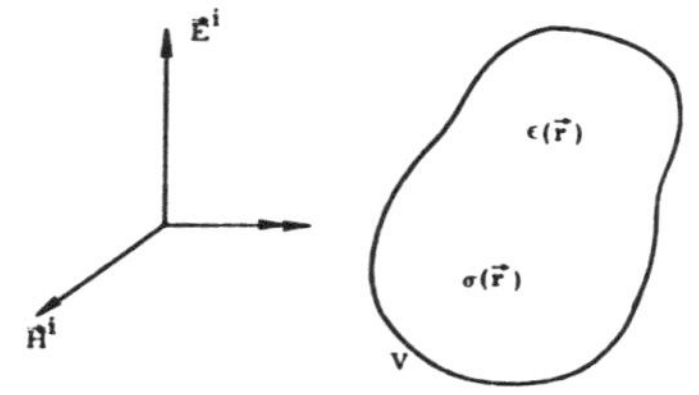

Fig. 1. An arbitrarily shaped biological body illuminated by a plane electromagnetic wave.

denotes the principal value of the integral as defined by Van Bladel [6].

We may write the total electric field $\boldsymbol{E}(r)$ inside the body as the sum of the incident electric field $\boldsymbol{E}^i$ and the scattered field $\boldsymbol{E}^s$. Thus

$$\boldsymbol{E}(r) = \boldsymbol{E}^i(r) + \boldsymbol{E}^s(r). \tag{5}$$

Substituting (2) in (5) and rearranging terms gives the desired integral equation for $\boldsymbol{E}(r)$:

$$\left[1 + \frac{\tau(r)}{3j\omega\epsilon_0}\right]\boldsymbol{E}(r) - \text{PV}\int_V \tau(r')\boldsymbol{E}(r')\cdot\mathsf{G}(r,r')\,dV' = \boldsymbol{E}^i(r). \tag{6}$$

In (6), $\boldsymbol{E}^i(r)$ is the incident electric field and is a known quantity. $\boldsymbol{E}(r)$ is the unknown total electric field inside the body. $\boldsymbol{E}(r)$ can be determined from (6) by moment methods.

III. TRANSFORMATION TO MATRIX EQUATION

We may represent the inner product $\boldsymbol{E}(r')\cdot\mathsf{G}(r,r')$ as

$$\boldsymbol{E}(r')\cdot\mathsf{G}(r,r') = \begin{bmatrix} G_{xx}(r,r') & G_{xy}(r,r') & G_{xz}(r,r') \\ G_{yx}(r,r') & G_{yy}(r,r') & G_{yz}(r,r') \\ G_{zx}(r,r') & G_{zy}(r,r') & G_{zz}(r,r') \end{bmatrix}\begin{bmatrix} E_x(r') \\ E_y(r') \\ E_z(r') \end{bmatrix}. \tag{7}$$

Let

$$x_1 = x \qquad x_2 = y \qquad x_3 = z.$$

Then, $G_{x_px_q}(r,r')$ is given by

$$G_{x_px_q}(r,r') = -j\omega\mu_0\left[\delta_{pq} + \frac{1}{k_0^2}\frac{\partial^2}{\partial x_q \partial x_p}\right]\psi(r,r'), \qquad p,q = 1,2,3. \tag{8}$$

The matrix in (7) is symmetric. Each scalar component of (6) may be written as

$$\left[1 + \frac{\tau(r)}{3j\omega\epsilon_0}\right]E_{x_p}(r) - \text{PV} \cdot\int_V \tau(r')\left[\sum_{q=1}^{3} G_{x_px_q}(r,r')E_{x_q}(r')\right]dV' = E_{x_p}{}^i(r), \qquad p = 1,2,3. \tag{9}$$

We can transform (6) into a matrix equation by using the method of moments. We partition the body into N subvolumes and assume that $\boldsymbol{E}(r)$ and $\tau(r)$ are constant in each subvolume. We will denote the mth subvolume by V_m, and denote the position of a representative interior point of V_m by r_m. By requiring that (9) be satisfied at each r_m, we obtain, after some rearranging,

$$\left[1 + \frac{\tau(r_m)}{3j\omega\epsilon_0}\right]E_{x_p}(r_m) - \sum_{q=1}^{3}\sum_{n=1}^{N}\left[\tau(r_n)\ \text{PV} \cdot\int_{V_n} G_{x_px_q}(r_m,r')\,dV'\right]E_{x_q}(r_n) = E_{x_p}{}^i(r_m). \tag{10}$$

After defining the following quantity:

$$\bar{G}_{x_px_q}{}^{mn} = \tau(r_n)\ \text{PV}\int_{V_n} G_{x_px_q}(r_m,r')\,dV' \tag{11}$$

we rewrite (10) as

$$\sum_{q=1}^{3}\sum_{n=1}^{N}\left[\bar{G}_{x_px_q}{}^{mn} - \delta_{pq}\delta_{mn}\left(1 + \frac{\tau(r_m)}{3j\omega\epsilon_0}\right)\right]E_{x_q}(r_n) = -E_{x_p}{}^i(r_m), \qquad m = 1,2,\cdots,N, \quad p = 1,2,3. \tag{12}$$

Let $[G_{x_px_q}]$ be the $N \times N$ matrix whose elements are defined by

$$G_{x_px_q}{}^{mn} = \bar{G}_{x_px_q}{}^{mn} - \delta_{pq}\delta_{mn}\left[1 + \frac{\tau(r_m)}{3j\omega\epsilon_0}\right] \tag{13}$$

and let $[E_{x_p}]$ and $[E_{x_p}{}^i]$ be N-dimensional vectors given by

$$[E_{x_p}] = \begin{bmatrix} E_{x_p}(r_1) \\ \cdot \\ \cdot \\ \cdot \\ E_{x_p}(r_N) \end{bmatrix} \qquad [E_{x_p}{}^i] = \begin{bmatrix} E_{x_p}{}^i(r_1) \\ \cdot \\ \cdot \\ \cdot \\ E_{x_p}{}^i(r_N) \end{bmatrix}, \qquad p = 1,2,3. \tag{14}$$

As m and p range over all possible values in (12), we obtain the following matrix representation of (6):

$$\begin{bmatrix} [G_{xx}] & [G_{xy}] & [G_{xz}] \\ [G_{yx}] & [G_{yy}] & [G_{yz}] \\ [G_{zx}] & [G_{zy}] & [G_{zz}] \end{bmatrix}\begin{bmatrix} [E_x] \\ [E_y] \\ [E_z] \end{bmatrix} = -\begin{bmatrix} [E_x{}^i] \\ [E_y{}^i] \\ [E_z{}^i] \end{bmatrix}. \tag{15}$$

Symbolically, we may write (15) as

$$[G][E] = -[E^i]. \tag{16}$$

$[G]$ is a $3N \times 3N$ matrix, while $[E]$ and $[E^i]$ have dimension $3N$. We can determine the total electric field at each of the N chosen points by inverting $[G]$ in (16).

IV. EVALUATION OF MATRIX ELEMENTS

Equations (11) and (13) define the elements of $[G_{x_px_q}]$. We have

$$G_{x_px_q}{}^{mn} = \tau(r_n)\ \text{PV}\int_{V_n} G_{x_px_q}(r_m,r')\,dV' - \delta_{pq}\delta_{mn}\left[1 + \frac{\tau(r_m)}{3j\omega\epsilon_0}\right]. \tag{17}$$

Let us first consider the off-diagonal elements of $[G_{x_p x_q}]$. Clearly, $r_m \notin V_n$, so $G_{x_p x_q}(r_m, r')$ is continuous throughout V_n. Therefore, we may omit the principal value operation from our evaluation, so (17) becomes

$$G_{x_p x_q}{}^{mn} = \tau(r_n) \int_{V_n} G_{x_p x_q}(r_m, r')\, dV', \qquad m \neq n. \qquad (18)$$

As a first approximation, we have

$$G_{x_p x_q}{}^{mn} = \tau(r_n) G_{x_p x_q}(r_m, r_n) \Delta V_n, \qquad m \neq n \qquad (19)$$

where

$$\Delta V_n = \int_{V_n} dV'.$$

Using (8) to evaluate $G_{x_p x_q}(r_m, r_n)$ gives

$$G_{x_p x_q}{}^{mn} = \frac{-j\omega\mu_0 k_0 \tau(r_n) \Delta V_n \exp(-j\alpha_{mn})}{4\pi\alpha_{mn}{}^3} \cdot [(\alpha_{mn}{}^2 - 1 - j\alpha_{mn})\delta_{pq} + \cos\theta_{x_p}{}^{mn} \cdot \cos\theta_{x_q}{}^{mn}(3 - \alpha_{mn}{}^2 + 3j\alpha_{mn})], \qquad m \neq n \quad (20)$$

where

$$\alpha_{mn} = k_0 R_{mn} \qquad R_{mn} = |r_m - r_n|$$

$$\cos\theta_{x_p}{}^{mn} = \frac{(x_p{}^m - x_p{}^n)}{R_{mn}} \qquad \cos\theta_{x_q}{}^{mn} = \frac{(x_q{}^m - x_q{}^n)}{R_{mn}}$$

and we have written r_m and r_n as

$$r_m = (x_1{}^m, x_2{}^m, x_3{}^m) \qquad r_n = (x_1{}^n, x_2{}^n, x_3{}^n).$$

If N is sufficiently large, the approximation given by (19) and (20) may yield adequate results. For greater accuracy, (18) may be integrated numerically by any convenient method.

If the body's cross section, as seen by the incident wave, is elongated, with the longer dimension parallel to E^i, numerical integration of (18) may considerably improve the accuracy of the solution. The reason is illustrated in Fig. 2. The scattered field in subvolume V_1 contributed by a parallel unit current in subvolume V_2 has approximately twice the magnitude in Fig. 2(a) that it has in Fig. 2(b), provided $k_0 R_{12} \ll 1$, even though R_{12} is the same in both figures. In the case we are considering, most of the contributions to the scattered field are of the type shown in Fig. 2(a). Since these contributions are more significant, they should be evaluated more carefully to preserve accuracy.

For the diagonal elements of $[G_{x_p x_q}]$, (17) becomes

$$G_{x_p x_q}{}^{nn} = \tau(r_n)\, \mathrm{PV} \int_{V_n} G_{x_p x_q}(r_n, r')\, dV' - \delta_{pq}\left[1 + \frac{\tau(r_n)}{3j\omega\epsilon_0}\right]. \qquad (21)$$

We approximate V_n by a sphere of equal volume centered at r_n. This enables us to evaluate the integral in (21) exactly.

Let a_n be the radius of the sphere. After a lengthy calculation (see Appendix), we find

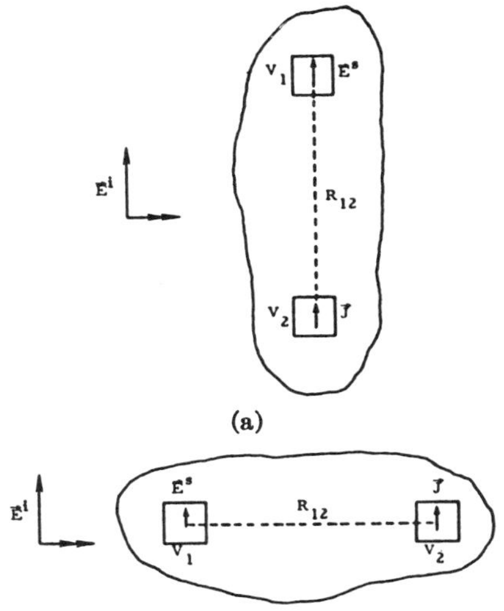

Fig. 2. (a) Scattered field in subvolume V_1 produced by unit current in subvolume V_2 when elongated body is parallel to incident field. (b) Scattered field in subvolume V_1 produced by unit current in subvolume V_2 when elongated body is perpendicular to incident field.

$$G_{x_p x_q}{}^{nn} = \delta_{pq}\left\{\frac{-2j\omega\mu_0 \tau(r_n)}{3k_0{}^2}[\exp(-jk_0 a_n)(1 + jk_0 a_n) - 1] - \left[1 + \frac{\tau(r_n)}{3j\omega\epsilon_0}\right]\right\} \qquad (22)$$

where

$$a_n = \left(\frac{3\Delta V_n}{4\pi}\right)^{1/3}.$$

If the actual shape of V_n differs appreciably from that of a sphere, our approximation may lead to poor numerical results. In such a case, the integration throughout a small sphere surrounding r_n can be performed as in the Appendix; the integration throughout the remainder of V_n can be done numerically.

V. SELECTED NUMERICAL RESULTS

The theory has been applied to some simple biological models. The first two examples evidence the accuracy of the numerical method.

Fig. 3 shows the electric field at the center of an electrically small dielectric cube (4 × 4 × 4 cm) illuminated by a plane wave, for various values of frequency and dielectric constant. Since the cube exhibits a symmetrical cross section to the incident wave, we only need to calculate the induced field in one quadrant, as illustrated in Fig. 3. As expected, the electric field intensity at the center of the cube is nearly equal to the electric field intensity at the center of a sphere with the same dielectric constant in a uniform electrostatic field. The field E in the sphere is given by

$$E = \left(\frac{3}{\epsilon_r + 2}\right) E^i \qquad (23)$$

where E^i is the externally applied field, and $\epsilon_r = \epsilon/\epsilon_0$.

For our second example, we calculated the electric field

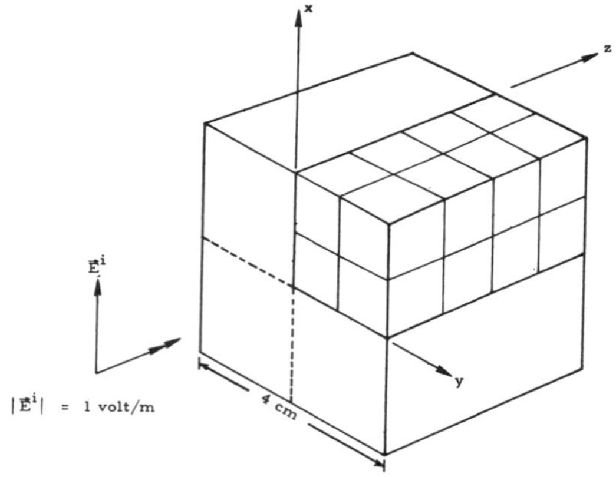

f (Hz)	λ_o (cm)	ϵ_r	λ_ϵ (cm)	$\frac{3}{\epsilon_r + 2}$	$\lvert E \rvert$ center (volts/m)
10^7	3×10^3	5.0	1.342×10^3	0.4286	0.4172
10^6	3×10^4	5.0	1.342×10^4	0.4286	0.4172
10^3	3×10^7	5.0	1.342×10^7	0.4286	0.4172
10^3	3×10^7	20.0	6.708×10^6	0.1364	0.1124
10^3	3×10^7	51.7	4.172×10^6	0.0559	0.0503

Fig. 3. Electric field at the center of a dielectric cube ($4 \times 4 \times 4$ cm) induced by plane waves of various frequencies.

inside a thin conducting layer ($\epsilon_r = 70$, $\sigma = 1$ mho/m) illuminated by a 300-MHz plane wave. Fig. 4 shows the results when the incident electric field is perpendicular to the plane of the layer. For this case the electric field inside the layer is approximately equal to E^i/ϵ_r, consistent with the boundary conditions. If the plane of the layer is now oriented parallel to the incident electric field, the electric field inside the layer increases about ten times, as shown in Fig. 5. This example shows that the intensity of the induced electric field inside a conducting body depends heavily on the body's orientation with respect to the incident wave.

The third example shows the induced electric field inside a tissue block ($16 \times 12 \times 4$ cm) comprising a fat layer and a muscle layer, illuminated by a 100-MHz plane wave. The distribution of the induced field is shown graphically in Fig. 6 and numerically in Fig. 7. Some interesting findings are as follows. 1) Inside the tissue block all three components of the electric field—E_x, E_y, and E_z—are induced, although the incident electric field has only an x component. We note that the induced E_z is comparable to the induced E_x in magnitude. 2) The distribution of the internal electric field is nonuniform, and its amplitude is quite different from that predicted by the plane slab model.

In the examples considered so far, the matrix elements were calculated using (20) and (22). For the remaining cases presented here, we computed the matrix elements using (22) and numerically integrating (18).

The final two examples illustrate the effects of an inhomogeneity in a thin tissue cylinder (10 cm $\times$ 1 mm $\times$ 1 mm) exposed to a 2.45-GHz plane wave. The dimensions

(only 1/2 of the volume is shown).
E_x is calculated at the centers of blocks.
E_y and E_z are very small.

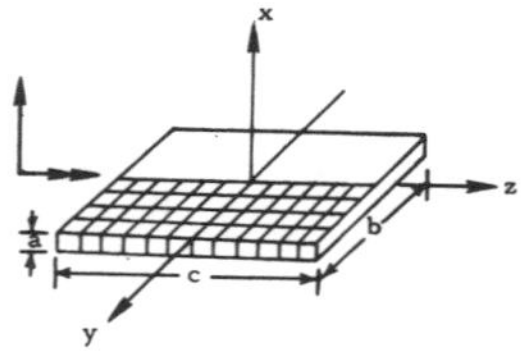

$\vec{E}^i = \hat{x}\, e^{-j2\pi(z/100)}$
freq = 300 MHz
σ = 1 mho/m
$\epsilon = 70\,\epsilon_o$
a = 0.5 cm
b = 4.0 cm
c = 6.0 cm

E_x/E^i (z vertical, y horizontal)

.0159	.0161	.0155	.0210
.0110	.0112	.0108	.0155
.0115	.0116	.0112	.0160
.0112	.0114	.0110	.0158
.0112	.0114	.0110	.0158
.0112	.0114	.0110	.0158
.0112	.0114	.0110	.0158
.0112	.0114	.0110	.0158
.0112	.0114	.0110	.0158
.0115	.0116	.0112	.0160
.0110	.0112	.0108	.0155
.0159	.0161	.0155	.0211

$\frac{\epsilon_o}{\epsilon} = 0.0143$

Fig. 4. Electric field inside of a layer of conducting medium with $\epsilon = 70\epsilon_0$, $\sigma = 1$ mho/m induced by a plane wave of 300 MHz with the incident electric field perpendicular to the flat surfaces of the layer.

$\vec{E}^i = \hat{x}\, e^{-j2\pi(z/100)}$
freq = 300 MHz
σ = 1 mho/m
$\epsilon = 70\,\epsilon_o$

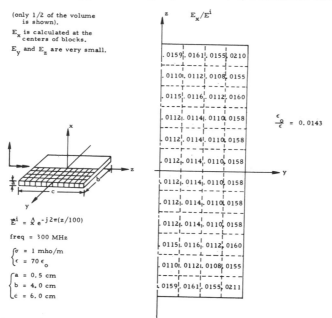

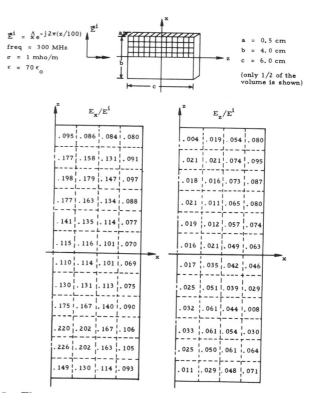

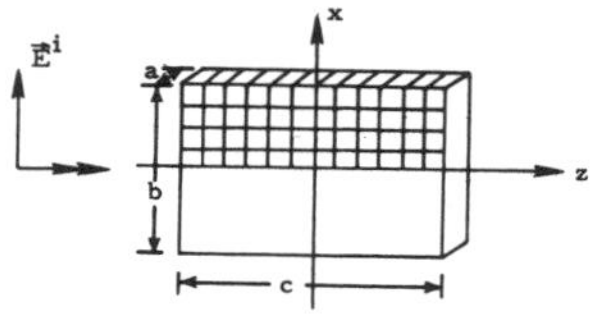

a = 0.5 cm
b = 4.0 cm
c = 6.0 cm
(only 1/2 of the volume is shown)

E_x/E^i (z vertical, x horizontal)

.095	.086	.084	.080
.177	.158	.131	.091
.198	.179	.147	.097
.177	.163	.134	.088
.141	.135	.114	.077
.115	.116	.101	.070
.110	.114	.101	.069
.130	.131	.113	.075
.175	.167	.140	.090
.220	.202	.167	.106
.226	.202	.163	.105
.149	.130	.114	.093

E_z/E^i (z vertical, x horizontal)

.004	.019	.054	.080
.021	.021	.074	.095
.018	.016	.073	.087
.021	.011	.065	.080
.019	.012	.057	.074
.016	.021	.049	.063
.017	.035	.042	.046
.025	.051	.039	.029
.032	.061	.044	.008
.033	.061	.054	.030
.025	.050	.061	.064
.011	.029	.048	.071

Fig. 5. Electric field inside of a layer of conducting medium with $\epsilon = 70\epsilon_0$, $\sigma = 1$ mho/m induced by a plane wave of 300 MHz with the incident electric field parallel to the flat surfaces of the layer.

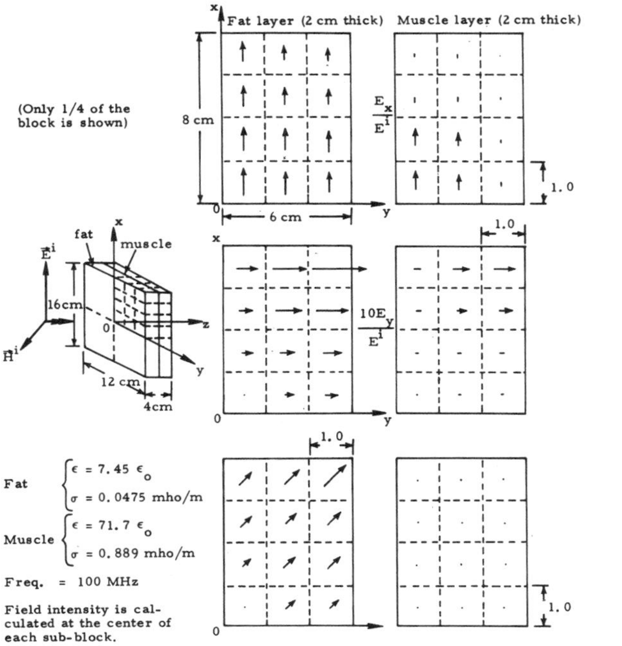

Fig. 6. Distribution of induced electric field inside of a tissue block (16 × 12 × 4 cm) consisting of a fat and a muscle layer. The incident plane wave is $\boldsymbol{E} = \boldsymbol{x}E^i$ with a frequency of 100 MHz. The field intensity in a corresponding infinite slab model is $E_x = 0.197E^i$ at the center of the fat layer and $E_x = 0.210E^i$ at the center of the muscle layer.

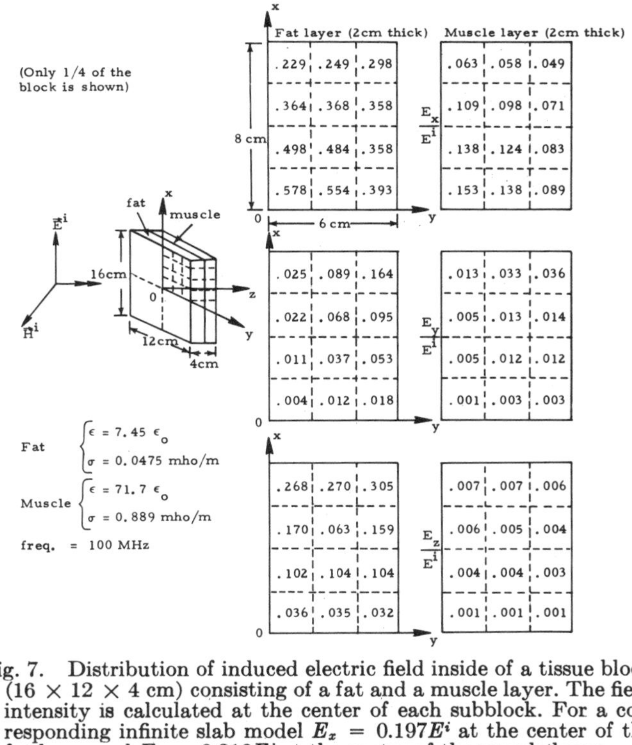

Fig. 7. Distribution of induced electric field inside of a tissue block (16 × 12 × 4 cm) consisting of a fat and a muscle layer. The field intensity is calculated at the center of each subblock. For a corresponding infinite slab model $E_x = 0.197E^i$ at the center of the fat layer and $E_x = 0.210E^i$ at the center of the muscle layer.

of the cylinder were chosen to help insure an accurate solution at this frequency. The incident electric field (x component only) is assumed to be parallel to the cylinder axis, so we may neglect the y and z components of the induced field.

Fig. 8(a) shows a homogeneous muscle cylinder, and Fig. 8(b) shows a muscle cylinder with a section of fat 1 cm long at its center. The field distribution in each cylinder is shown in Fig. 8(c), and the power density is plotted in Fig. 8(d). The discontinuity of the electric field at the muscle–fat interface in Fig. 8(c) obeys the relation

$$(\sigma_F + j\omega\epsilon_F)E_F = (\sigma_M + j\omega\epsilon_M)E_M \tag{24}$$

where the subscripts F and M refer to fat and muscle, respectively, and E_F and E_M are both normal to the interface. Equation (24) is the boundary condition for the normal component of the electric field at the boundary of two different biological tissues.

In Fig. 9(a) we have a homogeneous fat cylinder, and in Fig. 9(b) we have a fat cylinder with a 1-cm section of muscle at its center. The distributions of the induced field and the power density are shown in Fig. 9(c) and (d), respectively. Again, the discontinuity of the electric field in Fig. 9(c) obeys (24).

From Figs. 8(d) and 9(d), we see that the inhomogeneity creates a local hot spot on the fat side of the muscle–fat boundary because the electric field is normal to the boundary. We note that in the plane slab model, the electric field is tangent to the muscle–fat interface. In that case, a local hot spot occurs in the muscle.

Of particular interest is the cylinder in Fig. 9(b). Because the muscle section is relatively small, the maximum power density occurs in the fat region. The heat generation near the center of this cylinder is several times greater than the heat generation near the center of the homogeneous cylinder of Fig. 9(a). Therefore, the temperature of an irradiated cylindrical fat structure could be significantly increased by the presence of one or more small muscle segments.

VI. DISCUSSION OF NUMERICAL RESULTS

When using a pulse function expansion in the method of moments, it is important to establish an upper limit on the dimensions of the subvolumes. To arrive at the limit for our method, we have performed two convergence tests. In both tests, the incident electric field was parallel to the x axis and had a magnitude of 1 V/m.

In the first test, we investigated a muscle cylinder, shown in Fig. 10, illuminated by a 2.45-GHz plane wave. Expressed in wavelengths, the dimensions of the cylinder were 3 × 1/2 × 1/2. Actual computations supported our assumption that the induced electric field had only an x component. We partitioned the cylinder into a variable number of cubical cells, or subvolumes, and calculated the induced field for each configuration. The models for $N = 6$, $N = 48$, and $N = 162$ are shown in Fig. 10(a)–

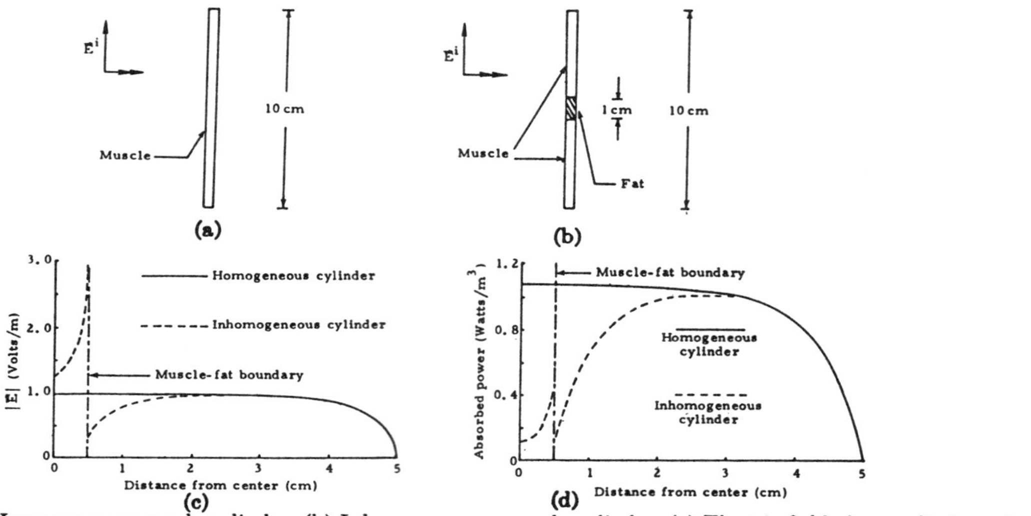

Fig. 8. (a) Homogeneous muscle cylinder. (b) Inhomogeneous muscle cylinder. (c) Electric field along cylinder axis. E^i is 1 V/m. (d) Power density ($1/2\sigma \mid E \mid^2$) along cylinder axis. Each cylinder was partitioned lengthwise into 100 subvolumes of equal size.

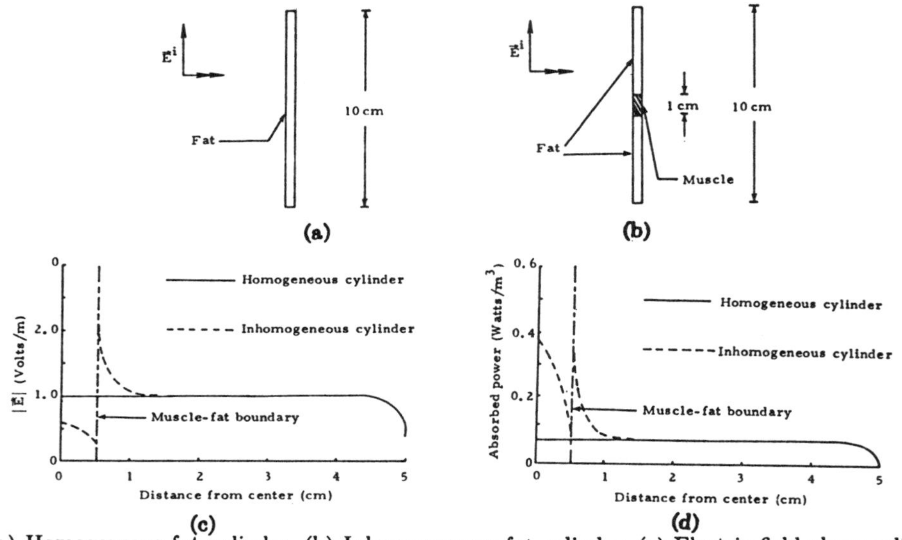

Fig. 9. (a) Homogeneous fat cylinder. (b) Inhomogeneous fat cylinder. (c) Electric field along cylinder axis. E^i is 1 V/m. (d) Power density ($1/2\sigma \mid E^2 \mid$) along cylinder axis. Each cylinder was partitioned lengthwise into 100 subvolumes of equal size.

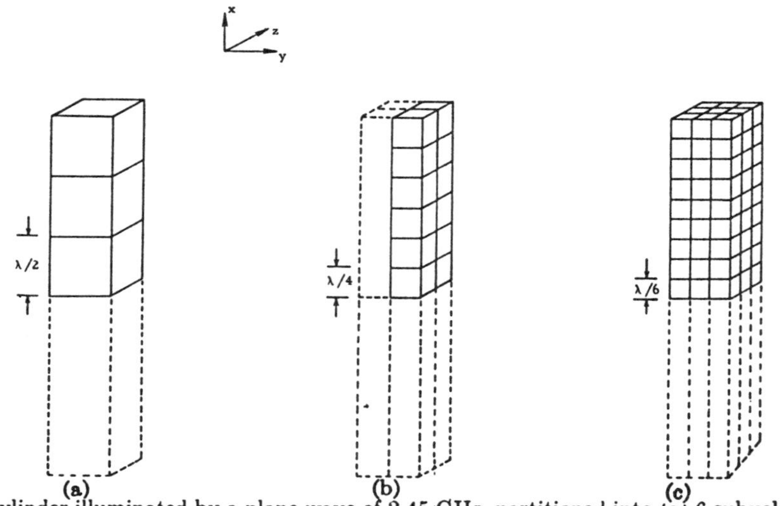

Fig. 10. A muscle cylinder illuminated by a plane wave of 2.45 GHz, partitioned into (a) 6 subvolumes, (b) 48 subvolumes, and (c) 162 subvolumes.

(c), respectively. The edges of each cell measured 1/2, 1/4, and 1/6 wavelength, respectively.

Fig. 11 shows the electric field intensity along the axis of the cylinder for each model. Since none of the subvolumes in Fig. 10(b) lie on the axis, we have plotted the average of the field intensities in the front and back of Fig. 10(b) to facilitate a comparison with Fig. 10(a) and (c). The results from all three models agree well, indicating that cells with dimensions as large as even 1/2 wavelength may produce useful data in some cases.

For the second test, we used a cube of muscle, again illuminated by a 2.45-GHz plane wave. The edges of our first sample measured one wavelength, as shown in Fig. 12(a). Treating the cube as a single cell, we calculated the induced electric field. Next, we partitioned the cube into 27 cubical subvolumes, as shown in Fig. 12(b), and again determined the induced field. We then compared the field intensity in the center cell with the value obtained from the first computation. Finally, we repeated the procedure for a 1/4-wavelength cube, shown in Fig. 12(c) and (d). The results are given in Table I. The two values agree well for the 1/4-wavelength cube. From these results, and from those of the first test, we conclude that subvolumes having edges of 1/4 wavelength or less should yield reliable data.

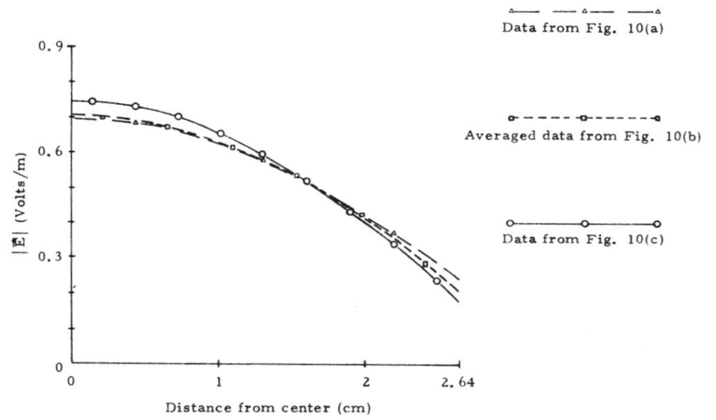

Fig. 11. Electric field intensity along the axis of the cylinder shown in Fig. 10, for models (a), (b), and (c). Frequency is 2.45 GHz. Incident field is 1 V/m. One wavelength in the cylinder is 1.76 cm.

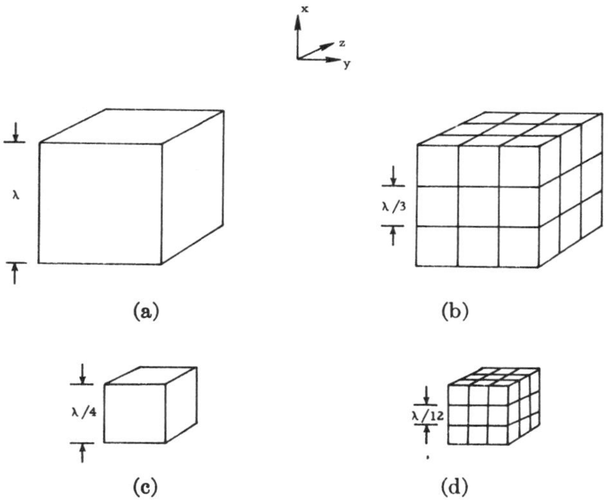

Fig. 12. Two cubes of muscle illuminated by a plane wave of 2.45 GHz, treated as single cells in (a) and (c), and partitioned into 27 subvolumes in (b) and (d). The edges of the cubes measure one wavelength and 1/4 wavelength, respectively.

TABLE I
INDUCED ELECTRIC FIELD AT THE CENTER OF THE MUSCLE CUBES SHOWN IN FIG. 12 FOR VARIOUS NUMBERS OF SUBVOLUMES

No. of cells	Size of each cell	$|\bar{E}|$ center (Volts/m)	Figure reference
1	λ	0.0789	12(a)
27	λ/3	0.0922	12(b)
1	λ/4	0.0592	12(c)
27	λ/12	0.0556	12(d)

Note: Frequency is 2.45 GHz; incident field is 1 V/m.

Concluding our discussion of the numerical results, we note the absence of supportive experimental data from our paper. Although we have not yet verified any of our numerical data experimentally, we are now beginning a research program in which we hope to do so.

APPENDIX
EVALUATION OF MATRIX DIAGONAL ELEMENTS

In (21), let us define the following:

$$I_{x_p x_q}{}^n = \mathrm{PV} \int_{V_n} G_{x_p x_q}(r_n, r')\, dV'. \tag{A1}$$

As outlined in Section IV, we approximate V_n by a sphere of equal volume centered at r_n. Hence, a_n, the radius of the sphere, is given by

$$a_n = \left(\frac{3\Delta V_n}{4\pi}\right)^{1/3}. \tag{A2}$$

It is easily verified that (8) may be rewritten as

$$G_{x_p x_q}(r,r') = -j\omega\mu_0 \left[\delta_{pq} + \frac{1}{k_0^2} \frac{\partial^2}{\partial x_q' \partial x_p'}\right] \psi(r,r'). \tag{A3}$$

Since the derivatives are taken with respect to the primed variables, we are free to take $r = r_n$ at the outset. $G_{x_p x_q}(r_n, r')$ is a function of $| r_n - r' |$ only, so we may define a spherical coordinate system centered at r_n, and set $r_n = \mathbf{0}$. Then

$$\psi(r_n, r') = \psi(r') = \frac{\exp(-jk_0 r')}{4\pi r'} \tag{A4}$$

where

$$r' = | r' |.$$

Now

$$\frac{\partial^2 \psi}{\partial x_q' \partial x_p'} = \frac{d^2\psi}{dr'^2} \frac{x_p'}{r'} \frac{x_q'}{r'} + \frac{1}{r'} \frac{d\psi}{dr'} \left[\delta_{pq} - \frac{x_p'}{r'} \frac{x_q'}{r'}\right]. \tag{A5}$$

In our spherical coordinate system

$$x_1'/r' = \sin\theta'\cos\phi' \qquad x_2'/r' = \sin\theta'\sin\phi'$$
$$x_3'/r' = \cos\theta' \qquad dV' = r'^2\sin\theta'\,d\theta'\,d\phi'\,dr' \tag{A6}$$

where θ' and ϕ' are the usual polar angles.

Since the primed variables are merely dummy variables, we shall omit the primes from now on. For convenience, we define the following:

$$f_{x_px_q}(\theta,\phi) = \frac{x_p}{r}\frac{x_q}{r}. \tag{A7}$$

Equation (A1) becomes

$$I_{x_px_q}{}^n = -j\omega\mu_0 \lim_{\eta\to 0}\int_\eta^{a_n} dr \int_0^{2\pi} d\phi \cdot \int_0^\pi \left\{\psi\delta_{pq} + \frac{1}{k_0^2}\left[\frac{d^2\psi}{dr^2} f_{x_px_q}(\theta,\phi) + \frac{1}{r}\frac{d\psi}{dr}\left(\delta_{pq} - f_{x_px_q}(\theta,\phi)\right)\right]\right\} r^2 \sin\theta\, d\theta. \tag{A8}$$

Integrating by parts gives

$$I_{x_px_q}{}^n = -j\omega\mu_0 \lim_{\eta\to 0}\left\{\int_\eta^{a_n} dr \int_0^{2\pi} d\phi \int_0^\pi \delta_{pq}\psi r^2 \sin\theta\, d\theta + \frac{1}{k_0^2}\left(r^2 \frac{d\psi}{dr}\bigg|_\eta^{a_n}\right)\int_0^{2\pi} d\phi \int_0^\pi f_{x_px_q}(\theta,\phi)\sin\theta\, d\theta + \frac{1}{k_0^2}\int_0^{2\pi} d\phi \int_0^\pi [\delta_{pq} - 3f_{x_px_q}(\theta,\phi)]\sin\theta\, d\theta \cdot \int_\eta^{a_n}\frac{d\psi}{dr} r\, dr\right\}. \tag{A9}$$

In the third term of (A9),

$$\int_\eta^{a_n}\frac{d\psi}{dr} r\, dr$$

becomes infinite as $\eta \to 0$. However, it is readily verified that

$$\int_0^{2\pi} d\phi \int_0^\pi [\delta_{pq} - 3f_{x_px_q}(\theta,\phi)]\sin\theta\, d\theta = 0, \qquad p,q = 1,2,3. \tag{A10}$$

Hence, the third term of (A9) is zero for all finite values of η, and contributes nothing as $\eta \to 0$.

It may also be readily verified that

$$\int_0^{2\pi} d\phi \int_0^\pi f_{x_px_q}(\theta,\phi)\sin\theta\, d\theta = \frac{4\pi}{3}\delta_{pq}. \tag{A11}$$

Equation (A9) then becomes

$$I_{x_px_q}{}^n = -j\omega\mu_0\delta_{pq} \lim_{\eta\to 0}\left\{\int_\eta^{a_n} dr \int_0^{2\pi} d\phi \int_0^\pi \psi r^2 \sin\theta\, d\theta + \frac{4\pi}{3k_0^2} r^2 \frac{d\psi}{dr}\bigg|_\eta^{a_n}\right\}. \tag{A12}$$

After a few simple steps, we obtain

$$I_{x_px_q}{}^n = \frac{-2j\omega\mu_0\delta_{pq}}{3k_0^2}[\exp(-jk_0a_n)(1 + jk_0a_n) - 1]. \tag{A13}$$

Substituting (A13) into (21) gives (22).

REFERENCES

[1] H. P. Schwan, "Radiation biology, medical applications, and radiation hazards," in *Microwave Power Engineering*, vol. 2, E. C. Okress, Ed. New York: Academic, 1968, pp. 215–232.

[2] J. F. Lehmann *et al.*, "Comparison of relative heating patterns produced in tissues by exposure to microwave energy at frequencies of 2450 and 900 megacycles," *Arch. Phys. Med. Rehabil.*, vol. 43, pp. 69–76, Feb. 1962.

[3] A. R. Shapiro, R. F. Lutomirski, and H. T. Yura, "Induced fields and heating within a cranial structure irradiated by an electromagnetic plane wave," *IEEE Trans. Microwave Theory Tech.* (*Special Issue on Biological Effects of Microwaves*), vol. MTT-19, pp. 187–196, Feb. 1971.

[4] H. N. Kritikos and H. P. Schwan, "Hot spots generated in conducting spheres by electromagnetic waves and biological implications," *IEEE Trans. Biomed. Eng.*, vol. BME-19, pp. 53–58, Jan. 1972.

[5] H. S. Ho, A. W. Guy, R. A. Sigelmann, and J. R. Lehmann, "Electromagnetic heating patterns in circular cylindrical models of human tissue," in *Proc. 8th Annu. Conf. Medical and Biological Engineering* (Chicago, Ill.), July 1969, p. 27.

[6] J. Van Bladel, "Some remarks on Green's dyadic for infinite space," *IRE Trans. Antennas Propagat.*, vol. AP-9, pp. 563–566, Nov. 1961.

[7] R. F. Harrington, *Field Computation by Moment Methods.* New York: Macmillan, 1968, ch. 1.

[8] J. H. Richmond, "Scattering by a dielectric cylinder of arbitrary cross-section shape," *IEEE Trans. Antennas Propagat.*, vol. AP-13, pp. 334–341, May 1965.

[9] ——, "TE-wave scattering by a dielectric cylinder of arbitrary cross-section shape," *IEEE Trans. Antennas Propagat.*, vol. AP-14, pp. 460–464, July 1966.

[10] R. E. Van Doeren, "An integral equation approach to scattering by dielectric rings," *IEEE Trans. Antennas Propagat.* (Communications), vol. AP-17, pp. 373–374, May 1969.

Dielectric coated wire antennas

J. H. Richmond and E. H. Newman

ElectroScience Laboratory, Department of Electrical Engineering, The Ohio State University, Columbus, Ohio 43212

(Received July 22, 1974.)

The problem considered is an electrically thin dielectric insulating shell on an antenna composed of electrically thin circular cylindrical wires. The solution is a moment method solution and the insulating shell is modeled by equivalent volume polarization currents. These polarization currents are related in a simple manner to the surface charge density on the wire antenna. In this way the insulating shell causes no new unknowns to be introduced, and the size of the impedance matrix is the same as for the uninsulated wires. The insulation is accounted for entirely through a modification of the symmetric impedance matrix. This modification influences the current distribution, impedance, efficiency, field patterns, and scattering properties. The theory will be compared with measurement for dielectric coated antennas in air.

1. INTRODUCTION

For a wire antenna in a conducting medium the radiation efficiency can often be improved by insulating all or part of the wire from the medium. This is accomplished with a thin dielectric layer coated on the wire surface. This paper considers the electromagnetic modeling of the dielectric layer or shell.

The problem of a circular insulating shell on a thin wire antenna has received considerable attention in the literature [*Wu et al.*, 1973; *King*, 1964; *Iizuka*, 1963; *King et al.*, 1974]. Unfortunately, much of this work is restricted as to the geometry of the antenna or the case where the complex permittivity of the exterior medium is much greater than that of the insulating layer. A simple theory is presented here which models the insulating shell by equivalent volume polarization currents, and which is restricted to electrically thin insulating layers covering electrically thin wire antennas or scatterers [*Richmond and Newman*, 1974; *Richmond*, 1974*a*]. Unlike previous theoretical and experimental work, the insulating shell is considered to terminate at the ends of the wire structure.

The solution presented is a modification of the piecewise-sinusoidal reaction formulation [*Richmond*, 1974*a*] for bare or uninsulated wires. Thus, a brief review of the theory for bare wires is presented first. Next it is shown how the theory for the bare wires can be modified to account for the insulating layer. Finally, a comparison is made between measured and calculated admittance.

2. BARE WIRE STRUCTURE

This section briefly reviews some aspects of the piecewise-sinusoidal reaction formulation for uninsulated thin-wire antennas or scatters in a homogeneous conducting medium. A more complete treatment is available elsewhere [*Richmond*, 1974*a*].

Let S denote the closed surface of the wire structure, which is composed of a number of straight segments, and let V denote the interior volumetric region. In the presence of the wire, an external source $(\mathbf{J}_i, \mathbf{M}_i)$ generates the field $(\mathbf{E}, \mathbf{H})$. When radiating in the homogeneous medium (μ, ϵ) without the wire, this source generates the incident field $(\mathbf{E}_i, \mathbf{H}_i)$. $(\mathbf{E}^m, \mathbf{H}^m)$ denotes the field of an electric test source located in V and radiating in the homogeneous medium (μ, ϵ). All sources and fields are considered to be time-harmonic with the same frequency. The time dependence $e^{j\omega t}$ is suppressed.

In the wire structure, let each segment have a circular cylindrical surface. At each point on the composite cylindrical surface of the wire, it is convenient to define a right-handed orthogonal coordinate system with unit vectors $(\hat{n}, \hat{\phi}, \hat{l})$ where $\hat{n}$ is the outward normal vector, $\hat{l}$ is directed along the wire axis and

$$\hat{\phi} = \hat{l} \times \hat{n} \qquad (1)$$

J. H. Richmond and E. H. Newman, *Radio Science*, vol. 11, no. 1, pp. 13–20, 1976, published by the National Bureau of Standards in cooperation with the U.S. National Committee of the International Scientific Radio Union (URSI).

Thus $(\hat{n}, \hat{\phi}, \hat{l})$ correspond directly with the unit vectors $(\hat{\rho}, \hat{\phi}, \hat{z})$ usually employed in the circular-cylindrical coordinate system.

To simplify the integral equation, we assume the wire radius a is much smaller than the wavelength λ, and the wire length is much greater than the radius. Furthermore, we shall neglect the integrations over the flat end surfaces of the wire, neglect the circumferential component J_ϕ of the surface-current density, and consider the axial component J_l to be independent of ϕ. (For thick wires, a more detailed treatment is essential for the ϕ-dependent current modes and the integrations over the junction regions and the open ends of the wire. A more elaborate formulation may also be required if one wire passes within a few diameters of another, or if a wire is bent to form a small acute angle.) In view of these approximations the reaction integral equation [*Rumsey*, 1954] reduces to

$$-\int_0^L I(l)(E_l^m - Z_s H_\phi^m)\, dl = V_m \tag{2}$$

where l is a metric coordinate measuring position along the wire axis, L denotes the overall wire length, Z_s is the surface impedance for exterior excitation, $I(l)$ is the total current (conduction plus displacement), and

$$V_m = \iiint (\mathbf{J}_i \cdot \mathbf{E}^m - \mathbf{M}_i \cdot \mathbf{H}^m)\, dv \tag{3}$$

$$E_l^m = \frac{1}{2\pi}\int_0^{2\pi} \hat{l} \cdot \mathbf{E}^m\, d\phi \tag{4}$$

$$H_\phi^m = \frac{1}{2\pi}\int_0^{2\pi} \hat{\phi} \cdot \mathbf{H}^m\, d\phi \tag{5}$$

The sinusoidal reaction formulation for thin wires is based on the integral equation (2). In this equation the known quantities are $\mathbf{E}^m, \mathbf{H}^m$, V_m, and Z_s. The current distribution $I(l)$ is regarded as an unknown function. To permit a solution for the current distribution, suitable test sources and expansion modes are now defined.

For a test source we choose a filamentary electric dipole with a sinusoidal current distribution. This is not a wire dipole, but merely an electric line source in the homogeneous medium. The sinusoidal dipole is probably the only finite line source with simple closed-form expressions for the near-zone fields. Furthermore, the mutual impedance between two sinusoidal dipoles is available in terms of

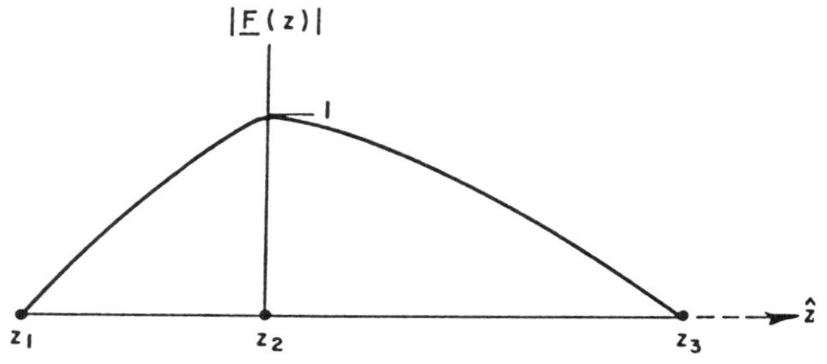

Fig. 1. A linear test dipole and its sinusoidal current distribution. The endpoints are at z_1 and z_3 with terminals at z_2.

exponential integrals, and the piecewise-sinusoidal function is evidently close to the natural current distribution on a perfectly conducting thin wire. These factors governed the choice of test sources. To simplify integrations in equations 4 and 5, we choose to locate the test dipole on the wire axis.

For the linear test dipole illustrated in Figure 1, the current distribution is $\mathbf{I}(z) = \mathbf{F}(z)$ where

$$\mathbf{F}(z) = [\hat{z}\, P_1 \sinh \gamma(z - z_1)]/\sinh \gamma d_1 + [\hat{z}\, P_2 \sinh \gamma(z_3 - z)]/\sinh \gamma d_2 \tag{6}$$

$P_1(z)$ is a pulse function with unit value for $z_1 < z < z_2$ and zero value elsewhere. The pulse function $P_2(z)$ has unit value for $z_2 < z < z_3$ and vanishes elsewhere. The segment lengths are $d_1 = z_2 - z_1$ and $d_2 = z_3 - z_2$. The current distribution on a V test dipole is

$$\mathbf{F}(l) = [\hat{l}_1\, P_1 \sinh \gamma(l - l_1)]/\sinh \gamma d_1 + [\hat{l}_2\, P_2 \sinh \gamma(l_3 - l)]/\sinh \gamma d_2 \tag{7}$$

In (6) and (7), γ denotes the complex propagation constant of the homogeneous exterior medium:

$$\gamma = j\omega(\mu\epsilon)^{1/2} \tag{8}$$

It is only with this value for γ that the sinusoidal test sources have the advantages mentioned earlier.

A typical problem requires not just one but several test dipoles located at different positions along the wire axis to form an overlapping array. Using N test dipoles, (2) is enforced for each one. Thus, (2) represents a system of N simultaneous integral equations with $m = 1, 2, \ldots, N$. In other words, (2) requires each test dipole in the array to have the correct reaction with the true source. Since the test dipoles and expansion modes both form an overlapping array, wire junctions require no special treatment.

The current distribution on the wire structure is expanded in a finite series as follows:

$$\mathbf{I}(l) = \sum_{n=1}^{N} I_n \mathbf{F}_n(l) \tag{9}$$

where the normalized expansion functions $\mathbf{F}_n(l)$ are the same as the test-dipole current distributions in (7). Since each expansion function extends over just a two-segment portion of the wire structure, these functions are subsectional bases. Since N is finite, (9) may be considered either as an expansion or an approximation, depending on the context. In (9), the coefficients I_n are complex constants which represent samples of the current function $I(l)$.

By inserting (9) into (2), we obtain the following system of simultaneous linear algebraic equations:

$$\sum_{n=1}^{N} I_n Z_{mn} = V_m \quad \text{where} \quad m = 1, 2, \ldots, N \tag{10}$$

$$Z_{mn} = -\int_n F_n(l)(E_l^m - Z_s H_\phi^m)\, dl \tag{11}$$

In (11), the integral extends over the two segments in the range of the expansion mode F_n. Equation 10 can be expressed in matrix form as $ZI = V$ where Z denotes the symmetric square impedance matrix, I is the current column and V is the voltage column. This matrix equation can be solved for the current column I. Inserting the components of I into (9) yields an approximation for the current on the wire structure. Once the current is known, it is straightforward to determine impedance, far-field patterns, or other quantities of interest.

3. INSULATED WIRE STRUCTURE

In this section it is shown how the matrix equation for the bare wire structure can be modified to account for the presence of a thin dielectric insulating layer. A typical cross section for an insulated wire is shown in Figure 2. Initially the insulating layer is considered to be homogeneous and with a circular cross section of radius b.

For simplicity, let the dielectric shell have the same permeability as the ambient medium. From the volume equivalence theorem, the dielectric shell may be replaced with ambient medium and an

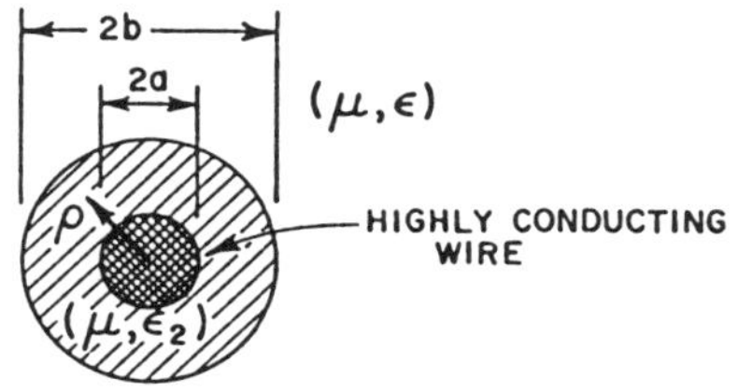

Fig. 2. A cross section of a dielectric coated wire.

equivalent source with electric current density

$$\mathbf{J} = j\omega(\epsilon_2 - \epsilon)\mathbf{E} \tag{12}$$

where $\mathbf{E}$ denotes the electric field intensity in the shell and ϵ_2 and ϵ are the complex permittivities of the shell and the ambient medium, respectively. From (12), the current $\mathbf{J}$ vanishes outside the region of the dielectric shell.

Let $(\mathbf{E},\mathbf{H})$ denote the field generated by $(\mathbf{J}_i,\mathbf{M}_i)$ in the presence of the insulated wire. Outside the wire, this field may also be generated by $(\mathbf{J}_i,\mathbf{M}_i)$, $(\mathbf{J}_s,\mathbf{M}_s)$, and $\mathbf{J}$, radiating in the homogeneous medium. These sources, radiating in the homogeneous medium, generate a null field in the interior region of the wire. The surface currents $(\mathbf{J}_s,\mathbf{M}_s)$ are located on the surface of the wire and are related to the field $(\mathbf{E},\mathbf{H})$ by [*Schelkunoff*, 1939]:

$$\mathbf{J}_s = \hat{n} \times \mathbf{H} \tag{13}$$

$$\mathbf{M}_s = \mathbf{E} \times \hat{n} \tag{14}$$

For the insulated wire, the reaction integral equation (2) is modified by replacing $\mathbf{J}_i$ with $\mathbf{J}_i + \mathbf{J}$. The current $\mathbf{J}$ may be regarded as an additional source which plays much the same role as the impressed source $\mathbf{J}_i$. However, $\mathbf{J}_i$ is considered to be a known source whereas $\mathbf{J}$ is unknown because $\mathbf{E}$ is unknown. If the dielectric shell is thin, $\mathbf{J}$ may be regarded as a dependent unknown function because it is simply related to the current distribution on the wire.

Using the equation of continuity [*Harrington*, 1961], the charge density on the wire surface is related to the current by

$$\rho_s = -I'(l)/j\omega 2\pi a \tag{15}$$

in which I' denotes dI/dl. For perfectly conducting wires, the ρ component of the electric flux density, D_ρ, at the surface of the wire is equal to ρ_s. We use the approximation that $D_\rho(\rho = a) = \rho_s$ for wires with high but not necessarily perfect conductivity. In this case the electric field on the surface of the wire is approximated by

$$\mathbf{E}(\rho = a) = \hat{\rho}\, \rho_s/\epsilon_2 = -\hat{\rho}\, I'(l)/j\omega 2\pi a \epsilon_2 \tag{16}$$

The $\hat{l}$ and $\hat{\phi}$ components of E are considered to be negligible in the dielectric shell. To extrapolate from this value at the wire surface to points in the insulating layer, we use a $1/\rho$ radial dependence and obtain from (16):

$$\mathbf{E} = -\hat{\rho}\, I'(l)/j\omega 2\pi\epsilon_2 \rho \tag{17}$$

From (12) and (17),

$$\mathbf{J} = -(\epsilon_2 - \epsilon)\,\hat{\rho}\, I'(l)/2\pi\epsilon_2 \rho \tag{18}$$

For an insulated wire, each expansion mode, $\mathbf{F}_n(l)$, has associated with it a shell of radial electric current J. Thus, from (9) and (18), the mutual impedance Z_{mn} between the filamentary test dipole m and the tubular expansion dipole n has an additional term given by

$$\Delta Z_{mn} = \iint_n \frac{(\epsilon_2 - \epsilon)}{\epsilon_2} F'_n(l)\, E^m_\rho(\rho, l)\, d\rho\, dl \tag{19}$$

where the integration extends through the dielectric shell in the range of the expansion dipole n. In deriving (19), the integration on ϕ was performed with the assumption that J and E^m_ρ are independent of ϕ. In integrating on ρ, the limits are a and b which denote the inner and outer radii of the dielectric shell.

In the dielectric shell, the test-dipole field E^m_ρ may be approximated by

$$E^m_\rho = -F'_m(l)/2\pi j\omega\epsilon\rho \tag{20}$$

In (19) and (20), ρ denotes distance from the axis of dipole n or m, respectively. Furthermore, the vector direction of the field component E^m_ρ in (19) generally differs from that in (20) unless dipoles m and n are colinear. If (20) is employed, (19) reduces to

$$\Delta Z_{mn} = -\frac{P}{2\pi j\omega\epsilon} \int_{m,n} F'_m(l)\, F'_n(l)\, dl \tag{21}$$

where (m, n) denotes the region of l shared by dipoles m and n, and the dimensionless parameter P is given by

$$P = \int_a^b \frac{(\epsilon_2 - \epsilon)}{\epsilon_2 \rho}\, d\rho = \frac{(\epsilon_2 - \epsilon)}{\epsilon_2} \ln(b/a) \tag{22}$$

Note that the elements of ΔZ_{mn} are very easy and fast to evaluate on a digital computer since the integration in (21) is available in closed form and in terms of simple functions (hyperbolic sines and cosines).

The current distribution on the wire in the presence of the insulating layer is found by solving the matrix equation

$$(Z + \Delta Z)\, I = V \tag{23}$$

where the elements of Z are given by (11) and the elements of ΔZ are given by (21). ΔZ_{mn} is zero if dipoles m and n do not share a segment. Also, that part of the integration in (21) which is over a portion of the wire which is not insulated is zero. Thus, (21) is sufficiently general to handle partly insulated and partly bare wires. The approximate result represented by (21) yields a matrix Z which is symmetric even when some of the wire segments are insulated and others are bare.

Equation 19 is sufficiently general to treat an inhomogeneous insulating layer where ϵ_2 varies as a function of ρ and l only.

If ϵ_2 is a function of ϕ, the analysis will be complicated because the current densities J and $\mathbf{J}_s$ will be functions of ϕ. Therefore this case is not considered.

If ϵ_2 is simply a function of ρ, then the only modification required is that ϵ_2 in (22) be considered as a function of ρ. If ϵ_2 has a sufficiently simple radial dependence, then P can still be obtained in closed form. For example, consider an L-layered insulating shell. If we use the notation that the ith layer extends from ρ_i to ρ_{i+1} and has uniform permittivity ϵ_i, then (22) becomes

$$P = \sum_{i=1}^{L} \frac{(\epsilon_i - \epsilon)}{\epsilon_i} \ln(\rho_{i+1}/\rho_i) \tag{24}$$

Thus by using (24) in (21) it is a simple matter to treat multilayered insulating shells.

4. NUMERICAL EXAMPLES

In this section numerical computations based on the theory presented above are compared with measurements. The computer program used for all computations and a user's manual describing these programs is available [*Richmond*, 1974*b*,*c*]. All

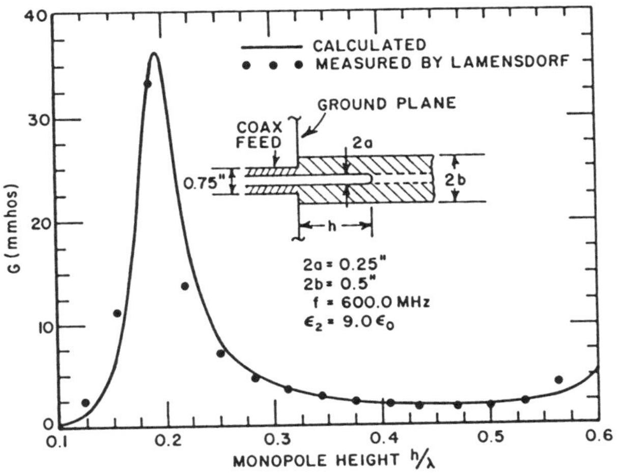

Fig 3. The conductance of an insulated monopole plotted versus its height in wavelengths.

antenna feeds are modeled by delta-gap generators. The ambient medium for all measurements and computations is free space. No more than N = seven current expansion modes were required for the computations to be presented.

The insert in Figure 3 illustrates the geometry for the dielectric coated monopole measured by *Lamensdorf* [1967]. His measurements were made at f = 600 MHz and with a monopole of diameter 0.25 inches. The dielectric sleeve extended beyond the monopole to simulate an extension to infinity.

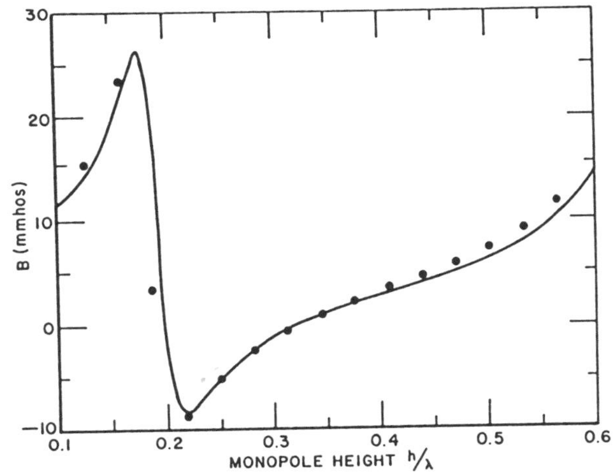

Fig. 4. The susceptance of the insulated monopole, shown in Figure 3, plotted versus its height in wavelengths.

Lamensdorf [1967] measured the monopole admittance for dielectric sleeves with diameters from 0.25 to 2.0 inches and relative dielectric constants from 3.2 to 15.0. Generally speaking, our computations agree well with *Lamensdorf's* [1967] measurements for the thinner dielectric sleeves and the smaller permittivities. The agreement deteriorates as the sleeve diameter increases or the permittivity increases. Figures 3 and 4 show the measured and calculated monopole admittance plotted versus the monopole height in wavelengths for the case $2b$ = 0.5 inches and $\epsilon_2 = 9.0\epsilon_0$. The agreement between experiment and theory is good for both the conductance, G, and the susceptance, B. As mentioned above, in the theoretical model the dielectric sleeve terminates at the tip of the monopole. *Lamensdorf* [1967] indicated that the extensions did affect the monopole's admittance, and its effect was greatest near resonance. Thus, it is possible that the discrepancies between our calculations and *Lamensdorf's* [1967] measurements for the thicker and denser dielectric layers resulted from the fact that our model does not include the extension of the dielectric.

The remaining measurements were made at The Ohio State University. The antennas were constructed by removing the outer insulation and copper braid from RG59/U coaxial cable. Then for the remaining examples the diameter of the copper wire is $2a$ = 0.025 inches, the diameter of the insulating layer is $2b$ = 0.146 inches, and the permittivity of the insulating layer is $\epsilon_2 = 2.3\epsilon_0$. The measurements were made with monopoles or half-loops mounted on a two-ft square aluminum ground plane whose edges were terminated in six-inch diameter cylinders to reduce edge reflections. The inner and outer diameter of the coaxial feed was 1/8 inch and 3/8 inch, respectively. The calculations were made for dipoles and complete loops, and it is the dipole or loop admittance which is plotted.

Figures 5 and 6 compare the measured and calculated admittance for a coated dipole of length L = 8 inches. Agreement between experiment and theory is good.

Figures 7 and 8 compare the measured and calculated admittance for a square loop of perimeter L = 8 inches. In Figures 9 and 10 the admittance of a loop identical to that shown in Figure 7, except that the insulation is removed from two arms, is shown. The good agreement between theory and experiment in Figures 7-10 illustrates the ability

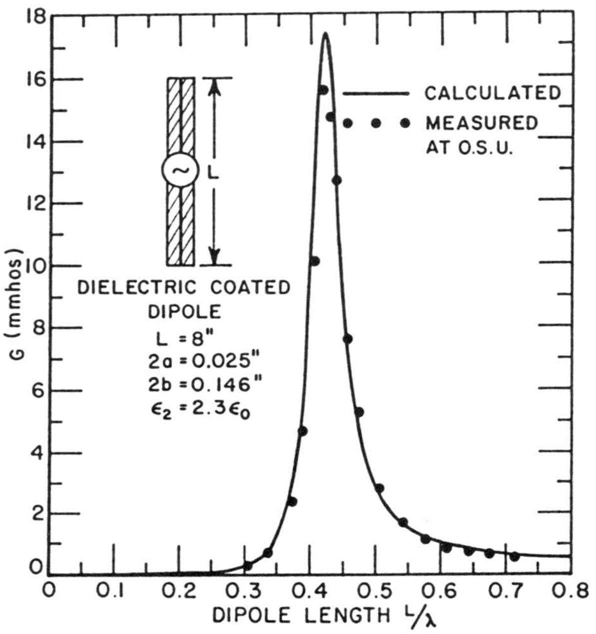

Fig. 5. The conductance of an insulated dipole plotted versus its length in wavelengths.

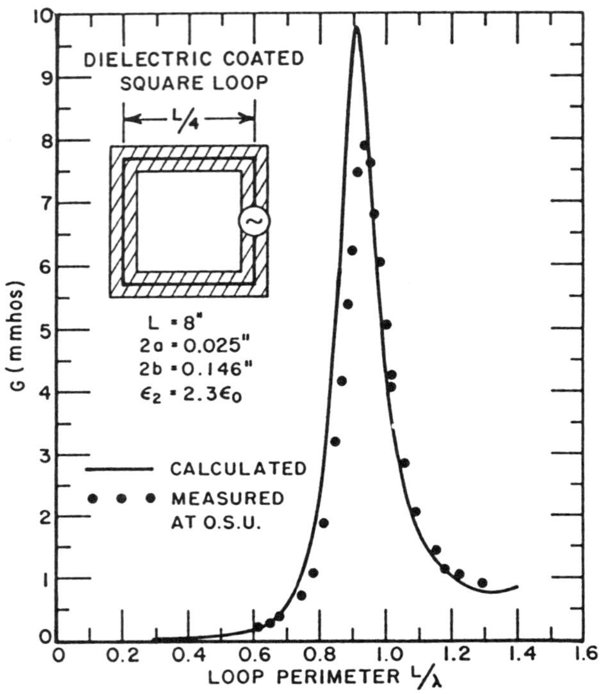

Fig. 7. The conductance of an insulated square loop plotted versus its perimeter in wavelengths.

of the computer programs to treat geometries other than simple dipoles as well as partially coated antennas.

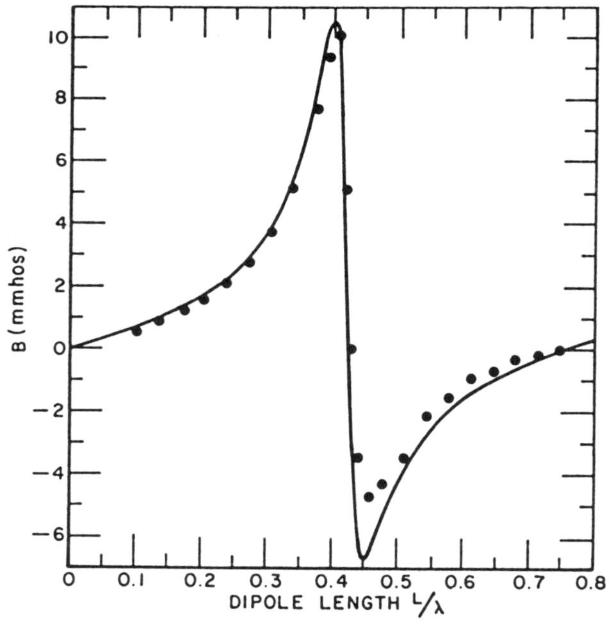

Fig. 6. The susceptance of the insulated dipole, shown in Figure 5, plotted versus its length in wavelengths.

Calculations not shown here were made to compare with measurements by *Iizuka* [1963] of insulated monopoles in water solutions. (In these calculations the feed was modeled by a magnetic frill current and not a delta-gap generator.) While the calculated and measured admittances generally followed the same locus, a significant resonant shift was noticeable. As was the case with *Lamensdorf's* [1967] measurements, the insulating sleeve in *Iizuka's* [1963] experiment extended beyond the monopole. Thus the discrepancies between our calculations and *Iizuka's* [1963] measurements could be explained in the same manner as was suggested above to account for differences between our calculations and some of *Lamensdorf's* [1967] measurements. A second possible explanation is that the theory presented here is inadequate to treat the case of insulating layers with permittivity vastly different from that of the ambient medium.

Comparing (21) and (22) or (24) it can be seen that the diameter and permittivity of the insulating layer affect ΔZ_{mn} only through the dimensionless parameter P. Thus, to the approximations being used here, several insulating shells with different

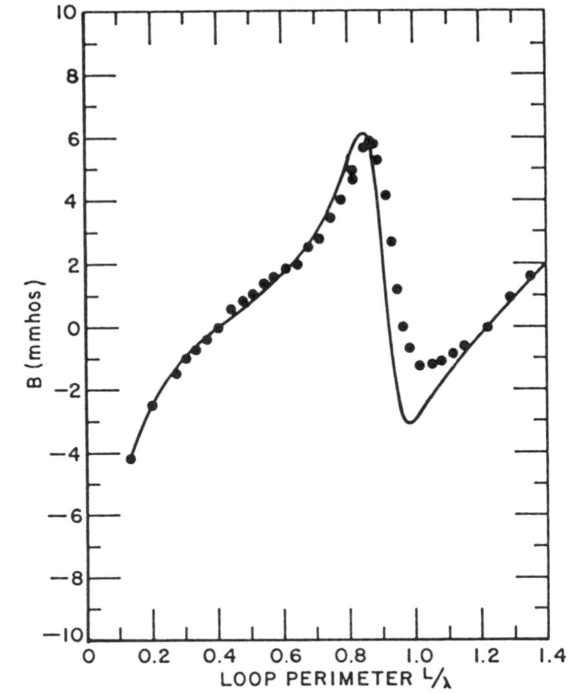

Fig. 8. The susceptance of the insulated square loop, shown in Figure 7, plotted versus its perimeter in wavelengths.

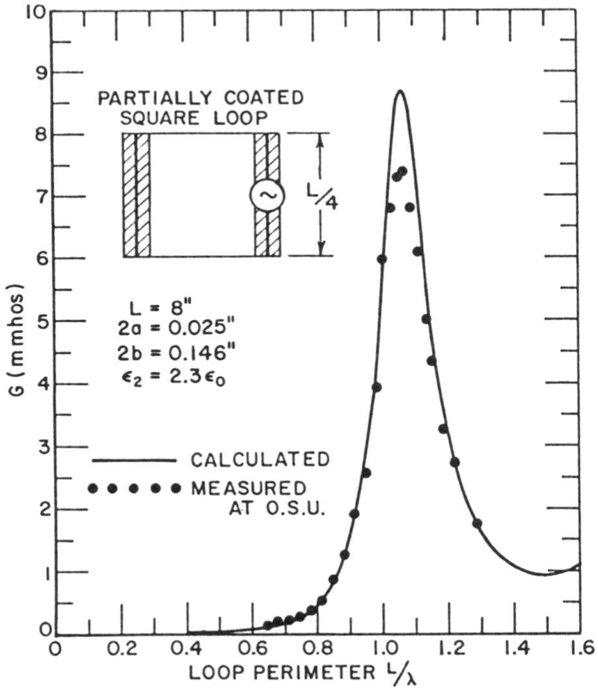

Fig. 9. The conductance of a partially insulated square loop plotted versus its perimeter in wavelengths.

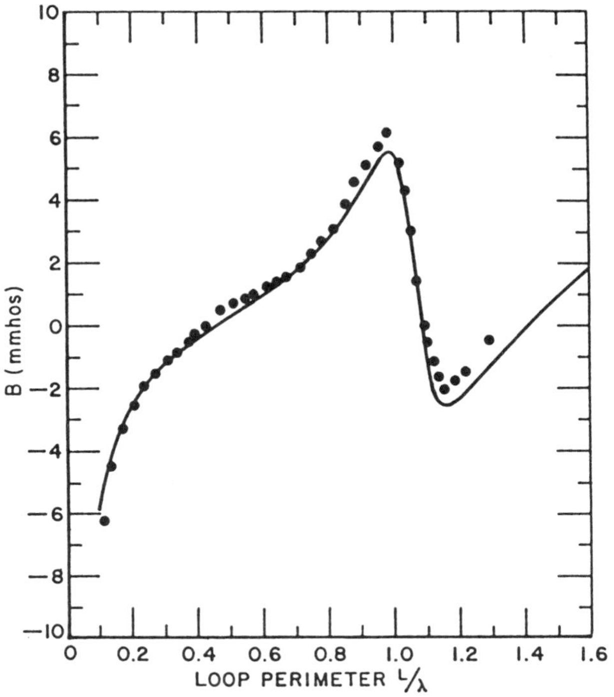

Fig. 10. The susceptance of the partially insulated square loop, shown in Figure 9, plotted versus its perimeter in wavelengths.

diameters and densities will have the same effect on an antenna if the value of P associated with each shell is identical. Figure 11 shows the resonant length and conductance of a dipole plotted versus P. Here it can be seen that the conductance at resonance increases almost linearly with increasing P, while the resonant length decreases almost linearly.

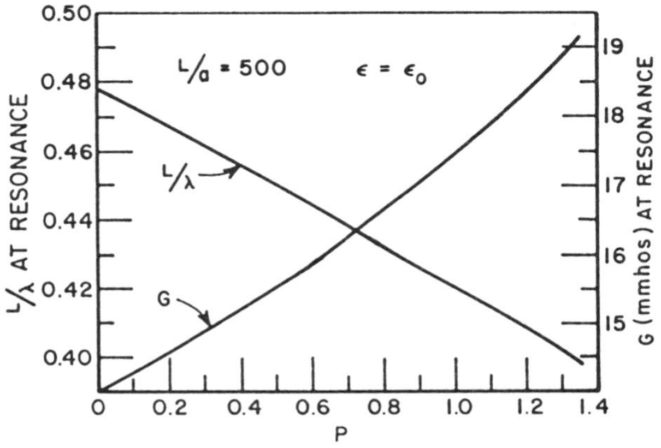

Fig. 11. The resonant length and conductance at resonance for an insulated dipole plotted vs. the dimensionless parameter P.

5. CONCLUSIONS

A simple method is presented to account for a thin insulating shell on a thin-wire antenna or scatterer. The shell is modeled by equivalent volume electric polarization currents. These polarization currents are simply related to the current on the wire structure. The insulating layer is accounted for by modifying the impedance matrix for the bare wire. Thus no increase in computer storage is required except that which is required to describe the insulating layer and to evaluate (21). The modification is obtained in terms of simple functions and is therefore easy to program on a digital computer. If no advantage is taken of symmetries, then (21) must be evaluated only N times for a given problem. Thus the modifications require very little computer time. Because of the simple approximation used for the electric field in the insulating layer, the theory was shown to be applicable to certain inhomogeneous insulating layers. Two important examples of inhomogeneous insulation which were treated are multilayered coatings and partially coated antennas.

The qualitative effect of placing an insulating layer, with permittivity ϵ_2, on an antenna is to shift its admittance (or impedance) so that it is between the admittance of the bare antenna and the admittance of the antenna in a homogeneous medium with permittivity ϵ_2. Specifically, the effects of placing an insulating layer, with permittivity greater than that of the ambient medium, on an antenna are to (a) lower the resonant frequency, (b) increase the peak admittance, and (c) narrow the bandwidth. If the insulating layer has permittivity less than that of the ambient medium, the effects are opposite to those indicated above.

An important advantage of the method presented is that it can be applied to any antenna geometry which could be treated in the absence of the insulating layer. Thus one can treat insulated dipoles, loops, crossed wires, spirals, etc. Further, the wires may have finite conductivity or contain lumped loads.

Acknowledgments. This work was supported in part by grant NGL 36-008-138 between National Aeronautics and Space Administration, Langley Research Center, Hampton, Virginia, and The Ohio State University Research Foundation, Columbus, Ohio.

REFERENCES

Harrington, R. F. (1961), *Time-Harmonic Electromagnetic Fields*, p. 2, McGraw-Hill, New York.

Iizuka, K. (1963), An experimental study of the insulated dipole antenna in a conducting medium, *IEEE Trans. Antennas Propagat.*, *AP-11*, 518-532.

King, R. W. P. (1964), Theory of terminated insulated antenna in a conducting medium, *IEEE Trans. Antennas Propagat.*, *AP-12*, 305-318.

King, R. W. P., K. M. Lee, S. H. Mishra, and G. S. Smith (1974), Insulated linear antenna: Theory and experiment, *J. Appl. Phys.*, *45*, 1688-1697.

Lamensdorf, D. (1967), An experimental investigation of dielectric coated antennas, *IEEE Trans. Antennas Propagat.*, *AP-15*, 767-771.

Richmond, J. H. (1974*a*), Radiation and scattering by thin-wire structures in the complex frequency domain, *NASA Contr. Rep. CR-2396*, available from National Technical Information Service, Springfield, Virginia 22151.

Richmond, J. H. (1974*b*), Computer program for thin-wire structures in a homogeneous conducting medium, *NASA Contr. Rep. CR-2399*, available from National Technical Information Service, Springfield, Virginia 22151.

Richmond, J. H. (1974*c*), Radiation and scattering by thin-wire structures in a homogeneous conducting medium, *IEEE Trans. Antennas Propagat.*, *AP-22*, 365.

Richmond, J. H., and E. H. Newman (1974), Dielectric coated wire antennas, paper presented at the 1974 Spring Meeting of URSI, Atlanta, Georgia.

Rumsey, V. H. (1954), Reaction concept in electromagnetic theory, *Phys. Rev.*, *94*, 1483-1491.

Schelkunoff, S. A. (1939), On diffraction and radiation of electromagnetic waves, *Phys. Rev.*, *56*, 308-316.

Wu, T. T., R. W. P. King, and D. V. Giri (1973), The insulated dipole antenna in a relatively dense medium, *Radio Sci.*, *8*, 699-709.

Electromagnetic Scattering from a Homogeneous Material Body of Revolution

by Joseph R. Mautz * and Roger F. Harrington *

This paper considers plane-wave scattering by a homogeneous material body of revolution. The problem is formulated in terms of equivalent electric and magnetic currents over the surface which defines the body. Application of the boundary conditions leads to four simultaneous surface integral equations to be satisfied by the two unknown equivalent currents, electric and magnetic. The set of four equations is reduced to a coupled pair of equations by taking linear combinations of the original four equations. Because many pairs of linear combinations are possible, there are many surface integral equation formulations for the problem. Two formulations commonly encountered in the literature are discussed and solved by the method of moments. Examples of numerical computations are given for dielectric spheres and a finite dielectric cylinder. The computed results for the sphere are compared to the exact series solution obtained by separation of variables.

Elektromagnetische Streuung eines homogenen Rotationskörpers

Die Streuung einer ebenen Welle an einem homogenen Rotationskörper wird mit Hilfe von äquivalenten elektrischen und magnetischen Oberflächenströmen auf dem Streukörper behandelt. Aus den Randbedingungen ergeben sich vier simultane Flächenintegralgleichungen, welche die beiden unbekannten äquivalenten elektrischen und magnetischen Ströme erfüllen müssen. Durch Linearkombinationen dieser vier Gleichungen wird das System auf ein Paar verkoppelter Gleichungen zurückgeführt. Weil viele Paare von Linearkombinationen möglich sind, gibt es auch viele Integralgleichungsansätze für das Problem. Zwei Ansätze, die man gewöhnlich in der Literatur findet, werden behandelt und mit der Momentenmethode gelöst. Numerische Beispiele werden für dielektrische Kugeln und endliche dielektrische Zylinder gegeben. Die errechneten Ergebnisse für die Kugel werden mit exakten Reihenlösungen durch Trennung der Variablen verglichen.

1. Introduction

The problem of plane-wave scattering by a homogeneous material body of revolution is formulated in terms of equivalent electric and magnetic currents over the body surface. Application of boundary conditions leads to a set of four integral equations to be satisfied. Linear combinations of these four equations lead to a coupled pair of equations to be solved. One choice of combination constants gives the formulation described by Poggio and Miller [1]. This formulation has been applied to material cylinders by Chang and Harrington [2], and to material bodies of revolution by Wu and Tsai [3]. We will call this choice the PMCHW formulation (formed by the initials of the above cited investigators).

Another choice of combination constants gives the formulation obtained by Müller [4]. This formulation has been applied to dielectric cylinders by Solodukhov and Vasil'ev [5] and by Morita [6], and to bodies of revolution by Vasil'ev and Materikova [7]. We will call this choice the Müller formulation. Conditions for the uniqueness of solutions are established in terms of the combination constants. It is found that solutions to both the PMCHW formulation and to Müller's formulation are unique at all frequencies.

Numerical solutions to the coupled pair of equations are obtained by the method of moments [8]. It is relatively easy to obtain numerical solutions to these equations because the required operators are the same as those evaluated in earlier work [9]—[11]. An exemplary computer program capable of obtaining the solution to both the PMCHW formulation and the Müller formulation is described and listed in a research report [12]. This is a main program which uses subroutines similar to those in [10] to compute the equivalent electric and magnetic currents and the two principal plane scattering patterns for a loss-free homogeneous body of revolution excited by an axially incident electromagnetic plane wave. Computed results for the equivalent currents and principal plane scattering patterns of a dielectric sphere whose relative dielectric constant is four show reasonable agreement between our solution to the PMCHW formulation, our solution to the Müller formulation, and the "exact" series [13] solution in the resonance region. Computer program subroutines which calculate the exact series solution for perfectly conducting spheres as well as for loss-free homogeneous spheres are described and listed in another research report [14].

* Dr. J. R. Mautz, Prof. R. F. Harrington, Department of Electrical and Computer Engineering, 111 Link Hall, Syracuse University, Syracuse, New York 13210, USA.

Reprinted with permission from *Arch. Elektron. Übertragungstech.*, vol. 33, pp. 71-80, 1979. © S. Hirzel Verlag.

2. Surface Integral Equation Formulation

An electromagnetic field propagating in a homogeneous medium of permeability μ_e and permittivity ε_e is incident on the surface S of a homogeneous obstacle of permeability μ_d and permittivity ε_d. The subscript e denotes exterior medium and the subscript d denotes diffracting medium. We wish to calculate the scattered electromagnetic field $\boldsymbol{E}^s$, $\boldsymbol{H}^s$ outside S and the diffracted electromagnetic field $\boldsymbol{E}$, $\boldsymbol{H}$ inside S in terms of the electromagnetic field $\boldsymbol{E}^i$, $\boldsymbol{H}^i$ which would exist on S in the absence of the obstacle. This original problem is shown in Fig. 1a where $\boldsymbol{J}^i$, $\boldsymbol{M}^i$ are the electric and magnetic sources of $\boldsymbol{E}^i$, $\boldsymbol{H}^i$ and $\boldsymbol{n}$ is the unit normal vector which points outward from S.

The equivalence principle [13] is used to piece together an outside situation consisting of medium μ_e, ε_e and field $\boldsymbol{E}^s$, $\boldsymbol{H}^s$ outside S and an inside situation consisting of medium μ_e, ε_e and field $-\boldsymbol{E}^i$, $-\boldsymbol{H}^i$ inside S. This composite situation is shown in Fig. 1b. Since $\boldsymbol{E}^s$, $\boldsymbol{H}^s$ is source-free outside S and $\boldsymbol{E}^i$, $\boldsymbol{H}^i$ is source-free inside S, the only sources in Fig. 1b are the equivalent electric surface current $\boldsymbol{J}$ and the equivalent magnetic surface current $\boldsymbol{M}$ on S.

As a second application of the equivalence principle, we combine an outside situation consisting of medium μ_d, ε_d and zero field with an inside situation consisting of medium μ_d, ε_d and field $\boldsymbol{E}$, $\boldsymbol{H}$. This combination of situations is shown in Fig. 1c. Since $\boldsymbol{E}$, $\boldsymbol{H}$ is source-free inside S, the only sources in Fig. 1c are the equivalent electric surface current $-\boldsymbol{J}$ and the equivalent magnetic surface current $-\boldsymbol{M}$ on S. By expressing the surface currents in terms of the discontinuities of the tangential fields across S and by using

$$\boldsymbol{n} \times \boldsymbol{E} = \boldsymbol{n} \times (\boldsymbol{E}^s + \boldsymbol{E}^i), \tag{1}$$

$$\boldsymbol{n} \times \boldsymbol{H} = \boldsymbol{n} \times (\boldsymbol{H}^s + \boldsymbol{H}^i) \tag{2}$$

on S, the interested reader can verify that the surface currents in Fig. 1c are indeed the negatives of those in Fig. 1b. Eqs. (1) and (2) are the boundary conditions that the tangential components of the fields in the original problem as shown in Fig. 1a are continuous across S.

The scattered field $\boldsymbol{E}^s$, $\boldsymbol{H}^s$ outside S and the diffracted field $\boldsymbol{E}$, $\boldsymbol{H}$ inside S could easily be calculated if $\boldsymbol{J}$ and $\boldsymbol{M}$ were known because the media into which $\boldsymbol{J}$ and $\boldsymbol{M}$ radiate is homogeneous in Fig. 1b and c. We have to determine $\boldsymbol{J}$ and $\boldsymbol{M}$. The equivalence principle states that there exist $\boldsymbol{J}$ and $\boldsymbol{M}$ which radiate the fields in Fig. 1b and c, but the equivalence principle does not tell what $\boldsymbol{J}$ and $\boldsymbol{M}$ are. The equivalence principle does state that

$$\boldsymbol{J} = \boldsymbol{n} \times \boldsymbol{H}, \quad \boldsymbol{M} = \boldsymbol{E} \times \boldsymbol{n} \tag{3), (4}$$

but this is not very useful because $\boldsymbol{E}$ and $\boldsymbol{H}$ are unknown.

From Fig. 1b and c,

$$-\boldsymbol{n} \times \boldsymbol{E}_e^- = \boldsymbol{n} \times \boldsymbol{E}^i, \quad -\boldsymbol{n} \times \boldsymbol{H}_e^- = \boldsymbol{n} \times \boldsymbol{H}^i, \tag{5), (6}$$

$$-\boldsymbol{n} \times \boldsymbol{E}_d^+ = 0, \quad -\boldsymbol{n} \times \boldsymbol{H}_d^+ = 0 \tag{7), (8}$$

where

$\boldsymbol{E}_e^-$ is the electric field just inside S due to $\boldsymbol{J}$, $\boldsymbol{M}$, radiating in μ_e, ε_e,

$\boldsymbol{H}_e^-$ is the magnetic field just inside S due to $\boldsymbol{J}$, $\boldsymbol{M}$, radiating in μ_e, ε_e,

$\boldsymbol{E}_d^+$ is the electric field just outside S due to $\boldsymbol{J}$, $\boldsymbol{M}$, radiating in μ_d, ε_d,

$\boldsymbol{H}_d^+$ is the magnetic field just outside S due to $\boldsymbol{J}$, $\boldsymbol{M}$, radiating in μ_d, ε_d.

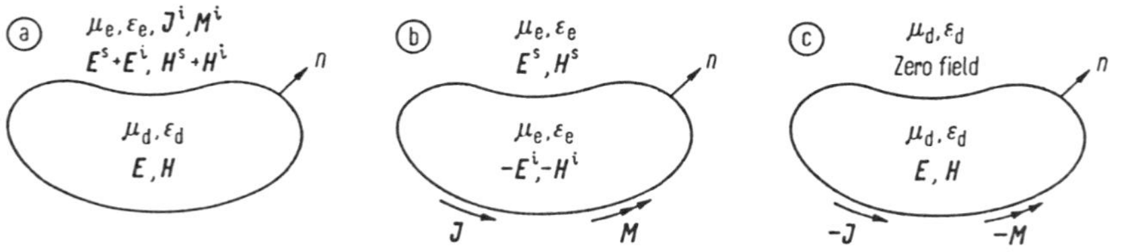

Fig. 1. (a) Original problem, (b) Outside equivalence, (c) Inside equivalence.

The equivalent currents $\boldsymbol{J}$, $\boldsymbol{M}$ which appear in Fig. 1b and c satisfy eqs. (5) to (8) because eqs. (5) to (8) were obtained from Fig. 1b and c. It is shown in the Appendix that the solution to eqs. (5) to (8) is unique. Therefore, eqs. (5) to (8) uniquely determine the equivalent currents $\boldsymbol{J}$, $\boldsymbol{M}$ of Fig. 1b and c.

Eqs. (5) to (8) form a set of four equations in the two unknowns $\boldsymbol{J}$ and $\boldsymbol{M}$. The usual methods of equation solving apply only when the number of equations is equal to the number of unknowns. We want to reduce the set of four equations (5) to (8) to two equations. One way to do this is to form the linear combination

$$-\boldsymbol{n} \times (\boldsymbol{E}_e^- + \alpha \boldsymbol{E}_d^+) = \boldsymbol{n} \times \boldsymbol{E}^i \tag{9}$$

of eqs. (5) and (7) and the linear combination

$$-\boldsymbol{n} \times (\boldsymbol{H}_e^- + \beta \boldsymbol{H}_d^+) = \boldsymbol{n} \times \boldsymbol{H}^i \tag{10}$$

of eqs. (6) and (8) where α and β are complex constants.

The solution $\boldsymbol{J}$, $\boldsymbol{M}$ to eqs. (5) to (8) satisfies eqs. (9) and (10). This $\boldsymbol{J}$, $\boldsymbol{M}$ will be the only solution to the pair of equations (9) and (10) if

$$-\boldsymbol{n} \times (\boldsymbol{E}_e^- + \alpha \boldsymbol{E}_d^+) = 0, \tag{11}$$

$$-\boldsymbol{n} \times (\boldsymbol{H}_e^- + \beta \boldsymbol{H}_d^+) = 0 \tag{12}$$

have only the trivial solution $\boldsymbol{J} = \boldsymbol{M} = 0$. From eqs. (11) and (12),

$$P_e = -\alpha \beta^* P_d \tag{13}$$

where P_e is the complex power flow of $\boldsymbol{E}_e^-$, $\boldsymbol{H}_e^-$ inside S and P_d is the complex power flow of $\boldsymbol{E}_d^+$, $\boldsymbol{H}_d^+$ outside S. The asterisk in eq. (13) denotes complex conjugate. If $\alpha\beta^*$ is real, then the real part of eq. (13) reduces to

$$\mathrm{Re}(P_e) = -\alpha \beta^* \,\mathrm{Re}(P_d). \tag{14}$$

If $\alpha\beta^*$ is not only real but also positive, then

$$\mathrm{Re}(P_\mathrm{d}) = 0 \tag{15}$$

because both $\mathrm{Re}(P_\mathrm{e})$ and $\mathrm{Re}(P_\mathrm{d})$ are greater than or equal to zero. Since there are no external resonances, eq. (15) implies that

$$\boldsymbol{n} \times \boldsymbol{E}_\mathrm{d}^+ = \boldsymbol{n} \times \boldsymbol{H}_\mathrm{d}^+ = 0\,. \tag{16}$$

Substitution of eq. (16) into eqs. (11) and (12) yields

$$\boldsymbol{n} \times \boldsymbol{E}_\mathrm{e}^- = \boldsymbol{n} \times \boldsymbol{H}_\mathrm{e}^- = 0\,. \tag{17}$$

The system of equations (16) and (17) is precisely the homogeneous system of equations associated with eqs. (5) to (8). It was shown in the Appendix that this homogeneous system of equations has only the trivial solution $\boldsymbol{J} = \boldsymbol{M} = 0$. Therefore, if $\alpha\beta^*$ is real and positive, then the coupled pair of equations (11) and (12) has only the trivial solution $\boldsymbol{J} = \boldsymbol{M} = 0$ so that the solution $\boldsymbol{J}, \boldsymbol{M}$ to eqs. (5) to (8) is the only solution to the coupled pair of equations (9) and (10).

If $\alpha = \beta = 1$, then eqs. (9) and (10) become

$$-\boldsymbol{n} \times (\boldsymbol{E}_\mathrm{e}^- + \boldsymbol{E}_\mathrm{d}^+) = \boldsymbol{n} \times \boldsymbol{E}^\mathrm{i}\,, \tag{18}$$

$$-\boldsymbol{n} \times (\boldsymbol{H}_\mathrm{e}^- + \boldsymbol{H}_\mathrm{d}^+) = \boldsymbol{n} \times \boldsymbol{H}^\mathrm{i}\,. \tag{19}$$

The set of equations (18) and (19) is the coupled pair of surface integral equations described by Poggio and Miller [1]. We call these equations the PMCHW equations. Since $\alpha = \beta = 1$ implies that $\alpha\beta^*$ is real and positive, the argument consisting of eqs. (11) to (17) and involving real power flow shows that eqs. (18) and (19) uniquely determine the desired $\boldsymbol{J}, \boldsymbol{M}$ of Fig. 1b and c.

That eqs. (18) and (19) uniquely determine $\boldsymbol{J}$, $\boldsymbol{M}$ of Fig. 1b and c can also be shown as follows. The desired $\boldsymbol{J}, \boldsymbol{M}$ of Fig. 1b and c satisfies eqs. (18) and (19) because eqs. (18) and (19) were obtained from Fig. 1b and c. This desired $\boldsymbol{J}, \boldsymbol{M}$ will be the only solution to eqs. (18) and (19) if the associated set of homogeneous equations

$$-\boldsymbol{n} \times (\boldsymbol{E}_\mathrm{e}^- + \boldsymbol{E}_\mathrm{d}^+) = 0\,, \tag{20}$$

$$-\boldsymbol{n} \times (\boldsymbol{H}_\mathrm{e}^- + \boldsymbol{H}_\mathrm{d}^+) = 0 \tag{21}$$

has only the trivial solution $\boldsymbol{J} = \boldsymbol{M} = 0$.

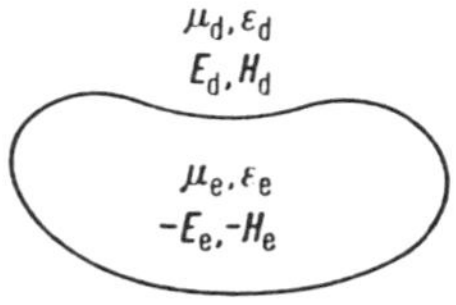

Fig. 2. Composite situation used to prove that eqs. (20) and (21) have only the trivial solution $\boldsymbol{J} = \boldsymbol{M} = 0$.

The following argument shows that eqs. (20) and (21) have only the trivial solution $\boldsymbol{J} = \boldsymbol{M} = 0$. Let $\boldsymbol{E}_\mathrm{d}$, $\boldsymbol{H}_\mathrm{d}$ be the electromagnetic field outside S due to $\boldsymbol{J}, \boldsymbol{M}$ radiating in μ_d, ε_d. Let $\boldsymbol{E}_\mathrm{e}$, $\boldsymbol{H}_\mathrm{e}$ be the electromagnetic field inside S due to $\boldsymbol{J}, \boldsymbol{M}$ radiating in μ_e, ε_e. Use the equivalence principle to form the composite situation consisting of medium μ_d, ε_d and field $\boldsymbol{E}_\mathrm{d}$, $\boldsymbol{H}_\mathrm{d}$ outside S and medium μ_e, ε_e and field $-\boldsymbol{E}_\mathrm{e}$, $-\boldsymbol{H}_\mathrm{e}$ inside S as shown in Fig. 2. In Fig. 2, $\boldsymbol{E}_\mathrm{d}$, $\boldsymbol{H}_\mathrm{d}$ is a source-free Maxwellian field outside S. Since $\boldsymbol{E}_\mathrm{e}, \boldsymbol{H}_\mathrm{e}$ is a source-free Maxwellian field inside S, the field $-\boldsymbol{E}_\mathrm{e}$, $-\boldsymbol{H}_\mathrm{e}$ appearing in Fig. 2 is also a source-free Maxwellian field inside S. Now, eqs. (20) and (21) state that the tangential components of the field in Fig. 2 are continuous across S. Thus, Fig. 2 is entirely source-free so that the field in Fig. 2 is zero everywhere in which case eqs. (A.1) to (A.4) are satisfied. But, as shown in the Appendix, eqs. (A.1) to (A.4) have only the trivial solution $\boldsymbol{J} = \boldsymbol{M} = 0$. Hence, eqs. (20) and (21) have only the trivial solution $\boldsymbol{J} = \boldsymbol{M} = 0$.

If

$$\alpha = -\,\varepsilon_\mathrm{d}/\varepsilon_\mathrm{e}\,, \quad \beta = -\,\mu_\mathrm{d}/\mu_\mathrm{e} \tag{22), (23}$$

then eqs. (9) and (10) become

$$-\boldsymbol{n} \times \left(\boldsymbol{E}_\mathrm{e}^- - \frac{\varepsilon_\mathrm{d}}{\varepsilon_\mathrm{e}} \boldsymbol{E}_\mathrm{d}^+\right) = \boldsymbol{n} \times \boldsymbol{E}^\mathrm{i}\,, \tag{24}$$

$$-\boldsymbol{n} \times \left(\boldsymbol{H}_\mathrm{e}^- - \frac{\mu_\mathrm{d}}{\mu_\mathrm{e}} \boldsymbol{H}_\mathrm{d}^+\right) = \boldsymbol{n} \times \boldsymbol{H}^\mathrm{i}\,. \tag{25}$$

The set of eqs. (24) and (25) is the coupled pair of surface integral equations obtained by Müller [4], and we call them the Müller equations. The static electric field contribution to the left-hand side of eq. (24) due to the electric charge associated with $\boldsymbol{J}$ is zero. Similarly, the static magnetic field contribution to the left-hand side of eq. (25) due to the magnetic charge associated with $\boldsymbol{M}$ vanishes. According to the last sentence on page 300 of [4], the singularity of the kernels in eqs. (24) and (25) due to the electric and magnetic charges is no more pronounced than the reciprocal of the distance between the source point and the field point. Hence, the singularity that the kernels of the integral equations (24) and (25) exhibit as the source point passes through the field point is not as pronounced as the singularity of the kernels of eqs. (18) and (19). If μ_e, ε_e, μ_d, and ε_d are real in eqs. (22) to (25), then $\alpha\beta^*$ is real and positive. In this case, the argument consisting of eqs. (11) to (17) shows that eqs. (24) and (25) uniquely determine the desired $\boldsymbol{J}, \boldsymbol{M}$, of Fig. 1b and c.

An alternate proof, valid for lossy media, that eqs. (24) and (25) uniquely determine the desired $\boldsymbol{J}, \boldsymbol{M}$ is presented. This proof is similar to the argument which used Fig. 2 to show that eqs. (18) and (19) uniquely determine the desired $\boldsymbol{J}, \boldsymbol{M}$ and is as follows. The desired $\boldsymbol{J}, \boldsymbol{M}$ of Fig. 1b and c satisfies eqs. (24) and (25) because eqs. (24) and (25) were obtained from Fig. 1b and c. This desired $\boldsymbol{J}, \boldsymbol{M}$ will be the only solution to eqs. (24) and (25) if the associated set of homogeneous equations

$$-\boldsymbol{n} \times \left(\boldsymbol{E}_\mathrm{e}^- - \frac{\varepsilon_\mathrm{d}}{\varepsilon_\mathrm{e}} \boldsymbol{E}_\mathrm{d}^+\right) = 0\,, \tag{26}$$

$$-\boldsymbol{n} \times \left(\boldsymbol{H}_\mathrm{e}^- - \frac{\mu_\mathrm{d}}{\mu_\mathrm{e}} \boldsymbol{H}_\mathrm{d}^+\right) = 0 \tag{27}$$

has only the trivial solution $\boldsymbol{J} = \boldsymbol{M} = 0$.

The following argument shows that eqs. (26) and (27) have only the trivial solution $\boldsymbol{J} = \boldsymbol{M} = 0$.

Let $\boldsymbol{E}_d$, $\boldsymbol{H}_d$ be the electromagnetic field outside S due to $\boldsymbol{J}$, $\boldsymbol{M}$ radiating in μ_d, ε_d. Because the electromagnetic field $\boldsymbol{E}_d$, $\boldsymbol{H}_d$ is a source-free Maxwellian field in μ_d, ε_d outside S, the dual electromagnetic field $\eta_d \boldsymbol{H}_d$, $-(1/\eta_d)\,\boldsymbol{E}_d$ where

$$\eta_d = \sqrt{\mu_d/\varepsilon_d} \tag{28}$$

is also a source-free Maxwellian field outside S. Let $\boldsymbol{E}_e$, $\boldsymbol{H}_e$ be the electromagnetic field inside S due to $\boldsymbol{J}$, $\boldsymbol{M}$ radiating in μ_e, ε_e. Because the electromagnetic field $\boldsymbol{E}_e$, $\boldsymbol{H}_e$ is a source-free Maxwellian field inside S, the dual electromagnetic field $\eta_e \boldsymbol{H}_e$, $-(1/\eta_e)\,\boldsymbol{E}_e$ where

$$\eta_e = \sqrt{\mu_e/\varepsilon_e} \tag{29}$$

is also a source-free Maxwellian field inside S. Use the equivalence principle to form the composite situation consisting of medium μ_d, ε_d and field $\eta_d \boldsymbol{H}_d$, $-(1/\eta_d)\boldsymbol{E}_d$ outside S and medium μ_e, ε_e and field

$$\sqrt{\frac{\mu_e\,\varepsilon_e}{\mu_d\,\varepsilon_d}}\left(\eta_e \boldsymbol{H}_e, -\frac{1}{\eta_e}\boldsymbol{E}_e\right)$$

inside S as shown in Fig. 3. Now, eqs. (26) and (27) state that the tangential components of the field

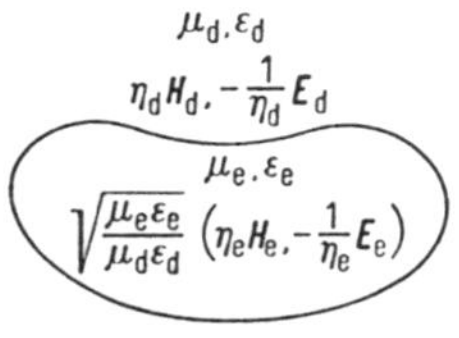

Fig. 3. Composite situation used to prove that eqs. (26) and (27) have only the trivial solution $\boldsymbol{J} = \boldsymbol{M} = 0$.

in Fig. 3 are continuous across S. Thus, Fig. 3 is entirely source-free so that the field in Fig. 3 is zero everywhere in which case eqs. (A.1) to (A.4) are satisfied. But, as shown in the Appendix, eqs. (A.1) to (A.4) have only the trivial solution $\boldsymbol{J} = \boldsymbol{M} = 0$. Hence, eqs. (26) and (27) have only the trivial solution $\boldsymbol{J} = \boldsymbol{M} = 0$.

3. Method of Moments Solution for a Body of Revolution

In this section, a method of moments solution to eqs. (9) and (10) is developed for a homogeneous loss-free body of revolution. Special cases of eqs. (9) and (10) are the PMCHW equations (18) and (19) and the Müller equations (24) and (25).

To give eq. (9) the dimensions of current, we rewrite it as

$$-\frac{1}{\eta_e}(\boldsymbol{E}_e^- + \alpha \boldsymbol{E}_d^+)_{\tan} = \frac{1}{\eta_e}\boldsymbol{E}_{\tan}^{i} \tag{30}$$

where tan denotes tangential components on S and η_e is given by eq. (29). The fields on the left-hand sides of eqs. (30) and (10) are written as the sum of fields due to $\boldsymbol{J}$ and fields due to $\boldsymbol{M}$. Advantage is taken of the fact that the operator which gives the electric field due to a magnetic current is the negative of the operator which gives the magnetic field due to an electric current and that the operator which gives the magnetic field due to a magnetic current is the square of the reciprocal of the intrinsic impedance times the operator which gives the electric field due to an electric current. In view of the above considerations, eqs. (30) and (10) become

$$\left[-\frac{1}{\eta_e}\boldsymbol{E}_e(\boldsymbol{J}) + \frac{1}{\eta_e}\boldsymbol{H}_e^-(\boldsymbol{M}) - \frac{\alpha}{\eta_e}\boldsymbol{E}_d(\boldsymbol{J}) + \frac{\alpha}{\eta_e}\boldsymbol{H}_d^+(\boldsymbol{M})\right]_{\tan} = \frac{1}{\eta_e}\boldsymbol{E}_{\tan}^{i}, \tag{31}$$

$$-\boldsymbol{n}\times\left[\boldsymbol{H}_e^-(\boldsymbol{J}) + \frac{1}{\eta_e^2}\boldsymbol{E}_e(\boldsymbol{M}) + \beta \boldsymbol{H}_d^+(\boldsymbol{J}) + \frac{\beta}{\eta_d^2}\boldsymbol{E}_d(\boldsymbol{M})\right] = \boldsymbol{n}\times\boldsymbol{H}^{i} \tag{32}$$

where $\boldsymbol{E}$ denotes the operator which gives the electric field due to an electric current. The subscript e or d on $\boldsymbol{E}$ denotes radiation in either μ_e, ε_e or μ_d, ε_d. The superscript + or −, if present on $\boldsymbol{H}$, denotes field evaluation either just outside S or just inside S. The $\boldsymbol{H}$'s in eqs. (31) and (32) are the corresponding magnetic field due to electric current operators. We stress that all $\boldsymbol{E}$'s and $\boldsymbol{H}$'s in eqs. (31) and (32) are, by definition, operators which give electric and magnetic fields due to electric currents, even though these operators act on both electric and magnetic currents $\boldsymbol{J}$ and $\boldsymbol{M}$ in eqs. (31) and (32).

Let

$$\boldsymbol{J} = \sum_{n=-\infty}^{\infty}\sum_{j=1}^{N}(I_{nj}^{t}\,\boldsymbol{J}_{nj}^{t} + I_{nj}^{\phi}\,\boldsymbol{J}_{nj}^{\phi}), \tag{33}$$

$$\boldsymbol{M} = \eta_e\sum_{n=-\infty}^{\infty}\sum_{j=1}^{N}(V_{nj}^{t}\,\boldsymbol{J}_{nj}^{t} + V_{nj}^{\phi}\,\boldsymbol{J}_{nj}^{\phi}) \tag{34}$$

where I_{nj}^{t}, I_{nj}^{ϕ}, V_{nj}^{t}, and V_{nj}^{ϕ} are coefficients to be determined and

$$\boldsymbol{J}_{nj}^{t} = \boldsymbol{u}_t\, f_j(t)\, e^{jn\phi}, \quad \boldsymbol{J}_{nj}^{\phi} = \boldsymbol{u}_\phi\, f_j(t)\, e^{jn\phi}. \tag{35), (36}$$

In eqs. (35) and (36), t is the arc length along the generating curve of the body of revolution and ϕ is the longitudinal angle. $\boldsymbol{u}_t$ and $\boldsymbol{u}_\phi$ are unit vectors in the t and ϕ directions respectively such that $\boldsymbol{u}_\phi \times \boldsymbol{u}_t = \boldsymbol{n}$ and $f_j(t)$ is a scalar function of t. The body of revolution and coordinate system are shown in Fig. 4. Substitution of eqs. (33) and (34) into eqs. (31) and (32) yields

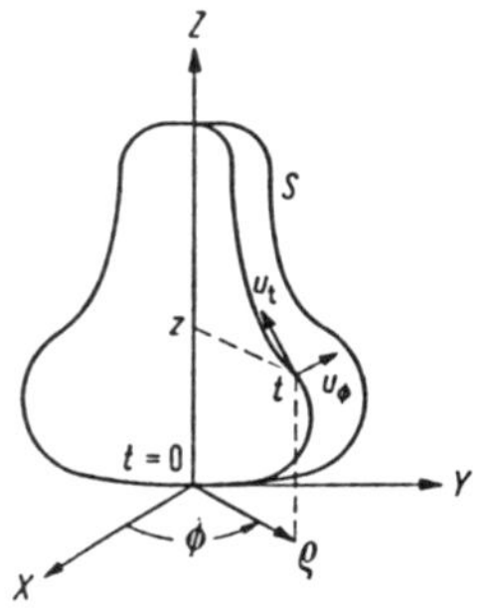

Fig. 4. Body of revolution and coordinate system.

$$\sum_{n=-\infty}^{\infty} \sum_{j=1}^{N} \Bigg\{ [\boldsymbol{H}_{\mathrm{e}}^{-}(\boldsymbol{J}_{nj}^{\mathrm{t}}) + \alpha \boldsymbol{H}_{\mathrm{d}}^{+}(\boldsymbol{J}_{nj}^{\mathrm{t}})]_{\tan} V_{nj}^{\mathrm{t}} + [\boldsymbol{H}_{\mathrm{e}}^{-}(\boldsymbol{J}_{nj}^{\phi}) + \alpha \boldsymbol{H}_{\mathrm{d}}^{+}(\boldsymbol{J}_{nj}^{\phi})]_{\tan} V_{nj}^{\phi} + \left[-\frac{\boldsymbol{E}_{\mathrm{e}}(\boldsymbol{J}_{nj}^{\mathrm{t}})}{\eta_{\mathrm{e}}} - \frac{\alpha \eta_{\mathrm{d}}}{\eta_{\mathrm{e}}} \frac{\boldsymbol{E}_{\mathrm{d}}(\boldsymbol{J}_{nj}^{\mathrm{t}})}{\eta_{\mathrm{d}}} \right]_{\tan} I_{nj}^{\mathrm{t}} + \left[-\frac{\boldsymbol{E}_{\mathrm{e}}(\boldsymbol{J}_{nj}^{\phi})}{\eta_{\mathrm{e}}} - \frac{\alpha \eta_{\mathrm{d}}}{\eta_{\mathrm{e}}} \frac{\boldsymbol{E}_{\mathrm{d}}(\boldsymbol{J}_{nj}^{\phi})}{\eta_{\mathrm{d}}} \right]_{\tan} I_{nj}^{\phi} \Bigg\} = \frac{1}{\eta_{\mathrm{e}}} \boldsymbol{E}_{\tan}^{\mathrm{i}}, \tag{37}$$

$$-\boldsymbol{n} \times \sum_{n=-\infty}^{\infty} \sum_{j=1}^{N} \Bigg\{ \left[\frac{\boldsymbol{E}_{\mathrm{e}}(\boldsymbol{J}_{nj}^{\mathrm{t}})}{\eta_{\mathrm{e}}} + \frac{\beta \eta_{\mathrm{e}}}{\eta_{\mathrm{d}}} \frac{\boldsymbol{E}_{\mathrm{d}}(\boldsymbol{J}_{nj}^{\mathrm{t}})}{\eta_{\mathrm{d}}} \right] V_{nj}^{\mathrm{t}} + \left[\frac{\boldsymbol{E}_{\mathrm{e}}(\boldsymbol{J}_{nj}^{\phi})}{\eta_{\mathrm{e}}} + \frac{\beta \eta_{\mathrm{e}}}{\eta_{\mathrm{d}}} \frac{\boldsymbol{E}_{\mathrm{d}}(\boldsymbol{J}_{nj}^{\phi})}{\eta_{\mathrm{d}}} \right] V_{nj}^{\phi} + [\boldsymbol{H}_{\mathrm{e}}^{-}(\boldsymbol{J}_{nj}^{\mathrm{t}}) + \beta \boldsymbol{H}_{\mathrm{d}}^{+}(\boldsymbol{J}_{nj}^{\mathrm{t}})] I_{nj}^{\mathrm{t}} + [\boldsymbol{H}_{\mathrm{e}}^{-}(\boldsymbol{J}_{nj}^{\phi}) + \beta \boldsymbol{H}_{\mathrm{d}}^{+}(\boldsymbol{J}_{nj}^{\phi})] I_{nj}^{\phi} \Bigg\} = \boldsymbol{n} \times \boldsymbol{H}^{\mathrm{i}}. \tag{38}$$

Define the inner product of two vector functions on S to be the integral over S of the dot product of these two vector functions. Because the field operators in eqs. (37) and (38) are the same as those considered in [9], only the nth term of the sum (37) or (38) contributes to the inner product of eqs. (37) or (38) with either $\boldsymbol{J}_{-ni}^{\mathrm{t}}$ or $\boldsymbol{J}_{-ni}^{\phi}$. Hence, the inner product of eq. (37) with $\boldsymbol{J}_{-ni}^{\mathrm{t}}, i=1,2,\ldots,N$, and J_{-ni}^{ϕ}, $i=1,2,\ldots,N$, successively and the inner product of eq. (38) with $\boldsymbol{J}_{-ni}^{\mathrm{t}}$, $i=1,2,\ldots,N$, and $\boldsymbol{J}_{-ni}^{\phi}$, $i=1,2,\ldots,N$, successively gives the matrix equation

$$\begin{bmatrix} (Y_{n\mathrm{e}}^{\phi\mathrm{t}} + \alpha Y_{n\mathrm{d}}^{\phi\mathrm{t}}) & (Y_{n\mathrm{e}}^{\phi\phi} + \alpha \hat{Y}_{n\mathrm{d}}^{\phi\phi}) & \left(Z_{n\mathrm{e}}^{\mathrm{tt}} + \frac{\alpha\eta_{\mathrm{d}}}{\eta_{\mathrm{e}}} Z_{n\mathrm{d}}^{\mathrm{tt}}\right) & \left(Z_{n\mathrm{e}}^{\mathrm{t}\phi} + \frac{\alpha\eta_{\mathrm{d}}}{\eta_{\mathrm{e}}} Z_{n\mathrm{d}}^{\mathrm{t}\phi}\right) \\ (-Y_{n\mathrm{e}}^{\mathrm{tt}} - \alpha \hat{Y}_{n\mathrm{d}}^{\mathrm{tt}}) & (-Y_{n\mathrm{e}}^{\mathrm{t}\phi} - \alpha Y_{n\mathrm{d}}^{\mathrm{t}\phi}) & \left(Z_{n\mathrm{e}}^{\phi\mathrm{t}} + \frac{\alpha\eta_{\mathrm{d}}}{\eta_{\mathrm{e}}} Z_{n\mathrm{d}}^{\phi\mathrm{t}}\right) & \left(Z_{n\mathrm{e}}^{\phi\phi} + \frac{\alpha\eta_{\mathrm{d}}}{\eta_{\mathrm{e}}} Z_{n\mathrm{d}}^{\phi\phi}\right) \\ \left(Z_{n\mathrm{e}}^{\phi\mathrm{t}} + \frac{\beta\eta_{\mathrm{e}}}{\eta_{\mathrm{d}}} Z_{n\mathrm{d}}^{\phi\mathrm{t}}\right) & \left(Z_{n\mathrm{e}}^{\phi\phi} + \frac{\beta\eta_{\mathrm{e}}}{\eta_{\mathrm{d}}} Z_{n\mathrm{d}}^{\phi\phi}\right) & (Y_{n\mathrm{e}}^{\mathrm{tt}} + \beta \hat{Y}_{n\mathrm{d}}^{\mathrm{tt}}) & (Y_{n\mathrm{e}}^{\mathrm{t}\phi} + \beta Y_{n\mathrm{d}}^{\mathrm{t}\phi}) \\ \left(-Z_{n\mathrm{e}}^{\mathrm{tt}} - \frac{\beta\eta_{\mathrm{e}}}{\eta_{\mathrm{d}}} Z_{n\mathrm{d}}^{\mathrm{tt}}\right) & \left(-Z_{n\mathrm{e}}^{\mathrm{t}\phi} - \frac{\beta\eta_{\mathrm{e}}}{\eta_{\mathrm{d}}} Z_{n\mathrm{d}}^{\mathrm{t}\phi}\right) & (Y_{n\mathrm{e}}^{\phi\mathrm{t}} + \beta Y_{n\mathrm{d}}^{\phi\mathrm{t}}) & (Y_{n\mathrm{e}}^{\phi\phi} + \beta \hat{Y}_{n\mathrm{d}}^{\phi\phi}) \end{bmatrix} \begin{bmatrix} \vec{V}_n^{\mathrm{t}} \\ \vec{V}_n^{\phi} \\ \vec{I}_n^{\mathrm{t}} \\ \vec{I}_n^{\phi} \end{bmatrix} = \begin{bmatrix} \vec{\hat{V}}_n^{\mathrm{t}} \\ \vec{\hat{V}}_n^{\phi} \\ \vec{\hat{I}}_n^{\mathrm{t}} \\ \vec{\hat{I}}_n^{\phi} \end{bmatrix} \tag{39}$$

for $n=0, \pm 1, \pm 2, \ldots$. In eq. (39), $\vec{V}_n^{\mathrm{t}}$, $\vec{V}_n^{\phi}$, $\vec{I}_n^{\mathrm{t}}$, and $\vec{I}_n^{\phi}$ are column vectors of the coefficients appearing in eqs. (33) and (34). Also,

$$(Y_{n\mathrm{f}}^{pq})_{ij} = -\iint_S \boldsymbol{J}_{-ni}^{p} \cdot \boldsymbol{n} \times \boldsymbol{H}_{\mathrm{f}}(\boldsymbol{J}_{nj}^{q})\, \mathrm{d}s, \tag{40}$$

$$(Z_{n\mathrm{f}}^{pq})_{ij} = -\frac{1}{\eta_{\mathrm{f}}} \iint_S \boldsymbol{J}_{-ni}^{p} \cdot \boldsymbol{E}_{\mathrm{f}}(\boldsymbol{J}_{nj}^{q})\, \mathrm{d}s, \tag{41}$$

$$\hat{V}_{ni}^{p} = \frac{1}{\eta_{\mathrm{e}}} \iint_S \boldsymbol{J}_{-ni}^{p} \cdot \boldsymbol{E}^{\mathrm{i}}\, \mathrm{d}s, \tag{42}$$

$$\hat{I}_{ni}^{p} = \iint_S \boldsymbol{J}_{-ni}^{p} \cdot \boldsymbol{n} \times \boldsymbol{H}^{\mathrm{i}}\, \mathrm{d}s \tag{43}$$

where p may be either t or ϕ, q may be either t or ϕ, and f may be either e or d. If $p=q$ in eq. (40), it matters whether the magnetic field $\boldsymbol{H}_{\mathrm{f}}(\boldsymbol{J}_{nj}^{q})$ is evaluated just outside or just inside S. The Y's without carets in eq. (39) are given by the right-hand side of eq. (40) in which the magnetic field is evaluated just inside S. The Y's with carets in eq. (39) are given by the right-hand side of eq. (40) with magnetic field evaluation just outside S.

The Y and Z submatrices on the left-hand side of eq. (39) are the same as in eq. (19) of [11] with the reservations that the caret on Y denotes magnetic field evaluation just outside S, and the extra subscript e or d denotes radiation in either μ_{e}, ε_{e} or μ_{d}, ε_{d}. The $\hat{I}$ column vectors on the right-hand side of eq. (39) are the same as in eq. (19) of [11] whereas the $\hat{V}$ column vectors in eq. (39) are the same as the V's without carets in eq. (19) of [11].

The solution $\vec{V}_n^{\mathrm{t}}$, $\vec{V}_n^{\phi}$, $\vec{I}_n^{\mathrm{t}}$, and $\vec{I}_n^{\phi}$ to the matrix equation (39) determines the equivalent electric and magnetic currents $\boldsymbol{J}$ and $\boldsymbol{M}$ according to eqs. (33) and (34). From Fig. 1b, these currents radiate in μ_{e}, ε_{e} to produce the scattered field outside S.

4. Far Field Measurement and Plane Wave Excitation

In this section, measurement vectors are used to obtain the far field of the equivalent surface currents $\boldsymbol{J}$ and $\boldsymbol{M}$ radiating in μ_{e}, ε_{e}. This far field is the far field scattered by the homogeneous body of revolution. For plane wave excitation, the composite vector on the right-hand side of eq. (39) is expressed in terms of these measurement vectors.

By reciprocity,

$$\boldsymbol{E}^{\mathrm{s}} \cdot I \boldsymbol{l}_{\mathrm{r}} = \iint_S [\boldsymbol{J}(\boldsymbol{r}) \cdot \boldsymbol{E}(I\boldsymbol{l}_{\mathrm{r}}) - \boldsymbol{M}(\boldsymbol{r}) \cdot \boldsymbol{H}(I\boldsymbol{l}_{\mathrm{r}})]\, \mathrm{d}s \tag{44}$$

where $\boldsymbol{E}^{\mathrm{s}}$ is the far electric field due to $\boldsymbol{J}$ and $\boldsymbol{M}$, $I\boldsymbol{l}_{\mathrm{r}}$ is a receiving electric dipole at the far field measurement point, $\boldsymbol{E}(I\boldsymbol{l}_{\mathrm{r}})$ is the electric field due to $I\boldsymbol{l}_{\mathrm{r}}$, and $\boldsymbol{H}(I\boldsymbol{l}_{\mathrm{r}})$ is the magnetic field due to $I\boldsymbol{l}_{\mathrm{r}}$. Both $\boldsymbol{E}(I\boldsymbol{l}_{\mathrm{r}})$ and $\boldsymbol{H}(I\boldsymbol{l}_{\mathrm{r}})$ are evaluated at point $\boldsymbol{r}$ on S where $\boldsymbol{r}$ is the point at which the differential portion of surface $\mathrm{d}s$ is located. If $\boldsymbol{l}_{\mathrm{r}}$ is tangent to the radiation sphere,

$$\boldsymbol{E}(I\boldsymbol{l}_{\mathrm{r}}) = \frac{-\mathrm{j} k \eta\, \mathrm{e}^{-\mathrm{j} k r_{\mathrm{r}}}}{4 \pi r_{\mathrm{r}}} I \boldsymbol{l}_{\mathrm{r}}\, \mathrm{e}^{-\mathrm{j} \boldsymbol{k}_{\mathrm{r}} \cdot \boldsymbol{r}}, \tag{45}$$

$$\boldsymbol{H}(I\boldsymbol{l}_{\mathrm{r}}) = \frac{-\mathrm{j}\, \mathrm{e}^{-\mathrm{j} k r_{\mathrm{r}}}}{4 \pi r_{\mathrm{r}}} (\boldsymbol{k}_{\mathrm{r}} \times I \boldsymbol{l}_{\mathrm{r}})\, \mathrm{e}^{-\mathrm{j} \boldsymbol{k}_{\mathrm{r}} \cdot \boldsymbol{r}} \tag{46}$$

where r_r is the distance between the measurement point and the origin in the vicinity of S. Also, $\boldsymbol{k}_r$ is the propagation vector of the plane wave coming from $I\boldsymbol{l}_r$, k is the propagation constant and η is the intrinsic impedance of the medium outside S. To simplify the notation in this section, we have omitted the subscript e from all parameters dependent on the medium. It is understood that all far field measurement vectors and plane wave excitation vectors depend only on the external medium μ_e, ε_e.

Substitution of eqs. (33), (34), (45), and (46) into eq. (44) gives

$$E_\theta^s = \frac{-\mathrm{j}\,\eta\,\mathrm{e}^{-\mathrm{j}kr_r}}{4\pi r_r} \cdot \sum_{n=-\infty}^{\infty} (\tilde{R}_n^{t\phi}\vec{V}_n^t + \tilde{R}_n^{\phi\phi}\vec{V}_n^\phi + \tilde{R}_n^{t\theta}\vec{I}_n^t + \tilde{R}_n^{\phi\theta}\vec{I}_n^\phi)\,\mathrm{e}^{\mathrm{j}n\phi_r} \tag{47}$$

for $I\boldsymbol{l}_r = \boldsymbol{u}_\theta^r$ and

$$E_\phi^s = \frac{-\mathrm{j}\,\eta\,\mathrm{e}^{-\mathrm{j}kr_r}}{4\pi r_r} \cdot \sum_{n=-\infty}^{\infty} (-\tilde{R}_n^{t\theta}\vec{V}_n^t - \tilde{R}_n^{\phi\theta}\vec{V}_n^\phi + \tilde{R}_n^{t\phi}\vec{I}_n^t + \tilde{R}_n^{\phi\phi}\vec{I}_n^\phi)\,\mathrm{e}^{\mathrm{j}n\phi_r} \tag{48}$$

for $I\boldsymbol{l}_r = \boldsymbol{u}_\phi^r$ where $\boldsymbol{u}_\theta^r$ and $\boldsymbol{u}_\phi^r$ are unit vectors in the θ_r and ϕ_r directions respectively. As shown in Fig. 5, θ_r and ϕ_r are the angular coordinates of the receiver location at which $I\boldsymbol{l}_r$ is placed. In eqs. (47) and (48), E_θ^s and E_ϕ^s are the θ_r and ϕ_r components of $\boldsymbol{E}^s$. Also, $\vec{V}_n^t$, $\vec{V}_n^\phi$, $\vec{I}_n^t$, and $\vec{I}_n^\phi$ are column vectors of the coefficients appearing in eqs. (33) and (34). Furthermore, $\tilde{R}_n^{pq}$ is a row vector whose jth element is given by

$$R_{nj}^{pq} = k\,\mathrm{e}^{-\mathrm{j}n\phi_r} \iint_S \boldsymbol{J}_{nj}^p \cdot \boldsymbol{u}_q^r\,\mathrm{e}^{-\mathrm{j}\boldsymbol{k}_r\cdot\boldsymbol{r}}\,\mathrm{d}s \tag{49}$$

where p may be either t or ϕ and q may be either θ or ϕ. In view of eqs. (35) and (36), eq. (49) is the same as eq. (92) on p. 26 of [9]. The right-hand side of eq. (49) does not depend on ϕ_r.

For plane wave incidence and expansion functions $\boldsymbol{J}_{nj}^t$ and $\boldsymbol{J}_{nj}^\phi$ given by eqs. (35) and (36), the equivalent currents (33) and (34) and the fields (47) and (48) have special forms. To obtain these forms, assume that the incident electromagnetic field $\boldsymbol{E}^i$, $\boldsymbol{H}^i$ is either a θ polarized field defined by

$$\boldsymbol{E}^i = k\,\eta\,\boldsymbol{u}_\theta^t\,\mathrm{e}^{-\mathrm{j}\boldsymbol{k}_t\cdot\boldsymbol{r}}, \quad \boldsymbol{H}^i = -k\,\boldsymbol{u}_y\,\mathrm{e}^{-\mathrm{j}\boldsymbol{k}_t\cdot\boldsymbol{r}} \qquad (50), (51)$$

or a ϕ polarized field defined by

$$\boldsymbol{E}^i = k\,\eta\,\boldsymbol{u}_y\,\mathrm{e}^{-\mathrm{j}\boldsymbol{k}_t\cdot\boldsymbol{r}}, \quad \boldsymbol{H}^i = k\,\boldsymbol{u}_\theta^t\,\mathrm{e}^{-\mathrm{j}\boldsymbol{k}_t\cdot\boldsymbol{r}} \qquad (52), (53)$$

where $\boldsymbol{k}_t$ is the propagation vector and, as shown in Fig. 5, $\boldsymbol{u}_\theta^t$ and $\boldsymbol{u}_y$ are unit vectors in the θ_t and y directions respectively. Here, θ_t is the colatitude of the direction from which the incident wave comes. $\boldsymbol{k}_t$ is in the x, z-plane. No generality is lost by putting $\boldsymbol{k}_t$ in the x, z-plane because if $\boldsymbol{k}_t$ were shifted out of the x, z-plane by an angle ϕ_t, the response would also be shifted by the same angle ϕ_t.

Substituting eqs. (50) and (51) into eqs. (42) and (43), then substituting eqs. (52) and (53) into eqs. (42) and (43), next taking advantage of the relationships

$$\boldsymbol{J}_{-ni}^t \times \boldsymbol{n} = \boldsymbol{J}_{-ni}^\phi, \quad \boldsymbol{J}_{-ni}^\phi \times \boldsymbol{n} = -\boldsymbol{J}_{-ni}^t \qquad (54), (55)$$

which are apparent from eqs. (35), (36) and Fig. 4, then comparing the results with eq. (49), and finally using eq. (104) on p. 29 of [9], we obtain

$$\begin{bmatrix} \vec{V}_n^{t\theta} & \vec{V}_n^{t\phi} \\ \vec{V}_n^{\phi\theta} & \vec{V}_n^{\phi\phi} \\ \vec{I}_n^{t\theta} & \vec{I}_n^{t\phi} \\ \vec{I}_n^{\phi\theta} & \vec{I}_n^{\phi\phi} \end{bmatrix} = \begin{bmatrix} \vec{R}_n^{t\theta} & -\vec{R}_n^{t\phi} \\ -\vec{R}_n^{\phi\theta} & \vec{R}_n^{\phi\phi} \\ -\vec{R}_n^{\phi\phi} & -\vec{R}_n^{\phi\theta} \\ -\vec{R}_n^{t\phi} & -\vec{R}_n^{t\theta} \end{bmatrix}. \tag{56}$$

The first superscript on $\vec{V}_n$ and $\vec{I}_n$ in eq. (56) is the superscript which appears on the right-hand side of eq. (39). The second superscript on $\vec{V}_n$ and $\vec{I}_n$ in eq. (56) denotes the polarization of the incident plane wave. If this second superscript is θ, the θ polarized field given by eqs. (50) and (51) is incident. If this second superscript is ϕ, the ϕ polarized field given by eqs. (52) and (53) is incident. The jth element of the column vector $\vec{R}_n^{pq}$ on the right-hand side of eq. (56) is given by eq. (49) with θ_r replaced by θ_t.

For plane wave incidence, the $+n$ and $-n$ terms in formulas (33) and (34) for the equivalent currents can be combined as follows. Because of even-odd properties with respect to n of the square submatrices on the left-hand side of eq. (39) and of the column vectors on the right-hand side of eq. (39), the solutions to eq. (39) satisfy

$$\begin{bmatrix} \vec{V}_{-n}^{t\theta} & \vec{V}_{-n}^{t\phi} \\ \vec{V}_{-n}^{\phi\theta} & \vec{V}_{-n}^{\phi\phi} \\ \vec{I}_{-n}^{t\theta} & \vec{I}_{-n}^{t\phi} \\ \vec{I}_{-n}^{\phi\theta} & \vec{I}_{-n}^{\phi\phi} \end{bmatrix} = \begin{bmatrix} -\vec{V}_n^{t\theta} & \vec{V}_n^{t\phi} \\ \vec{V}_n^{\phi\theta} & -\vec{V}_n^{\phi\phi} \\ \vec{I}_n^{t\theta} & -\vec{I}_n^{t\phi} \\ -\vec{I}_n^{\phi\theta} & \vec{I}_n^{\phi\phi} \end{bmatrix}. \tag{57}$$

The first superscript on the column vectors $\vec{V}_{\pm n}$ and $\vec{I}_{\pm n}$ in eq. (57) is that which appears on the column vectors $\vec{V}_n$ and $\vec{I}_n$ on the left-hand side of eq. (39). The second superscript on the column vectors in eq. (57) denotes either the θ or the ϕ polarized incident plane wave. Substitution of

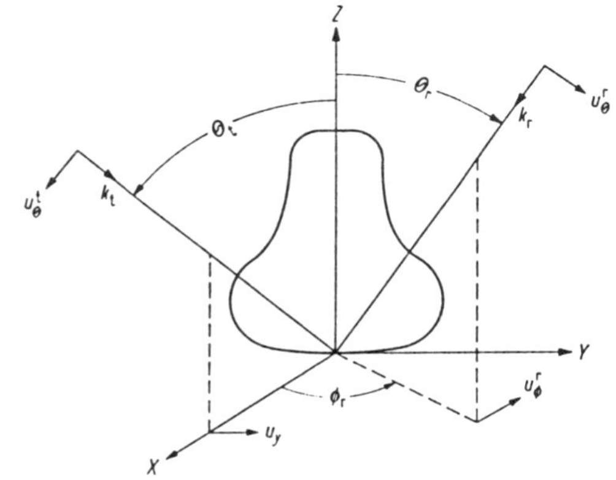

Fig. 5. Plane wave scattering by a dielectric body of revolution.

eqs. (57), (35), and (36) into eqs. (33) and (34) yields

$$J^\theta = (\tilde f \vec I_0^{t\theta})\, u_t + \sum_{n=1}^{\infty} [2(\tilde f \vec I_n^{t\theta})\, u_t \cos n\phi + 2\mathrm{j}(\tilde f \vec I_n^{\phi\theta})\, u_\phi \sin n\phi], \tag{58}$$

$$\frac{1}{\eta} M^\theta = (\tilde f \vec V_0^{\phi\theta})\, u_\phi + \sum_{n=1}^{\infty} [2\mathrm{j}(\tilde f \vec V_n^{t\theta})\, u_t \sin n\phi + 2(\tilde f \vec V_n^{\phi\theta})\, u_\phi \cos n\phi] \tag{59}$$

for the θ polarized incident wave and

$$J^\phi = (\tilde f \vec I_0^{\phi\phi})\, u_\phi + \sum_{n=1}^{\infty} [2\mathrm{j}(\tilde f \vec I_n^{t\phi})\, u_t \sin n\phi + 2(\tilde f \vec I_n^{\phi\phi})\, u_\phi \cos n\phi], \tag{60}$$

$$\frac{1}{\eta} M^\phi = (\tilde f \vec V_0^{t\phi})\, u_t + \sum_{n=1}^{\infty} [2(\tilde f \vec V_n^{t\phi})\, u_t \cos n\phi + 2\mathrm{j}(\tilde f \vec V_n^{\phi\phi})\, u_\phi \sin n\phi] \tag{61}$$

for the ϕ polarized incident wave. In eqs. (58) to (61), $\tilde f$ is a row vector of the $f_j(t)$. The superscript θ or ϕ on J or M in eqs. (58) to (61) differentiates the equivalent currents for the θ polarized incident wave from those for the ϕ polarized incident wave.

The far scattered fields (47) and (48) are specialized to the θ polarized incident plane wave by appending the additional subscript θ to E^s on the left-hand sides of eqs. (47) and (48) and the additional superscript θ to $\vec V_n^t$, $\vec V_n^\phi$, $\vec I_n^t$, and $\vec I_n^\phi$ on the right-hand sides of eqs. (47) and (48). Moreover, the $+n$ and $-n$ terms in eqs. (47) and (48) can be combined. As a result, eqs. (47) and (48) become

$$E^s_{\theta\theta} = \frac{-\mathrm{j}\eta\, e^{-\mathrm{j}kr_r}}{4\pi r_r} \Big[\tilde R_0^{\phi\phi} \vec V_0^{\phi\theta} + \tilde R_0^{t\theta} \vec I_0^{t\theta} + 2\sum_{n=1}^{\infty} (\tilde R_n^{t\phi} \vec V_n^{t\theta} + \tilde R_n^{\phi\phi} \vec V_n^{\phi\theta} + \tilde R_n^{t\theta} \vec I_n^{t\theta} + \tilde R_n^{\phi\theta} \vec I_n^{\phi\theta}) \cos n\phi_r \Big], \tag{62}$$

$$E^s_{\phi\theta} = \frac{\eta\, e^{-\mathrm{j}kr_r}}{2\pi r_r} \sum_{n=1}^{\infty} (-\tilde R_n^{t\theta} \vec V_n^{t\theta} - \tilde R_n^{\phi\theta} \vec V_n^{\phi\theta} + \tilde R_n^{t\phi} \vec I_n^{t\theta} + \tilde R_n^{\phi\phi} \vec I_n^{\phi\theta}) \sin n\phi_r \tag{63}$$

for the θ polarized incident plane wave. Similarly, eqs. (47) and (48) become

$$E^s_{\theta\phi} = \frac{\eta\, e^{-\mathrm{j}kr_r}}{2\pi r_r} \sum_{n=1}^{\infty} (\tilde R_n^{t\phi} \vec V_n^{t\phi} + \tilde R_n^{\phi\phi} \vec V_n^{\phi\phi} + \tilde R_n^{t\theta} \vec I_n^{t\phi} + \tilde R_n^{\phi\theta} \vec I_n^{\phi\phi}) \sin n\phi_r, \tag{64}$$

$$E^s_{\phi\phi} = \frac{-\mathrm{j}\eta\, e^{-\mathrm{j}kr_r}}{4\pi r_r} \Big\{ -\tilde R_0^{t\theta} \vec V_0^{t\phi} + \tilde R_0^{\phi\phi} \vec I_0^{\phi\phi} + 2\sum_{n=1}^{\infty} (-\tilde R_n^{t\theta} \vec V_n^{t\phi} - \tilde R_n^{\phi\theta} \vec V_n^{\phi\phi} + \tilde R_n^{t\phi} \vec I_n^{t\phi} + \tilde R_n^{\phi\phi} \vec I_n^{\phi\phi}) \cos n\phi_r \Big\} \tag{65}$$

for the ϕ polarized incident plane wave. The first subscript on E^s on the left-hand sides of eqs. (62) to (65) denotes the receiver polarization and the second subscript on E^s denotes the transmitter polarization.

The scattering cross section σ_{pq} is defined by

$$\sigma_{pq} = 4\pi r_r^2\, |E^s_{pq}|^2 / |E^i|^2 \tag{66}$$

where p is either θ or ϕ and q is either θ or ϕ. In eq. (66), E^s_{pq} is a component of the scattered field given by eqs. (62) to (65) and $|E^i|$ is the magnitude of the electric field of the incident plane wave. According to eqs. (50) and (52),

$$|E^i| = k\eta \tag{67}$$

for both polarizations so that

$$\sigma_{pq} = 4\pi r_r^2\, |E^s_{pq}|^2 / k^2 \eta^2 . \tag{68}$$

Normalized versions of eq. (68) are

$$\frac{\sigma_{pq}}{\pi a^2} = \frac{4 r_r^2\, |E^s_{pq}|^2}{k^2 a^2 \eta^2}, \qquad \frac{\sigma_{pq}}{\lambda^2} = \frac{r_r^2\, |E^s_{pq}|^2}{\pi \eta^2} \tag{69), (70}$$

where a is some characteristic length associated with the scatterer and λ is the wavelength in the external medium.

5. Examples

A computer program has been written to calculate the equivalent currents and scattering patterns for a dielectric body of revolution excited by an axially incident plane wave. This program is described and listed in [12]. For the program the functions $f_j(t)$ are pulse approximations to triangle functions divided by cylindrical radius. The integrations in ϕ were obtained using Gaussian quadrature formulas. Some computational results obtained with this program are given in this section.

Computations were made of the equivalent electric and magnetic currents using both the general body of revolution program and the exact Mie series solution [13]. The currents obtained from our PMCHW solution and from our Müller solution both agreed well with the exact solution when the dielectric constant was 4 and $ka = 3$. All integrations in ϕ were obtained by using a 20 point Gaussian quadrature formula. All integrations in $t = a(\pi - \theta)$ over the functions $f_j(t)$ were done by sampling each $f_j(t)$ four times. Fourteen overlapping triangle functions divided by cylindrical radius, equally spaced in θ, were used for each component of current. Plots of these equivalent currents can be found in [12].

Fig. 6a, b shows the scattering patterns radiated by the equivalent currents. The symbols $\times$ and $+$ denote $\sigma_{\theta\theta}/\pi a^2$ and $\sigma_{\phi\theta}/\pi a^2$ respectively. The solid curves are the exact patterns obtained from the Mie series solution [13]. The patterns $\sigma_{\theta\theta}$ and $\sigma_{\phi\theta}$ are given by eqs. (69), (62) and (63). Here, $\sigma_{\theta\theta}$ is the θ polarized pattern versus θ_r in the $\phi = 0$ plane and $\sigma_{\phi\theta}$ is the ϕ polarized pattern versus θ_r in the $\phi = 90°$ plane. For axial incidence, only the $n = 1$ terms are present in eqs. (62) and (63). Elsewhere [3], [15], the pattern $\sigma_{\theta\theta}$ is called the horizontal polarization because it is polarized parallel to the scattering plane. Similarly, the pattern $\sigma_{\phi\theta}$ is called the vertical polarization because it is polarized perpendicular to the scattering plane.

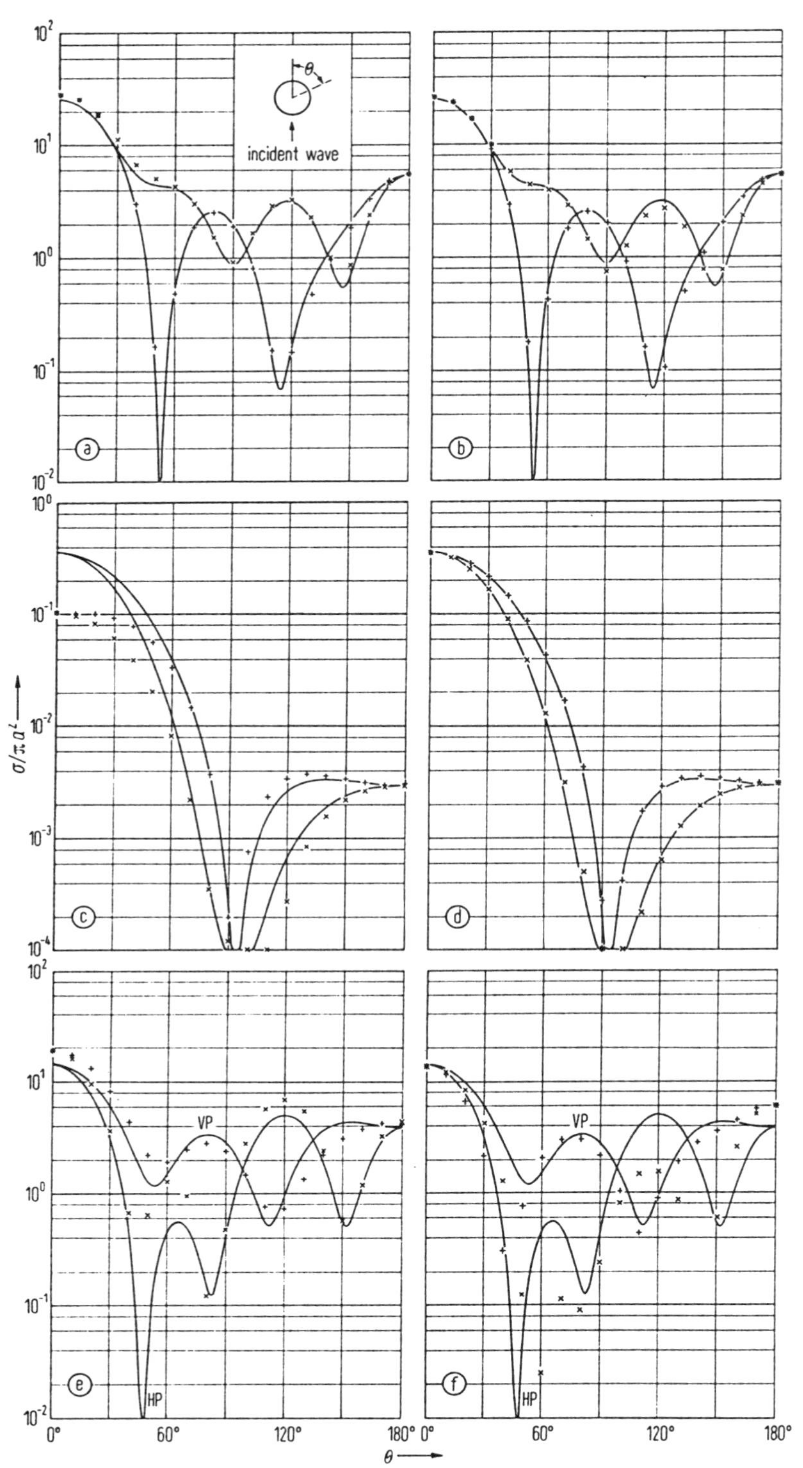

Fig. 6. Plane wave scattering patterns for dielectric sphere, $ka = 3$. Symbols × and + denote horizontal polarization and vertical polarization respectively. Solid line denotes exact solution.
(a) $\varepsilon_r = 4$, PMCHW solution,
(b) $\varepsilon_r = 4$, Müller solution,
(c) $\varepsilon_r = 1.1$, PMCHW solution,
(d) $\varepsilon_r = 1.1$, Müller solution,
(e) $\varepsilon_r = 20$, PMCHW solution,
(f) $\varepsilon_r = 20$, Müller solution.

Fig. 6c–f shows the scattering patterns for two other dielectric spheres. Fig. 6c, d is for relative dielectric constant $\varepsilon_r = 1.1$, and Fig. 6e, f for $\varepsilon_r = 20$. All other parameters in Fig. 6c–f are the same as in Fig. 6a, b. (In Fig. 6c, d, values less than 0.0001 are plotted at 0.0001.)

Fig. 7 shows the computed scattering patterns of a finite dielectric cylinder of radius a and height $2a$ when a is 0.25 free space wavelengths. The relative dielectric constant of the cylinder is $\varepsilon_r = 4$. The incident field is a plane wave traveling in the positive z direction, the same field which was incident upon the previous dielectric spheres. In Fig. 7, the patterns $\sigma_{\theta\theta}/\pi a^2$ and $\sigma_{\phi\theta}/\pi a^2$ as obtained from our solution of the PMCHW formulation are plotted with the symbols × and + respectively. The solid curves are $\sigma_{\theta\theta}/\pi a^2$ and $\sigma_{\phi\theta}/\pi a^2$ as obtained from our solution of the Müller formulation.

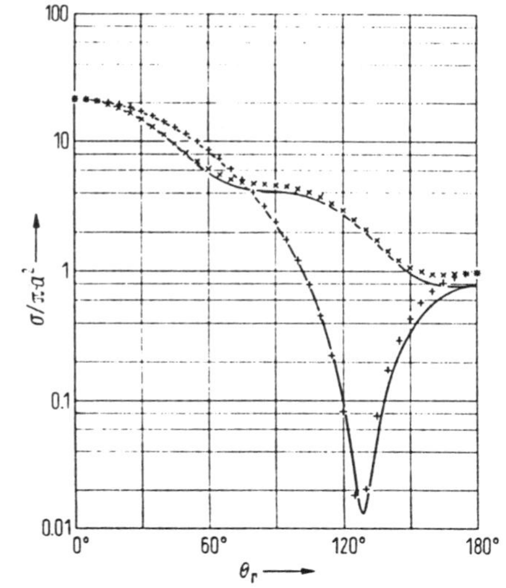

Fig. 7. Plane wave scattering patterns for finite dielectric cylinder of radius a and height $2a$. $a = 0.25$ free space wavelengths, $\varepsilon_r = 4$. Symbols $\times$ and $+$ denote horizontal polarization and vertical polarization respectively, PMCHW solution. Solid curves denote Müller solution.

The equivalent currents which radiate the patterns of Fig. 7 were obtained by using a 48 point Gaussian quadrature formula for all integrations in ϕ. All integrations in t over the functions $f_j(t)$ were done by sampling each $f_j(t)$ four times. Eleven overlapping triangle functions divided by the cylindrical coordinate radius equally spaced in t were used for each component of current.

6. Discussion

According to Fig. 6c, d, the scattering patterns obtained from our solution of the Müller formulation are more accurate than those obtained from our solution of the PMCHW formulation for the dielectric sphere with $ka = 3$ and $\varepsilon_r = 1.1$. A more accurate PMCHW solution for this case could, of course, be obtained at the expense of more accurate calculation of the moment matrix elements. From plots not included in this paper, we observed that both our PMCHW solution and our Müller solution for the equivalent currents on the dielectric sphere were reasonably accurate. However, the following argument shows that when ε_r is near one, a slight inaccuracy in the equivalent currents could affect the scattering patterns drastically. As ε_r approaches one, the equivalent electric and magnetic currents approach $\boldsymbol{n} \times \boldsymbol{H}^i$ and $\boldsymbol{E}^i \times \boldsymbol{n}$ respectively whereas the scattering patterns approach zero. This means that the equivalent electric and magnetic currents produce fields which nearly cancel each other. Hence, a slight inaccuracy in the equivalent currents could cause a large percentage inaccuracy in the scattering patterns.

We believe that our Müller solution is more accurate than our PMCHW solution whenever ε_r is close to one. When α and β are given by eqs. (22) and (23) as in the Müller formulation, the left-hand sides of eqs. (9) and (10) approach $-\boldsymbol{M}$ and $\boldsymbol{J}$ respectively as ε_r approaches one. In this case, the expected solution

$$\boldsymbol{J} = \boldsymbol{n} \times \boldsymbol{H}^i, \quad \boldsymbol{M} = \boldsymbol{E}^i \times \boldsymbol{n}$$

can be obtained by inspection of eqs. (9) and (10). However, if $\alpha = \beta = 1$ as in the PMCHW formulation, the solution to eqs. (9) and (10) is not obvious when $\varepsilon_r = 1$ because the field operators on the left-hand sides of eqs. (9) and (10) are not diagonal. With our Müller solution, the matrix on the left-hand side of eq. (39) would become tridiagonal for $\varepsilon_r = 1$ if its first two rows of submatrices were interchanged. With our PMCHW solution, no such simplification of this matrix is possible for $\varepsilon_r = 1$.

We recommend at least 10 expansion functions per free space wavelength per component of current along the generating curve of the dielectric body of revolution. For example, if the generating curve were one wavelength long, the order of the square matrix on the left-hand side of eq. (39) should be at least 36. The number 36 is arrived at as follows. There should be at least 9 expansion functions per component of current. We say 9 expansion functions rather than 10 because we are using overlapping triangle functions with no peak of triangle function at either ends of the generating curve. There are two components of electric current and two components of magnetic current.

Each of the elements of Y_{ne}, Y_{nd}, Z_{ne}, and Z_{nd} appearing in eq. (39) is a triple integral consisting of one integration with respect to ϕ and two integrations with respect to t. The ϕ integral is evaluated by using a Gaussian quadrature formula. Each t integration is done by crude sampling akin to the trapezoid rule. For any element of Y_{ne}, Y_{nd}, Z_{ne}, or Z_{nd} which contributes substantially to an element of eq. (39), the pertinent triple integral should be evaluated by using at least 10 sample points per wavelength in the media in question. Loss of accuracy in the computed patterns of Fig. 6e, f may be due to the fact that this condition was violated.

Both the PMCHW solution and the Müller solution are obtained by taking a linear combination of eqs. (5) and (7) and a linear combination of eqs. (6) and (8). There are two other possibilities which are

(1) A linear combination of eqs. (5) and (6) and a linear combination of eqs. (7) and (8).

(2) A linear combination of eqs. (5) and (8) and a linear combination of eqs. (6) and (7).

These other two possibilities give rise to alternative numerical solutions which may compare favorably with the PMCHW solution and the Müller solution. We are currently considering such alternative solutions.

Appendix

Proof that the solution to eqs. (5) to (8) is unique

The solution $\boldsymbol{J}, \boldsymbol{M}$ to eqs. (5) to (8) will be unique if the associated set of homogeneous equations

$$-\boldsymbol{n} \times \boldsymbol{E}_e^- = 0\,, \quad \text{(A.1)}$$

$$-\boldsymbol{n} \times \boldsymbol{H}_e^- = 0\,, \quad \text{(A.2)}$$

$$-\boldsymbol{n} \times \boldsymbol{E}_d^+ = 0\,, \quad \text{(A.3)}$$

$$-\boldsymbol{n} \times \boldsymbol{H}_d^+ = 0 \quad \text{(A.4)}$$

has only the trivial solution $\boldsymbol{J} = \boldsymbol{M} = 0$.

From eqs. (A.1) and (A.2), $\boldsymbol{J}, \boldsymbol{M}$ radiate in μ_e, ε_e to produce a field whose tangential components are zero just inside S. Hence, according to the relation between $\boldsymbol{J}, \boldsymbol{M}$ and the discontinuity of tangential field across S, the field $\boldsymbol{E}_e$, $\boldsymbol{H}_e$ radiated by $\boldsymbol{J}, \boldsymbol{M}$ in μ_e, ε_e outside S satisfies

$$\boldsymbol{n} \times \boldsymbol{H}_e = \boldsymbol{J}\,, \quad \text{(A.5)}$$

$$\boldsymbol{E}_e \times \boldsymbol{n} = \boldsymbol{M} \quad \text{(A.6)}$$

just outside S. See Fig. A.1a.

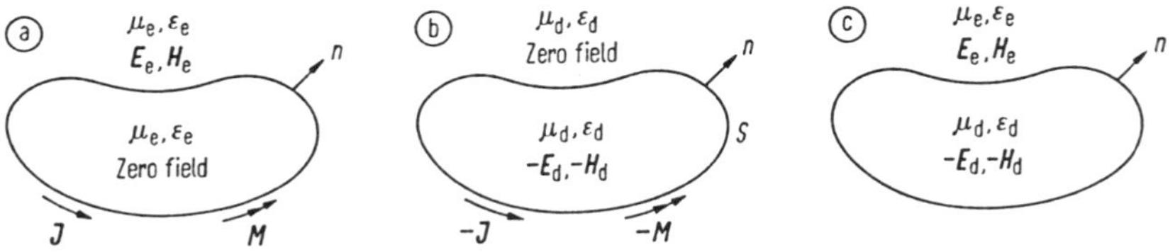

Fig. A.1 (a) Radiation of $\boldsymbol{J}, \boldsymbol{M}$ according to eqs. (A.1) and (A.2), (b) Radiation of $-\boldsymbol{J}$, $-\boldsymbol{M}$ according to eqs. (A.3) and (A.4), (c) Composite situation.

From eqs. (A.3) and (A.4) the electric and magnetic currents $-\boldsymbol{J}$, $-\boldsymbol{M}$ radiate in μ_d, ε_d to produce a field whose tangential components are zero just outside S. Hence, according to the relation between $-\boldsymbol{J}$, $-\boldsymbol{M}$ and the discontinuity of tangential field across S, the field $-\boldsymbol{E}_d$, $-\boldsymbol{H}_d$ radiated by $-\boldsymbol{J}$, $-\boldsymbol{M}$ in μ_d, ε_d satisfies

$$\boldsymbol{n} \times (-\boldsymbol{H}_d) = \boldsymbol{J}\,, \quad \text{(A.7)}$$

$$(-\boldsymbol{E}_d) \times \boldsymbol{n} = \boldsymbol{M} \quad \text{(A.8)}$$

just inside S. See Fig. A.1b.

The equivalence principle is used to combine the outside situation in Fig. A.1a with the inside situation in Fig. A.1b to obtain the composite situation shown in Fig. A.1c. Because of eqs. (A.5) to (A.8), the composite situation in Fig. A.1c is source-free. Therefore the field in Fig. A.1c is zero everywhere. Hence, the fields in Fig. A.1a, b are zero everywhere which implies that $\boldsymbol{J} = \boldsymbol{M} = 0$.

Thus, the solution to eqs. (5) to (8) is unique because the associated set (A.1) to (A.4) of homogeneous equations has only the trivial solution.

Acknowledgement

This work was supported by the United States Air Force under Contracts No. F19628-76-C-0300 and No. F30602-75-0121.

(Received February 2nd, 1978.)

References

[1] Poggio, A. J. and Miller, E. K., Integral equation solutions of three-dimensional scattering problems. Ch. 4 of R. Mittra (ed.), Computer techniques for electromagnetics. Pergamon Press, Oxford 1973, eq. (4.17).

[2] Yu Chang and Harrington, R. F., A surface formulation for characteristic modes of material bodies. Transact. IEEE AP-25 [1977], 789—795.

[3] T.-K. Wu and Tsai, L. L., Scattering from arbitrarily-shaped lossy dielectric bodies of revolution. Radio Sci. 12 [1977], 709—718.

[4] Müller, C., Foundations of the mathematical theory of electromagnetic waves. Springer-Verlag, Berlin 1969, p. 301, eqs. (40) and (41). (There are some sign errors in these equations.)

[5] Solodukhov, V. V. and Vasil'ev, E. N., Diffraction of a plane electromagnetic wave by a dielectric cylinder of arbitrary cross section. Soviet Physics — Tech. Physics 15 [1970], 32—36.

[6] Morita, N., Analysis of scattering by a dielectric rectangular cylinder by means of integral equation formulation. Electron. and Commun. in Japan 57-B [1974], Oct., 72—80.

[7] Vasil'ev, E. N. and Materikova, L. B., Excitation of dielectric bodies of revolution. Soviet Physics — Tech. Physics 10 [1966], 1401—1406.

[8] Harrington, R. F., Field computation by moment methods. Macmillan Co., New York 1968.

[9] Mautz, J. R. and Harrington, R. F, H-field, E-field, and combined field solutions for bodies of revolution. Interim Tech. Rep. RADC-TR-77-109, Rome Air Development Center, Griffiss Air Force Base, New York, March 1977.

[10] Mautz, J. R. and Harrington, R. F., Computer programs for H-field, E-field, and combined field solutions for bodies of revolution. Interim Tech. Rep. RADC-TR-77-215, Rome Air Development Center, Griffiss Air Force Base, New York, June 1977.

[11] Mautz, J. R. and Harrington, R. F., H-field, E-field, and combined-field solutions for conducting bodies of revolution. AEÜ 32 [1978], 157—164.

[12] Mautz, J. R. and Harrington, R. F., Electromagnetic scattering from a homogeneous body of revolution. Rep. TR-77-10, Dept. of Electrical and Computer Engrg., Syracuse University, N.Y., Nov. 1977.

[13] Harrington, R. F., Time-harmonic electromagnetic fields. McGraw-Hill Book Co., New York 1961, Sections 3—5 and 6—9.

[14] Mautz, J. R., Computer program for the Mie series solution for a sphere. Rep. TR-77-12, Dept. of Electrical and Computer Engrg., Syracuse University, N.Y., Dec. 1977.

[15] Barber, P. and Yeh, C., Scattering of electromagnetic waves by arbitrarily shaped dielectric bodies. Appl. Optics 14 [1975], 2864—2872.

Electromagnetic Scattering from Axially Inhomogeneous Bodies of Revolution

LOUIS N. MEDGYESI-MITSCHANG, MEMBER, IEEE, AND JOHN M. PUTNAM

***Abstract*—The electromagnetic scattering from partially or totally penetrable bodies of revolution (BOR) is formulated in terms of coupled Fredholm integral equations, solved by the method of moments (MM). The scatterers can have axial inhomogeneities, formed by dissimilar dielectric materials. The case of conducting bodies with axially discontinuous coatings is also treated. The penetrable regions can be lossy, characterized by complex permeability and permittivity. Boundary conditions are rigorously treated everywhere including the intersection of the various regions. The solutions are expressed in terms of combinations of two special matrices arising from the Galerkin technique. These solutions are implemented numerically for a class of generic axially inhomogeneous BOR scatterers. Numerical results given for various conducting/dielectric cylinder combinations using this formulation are compared with experimental data. For special cases where comparisons are possible, the present analysis replicates the results of the Mie theory.**

I. INTRODUCTION

THE SCATTERING from homogeneous, penetrable scatterers has been the subject of intense investigation. As a pioneer, Mie developed solutions for homogeneous spheres of arbitrary size [1]. Möglich constructed formal solutions for the scattered and internal fields of homogeneous ellipsoids using ellipsoidal harmonics [2]. Certain canonic scatterers with at most one-dimensional inhomogeneity have been treated by the Bromwich solution using transverse electric (TE) and transverse magnetic (TM) potentials [3]. Raleigh [4] and Wait [5] examined the scattering from infinitely long cylinders at normal and oblique illumination, respectively. Generalizations of these solutions for radial layered inhomogeneities as well as coated conductors have been obtained by a number of authors [6]-[7]. Recent investigations based on the method of moments (MM), the extended boundary condition method (EBCM) and the finite element method (FEM) extend the class of penetrable scatterers that can be treated, particularly those defined by nonseparable coordinate geometries such as general axisymmetric scatters, i.e., bodies of revolution (BOR). The penetrable region can be inhomogeneous in both the radial and axial dimension. Specific results for layered and radially inhomogeneous BOR's using the MM, EBCM, and FEM solutions are described [8]-[14].[1]

Scattering from BOR, characterized by various generic axial inhomogeneities, shown in Fig. 1, has not been treated. Morgan *et al.* [15] examined a related problem using an approach based on the coupled azimuthal potential (CAP) formulation. This analysis leads to a self-adjoint system of second order partial differential equations for the potentials or equivalently determined by an Euler-Lagrange variation criterion. This technique

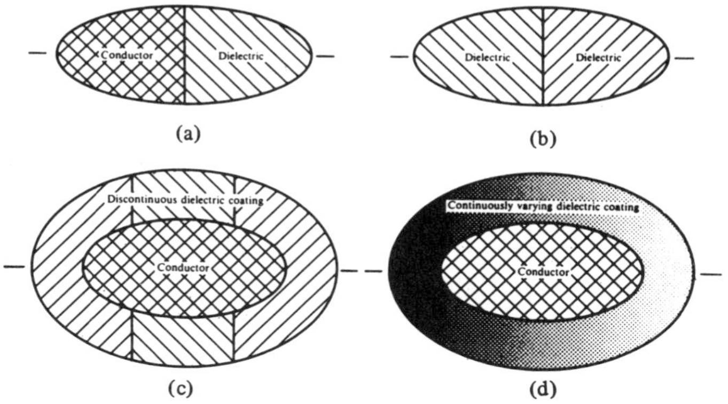

Fig. 1. Generic axially inhomogeneous bodies of revolution. (a) Conducting-dielectric. b) Dissimilar dielectric. (c) Discontinuously coated conducting. (d) Continuously varying coated conducting body.

has only been demonstrated for radial inhomogeneities although in principle it applies to axial inhomogeneities as well. The present analysis, based on the integral formulation of Maxwell's equations and the MM (Galerkin) technique, is computationally less restrictive and can treat larger scatterers than the CAP formulation. In a recent investigation, Mautz and Harrington [16] considered the problem of electromagnetic coupling to a perfectly conducting BOR with a homogeneous interior region. This formulation was solved with the MM technique.

To construct solutions for the scatterers in Fig. 1, it is evident that one requires valid solutions at the interface between two dielectrics and a conductor (Fig. 1(a)) or two or more dielectrics (Fig. 1(b)). Meixner [17] first examined EM solutions in the presence of two related but much simpler, planar two-dimensional boundary conditions depicted in Fig. 2: a region composed of a conducting wedge and two different dielectric wedges, and a region containing two dielectric wedges with a common edge (i.e., a dielectric wedge immersed in a different dielectric media). The formulation developed by Meixner expands the fields (electric and magnetic) in terms of radial harmonics. Subsequent generalizations of Meixner's approach were developed by Mittra and Lee [18], Hurd [19], and Lang [20]. All these solutions are restricted to two-dimensional geometries. The implications of Meixner's expansion are examined by Andersen *et al.* [21].

The scatterers in Fig. 1 can be visualized as combinations of the generalized forms of the two Meixner cases. For the inhomogeneous scatterers considered here, the intersecting planes forming the region boundaries are curved and thus define a nonlinear edge. A partial differential equation formulation of this problem, subject to the foregoing boundary conditions, leads to formidable analytical problems.

A numerical solution for the dielectric wedge problem (Fig. 2(b)) has been obtained by Wu and Tsai for TE plane wave illumination [22]. A coupled set of integral equations was solved

Manuscript received July 12, 1983; revised February 29, 1984. This work was supported by the McDonnell Douglas Independent Research and Development program.

The authors are with the McDonnell Douglas Research Laboratories, St. Louis, MO 63166.

[1] A space-time integral equation formulation for homogeneous BOR scatterers has been solved by Mieras and Bennett (see *IEEE Trans. Antennas Propagat.*, vol. AP-30, no. 1, pp. 2-9, January 1982).

Reprinted from *IEEE Trans. Antennas Propagat.*, vol. AP-32, no. 8, pp. 797-806, August 1984.

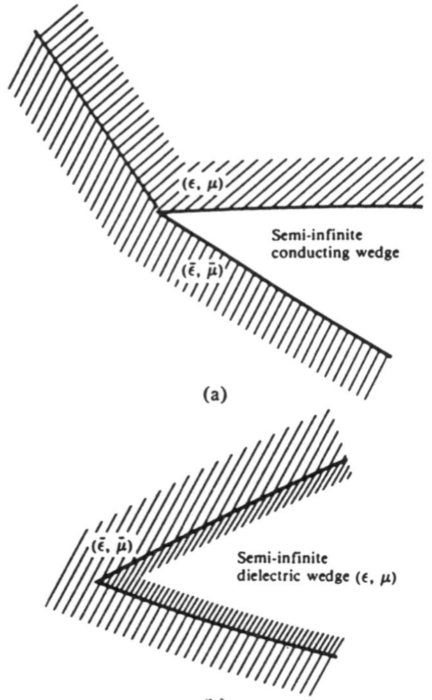

Fig. 2. Meixner boundaries. (a) Conducting wedge and two different dielectric wedges with a common edge. (b) Two different dielectric wedges with common edge.

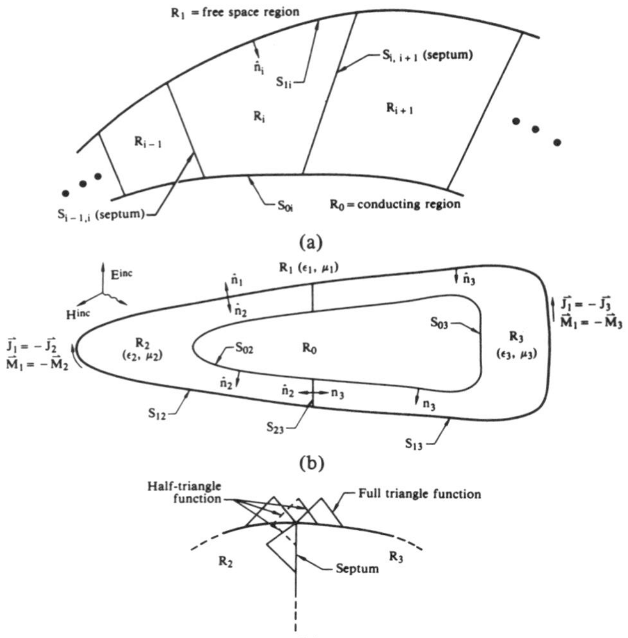

Fig. 3. Discontinuously coated conducting scatterer. (a) General case. (b) Two-region coated case. (c) Detail of expansion functions near intersection of regions.

using flat-pulse expansion functions for the fields near the wedge tip and physical optics for fields distant from the tip. Point matching was used to determine the coefficients of the expansion functions. The solution specifically neglects multiple reflections internal to the wedge which is a severe limitation and is avoided.

The present approach circumvents these analytical difficulties. The boundary conditions are rigorously imposed at the boundaries of the various regions. The resulting equations are rewritten in terms of equivalent surface currents. While the general analysis includes the septum (the boundary between internal regions, see Fig. 3) in a number of interesting and practical cases, the septum effects on the scattering cross section are negligible and the general formulation can be simplified.

In Section II, a general formulation for the penetrable scatterer with discontinuities including the septum effects is developed. In Section III, the resulting system of integral equations is specialized to a BOR and solved by the MM technique. In Section IV, the following special subcases of the preceding formulation are considered: 1) partially penetrable bodies (dielectric-conductor case), 2) totally penetrable bodies (dielectric-dielectric case), and 3) coated bodies with and without septum effects. The scattering cross sections for these cases are computed and compared with experimental data.

II. GENERAL FORMULATION

Consider a perfectly conducting body, having a nonhomogeneous penetrable coating, depicted in Fig. 3(a). Let the coating be composed of $(n-1)$ homogeneous contiguous subregions R_i having as the respective permittivity and permeability, ϵ_i and μ_i, $(i = 2 \cdots n)$ both of which may be complex, representing a lossy material. Let R_0 and R_1 denote the conducting and the free space region, respectively. Let $S_{ij} = R_i \cap R_j$, $(i, j = 0, 1, \cdots n)$ be the surface boundary between regions i and j, if $i \neq j$. Thus $S = \{S_{ij} | S_{ij} \text{ is a surface}\}$ is a set containing all the region interfaces, noting that $S_{ij} = S_{ji}$ and $S_{ii} \notin S$ for all i and j. Let ∂R_i denote the boundary of region R_i, with $\hat{n}_i$ being the surface normal on ∂R_i, pointing into the region R_i. Note $\partial R_i = \bigcup_{j \neq i} S_{ij}$.

Starting with the symmetric form of Maxwell's equations, incorporating electric and magnetic charges and currents for a homogeneous region with electric and magnetic sources, the fields in each of the regions in Fig. 3(a) can be written as follows. In region R_1, the electric and magnetic fields E_1 and H_1 in terms of the incident fields $\vec{E}^i$ and $\vec{H}^i$, can be written as

$$\theta(\vec{r})\vec{E}_1 = \vec{E}^i(\vec{r}) - L_1\vec{J}_1(\vec{r}) + K_1\vec{M}_1(\vec{r}) \tag{1}$$

$$\theta(\vec{r})\vec{H}_1 = \vec{H}^i(\vec{r}) - K_1\vec{J}_1(\vec{r}) - \frac{1}{\eta_1^2} L_1\vec{M}_1(\vec{r}). \tag{2}$$

The electric and magnetic surface currents along the region boundaries in general are

$$\vec{J}_i = \hat{n}_i \times \vec{H}_i|_{\partial R_i}, \quad \vec{M}_i = -\hat{n}_i \times \vec{E}_i|_{\partial R_i},$$

and

$$\eta_i = \eta_0 \sqrt{\frac{\mu_{ir}}{\epsilon_{ir}}}, \quad \epsilon_1 = \epsilon_0 \epsilon_{ir}, \quad \mu_i = \mu_0 \mu_{ir}, \quad \eta_0 = 377\ \Omega.$$

Time variation of $e^{j\omega t}$ is implied and suppressed in this discussion. The integro-differential operators L_i and K_i are defined as

$$L_i\vec{X}(\vec{r}) = j\omega\mu \int_{\partial R_i} \left(\vec{X}(\vec{r}\,') + \frac{1}{\omega^2 \mu\epsilon} \nabla\nabla' \cdot \vec{X}(\vec{r}\,')\right) \Phi(\vec{r} - \vec{r}\,')\, ds', \tag{3}$$

$$K_i\vec{X}(\vec{r}) = \int_{\partial R_i} \vec{X}(\vec{r}\,') \times \nabla\Phi(\vec{r} - \vec{r}\,')\, ds', \tag{4}$$

where for $\vec{r} = \vec{r}\,'$, the operators are interpreted as Cauchy principal value integrals. Note

$$\int_{\partial R_i} = \sum_{j \neq 1} \int_{S_{ij}}$$

and the vector function $\vec{X}$ is in the domain of L_i and K_i, spanning ∂R_i and defined for the kernel Φ which is the Green's function; ϵ, μ, and Φ are defined in region R_i with boundary ∂R_i, and

$$\theta(\vec{r}) = \begin{cases} 1, & \text{for } \vec{r} \in R_i, \quad i = 1, \cdots, n \\ 1/2, & \text{for } \vec{r} \in \partial R_i, \quad i = 1, \cdots, n \\ 0, & \text{otherwise.} \end{cases} \tag{5}$$

The fields in region $R_i (i = 2 \cdots n)$ can be expressed likewise as

$$\theta(\vec{r})\vec{E}_i = -L_i\vec{J}_i(\vec{r}) + K_i\vec{M}_i(\vec{r}) \tag{6}$$

$$\theta(\vec{r})\vec{H}_i = -K_i\vec{J}_i(\vec{r}) - \frac{1}{\eta_i^2} L_i\vec{M}_i(\vec{r}). \tag{7}$$

Applying the boundary conditions on each $S_{ij} \in S$ yields a set of coupled integral equations for the unknown electric and magnetic currents on these surfaces. Imposing continuity of the tangential fields at the interfaces, the boundary conditions on $S_{ij} \in S$ for i and $j \neq 0$ (i.e., dielectric boundary) are

$$(\hat{n}_i \times \vec{E}_i + \hat{n}_j \times \vec{E}_j)|_{S_{ij}} = 0 \tag{8}$$

$$(\hat{n}_i \times \vec{H}_i + \hat{n}_j \times \vec{H}_j)|_{S_{ij}} = 0. \tag{9}$$

Similarily, since the tangential electric fields vanish on a conducting boundary S_{ij} where i or $j = 0$,

$$\hat{n}_i \times \vec{E}_i|_{S_{ij}} = 0. \tag{10}$$

Substituting (1), (2) and (6), (7) into the above boundary conditions on each $S_{ij} \in S$, for i and $j \neq 0$ (i.e., dielectric interior or exterior boundary), one obtains

$$\{\hat{n}_i \times [L_i\vec{J}_i(\vec{r}) - K_i\vec{M}_i(\vec{r})] + \hat{n}_j \times [L_j\vec{J}_j(\vec{r}) - K_j\vec{M}_j(\vec{r})]\}|_{S_{ij}} = \delta\hat{n}_1 \times \vec{E}^i|_{S_{ij}} \tag{8a}$$

$$\left\{\hat{n}_i \times \left[K_i\vec{J}_i(\vec{r}) + \frac{1}{\eta_i^2} L_i\vec{M}_i(\vec{r})\right] + \hat{n}_j \times \left[K_j\vec{J}_j(\vec{r}) + \frac{1}{\eta_j^2} L_j\vec{M}_j(\vec{r})\right]\right\}\Bigg|_{S_{ij}} = \delta\hat{n}_1 \times \vec{H}^i\Big|_{S_{ij}}, \tag{9a}$$

and for i or $j = 0$ (i.e., conducting interior or exterior boundary):

$$\{\hat{n}_i \times L_i\vec{J}_i(\vec{r})\}|_{S_{ij}} = \delta\hat{n}_1 \times \vec{E}^i|_{S_{ij}} \tag{10a}$$

where δ is either 0 or 1 depending on whether S_{ij} is an interior or exterior boundary, respectively. In general, the coating need not be uniformly thick along the body and thus the outer coating surface and inner conducting body need not be congruent.

Imposing the foregoing boundary conditions yields a coupled system of integral equations, expressed compactly in matrix form as

$$\begin{bmatrix} L_E & K_E \\ K_H & L_H \end{bmatrix} \begin{bmatrix} \vec{J} \\ \vec{M} \end{bmatrix} = \begin{bmatrix} \vec{E} \\ \vec{H} \end{bmatrix} \tag{11}$$

where the submatrices are defined in terms of combinations of the L_i and K_i operators. As an example, consider the conducting scatterer, coated with a two-region dielectric coating, depicted in Fig. 3(b). The submatrices in (11) written explicitly are

$$L_E = \begin{bmatrix} \hat{n}_1 \times L_1|_{S_{12}} & \hat{n}_2 \times L_2|_{S_{12}} & 0 \\ \hat{n}_1 \times L_1|_{S_{13}} & 0 & \hat{n}_3 \times L_3|_{S_{13}} \\ 0 & \hat{n}_2 \times L_2|_{S_{23}} & \hat{n}_3 \times L_3|_{S_{23}} \\ 0 & \hat{n}_2 \times L_2|_{S_{02}} & 0 \\ 0 & 0 & \hat{n}_3 \times L_3|_{S_{03}} \end{bmatrix} \tag{11a}$$

and

$$K_H = \begin{bmatrix} \hat{n}_1 \times K_1|_{S_{12}} & \hat{n}_2 \times K_2|_{S_{12}} & 0 \\ \hat{n}_1 \times K_1|_{S_{13}} & 0 & \hat{n}_3 \times K_3|_{S_{13}} \\ 0 & \hat{n}_2 \times K_2|_{S_{23}} & \hat{n}_3 \times K_3|_{S_{23}} \end{bmatrix} \tag{11b}$$

where $L_E \to K_E$ when L_i is replaced by $-K_i$; similarly, $K_H \to L_H$ with K_i replaced by $(1/\eta_i^2)L_i$. The column vectors in (11) are

$$\vec{J} = \begin{bmatrix} \vec{J}_1 \\ \vec{J}_2 \\ \vec{J}_3 \end{bmatrix} \tag{11c}$$

$$\vec{M} = \begin{bmatrix} \vec{M}_1 \\ \vec{M}_2 \\ \vec{M}_3 \end{bmatrix} \tag{11d}$$

$$\vec{E} = \begin{bmatrix} \hat{n}_1 \times \vec{E}^i|_{S_{12}} \\ \hat{n}_1 \times \vec{E}^i|_{S_{13}} \\ 0 \\ 0 \\ 0 \end{bmatrix} \tag{11e}$$

$$\vec{H} = \begin{bmatrix} \hat{n}_1 \times \vec{H}^i|_{S_{12}} \\ \hat{n}_1 \times \vec{H}^i|_{S_{13}} \\ 0 \end{bmatrix} \tag{11f}$$

The rows of the system matrix in (11) are associated with the individual interfaces (i.e., S_{ij}) on which the boundary conditions are enforced. The number of columns of this matrix is determined by the number of regions (and currents along ∂R_i) defining the scatterer. To this point, no restrictions have been placed on the exact shape of the scatterer. In the next section, the results are specialized to the important case of an arbitrary BOR with a discontinuous (inhomogeneous) coating.

III. METHOD OF MOMENTS SOLUTION

All the physics of the scattering interactions is rigorously contained in the coupled set of Fredholm integral equations (11). The submatrices L_E, L_H, K_E, and K_H are combination of the integral operators L and K arising from the Stratton–Chu formulation of Maxwell's equations. Because of its complexity, no general analytical solution is available for (11). However, one may reduce the infinite dimensional problem for the unknown vector functions (i.e. the electric and magnetic currents) to a finite dimensional one using the method of moments (MM). A set of algebraic equations then results for the unknown expansion coefficients associated with the original vector functions. An outline of the MM solution is given below. A detailed discussion of the mathematical foundations of the MM technique may be found in [23]-[25].

The unknown currents in (11) are expanded in a finite sequence of basis functions spanning the various interfaces formed by the discontinuities of the dielectric regions and the surface of the scatterer. Subdividing each boundary ∂R_i into annular segments, the surface currents along ∂R_i can be expanded as

$$\vec{J}_i = \sum_{n,k} (a^t_{ink}\vec{J}^t_{ink} - a^\phi_{ink}\vec{J}^\phi_{ink}) \tag{12}$$

and

$$\vec{M}_i = \sum_{n,k} (b^t_{ink}\vec{M}^t_{ink} - b^\phi_{ink}\vec{M}^\phi_{ink}) \tag{13}$$

where a^α_{ink} and b^α_{ink} ($\alpha = t$ or ϕ) are the unknown expansion coefficients. The indices n and k are associated with the expansion along the principal surface coordinates of the boundary ∂R_i namely, ϕ and t, respectively. Introducing the minus sign in the current expansion leads to a symmetric system of equations.

The expansion functions in (12), (13) are chosen to form a linearly independent basis set with suitable smoothness properties so that the differentiations in the integral operator L_i can be carried out. Superposition of the basis functions should optimally closely approximate the unknown currents. Clearly, a plethora of choices exist. Here we pick for the currents on ∂R_i, expansion functions suggested in [26]:

$$\vec{J}^\alpha_{ink} = \vec{u}_\alpha f_{ik}(t)e^{jn\phi} \tag{14a}$$

$$\vec{M}^\alpha_{ink} = \eta_0 \vec{u}_\alpha f_{ik}(t)e^{jn\phi} \tag{14b}$$

where $\vec{u}_\alpha$ is a unit vector directed along $\alpha(=t$ or $\phi)$ on the surface of the BOR; $f_{ik}(t) = \rho_i(t)T_{ik}(t)$ where $T_{ik}(t)$ is a triangle function spanning the k annulus into which the surface ∂R_i is subdivided (see Fig. 3(c)). The triangle functions used in this analysis are defined as

$$T_{ik}(t) = \begin{cases} 1 - |t'|; & |t'| \leq 1 \\ 0; & |t'| > 1 \end{cases} \tag{14c}$$

where $t' = t - t_{ik}$ and $t_{ik} \in \partial R_i$. Introducing the factor η_0 in the magnetic current expansion leads to a well-conditioned system of equations. The current expansions are substituted into (11) and the inner products are formed with a set of testing functions $\vec{W}^\alpha_{ink}$ over the surface of the BOR. In the Galerkin form of the MM solution, $\vec{W}^\alpha_{ink} = (\vec{J}^\alpha_{ink})^*$ for the E-field equations and $\vec{W}^\alpha_{ink} = (\vec{M}^\alpha_{ink})^*$ for the H-field equations. The original integral equations are thus transformed into a linear set of algebraic equations. In the intersection neighborhood, where three or more boundaries intersect, half-triangle functions are used for $T_{ik}(t)$ in the expansion of the magnetic and electric currents (see Fig. 3(c)). Conceptually, the coefficients of these $T_{ik}(t)$ are treated as unknown and different, and matrix elements are formally obtained for each of these terms. The relationship between the currents along the region boundary at a given interface S_{ij} is $\vec{J}_i|_{S_{ij}} = -\vec{J}_j|_{S_{ij}}$ and $\vec{M}_i|_{S_{ij}} = -\vec{M}_j|_{S_{ij}}$. Hence some of the unknown expansion coefficients are equal and similar arguments apply for the testing functions. Although these simplifications are incorporated in the numerical implementation of the MM analysis (reducing the order of the MM system matrix), in the following discussion they are not invoked. The original integral equations are thus transformed into the linear set of algebraic equations:

$$\begin{bmatrix} S_{EE} & S_{EH} \\ S_{HE} & S_{HH} \end{bmatrix}_n \cdot \begin{bmatrix} A \\ B \end{bmatrix}_n = \begin{bmatrix} \bar{E} \\ \bar{H} \end{bmatrix}_n . \tag{15}$$

The column vectors A and B have as respective elements a_{ink} and b_{ink}. The elements of the submatrices in (15) are expressed in terms of the inner products of the testing functions $\vec{W}^\alpha(s)$ and the integral operators $L_i\vec{J}^\beta(s)$ and $K_i\vec{J}^\beta(s)$ for region R_i where $s \in S_{ij}$, i.e.,

$$(S^{\alpha\beta}_{EE})_{S_{ij},R_i} = \int_{S_{ij}} ds\, \vec{W}^\alpha(s) \cdot L_i\vec{J}^\beta(s)|_{S_{ij}} \tag{16a}$$

$$(S^{\alpha\beta}_{HH})_{S_{ij},R_i} = \frac{1}{\eta_i^2}\int_{S_{ij}} ds\, \vec{W}^\alpha(s) \cdot L_i\vec{M}^\beta(s)|_{S_{ij}} \tag{16b}$$

$$(S^{\alpha\beta}_{EH})_{S_{ij},R_i} = -(S^{\alpha\beta}_{HE})_{S_{ij},R_i} = -\int_{S_{ij}} ds\, \vec{W}^\alpha(s) \cdot K_i\vec{J}^\beta(s)|_{S_{ij}} \tag{16c}$$

where L_i and K_i are defined in (3), (4). The column vectors on the right side of (15) are defined similarly,

$$\bar{E}^\alpha = \int_{S_{ij}} ds\, \vec{W}^\alpha(s) \cdot \vec{E}^i \tag{16d}$$

and likewise for $\bar{H}^\alpha$. Note each of the S submatrices above is partitioned into (α, β) components. Each of the boundaries ∂R_i is segmented into annuli along the generating curve and are spanned by the triangle functions of (14c). One can rewrite (16a)-(16c) explicitly in terms of two operators L and K (on these annuli), to be defined subsequently that generate two matrices with special properties. The matrix elements of $S^{\alpha\beta}_{EE}$, $S^{\alpha\beta}_{HH}$, and $S^{\alpha\beta}_{EH}$, are formed by the integrations over the unprimed and primed surface coordinates. The interactions of the k and lth annular segments where $s_k \in S_{ij}$ and $s_l \in \partial R_i$, can be written in terms of $K^{\alpha\beta}_{kl}(s_k, s_l, R_i)$; $1/\eta_i^2 K^{\alpha\beta}_{kl}(s_k, s_l; R_i)$, and $K^{\alpha\beta}_{kl}(s_k, s_l; R_i)$. Note, the integration of the primed coordinates is over ∂R_i (see (3) and (4)); and that of the primed coordinates is over the intersection of R_i and R_j (i.e., S_{ij}).

The two-region dielectric coated case (Fig. 3(b)) will be considered as an example. The elements of the S matrices in (15) can be written as

$$S_{EE} = \begin{bmatrix} L(S_{12}, \partial R_1; R_1) & L(S_{12}, \partial R_2; R_2) & 0 \\ L(S_{13}, \partial R_1; R_1) & 0 & L(S_{13}, \partial R_3; R_3) \\ 0 & L(S_{23}, \partial R_2; R_2) & L(S_{23}, \partial R_3; R_3) \\ 0 & L(S_{02}, \partial R_2; R_2) & 0 \\ 0 & 0 & L(S_{03}, \partial R_3; R_3) \end{bmatrix}_n$$

$$S_{HH} = \begin{bmatrix} \eta_1^{-2} L(S_{12}, \partial R_1; R_1) & \eta_2^{-2} L(S_{12}, \partial R_2; R_2) & 0 \\ \eta_1^{-2} L(S_{13}, \partial R_1; R_1) & 0 & \eta_3^{-2} L(S_{13}, \partial R_3; R_3) \\ 0 & \eta_2^{-2} L(S_{23}, \partial R_2; R_2) & \eta_3^{-2} L(S_{23}, \partial R_3; R_3) \end{bmatrix}_n$$

and

$S_{EE} \to S_{EH}$ by replacing $L(S_{ij}, \partial R_i; R_i) \to -K(\cdot)$;

and

$S_{HH} \to S_{HE}$, letting $\eta_i^{-2} L(S_{ij}, \partial R_i; R_i) \to K(\cdot)$.

For notational simplicity, in the above expressions the superscripts (α, β) and subscripts (k, l) on $L(\cdot)$ and $K(\cdot)$ are suppressed, and the modal dependence on n is implied. Note

$$L(S_{12}, \partial R_1; R_1) \equiv L(S_{12}, S_{12}; R_1) + L(S_{12}, S_{13}, R_1),$$

$$L(S_{12}, \partial R_2; R_2) \equiv L(S_{12}, S_{12}; R_2) + L(S_{12}, S_{02}; R_2) + L(S_{12}, S_{23}; R_2)$$

and likewise for the other expressions.

The operators L and K arise from the MM(Galerkin) procedure applied to the Fredholm integral equations of the first and second kind (i.e., the electrical field integral equation (EFIE) and magnetic field integral equation (MFIE) formulations for a perfectly conducting scatterer). Hence L and K are denoted here as the Galerkin operators of the first and second type, respectively. Their general definition for an arbitrary vector $\vec{X}$ is

$$L_{kl}^{\alpha\beta}(s_k, s_l; R_i)\vec{X} = \langle \vec{W}_{ink}^{\alpha}, L_i \vec{X}_{inl}^{\beta} \rangle = \pm \int_{s_k} ds \int_{s_l} ds' \cdot \left[j\omega\mu \vec{W}_{ink}^{\alpha} \cdot X_{inl}^{\beta} + \frac{1}{j\omega\epsilon} (\nabla \cdot \vec{W}_{ink}^{\alpha})(\nabla' \cdot X_{inl}^{\beta}) \Phi(\vec{r} - \vec{r}') \right] \quad (17)$$

and

$$K_{kl}^{\alpha\beta}(s_k, s_l; R_i)\vec{X} = \langle \vec{W}_{ink}^{\alpha}, K_i \vec{X}_{inl}^{\beta} \rangle = \pm \int_{s_k} ds \int_{s_l} ds' \, \vec{W}_{ink}^{\alpha} \, (\vec{X}_{inl}^{\beta} \times \nabla \Phi(\vec{r} - \vec{r}')) \quad (18)$$

where s_k and s_l refer to the surfaces of integration in the operator (i.e., k and lth annuli on ∂R_i) and R_i refers to the region in which the Green's function Φ is evaluated; $\vec{W}_{ink}^{\alpha}$ and $\vec{J}_{inl}^{\beta}$ are the testing and expansion functions on the k and lth annuli, corresponding to mode number n. In (17), (18), the sign equals + or − for $\beta = t$ or $\beta = \phi$. All the submatrices in (15) are generated from the L and K operators. In principle these operators can be formed and evaluated for any arbitrary scatterer with suitable expansion and testing functions. Because of the orthogonality of $\vec{W}_{ink}^{\alpha}$ and the integral operators in (3) and (4) for a BOR, modal decoupling occurs, i.e., (15) is a function of only one given mode number n. Restricting the discussion to a BOR, the following computationally tractable forms for L and K are obtained. Specifically, $L_{kl}^{\alpha\beta}(\cdot)$ is given as

$$L_{kl}^{tt}(\cdot) = \sum_{p,q=1}^{4} \left\{ j\omega\mu T_p T_q (\sin \nu_p \sin \nu_q G_{cn} + \cos \nu_p \cos \nu_q G_n) - \frac{j}{\omega\epsilon} \dot{T}_p \dot{T}_q G_n \right\}, \quad (19a)$$

$$L_{kl}^{t\phi}(\cdot) = -\sum_{p,q=1}^{4} \left\{ \omega\mu \sin \nu_p T_p G_{sn} + \frac{n}{\omega\epsilon\rho_q} \dot{T}_p T_q G_n \right\}, \quad (19b)$$

$$L_{kl}^{\phi t}(\cdot) = L_{kl}^{t\phi}(\cdot), \quad (19c)$$

where p and q are interchanged, and

$$L_{kl}^{\phi\phi}(\cdot) = -\sum_{p,q=1}^{4} \left\{ j\omega\mu G_{cn} + \frac{n^2 G_n}{j\omega\epsilon\rho_p\rho_q} \right\} T_p T_q, \quad (19d)$$

where ρ_p and ρ_q are evaluated at points corresponding to the kth and lth annular surface segments s_k and s_l into which the BOR is subdivided; ν_p and ν_q are the angles subtended between a tangent to s_k and s_l and the z-axis of the BOR. The terms T_p and $\dot{T}_p$ are the pulse approximations for the triangle functions and their derivatives, i.e., $T_p = \{1/4, 3/4, 3/4, 1/4\}$ and $\dot{T}_p = \{1, 1, -1, -1\}$ for uniform segmentation. (The computer implementation of this analysis allows nonuniform segmentation.) The terms G_n, G_{cn}, and G_{sn} are the integrals of the Green's function Φ, and are given as

$$G_n = \Delta t_p \Delta t_q \int_0^{\pi} d\phi \cos n\phi \, f(\phi), \quad (20a)$$

$$G_{cn} = \Delta t_p \Delta t_q \int_0^{\pi} d\phi \cos \phi \cos n\phi \, f(\phi) \quad (20b)$$

$$G_{sn} = \Delta t_p \Delta t_q \int_0^{\pi} d\phi \sin \phi \sin n\phi \, f(\phi) \quad (20c)$$

where

$$f(\phi) = \frac{e^{-jkR_{pq}}}{R_{pq}} \quad (21a)$$

$$R_{pq} = \begin{cases} [(\rho_p - \rho_q)^2 + (z_p - z_q)^2 + 2\rho_p\rho_q(1 - \cos\phi)]^{1/2}; & \text{if } |\vec{r}_p - \vec{r}_q| \neq 0 \\ \left[\left(\frac{\Delta t_p}{4}\right)^2 + 2\rho_p^2(1 - \cos\phi)\right]^{1/2}; & \text{if } |\vec{r}_p - \vec{r}_q| = 0. \end{cases} \tag{21b}$$

Similarly, the expressions for $K_{kl}^{\alpha\beta}(\cdot)$ are

$$K_{kl}^{tt}(\cdot) = -j\eta_0 \sum_{p,q=1}^{4} \{(z_p - z_q)\sin\nu_p \sin\nu_q + \rho_q \sin\nu_p \cos\nu_p - \rho_p \cos\nu_p \sin\nu_q\} T_p T_q H_{sn} \tag{22a}$$

$$K_{kl}^{t\phi}(\cdot) = \eta_0 \sum_{p,q=1}^{4} \{(z_p - z_q)\sin\nu_p H_{cn} + \cos\nu_p(\rho_q H_n - \rho_p H_{cn})\} T_p T_q \tag{22b}$$

$$K_{kl}^{\phi t}(\cdot) = -K_{lk}^{t\phi}(\cdot) \tag{22c}$$

where p and q are interchanged, and

$$K_{kl}^{\phi\phi}(\cdot) = j\eta_0 \sum_{p,q=1}^{4} (z_p - z_q) T_p T_q H_{sn}. \tag{22d}$$

The terms H_n, H_{cn}, and H_{sn} are given by

$$H_n = \Delta t_p \Delta t_q \int_0^\pi d\phi \cos n\phi \, h(\phi) \tag{23a}$$

$$H_{cn} = \Delta t_p \Delta t_q \int_0^\pi d\phi \cos\phi \cos n\phi \, h(\phi) \tag{23b}$$

$$H_{sn} = \Delta t_p \Delta t_q \int_0^\pi d\phi \sin\phi \sin n\phi \, h(\phi) \tag{23c}$$

$$h(\phi) = \frac{(1 + jkR_{pq})}{R_{pq}^3} e^{-jkR_{pq}} \tag{23d}$$

and R_{pq} is given by (21b).

The following symmetry relations can be established and are used to enhance the computational efficiency of the analysis:

$$L_{kl}^{tt}(\cdot) = L_{lk}^{tt}(\cdot) \qquad K_{kl}^{tt}(\cdot) = -K_{lk}^{tt}(\cdot)$$

$$L_{kl}^{\phi t}(\cdot) = L_{lk}^{t\phi}(\cdot) \qquad K_{kl}^{\phi t}(\cdot) = -K_{lk}^{t\phi}(\cdot)$$

$$L_{kl}^{\phi\phi}(\cdot) = L_{lk}^{\phi\phi}(\cdot) \qquad K_{kl}^{\phi\phi}(\cdot) = -K_{lk}^{\phi\phi}(\cdot).$$

Note the L operator is symmetric (i.e., $L_{kl}(\cdot) = L_{lk}(\cdot)$) and the K operator is skew-symmetric (i.e., $K_{kl}(\cdot) = -K_{lk}(\cdot)$). It can also be shown that the system matrix in (15) is symmetric, and hence only the upper triangular portion of this matrix needs to be stored.

In summary, the MM (Galerkin) method transforms the original integral equations (11) into an algebraic set of equations defined in terms of the two Galerkin operators L and K. All the matrix elements of (15) are thus determined. The unknown expansion coefficients A and B are obtained in terms of the incident field by solving this matrix equation. Substituting these coefficients into (12) and (13), the currents on the scatterer are obtained and from these, the scattered field can be obtained everywhere, completing the solution.

IV. RESULTS

Using the foregoing formulation, the bistatic and monostatic (backscatter) cross sections for various axially inhomogeneous scatterers are computed and compared with other analytical results and experimental data. In Fig. 4, the computed vertical and horizontal bistatic cross sections of a closed conducting half-sphere ($ka = 3$) are compared with that for a sphere ($ka = 3$) composed of conducting and dielectric ($\epsilon_r = 1$) halves. The results for the conducting half-sphere are obtained using a derivative of the MM/BOR code developed by Mautz and Harrington for perfectly conducting bodies [26]. The conducting-dielectric body is treated using the present formulation with and without the septum effects. Omitting the septum, leads to erroneous results. With the septum, the agreement is excellent. Note the dielectric half-sphere-air interface is an "artificial" boundary in the calculations. In the MM/BOR calculation, the closed half-sphere is represented with 15 triangle (expansion) functions; in the present formulation 24 triangle functions are used, including five on the septum. Because of axial incidence, only modes $n = \pm 1$ are used. The corresponding results for a dielectric half-sphere ($ka = 3$) with $\epsilon_r = 4$, and a sphere composed of two dielectric halves, $\epsilon_r = 1$ and $\epsilon_r = 4$, are depicted in Fig. 5. For the former calculations 15 triangle functions are used; in the latter 24 triangle functions are employed, including five on the septum; in both analyses, $n = \pm 1$. Again the necessity of incorporating the septum effects is illustrated.

The scattering from a homogeneous dielectric sphere ($\epsilon_r = 4$, $ka_1 = 3.0$), a bisected dielectric sphere ($\epsilon_r = 4$, $ka_1 = 3.0$) and a layered dielectric sphere ($\epsilon_r = 4$, $ka_1 = 3.0$, $ka_2 = 0.75$) are compared in Fig. 6. The homogeneous sphere results are obtained from the Mie theory. The other two are treated with the present analysis. Since all three scatterers are physically identical, this case provides an excellent test of the formulation and the accuracy of its computer implementation. Indeed, the results of the calculations are indistinguishable. In Fig. 7, the computed bistatic cross section for a homogeneously coated ($\epsilon_r = 4$) perfectly conducting sphere ($ka = 0.75$) with a septum is compared with the Mie solution; the agreement between the two solutions is excellent.

The calculated cross sections for cylindrical bodies are compared with experimental data. Backscatter measurements were carried out in a large anechoic chamber measuring 36 × 9 × 9 m using a bistatic nulled CW system. Typical nulls of < -50 dBsm were attained. Absolute calibration of the backscatterer data was obtained using a -20 dBsm test sphere. The chamber quiet zone, wherein spurious reflections are attenuated by 40 dB or more, was 2.4 m in diameter. The test articles were mounted on a Styrofoam tower 11 m from the source. First, a metal-dielectric cylindrical scatterer was tested. The dielectric is plexiglass ($\epsilon_r = 2.6$); the metallic part is aluminum. The object is 7.62 cm in diameter and 10.16 cm in length. The dielectric segment is 5.08 cm long. The computed and measured backscatter cross sections for this case (in polar plot) are depicted in Fig. 8. The computations used 23 triangle functions, with four on the septum. Convergent answers are obtained with $N = 4$ circumferential modes. The corresponding results for a metal-dielectric cylinder with a diameter of 7.62 cm and a total length

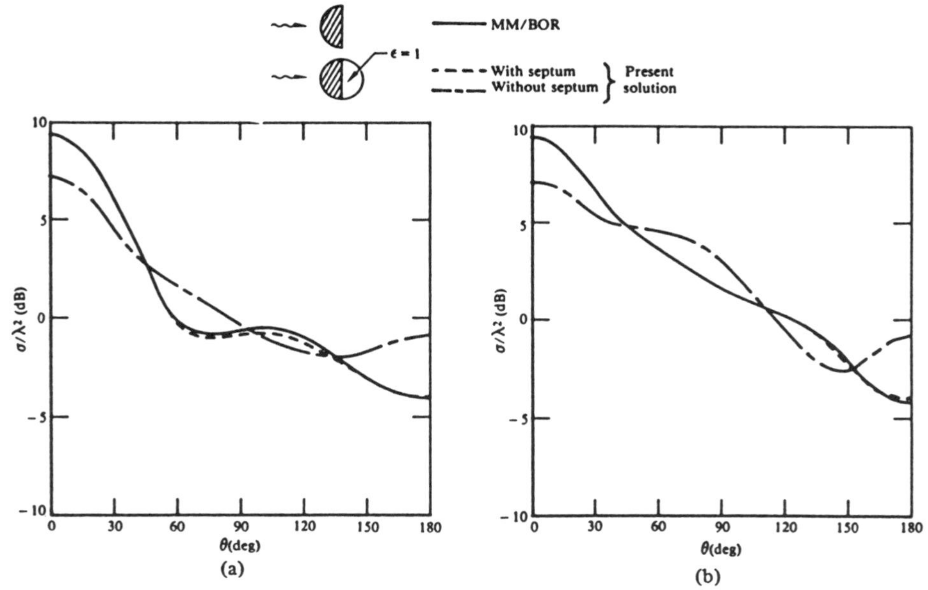

Fig. 4. Computed bistatic scattering cross sections for a conducting hemisphere–air hemisphere body and a conducting hemisphere. (a) $\phi\phi$-polarization. (b) $\theta\theta$-polarization.

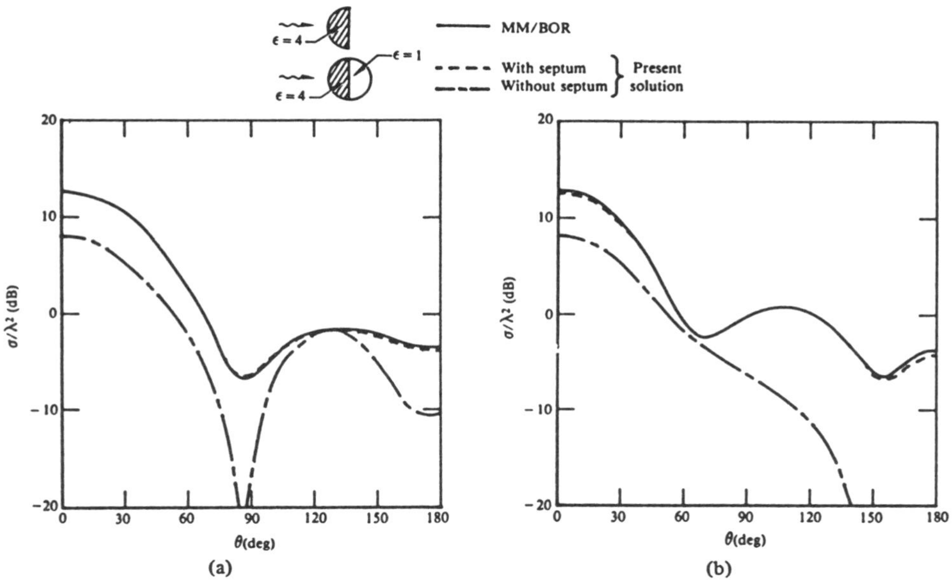

Fig. 5. Computed bistatic scattering cross sections for a dielectric ($\epsilon = 4$) hemisphere–air hemisphere body and a dielectric ($\epsilon = 4$) hemisphere. (a) $\phi\phi$-polarization. (b) $\theta\theta$-polarization.

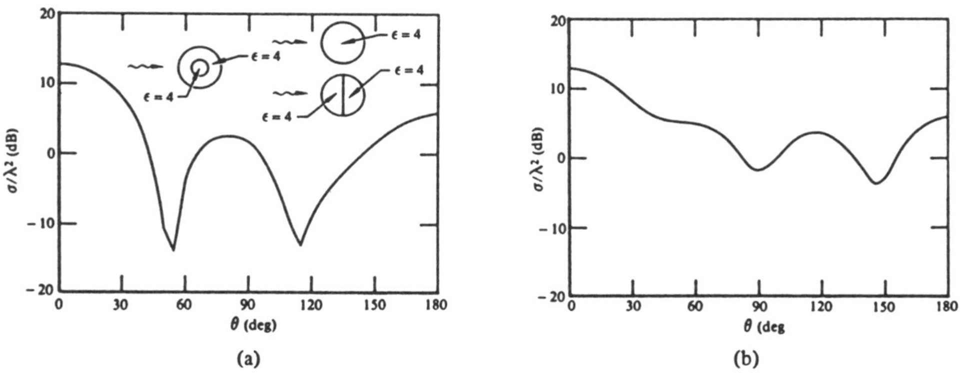

Fig. 6. Computed bistatic scattering cross sections for a homogeneous dielectric sphere ($\epsilon = 4$, $ka = 3$) using Mie solution; a layered sphere ($\epsilon = 4$, $ka_1 = 3.0$, $ka_2 = 0.75$) and a bisected dielectric ($\epsilon = 4$) sphere using present solution. (a) $\phi\phi$-polarization. (b) $\theta\theta$-polarization.

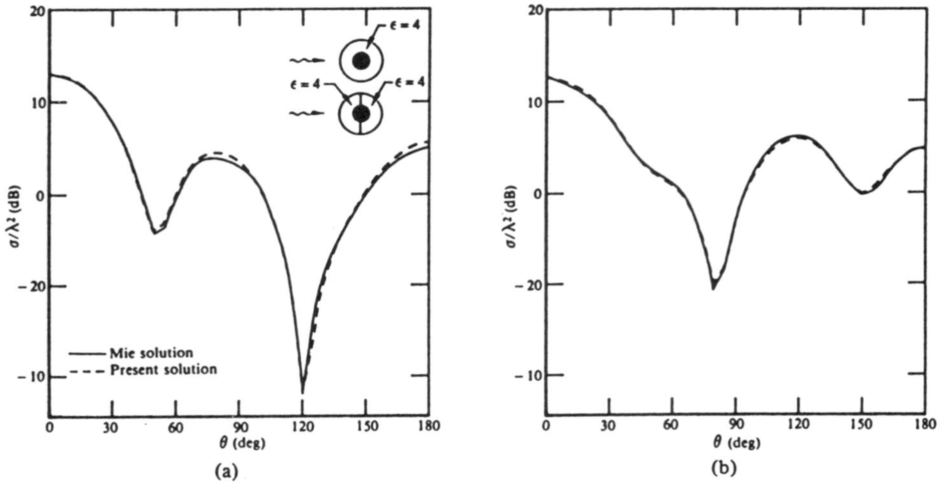

Fig. 7. Computed bistatic cross sections for: a conducting sphere (ka_2 = 0.75) coated with a homogeneous dielectric (ϵ = 4, ka_1 = 3.0) spherical shell using Mie solution, and same geometry bisected by a septum boundary using present solution. (a) $\phi\phi$-polarization. (b) $\theta\theta$-polarization.

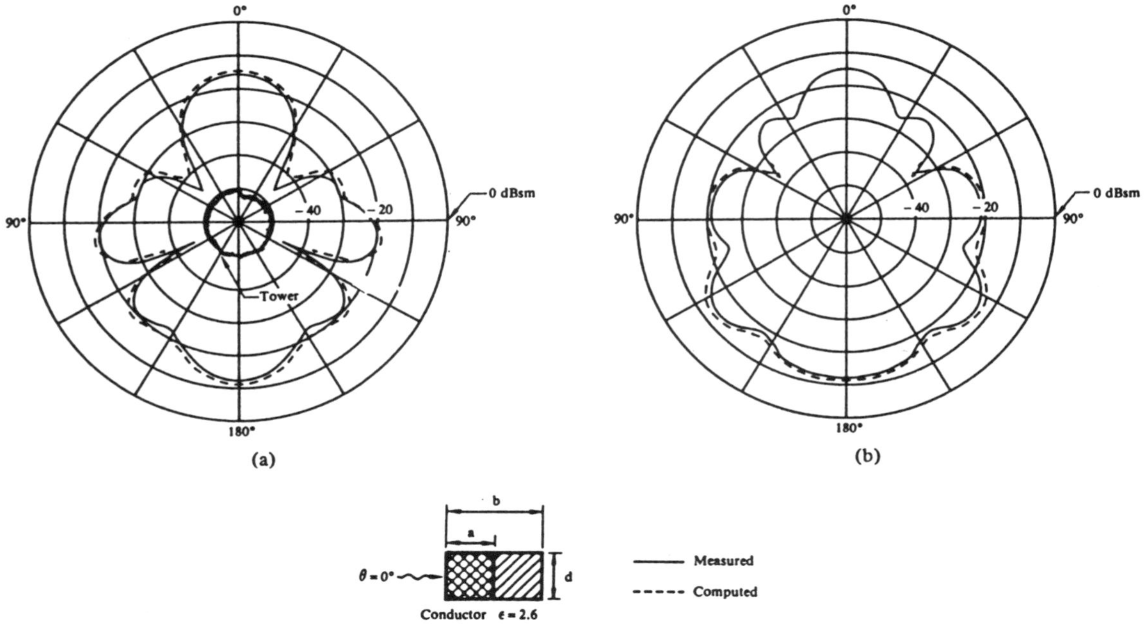

Fig. 8. Computed and measured backscatter cross sections at 3 GHz for an inhomogeneous conducting-dielectric cylinder (a = 5.08 cm, b = 10.16 cm, d = 7.62 cm). (a) $\phi\phi$-polarization. (b) $\theta\theta$-polarization.

of 15.24 cm where the metal region is 10.16 cm long, are depicted in Fig. 9. The foregoing calculations and measurements were at 3 GHz.

The scattering cross section from a dielectric cylinder, half nylon (ϵ_r = 3.4) and half plexiglass (ϵ_r = 2.6), is shown in Fig. 10. The cylinder length and diameter are 10.16 cm and 7.62 cm, respectively. The measurement frequency in this case was 3.5 GHz. In the calculations 23 triangle functions, including four on the septum, and N = 6 modes are used. The discrepancy between the measured and computed results are due to the uncertainty in the electrical properties (loss tangent) of the nylon.

In Figs. 4 and 5, we showed that the septum effects must be included to obtain correct results. For coated scatterers where the coatings are electrically thin or the material discontinuity is small, the septum effects contribute negligibly to the scattering cross sections as illustrated in Fig. 11 where the bistatic cross sections for a conducting sphere (ka_2 = 2.6858) coated with two different dielectric hemispherical shells (ka_1 = 3.0, with ϵ_r = 3 and 4) are computed with and without septum effects. Similar results were obtained at other different incident angles.

In the computer implementation of the MM solution in the present formulation, the scatterer is subdivided into narrow

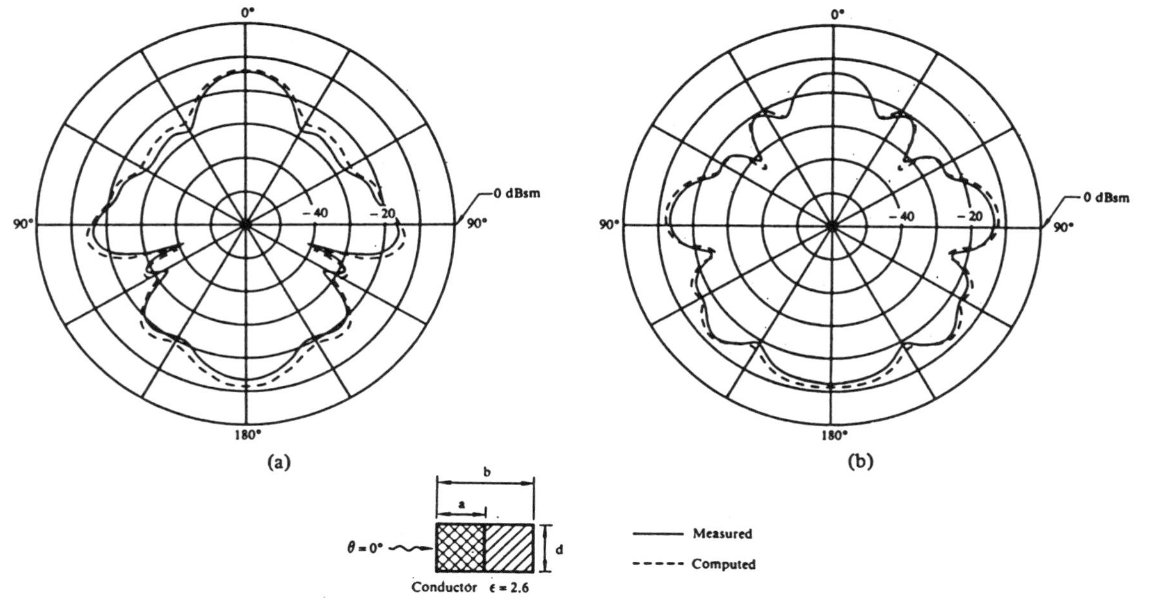

Fig. 9. Computed and measured backscatter cross sections at 3 GHz for an inhomogeneous conducting-dielectric cylinder (a = 10.16 cm, b = 15.24 cm, d = 7.62 cm). (a) $\phi\phi$-polarization. (b) $\theta\theta$-polarization.

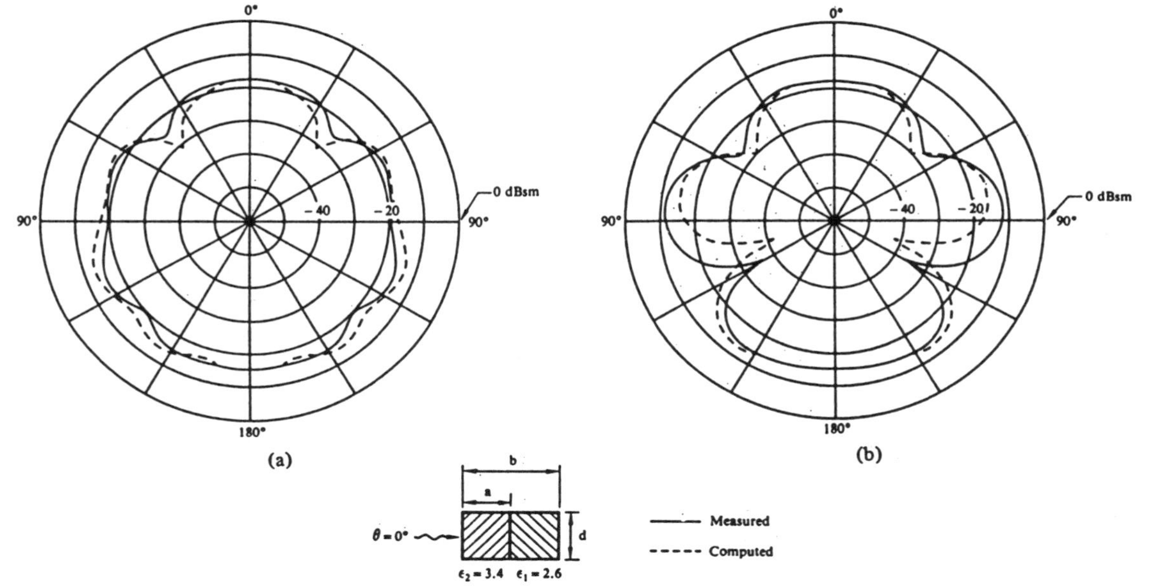

Fig. 10. Computed and measured backscatter cross sections at 3.5 GHz for an inhomogeneous cylinder (a = 5.08 cm, b = 10.16 cm, d = 7.62 cm. (a) $\phi\phi$-polarization. (b) $\theta\theta$-polarization.

($\lesssim$0.1 λ wide) annular segments and in each segment the coating properties can be different. In many practical applications using graded absorber coatings, the material properties change slowly and can be effectively discretized at 0.1 λ or larger intervals. Thus, in its numerically implemented form this formulation provides EM scattering solutions for coated conducting BOR with arbitrary variation in the coating properties. These solutions can be obtained for both monostatic and bistatic illumination and for arbitrary linear or circular polarization.

SUMMARY

The electromagnetic scattering from partially or totally penetrable BOR with axial inhomogeneities was formulated in terms of coupled integral equations, solved by the MM technique. The solutions account for all the EM interactions at the intersection between the various regions of the scatterer. The MM solutions are obtained in terms of combinations of two operators which are readily computed, and lead to a symmetric system matrix. Numerical implementation of the formulation provides solutions for coated conducting BOR's having arbitrary axial variations in the coating thickness and in permeability and permittivity. The formulation was tested for spherical, hemispherical and cylindrical scatterers. Predictions for various limiting cases agreed with published results. Calculations for other cases treated by this formulation were confirmed by experimental measurements.

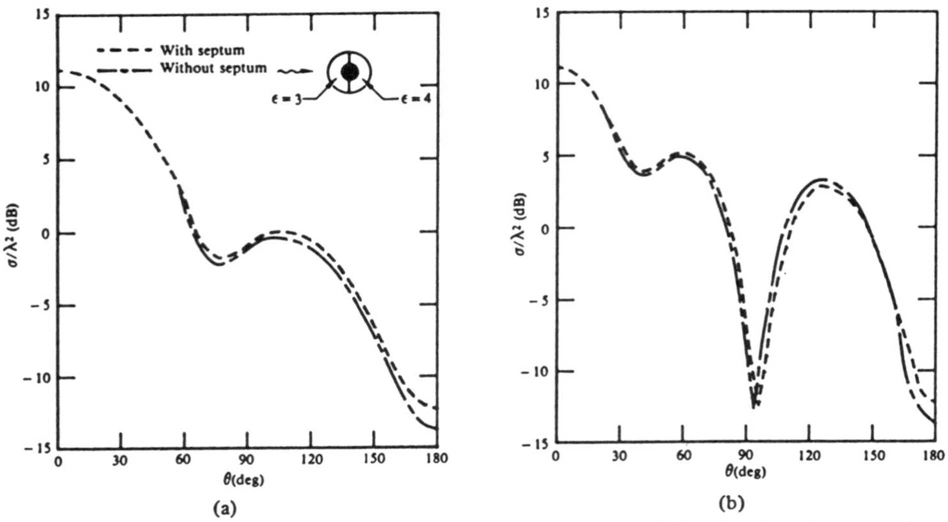

Fig. 11. Computed bistatic scattering cross sections for a conducting sphere ($ka_2 = 2.6858$) with a discontinuous coating ($ka_1 = 3.0$, $\epsilon = 3$ and 4) with and without septum effects. (a) $\phi\phi$-polarization. (b) $\theta\theta$-polarization.

ACKNOWLEDGMENT

The authors thank Mr. V. R. Ditton, Jr. of the McDonnell Aircraft Company for assistance with the experimental measurements.

REFERENCES

[1] G. Mie, "Beiträge zur Optik trüber Medien, speziell kolloider Metallösungen." *Ann. Phys.*, vol. 25, p. 377, 1908.

[2] F. Möglich, "Beugungerscheinungen an Körpern von Ellipsoidischer Gestalt," *Ann. Phys.*, vol. 83, p. 609, 1927.

[3] T. J. I'a. Bromwich, "Electromagnetic waves," *Phil. Mag.*, S.6., vol. 38, p. 223, 1919.

[4] Lord Rayleigh, "The dispersal of light by a dielectric cylinder," *Phil. Mag.*, vol. 36, p. 365, 1918. (Alternate reference: *Scientific Papers*, vol. VI. New York: Dover, 1964, p. 554.)

[5] J. R. Wait, "Scattering of a plane wave from a circular dielectric cylinder at oblique incidence," *Can. J. Phys.*, vol. 33, p. 189, 1955. (See also "The long wavelength limit in scattering from a dielectric cylinder a oblique incidence," *Can. J. Phys.*, vol. 43, p. 2212, 1965.)

[6] A. Güttler, "Die Miesche Theorie der Beugung durch dielektrische Kugeln mit absorbierendem Kern und ihre Bedeutung für Probleme der interstellaren Materie und des atmosphärischen Aerosols," *Ann. Phys.*, vol. 6, series 11, p. 65, 1952.

[7] G. T. Ruck, Ed., *Radar Cross Section Handbook*, vol. I. New York: Plenum, 1970, chs. 3 and 4.

[8] L. N. Medgyesi-Mitschang and C. Eftimiu, "Scattering from axisymmetric obstacles embedded in axisymmetric dielectrics: The method of moments solution," *Appl. Phys.*, vol. 19, 275, 1979.

[9] L. N. Medgyesi-Mitschang, "Method of moments solutions for resonant and near-resonant scatterers," *Proc. SPIE*, vol 358, p. 64, 1982.

[10] V. N. Bringi and T. A. Seliga, "Scattering from axisymmetric dielectrics or perfect conductors imbedded in an axisymmetric dielectric," *IEEE Trans. Antennas Propagat.*, vol. AP-25, p. 575, 1977.

[11] D.-S. Wang and P. W. Barber, "Scattering by inhomogeneous nonspheroidal objects," *Appl. Opt.*, vol. 18, p. 1190, 1979.

[12] D.-S. Wang *et al.*, "Light scattering by polydisperse suspensions of inhomogeneous nonspherical particles," *Appl. Opt.*, vol. 18, p. 2672, 1979.

[13] V. K. Varadan and V. V. Varadan, Eds., *Acoustic, Electromagnetic and Elastic Wave Scattering—Focus on the T-Matrix Approach.* New York: Pergamon, 1979.

[14] M. A. Morgan and K. K. Mei, "Finite-element computation of scattering by inhomogeneous penetrable bodies of revolution," *IEEE Trans. Antennas Propagat.*, vol. AP-27, p. 202, 1979.

[15] M. A. Morgan, S. K. Chang, and K. K. Mei, "Coupled azimuthal potentials for electromagnetic field problems in inhomogeneous axially-symmetric media," *IEEE Trans. Antennas Propagat.*, vol. AP-25, p. 413, 1977.

[16] J. R. Mautz and R. F. Harrington, "Electromagnetic coupling to a conducting body of revolution with a homogeneous material region," *Electromagn.*, vol. 2, p. 257, 1982.

[17] J. Meixner, "The behavior of electromagnetic fields at edges," *IEEE Trans. Antennas Propagat.*, vol. AP-20, p. 442, 1972.

[18] R. Mittra and S. W. Lee, *Analytical Techniques in the Theory of Guided Waves.* New York: MacMillan, 1971, ch. 1.

[19] R. A. Hurd, "The edge condition in electromagnetics," *IEEE Trans. Antennas Propagat.*, vol. AP-24, p. 70, 1976. (See also vol. AP-24, p. 904, 1976 for corrections.)

[20] K. C. Lang, "Edge conditions of a perfectly conducting wedge with its exterior region divided by a resistive sheet," *IEEE Trans. Antennas Propagat.*, vol. AP-21, p. 237, 1973.

[21] J. B. Andersen and V. V. Solodukhov, "Field behavior near a dielctric wedge," *IEEE Trans. Antennas Propagat.*, vol. AP-26, p. 598, 1978.

[22] T. K. Wu and L. L. Tsai, "Scattering by a dielectric wedge: A numerical solution," *IEEE Trans. Antennas Propagat.*, vol. AP-25, p. 570, 1977.

[23] L. V. Kantorovich and V. I. Krylov, *Approximate Methods of Higher Analysis*, translated by C. D. Benster. New York: Wiley, 1964.

[24] L. V. Kantorovich and G. P. Akilov, *Functional Analysis in Normed Spaces*, translated by D. E. Brown. Oxford: Pergamon, 1964.

[25] R. F. Harrington, *Field Computation by Method of Moments.* New York: MacMillian, 1968.

[26] J. R. Mautz and R. F. Harrington, "Radiation and scattering from bodies of revolution," *Appl. Sci. Res.*, vol. 20, p. 405, 1969.

A Tetrahedral Modeling Method for Electromagnetic Scattering by Arbitrarily Shaped Inhomogeneous Dielectric Bodies

DANIEL H. SCHAUBERT, SENIOR MEMBER, IEEE, DONALD R. WILTON, SENIOR MEMBER, IEEE, AND ALLEN W. GLISSON, MEMBER, IEEE

***Abstract*—A method for calculating the electromagnetic scattering from and internal field distribution of arbitrarily shaped, inhomogeneous, dielectric bodies is presented. A volume integral equation is formulated and solved by using the method of moments. Tetrahedral volume elements are used to model a scattering body in which the electrical parameters are assumed constant in each tetrahedron. Special basis functions are defined within the tetrahedral volume elements to insure that the normal electric field satisfies the correct jump condition at interfaces between different dielectric media. An approximate Galerkin testing procedure is used, with special care taken to correctly treat the derivatives in the scalar potential term. Calculated internal field distributions and scattering cross sections of dielectric spheres and rods are compared to and found in agreement with other calculations. The accuracy of the fields calculated by using the tetrahedral cell method is found to be comparable to that of cubical cell methods presently used for modeling arbitrarily shaped bodies, while the modeling flexibility is considerably greater.**

I. INTRODUCTION

THE INTERACTION of electromagnetic waves with dielectric bodies has been extensively studied because of its importance to problems including propagation through rain or snow, scattering by and detection of airborne particulates, medical diagnostics and power absorption in biological bodies, coupling to missiles with plasma plumes or dielectric-filled apertures, and performance of communication antennas in the presence of dielectric and magnetic inhomogeneities. When the size of the dielectric body is neither large nor small compared to the wavelength of the excitation, asymptotic methods cannot be used to solve the interaction problem. In the intermediate size region, often called the resonance region, a rigorous solution of Maxwell's equations is required.

The work of Richmond [1], [2] heralded the use of modern computational methods to solve dielectric interaction problems. A number of methods have been developed and applied to two- and three-dimensional scattering problems [3]-[12]. Some of these methods have been demonstrated only for homogeneous bodies while others are restricted to bodies of revolution. The methods of [6] and [7] are based on time-domain formulations and may not be well-suited to single frequency applications. The methods of [8]-[10] utilize cubical or rectangular cells to model the body and they are applicable to arbitrarily shaped bodies.

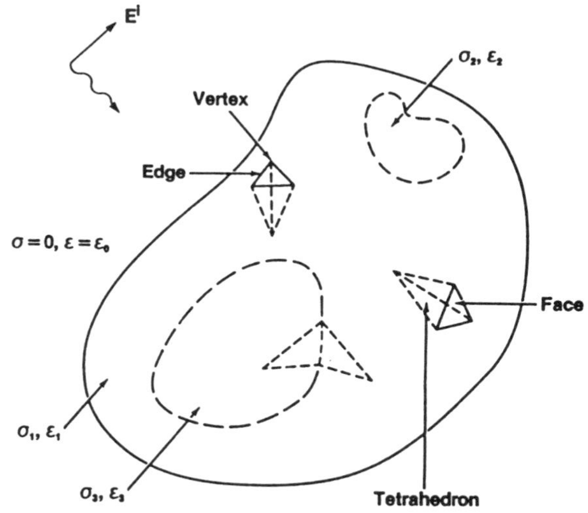

Fig. 1. Piecewise homogeneous dielectric body modeled by tetrahedral volume elements.

However, in order to accurately model arbitrarily curved body surfaces, many cells may be required. Also, in these approaches pulse basis functions were used to expand the unknown field within the body. This can lead to a divergence of the numerical solutions when the cells are subdivided to obtain a better representation of the body [13, p. 59].

An arbitrarily shaped, inhomogeneous body with piecewise constant electrical parameters is shown in Fig. 1. The simplest volume element that can be used to model each homogeneous region of the body is the tetrahedron, which is defined by four vertices and is bounded by four triangular faces. Any volume that is bounded by a polygonal surface or specified by discrete points on the boundary surface can be decomposed into tetrahedral elements. Higher order elements, such as cubes and parallelepipeds, can be formed by combining tetrahedral elements. The advantages of tetrahedral elements for volume modeling are analogous to the advantages of triangles for surface modeling [14], [15].

Special vector basis functions are defined for use in conjunction with the tetrahedral elements. These functions are analogous to the rooftop functions that are used for rectangular surface patches in conductor scattering [16] and are constructed to have continuous flux density across the faces of the tetrahedrons.

In the next section, the volume integral equation for the scattering problem is developed and converted to a matrix equation by the method of moments [13]. In Section III, numerical results are presented. Only results for relatively simple bodies are presented because the accurate numerical and experimental data needed to assess the accuracy of the tetrahedral calculations is not presently available for complex, three-dimensional bodies.

Manuscript received March 28, 1983; revised August 26, 1983.

D. H. Schaubert was with the Division of Electronic Products, Bureau of Radiological Health, Rockville, MD. He is now with the Department of Electrical and Computer Engineering, University of Massachusetts, Amherst, MA 01003.

D. R. Wilton was with the Department of Electrical Engineering, University of Mississippi, University, MS. He is now with the Department of Engineering, University of Houston, Houston, TX 77004.

A. W. Glisson is with the Department of Electrical Engineering, University of Mississippi, University, MS 38677.

Reprinted from *IEEE Trans. Antennas Propagat.*, vol. AP-32, no. 1, pp. 77–85, January 1984.

II. FORMULATION

Derivation of Volume Integral Equation

Let V denote the volume of a lossy, inhomogeneous, dielectric body with complex dielectric constant $\hat{\epsilon}(\mathbf{r}) = \epsilon(\mathbf{r}) - j\sigma(\mathbf{r})/\omega$ where ϵ and σ are the medium permittivity and conductivity at $\mathbf{r}$. An electric field $\mathbf{E}^i$, defined to be the field due to an impressed source $\mathbf{J}^i$ in the absence of the body, is incident on the body.

Decomposing the total electric field into an incident and a scattered field where the scattered field is due to a volume polarization current $\mathbf{J}$,

$$\mathbf{J}(\mathbf{r}) = j\omega[\hat{\epsilon}(\mathbf{r}) - \epsilon_0]\,\mathbf{E}(\mathbf{r}) \tag{1}$$

leads to

$$\mathbf{E}(\mathbf{r}) = \mathbf{E}^i(\mathbf{r}) + \mathbf{E}^s(\mathbf{r}) \tag{2}$$

where

$$\mathbf{E}^s(\mathbf{r}) = -j\omega\mathbf{A}(\mathbf{r}) - \nabla\Phi(\mathbf{r}) \tag{3}$$

$$\mathbf{A}(\mathbf{r}) = \frac{\mu_0}{4\pi}\int_V \mathbf{J}(\mathbf{r}')\,\frac{e^{-jk_0|\mathbf{r}-\mathbf{r}'|}}{|\mathbf{r}-\mathbf{r}'|}\,dv' \tag{4}$$

$$\Phi(\mathbf{r}) = \frac{1}{4\pi\epsilon_0}\int_V \rho(\mathbf{r}')\,\frac{e^{-jk_0|\mathbf{r}-\mathbf{r}'|}}{|\mathbf{r}-\mathbf{r}'|}\,dv' \tag{5}$$

and $k_0 = \omega\sqrt{\mu_0\epsilon_0} = 2\pi/\lambda_0$. The charge density $\rho(\mathbf{r})$ is related to the polarization current in (1) by

$$\nabla \cdot \mathbf{J}(r) = -j\omega\rho(r). \tag{6}$$

Equation (2), taken with (1) and (3)-(6) constitutes an integro-differential equation for the polarization current $\mathbf{J}$. In the following, however, it is found convenient to express $\mathbf{J}$ in terms of the intermediate quantity

$$\mathbf{D} = \hat{\epsilon}\mathbf{E} \tag{7}$$

which has a continuous normal component at media interfaces. From (7) and (1) it follows that $\mathbf{J}$ can be written

$$\mathbf{J}(\mathbf{r}) = j\omega\kappa(\mathbf{r})\mathbf{D}(\mathbf{r}) \tag{8}$$

where the *contrast ratio*

$$\kappa(\mathbf{r}) = \frac{\hat{\epsilon}(\mathbf{r}) - \epsilon_0}{\hat{\epsilon}(\mathbf{r})} \tag{9}$$

now accounts for all discontinuities in the normal component of $\mathbf{J}$ at media interfaces.

Basis Functions

The volume V is assumed to be subdivided into a number of tetrahedral elements such that an inhomogeneous dielectric region is approximated by a number of tetrahedrons, in each of which the dielectric properties are approximated as constant. A homogeneous dielectric region is bounded by a surface that is approximated by triangular faces assigned so as to fit the shape of the surface well and also meet the modeling guidelines discussed in Section III.

Once the volume of the scatterer has been appropriately modeled by tetrahedral volumes, the faces of the tetrahedrons are of primary importance for the development of the basis functions. Fig. 2 shows two tetrahedrons, T_n^+ and T_n^-, associated with the nth face of the subdivided region V modeling a scatterer. Points in T_n^+ may be designated either by the position vector $\mathbf{r}$ defined with respect to O, or by the position vector ρ_n^+ defined

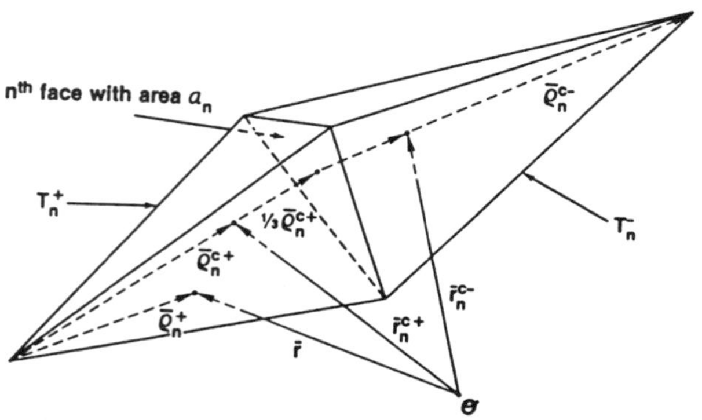

Fig. 2. Pair of tetrahedrons and geometrical parameters associated with nth face.

with respect to the free vertex of T_n^+. Similar remarks apply to points in T_n^- and the position vector ρ_n^-, except that the latter is directed toward the free vertex of T_n^-. The plus or minus designation of a tetrahedron is determined by the choice of a positive flux reference direction, which is assumed to be from T_n^+ to T_n^-. The basis functions that are used for the moment method solution of (2) are such that one basis function is associated with each face of the tetrahedral model of V. The basis function associated with the nth face is

$$\mathbf{f}_n(\mathbf{r}) = \begin{cases} \dfrac{a_n}{3V_n^3}\,\rho_n^+, & \mathbf{r} \in T_n^+ \\[2ex] \dfrac{a_n}{3V_n^-}\,\rho_n^-, & \mathbf{r} \in T_n^- \\[2ex] 0, & \text{otherwise} \end{cases} \tag{10}$$

where a_n is the area of the face and $V_n^\pm$ is the volume of $T_n^\pm$. (Note the convention that subscripts refer to faces while superscripts refer to tetrahedrons.)

The unknown electric flux density that enters (2) through (3)-(6) can be represented throughout V by

$$\mathbf{D}(\mathbf{r}) = \sum_{n=1}^{N} D_n\mathbf{f}_n(\mathbf{r}). \tag{11}$$

The summation is over the N faces that make up the tetrahedral model of V.

The basis functions $\mathbf{f}_n(\mathbf{r})$ have several properties that make them useful for representing $\mathbf{D}(\mathbf{r})$.

1) Within each tetrahedron $\mathbf{D}(\mathbf{r})$ is the sum of four linearly independent basis functions (one associated with each face) which can be combined to represent a constant vector in any direction.

2) $\mathbf{f}_n(\mathbf{r})$ has no component normal to any face except the common face of the conjoined pair T_n^+ and T_n^-.

3) The component of $\mathbf{f}_n$ normal to the nth face is constant and continuous across the face because the normal component of $\rho_n^\pm$ along face n is just the height of $T_n^\pm$ with face n as the base and the height expressed as $3\,V_n^\pm/a_n$. This latter factor normalizes $\mathbf{f}_n$ in (10) such that its flux density normal to face n is unity, ensuring continuity of the component of $\mathbf{f}_n$ normal to the face.

4) The divergence of the basis function is

$$\nabla \cdot \mathbf{f}_n(\mathbf{r}) = \begin{cases} \dfrac{a_n}{V_n^+}, & \mathbf{r} \in T_n^+ \\[2ex] \dfrac{-a_n}{V_n^-}, & r \in T_n^- \\[2ex] 0, & \text{otherwise} \end{cases} \tag{12}$$

where the divergence in $T_n^\pm$ is given by $(\rho_n^\pm)^{-2}\partial[\pm(\rho_n^\pm)^2 f_n]/\partial\rho_n^\pm$. The charge density, which is proportional to $\nabla\cdot\mathbf{f}_n$ through (6), (8), and (11), is constant within each tetrahedron.

5) The moment of $\mathbf{f}_n$ over $T_n^\pm$ is

$$\int_{T_n^\pm} \mathbf{f}_n\, dv = \frac{a_n}{3}\rho_n^{c\pm} \tag{13}$$

where $\rho_n^{c\pm}$ is the vector between the free vertex and the centroid of $T_n^\pm$ with ρ_n^{c-} directed toward and ρ_n^{c+} directed away from the vertex. Equation (13) may be most easily derived by expressing the integral in terms of volume coordinates [17].

6) If face n is on the boundary of V, then only one of the tetrahedrons, T_n^+ or T_n^-, is interior to V. In this case it is assumed that $\mathbf{f}_n$ is defined only over the interior tetrahedron and that the exterior tetrahedron is not defined.

Some of the less obvious of these properties can be verified by arguments similar to those in [15]. An important interpretation of (11) that follows from 1) and 3) is that the expansion coefficient D_n represents the normal component of $\mathbf{D}(\mathbf{r})$ at the nth face.

The modeling properties of the basis functions $\mathbf{f}_n$ are similar to those of the rooftop functions [16] that are used to model surface current distributions. Property 1) guarantees that the expansion (11) can provide at least a piecewise constant approximation to the flux density. Furthermore, each function varies linearly with distance from its defining vertex. This provides some capability to model linear field variations. However, like the rooftop functions, these basis functions are constant in the transverse direction so they cannot accurately represent a vector field with arbitrary linear variation; they can reproduce linear variations of the field only if the variation is parallel to the direction of the field defined by the basis function.

In order to solve (2) with $\mathbf{D}(\mathbf{r})$ represented by (11), it is necessary to find expressions for $\mathbf{J}(\mathbf{r})$ and $\rho(\mathbf{r})$ in terms of the basis functions and expansion coefficients. From (11) and (8)

$$\mathbf{J}(\mathbf{r}) = j\omega \sum_{n=1}^{N} D_n \kappa(\mathbf{r})\mathbf{f}_n(\mathbf{r}). \tag{14}$$

In (14), the parameter $\kappa(\mathbf{r})$ is taken inside the summation to emphasize that the polarization or contrast current at each point in the body is a weighted sum of the basis functions at that point times a factor that describes the medium at that point. By substituting (14) into (6) and using (12), the charge density is found to be represented by

$$\rho(\mathbf{r}) = -\sum_{n=1}^{N} D_n \kappa(\mathbf{r})\nabla\cdot\mathbf{f}_n(\mathbf{r}) - \sum_{n=1}^{N} D_n \mathbf{f}_n(r)\cdot\nabla\kappa(\mathbf{r}). \tag{15}$$

The first summation is the induced volume charge densities ρ_{vn} associated with the basis functions and can be evaluated with the aid of (12)

$$\rho_{vn}(\mathbf{r}) = \begin{cases} -D_n\kappa_n^+ \dfrac{a_n}{V_n^+}, & \mathbf{r}\in T_n^+ \\ D_n\kappa_n^- \dfrac{a_n}{V_n^-}, & \mathbf{r}\in T_n^- \\ 0, & \text{otherwise} \end{cases} \tag{16}$$

where $\kappa_n^\pm$ is the constant value of $\kappa(\mathbf{r})$ in $T_n^\pm$. The second summation in (15) represents the induced surface charge densities associated with the basis functions. When the nth face separates dissimilar media, κ is discontinuous and its gradient is a generalized function representing a surface charge density ρ_{sn}, residing on a_n,

$$\rho_{sn}(\mathbf{r}) = \begin{cases} D_n(\kappa_n^+ - \kappa_n^-), & \mathbf{r}\in a_n \\ 0, & \text{otherwise.} \end{cases} \tag{17}$$

An induced volume charge associated with each basis function exists in each tetrahedron for which the medium parameters differ from those of free space. It can be shown that the formulation of the problem in (1)-(6) guarantees that the total volume charge in a homogeneous region is zero. In the numerical procedure, the total volume charge, which is the sum of the volume charges associated with the four basis functions in each tetrahedron, is not explicitly constrained to be zero, but should be small for any solution that is close to the true solution. An induced surface charge exists only on faces that separate dissimilar media, and the charge density is constant on such faces. The correct surface charge representation is obtained at the faces because of the careful construction of the representation of $\mathbf{J}$.[1] Equations (11), (14), and (15) can be used in (2)-(7) to obtain an equation involving only the unknown coefficients $\{D_n\}$ and known vector functions.

Testing Procedure

The next step in applying the method of moments is to select a testing procedure that will generate N independent equations for the unknown expansion coefficients. The expansion functions $\mathbf{f}_m$ developed in the previous section are chosen as testing functions. With a symmetric product defined as

$$\langle \mathbf{f}, \mathbf{g}\rangle \equiv \int_V \mathbf{f}\cdot\mathbf{g}\, dv \tag{18}$$

(2) is tested with $\mathbf{f}_m$, yielding

$$\langle \mathbf{D}/\hat{\epsilon}, \mathbf{f}_m\rangle + j\omega\langle \mathbf{A}, \mathbf{f}_m\rangle + \langle\nabla\Phi, \mathbf{f}_m\rangle = \langle \mathbf{E}^i, \mathbf{f}_m\rangle, \quad m = 1, 2, \cdots, N. \tag{19}$$

This represents N equations for the N unknown coefficients $\{D_n\}$ and can be written in matrix form as

$$[S_{mn}][D_n] = [E_m] \tag{20}$$

where $[S_{mn}]$ is an $N\times N$ matrix with dimensions of m^4/F, and $[D_n]$ and $[E_m]$ are column vectors of length N. Since the elements of $[S_{mn}]$ and $[E_m]$ cannot be evaluated in closed form, it is necessary to derive accurate and efficient numerical approximations for these quantities. Formulas for the matrix elements and some considerations relevant to their computation are given in the Appendix.

III. NUMERICAL RESULTS

In order to test the accuracy of the solutions obtained by using the present method, it is necessary to apply the method to

[1] The representation is achieved by expressing $\mathbf{J}$ in terms of $\mathbf{D}$. However, $\nabla\cdot(\hat{\epsilon}\mathbf{E}) = 0$ so that if $\mathbf{D} = \hat{\epsilon}\mathbf{E}$ had been chosen as the unknown it would have been possible to impose on the basis set the condition that $\mathbf{D}$ is divergenceless. This would result in fewer unknowns than the present method, which imposes the condition numerically, but it greatly complicates the procedure of generating the basis functions. This alternative formulation, which apparently has not been previously explored, is worthy of further investigation because the savings that result from fewer unknowns may outweigh the additional complexity of the basis functions.

problems for which accurate solutions are available. Unfortunately, very few three-dimensional dielectric scattering problems can be solved analytically. Since the homogeneous sphere and layered, inhomogeneous sphere are examples where the scattered fields can be analytically determined, these examples are considered below. The radar cross section of a dielectric rod is also calculated and compared to the results of other numerical methods and to available experimental data.

Modeling Considerations

The most arduous task required for the solution of three-dimensional scattering problems by the tetrahedral cell method is the development of a multicell model that accurately represents the physical body. Although some automatic grid generation schemes may be helpful in this task, the models for the examples shown below were developed manually with the aid of interactive stereo graphics. The procedure that was used for model development consisted of three basic steps.

1) Model the surfaces of each homogeneous dielectric subregion with triangular faces. This step ensures proper modeling of the boundary shape of each subregion. The approximation of the surfaces by triangular faces is essentially the same as that required by the method of Rao, Wilton, and Glisson [15] for conducting bodies. Experience indicates that useful results for scattered field quantities often can be obtained when the edges of the triangular faces are longer than 0.25 wavelengths in the dielectric medium, but accurate determination of the internal fields usually requires that the edge lengths be about one half this value. Best results have been obtained when the triangular faces are approximately equilateral.

2) Divide each dielectric subregion into tetrahedral volume elements. For best results, tetrahedrons should be approximately equilateral.

3) Electrical parameters are assigned to each tetrahedron. For the examples shown below this was accomplished by interactively using computer generated stereo graphics.

Tetrahedral models usually require fewer unknowns per volume element than cubical models. This is because the tetrahedral models require one unknown per triangular face. For interior tetrahedrons where each face is common to two tetrahedrons, this results in only two unknowns per tetrahedron compared to the three unknowns that are required for each cubical element. As an example of tetrahedral modeling, consider the sphere shown in Fig. 3. This model for the sphere was developed as described above and consists of three concentric layers of tetrahedral cells comprising a total of 216 cells in the entire sphere. The diameter of the model is such that the total volume of the 216 tetrahedrons is equal to the actual volume of the sphere that is being modeled. The model consists of 468 faces, which means that 468 unknowns are required if the symmetry is not exploited. However, the number of unknowns can be reduced to 135 if one makes use of two planes of symmetry.

Dielectric Sphere

The 216-tetrahedron sphere model has been used to calculate the radar cross section of an electrically small sphere. The agreement of the results with the Rayleigh theory [18] approximation, which is asymptotically correct at low frequencies, is excellent. However, the radar cross section is not sensitive to minor errors in the internal field distribution. Therefore, a more critical check on the accuracy of the calculations is the accuracy of the internal field. At low frequencies the internal electric field is constant and can be shown to be equal to $3/(\epsilon_r + 2)$ times the incident field. The calculated field is within 10 percent of this value.

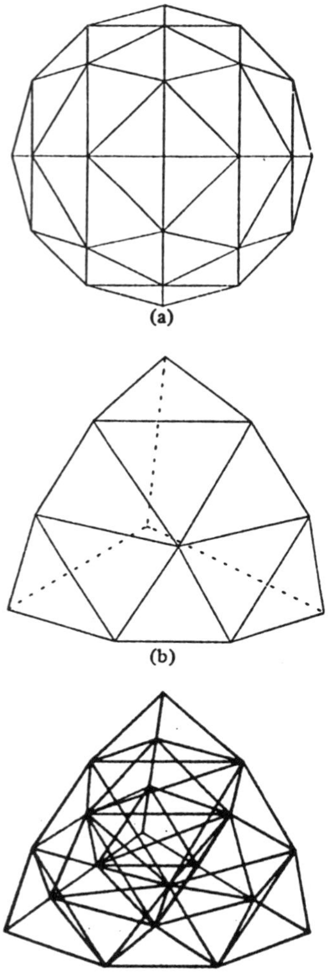

(a)

(b)

(c)

Fig. 3. Model of sphere consisting of three concentric layers of tetrahedrons. (a) Top view. (b) Outer surface of one-eighth of sphere. (c) Interior subdivision of one-eighth of sphere into 27 tetrahedrons.

At higher frequencies, the internal field exhibits a standing wave behavior. True and calculated electric field distributions inside a sphere of moderate size are shown in Fig. 4. A 512-tetrahedron model was used for the sphere, which has a radius of 0.39 wavelength relative to the dielectric medium. The agreement between the field calculated via the tetrahedral method and the Mie series solution [19] is very good. As already mentioned, the representation (11) provides at least a piecewise constant approximation to the internal fields.[2]

[2] All plots of internal field that are shown in this paper display the interpolated field values as obtained from the basis functions that are used in the moment method. This convention was chosen instead of plotting the calculated field values at discrete points for two reasons: 1) the centroids of the faces and tetrahedrons, which are the logical and conventional points to use for numerical evaluations, do not lie along a line or in a plane and cannot be readily displayed in a two-dimensional graph, and 2) for applications where internal field distribution is important the ability of the basis functions to accurately represent the true field is of interest.

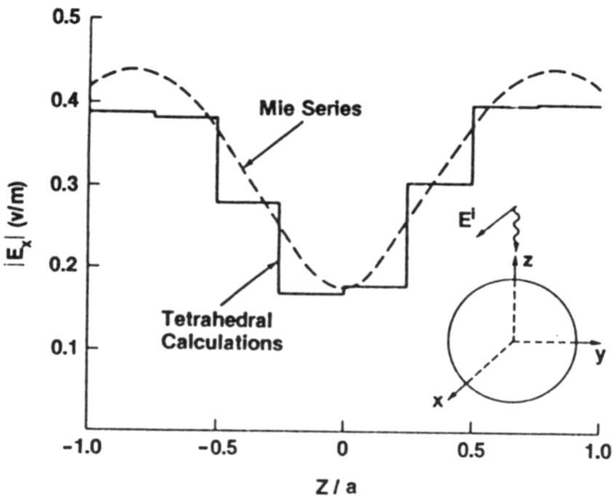

Fig. 4. Field along z axis inside lossless dielectric sphere; $\epsilon_r = 36$, $k_0 a = 0.408$.

As a further check on the results of the case shown in Fig. 4, the values of the expansion coefficients D_n in (11), which are equal to the electric flux density normal to the associated face of the tetrahedral model, were compared to the true values at the centroid of each face. The worst-case error of these coefficients was found not to exceed 10 percent of the largest coefficient, and for most points the error was less than 5 percent. The tetrahedral calculations utilized two planes of symmetry so that the computations could be performed with 304 unknowns.

The fields inside a lossy sphere have also been calculated. These fields agree well with the true fields as indicated in Fig. 5, which is a plot of the absorbed power distribution along the z axis. Also shown are the results obtained by Rukspollmuang and Chen [20] who used cubical cells with pulse basis functions and point matching.

One of the primary advantages of tetrahedral cell modeling is the ability to model arbitrarily shaped inhomogeneities. The tetrahedral and cubical models that were used to obtain the results in Fig. 5 are approximately equivalent in terms of the number of unknowns that are used to represent the field, but not in terms of modeling capability. The 512-tetrahedron model has 1088 unknowns. The 320-cube model has three unknowns per cube, so it results in 960 unknowns for the entire sphere. Stated differently, with only 13 percent more unknowns, the tetrahedral model has 60 percent more volume elements with which to represent inhomogeneous regions in the sphere. To demonstrate the usefulness of this capability it would be necessary to calculate the field inside a body with one or more irregularly shaped inhomogeneities. However, the only such bodies for which analytical solutions are known to the authors are spheres with spherically symmetric inhomogeneities. Therefore, the fields inside a two-layered sphere have been calculated by using the 512-tetrahedron model and are compared to the analytical solution in Fig. 6. The tetrahedral solution agrees well with the analytical solution except in the vicinity of the discontinuity in E_x at the dielectric interface along the x axis. The tetrahedral solution appears to violate the boundary condition,

$$\hat{\epsilon}_1 \mathbf{E}_1^{\text{normal}} = \hat{\epsilon}_2 \mathbf{E}_2^{\text{normal}}. \tag{21}$$

However, the apparent discrepancy results from the particular model that was used for the sphere. In the actual sphere, E_x is normal to the dielectric interface at the x axis, but in the tetrahedral model, the dielectric interface coincides with a triangular

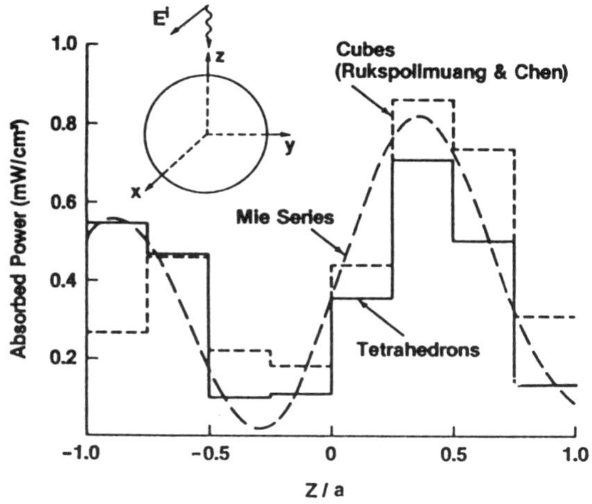

Fig. 5. Absorbed power distribution along z axis of lossy dielectric sphere; $\epsilon_r = 35$, $\tan\delta = 0.39$, $k_0 a = 0.577$.

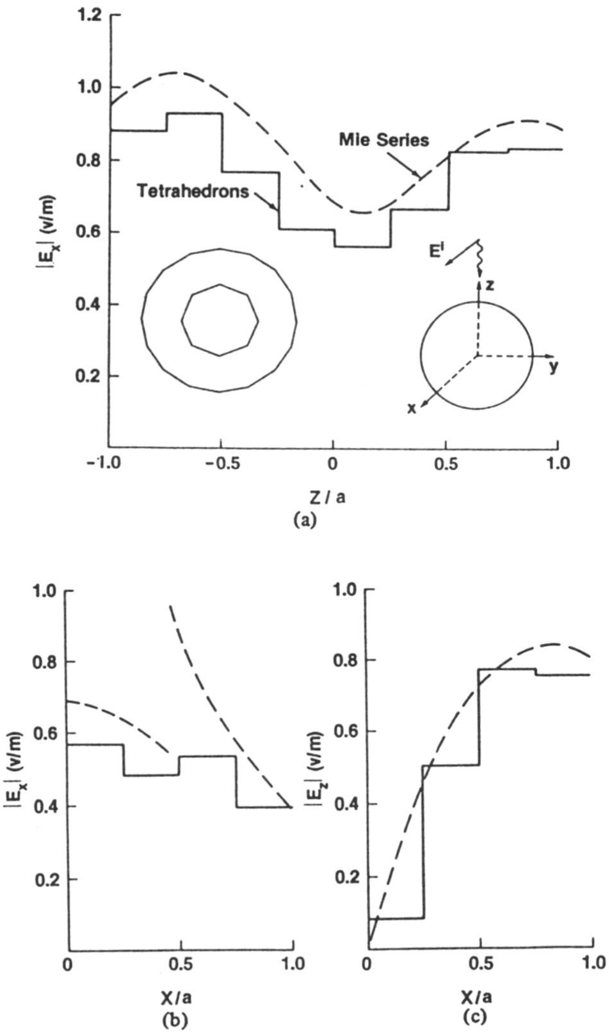

Fig. 6. Fields inside lossless, inhomogeneous sphere; $\epsilon_{r1} = 16$, $k_0 a_1 = 0.0595$, $\epsilon_{r2} = 9$, $k_0 a_2 = 0.13$. (a) Magnitude of E_x along z axis. (b) Magnitude of E_x along x axis. (c) Magnitude of E_z along x axis.

face that is not perpendicular to the axis (see, for example, Fig. 3(b)). Therefore, the x component of the electric field for the tetrahedral solution is not the component normal to the interface. The tetrahedral solution does satisfy (21) exactly at faces because the basis functions (10) and the field expansion (11) were constructed to insure this.

Scattering Calculations

Although the accuracy of internal field values is a more stringent test of a computational method than is the accuracy of scattered fields, it is difficult to accurately measure the fields inside most dielectric objects. Furthermore, many applications require only scattering data and not internal field data. Consequently, the scattering cross sections of dielectric objects are calculated and compared to published data. The comparison of these scattering data, which are obtained by different methods and by measurements, serve as an indicator of the overall accuracy of the tetrahedral method.

The radar cross section of a thin dielectric rod (radius = 0.16 cm) has been computed and compared to the results of Richmond [21] and Wang and Papanicolopulos [22]. These data are shown in Fig. 7, where the agreement is seen to be excellent. The results of Wang and Papanicolopulos were obtained with the cubical cell method of Livesay and Chen [9]. The thin rod was modeled by a single column of cubes with cross-sectional area equal to the circular rod. The tetrahedral model was a triangular column with the same cross-sectional area as the circular rod.

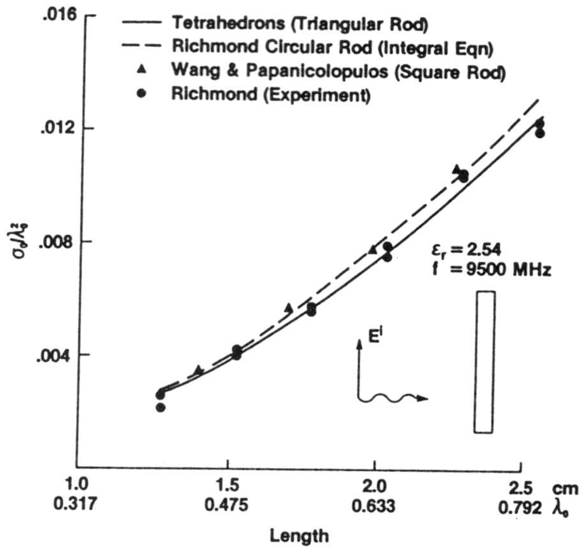

Fig. 7. Radar cross section of a thin dielectric rod; $a = 0.16$ cm $= 0.05\ \lambda_0$ $\epsilon_r = 2.54$.

The scattering cross section of a dielectric sphere can be calculated analytically from the Mie series expansion for the internal fields. In Fig. 8, the results of the tetrahedral model calculations are compared to those of Barber, Owen, and Chang [23], who employed the series expansion. The calculations used the 512-tetrahedron model for the sphere. The resonant frequency is predicted within 1 percent. Similar calculations with the 216-tetrahedron model predict the resonant frequency within 2 percent, but the scattering cross section is 35 percent below the true value. A slightly lossy sphere with the index of refraction equal to $6\text{-}j.01$ was also analyzed with the two models, and the predicted cross sections were slightly more accurate than for the lossless sphere.

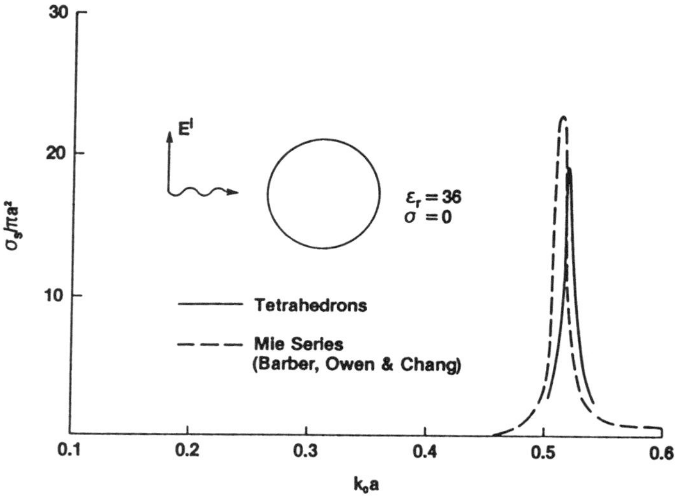

Fig. 8. Total scattering cross section of dielectric sphere.

These calculations illustrate the improvement in the computed solutions that results from using smaller volume elements and more unknowns to model the sphere. Table I shows the edge lengths and tetrahedron volumes for the above calculations at the peak of the scattering cross section. The parameters for the model that was used to obtain the results shown in Fig. 4 are also shown. It is evident from these data that the best results are obtained when the average edge length is somewhat less than 0.25 wavelengths in the dielectric medium.

Computation Time

An important aspect of any numerical solution is the amount of computer time required to obtain the desired accuracy in the solution. The accuracies that have been achieved with the tetrahedral modeling program are illustrated by the examples above. The computation times required for some of these solutions are presented in Table II. These times are for a VAX 11/780 computer running under the VMS operating system. The matrix fill time is quite long because of a large number of numerical integrations that are required to evaluate the vector and scalar potential contributions. Considerable savings in computation time are achieved by using the computed integrals in as many matrix elements as possible (similar to the procedure described by Rao, Wilton, and Glisson [15]). All volume and surface integrations were evaluated by using special numerical integration formulas for tetrahedrons and triangles [24] requiring five and seven points, respectively. The $1/R$ singularity in the self-terms was extracted and integrated analytically [25]. The table also contains a comparison of the tetrahedral method with a cubical cell method (pulse basis functions and point matching) developed by Hagmann, Gandhi and Durney [10]. The time required to fill the matrix for the tetrahedral model was about four times that required for the cubical model. This difference in filling times would be reduced or eliminated by decreasing the order of the numerical quadrature scheme when the integrand is slowly varying. The fill time per element is greater for the problems with two symmetry planes because there are four portions of the body (one real and three images) that contribute to each matrix element.

TABLE I
PARAMETERS OF SPHERICAL MODELS

MODEL	NUMBER OF UNKNOWNS	EDGE LENGTHS/λ_d* min.	max.	avg.	TETRAHEDRON VOLUMES/λ_d^3 min.	max.	avg.
512-Tetrahedron Sphere (Fig. 8) $k_o a$ =0.5177	1088	0.127	0.315	0.214	3.41×10^{-4}	1.67×10^{-3}	9.85×10^{-4}
216-Tetrahedron Sphere $k_o a$ = 0.5207	468	0.175	0.416	0.283	8.92×10^{-3}	4.18×10^{-3}	2.38×10^{-3}
512-Tetrahedron Sphere (Fig. 4) $k_o a$ = 0.408	1088	0.100	0.248	0.169	1.69×10^{-4}	8.22×10^{-4}	4.83×10^{-4}

*λ_d is the wavelength in the dielectric medium

TABLE II
COMPUTATION TIMES

MODEL	NUMBER OF UNKNOWNS	NUMBER OF SYMMETRY PLANES	MATRIX FILL TIME	SOLUTION TIME (by LU-factorization)
3-Layer Sphere $k_o a$ = 0.500 (Fig. 8)	135	2	283.67 sec (15.6 msec/element)	64.87 sec
4-Layer Sphere $k_o a$ = 0.408 (Fig. 4)	304	2	1372.48 sec (14.8 msec/element)	734.64 sec
Rectangular Box Modeled by Tetrahedrons	200	0	179.10 sec (4.5 msec/element)	209.41 sec
Rectangular Box Modeled by Cubes	192	0	42.96 sec (1.2 msec/element)	178.55 sec

IV. SUMMARY

A new method for computing the electromagnetic fields scattered from and penetrating into arbitrarily shaped, three-dimensional, inhomogeneous dielectric bodies has been presented. The method is based upon the use of tetrahedral volume elements, which are the simplest polyhedral volume elements that can be used to model three-dimensional bodies. Special basis functions have been developed to represent the field within each tetrahedron. These basis functions automatically incorporate the boundary conditions of the normal component of electric flux density, which is related to the surface polarization charge density.

The testing procedure that is used for the method of moments solution is an approximate Galerkin procedure. The formulation is exactly Galerkin, but the required volume integration is approximated by recognizing that the vector potential **A** and the scalar potential Φ are slowly varying and approximately constant within each tetrahedron.

Numerical results indicate that the tetrahedral modeling method predicts the internal field distribution and scattered field at least as accurately as existing cubical modeling methods. This accuracy is achieved with approximately the same number of unknowns as the cubical modeling methods, but with much greater modeling flexibility. In particular, the tetrahedral models have almost 50 percent more volume elements for the same number of unknowns. Furthermore, arbitrarily shaped boundaries of homogeneous subregions can be modeled more accurately by the triangular faces, which are not restricted in their spatial orientation.

The new tetrahedral modeling method also overcomes the fundamental limitation of pulse basis functions that are used for the existing cubical modeling methods. The basis functions for the tetrahedral elements are constructed to produce the correct jump in the normal component of **E** without fictitious surface charges. This should ensure convergence with no numerical instabilities when the model is successively subdivided to achieve greater accuracy.

APPENDIX

ELEMENTS OF THE MATRIX AND EXCITATION VECTOR

The formulas that are used to fill the matrix $[S_{mn}]$ and the excitation vector $[E_m]$ are derived from (19) as follows. The first term of (19) can be written as

$$\langle \mathbf{D}/\hat{\epsilon}, \mathbf{f}_m \rangle = \sum_{n=1}^{N} D_n \langle \mathbf{f}_n/\hat{\epsilon}, \mathbf{f}_m \rangle$$

$$= \sum_{n=1}^{N} D_n \left[\frac{1}{\hat{\epsilon}_n^+} \int_{T_m^+} \mathbf{f}_n \cdot \mathbf{f}_m \, dv + \frac{1}{\hat{\epsilon}_n^-} \int_{T_m^-} \mathbf{f}_n \cdot \mathbf{f}_m \, dv \right]. \tag{22}$$

(Recall that $\mathbf{f}_m$ is zero outside of $T_m^{\pm}$.) These integrals can be evaluated by utilizing normalized volume coordinates [17]. By using the volume coordinate representation of the basis functions and formulas for integration in volume coordinates [17, appendix 4], the integral of $\mathbf{f}_i \cdot \mathbf{f}_j$ over T^p can be evaluated as

$$\int_{T^p} \mathbf{f}_i \cdot \mathbf{f}_j \, dv = \frac{a_i a_j}{9V^p} \Big[\tfrac{8}{5} |\mathbf{r}^{cp}|^2 - (\mathbf{r}_i + \mathbf{r}_j) \cdot \mathbf{r}^{cp} + (\mathbf{r}_i \cdot \mathbf{r}_j) - \tfrac{1}{10} (\mathbf{r}_1 \cdot \mathbf{r}_2 + \mathbf{r}_1 \cdot \mathbf{r}_3 + \mathbf{r}_1 \cdot \mathbf{r}_4 + \mathbf{r}_2 \cdot \mathbf{r}_3 + \mathbf{r}_2 \cdot \mathbf{r}_4 + \mathbf{r}_3 \cdot \mathbf{r}_4) \Big] \tag{23}$$

where $\mathbf{r}^{cp}$ is the vector from O to the centroid of T^p and $\mathbf{r}_1, \mathbf{r}_2, \mathbf{r}_3, \mathbf{r}_4$ are vectors from O to the vertices of T. This result can be used to complete the evaluation of (22).

The second term and the right side of (19) both involve an integration over a vector field, and they can be treated similarly. In particular, $\mathbf{A}$ and $\mathbf{E}^i$ are assumed to vary slowly enough to be approximated within each tetrahedron by their values at the centroid of the tetrahedron. Then, by (13),

$$\left\langle \begin{pmatrix} \mathbf{A} \\ \mathbf{E}^i \end{pmatrix}, \mathbf{f}_m \right\rangle \approx \begin{pmatrix} \mathbf{A}(\mathbf{r}_m^{c+}) \\ \mathbf{E}^i(\mathbf{r}_m^{c+}) \end{pmatrix} \cdot \frac{a_m \boldsymbol{\rho}_m^{c+}}{3} + \begin{pmatrix} \mathbf{A}(\mathbf{r}_m^{c-}) \\ \mathbf{E}^i(\mathbf{r}_m^{c-}) \end{pmatrix} \cdot \frac{a_m \boldsymbol{\rho}_m^{c-}}{3}. \tag{24}$$

If T_m^+ or T_m^- is not in $V (T_m^{\pm} \notin V)$, then the corresponding term is absent in (24). In (24), $\mathbf{r}_m^{c\pm}$ is the vector from O to the centroid of $T_m^{\pm}$ and $\boldsymbol{\rho}_m^{c\pm}$ is the vector from the vertex opposite a_m in $T_m^{\pm}$ to the centroid of $T_m^{\pm}$.

The scalar potential term of (19) can be written as

$$\langle \nabla \Phi, \mathbf{f}_m \rangle = \int_S \Phi \mathbf{f}_m \cdot \hat{\mathbf{n}} \, ds - \int_V \Phi \nabla \cdot \mathbf{f}_m \, dv \tag{25}$$

where S is the boundary of V. Since the tangential component of $\mathbf{f}_m$ is discontinuous at tetrahedral boundaries, verification of (25) requires that the contributions from individual tetrahedral volume elements and their boundaries be summed. This leads directly to the last term of (25). The surface integral follows from the fact that the normal component of $\mathbf{f}_m$ is continuous or vanishes at all boundaries internal to V. Then, with the assumption that Φ is sufficiently slowly varying that it may be replaced by its value at the centroid of a tetrahedron or face, (25) becomes

$$\langle \nabla \Phi, \mathbf{f}_m \rangle = \begin{cases} a_m [\Phi(\mathbf{r}_m^{c-}) - \Phi(\mathbf{r}_m^{c+})], & T_m^+ \text{ and } T_m^- \in V \\ \pm a_m [\Phi(\mathbf{r}_m^{c\pm} \pm 1/3 \boldsymbol{\rho}_m^{c\pm}) - \Phi(\mathbf{r}_m^{c\pm})], & T_m^{\mp} \notin V. \end{cases} \tag{26}$$

The argument $(\mathbf{r}_m^{c\pm} \pm 1/3 \boldsymbol{\rho}_m^{c\pm})$ denotes that Φ is evaluated at the centroid of face m when $T_m^{\pm} \notin V$. When the elements of $[S_{mn}]$ are calculated, the contributions to $\mathbf{A}$ and Φ from a single basis function are needed. These are given by

$$\mathbf{A}_n(\mathbf{r}) = \frac{\mu_0 a_n}{12\pi} \left[\frac{\kappa_n^+}{V_n^+} \int_{T_n^+} \boldsymbol{\rho}_n^+ \frac{e^{-jkR}}{R} \, dv' + \frac{\kappa_n^-}{V_n^-} \int_{T_n^-} \boldsymbol{\rho}_n^- \frac{e^{-jkR}}{R} \, dv' \right] \tag{27}$$

$$\Phi_n(\mathbf{r}) = \frac{-a_n}{j\omega 4\pi\epsilon_0} \left[\frac{\kappa_n^+}{V_n^+} \int_{T_n^+} \frac{e^{-jkR}}{R} \, dv' - \frac{\kappa_n^-}{V_n^-} \int_{T_n^-} \frac{e^{-jkR}}{R} \, dv' - \frac{(\kappa_n^+ - \kappa_n^-)}{a_n} \int_{a_n} \frac{e^{-jkR}}{R} \, ds' \right], \tag{28}$$

where $\kappa_n^{\pm}$ is taken to be zero if $T_n^{\pm} \notin V$.

By expressing them in terms of normalized volume coordinates, the integrals in (27) and (28) may be readily evaluated numerically. They can, in fact, be expressed in terms of four independent scalar integrals over each tetrahedron. These four integrals contribute to 16 different elements of $[S_{mn}]$. Considerable savings in computation time is achieved by making use of these multiple contributions.

REFERENCES

[1] J. H. Richmond, "Scattering by a dielectric cylinder of arbitrary cross section shape," *IEEE Trans. Antennas Propagat.*, vol. AP-13, pp. 334–341, May 1965.

[2] J. H. Richmond, "TE-wave scattering by a dielectric cylinder of arbitrary cross-section shape," *IEEE Trans. Antennas Propagat.*, vol. AP-14, pp. 460–464, July 1966.

[3] T. K. Wu and L. L. Tsai, "Scattering from arbitrarily-shaped lossy dielectric bodies of revolution," *Radio Sci.* vol. 12, pp. 709–718, Sept. 1977.

[4] P. W. Barber and C. Yeh, "Scattering of EM waves by arbitrarily shaped dielectric bodies," *Appl. Opt.*, vol. 14, pp. 2864–2872, Dec. 1975.

[5] M. A. Morgan and K. K. Mei, "Finite element computation of scattering by inhomogeneous penetrable bodies of revolution," *IEEE Trans. Antennas Propagat.*, vol. AP-27, pp. 202–214, Mar. 1979.

[6] A. Taflove and M. E. Brodwin, "Computation of the electromagnetic fields and induced temperatures within a model of the microwave-irradiated human eye," *IEEE Trans. Microwave Theory Tech.*, vol. MTT-23, pp. 888–896, Nov. 1975.

[7] R. Holland, L. Simpson, and K. S. Kunz, "Finite-difference analysis of EMP coupling to lossy dielectric structures," *IEEE Trans. Electromagn. Compat.*, vol. EMC-22, pp. 203–209, Aug. 1980.

[8] E. H. Newman and P. Tulyathan, "Wire antennas in the presence of a dielectric/ferrite inhomogeneity," *IEEE Trans. Antennas Propagat.*, vol. AP-26, pp. 587–593, July 1978.

[9] D. E. Livesay and K-M Chen, "Electromagnetic fields induced inside arbitrarily shaped biological bodies," *IEEE Trans. Microwave Theory Tech.*, vol. MTT-22, pp. 1273–1280, Dec. 1974.

[10] M. J. Hagmann, O. P. Ghandhi, and C. H. Durney, "Numerical calculation of electromagnetic energy deposition for a realistic model of man," *IEEE Trans. Microwave Theory Tech.*, vol. MTT-27, pp. 804–809, Sept. 1979.

[11] J. J. H. Wang, "Numerical analysis of three-dimensional arbitrarily-shaped conducting scatterers by trilateral surface cell modelling," *Radio Sci.*, vol. 13, no. 6, pp. 947–952, Nov.-Dec. 1978.

[12] J. R. Mautz and R. F. Harrington, "Electromagnetic scattering from a homogeneous material body of revolution," *Arch. Elec. Ubertragung.*, vol. 33, pp. 71–80, Feb. 1979.

[13] R. F. Harrington, *Field Computation by Moment Methods.* New York: Macmillan, 1968.

[14] A. Sankar and T. C. Tong, "Current computation on complex struc-

tures by finite element method," *Electron. Lett.*, vol. 11, no. 20, pp. 481–482, Oct. 1975.

[15] S. M. Rao, D. R. Wilton, and A. W. Glisson, "Electromagnetic scattering by surfaces of arbitrary shape," *IEEE Trans. Antennas Propagat.*, vol. AP-30, pp. 409–418, May 1982.

[16] A. W. Glisson and D. R. Wilton, "Simple and efficient numerical methods for problems of electromagnetic radiation and scattering from surfaces," *IEEE Trans. Antennas Propagat.*, vol. AP-28, no. 5, pp. 593–603, Sept. 1980.

[17] O. C. Zienkiewicz, *The Finite Element Method in Engineering Science.* New York: McGraw-Hill, 1971.

[18] R. F. Harrington, *Time-Harmonic Electromagnetic Fields.* New York: McGraw-Hill, 1961, pp. 297–298.

[19] S. M. Neuder, "Electromagnetic fields in biological media, Part II—The SCAT program, multilayered spheres, theory and applications," HEW Pub. (FDA) 79-8072, Aug. 1979.

[20] S. Rukspollmuang and K-M Chen, "Heating of spherical versus realistic models of human and infrahuman heads by electromagnetic waves," *Radio Sci.*, vol. 14, no. 6S, pp. 51–62, Nov.-Dec. 1979.

[21] J. H. Richmond, "Digital computer solutions of the rigorous equations for scattering problems," *Proc. IEEE*, vol. 53, pp. 796–804, Aug. 1965.

[22] J. J. H. Wang and C. Papanicolopulos, "A study of the analysis and measurements of three-dimensional arbitrarily-shaped dielectric scatterers," Rome Air Development Center, Rep. RADC-TR-30-372, Dec. 1980.

[23] P. W. Barber, J. F. Owen, and R. K. Chang, "Resonant scattering for characterization of axisymmetric dielectric objects," *IEEE Trans. Antennas Propagat.*, vol. AP-30, pp. 168–172, Mar. 1982.

[24] P. C. Hammer, O. P. Marlowe, and A. H. Stroud, "Numerical integration over simplexes and cones," *Math Tables Aids Comp.*, vol. 10, pp. 130–137, 1956.

[25] D. R. Wilton, S. M. Rao, A. W. Glisson, D. H. Schaubert, O. Al-Bundak, and C. M. Butler, "Potential integrals of uniform and linear source distributions on polygonal and polyhedral domains," to appear in *IEEE Trans. Antennas Propagat.*, Mar. 1984.

Integral Equation Formulations for Imperfectly Conducting Scatterers

LOUIS N. MEDGYESI-MITSCHANG, MEMBER, IEEE, AND JOHN M. PUTNAM

Abstract—**Integral equation formulations are presented for characterizing the electromagnetic (EM) scattering interaction for nonmetallic surfaced bodies. Three different boundary conditions are considered for the surfaces: namely, the impedance (Leontovich), the resistive sheet, and its dual, the magnetically conducting sheet boundary. The integral equation formulations presented for a general geometry are specialized for bodies of revolution and solved with the method of moments (MM). The current expansion functions, which are chosen, result in a symmetric system of equations. This system is expressed in terms of two Galerkin matrix operators that have special properties. The solutions of the integral equation for the impedance boundary at internal resonances of the associated perfectly conducting scatterer are examined. The results are compared with the Mie solution for impedance-coated spheres and with the MM solutions of the electric, magnetic, and combined field formulations for impedance-coated bodies.**

I. INTRODUCTION

An UNDERSTANDING of the scattering from bodies with imperfectly conducting surfaces is central to many problems in optics and electroscience. Imperfectly conducting surfaces characterize many nonmetallic scatterers. To study the optical or electromagnetic (EM) phenomena of such surfaces, two approaches suggest themselves. Either a detailed investigation of material properties and boundary conditions (BC) can be undertaken; or alternately, a study of the net effect of these properties can be made. The latter approach is often more fruitful and expositive, leading to global understanding of the physical processes. This approach proved seminal in the study of a wide range of long-wave radio propagation problems.

The pioneering investigations of Leontovich and co-workers [1]-[3] showed that the observed phenomena in these problems could be represented by imposing the impedance (Leontovich) boundary condition (IBC) on the EM fields. This formulation obviated detailed computations that were impractical in the predigital computer era. Subsequent investigations showed that this approach leads to physically correct solutions at interfaces where the refractive index of the (nonfree space) dielectric media is $\gg 1$ and where the surface impedance can be expressed in terms of the permeability and permittivity of the surface material.

In the present investigation, we examine three integral representations for the boundary conditions at the interface of nonmetallic and coated bodies. They are the impedance (Leontovich), resistive sheet, and its dual, the magnetically conducting sheet boundary conditions, abbreviated here as IBC, RBC, and MBC, respectively. These representations model classes of scatterers with composite, artificial dielectric, or coated conducting surfaces. The details of the relationship between the physical and chemical properties of a material (or material systems) and its electrical characteristics are outside the scope of the present discussion. This paper focuses on the EM formulations and their solutions associated with these boundary conditions.

A detailed study of the IBC was made by Senior [4] in which the fundamental assumptions inherent in the IBC are examined for planar (flat) interfaces between a homogeneous (isotropic) dielectric and free space. Generalizations to curved interfaces with a large radius of curvature as well as to surfaces having variable properties are discussed. The relationships between surface curvature, surface properties, and assumptions on surface field variations are examined. The foregoing results are obtained from *ab initio* considerations of the partial differential equation formulations of Maxwell's equations.

In a companion investigation, Senior examined the use of the IBC for modeling optically rough surfaces [5]. In that study, the relationship of surface roughness to surface impedance was used to obtain the incoherent scattered fields from such bodies. As pointed out by Wait [6], the IBC representation also holds for finely corrugated surfaces and coated conductors.

A formulation based on the magnetic field integral equation (MFIE) for bodies satisfying the IBC was presented in theoretical terms by Mitzner [7]. Numerical results were obtained by Oshiro and co-workers [8]. Lossy dielectric spheres subject to the IBC were investigated by Garbacz using the classical modal solution derived from Mie's original studies [9]. The explicit assumption in these solutions is that the bodies are sufficiently lossy that each modal surface impedance in the Mie formulation can be replaced by a single impedance intrinsic to the lossy medium. The case of acoustic scattering by an impedance surface was treated by Angell and Kleinman [10].

Recent studies by Heath [11] and Rogers [12] examined the well-posedness of the combined field integral equation (CFIE) extended for scatterers satisfying the IBC. The method of moments (MM) solution of the IBC-CFIE formulation is shown to be stable at internal resonances of the associated perfectly conducting body. By contrast the IBC-EFIE (electric field integral equation) and IBC-MFIE formulations are shown to be ill-posed and lead to ill-conditioned matrices when solved by the MM technique. As in the earlier work on the perfectly conducting scatterers [13], the IBC-CFIE formulation requires the introduction of an arbitrary weighting constant. As noted in [12], this constant is a sensitive function of the surface impedance of the body. While the accuracy of the IBC-CFIE formulation depends upon the constant used, no general *a priori* procedure for its choice exists. Possible choices are suggested in [12]. A related investigation was made by Kress and Spassov [14].

In a recent investigation, Graglia and Uslenghi examined impedance-coated bodies of revolution (BOR's) using the generalized MFIE formulation due to Maue [15]. Their results are restricted to axial illumination of the body and implicitly are invalid for frequencies where internal resonances occur. The issue of ill-posedness of the IBC-MFIE is not addressed.

In this paper we present an integral formulation for bodies where their surface properties satisfy the IBC. Extensive compu-

Manuscript received May 2, 1984; revised September 26, 1984. This work was conducted under the McDonnell Douglas Independent Research and Development program.

The authors are with the McDonnell Douglas Research Laboratories, St. Louis, MO 63166.

Reprinted from *IEEE Trans. Antennas Propagat.*, vol. AP-33, no. 2, pp. 206–214, February 1985.

tations suggest that the resulting formulation is apparently well-posed at all frequencies for a range of surface impedances. This formulation obviates the need to introduce a weighting coefficient. As discussed later, the IBC-CFIE of Jones [16] coalesces to the present case for a particular choice of coupling constant. If a MM procedure is used to solve the present formulation, the solution is expressible in terms of two Galerkin matrix operators for which an efficient computational algorithm exists [17]. Furthermore, the present approach leads to special matrix symmetries permitting even electrically relatively large bodies to be treated. The analysis is valid for all angles of illumination including axial incidence. An analogous development is outlined for the RBC and MBC cases.

II. FORMULATION

The integral equation formulations for scatterers with surfaces satisfying the IBC, RBC, and MBC representation are considered. These equations are solved by the MM technique in terms of Galerkin matrix operators introduced and discussed in detail in [17], specialized for BOR geometries.

Case 1: Surfaces Satisfying the IBC

The IBC implies that only the electric and magnetic fields external to the scatterer are relevant and their relationship is a function of the material constitution (i.e., surface impedance) or surface characteristics (i.e., roughness) of the scatterer. In most practical situations, this is an approximation of the true physics of the problem. However, it is a suitable approach that allows highly complex effects in surface wave phenomena to be predicted for imperfectly conducting bodies involving only an incremental analytical/computational penalty over that of treating the scattering from perfectly conducting bodies as illustrated subsequently.

Given that the IBC is operative for a given scatterer (Fig. 1(a)), the electric and magnetic fields external to the body $\vec{E}_1$ and $\vec{H}_1$ are related as [4]

$$\vec{E}_1 - (\hat{n}_1 \cdot \vec{E}_1)\hat{n}_1 = \eta_s\eta_0(\hat{n}_1 \times \vec{H}_1) \tag{1}$$

or

$$\vec{E}_1\,|_{\tan} = \eta_s\eta_0(\hat{n}_1 \times \vec{H}_1). \tag{1a}$$

The dual form of the IBC is

$$\vec{H}_1 - (\hat{n}_1 \cdot \vec{H}_1)\hat{n}_1 = \frac{-1}{\eta_s\eta_0}(\hat{n}_1 \times \vec{E}_1) \tag{2}$$

or

$$\vec{H}_1\,|_{\tan} = \frac{-1}{\eta_s\eta_0}(\hat{n}_1 \times \vec{E}_1) \tag{2a}$$

where $\eta_0 = \sqrt{\mu_0/\epsilon_0} = 377\,\Omega$; η_s is the relative surface impedance; and $\hat{n}_1$ is the unit surface normal on the boundary ∂R_1 of region R_1 and directed into R_1, where R_1 is external to the scatterer. In this discussion, η_s is assumed constant on ∂R_1. If η_s varies along the axial direction of a BOR, modification of the present development is straightforward.

We start with the symmetric form of Maxwell's equations, written in terms of electric and magnetic charges and currents for a homogeneous region with electric and magnetic sources [18]. In region R_1 (external to ∂R_1) at a field point $\vec{r}$, the total electric and magnetic fields $\vec{E}_1$ and $\vec{H}_1$ in terms of the incident fields $\vec{E}^i$ and $\vec{H}^i$, can be written as

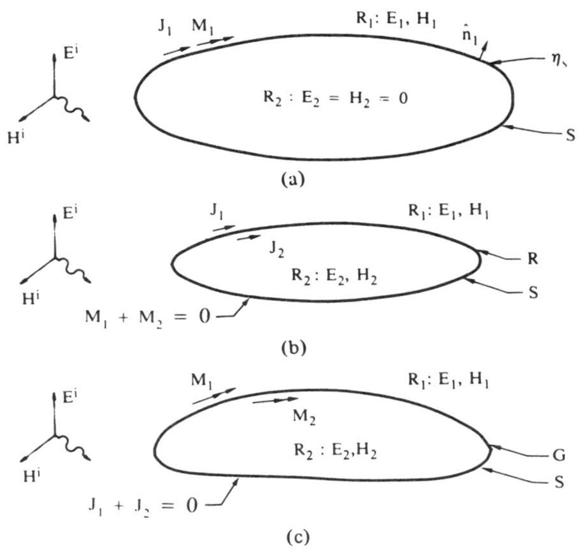

Fig. 1. Generic scatterers with (a) impedance, (b) resistive sheet, and (c) magnetically conducting sheet boundary conditions.

$$\theta(\vec{r})\vec{E}_1(\vec{r}) = \vec{E}^i(\vec{r}) - L_1\vec{J}_1(\vec{r}) + K_1\vec{M}_1(\vec{r}) \tag{3}$$

$$\theta(\vec{r})\vec{H}_1(\vec{r}) = \vec{H}^i(\vec{r}) - K_1\vec{J}_1(\vec{r}) - \frac{1}{\eta_0^2}\,L_1\vec{M}_1(\vec{r}). \tag{4}$$

Monochromatic time variation $e^{j\omega t}$ is assumed. The electric and magnetic surface currents (densities) on an arbitrary ∂R_i in general are

$$\vec{J}_i(\vec{r}) = \hat{n}_i \times \vec{H}_i(\vec{r})|_{\partial R_i}, \tag{5a}$$

$$\vec{M}_i(\vec{r}) = -\hat{n}_i \times \vec{E}_i(\vec{r})|_{\partial R_i}. \tag{5b}$$

(The general definitions for $\vec{J}_i$ and $\vec{M}_i$ introduced here for an arbitrary ∂R_i are needed later.) The integrodifferential operators L_i and K_i in the foregoing equations are defined as

$$L_i\vec{X}(\vec{r}) = j\omega\mu \int_{\partial R_i} \left(\vec{X}(\vec{r}\,') + \frac{1}{\omega^2\mu\epsilon}\,\nabla\nabla' \cdot \vec{X}(\vec{r}\,')\right) \cdot \Phi(\vec{r} - \vec{r}\,')\,ds' \tag{6a}$$

$$K_i\vec{X}(\vec{r}) = \int_{\partial R_i} \vec{X}(\vec{r}\,') \times \nabla\Phi(\vec{r} - \vec{r}\,')\,ds' \tag{6b}$$

where the vector function $\vec{X}$ is in the domain of L_i and K_i, spanning ∂R_i and defined for the kernel $\Phi = (4\pi|\vec{r} - \vec{r}\,'|)^{-1} \exp[-jk|\vec{r} - \vec{r}\,'|]$ which is the Green's function; ∇' is the surface gradient defined on the primed coordinates; ϵ, μ, and Φ are defined in region R_i with boundary ∂R_i; region R_1 is assumed lossless; and $\theta(\vec{r})$ is the Heaviside function defined as

$$\theta(\vec{r}) = \begin{cases} 1, & \text{for } \vec{r} \in R_i \\ 1/2, & \text{for } \vec{r} \in \partial R_i \\ 0, & \text{otherwise.} \end{cases} \tag{7}$$

When $\vec{r} = \vec{r}\,'$ (i.e., field and source points overlap), the integrals in (6a) and (6b) are interpreted in the Cauchy principal value sense. Equations (1) and (2) imply the following relationship between the electric and magnetic currents:

$$\vec{M}_1(\vec{r}) = -\eta_s\eta_0(\hat{n}_1 \times \vec{J}_1(\vec{r})). \tag{8}$$

Substituting (3) and (4) for $\vec{E}_1$ and $\vec{H}_1$ into (1) yields the following integral equation, denoted as IBC-IE:

$$\vec{E}^i(\vec{r})|_{\tan} - \eta_s\eta_0\hat{n}_1 \times \vec{H}^i(\vec{r}) = \{L_1\vec{J}_1(\vec{r}) - K_1\vec{M}_1(\vec{r})\}|_{\tan} - \eta_s\eta_0\hat{n}_1 \times \left\{K_1\vec{J}_1(\vec{r}) + \frac{1}{\eta_0^2} L_1\vec{M}_1(\vec{r})\right\}. \quad (9)$$

Substituting (8) for $\vec{M}_1$ into (9) yields

$$\vec{E}^i(\vec{r})|_{\tan} - \eta_s\eta_0\hat{n}_1 \times \vec{H}^i(\vec{r}) = L_1\vec{J}_1(\vec{r})|_{\tan} + \eta_s\eta_0K_1[\hat{n}_1' \times \vec{J}_1](\vec{r})|_{\tan} - \eta_s\eta_0\hat{n}_1 \times K_1\vec{J}_1(\vec{r}) + \eta_s^2\hat{n}_1 \times L_1[\hat{n}_1' \times \vec{J}_1](\vec{r}). \quad (10)$$

It can be shown that the IBC-CFIE of Jones [16, eq. (6.132)] reduces to (10) when the coupling constant β (in Jones' notation) is zero. As pointed out by Jones, the IBC-CFIE is a well-posed problem if $\beta - \eta_s \neq 0$ or pure imaginary. If one proceeds from the dual form of the IBC (2), the corresponding integral equation in terms of the magnetic currents is obtained:

$$\vec{H}^i(\vec{r})|_{\tan} + \frac{1}{\eta_s\eta_0}(\hat{n}_1 \times \vec{E}_1^i(\vec{r})) = \left\{\frac{1}{\eta_0^2} L_1\vec{M}_1(\vec{r})| + \frac{1}{\eta_s\eta_0} K_1[\hat{n}_1' \times \vec{M}_1](\vec{r})\right\}\Bigg|_{\tan} - \frac{1}{\eta_s\eta_0}\hat{n}_1 \times K_1\vec{M}_1(\vec{r}) + \left(\frac{1}{\eta_s\eta_0}\right)^2 \hat{n}_1 \times L_1[\hat{n}_1' \times \vec{M}_1](\vec{r}). \quad (11)$$

Galerkin (MM) Solution

The foregoing integral equations ((10) and (11)) are solved with the Galerkin (MM) technique. The unknown currents in (10) are expanded in a finite series of basis functions on the surface $\partial R_1(\equiv S)$. For a BOR, the surface S can be parameterized in circular cylindrical coordinates by $\rho(t)$, $z(t)$, and ϕ, where t is the distance measured along the BOR generating curve and ϕ is the azimuthal angle. The surface currents are expanded as

$$\vec{J}_1 = \sum_{n,k} (a_{nk}^t \vec{J}_{nk}^t - a_{nk}^\phi \vec{J}_{nk}^\phi) \quad (12)$$

where $a_{nk}^\alpha(\alpha = t \text{ or } \phi)$ are the unknown expansion coefficients. (Note the negative sign on a_{nk}^ϕ.) The indices n and k are associated with the expansion along the principal surface coordinates of the BOR, namely, ϕ and t, respectively. The expansion functions in (12) are

$$\vec{J}_{nk}^\alpha = \vec{u}_\alpha \frac{T_k(t)}{\rho(t)} e^{jn\phi} \quad (13)$$

where $T_k(t)$ is a triangle function spanning the kth annulus into which the surface S is subdivided. These overlapping triangle functions result in an approximation to $\rho\vec{J}_1$ which is piecewise linear in t and is a finite Fourier series in ϕ.

The current expansion in (12) is substituted into (10), and inner products are formed with a set of testing functions $\vec{W}_{nk}^\alpha$ over the BOR surface. In the Galerkin form of the MM solution, $\vec{W}_{nk}^\alpha = (\vec{J}_{nk}^\alpha)^*$. The original integral equation is thus transformed into a set of linear equations for each of the Fourier modes n:

$$\begin{bmatrix} Z^{tt} & Z^{t\phi} \\ Z^{\phi t} & Z^{\phi\phi} \end{bmatrix}_n \begin{bmatrix} A^t \\ A^\phi \end{bmatrix}_n = \begin{bmatrix} \bar{E}^t \\ \bar{E}^\phi \end{bmatrix}_n - \eta_s \begin{bmatrix} \bar{H}^\phi \\ -\bar{H}^t \end{bmatrix}_n \quad (14)$$

where the system matrix Z is given in terms of the $\mathbf{L} = L(S, S; R_1; n)$ and $\mathbf{K} = K(S, S; R_1; n)$ matrices derived from the conducting BOR case (see [19, Appendix]), i.e.,

$$Z^{tt} = \mathbf{L}^{tt} + \eta_s(\mathbf{K}^{t\phi} - \mathbf{K}^{\phi t}) + \eta_s^2\mathbf{L}^{\phi\phi} \quad (15a)$$

$$Z^{\phi t} = \mathbf{L}^{\phi t} + \eta_s(\mathbf{K}^{\phi\phi} + \mathbf{K}^{tt}) - \eta_s^2\mathbf{L}^{t\phi} \quad (15b)$$

$$Z^{t\phi} = \mathbf{L}^{t\phi} - \eta_s(\mathbf{K}^{tt} + \mathbf{K}^{\phi\phi}) - \eta_s^2\mathbf{L}^{\phi t} \quad (15c)$$

$$Z^{\phi\phi} = \mathbf{L}^{\phi\phi} - \eta_s(\mathbf{K}^{\phi t} - \mathbf{K}^{t\phi}) + \eta_s^2\mathbf{L}^{tt}. \quad (15d)$$

It can be shown that the Z matrix in (14) is symmetric since $\mathbf{L}$ is symmetric and $\mathbf{K}$ is skew-symmetric. Hence, in the computations only the upper triangular part needs to be stored. The kth elements of $\bar{E}^\alpha$ and $\bar{H}^\alpha$ in (14) are given by

$$\bar{E}_k^\alpha = e^{-jn\phi_i}R_k^{\alpha\beta}(\theta_i), \qquad \alpha = t \text{ or } \phi \quad (16a)$$

where $\beta = \theta$ for a θ-polarized incident electric field, and $\beta = \phi$ for a ϕ-polarized incident field. Similarly,

$$\bar{H}_k^\alpha = -e^{-jn\phi_i}R_k^{\alpha\phi}(\theta_i), \qquad \alpha = t \text{ or } \phi \quad (16b)$$

for a θ-polarized incident electric field, and

$$\bar{H}_k^\alpha = e^{-jn\phi_i}R_k^{\alpha\theta}(\theta_i), \qquad \alpha = t \text{ or } \phi \quad (16c)$$

for a ϕ-polarized incident electric field.

The angles ϕ_i and θ_i specify the direction of incidence. The elements $R_k^{\alpha\beta}$ are

$$R_k^{t\theta}(\theta_i) = \pi j^{n+1} \sum_{p=1}^{4} \Delta t_p T_p e^{jkz_p\cos\theta_i}\{\cos\theta_i \cdot \sin\nu_p (J_{n+1} - J_{n-1}) + 2j\sin\theta_i \cos\nu_p J_n\} \quad (17a)$$

$$R_k^{\phi\theta}(\theta_i) = \pi j^n \cos\theta_i \sum_{p=1}^{4} \Delta t_p T_p e^{jkz_p\cos\theta_i}(J_{n+1} + J_{n-1}) \quad (17b)$$

$$R_k^{t\phi}(\theta_i) = -\pi j^n \sum_{p=1}^{4} \Delta t_p T_p \sin\nu_p e^{jkz_p\cos\theta_i}(J_{n+1} + J_{n-1}) \quad (17c)$$

and

$$R_k^{\phi\phi}(\theta_i) = \pi j^{n+1} \sum_{p=1}^{4} \Delta t_p T_p e^{jkz_p\cos\theta_i}(J_{n+1} - J_{n-1}) \quad (17d)$$

where $J_n = J_n(k\rho_p \sin\theta_i)$ are Bessel functions of order n. For a given incident electric field, the column vector for the unknown current coefficients for mode n in (14) are denoted by

$$\begin{bmatrix} A^t \\ A^\phi \end{bmatrix}_n^\beta$$

where $\beta = \theta$ or ϕ for a θ-polarized or ϕ-polarized incident electric field, respectively.

Having specified the system matrix and solving for the current coefficients in (14), the scattering cross section can be computed. Specifically, the cross section for θ-polarized, scattered fields and β-polarized incident fields, is given by

$$\sigma^{\theta\beta} = \frac{(\omega\mu)^2}{4\pi}\left| \sum_{n=0,\pm 1,\cdots} e^{jn\phi_s}[R^{t\theta} + \eta_s R^{\phi\phi}, -R^{\phi\theta} + \eta_s R^{t\phi}]_{-n} \cdot \begin{bmatrix} A^t \\ A^\phi \end{bmatrix}_n^\beta \right|^2, \qquad \beta = \theta \text{ or } \phi. \tag{18}$$

Similarly, for ϕ-polarization, the result is

$$\sigma^{\phi\beta} = \frac{(\omega\mu)^2}{4\pi}\left| \sum_{n=0,\pm 1,\cdots} e^{jn\phi_s}[R^{t\phi} - \eta_s R^{\phi\theta}, -R^{\phi\phi} - \eta_s R^{t\theta}]_{-n} \cdot \begin{bmatrix} A^t \\ A^\phi \end{bmatrix}_n^\beta \right|^2, \qquad \beta = \theta \text{ or } \phi. \tag{19}$$

When $\eta_s = 1$ for axial illumination ($\theta_i = 0^\circ$ or 180°), the backscatter cross section (for $\theta_i = \theta_s$) computed from (18) or (19) can be shown to be zero. Proceeding from (18), it can be shown that for $n = \pm 1$,

$$\sigma^{\theta\theta}(\theta_i = \theta_s = 0 \text{ or } 180^\circ) \alpha \left| [\tilde{R}^{t\theta}, \pm j\tilde{R}^{t\theta}]_n \cdot \begin{bmatrix} Y^{tt} & Y^{t\phi} \\ -Y^{t\phi} & Y^{tt} \end{bmatrix}_n \begin{bmatrix} \tilde{R}^{t\theta} \\ \pm j\tilde{R}^{t\theta} \end{bmatrix}_n \right|^2 \tag{18a}$$

where the Y matrix is the inverse of (15),

$$\begin{bmatrix} \tilde{R}^{t\theta} \\ \tilde{R}^{\phi\theta} \end{bmatrix}_n \equiv \begin{bmatrix} R^{t\theta} + R^{\phi\phi} \\ R^{\phi\theta} - R^{t\phi} \end{bmatrix}_n \tag{18b}$$

and

$$\begin{bmatrix} -\tilde{R}^{\phi\theta} \\ \tilde{R}^{t\theta} \end{bmatrix}_n \equiv \begin{bmatrix} R^{t\phi} - R^{\phi\theta} \\ R^{\phi\phi} + R^{t\theta} \end{bmatrix}_n. \tag{18c}$$

Since $Y_n^{tt} = {}_tY_n^{tt}$ and $Y_n^{t\phi} = -{}_tY_n^{t\phi}$ (t denotes the transpose) and for $\theta_i = 0^\circ$ or 180° and $n = \pm 1$, $R_n^{t\theta} = \pm jR_n^{t\phi}$ and $R_n^{\phi\theta} = \pm jR_n^{\phi\phi}$, using these relationships, (18a) becomes

$$\sigma^{\theta\theta} \alpha |\{Y_n^{tt} \pm jY_n^{t\phi} \mp jY_n^{t\phi} - Y_n^{tt}\}[\tilde{R}_n^{t\theta}]^2|^2 = 0.$$

Similar results can be obtained from (19). The above result is frequency independent and is an alternate proof of cases treated by two general theorems in electromagnetics due to Weston [20] and Wagner and Lynch [21]. (The corresponding general result for an axisymmetric dielectric scatterer with $\mu_r = \epsilon_r$ can be proven likewise from an MM formulation.)

Case 2: Surfaces with Resistive Sheet Boundary Condition

Consider a body formed by an infinitesimally thin resistive sheet S having a resistance of $R\eta_0$ (ohms) (Fig. 1(b)). At the surface S, the tangential electric field is continuous, i.e.,

$$(\hat{n}_1 \times \vec{E}_1 + \hat{n}_2 \times \vec{E}_2)|_S = 0 \tag{20}$$

and a discontinuity occurs in the tangential magnetic field components, such that the total electric current on S is [22], [23]

$$\vec{J} = -\frac{1}{R\eta_0}\, \hat{n}_1 \times (\hat{n}_1 \times \vec{E}_1) \tag{21}$$

where $\vec{E}_1$ is the total electric field external to S defined by (3), subject to the condition that the total magnetic current $\vec{M} = 0$ on S. Summing the fields inside and outside S, assuming the permeability and permittivity of R_1 and R_2 are the same, yields

$$\vec{E}^i(\vec{r})|_{\tan} = L_1\vec{J}(\vec{r})|_{\tan} + R\eta_0\vec{J}(\vec{r})|_{\tan}. \tag{22}$$

If the composition of the interior region R_2 differs from that of R_1, the foregoing result is easily generalized. If the interior of the body is inhomogeneous, i.e., layered, the interior fields of the various regions can be related in a direct manner. (See [24] for the coated BOR.)

The integral equation above can be solved using the Galerkin technique with the expansion of the surface current, as in (12). A matrix equation results having the form:

$$\begin{bmatrix} L^{tt} + R\tilde{L} & L^{t\phi} \\ L^{\phi t} & L^{\phi\phi} - R\tilde{L} \end{bmatrix}_n \begin{bmatrix} A^t \\ A^\phi \end{bmatrix}_n = \begin{bmatrix} \bar{E}^t \\ \bar{E}^\phi \end{bmatrix}_n \tag{23}$$

where L is defined as before and the (k, l)th elements of the tridiagonal matrix $\tilde{L}$ are given by

$$\tilde{L}_{kl} = 2\pi\eta_0 \sum_{p,q=1}^{4} \Delta t_p \frac{T_p T_q}{\rho_q} \delta_{\vec{r}_p, \vec{r}_p}.$$

When $R \to 0$, (23) reduces to the well-known MM system matrix of the EFIE formulation. The scattering cross section is computed analogously to case 1.

Case 3: Surfaces with the Magnetically Conducting Sheet Boundary Condition

The boundary condition here (Fig. 1(c)) is the dual of that for case 2 [22], [23]. A magnetically conducting sheet with conductivity G/η_0 (mhos) implies continuity of the tangential magnetic fields, i.e.,

$$(\hat{n}_1 \times \vec{H}_1 + \hat{n}_2 \times \vec{H}_2)|_S = 0 \tag{24}$$

and a discontinuity of the tangential electric fields across S such that

$$\vec{M} = \frac{-\eta_0}{G}\, n_1 \times (n_1 \times \vec{H}_1). \tag{25}$$

Substituting (4) with the total electric current $\vec{J} = 0$ and adding the fields exterior and interior to S, assuming the constitutive properties of R_1 and R_2 are the same, yields:

$$\vec{H}^i(\vec{r})|_{\tan} = \frac{1}{\eta_0^2}\, L_1\vec{M}(\vec{r})|_{\tan} + \frac{G}{\eta_0}\, \vec{M}(\vec{r})|_{\tan}. \tag{26}$$

We solve (26) with the Galerkin technique, expanding the magnetic currents $\vec{M}_1(\vec{r})$ as

$$\vec{M}_1 = \eta_0 \sum_{n,k} (b_{nk}^t \vec{M}_{nk}^t - b_{nk}^\phi \vec{M}_{nk}^\phi), \tag{27}$$

where $\vec{M}_{nk}^\alpha = \vec{J}_{nk}^\alpha$, and $\vec{W}_{nk}^\alpha = \eta_0(\vec{J}_{nk}^\alpha)^*$. A matrix equation identical to (23) results by letting $R \to G$, $a_{nk}^\alpha \to b_{nk}^\alpha$ and $\bar{E}^\alpha \to \bar{H}^\alpha$. The scattering cross section is computed similarly to case 1.

III. RESULTS

The integral equations in Section II were solved with the MM technique as outlined above. Representative examples of calculations for impedance-coated BOR's follow. The bistatic cross sections for lossy dielectric spheres, satisfying the IBC for various

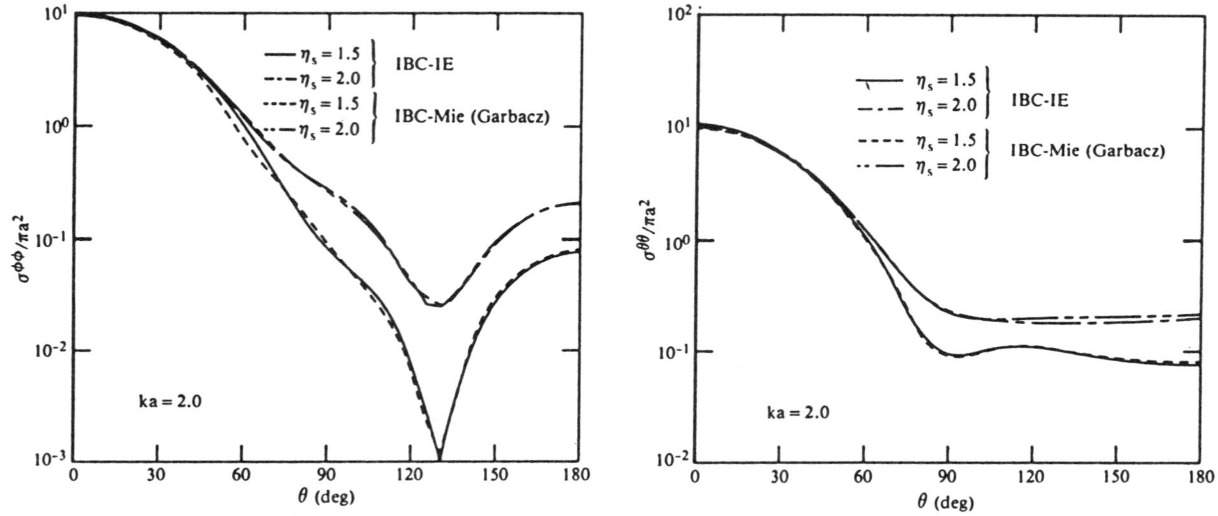

(a) (b)

Fig. 2. Bistatic scattering cross sections of impedance-coated sphere ($ka = 2$) as a function of η_s. (a) $\phi\phi$-polarization. (b) $\theta\theta$-polarization.

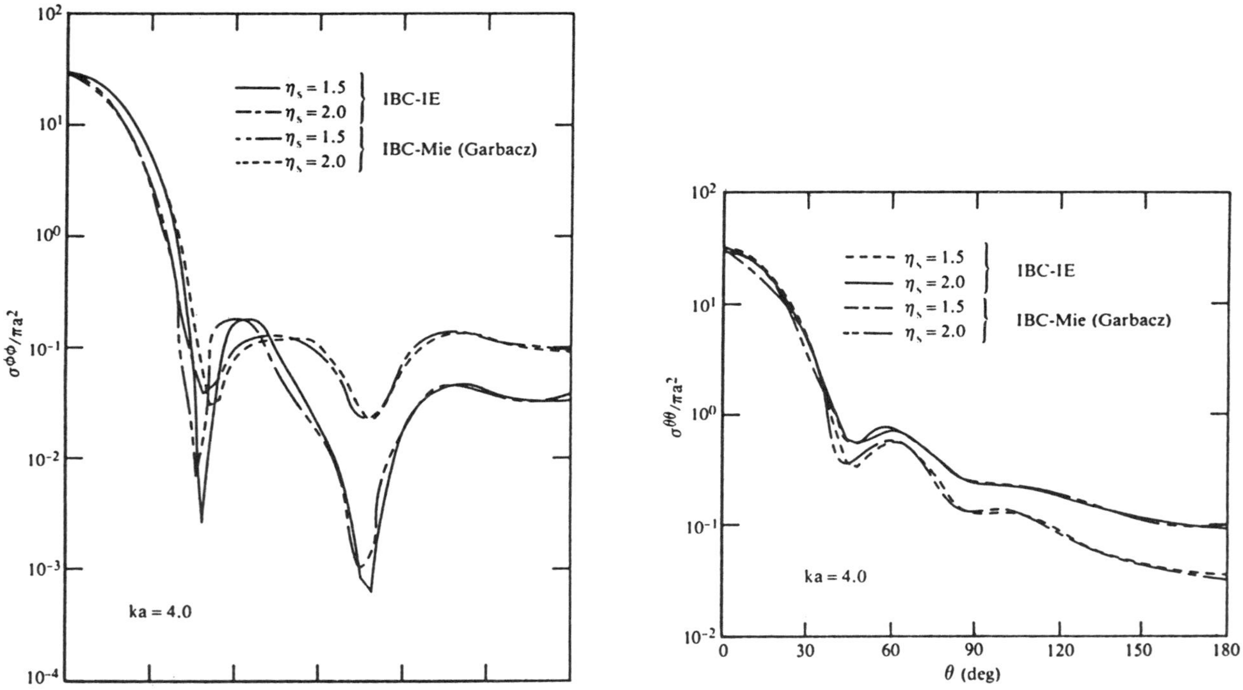

(a) (b)

Fig. 3. Bistatic scattering cross sections of impedance-coated sphere ($ka = 4$) as a function of η_s. (a) $\phi\phi$-polarization. (b) $\theta\theta$-polarization.

values of the surface impedance η_s, are depicted in Figs. 2 and 3. Comparison of the IBC-IE formulation's solutions is made with those of the Mie-based formulation (IBC-Mie) in [9]. In the IBC-IE calculations, $n = \pm 1$ (because of axial incidence), with nine and 19 triangle functions used to span the spheres in Figs. 2 and 3, respectively. The corresponding results for spheres with $ka = 10$ and 20 with complex surface impedance, $\eta_s = 0.1 + j\,0.1$, are summarized in Figs. 4 and 5, respectively. For the IBC-IE calculations 40 and 80 triangle functions were used, respectively. The IBC-Mie results are due to Rogers [25].

The present formulation was examined for the limiting case when $\eta_s = 1$. For this case, the submatrices in (15) simplify such that $Z_n^{tt} = Z_n^{\phi\phi}$ and $Z_n^{\phi t} = -Z_n^{t\phi}$. In computing the individual elements of these submatrices, we observe that for elements sufficiently removed from the diagonal, $Z_n^{tt} \cong \pm jZ_n^{t\phi}$. Our computations show that for $\eta_s = 1$, the Z matrix becomes ill-conditioned. The reasons underlying this pathology are unclear. For $\eta_s \approx 1$, the system matrices arising from the IBC-IE formulation are well conditioned. As an example, consider a sphere with $ka = 4.5$ and $\eta_s = 1.005$. The bistatic scattering cross sections computed from the IBC-IE and IBC-Mie formulations depicted in Fig. 6 are practically indistinguishable. Similar comparisons hold for nonspheri-

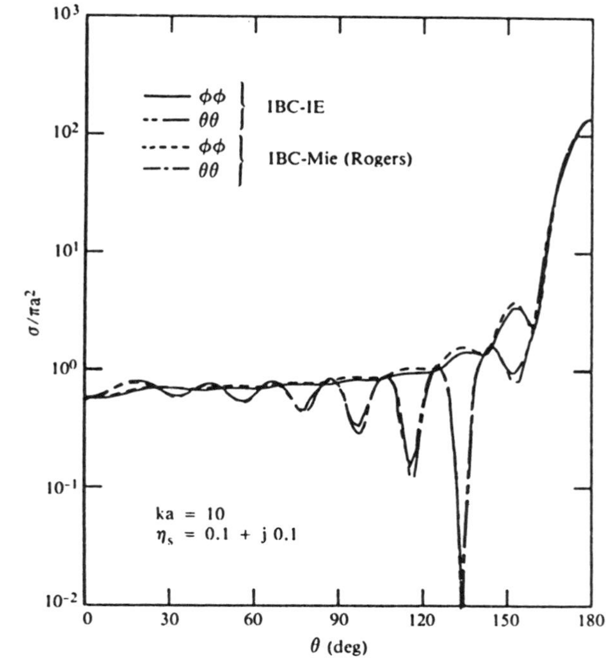

Fig. 4. Bistatic scattering cross sections of impedance-coated sphere (ka = 10) and η_s = 0.1 + 0.1j.

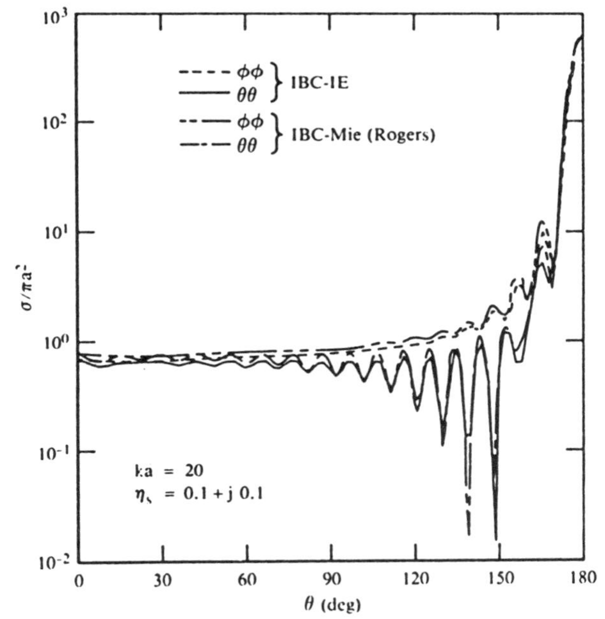

Fig. 5. Bistatic scattering cross sections of impedance-coated sphere: ka = 20 and η_s = 0.1 + 0.1j.

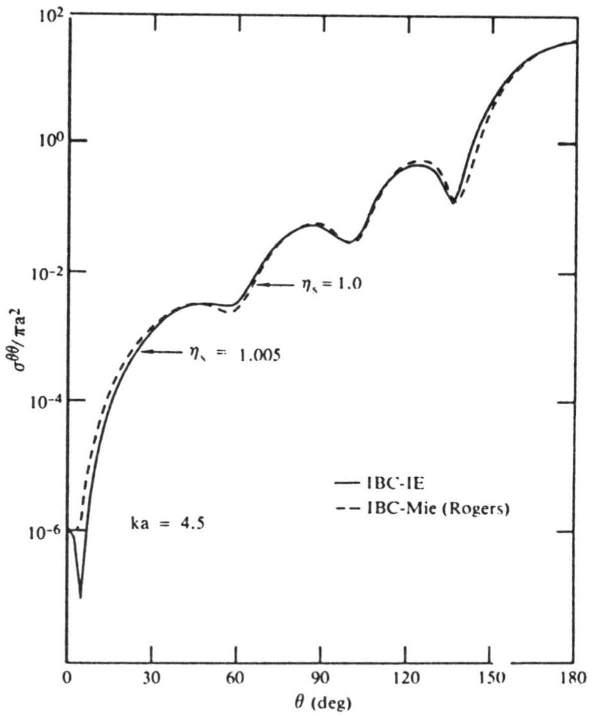

Fig. 6. Bistatic scattering cross section of impedance-coated spheres: ka = 4.5, η_s = 1.0 and 1.005.

cal BOR scatterers. Since $\eta_s = 1$ is physically unrealizable, the ill-conditioning of the present formulation for this case is not severely debilitating.

The orthogonal components of the currents on a BOR have an interesting property at $\eta_s \cong 1$. This is depicted in Fig. 7. For axial illumination of a sphere (or any BOR), the amplitude of the t- and ϕ-directed currents are equal at each point on the surface, and the phases are 180° apart. (The discrepancy in amplitude and phase for triangle function 1 in Fig. 7 is due to the approximate numerical representation of the sphere near the illuminated pole.) This same property can be deduced analytically from (1) and (2) and its associated relationships in [6]. The consequence of this property of the currents on an axisymmetric scatterer leads to the well-known null axial cross section stated by Weston's theorem.

The backscatter (monostatic) cross section predicted by the present formulation for an impedance-coated conesphere with a cone half-angle of 15° and a spherical cap with $ka = 8$ is depicted

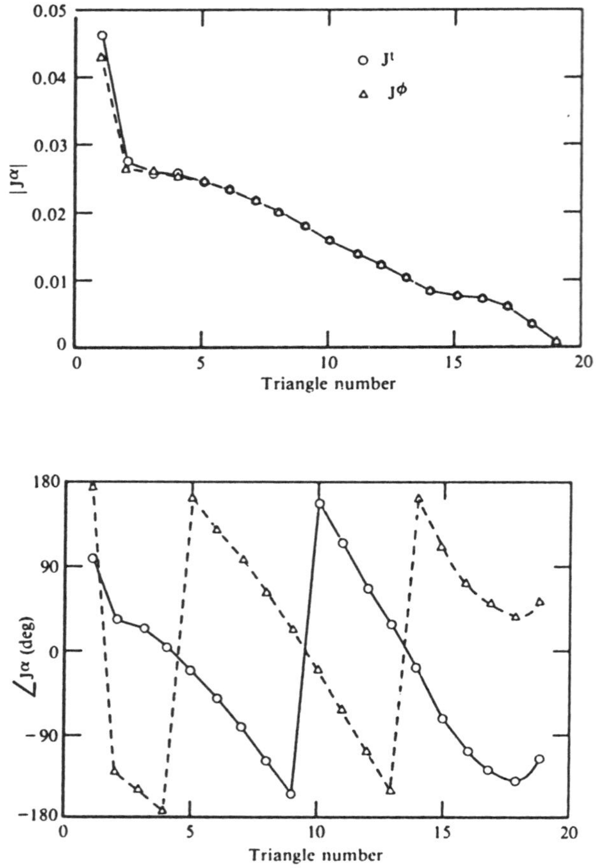

(a)

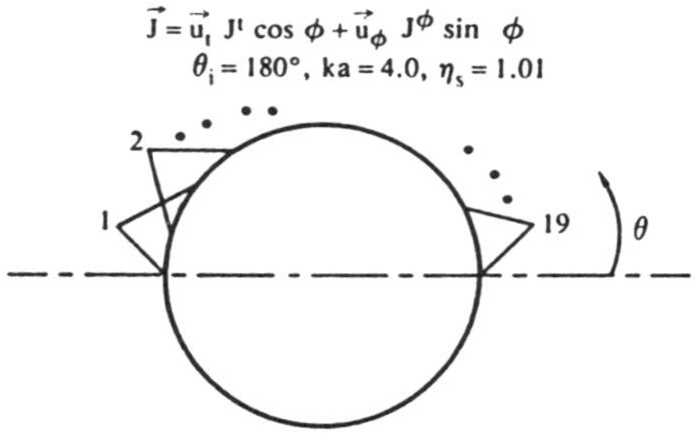

(b)

Fig. 7. Normalized currents on sphere (ka = 4.0, η_s = 1.01) as a function of annular segment position (triangle numbers). (a) Amplitude. (b) Phase. (Triangle number 1 is located near the illuminated pole of the sphere.)

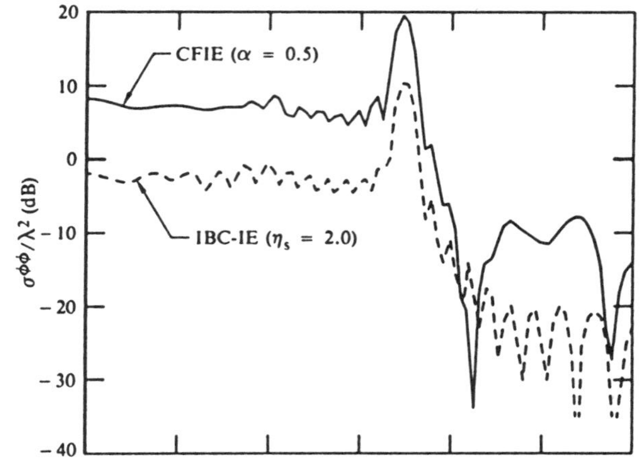

(a)

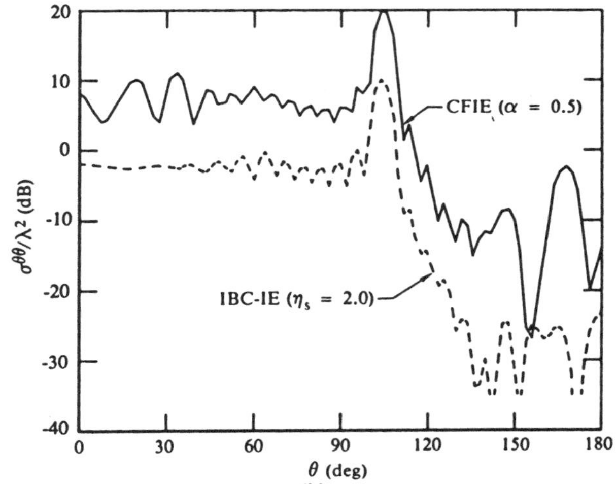

(b)

Fig. 8. Backscatter cross sections for conducting and impedance-coated (η_s = 2.0) conespheres. (a) $\phi\phi$-polarization. (b) $\theta\theta$-polarization.

in Fig. 8. The surface impedance is $\eta_s = 2.0$. For comparison, the corresponding results for a perfectly conducting conesphere using the CFIE (with α = 1.0) is also given. For both sets of calculations the highest mode number was nine, and 34 triangle functions were used to span the body. The condition numbers of the system matrices from the IBC-IE and CFIE calculations were comparable for each of the modes. As seen, the impedance coating dramatically alters the levels and features of the cross section.

Finally, we considered the case of impedance-coated bodies near internal resonances. As noted in [11], [12] the IBC-EFIE and IBC-MFIE formulations lead to spurious results at frequencies where the associated conducting scatterer exhibits cavity resonances. The ill-conditioning of these formulations is discussed at length in the cited works. The IBC-CFIE formulation with a proper weighting coefficient (α) circumvents this problem. However, as noted before, in the IBC-CFIE approach the proper choice of α must be made. In Fig. 9, the IBC-CFIE (with the optimum α) and the IBC-IE results are compared. In Fig. 9(a), the cross sections of impedance-coated spheres ($\eta_s = 0.1 + j\ 0.1$) as a function of ka are depicted, computed using the various integral equation formulations as well as the IBC-Mie solution. At the resonant frequency corresponding to ka = 2.75, the IBC-CFIE (with $\alpha = 0.375$), the IBC-IE, and the Mie (exact) solutions are in excellent agreement. The IBC-EFIE and IBC-MFIE solutions are invalid. In Fig. 9(b), the matrix condition numbers for each of

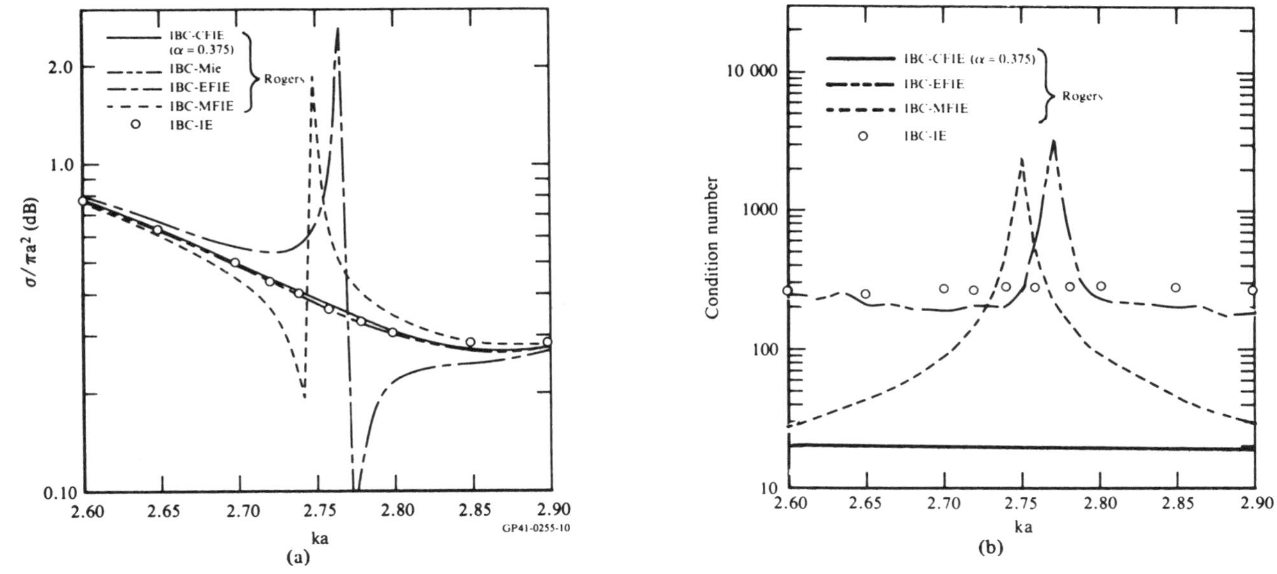

Fig. 9. Comparison of formulations for impedance-coated (η_s = 0.1 + 0.1j) spherical scatterers as a function of ka. (a) Scattering cross section. (b) Matrix condition number for mode n = 1.

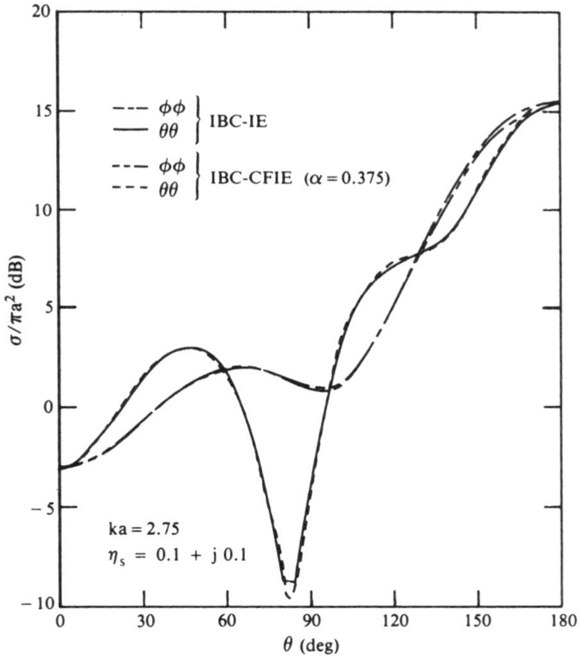

Fig. 10. Bistatic scattering cross sections of impedance-coated η_s = 0.1 + 0.1j) sphere: ka = 2.75 (resonant case).

the foregoing methods are given. While the condition number for the present formulation is higher than that for the IBC-CFIE when $\alpha = 0.375$ is chosen it is nevertheless bounded throughout the range. The bistatic cross section of an impedance-coated resonant sphere (ka = 2.75 and $\eta_s = 0.1 + j\ 0.1$) compared with the exact solution is shown in Fig. 10. The two sets of results overlay each other. Subsequent computations for $\eta_s = \pm j\ 0.1$ have also led to stable solutions. For $\eta_s \leqslant j\ 0.01$, the condition number of the system matrix becomes large, and the results are unstable as is expected since the IBC-IE approaches effectively the EFIT formulation in these cases.

IV. CONCLUSION

Integral equation formulations are developed for scatterers satisfying the impedance, resistive, and magnetically conducting sheet boundary conditions. The MM procedure is applied, yielding a solution in terms of two Galerkin matrix operators for BOR's. Numerical results presented for impedance-coated bodies are shown to be in excellent agreement with CFIE and the Mie solutions modified for the impedance boundary conditions. The result of the Weston theorem for the axial backscatter cross section of BOR's with η_s = 1 is satisfied. The present formulation applied to impedance-coated, internally resonant scatterers appears to be stable for the range of surface impedances described. The results apply equally for all angles of illumination and arbitrary polarization.

REFERENCES

[1] M. A. Leontovich, *Investigation of Propagation of Radiowaves,* pt. II. Moscow, 1948.

[2] ——, *Appendix of Diffraction, Refraction, and Reflection of Radio Waves,* (papers by V. A. Fock), N. Logan and P. Blacksmith, Eds. Washington, DC: U.S. Gov. Printing Office, 1957, DDC AD-117276.

[3] V. A. Fock, *J. Phys. USSR,* vol. 10, p. 13, 1946.

[4] T. B. A. Senior, "Impedance boundary conditions for imperfectly conducting surfaces," *Appl. Sci. Res.,* sec. B, vol. 8, p. 418, 1960. (See also "A note on impedance boundary condition," *Can. J. Phys.,* vol. 40, p. 663, 1962.)

[5] T. B. A. Senior, "Impedance boundary conditions for statistically rough surfaces," *Appl. Sci. Res.,* sec. B, vol. 8, p. 437, 1960.

[6] J. R. Wait and C. M. Jackson, "Calculations of the bistatic scattering cross section of a sphere with an impedance boundary condition," *Radio Sci. J. Res.* NBS/USNC-URSI, vol. 69D, p. 299, 1965.

[7] K. M. Mitzner, "An integral equation approach to scattering from a body of finite conductivity," *Radio Sci.,* vol. 2, p. 1459, 1967.

[8] F. K. Oshiro *et al.,* "Calculation of Radar Cross Section," pt. I, vol. I, Tech. Rep. AFAL-TR-67-308 Wright-Patterson AFB, OH, Dec. 1967. (Also Part II. Analytical Rep., Tech. Rep. AFAL-TR-70-21 Wright-Patterson AFB, OH, Apr. 1970, DDC AD-867969.)

[9] R. J. Garbacz, "Bistatic scattering from a class of lossy dielectric spheres with surface impedance boundary conditions," *Phys. Rev.,* vol. 133, p. A14, 1964.

[10] T. S. Angell and R. E. Kleinman, "Scattering of acoustic waves by impedance surfaces," in *Wave Phenomena 83,* B. Moodie and C. Rogers, Eds. Amsterdam: North Holland, (in press).

[11] G. E. Heath, "Impedance boundary condition integral equations," in *1984 Int. Symp. Antennas Propagat. Symp. Digest,* vol. II, p. 697.

[12] J. R. Rogers, "Numerical solutions to ill-posed and well-posed impedance boundary condition integral equations," Lincoln Lab. (MIT), Tech. Rep. 641, Nov. 1983.

[13] J. R. Mautz and R. F. Harrington, "*H*-field, *E*-field, and combined-field solutions for conducting bodies of revolution," *Arch. Elektron. Übertragungstech.* (Electron. Commun.) vol. 32, no. 4, pp. 159-164, 1978.

[14] R. Kress and W. T. Spassov, "On the condition number of boundary integral operators for the exterior Dirichlet problem for the Helmholtz equation," *Numer. Math.,* vol. 42, p. 77, 1983.

[15] R. D. Graglia and P. L. E. Uslenghi, "Electromagnetic scattering by impedance bodies of revolution," Nat. Radio Sci. Meeting, session URSI/B-6-3, paper 4, Univ. Houston, Houston, TX, 23-26 May 1983.

[16] D. S. Jones, *Methods in Electromagnetic Wave Propagation.* Oxford: Clarendon, 1979.

[17] L. N. Medgyesi-Mitschang and J. M. Putnam, "Electromagnetic scattering from axially inhomogeneous bodies of revolution," *IEEE Trans. Antennas Propagat.,* vol. AP-32, p. 797, 1984.

[18] R. F. Harrington, *Time Harmonic Electromagnetic Fields.* New York: McGraw-Hill, 1961, ch. 3.

[19] L. N. Medgyesi-Mitschang and D.-S. Y. Wang, "Hybrid solutions for scattering from large bodies of revolution with material discontinuities and coatings," *IEEE Trans. Antennas Propagat.,* vol. AP-32, p. 717, 1984.

[20] V. H. Weston, "Theory of absorbers in scattering," *IEEE Trans. Antennas Propagat.,* vol. AP-11, p. 578, 1963.

[21] R. J. Wagner and P. J. Lynch, "Theorem on electromagnetic backscatter," *Phys. Rev.,* vol. 131, p. 21, 1963.

[22] T. B. A. Senior, "Scattering by resistive strips," *Radio Sci.,* vol. 14, p. 911, 1979.

[23] ——, "Some extensions of Babinet's principle in electromagnetic theory," *IEEE Trans. Antennas Propagat.,* vol. AP-25, p. 417, 1977.

[24] L. N. Medgyesi-Mitschang and C. Eftimiu, "Scattering from axisymmetric obstacles embedded in axisymmetric dielectrics: The method of moment solution," *Appl. Phys.,* vol. 19, p. 275, 1979.

[25] J. R. Rogers, private communication, Feb. 1984.

Electromagnetic Scattering by Arbitrary Shaped Three-Dimensional Homogeneous Lossy Dielectric Objects

KORADA UMASHANKAR, SENIOR MEMBER, IEEE, ALLEN TAFLOVE, SENIOR MEMBER, IEEE, AND SADASIVA M. RAO

***Abstract*—The recent development and extension of the method of moments technique for analyzing electromagnetic scattering by arbitrary shaped three-dimensional homogeneous lossy dielectric objects is presented based on the combined field integral equations. The surfaces of the homogeneous three-dimensional arbitrary geometrical shapes are modeled using surface triangular patches, similar to the case of arbitrary shaped conducting objects. Further, the development and extensions required to treat efficiently three-dimensional lossy dielectric objects are reported. Numerical results and their comparisons are also presented for two canonical dielectric scatterers—a sphere and a finite circular cylinder.**

I. Introduction

THIS PAPER DEALS with the use of analytical and numerical methods to analyze electromagnetic scattering and corresponding radar cross section of three-dimensional arbitrary shaped homogeneous lossy dielectric objects. There is, in fact, a variety of frequency domain analytical and/or numerical methods applied for studying electromagnetic scattering by homogeneous dielectric objects. For those objects whose boundary surface just coincides with a given coordinate system, separation of variables [1] can be applied. For these cases, analytically exact solutions have been obtained only for simple scatterers, such as a sphere [2] and circular and elliptical cylinders [3]–[5]. For an object not much different from the sphere, namely a prolate spheroid, the perturbation technique [6] has been applied.

Also for objects which are arbitrary in shape, either the volume or the surface integral equation has been applied. The volume integral equation [7], [8] is principally based on relating the induced polarization currents to the corresponding total fields consisting of the scattering and incident fields. By associating an unknown polarization current coefficient either with a cubic cell or with a tetrahedral cell inside the scatterer [8], the operator form of the integral equation is converted into an equivalent matrix equation. To a limited extent, the recent advancements [9], [10] in the area of the fast Fourier transform and also iterative methods have extended the usefulness of the volume integral equation approaches. Employing the volume equivalence principle and method of moments, scattering results are reported for dielectric [3], [7] cylinders and also for plane slabs and biological tissue cylinders [11]. Some work is also reported in the area of the extended boundary condition approach [12] and also in the area of the unimoment method [13], [14] related to scattering by inhomogeneous objects. These approaches express fields in terms of integrals over surfaces separating one or more homogeneous regions surrounding a given scatterer [15], [16]. For detailed discussion on the various techniques and their limitations, the reader may refer to the recent technical report [17] by the authors of the present paper.

The surface integral equation approach is very well suited to analyzing homogeneous dielectric objects or to objects modeled by or made up of homogeneous layers [13], [18]. The usual procedure in this method is to set up coupled integral equations in terms of equivalent electric and magnetic currents on the surfaces of the homogeneous regions. For an object made up of a large number of layers, fields induced in any region are expressed in terms of the equivalent currents on the adjacent interfaces. An iterative procedure [18] has been utilized for solving currents on the outermost surface in terms of the currents on inner interfaces. Particularly, for the case of simple objects such as dielectric cylinders [5] and bodies of revolution [19], the surface coupled integral equations method has been extensively applied. But, when the surface of the scatterer takes on arbitrary shape, an efficient modeling of the surface geometry and also the surface electric and magnetic fields become complicated. A simple and an efficient modeling scheme is presented here and is the subject of discussion in this paper in the context of scattering by arbitrarily shaped objects [17], [20].

II. Triangular Surface Patch Modeling

General arbitrarily shaped scattering objects can be analyzed based on integral equations and the method of moments (MM) approach [17]. This method is suited for low-frequency scattering problems, but can be extended to bodies spanning approximately one to two wavelengths in three dimensions, Fig. 1. The following formulations of the MM have been found to be generally suited for certain scattering problems

Manuscript received November 9, 1984. This work was sponsored by the Electromagnetic Sciences Division, the Rome Air Development Center, Hanscom Air Force Base, MA, under Contract F19628-82-C-0140 to IIT Research Institute, Chicago, IL.

K. Umashankar is with the Communications Laboratory, Department of Electrical Engineering and Computer Science, University of Illinois at Chicago, IL 60680.

A. Taflove is with the Department of Electrical Engineering and Computer Science, Northwestern University, Evanston, IL 60201.

S. M. Rao is with the Department of Electrical Engineering, Rochester Institute of Technology, Rochester, NY 14623.

IEEE Log Number 8608070.

Reprinted from *IEEE Trans. Antennas Propagat.*, vol. AP-34, no. 6, pp. 758–766, June 1986.

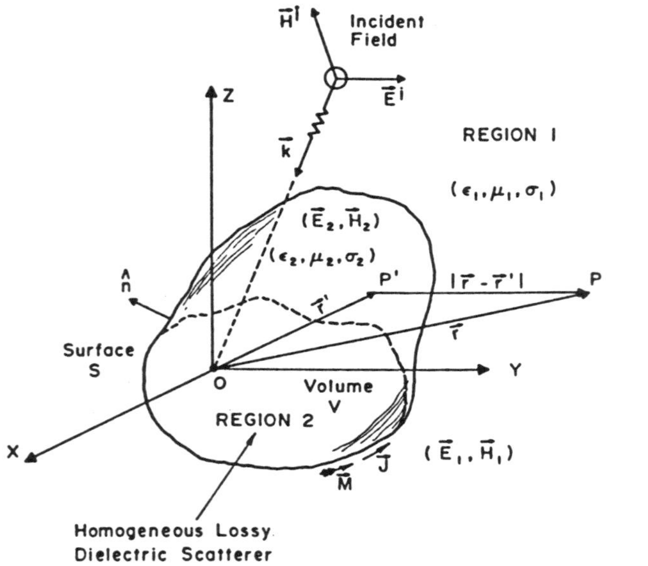

Fig. 1. Geometry of a homogeneous lossy dielectric scatterer in an isotropic free space medium.

based upon their geometry and material characteristics:

1) *Conducting Scatterers* (homogeneous, isotropic)
   a) Electric field integration formulation (EFIE) for closed and open bodies
   b) Magnetic field integral equation formulation (MFIE) for closed bodies
2) *Dielectric Scatterers* (homogeneous, isotropic)
   Combined field integral equation formulation (CFIE)
3) *Anisotropic Scatterers* (homogeneous)
   Combined field integral equation formulation (CFIE) modified for material characteristics.

Recent work has concentrated on case 1, especially in the development of the triangular patch model for arbitrary scatterers [20]. Recent further work is reported [17], using the combined field formulation for cases 2 and 3; only case 2 is discussed in this paper. There are several approaches by which one can efficiently model a given arbitrary shaped surface. For planar surface, one can conveniently use either rectangular or square patches [19]. Difficulties arise in case of nonplanar surfaces and even in case of planar surfaces with irregular boundary edges. An elaborate discussion on the triangular surface patch techniques, orientation of patches and their applicabilities to various types of integrodifferential equations is well discussed in [21]. Since the analysis of both EFIE and MFIE are already studied for conducting bodies [21], necessary analytical and numerical developments can be conveniently developed and extended [17] to surface patch model any arbitrary shaped dielectric objects, for any given excitation. Using MM to solve these equations, the equivalent electric and magnetic surface currents are expanded in terms of triangular surface patch currents; and tested on both sides with respect to the same surface patch basis functions to yield an efficient matrix expression. Once the surface equivalent currents are known, the near scattered fields, scattered far fields, and even penetrated fields are directly evaluated. To demonstrate validity, numerical results and their comparisons are presented for two canonical dielectric scatterers, namely a homogeneous dielectric sphere and a finite circular cylinder.

III. Summary of Combined Field Integral Equations

The detail derivation of the combined field integral equations can be found in [17] and [22], but for completeness and further numerical development only a summary of the CFIE equations is given below. Referring to Fig. 1, S denotes the surface of a homogeneous, lossy dielectric scatterer having a volume V contained in region 2 and bounded by the surface S. The scatterer is located in region 1 representing an isotropic, lossless free space medium. Let

$(\vec{E}_1^s, \vec{H}_1^s)$ = scattered electric and magnetic fields in region 1

$(\vec{E}_2^s, \vec{H}_2^s)$ = scattered electric and magnetic fields in region 2.

Then, referring to the electromagnetic equivalent principle, various scattered electric and magnetic fields in regions 1 and 2 are given by

$$\vec{E}_1^s(\vec{r}) = -j\omega\vec{A}_1(\vec{r}) - \nabla V_1(\vec{r}) - \frac{1}{\epsilon_1'}\nabla\times\vec{F}_1(\vec{r}) \tag{1a}$$

$$\vec{H}_1^s(\vec{r}) = -j\omega\vec{F}_1(\vec{r}) - \nabla U_1(\vec{r}) + \frac{1}{\mu_1}\nabla\times\vec{A}_1(\vec{r}), \quad \text{for } \vec{r} \text{ on or outside } S \tag{1b}$$

$$\vec{E}_2^s(\vec{r}) = j\omega\vec{A}_2(\vec{r}) + \nabla V_2(\vec{r}) + \frac{1}{\epsilon_2'}\nabla\times\vec{F}_2(\vec{r}) \tag{2a}$$

$$\vec{H}_2^s(\vec{r}) = j\omega\vec{F}_2(\vec{r}) + \nabla U_2(\vec{r}) - \frac{1}{\mu_2}\nabla\times\vec{A}_2(\vec{r}), \quad \text{for } \vec{r} \text{ on or inside } S \tag{2b}$$

where the various vector potentials $\vec{A}_i$ and $\vec{F}_i$ and the scalar potentials V_i and U_i, for $i = 1, 2$ are given by

$$\vec{A}_i(\vec{r}) = \frac{\mu_i}{4\pi}\iint_S \vec{J}(\vec{r}')G_i(\vec{r}, \vec{r}')\, dS(\vec{r}') \tag{3a}$$

$$\vec{F}_i(\vec{r}) = \frac{\epsilon_i'}{4\pi}\iint_S \vec{M}(\vec{r}')G_i(\vec{r}, \vec{r}')\, dS(\vec{r}') \tag{3b}$$

$$V_i(\vec{r}) = \frac{1}{4\pi\epsilon_i'}\iint_S \rho^e(\vec{r}')G_i(\vec{r}, \vec{r}')\, dS(\vec{r}') \tag{3c}$$

$$U_i(\vec{r}) = \frac{1}{4\pi\mu_i}\iint_S \rho^m(\vec{r}')G_i(\vec{r}, \vec{r}')\, dS(\vec{r}') \tag{3d}$$

$$\epsilon_i' = \epsilon_i\left[1 - j\frac{\sigma_i}{\omega\epsilon_i}\right] \tag{3e}$$

and

$$\rho^e(\vec{r}\,')=\frac{-1}{j\omega}\,[\nabla_S'\cdot\vec{J}(\vec{r}\,')] \tag{4a}$$

$$\rho^m(\vec{r}\,')=\frac{-1}{j\omega}\,[\nabla_S'\cdot\vec{M}(\vec{r}\,')]. \tag{4b}$$

In obtaining the above expressions, $e^{j\omega t}$ time dependence is assumed for various field quantities and ω is the frequency in radians per second. The Green's function defined in (3a)–(3d) for $i = 1, 2$ is given by

$$G_i(\vec{r},\vec{r}\,')=\frac{e^{-jk_iR}}{R} \tag{5a}$$

$$R=|\vec{r}-\vec{r}\,'| \tag{5b}$$

and the propagation constant is

$$k_i=[\omega^2\mu_i\epsilon_i']^{1/2}. \tag{6}$$

In (3a) and (3b), $\vec{J}$ is the equivalent electric current and $\vec{M}$ is the equivalent magnetic current on the surface of the dielectric scatterer. The equivalent electric and magnetic currents are, in fact, related to the surface total magnetic and electric fields tangential to the surface S:

$$\vec{J}(\vec{r}\,')=\hat{n}\times\vec{H}(\vec{r}\,') \tag{7a}$$

$$\vec{M}(\vec{r}\,')=\vec{E}(\vec{r}\,')\times\hat{n}, \qquad \vec{r}\,' \text{ on the surface } S \tag{7b}$$

where $\hat{n}$ is an outward unit normal on S shown in Fig. 1. Further, in the above expressions, $(\epsilon_1, \mu_1, \sigma_1 = 0)$ and $(\epsilon_2, \mu_2, \sigma_2)$ are the permittivity, permeability, and conductivity for the regions 1 and 2.

On enforcing the boundary condition that the total tangential electric field and the total tangential magnetic field should be continuous across the surface of the arbitrary dielectric scatterer, the following combined field integral equations are obtained in terms of the unknown surface equivalent electric and magnetic currents:

$$\vec{E}^i(\vec{r})|_{\tan}=\left\{j\omega[\vec{A}_1(\vec{r})+\vec{A}_2(\vec{r})]+[\nabla V_1(\vec{r})+\nabla V_2(\vec{r})] +\nabla\times\left[\frac{\vec{F}_1(\vec{r})}{\epsilon_1'}+\frac{\vec{F}_2(\vec{r})}{\epsilon_2'}\right]\right\}\Bigg|_{\tan} \tag{8a}$$

$$\vec{H}^i(\vec{r})|_{\tan}=\left\{j\omega[\vec{F}_1(\vec{r})+\vec{F}_2(\vec{r})]+[\nabla U_1(\vec{r})+\nabla U_2(\vec{r})] -\nabla\times\left[\frac{\vec{A}_1(\vec{r})}{\mu_1}+\frac{\vec{A}_2(\vec{r})}{\mu_2}\right]\right\}\Bigg|_{\tan}, \qquad \vec{r} \text{ on surface } S \tag{8b}$$

where $\vec{E}^i$ and $\vec{H}^i$ are the incident electric and magnetic fields in the region 1 and the subscript "tan" refers to tangential component only. A detailed numerical approach is discussed in the following based on the method of moment technique to reduce the coupled integrodifferential equations (8a) and (8b) to the corresponding partitioned matrix equation for the unknown electric and magnetic currents on the surface of the scatterer. Especially to treat arbitrary shaped bodies, the surface S of the dielectric scatterer should be efficiently modeled as proposed [21] by dividing the surface of the scatterer into a number of triangular shaped surface patches. An example is shown in Fig. 2 indicating how one can use triangles to patch model scatterer surfaces efficiently. Fig. 2 shows the case of a finite circular dielectric cylinder scatterer modeled in terms of triangular surface patches. In each of the triangular surface patch, the electric and magnetic currents are represented in terms of a known triangular basis function. The complete development of current expansion or basis functions for triangular surface patches suitable for homogeneous dielectric objects is discussed in detail in [17].

IV. Basis Functions and Current Representation

Given a closed surface S the surface is first approximated by a number of triangles, Fig. 2. Each triangle is defined by an appropriate set of faces, edges, and vertices. Fig. 3(a) shows two triangles T_n^+ and T_n^- with the nth common edge. The electric and magnetic currents flow along radial direction $\hat{\rho}_n^+$ in triangle T_n^+ and similarly flow along radial direction $\hat{\rho}_n^-$ in triangle T_n^-. Referring to Fig. 3(b), if l_n is the base length of common edge, then height lengths of the two triangles T_n^+ and T_n^- are, respectively, given by $2A_n^+/l_n$ and $2A_n^-/l_n$, where $A_n^\pm$ represents the area of $T_n^\pm$. Any point in triangles $T_n^\pm$ can be defined either with respect to global origin, O, or with respect to the triangle vertices $O_n^\pm$. In Fig. 3, the superscripts plus and minus signs designation of the triangles is determined by choice of a positive current reference direction [21] for the nth edge, which is always assumed to be from T_n^+ to T_n^-. Hence, a vector basis function associated with nth edge is

$$\vec{f}_n(\vec{r})=\begin{cases}\dfrac{l_n}{2A_n^+}\vec{\rho}_n^{\,+}, & \vec{r} \text{ in } T_n^+ \quad (9a)\\[2mm] \dfrac{l_n}{2A_n^-}\vec{\rho}_n^{\,-}, & \vec{r} \text{ in } T_n^- \quad (9b)\\[2mm] \vec{0}, & \text{otherwise.} \quad (9c)\end{cases}$$

The vector basis function stated in (9) is ideally suited for representing surface electric current $\vec{J}$ and the surface magnetic current $\vec{M}$ on the triangulated surface S of the given dielectric scatterer. The reader may refer to [17] and [21] for the detailed discussion on various mathematical properties of the vector basis functions. In fact the surface integral of basis function over adjacent triangles represents moment given by

$$\iint_{T_n^+ + T_n^-} \vec{f}_n\, ds=\frac{l_n}{2}\,[\vec{\rho}_n^{\,c+} + +\vec{\rho}_n^{\,c-}] \tag{10a}$$

$$=l_n(\vec{r}_n^{\,c+}-\vec{r}_n^{\,c-}) \tag{10b}$$

$\vec{\rho}_n^{\,c+}$ = vector between O_n^+ and centroid of T_n^+

$\vec{\rho}_n^{\,c-}$ = vector between centroid T_n^- and O_n^-

and referring to Fig. 4, $\vec{r}_n^{\,c+}$ and $\vec{r}_n^{\,c-}$ are the distances to centroids of triangles from the arbitrary reference point.

Referring to the dielectric scatterer shown in Fig. 2, the

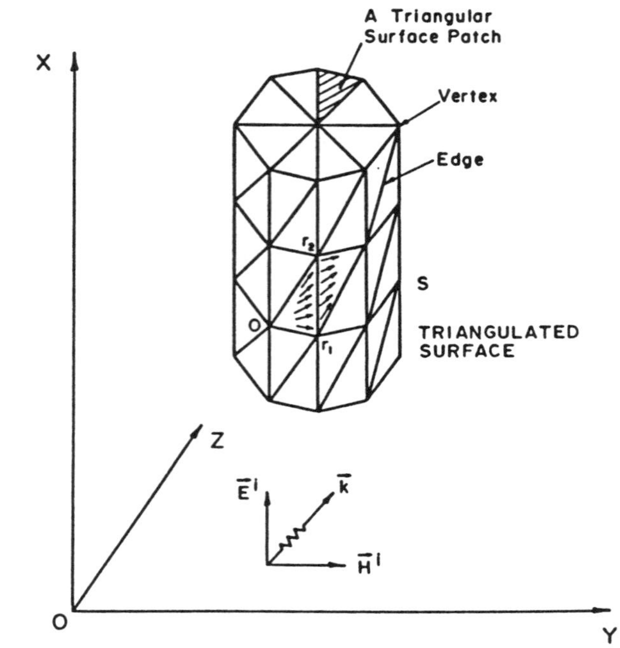

Fig. 2. Finite circular cylinder dielectric scatterer.

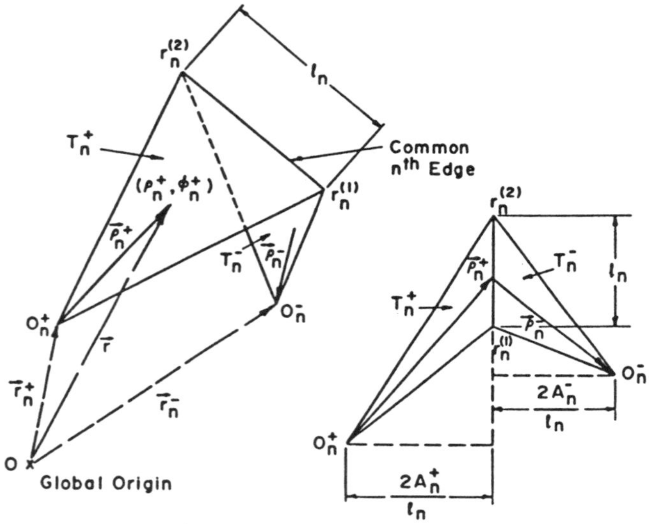

(a) (b)

Fig. 3. (a) Coordinates of common edge associated with two triangles. (b) Geometry for normal component of basis function at common edge.

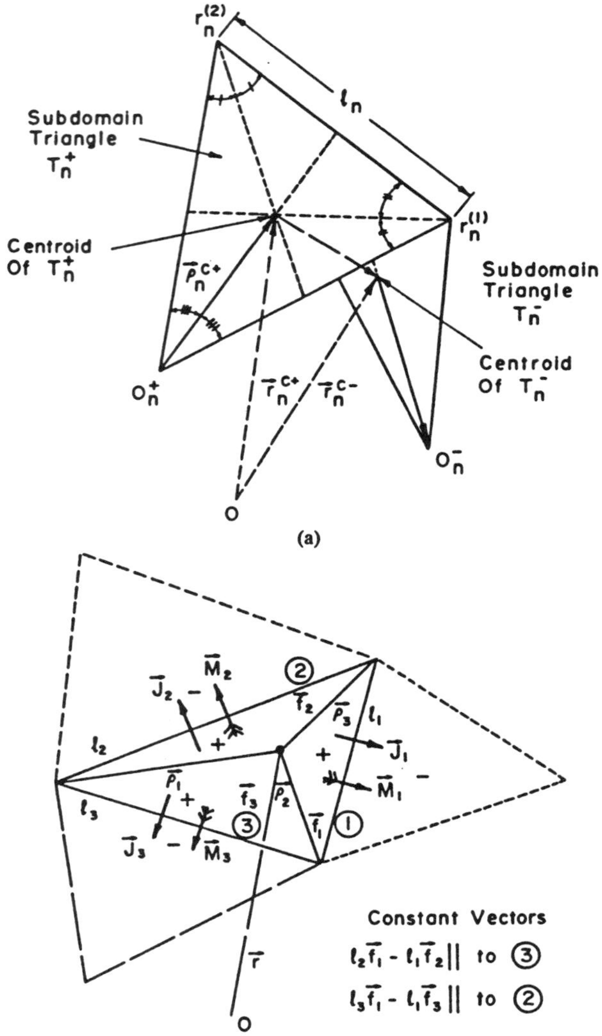

(a)

(b)

Fig. 4. (a) Coordinate for calculating centroids and moment of basis vectors. (b) Three edge currents associated with a triangle.

surface electric and magnetic current, $\vec{J}$ and $\vec{M}$, distributions are expanded in terms of vector basis functions defined in (9). Let N represents the total number of edges. Then,

$$\vec{J}(\vec{r}\,')=\sum_{n=1}^{N} I_n\vec{f}_n(\vec{r}\,') \tag{11a}$$

$$\vec{M}(\vec{r}\,')=\sum_{n=1}^{N} M_n\vec{f}_n(\vec{r}\,') \tag{11b}$$

where I_n and M_n are constants yet to be determined. Since the normal component of $\vec{f}_n$ at the nth common edge connecting T_n^+ and T_n^- is unity, each coefficient of I_n and M_n can be interpreted as the normal components of the electric and magnetic current density flowing past the nth common edge. Further, we note, for a given triangular face, that there are three edges and correspondingly there exists three vector basis functions, Fig. 4. It is also clearly pointed out in basis vectors development [17], [21] that the superposition of the basis functions with a triangle conveniently represent a constant current flowing in an arbitrary direction within the triangle.

V. Testing of CFIE

In order to find the current coefficients, the combined field integral equations (8a) and (8b) are tested with respect to testing functions. One suitable choice is to pick testing functions identically same as basis functions, given by (9a)–(9c), and test the equations based on the following symmetric product to reduce the operator type integral equations to the corresponding functional type,

$$\langle \vec{f},\, \vec{g}\rangle \equiv \iint_S \vec{f}\cdot\vec{g}\; dS. \tag{12}$$

Hence, testing (8a) and (8b), we obtain

$$\langle \vec{E}^i, \vec{J}_m\rangle = \langle j\omega(\vec{A}_1+\vec{A}_2), \vec{J}_m\rangle + \langle(\nabla V_1+\nabla V_2), \vec{J}_m\rangle + \left\langle \nabla\times\left(\frac{\vec{F}_1}{\epsilon_1'}+\frac{\vec{F}_2}{\epsilon_2'}\right), \vec{J}_m\right\rangle, \quad \text{on surface } S \tag{13a}$$

and

$$\langle \vec{H}^i, \vec{J}_m\rangle = \langle j\omega(\vec{F}_1+\vec{F}_2), \vec{J}_m\rangle + \langle(\nabla U_1+\nabla U_2), \vec{J}_m\rangle - \left\langle \nabla\times\left(\frac{\vec{A}_1}{\mu_1}+\frac{\vec{A}_2}{\mu_2}\right), \vec{J}_m\right\rangle, \quad \text{on surface } S \tag{13b}$$

where the subscript m denotes an edge formed by two triangles T_m^+ and T_m^-. The first-term in (13a) and (13b) can be simplified by evaluating the vector potentials at centroids of respective triangles. Further, the second-term with gradient in (13a) and (13b) can be simplified as

$$\langle \nabla V, \vec{J}_m\rangle = -\iint_S V(\nabla\cdot\vec{J}_m)\, dS \tag{14a}$$

$$= l_m\left[\frac{1}{A_m^-}\iint_{T_m^-} V\, dS - \frac{1}{A_m^+}\iint_{T_m^+} V\, dS\right] \tag{14b}$$

$$\simeq l_m[V(r_m^{c-}) - V(r_m^{+c})]. \tag{14c}$$

We note that the integrals in (14b) have been approximated by evaluating the scalar potentials at the respective centroids of the two $T_m^\pm$ triangles. Similarly, the third-term with "curl" in the above tested equations (13a) and (13b) can be simplified as

$$\langle \nabla\times\vec{A}, \vec{J}_m\rangle = \iint_S (\nabla\times\vec{A})\cdot\vec{J}_m\, dS \tag{15a}$$

$$= \frac{l_m}{2A_m^+}\iint_{T_m^+} \vec{\rho}_m^{\,+}\cdot[(\nabla\times\vec{A})]^+\, dS + \frac{l_m}{2A_m^-}\iint_{T_m^-} \vec{\rho}_m^{\,-}\cdot[(\nabla\times\vec{A})]^-\, ds. \tag{15b}$$

On substituting the relationships (14) and (15) into the tested form of (13a) and (13b), the following functional form of equations are obtained:

$$j\omega l_m\left[\frac{\vec{\rho}_m^{\,c+}}{2}\cdot\{\vec{A}_1(\vec{r}_m^{\,c+})+\vec{A}_2(\vec{r}_m^{\,c+})\} + \frac{\vec{\rho}_m^{\,c-}}{2}\cdot\{\vec{A}_1(\vec{r}_m^{\,c-})+\vec{A}_2(\vec{r}_m^{\,c-})\}\right]$$
$$+ l_m\left[\{V_1(\vec{r}_m^{\,c-})-V_1(\vec{r}_m^{\,c+})\}+\{V_2(\vec{r}_m^{\,c-})-V_2(\vec{r}_m^{\,c+})\}\right]$$
$$+\left[\left\{\frac{\vec{P}_1(\vec{r}_m^{\,+})}{\epsilon_1'}+\frac{\vec{P}_2(\vec{r}_m^{\,+})}{\epsilon_2'}\right\}+\left\{\frac{\vec{P}_1(\vec{r}_m^{\,-})}{\epsilon_1'}+\frac{\vec{P}_2(\vec{r}_m^{\,-})}{\epsilon_2'}\right\}\right]$$
$$= l_m\left[\frac{\vec{\rho}_m^{\,c+}}{2}\cdot\vec{E}^{i+}(\vec{r}_m^{\,c+})+\frac{\vec{\rho}_m^{\,c-}}{2}\cdot\vec{E}^{i-}(\vec{r}_m^{\,c-})\right],$$
$$\text{on scatter surface } S, \quad m=1, 2, 3, \cdots, N \text{ edges} \tag{16a}$$

and

$$j\omega l_m\left[\frac{\vec{\rho}_m^{\,c+}}{2}\cdot\{\vec{F}_1(\vec{r}_m^{\,c+})+\vec{F}_2(\vec{r}_m^{\,c+})\} + \frac{\vec{\rho}_m^{\,c-}}{2}\cdot\{\vec{F}_1(\vec{r}_m^{\,c-})+\vec{F}_2(\vec{r}_m^{\,c-})\}\right]$$
$$+ l_m\left[\{U_1(\vec{r}_m^{\,c-})-U_1(\vec{r}_m^{\,c+})\}+\{U_2(\vec{r}_m^{\,c-})-U_2(\vec{r}_m^{\,c+})\}\right]$$
$$-\left[\left\{\frac{\vec{Q}_1(\vec{r}_m^{\,+})}{\mu_1}+\frac{\vec{Q}_2(\vec{r}_m^{\,+})}{\mu_2}\right\}+\left\{\frac{\vec{Q}_1(\vec{r}_m^{\,-})}{\mu_1}+\frac{\vec{Q}_2(\vec{r}_m^{\,-})}{\mu_2}\right\}\right]$$
$$= l_m\left[\frac{\vec{\rho}_m^{\,c+}}{2}\cdot\vec{H}^{i+}(\vec{r}_m^{\,c+})+\frac{\vec{\rho}_m^{\,c-}}{2}\cdot\vec{H}^{i-}(\vec{r}_m^{\,c-})\right],$$
$$\text{on scatter surface } S, \quad m=1, 2, 3, \cdots, N \text{ edges} \tag{16b}$$

where in the above tested equations,

$$\vec{P}_{1,2}(\vec{r}_m^{\,\pm}) = \frac{l_m}{2A_m^\pm}\iint_{T_m^\pm} \vec{\rho}_m^{\,\pm}\cdot[\nabla\times\vec{F}_{1,2}(\vec{r}_m^{\,\pm})]\, dS \tag{17a}$$

$$\vec{E}^{i+}(\vec{r}_m^{\,c+}) = \vec{E}^i(\vec{r}_m^{\,c+}) \tag{17b}$$

$$\vec{E}^{i-}(\vec{r}_m^{\,c-}) = \vec{E}^i(\vec{r}_m^{\,c-}) \tag{17c}$$

$$\vec{Q}_{1,2}(\vec{r}_m^{\,\pm}) = \frac{l_m}{2A_m^\pm}\iint_{T_m^\pm} \vec{\rho}_m^{\,\pm}\cdot[\nabla\times\vec{A}_{1,2}(\vec{r}_m^{\,\pm})]\, dS \tag{18a}$$

$$\vec{H}^{i+}(\vec{r}_m^{\,c+}) = \vec{H}^i(\vec{r}_m^{\,c+}) \tag{18b}$$

$$\vec{H}^{i-}(\vec{r}_m^{\,c-}) = \vec{H}^i(\vec{r}_m^{\,c-}). \tag{18c}$$

In the above functional equations (16a) and (16b), the vector and the scalar potentials $\vec{A}_i$, $\vec{F}_i$, V_i, U_i are given by (3a)–(3d). The $\vec{P}_i$ and $\vec{Q}_i$ terms containing the curl operations can be further simplified as, for $i = 1, 2$ regions,

$$[\nabla\times\vec{F}_i(\vec{r})] = \frac{\epsilon_i'}{4\pi} ⨌_S \vec{M}(\vec{r}\,')\times\nabla' G_i(\vec{r}, \vec{r}\,')\, dS(\vec{r}\,') \tag{19a}$$

$$[\nabla\times\vec{A}_i(\vec{r})] = \frac{\mu_i}{4\pi} ⨌_S \vec{J}(\vec{r}\,')\times\nabla' G_i(\vec{r}, \vec{r}\,')\, dS(\vec{r}\,') \tag{19b}$$

where the symbol ⨌ represents Cauchy principle value of the integral. In the numerical development, the integrals defined in (17a) and (18a) will be evaluated numerically [17], [21] using seven point integration method principally applicable to triangular distributions.

VI. Matrix Equation (CFIE)

The electric current and the magnetic current expansion terms defined in (11a) and (11b) are now substituted into the CFIE tested (16a) and (16b) to reduce the functional form of the equation to a corresponding partitioned matrix equation [17],

[21]:

$$\begin{bmatrix} [Z_{mn}^{JJ}] & [C_{mn}^{JM}] \\ [D_{mn}^{MJ}] & [Y_{mn}^{MM}] \end{bmatrix} \begin{bmatrix} [I_n] \\ [M_n] \end{bmatrix} = \begin{bmatrix} [V_m] \\ [H_m] \end{bmatrix} \quad (20)$$

where the various matrix elements are given by the following, for $m = 1, 2, 3, \cdots, N$ edges and $n = 1, 2, 3, \cdots, N$ edges.

Elements of diagonal submatrix for electric current:

$$Z_{mn}^{JJ} = l_m \left[\frac{\vec{\rho}_m^{c+}}{2} \cdot \sum_{i=1}^{2} (jk_i\eta_i)\vec{A}_{imn}^{+} + \frac{\vec{\rho}_m^{c-}}{2} \cdot \sum_{i=1}^{2} (jk_i\eta_i)\vec{A}_{imn}^{-} + \sum_{i=1}^{2} \left(\frac{-\eta_i}{jk_i} \right) \{ \Phi_{imn}^{-} - \Phi_{imn}^{+} \} \right]. \quad (21a)$$

Elements of diagonal submatrix for magnetic current:

$$Y_{mn}^{MM} = l_m \left[\frac{\vec{\rho}_m^{c+}}{2} \cdot \sum_{i=1}^{2} \left(\frac{jk_i}{\eta_i} \right) \vec{F}_{imn}^{+} + \frac{\vec{\rho}_m^{c-}}{2} \cdot \sum_{i=1}^{2} \left(\frac{jk_i}{\eta_i} \right) \vec{F}_{imn}^{-} + \sum_{i=1}^{2} \left(\frac{-1}{jk_i\eta_i} \right) \{ \Psi_{imn}^{-} - \Psi_{imn}^{+} \} \right]. \quad (21b)$$

Elements of off-diagonal submatrix:

$$C_{mn}^{JM} = \left[\sum_{i=1}^{2} P_{imn}^{+} + \sum_{i=1}^{2} P_{imn}^{-} \right]. \quad (21c)$$

$$D_{mn}^{mJ} = \left[\sum_{i=1}^{2} Q_{imn}^{+} + \sum_{i=1}^{2} Q_{imn}^{-} \right]. \quad (21d)$$

Elements of electric and magnetic field excitation:

$$V_m = l_m \left[\frac{\vec{\rho}_m^{c+}}{2} \cdot \vec{E}^{i+}(\vec{r}_m^{c+}) + \frac{\vec{\rho}_m^{c-}}{2} \cdot \vec{E}^{i-}(\vec{r}_m^{c-}) \right] \quad (21e)$$

$$H_m = l_m \left[\frac{\vec{\rho}_m^{c+}}{2} \cdot \vec{H}^{i+}(\vec{r}_m^{c+}) + \frac{\vec{\rho}_m^{c-}}{2} \cdot \vec{H}^{i-}(\vec{r}_m^{c-}) \right] \quad (21f)$$

and, the vector and the scalar potential integrals take the following form [17], [21], for $i = 1, 2$:

$$\vec{A}_{imn}^{\pm} = \frac{1}{4\pi} \iint_{(T_n^+ + T_n^-)} \vec{J}_n(\vec{r}\,')G_i(\vec{r}_m^{c\pm}, \vec{r}\,')\, dS(\vec{r}\,') \quad (22a)$$

$$= \vec{F}_{imn}^{\pm} \quad (22b)$$

$$\Phi_{imn}^{\pm} = \frac{1}{4\pi} \iint_{(T_n^+ + T_n^-)} [\nabla_S' \cdot \vec{J}_n(\vec{r}\,')]G_i(\vec{r}_m^{c\pm}, \vec{r}\,')\, dS(\vec{r}\,') \quad (22c)$$

$$= \Psi_{imn}^{\pm} \quad (22d)$$

$$P_{imn}^{\pm} = \frac{l_m}{2A_m^{\pm}} \iint_{T_m^{\pm}} \vec{\rho}_m^{\pm} \cdot \left[\frac{1}{4\pi} \iint_{(T_n^+ + T_n^-)} \vec{J}_n \times \nabla' G_i(\vec{r}_m^{\pm}, \vec{r}\,')\, dS(\vec{r}\,') \right] dS(\vec{r}) \quad (23a)$$

$$= Q_{imn}^{\pm} \quad (23b)$$

and

$$G_i(\vec{r}_m^{\pm}, \vec{r}\,') = \frac{e^{-jk_iR^{\pm}}}{R^{\pm}} \quad (23c)$$

$$R^{\pm} = |\vec{r}_m^{\pm} - \vec{r}\,'| \quad (23d)$$

$$\nabla' G_i(\vec{r}_m^{\pm}, \vec{r}\,') = (\vec{r}_m^{\pm} - \vec{r}\,')(1 + jk_iR^{\pm}) \frac{e^{-jk_iR^{\pm}}}{(R^{\pm})^3}. \quad (23e)$$

VII. Efficient Numerical Algorithm Development

The integrals (22a), (22c) and (23a) are in a convenient form for numerical evaluation. The simplification of these integrals are discussed in detail in [21], which are useful for numerical algorithm development. The matrix equation (20) can be inverted to obtain the electric and magnetic current coefficients, I_n and M_n.

We note that the various matrix elements can be easily generated by considering faces rather than edges. This cuts down by approximately ninefold, computer time required to generate matrix elements [21]. We further note that the matrix elements Z_{mn} and Y_{mn} are similar except for floating constants which can be conveniently incorporated while filling matrix elements; so that one has to generate only Z_{mn} matrix elements and the Y_{mn} elements are obtained by changing multiplying constants while filling in matrix elements. Similarly C_{mn} and D_{mn} matrix elements differ by just a multiplying constant; so that one is to generate just C_{mn} elements only and the D_{mn} elements are obtained directly from the C_{mn} elements. Also referring to the expression (21), the elements of the submatrices contain terms belonging to both regions one and two. The various potential integral expressions for the regions 1 and 2 are in fact identical except for the electrical characteristics which appear in the propagation constants and in the multiplication constants. For an efficient numerical algorithm development and to save computer time, same subroutines are simultaneously utilized for calculating region 1 and region 2 integral terms which principally make up various matrix elements.

To obtain the electric and the magnetic current distributions one can either directly invert the matrix equation (20), or the matrix equation (20) can be rearranged so as to eliminate one unknown, and resubstituted back to obtain the second unknown. This takes less computer time than directly inverting a large composite matrix equation. Another possible method seems to be application of the iterative methods [10]. Detail studies are still underway in this specific area to utilize iterative schemes.

VIII. Numerical Results—Homogeneous Dielectric Sphere and Finite Circular Cylinder

To demonstrate applicability of the above formulation and to validate computer algorithms, numerical results are presented for the case of a homogeneous dielectric sphere and for the case of a homogeneous dielectric finite circular cylinder located in free space and excited by a plane wave.

The surface of the sphere is first modeled in terms of triangles having arbitrary edges and vertices arranged to depict the shape of a sphere. Fig. 5(a) shows the plan view (top

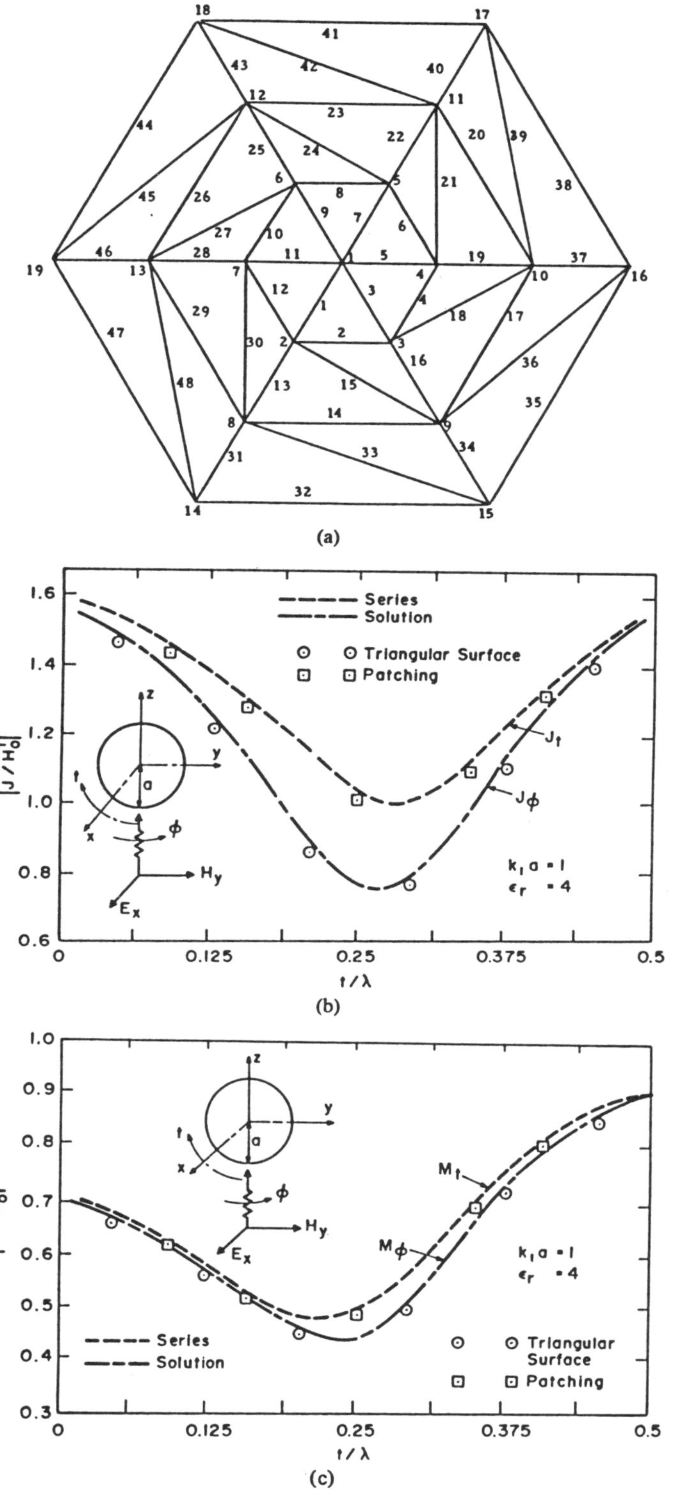

Fig. 5. (a) Triangular surface patching for top half of sphere (plan view). (b) Equivalent surface electric current distribution on a homogeneous dielectric sphere. (c) Equivalent surface magnetic current distribution on a homogeneous dielectric sphere.

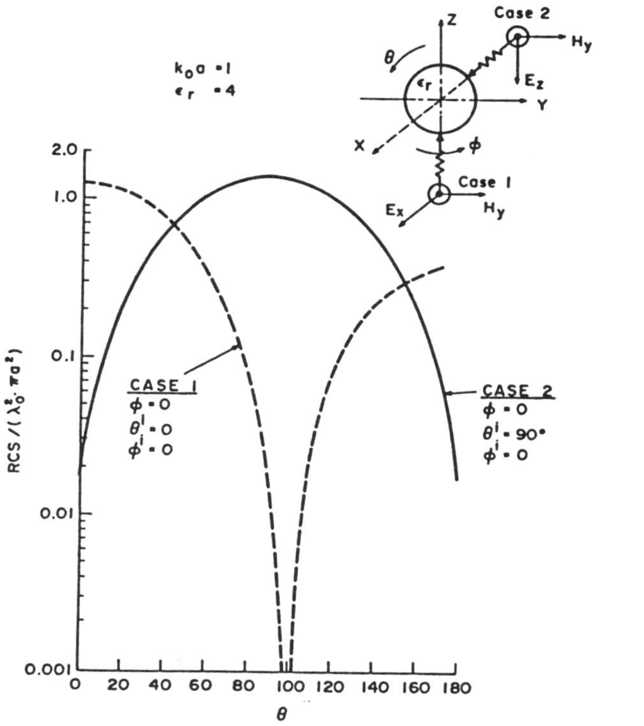

Fig. 6. Bistatic radar cross section of homogeneous dielectric sphere.

hemisphere only) of the triangular scheme adopted. There are in total 60 triangular faces consisting of 90 edges at which the unknown normal components of the electric and magnetic currents solved. The matrix size adopted to check the accuracy is 180 × 180. The electrical size of the sphere is $k_1a = 1$ where the free space propagation constant $k_1 = 2\pi/\lambda_0$ and the radius of the sphere is a. The relative dielectric constant of the sphere is $\epsilon_r = 4$. The sphere is located in free space and is excited by an axial incident plane wave. In Figs. 5(b) and 5(c) are shown the induced electric and the induced magnetic currents on the surface of the sphere along a circumferential arc in xz plane. Along the arc, there are two components of the electric currents J_t and J_ϕ; and magnetic currents M_t and M_ϕ. The results of the induced electric currents are shown normalized with respect incident magnetic field and similarly the induced magnetic currents are shown normalized with respect to incident electric field. The induced surface fields for a sphere problem can also be obtained by the eigenfunction analysis using spherical harmonic functions. The results of this approach are also shown in Figs. 5(b) and 5(c). The CFIE/MM solution based on triangular surface patching has good agreement with the eigenfunction series solution [2].

Fig. 6 gives the computed bistatic radar cross section for the homogeneous dielectric sphere based on the equivalent surface electric and magnetic currents. The radar cross section results are shown as a function of θ in the vertical plane cut $\phi = 0$ for two different angles of incidence; one along axial excitation and the other along broad excitation. The results check very well with the results based on body of revolution treatment [5].

Figs. 7(a) and 7(b) show the distribution of surface electric and magnetic currents for the case of homogeneous dielectric

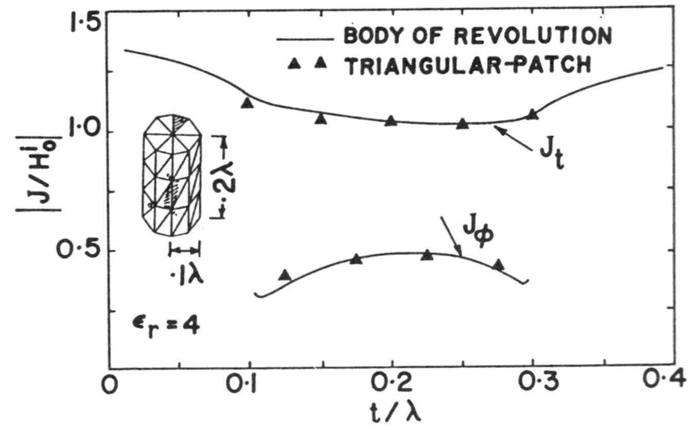

(a)

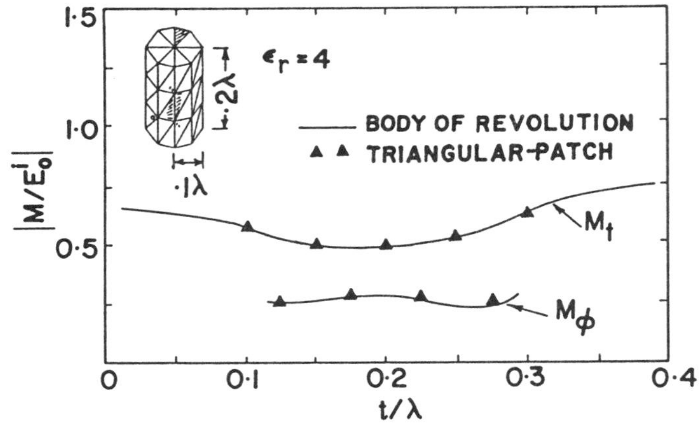

(b)

Fig. 7. (a) Electric surface current distribution on a dielectric finite circular cylinder due to axially incident plane wave. (b) Magnetic surface current distribution on a dielectric finite circular cylinder due to axially incident plane wave.

finite circular cylinder located in free space and externally excited by a plane wave. The results of the surface induced currents agree very well with those obtained based on the body of revolution [19] treatment.

IX. Conclusion

Based on the method of moments technique and the combined field integral equations, this paper reported recent developments and extensions to analyze electromagnetic scattering by arbitrary shaped three dimensional homogeneous lossy and lossless objects. Similar to the case of conducting objects, the arbitrary geometrical shapes have been modeled using surface triangular patches. Efficient and simple numerical algorithms are also developed; and validations reported here for two canonical dielectric scatterers namely a sphere and a finite circular cylinder. The numerical technique discussed here can be conveniently extended to analyze electromagnetic scattering by objects which are homogeneous, lossy, but having diagonizable tensor material characteristics [17]. The results of this study and their verifications will be reported in an upcoming paper.

References

[1] J. J. Bowman, T. B. A. Senior, and P. L. E. Uslenghi, *Electromagnetic and Acoustic Scattering by Simple Shapes*. Amsterdam: North-Holland, 1969.

[2] R. W. P. King and C. W. Harrison, "Scattering by imperfectly conducting spheres," AFCRL-70-0483, Cruft Lab. Rep. by Harvard Univ., 1970.

[3] M. G. Andreasen, "Scattering from bodies of revolution," *IEEE Trans. Antennas Propagat.*, vol. AP-13, pp. 303–310, Mar. 1965.

[4] V. A. Erma, "Exact solution for the scattering of electromagnetic waves from bodies of arbitrary shape, III, Obstacles with arbitrary electromagnetic properties," *Phys. Rev.*, vol. 179, pp. 1238–1246, Mar. 1969.

[5] T. K. Wu, "Electromagnetic scattering from arbitrarily-shaped lossy dielectric bodies," Ph.D. dissertation, Univ. Mississippi, University, MS, May 1976.

[6] C. Yeh, "Perturbation approach to the diffraction of electromagnetic waves by arbitrarily shaped dielectric obstacles," *Phys. Rev.*, vol. 135, no. 5A, pp. A1193–A1201, Aug. 1964.

[7] J. H. Richmond, "Scattering by dielectric cylinder of arbitrary cross section shape," *IEEE Trans. Antennas Propagat.*, vol. AP-13, pp. 334–341, May 1965.

[8] D. H. Schaubert, D. R. Wilton, and A. W. Glisson, "A tetrahedral modeling for electromagnetic scattering by arbitrary shaped homogeneous dielectric bodies," *IEEE Trans. Antennas Propagat.*, vol. AP-32, no. 1, pp. 77–85, Jan. 1984.

[9] D. T. Borup and O. P. Gandhi, "Fast-Fourier transform method for calculation of SAR distributions in finely discretized inhomogeneous models of biological bodies," *IEEE Trans. Microwave Theory Tech.*, vol. MTT-32, no. 4, pp. 355–360, Apr. 1984.

[10] T. K. Sarkar and S. M. Rao, "An iterative method for solving electrostatic problems," *IEEE Trans. Antennas Propagat.*, vol. AP-30, pp. 611–616, July 1982.

[11] D. E. Livesay and K. M. Chen, "Electromagnetic fields induced inside arbitrarily shaped biological bodies," *IEEE Trans. Microwave Theory Tech.*, vol. MTT-22, no. 12, pp. 1273–1280, Dec. 1974.

[12] P. C. Waterman, "Matrix formulation of electromagnetic scattering," *Proc. IEEE*, vol. 53, no. 8, pp. 805–812, Aug. 1965.

[13] S. Govind, "Numerical computation of electromagnetic scattering by inhomogeneous penetrable bodies," Ph.D. dissertation, Univ. Mississippi, University, MS, Dec. 1978.

[14] K. K. Mei, "Unimoment method of solving antenna and scattering problems," *IEEE Trans. Antennas Propagat.*, vol. AP-22, pp. 760–766, Nov. 1974.

[15] P. C. Waterman, "Scattering by dielectric obstacles," *Alta. Freq.*, vol. 38, (Speciale), p. 348, 1969.

[16] S. Strom, "T-Matrix for electromagnetic scattering from an arbitrary number of scatterers with continuously varying electromagnetic properties," *Phys. Rev. D*, vol. 10, pp. 2685–2690, Oct. 1974.

[17] K. R. Umashankar and A. Taflove, "Analytical models for electromagnetic scattering," Part-I, Final Rep., Contract F19628-82-C-0140 to RADC/ESD, Hanscom Air Force Base, MA, June 1984.

[18] R. J. Pogorzelski, "On the numerical computation of scattering from inhomogeneous penetrable objects," *IEEE Trans. Antennas Propagat.*, vol. AP-26, pp. 616–618, July 1978.

[19] A. W. Glisson, "On the development of numerical techniques for treating arbitrarily-shaped surfaces," Ph.D. dissertation, Univ. Mississippi, University, MS, 1978.

[20] S. M. Rao, D. R. Wilton, and A. W. Glisson, "Electromagnetic scattering by surfaces of arbitrary shape," *IEEE Trans. Antennas Propagat.*, vol. AP-30, pp. 409–418, May 1982.

[21] S. M. Rao, "Electromagnetic scattering and radiation of arbitrarily-shaped surfaces by triangular patch modeling, "Ph.D. dissertation, Univ. Mississippi, University, MS, Aug. 1980.

[22] J. R. Mautz and R. F. Harrington, "Electromagnetic scattering from a homogeneous material body of revolution," *Arch. Elek. Ubertragung*, vol. 33, no. 4, pp. 71–80, Apr. 1979.

Part 5
Apertures

AN APERTURE is a hole, crack, or gap in a perfectly conducting surface that divides space into two regions. Apertures are important in that they provide a means for the transmission of either desired or undesired energy from one region of space to the other. Apertures can also significantly increase the scattering from an otherwise smooth conducting surface with no holes. Aperture problems can be formulated in terms of the electric surface currents on the conducting surface separating the two regions of space. However, since the area of the aperture is usually much smaller than that of the conducting surface, it is more efficient to employ the MM to solve for either the tangential fields or equivalent surface currents in the aperture.

Van Bladel showed that aperture problems can be conveniently analyzed in terms of the tangential electric fields in the aperture in his paper, "The matrix formulation of scattering problems." He formulated the MM solution for cavity and waveguide apertures, but provided no numerical results. He also considered the internal cavity resonance problem and developed a Norton equivalent circuit for an aperture radiating into a region of space. Wallenberg and Harrington, in their paper "Radiation from apertures in conducting cylinders of arbitrary cross section," illustrated that the surface equivalence theorem could be used to formulate the aperture problem in terms of either equivalent electric or magnetic currents. The magnetic current formulation for the TE aperture in a cylinder was carried out in detail. The problem of a thin wire in the vicinity of an aperture was treated by Butler and Umashankar in "Electromagnetic excitation of a wire through an aperture-perforated conducting screen." They formulated the problem in terms of a pair of coupled integro-differential equations for the equivalent magnetic currents in the aperture and the equivalent electric currents on the surface of the thin wire. These coupled equations enforce continuity of the tangential magnetic field across the aperture and zero tangential electric field on the wire, and are solved by the MM. A tutorial review of methods for the analysis of apertures was presented by Butler *et al.* in "Electromagnetic penetration through apertures in conducting surfaces." This included an aperture in a plane screen, cavity apertures, and small aperture theory. In addition to being an excellent review of MM solutions for apertures, this paper includes an extensive bibliography of aperture research. At the same time, Schuman and Warren solved the problem of an aperture in a body of revolution (BOR). In their paper, "Aperture coupling in bodies of revolution," they basically showed how the equivalence theorems can be used to modify a conducting body BOR code so that it can treat aperture coupling. They presented data for aperture-coupled electric fields, and also presented a Norton equivalent network for apertures. Rahmat-Samii, in "Electromagnetic pulse coupling through an aperture into a two-parallel-plate region," used the fast Fourier transform to determine the penetration of an electromagnetic pulse through an aperture into a parallel plate region.

The Matrix Formulation of Scattering Problems

J. VAN BLADEL, SENIOR MEMBER, IEEE

Abstract—Two regions in space are coupled through an opening in a perfectly conducting surface. By using a complete set of eigenvectors in the opening, each region can be represented by an equivalent Norton circuit involving a short-circuit current (a vector) and a generator admittance (a matrix). The particular case of a cavity at resonance is investigated. Application to a cavity terminated in a waveguide is considered, and the transformation of the equivalent circuit resulting from the shift of the terminal plane is analyzed. After solving the example of a slotted waveguide, a possible set of eigenvectors for an arbitrary opening is proposed.

I. Introduction

A TYPICAL "coupled regions" configuration is shown in Fig. 1, where a field $\bar{e}_i$, $\bar{h}_i$ is incident on a metallic cavity I bounded by an infinitely thin conducting wall S. The wall is provided with an aperture S'. The fields in Regions I and II can be computed (in principle at least) once the tangential component $\bar{e}_{\text{tang}}$ of the electric field in S' is known. Suitable assumptions can be made concerning $\bar{e}_{\text{tang}}$ in certain particular cases (e.g., for small holes and for slots). In general, however, $\bar{e}_{\text{tang}}$ must be determined by:[1]

1) expressing $\bar{h}_{\text{tang}}$ on the cavity side in terms of $\bar{e}_{\text{tang}}$
2) expressing $\bar{h}_{\text{tang}}$ in Region II in terms of $\bar{e}_{\text{tang}}$, and
3) equating the two values of $\bar{h}_{\text{tang}}$ in S', and solving the resulting integral equation for $\bar{e}_{\text{tang}}$.

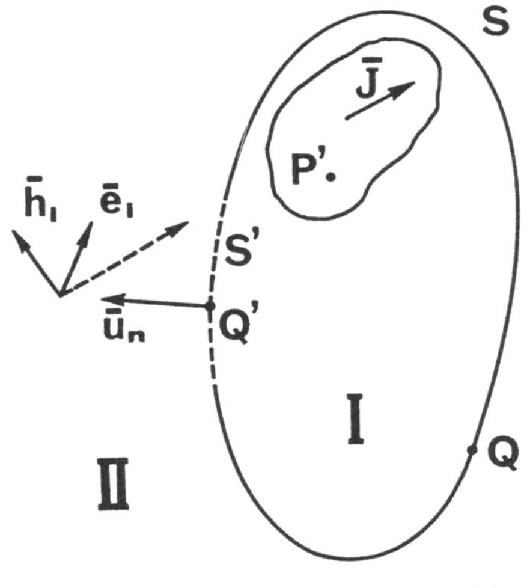

Fig. 1. A typical "coupled regions" problem.

In this paper, we seek to formulate the problem in terms of an equivalent network problem. Truly, the computational work is not simplified by this approach, but we believe that some conceptual clarity can be achieved by showing the connection between the electromagnetic problem and the (perhaps) more familiar network structure. Our treatment remains very general, and we leave for future reports the application of the network formulation to cavity filters, periodic structures, etc. Basically, our method rests on the use of an old workhorse—the eigenvector method—to the determination of the fields in the aperture.

II. Scattering of an Incident Field by a Cavity with an Opening

The cavity shown in Fig. 1 is excited by the volume currents $\bar{J}$ and by the aperture fields in S'. We shall first assume that the frequency does not coincide with one of its resonant values, leaving for Section III a discussion of the phenomena at resonance. Under those circumstances, the electromagnetic field in the cavity is uniquely determined by the values of $\bar{J}$ and $\bar{u}_n \times \bar{E}$ on S'. The two contributions are additive. For an evacuated cavity, for example, the tangential magnetic field due to $\bar{J}$ is[2]

$$\bar{H}_g{}^{\text{I}}(Q) = \iiint \Big[\sum_m \frac{k_m \bar{h}_m(Q)\bar{e}_m(P')}{(k_m{}^2 - k^2)} \Big] \cdot \bar{J}(P')dV' \quad (1)$$

where k_m is one of the resonant wave numbers, and $\bar{e}_m$ and $\bar{h}_m$ are the normalized solenoidal eigenvectors. These are connected by the relationships

$$\bar{h}_m = \frac{1}{k_m} \text{curl}\, \bar{e}_m \quad \text{and} \quad \bar{e}_m = \frac{1}{k_m} \text{curl}\, \bar{h}_m.$$

Clearly, $\bar{H}_g{}^{\text{I}}$ is the field which exists on S' when the latter surface is short circuited. The contribution from $(\bar{u}_n \times \bar{E})$ can be written as

$$\bar{H}_b{}^{\text{I}}(Q) = \iint_S \Big[-\frac{1}{j\omega\mu} \sum \bar{g}_m(Q)\bar{g}_m(Q') - j\omega\epsilon \sum \frac{\bar{h}_m(Q)\bar{h}_m(Q')}{k_m{}^2 - k^2} \Big] \cdot \bar{u}_n \times \bar{E}(Q')dS'$$

$$= \iint_{S'} \mathcal{G}^{\text{I}}(Q \mid Q') \cdot \bar{u}_n \times \bar{E}(Q')dS' \quad (2)$$

Manuscript received May 11, 1965; revised September 20, 1965.
The author is with the Laboratory for Electromagnetism and Acoustics, University of Ghent, Ghent, Belgium.

[1] For the solution of an actual problem, see F. J. Kriegler, F. E. Mills, and J. Van Bladel, "Fields excited by periodic beam currents in a cavity-loaded tube," *J. Appl. Phys.*, vol. 35, pp. 1721–6, June 1964.

[2] J. Van Bladel, *Electromagnetic Fields*. New York: McGraw-Hill, 1964, pp. 299, 415, 500, and 504.

Reprinted from *IEEE Trans. Microwave Theory Tech.*, vol. MTT-14, no. 3, pp. 130–135, March 1966.

where $\bar{g}_m$ denotes a normalized irrotational magnetic eigenvector.

The field $\bar{H}_{tang}$ on the outer side of S' is similarly given by the sum of a "short-circuit" component $\bar{H}_g^{II}(Q)$ and a contribution due to $(-\bar{u}_n)\times\bar{E}$ (we write $-\bar{u}_n$ because it is the unit vector along the *outer* normal of Region II which should be used). This contribution can be written as

$$\bar{H}_b^{II}(Q) = -\iint_{S'} \mathcal{G}^{II}(Q\mid Q')\cdot\bar{u}_n\times\bar{E}(Q')dS'. \quad (3)$$

Equating the two values of $\bar{H}_{tang}$ on both sides of S' leads to the integral equation

$$\iint_S [\mathcal{G}^I(Q\mid Q') + \mathcal{G}^{II}(Q\mid Q')]\cdot\bar{u}_n\times\bar{E}(Q')dS' = \bar{H}_g^{II}(Q) - \bar{H}_g^{I}(Q). \quad (4)$$

At this point we introduce[3] a set of vectors $\bar{\alpha}_m$, complete in S', and satisfying the orthonormality property

$$\iint_{S'} \bar{\alpha}_m\cdot\bar{\alpha}_k^* dS' = \delta_{mk}. \quad (5)$$

Utilizing the expansions

$$\bar{E} = \sum_m V_m\bar{\alpha}_m$$

$$\bar{J}_s^I = \bar{H}_g^I\times\bar{u}_n = \sum_m I_{gm}^I\bar{\alpha}_m$$

$$\bar{J}_s^{II} = \bar{H}_g^{II}\times(-\bar{u}_n) = \sum_m I_{gm}^{II}\bar{\alpha}_m$$

$$\mathcal{G}^I(Q\mid Q') = -\sum_n\sum_p Y_{np}^I\bar{u}_n\times\bar{\alpha}_n(Q)\bar{u}_n\times\bar{\alpha}_p^*(Q')$$

$$\mathcal{G}^{II}(Q\mid Q') = -\sum_n\sum_p Y_{np}^{II}(-\bar{u}_n\times\bar{\alpha}_n(Q))\cdot(-\bar{u}_n\times\bar{\alpha}_p^*(Q')) \quad (6)$$

and equating the values of $\bar{H}\times\bar{u}_n$ on both sides of S' leads to the following network equations

$$I_{g1}^I - Y_{11}^IV_1 - Y_{12}^IV_2\cdots = -(I_{g1}^{II} - Y_{11}^{II}V_1 - Y_{12}^{II}V_2 - \cdots)$$

$$I_{g2}^I - Y_{21}^IV_1 - Y_{22}^IV_2\cdots = -(I_{g2}^{II} - Y_{21}^{II}V_1 - Y_{22}^{II}V_2 - \cdots) \quad (7)$$

$$\text{etc.}\cdots.$$

which can be written more concisely as

$$\bar{I}_g^{II} + \bar{I}_g^I = (\mathcal{Y}^I + \mathcal{Y}^{II})\cdot\bar{V}. \quad (8)$$

It will be noticed that $\bar{J}_S^I$ and $\bar{J}_S^{II}$ are the surface current densities on the short-circuited surface S'. A matrix such as $\mathcal{Y}^I$ has a simple physical interpretation. From (7), indeed, it is the linear relationship which exists between the tangential electric field on S' and the resulting field $\bar{u}_n\times\bar{H}_b$ on the cavity side of S'. Thus,

"Projection $\bar{u}_n\times\bar{H}_b^I$ on the $\bar{\alpha}$ space
$=\mathcal{Y}^I\times$projection of $\bar{E}$ on the $\bar{\alpha}$ space."

In circuit terms, $\mathcal{Y}^I$ is the admittance looking into I, i.e., the ratio between $\bar{H}\times\bar{u}_n$, and $\bar{E}_{tang}$, where $\bar{u}_n$ is the unit vector pointing inside I. The sign conventions embodied in (6) ensure that the admittance looking into a matched waveguide load, for instance, is equal to the *positive* characteristic resistance R_c (and not to $-R_c$). This point will be belabored in Section IV.

The circuit equation (8) can be represented schematically as in Fig. 2. It is seen that the field problem reduces to the determination of $\bar{V}$, i.e., of the inverse of the $\mathcal{Y}^I+\mathcal{Y}^{II}$ matrix. The $\mathcal{Y}$ matrix has the usual properties of an admittance matrix. Assume, for example, that the cavity contains an anisotropic medium whose ϵ and μ are Hermitian tensors ($\tilde{\epsilon}=\epsilon^*$ and $\tilde{\mu}=\mu^*$). It is easy to

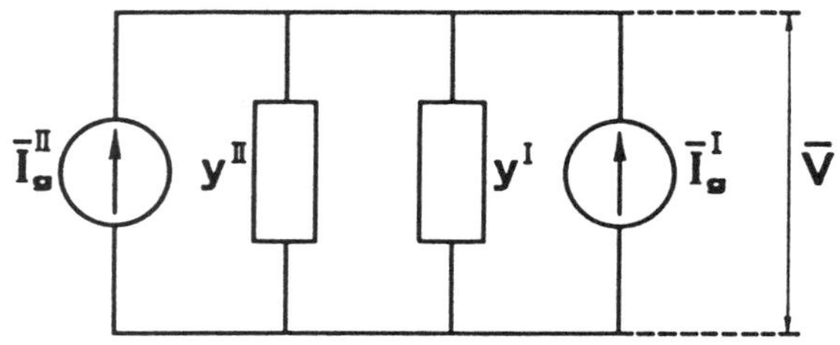

Fig. 2. Equivalent network for a scattering problem.

show, by classical methods,[2,4] that

$$Y_{ik} = -Y_{ki}^* \quad (9)$$

i.e., that the admittance matrix is skew-Hermitian. For symmetric tensors and real base vectors $\bar{\alpha}$, the matrix becomes symmetric, i.e.,

$$Y_{ik} = Y_{ki}. \quad (10)$$

III. Cavities at Resonance, Coupled Cavities, and Quadrupoles

Assume for simplicity that Cavity I does not contain any volume sources and is filled with a nondissipative medium. When k coincides with one of the resonant values k_ν, the Green's dyadic becomes infinite. The fields in the aperture, however, must remain finite because of the finite value of Q of the loaded cavity. It is therefore necessary [see (2)] that the tangential field satisfy the relationship.

$$\lim_{k\to k_\nu}\iint_{S'}(\bar{u}_n\times\bar{E})\cdot\bar{h}_\nu dS = 0. \quad (11)$$

[3] The idea of expanding the tangential fields in a complete set seems to be due to A. Tonning, "On the network description of electromagnetic field problems," Rept. AFCRL-62-967, 1962. Tonning utilizes two biorthogonal sets, one for $\bar{E}_{tang}$ and one for $\bar{H}_{tang}$.

[4] See, e.g., Tonning,[3] or R. F. Harrington and A. T. Villeneuve, "Reciprocity relationships for gyrotropic media," *IEEE Trans. on Microwave Theory and Techniques*, vol. MTT-6, pp. 308–310, July 1958.

The situation is not unlike that of Fig. 3, where the voltage across LC becomes zero at resonance (while the current remains finite) even though the admittance becomes infinite. For the cavity, similarly, the coefficient of excitation of the resonant mode $\bar{e}_\nu$, $\bar{h}_\nu$ must remain finite. It follows that the expression

$$\lim_{k\to k_\nu}\left[\iint_{S'} \frac{(\bar{u}_n \times \bar{E})\cdot \bar{h}_\nu dS}{k_\nu^2 - k^2}\right] \tag{12}$$

must remain finite. To further examine this limit, notice that the real eigenvector $\bar{h}_m$ can be represented by the following expansion, valid in S':

$$\bar{h}_m(Q') = \sum_n B_{nm}\bar{u}_n \times \bar{\alpha}_n(Q') = \sum_n B_{nm}{}^*\bar{u}_n \times \bar{\alpha}_n{}^*(Q').$$

It is then a simple matter to show that each resonant mode contributes a term

$$\frac{j\omega\epsilon}{k_\nu^2 - k^2} B_{m\nu}B_{n\nu}{}^*$$

to Y_{mn}. Let $\mathcal{Y}'$ be the matrix obtained by deleting the contribution of mode ν from $\mathcal{Y}^{\rm I}$. The circuit equations can now be written in the following way to emphasize the contribution of mode ν:

$$I_{g1}{}^{\rm II} = (Y_{11}' + Y_{11}{}^{\rm II})V_1 + (Y_{12}' + Y_{12}{}^{\rm II})V_2 + \cdots + \frac{j\omega\epsilon}{k^2 - k_\nu^2} B_{1\nu}[B_{1\nu}{}^*V_1 + B_{2\nu}{}^*V_2 + \cdots]$$

$$I_{g2}{}^{\rm II} = (Y_{21}' + Y_{21}{}^{\rm II})V_1 + (Y_{22}' + Y_{22}{}^{\rm II})V_2 + \cdots + \frac{j\omega\epsilon}{k^2 - k_\nu^2} B_{2\nu}[B_{1\nu}{}^*V_1 + B_{2\nu}{}^*V_2 + \cdots]. \tag{13}$$

etc. $\cdots$.

As k approaches k_ν, the quantity $D_\nu = B_{1\nu}{}^*V_1 + B_{2\nu}{}^*V_2 + \cdots$ must approach zero in such a manner that

$$\lim_{k\to k_\nu} j\omega\epsilon(B_{1\nu}{}^*V_1 + \cdots) = C(k^2 - k_\nu^2)$$

where C should be determined in order to fully evaluate the fields in the cavity. The formal solution proceeds by rewriting (13) in terms of C, and adding the condition $D_\nu = 0$ (the uncoupling condition) to the equations, viz.

$$(Y_{11}' + Y_{11}{}^{\rm II})V_1 + (Y_{12}' + Y_{12}{}^{\rm II})V_2 + \cdots + B_{1\nu}C = I_{g1}{}^{\rm II}$$

$$(Y_{21}' + Y_{21}{}^{\rm II})V_1 + (Y_{22}' + Y_{22}{}^{\rm II})V_2 + \cdots + B_{2\nu}C = I_{g2}{}^{\rm II}$$

$$B_{1\nu}{}^*V_1 + B_{2\nu}{}^*V_2 + \cdots = 0. \tag{14}$$

This is a system of equations with as many equations as unknowns, and out of which $V_1 V_2, \cdots, C$ can be determined.

The schematic diagram of Fig. 2 clearly shows that the region situated on a given side of the opening can be represented by a current generator in parallel with an admittance matrix. In this extension of Norton's theorem, generator and matrix have poles at the eventual resonant frequencies of the region. When two cavities are coupled together through a common aperture, the resonant frequencies of the total structure are determined by the condition that $(\mathcal{Y}^{\rm I} + \mathcal{Y}^{\rm II})\cdot\bar{V} = 0$ admits a nonzero solution for $\bar{V}$. This condition implies that the determinant of the (infinite) system vanishes,[5] which is the desired equation for the eigenfrequencies.

The equivalent network method also can be used to analyze the composite structure shown in Fig. 4. We now introduce two sets of vectors, $\bar{\alpha}_m'$ and $\bar{\alpha}_m''$, respectively complete and orthonormal on S' and S''. The network equations take the form

$$\bar{I}_g{}^{\rm II} + \bar{I}_g' = (\mathcal{Y}' + \mathcal{Y}^{\rm II})\cdot\bar{V}' + \mathcal{Y}^m\cdot\bar{V}''$$

$$\bar{I}_g{}^{\rm III} + \bar{I}_g'' = \mathcal{Y}^p\cdot\bar{V}' + (\mathcal{Y}'' + \mathcal{Y}^{\rm III})\cdot\bar{V}''. \tag{15}$$

As before, a matrix such as $\mathcal{Y}'$ represents the linear relationship between $\bar{E}_{\rm tang}$ and the resulting tangential field $\bar{\mu}_{n'}\times\bar{H}_b$ on S'. The currents $\bar{I}_g'$ and $\bar{I}_g''$ are the short-circuit currents produced by the volume sources of I on the short-circuited surfaces S' and S'', respectively. In the quadrupole equations (15), symmetry properties exist for the mutual admittance matrices $\mathcal{Y}^m$ and $\mathcal{Y}^p$. Remembering that $\mathcal{Y}^m\cdot\bar{V}''$, for example, represents the field, $\bar{\mu}_n'\times\bar{H}$ produced on the short-circuited surface S' by the electric field $\bar{E} = \Sigma V_m''\bar{\alpha}_m''$ on S'', it is easy to show[4] that

$$Y_{12}{}^p = -(Y_{21}{}^m)^* \tag{16}$$

[5] For an early application of this method to the nosed-in klystron cavity, see W. C. Hahn, "A new method for the calculation of cavity resonators." *J. Appl. Phys.*, vol. 12, pp. 62–68, January 1941.

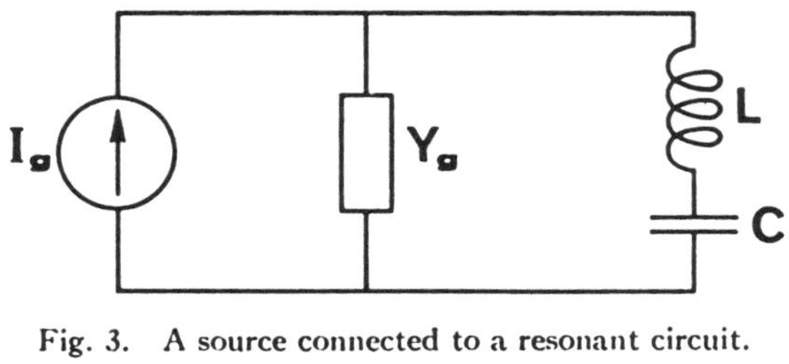

Fig. 3. A source connected to a resonant circuit.

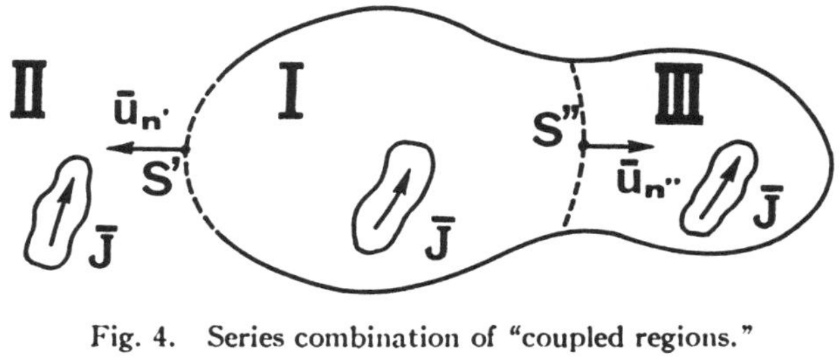

Fig. 4. Series combination of "coupled regions."

when the medium inside I is Hermitian, and

$$Y_{12}{}^{p} = Y_{21}{}^{m} \tag{17}$$

when the medium is symmetric and the $\bar{\alpha}$'s are real.

The quadrupole equations (15) allow one to apply the whole body of network theory to the cascade connection of electromagnetic structures. The structure to the left of S'' in Fig. 4, for example, can be replaced by a short-circuit current

$$\bar{I}_g{}'' - \mathcal{Y}^p\cdot(\mathcal{Y}' + \mathcal{Y}^{\text{II}})^{-1}\cdot(\bar{I}_g{}^{\text{II}} + \bar{I}_g{}') \tag{18}$$

in parallel with an admittance

$$\mathcal{Y}'' + \mathcal{Y}^p\cdot(\mathcal{Y}' + \mathcal{Y}^{\text{II}})^{-1}\cdot\mathcal{Y}^m. \tag{19}$$

IV. Application to a Waveguide Problem

Figure 5 shows a cavity I terminated by a waveguide arm. Seen from the cross section S', the cavity has an (assumedly given) admittance matrix $\mathcal{Y}'$. It is our purpose to determine the admittance $\mathcal{Y}''$ in S''.

The eigenvectors $\bar{\alpha}_n$ suitable for the present problem are, from classical waveguide theory,[2]

$$\text{grad } \phi_{mp}$$

$$\bar{u}_n \times \text{grad } \Psi_{ns} = \bar{u}_z \times \text{grad } \Psi_{ns}$$

where

$$\nabla_{xy}{}^2\Phi_{mp} + \mu_{mp}{}^2\Phi_{mp} = 0 \qquad \Phi_{mp} = 0 \text{ on contour } C \text{ of } S'$$

$$\nabla_{xy}{}^2\Psi_{ns} + \nu_{ns}{}^2\Psi_{ns} = 0 \qquad \frac{\partial\Psi_{ns}}{\partial n} = 0 \text{ on contour } C \text{ of } S'.$$

These eigenvectors are normalized in such a way that

$$\iint_S (\text{grad } \Phi_{mp})^2 dS = \iint_S (\text{grad } \Psi_{ns})^2 dS = 1.$$

The operator ∇^2 and the eigenvalues $\mu_{mp}{}^2$ and $\nu_{ns}{}^2$ are real. The eigenvectors can therefore be taken as real. We shall assume that all modes are damped except the lowest TE mode (subscript 11), and set

$$\gamma_{mp}{}^2 = \mu_{mp}{}^2 - k^2$$
$$\delta_{ns}{}^2 = \nu_{ns}{}^2 - k^2 \qquad (n, s \neq 1, 1)$$
$$k_{11}{}^2 = k^2 - \nu_{11}{}^2$$

with $k^2 = \omega^2\epsilon_0\mu_0$. The fields on S' can be expanded as

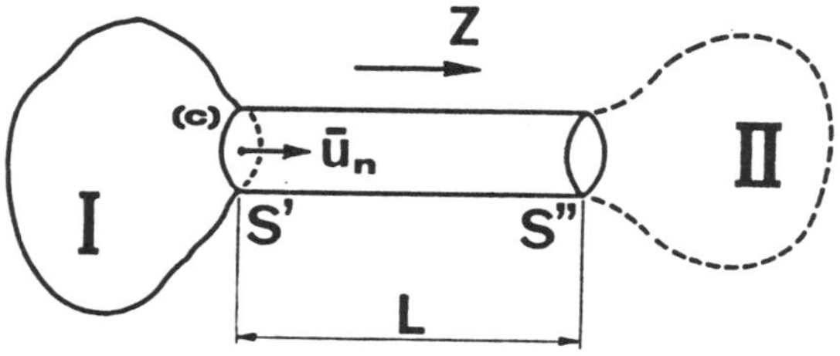

Fig. 5. Cavity with a waveguide output.

$$\bar{E} = \sum_{mp} V_{mp}{}' \text{ grad } \Phi_{mp} + \mathcal{V}_{11}{}' \text{ grad } \Psi_{11} \times \bar{u}_z + \sum_{n,s\neq 1,1} \mathcal{V}_{ns} \text{ grad } \Psi_{ns} \times \bar{u}_z$$

$$\bar{H} \times \bar{u}_z = \bar{H} \times \bar{u}_n = \sum_{mp} I_{mp}{}' \text{ grad } \Phi_{mp} + \mathcal{I}_{11}{}' \text{ grad } \Psi_{11} \times \bar{u}_z + \sum_{n,s\neq 1,1} \mathcal{I}_{ns}{}' \text{ grad } \Psi_{ns} \times \bar{u}_z. \tag{20}$$

Similar equations can be written for the fields on S'', provided the primes are replaced by double primes. Solution of the waveguide equations gives the following relationship between "voltages" and "currents." For the propagated mode:

$$\mathcal{V}_{11}{}' = \mathcal{V}_{11}{}'' \cos k_{11}L + jR_c\mathcal{I}_{11}{}'' \sin k_{11}L$$
$$\mathcal{I}_{11}{}' = \frac{j}{R_c}\mathcal{V}_1{}'' \sin k_{11}L + \mathcal{I}_{11}{}'' \cos k_{11}L \tag{21}$$

where $R_c = \omega\mu_0/R_{11}$. For the damped modes:

$$V' = V'' \cosh \gamma L + Z_c I'' \sinh \gamma L \tag{22}$$
$$I' = \frac{1}{Z_c} V'' \sinh \gamma L + I'' \cosh \gamma L$$

and

$$\mathcal{V}' = \gamma'' \cosh \delta L + \mathcal{Z}_c\mathcal{I}'' \sinh \delta L$$
$$\mathcal{I}' = \frac{1}{\mathcal{Z}_c}\mathcal{V}'' \sinh \delta L + \mathcal{I}'' \cosh \delta L \tag{23}$$

where

$$Z_c = \frac{\gamma}{j\omega\epsilon_0} \quad \text{and} \quad \mathcal{Z}_c = \frac{j\omega\mu_0}{\delta}.$$

The primed voltages and currents which appear in (21) through (23) are relative to the fields to the right of S'. The magnetic field to the left of S' is given by the vector

$$\bar{I}' = \bar{I}_g{}' - \mathcal{Y}'\cdot\bar{V}' \tag{24}$$

where $\bar{I}_g{}'$, for example, is the column vector representing the short-circuit field

$$\bar{H}_g{}' \times \bar{u}_n = \sum_{mp} u_{mp}{}' \text{ grad } \Phi_{mp} + \mathcal{U}_{11}{}' \text{ grad } \Psi_{11} \times \bar{u}_z + \sum_{n,s\neq 1,1} \mathcal{U}_{ns}{}' \text{ grad } \Psi_{ns} \times \bar{u}_z.$$

The components of $\bar{I}_g{}'$ are, therefore, $u_{11}{}'$, $u_{12}{}'$, $\cdots$, $\mathcal{U}_{11}{}'$, $\mathcal{U}_{12}{}'$, $\cdots$. Introducing the values of $\bar{I}'$ and $\bar{V}'$ obtained from (21) through (23) into (24) leads to an equation of the type

$$\bar{I}'' = \bar{I}_g{}'' - \mathcal{Y}''\cdot\bar{V}'' \tag{25}$$

where the desired Norton $\bar{I}_g{}''$ and $\mathcal{Y}''$ can easily be found after some cumbersome algebraic manipulation.

One obtains

$$\bar{I}_g'' = [C + \mathcal{Y}'\cdot\mathcal{Y}_c^{-1}\cdot S]^{-1}\cdot\bar{I}_g' \qquad (26)$$

$$\mathcal{Y}'' = [C + \mathcal{Y}'\cdot\mathcal{Y}_c^{-1}\cdot S]^{-1}\cdot[\mathcal{Y}'\cdot C + \mathcal{Y}_c\cdot S] \qquad (27)$$

where the diagonal matrices $\mathcal{Y}_c$, C, and S are given by

$$\mathcal{Y}_c = \begin{pmatrix} \frac{1}{Z_{c11}} & 0 & 0 & \cdots \\ 0 & \frac{1}{Z_{c12}} & & \\ & & \ddots & \\ 0 & & \frac{1}{R_c} & \\ 0 & & & \frac{1}{Z_{c12}} \\ \vdots & & & & \ddots \end{pmatrix}$$

$$C = \begin{pmatrix} \cosh\gamma_{11}L & 0 & 0 & \cdots \\ 0 & \cosh\gamma_{12}L & & \\ & & \ddots & \\ 0 & & \cos k_{11}L & \\ \vdots & & & \cosh\delta_{12}L \\ \vdots & & & & \ddots \end{pmatrix}$$

$$S = \begin{pmatrix} \sinh\gamma_{11}L & 0 & \cdots & \cdots \\ 0 & \sinh\gamma_{12}L & & \\ & & \ddots & \\ 0 & & \sin k_{11}L & \\ \vdots & & & \sinh\delta_{12}L \\ \vdots & & & & \ddots \end{pmatrix}.$$

Equations (25) through (27) allow one to see how the Norton terms vary with the distance L. For very large L, the contribution from the damped modes tends to vanish, and (25) takes the form

$$\begin{pmatrix} I_{11}'' \\ I_{12}'' \\ \vdots \\ \mathfrak{I}_{11}'' \\ \\ \mathfrak{I}_{12}'' \\ \vdots \end{pmatrix} = \begin{pmatrix} 0 \\ 0 \\ \vdots \\ \mathfrak{I}''_{g11} \\ \\ 0 \\ \vdots \end{pmatrix} - \begin{pmatrix} \frac{1}{Z_{c11}} & 0 & & \cdots \\ 0 & \frac{1}{Z_{c12}} & & \\ & & \ddots & \\ 0 & & \mathcal{Y}_{11}'' & \\ \vdots & & & \frac{1}{Z_{c12}} \end{pmatrix} \cdot \begin{pmatrix} V_{11}'' \\ V_{12}'' \\ \vdots \\ \mathcal{V}_{11}'' \\ \\ \mathcal{V}_{12}'' \\ \vdots \end{pmatrix} \qquad (28)$$

where the contribution from the propagated mode has the value

$$\mathfrak{I}_{g11}'' = \frac{\mathfrak{I}_{g11}'}{\cos k_{11}L + jR_c\mathcal{Y}_{11}'\sin k_{11}L}$$

$$\mathcal{Y}_{11}'' = \frac{1}{R_c}\,\frac{R_c\mathcal{Y}_{11}' + j\tan k_{11}L}{1 + jR_c\mathcal{Y}_{11}'\tan k_{11}L}. \qquad (29)$$

The actual value of the voltage vector $\bar{V}''$, i.e. of the coefficient of excitation of the various modes at the level of cross section S'', cannot be determined unless the input admittance of Volume II is given (Fig. 5). If II is an infinitely long waveguide, for instance, one has, in addition to (28), the relationship

$$\bar{I}'' = \mathcal{Y}_c\cdot\bar{V}''.$$

Comparison of the two equations leads to the expected result that all V's vanish except $\mathcal{V}_{11}''$, i.e., that no damped modes exist at the "junction" S''.

To illustrate the actual calculation of an equivalent circuit, we consider the slot antenna shown in Fig. 6, where the contour of the cross section is arbitrary. It is assumed that the slot is resonant, so that the voltage across the slot varies according to the relationship

$$v = V_0\cos\frac{\pi}{L}(Z - Z_0)$$

where Z_0 refers to the position of the center of the slot, and L (approximately equal to $\lambda/2$) is the length of the slot. Let c_0 be the value of the contour coordinate at the center of the slot. We assume that the voltage across the slot is a constant, independent of the loading conditions, so that the tangential electric field in the slot is of the form

$$\bar{E}_{\text{tang}} = V_0\cos\frac{\pi}{L}(Z - Z_0)\bar{u}_c\delta(c - c_0). \qquad (30)$$

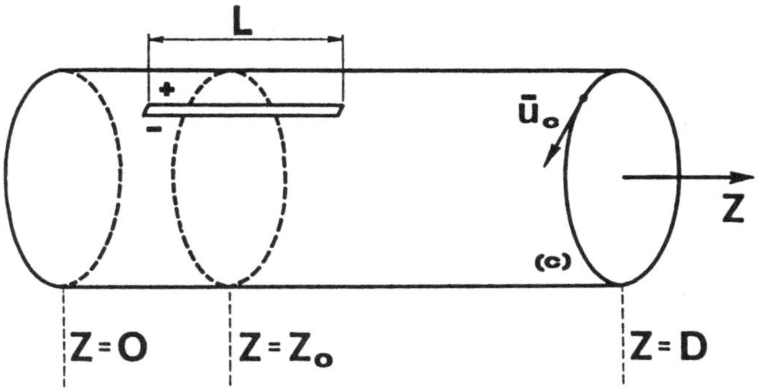

Fig. 6. Waveguide with a slot.

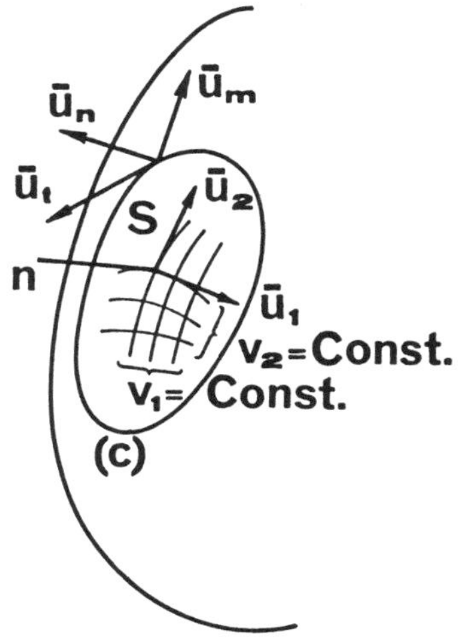

Fig. 7. Opening of arbitrary shape.

Utilizing this value of the boundary excitation in the waveguide equations[2] allows determination of the cross-sectional fields at $z'=D$, i.e., of $\mathfrak{J}_{e11}''$ and $\mathcal{Y}_{11}''$. With a short circuit at $z=0$, one finds, for example,

$$\bar{E} = \mathcal{V}_{11} \operatorname{grad} \psi_{11} \times \bar{u}_z$$

$$\bar{H} \times \bar{u}_z = \Bigg[\underbrace{\frac{2\pi\nu_{11}^2\psi_{11}(c_0)}{j\omega\mu_0 L N_{11}^2\left(k_{11}^2 - \dfrac{\pi^2}{L^2}\right)} \cos\frac{k_{11}L}{2} \, \frac{\sin(k_{11}Z_0)}{\sin(k_{11}D)} V_0}_{\mathfrak{J}_{e11}''} - \underbrace{\frac{k_{11}}{j\omega\mu_0 \tan k_{11}D}}_{\mathcal{Y}_{11}''} \mathcal{V}_{11}'' \Bigg] \operatorname{grad} \Psi_{11} \times \bar{u}_z$$

in which the normalization factor

$$N_{11}^2 = \iint (\operatorname{grad} \Psi_{11})^2 dS$$

has been left unspecified, and where $\psi_{11}(c_0)$ is the value of the eigenfunction at the location of the slot.

The choice of the most suitable eigenvectors $\bar{\alpha}$ is simple for a waveguide cross section. For an opening S of arbitrary shape, we propose the following generalization for the waveguide eigenvectors.[6] We choose orthogonal coordinates v_1, v_2, taken for instance along the lines of curvature of the surface[2], and such that the direction of increasing v_1, the direction of increasing v_2, and the positive normal n form a right-handed system of axes (see Fig. 7). An increase dv_1, dv_2 in the value of the coordinates results in a displacement dl of magnitude.

$$dl = (h_1^2 dv_1^2 + h_2^2 dv_2^2)^{1/2}.$$

Some important surface differential operators for a scalar function $f(v_1, v_2)$ are

$$\operatorname{grad}_s f = \frac{1}{h_1}\frac{\partial f}{\partial v_1}\bar{u}_1 + \frac{1}{h_2}\frac{\partial f}{\partial v_2}\bar{u}_2$$

$$\nabla_S^2 f = \frac{1}{h_1 h_2}\frac{\partial}{\partial v_1}\left(\frac{h_2}{h_1}\frac{\partial f}{\partial v_1}\right) + \frac{1}{h_1 h_2}\frac{\partial}{\partial v_2}\left(\frac{h_1}{h_2}\frac{\partial f}{\partial v_2}\right)$$

where $\bar{u}_1$ and $\bar{u}_2$ are vectors of unit length and tangent, respectively, to the curves of constant v_2 and constant v_1, and directed to increasing coordinates. The Dirichlet functions of interest are now defined by the relationships

$$\nabla_S^2 \Phi_{mp} + \mu_{mp}^2 \Phi_{mp} = 0$$
$$\Phi_{mp} = 0 \text{ on } (C).$$

The Neumann eigenfunctions are defined by

$$\nabla_S^2 \Psi_{ns} + \nu_{ns}^2 \Psi_{ns} = 0$$
$$\frac{\partial \Psi_{ns}}{\partial m} = 0 \text{ on } (C)$$

where m is a direction in the tangent plane, perpendicular to (C), and directed outward from the region enclosed by (C). Utilizing the basic relationships[2]

$$\iint_S (A\nabla_S^2 B - B\nabla_S^2 A)\, dS = \int_C (A \operatorname{grad}_s B - B \operatorname{grad}_s A)\cdot \bar{u}_m \, dC \quad (31)$$

$$\iint_S (A\nabla_S^2 B + \operatorname{grad}_s A \cdot \operatorname{grad}_s B)\, dS = \int_C A(\bar{u}_m \cdot \operatorname{grad} B)\, dS \quad (32)$$

it is easy to show that the eigenvalues μ^2 and ν^2 are real and non-negative (and hence, that the eigenfunctions can be chosen real), that the Φ's form an orthogonal system and the Ψ's another orthogonal system, and that the vectors

$$\operatorname{grad}_s \Phi_{mp}$$
$$\operatorname{grad}_s \Psi_{ns} \times \bar{u}_n$$

form an orthogonal system on S. These vectors are normal to the contour (C) of the aperture, and are therefore particularly well suited for expanding the vectors $\bar{E}_{\text{tang}}$ and $\bar{H}\times\bar{u}_n$, which are also normal to (C).

[6] An actual example of application of these generalized eigenvectors can be found in Van Bladel,[2] p. 467, where the aperture S' is a spherical cap.

Radiation from Apertures in Conducting Cylinders of Arbitrary Cross Section

ROBERT F. WALLENBERG, MEMBER, IEEE, AND ROGER F. HARRINGTON, FELLOW, IEEE

Abstract—An analysis of two-dimensional radiation from apertures in perfectly conducting cylinders of arbitrary cross section is given. Solutions are expressed in terms of generalized network parameters, obtained by applying moment methods to the superposition integral equation. Formulas are given for current distributions, self- and mutual admittances, and radiation patterns. Representative computations are included to illustrate the theory.

I. Introduction

A UNIFIED analysis of the problem of two-dimensional radiation from a perfectly conducting cylinder of arbitrary shape is presented. The analysis is similar to that for the scattering problem [1], [2]. Integral equation formulations are given for both transverse magnetic (TM) and transverse electric (TE) polarizations. Two TE formulations are presented, one based on computing the electric field due to equivalent magnetic surface currents, and the other based on the retarded scalar and vector potential integral formulas.

The solution is described in terms of the method of moments, by which the integral equations are reduced to matrix equations. The solution is basically one which treats the cylinder as a collection of current elements, and expresses the interaction between every pair of elements by an impedance matrix. The excitation of the cylinder is due to an impressed electric field and is represented by a voltage matrix. The current on the cylinder is found by inverting the impedance matrix and multiplying the resulting admittance matrix by the voltage matrix. Computation of scattering and radiation patterns correspond to an additional matrix multiplication with a measurement matrix. The radiation admittance of an aperture and mutual admittance between pairs of apertures are computed from the matrix solution for the tangential electric current, given by $n \times H$.

The impedance and the $n \times H$ matrices are computed using a point matching approximation; that is, the field equations are satisfied at discrete points along the conducting boundary. Numerical results for current distributions and radiation patterns of various cylindrical antennas are also presented. Using the equivalence principle and duality [3], the results may be directly extended to give a surface formulation of the problem of scattering and radiation from a dielectric cylinder of arbitrary cross section [4].

Consider an aperture antenna consisting of a conducting cylinder and one or more apertures through which electromagnetic energy may pass. A knowledge of the tangential components of E over a closed surface S is sufficient to uniquely determine the field external to S. Over the conductor tangential E is zero, and over the aperture tangential $E = E^a$ is assumed known. According to the equivalence principle [3], many different source distributions on or within S may produce the correct field external to S. Hence there are many equivalent formulations for the problem. These will all be of the form

$$L\begin{pmatrix} J \\ M \end{pmatrix} = -E_{\text{tang}} \tag{1}$$

where

$$E_{\text{tang}} = \begin{cases} E^a, & \text{in the aperture} \\ 0, & \text{on the conductor.} \end{cases} \tag{2}$$

In (1) J and M represent electric and magnetic currents on or within S, and L represents the integral operator relating J, or M, or both, to the tangential E on S.

Fig. 1 illustrates some of the possible equivalent problems for the region external to C, the contour bounding the cylinder. Fig. 1(a) shows the equivalent currents J and M on C, which produce the correct field E, H external to C, and zero field internal to C. They are given by

$$J = n \times H, \quad M = E \times n \tag{3}$$

where n is the unit outward normal to C. Note that M is known because E_{tang} is known, but J is unknown. Equation (1) therefore represents an operator equation for the unknown J on C. Fig. 1(c) shows an equivalent electric current J^e on C, which produces the correct field E, H external to C, and a field E', H' internal to C. Note that the current J of Fig. 1(c) is not equal to that of Fig. 1(b), but

$$J^e = n \times (H - H'). \tag{4}$$

Equation (1) now reduces to $L(J^e) = -E_{\text{tang}}$, an operator equation for the unknown J^e on C. Finally, Fig. 1(d) shows an equivalent magnetic current M^e on C, which produces the correct field E, H external to C, and a field E'', H'' internal to C. Again M^e of Fig. 1(d) is not equal to that of Fig. 1(b), but

$$M^e = (E - E'') \times n. \tag{5}$$

Manuscript received July 1, 1968; revised September 19, 1968. This work was supported in part by the Rome Air Development Center under Contract AF 30(602)-3724 and in part by NSF under Grant GK-1233.

R. F. Wallenberg is with the Special Projects Laboratory, Syracuse University Research Corporation, Syracuse, N. Y. 13210.

R. F. Harrington is with the Department of Electrical Engineering, Syracuse University, Syracuse, N. Y. 13210.

Reprinted from *IEEE Trans. Antennas Propagat.*, vol. AP-17, no. 1, pp. 56–62, January 1969.

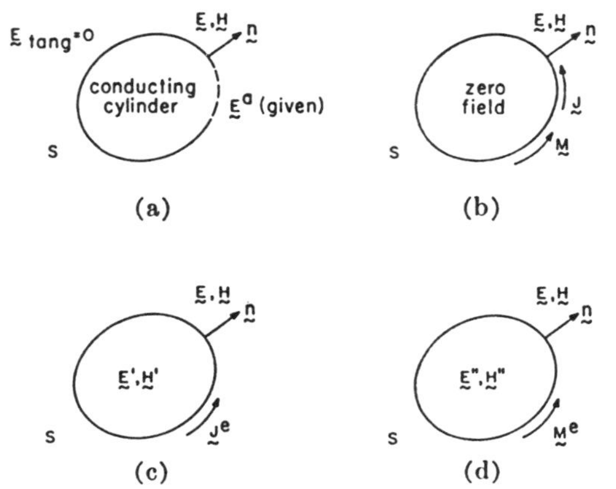

Fig. 1. (a) Original problem. (b) Equivalent $\boldsymbol{J}$ and $\boldsymbol{M}$ on S. (c) Equivalent $\boldsymbol{J}^e$ only on S. (d) Equivalent $\boldsymbol{M}^e$ only on S.

Now (1) reduces to $L(\boldsymbol{M}^e) = -\boldsymbol{E}_{\text{tang}}$, an operator equation for the unknown $\boldsymbol{M}^e$ on C.

In the equivalent problem of Fig. 1(b), the current $\boldsymbol{J}$ on C outside of the aperture is the actual current on the conductor of Fig. 1(a). The currents $\boldsymbol{J}$ and $\boldsymbol{M}$ in the aperture are equivalent currents. The equivalent problem of Fig. 1(c) can be viewed as the sum of two problems, one external to C and the other internal to C. The current associated with the external problem is

$$\boldsymbol{J} = \boldsymbol{n} \times \boldsymbol{H}|_{C+} \qquad (6)$$

where the subscript $C+$ denotes that $\boldsymbol{H}$ is evaluated just outside the contour C. The interior current density $\boldsymbol{J}^{\text{int}}$ may be found by a similar formula or by subtracting the external current from $\boldsymbol{J}$. Hence

$$\boldsymbol{J}^{\text{int}} = -\boldsymbol{n} \times \boldsymbol{H}|_{C-} = \boldsymbol{J}^e - \boldsymbol{n} \times \boldsymbol{H}|_{C+} \qquad (7)$$

where the subscript $C-$ denotes that $\boldsymbol{H}$ is evaluated just inside the contour C. Note that, if only the fields external to C are of interest, one need not calculate $\boldsymbol{J}$ or $\boldsymbol{J}^{\text{int}}$. The internal current produces no field external to C, and $\boldsymbol{J}^e$ produces the same field as $\boldsymbol{J}$ external to C. However, a knowledge of $\boldsymbol{J}$ is convenient for computing aperture admittances.

The equivalent problem of Fig. 1(d) has an interpretation analogous to that for Fig. 1(c). Again, the equivalent problem of Fig. 1(d) can be viewed as the sum of two problems, one external to C and the other internal to C. The magnetic currents associated with the external and internal problems are given by equations dual to (6) and (7), but they are of no direct interest. The electric current on C in the original problem is again given by (6), where $\boldsymbol{H}$ is now the magnetic field from $\boldsymbol{M}$. If only fields external to C are desired, one need not compute the current $\boldsymbol{J}$, since $\boldsymbol{M}$ is equivalent to $\boldsymbol{J}$ in the external region.

II. Integral Equation Formulations

A. Transverse Magnetic Formulation

For TM fields the electric field E_z and equivalent surface current J_z are in the z direction. The field at $\boldsymbol{\rho} = \boldsymbol{u}_x x + \boldsymbol{u}_y y$ is given by

$$E_z(\boldsymbol{\rho}) = -\frac{k\eta}{4}\int_C J_z(\rho')H_0^{(2)}(kR)\,dl' \qquad (8)$$

where the integration is over the contour C of the cylinder, l' is the length variable around C, and

$$R = |\boldsymbol{\rho} - \boldsymbol{\rho}'| = \sqrt{(x - x')^2 + (y - y')^2}. \qquad (9)$$

Here, $k = \omega\sqrt{\mu\epsilon} = 2\pi/\lambda$ (λ = wavelength), $\eta = \sqrt{\mu/\epsilon}$ is the impedance of free space, and $H_0^{(2)}$ is the Hankel function of the second kind and zero order. The boundary condition is given by

$$E_z(\boldsymbol{\rho})|_C = \begin{cases} E_z^a, & \text{in the aperture} \\ 0, & \text{on the conductor} \end{cases} \qquad (10)$$

where $|_C$ indicates that the field point $\boldsymbol{\rho}$ is on C. In operator form the integral equation to be solved is then

$$-E_z(\boldsymbol{\rho})|_C = L(J_z) \qquad (11)$$

where

$$L(J_z) = \frac{k\eta}{4}\int_C J_z(\boldsymbol{\rho}')H_0^{(2)}(kR)\,dl'|_C. \qquad (12)$$

The field $E_z(\rho)|_C$ is known, and J_z is the unknown to be determined. This is the formulation corresponding to Fig. 1(c).

B. Magnetic Current Formulation of the TE Problem

The formulation for the TE problem dual to the above TM formulation is obtained from the magnetic current representation of Fig. 1(d). It is in some respects similar to the $\boldsymbol{H}$-field formulation used for scattering problems in the literature [1], [2]. However, the $\boldsymbol{H}$-field method requires a specification of H_z on $C+$, which is not readily available in the aperture problem.

The TE field has only a z component of $\boldsymbol{H}$ and a transverse component of $\boldsymbol{E}$ and $\boldsymbol{J}$. It may be related to an equivalent z-directed magnetic surface current M_z by the electric vector potential $\boldsymbol{F}$ as

$$\boldsymbol{E} = -\nabla \times \boldsymbol{F} \qquad (13)$$

where

$$\boldsymbol{F}(\boldsymbol{\rho}) = \frac{1}{4j}\int_C \boldsymbol{u}_z M_z(\boldsymbol{\rho}')H_0^{(2)}(kR)\,dl'. \qquad (14)$$

Forming the tangential component of the electric field on $C+$, interchanging the order of integration and differentiation, and using the relationship between unit normal and unit tangent on C, one obtains the integral equation

$$-\boldsymbol{E}(\boldsymbol{\rho})|_{\text{tang}} = L(M_z) \qquad (15)$$

where

$$L(M_z) = \frac{k}{4j}\int_C M_z(\rho')\left(\frac{\boldsymbol{n}\cdot\boldsymbol{R}}{R}\right)H_1^{(2)}(kR)\,dl'|_{C+}. \qquad (16)$$

Here $\boldsymbol{E}_{\text{tang}}$ is the known field (2), $\boldsymbol{n}$ is the unit normal to C at $\boldsymbol{\rho}$, and M_z is the unknown to be determined. Once $\boldsymbol{M}$ is found, the current on C can be obtained from (6).

C. Electric Current Formulation of the TE Problem

The TE problem can also be formulated directly in terms of electric current on C, using the equivalence of Fig. 1(c). However, there is now a charge accumulation σ due to the tangentially directed electric currents given by

$$\sigma = \frac{-1}{j\omega} \nabla \cdot J. \tag{17}$$

This results in an electric scalar potential Φ as well as a magnetic vector potential $\boldsymbol{A}$. The electric field is related to the potentials by

$$\boldsymbol{E} = -j\omega A - \nabla\Phi. \tag{18}$$

The derivatives in (17) and (18) result in an integro-differential equation, which is more difficult to solve approximately than a purely integral equation. One solution involves the use of finite difference approximations to the derivatives, as used in the analogous three-dimensional problem of wire antennas and scatterers [5], [6]. However, better results can be obtained by using actual derivatives of J and Φ if sufficient care is exercised.

The derivation of the operator equation is straightforward, resulting in

$$-\boldsymbol{E}(\boldsymbol{\varrho})_{\text{tang}} = L(J^e) \tag{19}$$

where

$$L(J^e) = \frac{k\eta}{4} \int_C \left[J^e(\boldsymbol{\varrho}')(\boldsymbol{u}_l \cdot \boldsymbol{u}_{l'}) H_0^{(2)}(kR) - \left(\frac{\boldsymbol{u}_l \cdot \boldsymbol{R}}{R}\right) H_1^{(2)}(kR) \frac{dJ^e(\rho')}{dkl'} \right] dl' \Big|_C. \tag{20}$$

Again $\boldsymbol{E}(\boldsymbol{\varrho})_{\text{tang}}$ is specified on C by (2), $\boldsymbol{u}_l$ is the unit tangent at $\boldsymbol{\varrho}$, and $J^e(\rho')$ is the unknown to be determined. Note that in this formulation both $J^e(\rho')$ and its derivative must be approximated. In contrast to the magnetic current formulation, the electric current formulation applies to open-ended infinitesimally thin surfaces because the electric field is continuous across an electric current [7]. The true electric current J on C in the aperture problem is given by (6).

D. Other Formulations

One can, of course, use the magnetic current formulation for the TM problem, but the resultant equation is more complicated than that for the electric current formulation of the preceding section. The mixed $\boldsymbol{J},\boldsymbol{M}$ formulations represented by Fig. 1(b) can also be used in either the TE or TM problems, but there is no advantage in using them as long as conducting cylinders are used. However, for problems involving homogeneous dielectric or magnetic cylinders, an equivalence similar to Fig. 1(b) leads to a surface integral equation [4]. This is in contrast to the volume integral equation used previously [8], [9].

III. Matrix Solutions

The solution is obtained by using the method of moments to approximate the previous integral equations by a set of matrix equations [6]. The matrix problem to be solved is then of the form

$$[Z_{mn}][I_n] = [V_m]. \tag{21}$$

The solution for the current is given by the inversion of (21), which is

$$[I_n] = [Y_{nm}][V_m] \tag{22}$$

where $[Y_{nm}] = [Z_{mn}]^{-1}$ is the generalized admittance matrix for the cylinder.

The integral equations to be solved are all of the form

$$L(f) = g \tag{23}$$

where L is a linear operator, g is the applied tangential electric field, and f is the electric or magnetic current to be determined. The current f is not denoted as a vector because its direction has been incorporated into the operator L. The method of moments requires that f be expanded in a set of expansion functions $f_1, \cdots, f_n$ in the domain of L, as

$$f = \sum_{n=1}^{N} I_n f_n \tag{24}$$

where I_n are, in general, complex constants to be determined. Substituting (24) into (23), using the linearity of L, and denoting a finite sum by the usual method, one has

$$\sum_{n}^{N} I_n L(f_n) = g. \tag{25}$$

Using the point-matching technique, one requires (25) to be satisfied at each point $l' = m$ in the middle of each subsection of the cylinder contour. This yields a matrix equation of the form (21). The elements of $[I_n]$ are I_n, and those of $[V_m]$ are $g(m)$, the tangential electric field at the center of the mth subsection. Examples of V_m and f_n are given later.

To express the approximate current f over the entire contour, each expansion function is multiplied by a vector $\boldsymbol{u}_n$, which is either z directed or in the direction of the cylinder contour, depending on the current direction the f_n represent. On defining the row matrix

$$[\tilde{f}_n] = [\boldsymbol{u}_1 f_1, \boldsymbol{u}_2 f_2, \cdots, \boldsymbol{u}_N f_N] \tag{26}$$

one can write the solution as

$$f = [\tilde{f}_n][I_n] = [\tilde{f}_n][Y_{nm}][V_m]. \tag{27}$$

A substantial amount of computer time can be saved when the cylinder is symmetrical about an axis. This is accomplished by expanding the excitation and current into even and odd functions of ϕ and by solving the two resulting even and odd matrix equations independently [2], [5].

The integral equations for $\boldsymbol{n} \times \boldsymbol{H}$ are of the same form as (23). However, now f is approximated as the expansion (24) whose coefficients I_n have been determined, and g is the unknown surface current $\boldsymbol{n} \times \boldsymbol{H}$ on a two-dimensional obstacle. The method of moments leads to a set of matrix equations analogous to (21) which are written as

$$[H_m] = [M_{mn}][I_n] = [M_{mn}][Y_{nm}][V_m]. \tag{28}$$

Here $[M_{mn}]$ is a square matrix whose elements are defined exactly as those of the impedance matrix. The column matrix $[I_n]$ is that of (21), and when point matching is used, the element H_m becomes the value $\boldsymbol{n} \times \boldsymbol{H}$ at the mth point. No matrix inversion is required, so (28) may be solved directly for the elements of $[H_m]$. Using the vector expansion functions defined by (26) the solution for $\boldsymbol{n} \times \boldsymbol{H}$ on the contour may be written as

$$\boldsymbol{n} \times \boldsymbol{H} = [\hat{f}_m][H_m] = [\tilde{f}_m][M_{mn}][Y_{nm}][V_m]. \quad (29)$$

The current approximation employed in the computations amounts to replacing the flat top of a pulse centered at the nth point with a second-order polynomial. The current approximation is continuous in an interval but is discontinuous at the end points. The term dJ/dl in (20) is evaluated by differentiating the current approximation using the chain rule. The delta functions which occur at the ends of an interval account for charge accumulation where the current is discontinuous. The method of dividing the cylinder contour into segments and integrating over singular and nonsingular terms has been treated elsewhere [5], [7], [11].

IV. Aperture Antenna Parameters

A. Radiation Field

The radiation field from the cylinder can be obtained by numerically evaluating the far-field integrals or by using reciprocity as suggested by Harrington [6]. Fig. 2 represents a distant z-directed electric or magnetic current filament at $\boldsymbol{\rho}_0$ of strength I_r or M_r (subscripts r denote receiver), adjusted to produce the unit plane wave incident on the obstacle,

$$\left.\begin{matrix} \boldsymbol{E}^i \\ \boldsymbol{H}^i \end{matrix}\right\} = \boldsymbol{u}_r \exp^{(j\boldsymbol{k}_r \cdot \boldsymbol{\rho}_n)}. \quad (30)$$

For the TM case it is convenient to consider the filament as an electric current producing a z-directed $\boldsymbol{E}^i$. For the TE case the filament is a magnetic current producing ϕ-directed $\boldsymbol{E}^i$ or a z-directed $\boldsymbol{H}^i$, depending on whether the induced surface currents are formulated as electric or magnetic. The unit vector $\boldsymbol{u}_r$ specifies the polarization of the wave. The wavenumber vector $\boldsymbol{k}_r$ points in the direction of wave travel, and $\boldsymbol{\rho}_n$ is the radius vector from the origin to a point n on the antenna. By reciprocity [3]

$$E_r = \frac{1}{I_r} \int_C \boldsymbol{E}^i \cdot \boldsymbol{J}\, dl \qquad \text{TM} \quad (31)$$

or

$$-H_r = \frac{1}{M_r} \int_C \left\{ \begin{matrix} \boldsymbol{E}^i \cdot \boldsymbol{J} \\ -\boldsymbol{H}^i \cdot \boldsymbol{M} \end{matrix} \right\} dl \quad \text{TE}. \quad (32)$$

The fields E_r and H_r are z directed (as I_r or M_r) and are due to the induced surface current $\boldsymbol{J}$ or $\boldsymbol{M}$ on the conductor. For the electric current formulations the integrand is $\boldsymbol{E}^i \cdot \boldsymbol{J}$, and for the magnetic current formulation it is $-\boldsymbol{H}^i \cdot \boldsymbol{M}$. Using the solution for $\boldsymbol{J}$ or $\boldsymbol{M}$ given by (27), the fields outside the cylinder are given by [5]

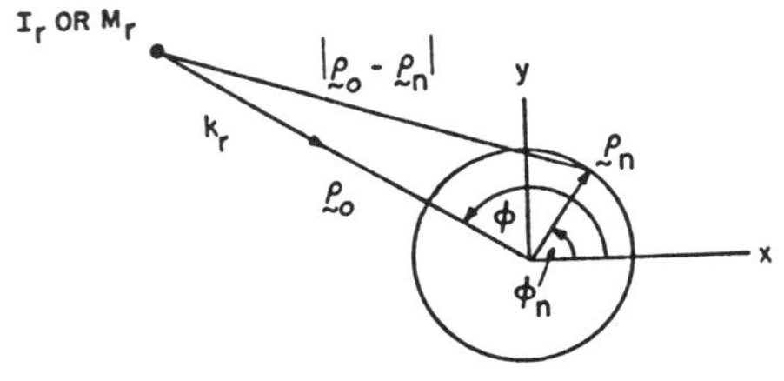

Fig. 2. Two-dimensional conductor and distant z-directed current filament.

$$E_r = -\frac{k\eta}{4} H_0^{(2)}(k\rho_0)[\hat{V}_n^r][Y_{nm}][V_m] \quad \text{TM} \quad (33)$$

$$H_r = \frac{k}{4\eta} H_0^{(2)}(k\rho_0)[\hat{V}_n^r][Y_{nm}][V_m] \quad \text{TE} \quad (34)$$

where the elements of the measurement matrix $[\hat{V}_n^r]$ are given by

$$V_n^r = \int_C \left\{ \begin{matrix} E_z^i J_n \\ \boldsymbol{u}_\phi E_\phi^i \cdot \boldsymbol{u}_{t_n} J_n \\ -H_z^i M_n \end{matrix} \right\} dl, \quad n = 1, \cdots, N \quad (35)$$

and depend on the formulation used.

The power gain pattern of the antenna is given by

$$g(\phi) = \lim_{\rho_0 \to \infty} \frac{2\pi\rho_0}{\eta} \frac{|E_r(\phi)|^2}{P_{\text{in}}} = \lim_{\rho_0 \to \infty} 2\pi\rho_0\eta \frac{|H_r(\phi)|^2}{P_{\text{in}}} \quad (36)$$

where P_{in} is the time average power input to the antenna. Using the matrix solution for the fields, given by (33) or (34), one obtains for the antenna gain

$$g(\phi) = K \frac{|[\hat{V}_n^r][Y_{nm}][V_m]|^2}{P_{\text{in}}} \quad (37)$$

where $K = k\eta/4$ for TM polarization and $K = k/4\eta$ for TE polarization. A convenient method for calculating the input power is to integrate over the far-field pattern. The result also gives the total radiation conductance of an aperture system [5].

The complex power supplied by an aperture in a cylinder may also be obtained using the moment solution (27) and integrating over the aperture surface [6], [12]. If the complex power is calculated using (27), the real part will be correct, but the imaginary part will include the energy radiated both external and internal to the conductor surface. The correct solution for the current due to an aperture in a solid conductor is given by (29).

B. Excitation and Measurement Matrix Elements

When the incident field is z directed, as in the TM formulation, the matrix element V_n^r is approximated as

$$V_n^r \approx \Delta l_n \exp[jk(x_n \cos\phi + y_n \sin\phi)]. \quad (38)$$

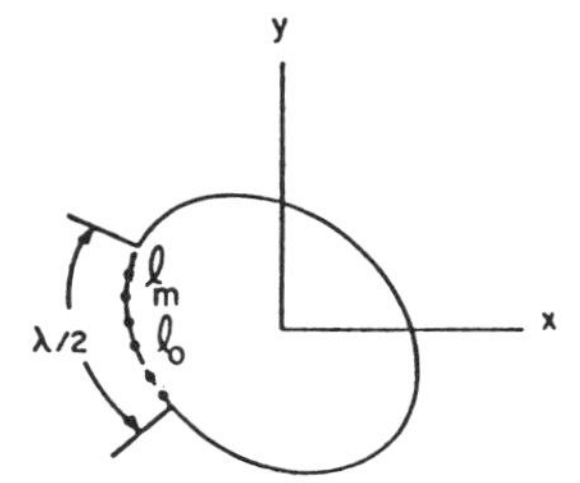

Fig. 3. Arbitrary cylinder with λ/2 transverse aperture.

For the TE formulations, the element V_n^r is given by

$$V_n^r \approx \Delta l_n u_{l_n} \cdot u_\phi e^{jk\rho_n}$$

$$\approx \Delta l_n(-l_{x_n} \sin\phi + l_{y_n} \cos\phi) \exp[jk(x_n \cos\phi + y_n \sin\phi)] \quad (39)$$

where l_{x_n} and l_{y_n} are the rectangular components of the unit tangent vector u_{l_n} at the nth point.

The elements of the excitation matrix for aperture excitation are evaluated using point matching. The elements are all zero except those corresponding to intervals within the aperture. For these intervals the elements are the negative of the aperture field evaluated at the midpoint of each. The nature of the nonzero elements will be illustrated by several examples.

For TM polarization consider a dominant mode cosine field distribution in an aperture of λ/2 width in an arbitrary shaped cylinder as shown in Fig. 3. The assumed aperture field is

$$E_z^{\text{apert}} = \cos k(l - l_0) \quad (40)$$

and hence the excitation element is

$$V_m = \begin{cases} -\cos k(l_m - l_0), & l_m \text{ in aperture} \\ 0, & l_m \text{ on conductor} \end{cases}$$

$$m = 1,2,\cdots,N. \quad (41)$$

For TE polarization the aperture is a narrow axial slot with a voltage V assumed constant across the slot. When the slot is assumed one interval wide, the field may be approximated as $V/\Delta l_m$, so the excitation element becomes

$$V_m = -\frac{V}{\Delta l_m}. \quad (42)$$

For a one-interval side axial slot in the mth interval, the complex current coefficients of (22) are just the elements of the mth column Y_{nm} multiplied by $-V/\Delta l_m$.

C. Aperture Admittance from Integration over Apertures

Consider one or more conducting bodies on which a system of apertures is defined. For the two-dimensional formulations given, where E is specified over a perfectly conducting surface in terms of the current J, the admittance element y_{ab} is given by

$$y_{ab} = \frac{1}{v_a v_b} \int_C E^a \cdot J^b \, dl. \quad (43)$$

The field E^a is due to voltage source v_a applied to aperture a, with all other ports short circuited, and J^b is the current due to v_b applied to aperture b, with all other ports short circuited. A solution in terms of the generalized network parameters proceeds in the same fashion as for power [13]. The admittance of apertures in solid conductors is properly obtained by using the current given by $n \times H$. The result for y_{ab} is

$$y_{ab} = \frac{1}{v_a v_b} [\hat{V}_m^a][M_{mn}][Y_{nm}][V_m^b] \quad (44)$$

where $[V_m^b]$ is the voltage excitation matrix with only aperture b excited. The elements of the aperture matrix $[\hat{V}_m^a]$ are given by

$$\hat{V}_m^a = \begin{cases} \int_{\text{apert } a} E^{\text{apert}} \cdot u_n J_n^b \, dl, & J_n^b \text{ in aperture } a \\ 0, & \text{outside aperture.} \end{cases} \quad (45)$$

V. Numerical Results

Computer subroutines were written which calculate the parameters of the previous sections. The field patterns remained relatively unchanged as the number of subsections is increased, indicating that the convergence of far-field quantities is good. Near-field quantities, such as surface current, are more slowly convergent and must be calculated with greater precision.

As a check, the scattering width versus azimuth angle ϕ was computed for several scatterers previously appearing in the literature [1], [2]. The same geometries were also used as aperture antennas. Some results are shown, and more are available in Wallenberg's dissertation [5].

Figs. 4 to 9 are plots of the square root of the power gain versus ϕ for several different aperture excitations on a square cylinder of side $3\lambda/\pi$. These curves are normalized such that each niche on the axes represents $\sqrt{g} = 0.5$. Figs. 4, 6, 8, and 9 illustrate the superposition of two apertures when the excitation is symmetric about the x axis. In this case only the even impedance matrix was inverted, and its excitation matrix is the superposition of the individual aperture excitation matrices; that is,

$$V_m^e = [V_m^{ea} + V_m^{eb}]. \quad (46)$$

The matrix $[V_m^{ea}]$ is the even excitation matrix with aperture a excited, and similarly for $[V_m^{eb}]$. The equal and in phase patterns are obtained when $V_m^{ea} = V_m^{eb}$, and the equal and out of phase patterns are obtained when $V_m^{ea} = -V_m^{eb}$. The use of both even and odd impedance matrices is illustrated in Figs. 5 and 7. The pattern for asymmetric excitation is obtained by superposition of the patterns for even and odd excitation.

Fig. 10 shows some calculated equivalent currents due to aperture excitation. Note that for the TM polarization a large current density is found just outside the aperture, whereas for the TE polarization a large electric or magnetic current is calculated within the interval coinciding with the aperture excitation. The electric current density, in general, seems to fall off more slowly than the magnetic current density, which is large only within the aperture slit.

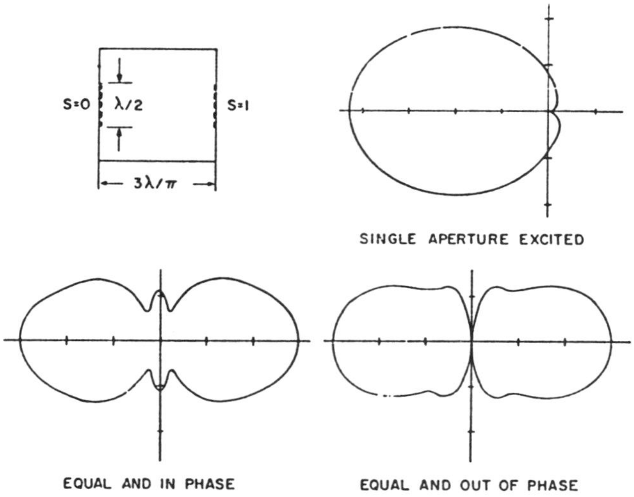

Fig. 4. TM radiation patterns, $|E_z|$.

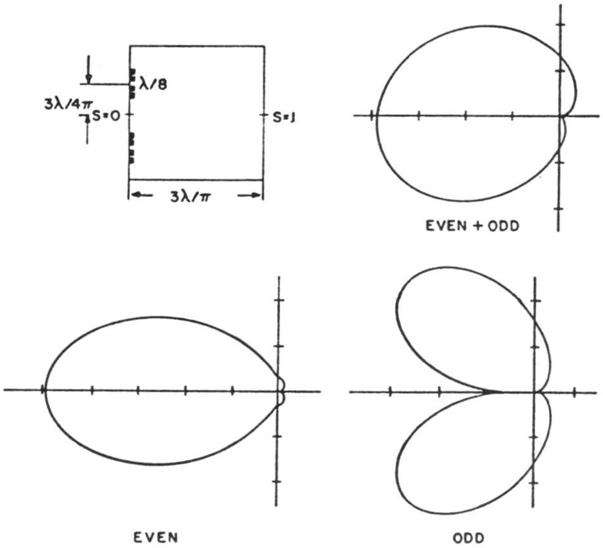

Fig. 5. TM radiation patterns, $|E_z|$.

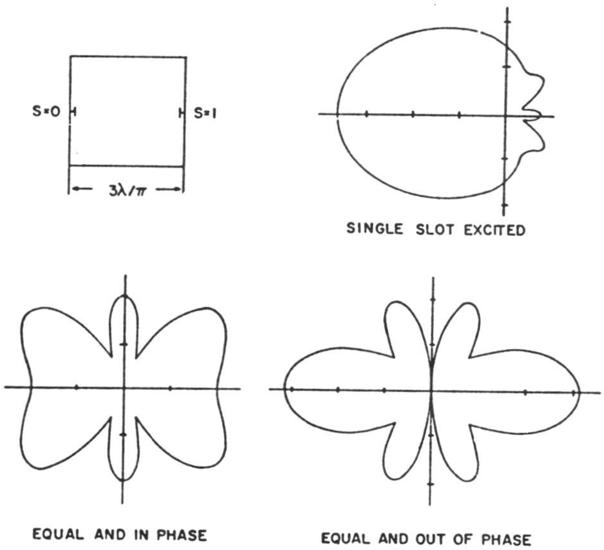

Fig. 6. TE radiation patterns, $|H_z|$.

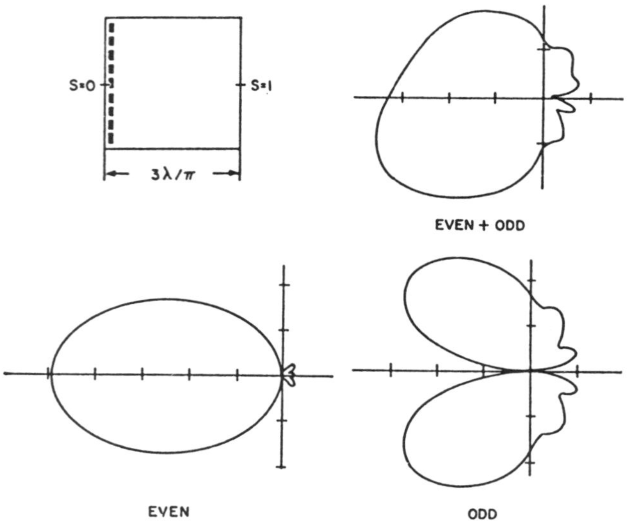

Fig. 7. TE radiation patterns, $|H_z|$.

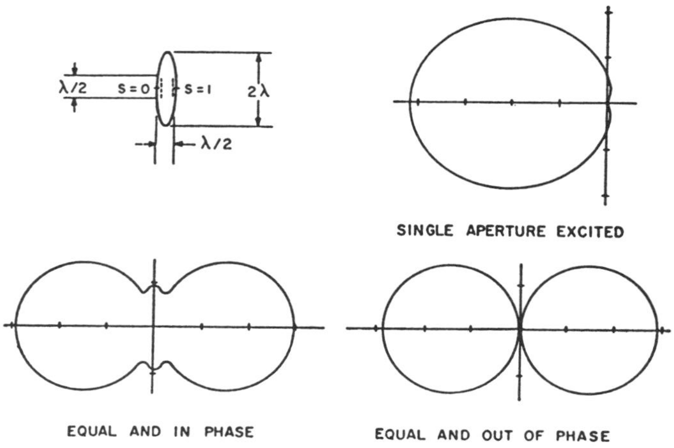

Fig. 8. TM radiation patterns, $|E_z|$.

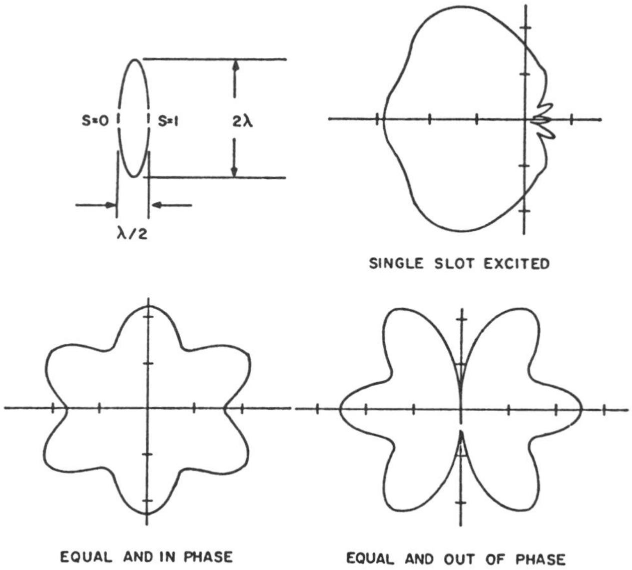

Fig. 9. TE radiation patterns, $|H_z|$.

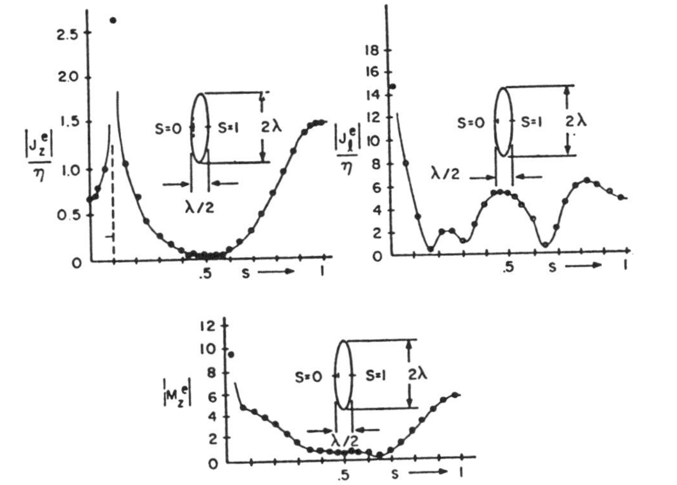

Fig. 10. Magnitude of equivalent current density distribution on elliptical cylinder (aperture defined by dotted lines).

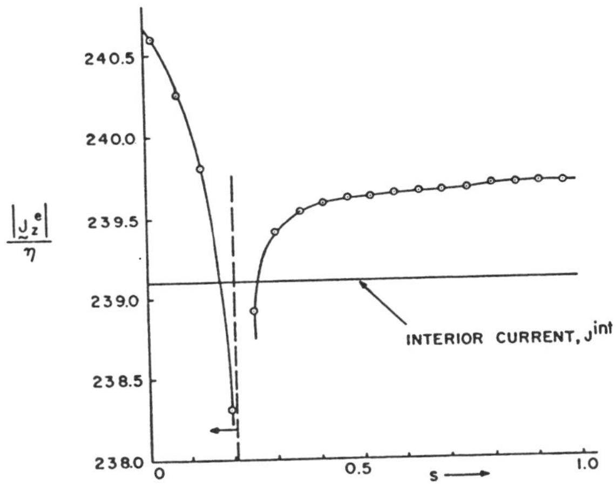

Fig. 11. Equivalent TM current on circular cylinder (ka = 2.405) excited by $\lambda/2$ aperture.

The equivalent TM current on a resonant circular cylinder excited with a half-wavelength aperture is plotted in Fig. 11. The TM_{01} mode current has been excited, and its amplitude, calculated by (6), is shown by the straight line. The current $\boldsymbol{n} \times \boldsymbol{H}$ which radiates external to the cylinder is plotted in Fig. 12. It was found that TM aperture excitation even excites significant mode current on a circular cylinder of nonresonant dimension [5].

VI. Discussion

The electromagnetic behavior of a perfectly conducting cylinder of arbitrary shape is described by a set of integral equations. The discussion has been given in terms of aperture excitation, but the excitation can be arbitrary. Solutions for plane-wave incidence are given in the literature [1], [2].

The integral equations with aperture excitation involve an unknown equivalent electric or magnetic current which

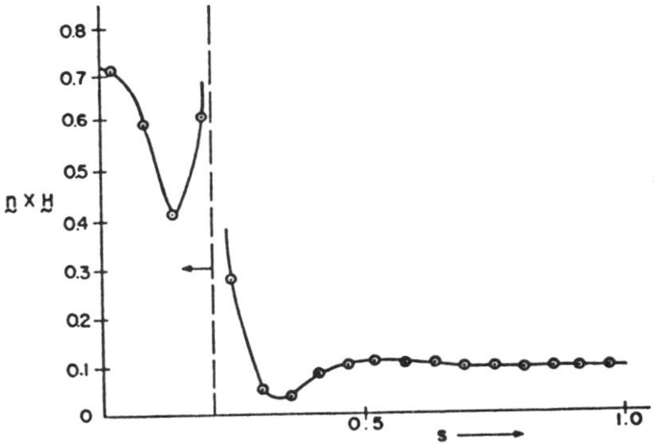

Fig. 12. External current $\boldsymbol{n} \times \boldsymbol{H}$ induced on a resonant circular cylinder (ka = 2.405) by $\lambda/2$ aperture.

give rise to the correct fields external to S but also give rise to fields internal to the conductor. Once an equivalent source distribution has been computed, the current associated with the external problem can be calculated from (6). This current is necessary when calculating aperture admittances. However, when only field patterns are needed, the equivalent currents are sufficient.

The principal limitation of the approach is set by the number of subdivisions and hence by the order of the matrix necessary to set up and invert. The current associated with the external problem varies rapidly within and near the aperture. Special treatment of the solution may be required in this region to obtain accurate current solutions. However, far-field quantities are relatively insensitive to small errors in the current distribution.

References

[1] K. K. Mei and J. G. Van Bladel, "Scattering by perfectly-conducting rectangular cylinders," *IEEE Trans. Antennas and Propagation*, vol. AP-11, pp. 185–192, March 1963.
[2] M. G. Andreasen, "Scattering from parallel metallic cylinders with arbitrary cross sections," *IEEE Trans. Antennas and Propagation*, vol. AP-12, pp. 746–754, November 1964.
[3] R. F. Harrington, *Time-Harmonic Electromagnetic Fields*. New York: McGraw-Hill, 1961.
[4] R. F. Harrington *et al.*, "Matrix methods for solving field problems," Rome Air Development Center, Griffiss AFB, Rome, N. Y., Final Rept., Contract AF30(602)-3724, March 1966.
[5] R. F. Wallenberg, "Two dimensional scattering and radiation from perfectly conducting cylinders of arbitrary shape," Ph.D. dissertation, Syracuse University, Syracuse, N. Y., Tech. Rept. 68-2, March 1968.
[6] R. F. Harrington, "Matrix methods for field problems," *Proc. IEEE*, vol. 55, pp. 136–149, February 1967.
[7] E. D. Sharp and M. G. Andreasen, "Cylindrical scatterers and antennas having open-ended surfaces of arbitrary surface impedance," unpublished paper.
[8] J. H. Richmond, "Scattering by a dielectric cylinder of arbitrary cross section shape," *IEEE Trans. Antennas and Propagation*, vol. AP-13, pp. 334–341, May 1965.
[9] J. H. Richmond, "TE-wave scattering by a dielectric cylinder of arbitrary cross-section shape," *IEEE Trans. Antennas and Propagation*, vol. AP-14, pp. 460–464, July 1966.
[10] T. W. Bristol, "Waveguides of arbitrary cross section by moment methods," Ph.D. dissertation, Syracuse University, Syracuse, N. Y., Tech. Rept. 2-68-4, November 1967.
[11] F. B. Hildebrand, *Introduction to Numerical Analysis*. New York: McGraw-Hill, 1956, pp. 75–76.
[12] R. F. Harrington, *Field Computation by Moment Methods*. New York: Macmillan, 1968, pp. 82–110.
[13] R. F. Harrington, "Generalized network parameters in field theory," *Proc. Symp. on Generalized Networks*, MRIS ser., vol. 16. Brooklyn, N. Y.: Polytechnic Press, 1966.

Electromagnetic Excitation of a Wire Through an Aperture-Perforated Conducting Screen

CHALMERS M. BUTLER, SENIOR MEMBER, IEEE, AND KORADA R. UMASHANKAR, MEMBER, IEEE

Abstract—Integro-differential equations are formulated for the general problem of a finite-length wire excited through an arbitrarily shaped aperture in a conducting screen. The wire is assumed to be electrically thin and perfectly conducting, and it is arbitrarily oriented behind the perfectly conducting screen of infinite extent. A known, specified incident field illuminates the perforated-screen/wire structure. The integro-differential equations fully account for the coupling between the wire and the aperture/screen. They are specialized to the case of the wire parallel to the screen with the aperture a narrow slot of general length. These special equations are solved numerically and data are presented for wire currents and aperture fields under selected conditions of wire/slot lengths and orientation. Data indicative of the coupling between the wire and slot are presented.

INTRODUCTION

THE PURPOSE of this paper is to present an analysis of the problem of a wire excited by an electromagnetic field which penetrates an aperture-perforated, conducting screen. General, coupled integro-differential equations are formulated in the paper with the wire current and aperture electric field (or equivalent aperture magnetic current) as unknowns. The coupling between the aperture and the wire is accounted for fully in the derived equations. These equations are specialized to the case in which the aperture is a narrow slot and the wire is parallel to the screen. Under these restricted but practical conditions, the appropriate equations are solved numerically and sample data are presented. Based upon the calculated results, the slot/wire problem is discussed in detail.

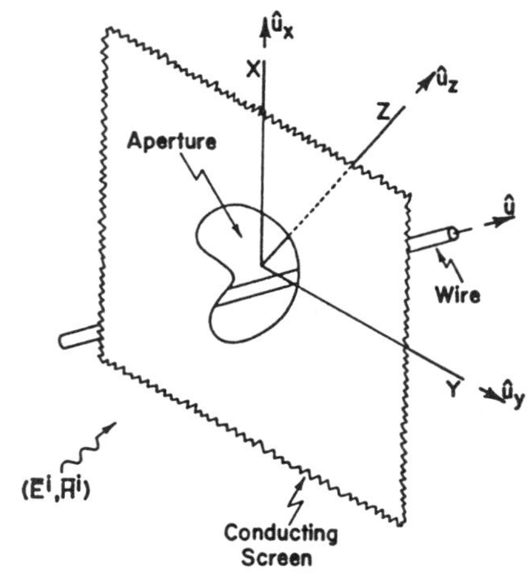

Fig. 1. Wire illuminated through aperture-perforated conducting screen.

The general problem to be considered here is illustrated in Fig. 1 where one sees a finite-length wire which is arbitrarily oriented behind a planar conducting screen. The screen is assumed to be perfectly conducting, vanishingly thin, and of infinite extent. The wire is also perfectly conducting and is thin relative to its length as well as to the wavelength of the electromagnetic field. There is an aperture A of general shape in the screen of Fig. 1, and an incident

Manuscript received July 22, 1975; revised January 19, 1976.
The authors are with the Department of Electrical Engineering, University of Mississippi, University, MS 38677.

Reprinted from *IEEE Trans. Antennas Propagat.*, vol. AP-24, no. 4, pp. 456–462, July 1976.

field $(\bar{E}^i, \bar{H}^i)$ is seen impinging upon the structure. The wire is excited by the electromagnetic field which penetrates the aperture.

The only previous literature which has a bearing on the present work comprises one moderately related paper by King and Owyang [1], one by Lin *et al.* [2], and a preliminary version of the present work by Butler [3]. In the first paper above, the authors treat an array consisting of two driven dipoles symmetrically-located on either side of a slotted screen. Lin *et al.* consider a wire excited through an aperture but, since they calculate aperture fields in the absence of the wire, their theory does not include the complete coupling between the wire and aperture. As substantiated in the discussion below, the energy scattered back into the aperture by the wire, which is ignored by Lin *et al.*, can be quite significant and can strongly influence the aperture fields.

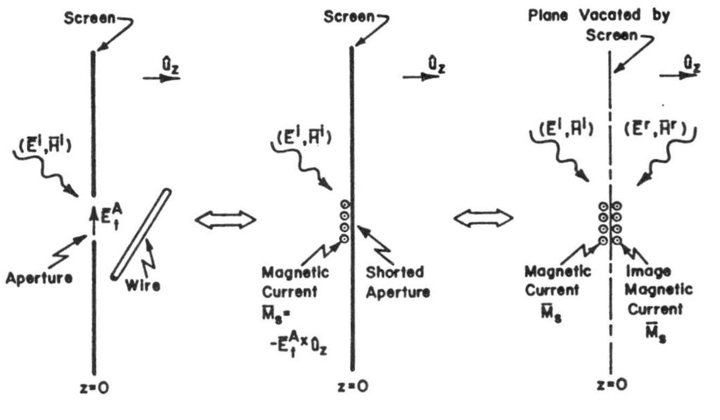

Fig. 2. Illuminated half-space equivalences.

Formulation

In the boundary value problem under consideration here, one observes that there exists some total electric field $\bar{E}_t^A$ in the aperture A and tangential to the aperture/screen plane. This electric field is, of course, unknown *a priori* and, in fact, is the quantity which is to be determined as a solution to the problem. In terms of this aperture field $\bar{E}_t^A$, the incident field $(\bar{E}^i, \bar{H}^i)$, and the geometry of the structure, as depicted in Fig. 1, one can express the magnetic field in each half-space in such a way that Maxwell's equations are satisfied in each half-space and boundary conditions are satisfied on the screen and on the wire. The magnetic field in each half-space is written as a function of $\bar{E}_t^A$, which is common to both half-spaces, ensuring continuity of electric field through A. Equating the transverse components of these two expressions for magnetic field enforces continuity of magnetic field through the aperture and leads directly to desired equations for the problem under investigation.

Fig. 2 depicts schematically[1] a step-by-step reduction of the left half-space $(z < 0)$ problem to a simple equivalent problem in a form which readily suggests how one may develop an expression for the left half-space total magnetic field $\bar{H}^-$. In Fig. 2(a) is seen the original problem while in Fig. 2(b) the aperture/screen is replaced by a perfectly conducting, continuous plane (aperture shorted) with the original tangential electric field $\bar{E}_t^A$ in the aperture restored at $z = 0^-$, $(x,y) \in A$, by an appropriate magnetic surface current $\bar{M}_s$ which is specified to have a value $-\bar{E}_t^A \times \hat{u}_z$, i.e., $\bar{M}_s = -\bar{E}_t^A \times \hat{u}_z$. Notice that this as-of-yet undetermined magnetic current, which resides on the illuminated side of the shorted screen, radiates in the presence of the screen and that $(\bar{E}^i, \bar{H}^i)$ of the original problem in Fig. 2(a) illuminates the shorted screen. Next, in Fig. 2(c), one appeals to image theory which enables him to remove the conducting screen entirely and which requires that he include the image magnetic current plus $(\bar{E}^r, \bar{H}^r)$, the field reflected from the shorted screen, so that the total electromagnetic field in the left half-space is unchanged from its value in the original problem (Fig. 2(a)). These last modifications leading to Fig. 2(c) simply serve to preserve the boundary conditions on the xy plane in the absence of the conducting screen, which are guaranteed by the presence of this screen in Fig. 2(b). The equivalent problems illustrated in Figs. 2(b) and 2(c) are valid only in the left half-space so the fields $(\bar{E}^i, \bar{H}^i)$ and $(\bar{E}^r, \bar{H}^r)$, plus that radiated by the magnetic current and its image, are to be calculated only for $z < 0$. Furthermore, since in the final equivalence of Fig. 2(c) all currents and fields exist in a homogeneous medium of infinite extent, one may use only the particular integral solutions of the wave equation for vector potential to calculate fields radiated by the magnetic current.

The total magnetic field $\bar{H}^-$ in the region $z < 0$ of Fig. 2(c) is the sum of that radiated by $2\bar{M}_s$, the incident field $\bar{H}^i$, and $\bar{H}^r$. If one defines the so-called short-circuit field $(\bar{E}^{sc}, \bar{H}^{sc})$ to be the sum of the incident and reflected fields, $\bar{E}^{sc} = \bar{E}^i + \bar{E}^r$ and $\bar{H}^{sc} = \bar{H}^i + \bar{H}^r$, in the half-space $z < 0$ with the aperture shorted, he may construct the total magnetic field at a point $\bar{r}$ in the left half-space as

$$\bar{H}^-(\bar{r}) = \bar{H}^{sc}(\bar{r}) - j\frac{\omega}{k^2}\left[k^2 F(\bar{r}) + \text{grad}\,(\text{div}\,\bar{F}(\bar{r}))\right], \qquad z < 0. \quad (1)$$

In (1), ω is the angular frequency of the suppressed harmonic variation in time, $e^{j\omega t}$, k is 2π/wavelength λ, and $\bar{F}$ is the electric vector potential

$$\bar{F}(\bar{r}) = \frac{\varepsilon}{4\pi}\iint_A 2\bar{M}_s(\bar{r}')G(\bar{r},\bar{r}')\,ds' \quad (2)$$

where ε is the permittivity of the medium of Fig. 1 and where

$$G(\bar{r},\bar{r}') = \frac{e^{-jk|\bar{r}-\bar{r}'|}}{|\bar{r}-\bar{r}'|} \quad (3)$$

[1] Directions of vector quantities shown in Figs. 2 and 3 are adopted for illustrative purposes and are not intended to indicate properties of final results.

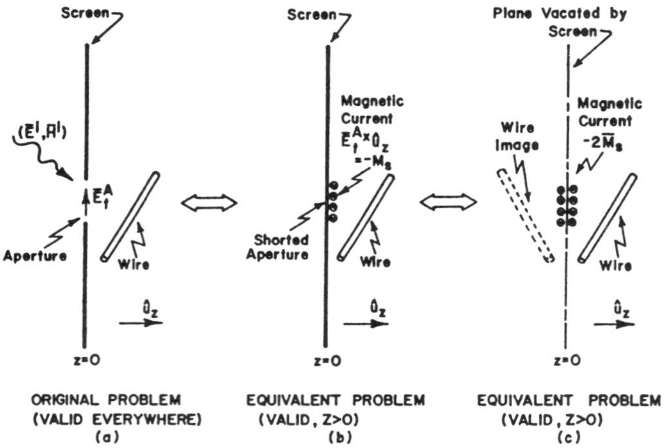

Fig. 3. Shadow half-space equivalences.

with

$$|\bar{r} - \bar{r}'| = [(x - x')^2 + (y - y')^2 + z^2]^{1/2}, \quad (x', y') \in A, \quad z < 0. \quad (4)$$

Next, attention is turned to the equivalences illustrated in Fig. 3 and the formulation of equations peculiar to the right half-space ($z > 0$). The unknown aperture electric field $\bar{E}_t^A$, common to the left and right half-space problems, is again postulated as illustrated in Fig. 3(a). Fig. 3(b) depicts the conducting screen with the aperture short-circuited over which is impressed a surface magnetic current $-\bar{M}_s (= \bar{E}_t^A \times \hat{u}_z)$ that serves to maintain the original value of tangential electric field by supporting a discontinuity from zero at $z = 0$ to $\bar{E}_t^A$ at $z = 0^+$ in A. In the equivalent problem depicted in Fig. 3(b), the wire remains, and both the wire current and the magnetic current radiate in the presence of the conducting screen. The final equivalence is the model shown in Fig. 3(c) from which the conducting screen has been removed. The boundary conditions on the electromagnetic field due to the presence of the conducting screen are maintained after the screen has been removed in Fig. 3(c) by the inclusion of the images of the wire and of the magnetic current. One may adopt the viewpoint that the conducting wire is replaced by an imaginary cylinder bearing an equivalent electric current, on whose surface one requires the total tangential electric field to be zero. For a wire satisfying the traditional assumptions of thin-wire theory, as is the case considered here, one neglects all but the axially directed surface current, which is assumed to be circumferentially independent, and accounts for it by a total axial current I. Furthermore, thin-wire assumptions permit one to require only that the axial component of electric field be zero on the wire surface. Subject to the qualifications above and the equivalence of Fig. 3(c), one has achieved in the region $z > 0$ a model having currents (both electric and magnetic) radiating in a homogeneous space of infinite extent. Again, such a situation permits one to employ only the particular integral solutions of the wave equations for the magnetic and electric vector potentials.

The total magnetic field in the region $z > 0$ is denoted $\bar{H}^+$ and can be written

$$\bar{H}^+ = +j \frac{\omega}{k^2} [k^2 \bar{F} + \text{grad}(\text{div}\, \bar{F})] + \frac{1}{\mu} \text{curl}\, \bar{A}, \qquad z > 0 \quad (5)$$

where $\bar{F}$ is defined in (2), and where μ is the permeability of the medium. $\bar{A}$ is the magnetic vector potential due to the wire current plus its image:

$$\bar{A}(\bar{r}) = \frac{\mu}{4\pi} \hat{u} \int_L I(\bar{r}_w')G(\bar{r},\bar{r}_w')\, dl' + \frac{\mu}{4\pi} \hat{u}_i \int_L I(\bar{r}_i')G(\bar{r},\bar{r}_i')\, dl' \quad (6)$$

where $\hat{u}$ and $\hat{u}_i$ are unit vectors in the wire and wire image directions, respectively. The wire is of length L and radius a, and the vector $\bar{r}_w'$ locates a source point on the wire while $\bar{r}_i'$ locates a source point on the wire image.

As a part of the right half-space problem, one must enforce the appropriate boundary condition on the wire by requiring that the total electric field in the axial direction on the wire's surface be zero. To this end, one requires

$$[\bar{E}_w^A + \bar{E}_w^i + \bar{E}_w^s] \cdot \hat{u} = 0, \qquad \text{on wire} \quad (7)$$

where $\bar{E}_w^A$ is the field due to the aperture/screen, $\bar{E}_w^i$ is that due to the wire image, and $\bar{E}_w^s$ is that due to the equivalent sources on the wire. The terms in (7) can be written

$$\bar{E}_w^A = +\frac{1}{\varepsilon} \text{curl}\, \bar{F}, \qquad z > 0 \quad (8)$$

$$\bar{E}_w^s + \bar{E}_w^i = -j \frac{\omega}{k^2} [k^2 \bar{A} + \text{grad}(\text{div}\, \bar{A})], \qquad z > 0 \quad (9)$$

so that (7) becomes

$$\frac{1}{\varepsilon} \hat{u} \cdot \text{curl}\, \bar{F} - j \frac{\omega}{k^2} \hat{u} \cdot [k^2 \bar{A} + \text{grad}(\text{div}\, \bar{A})] = 0, \qquad \text{on wire} \quad (10)$$

an equation which demands that the wire boundary condition be honored.

One observes that the formulations of (1), (5), and (10) are based on the magnetic and electric vector potentials, $\bar{A}$ and $\bar{F}$, which implies that $\bar{H}^+$ and $\bar{H}^-$ satisfy Maxwell's equations and the radiation condition in the appropriate half-spaces. Boundary conditions on the screen are satisfied because (1) and (5) are based upon the models of Figs. 2(c) and 3(c), and (10) ensures that the tangential electric field along the wire surface be zero. $\bar{E}_t^A$ is common to both half-space formulations (through $\bar{M}_s$ and $\bar{F}$) so the final remaining condition that must be enforced is continuity of the total transverse magnetic field through the aperture, which serves to couple the two individual half-space formulations.

The continuity requirement on tangential magnetic field is

$$\lim_{z\uparrow 0} \bar{H}^- \times \hat{u}_z = \lim_{z\downarrow 0} \bar{H}^+ \times \hat{u}_z$$

which, in view of (1) and (5), becomes

$$j2\frac{\omega}{k^2}[k^2\bar{F} + \text{grad}(\text{div}\,\bar{F})] \times \hat{u}_z + \frac{1}{\mu}(\text{curl}\,\bar{A}) \times \hat{u}_z = \bar{H}^{sc} \times \hat{u}_z, \quad \begin{cases}\text{in } A \\ z = 0.\end{cases} \tag{11}$$

Since $\bar{H}^{sc} \times \hat{u}_z = 2\bar{H}^i \times \hat{u}_z$ at $z = 0$ on the shorted screen, (11) reduces to

$$j\frac{\omega}{k^2}[k^2\bar{F} + \text{grad}(\text{div}\,\bar{F})] \times \hat{u}_z + \frac{1}{2\mu}(\text{curl}\,\bar{A}) \times \hat{u}_z = \bar{H}^i \times \hat{u}_z, \quad \begin{cases}\text{in } A \\ z = 0\end{cases} \tag{12}$$

which, in terms of the components of $\bar{F}$ and $\bar{A}$, can be written in scalar form,

$$\left(\frac{\partial^2}{\partial x^2} + k^2\right)F_x + \frac{\partial^2}{\partial x\,\partial y}F_y + j\frac{\omega\varepsilon}{2}\left(\frac{\partial}{\partial z}A_y - \frac{\partial}{\partial y}A_z\right) = -j\frac{k^2}{\omega}H_x^i, \quad z = 0, (x,y) \in A \tag{13a}$$

$$\left(\frac{\partial^2}{\partial y^2} + k^2\right)F_y + \frac{\partial^2}{\partial y\,\partial x}F_x + j\frac{\omega\varepsilon}{2}\left(\frac{\partial}{\partial x}A_z - \frac{\partial}{\partial z}A_x\right) = -j\frac{k^2}{\omega}H_y^i, \quad z = 0, (x,y) \in A. \tag{13b}$$

Also, (10) can be written as a scalar equation:

$$\left(\frac{\partial^2}{\partial s^2} + k^2\right)\int_{s'=-L/2}^{L/2} I(s')\frac{e^{-jk[a^2+(s-s')^2]^{1/2}}}{[a^2 + (s-s')^2]^{1/2}}\,ds' + k^2(2\cos^2\gamma - 1)\int_{s'=-L/2}^{L/2} I(s')g(s,s')\,ds' - \frac{\partial}{\partial s}\int_{s'=-L/2}^{L/2}\frac{d}{ds'}I(s')g(s,s')\,ds'$$
$$= -j4\pi\omega\left[\frac{\partial}{\partial z}F_x\cos\beta - \frac{\partial}{\partial z}F_y\cos\alpha + \left(\frac{\partial}{\partial x}F_y - \frac{\partial}{\partial y}F_x\right)\cos\gamma\right], \quad \text{on the wire} \tag{14}$$

in which

$$g(s,s') = \frac{e^{-jkR(s,s')}}{R(s,s')} \tag{15a}$$

with

$$R(s,s') = [4z_c^2 + 4z_c(s + s')\cos\gamma + 4ss'\cos^2\gamma + (s - s')^2]^{1/2}. \tag{15b}$$

The geometric quantities which are pertinent to the wire/aperture and which appear in (14) and (15) are listed below:

center of wire: (x_c, y_c, z_c)

unit vector along wire (with sense of I):

$$\hat{u} = \cos\alpha\hat{u}_x + \cos\beta\hat{u}_y + \cos\gamma\hat{u}_z$$

direction cosines of wire:

$$\cos\alpha = \hat{u}\cdot\hat{u}_x,\ \cos\beta = \hat{u}\cdot\hat{u}_y,\ \cos\gamma = \hat{u}\cdot\hat{u}_z.$$

Specialization of Aperture to Narrow Slot of Finite Length

Without the wire, Fig. 1 would depict the traditional, general aperture/screen problem, and (13) with terms involving components of $\bar{A}$ deleted would be the appropriate set of integro-differential equations. Recently, a few researchers [4], [5] have undertaken to solve numerically various noncircular aperture problems and have found computer storage and run times to be very high; the problem under consideration here of a wire excited through a general aperture is even more demanding of storage and time. Therefore, in order to reduce the present problem to a feasible size computer-wise and yet to retain the fundamental features of practical interest, totally accounting for the coupling between the wire and the aperture, we next specialize the aperture geometry to be an electrically narrow slot of width w and length l. The slot is centered at (0,0,0) and its axis is along the x axis.

Narrow-slot assumptions, similar in principle to those invoked in thin-wire theory, can be employed here to simplify the analysis. When the slot is very narrow relative to the wavelength and long compared to its width, the electric field in the slot is principally transverse to the longer slot dimension (transverse to x) and has a known transverse variation. This transverse variation of electric field, or equivalent magnetic current, can be determined via electrostatics to be simply

$$\xi(y) = \frac{1/\pi}{\sqrt{\left(\frac{w}{2}\right)^2 - y^2}} \tag{16}$$

provided the slot excitation does not possess an appreciable component which is an odd function with respect to y. Ignoring the axial (x-component) electric field, or, transverse equivalent magnetic current, in the slot, which, of course, implies $F_y = 0$, and evaluating quantities of interest along the slot axis ($y = 0$, $z = 0$), one reduces the surviving component of the electric vector potential F_x to

$$F_x(x,0,0) = \frac{\varepsilon}{2\pi}\int_{x'=-l/2}^{l/2} m(x')\int_{y'=-w/2}^{w/2}\xi(y')\cdot\frac{e^{-jk[(x-x')^2+y'^2]^{1/2}}}{[(x-x')^2 + y'^2]^{1/2}}\,dy'\,dx' \tag{17}$$

where $m(x)$ and $\xi(y)$ are the axial and transverse variations, respectively, of the slot magnetic current:

$$M_{s_x}(x,y) = m(x)\xi(y). \tag{18}$$

Subject to the above expression for M_{s_x} and the integration variable transformation

$$y' = \frac{w}{2} \sin \frac{\alpha}{2}$$

(17) becomes

$$F_x(x,0,0) = \frac{\varepsilon}{2\pi} \int_{x'=-l/2}^{l/2} m(x') K\left(x - x', \frac{w}{4}\right) dx' \quad (19)$$

where the kernel is

$$K(\zeta,a) = \frac{1}{2\pi} \int_{\alpha=-\pi}^{\pi} \frac{\exp(-jk[\zeta^2 + 4a^2 \sin^2 \alpha/2]^{1/2})}{[\zeta^2 + 4a^2 \sin^2 \alpha/2]^{1/2}} d\alpha \quad (20a)$$

$$\doteq \frac{\exp(-jk[\zeta^2 + a^2]^{1/2})}{[\zeta^2 + a^2]^{1/2}}. \quad (20b)$$

F_x of (19) with K of (20) is recognized to have the form of the magnetic vector potential associated with a thin wire of radius $w/4$. The same result relating the width to an equivalent radius may be obtained in a different way [6].

The above simplifications enable one to reduce (13) to

$$\left(\frac{\partial^2}{\partial x^2} + k^2\right) F_x + j\frac{\omega\varepsilon}{2}\left(\frac{\partial}{\partial z} A_y - \frac{\partial}{\partial y} A_z\right) = -j\frac{k^2}{\omega} H_x^i. \quad (21)$$

Equation (14), with $F_y = 0$, applies to the present case. Finally, under the further specialization that the wire be parallel to the screen, the particularly simple coupled integro-differential equations below are obtained:

$$\left(\frac{\partial^2}{\partial x^2} + k^2\right) \int_{x'=-l/2}^{l/2} m(x') K\left(x - x', \frac{w}{4}\right) dx' + j\pi\omega \frac{\partial}{\partial z} A_y = -j2\pi\omega\mu H_x^i, \quad \text{on slot} \quad (22a)$$

$$\left(\frac{\partial^2}{\partial s^2} + k^2\right) \int_{s'=-L/2}^{L/2} I(s')[K(s - s', a) - K(s - s', 2z_c)] ds' + j4\pi\omega \cos\beta \frac{\partial}{\partial z} F_x = 0, \quad \text{on wire.} \quad (22b)$$

Results and Conclusions

Equations (22) have been solved numerically by the method of moments [7] with pulses used for representation of both $m(x)$ and $I(s)$ and triangles used for testing, and selected results are presented and discussed below. In the following explanations of results, an effort is made to outline how quantities of interest depend upon important geometric features of the slotted-screen/wire problem. Also, the influence of the properties of the incident field is mentioned.

In all cases for which results are given below, the wire is parallel to the screen and its center is designated (x_c, y_c, z_c). In the following discussion of data and results, the angular rotation of the wire about its center is in a plane parallel

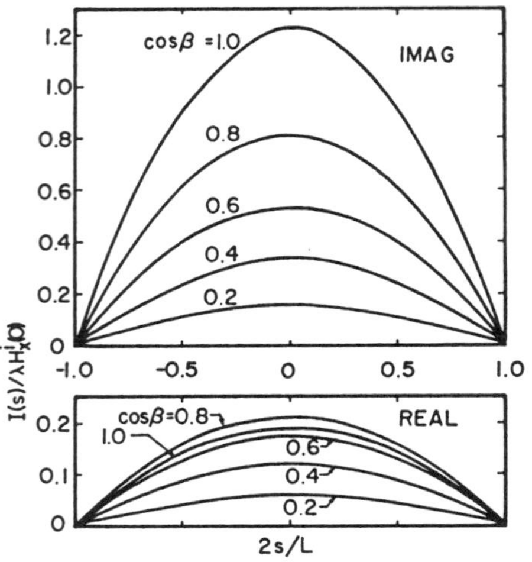

Fig. 4. Current on wire illuminated through slotted screen ($w/\lambda = 0.05$, $l/\lambda = 0.5$; $a/\lambda = 0.001$, $L/\lambda = 0.5$; $x_c/\lambda = 0$, $y_c/\lambda = 0$, $z_c/\lambda = 0.25$; normal incidence).

to the screen and is measured by the angle β defined as the angular displacement from the y axis to the axis of the wire. When the narrow slot is viewed as the dual of a thin wire, subject to the usual thin-wire/narrow-slot assumptions, one readily sees that the wire is not excited by the fields which penetrate the slot whenever the two are parallel and, also, that the coupling between wire and slot is maximum when they are perpendicular.

It can be shown easily that, if the excitation of a thin wire or a narrow slot is an even function of axial displacement measured with respect to the center of the element, the current must be even too and, also, if the excitation is odd, so must be the current. In the case of a half-wavelength wire or slot, the resonant current is nearly cosinusoidal whereas, in a one-wavelength element, the antiresonant current is approximately a sine function. Hence, an odd-function excitation cannot excite a significant resonant component of current in a half-wavelength wire or slot, and an even-function excitation cannot excite an appreciable antiresonant component of current in a one-wavelength element. These observations, made relative to a wire in free space or a single slot in an infinite screen, also hold for a thin wire parallel to an infinite screen.

Half-Wavelength Slot, Half-Wavelength Wire

Fig. 4 shows the current I on a half-wavelength wire, as a function of position along the wire, induced by the field which penetrates a half-wavelength slot; this current is given for selected values of the angle β. When $\cos\beta = 1$, the wire and slot are perpendicular, and the coupling is seen to be maximum as expected, while there is no coupling (hence the wire current is zero) when the wire and slot are parallel ($\cos\beta = 0$). The wire current distribution is essentially a cosine function as one would expect for resonant length. One can demonstrate that there is a small component of odd-function current on the resonant wire due to asymmetric coupling to the slot caused by displacement of the wire center from the z axis. With the wire center above

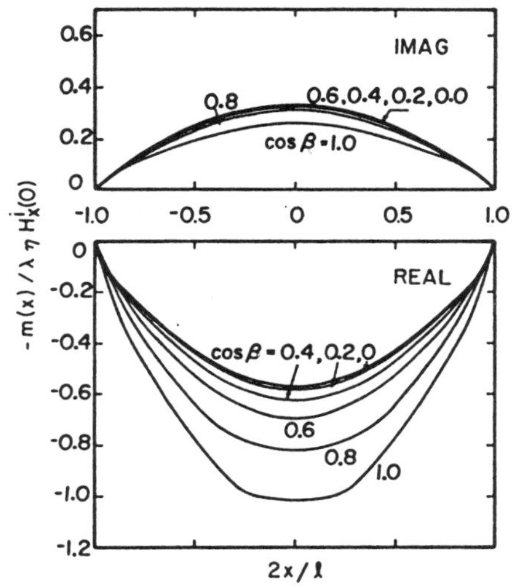

Fig. 5. Axial distribution of slot magnetic current ($w/\lambda = 0.05$, $l/\lambda = 0.5$; $a/\lambda = 0.001$, $L/\lambda = 0.5$; $x_c/\lambda = 0$, $y_c/\lambda = 0$, $z_c/\lambda = 0.25$; normal incidence).

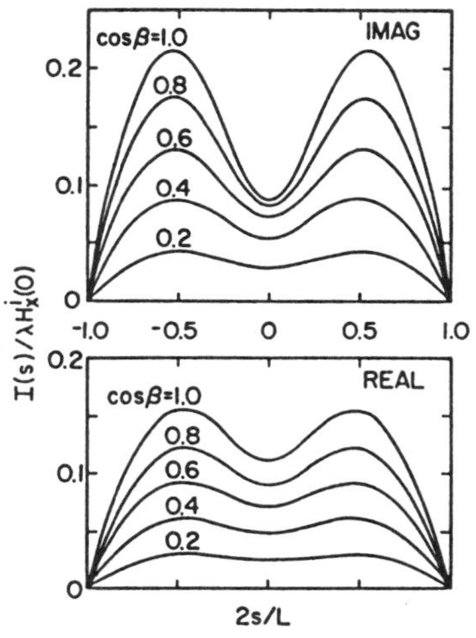

Fig. 6. Current on wire illuminated through slotted screen ($w/\lambda = 0.05$, $l/\lambda = 0.5$; $a/\lambda = 0.001$, $L/\lambda = 1.0$; $x_c/\lambda = 0$, $y_c/\lambda = 0$, $z_c/\lambda = 0.125$; normal incidence).

the slot axis but not on the z axis, the illumination of the wire is always an even function with respect to its center whenever $\cos \beta = 1$ but not for other angles.

In Fig. 5 is shown the axial distribution of magnetic current $m(x)$, or, equivalently, electric field, in the slot for several values of $\cos \beta$. Here we note that, when the wire and the slot are perpendicular ($\cos \beta = 1$), the magnitude of the real part of the magnetic current is approximately twice that for the $\cos \beta = 0$ case. The wire and slot are uncoupled when $\cos \beta = 0$, and the curve so-designated represents the slot magnetic current in the absence of the wire. One readily appreciates from Fig. 5 how serious the errors may become in a two-step analysis in which, first, the slot field or magnetic current is calculated in the absence of the obstacle behind the screen and, second, induced current on the obstacle is calculated on the basis of illumination determined from the slot field of the first step.

Half-Wavelength Slot, One-Wavelength Wire

The current in Fig. 6 is on a wire whose center falls on the z axis $(0,0,0.125\lambda)$ so for any value of β the slot radiation gives rise to an even-function excitation of the wire which, in turn, causes an even-function current. If the wire center is displaced from the z axis along the slot axis to the point $(0.25\lambda,0,0.125\lambda)$, the current is seen in Fig. 7 to be quite different. Still, for $\cos \beta = 1$, the excitation and, hence, the current are even, but, for any other value of $\cos \beta$, the excitation is not entirely an even function and, thus, the odd function antiresonant current is strongly excited on the one-wavelength wire. With the wire axis not above the slot axis, the wire excitation is never an even function and a strong antiresonant current is excited for all angles β. In the half-wavelength slot, $m(x)$ is predominantly resonant.

One-Wavelength Slot, One-Wavelength Wire

Distributions of wire current and slot magnetic current are very sensitive to the location of the wire center and to β when the length of both the slot and the wire is one wave-

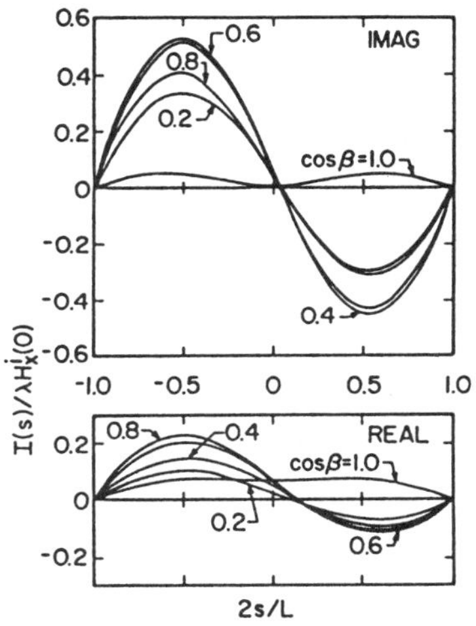

Fig. 7. Current on wire illuminated through slotted screen ($w/\lambda = 0.05$, $l/\lambda = 0.5$; $a/\lambda = 0.001$, $L/\lambda = 1.0$; $x_c/\lambda = 0.25$, $y_c/\lambda = 0$, $z_c/\lambda = 0.125$; normal incidence).

length (Figs. 8–11). If the wire center is above the slot axis but not above its center, the one-wavelength antiresonant current is excited whenever $\cos \beta \neq 1$ (Fig. 8). Fig. 9 displays $m(x)$ pertaining to the same configuration as does Fig. 8. The normally incident illumination is an entirely even-function excitation of the slot and, thus, gives rise to a slot magnetic current having a shifted cosine (forced response) distribution and having no odd-function antiresonant component. For $\cos \beta = 0$, there is no coupling between the wire and slot, so the anticipated shifted cosine due to the incident field is observed in Fig. 9. For $\cos \beta = 1$, however, one observes a large forced component of $m(x)$ due to the incident field plus a smaller antiresonant part due to the unsymmetric backscatter from the wire.

All results discussed above pertain to cases where the plane wave illumination is normally incident upon the slotted screen. An antiresonant magnetic current can be excited in a one-wavelength slot by an obliquely incident plane wave as seen in Fig. 10. The wire current is very

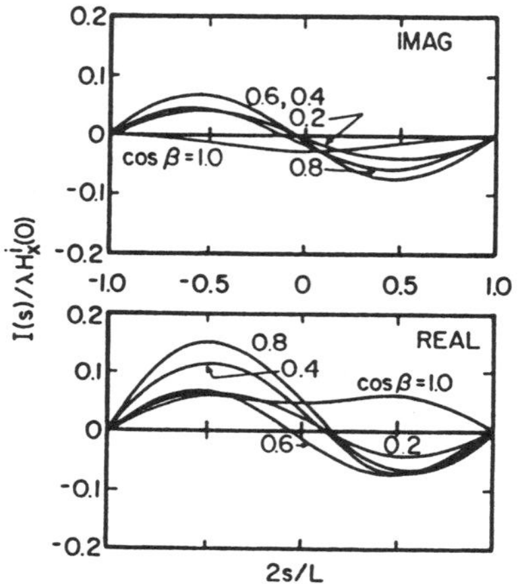

Fig. 8. Current on wire illuminated through slotted screen ($w/\lambda = 0.05$, $l/\lambda = 1.0$; $a/\lambda = 0.001$, $L/\lambda = 1.0$; $x_c/\lambda = 0.25$, $y_c/\lambda = 0$, $z_c/\lambda = 0.25$; normal incidence).

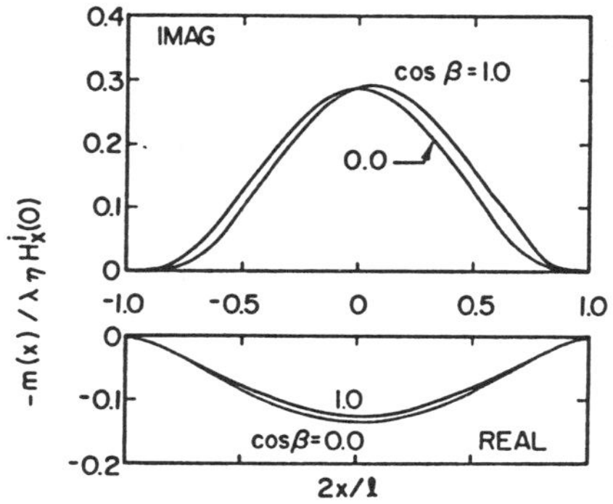

Fig. 9. Axial distribution of slot magnetic current ($w/\lambda = 0.05$, $l/\lambda = 1.0$; $a/\lambda = 0.001$, $L/\lambda = 1.0$; $x_c/\lambda = 0.25$, $y_c/\lambda = 0$, $z_c/\lambda = 0.25$; normal incidence).

sensitive to wire/slot coupling in this orientation as can be seen from the curves of Fig. 11.

Summary

From the data presented above, one appreciates the fact that the current on a wire excited through a slotted screen can vary markedly with wire and slot lengths as well as with wire/slot orientation. Also, one sees that coupling between the wire and slot can be quite strong and that it cannot be ignored in general. The *distribution* of current on an anti-resonant-length wire or slot is highly dependent upon wire/slot coupling and incident field, whereas near resonant length the influence is minimal; however, the *magnitude* can be strongly dependent upon coupling for any lengths.

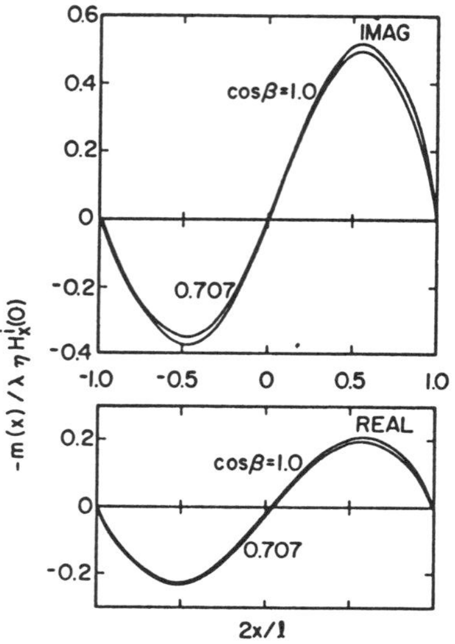

Fig. 10. Axial distribution of slot magnetic current ($w/\lambda = 0.05$, $l/\lambda = 1.0$, $a/\lambda = 0.001$, $L/\lambda = 1.0$; $x_c/\lambda = 0.25$, $y_c/\lambda = 0$, $z_c/\lambda = 0.25$; 60°-incidence angle).

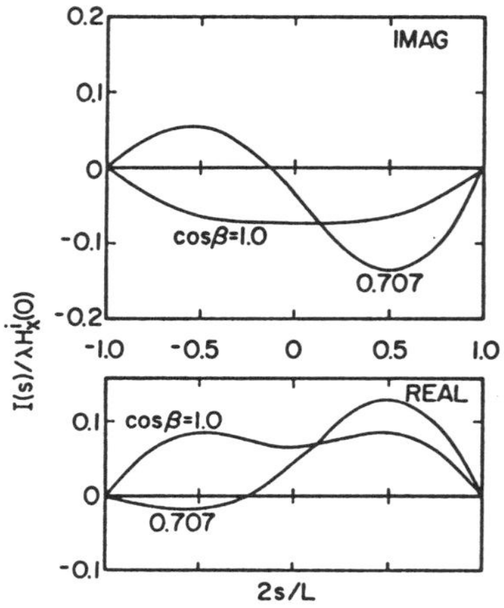

Fig. 11. Current on wire illuminated through slotted screen ($w/\lambda = 0.05$, $l/\lambda = 1.0$, $a/\lambda = 0.001$, $L/\lambda = 1.0$, $x_c/\lambda = 0.25$, $y_c/\lambda = 0$, $z_c/\lambda = 0.25$; 60°-incidence angle).

References

[1] R. W. P. King and G. H. Owyang, "The slot antenna with coupled dipoles," *IRE Trans. on Antennas and Propagation*, vol. AP-8, no. 2, pp. 136–143, March 1960.

[2] J. L. Lin, W. L. Curtis, and M. C. Vincent, "Electromagnetic coupling to a cable through apertures," 1974 *International IEEE/AP-S Symposium Digest*, Atlanta, Georgia, pp. 196–199, June 1974.

[3] C. M. Butler, "Analysis of a wire excited through an aperture perforated conducting screen," 1974 URSI Annual Meeting, Boulder, Colorado, October 1974.

[4] Y. Rahmat-Samii and R. Mittra, "Integral equation solution and RCS computation of a thin rectangular plate," *IEEE Trans. on Antennas and Propagation*, vol. AP-22, no. 4, pp. 608–610, July 1974.

[5] J. L. Lin, W. L. Curtis, and M. C. Vincent, "On the field distribution of an aperture," *IEEE Trans. on Antennas and Propagation*, vol. AP-22, no. 3, pp. 467–471, May 1974.

[6] R. W. P. King, *The Theory of Linear Antennas*. Cambridge, Mass.: Harvard University Press, 1956.

[7] R. F. Harrington, *Field Computation by Moment Methods*. New York: MacMillan, 1968.

Electromagnetic Penetration Through Apertures in Conducting Surfaces

CHALMERS M. BUTLER, SENIOR MEMBER, IEEE, YAHYA RAHMAT-SAMII, MEMBER, IEEE, AND RAJ MITTRA, FELLOW, IEEE

Abstract—In designing hardened systems, one must be able to characterize as well as quantitatively determine the penetration of EMP signals through apertures of general shapes in structures of varying configurations. In this paper a tutorial review of a number of methods for analyzing such aperture problems is presented with an emphasis on techniques. The discussion presented herein is reasonably self-contained and is supplemented by references to classical as well as current approaches to the aperture problem. An extensive set of representative numerical results is included.

I. INTRODUCTION

IN EMP STUDIES, it is desirable to quantify electromagnetic penetration through apertures in conducting surfaces. Apertures of interest usually are electromagnetically small over the spectrum of the EMP, and their existence may be intentional, e.g., windows, open access holes, and bombay doors, or they may be inadvertent as in the case of cracks around doors and plates covering access ports or poor electrical seams in outer skins.

Even though the classic problem of penetration of time-harmonic electromagnetic fields through an aperture in a planar conducting screen of infinite extent has been the subject of intensive research for many years [1]-[3], still the body of theory pertaining to this simplest of aperture problems remains a rather complicated subject, and only in the special case of diffraction by a circular aperture are analytical results available. Greater progress has been made for small[1] apertures in planar screens, where, in the present context, small implies that the maximum dimension across the aperture is small relative to the wavelength of the time-harmonic electromagnetic field.

The purpose of the present paper is to provide a tutorial review of aperture theory in its present state of maturity with emphasis upon those facets of the theory which lead to a better understanding of EMP penetration. A brief discussion is given of the boundary value problem involving an aperture-perforated, planar screen separating two homogeneous half spaces having the same electromagnetic properties, and integro-differential equations for this problem are formulated. Due to the importance of this fundamental but complex problem, these introductory developments are presented at an elementary level. These preliminary concepts are generalized and equations are derived for the problem of diffraction by a closed conducting surface in which an aperture has been cut. Because large apertures are of no interest in EMP studies, they are not considered here.

Due to space limitations, no attempt is made to include an extensive literature review in this paper. The interested reader is referred to the papers of Bouwkamp [1] and Eggimann [2]. However, the authors do provide a classified bibliography comprising primarily papers published since [1] and [2].

II. GENERAL APERTURE/SCREEN EQUATIONS

The fundamental problem considered here is the electromagnetic interaction of the field due to impressed (specified) sources and a planar conducting screen having a hole (aperture) cut in it (Fig. 1). The screen is in an infinite homogeneous medium characterized by (μ, ϵ),[2] it is assumed to be perfectly conducting, vanishing thin, and of infinite extent, and it is located, for convenience, in the xy plane of a Cartesian coordinate system (Fig. 1). As usual, the sources $(\bar{J}^{i-}, \bar{M}^{i-})$ and $(\bar{J}^{i+}, \bar{M}^{i+})$, located in the left and right half spaces, respectively, as shown in Fig. 1, vary harmonically in time according to $e^{j\omega t}$, which factor is suppressed in subsequent equations.

The general procedure which leads to the desired aperture/screen integro-differential equations is briefly outlined below. First, we identify the transverse aperture electric field $\bar{E}_t^a$, the component of total electric field in the aperture parallel to the screen, as the unknown to be determined. Next, we derive individual expressions for the magnetic field on both sides of the screen in terms of E_t^a (or, in terms of an equivalent magnetic current). The expressions for the magnetic field are formulated in terms of the electric vector potential, which ensures us that Maxwell's equations and the radiation condition are satisfied in both half spaces, and they are based upon image theory which ensures us that the boundary conditions on the screen itself are satisfied. Moreover, the magnetic field in each half space is written as a function of $\bar{E}_t^a$, which is common to the two half-space problems; thus the continuity of electric field through A is automatically ensured. The last remaining requirement of electromagnetics is that the mag-

Manuscript received September 13; revised June 2, 1977. A more extensive version of the present paper is available as *AFWL Interaction Note 308* from the Air Force Weapons Laboratory, Kirtland Air Force Base, NM.

C. M. Butler is with the Department of Electrical Engineering, School of Engineering, University of Mississippi, University, MS.

Y. Rahmat–Samii and R. Mittra are with the University of Illinois, Urbana, IL.

[1] See references under appropriate heading in classified bibliography.

[2] If desired, the medium may be lossy which condition may be accounted for by replacing ϵ by $\epsilon - j(\sigma/\omega)$.

Reprinted from *IEEE Trans. Antennas Propagat.*, vol. AP-26, no. 1, pp. 82–93, January 1978.

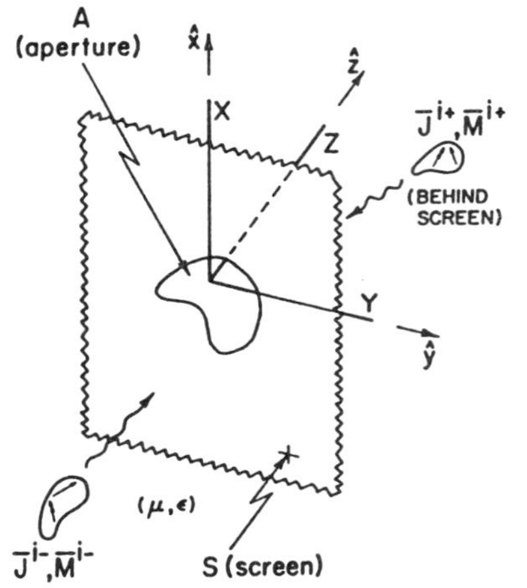

Fig. 1. Aperture in planar conducting screen of infinite extent.

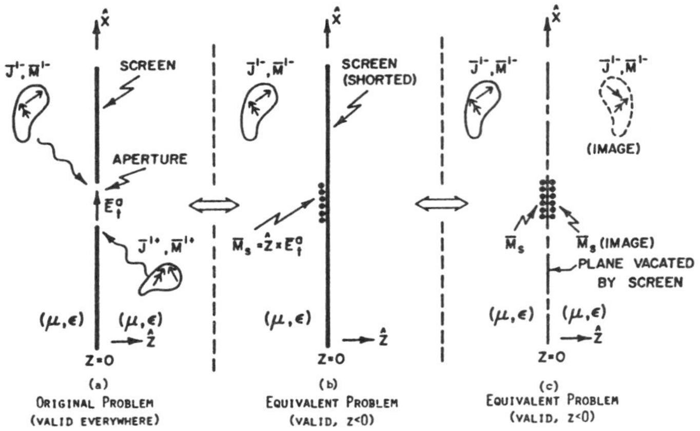

Fig. 2. Left half-space equivalences.

netic field must be continuous along any path through the aperture. Enforcement of this last condition leads to the desired equations.

Fig. 2 depicts a sequential procedure for deriving the expression for $\bar{H}^-$, the total magnetic field in the left half space. The original problem is seen in Fig. 2(a), while in Fig. 2(b) the aperture is short circuited, i.e., the conducting screen is made continuous, and the electric field is restored to its original value $\bar{E}_t^a$ at $z = 0^-$ by the equivalent surface magnetic current $\bar{M}_s$ $(= \hat{z} \times \bar{E}_t^a)$ placed over the region A on the short-circuited screen. From Fig. 2(b) one obtains Fig. 2(c) directly by use of image theory.[3] Now, since all currents in Fig. 2(c) reside in a homogeneous space of infinite extent, one can write $\bar{H}^-$ in terms of the particular integral solution of the wave equation for the electric vector potential:

$$\bar{H}^-(\bar{r}) = \bar{H}^{sc-}(\bar{r}) - j\frac{\omega}{k^2}\,[k^2\bar{F}(\bar{r}) + \nabla(\nabla \cdot \bar{F}(\bar{r}))], \quad z < 0 \tag{1}$$

where $\bar{r}$ is the point of observation and where $k = 2\pi/\lambda$. $\bar{H}^{sc-}$ is the so-called short-circuit magnetic field [4] and is that field due to the sources $(\bar{J}^{i-}, \bar{M}^{i-})$ which would exist in the left half space with the aperture shorted. The remaining terms are the contributions from the equivalent magnetic current plus its image, and they account for the presence of the hole in the screen. The vector potential $\bar{F}$ is given by

$$\bar{F}(\bar{r}) = \frac{\epsilon}{2\pi}\iint_A \bar{M}_s(\bar{r}')\,\frac{e^{-jk|\bar{r}-\bar{r}'|}}{|\bar{r}-\bar{r}'|}\,dS'. \tag{2}$$

In an analogous manner, the right half-space magnetic field $\bar{H}^+$ is

$$\bar{H}^+(\bar{r}) = \bar{H}^{sc+}(\bar{r}) + j\frac{\omega}{k^2}\,[k^2\bar{F}(\bar{r}) + \nabla(\nabla \cdot \bar{F}(\bar{r}))], \quad z > 0 \tag{3}$$

where, of course, $\bar{H}^{sc+}$ is the right half-space short-circuit magnetic field due to $(\bar{J}^{i+}, \bar{M}^{i+})$. The equivalent magnetic current for the right half-space problem is $-\bar{M}_s$, which accounts for the positive sign of the contribution from vector potential terms in (3).

Enforcement of continuity of magnetic field through A is achieved by

$$\lim_{z \uparrow 0}\,(\bar{H}^-(\bar{r}) \times \hat{z}) = \lim_{z \downarrow 0}\,(\bar{H}^+(\bar{r}) \times \hat{z}), \qquad \bar{r} \in A \tag{4}$$

which, in view of (1) and (3), becomes

$$j\frac{\omega}{k^2}\,[k^2\bar{F} + \nabla_t\nabla_t \cdot \bar{F}] \times \hat{z} = \begin{cases} \frac{1}{2}(\bar{H}^{sc-} - \bar{H}^{sc+}) \times \hat{z}, & \text{in } A \\ \text{or} \\ (\bar{H}^{i-} - \bar{H}^{i+}) \times \hat{z} \end{cases} \tag{5}$$

where ∇_t is the transverse (to z) gradient operator and where one interprets (5) in the limiting sense of (4). $\bar{H}^{i-}$ and $\bar{H}^{i+}$ in (5) are the incident magnetic fields in the left and right space, respectively, due to the specified sources radiating in the absence of the screen. Implicit in (5) is the fact that $\bar{H}^{sc} \times \hat{z} = 2\bar{H}^i \times \hat{z}$ on a planar conducting screen. Note that (5) embodies two scalar, coupled integro-differential equations with the two transverse (to z) components of $\bar{M}_s$, or, equivalently, of $\bar{E}_t^a$, as the unknown quantities.

In view of the well-known behavior of electric fields near edges [5], together with the relationship $\bar{M}_s = \hat{z} \times \bar{E}_t^a$, the component of $\bar{M}_s$ normal to the aperture/screen edge must approach zero at a point in A as the square root of the distance from this point to the screen, and the tangential component of $\bar{M}_s$ must be singular as the reciprocal of this square root of distance.

When $\bar{M}_s$ or $\bar{E}_t^a$ is available from the solution of (5) for a specified aperture problem, the magnetic fields on the two sides of the screen can be determined from (1) and (3) and the electric fields can be calculated from

$$\bar{E}^{\pm}(\bar{r}) = \bar{E}^{sc\pm}(\bar{r}) \pm \frac{1}{\epsilon}\nabla \times \bar{F}(\bar{r}) \tag{6}$$

[3] Fig. 2 is only schematic and so the vector directions shown should not be interpreted as actual directions of quantities.

where $\bar{E}^{sc\pm}$ represents the short-circuit electric fields on the two sides of the screen.

Field Properties

From (1), (3), and (5), one can show that

$$\bar{H}^{\pm} \times \hat{z} = \frac{1}{2}(\bar{H}^{sc+} + \bar{H}^{sc-}) \times \hat{z} = (\bar{H}^{i+} + \bar{H}^{i-}) \times \hat{z} \quad \text{in } A. \tag{7}$$

Also, since the normal component of the total electric field must be continuous along any path through A, it follows from (6) that

$$\bar{E}^{\pm} \cdot \hat{z} = \frac{1}{2}(\bar{E}^{sc+} + \bar{E}^{sc-}) \cdot \hat{z} = (\bar{E}^{i+} + \bar{E}^{i-}) \cdot \hat{z} \quad \text{in } A. \tag{8}$$

An interpretation of (7) and (8) is that the transverse component of $\bar{H}$ and the normal component of $\bar{E}$ are the same in A as what they would be if the screen were not present. In addition, the field components exhibit interesting symmetry properties, with respect to the xy plane, which can be expressed as

$$\hat{z} \times [\bar{E}^{-}(x, y, z) - \bar{E}^{sc-}(x, y, z)] = \hat{z} \times [\bar{E}^{+}(x, y, -z) - \bar{E}^{sc+}(x, y, -z)] \tag{9a}$$

and

$$\hat{z} \cdot [\bar{E}^{-}(x, y, z) - \bar{E}^{sc-}(x, y, z)] = -\hat{z} \cdot [\bar{E}^{+}(x, y, -z) - \bar{E}^{sc+}(x, y, -z)] \tag{9b}$$

plus

$$\hat{z} \times [\bar{H}^{-}(x, y, z) - \bar{H}^{sc-}(x, y, z)] = -\hat{z} \times [\bar{H}^{+}(x, y, -z) - \bar{H}^{sc+}(x, y, -z)] \tag{10a}$$

and

$$\hat{z} \cdot [\bar{H}^{-}(x, y, z) - \bar{H}^{sc-}(x, y, z)] = \hat{z} \cdot [\bar{H}^{+}(x, y, -z) - \bar{H}^{sc+}(x, y, -z)] \tag{10b}$$

where we emphasize that $(\bar{E}^{\pm} - \bar{E}^{sc\pm}, \bar{H}^{\pm} - \bar{H}^{sc\pm})$ is that part of the field which is due to the presence of the hole in the screen.

III. GENERAL APERTURE/CAVITY-WALL EQUATIONS

All surfaces of practical interest are not planar and the regions separated by them are not always empty half spaces. Therefore, it is appropriate here to outline a formulation which is generally applicable to a class of problems involving an aperture in a curved surface. Since a vector source radiates a vector field whose boundary conditions are vector in nature too, dyadic formalism is employed as a convenience [6].

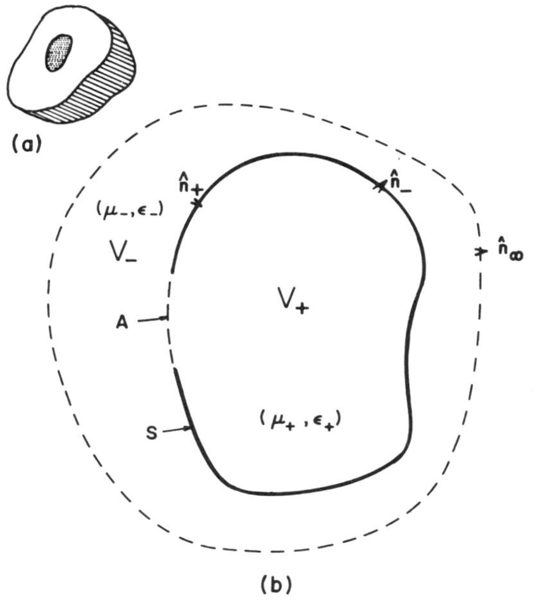

Fig. 3. (a) Aperture in a cavity. (b) Cross sectional view of (a) where V_- and V_+ are exterior and interior regions, respectively; A is the aperture and S is the surface of the cavity.

The geometry under consideration is a cavity with an aperture A in its shell S as shown in Fig. 3 where V_- and V_+ denote the exterior and interior regions of the cavity, respectively. It is assumed that V_- and V_+ are filled with homogeneous and isotropic materials (μ_-, ϵ_-) and (μ_+, ϵ_+), respectively, and that S is a perfectly conducting, vanishingly thin shell. The starting point is Maxwell's equations

$$\begin{bmatrix} \nabla\times & j\omega\mu \\ -j\omega\epsilon & \nabla\times \end{bmatrix} \begin{bmatrix} \bar{E} \\ \bar{H} \end{bmatrix} = \begin{bmatrix} -\bar{M}^i \\ \bar{J}^i \end{bmatrix} \tag{11}$$

which apply to V_- and V_+ individually. From (11) one arrives at the inhomogeneous vector wave equation for the electric field $\bar{E}$ in each region

$$(\nabla \times \nabla \times - k^2)\bar{E} = -\nabla \times \bar{M}^i - j\omega\mu\bar{J}^i. \tag{12}$$

Next, the dyadic Green's function $\bar{\bar{G}}(\bar{r} \mid \bar{r}')$ is defined in the usual way as the solution of

$$(\nabla \times \nabla \times - k^2)\bar{\bar{G}}(\bar{r} \mid \bar{r}') = \hat{I}\delta(\bar{r} - \bar{r}') \tag{13}$$

subject to boundary conditions discussed subsequently, and where $\hat{I}$ is the unit dyadic and δ is the Dirac delta distribution. Note from (13) that $\nabla\cdot\bar{\bar{G}}(\bar{r} \mid \bar{r}') \neq 0$. To establish a relationship between $\bar{E}$ and $\bar{\bar{G}}$, Green's Theorem in dyadic form [6] is used

$$\iiint_V [\bar{E} \cdot \nabla \times \nabla \times \bar{\bar{G}} - (\nabla \times \nabla \times \bar{E}) \cdot \bar{\bar{G}}]\, dv = -\iint_{\partial V} \hat{n} \cdot [\bar{E} \times \nabla \times \bar{\bar{G}} + (\nabla \times \bar{E}) \times \bar{\bar{G}}]\, dS \tag{14}$$

where V denotes the domain of the volume integration, ∂V designates the closed surface surrounding the volume V; and $\hat{n}$ is the outward unit normal to the boundary ∂V.

Since the shell S is assumed to be perfectly conducting, the electric field satisfies

$$\hat{n}_{\mp} \times \bar{E}^{\mp} = \bar{0}, \qquad \bar{r} \in S \tag{15}$$

where "−" and "+" signify quantities in V_- and V_+, respectively. The proper boundary condition for the interior and exterior dyadic Green's functions is

$$\hat{n}_{\mp} \times \bar{\bar{G}}^{\mp}(\bar{r} \mid \bar{r}') = \bar{\bar{0}}, \qquad \bar{r} \in S \cup A. \tag{16}$$

and, for the exterior region, one also imposes the radiation condition, viz.,

$$\lim_{r \to \infty} r \left(\nabla \times \begin{Bmatrix} \bar{E}^- \\ \bar{\bar{G}}^- \end{Bmatrix} + jk_- \hat{r} \times \begin{Bmatrix} \bar{E}^- \\ \bar{\bar{G}}^- \end{Bmatrix} \right) = \begin{Bmatrix} \bar{0} \\ \bar{\bar{0}} \end{Bmatrix}, \qquad \bar{r} \in V_-. \tag{17}$$

Employing (12)-(14) in regions V_- and V_+ and using (15)-(17), we obtain the following representation for the exterior and interior fields

$$\bar{E}^{\mp}(\bar{r}) = \iiint_{V_{\mp}} (-\nabla' \times \bar{M}^{i\mp} - j\omega\mu_{\mp}\bar{J}^{i\mp}) \cdot \bar{\bar{G}}^{\mp}(\bar{r}' \mid \bar{r})\, dv'$$

$$- \iint_A \hat{n}_{\mp} \times \bar{E}^a \cdot \nabla' \times \bar{\bar{G}}^{\mp}(\bar{r}' \mid \bar{r})\, dS', \qquad \bar{r} \in V_{\mp} \tag{18}$$

where, clearly, $\hat{n}_+ = -\hat{n}_-$ and $\bar{E}^-(\bar{r}) = \bar{E}^+(\bar{r}) = \bar{E}^a$ for $\bar{r} \in A$. It should be mentioned that the exterior and interior Green's functions, respectively, $\bar{\bar{G}}^-$ and $\bar{\bar{G}}^+$, generally are different in form. Equation (18) also reveals that, from knowledge of the tangential electric field in the aperture and the Green's functions, one can construct the field everywhere. Our goal now is to construct an integral equation for the unknown tangential electric field in the aperture, i.e., $\hat{n} \times \bar{E}^a$. This is done by first deriving the proper Green's functions for the geometry of interest and then enforcing the condition,

$$\hat{n} \times \bar{H}^- = \hat{n} \times \bar{H}^+, \qquad \text{for } \bar{r} \in A \tag{19}$$

where $\bar{H}$ is determined by substituting (18) into (11). Applications of this procedure are found in [7].

IV. SMALL APERTURES

In many applications, apertures of interest are electromagnetically small, a property which leads to very useful simplifications in computations. Diffraction by small circular and elliptic apertures has been investigated by numerous workers employing a wide variety of different approaches. A comprehensive review of articles pertaining to aperture diffraction in general is given in [3], and an extensive bibliography is accumulated in [2]. Recently, attempts have been made to use an integral equation approach for aperture diffraction problems with the goal of attacking non-separable geometries, e.g., rectangular apertures [8]-[13].

One can utilize the results given in [13] to analyze the problem of electromagnetic diffraction by small apertures. Unlike some of the earlier work referred to above, the procedure discussed below is not restricted in its application to separable geometries only, and its use is particularly suitable when numerical techniques are considered. Here we mention two closely related procedures, both based upon Stevenson's method [14], [15] and both involving an expansion of the unknown magnetic current $\bar{M}_s$ in a so-called Rayleigh series

$$\bar{M}_s(\bar{r}) = \sum_{m=0}^{\infty} \bar{M}_s^{(m)}(\bar{r})k^m. \tag{20}$$

One procedure, advanced by Rahmat-Samii and Mittra [13], is based upon an expansion of both sides of an integral equation in Rayleigh series and an incorporation of constraints on constants in the homogeneous solution. An individual integral equation for each $\bar{M}_s^{(m)}$ is obtained by equating coefficients of like powers of k on the two sides of the expanded version of the integral equation.

In a related procedure developed by Butler [8], [9], (20) is substituted into (5), which is expanded in Rayleigh series. Equating like powers of k, he achieves directly a set of integro-differential equations for the coefficients $\bar{M}_s^{(m)}$. By making use of the concepts of potential theory, he obtains homogeneous and particular solutions to the differential operator equations and converts the integro-differential equations to integral equations in a manner reminiscent of the way one can convert Pocklington's thin-wire equation to Hallen's equation.

Both procedures [8], [9], [13] alluded to above lead to integral equations which are well-suited for numerical methods and from which, in principle, the coefficients $\bar{M}_s^{(m)}$ for all m can be determined for any aperture whose maximum dimension is less than $\lambda/2$. To solve the integral equation for $\bar{M}_s^{(m)}$, one must know $\bar{M}_s^{(m-1)}$; $\bar{M}_s^{(0)}$ can be determined directly. The integral equation for each $\bar{M}_s^{(m)}$ is relatively simple since its kernel is of the electrostatic type and since no differential operators are involved. Another interesting feature of the equations of both procedures is that coupling between the two vector components of $\bar{M}_s^{(m)}$ is realized through relationships among constants in the homogeneous solutions of the differential operator equations and not through the operators themselves.

For a circular aperture of radius a, leading coefficients $\bar{M}_s^{(m)}$ can be determined exactly from solutions of the equations discussed above. With the incident plane wave $E^i = (\hat{x}E_{0x}{}^i + \hat{y}E_{0y}{}^i + \hat{z}E_{0z}{}^i) \exp[-jk(\alpha x + \beta y + \gamma z)]$, where α, β, and γ are the usual direction cosines, these coefficients are found to be [13]

$$M_{s_\rho}{}^{(0)} = 0 \tag{21a}$$

$$M_{s_\phi}{}^{(0)} = \frac{2\rho}{\pi(a^2 - \rho^2)^{1/2}} E_{0z}{}^i \tag{21b}$$

where $\bar{M}_s^{(0)} = M_{s_\rho}^{(0)}\hat{\rho} + M_{s_\phi}^{(0)}\hat{\phi}$. Similarly, the next higher order terms are

$$M_{s_\rho}^{(1)} = -\frac{8j}{3\pi}\gamma(E_{0y}^i \cos\phi - E_{0x}^i \sin\phi)(a^2-\rho^2)^{1/2} - \frac{4j}{3\pi}(\alpha\sin\phi - \beta\cos\phi)(a^2-\rho^2)^{1/2}E_{0z}^i \quad (22a)$$

$$M_{s_\phi}^{(1)} = \frac{2j}{3\pi}\gamma(E_{0x}^i\cos\phi + E_{0y}^i\sin\phi)\left[4(a^2-\rho^2)^{1/2} + \frac{2\rho^2}{(a^2-\rho^2)^{1/2}}\right] - \frac{4j}{3\pi}(\alpha\cos\phi + \beta\sin\phi) \cdot \frac{\rho^2+a^2}{(a^2-\rho^2)^{1/2}}E_{0z}^i. \quad (22b)$$

For normally incident illumination, one can show that $M_\phi^{(2)} = M_\rho^{(2)} = 0$; for this case, the low-frequency expansion (20), correct up to the k^2 term, is

$$M_{s_\rho} = -\frac{8j}{3\pi}(E_{0y}^i\cos\phi - E_{0x}^i\sin\phi)(a^2-\rho^2)^{1/2}k + 0(k^3) \quad (23a)$$

$$M_{s_\phi} = \frac{2j}{3\pi}(E_{0x}^i\cos\phi + E_{0y}^i\sin\phi)\left[4(a^2-\rho^2)^{1/2} + \frac{2\rho^2}{(a^2-\rho^2)^{1/2}}\right]k + 0(k^3). \quad (23b)$$

Equivalent Dipole Moments and Polarizabilities of Small Apertures

Preliminary to the introduction of an aperture-perforated screen, we consider a magnetic surface current density $\bar{M}_s$ in a small planar region R_s about the origin in the xy plane and residing in a homogeneous medium with properties (μ, ϵ). In R_s the magnetic surface charge density m_s is

$$m_s = \frac{j}{\omega}\nabla_t \cdot \bar{M}_s. \quad (24)$$

For R_s sufficiently small and at a point $\bar{r}$ sufficiently remote from this source region, the electromagnetic field due to this magnetic source can be approximated [1], [16] by the radiation from an electric dipole moment $\bar{p}_e$ and a magnetic dipole of moment $\bar{p}_m$, both located at (0, 0, 0), where

$$\bar{p}_e = -\frac{\epsilon}{2}\iint_{R_s}\bar{r}' \times \bar{M}_s(\bar{r}')\,dS' \quad (25a)$$

and

$$\bar{p}_m = \frac{1}{\mu}\iint_{R_s}\bar{r}'m_s(\bar{r}')\,dS' = -\frac{j}{\omega\mu}\iint_{R_s}\bar{M}_s(\bar{r}')\,dS' \quad (25b)$$

with $\bar{r}'$ in R_s. If the dipole moments (25) are known for a magnetic source, one can readily compute the approximate field by means of simple formulas for dipole radiation. If the moments are known for the magnetic sources residing on a conducting surface, the presence of the surface must be accounted for in the calculation of the field from the equivalent dipoles. Returning now to the aperture/screen problem, we note that the equivalent dipole moments of a given small aperture in a screen are related to the specified excitation by the so-called aperture polarizabilities [1], [4], [16], [17]. Knowing the polarizabilities for an aperture and the illumination of the perforated screen, one can determine dipole moments and, subsequently, the diffraction caused by the presence of a small aperture in the screen. For a small aperture A in a screen, the electric polarizability α_e is defined by

$$p_e^{\pm}\hat{z} = \pm\epsilon\alpha_e(E_z^{sc-}(\bar{0}) - E_z^{sc+}(\bar{0}))\hat{z} \quad (26a)$$

and the magnetic polarizability $\bar{\bar{\alpha}}_m$, a dyadic, by

$$\bar{p}_m^{\pm} = \mp\bar{\bar{\alpha}}_m \cdot (\bar{H}^{sc-}(\bar{0}) - \bar{H}^{sc+}(\bar{0})). \quad (26b)$$

The polarizabilities above are defined in such a way that $p_e^+\hat{z}$ and $\bar{p}_m^+$ are the moments for the equivalent dipoles *in the presence of the screen* for the right half-space, i.e., the dipoles are located at $(0, 0, 0^+)$ on the short-circuited screen, while $p_e^-\hat{z}$ and $\bar{p}_m^-$ are the equivalent moments for the left half space. Polarizabilities are available in the literature for several small apertures and values are listed in Table I for three shapes which are of practical interest and for which values are expressible in closed form.

At points far from A, relative to the maximum dimension across the small aperture, the electric fields in the two half spaces are approximately

$$\bar{E}^{\pm} = \bar{E}^{sc\pm} + \bar{E}_e^{\pm} + \bar{E}_m^{\pm} \quad (27)$$

where $\bar{E}_e^{\pm}$ and $\bar{E}_m^{\pm}$ are due to the equivalent electric and magnetic dipoles, respectively, and are given by

$$\bar{E}_e^{\pm}(\bar{r}) = \frac{p_e^{\pm}}{2\pi\epsilon}\left(k^2\hat{z} + \nabla\frac{\partial}{\partial z}\right)g(\bar{r}) \quad (28a)$$

and

$$\bar{E}_m^{\pm}(\bar{r}) = -j\frac{k\eta}{2\pi}\nabla g(\bar{r}) \times \bar{p}_m^{\pm} \quad (28b)$$

with $g(\bar{r}) = \exp(-jk|\bar{r}|)/|\bar{r}|$. Of course, corresponding expressions for the magnetic fields are available from the dipole moments too.

V. EXCITATION OF AN OBJECT THROUGH AN APERTURE IN A SCREEN

In this section, attention is turned to the important problem of calculating the current induced on an object by the field which penetrates a screen through an aperture [18],

TABLE I
APERTURE POLARIZABILITIES*

Shape	α_e	$\alpha_{m_{xx}}$	$\alpha_{m_{yy}}$
Circle (Radius = R)	$\frac{2}{3} R^3$	$\frac{4}{3} R^3$	$\frac{4}{3} R^3$
Ellipse	$\frac{1}{3} \frac{\pi w^2 \ell}{E(\epsilon)}$	$\frac{1}{3} \frac{\pi \ell^3 \epsilon^2}{K(\epsilon) - E(\epsilon)}$	$\frac{1}{3} \frac{\pi \ell^3 \epsilon^2}{\left(\frac{\ell}{w}\right)^2 E(\epsilon) - K(\epsilon)}$
Narrow Ellipse ($w << \ell$)	$\frac{1}{3} \pi w^2 \ell$	$\frac{1}{3} \frac{\pi \ell^3}{\ln\left(\frac{4\ell}{w}\right) - 1}$	$\frac{1}{3} \pi w^2 \ell$

*See [1,16,17]

Notes: (1) $\bar{\bar{\alpha}}_m = \alpha_{m_{xx}} \hat{x}\hat{x} + \alpha_{m_{yy}} \hat{y}\hat{y}$ ($\bar{\bar{\alpha}}_m$ is diagonal for symmetric shapes given.)

(2) Ellipse Eccentricity $\epsilon = \sqrt{1 - \left(\frac{w}{\ell}\right)^2}$

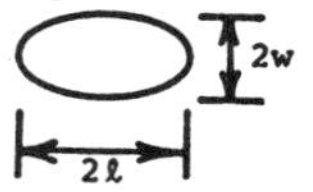

(3) K and E are the complete elliptic integrals of the first and second kind, respectively, as defined in [27].

[19]. The object or scatterer is perfectly conducting and of general shape. The only sources are in the left half space, i.e., $(\bar{J}^{i+}, \bar{M}^{i+}) = (\bar{0}, \bar{0})$, and the scatterer is in the right half space. Desired equations for this problem are achieved by modifying both (5) and the equation for the scatterer in the presence of a continuous (shorted aperture) screen.

We extend (5) to include the effect of the scatterer by treating the field scattered back to the aperture by the object as part of the forcing function of (5):

$$j \frac{\omega}{k^2} (k^2 \bar{F} + \nabla_t \nabla_t \cdot \bar{F}) \times \hat{z} = \begin{cases} \tfrac{1}{2}[\bar{H}^{sc-} - \bar{h}^{sc}] \times \hat{z}, & \text{in } A \\ \text{or,} \\ [\bar{H}^{i-} - \bar{h}^{i}] \times \hat{z} \end{cases} \quad (29)$$

where $(\bar{e}^{sc}, \bar{h}^{sc})$ and $(\bar{e}^{i}, \bar{h}^{i})$ are, respectively, the short-circuit and "incident" fields in the right half space due to scattering by the object and where $\bar{F}$ of (29) is given in (2). Since the current $\bar{J}_s$ induced on the object is due to fields which penetrate the aperture and since $\bar{h}^{sc}$ (or $\bar{h}^{i}$) is entirely due to $\bar{J}_s$, we view this portion of the forcing function of (29) as excitation from a dependent "generator":

$$\bar{h}^{i}(\bar{r}) = \frac{1}{4\pi} \nabla \times \iint_{S_B} \bar{J}_s(\bar{s}') \frac{e^{-jk|\bar{r}-\bar{s}'|}}{|\bar{r} - \bar{s}'|} dS' \quad (30)$$

where S_B represents the surface of the object and $\bar{s}'$ locates a point on S_B.

In addition to (29), an equation must be available which characterizes the scatterer in the presence of the shorted screen subject to illumination from the aperture equivalent magnetic current $\bar{M}_s$. The electric field in the right half space due to $\bar{M}_s$ is given by (6) (with $\bar{E}^{sc+} = 0$) so, for the perfectly conducting scatterer, one requires the tangential electric field on the surface S_B of the object to be zero:

$$\left(j \frac{\omega}{k^2} [k^2 \bar{A} + \nabla\nabla \cdot \bar{A}] + \frac{1}{\epsilon} \nabla \times \bar{F}\right) \times \hat{n} = \bar{0} \text{ on } S_B \quad (31)$$

where $\hat{n}$ is the outward unit normal at a point $\bar{r}$ on S_B and where the magnetic vector potential due to the surface current $\bar{J}_s$ on the object in the presence of the screen is

$$\bar{A}(\bar{r}) = \frac{\mu}{4\pi} \iint_{S_B} \bar{J}_s(\bar{s}') \cdot \bar{\bar{g}}(\bar{r}, \bar{s}')\, dS' \quad (32a)$$

with

$$\bar{\bar{g}}(\bar{r}, \bar{s}') = \bar{\bar{I}} \frac{e^{-jk|\bar{r}-\bar{s}'|}}{|\bar{r} - \bar{s}'|} + (2\hat{z}\hat{z} - \bar{\bar{I}}) \cdot \frac{e^{-jk|\bar{r}-\bar{s}'+2(\bar{s}'\cdot\hat{z})\hat{z}|}}{|\bar{r} - \bar{s}' + 2(\bar{s}' \cdot \hat{z})\hat{z}|} \quad (32b)$$

in which $\bar{\bar{I}}$ is the unit dyadic.

For small apertures of shapes whose polarizabilities are known, a simple formulation for an equation characterizing $\bar{J}_s$ has been devised [18]. The resulting equation is of the order of difficulty to solve as would be the equation for the same scatterer in the presence of the screen with no aperture.

VI. EXAMPLE FREQUENCY-DOMAIN DATA

Ultimate interest in EMP investigations lies, of course, in the time history of the electromagnetic field at critical points in a system under evaluation. Usually such a time history is computed (via Fourier inversion) from knowledge of the corresponding time-harmonic field over a frequency spectrum of practical limits. For this reason and, also, due to the present

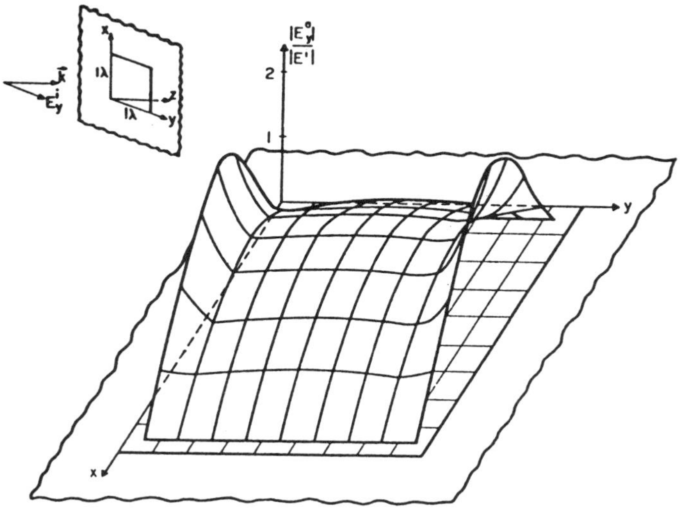
Fig. 4. Three-dimensional representation of the amplitude distribution of the E_y-field in the aperture.

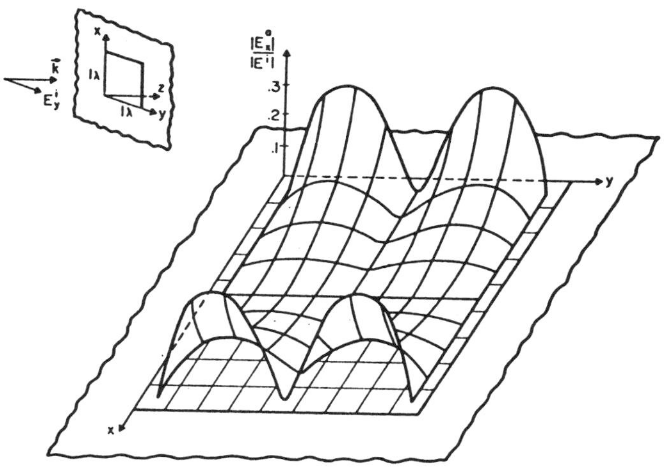
Fig. 5. Three-dimensional representation of the amplitude distribution of the E_x-field in the aperture.

paucity of time-domain electromagnetic field data that are of utility in EMP studies, it is of value to become familiar with available information in the frequency domain. To this end, a summary of time-harmonic results is provided in this section.

For the most part, data presented below were calculated from numerical solutions of the equations discussed in Sections II–V, but, even though such numerical methods are important in present-day work, space limitations do not allow coverage of this subject here. Original sources are cited and should be consulted by those interested in numerical techniques appropriate for a given type problem.

Square and Circular Apertures in Planar Screens

In Figs. 4 and 5 are displayed the magnitudes of the components of the transverse electric field in a $1\lambda \times 1\lambda$ square aperture in an infinite conducting screen, excited by a normally incident plane wave in the left half space with $\bar{E}^{i-} = E^i\hat{y}$. The singularity in the field at the aperture/screen edges is clearly exhibited, and one should note that the field components are different in both peak magnitude and distribution across the aperture. These data were computed by Rahmat-Samii and Mittra [7], and we wish to point out that Wilton and Glisson [20] have devised a clever scheme for solving (5) from which these results are obtainable too. Others [21], [22] have employed moderately reliable solution techniques which are founded upon an approximate wire-grid model of a conducting plate, the Babinet equivalent of an aperture in a planar screen.

Fig. 6 depicts penetrated fields obtained from integral equation solutions, from the Kirchhoff approximation, and from measurements made by Andrews [23] for a circular aperture. The three curves exhibit essentially the same behavior for $z/\lambda > 1.5\lambda$.

Small Apertures

Over the practical spectrum of the EMP, many apertures of interest are electromagnetically small, while a majority of others fall into this category over a significant portion of the spectrum, and, fortunately, far more data are available for small apertures than for those whose maximum dimension is a sizeable fraction of the wavelength.

In Fig. 7 are found plots of the dominant component of electric field in small square and circular apertures, subject to a normally incident plane wave ($\bar{E}^{i-} = E^i\hat{y}$). The square-aperture data were obtained from numerical methods [13], while those for the circular case were computed from (23) directly. We observe from this set of curves that the electric field along the principal axes of the small square aperture is quite close to that in the (inscribed) circular aperture.

For an electrically small aperture one may approximate the aperture-produced fields by making use of (28) and the dipole moments of the given aperture. The closeness of the dipole moment approximation to the actual fields depends upon the electrical size of the aperture, the distance from the aperture to the point at which the field is evaluated, and the choice of the coördinate origin with respect to which the dipole moments are calculated. Therefore, due to the attractiveness afforded by the simplicity of the dipole moment approximation, it is of interest to determine fields directly from the numerically calculated magnetic current $\bar{M}_s$ as well as from the moments and then to compare values so obtained. Such comparisons enable one to assess the accuracy of fields calculated from moments.

Fig. 8 shows the electric field which penetrates a small square aperture ($2a = 2b = 0.15\lambda$) subject to edge-on incident illumination with $E_z{}^i(0) = 1$ V/m and with the direction of propagation along either the x axis or the y axis. These approximate values of fields together with exact values determined from computed $\bar{M}_s$ are displayed for comparison. One sees good agreement at a radial distance $r = 10a$ but sees significant differences at $3a$ and $2a$. The primary reason for the departure of the two results is the approximate nature of the dipole moment calculation, which incorrectly predicts an infinite field for $z/\lambda \to 0$. Thus one must exercise caution in using the dipole moment approach to compute the diffracted field close to an aperture.

Now we return to observations of similarities in small circular and square apertures. In particular we compare diffracted

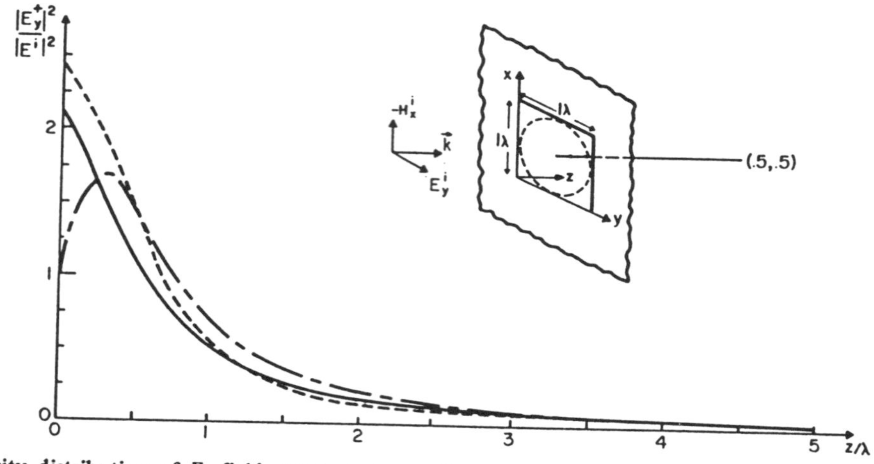

Fig. 6. Intensity distribution of E_y-field sampled along a line parallel to the z-axis and passing through the center of square and circular apertures. Integral equation (—) solution for square aperture. Kirchhoff approximation for square aperture (— -). Experimental result (- - -) for circular aperture from Andrews [23].

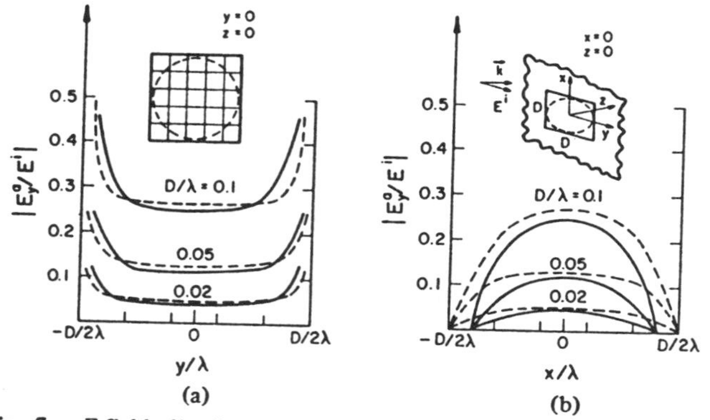

Fig. 7. E-field distribution in electrically small square and circular apertures. (a) E_y-field sampled along y-axis. (b) E_y-field sampled along x-axis. Integral equation (—) solution for square apertures. First-order (- - -), low frequency, from (23) for the circular apertures with diameter D.

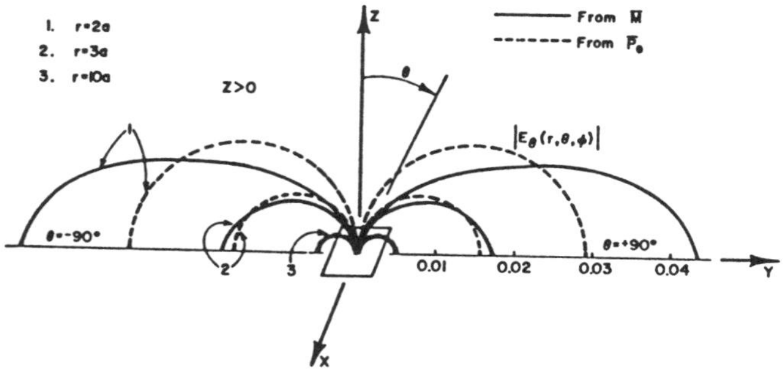

Fig. 8. Electric field on shadow side of square aperture ($2a = 2b = 0.15\lambda$, $E_z^{i-} = 1$ V/m, edge-on incidence).

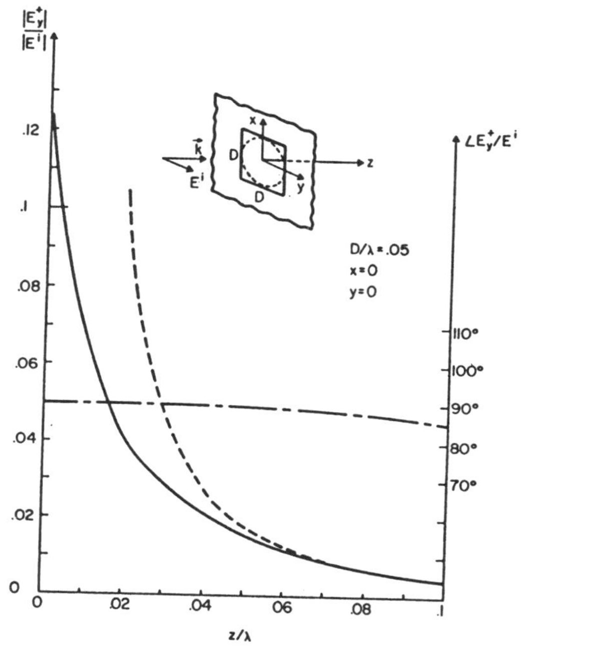

Fig. 9. E_y-field distribution sampled along the z-axis. Integral equation solution (—) for square aperture. Dipole moment results (- - -) for circular aperture. Phase curve (— -) obtained from integral equation and dipole moment techniques.

fields computed for the square aperture from integral equation solutions with those for the circular hole from dipole moments. We note from Fig. 9, which displays the behavior of the E_y field as a function of z/λ, that the phases coincide and that the amplitude curves also agree well for $z/\lambda > 0.08$. However, for closer points, i.e., $0 < z/\lambda < 0.08$, the dipole moment approximation for the circular aperture field deviates substantially from the numerically determined solution for the square aperture.

Wire Scatterer Behind a Slotted Screen

As a special but practically important case of the type problem discussed in Section V, we let the scatterer be a thin wire of length L parallel to the screen and the aperture be a narrow slot of length l, as depicted in the insert of Fig. 10. The wire center is designated (x_c, y_c, z_c), the slot center is at $(0, 0, 0)$, and the orientation is specified by the angle β defined as the angular displacement of the wire axis from the y axis. The excitation is a plane wave normally incident upon the screen, the wire radius a is 0.001λ, and the slot width w is 0.05λ.

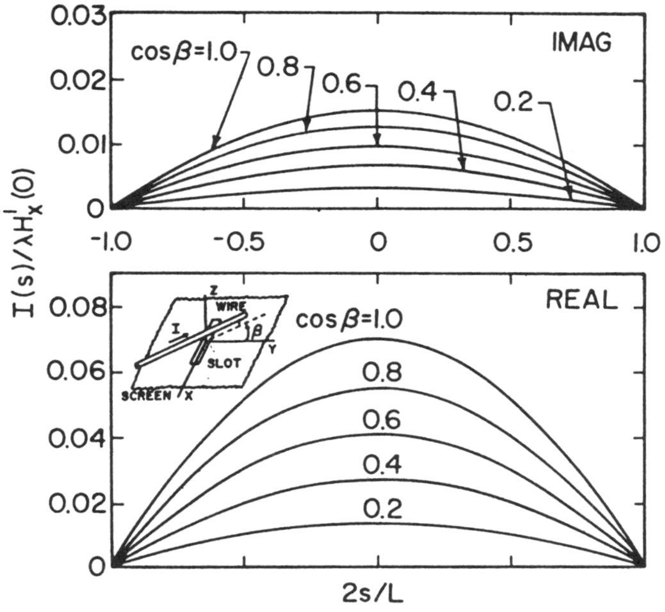

Fig. 10. Current on wire illuminated through slotted screen (w/λ = 0.05, l/λ = 0.25; a/λ = 0.001, L/λ = 0.5; x_c/λ = 0, y_c/λ = 0, z_c/λ = 0.25; normal incidence).

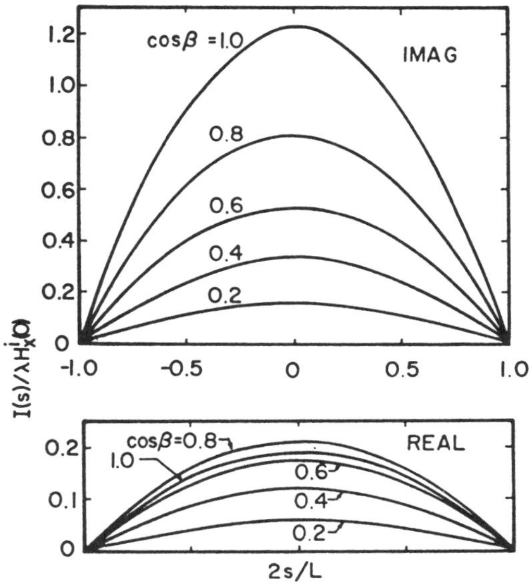

Fig. 11. Current on wire illuminated through slotted screen (w/λ = 0.05, l/λ = 0.5; a/λ = 0.001, L/λ = 0.5; x_c/λ = 0, y_c/λ = 0, z_c/λ = 0.25; normal incidence).

Fig. 10 shows the current on a half-wavelength wire, as a function of position along the wire, induced by the field which penetrates a quarter-wavelength slot; the center of the wire is on the z axis, $\lambda/4$ behind the screen, and I is given for selected values of the angle β. When $\cos\beta = 1$, the wire and slot are perpendicular, and the coupling is seen to be maximum as expected, while there is no coupling when the wire and slot are parallel ($\cos\beta = 0$). For resonant length ($L = \lambda/2$), the wire current distribution is essentially a cosine function as one would expect. With the slot length increased to $\lambda/2$, the wire current is far greater (Fig. 11) due to the fact that the penetration through a half-wavelength slot is much greater than that through a quarter-wavelength slot.

As pointed out by Butler and Umashankar [19], the energy scattered back into the aperture from the wire can be quite significant and can strongly influence the aperture fields. Calculation of aperture fields under the assumption that the wire is not present can lead to serious errors in certain cases.

VII. EXAMPLE TIME-DOMAIN DATA

As an example of the response of an aperture to an EMP, we compute the time history $\bar{e}^+(t, \bar{r})$ of the field which passes through a 115 cm × 1.3 cm rectangular aperture in a planar screen. The excitation of the aperture/screen is a normally incident, double-exponential, plane-wave EMP represented by

$$\bar{e}^{i}(t,\bar{r}) = \hat{y}E_0\{(e^{-\alpha(t-z/c)} - e^{-\beta(t-z/c)})\theta(t-z/c) - (e^{-\alpha(t-\tau-z/c)} - e^{-\beta(t-\tau-z/c)}) \cdot \theta(t-\tau-z/c)\} \tag{33}$$

where $c = 3.0 \times 10^8$ m/s, $\alpha = 6.0 \times 10^6$ s^{-1}, $\beta = 2.0 \times 10^8$ s^{-1}, $\tau = 2.04189 \times 10^{-9}$ s, $E_0 = 10^3$ V/m, and θ is the unit step function. One computes the spectrum of (33) and, subject to this excitation, determines frequency-domain quantities for the aperture problem; then the time history of each quantity is available from Fourier inversion. The time domain response $\bar{e}^+(t, \bar{r})$ is computed from knowledge of $\bar{E}^+(f, \bar{r})$ via standard transform means and the integrals involved in this process may be handled by the FFT algorithm [24] which is known to be efficient for such purposes. Also, using the fact that $\bar{e}^+(t, \bar{r})$ is real, one can show [25] that either the real or the imaginary part of $\bar{E}^+(f, \bar{r})$ is sufficient to determine $\bar{e}^+(t, \bar{r})$, an observation which lessens computational labor. For an accurate evaluation of the integrals, $\bar{E}^+(f, \bar{r})$ must be adequately sampled in the frequency range of interest so that the criterion suggested by the sampling theorem [26] is not violated. According to this criterion, it is necessary to evaluate $\bar{E}^+(f, \bar{r})$ at no fewer than 1024 points in the interval 0-10^9 Hz, beyond which $|\bar{E}^+(f, \bar{r})|$ decreases to 60 dB below its maximum value. However, rather than sampling $\bar{E}^+(f, \bar{r})$ uniformly in the frequency range 0-10^9 Hz, one finds it more desirable to compute $\bar{E}^+(f, \bar{r})$ relatively densely in the neighborhood of the aperture's resonant frequencies, where $\bar{E}^+(f, \bar{r})$ varies rapidly, and sparsely at the frequencies where the response is smooth. Then one can compute $\bar{E}^+(f, \bar{r})$ at the needed 1024 points by applying an interpolation scheme to the values of $\bar{E}^+(f, \bar{r})$ at the fewer (seventy in this example), nonuniformly spaced points in the spectrum.

In Fig. 12 is found the frequency response of the magnitude of the electric field component $E_y{}^+$ evaluated at a point on the z axis 2 meters behind the aperture/screen. The time-domain electric field $e_y{}^+(t, \bar{r})$, calculated at $\bar{r} = 2\hat{z}$ by the procedure outlined above, is displayed in Fig. 13.

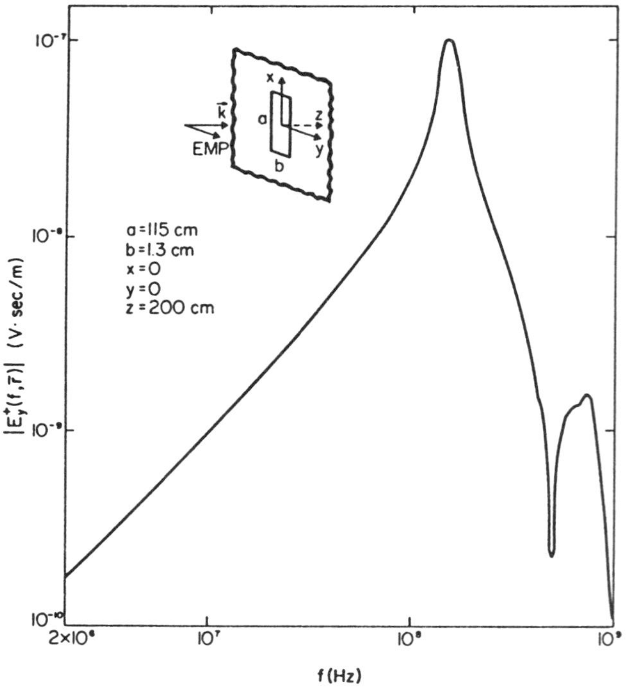

Fig. 12. Frequency domain behaviour of $|E_y{}^+(f,\bar{r})|$ sampled at a point 2 m behind the aperture on the z-axis.

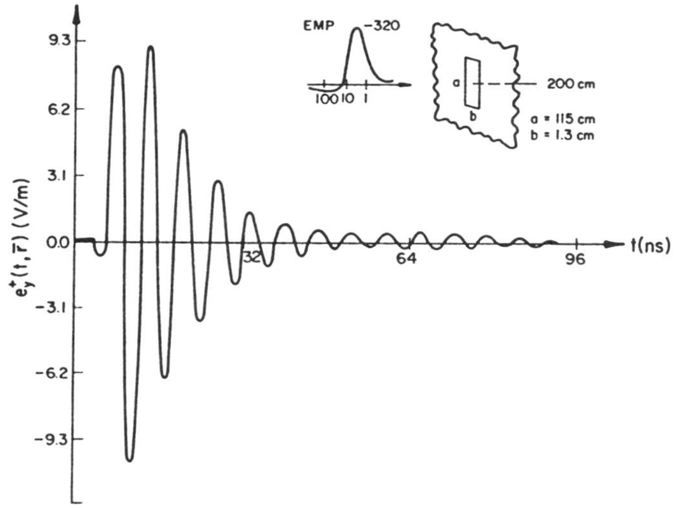

Fig. 13. Time domain behavior of the $e_y{}^+(t,\bar{r})$ field sampled at a point 2 m behind a single aperture.

REFERENCES

Cited

[1] C. J. Bouwkamp, "Diffraction theory," *Rep. Prog. Phys.*, vol. 17, pp. 35–100, 1954.

[2] W. H. Eggimann, "Higher-order evaluation of electromagnetic diffraction by circular disks," *IRE Trans. Microwave Theory Tech.*, vol. MTT-9, pp. 408–418, Sept. 1961.

[3] C. J. Bouwkamp, "Theoretical and numerical treatment of diffraction through a circular aperture," *IEEE Trans. Antennas Propagation*, vol. AP-18, pp. 152–176, Mar. 1970.

[4] J. Van Bladel, *Electromagnetic Fields.* New York: McGraw-Hill, 1964.

[5] J. Meixner, "The behavior of electromagnetic fields at edges," *IEEE Trans. Antennas Propagat.*, vol. AP-21, pp. 442–446, July 1972.

[6] P. M. Morse and H. Feshbach, *Methods of Theoretical Physics*, Part I. New York: McGraw-Hill, 1953.

[7] Y. Rahmat-Samii and R. Mittra, "A new integral equation solution of electromagnetic aperture coupling and thin plate scattering problems," *AFWL Interaction Note 224*, Feb. 1975.

[8] C. M. Butler, "Formulation of integral equations for an electrically small aperture in a conducting screen," *AFWL Interaction Note 149*, Dec. 1973.

[9] —, "Formulation of integral equations for an electrically small aperture in a conducting screen," presented at the 1974 IEEE Int. Symp. Antennas and Propagation, Atlanta, GA, June 1974.

[10] K. R. Umashankar and C. M. Butler, "A numerical solution procedure for small aperture integral equations," *AFWL Interaction Note 212*, July 1974.

[11] D. R. Wilton, C. M. Butler, and K. R. Umashankar, "Penetration of EM fields through small apertures in planar screens: Selected data," *AFWL Interaction Note 213*, Sept. 1974.

[12] D. R. Wilton and O. C. Dunaway, "Electromagnetic Penetration through apertures of arbitrary shape: Formulation and numerical solution procedure," Air Force Weapons Lab., Albuquerque, NM, Tech. Rep. AFWL-TR-74-192, Jan. 1975 (also *AFWL Interaction Note 214*, July 1974).

[13] Y. Rahmat-Samii and R. Mittra, "Electromagnetic coupling through small apertures in a conducting screen," *IEEE Trans. Antennas Propagat.*, vol. AP-25, pp. 180–187, Mar. 1977.

[14] A. F. Stevenson, "Solution of electromagnetic scattering problems as power series in the ratio (dimensions scatterer)/wavelength," *J. Appl. Phys.*, vol. 24, pp. 1134–1142, 1953.

[15] A. F. Kleinman, "Low frequency solutions of electromagnetic scattering problems, " in *Electromagnetic Wave Theory* (Delft Symposium). New York: Pergamon, 1967.

[16] R. E. Collin, *Field Theory of Guided Waves. New York:* McGraw-Hill, 1960.

[17] H. A. Bethe, "Theory of diffraction by small holes," *Phys. Rev.*, vol. 66, pp. 163–182, Oct. 1944.

[18] C. M. Butler, "Excitation of a general scatterer in the vicinity of an aperture in a planar, conducting screen," in preparation.

[19] C. M. Butler and K. R. Umashankar, "Electromagnetic Excitation of a wire through an aperture-perforated, conducting screen," *IEEE Trans. Antennas Propagat.*, vol. AP-25, pp. 456–462, July 1976.

[20] D. R. Wilton and A. W. Glisson, "Toward simple, efficient numerical techniques for scattering by surfaces," presented at the 1976 IEEE/AP-S Symp., Amherst, MA, Oct. 1976.

[21] J. L. Lin, W. L. Curtis, and M. C. Vincent, "On the field distribution of an aperture," *IEEE Trans. Antennas Propagat.*, vol. AP-22, pp. 467–471, May 1974.

[22] J. H. Richmond, "A wire-grid model for scattering by conducting bodies," *IEEE Trans. Antennas Propagat.*, vol. AP-14, pp. 782–786, Nov. 1966.

[23] C. L. Andrews, "Diffraction pattern in a circular aperture measured in the microwave region," *J. Appl. Phys.*, vol. 22, pp. 761–767, 1950.

[24] E. O. Bringham, *The Fast Fourier Transform.* Englewood Cliffs, NJ: Prentice-Hall, 1974.

[25] A. Papoulis, *The Fourier Transform and its Application.* New York: McGraw-Hill, 1962.

[26] P. L. Ransom, "Evaluation of the Fourier integral using the fast fourier transform," Univ. of Illinois, Urbana, Antenna Lab. Rep. 72-9.

[27] I. S. Gradshteyn and I. M. Ryzhik, *Tables of Integrals, Series, and Products.* New York: Academic, 1965.

Electrically Small Apertures

[28] S. B. Cohn, "Determination of aperture parameters by electrolytic-tank measurements," *Proc. IRE*, vol. 39, pp. 1416–1421, Nov. 1951. (See Correction: *ibid.*, vol. 40, p. 33, Jan. 1952.)

[29] —, "The electric polarizability of apertures of arbitrary shape," *Proc. IRE*, vol. 40, pp. 1069–1071, Sept. 1952.

[30] F. De Meulenaere, and J. Van Bladel, "Polarizability of some small apertures," *IEEE Trans. Antennas Propagat.*, vol. AP-25, pp. 198–205, Mar. 1977.

[31] Lord Rayleigh, "On the Incidence of aerial and electric waves upon small obstacles in the form of ellipsoids or elliptic cylinders, and on the passage of electric waves through a circular aperture in a conducting screen," *Phil. Mag.*, series 5, vol. 44, pp. 28–52, July 1897.

[32] C. D. Taylor, "Electromagnetic pulse penetration through small apertures," *IEEE Trans. Electromagn. Compat.*, vol. EMC-15, pp. 17–23, Feb. 1973.

[33] R. F. Fikhmanas and P. Sh. Fridberg, "Theory of diffraction at small apertures. Computation of upper and lower boundaries of the polarizability coefficients," *Radio Eng. Electron. Phys. (USSR)*, vol. 18, pp. 824–829, 1973.

[34] J. S. Asvestas and R. E. Kleinman, "Low frequency scattering by spheroids and disks, I: Dirichlet problem for a prolate spheroid," *J. Inst. Math. Appl., 6*, pp. 42–56, 1969.

[35] —, "Low frequency scattering by spheroids and disks, II: Newmann problem for a prolate spheroid," *J. Inst. Math. Appl. 6*, pp. 57–75, 1969.

[36] —, "Low frequency scattering by spheroids and disks, III. Oblate spheroids and disks," *J. Inst. Math. Appl. 6*, pp. 157–163, 1970.

[37] R. E. Kleinman, "The Rayleigh region," *Proc. IEEE*, vol. 53, pp. 848–856, 1965.

Electrically Large Apertures

[38] T. E. Cherot, Jr., "Calculation of near field of circular aperture antenna using geometrical theory of diffraction," *IEEE Trans. Electromagn. Compat.*, vol. EMC-13, pp. 29–34, May 1971.

[39] G. F. Koch and K. S. Kolbig, "The transmission coefficient of elliptical and rectangular apertures for electromagnetic waves," *IEEE trans. Antennas propagat.*, vol. AP-16, Jan. 1968.

[40] H. H. Snyder, "On certain wave transmission coefficients for elliptical and rectangular apertures," *IEEE Trans. Antennas Propagat.*, vol. AP-17, pp. 107–109, Jan. 1969.

[41] F. Bekefi, "Diffraction of electromagnetic waves by an aperture in a large screen," *J. Appl. Phys.*, vol. 24, pp. 1123–1130, Sept. 1953.

[42] J. B. Keller, R. M. Lewis, and B. D. Seckler, "Diffraction by an aperture II," *J. Appl. Phys.*, vol. 28, pp. 570–579, May 1957.

[43] J. B. Keller, "Geometrical theory of diffraction," *J. Appl. Phys.*, vol. 28, pp. 426–444, 1962.

[44] R. Mittra, Y. Rahmat-Samii, and W. Ko, "Spectral theory of diffraction," *Appl. Phys.*, vol. 10, pp. 1–13, 1976.

[45] Y. Rahmat-Samii and R. Mittra, "A Spectral domain interpretation of high frequency diffraction phenomena," *IEEE Trans. Antennas Propagat.*, vol. AP-25, pp. 676–687, Sept. 1977.

Slots

[46] J. L. Guiraud, "Lignes de propagation de i'energie dans diffraction d'une onde electromagnetique plane par une fente dans un plan conducteur," *C. R. Acad. Sc. Paris*, t. 277, July 1973.

[47] R. F. Millar, "Radiation and reception properties of a wide slot in a parallel-plate transmission line, Parts I and II," *Can. J. Phys.*, vol. 37, pp. 144–169, 1959.

[48] —, "Diffraction by a wide slit and complementary strip I," *Proc. Camb. Phil. Soc.*, vol. 54, pp. 479–496, 1958.

[49] —, "Diffraction by a wide slit and complementary strip, II," *Proc. Camb. Phil. Soc.*, vol. 54, pp. 497–511, 1958.

[50] S. N. Karp and A. Russek, "Diffraction by a wide slit," *J. Appl. Phys.*, vol. 27, pp. 886–894, Aug. 1956.

[51] C. M. Butler and E. Yung, "Properties of a slotted parallel-plate waveguide excited by an incident TEM wave, presented at the 1976 IEEE/AP-S Symp., Amherst, MA, Oct. 1976.

[52] R. F. Millar, "A note on the diffraction by a wide slit," *Can. J. Phys.*, vol. 38, pp. 38–47, 1960.

[53] H. G. Booker, "Slot aerials and their relation to complementary wires aerials (Babinet's Principle)," *J. Inst. Elec. Eng.*, vol. III-A, pp. 620–626, 1946.

[54] R. W. P. King and G. H. Owyang, "The slot antenna with coupled dipoles," *IRE Trans. Antennas Propagat.*, vol. AP-8, pp. 136–143, Mar. 1960.

[55] Y. Nomura and S. Katsura, "Diffraction of electromagnetic waves by ribbon and slit, I," *J. Phys. Soc. Japan*, vol. 12, pp. 190–200, Feb. 1957.

Aperture in Screen Separating Different Media

[56] C. M. Butler and K. R. Umashankar, "Electromagnetic penetration through an aperture in an infinite, planar screen separating two half-spaces of different electromagnetic properties," *Radio Sci.*, vol. 11, pp. 611–619, July 1976.

[57] R. Barakat, "Diffraction of plane waves by a slit between two different media," *J. Opt. Soc. Amer.*, vol. 53, pp. 1231–1243, 1963.

[58] K. Houlberg, "Diffraction by a narrow slit in the interface between two media," *Can. J. Phys.*, vol. 45, pp. 57–81, 1967.

[59] D. P. Thomas, "Electromagnetic Diffraction by a circular aperture in a plane screen between different media." *Can. J. Phys.*, vol. 47, pp. 921–930, 1969.

Loaded Apertures

[60] K. C. Lang, "Babinet's principle for a perfectly conducting screen with aperture covered by resistive sheet," *IEEE Trans. Antennas Propagat.*, vol. AP-21, pp. 738–740, Sept. 1973.

[61] R. F. Harrington and J. R. Mautz, "Comments on Babinet's principle for a perfectly conducting screen with aperture covered by resistive sheet," *IEEE Trans. Antennas Propagat.*, vol. AP-22, p. 842, Nov. 1974.

[62] T. B. A. Senior, "Some extensions of Babinet's principle," *J. Acoust. Soc. Amer.*, vol. 58, pp. 501–503, Aug. 1975.

[63] C. M. Butler and G. C. Lewis, Jr., "Electromagnetic penetration through a dielectric-sheet-covered slotted screen," presented at the *1975 URSI Ann. Meeting*, Bolder, CO, Oct. 1975.

[64] R. A. Hurd and B. K. Sachdeva, "Scattering by a dielectric-loaded slit in a conducting plane," *Radio Sci.*, vol. 10, pp. 565–572, May 1975.

[65] R. W. Latham and M. S. H. Lee, "Magnetic field leakage into a semi-infinite pipe," *Can. J. Phys.*, vol. 46, pp. 1455–1462, 1968.

Apertures in Thick Screen

[66] F. L. Neerhoff and G. Mur, "Diffraction of a plane electromagnetic wave by a slit in a thick screen placed between two different media," *Appl. Sci. Res.*, vol. 28, pp. 73–88, July 1973.

[67] A. N. Akhiezer, "On the inclusion of the effect of the thickness of the screen in certain diffraction problems," *Sov. Phys.-Tech. Phys.*, vol. 2, pp. 1190–1196, 1957.

[68] M. A. Allen and G. S. Kino, "On the theory of strongly coupled cavity chains," *IRE Trans. Microwave Theory Tech.*, vol. MTT-8, pp. 362–372, May 1960.

[69] Kh. L. Garb, I. B. Levinson, and P. Sh. Fridberg, "Effect of wall thickness in slot problems of electrodynamics," *Radio Eng. Electron. Phys., (USSR)*, vol. 13, pp. 1888–1896, 1968.

[70] S. C. Kashyap, M. A. K. Hamid, and N. J. Mostowy, "Diffraction pattern of a slit in a thick conducting screen," *J. Appl. Phys.*, pp. 894–895, Feb. 1971.

[71] S. C. Kashyap, and M. A. K. Hamid, "Diffraction characteristics of a slit in a thick conducting screen," *IEEE Trans. Antennas Propagat.*, vol. AP-19, pp. 499–507, July 1971.

[72] G. W. Lehman, "Diffraction of electromagnetic waves by planar dielectric structures, I: Transverse electric excitation," *J. Math. Phys.*, vol. 11, pp. 1522–1535, May 1970.

[73] N. A. McDonald, "Electric and magnetic coupling through small apertures in shield walls of any thickness," *IEEE Trans. Microwave Theory Tech.*, vol MTT-20, pp. 689–695, Oct. 1972.

[74] A. A. Oliner, "The impedance properties of narrow radiating slots in the broad face of rectangular waveguide, Part I: Theory," *IRE Trans. Antennas Propagat.*, vol. AP-5, pp. 4–11, Jan. 1957.

[75] —, "The impedance properties of narrow radiating slots in the broad face of rectangular waveguide, Part II: Comparison with measurement," *IRE Trans. Antennas Propagat.*, vol. AP-5, pp 12–20, Jan. 1957.

Multiple Apertures

[76] C. C. Chen, "Transmission through a conducting screen perforated periodically with apertures," *IEEE Trans. Microwave Theory Tech.*, vol. MTT-18, Sept. 1970.

[77] —, "Transmission of microwave through perforated flat plates of finite thickness," *IEEE Trans. Microwave Theory Tech.*, vol. MTT-21, pp. 1-6, Jan. 1973.

[78] R. B. Kieburtz and A. Ishimaru, "Aperture fields of an array of rectangular apertures," *IRE Trans. Antennas Propagat.*, vol. AP-10, pp. 663-671, Nov. 1962.

Object Excited through Aperture in Screen

[79] C. M. Butler and K. R. Umashankar, "Electromagnetic excitation of a wire through an aperture-perforated, conducting screen," *IEEE Trans. Antennas Propagat.* vol. AP-24, 456-462, July 1976.

[80] J. L. Lin, W. L. Curtis, and M. C. Vincent, "Electromagnetic coupling to a cable through apertures," *IEEE Trans. Antennas Propagat.*, vol. AP-24, pp. 198-203, Mar. 1976.

[81] D. Kajfez, "Excitation of a terminated TEM transmission line through a small aperture," (Air Force Weapons Lab., Albuquerque, NM, Tech. Rep. AFWL-TR-74-195, Aug. 1975 (also, *AFWL Interaction Note 215*, July 1974).

[82] D. Kajfez, C. E. Smith, and D. R. Wilton, "Equivalent circuits of quasi-TEM transmission lines in the presence of small apertures," Report currently in preparation for AFWL.

Apertures in Cylinder (Infinite) Wall

[83] J. N. Bombardt, Jr., "Magnetic field shielding degradation due to circular apertures in long hollow cylinders," *AFWL Interaction Note 3*, Sept. 1966.

[84] —, "Quasistatic magnetic field transmission through circular apertures," *AFWL Interaction Note 9*, July 1967.

[85] R. W. Latham, "Small holes in cable shields," *AFWL Interaction Note 118*, Sept. 1972.

[86] J. Van Bladel, "Small holes in a waveguide wall," *Proc. Inst. Elec. Eng.*, vol. 118, pp. 43-50, Jan. 1971.

[87] T. B. A. Senior, "Electromagnetic field penetration into a cylindrical cavity," *IEEE Trans. Electromagn. Compat.*, vol. EMC-18, pp. 71-73, May 1976.

Cavity-Backed Apertures

[88] D. K. Cheng, and C. A. Chen, "On transient electromagnetic excitation of a rectangular cavity through an aperture," *AFWL Interaction Note 237*, Feb. 1975.

[89] C. W. Harrison, Jr., and R. W. P. King, "Excitation of a coaxial line through a transverse slot," *IEEE Trans. electromagn. Compat.*, vol. EMC-14, pp. 107-112, Nov. 1972.

[90] K. Kurokawa, "The expansions of electromagnetic fields in cavities," *IRE Trans. Microwave Theory Tech.*, vol. MTT-6, pp. 178-187, Apr. 1958.

[91] M. I. Sancer and A. D. Varvatsis, "Electromagnetic penetrability of perfectly conducting bodies containing an aperture," *AFWL Interaction Note 49*, Aug. 1970.

[92] T. B. A. Senior, "Electromagnetic field penetration into a cylindrical cavity," *AFWL Interaction Note 221*, Jan. 1975.

[93] T. B. A. Senior and G. A. Desjardins, "Electromagnetic field penetration into a spherical cavity," *IEEE Trans. Electromagn. Compat.* vol. EMC-16, pp. 205-208, Nov. 1974 (also *AFWL Interaction Note 142*, Aug. 1973).

[94] C. D. Taylor and C. W. Harrison, Jr., "On the excitation of a coaxial line by an incident field propagating through a small aperture in the sheath," *IEEE Trans. Electromagn. Compat.*, vol. EMC-15, pp. 127-131, Aug. 1973.

[95] T. Teichmann and E. P. Wigner, "Electromagnetic field expansions in loss-free cavities excited through holes," *J. Appl. Phys.*, vol. 24, pp. 262-267, Mar. 1953.

[96] J. R. Wait and D. A. Hill, "Electromagnetic fields of a dielectric coated coaxial cable with an interrupted shield–Quasistatic approach," *IEEE Trans. Antennas Propagat.*, vol. AP-23, pp. 679-682, Sept. 1975.

[97] J. R. Wait, "Electromagnetic field analysis for a coaxial cable with periodic slots," *AFWL Interaction Note 265*, Feb. 1976.

[98] D. C. Chang, "Equivalent-circuit representation and characteristics of a radiating cylinder driven through a circumferential slot," *IEEE Trans. Antennas Propagat.*, vol. AP-21, pp. 792-796, Nov. 1973.

[99] —, "A general theory on small, radiating apertures in the outer sheath of a coaxial cable," EM Laboratory, Univ. of Colorado, Boulder, Scientific Rep. 19, 1976.

[100] J. Van Bladel, "Small apertures in cavities at low frequencies," *Archiv fur Elektronik und Ubertragungstechnik (AEU)-Electronics and Communication*, Band 26, pp. 481-486, 1972.

[101] N. N. Voytovich, B. E. Katsenelenbaum, and A. N. Sivov, "Excitation of a two-dimensional metal cavity with a small opening (Slotted cylinder)," *Radio Eng. Electron. Phys. (USSR)*, vol. 19, pp. 8-17, 1974.

[102] K. C. Chen and C. E. Baum, "On EMP excitations of cavities with small openings," *AFWL Interaction Note 170*, Jan. 1974.

Numerical Methods

[103] B. D. Graves, T. T. Crow, and C. D. Taylor, "On the electromagnetic field penetration through apertures," *AFWL Interaction Note 199*, Aug. 1974.

[104] R. Mittra, Y. Rahmat-Samii, D. V. Jamnejad, and W. A. Davis, "A new look at the thin-plate scattering problem," *Radio Sci.*, vol. 8, pp. 869-875, 1973.

[105] Y. Rahmat-Samii, and R. Mittra, "A new integral equation solution of electromagnetic aperture coupling and thin plate scattering problems," *AFWL Interaction Note 224*, Feb. 1975.

[106] J. R. Mautz and R. F. Harrington, "Electromagnetic transmission through a rectangular aperture in a perfectly conducting plane," Scientific Rep. 10 on Contract F19628-73-C-0047, with AF Cambridge Res. Lab. Rep. AFCRL-TR-76-0056, Feb. 1976.

[107] D. R. Wilton and S. Govind, "Incorporation of edge conditions into moment method solutions," *IEEE Trans. Antennas Propagat.*, vol. AP-25, pp. 845-850, Nov. 1977.

Other

[108] S. J. Buchsbaum, A. R. Milne, D. C. Hogg, G. Berkefi, and G. A. Woonton, "Microwave diffraction by apertures of various shapes," *J. Appl. Phys.*, vol. 26, pp. 706-715, June 1955.

[109] C. Huang, R. D. Kodis, and H. Levine, "Diffraction by apertures," *J. Appl. Phys.*, vol. 26, pp. 151-165, Feb. 1954.

[110] M. Suzuki, "Diffraction of plane electromagnetic waves by a rectangular aperture," *IRE Trans. Antennas Propagat.*, vol. AP-4, pp. 149-156, Apr. 1956.

[111] H. L. Robinson, "Diffraction patterns in circular apertures less than one wavelength in diameter," *J. Appl. Phys.*, vol. 24, pp. 35-38, 1953.

[112] J. Van Bladel, "Small hole coupling of resonant cavities and waveguides," *Proc. Inst. Elec. Eng.*, vol. 117, pp. 1098-1104, June 1970.

[113] R. W. P. King and T. T. Wu, *Scattering and Diffraction of Waves.* Cambridge, MA: Harvard Univ. Press, 1959.

[114] J. J. Bowman, T. B. A. Senior, and P. L. E. Uslenghi, *Electromagnetic and Acoustic Scattering by Simple Shapes.* New York: Wiley-Interscience, 1969.

[115] D. B. Seidel and C. M. Butler, "Determination of current on a wire which passes through a hole in a planar screen," presented at the 1976 USNC/URSI Meeting, Amherst, MA, Oct. 1976.

[116] L. Marin, "Dielectric effects on the electric polarizability of an aperture," *AEU*, vol. 30, pp. 8-12, 1976.

Aperture Coupling in Bodies of Revolution

HARVEY K. SCHUMAN, MEMBER, IEEE, AND D. E. WARREN

Abstract—A method for predicting the field penetrating a circumferential opening in a body of revolution is described here. This method employs the method of moments and an aperture equivalence theorem. The former permits rotationally symmetric cavities with otherwise arbitrary contours. The latter improves the sensitivity of field computation when the aperture height is tiny and/or points deep within the cavity are sought. This method is evaluated with comparison to classical theory and experiment via application to spherical and cylindrical cavities. Results for a missile-like cavity irradiated by an obliquely incident plane wave are given. A computer program and user manual is available.

I. INTRODUCTION

AN EQUIVALENT aperture excitation method (referred to as BOR3 in this paper) for predicting the fields coupled through a rotationally symmetric aperture in a body (cavity) of revolution (BOR) is described here. BOR3 has an advantage over a conventional near-field scattering method [2] (referred to as BOR2 in this paper) in that BOR3 is characterized by greater sensitivity; tiny field strengths are determined more accurately. In BOR2, first the current induced on the metallic portion of the body and its corresponding scattered field are determined. Then the cavity field is found by adding the impressed and scattered fields. For small field levels, such as those associated with tiny apertures or those at locations deep within a cavity, this computation involves a subtraction of almost equal numbers with a corresponding loss in accuracy. In BOR3, this difficulty is avoided because the cavity field is computed as a radiation problem once an equivalent aperture excitation is determined. The improvement in accuracy of BOR3 over BOR2 is demonstrated with the open-ended cylinder problem discussed in Section III.

Based on a moment method solution to an E-field formulation, BOR3 is an extension of the BOR work originating at Syracuse University [1], [3]. (A computer program implementation of BOR3 [7] also employs a near-field subroutine developed at Lawrence Livermore Laboratory [2].) There are two principal reasons for restricting a study to BOR's. First, many structures of interest can be approximated by BOR's. These include streamlined bodies such as aircraft fuselages and missiles. Second, the electromagnetic characteristics of a BOR conveniently decouple. That is, a single Fourier mode, or circumferentially varying sinusoid, of excitation (impressed field, aperture field, etc.) excites only the corresponding mode of response (far or near field, current density, etc.). This permits significant savings in computer time and memory. With modal decoupling, a matrix formulation of the problem results in a number of small matrices to deal with rather than one huge matrix. The latter often arises with use of an arbitrary surface formulation.

In Section II, the theory behind BOR3 is briefly described. The details are available in [8], where BOR3 is presented as an extension of the Syracuse BOR method. In Section III, BOR3 is evaluated by comparison with BOR2, classical methods and experiment. These methods are applied to both spherical and finite-cylindrical cavities with rotationally symmetric apertures. Finally, in Section IV, BOR3 is applied to the problem of a thin circumferential slot in a missile-shaped body immersed in an obliquely incident plane wave.

A user's description of a BOR3 computer code and a Fortran source listing with example printout are available [7]. This version of BOR3 is limited to $\hat{\theta}$-polarized plane-wave excitation.

II. THEORY

Fig. 1 shows a plane wave incident upon a perfectly conducting BOR with a rotationally symmetric aperture. The surface of a BOR may be formed by constructing any contour in the y-z plane, called the generating curve, and rotating it about the z-axis. The cylinder and conesphere are typical BOR's. The basic treatment of BOR's by the method of moments has been set forth by Harrington and Mautz [3] for radiation and scattering problems. They developed a computer code for this method based on an E-field formulation, which has proven useful in solving a large class of problems [1], [3].

Manuscript received August 3, 1977; revised February 23, 1978. This work was supported in part by the Rome Air Development Center Post-Doctoral Program under Contract no. F30602-75-C-0121.

H. K. Schuman was with the Department of Electrical and Computer Engineering, Syracuse University, Syracuse, NY 13210. He is now with Atlantic Research Corp., Rome, NY 13340.

D. E. Warren is with the Rome Air Development Center, Griffiss Air Force Base, NY 13441.

Reprinted from *IEEE Trans. Antennas Propagat.*, vol. AP-26, no. 6, pp. 778–783, November 1978.

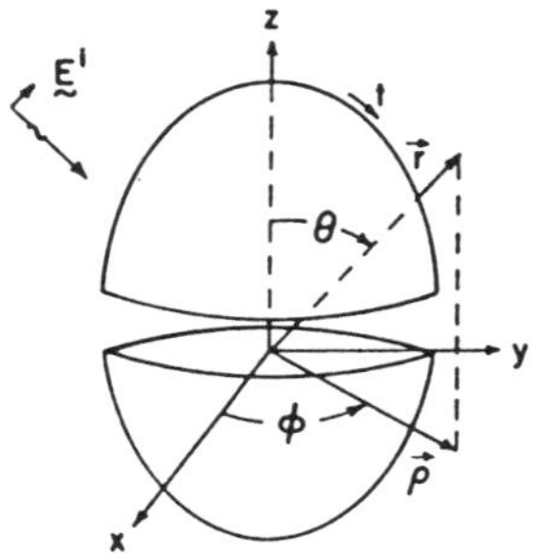

Fig. 1. Body of revolution with rotationally symmetric aperture.

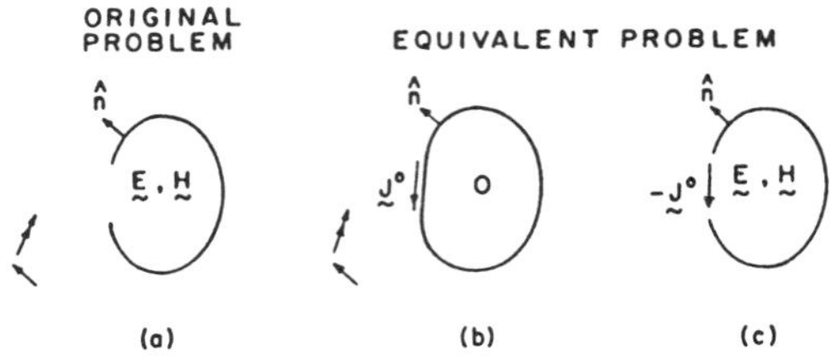

Fig. 2. Equivalence theorem.

Essentially, the task is reduced to inverting a set of matrices each of order equal to the number of expansion functions applied to the BOR generating curve. Each matrix corresponds to a different Fourier component or mode n of circumferential variation of fields and current densities.

The basis for the adaptation of the Harrington-Mautz (Syracuse) BOR method to aperture coupling via the equivalent aperture excitation method, BOR3, is an equivalence theorem due to Schelkunoff [4] which states that the original problem containing sources external to the body can be replaced with an equivalent problem having only aperture current sources for excitation. Fig. 2 illustrates this where the original problem depicted in (a) is equivalent to the sum of the solutions of the two problems (b) and (c). In (b), the induced surface current density $\mathbf{J}^0$ in the vicinity of the aperture is found for the apertureless (replace aperture with conductor) body in the presence of the external sources. Then, as shown in (c), $\mathbf{J}^A = -\mathbf{J}^0$ becomes an aperture excitation resulting in the internal fields of the original problem. For BOR's both problems (b) and (c) can be solved by proper application of the Syracuse method. It is shown in [5] that this procedure is also applicable to nonrotationally symmetric-aperture BOR problems, with all the computational advantages of BOR modeling maintained, although the details are then considerably more complex.

The method for solving the problems of (b) and (c) (Fig. 2) when dealing with cavities of revolution with rotationally symmetric apertures is outlined as follows.

The problem depicted by Fig. 2(b) is a straightforward BOR scattering problem. Once $\mathbf{J}^A = -\mathbf{J}^0$ is determined, the problem of Fig. 2(c) is then solved by first replacing the current excitation $\mathbf{J}^A$ with an equivalent E-field excitation. The latter is found by dividing the aperture into thin ribbon-like cells or ports along the generating curve, short-circuiting them, and exciting each one in turn with a bi-directed (since the surface is two dimensional) E-field, each component of which is unity across the excited port and zero elsewhere. The number of cells is equal to the number of expansion functions for $\mathbf{J}^0$ that reside in the aperture. The circumferential variation of each port E-field excitation is, of course, the Fourier component or mode under consideration. Thus for each port E-field, the bi-directed current density induced at every port is easily obtained by solving a straightforward BOR radiation problem. This results in an "aperture admittance matrix" $[y^A]$ and the matrix equation

$$\vec{j} = [y^A]\,\vec{e}. \tag{1}$$

The j and e in (1) are, respectively, column matrix representations of $\mathbf{J}^A$ and the aperture tangential E-field of the original problem $\mathbf{E}^A$. The elements of e are the coefficients of the expansion for $\mathbf{E}^A$ comprised of port E-field excitations referred to above. These functions are pulse-like with pulses terminating half intervals prior to aperture edges for ϕ-components and at these edges for t-components [8]. Each element of j is $-\mathbf{J}^0$ evaluated at a point within the corresponding port. The solution of (1) for e can then be applied as an aperture excitation in a manner that is, again, a straightforward BOR radiation problem. This provides the surface current which in turn yields the internal field.

III. EVALUATION OF BOR3

Cavities of convenient shape were chosen in order to compare BOR3 with classical techniques. Results with spherical and cylindrical cavities are presented in this section. Also included are results obtained from both the conventional application of a BOR code, BOR2, and experiment. The purpose of these comparisons was twofold. First was the determination of the number of triangle[1] functions per aperture, "NAM", required for reasonable accuracy in field (~80 percent). Second was the evaluation of the effect of severe size variation in adjacent triangle segments (four segments per triangle extent [1]). These considerations are pertinent if the height of the circumferential aperture is only a tiny fraction of the cavity length. For a given NAM and a reasonable number of triangle functions on the cavity generating curve, "NM", a large variation, e.g., 10:1, in adjacent segment size may result.

A. Spherical Cavity

The equivalent aperture excitation method, BOR3, was applied to the spherical cavity aperture coupling problem studied by Senior and Desjardins [6]. The latter employed a spherical mode matching method. The case of a $ka = 2.5$ sphere with 30° half-angle front-end aperture and excited by a unit amplitude, linearly polarized, E-field incident along the axis and toward the aperture side of the body is shown in Fig. 3. The magnitude of the internal E-field along the axis is plotted. Only the $n = 1$ mode ($\sin\phi$, $\cos\phi$) was excited because the impressed plane wave was axially incident. The BOR3 result for a total of NM = 15 expansion functions along the generating curve of the apertureless sphere is in good agreement with the spherical mode matching result. Five of these triangles traversed the aperture (NAM = 5). This necessitated a 3:1 variation in segment size with no appreciable loss

[1] Each triangle function is the variation along the generating curve of an expansion function of current density. Triangle functions were chosen by Harrington and Mautz to achieve a piecewise linear variation to both the $\hat{t}$-directed and $\hat{\phi}$-directed components of current density in accordance with the moment method [1], [3].

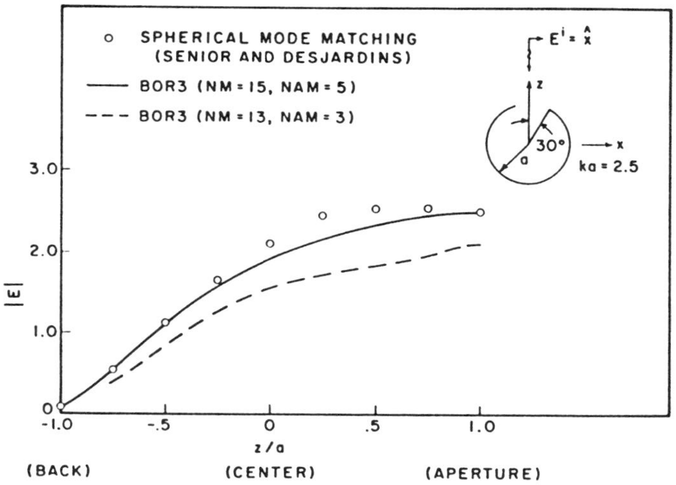

Fig. 3. Amplitude of E-field along axis of sphere, $ka = 2.5$.

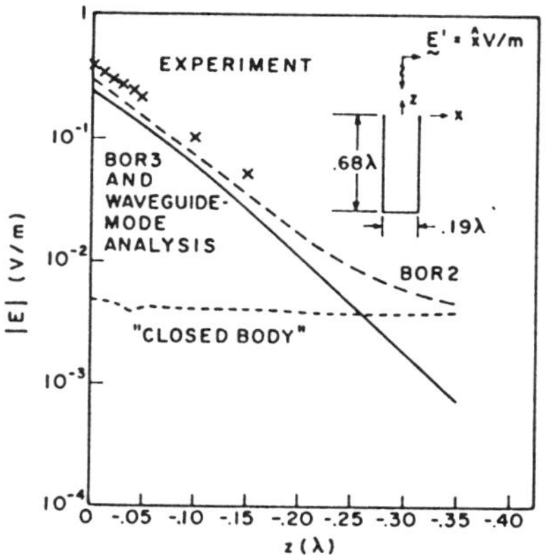

Fig. 4. Amplitude of E-field along axis of open-ended finite cylinder.

in accuracy. The restriction of uniform segment size and NM = 11 (not plotted) resulted in only one triangle peak lying in the aperture. However, the aperture field then differed from the NM = 15 case by about 6 dB. The result with NM = 13, NAM = 3 (Fig. 3) differs by less than 20 percent. Thus a minimum of three triangle functions should "excite" the aperture. It is not surprising that one aperture expansion function is inadequate, even for tiny aperture heights. The small-hole theory predicts that two modes—a "magnetic" (proportional to $\mathbf{J}^A$) and an "electric" (proportional to $\nabla \cdot \mathbf{J}^A$)—are generally significant.

B. Cylindrical Cavity

Further evaluation of BOR3 was obtained with application to an open-ended cylinder problem. The comparison was with BOR2, classical theory, and experiment. The magnitude of the E-field along the axis of the cylinder is plotted in Fig. 4. The excitation was a linearly polarized unit plane wave axially incident on the open-ended side. Again, only the $n = 1$ mode was excited due to the axial incidence. The conventional application of the BOR code, BOR2, and the equivalent aperture excitation method, BOR3, agree well to a distance of about 0.2λ into the cylinder. The scale factor difference between them is attributed to the different ways in which each models the aperture; an increase in the number of expansion functions should bring both methods into better agreement. Beyond 0.2λ into the cylinder, the BOR2 result tends to level off and approach the "closed body" (apertureless) result which is also plotted. This is to be expected since the "closed body" field is essentially the error field (ideally zero) resulting from the approximation in moment method modeling, and the errors associated with the aforementioned (Section I) addition of impressed and scattered fields. BOR3 is not limited by the "closed body" field. Thus it demonstrates a minimum of perhaps a 20-dB improvement in sensitivity over BOR2.

A check on the BOR3 result was obtained from a classical waveguide-mode analysis. The modal components were determined by requiring zero tangential E-field on the cylinder surface except in the aperture where the E-field predicted by BOR3 was prescribed. Five electric and five magnitic modes were used. Due to the thinness of the cylinder (the frequency was well below cutoff) the dominant mode alone essentially resulted in the fields beyond the aperture. As shown in Fig. 4, there is no observable difference between the cylindrical mode matching method and BOR3.

A further check on the computational methods was obtained with an anechoic chamber experiment. An electrically short diode-loaded probe and a high resistance transmission line were used to minimize measurement apparatus interference. The probe was mounted on a nonmetallic "plunger" with placement holes radially spaced. The plunger was mounted on a calibrated dielectric rod. In this way, all positions of the probe could be determined accurately. A sampling of the results are plotted in Fig. 4. They are adjusted to account for the dielectric support which, experiment showed, increases the observable field by about 20 percent. There is good agreement between theory and experiment for distances of about .15λ or less into the cylinder. For greater distances, the experimental results were not reliable, and, are therefore omitted from the graph. The reason for this is that the 0.15λ-point, the dominant-mode field was approximately 30 dB below the impressed field, and harmonics of the fundamental, which were closer to or above the cutoff frequency of the "guide," became significant. Attempts at filtering the transmitter to suppress these harmonics could not appreciably extend the curve due to the severe attenuation of the dominant mode at the fundamental frequency.

IV. THIN CIRCUMFERENTIAL APERTURE

The equivalent aperture excitation method, BOR3, was applied to the missile-like structure shown in Fig. 5. The aperture is modeled simply as a small separation between two sections. It is noted, however, that actual apertures are often considerably more complex since the section walls can have odd shapes at the junction, or the aperture can be filled with a dielectric material or sealed with a conductive putty. An extension of the foregoing technique is available, which can account for the aperture complexity. This is discussed later in this section although computations have not as yet been performed.

A. Oblique Incidence Plane Wave Excitation

Figs. 6-10 show the E-field in the vicinity of the aperture for a unit, θ-polarized impressed field arriving from the $\theta = 45°$ direction in the x-z plane. Each figure plots three lines. Each line corresponds to a path indicated by one of the dashed lines shown in Fig. 5. The azimuth orientation of each path

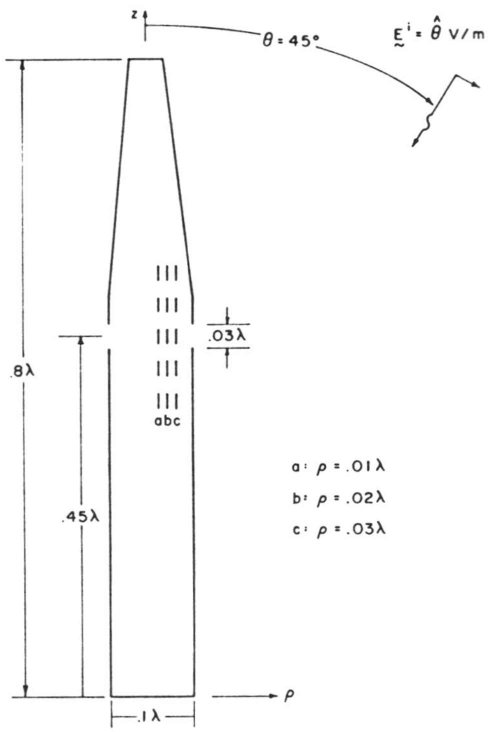

Fig. 5. Circumferential opening in thin cavity.

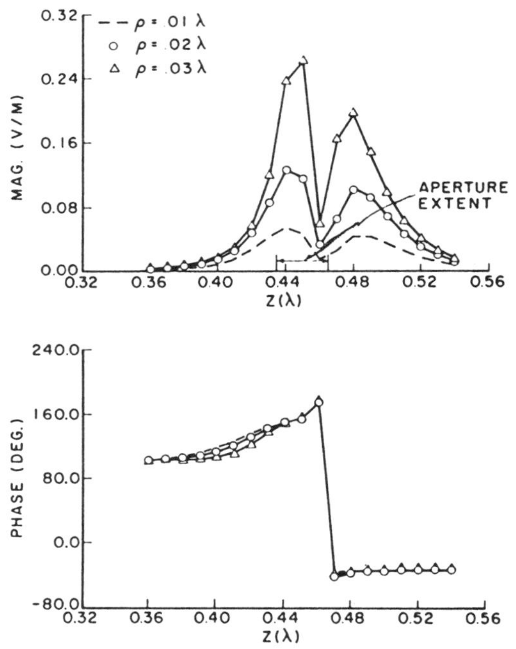

Fig. 7. $n = 0$ mode (uniform in ϕ) ρ component of aperture coupled E-field $E_\rho{}^0$.

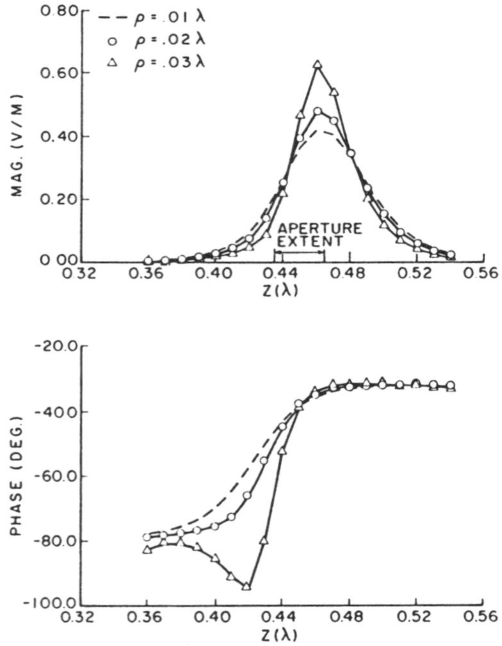

Fig. 6. $n = 0$ mode (uniform in ϕ) z component of aperture coupled E-field $E_z{}^0$.

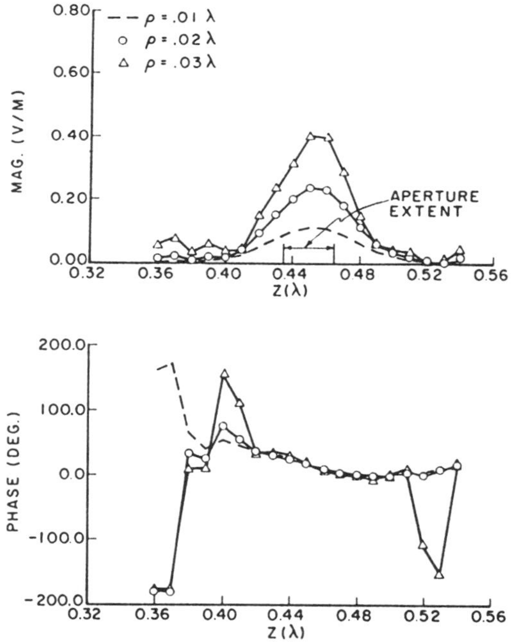

Fig. 8. $n = 1$ mode (cos ϕ varying) z component of aperture coupled E-field $E_z{}^1$.

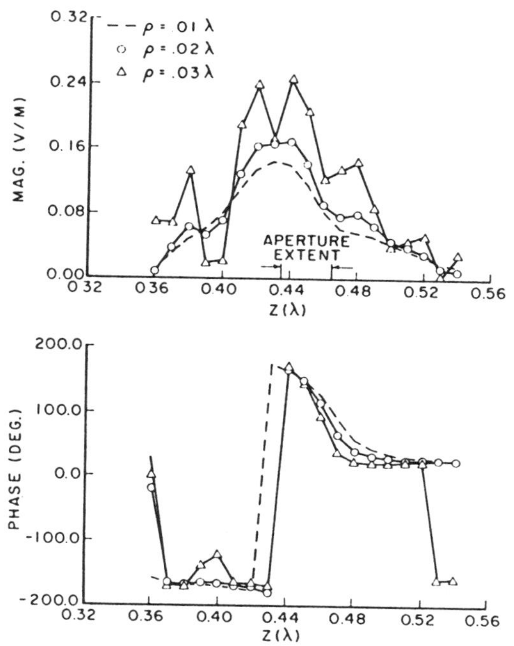

Fig. 9. $n = 1$ mode ($\cos\phi$ varying) ρ component of aperture coupled E-field $E_\rho{}^1$.

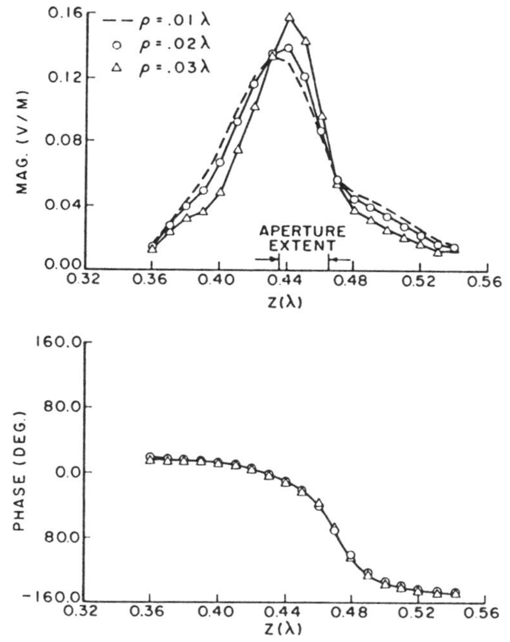

Fig. 10. $n = 1$ mode ($\sin\phi$ varying) ϕ component of aperture coupled E-field $E_\phi{}^1$.

corresponds to that of peak amplitude. The ρ– and z– components of the E-field ($E_\rho{}^n$, $E_z{}^n$) each vary as $\cos n\phi$, and the ϕ-component, $E_\phi{}^n$, varies as $\sin n\phi$. Plots of the three E-field components are shown for the $n = 0$ and 1 modes except for $E_\phi{}^0$ which is identically zero due to the θ-polarization of impressed field. Other modes are not presented since the $n = 2$ modal fields were found to be more than 20 dB below the $n = 0$, 1 fields. This is not surprising since the number of significant modes is limited to all n such that $|n| \leqslant 1 + C_\lambda$, where C_λ is the largest circumference of the body in wavelengths [1]. However, it is interesting to note that the $n = 1$

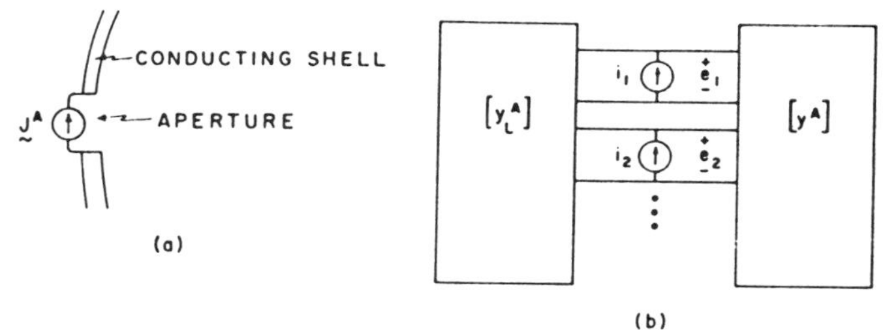

Fig. 11. Aperture excitation. (a) current density excitation. (b) network representation.

mode is significant even though the structure is quite thin. Although the sinusoidal circumferential variation of the $n = 1$ mode would make it insignificant for far-field scattering, it is certainly of significance in aperture coupling.

As expected, $E_\rho{}^0$ exhibits a dipole-like behavior (Fig. 7). For ρ close to an aperture edge, where the charge density is infinite, $E_\rho{}^0$ becomes large. Also, the phase of $E_\rho{}^0$ changes rapidly in passing by the aperture. The behavior of $E_z{}^0$ and $E_z{}^1$ are also as expected in that close to the cavity surface they are small (except at the aperture) (Figs. 6 and 10), and close to the axis, $E_z{}^1$ is small. (From symmetry, $E_z{}^1$ should vanish along the axis.) The $\rho = .03\lambda$ curve of $|E_\rho{}^1|$ in Fig. 9 exhibits unexpected oscillation away from the aperture. Numerical inaccuracies may be significant here since the field point-to-body surface separation is not much larger than an expansion function segment.

The number of triangle functions used in computing these results was NM = 44. (The corresponding generalized impedance matrix [1], [3] was of order 88 since the current density is bi-directed.) Also, as suggested by the results of Section III, three triangle peaks were chosen to lie in the thin aperture. This resulted in a 3:1 triangle segment size variation near the aperture.

B. Complex Aperture as Equivalent Load

In the beginning of this section, it was pointed out that in practice, apertures are usually considerably more complex than simple circumferential openings. A method for including the effect of the aperture shape and material composition in computing cavity fields is presented here. Essentially, it involves the determination of an "aperture-load" matrix $[y_L{}^A]$ which can be combined with $[y^A]$ such that e, the column vector representation of $\mathbf{E}^A$, can readily be determined.

As discussed in Section II, both j, the current vector representation of $\mathbf{J}^A$, and also $[y^A]$ are computed with the aperture "short circuited." Hence, j and $[y^A]$ are independent of the aperture complexity (or "loading"). With the aperture present and excited with $\mathbf{J}^A$ (Fig. 11(a)) the problem can be represented in network form as shown in Fig. 11(b). The generalization of (1) to include the aperture complexity is then

$$\vec{j} = ([y^A] + [y_L{}^A])\vec{e}. \qquad (2)$$

Once $[y_L{}^A]$ is known, e and thus $\mathbf{E}^A$, can be determined from (2). With $\mathbf{E}^A$ known, the internal fields can be computed, as before, from a straightforward aperture radiation problem.

For thin apertures, the load matrix elements relating the t-directed components of current and E-field can be approximated with quasi-static solutions. The shape of the conducting shell at the aperture presents essentially a capacitance, the value of which can be approximated by contour mapping. Also, the conductance and susceptance of material (other than free space) filling the aperture can be determined by mapping methods. The ϕ-directed components of current can easily skirt the thin aperture. Thus only the t-directed components of loading need be considered.

V. CONCLUSIONS

A moment method for predicting aperture-coupled fields within a rotationally symmetric cavity has been presented. This method, BOR3, employs an equivalent aperture excitation concept which demonstrates greater sensitivity over more conventional procedures if the aperture is thin. This improvement is important, for example, in analyzing the fields coupled through thin circumferential apertures in missile-like structures. It was found that a 3:1 adjacent segment size variation can be tolerated. Also, a minimum of three current expansion functions should "reside" within the aperture. A computer program with user manual is available.

ACKNOWLEDGMENT

The authors are indebted to Prof. B. J. Strait of Syracuse University for valuable contributions to all aspects of this work. Thanks is also due to T. Baustert of the Rome Air Development Center for Assistance in conducting the experiment.

REFERENCES

[1] R. F. Harrington and J. R. Mautz, "Radiation and scattering from bodies of revolution," Final Rep., prepared for Air Force Cambridge Research Laboratories, Bedford, MA, under Contract No. F19628-67-C-0233, July 1969.

[2] R. M. Bevensee, "S3F - SYR/LLL1 The Syracuse computer code for radiation and scattering from bodies of revolution, extended for near-field computations," Rep. no. UCRL-51622, prepared for U.S. Atomic Energy Commission under Contract No. W-7405-Eng-48 with the Lawrence Livermore Laboratory, Univ. California, Livermore, CA, May 1974.

[3] J. R. Mautz and R. F. Harrington, "Generalized network parameters for bodies of revolution," Scientific Rep. no. 1, Contract no. F-19628-67-C-0233, Air Force Cambridge Research Laboratories, Bedford, MA, with Syracuse University, Syracuse, NY, May 1968. Or see instead: J. R. Mautz and R. F. Harrington, "Radiation and Scattering from Bodies of Revolution," *Applied Sci. Res.*, vol. 20, 1969.

[4] S. A. Schelkunoff, "Field equivalence theorems," *Comm. Pure Appl. Math.*, vol. 4, pp. 43-59, June 1951.

[5] H. K. Schuman and B. J. Strait, "Coupling through apertures of arbitrary shape in bodies of revolution," *IEEE AP-S International Symposium*, Univ. Illinois, Urbana-Champaign, June 1975. See also, A. T. Adams, *et al.*, "On the coupling of electromagnetic energy through apertures," *IEEE EMC Symposium Record*, Montreaux, Switzerland, May 1975. For greater detail, see H. K. Schuman, "Coupling through non-azimuthally symmetric apertures in a body-of-revolution", Internal Memorandum, Dept. of Electrical and Computer Engineering, Syracuse University, July 1974.

[6] T. B. A. Senior and G. A. Desjardins, "Electromagnetic field penetration into a spherical cavity," *IEEE Trans. on Electromagn. Compat.*, Vol. EMC-16, pp. 205-208, Nov. 1974.

[7] H. K. Schuman, "Coupling through rotationally symmetric apertures in cavities of revolution - Computer Code (BOR3) Description", Tech. Memo, Electrical and Computer Engineering Dept., Syracuse University, Syracuse, NY 13210, Sept. 1976.

[8] H. K. Schuman and D. E. Warren, "Coupling through rotationally symmetric apertures in cavities of revolution", TR-77-214, Rome Air Development Center, Griffiss AFB, Rome, NY, June 1977.

Electromagnetic Pulse Coupling Through an Aperture into a Two-Parallel-Plate Region

YAHYA RAHMAT-SAMII, MEMBER, IEEE

Abstract—Analysis of electromagnetic-pulse (EMP) penetration via apertures into cavities is an important study in designing hardened systems. In this paper, an integral equation procedure is developed for determining the frequency and consequently the time behavior of the field inside a two-parallel-plate region excited through an aperture by an EMP. Some discussion of the numerical results is also included in the paper for completeness.

Key Words: EMP, coupling, aperture, two-parallel-plates.

I. INTRODUCTION

RECENTLY, the problem of electromagnetic compatibility of electronic systems subject to an electromagnetic pulse (EMP) has revived researchers' interests in analyzing the coupling phenomenon through apertures into cavities. The application of this study is vitally important in solving many typical problems. For instance, apertures may occur due to an improperly seated shielded door, defects in welding, etc., all of which may provide points of entry via the aperture coupling.

Though the problem of electromagnetic coupling through small circular or elliptical apertures into a cavity has received considerable attention, little is known about this problem when the aperture size is approximately that of the wavelength or when the aperture has a more complex shape. The reader is referred to the works of Muller [1], Liu [2], Chen [3], Sancer and Varvatsis [4], Van Bladel [5], Taylor [6], Chen and Baum [7], and others who have investigated small aperture-cavity coupling problems. The basic idea behind their approaches has been that the aperture field could be approximated as though the aperture were perforated in a single infinite screen, using Bethe's [8] approximation. This approach, however, fails when the size of the aperture is on the order of the wavelength. Some limitations of this approach for computing the field behind the aperture have been discussed by Rahmat-Samii and Mittra [9]. A rather complete listing of classical and recent works concerning penetration of time harmonic fields through apertures in a planar screen is given in [10]-[12].

In this paper, a new integral equation formulation is constructed for the problem of electromagnetic wave penetration through apertures, not necessarily circular or elliptical, into a two-parallel-plate region. Special attention is given to expressing the integral equation in the form successfully used for the aperture in planar screen problems [9]. Some discussion regarding the numerical behavior of the kernel is included. Finally, a time-domain analysis is presented, based on the application of fast Fourier transform (FFT) to determine the time history of the field inside the cavity.

Manuscript received August 16, 1977; revised April 11, 1978.

The author was with the Electromagnetics Laboratory, Department of Electrical Engineering, University of Illinois, Urbana. He is now with the Jet Propulsion Laboratory, Pasadena, CA 91103.

II. GENERAL FORMULATION

All surfaces of practical interest are not planar and the regions separated by such surfaces are not always empty spaces. In this section, a procedure is outlined which is generally applicable to a class of problems involving an aperture in a curved surface. Since a vector source radiates a vector field whose boundary conditions are also vector in nature, for convenience dyadic formulation is employed as the mathematical tool [13], [14].

The geometry under consideration in this section is a cavity with an aperture A in its shell S where V_- and V_+ are used to indicate the exterior and interior regions of the cavity, respectively. It is assumed that V_- and V_+ are filled with homogenous and isotropic materials (μ_-, ϵ_-) and (μ_+, ϵ_+), respectively, and that S is a perfectly conducting and vanishingly thin shell. The starting point is Maxwell's equations [time convention $\exp(j\omega t)$]

$$\begin{bmatrix} \nabla\times & j\omega\mu \\ -j\omega\epsilon & \nabla\times \end{bmatrix} \begin{bmatrix} \bar{E} \\ \bar{H} \end{bmatrix} = \begin{bmatrix} -\bar{M}^i \\ \bar{J}^i \end{bmatrix} \tag{1}$$

which apply to V_- and V_+ individually. From (1), one arrives at the inhomogeneous vector wave equation for the electric field $\bar{E}$ in each region

$$(\nabla\times\nabla\times -k^2)\bar{E} = -\nabla\times\bar{M}^i - j\omega\mu\bar{J}^i. \tag{2}$$

Next we define the dyadic Green's function $\bar{\bar{G}}(\bar{r}|\bar{r}')$ in the usual way as the solution of

$$(\nabla\times\nabla\times -k^2)\bar{\bar{G}}(\bar{r}|\bar{r}') = \bar{\bar{I}}\delta(\bar{r}-\bar{r}') \tag{3}$$

subject to boundary conditions discussed subsequently, where $\bar{\bar{I}}$ is the unit dyadic and δ is the Dirac delta distribution. It is noted from (3) that $\nabla\cdot\bar{\bar{G}}(\bar{r}|\bar{r}') \neq 0$. In order to establish a relationship between $\bar{E}$ and $\bar{\bar{G}}$, Green's theorem in dyadic form

Reprinted from *IEEE Trans. Electromagn. Compat.*, vol. EMC-20, no. 3, pp. 436–442, August 1978.

[14] is used

$$\iiint_V [\bar{E}\cdot\nabla\times\nabla\times\bar{\bar{G}}-(\nabla\times\nabla\times\bar{E})\cdot\bar{\bar{G}}]\,dv$$
$$=-\iint_{\partial V}\hat{n}\cdot[\bar{E}\times\nabla\times\bar{\bar{G}}+(\nabla\times\bar{E})\times\bar{\bar{G}}]\,dS \qquad (4)$$

where V denotes the domain of the volume integration, ∂V designates the closed surface surrounding the volume V, and $\hat{n}$ is the outward unit normal to the boundary ∂V.

Since the shell S is assumed to be a perfectly conducting material, the electric field satisfies the following boundary condition

$$\hat{n}_{\mp}\times\bar{E}^{\mp}=\bar{0}, \qquad \bar{r}\in S \qquad (5)$$

where "−" and "+" are used to denote the quantities in V_- and V_+, respectively. The proper boundary condition to be imposed on the interior and exterior dyadic Green's functions is

$$\hat{n}_{\mp}\times\bar{\bar{G}}^{\mp}(\bar{r}|\bar{r}')=\bar{\bar{0}}, \qquad \bar{r}\in S\cup A. \qquad (6)$$

Furthermore, for the exterior region, the satisfaction of the radiation condition is required, viz.,

$$\lim_{r\to\infty} r\left(\nabla\times\begin{Bmatrix}\bar{E}^-\\ \bar{\bar{G}}^-\end{Bmatrix}+jk_-\hat{r}\times\begin{Bmatrix}\bar{E}^-\\ \bar{\bar{G}}^-\end{Bmatrix}\right)=\begin{Bmatrix}\bar{0}\\ \bar{\bar{0}}\end{Bmatrix}, \qquad \bar{r}\in V_-. \qquad (7)$$

Employing (2)-(4) in regions V_- and V_+, and using (5)-(7), one finds the following representation for the interior and exterior fields:

$$\bar{E}^{\mp}(\bar{r})=\iiint_{V_{\mp}}\left(-\nabla'\times\bar{M}^{i\mp}-j\omega\mu_{\mp}\bar{J}^{i\mp}\right)\cdot\bar{\bar{G}}^{\mp}(\bar{r}'|\bar{r})\,dv'$$
$$-\iint_A \hat{n}_{\mp}\times\bar{E}^a\cdot\nabla'\times\bar{\bar{G}}^{\mp}(\bar{r}'|\bar{r})\,dS', \qquad \bar{r}\in V_{\mp} \qquad (8)$$

where, clearly, $\hat{n}=\hat{n}_+=-\hat{n}_-$, and $\bar{E}^-(\bar{r})=\bar{E}^+(\bar{r})=\bar{E}^a$, for $\bar{r}\in A$. It should be mentioned that the exterior and interior Green's functions, $\bar{\bar{G}}^-$ and $\bar{\bar{G}}^+$, respectively, generally speaking are different in form. Equation (8) also reveals the fact that, from knowledge of the tangential electric field in the aperture and the Green's functions, one can construct the field everywhere else. The goal is therefore to construct an integral equation for the unknown tangential electric field in the aperture, i.e., $\hat{n}\times\bar{E}^a$. This construction is done by first deriving the proper Green's functions for the geometry of interest and then enforcing the condition

$$\hat{n}\times\bar{H}^-=\hat{n}\times\bar{H}^+, \qquad \text{for } \bar{r}\in A \qquad (9)$$

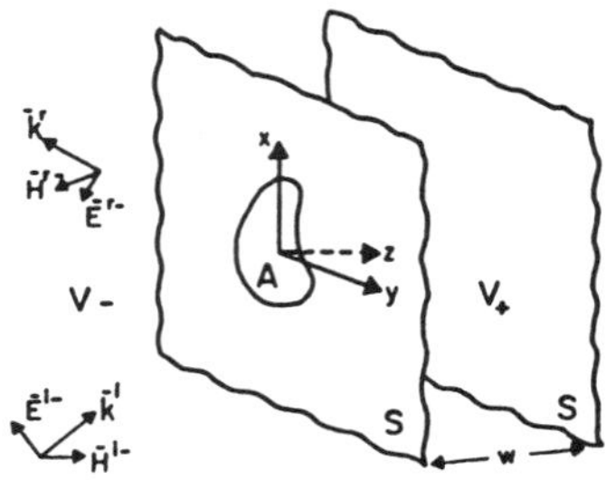

Fig. 1. Aperture in a perfectly conducting screen with a backplate.

where $\bar{H}$ is determined by substituting (8) into (1). In the following section, the above results will be employed and simplified for the case where the shell S is a two-parallel-plate structure immersed in a homogeneous medium.

III. CONSTRUCTION OF THE INTEGRAL EQUATION

In this section, attention is focused on the problem of penetration into a two-parallel-plate region and the appropriate integral equation for this structure is constructed. The geometry of the two-parallel-plate structure is shown in Fig. 1 where one sees two perfectly conducting, parallel plates separated by a distance w. A Cartesian coordinate system, with its z axis normal to, and its xy plane parallel to the plates is erected as shown in the figure. The plate at $z=0$ is perforated by an arbitrarily shaped aperture and it is assumed that a monochromatic wave $\bar{E}^{i-}$ and $\bar{H}^{i-}$ originating from a source situated in the half space V_- is incident on the structure.

The total electromagnetic field $(\bar{E}^{\mp}, \bar{H}^{\mp})$ at any point in either space is partitioned into an incident field $(\bar{E}^{i-}, \bar{H}^{i-})$, a reflected field $(\bar{E}^{r-}, \bar{H}^{r-})$ associated with the reflected wave which exists when the aperture is closed, and a diffracted field $(\bar{E}^{d\mp}, \bar{H}^{d\mp})$ due to the aperture. In V_-, the total electric field can be written

$$\bar{E}^-=\bar{E}^{i-}+\bar{E}^{r-}+\bar{E}^{d-} \qquad (10a)$$

and in V_+

$$\bar{E}^+=\bar{E}^{d+} \qquad (10b)$$

with similar expressions for the magnetic fields. The reflected field can, in general, be constructed from knowledge of the incident field and the reflecting surface. For an incident plane wave

$$\bar{E}^{i-}=(\hat{x}E_{0x}{}^i+\hat{y}E_{0y}{}^i+\hat{z}E_{0z}{}^i)e^{-jk(\alpha x+\beta y+\gamma z)} \qquad (11a)$$

the field reflected from the plate at $z=0$ is

$$\bar{E}^{r-}=(-\hat{x}E_{0x}{}^i-\hat{y}E_{0y}{}^i+\hat{z}E_{0z}{}^i)e^{-jk(\alpha x+\beta y-\gamma z)} \qquad (11b)$$

where $\alpha=\sin\theta^i\cos\phi^i$, $\beta=\sin\theta^i\sin\phi^i$, and $\gamma=\cos\theta^i$ are the direction cosines of the incident wave vector, and θ^i and ϕ^i are the corresponding elevation and azimuthal angles. The diffracted fields $\bar{E}^{d-}$ and $\bar{E}^{d+}$ are obtained from (8)

$$\bar{E}^{d\mp} = \mp \iint_A \hat{z} \times \bar{E}^a \cdot \nabla' \times \bar{\bar{G}}^{\mp}(\bar{r}'|\bar{r})\, dS'. \tag{12}$$

In (12), $\bar{\bar{G}}^+$ is the dyadic Green's function of the parallel-plate region and $\bar{\bar{G}}^-$ is the dyadic Green's function of the half-space $z < 0$. These dyadic Green's functions take the following form [15]

$$\bar{\bar{G}}^{\mp}(\bar{r}'|\bar{r}) = \left(\bar{\bar{I}} - \frac{1}{k^2}\nabla'\nabla\right)\begin{Bmatrix} g^- - g^+ \\ G^+ - G^- \end{Bmatrix} + 2\hat{z}\hat{z}\begin{Bmatrix} g^- \\ G^- \end{Bmatrix} \tag{13}$$

where

$$G^{\mp} = \sum_{n=-\infty}^{\infty} \frac{e^{-jk\sqrt{(x-x')^2+(y-y')^2+(z\pm z'+2nw)^2}}}{4\pi\sqrt{(x-x')^2+(y-y')^2+(z\pm z'+2nw)^2}} \tag{14a}$$

and

$$g^{\mp} = \frac{e^{-jk\sqrt{(x-x')^2+(y-y')^2+(z\pm z')^2}}}{4\pi\sqrt{(x-x')^2+(y-y')^2+(z\pm z')^2}}. \tag{14b}$$

Substituting (13) into (12) and using the fact that

$$\nabla' \times \bar{\bar{G}}^{\mp}(\bar{r}'|\bar{r}) = \nabla'\begin{Bmatrix} g^- - g^+ \\ G^+ - G^- \end{Bmatrix} \times \bar{\bar{I}} + 2\nabla'\begin{Bmatrix} g^- \\ G^- \end{Bmatrix} \times \hat{z}\hat{z} \tag{15}$$

and that

$$G^-|_{z'=0} = G^+|_{z'=0}, \qquad \frac{\partial}{\partial z'}G^{\mp}\Big|_{z'=0} = \mp\frac{\partial}{\partial z}G^{\mp}\Big|_{z'=0} \tag{16}$$

one finally obtains the simplified form of (12), viz.,

$$\bar{E}^{d\mp} = \pm 2\frac{\partial}{\partial z}\iint_A [E_x{}^a\hat{x} + E_y{}^a\hat{y}]\, g^{\overset{0}{\infty}}(\bar{r}'|\bar{r})\, dS'$$
$$\mp 2\hat{z}\left[\frac{\partial}{\partial x}\iint_A E_x{}^a g^{\overset{0}{\infty}}(\bar{r}'|\bar{r})\, dS' + \frac{\partial}{\partial y}\iint_A E_y{}^a g^{\overset{0}{\infty}}(\bar{r}'|\bar{r})\, dS'\right] \tag{17}$$

where

$$g^{\infty}(\bar{r}'|\bar{r}) = G^-|_{z'=0} = G^+|_{z'=0}$$
$$g^{0}(\bar{r}'|\bar{r}) = g^-|_{z'=0} = g^+|_{z'=0}. \tag{18}$$

Defining the equivalent surface magnetic current in the aperture as

$$\bar{M} = \hat{z} \times \bar{E}^a \tag{19}$$

and introducing the vector potential

$$\bar{F}^{\mp} = 2\epsilon\iint_A \bar{M}(\bar{r}')g^{\overset{0}{\infty}}(\bar{r}'|\bar{r})\, dS' \tag{20}$$

one may then readily express the total field (10) in the following form with the help of (17)

$$\bar{E}^{\mp} = \bar{E}^{sc\mp} \mp \frac{1}{\epsilon}\nabla \times \bar{F}^{\mp} \tag{21}$$

where $\bar{E}^{sc\mp} = \bar{E}^{i\mp} + \bar{E}^{r\mp}$. From (1) and (21), the magnetic field is determined as

$$\bar{H}^{\mp} = \bar{H}^{sc\mp} \mp j\frac{\omega}{k^2}[k^2\bar{F}^{\mp} + \nabla(\nabla \cdot \bar{F}^{\mp})]. \tag{22}$$

To establish an integral equation for the unknown aperture field, or equivalently for $\bar{M}$, the continuity of the tangential magnetic field is enforced in A. It is noticed that the continuity of the tangential electric field is guaranteed by (17). Using (22) and simplifying the result, one finally arrives at the following conventional integro-differential equation for $\bar{M}$,

$$j\frac{\omega}{k^2}[k^2\bar{F} + \nabla_t\nabla_t \cdot \bar{F}] \times \hat{z} = (\bar{H}^{sc-} - \bar{H}^{sc+}) \times \hat{z} \tag{23}$$

where $\bar{F}$, which is entirely transverse, is defined as

$$\bar{F} = (\bar{F}^- + \bar{F}^+)|_{z=0}. \tag{24}$$

Integral equation (23) can be solved numerically for determining $\bar{M}$; however, due to its highly singular integrand, special care must be exercised. Some discussion on the application of the finite-difference technique for equations similar to (23) is given in [16].

An Alternate Integral Equation

For the case of plane-wave excitation described in (11a), one can construct an alternate integral equation which has many desirable features when numerical techniques are considered. To this end, the continuity condition (9) is enforced in an indirect fashion. The continuity of the normal component of the electric field is employed together with the continuity of the normal derivative of the tangential electric field in the aperture to ensure the continuity of the tangential magnetic field in A. In other words, one requires that the following conditions be satisfied:

$$\lim_{z\uparrow 0} (\bar{E}^- \cdot \hat{z}) = \lim_{z\downarrow 0} (\bar{E}^+ \cdot \hat{z}) \qquad \text{and}$$

$$\lim_{z\uparrow 0} \frac{\partial}{\partial z}(\bar{E}^- \times \hat{z}) = \lim_{z\downarrow 0} \frac{\partial}{\partial z}(\bar{E}^+ \times \hat{z}), \qquad \bar{r}\in A. \tag{25}$$

The continuity of the normal component of the electric field can be expressed as

$$\left[\frac{\partial}{\partial x}F_x - \frac{\partial}{\partial y}F_y\right] = \epsilon\hat{z}\cdot(\bar{E}^{sc-} - \bar{E}^{sc+}), \qquad \bar{r}\in A \tag{26}$$

where $\bar{F}$ is defined in (24). The continuity of the normal derivative leads to

$$\left[\frac{\partial^2}{\partial x^2} + \frac{\partial^2}{\partial y^2} + k^2\right]\bar{F} = \epsilon\left[\hat{z}\times\frac{\partial}{\partial z}(\bar{E}^{sc-} - \bar{E}^{sc+})\right]_{z=0}, \qquad \bar{r}\in A \tag{27}$$

where (21) and (25) are employed for the derivation of (27).

For the case of a plane wave incident from the sources in V_-, i.e., $\bar{E}^{sc-} = \bar{E}^{i-} + \bar{E}^{r-}$ defined in (11), and $\bar{E}^{sc+} = \bar{0}$, a solution of (27) subject to (26) may be written as follows [9], [15]:

$$\iint_A \bar{M}(\bar{r}')g(\bar{r}'|\bar{\rho})\,dS' = \frac{1}{jk\gamma}\hat{z}\times\bar{E}_0{}^i e^{-jk(\alpha x+\beta y)} + \bar{h}, \qquad \bar{\rho}\in A \tag{28}$$

where $\gamma \neq 0$, $\bar{h} = h_x\hat{x} + h_y\hat{y}$ is the homogeneous solution of the operator

$$\left(\frac{\partial^2}{\partial x^2} + \frac{\partial^2}{\partial y^2} + k^2\right)$$

and

$$g(\bar{r}'|\bar{r}) = [g^0(\bar{r}'|\bar{r}) + g^\infty(\bar{r}'|\bar{r})]_{z=0} = \sum_{n=0}^{\infty} \frac{e^{-jk|\bar{r}-\bar{r}'+(2nw)\hat{z}|}}{2\pi|\bar{r}-\bar{r}'+(2nw)\hat{z}|}. \tag{29}$$

The components of $\bar{h}$, subject to (26), can be expressed as

$$h_{\substack{x\\y}} = \frac{\pi}{k}\begin{Bmatrix} j \\ -1\end{Bmatrix}\sum_{n=-\infty}^{\infty} C_n\left[j^{n+1}e^{j(n+1)\phi}J_{n+1}(k\rho) + \begin{Bmatrix} -1 \\ 1\end{Bmatrix} j^{n-1}e^{j(n-1)\phi}J_{n-1}(k\rho)\right] \tag{30}$$

where (ρ, ϕ) are the cylindrical coordinates, J_n is the Bessel function, and C_n's are newly introduced unknown constants yet to be determined. Equation (28) is an integral equation for the unknown $\bar{M}(\bar{r}')$ or, equivalently, the tangential component of the electric field in A. This equation is solved in conjunction with the following condition for the determination of the C_n's:

$$\hat{c}\cdot\bar{M}(\bar{r}) = 0 \quad \text{or} \quad \hat{\tau}\cdot\bar{E}^a(\bar{r}) = 0, \qquad \bar{r}\in C \tag{31}$$

where C is the rim of the aperture, and $\hat{c}$ and $\hat{\tau}$ are the unit normal and tangent vectors, respectively, to the rim. Comparing (23) and (28) with equations appearing in linear antenna theory, one finds that (23) and (29) are the counterparts of Pocklington's and Hallen's equations, respectively. Although (23) and (29) apply to the problem represented by Fig. 1, they can be used for the single-screen problem by simply replacing g with g^0 in (28). Some numerical results for the latter case are given in [9], [15].

IV. NUMERICAL RESULTS AND DISCUSSION

This section is concerned with extracting the numerical solution of the integral equation (28) by applying the method of moments. Because of the almost identical structures of (28) and the equation used in [9], i.e., (4), the numerical procedure developed in [9] can be readily employed here; the interested reader is referred to this reference.

Kernel Approximation

Since the kernel of integral equation (28), i.e., (29), is more complicated than that of [9], it should be analyzed in more detail. An analysis of the kernel shows that the singular term arises when $n = 0$ in the summation appearing in the kernel. This, as a result, allows the self-patch integration as obtained in (4) of [9] for the singular term to be used. Since an infinite summation cannot be handled directly by the computer, it must be truncated in a manner that provides an acceptable numerical approximation to the summation. For the infinite summation given in (29), one first truncates the summation and then approximates the residual terms.

Truncating (29) results in the following equation:

$$g(\bar{r}'|\bar{r}) = \sum_{n=0}^{L} \frac{e^{-jk|\bar{r}-\bar{r}'+(2nw)\hat{z}|}}{2\pi|\bar{r}-\bar{r}'+(2nw)\hat{z}|} + R_L \tag{32}$$

where R_L denotes the residual term. For the index L such that

$$\frac{|\bar{r}-\bar{r}'|}{2Lw} \ll 1 \tag{33}$$

the residual term R_L may be approximated as follows:

$$R_L = \sum_{n=L+1}^{\infty} \frac{e^{-jk|\bar{r}-\bar{r}'+(2nw)\hat{z}|}}{2\pi|\bar{r}-\bar{r}'+(2nw)\hat{z}|} \approx \left[\sum_{n=1}^{\infty} \frac{e^{-jk(2nw)}}{2\pi(2nw)} - \sum_{n=1}^{L} \frac{e^{-jk(2nw)}}{2\pi(2nw)}\right]. \tag{34}$$

Using the following formula [17]:

$$\sum_{n=1}^{\infty} \frac{e^{-jnx}}{n} = -\ln(1-e^{-jx}) = -\ln\left(2\sin\frac{x}{2}\right) + j\frac{\pi - x}{2}, \quad 0<x<2\pi \tag{35}$$

one may finally approximate (32) in the following form:

$$g(\bar{r}'|\bar{r}) \approx \sum_{n=0}^{L}\left[\frac{e^{-jk|\bar{r}-\bar{r}'+(2nw)\hat{z}|}}{2\pi|\bar{r}-\bar{r}'+(2nw)\hat{z}|} - \frac{e^{-jk(2nw)}}{4\pi nw}\right] + \frac{1}{4\pi w}\ln(1-e^{-j2kw}). \tag{36}$$

It is noticed that (36) is exact for $L \to \infty$. Furthermore, the last two terms on the right-hand side of (36) are independent of the location of observation points ($\bar{r}$) and source points ($\bar{r}'$); hence, they can be evaluated once for all the points. This approximation introduces only a small loss of accuracy but produces a substantial savings in computing time. It is further noticed that the aforementioned infinite series is not uniformly convergent and, in fact, is divergent for the values of $w = \lambda/2$, λ, $3\lambda/2$, $\cdots$. These values are related to the resonances of the two-parallel-plate structures. Numerical results have revealed that for values of $|\rho - \rho'|/\lambda \leqslant \sqrt{2}$, $w/\lambda > 2$, and w slightly away from the resonant separations, $L = 20$ would provide an adequate approximation in (36). The validity of using (36) is studied in detail in Appendix F of [15] and the interested reader is referred to this appendix.

Numerical Results of the Aperture Field

Numerical results for the aperture field are obtained by solving (28) in the manner discussed in the previous sections. Although the method of solution is almost the same for any aperture configuration (as an example here rectangular apertures are considered), the basic difference is the design of a proper algorithm for enforcing the condition $\bar{M}\cdot\hat{c} = 0$ (or $\bar{E}\cdot\hat{\tau} = 0$) at the rim, which is done very easily for rectangular apertures.

Fig. 2 shows the amplitude distribution of the dominant component of the E field along the principal axes of a $0.5\lambda \times 0.5\lambda$ aperture. Note that since pulse expansion is used, the field is zero at the center of the edge patches. The result of Fig. 2, which has been compared with that of a single aperture (no back plate), shows that away from resonant separations (in this case $w = 2.8\lambda$) the aperture fields are almost the same. This observation is particularly accurate for small apertures

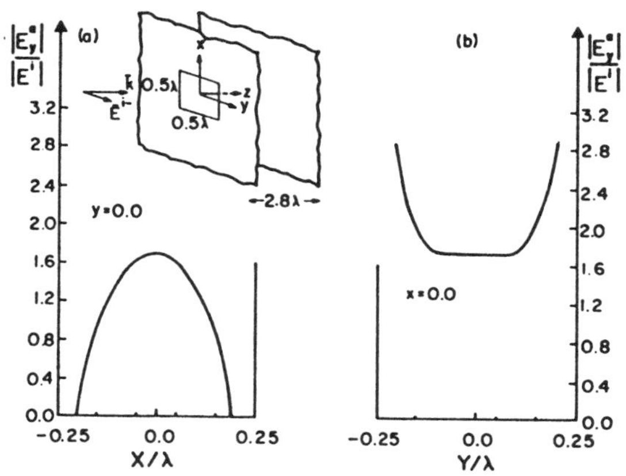

Fig. 2. $|E_y^a|$ field in a $0.5\lambda \times 0.5\lambda$ aperture with backplate. (a) $|E_y^a|$ sampled along the x axis. (b) $|E_y^a|$ sampled along y axis.

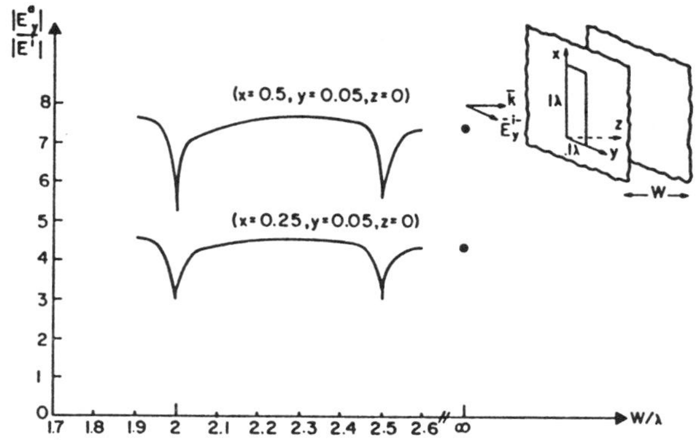

Fig. 3. $|E_y^a|$ sampled at two different points in the aperture as a function of plates' separation.

and is the basis of applying Bethe's approximation to the cavities that are not in their resonant regime [6]. The effect of the resonant separation is studied in Fig. 3. This figure displays the behavior of the amplitude of the E_y field at two different aperture points as a function of the separation distance w/λ. Except around the resonant separation, this field varies very slowly as a function of w/λ, and does not show substantial change when compared with the case of $w/\lambda \to \infty$ (no backplate).

Time-Domain Response Due to an EMP

As an example of an aperture response to an EMP, the time history of the electric field $\bar{e}(t, \bar{r})$ which passes through a 115 cm × 1.3 cm rectangular aperture with backplate separation $w = 348$ cm is computed. The excitation field is a normally incident double-exponential plane-wave EMP represented by

$$\bar{e}^{i-}(t,\bar{r}) = \hat{y}E_0\{[e^{-\alpha(t-z/c)} - e^{-\beta(t-z/c)}]H(t-z/c) - [e^{-\alpha(t-\tau-z/c)} - e^{-\beta(t-\tau-z/c)}] \cdot H(t-\tau-z/c)\} \tag{37}$$

where $H(\cdot)$ is the unit step function, $c = 3.0 \times 10^8$ m/s, $\alpha = 6.0 \times 10^6$ s^{-1}, $\beta = 2.0 \times 10^8$ s^{-1}, $\tau = 2.04189 \times 10^{-9}$ s,

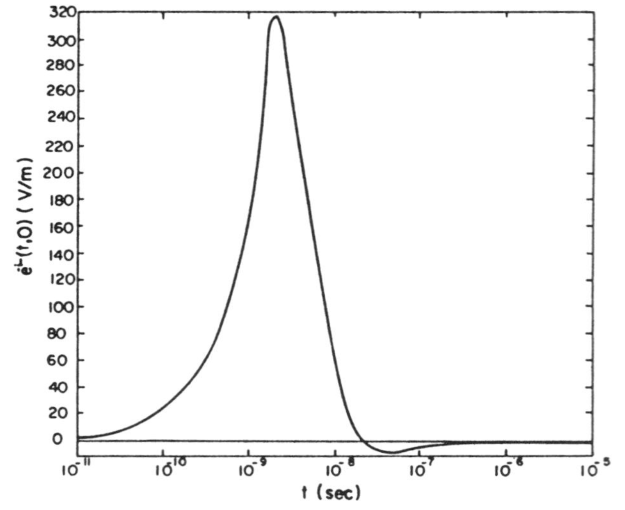

Fig. 4. Time-domain plot of the incident field.

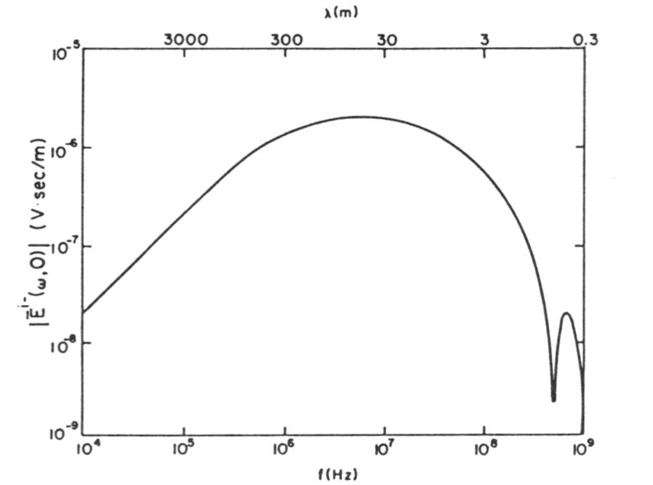

Fig. 5. Frequency-domain behavior of the magnitude of the incident field.

and $E_0 = 10^3$ V/m. The spectrum of $\bar{e}^i(t, \bar{r})$ can be readily determined by Fourier transforming (37), which yields

$$\bar{E}^{i-}(\omega, \bar{r}) = \hat{y}E_0\left(\frac{1}{\alpha + j\omega} - \frac{1}{\beta + j\omega}\right)(1 - e^{-j\omega\tau})e^{-jkz}. \tag{38}$$

Figs. 4 and 5 show the behavior of $\bar{e}^{i-}(t, 0)$ as a function of time and the behavior of $\bar{E}^{i-}(\omega, 0)$ as a function of frequency ($f = \omega/2\pi$), respectively.

Since the convolution in the time domain is transformed into a multiplication in the Fourier domain, one first needs to determine the transfer function (delta response) of the system. This is done by finding the field values at a given point due to incident plane waves of different frequencies. Having obtained this transfer function, one then multiplies it by the transform of the EMP, i.e., (38), to obtain the Fourier transform of the response $\bar{E}(\omega, \bar{r})$. Fig. 6 displays this transform for the dominant component of the field sampled at the point ($x = 0, y = 0$, $z = 200$ cm) inside the cavity. The dominant feature in Fig. 6 is governed by the first resonance of the aperture. Other peaks in Fig. 6 are due to the back-plate resonances and the aperture secondary resonances.

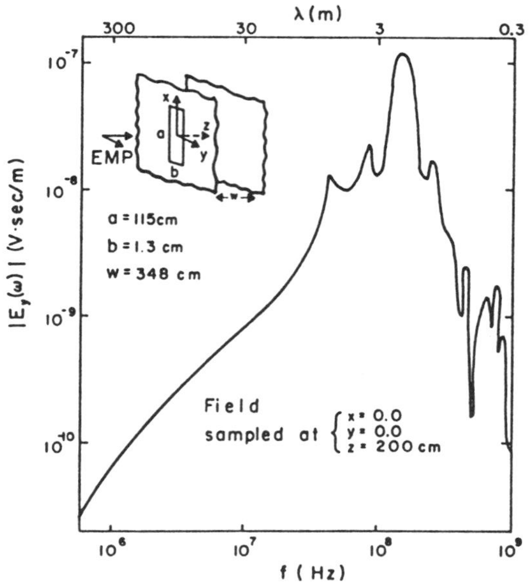

Fig. 6. Frequency-domain behavior of the amplitude of $E_y(\omega)$ field sampled at a point 2 m behind the aperture on the z axis.

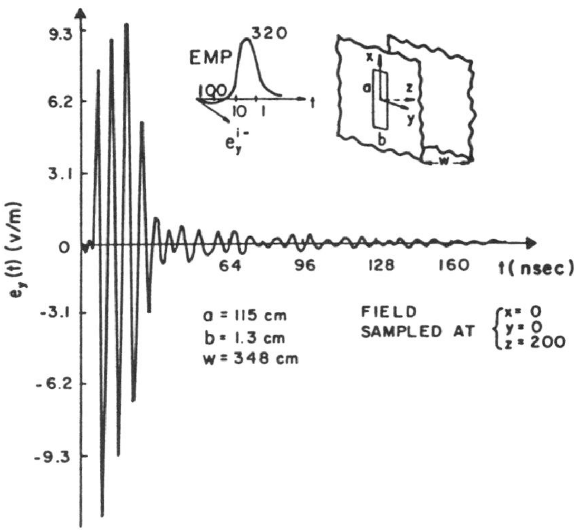

Fig. 7. Time-domain behavior of $e_y(t)$ sampled at a point 2 m behind the aperture on the z axis.

The time-domain response $\bar{e}(t, \bar{r})$ is computed from the knowledge of $\bar{E}(\omega, \bar{r})$ via standard transform means, and the integrals involved in this process may be handled using the FFT algorithm [18], which is known to be efficient in such cases. Employing the fact that $\bar{e}(t, \bar{r})$ is a real and causal function, one can show [19] that either the real or the imaginary part of $\bar{E}(\omega, \bar{r})$ is sufficient to determine $\bar{e}(t, \bar{r})$. For an accurate evaluation of the integral, $\bar{E}(\omega, \bar{r})$ must be adequately sampled in the frequency range of interest so that the criterion suggested by the sampling theorem [20] is not violated. According to this criterion, it is necessary to evaluate $\bar{E}(\omega, \bar{r})$

at no fewer than 1024 points in the interval 0-10^9 Hz, beyond which $|\bar{E}(\omega, \bar{r})|$ decreases to 60 dB below its maximum value. However, rather than sampling $\bar{E}(\omega, \bar{r})$ uniformly in the above frequency range, it is more efficient to compute $\bar{E}(\omega, \bar{r})$ at relatively dense samples in the neighborhood of the resonant frequencies, where the function varies rapidly and sparsely at the frequencies where the response is smooth. Then $\bar{E}(\omega, \bar{r})$ is computed at the uniformly spaced needed points by employing an interpolation scheme through the values of $\bar{E}(\omega, \bar{r})$ constructed at the fewer nonuniformly spaced points in the spectrum.

Fig. 7 displays the time-domain electric field $\bar{e}(t, \bar{r})$, sampled at $\bar{r} = 200\hat{z}$ using the procedure outlined above. It has been found that: i) the rise time of the exciting EMP has a profound effect on the intensity of the time response, ii) the time-domain response oscillates with the resonant frequency of the aperture, and iii) the cavity resonances prolong the time duration of the response.

REFERENCES

[1] R. Muller, "Theory of cavity resonators," in *Electromagnetic Waveguides and Cavities*, G. Goubau, Ed. New York: Pergamon Press, 1961, chap. 2.

[2] Y. P. Liu, "Penetration of electromagnetic fields through small apertures into closed shields," Interaction Note 48, Air Force Weapons Lab., Albuquerque, NM, 1969.

[3] L. W. Chen, "On cavity excitation through small apertures," Interaction Note 45, Air Force Weapons Lab., Albuquerque, NM, 1970.

[4] M. I. Sancer and A. D. Varvatsis, "Electromagnetic penetrability of perfectly conducting bodies containing an aperture," Interaction Note 49, Air Force Weapons Lab., Albuquerque, NM, 1970.

[5] J. Van Bladel, "Small apertures in cavities at low frequencies," *AEÜ*, pp. 481-486, 1972.

[6] C. D. Taylor, "Electromagnetic pulse penetration through small apertures," *IEEE Trans. Electromagn. Compat.*, vol. EMC-15, no. 1, 1973.

[7] K. C. Chen and C. E. Baum, "On EMP excitations of cavities with small openings," Interaction Note 170, Air Force Weapons Lab., Albuquerque, NM, 1974.

[8] H. A. Bethe, "Theory of diffraction by small holes," *Phys. Rev.*, vol. 66, pp. 163-182, 1944.

[9] Y. Rahmat-Samii and R. Mittra, "Electromagnetic coupling through small apertures in a conducting screen," *IEEE Trans. Antennas Propagat.*, vol. AP-25, no. 2, pp. 180-187, Mar. 1977.

[10] C. J. Bouwkamp, "Diffraction theory," *Repts. Prog. in Phys.*, vol. 17, pp. 35-100, 1954.

[11] W. H. Eggimann, "Higher-order evaluation of electromagnetic diffraction by circular disks," *IRE Trans. Microwave Theory Tech.*, vol. MTT-9, no. 5, pp. 408-418, Sept. 1961.

[12] C. Butler, Y. Rahmat-Samii and R. Mittra, "Electromagnetic penetration through apertures in conducting surfaces," *IEEE Trans. Antennas Propagat.*, AP-26, PP. 82-93, Jan. 1978; also, *IEEE Tran. Electromagn. Compat.*, vol. EMC-20, pp. 82-93, Feb. 1978.

[13] M. Morse and H. Feshbach, *Methods of Theoretical Physics, Part I.* New York: McGraw-Hill, 1953.

[14] C. T. Tai, *Dyadic Green's Functions in Electromagnetic Theory.* Scranton, PA: Intext Educational Publishers, 1971.

[15] Y. Rahmat-Samii and R. Mittra, "A new integral equation solution of electromagnetic aperture coupling and thin plate scattering problems," Interaction Note 224, Air Force Weapons Lab., Albuquerque, NM, Feb. 1975.

[16] P. Parhami, Y. Rahmat-Samii, and R. Mittra, "Technique for calculating the radiation and scattering characteristics of antennas mounted on a finite ground plane," *Proc. IEE*, vol. 124, pp. 1009-1016, Nov. 1977.

[17] R. E. Collin, *Field Theory of Guided Waves.* New York: McGraw-Hill, 1960.

[18] E. O. Bringham, *The Fast Fourier Transform.* Englewood Cliffs, NJ: Prentice-Hall, 1974.

[19] A. Papoulis, *The Fourier Transform and Its Applications.* New York: McGraw-Hill, 1962.

[20] P. L. Ransom, "Evaluation of the Fourier integral using the Fast Fourier Transform," Antenna Laboratory Report No. 72-9, Univ. Illinois, Urbana, IL, 1972.

Part 6
Hybrid MM/Alternate Green's Functions

HYBRID techniques are those that involve use of different kinds of field propagators to model separate parts of an overall problem, such as combining the MM for a wire with geometrical theory of diffraction (GTD) for scattering from nearby edges. Alternate Green's functions go beyond the one commonly used for an infinite medium by incorporating a special Green's function that satisfies field boundary conditions over special surfaces such as infinite plane, circular cylinder, or sphere. In either case, the goal is to reduce the number of unknowns needing solution in the MM model. Problems discussed in this section include a monopole attached to a sphere, use of the MM with GTD in various ways, and the problem of an infinite interface that arises when modeling antennas near the earth's surface.

We begin with the paper by Tesche and Neureuther, "Radiation patterns for two monopoles on a perfectly conducting sphere." This was one of the first MM models to employ a specialized Green's function, in this case for a sphere, to reduce a problem involving a sphere–wire combination to an MM solution over only the wire. Albertsen *et al.*, in "Computation of radiation from wire antennas on conducting bodies," present a hybrid treatment based on the MM magnetic-field integral equation (MFIE) solution for a conducting surface, combined with the EFIE solution for thin wires and a wire–surface attachment mode. Still another hybrid approach is described by Thiele and Newhouse, "A hybrid technique for combining moment methods with the geometrical theory of diffraction." This paper was the first to combine high-frequency, or GTD, approximation with an MM model, exploiting the separate advantages of each approach to achieve a capability not feasible using either alone. Newman and Tulyathan, in "Wire antennas in the presence of a dielectric/ferrite inhomogeneity," present the volume-current approach for arbitrary 3D material body in the presence of a thin wire. The paper shows how to treat a material body with permittivity and permeability different from that of free space, as well as how to treat a thin wire in the presence of the penetrable body. In "A summary of hybrid solutions involving moment methods and GTD," Burnside and Pathak survey a variety of hybrid MM–GTD methods. A different kind of hybrid approach is described by Medgyesi-Mitschang and Wang in "Hybrid solutions for scattering from perfectly conducting bodies of revolution." They develop a current-based hybrid MM formulation, incorporating the Fock solution for the surface currents into the MM model. In "Modeling antennas near to and penetrating a lossy interface," Burke and Miller present an MM/Green's function solution for wire antennas in the presence of a flat earth. The accurate and efficient evaluation of the flat-earth Green's function, the Sommerfeld integrals, is considered, as is the important case where the wire penetrates the air/earth interface. Mosig and Gardiol describe the basic integral equation and MM solution for a microstrip antenna in "General integral equation formulation for microstrip antennas and scatterers." The problem of modeling microstrip structures is related to, and a generalization of, the flat-earth problem just mentioned, and is one of increasing importance. Microstrip structures are also the subject of the paper "Analysis of planar strip geometries in a substrate-superstrate configuration," by Jackson and Alexópoulos. This paper includes examples of the microstrip transmission line and coupled dipoles, and presents a unified treatment of MM models for such structures in layered media. The next two papers in this section concern hybrid approaches in which integral- and differential-equation models are combined, one reason being to rigorously "close" the differential-equation mesh by using an integral equation. Another is to employ integral equations for homogeneous regions for which only surface sampling is needed, while using differential equations for inhomogeneous regions. In "A note on hybrid finite element method for solving scattering problems," Jin and Liepa show how modifying a hybrid finite-element, integral-equation or modal-expansion to produce matrices that are easier to solve can improve computational efficiency. The paper by Yuan *et al.*, "Coupling of finite element and moment methods for electromagnetic scattering from inhomogeneous objects" elaborates formulation of a hybrid finite-element, integral-equation model. They demonstrate application of an integral equation to close the differential-equation mesh used to solve an interior, inhomogeneous region. The concluding hybrid paper by Olsen and Mannikko, "Validation of the hybrid quasi-static/full-wave method for capacitively loaded thin-wire antennas," discusses quite a different kind of hybrid model, where a quasistatic approach is used to avoid problems that can occur in regions very much smaller than a wavelength.

Radiation Patterns for Two Monopoles on a Perfectly Conducting Sphere

Abstract—**A general numerical technique for analyzing monopole antennas protruding from a conducting sphere is suggested. Computed results for special cases of antenna locations are compared to known and approximate solutions showing that the method is practical and accurate.**

I. Introduction

During the past few years there has been considerable interest in the use of numerical techniques for the solution of problems in electromagnetic theory that are difficult to solve exactly by the classical methods. One such problem is that of determining the behavior of a thin-wire antenna which protrudes from a perfectly conducting sphere having a radius of the order of magnitude of the wavelength. An example of such a radiating system would be a conducting vehicle, perhaps a satellite, with antennas mounted on it. The special case of antenna wires mounted on a conducting sphere would permit the estimation of the effect of vehicle shape on the antenna performance by approximating the actual vehicle by the sphere.

The spherical shape of the vehicle will affect the current distribution, input impedance, and radiation pattern of the antenna. The major part of the work to date, however, has been only on the behavior of the pattern. There are several different methods which may be employed to determine the radiation pattern from such a spherical antenna, but little appears to be known about the accuracy of the resulting solution. One approximate method is to replace the sphere by a thin wire and then solve for the current on the resulting structure which consists of the original monopoles plus the new wire. The results of this approximate analysis are thought to be fairly good for small sphere sizes, but inaccurate for large spheres. To account for the effect of sphere exactly, it is possible to represent the fields produced by a point dipole near the sphere by an infinite series of Hankel functions and spherical harmonics. For spheres having a radius on the order of several wavelengths, it is possible to sum the series directly to find the fields due to the dipole source. This was done by Du and Tai [1] for the case of four symmetrically located point dipoles on the surface of a conducting sphere. The resulting radiation patterns, although exact for the point dipole, are not correct for antennas with a length comparable to the wavelength. Assuming that the fields produced by a point dipole located at an arbitrary distance from the sphere surface are known, the principle of superposition may be used to calculate the radiation pattern of a long antenna. This method, however, requires a prior knowledge of the current distribution on the wire antenna, and if only an approximate distribution is used, the far-field components are in error.

Recently Bolle and Morganstern [2] numerically solved the classically formulated case of a single monopole antenna protruding from a small sphere by considering it to be the limiting case of a conical antenna. They were able to obtain the input impedance and far-field radiation pattern, but again the actual current distribution on the antenna was not found.

In this communication a more general and convenient numerical approach is considered which provides a solution not only for the input impedances and radiation pattern, but also for the current distributions on two monopole antennas which lie along the polar axis $\theta = 0,\pi$ of a perfectly conducting sphere. In addition, it is possible to determine the current distribution on the sphere itself. By a simple change of the forcing function of the matrix equation, one monopole can be considered to be a scatterer. This approach is more general in that it can be easily extended to consider monopoles at any angle on the sphere, curved wires, impedance loading of the wires, and finite conductivity of the sphere. Due to the limitations on space, the main emphasis in this communication will be on the radiation patterns. The behavior of the current and input impedance for some of these more general structures will be published at a later data.

Manuscript received December 10, 1969; revised March 10, 1970. This research was sponsored by the Joint Services Electronics Program under Grant AFOSR-68-1488.

II. Theory

By using the Green's tensor [3] $\mathbf{\Gamma}(\bar{x}',\bar{x})$ for the electric field produced by a point current element in the presence of a perfectly conducting sphere, an integral equation for the current on the monopoles can be formulated. The electric field produced by a primary current density $\bar{J}(\bar{x}')$ is given by

$$\bar{E}(\bar{x}) = \int_{\text{sources}} \bar{J}(\bar{x}')\cdot\mathbf{\Gamma}(\bar{x}',\bar{x})\,d\bar{x}'. \tag{1}$$

Letting $\bar{J}(\bar{x}')$ be constrained to flow only along the wires of the antenna and looking only at the radial component of the electric field, (1) gives the following integral equation for the current $I(r')$ as the observation point $\bar{x}$ approaches the perfectly conducting thin-wire antenna having $E_{\tan} = 0$:

$$-E_r^{\text{inc}}(r) = \int_{\text{antenna wires}} I(r')\Gamma_{rr}(r',r)\,dr'. \tag{2}$$

Here E_r^{inc} is the radial component of the electric field incident on the antenna from the driving sources located at the points where the monopoles meet the surface of the sphere.

Upon analysis of Γ_{rr}, it is seen that there is a triplet-like singularity near the point where r and r' coincide. This behavior is like that encountered in the kernal of Pocklington's integral equation, and in fact it can be shown that (2) can be rewritten as

$$E_r^{\text{inc}}(r) = \int_{\text{antenna wires}} I(r')\left(k^2 + \frac{d^2}{dr^2}\right)G_T(r',r)\,dr' \tag{3}$$

where

$$G_T(r',r) = \frac{1}{4\pi}\frac{r}{r'}\left(\frac{\mu_0}{\epsilon_0}\right)^{1/2}\left[j\,\frac{\exp(-jk\,|r-r'|)}{k\,|r-r'|} - j_0(kr_<)h_0^{(2)}(kr_>) - \sum_{n=1}^{\infty}\frac{(d/da)aj_n(ka)}{(d/da)ah_n^{(2)}(ka)}(2n+1)(\pm1)^n h_n^{(2)}(kr)h_n^{(2)}(kr')\right]. \tag{4}$$

Here the parameter a is the sphere radius and the (±1) term is chosen according to whether the source point r' and observation point r are on the same monopole $(+1)$ or on opposite monopoles (-1). The resulting integral equation in (3) is then solved numerically for the current distribution by a method of moments similar to that used by Harrington [4].

In calculating the radiation pattern from the current $I(r')$ on the antenna, (1) is again used, but with a slight change in $\mathbf{\Gamma}(\bar{x}',\bar{x})$ to give only the far-field components. The use of the exact current distribution in calculating the radiation pattern will clearly give a more accurate solution than if an approximate distribution is used. It is recognized, however, that the far fields are not extremely sensitive to slight variations in this distribution.

Once the current distribution on the monopoles is known, the induced currents on the sphere surface can then be determined by using the Green's tensor $\mathbf{\Gamma}$. The magnetic field intensity at an

Reprinted from *IEEE Trans. Antennas Propagat.*, vol. AP-18, no. 5, pp. 692–694, September 1970.

observation point $\bar{x}$ exterior to the sphere is given by the following [3]:

$$\bar{H}(\bar{x}) = -\frac{1}{j\omega\mu}\int_{\text{sources}} \nabla \times \mathbf{\Gamma}(\bar{x},\bar{x}')\cdot\bar{J}(\bar{x}')\, d\bar{x}'. \quad (5)$$

By letting the observation point $\bar{x}$ approach the sphere and noting that the tangential $\bar{H}$ field gives the induced surface currents, it is possible to obtain the values of J_ϕ and J_θ on the sphere surface. The results are given by

$$J_\theta(a,\theta) = \int_{\text{antenna wires}} I(r')K(a,\theta;r')\, dr' \quad (6)$$

$$J_\phi = 0$$

where the kernal K is given by

$$K(a,\theta;r') = -\frac{jk}{4\pi}\sum_{n=1}^{\infty}(2n+1)\frac{h_n^{(2)}(kr')}{r'}(\pm 1)^n$$

$$\cdot\left(j_n(ka) - \frac{(d/da)aj_n(ka)}{(d/da)ah_n^{(2)}(ka)}h_n^{(2)}(ka)\right)P_n^1(\cos\theta). \quad (7)$$

In this symmetric case of monopoles at $\theta = 0$ and $\theta = \pi/2$, there is no ϕ variation of the current.

III. Numerical Results

With a computer program which embodied the preceding method of solution, the radiation patterns for a number of antenna configurations were computed. Typical computational times on a CDC 6400 computer for calculating currents, input impedance, and pattern for each structure were 5 seconds for 10 cells on each monopole. In this preliminary study, the current distribution on the sphere has not been included.

In Fig. 1(a) the radiation pattern for a single monopole of length $L = \lambda/4$ and located at $\theta = 0°$ is presented for two values of sphere radius. The solid lines represent the authors' results, and those of Bolle and Morganstern are represented by dashed lines. As is evident, the two different approaches given patterns which agree closely. The probable cause of the slight difference between the curves in Fig. 1 is that it was difficult to reproduce accurately the patterns of Bolle and Morganstern due to the small size of the polar plot in their paper. The parameter $\Omega = 2\ln(2L/b)$, b being the wire radius and L the wire length, was chosen to be 9.6, a value which corresponds roughly to the size of their conical antenna.

A plot of the input impedance of a single monopole of length $L = a$ mounted on the sphere as a function of the sphere radius a is shown in Fig. 1(b). The solid lines represent the input resistance and reactance obtained by the integral equation approach while the dotted curves are those presented by Bolle and Morganstern. The observed differences in the curves can possibly be attributed to the fact that the authors considered a cylindrical wire while that used by Bolle and Morganstern was conical.

Fig. 2 presents the radiation pattern for two monopoles each of length $L = \lambda/4$, located at $\theta = 0,\pi$ on the sphere. The antennas are excited in such a manner that if the sphere would shrink to zero, the exciting voltages would add, giving an equivalent dipole. The solid lines are the results of the rigorous numerical approach, and the dotted lines represent the pattern by approximating the sphere by a conducting wire as is shown in the diagram. As expected, the radiation pattern in the approximate analysis is very good for small spheres, having a radius less than about $\lambda/8$. However, the accuracy is seen to deteriorate as the sphere radius increases.

The data given in Fig. 3 is for one active monopole at $\theta = 0$ and one passive monopole at $\theta = \pi$. As in the previous cases, both lengths are $\lambda/4$, and the sphere radius is varied. This case would be of special interest in modifying the radiation pattern by the use of a

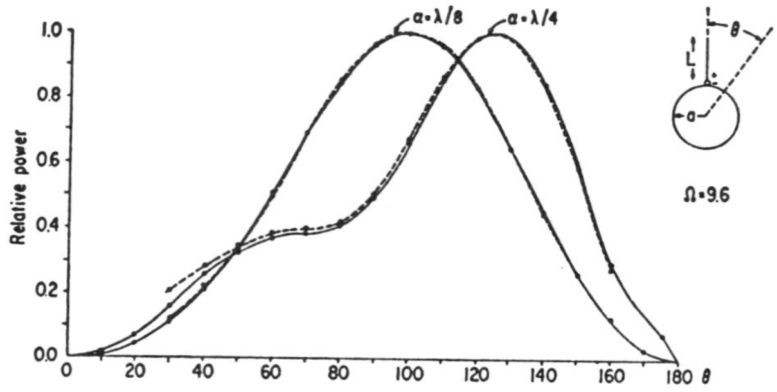

(a)

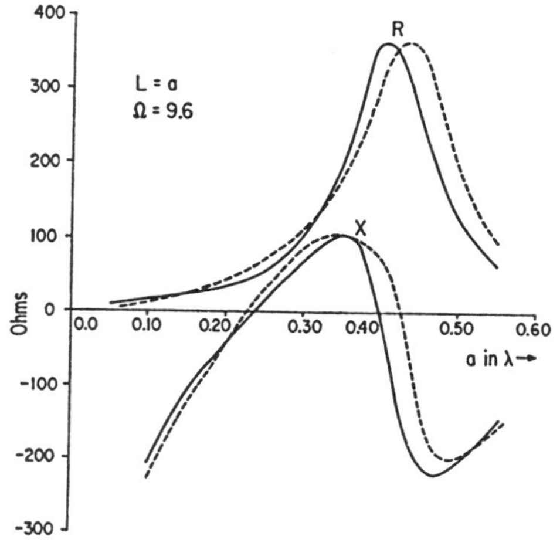

(b)

Fig. 1. (a) Normalized radiation patterns for single monopole of length $L = \lambda/4$ on conducting sphere of radius a. Solid lines represent authors' results and dotted lines are those of Bolle and Morganstern [2]. (b) Comparison of input impedance of single monopole of length $L = a$ mounted over conducting sphere of radius a. Authors' results are represented by solid lines and those of Bolle and Morganstern [2] are dotted lines.

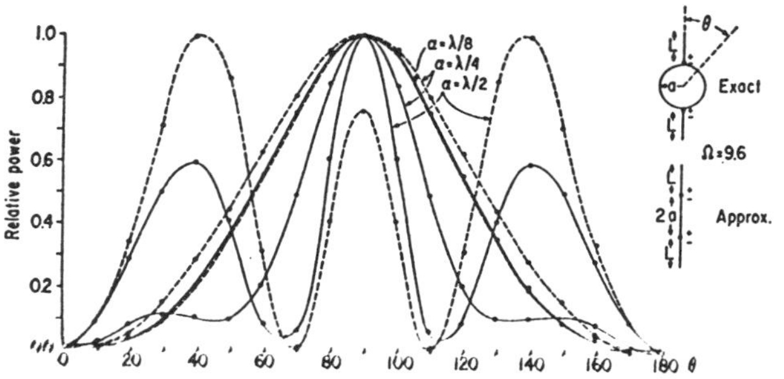

Fig. 2. Comparison between exact and approximate methods for finding radiation pattern of two monopoles of equal lengths $L = \lambda/4$ located at poles of sphere of radius a. Solid lines are exact results and dotted lines are approximate.

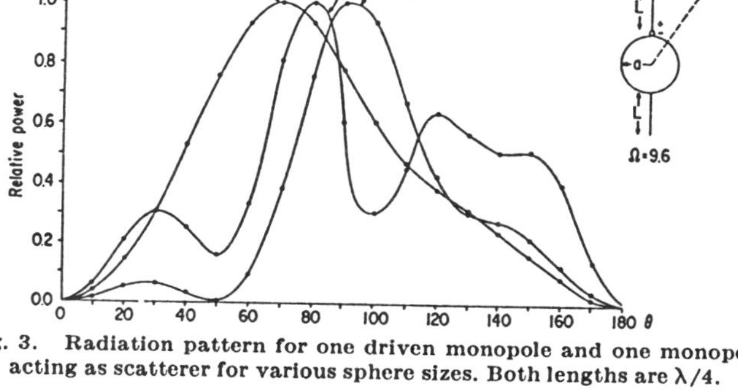

Fig. 3. Radiation pattern for one driven monopole and one monopole acting as scatterer for various sphere sizes. Both lengths are $\lambda/4$.

scattering element or in evaluating the interference of one monopole on another. It is seen that the effect of the additional monopole is to significantly change the position of the main lobe. As might be expected, sidelobes begin appearing as the sphere radius is increased.

IV. Conclusions

Using rigorous numerical methods, the behavior of one and two thin-wire monopoles mounted on a sphere has been studied. The radiation patterns obtained by the numerical approach were shown to agree closely with similar patterns for a single conical antenna on a sphere, and a comparison of input impedances was made. An approximate analysis of the antenna system with two monopoles indicates that replacing the sphere by a wire is not adequate for relatively large spheres. Finally, it was shown that the presence of a passive monopole on the sphere modifies the radiation pattern. Since the coupling of elements in an antenna system is important and can be most adequately treated by numerical techniques, the authors are pursuing the more general aspects of this problem.

F. M. Tesche
A. R. Neureuther
Dept. of Elec. Eng. and Comput. Sci.
Electron. Res. Lab.
University of California,
Berkeley, Calif. 94720

References

[1] L. Du and C. T. Tai, "Radiation patterns of four symmetrically located sources on a perfectly conducting sphere," Ohio State University, Research Foundation Rept. 169-10, December 15, 1964.
[2] D. M. Bolle and M. D. Morganstern, "Monopole and conic antennas on spherical vehicles," *IEEE Trans. Antennas Propagat.*, vol. AP-17, pp. 477–484, July 1969.
[3] D. S. Jones, *The Theory of Electromagnetism.* London: Pergamon, 1964, pp. 495–498.
[4] R. F. Harrington, *Field Computation by Moment Methods.* New York: Macmillian, 1968.

Computation of Radiation from Wire Antennas on Conducting Bodies

N. CHRISTIAN ALBERTSEN, JESPER E. HANSEN, AND NIELS E. JENSEN

Abstract—**A theoretical formulation, in terms of combined magnetic and electric field integral equations, is presented for the class of electromagnetic problems in which one or more wire antennas are connected to a conducting body of arbitrary shape. The formulation is suitable for numerical computation provided that the overall dimensions of the structure are not large compared to the wavelength. A computer program is described, and test runs on various configurations involving a cylindrical body with one or more straight wires are presented. The results obtained agree well with experimental data.**

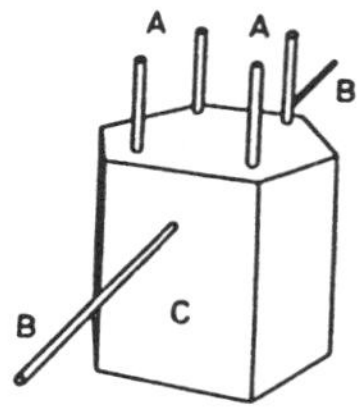

Fig. 1. Example of structure.

I. Introduction

FOR RADIATING systems that are not large compared to the wavelength, several methods are available for the determination of radiation patterns. For antennas mounted on structures which closely approximate simple geometrical shapes, the problem may well be attacked analytically [1]–[3], or by combined analytical and numerical methods [4]. However, when more complex geometries are involved, as in the case of antennas on spacecrafts, resort has to be taken to purely numerical techniques, in which solutions to Maxwell's equations or integral equations derived from these, are obtained with the aid of a fast computer. A computer program for solving complicated radiation or scattering problems of the kind indicated is in many cases equally as accurate as experimental procedures, and more convenient. Various examples of such programs have been described in the past few years [5]–[10].

The present work deals with a specific class of radiating structures, the typical features of which are illustrated in Fig. 1. Here, a number of wire antennas A and a number of booms (passive antennas) B are attached to the surface of a metallic body C which may, at least in principle, be arbitrarily shaped. The complete structure is assumed to have a maximum linear dimension which is not large compared to the wavelength. Fig. 1 is actually an idealization of the shape of a small scientific satellite with four antennas operating in a turnstile mode, and two booms for support of measurement equipment. Although this particular interpretation of Fig. 1 formed the primary motivation of the work reported here, the geometry is common to other types of radiating systems of interest, e.g., antennas on ships, cars, etc.

For the class of structures described it seems a natural choice to attempt a formulation in terms of integral equations. Several forms of integral equations suitable for solid bodies and thin wires exist [6]. Although the electric-field integral equation (EFIE) has been used as a basis for the numerical solution of scattering or antenna problems involving (large) bodies of revolution [11], the magnetic-field integral equation (MFIE) is more attractive for generally shaped voluminous structures. For thin wires or plates, however, the MFIE is known to fail, and one is forced to use the EFIE equation in such cases, or to modify the MFIE equation in a suitable way [12], [13].

In view of the properties of the MFIE and the EFIE, the authors have used the MFIE for the solid body, whereas the EFIE in its thin-wire approximation has been applied to the wire parts of the structure. The mutual coupling between wires and body, and vice versa, is taken account of through the source terms in the integral equations [14]. A computer program based on this technique has been developed.

In Section II the important features of the MFIE and EFIE are summarized. In Section III it is explained how the coupling between body and wires may be computed using a surface current interpolation scheme. A formulation of the combined MFIE and EFIE equations particularly suitable for computer solution is given in Section IV. A short description of the structure of the computer program and examples of computations with experimental verifications are given in Section V.

Manuscript received May 1, 1973; revised October 1, 1973. This work was supported by the European Space Research and Technology Centre, Noordwijk, The Netherlands, under Contract 1340-71.

N. C. Albertsen and J. E. Hansen are with the Laboratory of Electromagnetic Theory, The Technical University of Denmark, Lyngby, Denmark.

N. E. Jensen was with the Laboratory of Electromagnetic Theory, The Technical University of Denmark, Lyngby, Denmark. He is now with the European Space Research and Technology Centre, Noordwijk, The Netherlands.

Reprinted from *IEEE Trans. Antennas Propagat.*, vol. AP-22, no. 2, pp. 200–206, March 1974.

II. Magnetic and Electric Field Integral Equations

From Maxwell's equations, the following MFIE for the surface current density $\bar{J}$ on the surface S of a perfectly conducting body in an incident field with magnetic field strength $\bar{H}^{\text{inc}}$ may be derived [6]:

$$-\tfrac{1}{2}\bar{J}(\xi,\eta) + \hat{n} \times \int_S^* \bar{J}(\xi',\eta') \times \nabla'\phi \, dS' = -\hat{n} \times \bar{H}^{\text{inc}}(\xi,\eta). \quad (2.1)$$

Here, $\hat{n}$ is an outward unit normal vector to the surface. Furthermore, $\phi = \exp(-jkR)/4\pi R$, where R is the distance between the observation point (ξ,η) and source point (ξ',η') on area element dS', and where k is the wavenumber. The asterisk denotes that the integration is carried out on the surface punctured at (ξ,η). The symbol ∇' denotes a gradient with respect to the primed coordinates. The time factor used is $\exp(j\omega t)$. In operator notation (2.1) may be written

$$L_M\bar{J} = -\hat{n} \times \bar{H}^{\text{inc}}. \quad (2.2)$$

For a perfectly conducting thin straight wire, the following EFIE may be derived [6]:

$$\frac{1}{j\omega\epsilon}\int_L \left(\frac{\partial^2\phi}{\partial s^2} + k^2\phi\right) I(s')\, ds' = -E_s^{\text{inc}}(s) \quad (2.3)$$

where it is assumed that the current I has an axial component only (thin wire approximation). Here, L denotes the interval for the coordinate s' of a source point on line element ds' along the wire, while

$$\phi = \exp\{-jk[(s - s')^2 + a^2]^{1/2}\}/[(s - s')^2 + a^2]^{1/2}$$

where a is the wire radius. E_s^{inc} is the component of the electric field strength of the incident field in the direction of the wire, and ϵ the dielectric constant. In operator notation we have

$$L_E I = -E_s^{\text{inc}}. \quad (2.4)$$

Using pulse functions for expansion and point-matching, the moment method formulations [15] of (2.2) and (2.4) become, respectively

$$[\langle\delta_p, L_M W_q\rangle][\bar{J}_q] = -[\langle\delta_p, \hat{n} \times \bar{H}^{\text{inc}}\rangle],$$
$$p = 1,2,\cdots,q_m; \quad q = 1,2,\cdots,q_m \quad (2.5)$$

and

$$[\langle\delta_i, L_E P_j\rangle][I_j] = -[\langle\delta_i, E_s^{\text{inc}}\rangle],$$
$$i = 1,2,\cdots,j_m; \quad j = 1,2,\cdots,j_m. \quad (2.6)$$

In these formulas δ_p and δ_i are Dirac delta functions on S and L, respectively. The centers of these (match points) are on the pth of the q_m patches into which the surface is divided, and on the ith of the j_m segments into which the wire is divided, respectively. W_q and P_j are pulse functions with amplitude unity on the qth patch, and on the jth segment, respectively, and zero elsewhere. Long-hand expansion of (2.6) may be found in [16].

While (2.5) is of a vector character, and may be split into two scalar systems of q_m linear equations in the $2q_m$ unknown components of the surface current density coefficients $\bar{J}_q$, (2.6) represents a system of j_m linear equations in the j_m unknown current coefficients I_j.

When implemented on a computer both the total number of surface patches q_m at the MFIE and the total number of wire segments j_m at the EFIE must be chosen large enough in order to obtain sufficiently accurate solutions [6].

III. Coupling Between Body and Wire

In order to take the wire–body coupling into account, as described in Section I, we need formulas for the magnetic and electric fields from known current distributions. These may be found in standard texts, e.g., [17]. At a patch with unit normal vector $\hat{n}$ and a segment with unit tangential vector $\hat{s}$ we get the following contributions to the tangential magnetic and electric fields, respectively

$$\hat{n} \times \bar{H}(I) = \hat{n} \times \int_L I(s')\hat{s}' \times \nabla'\phi \, ds' \doteq L_M' I \quad (3.1)$$

and

$$E_s(\bar{J}) = \frac{1}{j\omega\epsilon}\hat{s}\cdot\int_S \nabla \times (\nabla \times (\bar{J}\phi))\, dS \doteq L_E'\bar{J} \quad (3.2)$$

where we have introduced an operator notation along the same lines as above.

After having expanded the currents in (3.1) and (3.2) into pulse functions with coefficients I_j and $\bar{J}_q$, in a similar manner as in Section II, we obtain the following matrix equation for the unknown currents in the mixed problem

$$\left[\begin{array}{c:c} \langle\delta_p, L_M W_q\rangle & \langle\delta_p, L_M' P_j\rangle \\ \hline \langle\delta_i, L_E' W_q\rangle & \langle\delta_i, L_E P_j\rangle \end{array}\right] \left[\begin{array}{c} \bar{J}_q \\ \hline I_j \end{array}\right] = -\left[\begin{array}{c} \langle\delta_p, \hat{n} \times \bar{H}^{\text{inc}}\rangle \\ \hline \langle\delta_i, E_s^{\text{inc}}\rangle \end{array}\right] \quad (3.3)$$

with $p = 1,2,\cdots,q_m$; $q = 1,2,\cdots,q_m$; $i = 1,2,\cdots,j_m$; and $j = 1,2,\cdots,j_m$.

This equation is of mixed scalar and vector form, and may be expanded into scalar form alone by introducing the proper decompositions of the vector quantities.

In order to compute the off-diagonal (coupling) coefficients in (3.3) a careful treatment of the currents close to the wire attachment area is necessary. The operator L_M' is well behaved, hence it suffices to use pulse functions in (3.1) for all segments including the attachment segment. The operator L_E', on the other hand, has a singularity of third order, which makes it necessary to use at patches on which wires are attached, a more accurate expression for the surface current density $\bar{J}$ than pulse functions. The technique used is described below for the case of quadrilateral patches.

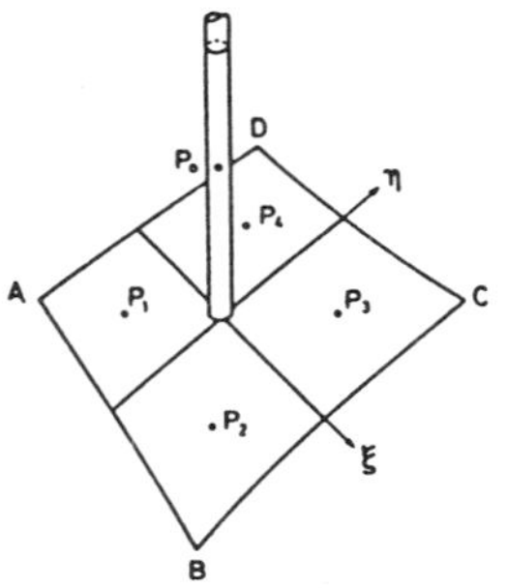

Fig. 2. Details of wire attachment area.

In Fig. 2, $ABCD$ is a quadrilateral patch in a body segmentation scheme set up before the wires are brought into the picture. In order to introduce an interpolation of the surface current density on $ABCD$ which takes into account the attachment of a wire, $ABCD$ is divided into four patches as shown. A local coordinate system (ξ,η) is introduced in which A, B, C, and D have the coordinates (ξ_1,η_1), (ξ_2,η_2), (ξ_3,η_3), and (ξ_4,η_4), respectively. The match points of the four patches are denoted $P_1,\cdots,P_4$, and the unknown surface current densities in these points $\bar{J}_1,\cdots,\bar{J}_4$. The match point on the first wire segment is P_0 and the corresponding unknown current I_0. The coordinates for $P_1,\cdots,P_4$ are $(\xi_1/2,\eta_1/2),\cdots,(\xi_4/2,\eta_4/2)$, respectively.

A suitable expansion of the surface current density $\bar{J}(\xi,\eta)$ over $ABCD$ should meet the following requirements:

$$\bar{J}(\xi_i/2,\eta_i/2) = \bar{J}_i, \qquad i = 1,\cdots,4 \tag{3.4}$$

and

$$\nabla_s\cdot\bar{J}(\xi,\eta) = J_0(\xi,\eta) - I_0\delta(\xi,\eta) \tag{3.5}$$

where $\nabla_s\cdot$ denotes surface divergence, and where $J_0(\xi,\eta)$ is continuous across $ABCD$. One such expansion is

$$\bar{J}(\xi,\eta) = I_0\frac{\xi\hat{\xi}+\eta\hat{\eta}}{2\pi(\xi^2+\eta^2)} + \sum_{i=1}^{4}\bar{J}_i'g_i(\xi,\eta). \tag{3.6}$$

The interpolation functions, $g_1(\xi,\eta),\cdots,g_4(\xi,\eta)$, are required to have the following properties across $ABCD$: 1) $g_i(\xi,\eta)$ is differentiable; 2) $g_i(\xi_j,\eta_j) = \delta_{ij}$, for $i = 1,\cdots,4$ and $j = 1,\cdots,4$, where δ_{ij} is the Kronecker delta; and 3) $g_1(\xi,\eta) + \cdots + g_4(\xi,\eta) = 1$.

These requirements are fulfilled by the hyperbolic paraboloids used in this study. Hyperbolic paraboloids are particularly well suited for quadrilateral surface patches, and are merely specific examples of the more general "wedge functions" introduced by Wachspress [18] as interpolation functions for surface patches of a general shape.

The fields due to the currents on the attachment patches and segment may be expressed in terms of $\bar{J}_1',\cdots,\bar{J}_4'$, and I_0 as the moments

$$[\langle\delta_p,\hat{n}\times\bar{H}\rangle] = [B][\bar{J}'] + [W_B]I_0 \tag{3.7}$$

$$\langle\delta_0,E_s\rangle = [B_W][\bar{J}'] + WI_0 \tag{3.8}$$

where $[B]$ is a matrix derived from (2.5), W is a coefficient derived from (2.6) and $[W_B]$ and $[B_W]$ are matrices derived as inner products between (3.1) and δ_p, and between (3.2) and δ_0, respectively. Here δ_p and δ_0 are delta functions defined in Section II. In the computations $[B]$, $[W_B]$, and W are evaluated using the simple technique used for all other elements of (3.3), i.e., the unknown current densities are approximated by pulse functions, while $[B_W]$ is computed as the moments of δ_0 (at P_0) and (3.2) with (3.6) inserted.

If (3.6) is inserted into (3.4), a set of linear equations is obtained which may be solved for $\bar{J}_1',\cdots,\bar{J}_4'$. In this manner the fields at $P_0,P_1,\cdots,P_4$ due to currents on the attachment patches and segment are expressed in terms of the unknowns $\bar{J}_1,\cdots,\bar{J}_4$ and I_0 occurring in (3.3).

It is noted that the interpolation technique described permits continuity of the current at the attachment point of the wire without introduction of new unknown quantities.

IV. Formulation for Computer Solution

In this section we shall show how (3.3) may be rewritten in order to cut down computer storage requirements. From (3.3) we have by multiplying out formally

$$[\langle\delta_p,L_MW_q\rangle][\bar{J}_q] + [\langle\delta_p,L_M'P_j\rangle][I_j] = -[\langle\delta_p,\hat{n}\times\bar{H}^{\text{inc}}\rangle] \tag{4.1}$$

and

$$[\langle\delta_i,L_E'W_q\rangle][\bar{J}_q] + [\langle\delta_i,L_EP_j\rangle][I_j] = -[\langle\delta_i,E_s^{\text{inc}}\rangle]. \tag{4.2}$$

Equation (4.1) may now be solved for $[\bar{J}_q]$.

$$[\bar{J}_q] = -[\langle\delta_p,L_MW_q\rangle]^{-1}[\langle\delta_p,\hat{n}\times\bar{H}^{\text{inc}}\rangle] - [\langle\delta_p,L_MW_q\rangle]^{-1}[\langle\delta_p,L_M'P_j\rangle][I_j] \tag{4.3}$$

which by insertion in (4.2) yields

$$(-[\langle\delta_i,L_E'W_q\rangle][\langle\delta_p,L_MW_q\rangle]^{-1}[\langle\delta_p,L_M'P_j\rangle] + [\langle\delta_i,L_EP_j\rangle])[I_j] = [\langle\delta_i,L_E'W_q\rangle][\langle\delta_p,L_MW_q\rangle]^{-1}[\langle\delta_p,\hat{n}\times\bar{H}^{\text{inc}}\rangle] - [\langle\delta_i,E_s^{\text{inc}}\rangle]. \tag{4.4}$$

When this equation has been solved for $[I_j]$, $[\bar{J}_q]$ may be found from (4.3).

Equation (4.4) is of a form similar to the moment solution (2.6) of the EFIE for a wire in free space. However, extra terms occur on both sides of the equality sign. These precisely account for the presence of the conducting body, and (4.4) may be considered to be the moment method formulation of an electric field integral equation similar to (2.3), with the (free space) Green's function ϕ modified to take the conducting body into account. This is analogous to the integral equation formulation given in [4] for the particular case of a monopole on a sphere, where the modified Green's function is available in analytical form.

An equation for $[\bar{J}_q]$ analogous to (4.4) may also be derived. The result is

$$\begin{aligned}
&([\langle\delta_p, L_M' P_j\rangle][\langle\delta_i, L_E P_j\rangle]^{-1}[\langle\delta_i, L_E' W_q\rangle] \\
&\quad + [\langle\delta_p, L_M W_q\rangle])[\bar{J}_q] \\
&= [\langle\delta_p, L_M' P_j\rangle][\langle\delta_i, L_E P_j\rangle]^{-1} \; [\langle\delta_i, E_s^{\mathrm{inc}}\rangle] \\
&\quad - [\langle\delta_p, \hat{n} \times \bar{H}^{\mathrm{inc}}\rangle]. \qquad (4.5)
\end{aligned}$$

This equation may be considered to be the moment method formulation of an MFIE similar to (2.1), with a kernel that takes the conducting wire into account.

For a problem involving j_m segments and q_m patches the original matrix in (3.3) has $(2q_m + j_m)^2$ elements. With the preceding approach totally no more than two matrices with a total number of elements not exceeding $2q_m(2q_m + j_m)$ or $j_m(j_m + 2q_m)$, any of which is the largest figure, must be in core storage in the computer, simultaneously.

In the present work, (4.4) forms the basis of computations. In the computer program, the operators L_M and L_E have been evaluated using rectangular rule for the integration. The coupling between the attachment segment and the attachment patch is computed using Romberg integration in L_M' and a rectangular rule with 100 subdivisions in L_E'. All other couplings are computed using a simple rectangular rule in both L_M' and L_E'.

The computation of the radiation pattern from $[I_j]$ and $[\bar{J}_q]$ is a standard procedure, which may be found in textbooks, e.g., [17]. For this reason details are omitted.

V. Computations and Comparisons with Experiments

The computer program constructed for the present investigation consists of a main program and 34 Fortran subroutines. The sequence of computations is divided into four distinct parts.

First, the complete geometry is specified. This includes all coordinates of the structure as well as the segmentation parameters for body, antennas, and passive booms. In the second part of the computer program, the left-hand side matrix in (4.4) is formed and stored in an array. In the third part of the program, (4.4) is solved for the currents on all antennas and other wires attached to the body. Equation (4.3) then yields the surface current densities. Finally, the radiation pattern is computed. A number of test runs of the program and comparisons with experiments have been made for a circular cylindrical body with straight antennas and passive booms.

The experimental model is shown in Fig. 3. The height and diameter of the cylindrical body are 220 mm and 200 mm, respectively. Inside the model, a battery-powered transmitter for the frequency range 0.5–1.0 GHz is placed. On the surface of the model, monopole antennas and passive booms may be mounted in a variety of positions. The antenna is 2 mm in diameter while booms (in metallic connection with the body) are 4 mm in diameter.

During the measurements the received signals were recorded digitally on paper tape. These tapes were used later as input data for a plotting program. Measured

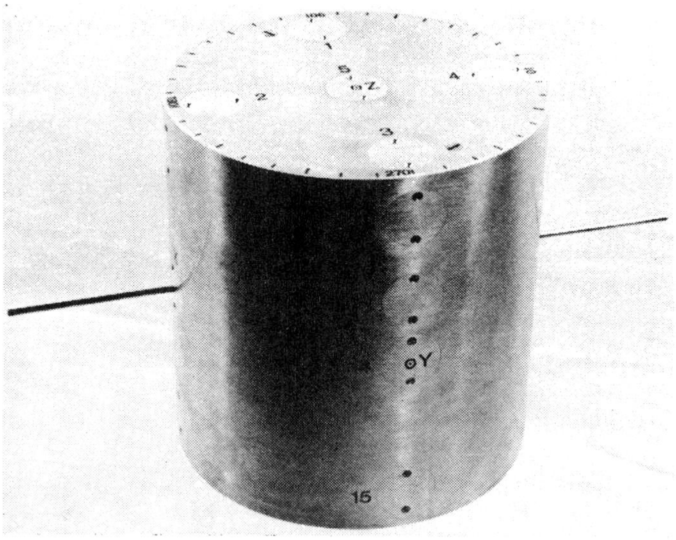

Fig. 3. Cylindrical model used for experiments.

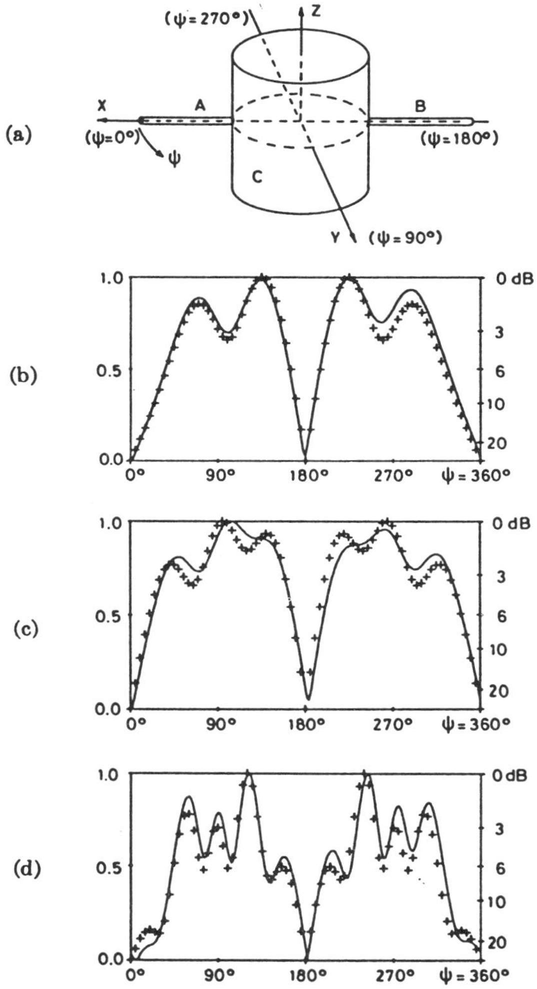

Fig. 4. xy-polarized radiation patterns in the xy-plane (field strength, normalized to maximum value of unity, as function of angle ψ) for monopole antenna (A) on metallic cylinder (C) with and without straight boom (B). Cylinder height = 0.22 m, cylinder diameter = 0.20 m, wavelength = 0.36 m. (a) Configuration, xyz-coordinate system (centered at cylinder midpoint), and definition of angular variable ψ. (b) Radiation pattern without boom; monopole length = 0.12 m. (c) Radiation pattern for boom length of 0.09 m; monopole length = 0.08 m. (d) Radiation pattern for boom length of 0.44 m; monopole length = 0.08 m.

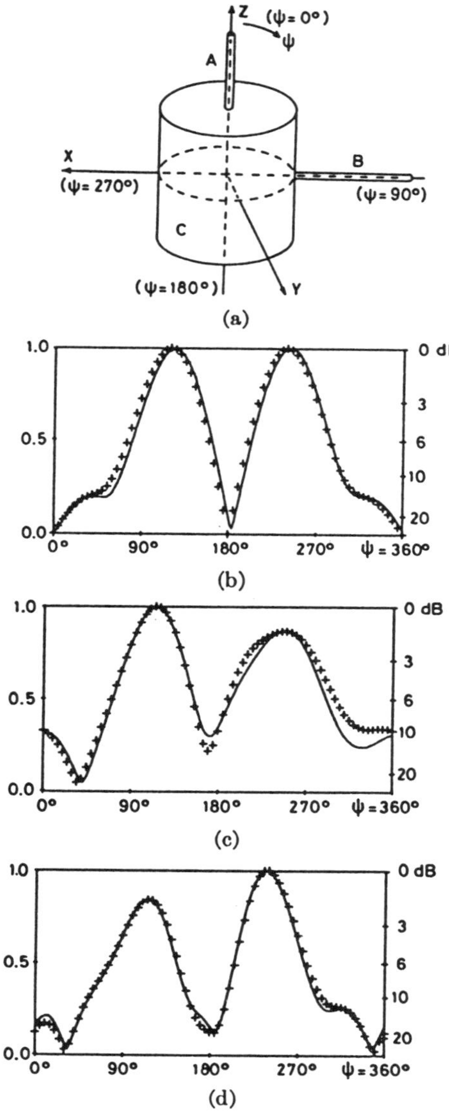

Fig. 5. *zx*-polarized patterns in the *zx* plane (field strength, normalized to maximum value of unity, as function of the angle ψ) for monopole antenna (A) on a metallic cylinder (C) with and without a straight boom (B). Cylinder height = 0.22 m, cylinder diameter = 0.20 m, monopole length = 0.12 m; wavelength = 0.48 m. (a) Configuration, *xyz*-coordinate system (centered at cylinder midpoint), and definition of angular variable ψ. (b) Radiation pattern without boom. (c) Radiation pattern for boom length of 0.18 m. (d) Radiation pattern for boom length of 0.44 m.

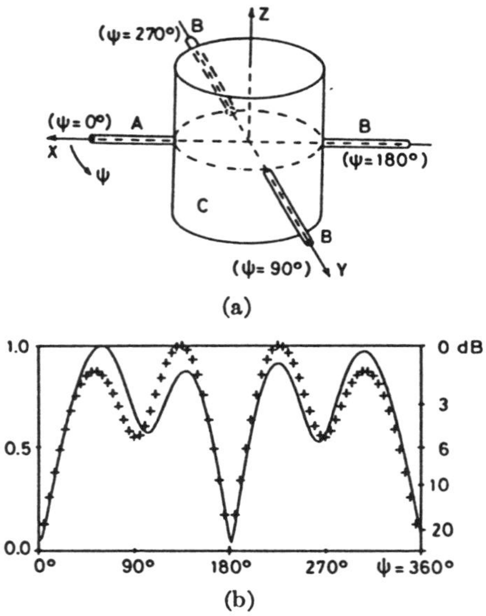

Fig. 6. *xy*-polarized radiation pattern in *xy* plane (field strength, normalized to maximum value of unity, as function of the angle ψ) for a monopole antenna (A) on metallic cylinder (C) with four straight booms (B). Cylinder height = 0.22 m, cylinder diameter = 0.20 m, monopole length = 0.12 m, boom length = 0.13 m, wavelength = 0.58 m. (a) Configuration, *xyz*-coordinate system (centered at cylinder midpoint), and definition of angular variable ψ. (b) Radiation pattern.

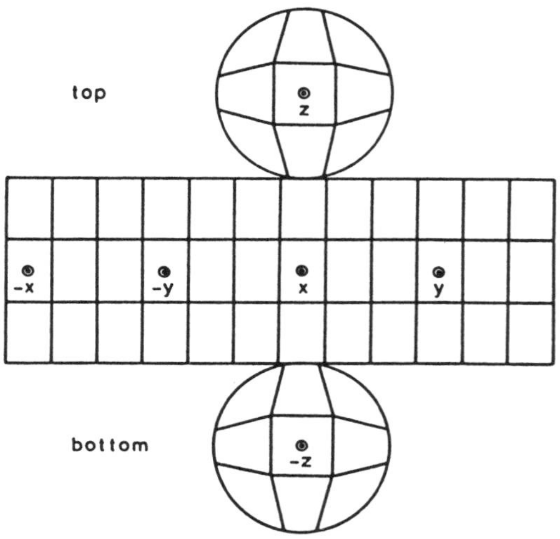

Fig. 7. Segmentation of cylinder.

patterns (solid line) for a number of representative configurations are shown together with the corresponding computer radiation patterns (crosses) in Figs. 4–6.

The small unsymmetries in the measured patterns in Figs. 4 and 6, where the configurations have complete symmetry with respect to the *xz* plane, give an impression of the role of the measurement inaccuracies. These were due to wall reflections in the Radio Anechoic Room where the patterns were taken.

In Fig. 4 the influence on the radiation pattern of adding booms diametrically opposite to an antenna on the curved surface of the cylinder, is shown. The undisturbed pattern is shown in Fig. 4(a), while patterns for booms of lengths $\lambda/4$ and $5\lambda/4$ are given in Figs. 4(b) and (c), respectively. It is seen that the radiation pattern in the *xy* plane is changed noticeably by the addition of booms. In all three cases the computations predict the measured behavior satisfactorily.

For the examples in Fig. 5 the antenna is placed at the top of the cylinder. The undisturbed pattern is shown in Fig. 5(a), while patterns for booms of lengths $\lambda/4$ and $3\lambda/4$ are given in Figs. 5(b) and (c), respectively. Also in these cases the measured results are satisfactorily predicted by the computations.

Fig. 6 demonstrates the numerical prediction of the radiation pattern of an antenna on a cylinder with three quarterwave booms.

No studies of the relative numerical convergence as a function of sampling density were made. The segmentation of body and wires were made with as many patches and

TABLE I

Configuration illustrated in figure	Number of surface patches	Number of wire segments	Run times (in seconds) on IBM 370/165				
			(T)[a]	(T)[b]	(T)[c]	(T)[d]	(T)[e]
4(a)	57	4	4	17	~0	21	10
4(b)	60	12	5	23	~0	28	19
4(c)	60	20	6	25	~0	31	22
5(a)	57	6	4	17	~0	21	10
5(b)	60	14	6	23	~0	29	18
5(c)	60	27	7	27	~0	34	25
6	66	24	8	34	~0	42	32

[a] Computation of L_M and L_M' for (4.1).
[b] Solution of (4.1).
[c] Computation of L_E and L_E' and solution of (4.4).
[d] Total time for computation of current distributions.
[e] Additional time for computation of radiation patterns.

segments as permitted by the available core storage in the computer (500k bytes) and the computations were verified by comparison with experiments. In all cases the number of surface patches was larger than the number used in [6] in an MFIE computation of surface currents on a sphere in a plane wave field.

The number of surface patches, wire segments, and run times for the seven examples described above were as given in Table I.

Other computational examples and a more detailed description of the computer program may be found in [19]. The location of booms and antennas in the cylinder version of the program is restricted to surface patches of quadrilateral shape. No boom or antenna should be placed directly on the edge of the cylinder or in the triangular shaped patches on the end surfaces of the cylinder shown in Fig. 7. No special treatment of the cylinder edges were attempted. Apparently, this is not of any practical importance for the computations, although the MFIE obviously is not valid for field points right on the edge.

VI. Conclusion

A theoretical formulation in terms of combined magnetic and electric field integral equations has been presented for antenna and scattering problems involving wires connected to a conducting body. Numerical computations using the method of moments for the solution have been made for a configuration consisting of a cylindrical body with straight wires attached to the surface. Comparison between the computations and experiments shows good agreement.

The formulation and computer program described in this paper seem suitable for solving electromagnetic problems in which thin wires are connected to a conducting body. Although the primary motivation for the work presented was the need for computerized prediction of radiation patterns for spacecraft VHF-monopole antennas, the technique may prove useful also for the examination of radiation from antennas on small aircrafts, portable radio sets, etc.

Acknowledgment

The authors wish to thank G. Paci of the European Space Research Organisation for suggesting the present investigation. The numerical computations were made at the NEUCC computing center at the Technical University of Denmark.

References

[1] F. V. Schultz, "The scattering of electromagnetic waves by perfectly reflecting objects of complex shape," School Elec. Eng., Purdue Univ., Indiana, Rep. TR-EE68-39, 1968.
[2] D. M. Bolle and M. D. Morganstern, "Monopole and conic antennas on spherical vehicles," *IEEE Trans. Antennas Propagat.*, vol. AP-17, pp. 477–84, July 1969.
[3] P. Bruscaglioni and A. Consortini, "Irraggiamento di un monopolo su una sfera conduttrice," *Alta Freq.*, vol. 36, pp. 224–26, 1967.
[4] F. M. Tesche and A. R. Neureuther, "Radiation patterns for two monopoles on a perfectly conducting sphere," *IEEE Trans. Antennas Propagat.*, vol. AP-18, pp. 692–94, Sept. 1970.
[5] E. K. Miller *et al.*, "Numerical analysis of aircraft antennas," in *Proc. Conf. Environmental Effects on Antenna Performance*, pp. 55–58, 1969.
[6] A. J. Poggio and E. K. Miller, "Integral equation solutions of three-dimensional scattering problems," in *Computer Techniques for Electromagnetics*, R. Mittra, Ed. Oxford: Pergamon, 1973, ch. 4.
[7] P. L. E. Uslenghi, "Computation of surface currents on bodies of revolution," *Alta Freq.*, vol. 39, pp. 213E–224E, Aug. 1970.
[8] G. A. Thiele and M. Travieso-Diaz, "Radiation of a monopole antenna on the base of a conical structure," in *Proc. Conf. Environmental Effects on Antenna Performance*, pp. 99–102, 1969.
[9] C. L. Bennett, A. M. Auckenthaler, and J. D. DeLorenzo, "Transient scattering by three-dimensional conducting surfaces with wires," in *1971 G-AP Int. Symp. Dig.*, pp. 349–351, Sept. 1971.
[10] J. E. Jones, J. H. Richmond, and T. G. Campbell, "A numerical method for obtaining radiation patterns of space shuttle annular slot antennas," *1972 G-AP Int. Symp. Dig.*, pp. 189–192, Dec. 1972.
[11] J. R. Mautz and R. F. Harrington, "Radiation and scattering from bodies of revolution," *Appl. Sci. Res.*, vol. 20, pp. 405–435, June 1969.
[12] R. Mittra, "How to use the H-integral equation for electrically thin antennas and scatterers," in *1972 USNC-URSI Spring Meeting Abst.*, 1972.
[13] R. Mittra, W. A. Davis, and D. V. Jamnejad, "An integral equation for plane wave scattering by thin plates," in *1972 USNC-URSI Fall Meeting Abst.*, pp. 93–94, Dec. 1972.
[14] N. C. Albertsen, J. E. Hansen, and N. E. Jensen, "Numerical prediction of radiation patterns for antennas mounted on spacecrafts," Conference on Aerospace Antennas, IEE Conference Publication Number 77, London, pp. 219–228, June 1971.
[15] R. F. Harrington, *Field Computation by Moment Methods.* New York: Macmillan, 1968.
[16] J. H. Richmond, "A wire-grid model for scattering by conducting bodies," *IEEE Trans. Antennas Propagat.*, vol. AP-14, pp. 782–86, Nov. 1966.
[17] R. E. Collin and F. J. Zucker, *Antenna Theory.* New York: McGraw-Hill, 1969, pt. I, ch. 2.
[18] E. L. Wachspress, "A rational basis for function approximation," in *Publ. Conf. Applications of Numerical Analysis.* Berlin, Germany: Springer-Verlag, 1971.
[19] N. C. Albertsen, J. E. Hansen, and N. E. Jensen, "Computation of spacecraft antenna radiation patterns," Lab. Electromag. Theory, Tech. Univ. Denmark, Lyngby, Denmark. Rep. R 108, June 1972; also Rep. ESRO CR-207, Sept. 1973.

A Hybrid Technique for Combining Moment Methods with the Geometrical Theory of Diffraction

GARY A. THIELE, SENIOR MEMBER, IEEE, AND THOMAS H. NEWHOUSE, STUDENT MEMBER, IEEE

Abstract—**A technique for combining moment methods with the geometrical theory of diffraction (GTD) is presented, which permits the application of the method of moments to a larger class of problems. The fundamental idea used to develop the hybrid technique is to modify the usual impedance matrix that characterizes, for example, a wire antenna such that a metallic body or discontinuity on that body is properly accounted for. It is shown in general that one can modify the impedance matrix for any basis and/or weighting functions if one can compute the correct modification to the impedance matrix element. The modification is readily accomplished using the geometrical theory of diffraction and/or geometrical optics. Several example problems are considered to illustrate the usefulness of the technique. First, the canonical problem of a monopole near a conducting wedge is investigated. Second, a monopole at the center of a four-sided and an eight-sided flat plate is considered. Impedance results for the latter case are in good agreement with measurements. Third, a monopole at the center of a circular disc is examined and compared with experimental measurements in the literature, and fourth, the problem of a monopole near a conducting step is solved and the dependence of the input impedance upon the step height shown.**

Manuscript received February 1, 1974; revised August 14, 1974. This work was supported in part by Contract N00014-67-A-0232-0018 between the Office of Naval Research and the Ohio State University Research Foundation. A portion of this work was presented at the 1973 URSI/G-AP Symposium, Boulder, Colo., August 21–24.

G. A. Thiele is with the Ohio State University ElectroScience Laboratory, Department of Electrical Engineering, Columbus, Ohio 43212.

T. H. Newhouse was with the Ohio State University ElectroScience Laboratory, Department of Electrical Engineering, Columbus, Ohio. He is now with the Department of Defense, Fort George, Md.

I. INTRODUCTION

THE ELECTROMAGNETIC characterization of wire antennas on or near a three-dimensional metallic surface by the method of moments is significantly limited by computer storage [1], [2]. This is true whether one uses a wire-grid [3] representation or a surface patch [4] representation of the surface. Thus we may refer to the method of moments as a low-frequency technique since its practical use is generally restricted to bodies that are not large in terms of the wavelength. On the other hand, geometrical optics and the geometrical theory of diffraction (GTD) [5], [6] are applicable to bodies that are arbitrarily large in the electrical sense and may thus be referred to as high frequency techniques. Both the method of moments (MM) and GTD are powerful computational methods in their own right. This is due in large measure to the flexibility inherent in each method that permits application to a wide range of radiation and scattering problems.

It is the purpose of this paper to present a new method, which will combine these two computational methods into a hybrid technique. The hybrid technique is developed by considering the general problem of how to extend moment methods to include a class of problems wherein a three-dimensional body, on or near which is located a radiating element, may be arbitrarily large. Thus in

Reprinted from *IEEE Trans. Antennas Propagat.*, vol. AP-23, no. 1, pp. 62–69, January 1975.

developing the hybrid technique in the next section, the approach is to modify the impedance matrix, which in the usual moment method approach characterizes only a portion of the problem (e.g., the radiating element), to properly account for the remainder of the problem to be solved. This is to be contrasted with other work being conducted at this time that considers the extension of GTD via moment methods [7].

II. THE HYBRID TECHNIQUE

In a moment method solution, we expand the current J in a series of basis functions $J_1, J_2, J_3, \cdots$ on the surface of interest and defined in the domain of an operator L. That is,

$$J = \sum_{n=1}^{N} I_n J_n. \tag{1}$$

Using the linearity of the operator, and forming the inner product with a set of weighting functions $W_1, W_2, W_3, \cdots$ in the domain of L, we write

$$\sum_{n=1}^{N} I_n \langle W_m, L(J_n) \rangle = \langle W_m, E^i \rangle. \tag{2}$$

In reality, the preceding expression represents the mth row of equations in a system of N such equations while the quantity $L(J_n)$ represents the electric field from the nth basis function of unit amplitude. In the usual moment method matrix notation, (2) is compactly represented by

$$[Z](I) = (V) \tag{3}$$

where the elements of the generalized impedance matrix are given by

$$Z_{mn} = \langle W_m, L(J_n) \rangle. \tag{4}$$

The symmetric inner product here is defined such that the following three axioms are satisfied:

$$\langle \alpha J_1 + \beta J_2, E \rangle = \alpha \langle J_1, E \rangle + \beta \langle J_2, E \rangle, \quad \alpha \text{ and } \beta \text{ are scalars} \tag{5}$$

$$\langle J, E \rangle = \langle E, J \rangle \tag{6}$$

$$\langle J^*, J \rangle > 0; \quad \text{but} \quad \langle J^*, J \rangle = 0, \quad \text{if and only if } J = 0. \tag{7}$$

Using the first two axioms (i.e., (5) and (6)) it is easily shown that

$$\langle J, aE_1 + bE_2 \rangle = a\langle J, E_1 \rangle + b\langle J, E_2 \rangle \tag{8}$$

where a and b are complex scalars.

Suppose that in (8) the quantity aE_1 represents $L(J_n)$ in (4). That is, $L(J_n)$ is the usual calculation made in a strictly moment method solution. The quantity bE_2 in (8) represents an additional field contribution to Z_{mn} in (4) that is also due to J_n but arrives at observation point (or region) m by a physical process not directly accounted for in the moment method formulation. Thus we may write for the new impedance matrix element

$$Z_{mn}' = \langle W_m, L(J_n) + bL(J_n) \rangle, \qquad a = 1,\ b = b(m,n) \tag{9}$$

or

$$Z_{mn}' = \langle W_m, L(J_n) \rangle + \langle W_m, bL(J_n) \rangle \tag{10}$$

or

$$Z_{mn}' = Z_{mn} + Z_{mn}^g \tag{11}$$

where the superscript g denotes that Z_{mn}^g is an additional term added to, in general, each impedance matrix element due to a physical process g that also directs energy from the nth basis current function to the mth observation point (or region). Thus we may modify (3) such that

$$[Z'](I') = (V) \tag{12}$$

where $[Z']$ is the generalized impedance matrix properly modified to account for physical processes not accounted for by the moment method formulation, which is assumed to represent only a portion of the problem. Note that modification of (V) is neglected since it has been found to be numerically insignificant in this paper. Solution of (12) can then be written as

$$(I') = [Z']^{-1}(V) \tag{13}$$

where (I') is the current on, for example, an antenna in the presence of other scattering mechanisms that may be accounted for by either geometrical optics techniques or GTD. In the next section we will briefly review some of the current advances in GTD before considering several problems in Section IV that have been treated successfully by the hybrid technique.

III. DIFFRACTION THEORY

Historically, the problem of straight edge diffraction by a perfectly conducting wedge was first solved by Sommerfeld [8]. Later, Pauli [9] introduced the V_B function as a practical formulation in the solution for the fields diffracted by a perfectly conducting wedge. The V_B function is related to the term $D(\beta)$, which Keller [11] has designated as a diffraction coefficient.

More recently, Hutchins and Kouyoumjian [11] have presented a formulation that yields better accuracy than does the Pauli solution in the regions near the incident and reflected shadow boundaries, particularly when $r < \lambda$. Their formulation can be conveniently found in [12, Sect. II] or in [6].

It is the introduction of the distance parameter L by Pathak and Kouyoumjian that permits near-zone to near-zone diffraction to be considered. Experience has thus far shown that accurate results may be obtained for $kL > 1$. Therefore, this is the formulation that will be used in the following section.

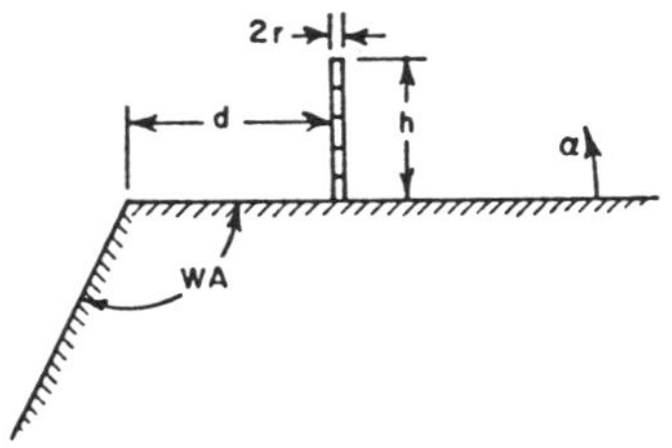

Fig. 1. Monopole on conducting wedge.

IV. EXAMPLES

A. Monopole Near a Wedge

Initially, to combine the method of moments and GTD into a hybrid technique, we have considered the canonical problem of a monopole near a perfectly conducting wedge as shown in Fig. 1. If we describe the monopole on an infinite ground plane strictly by the moment method matrix representation given in (3), then for the monopole near the conducting wedge we utilize (12) where in (11) the term Z_{mn}^{g} is obtained by considering that energy radiated by the nth basis function on the monopole that is diffracted by the wedge to the mth observation point or region. In our work here we have employed pulse basis functions and point-matching wherein the testing functions are delta functions [2]. However, as indicated in Section II, the choice of basis and testing functions is not so restricted.

To calculate Z_{mn}^{g} we compute the electric field from the nth pulse basis function incident upon the edge of the wedge at the stationary point. Taking that component of the electric field perpendicular to the edge and to the direction of propagation of the incident field, we then compute the energy diffracted to the observation point at the center of the mth segment on the monopole. The component of this field tangential to segment m is the term Z_{mn}^{g} of (11) since we are employing delta weighting functions. To compute the diffracted field, we use the formulation of Kouyoumjian and Pathak discussed in the previous section for the case of spherical wave incidence.

Shown in Fig. 2(a) is a calculated curve for the input resistance of a quarter wavelength monopole a distance d from the edge of a perfectly conducting wedge. We note that the resistance oscillates about the value for a quarter wavelength monopole on an infinite ground plane and also that the amount of variation is relatively small being only a few ohms. A similar curve is shown in Fig. 2(b) for the input reactance. Data for both curves was obtained directly from (13) without the need for any *a priori* knowledge of the current distribution or the terminal current value.

In an effort to find a case for which experimental confirmation could readily be obtained, a monopole at the center of a square plate was investigated. In this problem there are four stationary points, one on each edge, that contribute to the Z_{mn}^{g} term of (11). Diffraction by the four vertices was neglected. A calculated curve for the input resistance of a quarter wavelength monopole on a square plate is shown in Fig. 3. As in the previous case the input resistance oscillates about its value for the infinite ground plane case, but the amount of variation is substantially greater. A similar curve was obtained for the input reactance. While fair agreement between theory and experiment was obtained for the square plate case, it was decided that an octagonal plate would be better for two reasons. First, there would be either stationary points, one on each edge and thus the sensitivity of the input impedance to the plate size would be greater, and second the vertex diffraction would be less since the angle of the vertex would be rather large (i.e., 135° in this case) [13]. A comparison between theory and experiment for a monopole at the center of an octagonal plate showed good agreement for both the input resistance and input reactance.

One objective of the experimental verification for the octagonal plate was to determine how small d/λ can become before inaccurate results are obtained. It was apparent from the measurements that $d > 0.25\lambda$ will yield accurate results. We hypothesize that the hybrid technique will probably permit d to approach $\lambda/8$ but further work is necessary to verify this.

It was thought that perhaps the accuracy of the calculations could be increased if double diffractions were included. That is, energy incident upon an edge that is diffracted to an opposite edge and then to the monopole is also included in the calculation of the Z_{mn}^{g} term in (11). However, from the calculations it was concluded that, at least for $d > 0.25\lambda$, the effect of double diffractions is negligible.

Thus far the discussion has centered upon the calculation of input impedance. Obviously, if one can accurately compute the input impedance, then quite accurate far-field information can readily be obtained too. For example, in the case of a monopole near a single wedge, as in Fig. 1, there may be as many as three contributors to the far-field. First, there is direct source radiation except in the shadow region. Second, there is the reflected field, which is most conveniently accounted for by using the image in the horizontal surface. Third, there is the diffracted field that contributes in all regions and, of course, is the only source of radiation into the shadow region. A typical far-field pattern is shown in Fig. 4. Note that for $\alpha = 90°$ the field does not go to zero as would be the case if the wedge were not present.

B. Monopole on a Circular Disc

If we wish to investigate a circular ground plane, we obviously cannot extend the method employed for the octagonal plate to an n-sided polygon (n large) since the principle of stationary phase will be violated and the method will fail. The difficulty can be eliminated, however, by replacing the edge of the disc with an equivalent

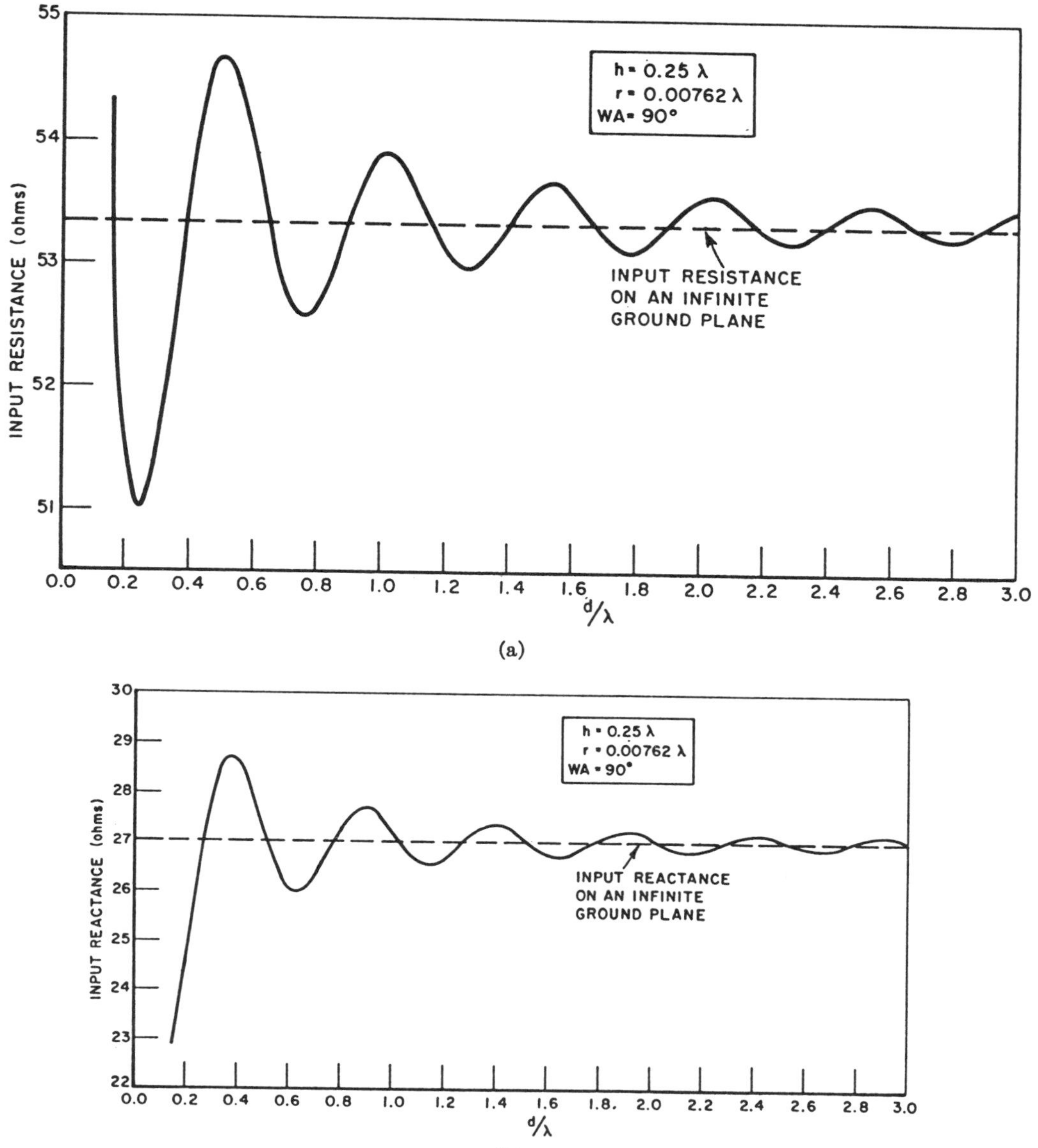

Fig. 2. (a) Input resistance of monopole on conducting wedge versus distance d as shown in Fig. 1. (b) Input reactance of monopole on conducting wedge versus distance d as shown in Fig. 1.

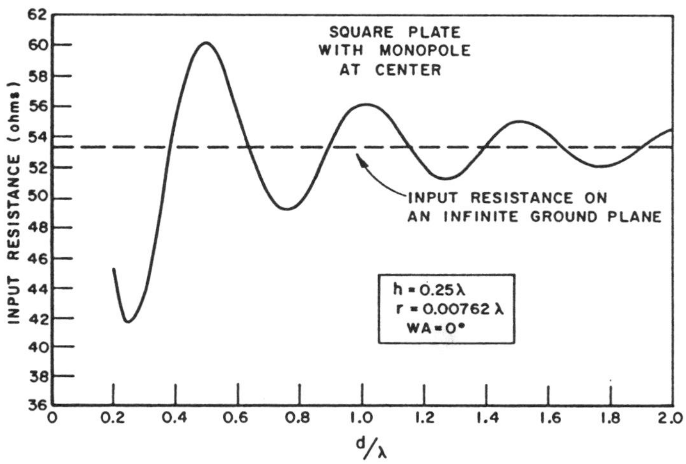

Fig. 3. Input resistance of monopole on square plate versus monopole distance to edges.

magnetic ring current M given by

$$M = -2E_\theta \left(\frac{ss'}{s+s'}\right)^{1/2} \exp\left[j\left(\frac{ss'}{s+s'}\frac{2\pi}{\lambda} - \frac{\pi}{4}\right)\right] \cdot [V_B(\phi - \phi') + V_B(\phi + \phi')]. \quad (14)$$

Equation (14) gives the equivalent magnetic current used to calculate the field at the segment at s due to the current at s' as indicated in Fig. 5. Note that an equivalent magnetic ring current must be calculated for each choice of s and s'.

It is useful for us to break up the equivalent magnetic ring current of Fig. 5 into differential elements dl so that the observation point is in the far-field of each element even though it may be in the near-field of the total ring current [14].

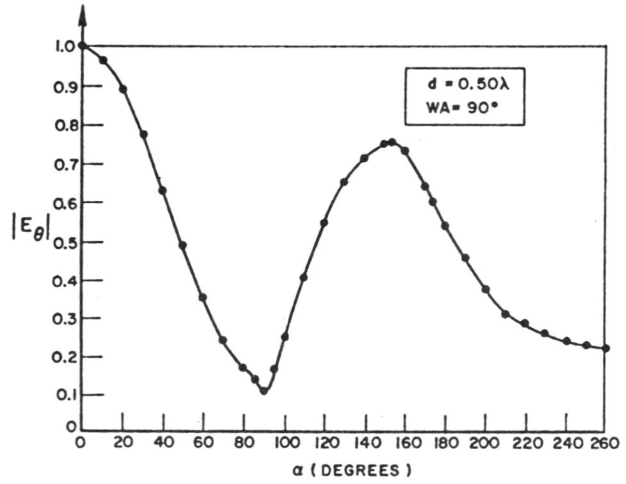

Fig. 4. Normalized far-field pattern of quarter wave monopole near conducting edge.

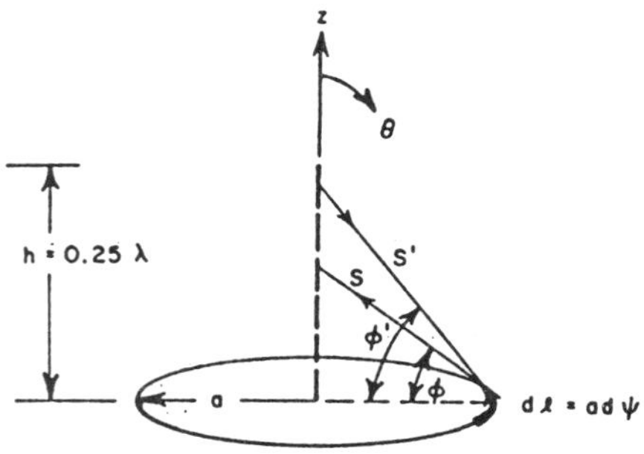

Fig. 5. Segmented monopole encircled by magnetic ring current.

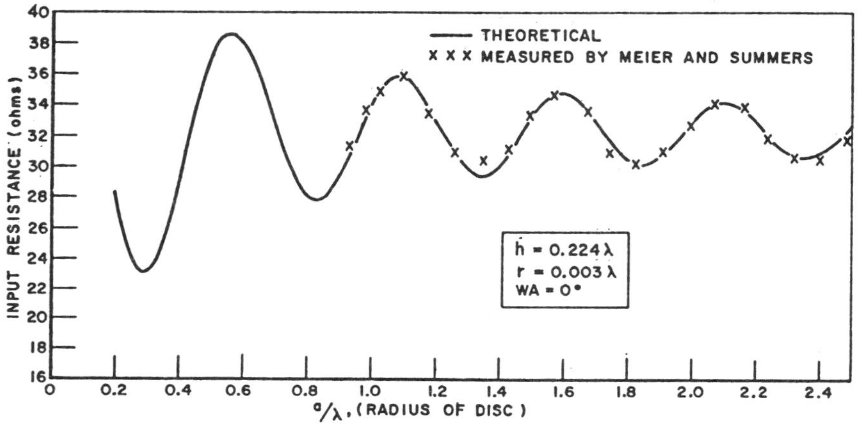

(a)

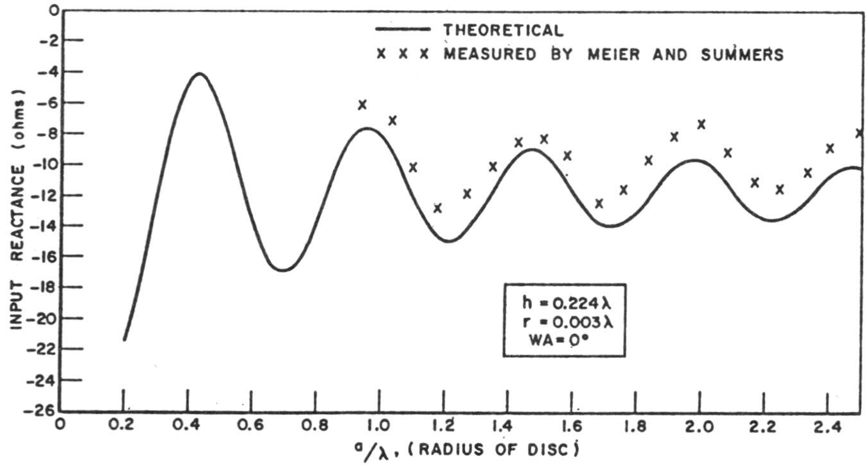

(b)

Fig. 6. (a) Theoretical and experimental input resistance of monopole at center of circular disc. (b) Theoretical and experimental input reactance of monopole at center of circular disc.

The electric field in a plane perpendicular to an element dl is given by [15]

$$dE = \frac{Mdl}{4\pi}\left[\frac{j\omega}{cr} + \frac{1}{r^2}\right] \exp\,(-jkr) \tag{15}$$

where $r = (a^2 + z^2)^{1/2}$. Letting $dl = a d\psi$, where ψ is the azimuth angle, taking only the z component at the monopole, and integrating over the range $\psi = 0$ to $\psi = 2\pi$ yields

$$E_z = \frac{Ma^2}{2r}\left[\frac{jk}{r} + \frac{1}{r^2}\right] \exp\,(-jkr). \tag{16}$$

The value for E_z is the term Z_{mn}^g that is added to the impedance element obtained for a monopole on an infinite ground plane. This process gives the modified impedance element needed to calculate the modified currents (and hence input impedance) of a monopole on the finite circular ground plane.

Equation (16), without the near-field term, is the one used by Green [16] in his work with a monopole on the base of a large cone. However, in our work we need the near-field term.

Figs. 6(a) and 6(b) show a comparison between calculations made with the equivalent magnetic ring current and measurements made by Meier and Summers [17], [18] for a monopole of length 0.224λ and radius 0.003λ on a circular ground plane of varying radius. It is apparent that the correct variation is accurately predicted for both the real and imaginary parts of the input impedance. For the input resistance the agreement between their measurements and the theory is excellent. For the input reactance the agreement is very good but there is a slight shift in the calculated curve when compared to the measurements. The amount of this shift is sufficiently small that it can be attributed to the usual problems associated with modeling the region in proximity to the driving point. The impedance predicted here is also in agreement with that predicted by Green [16] whose results are based on the measurements by Meier and Summers.

In a partial attempt to resolve the accuracy of our calculations for smaller values of a/λ, measurements were made with a quarter wavelength long monopole of 1/16-in radius on a circular plate. These results are shown in Figs. 7(a) and 7(b). The agreement between theory and experiment for the input resistance and input reactance is fairly good. Most of the difference is attributed to mismatches in the measuring system. In our calculations we use a magnetic frill representation for the coaxial aperture at the base of the monopole [19], which closely models the actual physical geometry and is known to permit prediction of the input impedance as accurately as is meaningful [2].

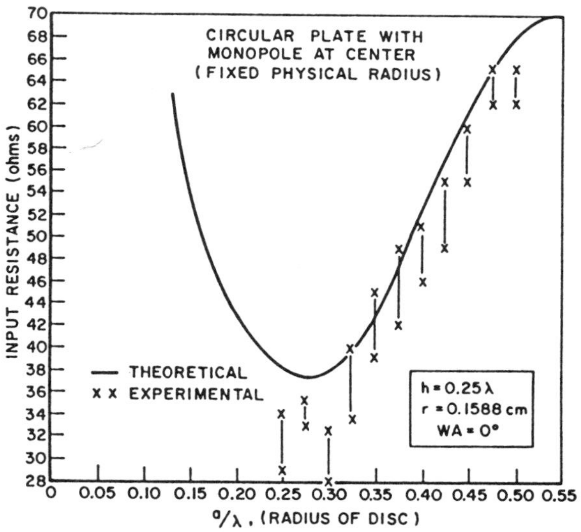

(a)

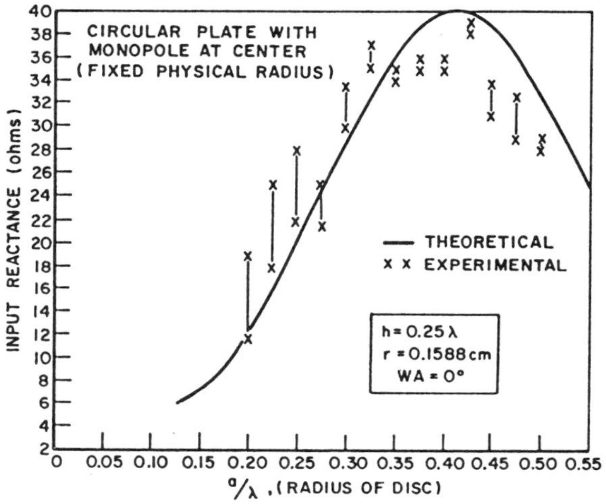

(b)

Fig. 7. (a) Theoretical and experimental input resistance of monopole at center of circular disc. (b) Theoretical and experimental input reactance of monopole at center of circular disc.

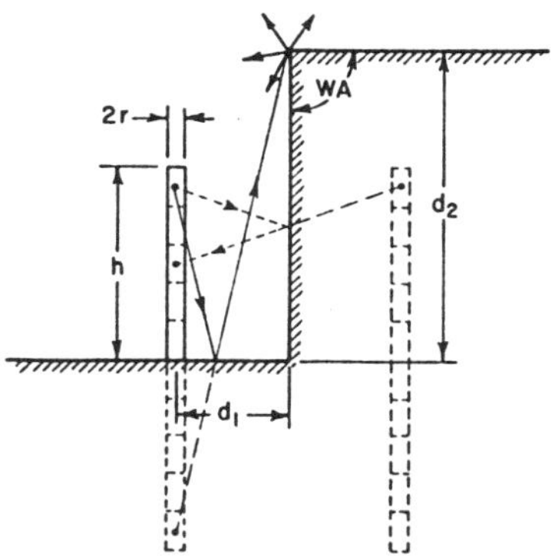

Fig. 8. Monopole near conducting step showing partial use of images.

C. *Monopole Near a Conducting Step*

Consider the situation shown in Fig. 8 where a monopole of height h is a distance d_1 away from a vertical conducting step. To properly determine the Z_{mn}^g term in (11), it is necessary to determine all the various combinations of reflections that can occur for rays emanating from the monopole and reflecting back to it as well as the diffraction from the top edge of the step. Since the vertical wall

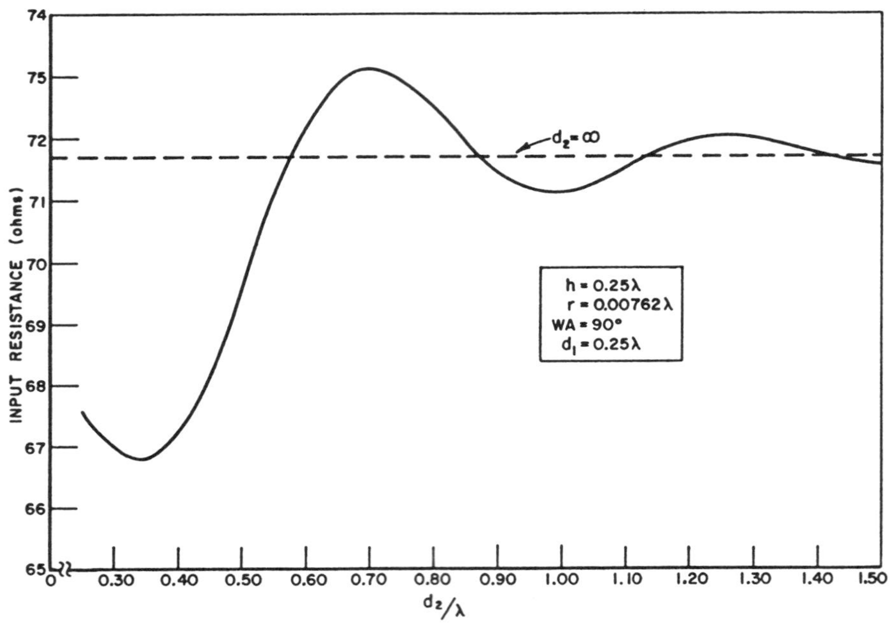

(a)

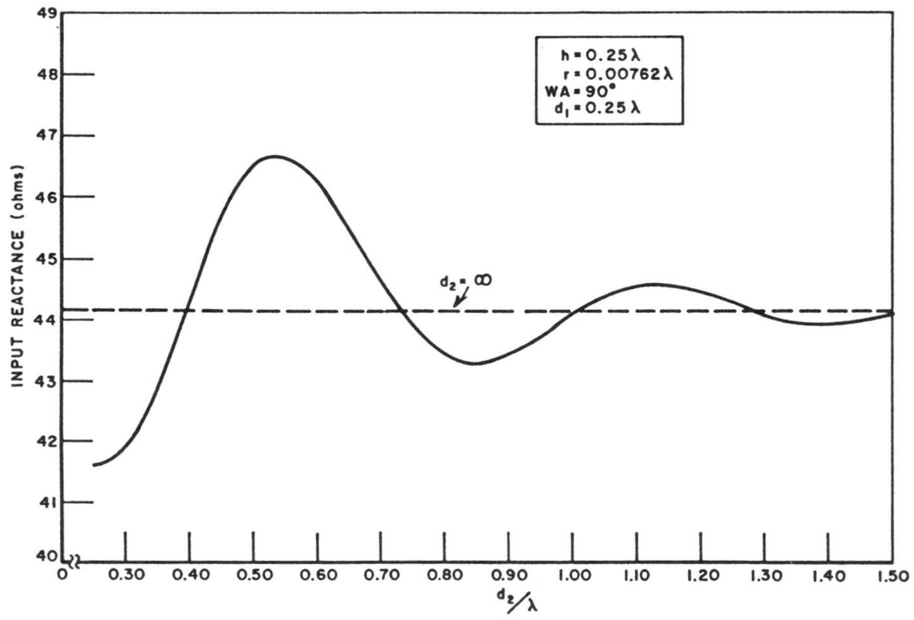

(b)

Fig. 9. (a) Input resistance of quarter wave monopole versus step height. (b) Input reactance of quarter wave monopole versus step height.

is at a right angle to the lower horizontal surface, there will be no diffraction from the interior wedge and all the reflections can most conveniently be accounted for by imaging the monopole into the horizontal ground plane and then imaging the resulting dipole into the plane of the vertical wall.

Shown in Fig. 8 are two example situations that depict the utilization of the images. Considering the uppermost segment of the monopole to be the source segment, one set of rays shows the use of the image in the horizontal surface to calculate reflected-diffracted energy reaching the segments of the monopole. The other set of rays shows the use of the image in the vertical wall to calculate singly reflected energy. In the calculated results that follow, all combinations of singly reflected, doubly reflected, diffracted, diffracted-reflected, reflected-diffracted, and reflected-diffracted-reflected rays are taken into account. All rays that involve combinations of double (or higher order) diffractions are negligible.

Fig. 9 shows the calculated input impedance for a quarter wavelength monopole a quarter wavelength away from a vertical wall whose height is $d_2 > 0.25\lambda$. As d_2 increases the impedance oscillates about the value for the case where $d_2 = \infty$. The value of the input impedance for the $d_2 = \infty$ case was verified by an independent computer program employing the piecewise sinusoidal basis [20]. The results of Figs. 9(a) and 9(b) show that as the diffracting edge recedes from the vicinity of the monopole, its effect upon the input impedance rapidly diminishes. Although we have not shown results for the case where the step height is less than the height of the monopole, the same method could be used to investigate such situations.

V. SUMMARY AND CONCLUSIONS

In presenting the hybrid technique we have attempted to show that the idea of modifying the generalized impedance matrix is a useful one. In combining moment methods with GTD we have proceeded from the philosophical viewpoint of extending the method of moments via GTD. In so doing we have shown that modifying the impedance matrix to account for diffraction effects (or geometrical optics effects) enables one to accurately treat a larger class of problems than could be treated by moment methods alone. A potential advantage of such an approach is that in many problems the impedance matrix may be small enough to calculate data at sufficiently many frequencies to obtain time domain solutions via Fourier transformation; whereas, attempting to do so using strictly moment methods may be economically prohibitive or impossible due to the increasing electrical size of the problem with increasing frequency.

An alternative interpretation of the hybrid method is also possible. That is, the procedure employed can be viewed as using GTD to obtain an approximation to the exact dyadic Green's function needed. For example, for the octagonal plate problem, the single and double diffraction terms represent the leading terms in a series expression for the Green's function of the plate. In this context, moment method solutions using other than free space Green's functions are not completely without precedent [21].

While the hybrid method possesses many of the advantages inherent in both MM and GTD, it also has some of the limitations peculiar to each. For example, as in the usual MM problem, one can treat arbitrary configurations of wire antennas (or slot antennas) taking into account lumped loading, finite conductivity, etc., and obtain accurate impedance data and current distributions. Naturally, one still must take the usual precaution of using a sufficient number of basis functions to assure convergence. On the other hand, as in the usual GTD problem, one must take care that the antenna is not too close to a source of diffraction (e.g., $d > 0.2\lambda$). Furthermore, the study described in this paper has mainly addressed itself to the determination of the input impedance of the antenna and hence the near-fields at its location. The determination of the currents and near-fields elsewhere, particularly near a source of diffraction, is a subject for future investigation.

Nevertheless, the examples considered in this paper illustrate well the advantages of combining MM and GTD in a hybrid method since the advantages intrinsic to each technique can be effectively utilized to treat new problems or to treat old ones in a more efficient and accurate manner.

ACKNOWLEDGMENT

The authors wish to express their gratitude to Prof. R. G. Kouyoumjian and R. J. Marhefka at the Ohio State University ElectroScience Laboratory for many stimulating discussions. Constructive reviews of the manuscript by Profs. C. H. Walter and R. J. Garbacz are also appreciated.

REFERENCES

[1] R. F. Harrington, *Field Computation by Moment Methods.* New York: Macmillan, 1968.

[2] G. A. Thiele, "Wire antennas," in ch. 2 of *Computer Techniques for Electromagnetics*, R. Mittra, Ed. London: Pergamon, 1973.

[3] J. H. Richmond, "A wire-grid model for scattering by conducting bodies," *IEEE Trans. Antennas Propagat.*, vol. AP-14, pp. 782–786, Nov. 1966.

[4] J. H. Richmond and N. N. Wang, "Sinusoidal reaction formulation for scattering by conducting bodies of arbitrary shape," presented at the 1973 URSI meeting, Boulder, Colo.

[5] R. G. Kouyoumjian, "Asymptotic high-frequency methods," *Proc. IEEE*, vol. 53, pp. 864–876, Aug. 1965.

[6] ——, Notes for a "Short course on application of GTD and numerical techniques to the analysis of electromagnetic and acoustic radiation and scattering," Ohio State University, Columbis, Sept. 1973.

[7] W. D. Burnside, C. L. Yu, and R. J. Marhefka, "A technique to combine the geometrical theory of diffraction and the moment method," presented at the 1973 URSI meeting, Boulder, Colo.

[8] A. Sommerfeld, *Optics.* New York: Academic, 1954, pp. 245–265.

[9] W. Pauli, "On asymptotic series for functions in the theory of diffraction of light," *Phys. Rev.*, vol. 54, pp. 924–931, Dec. 1, 1968.

[10] J. B. Keller, "Geometrical theory of diffraction," *J. Opt. Soc. Amer.*, vol. 52, pp. 116–130, Feb. 1962.

[11] D. L. Hutchins and R. G. Kouyoumjian, "A new asymptotic solution to the diffraction by a wedge," at *URSI 1967 Spring Meeting*, Ottawa, Ont., Canada, pp. 154–155.

[12] W. D. Burnside, R. J. Marhefka, and C. L. Yu, "Roll-plane analysis of on-aircraft antennas," *IEEE Trans. Antennas Propagat.*, vol. AP-21, pp. 780–786, Nov. 1973.

[13] R. G. Kouyoumjian, private communication.

[14] C. H. Walter, *Traveling Wave Antennas.* New York: Dover, 1970, p. 41.

[15] J. D. Kraus, *Antennas.* New York: McGraw-Hill, 1950, p. 159.

[16] H. E. Green, "Impedance of a monopole on the base of a large cone," *IEEE Trans. Antennas Propagat.*, vol. AP-17, pp. 703–706, Nov. 1969.

[17] A. S. Meier and W. P. Summers, "Measured impedance of vertical antennas over finite ground planes," *Proc. IRE*, vol. 37, pp. 609–616, June 1949.

[18] D. W. Little, D. R. Rhodes, W. P. Summers, and A. S. Meier, "Measured impedance of vertical antennas over finite ground planes," Antenna Laboratory, Department of Electrical Engineering, Ohio State University, Columbus, Rep. 233-3, Oct. 1, 1946.

[19] L. L. Tsai, "A numerical solution for the near and far fields of an annular ring of magnetic current," *IEEE Trans. Antennas Propagat.*, vol. AP-20, pp. 569–576, Sept. 1972.

[20] J. H. Richmond and N. H. Geary, "Mutual impedance between coplanar-skew dipoles," *IEEE Trans. Antennas Propagat.* (Commun.), vol. AP-18, pp. 414–416, May 1970.

[21] F. M. Tesche and A. R. Neureuther, "Radiation patterns for two monopoles on a perfectly conducting sphere," *IEEE Trans. Antennas Propagat.* (Commun.), vol. AP-18, pp. 692–694, Sept. 1970.

Wire Antennas in the Presence of a Dielectric/Ferrite Inhomogeneity

EDWARD H. NEWMAN, MEMBER, IEEE, AND P. TULYATHAN

Abstract—A moment method solution for treating thin-wire antennas in the presence of an arbitrary dielectric and/or ferrite inhomogeneity is presented. The wire is modeled by an equivalent surface current density, and the dielectric/ferrite inhomogeneity is modeled by equivalent volume polarization currents. The conduction currents on the wire and the polarization currents in the dielectric/ferrite inhomogeneity are treated as independent unknowns and determined in the moment method solution. The method is applied to the problem of a loop antenna loaded with dielectric or ferrite. Numerical results are presented, and are in good agreement with measurements and previous calculations.

I. INTRODUCTION

THE INTERACTION of electromagnetic fields with metallic as well as dielectric and/or ferrite inhomogeneities is a problem of practical interest. Numerous techniques have been employed by various authors to treat antenna or scattering problems with dielectric/ferrite inhomogeneities [1]–[5]. Each of the above analyses treated a specific or very limited class of antennas or inhomogeneities. Often the geometries were two-dimensional or coincided with a separable coordinate system. It is desirable to have a single method to analyze a general class of problems where the antennas, scatterers, and inhomogeneities can be of arbitrary shape, and the inhomogeneities can be dielectric and/or ferrite.

The purpose of this paper is to present a moment method solution [6] to the problem of thin-wire antennas and scatterers in the presence of a dielectric/ferrite body. The analysis has the advantage that it is applicable to a wide variety of antenna, scatterer, and dielectric/ferrite geometries. It is sufficiently general to treat isotropic, lossy, and inhomogeneous bodies. The dielectric/ferrite inhomogeneities are modeled by equivalent volume polarization currents, and the wire is modeled by equivalent surface currents. The complex magnitudes of these currents are determined in the course of the moment method solution. The electrical sizes of the antennas, scatterers, and dielectric/ferrite inhomogeneities are limited by the finite computer storage.

II. THEORY

Following the development of Richmond [7] and Newman [8], the moment method solution for thin-wire antennas and scatterers in the presence of a linear isotropic dielectric and/or ferrite inhomogeneity will now be developed. Fig. 1(a) shows the geometry to be considered. The impressed sources ($\mathbf{J}_i$, $\mathbf{M}_i$) are confined to the volume V_1, and radiate the field (E, H) in the presence of two inhomogeneities. The first inhomogeneity is a thin-wire structure in the volume V_w and enclosed by the surface S_w. The second inhomogeneity is the dielectric/ferrite body in the volume V_2. The permittivity and permeability of this inhomogeneity are defined by the parameters (μ, ϵ), which may be complex functions of position. Although for simplicity not shown here, μ and ϵ could be considered as tensor quantities. This would permit the treatment of anisotropic inhomogeneities. The ambient medium has parameters (μ_0, ϵ_0). All sources and fields are considered to be time harmonic, and the $e^{j\omega t}$ time dependence will be suppressed.

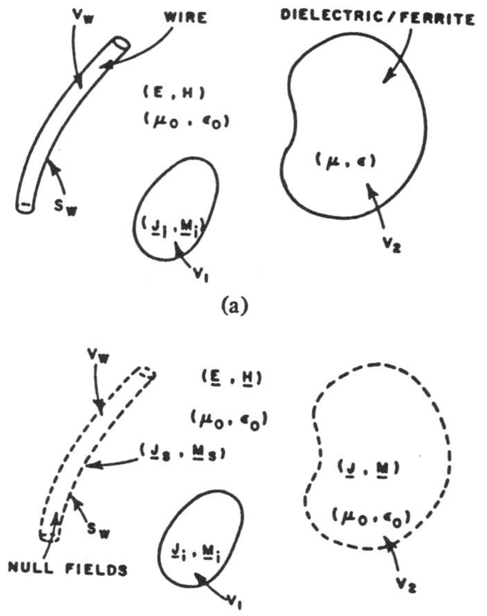

Fig. 1. (a) Wire antenna in presence of dielectric and/or ferrite inhomogeneity. (b) Equivalent problem.

The first step in the solution will be to replace the two inhomogeneities by equivalent sources. Employing the surface equivalence principle of Schelkunoff [9], the wire is removed and the following surface current densities are introduced on the surface S_w:

$$\mathbf{J}_s = \hat{n} \times \mathbf{H} \tag{1}$$

$$\mathbf{M}_s = \mathbf{E} \times \hat{n} \tag{2}$$

where $\hat{n}$ is the outward directed normal vector on S_w. By defining ($\mathbf{J}_s$, $\mathbf{M}_s$) as in (1) and (2), the total field inside the wire is zero. Next, the dielectric/ferrite inhomogeneity is replaced by the ambient medium and the equivalent volume polarization currents

$$\mathbf{J} = j\omega(\epsilon - \epsilon_0)\mathbf{E} \tag{3}$$

$$\mathbf{M} = j\omega(\mu - \mu_0)\mathbf{H} \tag{4}$$

which are confined to the volume V_2.

Manuscript received February 13, 1977; revised September 14, 1977. This work was supported in part between the U.S. Army Research Office, Research Triangle Park, NC, and the Ohio State University Research Foundation under Grant DAAG29-76-G-0067.

The authors are with the ElectroScience Laboratory, Department of Electrical Engineering, Ohio State University, Columbus, OH 43212.

Reprinted from *IEEE Trans. Antennas Propagat.*, vol. AP-26, no. 4, pp. 587–592, July 1978.

The equivalent problem is shown in Fig. 1(b). Here, in the homogeneous medium (μ_0, ϵ_0), the sources $(\mathbf{J}_i, \mathbf{M}_i)$, $(\mathbf{J}_s, \mathbf{M}_s)$, and $(\mathbf{J}, \mathbf{M})$ radiate the field $(\mathbf{E}, \mathbf{H})$ exterior to S_w and the field (0, 0) interior to S_w. We use the notation that the sources $(\mathbf{J}_i, \mathbf{M}_i)$, $(\mathbf{J}_s, \mathbf{M}_s)$, $(\mathbf{J})$, and $(\mathbf{M})$, radiating in the ambient medium (μ_0, ϵ_0), produce the fields $(\mathbf{E}_i, \mathbf{H}_i)$, $(\mathbf{E}_s, \mathbf{H}_s)$, $(\mathbf{E}^J, \mathbf{H}^J)$, and $(\mathbf{E}^H, \mathbf{H}^H)$, respectively. These relationships are summarized in Table I.

Let us consider the wire to have a circular cross section. At each point on the wire we define a local right-handed orthogonal coordinate system with unit vectors $(\hat{n}, \hat{\phi}, \hat{l})$ where $\hat{n}$ is the outward normal vector to S_w, $\hat{l}$ is directed along the wire axis, and

$$\hat{\phi} = \hat{l} \times \hat{n}. \tag{5}$$

If the wire radius $a \ll \lambda$, then the surface current density on the wire structure can be approximated by the "thin-wire approximation"

$$\mathbf{J}_s(l) = \frac{\hat{l}I(l)}{2\pi a} = \frac{\mathbf{I}(l)}{2\pi a} \tag{6}$$

where $I(l)$ is the total current (conduction plus displacement).

The reaction integral equation (RIE) for $I(l)$ is obtained by placing a test source with current $(\mathbf{J}_m, \mathbf{M}_m)$ in V_w, and noting that since a null field exists in V_w, $(\mathbf{J}_m, \mathbf{M}_m)$ will have zero reaction with the sources $(\mathbf{J}_i, \mathbf{M}_i)$, $(\mathbf{J}_s, \mathbf{M}_s)$, and $(\mathbf{J}, \mathbf{M})$. Denoting $(\mathbf{E}^m, \mathbf{H}^m)$ as the fields radiated by $(\mathbf{J}_m, \mathbf{M}_m)$ in the ambient medium (μ_0, ϵ_0) and Z_s as the wire surface impedance for exterior excitation, the RIE is

TABLE I

Source Current	Field of the source radiating in the ambient medium (μ_0, ϵ_0)
$(\underline{J}_s, \underline{M}_s)$	$(\underline{E}_s, \underline{H}_s)$
$\underline{J}$	$(\underline{E}^J, \underline{H}^J)$
$\underline{M}$	$(\underline{E}^H, \underline{H}^H)$
$(\underline{J}_i, \underline{M}_i)$	$(\underline{E}_i, \underline{H}_i)$
$(\underline{J}_m, \underline{M}_m)$	$(\underline{E}^m, \underline{H}^m)$
$\underline{F}_n$	$(\underline{E}_{sn}, \underline{H}_{sn})$
$\underline{G}_n$	$(\underline{E}_n^J, \underline{H}_n^J)$
$\underline{Q}_n$	$(\underline{E}_n^H, \underline{H}_n^H)$

$$-\int_0^L I(l)(E_l{}^m - Z_s H_\phi{}^m)\,dl - \iiint_{V_2} (\mathbf{J} \cdot \mathbf{E}^m - \mathbf{M} \cdot \mathbf{H}^m)\,dv = V_m \tag{7}$$

where L represents the overall wire length and

$$V_m = \iiint_{V_1} [(\mathbf{J}_i \cdot \mathbf{E}^m - \mathbf{M}_i \cdot \mathbf{H}^m)\,dv \tag{8}$$

$$E_l{}^m = \frac{1}{2\pi}\int_0^{2\pi} (\hat{l} \cdot \mathbf{E}^m)\,d\phi \tag{9}$$

$$H_\phi{}^m = \frac{1}{2\pi}\int_0^{2\pi} (\hat{\phi} \cdot \mathbf{H}^m)\,d\phi. \tag{10}$$

Equation (7) is an integral equation for the three unknown currents $I(l)$, $\mathbf{J}$, and $\mathbf{M}$. This equation insures that these currents have the proper reaction with a test source in the wire surface, but does not insure that the proper conditions are satisfied in the dielectric/ferrite inhomogeneity. In the dielectric/ferrite inhomogeneity it is required that $(\mathbf{E}, \mathbf{H})$ be related to $(\mathbf{J}, \mathbf{M})$ by (3) and (4). Thus we have

$$\mathbf{E}_s + \mathbf{E}^J + \mathbf{E}^H + \mathbf{E}_i = \mathbf{E} = \frac{\mathbf{J}}{j\omega(\epsilon - \epsilon_0)}, \quad \text{in } V_2 \tag{11}$$

and

$$\mathbf{H}_s + \mathbf{H}^J + \mathbf{H}^H + \mathbf{H}_i = \mathbf{H} = \frac{\mathbf{M}}{j\omega(\mu - \mu_0)}, \quad \text{in } V_2. \tag{12}$$

Multiplying (11) by the vector weighting function $\mathbf{w}_m$, (12) by the vector weighting function $\mathbf{w}_m{}'$, and integrating over the volume V_2 yields

$$\iiint_{V_2} \left(\mathbf{E}_s + \mathbf{E}^J + \mathbf{E}^H - \frac{\mathbf{J}}{j\omega(\epsilon - \epsilon_0)}\right) \cdot \mathbf{w}_m\,dv = -\iiint_{V_2} \mathbf{E}_i \cdot \mathbf{w}_m\,dv \tag{13}$$

$$\iiint_{V_2} \left(\mathbf{H}_s + \mathbf{H}^J + \mathbf{H}^H - \frac{\mathbf{M}}{j\omega(\mu - \mu_0)}\right) \cdot \mathbf{w}_m{}'\,dv = -\iiint_{V_2} \mathbf{H}_i \cdot \mathbf{w}_m{}'\,dv. \tag{14}$$

Equations (7), (13), and (14) are three coupled integral equations which can be solved for the unknown currents $I(l)$ and $(\mathbf{J}, \mathbf{M})$. The moment method solution to these equations is presented below. Strictly speaking, (7), (13), and (14) must be satisfied by an arbitrary $(\mathbf{J}_m, \mathbf{M}_m)$, $\mathbf{w}_m$, and $\mathbf{w}_m{}'$. However, in the moment method solution we only enforce these equations for N distinct $(\mathbf{J}_m, \mathbf{M}_m)$, M distinct $\mathbf{w}_m$, and P distinct $\mathbf{w}_m{}'$. In this case (7), (13), and (14) represent N, M, and P simultaneous linear integral equations, respectively.

The next step in the moment method solution will be to transform these simultaneous linear integral equations to simultaneous linear algebraic equations. This is accomplished by expanding the unknown currents $\mathbf{I}$, $\mathbf{J}$, and $\mathbf{M}$ as follows:

$$\mathbf{I}(l) = \sum_{n=1}^{N} I_n \mathbf{F}_n(l) \tag{15a}$$

$$\mathbf{J} = \sum_{n=N+1}^{N+M} I_n \mathbf{G}_n \tag{15b}$$

$$\mathbf{M} = \sum_{n=N+M+1}^{N+M+P} I_n \mathbf{Q}_n. \tag{15c}$$

In (15) the wire current is expanded in terms of the basis set $\mathbf{F}_n$, and the electric and magnetic volume polarization currents in terms of the basis sets $\mathbf{G}_n$ and $\mathbf{Q}_n$, respectively. To simplify the following equations we will denote $(\mathbf{E}_{sn}, \mathbf{H}_{sn})$, $(\mathbf{E}_n{}^J, \mathbf{H}_n{}^J)$, and ($\mathbf{E}_n{}^H$, $\mathbf{H}_n{}^H$) as the fields radiated by $\mathbf{F}_n$, $\mathbf{G}_n$, and $\mathbf{Q}_n$, respectively, in the homogeneous medium (μ_0, ϵ_0). This notation is summarized in Table I.

Substituting (15) into (7), (13), and (14), and changing the order of integration and summation yields

$$\sum_{n=1}^{N} I_n\left\{-\int_0^L F_n(l)(E_l{}^m - Z_s H_\phi{}^m)\,dl\right\} + \sum_{n=N+1}^{N+M} I_n\left\{-\iiint_{V_2} \mathbf{G}_n \cdot \mathbf{E}^m\,dv\right\} + \sum_{n=N+M+1}^{N+M+P} I_n\left\{+\iiint_{V_2} \mathbf{Q}_n \cdot \mathbf{H}^m\,dv\right\} = V_m, \qquad m = 1, 2, \cdots, N \tag{16a}$$

$$\sum_{n=1}^{N} I_n\left\{\iiint_{V_2} \mathbf{E}_{sn} \cdot \mathbf{w}_m\,dv\right\} + \sum_{n=N+1}^{N+M} I_n\left\{\iiint_{V_2}\left(\mathbf{E}_n{}^J - \frac{\mathbf{G}_n}{j\omega(\epsilon-\epsilon_0)}\right)\cdot \mathbf{w}_m\,dv\right\} + \sum_{n=N+M+1}^{N+M+P} I_n\left\{\iiint_{V_2} \mathbf{E}_n{}^H \cdot \mathbf{w}_m\,dv\right\} = -\iiint_{V_2} \mathbf{E}_i \cdot \mathbf{w}_m\,dv, \qquad m = N+1, N+2, \cdots, N+M \tag{16b}$$

$$\sum_{n=1}^{N} I_n\left\{\iiint_{V_2} \mathbf{H}_{sn} \cdot \mathbf{w}_n{}'\,dv\right\} + \sum_{n=N+1}^{N+M} I_n\left\{\iiint_{V_2} \mathbf{H}_n{}^J \cdot \mathbf{w}_m{}'\,dv\right\} + \sum_{n=N+M+1}^{N+M+P} I_n\left\{\iiint_{V_2}\left(\mathbf{H}_n{}^H - \frac{\mathbf{Q}_n}{j\omega(\mu-\mu_0)}\right)\cdot \mathbf{w}_m{}'\,dv\right\} = -\iiint_{V_2} \mathbf{H}_i \cdot \mathbf{w}_m{}'\,dv,$$

$$m = N+M+1, N+M+2, \cdots N+M+P. \tag{16c}$$

Equation (16) can be written compactly as

$$\sum_{n=1}^{N+M+P} I_n Z_{mn} = V_m, \qquad m = 1, 2, \cdots, N+M+P \tag{17}$$

or in the matrix form as

$$ZI = V \tag{18}$$

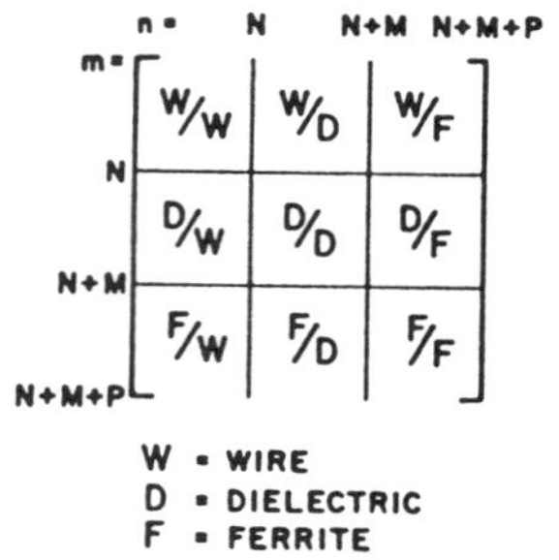

Fig. 2. Symbolic representation of impedance matrix Z.

where Z is the square impedance matrix, V is the excitation voltage column, and I is the current column which contains the unknown coefficients I_n, $n = 1, 2, \cdots, N + M + P$, as defined in (15). In Fig. 2, the impedance matrix Z is symbolically shown divided into nine regions, and the type of coupling accounted for in each region is indicated. In (16b) and (16c), when $m = n$, one must evaluate fields inside a current distribution. This problem is discussed in the Appendix.

In order to obtain numerical results from the above formulation, it is necessary to define specific expansion and weighting functions. The choices made for this study will now be presented. For $\mathbf{F}_n$ we choose the piecewise-sinusoidal function used by Richmond [7]. The unit magnitude vector volumetric pulse function is chosen for the expansion modes $\mathbf{G}_n$ and $\mathbf{Q}_n$. The volumetric pulses are the parallelepipeds obtained by dividing the dielectric/magnetic body into smaller rectangular volumes. Since the volume polarization current has an arbitrary polarization, it is necessary to have three orthogonal vector volumetric pulses occupying the same parallelepiped. Each will have a different polarization, i.e., $\hat{x}$, $\hat{y}$, and $\hat{z}$. There are no magnetic test sources and $\mathbf{M}_m = 0$. The piecewise-sinusoidal test function is chosen for $\mathbf{J}_m$. The choice of piecewise-sinusoidal functions for both the expansion and test modes enables us to use the computer program for thin-wire antennas and scatterers in a homogeneous medium, developed by Richmond [10], to calculate the wire/wire region of the impedance matrix Z. For the $\mathbf{w}_m$ or $\mathbf{w}_m{}'$ we choose a delta function which is located in the center of the corresponding volumetric pulse expansion $\mathbf{G}_m$ or $\mathbf{Q}_m$, and with the same polarization.

Since the test modes in the wire structure have the same current distribution as the expansion mode, this may be regarded as an application of Galerkin's method. The use of delta functions as the test modes for the dielectric/ferrite inhomogeneity results in (13) and (14) being satisfied at discrete points in V_2. This may be recognized as an application of the point-matching method.

III. EXPERIMENTAL AND NUMERICAL RESULTS

In this section, numerical results based on equations derived in the preceding section are presented for a loop antenna loaded with a dielectric and/or ferrite core. The calculated input admittance of the dielctric loaded loops are compared with experimental results. In the case of ferrite loaded loops, the computed radiation resistances are compared with the previous theoretical results by Stewart [1]. The loop geometry was chosen because it is a basic geometry and yet still required treating wire segments which are parallel or intersect at an angle.

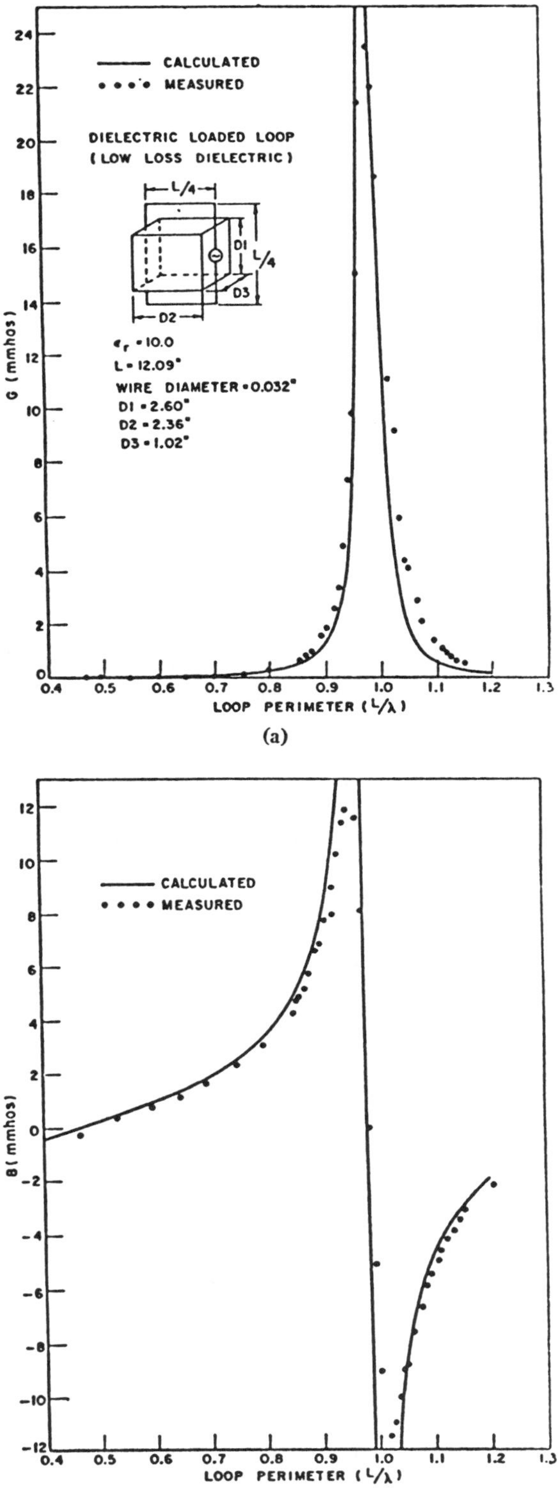

Fig. 3. (a) Comparison of calculated and measured conductance for dielectric (ϵ_r = 10.0) loaded loop. (b) Comparison of calculated and measured susceptance for dielectric (ϵ_r = 10.0) loaded loop.

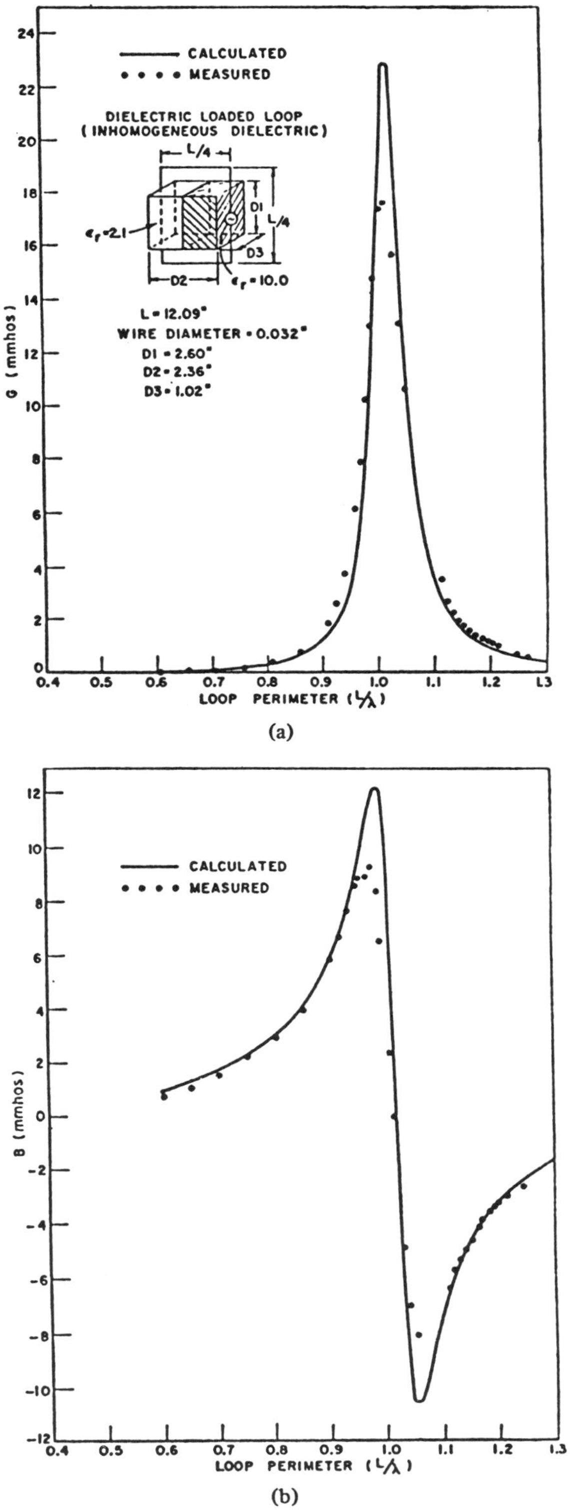

Fig. 4. (a) Comparison of calculated and measured conductance for dielectric (inhomogeneous) loaded loop. (b) Comparison of calculated and measured susceptance for dielectric (inhomogeneous) loaded loop.

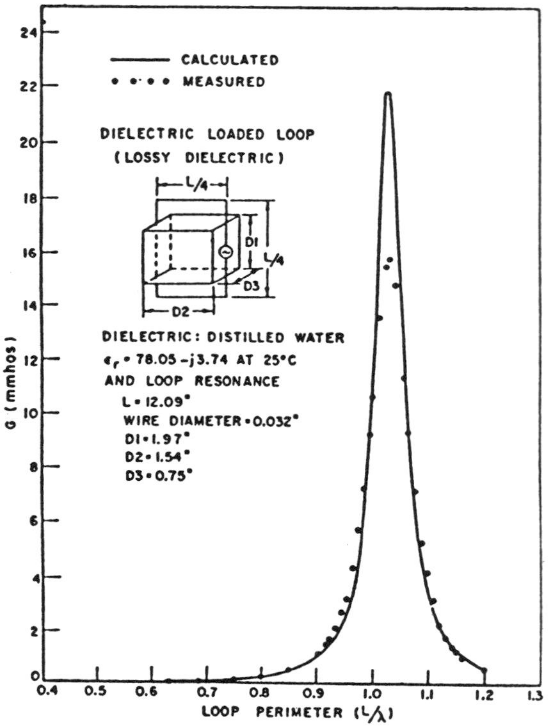

(a)

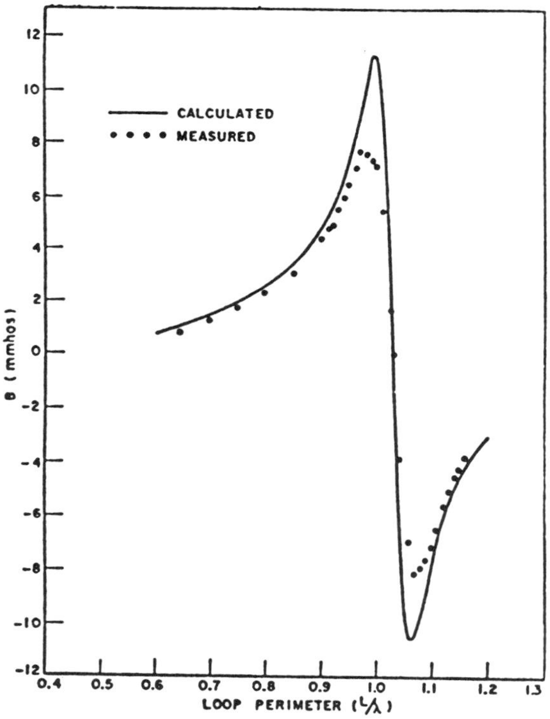

(b)

Fig. 5. (a) Comparison of calculated and measured conductance for dielectric (distilled water) loaded loop. (b) Comparison of calculated and measured susceptance for dielectric (distilled water) loaded loop.

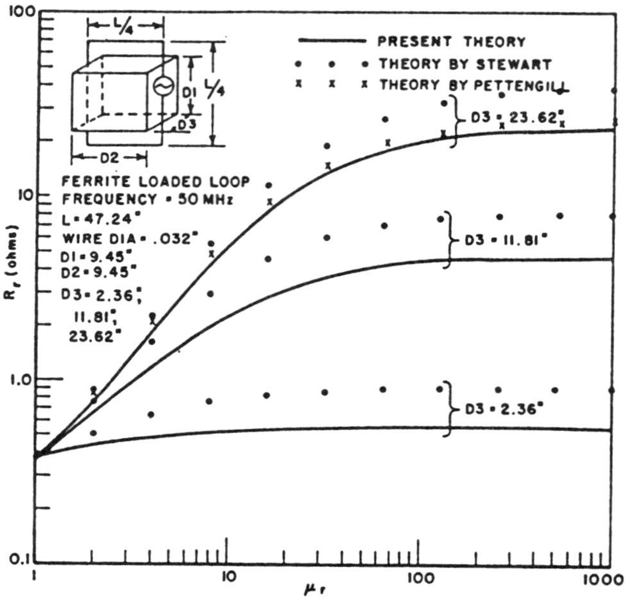

Fig. 6. Comparison of present theory with previous calculation for the radiation resistance of ferrite loaded loops.

The measurements were made with a half-loop mounted on a 4-ft square aluminum ground plane. The half-loop is made of copper wire coated with a thin layer of tin and has a total diameter of 0.032 in. The coaxial feed has an inner radius of 1/16 in and an outer radius of 3/8 in. The half-loop has a dimension of 3.02 in by 1.51 in. The dielectrics are rectangular parallelepipeds and have the dimensions $D1/2$, $D2$, and $D3$. The calculations were made for the complete loop, and it is the complete loop admittances which are plotted. The feed was modeled by a delta-gap generator.

Three examples will now be presented to demonstrate the ability of the theory to treat different types of dielectric inhomogeneities. Figs. 3(a) and (b) compare measured and calculated admittance of the loop loaded with Eccoflo-Hik ($\epsilon_r = 10.0$, produced by Emerson and Cuming). The dimensions of the block are $D1 = 2.60$ in, $D2 = 2.36$ in, and $D3 = 1.02$ in.

Next, the ability of the theory and computer program to treat problems with inhomogeneous dielectrics will be demonstrated. As seen in the insert in Fig. 4(a), the loop is loaded with an inhomogeneous dielectric of the same size as was considered above. The dielectric is half-filled with Teflon ($\epsilon_r = 2.1$) and half-filled with Eccoflo-Hik. Figs. 4(a) and (b) show the measured and calculated admittance of the loaded loop.

Distilled water ($\epsilon_r = 78.05 - j3.74$ at the resonance frequency of the loop) is used to demonstrate the ability to treat lossy dielectrics. Due to the large relative permittivity of water, the size of the inhomogeneity has been reduced to $D1 = 1.97$ in, $D2 = 1.54$ in, and $D3 = 0.75$ in. The measured and calculated admittance of the water loaded loop are shown in Figs. 5(a) and (b).

One of the most common uses of ferrites is to increase the radiation resistance of loop antennas. Fig. 6 shows the radiation resistance of a ferrite loaded loop versus μ_r. Curves are shown for various sizes of the ferrite cores. The loop has a perimeter $L = 47.24$ in, and the ferrite cores are of size $D1 = 9.45$ in, $D2 = 9.45$ in, and $D3 = 2.36$ in, 11.81 in, and 23.62 in. The computations are at 50 MHz where the loop circumference is about 0.20λ. The present results are compared with the previous approximate results of Stewart. These two results are seen to agree only qualitatively. It is felt that most of the quantitative differences are due to approximations given by Stewart for the demagnetization factor of a rectangular rod. More recently, Pettengill *et al.* [11] have presented different expressions for the demagnitization factor of a core with length to diameter ratio greater than two. The $D3 = 23.62$ case is compared to a computation using Pettengill's demagnitization factor, and the agreement is good.

IV. SUMMARY

The purpose of this study has been to develop a technique to analyze thin-wire antennas and scatterers in the presence of a dielectric and/or ferrite inhomogeneity. The method presented is a moment method solution and a modification of the piecewise-sinusoidal reaction formulation for thin-wire antennas and scatterers in a homogeneous medium. The technique is sufficiently general to be applicable to lossy and loaded thin-wire antennas and scatterers in the presence of isotropic, inhomogeneous, and lossy dielectric/ferrite inhomogeneities.

Numerical results have been presented and compared with measurements and previous calculations. These results show the ability of the theory and computer program to analyze antennas in the presence of dielectrict of varying sizes and permittivities, inhomogeneous dielectrics, lossy dielectrics, and ferrites. The method is, however, limited by the finite computer core to electrically small inhomogeneities.

Using the technique presented here for treating antennas and scatterers in the presence of a dielectric/ferrite inhomogeneity, one can analyze problems such as ferrite loaded loops, manpack transceiver antennas, scattering from rain or blood cells, effects of microwave radiation on biological tissues, and antennas covered by dielectric radomes.

APPENDIX

Consider the problem of finding the electric field at the center of a column of current with uniform current density $\mathbf{J} = \hat{z}J$, occupying a circular cylindrical volume. In numerically evaluating this field, one must exclude a small region about the field point. Unfortunately, the value of the resultant integral is dependent upon the shape of the volume excluded. If one excludes a circular cylinder of radius η and height $2h$ surrounding the singularity, Van Bladel [12] found that the result will be in error by

$$\delta = \left(1 - \frac{1}{\sqrt{1 + \frac{\eta^2}{h^2}}}\right) \frac{J}{j\omega\epsilon_0}. \tag{A-1}$$

Note that if one excludes a long thin cylinder where $\eta \ll h$, δ goes to zero and the correct result is obtained. This idea is used in determining the electric field inside a volumetric electric current expansion mode, and also for the dual problem of determining the magnetic field inside a volumetric magnetic current expansion mode.

REFERENCES

[1] J. L. Stewart, "Research in magnetic antennas," California Inst. Tech.; prepared under Contract No. DA-36-039 sc-73189 for Signal Corps., Dep. of the Army, Sept. 13, 1957.

[2] J. H. Richmond, "Scattering by a dielectric cylinder of arbitrary cross section shape," *IEEE Trans. Antennas Propagat.*, vol. AP-13, pp. 334-341, May 1965.

[3] —, "TE-wave scattering by a dielectric cylinder of arbitrary cross-section shape," *IEEE Trans. Antennas Propagat.*, vol. AP-14, pp. 460-464, July 1966.

[4] S. K. Chang and K. K. Mei, "Application of the unimoment method to electromagnetic scattering of dielectric cylinders," *IEEE Trans. Antennas Propagat.*, vol. AP-24, pp. 35-42, Jan. 1976.

[5] D. E. Livesay and K. M. Chen, "Electromagnetic fields induced inside arbitrarily shaped biological bodies," *IEEE Trans. Microwave Theory Tech.*, vol. MTT-22, pp. 1273-1280, Dec. 1974.

[6] R. F. Harrington, *Field Computations by Moment Methods.* New York: MacMillan, 1968.

[7] J. H. Richmond, "Radiation and scattering by thin-wire structures in the complex frequency domain," the Ohio State Univ. ElectroScience Lab., Dep. Elec. Eng., prepared under Grant No. NGL 36-008-138 for National Aeronautics and Space Administration, Rep. 2902-10, July 1973.

[8] E. H. Newman, "Analysis of strip antennas in the presence of a dielectric inhomogeneity," Ph.D. dissertation, the Ohio State University, 1974.

[9] S. A. Schelkunoff, "On diffraction and radiation of electromagnetic waves," *Phys. Rev.*, vol. 56, pp. 308-316, Aug. 15, 1939.

[10] J. H. Richmond, "Computer program for thin-wire structures in a homogeneous conducting medium," the Ohio State Univ. ElectroScience Lab., Dep. Elec. Eng., prepared under Grant No. NGL 36-008-138 for National Aeronautics and Space Administration, Rep. 2902-12, Aug. 1973.

[11] R. C. Pettengill, H. T. Garland, and J. D. Meindl, "Receiving antenna design for miniature receivers," *IEEE Trans Antennas Propagat.*, vol. AP-25, no. 4, pp. 528-530, July 1977.

[12] J. Van Bladel, "Some remarks on Green's dyadic for infinite space," *IRE Trans. Antennas Propagat.*, vol. AP-9, pp. 563-566, Nov. 1961.

A Summary of Hybrid Solutions Involving Moment Methods and GTD*

W. D. BURNSIDE AND P. H. PATHAK

THE OHIO STATE UNIVERSITY ELECTROSCIENCE LABORATORY, ELECTRICAL ENGINEERING DEPARTMENT, COLUMBUS, OHIO 43212

Abstract—**This paper attempts to summarize the recent advances associated with hybrid solutions which combine the Moment Method (MOM) and Geometrical Theory of Diffraction (GTD). This discussion of hybrid solutions is limited to combinations of MOM and GTD in that they potentially can provide some excellent analytical tools for the electromagnetic compatibility community. The essential thrust of the previous comment is based on the basic complementary nature of the two analyses. The MOM can be used to solve arbitrary shapes provided they are small in terms of the wavelength; whereas, the GTD is appropriate for large structures which can be simulated by simpler component parts. There are at least two basic methods [1], [2] which have successfully combined the MOM and GTD and have been extended to allow one to treat, for example, the coupling between antennas mounted on very large complex aircraft structure. Based on the great strides made in the recent development of uniform GTD solutions and their applications to complex electromagnetic problems, it is apparent that the combination of MOM and GTD can provide the EMC community with many new and useful solutions.**

I. Introduction

THERE seem to be as many hybrid electromagnetic solutions being developed as there are combinations of theories such as moment methods (MOM) and the Geometrical Theory of Diffraction (GTD) (together with its uniform version), MOM and Physical Optics, MOM and Eigenmode solutions, etc. Thus, summarizing the complete hybrid area is beyond the scope of this paper. In any event an attempt is made here to illustrate the advances associated with combining MOM and GTD. It is felt that the combination of MOM and GTD offers the EMC community some new and very useful analytic tools, especially in terms of handling structures which are large in terms of the wavelength. There are at least two basic methods that can be used to combine the MOM and GTD. The underlying principles of these two approaches will be discussed as well as the advances that have been made since their inception.

In order to illustrate the need for and usefulness of these hybrid techniques, let us consider some of the properties of the MOM and GTD as shown in Table 1. It is apparent from these properties that the MOM and GTD are complementary in several respects. As a result of these observations, it is conceivable that a combination of solutions could be used to supplement their shortcomings. For example, the hybrid solution originated by Burnside et al. [1] can be used to numerically generate new diffraction coefficients; whereas, the approach developed by Thiele et al. [2][1] can be used to eliminate the need of representing the scattered field using current samples around the complete structure. As a consequence of these analyses, one can use the MOM combined with the GTD to solve for the electromagnetic properties of very large structures, such as the coupling between a pair of antennas mounted on an aircraft even at microwave frequencies.

TABLE 1
PROPERTIES OF VARIOUS NUMERICAL SOLUTIONS

Geometrical Theory of Diffraction
Can analyze objects which are large in terms of wavelengths.
Can incorporate other solutions into its format.
Gives little information about antenna parameters (especially wire antennas).
Only have diffraction coefficients for a small number of structures.
Moment Method Solutions
Can analyze objects which are small in terms of wavelengths.
Gives information about antenna parameters.
Can treat arbitrarily shaped objects.

Before going into the details of these two basic hybrid techniques, it is appropriate to examine in some detail the various advances made in the MOM and GTD. In that this paper is surrounded in this volume by the modern developments in the MOM, it is not necessary to reiterate those achievements. However, the GTD has not been addressed, so an attempt is made here to describe some of the major GTD achievements.

II. The Geometrical Theory of Diffraction (GTD)

The GTD is a systematic extension of classical Geometrical Optics (GO) proposed by Keller and his associates to describe the phenomenon of diffraction at high frequencies in terms of diffracted rays. The diffracted rays are introduced in addition to the usual GO incident, reflected, and transmitted rays. In the case of an impenetrable obstacle, the GO rays do not exist in the shadow region behind the obstacle, i.e., in the region where the rays emanating from the source are blocked by the diffracting obstacle. Furthermore, the GO field is

*The work reported in this paper was supported in part by Contract No. N00014-78-C-0049 between the Department of the Navy, Office of Naval Research, and The Ohio State University Research Foundation.

[1] See pp. 328–335 of this book for a reprint of this paper.

The paper, "A summary of hybrid solutions involving moment methods and GTD," by W. D. Burnside and P. H. Pathak appears with 18 other papers on Moment Methods in a 1980 SCEEE Press book entitled, *Applications of the Method of Moments to Electromagnetic Fields*; the book is still in print and is available from the SCEEE Press, 1101 Massachusetts Avenue, St. Cloud, FL 34769.

incorrect near, and at the incident and reflection shadow boundaries where the GO incident and reflected fields vanish abruptly, thereby exhibiting a highly discontinuous field behavior. The failure of GO to provide a nonzero field in the shadow region is overcome by the presence of Keller's diffracted rays which entirely account for the field in the shadow zone. The diffracted rays are excited, for example, by an incident ray which strikes a geometrical (or even an electrical) discontinuity in the obstacle surface, or when it grazes a smooth convex surface; they are, in general, also present in the illuminated region where the source is directly visible.

These diffracted rays of the GTD obey the generalized Fermat's principle postulated by Keller. As a consequence of the generalized Fermat's principle, the rays diffracted by an edge type surface discontinuity form a cone of half angle β_0 where β_0 denotes the acute angle between the incident ray and the edge at the point of diffraction. Also, the rays launched on a smooth convex surface by an incident ray at grazing follow geodesic surface ray paths while continually shedding energy via surface diffracted rays (with the associated surface ray field decaying exponentially) along the forward tangents to the surface ray paths. Likewise, the rays diffracted by a tip emanate in all directions from that tip, etc. In the GTD, the initial values of the diffracted ray field is specified by a diffraction coefficient which is associated, for example, with an edge, tip, or a convex surface, in a manner analogous to the reflection and transmission coefficients of GO. Away from the diffracting point, the diffracted rays propagate according to the principles of GO. The diffraction coefficients and other relevant parameters of the GTD are independent of range, and they may be found from the asymptotic high frequency solutions to appropriate but simpler canonical problems, i.e. simpler problems whose geometrical and electrical properties locally model the essential features of the original diffraction problem. The generalization of the results of the canonical problems to treat the original problems is based on the principle of locality of high frequency radiation, propagation, and diffraction as postulated in the GTD. As a result of this local property, the GTD converts a complex electromagnetic boundary value problem into one of geometry and solving for the various ray paths, assuming that the relevant GTD parameters are known so that the fields associated with the ray paths can be calculated.

Since in its original form, the GTD is a purely ray optical technique, it fails at and near shadow boundaries and caustics. The modification of the GTD which is required to calculate the fields near the caustics* of diffracted rays is described in [5] in terms of the method of equivalent currents which indirectly employs the GTD to calculate these currents. On the other hand, the failure of the purely ray optical GTD field description at and near the shadow boundaries results from the fact that within these regions referred to as the shadow boundary transition regions, the "dominant" character of the true diffracted field must change rapidly but continuously from a purely ray optical field behavior outside the transition regions to a behavior which allows one to properly compensate for the discontinuities in the GO field at the shadow boundaries such that the total high frequency field remains continuous at these boundaries. To accomplish such a task, the diffraction coefficients and other GTD parameters cannot be allowed to remain range independent, i.e., the field description has to depart from a purely ray optical one. The uniform GTD [3], [4], [5] which has been developed recently accomplishes this task in a simple and accurate manner, and it employs the same ray paths as in the GTD. Thus, the uniform GTD field departs from the purely ray optical character to yield a bounded and continuous total field across the shadow boundary transition regions; whereas exterior to these transition regions, it automatically reduces to the purely ray optical GTD field description where the latter indeed becomes valid. Clearly, the development of the Uniform GTD (or UTD) represents a very significant step in extending the utility and power of the GTD to solve a wide variety of engineering problems.

It is noted that the polarization of the GTD (and therefore the UTD) field is to first order found to be perpendicular to the ray path. Further, the diffracted field appears to emanate from an isolated or set of isolated points in much the same way as ordinary reflection and transmission. This implies that the diffracted field is treated simply as another virtual source of scattered fields which are added to the incident and reflected fields to give a complete high frequency solution. Note that a much more complete overview of GTD and its uniform version can be found in the literature [3], [4], [5].

In order to illustrate the GTD solution approach, let us consider the wedge diffraction problem illustrated in Fig. 1. The GO solution consists of the superposition of the incident and reflected fields given respectively by

$$U^i = \begin{cases} K\dfrac{e^{-jk\rho_i}}{\sqrt{\rho_i}} & \text{in Regions I \& II, and} \\ 0 & \text{in Region III,} \end{cases} \tag{1}$$

and

$$U^r = \begin{cases} \pm K\dfrac{e^{-jk\rho_r}}{\sqrt{\rho_r}} & \text{in Region I, and} \\ 0 & \text{in Region II \& III,} \end{cases} \tag{2}$$

where K is a complex constant which is dependent on the line source excitation. Note that the $\pm$ sign in the reflection expression represents the reflection coefficient which is dictated by the polarization of the incident field. It is obvious from (1) and (2) that the GO solutions are discontinuous across both the incident and reflected shadow boundaries shown in Fig. 1. The edge diffracted field associated with this configuration is given by

$$U^d = K\frac{e^{-jk\rho'}}{\sqrt{\rho'}}\left[D\left(\frac{\rho'\rho}{\rho'+\rho}, \phi-\phi', n\right) \pm D\left(\frac{\rho'\rho}{\rho'+\rho}, \phi+\phi', n\right)\right]\frac{e^{-jk\rho}}{\sqrt{\rho}} \tag{3}$$

*Caustics are formed by ray congruences; in general, caustics are the envelope surfaces of a family of intersecting rays. These surfaces could degenerate to a line or a point (focus).

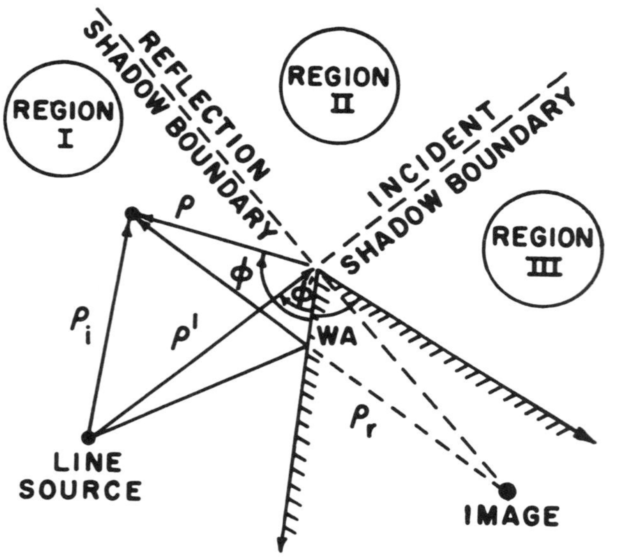

Fig. 1. A line source radiating in the presence of a wedge. Note the wedge angle is WA = (2 − n)π in the diffraction solution.

where $D()$ is a UTD edge diffraction coefficient [6]. Note that the $D(\frac{\rho'\rho}{\rho'+\rho}, \phi - \phi', n)$ term is discontinuous in just the proper manner that it plus the incident field is continuous across the incident shadow boundary; whereas, the $D(\frac{\rho'\rho}{\rho'+\rho}, \phi + \phi', n)$ term enforces continuity across the reflection shadow boundary. The total GTD (or more precisely the total UTD) solution is, then, given by

$$U^{\mathrm{GTD}} = U^i + U^r + U^d. \tag{4}$$

An example of the accuracy of this solution is illustrated in Fig. 2.

In that the first-order edge diffraction solution [6] is dependent on the incident field at the edge, one might think that a null field at the edge will eliminate the edge diffraction term. This is not the case. There is a slope diffraction term [3], [4], [5] which must also be included in the total solution; it depends on the spatial derivative (or slope) of the incident field which may be nonzero, even though the incident field itself may vanish at the edge.

As shown in [3], [4], [5], the edge diffracted field can be rather easily extended to the three-dimensional [3-D] case and still maintain its geometric interpretation. When the incident field strikes the edge at an angle (β_0), the diffracted field is contained within a cone of rays emanating from the diffraction point and making the same angle (β_0) with the edge as discussed earlier. In the 3-D case, the edge-diffracted field is discontinuous in that practical structures are composed of finite length edges. In order to compensate for these discontinuities, one must include a corner diffracted field [3], [4], [5], [7]. Thus, the GTD solution for a finite flat plate consists of incident, reflected, edge diffracted, and corner diffracted field components. Such geometries have been analyzed for the past decade using GTD [5], [8], [9].

There is a second major category of diffracted fields associated with curved surface diffraction. When an incident ray system emanating from a source strikes a smooth, perfectly-conducting convex surface, it produces a system of rays reflected from that surface; at grazing, the incident ray merges with the reflected ray giving rise to a surface ray which propagates into the shadow region along a geodesic on the convex surface according to the generalized Fermat's principle. The field associated with the surface ray exhibits an exponential decay in the deep shadow region due to the continual leakage of energy along the surface ray resulting from diffracted rays shedding tangentially from the surface ray, as mentioned earlier. One notes that surface rays can be excited by sources which are located either on or off a smooth convex surface; they can also be excited by the illumination of an edge, or other geometrical or electrical discontinuities in an otherwise smooth convex surface.

The recently obtained UTD solutions for diffraction by curved surfaces are based on the GTD ray format; however, they remain uniformly valid across the shadow boundary transition regions [12]–[14]. One should realize that these solutions will all shortly appear in the open literature and represent a significant advance in the GTD technology. Further, they are based on a consistent set of principles such that one can directly extend his GTD radiation solution, for example, to those of scattering, or coupling in conformal antennas (or conformal arrays).

The ray paths used to analyze the radiation patterns for an antenna mounted on a general convex curved surface is shown in Fig. 3. The observation point must be located at least a wavelength from the surface even though it may be in the near zone of the complete structure. The high-frequency electric field is given by

$$d\vec{E} = \hat{n}\, dE_n + \hat{b}\, dE_b,$$

where dE_n and dE_b have been deduced from a careful study of the cylinder and sphere canonical problems; in addition, experimental results for antennas on a spheroid were helpful in the generalization to general convex surfaces. The UTD solutions for an infinitesimal monopole are given in the lit region by [3], [4], [5], [10]

$$dE_n(P_L) = CZ_0\, dp_e(Q') \sin\theta^i \left[\frac{H^l + T_0^2 S^l \cos\theta^i}{1 + T_0^2\cos^2\theta^i}\right]\frac{e^{-jks}}{s}$$

and

$$dE_b(P_L) = CZ_0\, dp_e(Q') \sin\theta^i T_0 F \frac{e^{-jks}}{s}$$

and in the shadow region by

$$dE_n(P_S) = CZ_0\, dp_e(Q') H e^{-jkt} \cdot \left[\frac{\rho_g(Q')}{\rho_g(Q)}\right]^{-1/2} \sqrt{\frac{d\psi_0}{d\eta(Q)}}\sqrt{\frac{\rho_2^d}{s(\rho_2^d + s)}}\, e^{-jks}$$

and

$$dE_b(P_S) = CZ_0\, dp_e(Q') T_0 S e^{-jkt} \cdot \left[\frac{\rho_g(Q')}{\rho_g(Q)}\right]^{-1/6} \sqrt{\frac{d\psi_0}{d\eta(Q)}}\sqrt{\frac{\rho_2^d}{s(\rho_2^d + s)}}\, e^{-jks}.$$

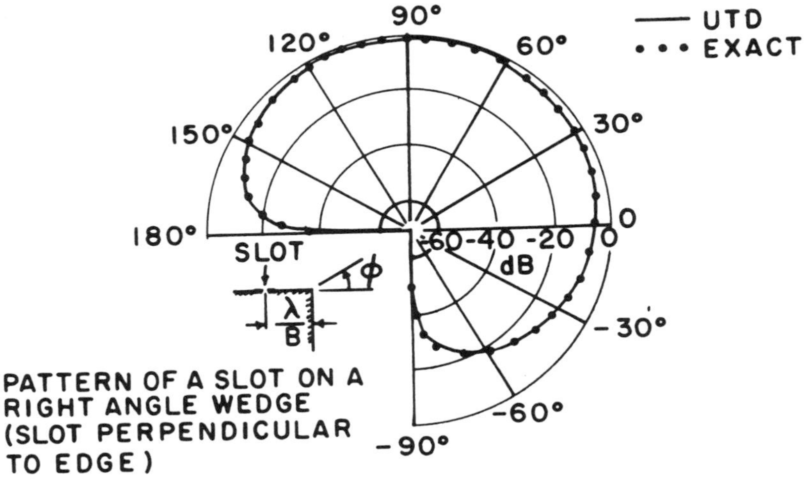

(a)

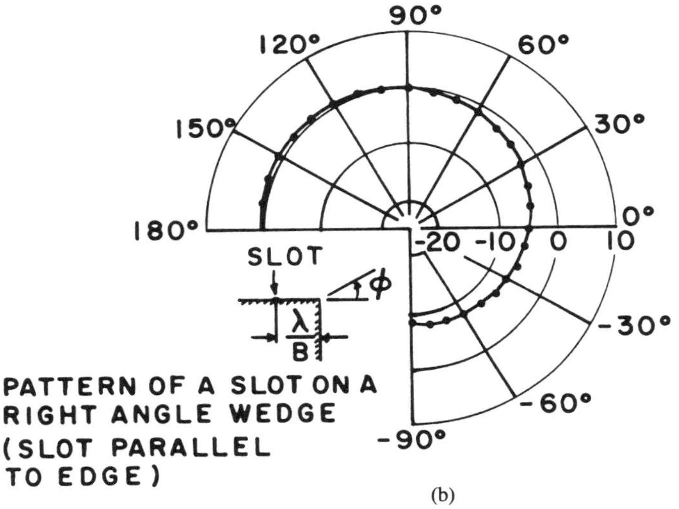

(b)

Fig. 2. Diffraction of the field of an infinitesimal slot by a right-angle edge.

Note that a complete description of the general convex curved surface radiation solution is given in [10]. The simplicity of the above solutions is very intriguing especially since the complex (Fock) functions H', H, S', and S can be easily computed based on standard interpolation schemes and well-tabulated data. A comparison between measured and calculated patterns for a monopole radiating from a prolate spheroid are shown in Fig. 4. Note that the E_b component in the previous shadow boundary pattern is due to the spheroidal surface in that E_b would vanish if the monopole were mounted on an infinite ground plane.

The UTD scattering and coupling solutions follow basically the same format as previously discussed for the radiation problem. The general curved surface scattering solutions are described in [3], [4], [5], [11], [12] with an example of the accuracy of the UTD analysis shown in Fig. 5. The UTD coupling solution is described in detail in [5], [13], [14] and an example of that analysis is demonstrated by the coupling comparisons shown in Fig. 6.

The UTD diffraction solutions for edges and curved surfaces have been applied for almost two decades to analyze various complex electromagnetic problems [5]. In fact, they have been applied in terms of the theoretical development applied in generating the three major computer codes described in Table 2. These codes [15] illustrate the very broad scope of the GTD analysis approach. Although these specific codes have not been applied to any great extent by the EMC community, they offer a definite potential in terms of directly solving numerous compatibility problems which involve large complex structures. For example, the Basic Scattering Code has already been interfaced with the NEC–Method of Moments Code [16] such that the NEC–MOM Code can be used to analyze the impedance and current distribution on a wire antenna, and the Basic Scattering Code can be used to code the radiation pattern for the same antenna in the presence of a complex structure.

III. Hybrid Solutions

As stated earlier, there are at least two basic ways to combine the MOM and GTD to solve complex electromagnetic problems that neither solution can treat alone. In that the hybrid solution of Burnside et al. [1] is used to extend the

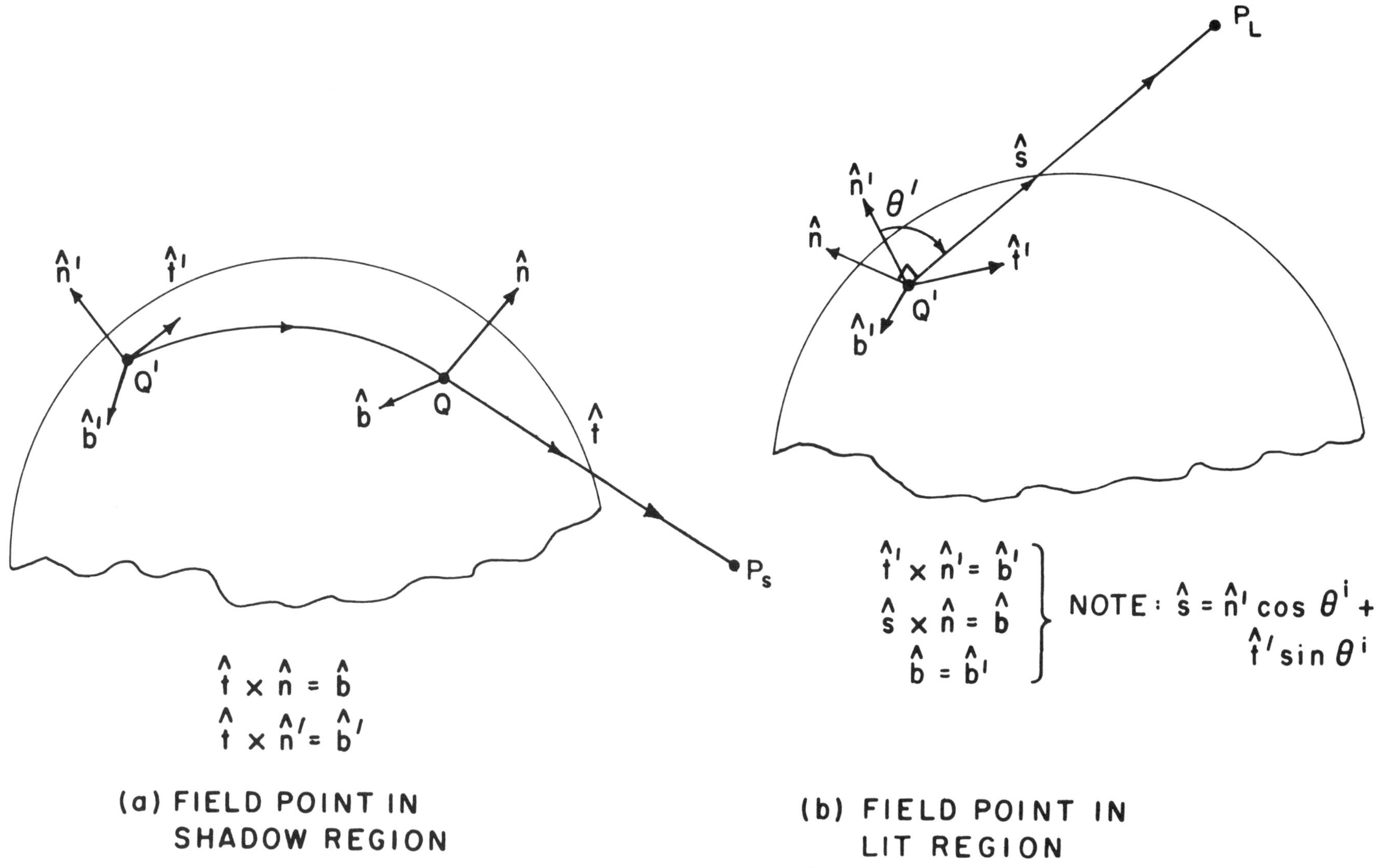

Fig. 3. Rays emanating from a source on a convex surface.

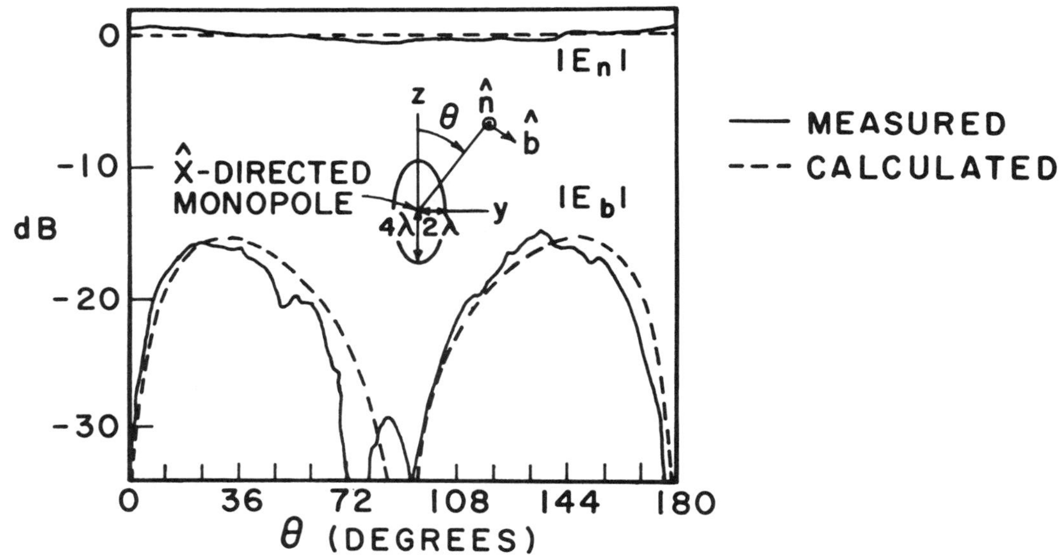

Fig. 4. Radiation patterns of a quarter-wavelength monopole on a prolate spheroid calculated and measured in the shadow boundary plane.

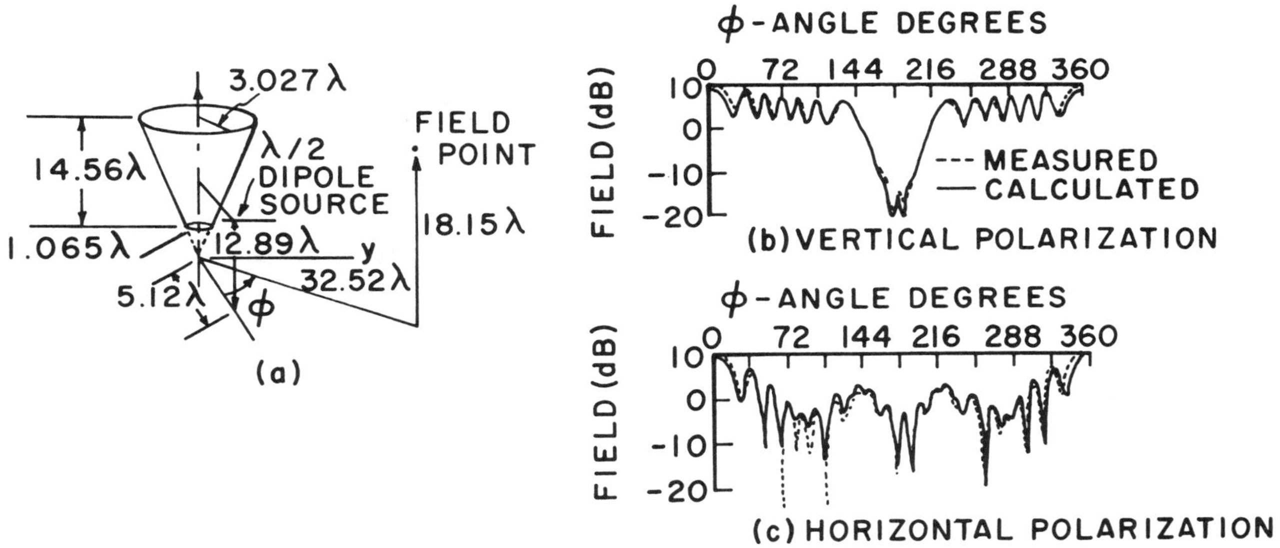

Fig. 5. Radiation patterns of an electric dipole near the frustum of a cone.

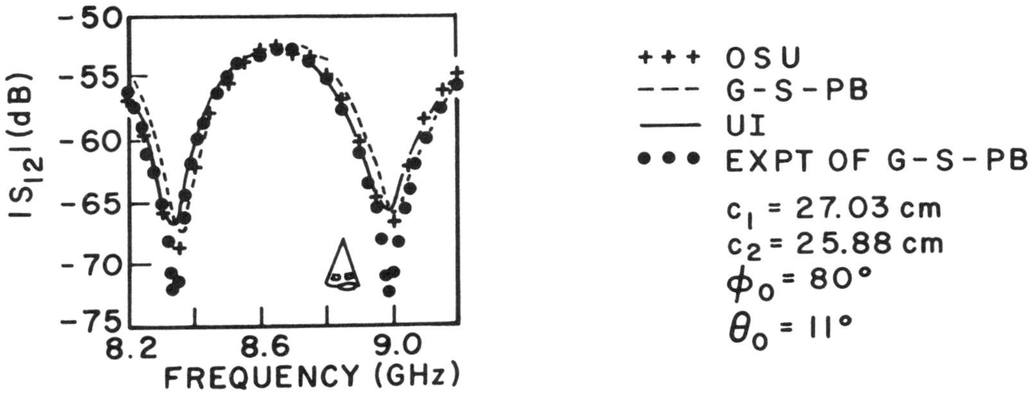

Fig. 6. Coupling coefficient S_{12} between two circumferential slots on a cone vs. frequency. The radial separation between the slots is $C_1 - C_2$ and angular separation is ϕ_0. The cone half angle is θ_0.

TABLE 2

CODE	CAPABILITIES
On-Aircraft Antenna Code	Near- or far-field patterns of antennas mounted on aircraft or similar structures. Can be used for slots, stubs, or arrays mounted on a fuselage or convex curved surface.
Basic Scattering Code	Far-field patterns of antennas in presence of conducting scattering structures. For example, this code can be used for antennas in shipboard environments, mounted on vehicles, etc.
Reflector Antenna Code	Near-field and far-field computations for parabolic reflector antennas, including offset reflectors. Can be used for calculations of near field radiation levels and EMC coupling to small antennas

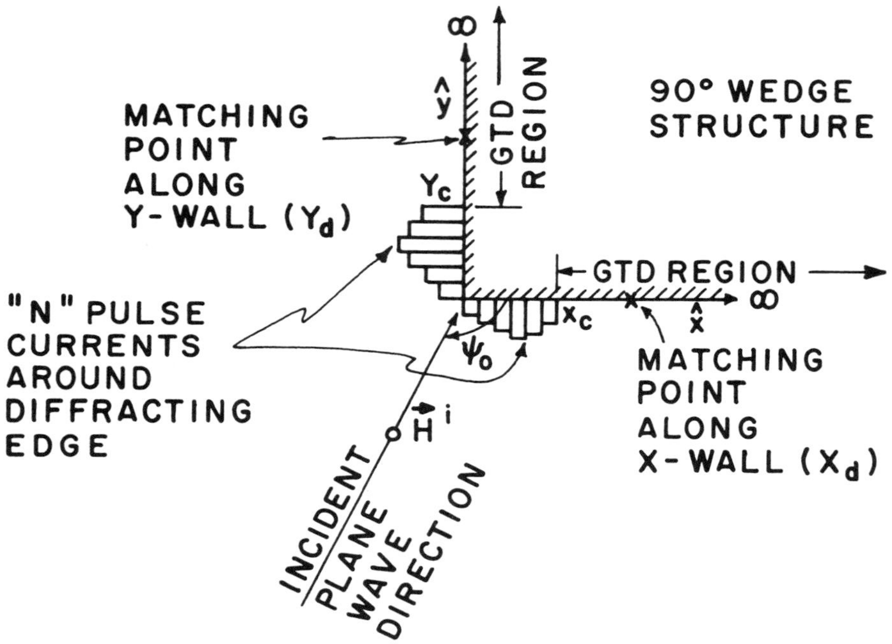

Fig. 7. Basic wedge diffraction geometry applied in MOM-GTD solution.

GTD by numerically providing new diffraction coefficients, let us continue the theme of the previous section and consider that solution approach first.

As shown in [1] the surface currents that exist on an infinite wedge can be defined in terms of MOM current samples around the diffracting edge and GTD currents otherwise as illustrated in Fig. 7. The currents in the GTD regions are given by

$$\bar{J} = \bar{J}^i + \bar{J}^r + \bar{J}^d \tag{5}$$

where

$\bar{J}^i$ = surface current associated with the incident field ($\bar{J}^i = \hat{n} \times \overline{H}^i$),

$\bar{J}^r$ = surface current associated with the reflected field ($\bar{J}^r = \hat{n} \times \overline{H}^r$), and

$\bar{J}^d$ = surface current associated with the diffracted field ($\bar{J} = \hat{n} \times \overline{H}^d$).

The diffracted field is given by

$$\overline{H}^d = \hat{Z} H^i(Q_E)\left[D(\rho, \phi - \phi', n) + D(\rho, \phi + \phi', n)\right] \frac{e^{-jk\rho}}{\sqrt{\rho}} \tag{6}$$

where the terms in the brackets represent the composite diffraction coefficient, $H^i(Q_E)$ is the $\hat{Z}$-directed field incident upon the edge, and ρ is the distance from the edge to the field point. Provided that the two faces making up the wedge are not illuminated by an incident TM plane wave at or near grazing, one can define $[D() + D()]$ such that it is independent of range (ρ) in the GTD regions. Using the procedure developed in [1], one can use the 90° wedge geometry of Fig. 7 in order to illustrate how to obtain a system of linear equations as represented by

$$\sum_{n=1}^{N} l_{mn}\alpha_n + D_x L_n\left(\frac{e^{-jkx}}{\sqrt{x}}\right) + D_y L_n\left(\frac{e^{-jky}}{\sqrt{y}}\right) = g_n \quad (7)$$

where the α_n's are the unknown MOM current samples, D_x and D_y are the unknown composite diffraction coefficients along the x and y faces, respectively, and g_n is the driving function which includes the known fields associated with the incident and reflected mechanisms. By satisfying the integral equation using, for example, point matching at the center of each pulse segment and in each of the two GTD regions, one can obtain a linear system of $(N + 2)$ equations which can be used to solve for the $N + 2$ unknowns.

Using this procedure, Burnside et al. [1] computed the surface current and scattering patterns for a 90° wedge using as few as six ($N = 4$) unknowns. Some examples of their solution are shown in Fig. 8 and compared with the exact eigenmode results. As illustrated in Fig. 8(c), one can also handle the case of grazing incidence by including a diffraction coefficient series of the form:

$$J^d = \sum_n C_n \frac{e^{-jk\rho}}{\left(\sqrt{\rho}\right)^n} \quad (8)$$

It was shown in [17], that one needs to include only three terms in (8) in order to obtain satisfactory results for all incidence angles.

In addition to the wedge problem, it was shown by Burnside et al. [1] that one can treat a rectangular cylinder of arbitrary dimensions using the same hybrid approach with as few as 24 unknowns. Further using the current description for the different regions shown in Fig. 9, they were able to solve for the complete surface current and scattered field for a TM plane wave illuminating a circular cylinder by including an unknown corresponding to the creeping wave excitation coefficient in the GTD shadow region. Some examples of this problem are shown in Fig. 9(b).

One should note that the previous solutions were obtained for problems which could have been analyzed directly using several methods. However, there are many practical examples for which the size of the structure doesn't allow one to attempt a MOM solution, and the geometry cannot be simulated by available GTD diffraction coefficients. For these problems, the hybrid solution just described is most appropriate. It was shown in [18] and [19] that one can analyze a wedge structure partially covered with a surface impedance material. In addition, they were successfully able to apply this hybrid approach to compute the scattering patterns for a plane wave illuminating an impedance loaded wing foil. An example of that solution is illustrated in Fig. 10. Their analysis was used to examine the backscatter reduction obtainable using impedance loaded wing structures. It was also shown in [19] that one can use this hybrid solution to analyze the TE polarization case.

More recently, this hybrid solution approach was successfully used in [20] to numerically compute the diffraction coefficient for a curved surface terminated ground plane as shown in Fig. 11. An example of that diffraction coefficient is shown in Fig. 12 and it was applied to compute the radiation pattern illustrated in Fig. 13. Note that this numerically derived diffraction coefficient can be computed with speeds approaching that of the analytic solutions such as for the wedge problem.

One should not be misled by the previous comments in that this hybrid solution approach is not necessarily limited to computing GTD diffraction coefficients. It can be used directly in a MOM sense to solve for the surface currents associated with a large structure and, indeed, that is what is done in terms of solving for the diffraction coefficients. However, there are many advantages associated with the diffraction coefficient approach:

1) One can solve for the scattered fields using asymptotic solutions for the various radiation integrals as done in [20].

2) Once a new diffraction coefficient is obtained, it can be applied to many structures which contain that geometry. Thus, a new MOM or hybrid solution need not be developed for the new geometry in that it can be solved in a GTD sense.

3) In many cases the 2-D diffraction coefficients can be directly extended to the 3-D case. This is illustrated in terms of the wedge problem where the 3-D solution is given by

$$\begin{bmatrix} E_{11}^d \\ E_{\perp}^d \end{bmatrix} = \begin{bmatrix} -D^s & 0 \\ 0 & -D^h \end{bmatrix} \begin{bmatrix} E_{11}^i \\ E_{\perp}^i \end{bmatrix} A(s)e^{-jks} \quad (9)$$

where D^s and D^h are basically 2-D solutions. The last point is most significant in that it is much easier to solve a 2-D problem as opposed to a 3-D one. Note that the planar/curved surface diffraction coefficient of [20] can be extended to 3-D problems using (9). For those individuals that might wish to solve the 3-D problem directly, they should refer to [21].

As shown by Thiele et al. [2], one can use the MOM to solve large structures by introducing GTD solutions to handle certain aspects of the structure. Consider the basic geometry illustrated in Fig. 14 in which structure (B) is too large to be analyzed.

In that the hybrid approach suggested by Thiele et al. [2] can be used to analyze large structures with relatively few unknowns, it is apparent that the resulting solutions will be extremely efficient. Recognizing this fact, it was shown in [24] that one could use a fast Fourier transform combined with this hybrid solution in order to obtain the time domain waveform response for a monopole mounted at the center of a circular disk. Of course, this methodology can be applied to obtain time domain waveforms for all types of geometries provided that the individual frequency domain hybrid solutions are efficient.

There has recently been a series of hybrid solutions developed using the approach outlined in [2]; however, these solutions are based on the GTD curved surface analyses

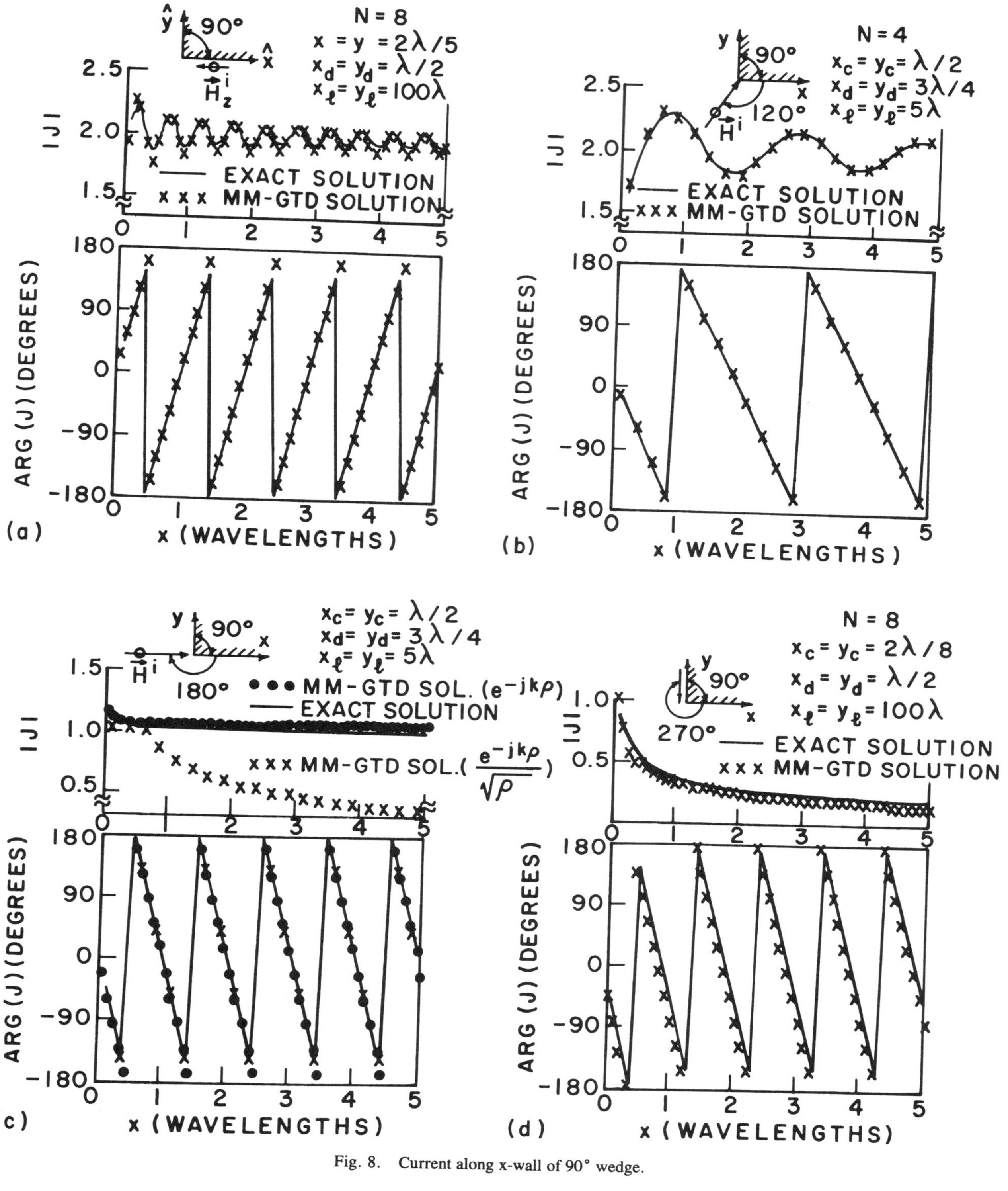

Fig. 8. Current along x-wall of 90° wedge.

presented earlier. The input impedance of a dipole antenna radiating in the presence of a curved surface is analyzed in [25]. The mutual coupling between a pair of dipoles positioned near but not on a curved surface is treated in [26]. Finally, the coupling between a pair of monopoles on large cylindrical surfaces is examined in [27]. It is very clear after reviewing the previous papers that this hybrid approach is a very powerful technique; however, it can only be applied provided the appropriate GTD solutions are available. For example, the hybrid solution of Thiele et al. [2] could not be used to solve for the input impedance for a monopole mounted on a ground plane terminated in a curved surface. However, if one used the previous hybrid solution [1], then one could numerically derive a diffraction coefficient which could be used in combination with the present hybrid solution [2] to solve for the input impedance.

There has been a great deal of effort applied to treat the near- and far-field radiation patterns for many structures. In

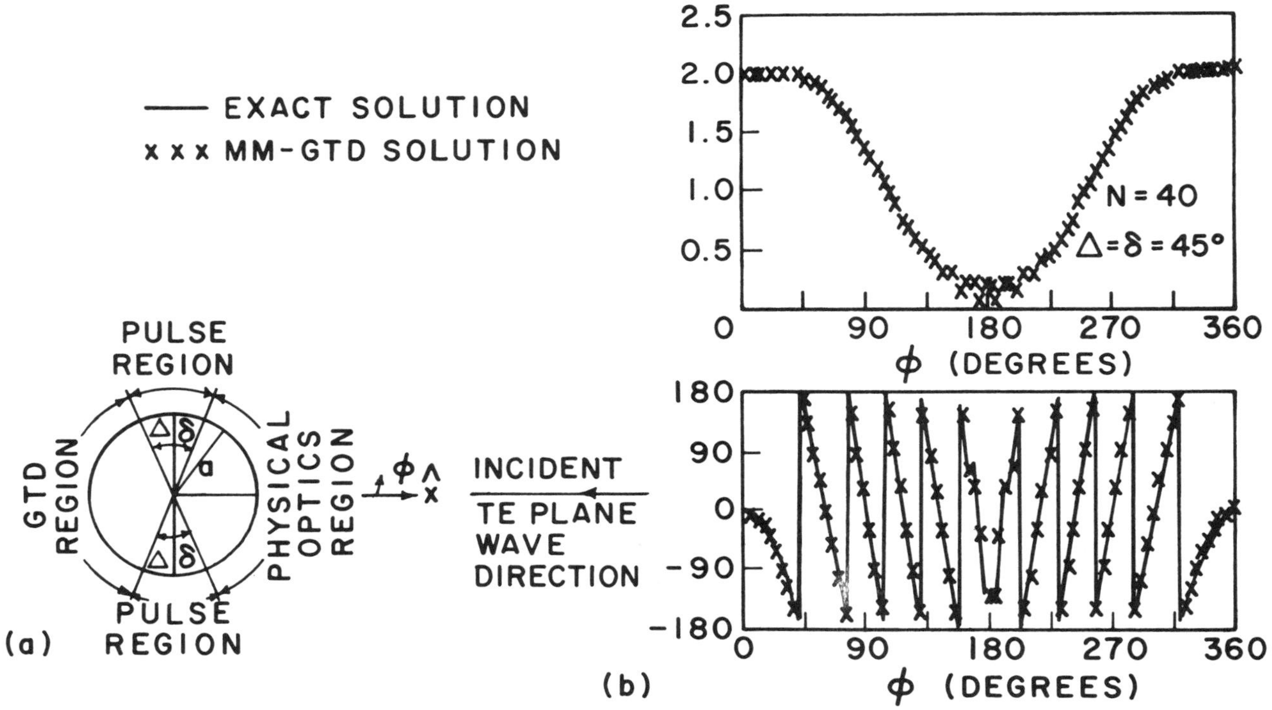

Fig. 9. (a) Various current regions used in MOM-GTD solution for circular cylinder of radius a. (b) Total current around circumference of circular cylinder with $a = 2\lambda$.

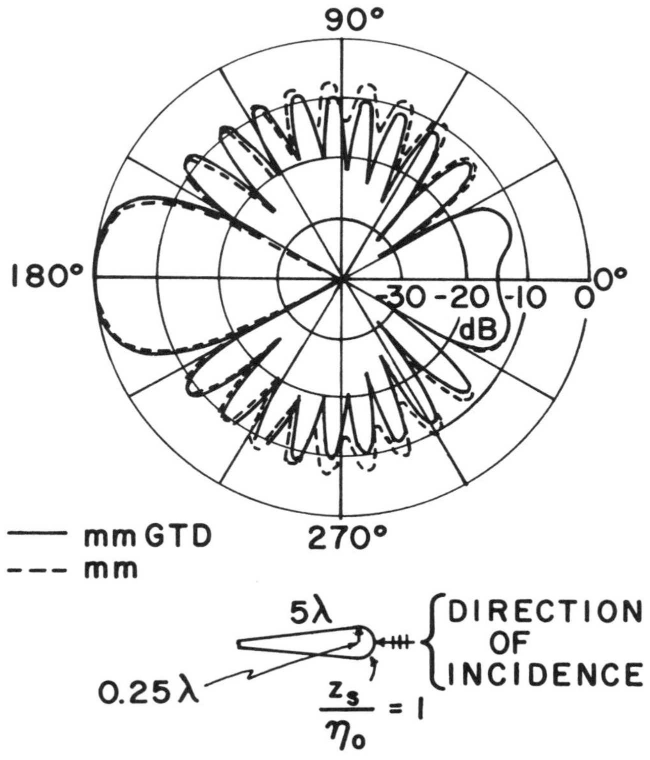

Fig. 10. Comparison of bistatic scattered fields obtained using MOM and MOMGTD solutions for a wing foil with impedance loading.

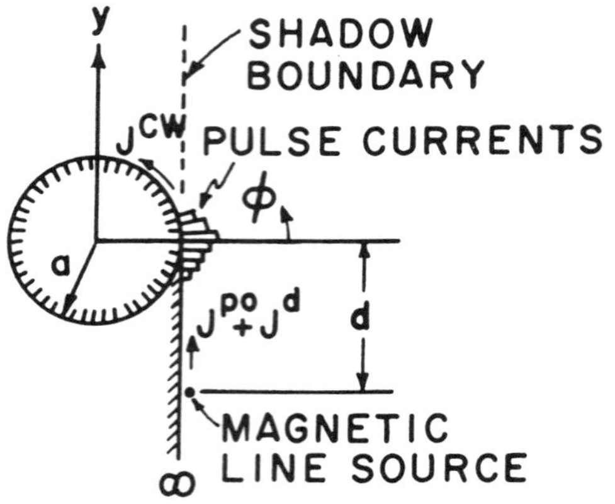

Fig. 11. Geometry associated with planar surface truncated by a circular cylinder.

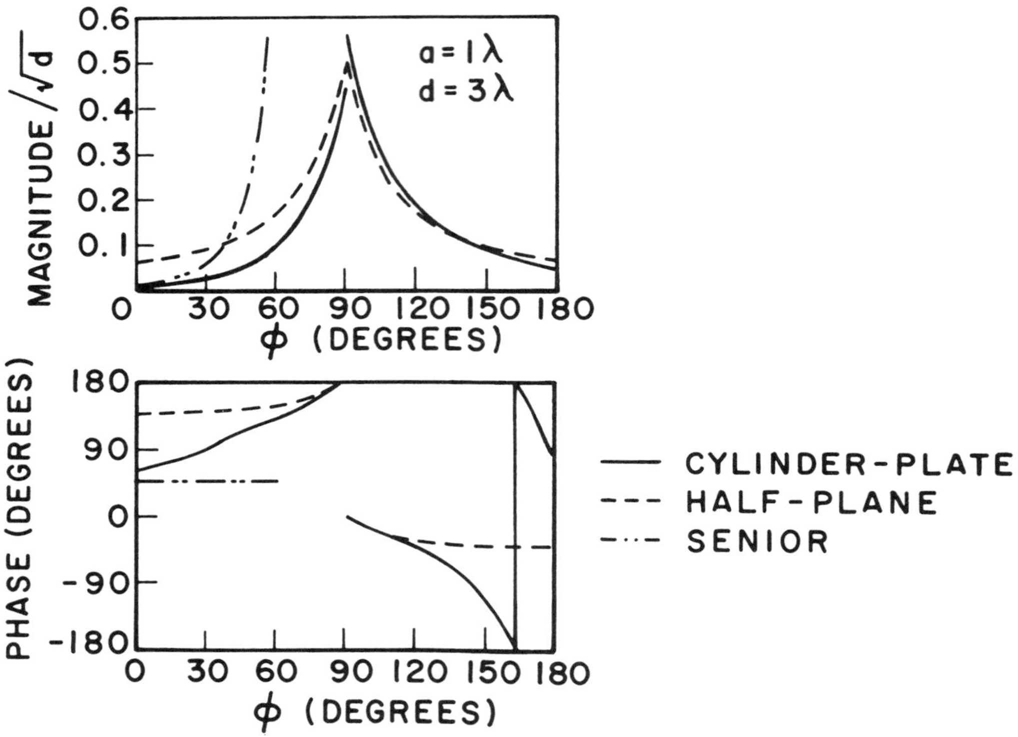

Fig. 12. Example of numerically derived diffraction coefficient compared against half-plane solution and Senior's solution.

fact, the three major GTD computer codes presented in the previous section could be modified using this hybrid approach to treat some very interesting EMC problems; one such example is to extend these using the MOM. However, structure (B) can be simulated by a simpler structure, if necessary, which can be analyzed completely using GTD. Using this concept, it is shown in [2] that the normal MOM solution given by

$$[Z_{mn}](I_n) = (V_m)$$

can be modified by using GTD solutions to treat the scattered field from structure (B) which results in giving

$$[Z'_{mn} + Z_{mn}^{\text{GTD}}](I'_n) = (V_m).$$

Note that Z_{mn}^{GTD} is the impedance matrix modification associated with the radiation from current element I_n, scattered by structure (B) in terms of GTD mechanisms, and received by current segment (m). Since the MOM current samples are only needed for the smaller structures, such as to represent a wire radiator, one can treat large objects using relatively few unknowns. This solution approach was used in [2] to analyze the input impedance of a monopole mounted on wedge structures, a square ground plane, and a circular ground plane. It is shown therein that one can obtain accurate input impedance results simply using a few unknowns to represent the current along the wire radiator. This method can be extremely useful for EMC applications which often require analysis of large structures. For example, a pair of UHF blade antennas mounted along the centerline of an aircraft

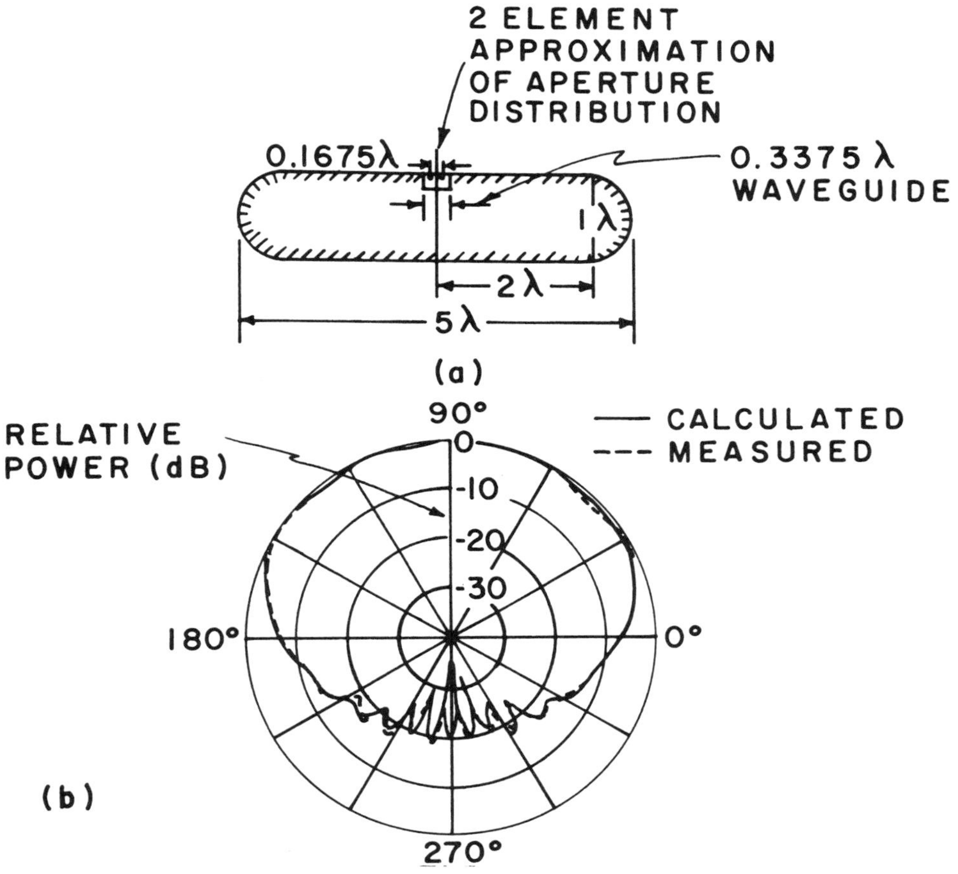

Fig. 13. (a) Axial waveguide mounted on a two-dimensional cylinder. (b) Principal plane radiation pattern.

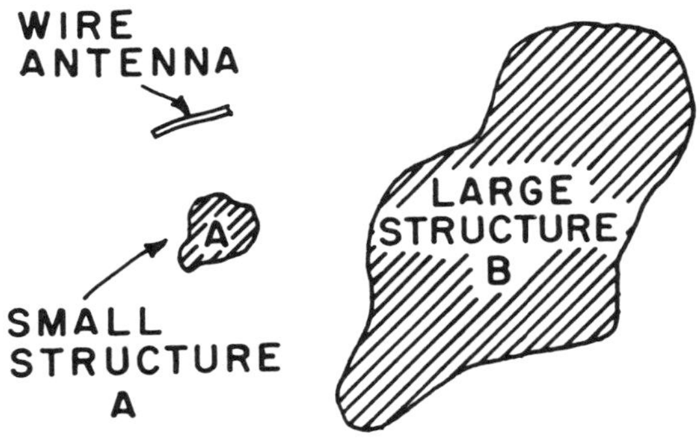

Fig. 14. Basic geometry treated using the hybrid method of [2].

can be represented by a few unknown MOM current samples; whereas, the aircraft scattered field can be represented by GTD solutions. Recall that the GTD has been used for about a decade to analyze complex airborne antenna patterns [7], [9], [22]. Even though the previous geometry is electrically very large, one should be able to obtain reasonable results for the coupling and radiation patterns using as few as three current samples to represent each UHF blade current distribution. Such an approach has been used in [23] to compute the radiation patterns for a 4-element UHF blade array mounted on an A-10 aircraft. Codes to analyze the coupling between a search radar and a UHF communications antenna mounted on a ship.

IV. Conclusions

This paper has attempted to present some of the recent advances in hybrid solutions involving Moment of Methods (MOM) and the Geometrical Theory of Diffraction (GTD). There are at least two basic ways to combine MOM and GTD. The one developed by Burnside et al. [1] can be used to numerically derive new diffraction coefficients. It can also be applied to analyze large structures for which GTD solutions do not exist. The hybrid solution approach originated by Thiele et al. [2] can be used in conjunction with existing GTD solutions to analyze large structures in which MOM current samples are only used to represent those portions of a structure which cannot be analyzed in terms of GTD alone. This hybrid solution approach has been successfully applied to analyze the input impedance and coupling for wire antennas near wedges, finite ground planes, and curved surfaces.

The EMC significance of these hybrid solutions is that one can use the MOM to represent the radiation or scattering from the antenna and complex subsections of a large structure which cannot be treated by GTD alone; whereas, the bulk of the configuration can be analyzed using GTD. Using these techniques one can, for example, analyze the input impedance and coupling for antennas mounted on an aircraft, ship, satellite, etc. where the supporting structure is large in terms of the wavelength but some critical detail of the structure is not.

References

[1] W. D. Burnside, C. L. Yu, and R. J. Marhefka, "A technique to combine the geometrical theory of diffraction and the moment method," *IEEE Trans. Antennas Propagat.*, vol. AP-23, no. 4, pp. 551–557, July 1975.

[2] G. A. Thiele and T. M. Newhouse, "A hybrid technique for combining moment methods with the geometrical theory of diffraction," *IEEE Trans. Antennas Propagat.*, vol. AP-23, no. 1, January 1975.

[3] *A Uniform GTD for the Diffraction by Edges, Vertices, and Convex Surfaces*, by R. C. Kouyoumjian, P. H. Pathak, and W. D. Burnside. Chapter to appear in: *Theoretical Methods for Determining the Interaction of Electromagnetic Waves with Structures*, edited by J. K. Skwirzynski, and being published by Sijthoff and Noordhoff, Netherlands.

[4] *A Uniform GTD and its Application to Electromagnetic Radiation and Scattering*, by R. G. Kouyoumjian, P. H. Pathak, and W. D. Burnside. Chapter to appear in: *Recent Developments in Classical Wave Scattering*, edited by V. K. Varadan and V. V. Varadan, and being published by the Pergamon Press, N.Y.

[5] "Modern Geometrical Theory of Diffraction," Ohio State University Short Course, Department of Electrical Engineering, September 1980.

[6] R. G. Kouyoumjian and P. H. Pathak, "A uniform geometrical theory of diffraction for an edge in a perfectly conducting surface," *Proc. IEEE*, vol. 62, pp. 1448–1461, 1975.

[7] W. D. Burnside, N. Wang, and E. L. Pelton, "Near field analysis of airborne antennas," *IEEE Trans. Antennas Propagat.*, vol. AP-28, no. 3, May 1980.

[8] W. D. Burnside, "Analysis of on-aircraft antenna patterns," Report 3390-1, The Ohio State University ElectroScience Lab., Dept. of Electrical Engineering: prepared under Contract N62269-72-C-0354 for Naval Air Development Center, August 1972.

[9] W. D. Burnside, R. J. Marhefka, and C. L. Yu, "Roll plane analysis of on-aircraft antennas," *IEEE Trans. Antennas Propagat.*, vol. AP-21, pp. 780–786, November 1973.

[10] P. H. Pathak, N. Wang, W. D. Burnside, and R. G. Kouyoumjian, "A uniform GTD solution for the radiation from sources on a perfectly-conducting convex surface," *IEEE Trans. Antennas Propagat.*, to appear.

[11] P. H. Pathak, "An asymptotic analysis of the scattering of plane waves by a smooth convex cylinder," *J. Radio Sci.*, vol. 14, no. 3, pp. 419–435, May–June 1979.

[12] P. H. Pathak, W. D. Burnside, and R. J. Marhefka, "A uniform GTD analysis of the diffraction of electromagnetic waves by a smooth convex surface," *IEEE Trans. Antennas Propagat.*, vol. AP-28, September 1980.

[13] P. H. Pathak and N. N. Wang, "An analysis of the mutual coupling between antennas on a smooth convex surface," Final Report 784583-7, The Ohio State University ElectroScience Lab., Dept. of Electrical Engineering; prepared under Contract N62269-76-C-0554 for Naval Air Development Center, Warminster, Pa., October 1978.

[14] P. H. Pathak and N. Wang, "Surface fields of sources on a perfectly-conducting convex surface," to appear in the *IEEE Trans. Antennas Propagat.*

[15] W. D. Burnside, R. C. Rudduck, and R. J. Marhefka, "Summary of GTD computer codes developed at the Ohio State University," *IEEE Trans. Electromagn. Compat.*, vol. EMC-22, no. 4, November 1980.

[16] G. J. Burke and A. J. Poggio, "Numerical electromagnetic code (NEC)—Method of moments," NOSC/TD116, Naval Oceans System Center, San Diego, Calif., July 1977.

[17] J. Sahalos and G. A. Thiele, "An improved formulation for extending the geometrical theory of diffraction by the moment method," Report 4372-2, The Ohio State University, ElectroScience Lab.; prepared under Contract N00014-76-C-0573, May 1977.

[18] S. Lee, "Control of electromagnetic scattering by antenna impedance loading," Report 3424-2, The Ohio State University ElectroScience Lab., Dept. of Electrical Engineering; prepared under Contract F19628-72-C-0202 (AFCRL-TR-74-0426), July 1974.

[19] J. A. Ass, "Control of electromagnetic scattering from wing profiles by impedance loading," Report 3424-4, The Ohio State University ElectroScience Lab., Dept. of Electrical Engineering; prepared under Contract F19628-72-C-0202 (AFCRL-TR-75-0463), August 1975.

[20] C. W. Chuang and W. D. Burnside, "A diffraction coefficient for a cylindrically truncated planar structure," *IEEE Trans. Antennas Propagat.*, vol. AP-28, no. 2, March 1980.

[21] J. N. Sahalos and G. A. Thiele, "On the application of the GTD–MOM technique and its limitations," Report 711353-1, The Ohio State University ElectroScience Lab.; prepared under Contract F19628-78-C-0198, October 1979.

[22] C. L. Yu, W. D. Burnside, and M. C. Gilreath, "Volumetric pattern analysis of airborne antennas," *IEEE Trans. Antennas Propagat.*, vol. AP-26, pp. 636–641, September 1978.

[23] W. D. Burnside and T. Chu, "Airborne antenna pattern code user's manual," Report 711679-2, The Ohio State University ElectroScience Lab.; prepared under Contract F30602-C-0068, March 1980.

[24] G. A. Thiele and G. K. Chan, "Application of the hybrid technique to time domain problems," *IEEE Trans. Antennas Propagat.*, vol. AP-26, no. 1, January 1978.

[25] E. P. Ekelman, Jr., "A hybrid technique for combining the moment method treatment of wire antennas with the GTD for curved surfaces," Ph.D. Dissertation, The Ohio State University, 1978.

[26] L. W. Henderson and G. A. Thiele, "A hybrid MOM–GTD technique for the treatment of wire antennas mounted on or near a curved surface," presented at the XX General Assembly of URSI; Munich, Germany, August 1980.

[27] S. A. Davidson and G. A. Thiele, "A hybrid method of moments–GTD technique for computing electromagnetic coupling between two monopole antennas on a large cylindrical surface," presented at the XX General Assembly of URSI; Munich, Germany, August 1980.

Hybrid Solutions for Scattering from Perfectly Conducting Bodies of Revolution

LOUIS N. MEDGYESI-MITSCHANG, MEMBER, IEEE, AND DAU-SING WANG

Abstract—A current-based hybrid formulation is developed for predicting the electromagnetic scattering from conducting bodies of revolution (BOR). The electric field integral equation (EFIE) formulation of the problem is solved by incorporating the Fock solution for the surface currents on the scatterer into the method of moments (MM) solution. To treat oblique illumination, the Fock results are extended to arbitrary surfaces with torsion. The formulation is illustrated for spheres and conespheres with smooth and discontinuous joins. The analysis includes nonspecular phenomena such as creeping wave effects. Application of the physical optics (PO) approximation in this hybrid formulation is discussed. The formulation is shown to be accurate even for scatterers in the near-resonance range (i.e., $ka \gtrsim 7.5$).

I. INTRODUCTION

RECENT treatments of electromagnetic radiation and scattering problems have followed two major avenues of analysis. The first is based on the generalization of classical optics through the geometrical and physical theories of diffraction (GTD and PTD) [1]-[3]. The second involves specialization of the concepts of linear spaces and orthogonal projection methods to solve Maxwell's equations. A particular attribute of the optic-derived methods, particularly GTD and its uniform extension UTD, is their conceptual simplicity and wide scope of applicability in treating certain nonspecular effects on electrically extended bodies. For bodies several wavelengths in extent (i.e., in the resonance range and beyond), the methods of Galerkin, Petrov, Kantorovich, and others [4], [5], often termed the method of moments (MM) or the method of projections, can yield results of great accuracy. Formally, this method is equivalent to the well-known Rayleigh-Ritz variational method [4], [6] and also to Rumsey's reaction concept in electromagnetics [7]. Harrington provided a systematic exposition of the use of MM to modern electromagnetics [8].

The MM approach is usually used with the integral formulation of an electromagnetic scattering problem. Such a formulation has the desirable property that all boundary and radiation conditions even for complex scatterers (and radiators) can be included. The unknown vector fields or currents are expanded in a sequence of basis functions, linearly independent and complete, spanning the entire domain of the integral operator. By appropriate testing of the domain of the operator, a linear algebraic set of equations results for the unknown vector quantities, from which all near-field and far-field quantities can be obtained. In principal, such a formulation incorporates the totality of all specular and nonspecular phenomena associated with the problem and hence is its great value.

While the MM approach is suitable for scattering problems involving complex geometries for a frequency range where the optic-derived methods are tenuous or inapplicable, the computational requirements of the method become prohibitive as the body dimensions approach ten wavelengths. The hybrid concept, initially proposed by Thiele and co-workers [9]-[12] provides an escape from this dilemma.[1] The hybrid approach is the logical confluence of the high- and low-frequency methods for treating electromagnetic problems where neither class of formulations by itself is adequate to the task. In its broadest context, the hybrid approach provides a unified theoretical framework into which the recent advances of the optic-derived and MM-based methods can be incorporated.

II. DISCUSSION OF CONCEPTS

A hybrid formulation can be formulated in various ways. In a field-based analysis, the GTD solutions for the fields, attributable to some physical process such as diffraction from an edge, are sampled by the MM testing functions, and these fields provide the effective source (voltage vector) for the MM solution. Alternately, in the current-based analysis developed here, the surface currents obtained from an optic-derived Ansatz, such as the Fock theory [15], are substituted into the original integral operators which are then sampled by the MM testing functions to provide the effective source for the MM solution. The relation between the field- and current-based approaches is analogous to that between geometrical and physical optics. The field-based approach is particularly useful for problems involving radiators on large convex surfaces for which canonical GTD solutions are available. The principal disadvantage of the field approach is that extensive ray tracing is often necessary. This requirement is particularly true in scattering problems. In the current-based approach, laborious ray tracing is avoided. Since surface currents (and fields) exhibit slower spatial variation than fields in the neighborhood of the scatterers, approximate forms for the surface currents are generally adequate. Since the integral operators provide a smoothing effect, sufficiently accurate results are often obtained in scattering problems even with approximate surface currents. A similar effect is observed when geometrical optic (GO) currents are used in the exact representations of scattered fields in physical optics (PO). While the GO currents are often demonstrably inaccurate (such as the currents near the vertex of a cone, for example), the PO results can be excellent. The principal disadvantage of the current-based hybrid formulation is that the currents must be integrated. However, stationary-phase approximations can be utilized (as in PTD) for the illuminated regions of the body.

[1] Recently, a hybrid-iterative method has been applied to the magnetic field integral equation (MFIE) formulation of Maxwell's equations. In this method, an asymptotic (i.e., physical optics) current is used in an iterative manner to solve the MFIE [13]. In another hybrid approach, Burnside *et al.* [14] demonstrated a combined MM-GTD technique to obtain numerically diffraction coefficients for noncanonical two-dimensional scatterers.

Manuscript received October 25, 1982; revised March 11, 1983. This work was supported under the McDonnell Douglas Independent Research and Development Program.

The authors are with the McDonnell Douglas Research Laboratories, St. Louis, MO 63166.

Reprinted from *IEEE Trans. Antennas Propagat.*, vol. AP-31, no. 4, pp. 570–583, July 1983.

In the current-based hybrid formulation, the integral equation describing the scattering physics is solved by incorporating the Galerkin expansion and local solutions of a parabolic-type partial differential equation (PDE). As shown by Fock, in diffraction problems involving smooth convex bodies, the latter equation can be solved in terms of the universal Fock functions for the fields (and currents) on the scatterer surface. Thus, in the present hybrid method, the integral equation is solved through the modality of coupling the solution of a PDE via the Galerkin technique. In this discussion we emphasize the use of the Fock-theory-based Ansatz solutions because of their generality. By the principle of locality, the Fock results hold for any convex body having at the point of illumination the same geometrical and material characteristics.

In this analysis, the scattering body is subdivided into smooth convex and irregular surfaces. The former refer to geometries for which the surface currents can be obtained from optic-derived methods such as the Fock theory or PO. Irregular surfaces refer to geometries not amenable to these methods, which include, but are not limited to, surfaces with various discontinuities caused by protrusions or concavities. (Surfaces with material discontinuities also fall into this category but are excluded from the present discussion.) These regions of the body are treated using a MM expansion for the surface currents. In some cases, it is also computationally convenient to treat the penumbra region with MM even though this part of the surface is smooth convex, since an optic-derived Ansatz requires calculation of higher order terms and complex geodesic lines.

The integro-differential operators with respect to the surface currents (or fields) in the EFIE or MFIE formulation can be separated into terms identified with the smooth convex and irregular surfaces of a body. In earlier work by Tew and Tsai [16] and Azarbar and Shafai [17], the full potential of this decomposition was not used since the Ansatz and the MM formulations spanned the entire body.

Since the present formulation is a current-based approach, the Ansatz solutions for the currents in the smooth convex regions are required. These currents are deduced from the Fock theory, which allows the surface currents (and fields) induced on a convex surface by a plane wave to be computed for surfaces with radii of curvature large with respect to λ. Furthermore, use of the Fock currents allows accurate representation of the fields generated by grazing illumination, which is important in many applications. While the original Fock-derived results were restricted to the penumbra region, these results can be extended to be valid everywhere on a canonical body [18]. Furthermore, the Fock-derived currents have the unique property that they provide a transition between the physical optics and creeping wave currents. This transition is important in a well-behaved hybrid analysis. The principle of locality in the Fock theory for the fields and currents in the penumbra region is implicitly used in constructing the Ansatz solutions.

In Section III we develop the hybrid formulation, using the electric field integral equation (EFIE). Specialized results needed for the Ansatz solutions are derived from the original Fock theory in Section IV. To treat the important case of oblique illumination, the Fock results are generalized to treat arbitrary convex surfaces with torsion. This generalization is carried out to include first- and second-order terms in the illuminated and shadow regions, respectively, although extension to higher orders is possible. In Section V analytical results are obtained

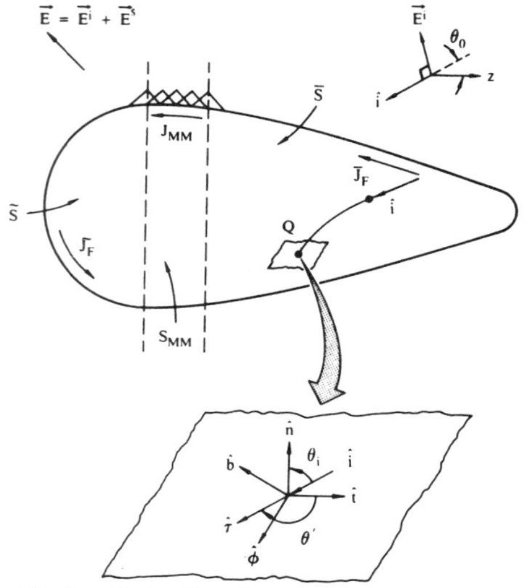

Fig. 1. Body of revolution scatterer geometry.

for axial and nonaxial illumination of bodies of revolution (BOR's) and are compared with MM solutions and experimental data for a number of configurations where surface wave phenomena dominate. These cases provide a sensitive indicator of the accuracy of the hybrid approach.

III. HYBRID GALERKIN FORMULATION

Consider a perfectly conducting body described in Fig. 1. The tangential scattered and incident fields on the surface, expressed in terms of the electric field integral equation are

$$\vec{E}_t^s = -\vec{E}_t^i = -L\vec{J}. \tag{1}$$

The integro-differential operator L is given as

$$L\vec{J} = jk\eta \int_S \left(\vec{J} + \frac{1}{k^2} \nabla\nabla' \cdot \vec{J}\right) \Phi \, ds', \tag{1a}$$

where the free space Green's function is $\Phi = (1/4\pi R) \exp(-jkR)$, R is the distance from the source to the field point, ∇ is the surface gradient on the body with respect to the unprimed variables, $k = 2\pi/\lambda$, and $\eta = \sqrt{\mu/\epsilon}$. The current density $\vec{J}$ is the unknown of the problem. (An alternative starting point, restricted to closed surfaces, is the MFIE formulation.) The domain of integration in the operator L can be decomposed to span parts of the illuminated and shadowed regions of the body $\bar{S}$ and $\tilde{S}$, respectively, for which an optic-derived Ansatz is available for the surface currents. The remainder of the surface is denoted as S_{MM} (i.e., the irregular part) treated with a MM representation. Since the MM expansion of the currents is the same for both the illuminated and shadowed parts, both regions are combined in S_{MM}. The MM region does not span the entire surface but is localized to those regions where the optic-derived solutions are inapplicable or must be corrected. One can rewrite (1) as

$$L_{S_{MM}}\vec{J} = E^i - L_{\bar{S}}\vec{J} - L_{\tilde{S}}\vec{J}, \tag{2}$$

where the domain of integration for each of the terms in (2) is indicated by the subscripts on L. The current density $\vec{J}$ is defined in the respective regions as

$$\vec{J} = \vec{J}_{MM} \equiv \sum_{n,j} (\hat{t}I_{nj}^t J_{nj}^t(s) + \hat{\phi}I_{nj}^{\phi} J_{nj}^{\phi}(s)), \quad s \in S_{MM}, \tag{3}$$

and the optic-derived (Fock) currents are

$$\vec{J} = \vec{J}_F \equiv \begin{cases} \hat{t}\bar{J}_F^t(s) + \hat{\phi}\bar{J}_F^\phi(s), & s \in \bar{S}; \\ \hat{t}\tilde{J}_F^t(s) + \hat{\phi}\tilde{J}_F^\phi(s), & s \in \tilde{S}, \end{cases} \tag{4}$$

where $\hat{t}$ and $\hat{\phi}$ are the surface unit vectors on the BOR. Explicit expressions for the currents in (4) are derived in Section IV. (Often as a first approximation, one can take $\bar{J}_F^t$ and $\bar{J}_F^\phi$ as the PO currents on the surface.)

A Galerkin solution procedure will be used to solve (2) in terms of the incident fields and the Ansatz currents. Let the basis set for the MM representation in (3) be defined as

$$\vec{J}_{nj}^\beta(s) = \hat{\beta} J_{nj}^\beta(s) = \hat{\beta} f_j^\beta(t) e^{jn\phi}, \qquad \beta = t \text{ or } \phi, \tag{5}$$

where $f_j^\beta(t)$ is a triangle function as defined in [19]. Taking the testing functions as $\vec{W}_{nj}^\beta = (\vec{J}_{nj}^\beta)^*$, the Galerkin procedure yields

$$Z_n I_n = V_n - \bar{V}_n - \tilde{V}_n, \tag{6}$$

where Z_n is a matrix containing the electromagnetic interactions of the MM-represented regions. The matrix Z_n, partitioned into four submatrices corresponding to the t and ϕ directed current components, is

$$Z_n = \begin{bmatrix} Z_n^{tt} & Z_n^{t\phi} \\ Z_n^{\phi t} & Z_n^{\phi\phi} \end{bmatrix}, \tag{7}$$

where the (i, j)th element of the submatrices is obtained from

$$Z_{n,ij}^{\alpha\beta} = \langle \vec{W}_{ni}^\alpha, L(\vec{J}_{nj}^\beta) \rangle, \tag{8}$$

where α, β are combinations of t and ϕ, n is the mode number, $\vec{W}_{ni}^\alpha$ is the ith BOR testing function, and $\vec{J}_{nj}^\beta$ is the jth BOR surface current basis function on the jth annulus into which the MM-represented part of the surface of the body is segmented. For a BOR, the inner product in (8) can be expressed as [19]

$$Z_{n,ij}^{\alpha\beta} = jk\eta \int_i dt \int_0^{2\pi} \rho\, d\phi \int_j dt' \int_0^{2\pi} \rho'\, d\phi' \cdot \left\{ \vec{W}_{ni}^\alpha \cdot \vec{J}_{nj}^\beta - \frac{1}{k^2} (\nabla \cdot \vec{W}_{ni}^\alpha)(\nabla' \cdot \vec{J}_{nj}^\beta) \right\} \Phi, \tag{9}$$

where

$$\nabla \cdot \vec{W}_{ni} = \frac{1}{\rho} \frac{\partial}{\partial t} (\rho W_{ni}^t) + \frac{1}{\rho} \frac{\partial}{\partial \phi} (W_{ni}^\phi) \tag{9a}$$

and similarly for $\nabla' \cdot \vec{J}_{nj}$.

The right side of (6) is a function of the illuminating field and of the Ansatz solutions for $\bar{S}$ and $\tilde{S}$, represented in V_n, $\bar{V}_n$, and $\tilde{V}_n$, respectively. Formally, the ith elements of these column vectors are

$$V_{ni}^\alpha = \langle \vec{W}_{ni}^\alpha, \vec{E}^i \rangle, \quad \alpha = t \text{ or } \phi, \tag{10a}$$

$$\bar{V}_{ni}^\alpha = \langle \vec{W}_{ni}^\alpha, L_{\bar{S}}(\vec{J}_F) \rangle \tag{10b}$$

and

$$\tilde{V}_{ni}^\alpha = \langle \vec{W}_{ni}^\alpha, L_{\tilde{S}}(\vec{J}_F) \rangle. \tag{10c}$$

These vectors are also partitioned into t and ϕ components and are evaluated by testing the incident fields and the fields obtained from Fock-derived solutions over the MM-represented part of the body. Physically, the right side of (6) represents an equivalent hybrid excitation of the MM region of the body consisting of a plane wave illumination and that resulting from the optic-derived Ansatz currents. Explicit expressions for $\bar{V}_n^\alpha$ and $\tilde{V}_n^\alpha$ for axial and nonaxial illumination are derived in Section V. Expressions for V_n^α are given in [19]. Equation (6) can be solved for the unknown currents in the MM-represented region of the body knowing the voltage column vectors.

IV. SURFACE FIELDS ON A CONVEX BODY WITH TORSION

In this section we develop the results for the surface fields on a general convex surface with torsion, resulting from plane wave illumination. Earlier, Pathak *et al.* [20] used uniform theory of diffraction (UTD) to obtain the surface fields due to electric and magnetic sources on a convex body. These results are not immediately applicable to our problem, necessitating a reformulation. A differently structured derivation of these results for spherical and cylindrical surfaces is given by Ruck [21].

Our formulation begins from the perspective of the Fock theory. We consider two canonical problems: a sphere and a general convex cylinder subject to plane wave illumination. To introduce torsion, the cylinder is illuminated obliquely. The expressions for the surface fields can be obtained from the Debye and Hertz potentials, expressed asymptotically in terms of the Fock functions. These two canonical solutions are then combined to construct the fields on the surface of a general convex torsional surface subject to plane wave illumination.

Case 1: Spherical Surface

Consider the canonical problem of a sphere of radius a, illuminated by a plane wave (Fig. 2(a)), i.e.,

$$\vec{E}^i = \hat{x} e^{-jkz},$$

where the time dependence $e^{-j\omega t}$ is implied. The magnetic fields (the η factor is suppressed) are obtained from

$$\vec{H} = \nabla \times \nabla \times \hat{r}\pi_1 + jk\nabla \times \hat{r}\pi_2, \tag{11}$$

where explicit expressions for the surface fields, in terms of the potentials π_1 and π_2, are

$$H_\theta = jk \frac{1}{r \sin\theta} \frac{\partial \pi_1}{\partial \phi} - \frac{1}{r} \frac{\partial^2 \pi_2}{\partial r \partial \theta} \tag{12}$$

$$H_\phi = -jk \frac{1}{r} \frac{\partial \pi_1}{\partial \theta} + \frac{1}{r \sin\theta} \frac{\partial^2 \pi_2}{\partial r \partial \phi}. \tag{13}$$

The potentials, approximated on the spherical surface at a point $(r = a, \phi, \theta)$, are

$$\pi_1 \cong \frac{1}{k^2} [e^{-jk\tau_1} g(\xi_1) - je^{-jk\tau_2} g(\xi_2)] D \cos\phi \tag{14}$$

and

$$\frac{\partial \pi_2}{\partial r} = (\hat{n} \cdot \nabla \pi_2) \cong \frac{1}{mk} [e^{-jk\tau_1} f(\xi_1) - je^{-jk\tau_2} f(\xi_2)] D \sin\phi. \tag{15}$$

The resulting magnetic fields from (12), (13) are [21, p. 90]

$$H_\theta = \left\{ \frac{j}{m} [c_0 e^{-jk\tau_1} f(\xi_1) + je^{-jk\tau_2} f(\xi_2)] - \frac{j}{ka \sin\theta} [e^{-jk\tau_1} g(\xi_1) - je^{-jk\tau_2} g(\xi_2)] \right\} D \sin\phi \tag{16}$$

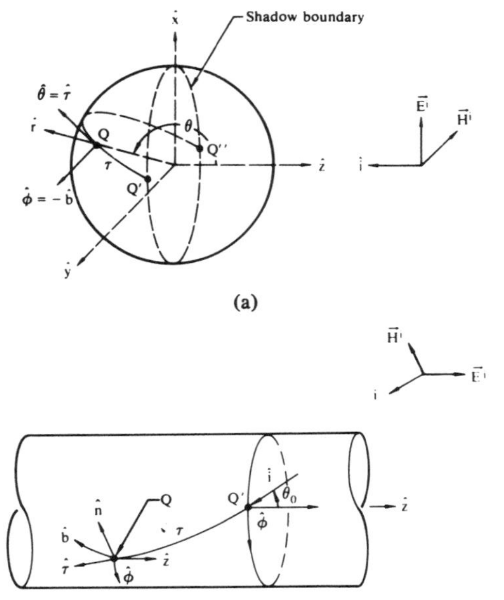

Fig. 2. (a) Coordinates for spherical scatterer. (b) Coordinates for cylindrical scatterer.

and

$$H_\phi = -\{c_0 e^{-jk\tau_1} g(\xi_1) + je^{-jk\tau_2} g(\xi_2) - \frac{1}{kam \sin\theta}[e^{-jk\tau_1} f(\xi_1) - je^{-jk\tau_2} f(\xi_2)]\} D \cos\phi. \quad (17)$$

The diffusion factor D is defined as

$$D = \frac{1}{\sqrt{\sin\theta}}.$$

The other parameters are

$$\tau_1 = \begin{cases} a(\theta - \pi/2), & \pi/2 \leqslant \theta \leqslant \pi \text{ (shadow region)}; \\ -a\cos\theta + \xi_1^3/3k, & 0 \leqslant \theta \leqslant \pi/2 \text{ (illuminated region)}, \end{cases}$$

$$\xi_1 = \begin{cases} m\tau_1/a, & \pi/2 \leqslant \theta \leqslant \pi; \\ -m\cos\theta, & 0 \leqslant \theta \leqslant \pi/2, \end{cases}$$

$$\tau_2 = a(3\pi/2 - \theta),$$
$$\xi_2 = m\tau_2/a,$$

$$c_0 = \begin{cases} 1, & \text{shadow region}, \\ D^{-1}, & \text{illuminated region}, \end{cases}$$

and

$$m = \left(\frac{ka}{2}\right)^{1/3},$$

where m^{-1} is the angular width of the penumbra region of the sphere. The Fock functions are defined as

$$f(\xi) = \frac{1}{\sqrt{\pi}} \int_{-\infty}^{\infty} \frac{e^{-j\xi\tau}}{w_2(\tau)} d\tau \quad (18)$$

and

$$g(\xi) = \frac{1}{\sqrt{\pi}} \int_{-\infty}^{\infty} \frac{e^{-j\xi\tau}}{w_2'(\tau)} d\tau, \quad (19)$$

where $w_2(\tau)$ and $w_2'(\tau)$ are Fock-type Airy functions, i.e., $w_2(\tau) = \sqrt{\pi}[\text{Bi}(\tau) - j\text{Ai}(\tau)]$ (see [22] and [23] for details).

The two values of τ and ξ arise from the fact that a given point in the shadow region is at the confluence of two different geodesic paths emanating from the shadow boundary on the sphere. Generally, in the deep illuminated region, the specular components of the fields dominate. Hence one can neglect the higher order terms in (16) and (17) in addition to those that are functions of τ_2 and ξ_2 and are associated with the creeping waves.

Case 2: Arbitrary Convex Cylindrical Surface

Consider an obliquely illuminated convex cylinder depicted in Fig. 2(b). The incident field forms an angle θ_0 with the z-axis, and the incident field direction $\hat{i}$ is in the xz plane. The magnetic fields are obtained in terms of the Hertz potentials, π_1 and π_2, defined below [24]. Thus,

$$\vec{H} = \nabla \times \nabla \times \hat{z}\pi_2 + jk\nabla \times \hat{z}\pi_1 \quad (20)$$

yielding

$$H_\phi = jk\frac{\partial \pi_1}{\partial \rho} + \frac{1}{\rho}\frac{\partial^2 \pi_2}{\partial z \partial \phi} \quad (21)$$

and

$$H_z = -\frac{jk}{\rho}\frac{\partial \pi_1}{\partial \phi} + \frac{1}{\rho}\left(\frac{\partial^2(\rho\pi_2)}{\partial \rho^2} - \frac{1}{\rho}\frac{\partial^2 \pi_2}{\partial \phi^2}\right). \quad (22)$$

In the vicinity of the cylindrical surface, the Hertz potentials can be expressed in terms of the Fock functions, defined in (18) and (19), as

$$\left.\begin{matrix}\pi_1 \\ \pi_2\end{matrix}\right\} \cong \frac{e^{-jk\tau}}{k^2 \sin\theta_0}\left(\frac{\rho_g(Q')}{\rho_g(Q)}\right)^{1/6} D \left\{\begin{matrix} df(\xi) \\ g(\xi)\end{matrix}\right., \quad (23)$$

where the diffusion factor $D = 1$ and ρ_g is the radius of curvature at Q' or Q along the geodesic path. The distance d above the surface is given in terms of the radius of the cylinder at point Q, $a(Q)$, i.e.,

$$d = \frac{k}{m(Q)}(\rho(Q) - a(Q)).$$

As a first approximation, creeping wave terms in (23) are deleted. The potentials π_1 and π_2 correspond to the perpendicularly polarized (TM) and parallel polarized (TE) cases, respectively. The surface magnetic field for the TM case is

$$H_\phi = \frac{-j}{m \sin\theta_0} e^{-jk\tau} f(\xi) \left(\frac{\rho_g(Q')}{\rho_g(Q)}\right)^{1/6} D. \quad (24)$$

Similarly for the TE case,the fields are

$$H_\phi = -e^{-jk\tau} g(\xi) \left(\frac{\rho_g(Q')}{\rho_g(Q)}\right)^{1/6} D \cos\theta_0 \quad (25)$$

and

$$H_z = -e^{-jk\tau} g(\xi) \left(\frac{\rho_g(Q')}{\rho_g(Q)}\right)^{1/6} D \sin\theta_0, \quad (26)$$

where

$$\tau = \begin{cases} p \sin\theta_0 + z\cos\theta_0, & \text{shadow region}; \\ \frac{1}{k}(\vec{k} \cdot \vec{r} + \xi^3/3), & \text{illuminated region}, \end{cases}$$

$$\xi = \begin{cases} \int_{Q'}^{Q} ds \, \dfrac{m(s)}{\rho_g(s)}, & \text{shadow region;} \\ -m_l(Q) \cos \theta_i, & \text{illuminated region,} \end{cases}$$

$$P = \int_{Q'}^{Q} a(s) \, d\phi, \quad a(s) = \text{radius at point } s,$$

$$m(s) = \left(\frac{k\rho_g(s)}{2} \right)^{1/3}, \quad \text{shadow region,}$$

and

$$m_l(Q) = \frac{m(Q)}{(1 + T_0^2 \cos^2 \theta_i)^{1/3}}, \quad \text{illuminated region,}$$

where $T_0 = \cot \theta_0$.

Case 3: Arbitrary Convex Surface with Torsion

The surface fields and currents for this case are obtained by combining the Fock-theory-based solutions for the sphere and cylinder discussed above. This generalization is based on the principle of locality, inherent in the Fock theory and the GTD (UTD) formulations, namely, that the high frequency solutions for the surface fields depend on the local geometrical characteristics of the surface. Two characteristics that distinguish an arbitrary convex surface are its two principal radii of curvature (R_1, R_2) and the varying torsion of the geodesic paths on its surface. In general, R_1 and R_2 are finite and unequal. Consider the two special cases of a sphere and a cylinder. A spherical surface has $R_1 \equiv R_2$, i.e., no torsion. On the other hand, a cylinder at oblique incidence has torsion (varying if the cylinder is noncircular) but has only one finite radius of curvature; the other radius of curvature is infinite. Thus the general solution for an arbitrary convex surface must encompass the features of both canonical solutions discussed before. As in the development in [20], the synthesis of the general solution is heuristic. Its validity is established by considering special cases, such as the conesphere at oblique illumination, where torsion factors play an important role in the surface fields and currents.

Because the surface fields on a general convex surface are most suitably expressed in terms of the tangent vector $\hat{\tau}$ along the geodesic path and its binormal vector $\hat{b} = \hat{\tau} \times \hat{n}$, we reexpress the fields in (16), (17), and (24)–(26) in terms of these components.

Consider the spherical case first. The geodesic paths are the great circles on the sphere; therefore $\hat{\tau} = \hat{\theta}$ and $\hat{b} = \hat{\tau} \times \hat{n} = \hat{\theta} \times \hat{r} = -\hat{\phi}$. Hence, $H_\theta \to H_\tau$ and $H_\phi \to -H_b$. Noting that $\tau = a\theta$, $b = -a\phi \sin \theta$, and $\rho_g = a$ everywhere on the sphere, for TM illumination along the geodesic path ($\phi = 0$), neglecting the creeping wave terms, (17) becomes

$$\vec{H} = \hat{b} \left[c_0 e^{-jk\tau} g(\xi) - \frac{e^{-jk\tau}}{kma \sin \theta} f(\xi) \right] \left(\frac{\rho_g(Q')}{\rho_g(Q)} \right)^{1/6} D, \tag{27}$$

and likewise for the TE case ($\phi = \pi/2$), (16) is

$$\vec{H} = \hat{\tau} \left[\frac{j}{m} c_0 e^{-jk\tau} f(\xi) - j \frac{e^{-jk\tau}}{ka \sin \theta} g(\xi) \right] \left(\frac{\rho_g(Q')}{\rho_g(Q)} \right)^{1/6} D. \tag{28}$$

The parameters τ and ξ have values as before depending upon whether the surface fields are in the shadow or illuminated regions of the sphere.

Next we consider the cylindrical case. Relating the cylindrical coordinates to the vectors $\hat{\tau}$ and $\hat{b}$ of the geodesic path, i.e.,

$$\hat{\phi} = \hat{\tau} \sin \theta_0 - \hat{b} \cos \theta_0,$$

$$\hat{z} = -\hat{\tau} \cos \theta_0 - \hat{b} \sin \theta_0,$$

then the magnetic fields for the TM case (24) become

$$\vec{H} = \hat{\phi} H_\phi = -je^{-jk\tau}(\hat{\tau} - \hat{b} T_0) \left(\frac{\rho_g(Q')}{\rho_g(Q)} \right)^{1/6} m^{-1}(Q) f(\xi) D \tag{29}$$

and for the TE case ((25), (26)),

$$\vec{H} = \hat{\phi} H_\phi + \hat{z} H_z = \hat{b} e^{-jk\tau} \left(\frac{\rho_g(Q')}{\rho_g(Q)} \right)^{1/6} g(\xi) D. \tag{30}$$

The functions $\rho_g(Q')$ and $\rho_g(Q)$ were defined before and

$$T_0 = T \rho_g(Q'),$$

where T is the surface torsion factor defined as

$$T = \frac{\sin \theta_0 \cos \theta_0}{a(Q')}.$$

The parameters τ and ξ in (29) and (30) are defined as before for the shadow and illuminated regions of the body.

The general solution is formed by combining the two sets of fields for the TE and TM polarizations obtained for the sphere and the cylinder ((27)–(30)). Using the nomenclature for TE and TM in cylindrical coordinates, then for the TE case,

$$\vec{H} = \hat{b} \left(c_0 F_1 + jF_2 D^2 \frac{\Lambda}{k\rho_g(Q)} \right), \tag{31}$$

and for the TM case,

$$\vec{H} = -\hat{\tau} \left(c_0 F_2 - jF_1 D^2 \frac{\Lambda}{k\rho_g(Q)} \right) + \hat{b} T_0 F_2, \tag{32}$$

where

$$\begin{rcases} F_1 \\ F_2 \end{rcases} = e^{-jk\tau} \left(\frac{\rho_g(Q')}{\rho_g(Q)} \right)^{1/6} D \begin{cases} g(\xi) \\ jm^{-1}(Q) f(\xi) \end{cases}. \tag{33}$$

F_1 and F_2 are defined in terms of the parameters for the sphere or the cylinder. The parameters in (33) for each of these canonical geometries are summarized in Table I. The shape factor Λ, which provides an interpolation between the sphere and cylinder solutions as given in [20], is defined as the ratio of the principal radii of curvature on a body, i.e.,

$$\Lambda = \begin{cases} 0, & \text{cylinder;} \\ 1, & \text{sphere;} \\ R_2(Q')/R_1(Q'), & \text{general convex surface;} \\ & R_1(Q') \geqslant R_2(Q'). \end{cases}$$

For arbitrary oblique illumination, the incident wave is a superposition of the TE and TM solutions. Define the vectors $\hat{e}$ and $\hat{h}$ as unit vectors along the incident fields, i.e.,

$$\hat{e} = \frac{\vec{E}^i}{|\vec{E}^i|}$$

TABLE I
GEOMETRY PARAMETERS

		Sphere	Cylinder	Cone*	Arbitrary convex surface**
D	diffusion factor	$(\sin\theta)^{-1/2}$	1	1	$\sqrt{\frac{\tau d\psi_0}{\rho_c d\psi}}$***
T(s)	torsion at s	0	$\frac{\sin 2\theta_0}{2a(s)}$	$\frac{\cos\alpha \sin 2A(s)}{2\rho(s)}$	$\frac{\sin 2\theta'}{2}\left[\frac{1}{R_2(s)} - \frac{1}{R_1(s)}\right]$
$\rho_g(s)$	radius of curvature at s	a	$\frac{a(s)}{\sin^2\theta_0}$	$\frac{\rho^3(s)}{C_1^2\cos\alpha}$	$\left[\frac{\cos^2\theta'}{R_1(s)} + \frac{\sin^2\theta'}{R_2(s)}\right]^{-1}$
$T_0(s)$	$= T(s)\rho_g(s)$	0	$\cot\theta_0$	$\tan A(s)$	—
$\Lambda(s)$	shape factor (ratio of principal radii of curvature)	1	0	0	$\frac{R_2(s)}{R_1(s)}$

* See Table II.
** $R_1(s) \geqslant R_2(s)$.
*** See [20].

TABLE II
GEOMETRY PARAMETERS FOR CONES

Parameters		Shadow region	Illuminated region
$\rho(s)$	radius at s along a given geodesic line	$C_1 \sec A(s)$ (*)	—
$R_1(s)$	principal radius along $\hat{t}$	∞	∞
$R_2(s)$	principal radius along $\hat{\phi}$	$\frac{\rho(s)}{\cos\alpha}$	$\frac{\rho(s)}{\cos\alpha}$
m(s)	(angular width)$^{-1}$ of penumbra region	$\left[\frac{k\rho^3(s)}{2C_1^2\cos\alpha}\right]^{1/3}$	$\left[\frac{k\rho(s)}{2\cos\alpha[1-\sin^2\theta_i\cos^2\theta']}\right]^{1/3}$
ξ_1	Fock parameter	$\left[\frac{kC_1}{2}\right]^{1/3} \mid \phi(Q') - \phi(Q) \mid \cos^{2/3}\alpha$	$-m(s)\cos\theta_i$
ξ_2	Fock parameter	$\left[\frac{kC_1}{2}\right]^{1/3} \mid \phi(Q'') - \phi(Q) \mid \cos^{2/3}\alpha$	—
τ_1	geodesic path length, Q′ to Q	$\left\lvert\frac{C_1}{\sin\alpha}[\tan A(Q') - \tan A(Q)]\right\rvert$	—
τ_2	geodesic path length, Q″ to Q	$\left\lvert\frac{C_1}{\sin\alpha}[\tan A(Q'') - \tan A(Q)]\right\rvert$	—

* Where $A(s) = [\phi(s) - C_2] \sin\alpha$; C_1 and C_2 are constants on a given geodesic line; θ_i and θ' are defined in Fig. 1; Q'' is opposite Q' on shadow boundary.

and

$$\hat{h} = \frac{\vec{H}^i}{|\vec{H}^i|}.$$

Thus, the total $\hat{b}$-directed magnetic field on an arbitrary convex surface in the shadow region is

$$H_b = (\hat{n}\cdot\hat{e})\left(F_1 + j\frac{\Lambda}{k\rho_g(Q)}D^2F_2\right) + (\hat{n}\cdot\hat{h})T_0F_2, \qquad (34)$$

and similarly, the $\hat{\tau}$-directed field is

$$H_\tau = -(\hat{n}\cdot\hat{h})\left(F_2 - j\frac{\Lambda}{k\rho_g(Q)}D^2F_1\right). \qquad (35)$$

For the illuminated region, neglecting second-order terms, the total field is

$$\vec{H} = [(\hat{e}\cdot\hat{n}')\hat{b}F_1 - (\hat{h}\cdot\hat{n}')\hat{\tau}F_2]D^{-1}, \qquad (36)$$

where $\hat{n}' = \hat{i} \times \hat{b}$ at the specular point and $\hat{i}$ is the incident field direction. The functions F_1 and F_2 are defined in (33) with the parameters in these expressions given following (26) and summarized in Table I. The corresponding parameters for conical surfaces are given in Table II.

V. SCATTERING CROSS SECTIONS FOR AXIAL AND OBLIQUE ILLUMINATION

To compute the scattering cross sections, the unknown current coefficients in the Galerkin expansion in (6) are solved in terms of the generalized voltage vectors in (10). Expressions for V_n^α are given in [19]. Specific analytical expressions for $\overline{V}_n^\alpha$ and $\tilde{V}_n^\alpha$ are obtained below. For example, the ith element of $\tilde{V}_n^\alpha$ is given by

$$\tilde{V}_{ni}^\alpha = j\eta k\int_0^{2\pi} d\phi \int_i \rho\, dt \cdot \int_{\tilde{S}}\left[\vec{W}_{ni}^\alpha\cdot\vec{J}_F - \frac{1}{k^2}(\nabla\cdot\vec{W}_{ni}^\alpha)(\nabla'\cdot\vec{J}_F)\right]\Phi\, ds', \qquad (37)$$

where from (34), (35), and (36) the Fock-Ansatz currents are

$$\vec{J}_F = \frac{1}{\eta}\left\{\hat{e}\cdot(\hat{n}\hat{\tau})\left[F_1 + \frac{j\Lambda}{k\rho_g(Q)}D^2F_2\right] + \hat{h}\cdot\left(\hat{n}\hat{b}\left[F_2 - \frac{j\Lambda}{k\rho_g(Q)}D^2F_1\right] + \hat{n}\hat{\tau}T_0F_2\right)\right\}. \qquad (38)$$

(Note, the η factor was suppressed in Section IV for simplicity, but is reinserted here.) One can relate the unit vectors of the geodesic line and its binormal to those of a BOR at a given point. If the geodesic line and the generating curve of the BOR subtend an angle θ', then as in Fig. 1

$$\hat{\tau} = \hat{t}\cos\theta' + \hat{\phi}\sin\theta'$$

and the binormal vector is

$$\hat{b} = \hat{\tau} \times (\hat{\phi} \times \hat{t}).$$

Equation (37) can be written explicitly as

$$\tilde{V}_{ni}^t = j\eta k\sum_j \int_0^{2\pi} d\phi \int_i \rho\, dt \int_0^{2\pi} d\phi' \int_j \rho'\, dt' \cdot \Big\{\{[(\sin v \sin v' \cos\psi + \cos v \cos v')\cos\theta' - \sin v' \sin\psi \sin\theta']J_F^\tau - [\sin v' \sin\psi \cos\theta' + (\sin v \sin v' \cos\psi + \cos v \cos v')\sin\theta']J_F^b\} \cdot f_i(t)e^{-jn\phi} - \frac{1}{k^2}\dot{f}_i(t)e^{-jn\phi}\nabla'\cdot\vec{J}_F\Big\}\Phi, \qquad (39)$$

where v and v' denote the angles subtended by t and t' with the z-axis, $\dot{f}_i(t)$ is the derivative of $f_i(t)$, $\psi = \phi - \phi'$, and J_F^τ and J_F^b denote the $\hat{\tau}$ and $\hat{b}$ directed components of $\vec{J}_F$. The integrals over t and t' span the ith annular segment of S_{MM} and the jth segment of $\tilde{S}$, respectively. An expression similar to (39) holds for $\tilde{V}_{ni}^\phi$. For the illuminated region, the expressions for $\overline{V}_{ni}^t$ and $\overline{V}_{ni}^\phi$ are obtained similarly, with

$$\vec{J}_F = \frac{1}{\eta}\{(\hat{e}\cdot\hat{n}')\hat{\tau}F_1 + (\hat{h}\cdot\hat{n}')\hat{b}F_2\}. \qquad (40)$$

In the general case, the Ansatz currents are complicated functions, and hence it is difficult to obtain a general expression for the divergence terms of $\vec{J}_F$. An alternative procedure to obtain the generalized voltage vectors is as follows. Expand the Ansatz currents in the shadow region as

$$\vec{J}_F = \sum_{nj}[\tilde{I}_{nj}^t\vec{J}_{nj}^t + \tilde{I}_{nj}^\phi\vec{J}_{nj}^\phi], \qquad (41)$$

where $\vec J^t_{nj}$ and $\vec J^{\phi}_{nj}$ are the same basis functions as used in the expansion of $\vec J_{MM}$ but are evaluated on $\tilde S_j$ (the jth annular section into which $\tilde S$ is divided). The current coefficients $\tilde I^t_{nj}$ and $\tilde I^{\phi}_{nj}$ can be calculated because the Ansatz currents are known. Substituting (41) into (37) gives

$$\tilde V^{\alpha}_{ni} = j\eta k \sum_j \int_0^{2\pi} d\phi \int_i \rho\, dt \int_{\tilde S_j} ds' \cdot \left\{ \tilde I^t_{nj}\left[\vec W^{\alpha}_{ni}\cdot \vec J^t_{nj} - \frac{1}{k^2}(\nabla\cdot \vec W^{\alpha}_{ni})(\nabla'\cdot \vec J^t_{nj})\right] + \tilde I^{\phi}_{nj}\left[\vec W^{\alpha}_{ni}\cdot\vec J^{\phi}_{nj} - \frac{1}{k^2}(\nabla\cdot\vec W^{\alpha}_{ni})(\nabla'\cdot\vec J^{\phi}_{nj})\right]\right\}\Phi. \tag{42}$$

$\tilde V^{\alpha}_{ni}$ in (42) can be rewritten in terms of the generalized impedance matrix $(\tilde Z^{\alpha\beta}_n)_{ij}$, i.e.,

$$\tilde V^{\alpha}_{ni} = \sum_j \left[(\tilde Z^{\alpha t}_n)_{ij}\tilde I^t_{nj} + (\tilde Z^{\alpha\phi}_n)_{ij}\tilde I^{\phi}_{nj}\right], \tag{43}$$

where

$$(\tilde Z^{\alpha\beta}_n)_{ij} = \langle \vec W^{\alpha}_{ni}, L_{\tilde S_j}(\vec J^{\beta}_{nj})\rangle. \tag{44}$$

The above procedure avoids the difficulty of evaluating the divergence of $\vec J_F$ since the general expression for $(\nabla'\cdot\vec J^{\beta}_{nj})$ is given by

$$\nabla'\cdot\vec J^{\beta}_{nj} = \begin{cases} \dot f_j(t')e^{jn\phi'}, & \text{for } \beta = t; \\ jnf_j(t')e^{jn\phi'}, & \text{for } \beta = \phi. \end{cases} \tag{45}$$

Similar expressions can be obtained for $\bar V^{\alpha}_{ni}$ in the illuminated region.

If the illuminated surface is large compared to λ and of slowly varying curvature, then the PO approximation can be used. In this case, in the illuminated region for an incident field $\vec H^i$ at a point $\vec r$,

$$\vec J_{PO} = 2\hat n \times \vec H^i = \frac{2}{\eta}(\hat\phi \sin v \sin\phi - \hat t\cos\phi)e^{-j\vec k\cdot\vec r}. \tag{46}$$

It can be shown following steps analogous to those used in deriving $\tilde V^{\alpha}_{ni}$ above that for horizontally polarized incidence,

$$\bar V^{\alpha}_{ni} = -\frac{e^{j\pi\left(\frac{n+1}{2}\right)}}{\eta}\sum_j e^{jkz_j\cos\theta_0}\{[J_{n+1}(\rho) - J_{n-1}(\rho)] \cdot (\bar Z^{\alpha t}_n)_{ij} - e^{j\pi/2}\sin v_j[J_{n-1}(\rho) + J_{n+1}(\rho)](\bar Z^{\alpha\phi}_n)_{ij}\} \tag{47}$$

and for vertically polarized incidence,

$$\bar V^{\alpha}_{ni} = \frac{-e^{j\pi\left(\frac{n+1}{2}\right)}}{\eta}\sum_j e^{jkz_j\cos\theta_0}\{[\sin v_j\cos\theta_0(J_{n+1}(\rho) - J_{n-1}(\rho)) - 2e^{j\pi/2}\cos v_j\sin\theta_0 J_n(\rho)](\bar Z^{\alpha\phi}_n)_{ij} - e^{j\pi/2}\cos\theta_0[J_{n+1}(\rho) + J_{n-1}(\rho)](\bar Z^{\alpha t}_n)_{ij}\}, \tag{48}$$

where $(\bar Z^{\alpha\beta}_n)_{ij}$ is defined analogously to (44). The argument of the Bessel functions is $\rho = k\rho_j \sin\theta_0$, where ρ_j and z_j are evaluated at the jth annulus in $\bar S$. Once the current coefficients in the MM region are solved in terms of V, $\bar V$, and $\tilde V$, the scattering cross sections can be evaluated.

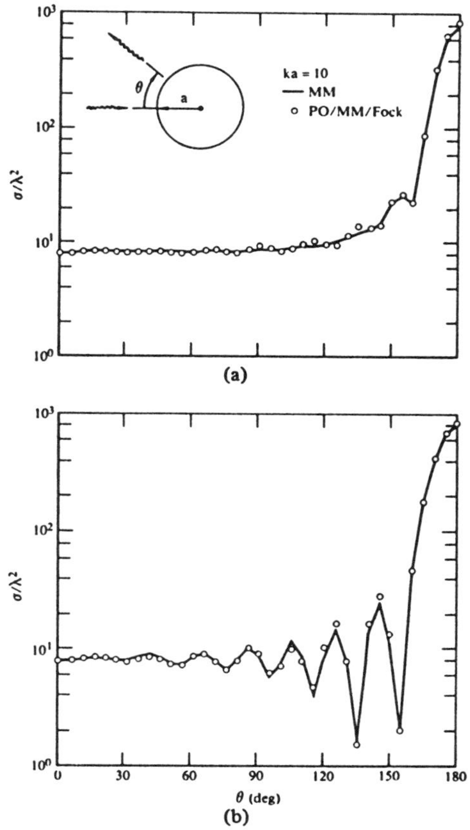

Fig. 3. Bistatic scattering cross sections of a sphere computed using an MM and a hybrid analysis. (a) Vertical polarization. (b) Horizontal polarization.

VI. COMPUTED RESULTS AND DISCUSSION

To provide a stringent test of the hybrid analysis, it was applied to three configurations: a sphere and two conespheres, one with a small cone half-angle ($\alpha = 10°$) and a relatively smooth conesphere join region and one with a large cone half-angle ($\alpha = 45°$) and a discontinuous join region. The scattering from these geometries at certain aspect angles is dominated by nonspecular effects such as creeping and traveling waves. In addition, the conespheres are surfaces with torsion and sharp discontinuities. To demonstrate the generality of the method, both axial and oblique illumination were considered. To provide a worst-case test of the hybrid formulation, the three bodies were chosen to be in the near-resonant range where the Ansatz solutions are least accurate. The results below clearly illustrate, however, that the total hybrid solution was accurate. In all cases presented, a MM solution for BOR geometries was used for comparison using a derivative of the Mautz-Harrington code for the numerical computations. The accuracy of this code has been verified in our laboratory for a variety of geometries. The computations were generally carried out at 2° or 5° increments in the azimuthal scattering angle θ.

Scattering cross-section predictions for the sphere ($ka = 10$) are shown in Fig. 3 for vertical and horizontal polarizations. The Fock and PO solutions were used in the hybrid analysis, in the shadow and the illuminated regions, respectively. In the penumbra region, subtending 1.34 λ, the currents were represented

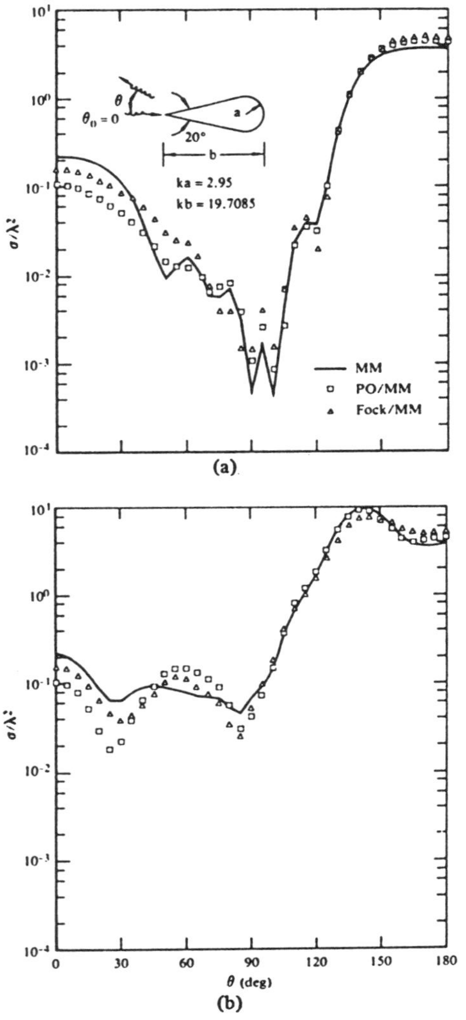

Fig. 4. Bistatic scattering cross sections for a conesphere with smooth join using MM analysis and two hybrid representations: axial incidence, $\theta_0 = 0°$. (a) Vertical polarization. (b) Horizontal polarization.

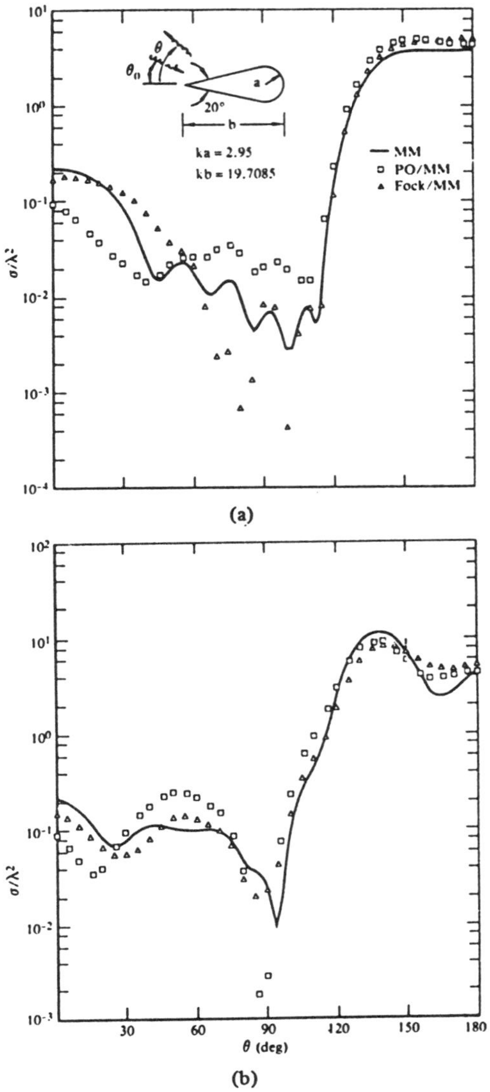

Fig. 5. Bistatic scattering cross sections for a conesphere with smooth join using MM analysis and two hybrid representations: oblique incidence, $\theta_0 = 5°$. (a) Vertical polarization. (b) Horizontal polarization.

by the Galerkin expansion, using 10 triangle functions and $n = \pm 1$ circumferential modes. Identical results were obtained by substituting the Fock Ansatz currents for the PO solution in the illuminated region. Obviously, the sphere is a canonical geometry for which the Fock solution can treat the entire body. The MM representation was introduced in the sphere case for the sole purpose of testing the hybrid formulation. The amplitude and phase of the surface fields on the sphere predicted by our analysis are in agreement with the Fock results given in [21, figs. 2-21 and 2-22]. (The legends for the "argument" in these figures are in error and should read $[-\exp(ik_0 a \cos\theta)]$.)

The analysis was also applied to the conesphere with $\alpha = 10°$. The length of the body was 3.136 λ (kb = 19.7085) and ka = 2.95. The scattering cross sections for vertical and horizontal polarizations are depicted in Fig. 4 for axial illumination (θ_0 = 0°). In the hybrid analysis of this problem, the MM representation is used for the spherical cap and join region and PO or Fock for the conical part. The discrepancy near nose-on is due to the fact the body is small and higher order terms are needed to calculate the generalized voltage vectors $\overline{V}_n$. In Fig. 5, the corresponding results are shown for oblique illumination (θ_0 = 5°) using two different hybrid formulations as shown in Fig. 4. The Fock/MM hybrid yields somewhat more accurate answers, i.e., the Fock Ansatz current is more accurate than the PO Ansatz. The MM region for the hybrid calculations in Figs. 4 and 5 spanned 1.09 λ of the body surface. For a larger body, the MM region and thus the matrix size will not increase. However, larger bodies require more execution time to evaluate the excitation vectors, i.e., $\overline{V}_n$ and $\tilde{V}_n$ in (6). But this is less restricting than the large matrices encountered in a pure MM analysis of an electrically extended body.

The conesphere with α = 45°, kb = 20, and ka = 10 was considered next. It had a discontinuous join region. The bistatic scattering cross sections are given in Fig. 6 for axial illumination using two hybrid formulations, i.e., Fock/MM/Fock and PO/MM/Fock. Here both hybrid formulations yield very accurate results. Apparently, the PO Ansatz is satisfactory for conical surfaces with large half-angles. The corresponding results for oblique illumination at θ_0 = 5, 10, 20, and 30° are depicted in Figs. 7-9 and 11, using the PO/MM/Fock hybrid solutions. These results are in good agreement with the MM solutions. Fig. 10 shows the

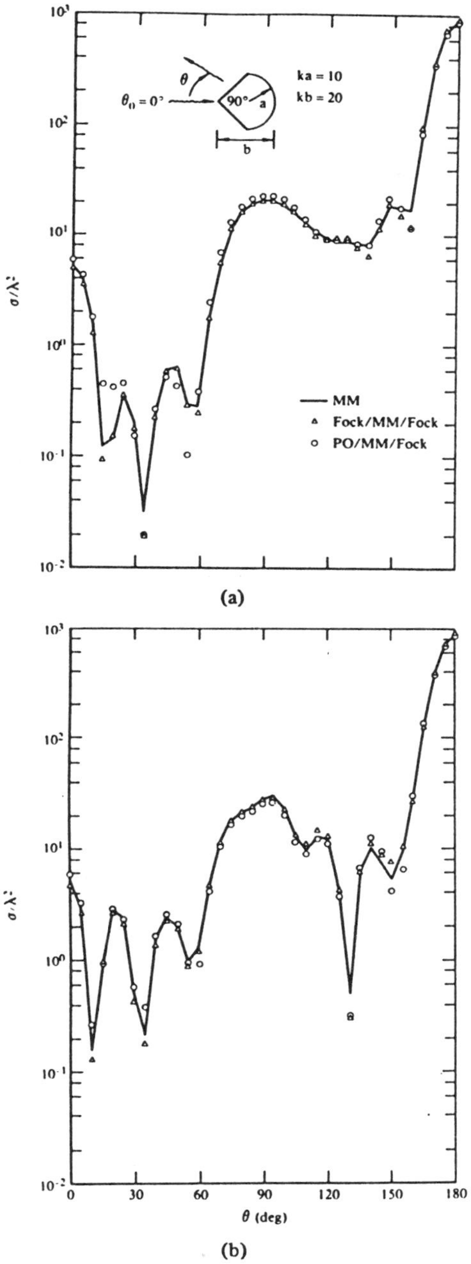

Fig. 6. Bistatic scattering cross section for a conesphere with a discontinuous join using MM analysis and two hybrid representations: axial illumination, $\theta_0 = 0°$. (a) Vertical polarization. (b) Horizontal polarization.

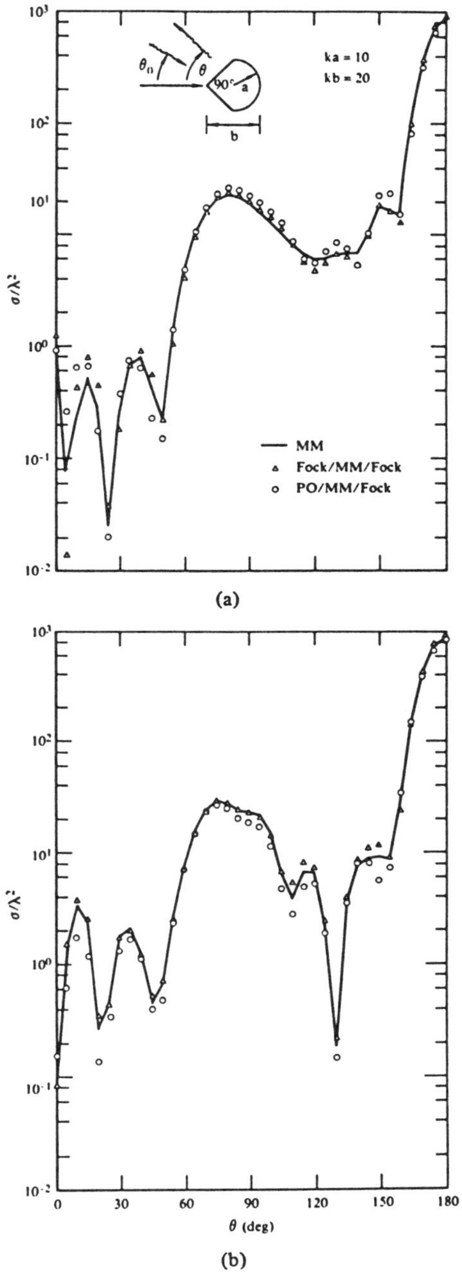

Fig. 7. Bistatic scattering cross sections for the conesphere in Fig. 6 using MM analysis and two hybrid representations: oblique incidence, $\theta_0 = 5°$. (a) Vertical polarization. (b) Horizontal polarization.

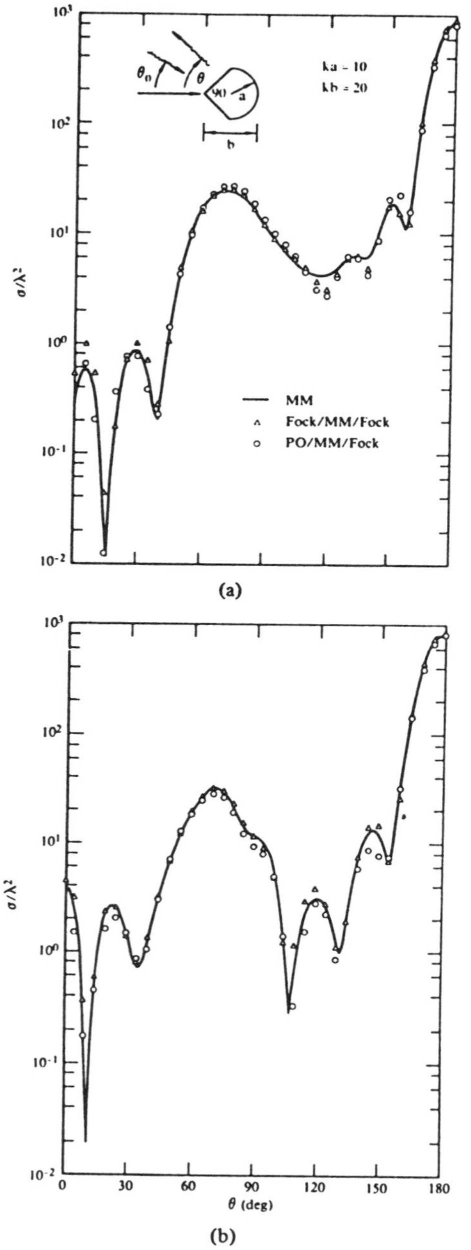

Fig. 8. Bistatic scattering cross sections for the conesphere in Fig. 6 using MM analysis and two hybrid representations: oblique incidence, $\theta_0 = 10°$. (a) Vertical polarization. (b) Horizontal polarization.

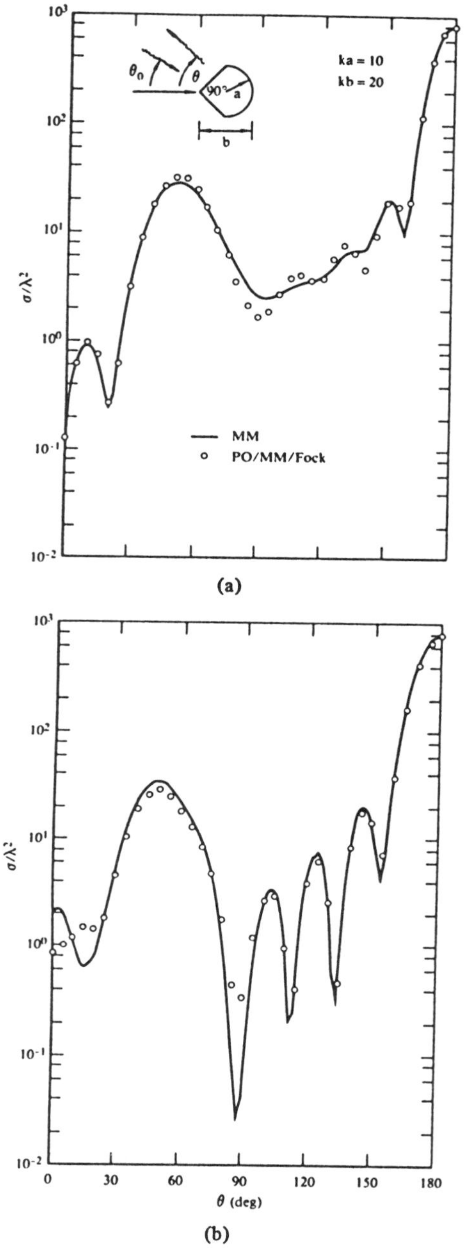

Fig. 9. Bistatic scattering cross sections for the conesphere in Fig. 6 using MM analysis and a hybrid representation: oblique incidence, $\theta_0 = 20°$. (a) Vertical polarization. (b) Horizontal polarization.

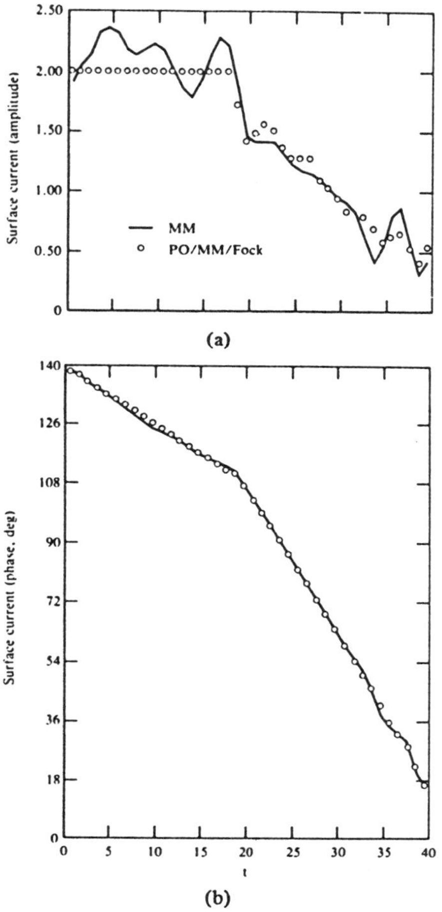

Fig. 10. Amplitude (a) and Phase (b) of t-directed surface currents on conesphere for illumination given in Fig. 9(b) as a function of t (t = 0 denotes the cone tip, t is in arbitrary units).

amplitude (normalized by the factor $|H^i|$) and phase of the t-directed surface currents along $\phi = 0$ on the conesphere, illuminated at $\theta_0 = 20^\circ$ in horizontal polarization, obtained from the MM and hybrid solutions. Apparently, the excellent agreement of the phases ensures that the cross-section results in Fig. 9 are satisfactory despite the discrepancy of the current magnitudes near the tip region of the body.

The bistatic cross sections for $\theta_0 = 50^\circ$ are shown in Fig. 12 using the two hybrid solutions that were employed to calculate Fig. 6. Setting the Fock currents equal to zero in the shadow region on the cone degrades the results as seen in Fig. 13. With the angle of illumination, $\theta_0 > \alpha$, the cross section is dominated by nonspecular effects. In this case, the PO solutions are known to be inaccurate, and yet the hybrid results are in good agreement with the MM calculation. The hybrid solution using the Fock Ansatz current everywhere except the join region yields better results than alternative combinations. The pure MM calculations were carried out with 123 annular segments on the body. In the hybrid analysis we used 23 segments in the MM region. The number of circumferential modes used for the oblique incidence cases was $n = 15$, and the same number was used in both hybrid calculations.

One of the benefits of using MM in the junction region is that exact knowledge of the shadow boundary (SB) in the neighborhood of the join was unnecessary. Although the conesphere is a relatively simple body, at $\theta_0 > \alpha$, the SB is discontinuous across the junction region and has a complicated form which must be recomputed for each angle of illumination. This problem can be overcome by choosing a sufficiently wide MM region.

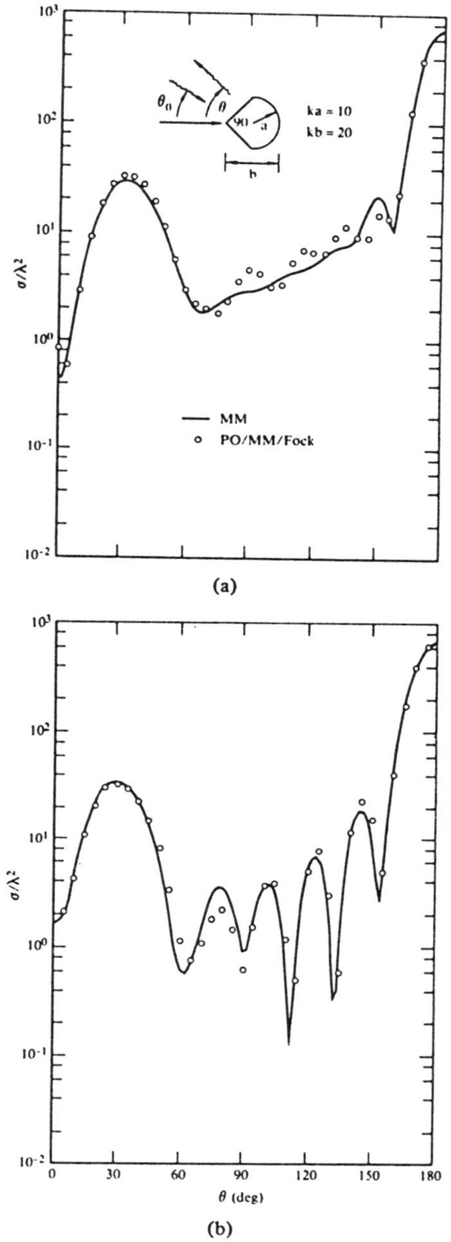

Fig. 11. Bistatic scattering cross sections for the conesphere in Fig. 6 using MM analysis and a hybrid representation: oblique incidence, $\theta_0 = 30^\circ$. (a) Vertical polarization. (b) Horizontal polarization.

We applied the hybrid theory to examine the creeping wave enhancement on an axially illuminated conesphere with a smooth join to provide a final test of the hybrid technique. Senior [25] studied this phenomenon in detail and developed an approximate theory to explain the data. In Fig. 14, we compare the results of the hybrid formulation with the Senior results (theory and experiment) for conespheres with $7.5 < ka < 10.5$ and a cone

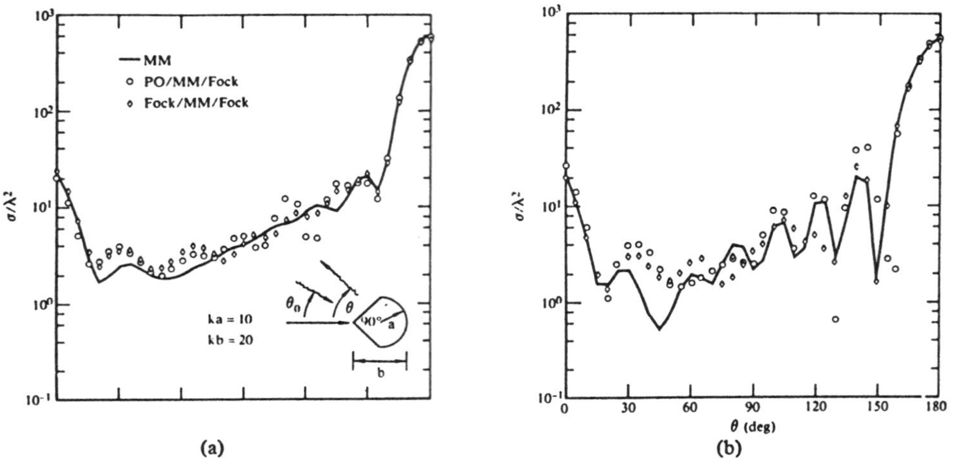

Fig. 12. Bistatic scattering cross sections for the conesphere in Fig. 6 using MM analysis and two hybrid representations: oblique incidence, $\theta_0 = 50°$. (a) Vertical polarization. (b) Horizontal polarization.

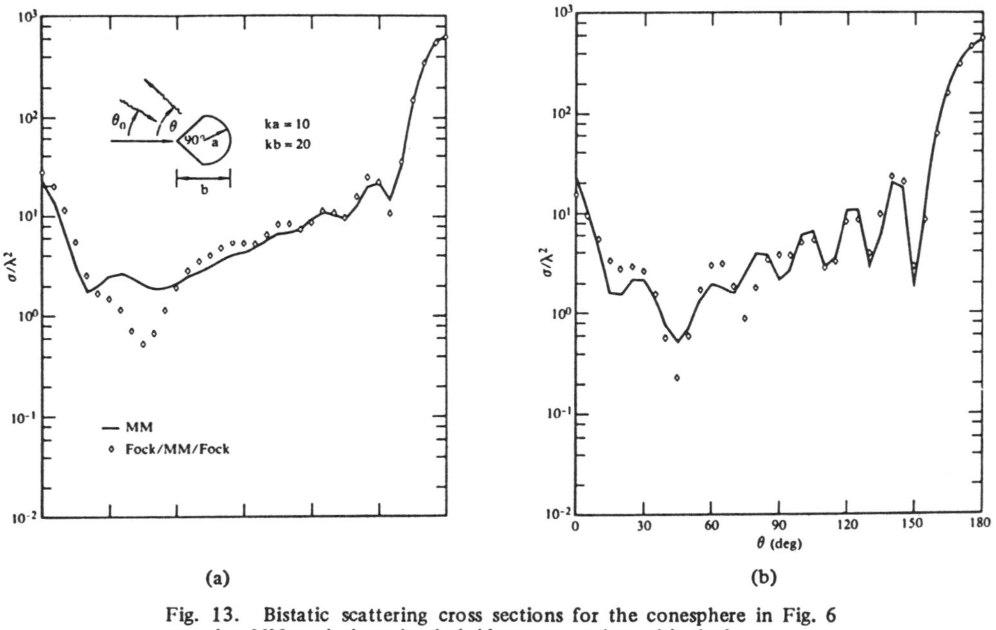

Fig. 13. Bistatic scattering cross sections for the conesphere in Fig. 6 using MM analysis and a hybrid representation with shadow currents on cone set to zero: oblique incidence, $\theta_0 = 50°$. (a) Vertical polarization. (b) Horizontal polarization.

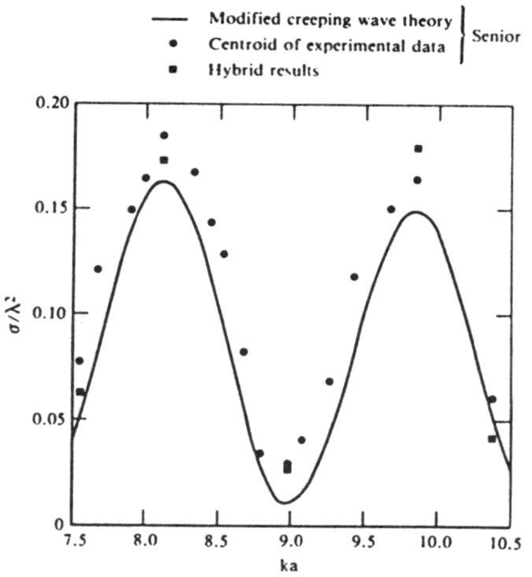

Fig. 14. Nose-on backscattering cross sections for a conesphere (half-angle is 12.5°) with smooth join: comparison of the hybrid analysis and the results of Senior as a function of *ka*.

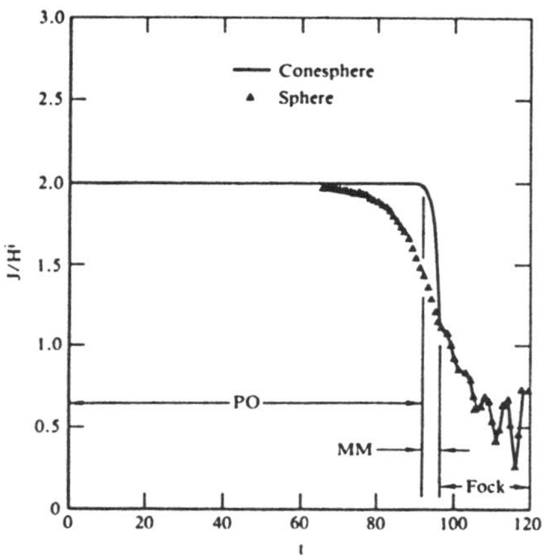

Fig. 15. Comparison of t-directed current amplitudes on a conesphere in Fig. 14 (ka = 9.852) and a sphere of same dimension as a function of surface parameter t (t = 0 denotes the cone tip, t is in arbitrary units).

half-angle, $\alpha = 12.5°$. The hybrid results agree well with the measurements. In the hybrid formulation, PO and Fock Ansatz currents were used for the cone and spherical cap surfaces; the MM representation was used in the SB region. Our calculations confirmed Senior's conclusion that the use of the PO currents on the sides of the cone near the join is valid for $ka \cos \alpha \ll 1$. The join, however, is not solely responsible for the creeping wave enhancement. Instead, the effect appears to be due to the perturbation of the currents on the spherical surface in the SB region caused by the presence of the cone. As shown in Fig. 15, in the SB region (where we used the MM expansion), there is a significant enhancement of the original spherical currents. The results depicted in Fig. 15 were computed for a conesphere with $ka = 9.852$ and $\alpha = 12.5°$. Senior noted similar enhancement effects experimentally.

Further observations from our analysis are as follows. Using PO currents near the SB leads to incorrect results as one would expect from Ufimtsev's observation about the presence of fringe currents near geometrical or shadow boundaries. The spherical Fock currents are also in error near the SB since the canonical sphere results are modified by the presence of the cone. Thus one must use the MM representation in this transition region. Finally, the Senior results in Fig. 14 are at the extreme limit of validity of the hybrid formulation since the Ansatz solutions (i.e., PO and Fock) are tenuous at best for $ka \gtrsim 10$. Importantly, although the Ansatz currents used in the hybrid formulation provide only approximations to the true currents because the body is not large enough or does not conform to a canonical shape for the asymptotic solutions to be strictly valid, the overall hybrid results were demonstrated to be quite accurate.

Finally, although the present discussion focused on bodies of revolution, this hybrid method was also applied to various two-dimensional geometries, namely flat and curved strips, dihedrals, and ogival cylinders. The MM representation was used in regions of surface discontinuities. For the ogival geometry at axial or near axial illumination, one MM treated discontinuity occurred in the deep shadow region. The calculated bistatic and monostatic scattering cross sections at grazing illumination, dominated by non-specular effects, agreed well with experimental data.

VI. CONCLUSION

A current-based hybrid formulation was developed in which we used as an Ansatz the results of the Fock theory and PO. Our hybrid formulation demonstrates that an integral equation, describing electromagnetic scattering, can be solved in a tractable manner by the MM technique using local solutions derived from the Fock theory or PO for the surface fields and currents on a scatterer. To treat scattering from BOR at large oblique angles, the Fock results were extended to arbitrary convex surfaces with torsion. The creeping wave enhancement for conespheres with smooth joins was predicted by this formulation and confirmed earlier experimental observations by Senior as to its genesis. It was shown that a MM representation for the currents in regions of surface discontinuities and also often in the shadow boundary region is preferred when the scattering on a body is primarily due to surface waves. In other regions of the scatterer, approximate expressions for the currents from the Fock theory or PO yield satisfactory results in the present hybrid framework.

The hybrid formulation was shown to yield accurate results for currents (phase and amplitude) and scattering cross sections even for objects near resonance where the Ansatz solutions are known to be least accurate. Conversely, as the scatterer size increases, the Ansatz solutions become more accurate and the overall hybrid analysis should yield results of even greater precision than obtained here.

REFERENCES

[1] G. L. James, *Geometrical Theory of Diffraction for Electromagnetic Waves*. London, England: Peter Peregrinus, 1976.
[2] R. C. Hansen, Ed., *Geometric Theory of Diffraction*. New York: IEEE Press, 1981.
[3] P. Ya Ufimtsev, "Method of edge waves in physical theory of diffraction," Air Force System Command, Foreign Tech. Div. Doc. ID FTD-HC-23-259-71, 1971.
[4] L. V. Kantorovich and V. I. Krylov, *Approximate Methods of Higher Analysis*, transl. by C. D. Benster. New York: Wiley, 1964.
[5] L. V. Kantorovich and G. P. Akilov, *Functional Analysis in Normed Spaces*, transl. by D. E. Brown, Oxford, England: Pergamon, 1964, pp. 586–587.

[6] D. S. Jones, "A critique of the variational method in scattering problems," *IRE Trans. Antennas Propagat.* vol. AP-4, p. 297, 1956.
[7] V. H. Rumsey, "The reaction concept in electromagnetic theory," *Phys. Rev., series 2*, vol. 94, pp. 1483–1491, June 1954.
[8] R. F. Harrington, *Field Computation by Moment Methods.* New York: Macmillan, 1968.
[9] G. A. Thiele and T. H. Newhouse, "A hybrid technique for combining moment methods with a geometrical theory of diffraction," *IEEE Trans. Antennas Propagat.*, vol. AP-23, pp. 62–69, 1975.
[10] E. P. Ekelman and G. A. Thiele, "A hybrid technique for combining the moment method treatment of wire antennas with the GTD for curved surfaces," *IEEE Trans. Antennas Propagat.*, vol. AP-28, pp. 831–839, 1980.
[11] L. W. Henderson and G. A. Thiele, "A hybrid MM-GTD technique for treatment of wire antennas near a curved surface," *Radio Sci.*, vol. 16, pp. 1125–1130, 1981. (See also, related paper in *IEEE Trans. Antennas Propagat.*, vol. AP-30, pp. 1257–1261, 1982).
[12] J. N. Sahalos and G. A. Thiele, "On the application of the GTD-MM technique and its limitation," *IEEE Trans. Antennas Propagat.*, vol. AP-29, pp. 780–786, 1981.
[13] T. J. Kim and G. A. Thiele, "A hybrid diffraction technique—General theory and applications," *IEEE Trans. Antennas Propagat.*, vol. AP-30, pp. 888–897, 1982.
[14] W. D. Burnside, C. L. Yu, and R. J. Marhefka, "A technique to combine the geometric theory of diffraction and the moment method," *IEEE Trans. Antennas Propagat.*, vol. AP-23, pp. 551–558, 1975.
[15] V. A. Fock, *Electromagnetic Diffraction and Propagation Problems.* New York: Pergamon, 1965.
[16] M. D. Tew and L. L. Tsai, "A method toward improved convergence of moment method solutions," *Proc. IEEE*, vol. 60, pp. 1436–1437, 1972.
[17] B. Azarbar and L. Shafai, "Application of moment method to large cylindrical reflector antennas," *IEEE Trans. Antennas Propagat.*, vol. AP-26, pp. 500–502, 1978.
[18] R. F. Goodrich, "Fock theory—An appraisal and exposition," *IRE Trans. Antennas Propagat.*, vol. AP-7, pp. S28–S36, 1959.
[19] J. R. Mautz and R. F. Harrington, "Radiation and scattering from bodies of revolution," *Appl. Sci. Res.*, vol. 20, pp. 405–435, 1969.
[20] P. H. Pathak and N. Wang, "Ray analysis of mutual coupling between antennas on a convex surface," *IEEE Trans. Antennas Propagat.*, vol. AP-29, pp. 911–922, 1981.
[21] G. T. Ruck, Ed., *Radar Cross Section Handbook*, Vol. I. New York: Plenum, 1970, pp. 76–94.
[22] P. H. Pathak, "An asymptotic analysis of the scattering of plane waves by a smooth convex cylinder," *Radio Sci.*, vol. 14, pp. 419–435, 1979.
[23] R. Mittra and S. Safavi-Naini, "Source radiation in presence of smooth convex bodies," *Radio Sci.*, vol. 14, pp. 217–237, 1979.
[24] G. T. Ruck, Ed., *Radar Cross Section Handbook*, Vol. I. New York: Plenum, 1970, p. 85.
[25] T. B. A. Senior, "The backscattering cross section of a cone sphere," *IEEE Trans. Antennas Propagat.*, vol. AP-13, p. 271, 1965.

Modeling Antennas Near to and Penetrating a Lossy Interface

GERALD J. BURKE, MEMBER, IEEE, AND EDMUND K. MILLER, FELLOW, IEEE

Abstract—**A technique for modeling wire objects interacting across or penetrating the planar interface which separates two half-spaces is described. The moment-method treatment is employed, based on the thin wire approximation to the electric-field integral equation, with the effect of the interface included via the Sommerfeld integrals. The computation time associated with evaluating the latter is substantially shortened by using an interpolation-based technique plus asymptotic field expressions. Although developed specifically for the wire problem, the procedure is also applicable, with slight modification, to modeling surface objects as well. Special care is taken to account for the charge discontinuity that occurs at the point a wire penetrates the interface. Example calculations are shown for a monopole antenna driven against ground stakes and simple ground screens, the fields of buried objects, and a simple electromagnetic pulse (EMP) simulator.**

I. INTRODUCTION

THE CAPABILITY of modeling antennas near to and penetrating a plane boundary such as the earth-air interface has long been needed. While the analytical solution of this problem is well established, its rigorous computational treatment is only now becoming practicable. In this presentation, we outline the basic problem, describe a numerical approach to modeling it, and present some representative results. For simplicity, our discussion is addressed to wire objects only, although the method used to include the interface is more general, being applicable to conducting surfaces and penetrable bodies as well. The computer model described below is incorporated in the Numerical Electromagnetic Code (NEC) [1], a widely used modeling code.

II. ANALYTICAL DEVELOPMENT

It is convenient to develop the treatment for a wire object penetrating an interface as a sequence of extensions to that of the same object located in free space. The discussion here (and in Section III on the numerical procedure) therefore considers a wire 1) in free space; 2) in a lossy infinite medium; 3) near an interface; and 4) penetrating an interface.

A. Wires in Free Space and Infinite Lossy Media

Our approach is to employ an integral equation to obtain the current distribution induced on an object by a monochromatic ($\epsilon^{j\omega t}$ time variation) source. For wires whose radius (a) is small with respect to the wavelength (λ), the Pocklington form of the electric field integral equation has been found to work well [2]. It can be written for a wire with contour C in free space in the form

$$\hat{s} \cdot \vec{E}^I_\infty(\vec{r}) = -\hat{s} \cdot \vec{E}^D_\infty(\vec{r}), \qquad \vec{r} \in C, \tag{1a}$$

where

$$\vec{E}^D_\infty(\vec{r}) = \int_{C'} I(s')\hat{s}' \cdot \bar{\bar{G}}^D_\infty(\vec{r},\vec{r}')\, ds'. \tag{1b}$$

In addition

$$\bar{\bar{G}}^D_\infty(\vec{r},\vec{r}') = C_1(\nabla\nabla/k_0^2 + \bar{\bar{I}})g_\infty, \tag{2}$$

$$g_\infty = \exp(-jk_0R)/R,$$

$$\bar{\bar{I}} = \hat{x}\hat{x} + \hat{y}\hat{y} + \hat{z}\hat{z},$$

$$R = |\vec{r} - \vec{r}'|,$$

$$k_0^2 = \omega^2\mu_0\epsilon_0,$$

and

$$C_1 = -j\omega\mu_0/4\pi,$$

where the subscript ∞ denotes an infinite medium quantity.

Also, I is the induced current, $\vec{E}^I_\infty$ is the exciting field, $\hat{s}$ and $\hat{s}'$ are unit vectors tangent to the wire at $\vec{s}$ and $\vec{s}'$ and $\vec{r}$ is a vector to the point s on the contour C on the wire surface while $\vec{r}'$ is the point s' on contour C' on the wire axis. $\bar{\bar{G}}^D_\infty$ is the dyadic Green's function for the electric field at $\vec{r}$ due to a current element at $\vec{r}'$.

The thin-wire approximation has been employed in reducing the original two-dimensional surface integration to the one-dimensional line integration shown in (1). This approximation involves the assumptions that 1) the longitudinal current has negligible circumferential variation; 2) the circumferential current is negligible; 3) the tube of current flowing on the wire can be replaced by a filament flowing on the wire axis, while the boundary condition is matched on the wire surface, so that $R \geqslant a$.

Typical grounds are lossy, so that our model of the general interface problem requires that it allow for the complex wavenumber

$$k = k_0\sqrt{\tilde{\epsilon}_\infty}, \qquad \text{where } \tilde{\epsilon}_\infty = \epsilon_r - j\frac{\sigma}{\omega\epsilon_0}$$

which replaces the free space wavenumber k_0 in (1) for an infinite lossy medium. The numerical impact of this change is nontrivial, but straightforward, as is discussed in Section III-A.

B. Wire Near an Interface

The free space (or infinite medium) integral equation (1) can be extended to handle a wire located near the boundary between two half-spaces by modifying the kernel to include the fields scattered from the interface. Various image approximations have been used for this purpose (e.g., see [3], [4]), but their limitations are so restrictive that the rigorous Sommerfeld integrals are required in general. Thus, we use in place of (1), for source and observation points above ground (+) or below ground (−), such that $z \cdot z' \geqslant 0$,

$$\hat{s} \cdot \vec{E}^I_\pm(\vec{r}) = -\hat{s} \cdot \vec{E}^D_\pm(\vec{r}) - \hat{s} \cdot \vec{E}^R_\pm(\vec{r}), \qquad \vec{r} \in C_\pm, \tag{3a}$$

where the direct field $\vec{E}^D_\pm$ is given by (1b) with $k_0 \to k_\pm = k_0\sqrt{\tilde{\epsilon}_\pm}$

Manuscript received October 20, 1983; revised April 26, 1984. This work was supported by the U.S. Department of Energy under Contract W-7405-Eng-48.

The authors are with the Lawrence Livermore National Laboratory, P.O. Box 5504, Livermore, CA 94550.

Reprinted from *IEEE Trans. Antennas Propagat.*, vol. AP-32, no. 10, pp. 1040–1049, October 1984.

and $\tilde{\epsilon}_\pm = \epsilon_\pm - j\sigma_\pm/\omega\epsilon_0$. The field component due to the interface is

$$\vec{E}_\pm^R(\vec{r}) = \int_{C'_\pm} I_\pm(s')\hat{s}' \cdot \bar{\bar{G}}_\pm^R(\vec{r},\vec{r}')\, ds', \tag{3b}$$

with

$$\bar{\bar{G}}_\pm^R(\vec{r},\vec{r}') = \frac{k_\mp^2 - k_\pm^2}{k_+^2 + k_-^2}\,\bar{\bar{G}}_\pm^I(\vec{r},\vec{r}') + \bar{\bar{R}}_\pm(\vec{r},\vec{r}'), \tag{4a}$$

$$\bar{\bar{G}}_\pm^I(\vec{r},\vec{r}') = -\bar{\bar{I}}_R \cdot \bar{\bar{G}}_\pm^D(\vec{r},\bar{\bar{I}}_R \cdot \vec{r}'), \tag{4b}$$

$$\bar{\bar{I}}_R = \hat{x}\hat{x} + \hat{y}\hat{y} - \hat{z}\hat{z}.$$

$\bar{\bar{R}}_\pm$ involves the Sommerfeld integrals and is constructed from the following vector components for horizontally and vertically oriented dipoles

$$R_{\pm\rho}^V = \frac{C_1}{k_\pm^2}\frac{\partial^2}{\partial\rho\partial z}\, k_\mp^2 V_\pm^R,$$

$$R_{\pm z}^V = \frac{C_1}{k_\pm^2}\left(\frac{\partial^2}{\partial z^2} + k_\pm^2\right) k_\mp^2 V_\pm^R,$$

$$R_{\pm\rho}^H = \frac{C_1}{k_\pm^2}\cos\phi\left(\frac{\partial^2}{\partial\rho^2}\, k_\pm^2 V_\pm^R + k_\pm^2 U_\pm^R\right),$$

$$R_{\pm\phi}^H = \frac{-C_1}{k_\pm^2}\sin\phi\left(\frac{1}{\rho}\frac{\partial}{\partial\rho}\, k_\pm^2 V_\pm^R + k_\pm^2 U_\pm^R\right),$$

$$R_{\pm z}^H = -\cos\phi R_{\pm\rho}^V,$$

where the source is at $\vec{r}' = z'\hat{z}$ and the field is evaluated at $\vec{r} = \rho(\cos\phi\hat{x} + \sin\phi\hat{y}) + z\hat{z}$.

The superscript on R indicates a vertical (V) or horizontal (H) current element and the subscript indicates the cylindrical component of the field vector. The horizontal current element is along the x axis.

The Sommerfeld integral terms are

$$U_\pm^R = \int_0^\infty D_1(\lambda) \exp\,[-\gamma_\pm|z + z'|]\, J_0(\lambda\rho)\, \lambda d\lambda, \tag{5e}$$

$$V_\pm^R = \int_0^\infty D_2(\lambda) \exp\,[-\gamma_\pm|z + z'|]\, J_0(\lambda\rho)\, \lambda d\lambda, \tag{5f}$$

with

$$D_1(\lambda) = \frac{2}{\gamma_+ + \gamma_-} - \frac{2k_\pm^2}{\gamma_\pm(k_+^2 + k_-^2)},$$

$$D_2(\lambda) = \frac{2}{k_-^2\gamma_+ + k_+^2\gamma_-} - \frac{2}{\gamma_\pm(k_+^2 + k_-^2)},$$

and

$$\gamma_\pm = (\lambda^2 - k_\pm^2)^{1/2}.$$

The terms $U_\pm^R$ and $V_\pm^R$ differ from those commonly used [5] by having static terms subtracted. The term subtracted from $V_\pm^R$ lets that integral converge as ρ and $z + z'$ approach zero with only a R^{-1} singularity in the second derivatives of $V_\pm^R$. The term subtracted from $U_\pm^R$ does not alter the singularity but completes the $\bar{\bar{G}}^I$ term in (4).

C. Wire Penetrating an Interface

Besides reflecting from it, fields are transmitted through the interface. As long as the object is located wholly in one half-space, the transmitted fields need not be considered in the solution for current. But, when the object penetrates the interface, or there are two (or more) objects located on opposite sides of the interface, the transmitted fields must then be included. Upon adding the transmitted fields, the integral equation (3a) becomes

$$\hat{s} \cdot \vec{E}_\pm^I(\vec{r}) = -\hat{s} \cdot \vec{E}_\pm^D(\vec{r}) - \hat{s} \cdot \vec{E}_\pm^R(\vec{r}) - \hat{s} \cdot \vec{E}_\pm^T(\vec{r}), \quad \vec{r} \in C_\pm \tag{6a}$$

where

$$\vec{E}_\pm^T(\vec{r}) = \int_{C'_\mp} I_\mp(\vec{s}')\hat{s}' \cdot \bar{\bar{T}}_\pm(\vec{r},\vec{r}')\, ds' \tag{6b}$$

and $\bar{\bar{T}}_\pm(\vec{r},\vec{r}')$ accounts for the Sommerfeld integral contributions to the transmitted field with the sign subscript chosen for the observation medium (with the source in the other as indicated by the $\mp$ sign on I and C'). It is constructed from the following expressions:

$$T_{\pm\rho}^V = C_1 \frac{\partial^2}{\partial\rho\partial z} V_\pm^T, \tag{7a}$$

$$T_{\pm z}^V = C_1\left(\frac{\partial^2}{\partial z^2} + k_\pm^2\right) V_\pm^T, \tag{7b}$$

$$T_{\pm\rho}^H = C_1 \cos\phi\left(\frac{\partial^2}{\partial\rho^2} V_\pm^T + U_\pm^T\right), \tag{7c}$$

$$T_{\pm\phi}^H = -C_1 \sin\phi\left(\frac{1}{\rho}\frac{\partial}{\partial\rho} V_\pm^T + U_\pm^T\right), \tag{7d}$$

$$T_{\pm z}^H = -C_1 \cos\phi \frac{\partial^2}{\partial\rho\partial z'} V_\pm^T, \tag{7e}$$

where

$$V_\pm^T = 2\int_0^\infty \frac{e^{-\gamma_\mp|z'| - \gamma_\pm|z|}}{k_-^2\gamma_+ + k_+^2\gamma_-} J_0(\lambda\rho)\, \lambda d\lambda \tag{7f}$$

$$U_\pm^T = 2\int_0^\infty \frac{e^{-\gamma_\mp|z'| - \gamma_\pm|z|}}{\gamma_- + \gamma_+} J_0(\lambda\rho)\, \lambda d\lambda. \tag{7g}$$

III. NUMERICAL TREATMENT AND VALIDATION

The numerical solution of integral equations of the kind presented above is now fairly standard in electromagnetic computer modeling. Only a brief summary is included here with detailed discussion limited, for the most part, to those aspects of the solution procedure that are new. Validation results similarly are given only where the computational procedure is significantly different from the basic case of the free space medium or where they round out the discussion.

A. Wires in Free Space and Infinite Lossy Media

The method of moments [6] provides a way to reduce an integral equation to a linear system whose solution can be obtained using standard matrix techniques. Details of the particular approach used here can be found elsewhere [1], [7]. The correctness of this procedure for the free space enviroment has been widely demonstrated (e.g., see [6]) so no specific validation results are given here.

When the medium is lossy, the numerical treatment follows the steps for free space, with k_0 replaced by k. Introduction of a complex wavenumber results in a minimal increase in computation time since the field quantities were already complex.

A comparison of the results obtained from the numerical treatment discussed here with those from an analytic closed-form solution due to Balmain [8] for a dipole in a lossy plasma was found to yield close agreement. Another kind of check on the numerical model for complex wave number is to obtain the resonance frequencies of a straight wire in the complex-frequency ($s = j\omega - \sigma$) plane. Resonance frequencies from an independent computation [9] are also found to agree closely with those obtained using the present treatment, providing further validation of the extension to a lossy medium. The resonance frequencies in both cases were obtained from a search of the complex frequency plane.

B. Wire Near an Interface

Addition of the Sommerfeld integrals to account for the fields reflected from the interface has a substantial impact on the numerical treatment. Straightforward numerical evaluation of the Sommerfeld contributions can be time consuming, increasing the overall computer time by a factor of 100 or more relative to the same computation performed for an infinite medium [2], [3].

An alternate approach, obtaining the Sommerfeld integral values by interpolation in a two-dimensional ($\rho, z + z'$) grid of prestored values, has been shown to be as accurate and much more efficient [10]. Accurate interpolation for the Sommerfeld integral terms ($\bar{\bar{R}}_{\pm}$) in (4a) is aided by the grouping of the more singular terms with the image field $\bar{\bar{G}}^I_{\pm}$. This form also aids in the evaluation of the integral in (3b) since the image field can be integrated in closed form for the sinusoidal terms of the current expansion. With this method, the computer time for evaluating matrix elements is about four times that required for the same object in free space. Time to solve the matrix equation is of course unchanged. For comparison, modeling the same object located near a perfectly conducting ground plane requires about twice the computation time to evaluate matrix elements as for free space since both the direct and image fields must be computed.

Validation of this numerical procedure is provided by the internal consistency check of integrating the far-field power

$$P_F = \frac{1}{2} r^2 \lim_{r \to \infty} \mathrm{Re} \int_{4\pi} \vec{E}(\bar{r}) \times \vec{H}^*(\bar{r}) \cdot \hat{r}\, d\Omega \tag{9a}$$

for comparison with the input power

$$P_I = \tfrac{1}{2} \mathrm{Re}\, (V_I I_I^*) \tag{9b}$$

obtained from the antenna-source model. For lossless media, any difference between these two results can be interpreted as a measure solution inaccuracy. However, this is a necessary, but not sufficient, requirement of a valid solution. Application of this check typically yields agreement between P_F and P_I within one percent or so, giving confidence that for lossless half-spaces at least, the overall model is quite accurate.

C. Wire Penetrating an Interface

Extension of the model to a wire penetrating the interface requires evaluation of the transmitted field $E^T_{\pm}$ in (6a). A similar approach was taken as for the reflected field, although the implementation is more involved. Also, the basis functions in the current expansion must be modified for the condition on current and charge at the point where a wire passes through the interface as described in Section III-C-2.

1) *Transmitted Field Evaluation*: In contrast to the two-dimensional dependence of the reflected fields as can be seen in (5), the transmitted fields depend on three coordinates: ρ, z, and z'. This is because the source and observation distances from the interface are multiplied by different z-dependent wavenumbers in the Sommerfeld integrals (see (7)). Consequently, the number of values in the interpolation table is on the order of the 3/2 power of that for the reflected field. Interpolation still results in a large reduction in computation time since, in the method of moments solution, values over a limited range of coordinates tend to be used repeatedly.

With this dependence on three parameters, it is particularly important to use a numerical treatment that yields accurate field values with a minimum number of evaluations of the Sommerfeld integrals. Using interpolation, this is accomplished by removing the singularity and dividing out the dominant phase factor as was done for the reflected field above ground. The transmitted field, however, can in some cases have a complex phase behavior described asymptotically by one ray traveling principally in the upper medium, above the straight line from source to observer, and another ray on or below the straight-line path. Since this phase dependence cannot be removed by division, an approach of least squares approximation was developed which can be viewed as parameter estimation with a model chosen from the known behavior of the field.

At sufficiently large distances from the source, the field can be obtained directly from asymptotic approximations. Such approximations were used to permit truncating the interpolation and least squares approximation tables. These three methods interpolation, least squares approximation and asymptotic approximation are combined in the present code to cover the range of source and observation point coordinates for the transmitted field. The technique, described briefly below, represents a tradeoff between accuracy, complexity, and computation time, and could probably continue to be refined with as much additional work as one cared to spend.

In the following, as is done in the code, we will consider only the case of a buried source and elevated observer, since the fields for elevated source and buried observer can be obtained from the former case through reciprocity.

Three-dimensional interpolation: The transmitted field components in (7a)-(7e) are dominated at small distances in $R_T = [\rho^2 + (z - z')^2]^{1/2}$ by the derivatives of $V^T_{\pm}$ which have R_T^{-3} singularities. Hence, to smooth the interpolated functions and ease the requirement for interpolation accuracy, an analytically integrable term is subtracted from the integrand of $V^T_{\pm}$ to obtain

$$\begin{aligned}\tilde{V}^T_+ &= 2\int_0^\infty \left[\frac{e^{-\gamma_+|z| - \gamma_-|z'|}}{k_+^2\gamma_- + k_-^2\gamma_+} - \frac{e^{-\gamma_+|z-z'|}}{\gamma_+(k_+^2 + k_-^2)}\right] J_0(\lambda\rho)\,\lambda\, d\lambda \\ &= V^T_+ - \frac{2}{k_+^2 + k_-^2}\,\frac{e^{-jk_+R_T}}{R_T}.\end{aligned} \tag{10}$$

$\tilde{V}_+^T$ is nonsingular as R_T goes to zero and, when it is substituted for V_+^T in (7a) through (7e), the second derivatives are singular as R_T^{-1}. The R_T^{-1} terms can also be obtained in closed form and subtracted from the integrals. The remainder multiplied by R_T yields the following functions to which interpolation is applied:

$$I_\rho^V = C_1 \left\{ R_T \frac{\partial^2}{\partial\rho\partial z} \tilde{V}_+^T - \left(C_3 \frac{1 - \sin\theta}{\cos\theta} - C_2 S \cos\theta \right) \right\} \quad (11a)$$

$$I_z^V = C_1 \left\{ R_T \left(\frac{\partial^2}{\partial z^2} + k_+^2 \right) \tilde{V}_+^T - (C_3 - C_2 S \sin\theta) \right\} \quad (11b)$$

$$I_\rho^H = C_1 \left\{ R_T \left(\frac{\partial^2}{\partial\rho^2} \tilde{V}_+^T + U_+^T \right) - \left[C_3 \left(\frac{\sin^2\theta - \sin\theta}{\cos^2\theta} \right) + C_2 S \left(\frac{\sin\theta\,(1 + \cos^2\theta) - 1}{\cos^2\theta} \right) + 1 \right] \right\} \quad (11c)$$

$$I_\phi^H = -C_1 \left\{ R_T \left(\frac{1}{\rho} \frac{\partial}{\partial\rho} \tilde{V}_+^T + U_+^T \right) + \left[(C_2 S - C_3) \frac{1 - \sin\theta}{\cos^2\theta} + 1 \right] \right\} \quad (11d)$$

$$I_z^H = -C_1 \left\{ R_T \frac{\partial^2}{\partial\rho\partial z'} \tilde{V}_+^T + \left(k_-^2 C_3 \frac{1 - \sin\theta}{\cos\theta} + k_+^2 C_2 S \cos\theta \right) \right\} \quad (11e)$$

where

$$\theta = \tan^{-1} \frac{|z - z'|}{\rho},$$

$$S = z'/R_T,$$

$$C_2 = \frac{k_-^2 - k_+^2}{k_+^2 + k_-^2},$$

and

$$C_3 = k_+^2 C_2/(k_+^2 + k_-^2).$$

The functions in (11a)–(11e) are zero for $R_T = 0$.

When, for $|k_- R_T|$ less than about three, values for the quantities in (11a)–(11e) are obtained by interpolation and subsequently divided by R_T and combined with the subtracted terms, the resulting errors in the field are considerably smaller than when the total field is obtained by interpolation. For larger values of R_T, however, the subtracted terms can dominate, since they are not attenuated by loss in the medium, and the interpolation error can be magnified. Thus, for larger values of R_T the field components of (7a)–(7e) are obtained directly by interpolation.

In the treatment used to obtain the results presented here, interpolation is used for the region

$$0 \leqslant \rho \leqslant 2\pi/|k_-|$$

$$0 \leqslant z \leqslant 2\pi/|k_-|$$

$$0 \leqslant |z'| \leqslant 2\pi/k_+$$

with three subregions.

1) $0 \leqslant |z'| \leqslant 0.4(2\pi)/|k_-|$; interpolation for the quanties in (11)
2) $0.4(2\pi)/|k_-| < |z'| \leqslant 2\pi/|k_-|$; interpolation for the quantities in (7) divided by $\exp(-jk_-|z'|)/R_T^2$
3) $2\pi/|k_-| < |z'| \leqslant 2\pi/k_+$; interpolation for the quantities in (7) divided by $\exp[-jk_-(\rho^2 + z'^2)^{1/2}]/R_T^2$.

Three-dimensional linear interpolation is used in each region. Outside of this interpolation region, the field is evaluated by least squares or asymptotic approximation.

Asymptotic approximation: Many asymptotic approximations for Sommerfeld integrals have been developed in forms ranging from simple first-order approximations, known as reflection coefficient and transmission coefficient approximations [4] to higher order uniform asymptotic approximations. The forms used here were chosen to be valid on the interface and to cover transmitted field paths without limit on the maximum depth in the ground.

For asymptotic approximation, the integral for $V_\pm^T$ in (7f) (and similarly $U_\pm^T$) can be written in the form

$$V_\pm^T = \int_{-\infty}^{\infty} G(\lambda) e^{-F(\lambda)}\, d\lambda \quad (12a)$$

where

$$G(\lambda) = \frac{\lambda H_0^{(2)}(\lambda\rho) \exp(j\lambda\rho)}{k_-^2 \gamma_+ + k_+^2 \gamma_-} \quad (12b)$$

and

$$F(\lambda) = \gamma_\mp |z'| + \gamma_\pm |z| + j\lambda\rho. \quad (12c)$$

When ρ is sufficiently large $G(\lambda)$ can be considered slowly varying relative to the exponential in (12a) and an approximation for the integral can be derived by the method of steepest descent. The integral is thus approximated in terms of contributions from the neighborhood of saddle points which are solutions of the equation

$$F'(\lambda) = 0. \quad (13)$$

The procedure is complicated by the existence of two branch points, $\lambda = k_+$, and $\lambda = k_-$, and a pole at the zero of the denominator of G. In general, two saddle points must be considered. The first λ_1 represents a ray that travels at a steep angle up to the interface and then at a shallower angle in the upper medium. The second λ_2 represents a ray that travels at a shallow angle in the lower medium and then upward to the observer. The λ_2 ray attenuates exponentially above the interface (since $\mathrm{Re}(\lambda_2) > k_+$) and is strongly attenuated in a lossy medium. In a lossy medium both rays are complex.

Rather than develop the complete asymptotic approximation for an integral of this form, an approximation was derived considering the λ_1 saddle point together with the pole but isolated from the λ_2 saddle point. Higher order terms were included in the contribution of the λ_1 saddle point. This part of the approximation thus includes the usually dominant ray and the surface wave. A first-order approximation was then added for the λ_2 saddle point which can be important when z is small relative to ρ and $z'(z \geqslant 0., z' \leqslant 0.)$. When ρ is small relative to z the first-order approximation for the λ_1 saddle point is used with higher order terms obtained by interpolating between the higher terms for a larger value of ρ and higher order terms from an approximation for $\rho = 0$. When ρ and z are small relative to z', only the first order approximation for λ_1 is used since a more involved analysis would be required for higher order terms.

These asymptotic approximations yield errors less than a few percent for ρ greater than about 0.5 λ to 1 λ_0 or for z greater than about 1 λ_0 to 2 λ_0. For small ρ and z, errors are less than 10 to 20 percent for $|z'|$ greater than about 1 λ_0 to 2 λ_0.

Least squares approximation (parameter estimation): For efficient interpolation the interpolated functions were made as smooth as possible, by suppressing the singularity and phase factor, to better match the linear or higher order interpolation functions. As an alternative, we might consider a model-based procedure. The basic difference is to employ interpolation functions chosen on physical or mathematical grounds to be close to the expected field behavior. The function amplitudes are obtained by enforcing a match to Sommerfeld integral values. If suitable model-based interpolation functions can be found, the possibility exists that the number of stored data values can be greatly reduced.

A form for such functions is suggested by the asymptotic approximations. The rays corresponding to the saddle points λ_1 and λ_2 have phase factors of the form

$$P_1(\rho, z, z') = e^{-F(\lambda_1)},$$

$$P_2(\rho, z, z') = e^{-F(\lambda_2)},$$

where λ_1 and λ_2 are solutions to (13) for the particular ρ, z, and z'. In addition, the field will involve spreading factors, such as R_1^{-n} for P_1 and R_2^{-n} for P_2 with $n = 1, 2, 3$, where

$$R_1 = [\rho^2 + (z - |k_+/k_-|z')^2]^{1/2},$$

$$R_2 = [\rho^2 + z'^2]^{1/2},$$

and angle factors such as 1., z/R_T, z^2/R_T, z'/R_T, ρ/R_T, etc. The field may then be approximated by a sum of functions

$$\tilde{E}(\rho, z, z') = \sum_n A_n f_n(\rho, z, z')$$

where each f_n is a product of a phase factor, a spreading factor, and an angle factor. The coefficients A_n are determined to provide a least squares fit of $\tilde{E}$ to field values computed by evaluating Sommerfeld integrals. $\tilde{E}$ can then be used to interpolate or extrapolate from the computed points that were fit. If the f_n are good approximations to the field behavior, then considerably fewer computer points should be needed for a given accuracy than with polynomial interpolation.

This method is effective for extrapolation as well as interpolation, as illustrated in Fig. 1. Both low frequency (P_1) and high frequency (P_2) components are present and are matched by the approximation.

In the present code, this least squares approximation is used from the outer boundary of the interpolation region out to 3 λ_0 in ρ, 2 λ_0 in z and to a depth of about $2\pi/|k_-|$ in z'. The method is applied in three subregions. Two, with a border on the interface, involve a least squares fit of 32 terms while the third, for small ρ and z greater than $2\pi/|k_-|$ involves 28 terms with no P_2 factors. The Sommerfeld integrals are fit at a total of 196 points which is far fewer than would be needed for simple interpolation over the region.

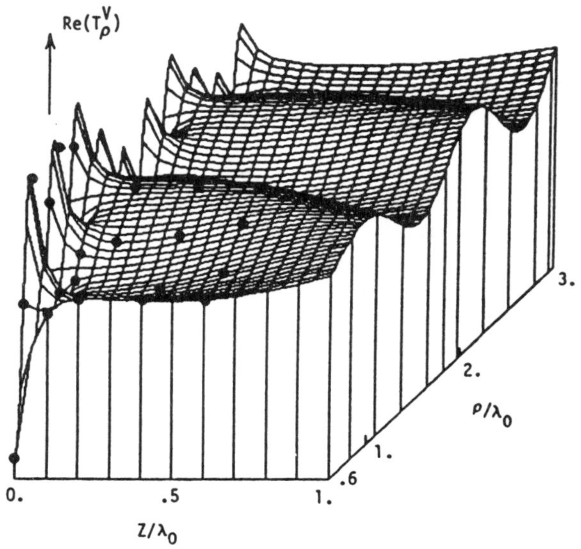

Fig. 1. Transmitted field, shown for T_ρ^V component with $\tilde{\epsilon}_- = 16 - j0$ and source depth of $z' = -0.1\ \lambda_0$, as modeled by least squares approximation. A total of 88 points are fit (36 in the plane of this figure with visible ones shown) to cover the three-dimensional region $0.6 \leqslant \rho/\lambda_0 \leqslant 3.0$, $0 \leqslant z/\lambda_0 \leqslant 2.0$, and $0 \leqslant |z'|/\lambda_0 \leqslant 0.25$ with less than 4 percent error.

2) *Current and Charge Conditions at the Interface*: To model a wire penetrating the interface the basis functions of the current expansion must be modified to conform to the appropriate conditions on current and charge at the point of penetration. These conditions can be derived from the continuity required of the tangential electric and magnetic fields at the interface.

The condition on current is determined by considering the azimuthal magnetic field of the current which, for a vertical wire, is tangent to the interface. Continuity of the magnetic field across the interface requires continuity of current (assuming $\mu_+ = \mu_- = \mu_0$), or

$$I_+ = I_- \tag{14}$$

where

$$I_\pm = I(z_\pm \to 0).$$

Similarly, a radial electric field is produced by the charge on the wire and is tangent to the interface when the wire is vertical. Continuity of the radial electric field across the interface requires that

$$\frac{q_+}{q_-} = \frac{\epsilon_+}{\epsilon_-}$$

where q_+ and q_- are linear charge densities on the wire, and ϵ_+ and ϵ_- are the real dielectric constants of the media. The continuity relation on the wire, taking account of the conduction current into the medium, is

$$I'_\pm = -j\omega q_\pm \tilde{\epsilon}_\pm/\epsilon_\pm$$

where

$$\tilde{\epsilon}_\pm = \left(\epsilon_\pm - j\frac{\sigma_\pm}{\omega\epsilon_0}\right).$$

Hence, the condition on the derivative of current, which is

needed to define the basis function, is

$$\frac{I'_+}{I'_-} = \frac{\bar{\epsilon}_+}{\bar{\epsilon}_-}. \tag{15}$$

These conditions are strictly valid only for the vertical wire. They have been used for oblique penetration, however, and appear to be the most reasonable conditions compatible with the thin wire approximation.

Two kinds of checks were made to assess the validity of the treatment of the penetrating conductor. One involved an independent formulation for a vertical circular cylinder penetrating the interface [11]. The cylinder was modeled with the electric field integral equation as a surface, without invoking the thin wire approximation. A Galerkin's technique was employed to solve for the current using piecewise linear basis and weight functions. This solution ensured continuity of current but placed no constraint on the derivative of current. Interface effects were included via the Sommerfeld integrals but these were evaluated independently of the routines used in NEC.

The solution for current obtained from this model revealed a variation in accord with (15). The overall current distribution was in close agreement with that obtained from NEC when the cylinder radius was small enough to satisfy the thin-wire approximation. It was found, however, that considerably more samples were needed in the vicinity of the interface to obtain convergence in the solution with the cylinder code than with NEC where the condition of (15) is built in.

The other check was to evaluate the electric field both radial and tangential to the wire due to the computed current. This was done for normal and oblique angles of penetration. The results, some of which are presented in Fig. 2, show no significant anomaly in the field until the penetration angle exceeds about 60° from normal. Even then, the perturbation in the field is small in an integral sense relative to the source field. Any perturbation in the condition (15) produced a noticeable increase in the field tangential to the wire at the interface.

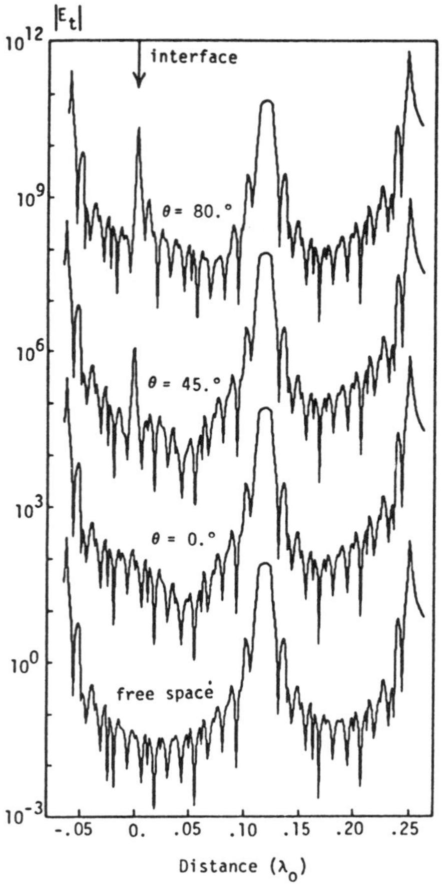

Fig. 2. Tangential electric field along surface of thin wire penetrating a ground at various angles, θ with respect to vertical, with free space shown for comparison. For $\theta = 0°$, the length of wire below the interface is 0.0625 λ_0 and the length above is 0.25 λ_0 with the source at 0.119 λ_0. The wire radius is 0.00025 λ_0 and $\bar{\epsilon}_- = 16 - j0$. This error field, which should be zero outside the source region, oscillates due to the numerical treatment. An increase is seen at the interface for θ of 45° and 80°; however, the integral over the region remains small relative to that for the source.

IV. REPRESENTATIVE APPLICATIONS

There are many practical problems to which a model of the kind discussed above can be productively applied. Some representative results for antenna, geophysical and electromagnetic pulse (EMP) simulation applications are given here.

One of the motivations for developing a capability for modeling a wire penetrating an interface was to model ground stakes, typically metal rods driven into the ground, that are used with many HF and VHF antennas. In Fig. 3 is shown the real part of the current distribution on a vertical quarterwave monopole driven against ground stakes of varying length. Input impedance is supplied directly by the model while radiation resistance R_R and ground-loss resistance R_G may be computed as

$$R_R = \frac{r^2}{|I_I|^2} \lim_{r \to \infty} \int_{2\pi} Re[\bar{E}(\bar{r}) \times \bar{H}^*(\bar{r})] \cdot \hat{r}\, d\Omega, \qquad \bar{r} \cdot \hat{z} > 0$$

for source current I_I and $R_G = R_I - R_R$. R_R and R_G are shown in Fig. 4 for variable ground stake length. Radiation efficiency η_R may be defined as

$$\eta_R \equiv \frac{R_R}{R_I}.$$

For low ground conductivity, η_R decreases sharply at resonances

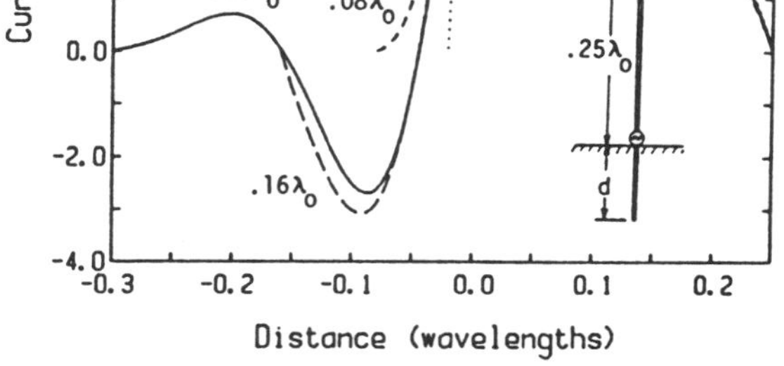

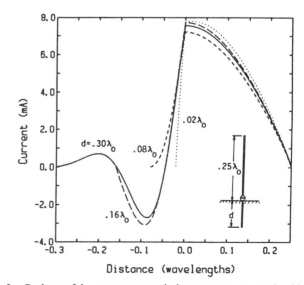

Fig. 3. Real part of the current on a vertical quarter wave monopole with 1 V source for various lengths of ground stake. The wire radius is 0.00025 λ_0 and $\bar{\epsilon}_- = 16 = j16$.

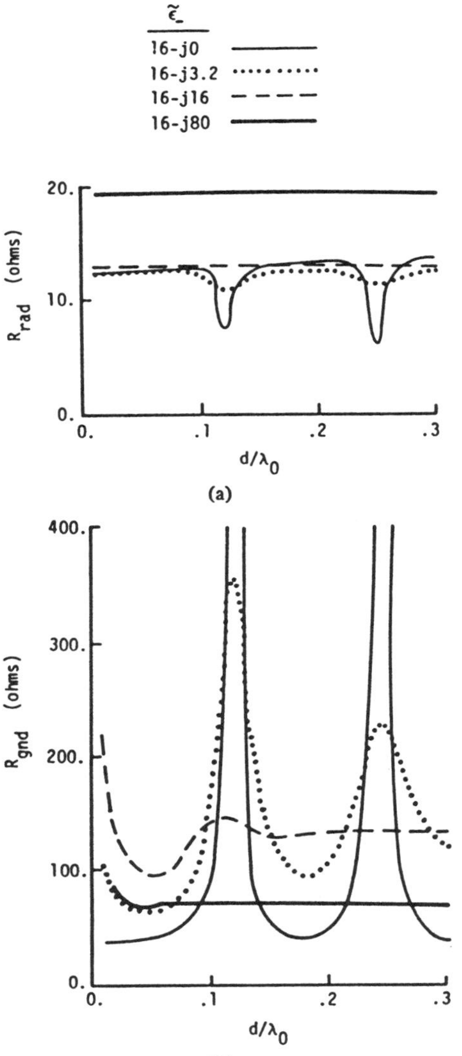

Fig. 4. Radiation (from upper half-space power) and ground loss resistances of the quarter-wave monopole as a function of ground-stake length with $\tilde{\epsilon}_-$ a parameter. Note the possibility, for low ground conductivity, of "tuning" the ground stake length to maximize the power radiated into the upper half space.

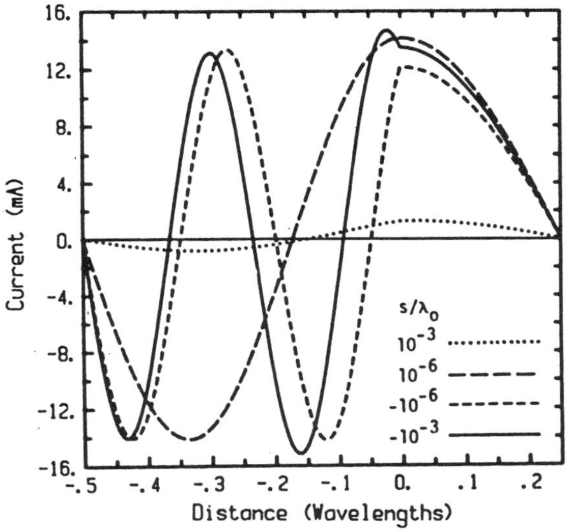

Fig. 5. Real part of the current on a quarter wave monopole driven against a ground screen consisting of six evenly spaced radial wires with a screen radius of 0.5 λ_0 and $\tilde{\epsilon}_- = 16 - j0$. The height of the ground screen above the interface is s. The sum of the radial currents is plotted to obtain a continuous current at the monopole-screen junction. It can be seen that transition from the upper to lower medium wavelength occurs largely within a vertical distance of $2(10^{-6})\ \lambda_0$.

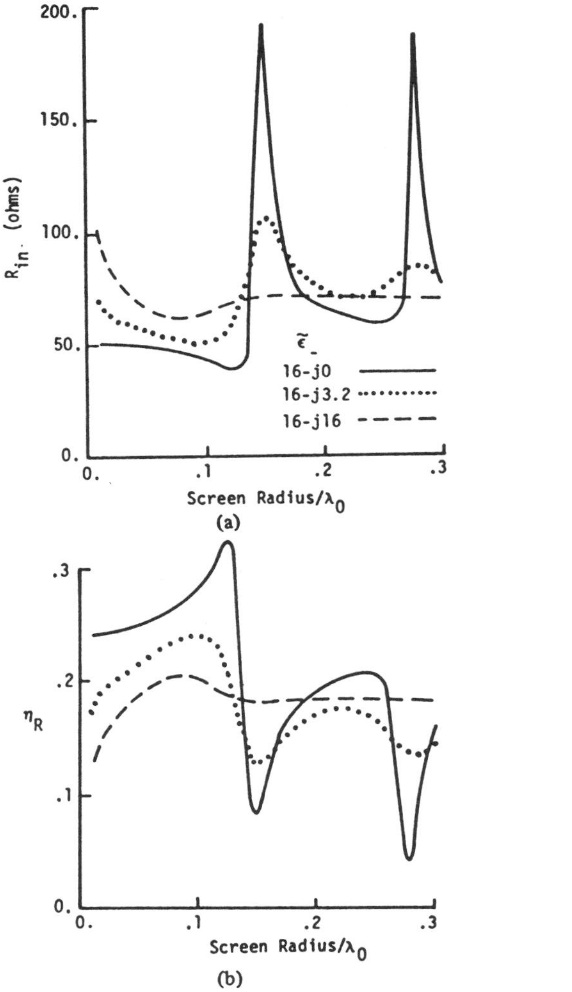

Fig. 6. (a) Input resistance and (b) radiation efficiency of monopole on a six-wire radial ground screen $10^{-3}\ \lambda_0$ below the interface with the screen-wire radius of $10^{-7}\ \lambda_0$, and $\tilde{\epsilon}_-$ a parameter. Strong resonance effects are seen for low ground conductivity showing the possibility of tuning the ground screen.

of the ground stake while for higher ground conductivity it increases with increasing wire radius.

Ground screens are used to stabilize antenna impedance, improve radiation efficiency, and control the radiation, besides providing a point against which to drive the antenna. Their effects are presently analyzed using various approximations such as the equivalent surface impedance and the compensation theorem. This situation is not surprising, since it has been difficult to treat a single wire near an interface, let alone the large collection of wires which may comprise a typical ground screen. Approximate analyses have, in the past, provided much useful data for dense ground screens [12]. The more rigorous analysis described here can provide a check on these approximations. Some preliminary results were previously presented for a sparce ground screen consisting of just a few (≤4) wires with the antenna and screen above the interface [13]. Here, we study a more general situation including the case of an elevated antenna driven against a buried ground, as illustrated in Figs. 5-7.

In Fig. 5 is shown the real part of the current distribution on a monopole and ground screen of six radial wires for the

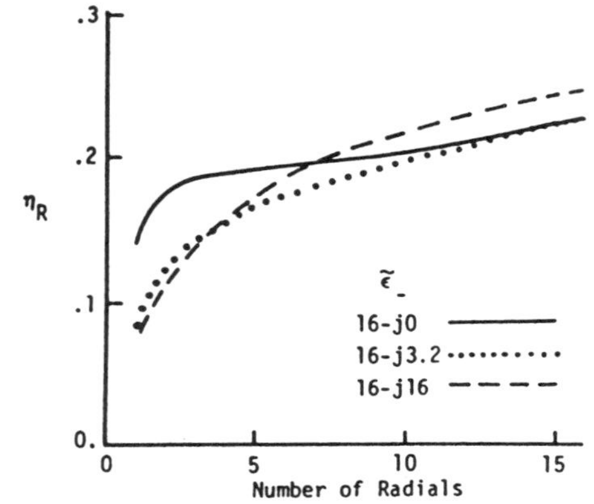

Fig. 7. Radiation efficiency of a monopole on a radial wire ground screen with screen radius of 0.2 λ_0, screen wire radius 10^{-7} λ_0 and screen depth of 10^{-3} λ_0 for varying number of radials. Increasing the number of radials beyond six or eight for this set of parameters yields diminishing benefit.

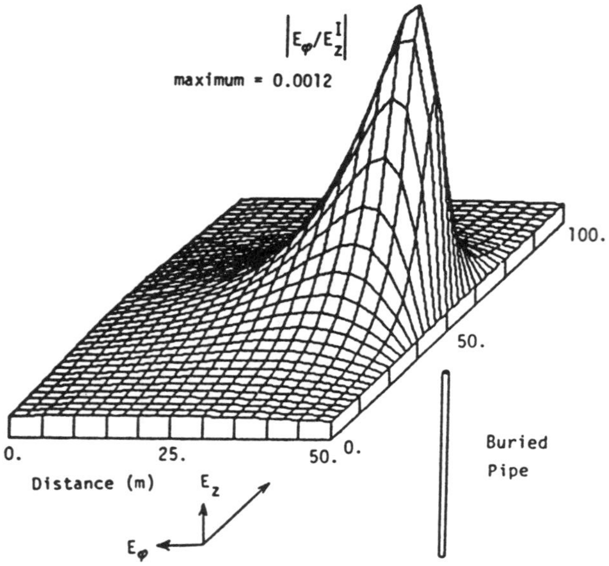

Fig. 8. Typical signature (defined by the magnitude of E_ϕ relative to E_z) of a buried vertical pipe (length L = 45 m, diameter d = 0.2 m) with upper end 5 m below the interface for $\tilde{\epsilon}_-$ = 16 − j80 and f = 200 kHz. The excitation is provided by a vertical electric source 10 km distant. Only the left half of the field, which is left-right symmetric, is plotted.

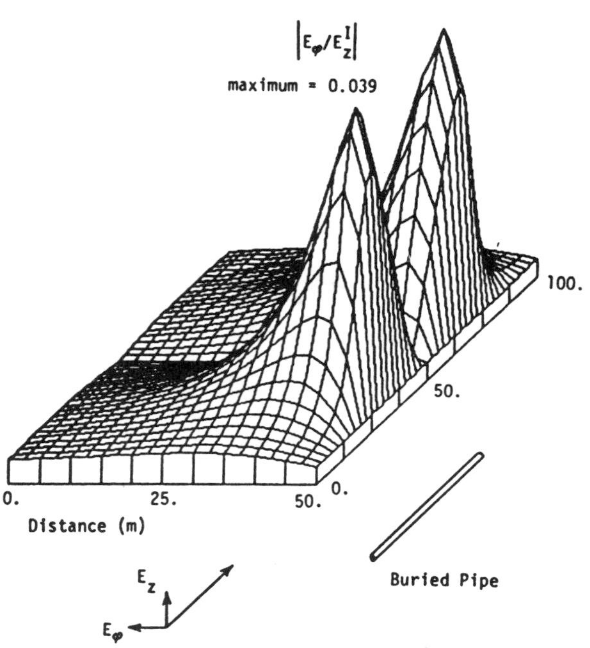

Fig. 9. Signature of a horizontal pipe of L = 50 m, d = 0.2 m, and buried 5 m below the interface with $\tilde{\epsilon}_-$ = 16 − j80 and f = 200 kHz, plotted for the same excitation as in Fig. 8. The two peaks in the field are associated with the ends of the pipe.

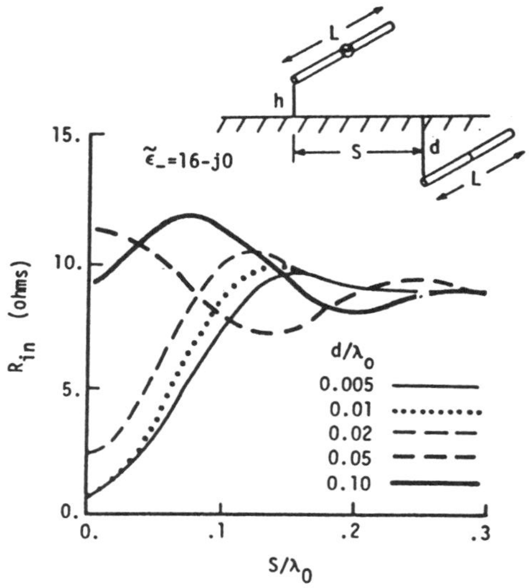

Fig. 10. Resistance of an elevated dipole due to a buried wire with burial depth a parameter. The wires are parallel to each other and the interface with their centers in a common vertical plane. Wire lengths are L = 0.125 λ_0, diameter 0.0002 λ_0, with the elevated dipole 0.02 λ_0 above the interface.

screen above and below the interface. The transition from a k_+ wavenumber for current to a k_- wavenumber occurs largely within a vertical distance of $2(10^{-6})\lambda_0$. The large change in amplitude from $s = 10^{-3}$ λ_0 to $s = 10^{-6}$ λ_0 is due to the choice of a screen radius of 0.5 λ_0 which probably would not be used in practice.

The input resistance and radiation efficiency are shown in Fig. 6 for varying screen radius. For low ground conductivity, the effect of the ground screen is strongly dependent on the lengths of the radial wires. The radiation efficiency as the number of radial wires is increased to a maximum of 16 is shown in Fig. 7. For 100 radials, the NEC results are in good agreement with those of Wait and Pope [12] which are expected to be accurate for dense screens.

An area in which there is a continuing need for the capability to predict the response of buried objects to various kinds of incident fields is that of geophysical exploration. Two approaches are discussed below, measurement of scattered field and measurement of impedance change in an active probe.

In Fig. 8 a buried vertical pipe is illuminated by the field of a distant vertical tower. The plot shows the magnitude of the scattered field E_ϕ ($\hat{\phi}$ is normal to the direction to the transmitting tower) relative to the incident field E_z and represents measurements made by moving a probe in a raster-like fashion over the search area. Fig. 9 shows the same result for a buried horizontal pipe. Other objects have been modeled, including a plate represented by a wire grid, for varying frequency, excitation, and ground parameters.

Another approach to geophysical probing is to measure the input impedance of a sensing antenna as it is moved over the search region. The change in input resistance of a horizontal dipole as it is moved over a buried horizontal wire is shown in Fig. 10 for a dielectric ground. Increasing ground conductivity

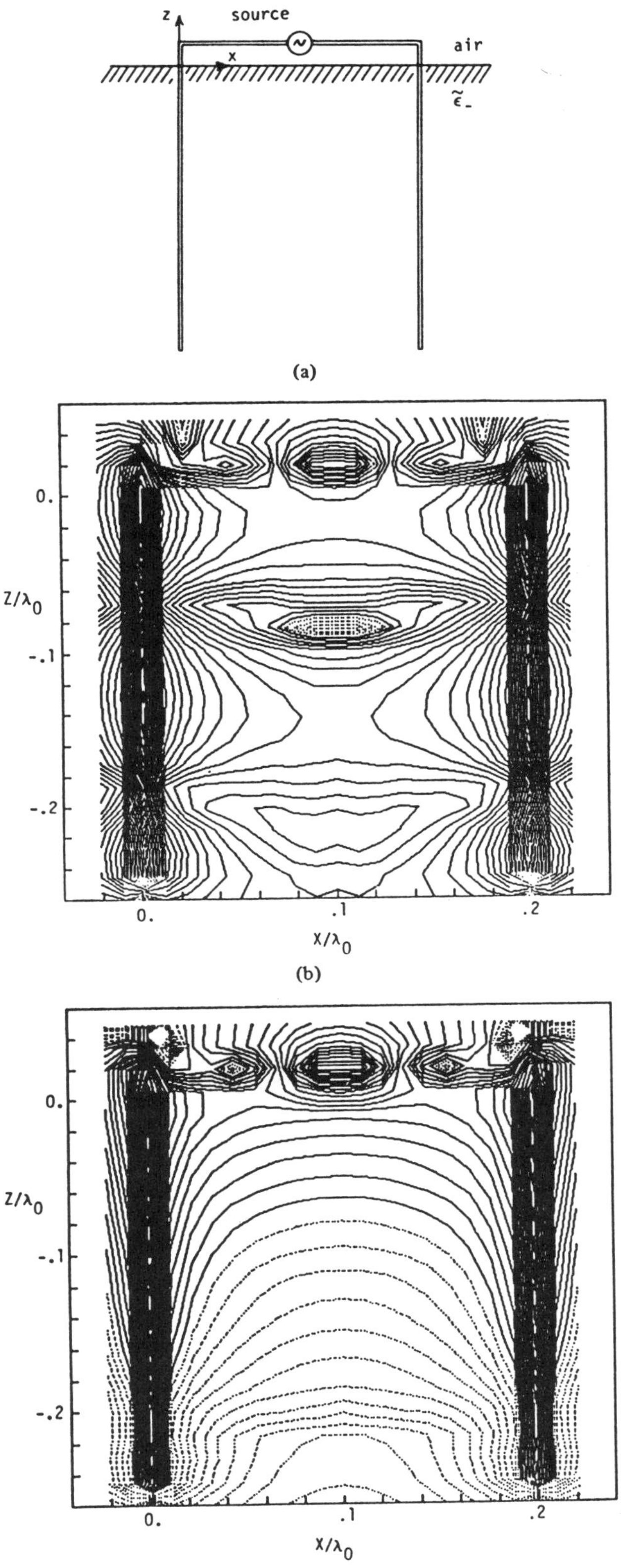

Fig. 11. Contours of the magnitude of the horizontal component of electric field produced by a single-wire model of a Seige EMP simulator. The source is 0.02 λ_0 above the interface, and $\tilde{\epsilon}_-$ is 16 − $j0$ in (b) and 16 − $j16$ in (c). A standing wave is apparent for zero loss. The contour interval is 2 dB in each case and the change from solid to dotted contours indicates the same field strength in each figure for a fixed voltage source.

would rapidly decrease the response. For either scattered field or impedance measurement, the response is relatively small, especially for the vertical pipe, so that noise could substantially reduce target detectability.

As a final example, contours of constant electric field strength are shown in Fig. 11 for a two wire transmission line penetrating into the ground. This structure represents a simple model of a Siege simulator for testing the effects of electromagnetic pulse on buried systems. Computer models of this kind can be helpful in evaluating design trade-offs.

V. CONCLUDING REMARKS

The development and application of a computer model for testing objects located near, or penetrating, the earth-air interface has been described in this paper. An integral equation provides the basis for the computer model, and its solution is obtained using the moment method.

The effect of the ground is included using the Sommerfeld integrals, which appear as part of the integral equation kernel. An innovative procedure is used to provide an accurate and efficient computation of these fields. It is based on a combination of interpolation, model-based parameter estimation and an asymptotic expansion. A charge discontinuity condition is employed to ensure continuity of the tangential electric field at the interface due to the wire current. The result is a computer model which can provide accurate results for problems involving penetrating conductors at a computer cost of only ~4-8 times that required for the same object in free space.

ACKNOWLEDGMENT

This work represents the culmination of much effort over many years by a number of people. The authors would like to acknowledge the contributions of R. M. Bevensee, J. N. Brittingham, F. J. Deadrick, W. A. Johnson, D. L. Lager, R. J. Lytle, J. T. Okada, A. J. Poggio, E. S. Selden, and R. W. Ziolkowski to various aspects of the developments reported here. We would also like to acknowledge the sponsorship of the Defense Advanced Research Projects Agency (D. E. Barrick); the Naval Ocean Systems Center, San Diego, CA (J. C. Logan and J. W. Rockway); the U.S. Army Communications Electronics Command, Ft. Monmouth, NJ (D. V. Campbell); and the U.S. Army Communication-Electronics Engineering Installation Agency, Ft. Huachuca, AZ (G. Lane and J. E. McDonald).

REFERENCES

[1] G. J. Burke and A. J. Poggio, "Numerical electromagnetics code (NEC)—method of moments, Part I: Theory, Part II: Code, Part III: User's manual," NOSC TD-116, Naval Ocean Syst. Center, San Diego, CA, July 18, 1977 (NEC-1), revised Jan. 2, 1980 (NEC-2).

[2] E. K. Miller and F. J. Deadrick, "Some computational aspects of thin-wire modeling," in *Numerical and Asymptotic Techniques in Electromagnetics*, R. Mittra, Ed. New York: Springer-Verlag, 1975, ch. 4.

[3] E. K. Miller, A. J. Poggio, G. J. Burke, and E. S. Selden, "Analysis of wire antennas in the presence of a conducting half space: Part I. The vertical antenna in free space," *Canadian J. Phys.*, vol. 50, pp. 879–888, 1972.

[4] ——, "Analysis of wire antennas in the presence of a conducting half space: Part II. The horizontal antenna in free space," *Canadian J. Phys.*, vol. 50, pp. 2614–2627, 1972.

[5] A. Banos, *Dipole Radiation in the Presence of a Conducting Half-Space.* New York: Pergamon, 1966.

[6] A. J. Poggio and E. K. Miller, "Integral equation solutions of three-dimensional scattering problems," in *Computer Techniques for Electromagnetics*, R. Mittra, Ed. New York: Pergamon, 1973, ch. IV.

[7] E. K. Miller, "The numerical electromagnetics code (NEC)," in

Applications of the Method of Moments to Electromagnetic Fields, B. J. Strait, Ed. SCEEE Press, 1980.

[8] K. G. Balmain, "Dipole admittance for magneto plasma diagnostics," *IEEE Trans. Antennas Propagat.*, vol. AP-17, pp.389-392, 1969.

[9] F. M. Tesche, "On the analysis of scattering and antenna problems using the singularity expansion technique," *IEEE Trans. Antennas Propagat.*, vol. AP-21, pp. 53-62, 1973.

[10] G. J. Burke, E. K. Miller, J. N. Brittingham, D. L. Lager, and R. J. Lytle, "Computer modeling of antennas near the ground," *Electromagn.*, vol. 1, pp. 29-49, 1981.

[11] W. A. Johnson, "Analysis of a vertical, tubular cylinder which penetrates an air-dielectric interface and which is excited by an azimuthally symmetric source," *Radio Sci.*,vol. 18, no. 6, pp. 1273-1281, Nov.-Dec. 1983.

[12] J. R: Wait and W. A. Pope, "The characteristics of a vertical antenna with a radial conductor ground system," *Appl. Sci. Res.*, section B, Vol. 4, 1954.

[13] E. K. Miller, J. N. Brittingham, and J. T. Okada, "Explicit modeling of antennas with sparse ground screens," *Electron. Lett.*, vol. 14, no. 19, pp. 627-629, Sept. 14, 1978.

[14] G. J. Burke, W. A. Johnson, and E. K. Miller, "Modeling of simple antennas near to and penetrating an interface," *Proc. IEEE*, vol. 71, no. 1, Jan. 1983.

General integral equation formulation for microstrip antennas and scatterers

J.R. Mosig, D.Appl.Sc., and Prof. F.E. Gardiol, M.Sc., D.Appl.Sc., Sen. Mem. IEEE

Indexing terms: *Antennas (Microstrip), Numerical Analysis*

Abstract: The paper deals with the dynamic analysis of microstrip structures. It is shown that the mixed-potential integral equation for stratified media, which was introduced in a previous publication, provides a rigorous and powerful approach. The Green's functions belonging to the kernel of the integral equation are expressed as Sommerfeld integrals, in which surface wave effects are automatically included. A two-dimensional moment's method using subsectional basis functions has been chosen. Thus, microstrip patches of any shape can be analysed at any frequency and for any substrate. Practical numerical aspects are carefully discussed, and special numerical devices are introduced to reduce computation time without loss of accuracy. Complete results for a rectangular patch and for a slotted patch are given and compared with measured values. Radiation patterns corresponding to the ideal situation of a substrate with infinite transverse dimensions are presented for a rectangular patch.

1 Introduction

Over the past decade, the range of application of microstrip structures has broadened considerably. In particular, microstrip antennas are used in an increasing numer of applications, ranging from biomedical diagnosis to satellite communications. Such a wide range of applications, coupled with the fact that microstrip structures are relatively simple to produce with good reproducibility, has turned microstrip analysis into a cornerstone problem, to which almost all the mathematical models developed in the field of electromagnetics have been applied. This is witnessed by the huge amount of technical literature and several monograph books published in recent years [1–3].

Models used to study microstrip patch antennas range from very simplified ones, such as the transmission-line model [4], through cavity models [5], segmentation techniques [6], full-wave analysis [7] and up to quite sophisticated approaches based on an integral formulation and numerical resolution on a computer [8–10]. Whereas simple approximations yield directly usable simplified formulas, the more complex approaches require increasingly lengthy calculations. Many models are directly linked to simple patch shapes (rectangular or circular), and in some instances an approximate distribution of the current is introduced, determined by an educated guess. Unfortunately, the range of validity of many of the assumptions made was not defined; as a matter of fact, this would be difficult to make in the absence of a rigorous solution as a basis for comparison. Also, some effects such as the presence of surface waves are lost in the approximation process (but not in the actual device!). Detailed surveys of the previously developed methods are available [11–12].

The purpose of the present study is to provide a rigorous treatment of the general problem, free from over-simplifying assumptions and applicable to arbitrarily shaped patches. The mixed-potential integral equation [13] was found to be better suited for numerical analysis than the previously used electric-field integral formulation [8–10]. Of particular interest is the fact that the numerical techniques presented by the authors in a previous publication can be taken full advantage of [14].

Paper 4095H (E11), first received 25th February and in revised form 10th June 1985

The authors are with the École Polytechnique Fédérale de Lausanne, Laboratoire d'Electromagnétisme et d'Acoustique, 16 ch. de Bellerive, CH-1007, Switzerland

The integral equation is solved by means by a moment's method using rooftop subsectional basis functions [15], which are much more flexible to use than sine waves defined over the entire domain. These choices, and a thorough treatment of the antenna's excitation, lead to a very general technique, suitable for the analysis of complex shaped patches with any combination of thickness and permittivity, taking dielectric and conducting losses into account. The frequency behaviour of a given structure can be determined from the quasistatic range up to its higher-order resonances.

2 Mixed-potential integral equation (MPIE)

This formulation was extensively used in the analysis of wire antennas by the moment's method [16]. Here, it will be applied to lossy microstrip structures. With reference to Fig. 1, the boundary conditions for the electric field on the

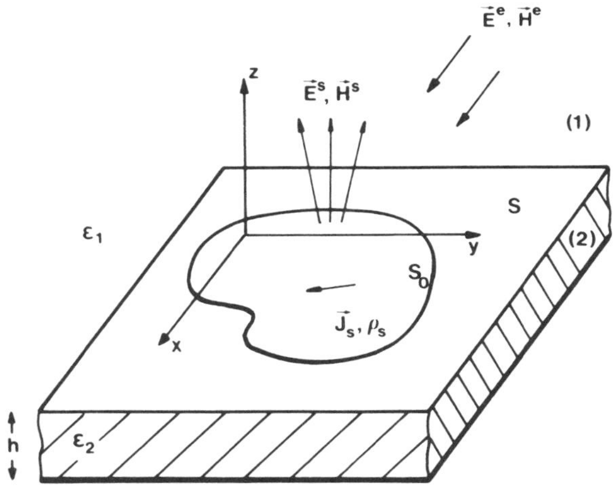

Fig. 1 *Arbitrarily shaped microstrip structure with dynamical excitation*
$E^{(e)}$: Excitation field (source)
$E^{(s)}$: Scattered field
J_s, ρ_s: Induced current and charge densities
S_0: Conducting patch (upper conductor)
S: air-substrate interface

surface of a patch of real conductor (non ideal) is

$$e_z \times [E^{(s)}(r) + E^{(e)}(r)] = Z_s[e_z \times J_s(r)] \quad (1)$$

$$r \in S_0$$

This equation simply expresses that the total electric field, sum of the excitation field $E^{(e)}$ and of the scattered field $E^{(s)}$ must be proportional to the electric surface current J_s. The proportionality factor Z_s has the dimensions of an impedance and depends on the metal conductivity σ, the thickness of the upper conductor t and the frequency f. In a perfect conductor, Z_s vanishes, whereas in most practical situations the metal skindepth is much smaller than the conductor thickness, so that Z_s becomes the classical plane-wave surface impedance $Z_s = (1 + j)\sqrt{\mu_0 f/\pi\sigma}$. As the upper conductor is always much thinner than the dielectric substrate, it can be replaced by a current sheet at all frequencies. The surface current density J_s in eqn. 1 is thus a total value, the sum of the actual surface currents flowing over both sides of the patch.

The scattered field derives from a scalar and a vector potential, which in turn are expressed in terms of superposition integrals of the corresponding Green's functions, weighted by the unknown distributions of surface electric charge and current [14]

$$A(\mathbf{r}) = \int_{S_0} \bar{G}_A(\mathbf{r} \mid \mathbf{r}') \cdot J_s(\mathbf{r}') \, dS';$$

$$V(\mathbf{r}) = \int_{S_0} G_V(\mathbf{r} \mid \mathbf{r}') \rho_s(\mathbf{r}') \, dS' \qquad (2)$$

The Green's functions $\bar{G}_A$ and G_V can be expressed in terms of Sommerfeld integrals. They are related to those introduced in Reference 10 to solve Pocklington's equation for printed wires. Their analytical and numerical properties have been extensively studied in a previous paper [14] for the lossless case. The modifications needed to account for a complex relative permittivity $\varepsilon_r = \varepsilon_r'(1 - j \tan \delta)$ have been outlined in Reference 17.

When the observer is very close to the source, the dominant term in the Green's functions is given by the static Green's functions corresponding to an homogeneous medium of permittivity $\varepsilon_0(\varepsilon_r + 1)/2$ and permeability μ_0

$$\frac{4\pi}{\mu_0} G_A^{xx}(\mathbf{r} \mid \mathbf{r}') = 2\pi(\varepsilon_r + 1)\varepsilon_0 G_V(\mathbf{r} \mid \mathbf{r}') = 1/\lvert \mathbf{r} - \mathbf{r}' \rvert \qquad (3)$$

This fact can be taken advantage of to expand any component of the Green's functions G in the following way:

$$G = G_H + (G - G_H) \qquad (4)$$

where H stands for homogeneous.

The term G_H, given by eqns. 3, exhibits at the origin a weak singularity of the r^{-1} type. Its integral over a rectangular domain can be performed analytically. Therefore, numerical techniques are only needed to evaluate the difference term, which is a regular well-behaved function at the origin.

3 Moment's method

To obtain an exact solution, one would have to satisfy the boundary condition, eqn. 1, at every point within the patch. This is clearly not feasible, as it would require the resolution of an infinite set of equations. Some kind of truncation of the set is an absolute requirement: the boundary condition, eqn. 1, will then be satisfied over a limited number of points, carefully chosen over the patch, using a method of moments.

3.1 Charge and current cells

The basis and test functions best suited to the study of arbitrary shaped patches at any frequency are selected. No *a priori* assumptions will be made for the distribution of currents on the patch: this actually eliminates the use of basis functions defined over the entire domain [8]. A comparison of available possibilities [11] led to the selection of rooftop functions for the surface current J_s, which were successfully used in similar problems [15]. To implement these functions, the patch's boundary is replaced by a Manhattan-type polygonal line (Fig. 2A). As most commonly used antennas exhibit this kind of geometry anyway, this requirement is easily satisfied.

The patch's surface is then divided into rectangular cells, called charge cells, which are all chosen of equal size, with dimensions $a \times b$ (Fig. 2A). This is not a basic requirement, but the use of different cell sizes would considerably increase the length of the computations.

Two adjacent charge cells, sharing a common border perpendicular to the x-direction (y-direction), will form an x-directed (y-directed) current cell (Fig. 2B). An automatic overlapping of current cells is obtained in this manner, in which a charge cell may belong to up to four different current cells. The number of charge cells is thus related to the number of current cells, although the relationship is not a simple one, depending as it does on the shape of the patch. For rectangular patches with $m \times n$ charge cells, the number of x-directed current cells is $M = (m - 1)n$, and that of y-directed current cells $N = m(n - 1)$.

Every current cell supports one rooftop basis function, to which is associated one test segment joining the centres of the two charge cells belonging to the current cell. The centre of the segment C_{xj} associated to the j-th x-directed current will be denoted by the vector $\mathbf{r}_{xj}$, its ends by $\mathbf{r}_{xj}^-$

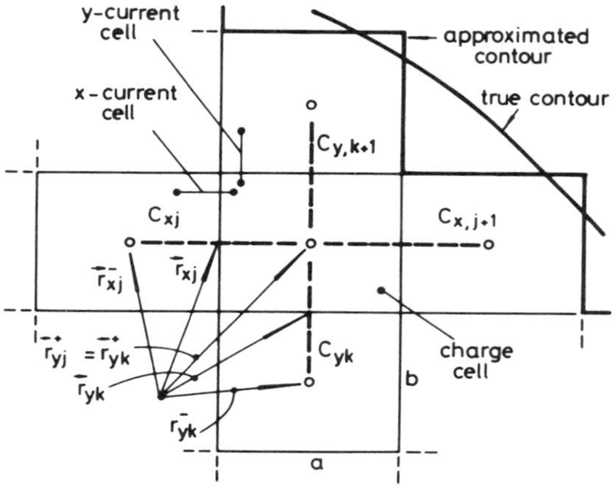

Fig. 2A *Segmentation of the patch in elementary charge and current cells, showing the network of test segments*

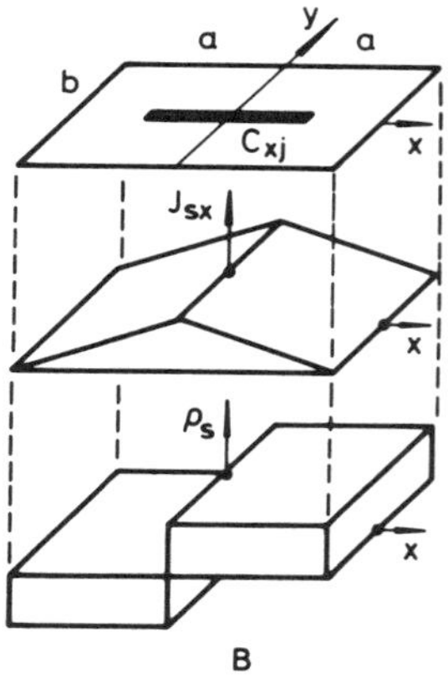

Fig. 2B *x-directed current cell centred at r = 0 and its associate surface current density $J_{sx} = T_x(\mathbf{r})$, and surface charge density $\rho_s = \Pi(\mathbf{r} - \mathbf{e}_x a/2) - \Pi(\mathbf{r} + \mathbf{e}_x a/2)$*

and r_{xj}^{+} (Fig. 2), with these three vectors related through

$$r_{xj}^{\pm} = r_{xj} \pm e_x(a/2) \qquad j = 1, 2, \dots M \tag{5}$$

A similar relationship is written for y-directed segments C_{yj} $(j = 1, 2 \dots, N)$.

3.2 Basis functions

The Cartesian components of the surface current are expanded over a set of basis functions T_x, T_y

$$J_{sx} = \frac{1}{b} \sum_{j=1}^{M} I_{xj} T_x(r - r_{xj})$$
$$J_{sy} = \frac{1}{a} \sum_{j=1}^{N} I_{yj} T_y(r - r_{yj}) \tag{6}$$

where the basis functions are of rooftop type defined as (Fig. 2)

$$T_x(r) = \begin{cases} 1 - |x|/a & |x| < a, |y| < b/2 \\ 0 & \text{elsewhere} \end{cases} \tag{7}$$

A similar expression is obtained for T_y by interchanging $a \leftrightarrow b$, $x \leftrightarrow y$ in eqn. 7.

The introduction of factors $1/a$ and $1/b$ in eqn. 6 yields unknown coefficients I_{xj} and I_{yj} having dimensions of a current. Moreover, every coefficient gives the total current flowing across the common boundary of two charge cells.

The associated surface charge density is obtained from eqn. 6 by using the continuity equation, yielding

$$\rho_s = \frac{1}{j\omega ab} \left\{ \sum_{j=1}^{M} I_{xj}[\Pi(r - r_{xj}^{+}) - \Pi(r - r_{xj}^{-})] + \sum_{j=1}^{N} I_{yj}[\Pi(r - r_{yj}^{+}) - \Pi(r - r_{yj}^{-})] \right\} \tag{8}$$

where $\Pi(r)$ is a two-dimensional unit pulse function defined over a rectangle of dimensions $a \times b$, centred at $r = 0$ (Fig. 2A).

The charge density within every elementary cell remains constant, justifying the appellation of charge cell. For the charge cell of Fig. 2B, with four test segments ending at its centre, the surface charge density is simply given by

$$\rho_s = \frac{1}{j\omega ab} [I_{x,j+1} - I_{x,j} + I_{y,k+1} - I_{y,k}] \tag{9}$$

The charge density is discontinuous on the borders between charge cells. The scalar potential remains bounded, while the electric field becomes singular, as ρ_s does not satisfy a Hölder condition [11]. This means that the test functions must be selected carefully, avoiding the locations where the electric field is singular.

3.3 Discrete Green's functions

The notation and the computational task can be simplified by introducing discrete Green's functions, which have as source a complete basis function, instead of the traditional elementary point source.

The vector potential $\bar{\Gamma}_A$ is created by a rooftop distribution of surface current, whereas Γ_V is the scalar potential resulting from a rectangular distribution of unit surface charge. It is convenient in practice to deal with dimensionless quantities, in a normalised space where physical lengths are replaced by electrical lengths. The following adimensional expressions are therefore introduced, defining the discrete Green's functions

$$\Gamma_A^{xx}(r \mid r_{xj}) = \int_{S_{xj}} \frac{1}{\mu_0 k_0} G_A^{xx}(r \mid r') T_x(r' - r_{xj})(k_0^2 \, dS') \tag{10a}$$

$$\Gamma_V(r \mid r_{0j}) = \int_{S_{0j}} \frac{\varepsilon_0}{k_0} G_V(r \mid r') \Pi(r' - r_{0j})(k_0^2 \, dS') \tag{10b}$$

A similar expression holds for Γ_A^{yy}. In these formulas, $r_{xj}(r_{0j})$ denotes the centre and $S_{xj}(S_{0j})$ the surface of a current (charge) cell.

The discrete Green's functions exhibit the same properties of translational invariance and of symmetry as the conventional ones do [14]. In the general case, the surface integrals in eqn. 10 must be evaluated numerically. When the observation point r belongs to the source cell, some difficulties arise in the integration process. It is then recommended to separate the Green's functions into their singular and regular parts, as indicated in eqn. 4, where the singular part can be integrated analytically. For an observation point at the centre of a charge cell, replacement of eqn. 3 in eqn $10b$ yields the singular part of the discrete Green's function as

$$2\pi(\varepsilon_r + 1)\Gamma_V(\mathbf{0} \mid \mathbf{0}) \simeq 2k_0 a \ln \tan\left(\frac{\alpha}{2} + \frac{\pi}{4}\right) - 2k_0 b \ln \tan(\alpha/2) \tag{11}$$

with $\tan \alpha = b/a$

When the observer is located many cells away from the sources, the latter can be concentrated at the centre of the cell. The following approximations may then be used:

$$\Gamma_A^{xx}(r \mid r_{xi}) \simeq \frac{1}{\mu_0 k_0} G_A^{xx}(r \mid r_{xi})(k_0 a)(k_0 b) \tag{12}$$
$$\Gamma_V(r \mid r_{0i}) \simeq \frac{\varepsilon_0}{k_0} G_V(r \mid r_{0i})(k_0 a)(k_0 b)$$

Discrete Green's functions provide a very compact notation for the potentials created by the whole structure. Introducing eqns. 6 and 8 in the definitions 2, and making use of eqn. 10, yields

$$A(r) = e_x \frac{\mu_0}{k_0 b} \sum_{j=1}^{M} I_{xj} \Gamma_A^{xx}(r \mid r_{xj}) + e_y \frac{\mu_0}{k_0 b} \sum_{j=1}^{N} I_{yj} \Gamma_A^{yy}(r \mid r_{yj}) \tag{13a}$$

$$V(r) = \frac{Z_0}{j(k_0 a)(k_0 b)} \left\{ \sum_{j=1}^{M} I_{xj}[\Gamma_V(r \mid r_{xj}^{+}) - \Gamma_V(r \mid r_{xj}^{-})] + \sum_{j=1}^{N} I_{yj}[\Gamma_V(r \mid r_{yj}^{+}) - \Gamma_V(r \mid r_{yj}^{-})] \right\} \tag{13b}$$

where Z_0 is the characteristic free-space impedance.

3.4 Test functions

The last step of the resolution with a moment's method is the selection of a suitable test function. Previous work [11] has shown that the most adequate choice, compatible with the basis functions selected, is the use of unidimensional rectangular pulses. This actually means that the boundary condition, eqn. 1, is integrated along all the test segments, yielding

$$j\omega \int_{C_{xi}} A_x \, dx + V(r_{xi}^{+}) - V(r_{xi}^{-}) + Z_s \int_{C_{xi}} J_{sx} \, dx = \int_{C_{xi}} E_x^{(e)} \, dx = -V_{xi}^{(e)} \tag{14}$$

where C_{xi} is the x-directed test segment extending from r_{xi}^{-} to r_{xi}^{+} and $V_{xi}^{(e)}$ is the excitation (impressed) voltage along the segment. A similar relationship is obtained for y-directed test segments. It is worth mentioning that this choice eliminates the need for computing field values near the edges, where field singularities can negatively affect the performances of the moment's method.

Eqns. 14 are well suited to a numerical treatment, as all derivatives have been removed. The integration of J_{sx} can be done easily by using expansion 6 with the result

$$\int_{C_{xi}} J_{sx}\, dx = \frac{a}{4b}\,[2I_{xi} + I_{xi+1} + I_{xi-1}] \simeq \frac{a}{b}\, I_{xi} \qquad (15)$$

The last approximation is valid for a reasonably smooth current distribution.

3.5 Matrix equation

Introducing expansions 13 into eqn. 14 yields the following matrix equation:

$$\begin{pmatrix} C^{xx} & C^{xy} \\ C^{yx} & C^{yy} \end{pmatrix} \begin{pmatrix} I_x \\ I_y \end{pmatrix} = \frac{1}{jZ_0} \begin{pmatrix} V_x^{(e)} \\ V_y^{(e)} \end{pmatrix} \qquad (16)$$

The elements in the submatrices are given by

$$\begin{aligned} C_{ij}^{xx} = {} & \frac{1}{k_0\, a k_0\, b}\,[-\Gamma_V(\boldsymbol{r}_{xi}^+ \mid \boldsymbol{r}_{xj}^-) - \Gamma_V(\boldsymbol{r}_{xi}^- \mid \boldsymbol{r}_{xj}^+) \\ & + \Gamma_V(\boldsymbol{r}_{xi}^+ \mid \boldsymbol{r}_{xj}^+) + \Gamma_V(\boldsymbol{r}_{xi}^- \mid \boldsymbol{r}_{xj}^-)] \\ & - \frac{1}{k_0\, b} \int_{C_{xi}} \Gamma_A^{xx}(\boldsymbol{r} \mid \boldsymbol{r}_{xj}) k_0\, d_x + j\,\frac{Z_s}{Z_0}\,\frac{a}{b}\,\delta_{ij} \\ & i = 1 \ldots M,\, j = 1 \ldots M \end{aligned} \qquad (17a)$$

$$\begin{aligned} C_{ij}^{xy} = {} & \frac{1}{k_0\, a k_0\, b}\,[-\Gamma_V(\boldsymbol{r}_{xi}^+ \mid \boldsymbol{r}_{yj}^-) - \Gamma_V(\boldsymbol{r}_{xi}^- \mid \boldsymbol{r}_{yj}^+) \\ & + \Gamma_V(\boldsymbol{r}_{xi}^+ \mid \boldsymbol{r}_{yj}^+) + \Gamma_V(\boldsymbol{r}_{xi}^- \mid \boldsymbol{r}_{yj}^-)] \\ & i = 1 \ldots M,\, j = 1 \ldots N \end{aligned} \qquad (17b)$$

where δ_{ij} is the Kronecker delta. The expression for C_{ij}^{yy} is obtained by interchanging the couples (x, y), (a, b) and (M, N) within eqn. 17*a*. Finally, it is easily shown that $C_{ij}^{yx} = C_{ji}^{xy}$.

For distances $|\boldsymbol{r}_{xi} - \boldsymbol{r}_{xj}|$ much greater than the dimensions of a cell, the integrals in eqn. 17*a* can be replaced by

$$\int_{C_{xi}} \Gamma_A^{xx}(\boldsymbol{r} \mid \boldsymbol{r}_{xj}) k_0\, dx \simeq k_0\, a \Gamma_A^{xx}(\boldsymbol{r}_{xi} \mid \boldsymbol{r}_{xj}) \qquad (18)$$

In principle, this approximation is not valid for short distances between cells. For these situations, however, the contribution of the vector potential to eqn. 17*a* is overshadowed by the one of the scalar potential, so that the approximation of eqn. 18 still suffices. As a matter of fact, eqn. 18 may be used everywhere but in the diagonal terms. This assumption was confirmed by extensive numerical tests.

A last point worth mentioning concerns the number of discrete Green's functions which must be calculated. For a rectangular patch with $m \times n$ charge cells, the number of matrix elements is $(M + N)^2$, with $M = (m - 1)n$ and $N = (n - 1)m$. When all the cells have identical sizes, only $m \times n$ values of Γ_V, M values of Γ_A^{xx} and N values of Γ_A^{yy} are needed in order to completely fill the matrix. This is the great advantage of using cells of equal size. It is generally more convenient to use a larger number of identical cells, rather than fewer cells of different sizes.

4 Numerical details

4.1 Interpolation among Green's functions

The evaluation of the matrix in eqn. 21 requires a large amount of computation. For a rectangular patch divided into 10×10 cells, the order of the matrix is 180, hence the number of elements in it is $180^2 = 32\,400$. Even when a simple 4×4 Gaussian quadrature is used to evaluate the discrete Green's functions, eqns. 16, the number of Sommerfeld integrals which should be evaluated would exceed half a million.

Fortunately, for a given structure these integrals only depend upon the distance from source to observer. It is thus possible to tabulate the integrals for a small number of distances, and then to interpolate between the tabulated values. The distances to be considered range from zero to the maximum linear dimension of the patch. Several interpolation schemes have been tried [11]. The best solution was obtained by separating the Green's functions according to eqn. 4, and then using a simple parabolic Lagrange interpolation for the regular part.

For a square patch with 10×10 cells, at frequencies for which the patch's length is less than a free-space wavelength, the error obtained when interpolating from 25 tabulated values is hardly noticeable: less than 0.5%, even though the computation time was reduced by a factor of 100!

4.2 Resolution of the linear system

The system of linear equations 16 is solved by standard Gaussian elimination. The $[C_{ij}]$ matrix is ill-conditioned, so that a careful evaluation of its elements is needed. This matrix is diagonal dominated, so that the accuracy requirements may be relaxed for the off-diagonal terms. The following approximation was therefore considered: the double numerical integration, eqn. 10, is replaced by its analytical approximation, eqn. 12, whenever the distance between cells i and j of a given element C_{ij} exceeds a certain critical distance D.

As the physical characteristics of the substrate itself do not significantly affect the numerical problem, the values $\varepsilon_r = 1$, $h \to \infty$ (isolated patch in free space) have been selected to determine the effect of the parameter D. A rectangular patch with 10×10 cells is again considered. When D becomes larger than the patch's diagonal ($D/a > 10\sqrt{2}$), no approximation is introduced, and the solution of the rigorous computation is obtained. When $D/a = 0$, the approximation is used everywhere but for the diagonal terms. The resulting relative error for the RMS value of the currents is about 25%. For $D/a = 4$, the error is reduced to 4%, and for $D/a = 8$, it further drops down to 0.1%. Hence, the use of this approximation can reduce considerably the length of computations, without significantly affecting the accuracy.

4.3 Relevance of surface waves and losses

The diagonal terms of eqn. 17*a*, which dominate the behaviour of the matrix, can be written as

$$\begin{aligned} C_{ii}^{xx} = {} & \frac{2}{k_0\, a k_0\, b}\,[\Gamma_V(\boldsymbol{0} \mid \boldsymbol{0}) - \Gamma_V(\boldsymbol{e}_x a \mid \boldsymbol{0})] \\ & - \frac{1}{k_0\, b} \int_{-a/2}^{+a/2} \Gamma_A^{xx}(\boldsymbol{r} \mid \boldsymbol{0}) k_0\, dx + j\,\frac{Z_s}{Z_0}\,\frac{a}{b} \end{aligned} \qquad (19)$$

For electrically small cells, the main contribution to C_{ii}^{xx} comes from the self term $\Gamma_V(\boldsymbol{0} \mid \boldsymbol{0})$, which gives the scalar potential produced by a cell on its own centre. In a homogeneous lossless case, this self term has a negative imaginary part. But it has been shown in previous works [11, 14] that in the microstrip case the imaginary part of $\Gamma_V(\mathbf{0}|\mathbf{0})$ is positive, due to the presence of a surface wave. The dielectric losses will also contribute a small positive imaginary term, due to the presence of the complex permittivity in the denominator of eqn. 11. Finally, it is apparent from eqn. 19 that conductor losses further add a positive imaginary contribution.

Summarising the above, surface waves and losses considerably affect the imaginary part of the diagonal terms. Still, their effect may go unnoticed, as the real part, which is not affected, is larger by several orders of magnitude. Nevertheless, at resonance, the currents are in quadrature with the excitation, in which case they are mainly determined by the imaginary part of the moment's matrix. This means that losses and surface waves play a significant role at resonance.

4.4 Resonant frequencies and matrix condition

The roots of the complex determinant of the moment's matrix yield the resonant frequencies, which are, in general, complex (they correspond to an open radiating structure). On the real frequency axis, the determinant does not vanish, but goes through sharp minima at the points closest to the complex roots. These minima, detected with standard numerical techniques, provide the real resonant frequencies of the antenna. The condition number of the matrix, which is often obtained as a by-product of resolution techniques for linear systems may also be used to locate the resonances.

As was shown in the previous Section, in electrically small cells the scalar potential is predominant in the matrix elements. As a result, when four test segments form a square loop, a test along one of them is practically equivalent to a test along the remainder of the loop. In other words, some rows of the moment's matrix are almost linear combinations of three other rows. It is for this reason that the matrix is severely ill-conditioned. In some extreme situations, some resonances may actually be missed, the numerical value of the determinant being masked by numerical noise and round-off errors. A considerable improvement can be obtained by systematically replacing every test segment closing a loop by the complete loop. Eqn. 14 for a test segment is now replaced by the loop's equation

$$+j\omega \oint_\Gamma A_x \, dx + Z_s \oint_\Gamma J_{sx} \, dx = + \oint_\Gamma E_x^{(e)} \, dx \qquad (20)$$

which is independent of the scalar potential [18].

4.5 Excitation and input impedance

The column vector $V^{(e)}$ in eqn. 16 is obtained by integrating the tangential excitation electric field. The simplest kind of excitation is provided by a plane wave impinging on the patch. In the previous numerical tests, this excitation was used.

For transmitting antennas, more complex excitations have to be considered. The numerical problems encountered with a coaxial probe excitation will be described here. A rigorous model would require the introduction of a frill of magnetic currents $\boldsymbol{M}_s$ within the ground plane (Fig. 3A). In practice, however, simpler models may be used. The inner coaxial conductor carries a total current $I = 1A$., supposed to be uniformly distributed on the surface, so that

$$\boldsymbol{J}_s = \boldsymbol{e}_z \frac{I}{2\pi r_c} \qquad (21)$$

where r_c is the radius of the inner conductor. The current then spreads radially across the patch, and several sophisticated attachment models have been devised to describe it [19]. A new, simpler model is introduced here, in which the excitation current spreads over a charge cell (Fig. 3A). The postulated current distribution is (Fig. 3B)

$$\boldsymbol{J}_s = \boldsymbol{e}_x \frac{I}{4b} \operatorname{sgn}(x)(1 - 2|x|/a) + \boldsymbol{e}_y \frac{I}{4a} \operatorname{sgn}(y)(1 - 2|y|/b) \qquad (22)$$

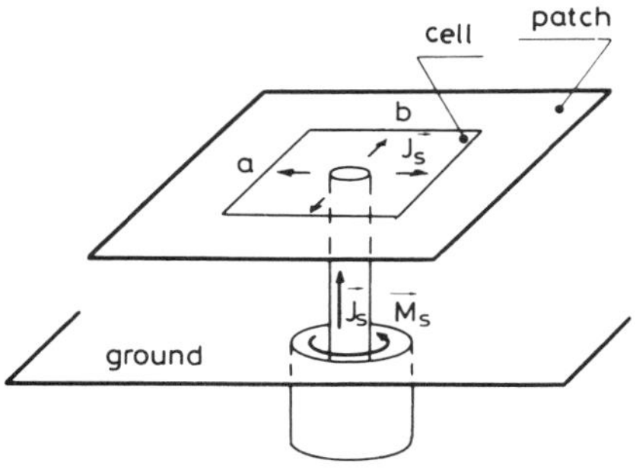

Fig. 3A *Coaxial-fed microstrip patch*

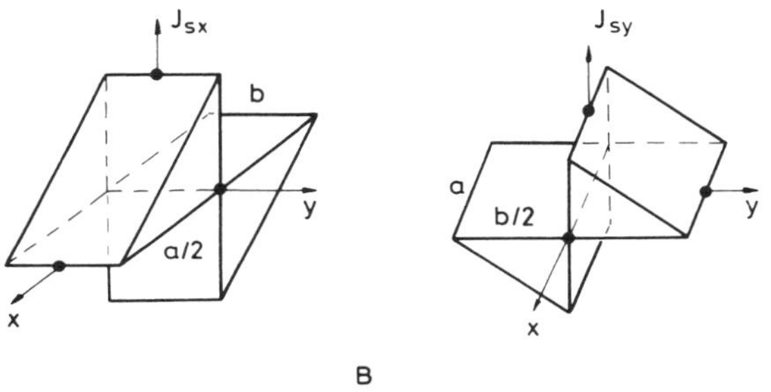

Fig. 3B *Electric surface current distribution associated with the coaxial excitation*

The surface current $\boldsymbol{J}_s$ thus defined is not a continuous function at the junction between coaxial line and patch. Still, the continuity equation is globally satisfied. The associated surface charge is constant over the spreading rectangle, having the value $\rho_s = I/j\omega ab$. Consequently, the total charge is $I/j\omega$, as required.

This model was developed to be compatible with the basis functions. For a coaxial excitation located at the centre of a charge cell, the excitation vector may be obtained from the matrix elements with little additional computation. When the surface currents have been determined over the whole patch, the antenna's input impedance is easily calculated as

$$Z_{IN} = -\frac{1}{I} \int_0^h (\boldsymbol{E}^{(e)} + \boldsymbol{E}^{(s)}) \cdot \boldsymbol{e}_z \, dz \qquad (23)$$

The field $\boldsymbol{E}^{(s)}$ results from the currents on the antenna J_{sx} and J_{sy}, whereas $\boldsymbol{E}^{(e)}$ is the excitation field, produced by the excitation currents (Fig. 3). Previous works [8] sometimes neglected $\boldsymbol{E}^{(e)}$ in the calculation of the input impedance. A somewhat artificial correction was then added to account for the 'inductive effect of the coaxial probe'. Even when $\boldsymbol{E}^{(e)}$ only represents a second-order contribution at resonances, it must be retained in the general theory, to obtain the correct input impedance at low frequencies, given by $Z_{in} = 1/(j\omega C_{stat})$ plus a small positive real term.

At low frequencies, in fact, the mixed potential integral

equation is formally identical to the scalar potential equation used to calculate microstrip capacitances [20]. This provides a useful way to check the computer implementation.

5 Results

5.1 Numerical convergence

As the number of charge cells m, n increases, the calculated values converge towards the true solution. The process was investigated for a rectangular patch having an aspect ratio of $B/A = \sqrt{2}$, excited by a normally incident plane wave with an x-polarised electric field. Relative errors were determined from comparison with extrapolated values (Fig. 4), for both the resonant frequency and the RMS

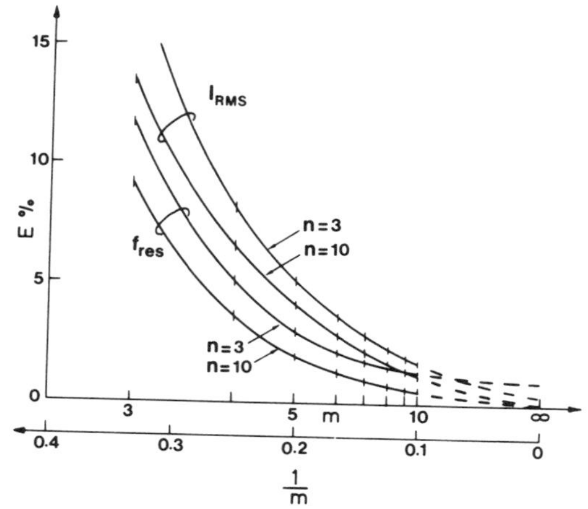

Fig. 4 *Relative error in the resonant frequency and the RMS value of the current at resonance as a function of the number of cells (m × n)*

current at the resonance. With 10 × 10 cells, the relative error on these two quantities is, respectively, 0.7% and 1.3% (Fig. 4). In this example, the number of cells n taken in the y-direction has little effect, as the transverse currents (along y) at resonance are quite small.

5.2 Rectangular patch

A rectangular patch of 60 × 40 mm was analysed theoretically and then measured. The substrate parameters are $\varepsilon_r = 4.34$, $\tan\delta = 2\ 10^{-3}$ and $h = 0.8$ mm. An effective conductivity $\sigma_e = \sigma_{Cu}/4$ was assumed, taking into account the surface roughness of the conductors. The number of cells taken is 9 × 6. To obtain adequate impedance levels at the first four resonances, the coaxial probe (a standard APC connector) was located at the centre of cell (2, 2). Computer-generated plots of the calculated surface currents are given in Fig. 5, at a low frequency (one half of the first resonant frequency), and at the first four resonances. The numerical values given correspond to the largest current value (longest arrow in every plot). The real part of $\boldsymbol{J}_s$ clearly depends on the position of the coaxial excitation, corresponding roughly to a total unit current spreading radially from the injection point. On the other hand, the imaginary part (in quadrature with the excitation current) is practically independent of the position of the excitation. This component is negligible out of the resonances, but becomes the dominant one at resonant frequencies. The current patterns are thus easily recognisable as those of the TM_{ij0} modes in the microstrip cavity. Fig. 6 provides the computed and the measured values of the input impedance near the two first resonances. A good agreement is observed in every case. Similar agreement was observed for the next two resonances and in the study of a more complicated structure, an L-shaped patch.

5.3 Slotted patch

To establish the accuracy limitations of the present approach, the slotted rectangular patch of Fig. 7 has been considered. This geometry provides a rather severe test, because a significant part of the structure is modelled with only one row of charge cells, over which transverse effects cannot be accounted for. For the slot dimensions of Fig. 7, two close resonant modes have been found at the frequencies of 1.28 and 1.32 GHz. The surface-current plots show that one mode is just the dominant mode, slightly perturbed by the slot, whereas the second one is an annular-like mode having dominant currents perpendicular to those of the first mode. Slotted patches may be used to generate circularly polarised radiation at some intermediate frequency, a fact which was recognised experimentally [21]. This work provides the first theoretical justification for these phenomena. Computed and measured values for the input impedance are presented in Fig. 7. The theoretical predictions are still qualitatively valid, but a shift in frequency and in impedance level is observed. A larger number of cells would be required to accurately study this structure. Also, the excitation point is here very close to the edge of the patch, so that a more sophisticated modelling of the excitation would be needed to describe the current distribution.

6 Radiation pattern

6.1 Asymptotic expressions for the radiated field

The dyadic Green's function formulation derived in Section 2 can still be used to determine the far field. As the fields are to be calculated far from the sources it is possible to define in a unambiguous way a dyadic Green's function associated with the electric field as

$$\bar{\boldsymbol{G}}_E = -j\omega\bar{\boldsymbol{G}}_A + \frac{1}{j\omega}\nabla\nabla G_V \tag{24}$$

Thus, G_E^{st} gives the s-component of the electric field created by a t-directed unit electric dipole located on the substrate.

The radiated field can be obtained by asymptotic evaluation of the Sommerfeld integrals appearing in eqn. 24. It is found [10, 11] that the far field is composed of a spherical and a cylindrical wave, currently termed spatial and surface wave. The surface wave is only relevant at grazing angles ($\theta \simeq \pi/2$). For a horizontal electric dipole, on a electrically thin substrate, the ratio between the power carried by the surface wave and the spatial wave is [22]

$$\eta = \frac{\text{Power surface wave}}{\text{Power spatial wave}} = \pi^2 \frac{(\varepsilon_r - 1)^3 h/\lambda_0}{\frac{2}{3}\varepsilon_r^2(\varepsilon_r - 1) + \frac{4}{15}\varepsilon_r} \tag{25}$$

The spatial wave is described by the spherical components G_E^{st} ($s = \theta, \phi, t = x, y$). Explicit expressions for these components can be found elsewhere [10, 11].

6.2 Radiation from a patch

In the previous Section the current density was obtained numerically as a set of discrete currents I_{xi} ($i = 1 \ldots M$) and I_{yj} ($j = 1 \ldots N$), each current related to a charge cell of dimensions a, b, with its centre defined by the vector

$$\boldsymbol{\rho}'_k = \boldsymbol{e}_x x_k + \boldsymbol{e}_y y_k \quad \text{with} \quad k = i, j \tag{26}$$

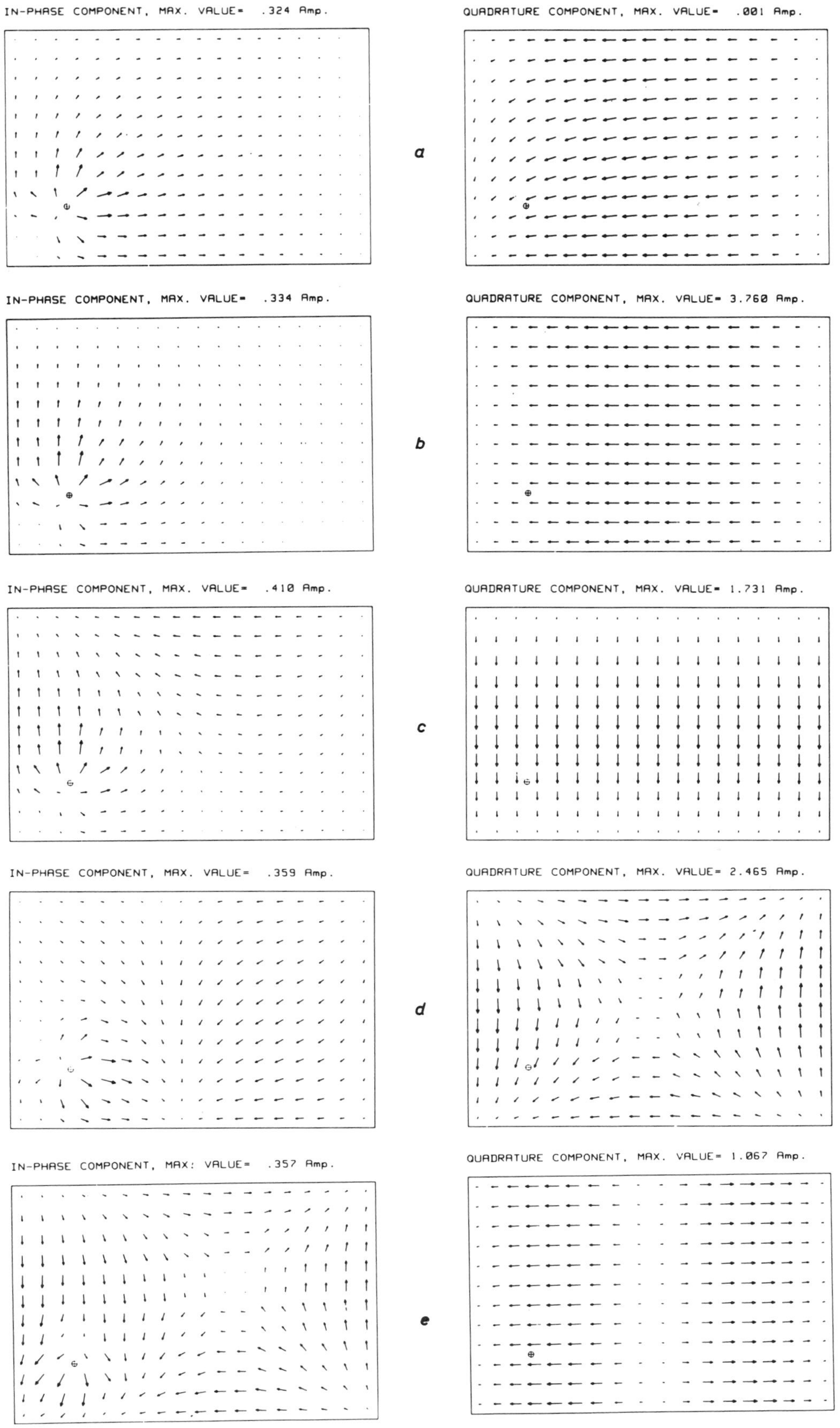

Fig. 5 *Real and imaginary parts of the surface current for a rectangular coaxial-fed microstrip antenna*

a Below resonance $f = f_{10}/2 = 0.603$ GHz
b TM_{10} resonance $f = f_{10} = 1.206$ GHz
c TM_{01} resonance $f = f_{01} = 1.783$ GHz
d TM_{11} resonance $f = f_{11} = 2.177$ GHz
e TM_{20} resonance $f = f_{20} = 2.405$ GHz

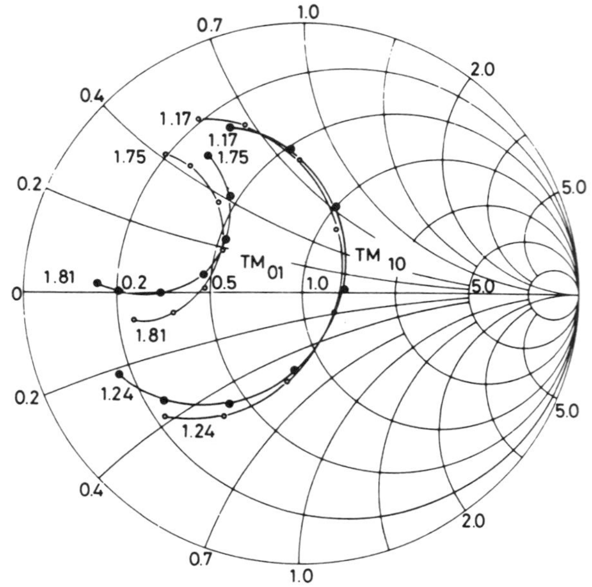

Fig. 6 *Input impedances near the TM_{10} and TM_{01} resonances*

Frequency increases clockwise by 0.01 GHz steps

○—○ theory

●—● measured

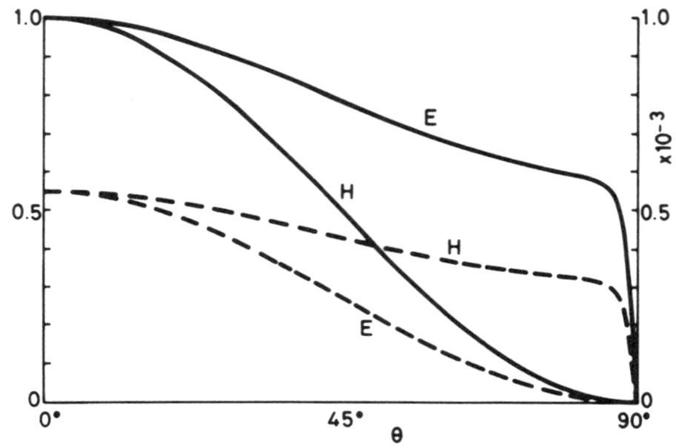

Fig. 8 *Theoretical predictions for the radiation pattern of a rectangular microstrip antenna*

E: E-plane ($\phi = 0°$)

H: H-plane ($\phi = 90°$)

——— copolar radiation (left vertical scale)

– – – – crosspolar radiation (right vertical scale)

a

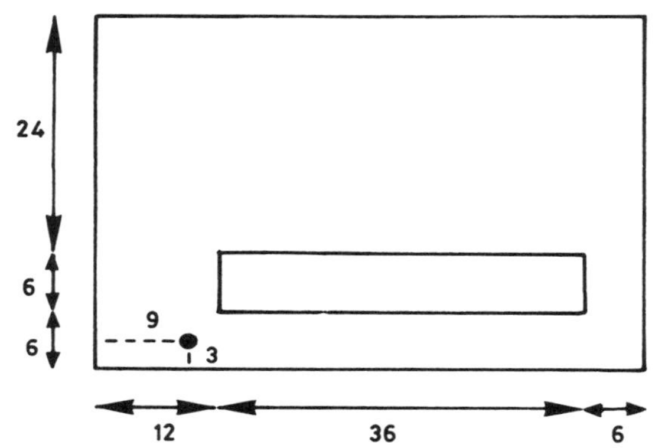

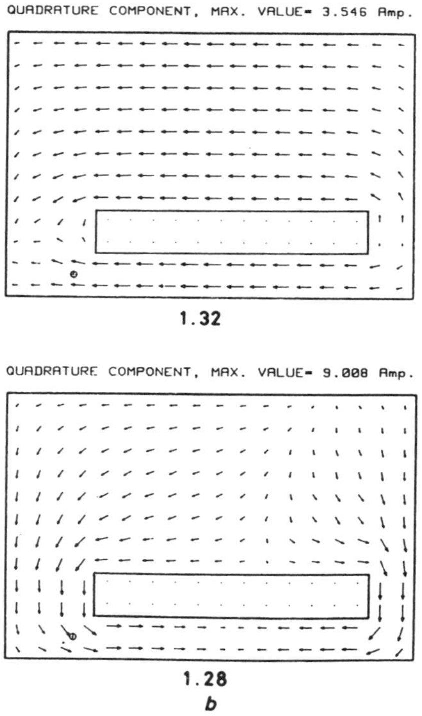

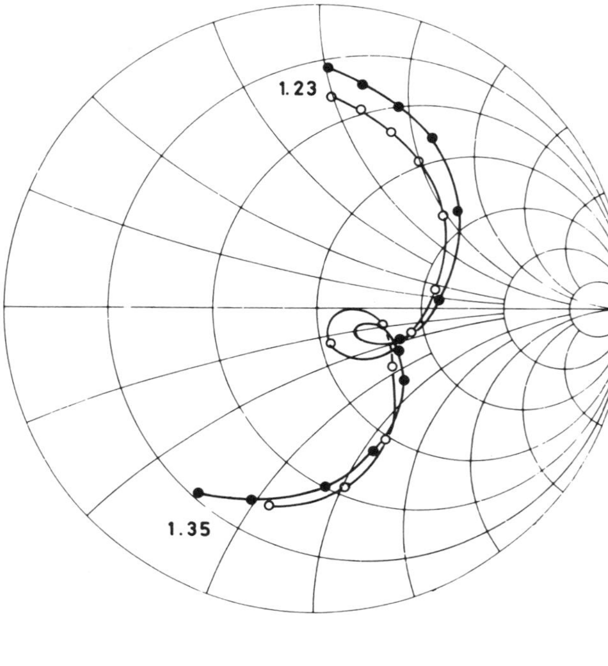

c

Fig. 7 *Slotted microstrip rectangular antenna $\varepsilon_r \doteq 4.34$, $h = 0.8$ mm*

a Dimensions in mm. and coaxial location

b Surface current distribution at two nearby resonant frequencies (1.28 and 1.32 GHz)

c Input impedance in the 1.23–1.35 GHz band. Frequency increases clockwise by 0.01 GHz steps

○—○ theory

●—● measured

The patch is then replaced by an array of Hertz dipoles, for which the radiated field is given by

$$E_\alpha = G_E^{\alpha x}(\boldsymbol{r}\,|\,\boldsymbol{0}) \sum_{i=1}^{M} aI_{xi} \exp(jk_0 \boldsymbol{e}_r \cdot \boldsymbol{\rho}_i')$$
$$+ G_E^{\alpha y}(\boldsymbol{r}\,|\,\boldsymbol{0}) \sum_{i=1}^{N} bI_{yj} \exp(jk_0 \boldsymbol{e}_r \cdot \boldsymbol{\rho}_j'),\ \alpha = \theta, \phi \qquad (27)$$

The radiated power density can then be determined, yielding the radiation pattern and related antenna properties.

As a typical example, the rectangular antenna of Section 5 was considered, at its first resonant frequency (1.206 GHz). The radiation patterns obtained, respectively, in the E-plane ($\phi = 0°$) and the H-plane ($\phi = 90°$) are represented in Fig. 8. In the E-plane, the radiation pattern is strongly affected by the substrate: radiation remains large even close to the substrate ($\theta \simeq \pi/2$). On the other hand, the H-plane pattern resembles the one of a half-wavelength dipole in free space. The polarisation is practically linear, with an electric field directed along x. The cross-polarisation component (dotted lines) is mostly due to currents in phase with the coaxial excitation.

7 Conclusion

The integral equation technique is a powerful tool for the analysis of planar microstrip antennas. Combined with the Green's function treatment introduced in a previous work [14], it provides a flexible and accurate numerical algorithm able to handle arbitrary microstrip shapes at any frequency and for any substrate parameters.

Standard feeds like coaxial probes and microstrip lines can be easily included in the model. In addition to the examples presented, other practical devices like coupled and short-circuited patches, parasitic elements and multiple-fed antennas can be studied without added complexity. The proposed model provides a good quantitative description of the electric surface currents on the patch. Hence, accurate theoretical predictions can be made for related quantities such as input impedance, near field values and polarisation purity.

In the theoretical developments, substrate and ground plane are assumed to be infinite. This departure from the real situation is not a drawback as far as near-field quantities (resonant frequencies, input impedances) are considered. On the other hand, the theoretical predictions for the radiation pattern can be considerably modified by a substrate having finite dimensions. As a surface wave reaches the antenna's edge, it is scattered, producing both a reflected surface wave and a radiated wave. The presence of secondary sources of radiation on the dielectric edges proved most troublesome in practice, as it contributes to secondary lobes and to cross-polarised radiation. Moreover, the spatial wave itself can no longer be considered separately from the surface wave.

The present technique allows one to determine the amplitude of the incident surface wave. Further study should look at the scattering of surface waves by edges and the resulting effect on the spatial radiation pattern.

8 References

1 BAHL, I.J., and BHARTIA, P.: 'Microstrip antennas' (Artech House, Dedham, MA, USA, 1980)

2 JAMES, J.R., HALL, P.S., and WOOD, C.: 'Microstrip antenna theory and design' (Peter Peregrinus, London, 1981)

3 DUBOST, G.: 'Flat radiating dipoles and application to arrays'. Research Studies Press, (John Wiley), New York, 1981)

4 LIER, L.: 'Improved formulas for input impedance of coax-fed microstrip patch antennas', *IEE Proc. H, Microwaves, Opt. & Antennas*, 1982, **129**, pp. 161–164

5 RICHARDS, W.F., LO, Y.T., and HARRISON, D.D.: 'An improved theory for microstrip antennas and applications', *IEEE Trans.*, 1981, **AP-29**, pp. 38–46

6 GUPTA, K.C., and SHARMA, P.C.: 'Segmentation and desegmentation techniques for the analysis of planar microstrip antennas', IEEE AP-S International Symposium, Los Angeles, 1981, pp. 19

7 ARAKI, K., and ITOH, T.: 'Hankel transform domain analysis of open circular microstrip radiating structures', *IEEE Trans.*, 1981, **AP-29**, pp. 84–89

8 POZAR, D.M.: 'Input impedance and mutual coupling of rectangular microstrip antennas', *ibid.*, 1982, **AP-30**, pp. 1191–1196

9 BAYLEY, M.C., and DESHPANDE, M.D.: 'Integral equations formulation of microstrip antennas', *ibid.*, 1982, **AP-30**, pp. 651–655

10 UZUNOGLU, N.K., ALEXOPOULOS, N.G., and FIKIORIS, J.G.: 'Radiation properties of microstrip dipoles'. *ibid.*, 1979, **AP-27**, pp. 853–858

11 MOSIG, J.R., and GARDIOL, F.E.: 'A dynamical radiation model for microstrip structures', *In* HAWKES, P. (Ed): 'Advances in electronics and electron physics' (Academic Press, New York, 1982), pp. 139–237

12 CARVER, K.R., and MINK, J.W.: 'Microstrip antenna technology', *IEEE Trans.*, 1981, **AP-29**, pp. 2–24

13 MILLER, E.K., and DEADRICK, F.: 'Some computational aspects of thin wire modeling'. *In* MITTRA, R. (Ed.): 'Numerical and asymptotical techniques in electromagnetics' (Springer Verlag, New York, 1975)

14 MOSIG, J.R., and GARDIOL, F.E.: 'Analytical and numerical techniques in the Green's function treatment of microstrip antennas and scatterers', *IEE Proc. Microwaves, Opt. & Antennas*, 1983, **130**, pp. 175–182

15 GLISSON, A.W., and WILTON, D.R.: 'Simple and efficient numerical methods for problems of electromagnetic radiation and scattering from surfaces', *IEEE Trans.*, 1980, **AP-28**, pp. 593–603

16 HARRINGTON, R.F.: 'Field computation by moment methods' (McMillan, New York, 1968)

17 MOSIG, J.R., and GARDIOL, F.E.: 'Dielectric losses, ohmic losses and surface wave effects in microstrip antennas', Int. U.R.S.I. Symposium, Santiago de Compostela, August 1983, pp. 425–428

18 WILTON, D.R., and GLISSON, A.W.: 'On improving the stability of the electric field integral equation at low frequency'. IEEE AP-S International Symposium., Los Angeles, June 1981

19 NEWMAN, E.M., and POZAR, D.M.: 'Electromagnetic modeling of composite wire and surface geometry'. *IEEE Trans.*, 1978, **AP-26**, pp. 784–789

20 SILVESTER, P., and BENEDEK, P.: 'Electrostatics of the microstrip revisited'. *ibid.*, 1972, **MTT-20**, pp. 756–758

21 KERR, J.L.: 'Microstrip polarization techniques'. Proceedings Antenna Application Symposium, Allerton Park, Illinois, USA, April 1977

22 MOSIG, J.R., and GARDIOL, F.E.: 'Radiation of an arbitrarily shaped microstrip antenna', *Ann. Telecommun.*, 1985, **40**, pp. 181–189

Analysis of Planar Strip Geometries in a Substrate-Superstrate Configuration

DAVID R. JACKSON, MEMBER, IEEE, AND NICÓLAOS G. ALEXÓPOULOS, SENIOR MEMBER, IEEE

***Abstract*—A moment method procedure is used to analyze the behavior of several different configurations consisting of planar strips in a substrate-superstrate geometry. These include the microstrip transmission line, center-fed dipole, the mutual impedance between two dipoles, and the transmission-line coupled dipole. In each case some of the basic superstrate effects are discussed.**

I. Introduction

RECENTLY, some of the basic effects of a superstrate (cover) layer on printed circuit antenna performance were investigated, based on the analysis of a Hertzian (infinitesimal) dipole. By properly choosing the layer thicknesses and material constants, it was observed that several different phenomena could occur, each of which could be used in a different way to improve the radiation characteristics of the dipole [1]–[4]. These results extend directly to apply to the radiation characteristics of full-size printed antennas. However, for the full-size antenna, input impedance and mutual coupling problems are also of concern, and these require calculation of the near-field together with a suitable numerical technique. The purpose of this investigation is the numerical analysis of four representative classes of full-size geometries in a substrate-superstrate configuration, each composed of planar strips. These four geometries, shown in Fig. 1, include the microstrip transmission line, the center-fed dipole, the mutual impedance between two dipoles, and the transmission-line coupled dipole. In each case the conductive strips of width w are embedded within a two-layered structure consisting of a grounded substrate of thickness B with relative permittivity and permeability ϵ_1, μ_1 with a superstrate of thickness t on top, having parameters ϵ_2, μ_2. The strip axes are taken to be in the x direction in all cases. An illustration of the layer geometry is shown in Fig. 2 for the case of the microstrip line. A convenient analysis for each of these cases can be provided by using a plane-wave spectrum formulation for the reaction between two arbitrary x-directed current sources, together with a Galerkin moment method formulation.

After the formulation of each problem is given, results are shown and the effects of a superstrate layer are discussed. For the microstrip transmission line, the effect of the superstrate on the line dispersion is demonstrated. For the center-fed dipole, the superstrate effect on the resonant length and input resistance is seen, and simple formulas are given to approximately calculate these quantities. The effect of a superstrate layer on the mutual impedance between two dipoles is then discussed, and it is demonstrated that a superstrate may be

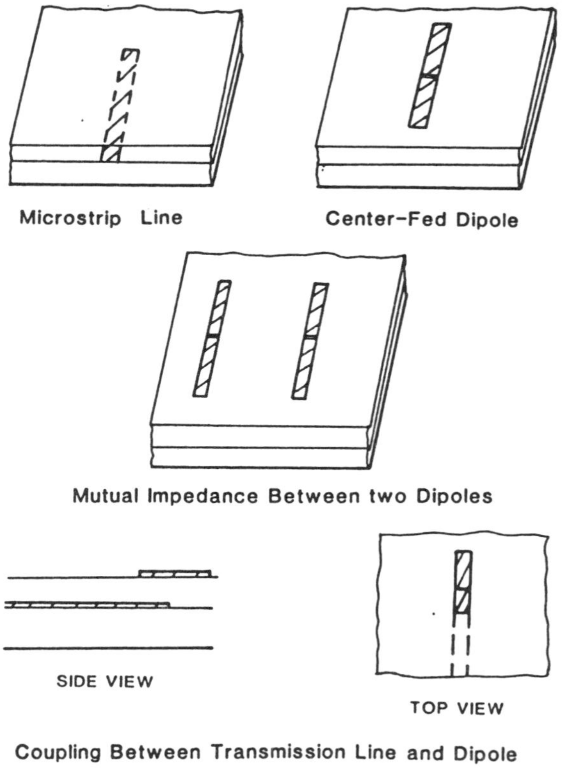

Fig. 1. Planar-strip geometries in a substrate-superstrate configuration.

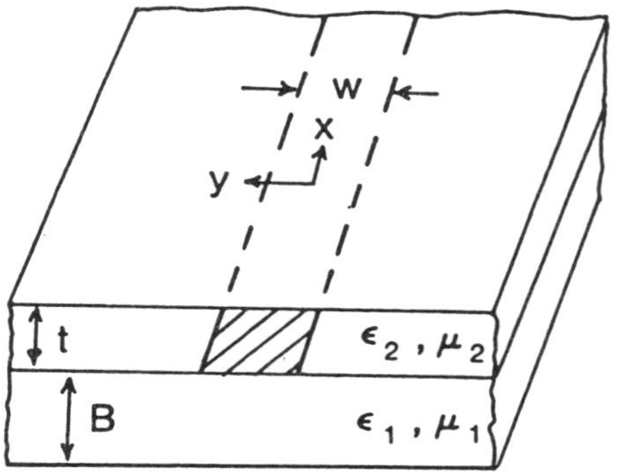

Fig. 2. Substrate-superstrate geometry.

Manuscript received February 7, 1986; revised May 29, 1986. This work was supported under U.S. Army Research Contract DAAG 29-83-K-0067 and Northrop Contract 82-110-1006.

D. R. Jackson is with the Electrical Engineering Department, University of Houston, Houston, TX 77004.

N. G. Alexópoulos is with the Electrical Engineering Department, University of California, Los Angeles, CA 90024.

IEEE Log Number 8610051.

Reprinted from *IEEE Trans. Antennas Propagat.*, vol. AP-34, no. 12, pp. 1430–1438, December 1986.

used to significantly reduce the mutual impedance for dipoles in an endfire configuration. Finally, the effect of a superstrate layer on the coupling between a dipole and transmission line is examined, and the capacitive nature of the coupling is demonstrated.

II. Formulation for Current Reaction

A foundation for the numerical analysis of full-size planar strip structures is the calculation of the reaction between two x-directed currents. Denoting $J_1(x, y)$ and $J_2(x, y)$ as two such currents located at $z = z_1$ and $z = z_2$, respectively, the reaction is defined as

$$r_{12} = \langle J_1, J_2 \rangle = \int_{-\infty}^{+\infty} \int_{-\infty}^{+\infty} E_{x1}(x, y) J_2(x, y)\, dx\, dy \quad (1)$$

where E_{x1} is the E_x-field from current $J_1(x, y)$ and the integration is performed in the plane z_2. A starting point for this calculation is the Sommerfeld solution for the magnetic Hertzian vector potential at (x, y, z) due to an electric Hertzian dipole source at (x', y', z') where $z' = z_1$. This can be written in cylindrical coordinates as

$$\Pi_x = \int_0^\infty f(\lambda) J_0(\lambda r)\, d\lambda \quad (2)$$

$$\Pi_z = \cos\phi \int_0^\infty g(\lambda) J_1(\lambda r)\, d\lambda \quad (3)$$

where

$$r = [(x - x')^2 + (y - y')^2]^{1/2}$$

and $f(\lambda)$, $g(\lambda)$ are complicated functions of λ which also depend on z, z', although this dependence is not explicitly shown here (the time dependence $e^{+j\omega t}$ is also being suppressed throughout the analysis). These functions are given explicitly in [2]. The E_x-field in the plane z_2 is then found from

$$E_{x1}(x, y, z_2) = \int_{-\infty}^{+\infty} \int_{-\infty}^{+\infty} J_1(x', y') \cdot \left(k^2 \Pi_x + \frac{\partial^2 \Pi_x}{\partial x^2} + \frac{\partial^2 \Pi_z}{\partial x \partial z} \right)_{z=z_2} dx'\, dy' \quad (4)$$

where the integration is performed in the plane z_1, and k is the wavenumber at $z = z_2$, given as $k = k_0 n_i$ when z is in the ith layer, with n_i being the index of refraction of the ith layer. Because of the differentiation appearing in (4), it is convenient to write

$$\frac{\partial \Pi_z}{\partial z} = \cos\phi \int_0^\infty h(\lambda) J_1(\lambda r) \lambda\, d\lambda \quad (5)$$

where

$$h(\lambda) = \frac{1}{\lambda} \frac{\partial g}{\partial z}. \quad (6)$$

Equation (5) may be put into a similar form as (2) by using

$$\frac{\partial}{\partial x} J_0(\lambda r) = -\lambda \cos\phi\, J_1(\lambda r). \quad (7)$$

Both (2) and (5) are then in the form of cylindrical (Fourier–Bessel) transforms, and a conversion to rectangular (Fourier) form may be accomplished. Substituting (4) into (1) then results in the plane-wave spectrum form

$$r_{12} = \frac{1}{2\pi} \int_{-\infty}^{+\infty} \int_{-\infty}^{+\infty} \frac{1}{\lambda} [(k^2 - \lambda_x^2) f(\lambda) + \lambda_x^2 h(\lambda)] \cdot \tilde{J}_1(\lambda_x, \lambda_y) \tilde{J}_2(-\lambda_x, -\lambda_y)\, d\lambda_x\, d\lambda_y \quad (8)$$

where

$$\tilde{J}(\lambda_x, \lambda_y) = \int_{-\infty}^{+\infty} \int_{-\infty}^{+\infty} J(x', y') e^{-j[\lambda_x x' + \lambda_y y']}\, dx'\, dy'$$

and

$$\lambda = \sqrt{\lambda_x^2 + \lambda_y^2}.$$

Defining new origins (x_1, y_1) and (x_2, y_2) for currents J_1, J_2, and assuming the currents are even functions in x and y about the new origins, (8) can be written as

$$r_{12} = \frac{2}{\pi} \int_0^{\pi/2} \int_C [(k^2 - \lambda_x^2) f(\lambda) + \lambda_x^2 h(\lambda)] \tilde{J}_1(\lambda_x, \lambda_y) \cdot \tilde{J}_2(\lambda_x, \lambda_y) \cdot \cos(\lambda_x \Delta x) \cos(\lambda_y \Delta y)\, d\lambda\, d\bar{\phi} \quad (9)$$

where $\Delta x = x_1 - x_2$, $\Delta y = y_1 - y_2$, and $\bar{\phi} = \tan^{-1}(\lambda_y/\lambda_x)$.

The contour C in the Sommerfeld integration is from 0 to ∞, above the surface-wave poles. A root extraction technique is used to remove the singular behavior of the integrand [5], [6], and other numerical techniques are used to improve computational efficiency. A matrix storage of the complicated part of the integrand in brackets is one such technique, and the use of a Filon-type integration to account for the oscillatory nature of the cosine terms is another. In the application of Galerkin's method to the antenna geometries, a convenient choice of basis functions is

$$b(x, y) = \xi(x) \eta(y) \quad (10)$$

where

$$\xi(x) = \frac{\sin k_e(d - |x|)}{\sin k_e d}, \qquad |x| \le d \quad (11)$$

$$\eta(y) = \frac{4}{5w} \left(1 + \left|\frac{2}{w} y\right|^3\right), \qquad |y| \le \frac{w}{2} \quad (12)$$

[7], [8]. An advantage of using (12) over the Maxwell current is faster convergence of the transform [8]. The planar strips of width w are then divided into N subsections of length d in the x direction, and the representation

$$J(x, y) = \eta(y) \sum_{n=1}^{N-1} I_n \xi(x - nd + L/2) \quad (13)$$

is used for each strip, with L the strip length and x measured from the strip center.

One of the difficulties in using (9) is that the cosine terms oscillate faster for increasing separation Δx, Δy between basis functions. Furthermore, convergence is provided by the Fourier transforms $\tilde{J}_1$, $\tilde{J}_2$ when $z_1 = z_2$ since $f(\lambda)$ and $h(\lambda)$ are $0(1)$ as $\lambda \to \infty$, so the integration in λ must extend to further values as the size of the basis functions decrease. This makes the necessary computation for mutual impedance very time consuming when computing the reaction between basis functions which are small and widely separated. In this case it is convenient to formulate the reaction in the real-space domain. If the basis functions are approximated as point dipoles of current this is equivalent to simply evaluating the E_x-field of a Hertzian dipole, and this may be formulated in an efficient manner [9].

III. Microstrip Transmission Line

Using (8), an analysis of the propagation on a microstrip line is straightforward. A current expansion function of the form

$$J_1(x, y) = \eta(y)e^{-j\beta x} \tag{14}$$

is assumed on the line, and a testing function

$$J_2(x, y) = \delta(x)\delta(y) \tag{15}$$

is used, which forces the E_x-field at the strip center to be zero. This is equivalent in principle to the original formulation of Denlinger [8], although the approach is based here on the three-dimensional Green's function. Using the integral representation of the δ-function, the result

$$\int_0^\infty \frac{1}{\lambda}\,\tilde{\eta}(\lambda_y)[(k^2-\beta^2)f(\lambda)+\beta^2 h(\lambda)]\;d\lambda_y = 0 \tag{16}$$

is obtained, where

$$\lambda = \sqrt{\beta^2 + \lambda_y^2}$$

and $\tilde{\eta}(\lambda_y)$ is the one-dimensional transform of $\eta(y)$. As pointed out in [11] the value of β must be greater than all surface-wave poles, since a residue contribution would result in a complex value for the integration, and a real-valued solution for β would not be possible.

Equation (16) allows for a convenient solution for the propagation on an infinitely thin strip. As $w \to 0$, $\tilde{\eta}(\lambda_y) \to 1$ and does not provide any convergence, so the term in brackets must vanish as $\lambda \to \infty$, resulting in

$$\beta^2 \sim k^2 \frac{f(\infty)}{f(\infty) - h(\infty)} \tag{17}$$

as $w \to 0$. From this formula the classical results for an infinitely thin wire may be obtained [12], [13], where the effective permittivity and permeability are

$$\epsilon_e = \frac{1}{2}(\epsilon_i + \epsilon_j), \quad \mu_e = \left[\frac{1}{2}(\mu_i^{-1} + \mu_j^{-1})\right]^{-1} \tag{18}$$

when the strip is at the interface between the ith and jth layers, and $\epsilon_e = \epsilon_i$, $\mu_e = \mu_i$ if the line is embedded within the ith layer. Equation (18) may be used as a basis for choosing k_e in (11), with $k_e = k_0\sqrt{\mu_e\epsilon_e}$ [6].

Results for a line at the interface ($z = B$) are shown in Fig. 3(a) for the case of a GaAs superstrate ($\epsilon_2 = 12.5$) over a teflon substrate ($\epsilon_1 = 2.1$). As the frequency increases, $\epsilon_e \to \epsilon_2$ since $\epsilon_2 > \epsilon_1$. The dispersion is less for the smaller line width at lower frequencies, but regardless of line width the solution is above the dominant TM_1 surface-wave mode here. As the superstrate thickness increases the amount of dispersion does as well, seen from Fig. 3(b). This is due to the fact that ϵ_2 is significantly larger than ϵ_1. If a superstrate with $\epsilon_2 = \epsilon_1$ is used instead, the dispersion would be less than without a superstrate.

IV. Center-Fed Dipole

For the three antenna geometries shown in Fig. 1, a moment solution using Galerkin's method is used. For the center-fed dipole the current distribution (13) is assumed, and a δ-gap excitation is taken at $x = 0$. The Z_{mn} coefficients of the Galerkin matrix then represent the reaction between basis functions centered at $x = md - L/2$ and $x = nd - L/2$ where L is the length of the dipole. The Galerkin set of matrix equations

$$[Z_{mn}][I_n] = [E_n] \tag{19}$$

is then formulated, where the excitation column vector E has components $E_n = \delta_{np}$ for N even, where $p = N/2$. A variational expression for the input impedance is [14]

$$Z_{\text{in}} = \frac{1}{I(0)^2}\sum_{m,n} Z_{mn} I_m I_n \tag{20}$$

where $I(0)$ is the current at $x = 0$. Using (19), this formula immediately reduces to the usual result $Z_{\text{in}} = 1/I(0)$. Once input impedance is found in this way, the resonant length may easily be determined. If the resonant length L_r is small compared to a wavelength, an approximate formula for the input resistance is [15]

$$R_r \simeq 120\left(\frac{L_r}{\lambda_0}\right)^2 \frac{P_T}{15\pi^2/\lambda_0^2}\ (\Omega) \tag{21}$$

where p_T is the total (radiated + surface wave) power (in watts) delivered by a Hertzian dipole of strength 1 A-m at the same z location.

In Fig. 4(a) the exact resonant length of a dipole at the interface is shown, and compared with the approximate value

$$L_r/\lambda_0 \simeq 1/\sqrt{2(\epsilon_1+\epsilon_2)} \tag{22}$$

as predicted by (18) when fringing fields are neglected in the infinitely narrow nonmagnetic case. When the thickness t becomes comparable to w the resonant length changes rapidly, becoming asymptotic to a limiting half-space value close to the value predicted by (22). In Fig. 4(b) the resonant input resistance is shown for the same case of Fig. 4(a), and compared with (21). From (21) it is seen that a superstrate affects resonant input resistance in two ways; by shortening the resonant length, and by changing P_T. The latter effect is discussed in [1].

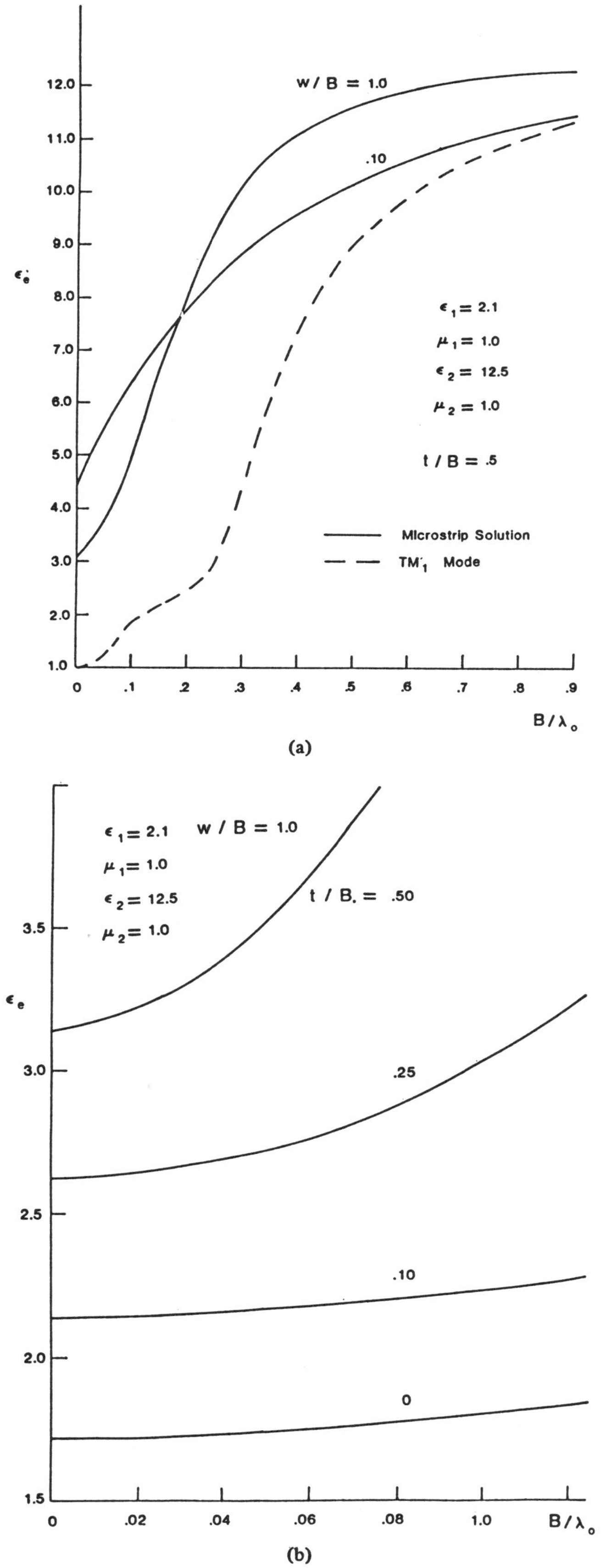

Fig. 3. (a) Effective dielectric constant versus substrate thickness for two different strip widths. (b) Effective dielectric constant versus substrate thickness for different superstrate thicknesses.

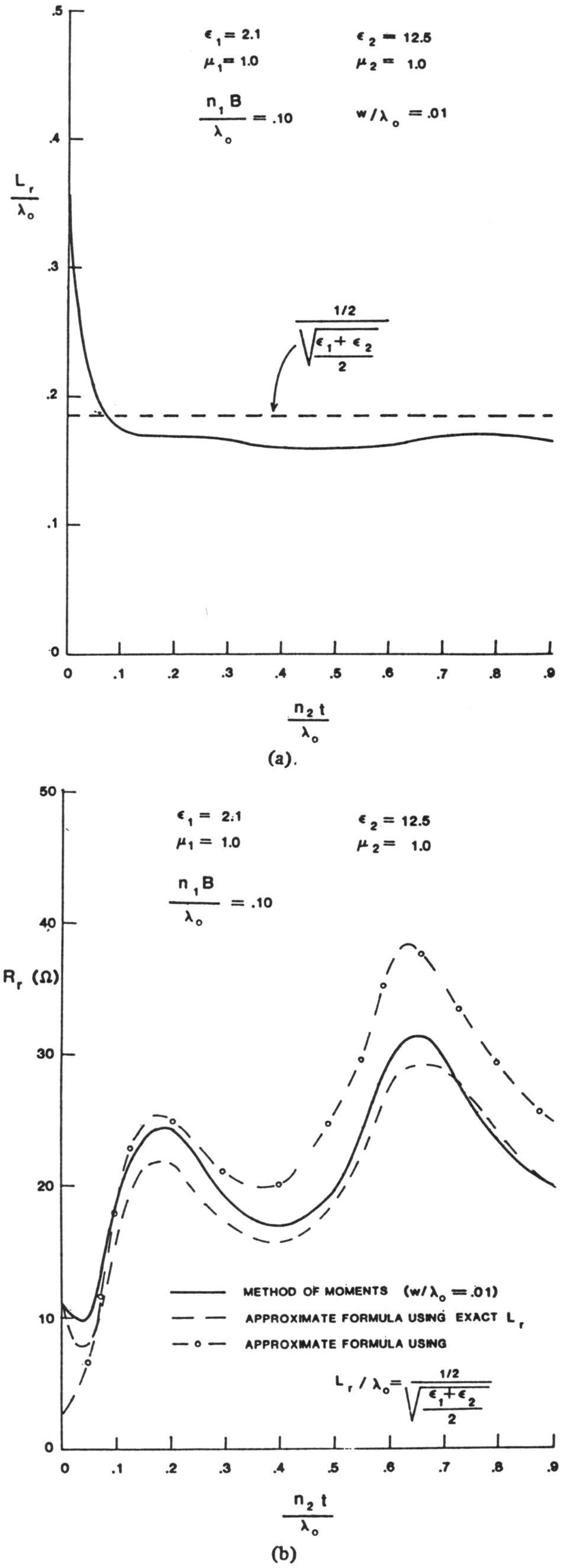

Fig. 4. (a) Resonant length versus superstrate thickness showing a comparison of exact and approximate values. (b) Resonant input resistance versus superstrate thickness showing a comparison of exact and approximate values.

V. Mutual Impedance Between Two Dipoles

The mutual impedance between two center-fed dipoles has been analyzed in two ways: by using the method of moments, and by using the electromotive force (emf) method. In the moment method approach basis functions are assumed on each dipole, with dipole 1 excited by a delta-gap source and dipole 2 short circuited. A partitioned Galerkin matrix set of equations is then solved to find the dipole currents. Using $[Z] = [Y]^{-1}$ where $[Y]$ is the admittance matrix results in the formula

$$Z_{12} = \frac{-1}{I_1(0)} \left[\frac{I_2(0)/I_1(0)}{1 - \left(\frac{I_2(0)}{I_1(0)} \right)^2} \right]. \tag{23}$$

Other formulas for Z_{12} may be derived, but an advantage of this form is the avoidance of numerical difficulties when $|I_2(0)| \ll |I_1(0)|$, which occurs for larger dipole separations.

In the emf method the mutual impedance is given by

$$Z_{12} = \frac{-1}{I_1(0) I_2(0)} \iint_{S_2} E_{x1}(x, y) J_2(x, y)\, dx\, dy \tag{24}$$

where E_{x1} is the field from the current J_1 on dipole 1 when excited in the absence of dipole 2, and J_2 is the current on dipole 2 (region denoted S_2) with dipole 1 open-circuited. Currents J_1 and J_2 are usually assumed, making the method approximate. In the case of infinitely narrow dipoles the currents become sinusoidal with an effective k_e as discussed in Section III however, allowing for an exact calculation. Since no self-term problems are encountered for separated dipoles, an efficient calculation of Z_{12} may be achieved using integration in real space. For narrow dipoles this approach is computationally more efficient, and has the advantage of being numerically stable as $w \rightarrow 0$ since there are no self-term problems.

For simplicity, results will be confined to the special case of both dipoles at $z = z_0$. First, a comparison of the moment solution with the emf method is shown in Fig. 5, for dipoles in an endfire configuration on a single teflon layer ($z_0 = B$). The real and imaginary parts of the mutual impedance, R_{12} and X_{12}, are plotted for several different strip widths. The moment method solution is used for the nonzero strip widths, and the emf method is used for the zero strip width. As can be seen from the figure, the moment solution converges toward the emf solution for narrower strip widths, supporting the validity of the assumed current using (18). Because all of the basic superstrate effects on mutual impedance can be seen from the infinitely narrow dipole case, the remainder of the results shown in this section will be for this case.

The basic properties of mutual impedance for dipoles in broadside or endfire configuration can be seen from Fig. 6, for dipoles on a single layer. The mutual impedance decays fairly rapidly with separation in the broadside case (Fig. 6(a)), but much slower in the endfire case (Fig. 6(b)). This is due to the fact that the far-field surface wave pattern of a transverse magnetic (TM)-mode wave is strongest in the endfire direction, having a null in the broadside direction [16]. The opposite is true for a transverse electric (TE)-mode wave, but for the layer thickness used here only the dominant TM mode exists. This strong endfire coupling is even more pronounced for dipoles on a magnetic layer, since the TM_1 mode is much more strongly excited. Such a case is shown in Fig. 7, where $\mu_1 = 10.0$. This strong endfire coupling can be reduced by using a superstrate layer to reduce or eliminate the surface waves. An example of this is shown in Fig. 8, in which a magnetic layer with $\mu_2 = 10.0$ is used as a superstrate over a teflon substrate, with $z_0 = B$. The superstrate thickness is chosen from

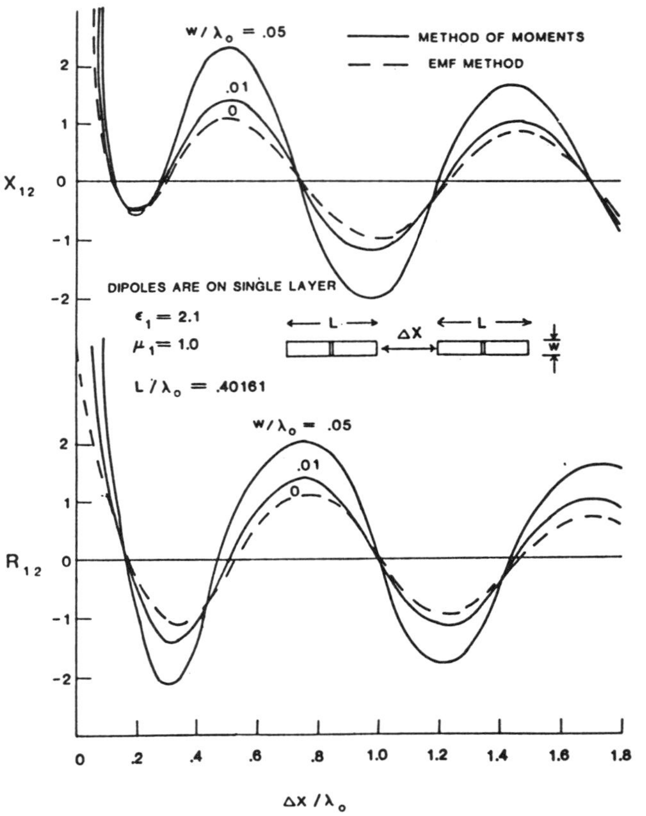

Fig. 5. Mutual impedance versus separation showing a comparison of the moment method solution for two different strip widths and the emf solution for zero strip width.

$$\frac{n_2 t_c}{\lambda_0} = \frac{n_2}{2\pi\sqrt{n_2^2 - n_1^2}} \tan^{-1} \left[\frac{\epsilon_2 \sqrt{n_1^2 - 1}}{\sqrt{n_2^2 - n_1^2}} \right] \tag{25}$$

which results in an elimination of all surface waves in this case. As discussed in [1], the elimination of surface waves is often difficult to achieve when using a nonmagnetic superstrate, because of the thin substrate limitation. However another way to reduce surface waves is by embedding the dipoles within the substrate layer and using a superstrate with a higher index of refraction. This reduces surface wave excitation due to the exponentially decaying nature of the surface-wave fields inside the substrate. An example showing mutual impedance for this configuration is presented in Fig. 9, where $z_0 = B/2$. Once the separation becomes sufficiently large in this case, the mutual impedance is very small. Also of interest

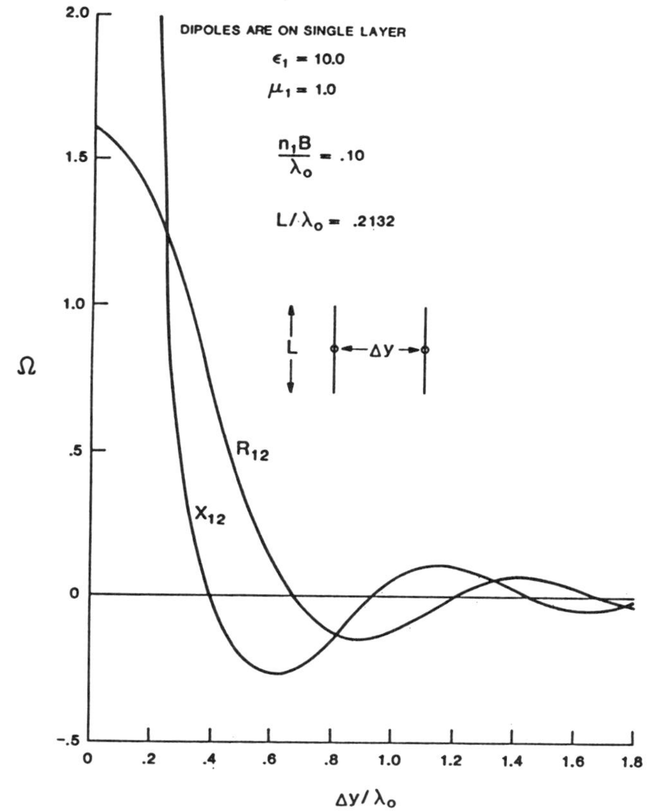

(a)

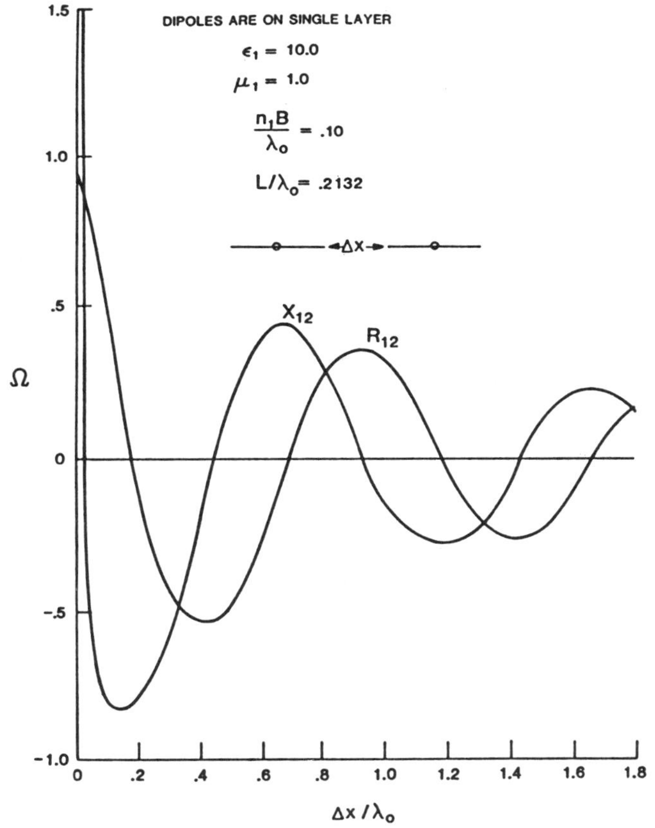

(b)

Fig. 6. (a) Mutual impedance versus separation for dipoles in broadside configuration on a single layer. (b) Mutual impedance versus separation for dipoles in endfire configuration on a single layer.

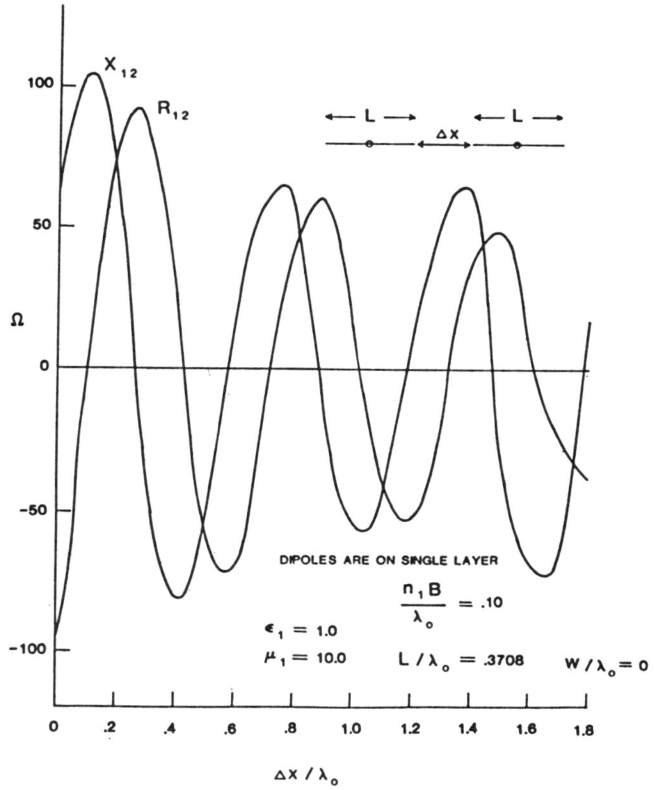

Fig. 7. Mutual impedance versus separation for dipoles in endfire configuration on a single magnetic layer.

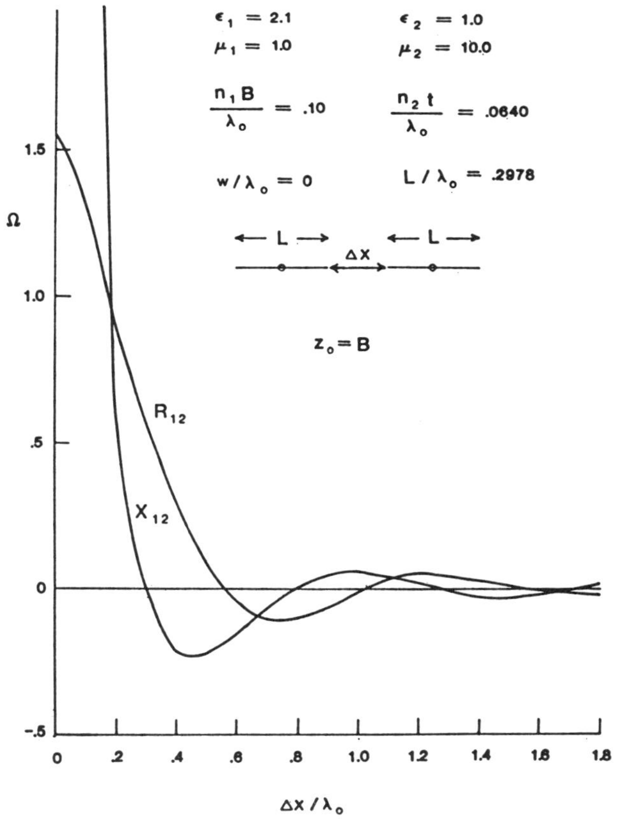

Fig. 8. Mutual impedance versus separation for dipoles in endfire configuration with a superstrate used to eliminate surface waves.

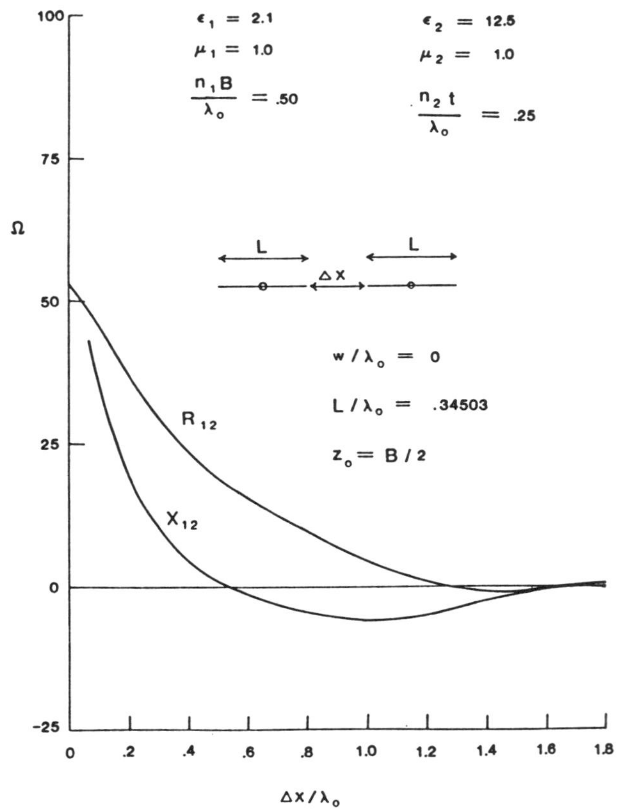

Fig. 9. Mutual impedance versus separation for dipoles in endfire configuration embedded in the middle of the substrate.

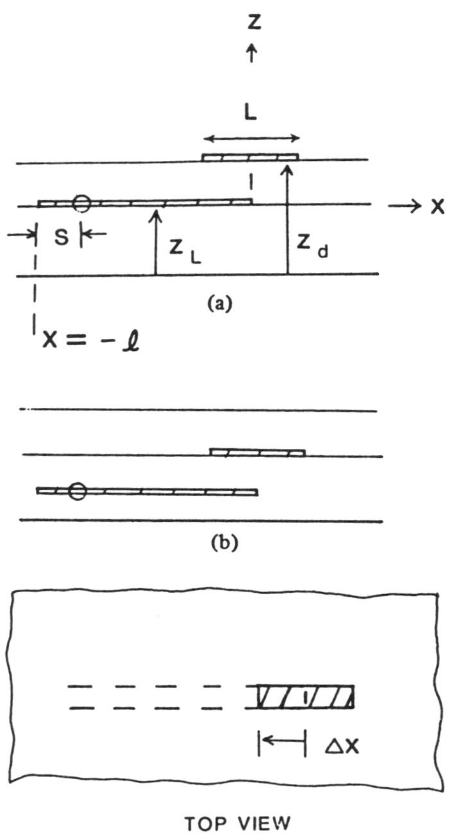

Fig. 10. Geometry of coupling between dipole and transmission line. (a) Dipole on top of superstrate. (b) Dipole at the interface.

is the almost monotonic nature of the curves in this case, which can be explained by assuming the superstrate acts as a conductor for the space-wave fields, and using image theory. The images will add in phase for z_0 and $B - z_0$ an odd number of electrical quarter-wavelengths in the substrate.

VI. Dipole Excitation by Transmission-Line Coupling

Another configuration which has been analyzed using the moment approach is the strip dipole excited by electromagnetic coupling to a microstrip transmission line, shown in Fig. 10. This problem has been successfully analyzed with the moment method in the past for the line and dipole within a single layer [17]. A δ-gap source is placed near the end of the line as shown, and the same procedure as for the mutual impedance is used to find the line and dipole currents. From this information the normalized admittance Y/Y_0 may be found, referred to the dipole-end of the line. For fixed line and dipole heights z_L and z_d, an input match may be achieved by varying the dipole length L and offset in either the x or y directions, provided z_L and z_d are sufficiently close. An offset variation in the x direction was chosen here in order to take advantage of an efficient matrix storage procedure for the reaction values. The offset is then taken in integral number of subsections, $\Delta x = N \cdot d$.

One of the main objectives in feeding a dipole in this manner is to minimize the radiation from the line while maintaining a significant radiation from the dipole. This implies keeping z_L small, while simultaneously keeping a prescribed level of coupling between the two. Also, to have a significant bandwidth, z_d should not be too small. However, decreasing z_L lowers the coupling for a fixed z_d, so it is important to investigate ways of increasing the coupling. One approach is to stack multiple dipoles above the line [18]. Another approach is to choose a different material between the line and dipole. Qualitatively, the coupling between the line and dipole may be regarded as largely capacitive, which suggests an improved coupling using $\epsilon_2 > \epsilon_1$ in the geometry of Fig. 10(a). To verify this, the end admittance is plotted versus offset for different dipole lengths in Figure 11(a) for the case of $\epsilon_1 = \epsilon_2 = 2.1$. As can be seen, the coupling is not sufficient to allow for an input match with the layer thicknesses used. In Fig. 11(b) the same layer electrical thicknesses are used, but with $\epsilon_2 = 10.0$. In this case the coupling is sufficiently strong to allow for an input match, for $L/\lambda_0 \simeq 0.186$ and $\Delta x/\lambda_0 \simeq 0.123$. In Fig. 11(c) a magnetic superstrate is used, with $\mu_2 = 10.0$. The coupling in this case is not significantly different than the case of Fig. 11(a), supporting the conclusion of a capacitive-like coupling mechanism. A similar improvement in coupling is possible when using stacked dipoles. The stacked dipole approach allows for an even greater improvement in bandwidth, but the total physical layer thickness will be larger when using $\epsilon_2 \simeq \epsilon_1$.

The dipole can also be placed at the interface and excited by a line in the substrate, as shown in Fig. 10(b). In this case the superstrate is not improving coupling, but may be used to increase the efficiency of the dipole or achieve other desired effects [1].

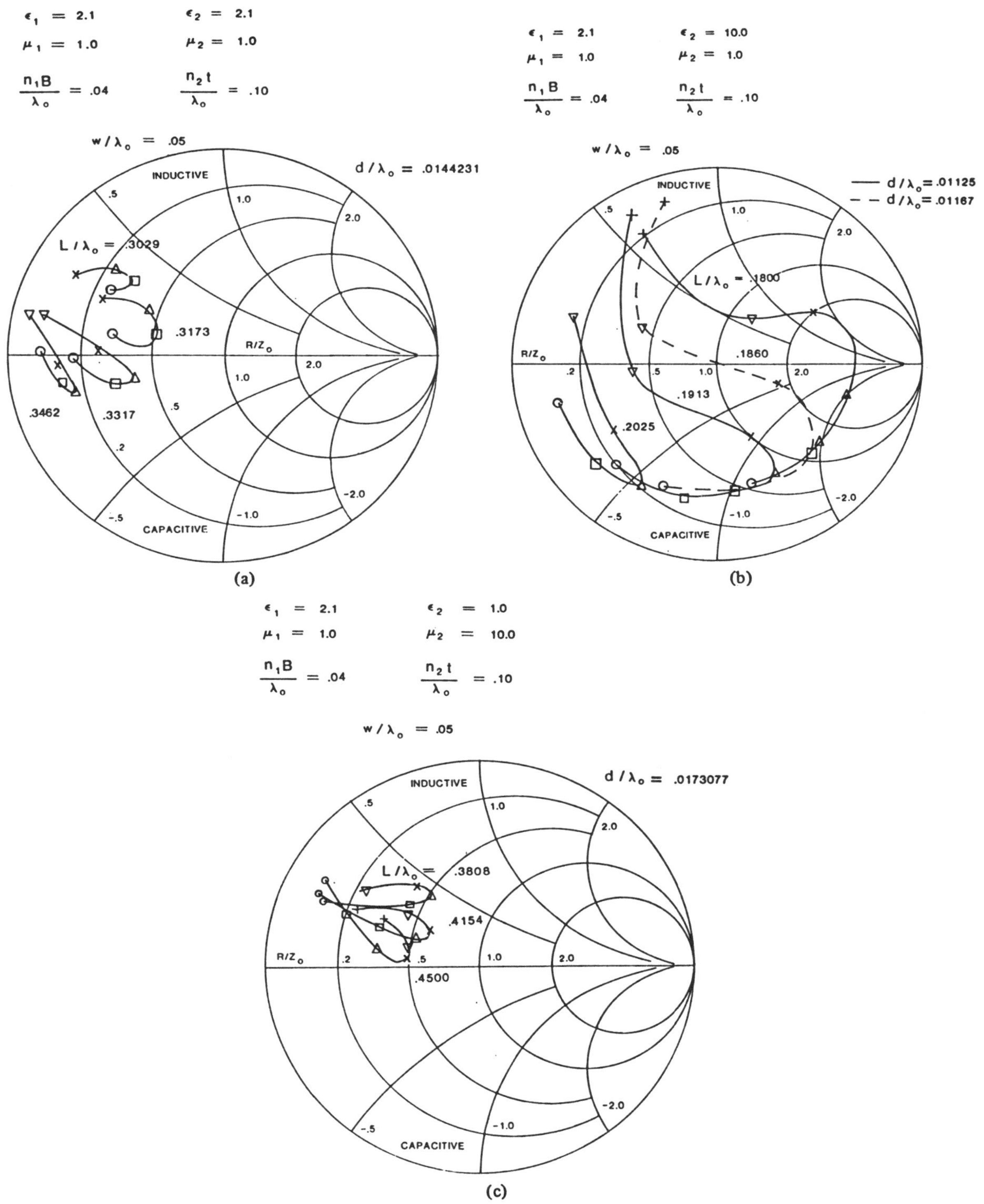

Fig. 11. (a) Normalized end admittance versus offset for different dipole lengths with $\epsilon_2 = \epsilon_1 = 2.1$. The offset along each curve is specified by $\Delta x = N \cdot d$ with $N = 4, 8, 12, 16, 20$ corresponding to points O, □, △, X, ▽, respectively. (b) Normalized end admittance versus offset for different dipole lengths with $\epsilon_2 = 10.0$, $\epsilon_1 = 2.1$. The dashed curve shows a dipole length which allows for an input match. The offset along each curve is specified by $\Delta x = N \cdot d$ with $N = 3, 5, 7, 10, 11, 12$ corresponding to points O, □, △, X, ▽, +, respectively. (c) Normalized end admittance versus offset for different dipole lengths with $\mu_2 = 10.0$, $\epsilon_1 = 2.1$. The offset along each curve is specified by $\Delta x = N \cdot d$ with $N = 8, 12, 14, 16, 18, 20$ corresponding to points O, □, △, X, ▽, +, respectively.

VII. Conclusion

Four geometries consisting of planar strips in a substrate-superstrate configuration have been analyzed, including the microstrip transmission line, the center-fed dipole, the mutual impedance between two dipoles, and the transmission-line coupled dipole. The starting point in the analysis of each of these was the plane-wave spectrum formulation for the reaction between two current sources. A real-space evaluation technique for finding the electric field was also used, which improved the efficiency of the moment-method calculation for mutual impedance, and also allowed for an easy determination of the mutual impedance between infinitely narrow dipoles.

For the microstrip transmission line, the amount of dispersion was seen to significantly increase as a superstrate is added when ϵ_2 is much larger than ϵ_1. The dispersion decreases as the line width decreases, although β must always be above the largest surface-wave pole. In the case of the center-fed dipole, the resonant length and input resistance were found from the moment method and compared with simple approximate formulas. These approximate formulas allow for an easy extension of Hertzian dipole results to the resonant strip dipole. The mutual impedance between dipoles was then investigated using both the moment method and the emf method. The emf method allows for an efficient computation for the case of very narrow dipoles, and gives accurate results when the proper value of k_e is used in the assumed current. It was demonstrated that a superstrate may be used to significantly decrease the mutual impedance between dipoles, especially for dipoles in an endfire configuration, by reducing or eliminating surface waves. Finally, results were shown for the case of a dipole excited by coupling to a transmission line, and it was verified that the coupling mechanism is predominantly capacitive. Hence the coupling is improved by using a material with a higher dielectric constant between the line and dipole.

References

[1] N. G. Alexopoulos and D. R. Jackson, "Fundamental superstrate (cover) effects on printed circuit antennas," *IEEE Trans. Antennas Propagat.*, vol. AP-32, pp. 807–816, 1984.

[2] ——, "Fundamental superstrate (cover) effects on printed circuit antennas," UCLA Rep. ENG-83-50, Oct. 1983.

[3] N. G. Alexopoulos, D. R. Jackson, and P. B. Katehi, "Criteria for nearly omnidirectional radiation patterns for printed antennas," *IEEE Trans. Antennas Propagat.*, vol. AP-33, pp. 195–205, Feb. 1985.

[4] D. R. Jackson and N. G. Alexopoulos, "Gain enhancement methods for printed circuit antennas," *IEEE Trans. Antennas Propagat.*, vol. AP-33, pp. 976–987, Sept. 1985.

[5] N. K. Uzunoglu, N. G. Alexopoulos, and J. G. Fikioris, "Radiation properties of microstrip dipoles," *IEEE Trans. Antennas Propagat.*, vol. AP-27, pp. 853–858, Nov. 1979. (See also correction, *IEEE Trans. Antennas Propagat.*, vol. AP-30, p. 526, May 1982.)

[6] D. M. Pozar, "Imput impedance and mutual coupling of rectangular microstrip antennas," *IEEE Trans. Antennas Propagat.*, vol. AP-30, pp. 1191–1196, Nov. 1982.

[7] I. E. Rana and N. G. Alexopoulos, "Current distribution and input impedance of printed dipoles," *IEEE Trans. Antennas Propagat.*, vol. AP-29, pp. 99–106, Jan. 1981.

[8] E. J. Denlinger, "A frequency dependent solution for microstrip transmission lines," *IEEE Trans. Microwave Theory Tech.*, vol. MTT-19, pp. 30–39, Jan. 1971.

[9] D. R. Jackson and N. G. Alexopoulos, "An asymptotic extraction technique for evaluating Sommerfeld-type integrals," *IEEE Trans. Antennas Propagat.*, pp. 1467–1470, this issue.

[10] D. M. Pozar, "Improved computational efficiency for the moment method solution of printed dipoles and patches," *Electromagn.*, vol. 3, no. 3–4, pp. 299–309, July-Dec. 1983.

[11] R. W. Jackson and D. M. Pozar, "Full-wave analysis of microstrip open-end and gap discontinuities," *IEEE Trans. Microwave Theory Tech.*, vol. MTT-33, pp. 1036–1042, Oct. 1985.

[12] J. R. Wait, "Theory of wave propagation along a thin wire parallel to an interface," *Radio Sci.*, vol. 7, pp. 675–679, June 1972.

[13] R. A. Pucel and D. J. Masse, "Microstrip propagation on magnetic substrates—Part I: design theory," *IEEE Trans. Microwave Theory Tech.*, vol. MTT-20, pp. 304–308, May 1972.

[14] V. H. Rumsey, "Reaction concept in electromagnetic theory," *Phy. Rev.*, vol. 94, pp. 1483–1491, June 1954.

[15] D. R. Jackson and N. G. Alexopoulos, "Microstrip dipoles on electrically thick substrates," *Int. J. Infrared and Millimeter Waves*, Jan. 1986.

[16] N. G. Alexopoulos and I. E. Rana, "Mutual impedance computation between printed dipoles," *IEEE Trans. Antennas Propagat.*, vol. AP-29, pp. 106–111, Jan. 1981.

[17] P. B. Katehi and N. G. Alexópoulos, "On the modeling of electromagnetically coupled microstrip antennas—The printed strip dipole," *IEEE Trans. Antennas Propagat.*, vol. AP-32, pp. 1179–1186, Nov. 1984.

[18] ——, "A bandwidth enhancement method for microstrip antennas," in *1985 IEEE Antennas Propagat. Soc. Symp. Dig.*, vol. 1, pp. 405–408.

[19] A. J. M. Soares, S. B. A. Fonseca, and A. J. Giarola, "The effect of a dielectric cover layer on the current distribution and input impedance of printed dipoles," *IEEE Trans. Antennas Propagat.*, vol. AP-32, pp. 1149–1153, Nov. 1984.

A Note on Hybrid Finite Element Method for Solving Scattering Problems

JIAN-MING JIN AND VALDIS V. LIEPA, MEMBER, IEEE

Abstract—**The hybrid finite element formulation is modified so that it results in a sparse banded symmetric matrix. This modification substantially improves the computational efficiency and enhances the capability of the method, which is demonstrated by numerical examples. A comparison with other numerical techniques is presented.**

I. Introduction

The hybrid finite element method (HFEM) was first introduced by Silvester and Hsieh [1] and McDonald and Wexler [2] in the early 1970's, and has been further improved and applied to solving various two-dimensional unbounded field problems, e.g., [3]-[9]. The method expresses the exterior fields using an eigenfunction series or an integral involving a Green's function. Such expressions are imposed on the variational equation or, more conveniently, on the matrix equation derived using the finite element method for the interior fields. As a result, HFEM produces a system matrix having a nonuniform block submatrix structure. Such a matrix, as pointed out by Mei and Morgan [10], [11], is not easily adaptable to banded matrix algorithms, and hence is numerically inefficient to solve using direct or iterative methods. As a consequence, HFEM in such a form is less efficient than the unimoment method [10]-[12] and its modification—the so-called hybrid finite element-extended boundary condition method (hybrid FEM-EBCM) [13], mainly because the unimoment method and the hybrid FEM-EBCM can take advantage of generating numerical basis functions by solving sparse or uniformly banded matrices.

In this communication, the HFEM formulation is modified in such a way that it also results in a sparse or uniformly banded matrix, rather than a partly full and partly sparse nonuniform matrix. The particular formulation used is that first proposed by McDonald and Wexler [2] and thereafter improved and used by others [5]-[9]. The modification is accomplished by changing the sequence of matrix substitutions, a procedure that has been discussed by McDonald and Wexler [6] for general HFEM formulations. The implementation is straightforward, but it substantially improves the computational efficiency and enchances the capability of the method. In the following we consider the technique applied to electromagnetic scattering, present sample computations, and compare with other methods.

II. Formulation

Consider a two-dimensional scattering problem illustrated in Fig. 1, where the contour Γ encloses the scatterer, and the contour Γ_A is an artificial boundary enclosing Γ. The details of the HFEM analysis of this problem can be found in [9]; however, for brevity, the notation used here is somewhat different. Application of the finite element analysis to the region enclosed by Γ_A results in the matrix equation

$$[K_{IA}]\{\phi_A\}+[K_{II}]\{\phi_I\}=0 \tag{1}$$

where the subscript A denotes the nodes on Γ_A, I the nodes interior to Γ_A, and $\{\phi\}$ is the discretized unknown field vector. There are no nodes between Γ_A and Γ. Equation (1) is referred to as the finite element equation.

To solve (1), another equation is needed to relate $\{\phi_A\}$ and $\{\phi_I\}$. Such an equation can be obtained by using a surface integral equation involving the free-space Green's function, which in discretized form is written as

$$[P_{AA}]\{\phi_A\}+[P_{AI}]\{\phi_I\}=\{\phi_A^{\text{inc}}\} \tag{2}$$

where $\{\phi_A^{\text{inc}}\}$ is the known incident field vector on boundary Γ_A. Equation (2) is referred to as the equivalent boundary constraint on Γ_A.

The solution for $\{\phi_A\}$ and $\{\phi_I\}$ can then be found by jointly solving (1) and (2). There are two approaches. The first, which is the one commonly used, imposes (2) on (1) and gives the final system equation

$$[K'_{II}]\{\phi_I\}=\{\psi_I\} \tag{3}$$

where

$$[K'_{II}]=[K_{II}]-[K_{IA}][P_{AA}]^{-1}[P_{AI}],\ \{\psi_I\}=-[K_{IA}][P_{AA}]^{-1}\{\phi_A^{\text{inc}}\}.$$

In this approach, two matrices have to be solved: one is complex and full matrix $[P_{AA}]$ having size of $N_A \times N_A$, and the other is a partly full and partly sparse complex matrix $[K'_{II}]$ having size of $N_I \times N_I$, where N_A is the total number of nodes on Γ_A and N_I is the total number of nodes interior to Γ_A. Usually, N_I is much larger than N_A, and hence the size of the scatterer to be treated by HFEM is mostly limited by the magnitude of N_I.

The second approach, which we present here, substitutes (1) into (2) and gives the equation

$$[P'_{AA}]\{\phi_A\}=\{\phi_A^{\text{inc}}\}$$

where

$$[P'_{AA}]=[P_{AA}]-[P_{AI}][K_{II}]^{-1}[K_{IA}].$$

Mathematically, this second approach is equivalent to the first one; however, computationally it is much more efficient. Here, one also needs to solve two matrices: one is a complex and full matrix $[P'_{AA}]$, but now the other is a symmetric and sparse matrix $[K_{II}]$, which becomes real-valued for lossless scatterers and can be narrowly banded if one numbers the nodes properly. A more obvious comparison is given in Table I. The difference between the first approach and the second approach is in the properties of the matrices $[K'_{II}]$ and $[K_{II}]$. Solving a symmetric, sparse or uniformly banded matrix $[K_{II}]$ is of course much easier and more efficient than solving a nonsymmetric matrix $[K'_{II}]$ with nonuniform block submatrix structure.

Three of the algorithms available for solving the matrix $[K_{II}]$ are briefly discussed here: the banded matrix algorithm, the frontal solution algorithm, and the Lanczos algorithm. The banded matrix algorithm decomposes the matrix into an upper triangular form, and then solves the equation by back substitution. In this process, neither storage nor operations are performed on zeros outside the band. This algorithm is used for computations presented in the next section. The frontal solution algorithm [14] applies Gaussian elimination during the assembly process of the matrices $[K_{II}]$ and $[K_{IA}]$ until all nodes have been processed. The unknowns are then solved by back substitution. This algorithm greatly reduces the storage requirement and avoids operations on zeros. The Lanczos algorithm, in contrast to the above two direct methods, solves the equation by iteration.

Manuscript received September 30, 1987; revised February 3, 1988.

The authors are with the Radiation Laboratory, Department of Electrical Engineering and Computer Science, The University of Michigan, Ann Arbor, MI 48109.

IEEE Log Number 8822586.

Reprinted from *IEEE Trans. Antennas Propagat.*, vol. 36, no. 10, pp. 1486-1490, October 1988.

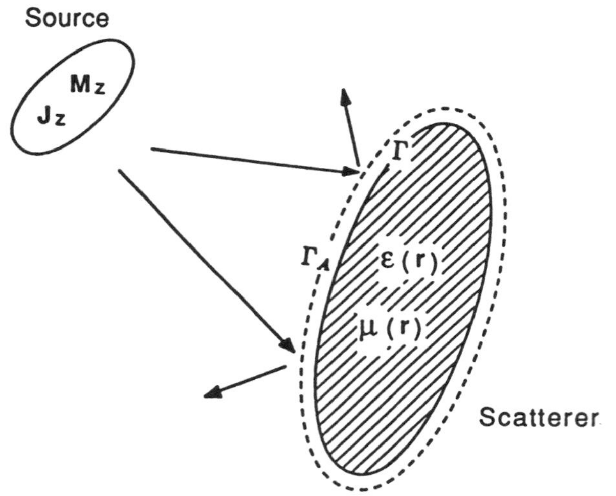

Fig. 1. Geometry of a general two-dimensional scattering problem.

TABLE I
COMPARISON OF THE TWO APPROACHES

Approach	Matrix to be solved	Matrix order	Properties of matrices
First approach	$[P_{AA}]$	N_A	Complex and full
	$[K'_{II}]$	N_I	Complex, partly full and partly sparse
Second approach	$[P'_{AA}]$	N_A	Complex and full
	$[K_{II}]$	N_I	Symmetric, sparse or banded. Real for lossless materials and complex for lossy materials

Convergence is guaranteed if $[K_{II}]$ is a positive definite matrix. This algorithm performs computations such as $[K_{II}]\{p_I\}$ by a summation $\Sigma_{e=1}^{M} K_{II}^{e}\{p_I\}$, where K_{II}^{e} is the element matrix for the element e, M is the total number of elements, and $\{p_I\}$ denotes a vector used in the algorithm. Hence, like the frontal solution algorithm, this algorithm does not actually assemble the matrix K_{II}, thus minimizing the storage requirement and avoiding operations on zeros. A more detailed discussion of the three algorithms can be found in [15, pp. 54-72]. Of the three algorithms, the banded matrix one is most efficient for a matrix with dimension up to a thousand. However, it requires larger storage than do the other two. The frontal solution and the Lanczos algorithms are better for solving large scatterer problems having thousands of unknowns. Even though their programming is complicated, some well developed programs are available.

III. Numerical Results

To show the capability of the modified HFEM formulation, we present here four numerical examples, each of which contains about two thousand unknowns. The first example is a coated circular cylinder with the radius of the conducting cylinder of 1.0 λ (free-space wavelength) and the coating thickness of 0.5 λ (with $\epsilon_r = 2.0$, $\mu_r = 2.0$). The numerical results for bistatic scattering, along with the exact solutions, are shown in Fig. 2. Considering the subdivision information given below and the fact that the coating is lossless, this is good agreement. The results were obtained with $N_A = 188$ and $N_I = 2068$. The area of the coating was subdivided into 3760 triangular elements with the largest edge length one-tenth wavelength in the material.

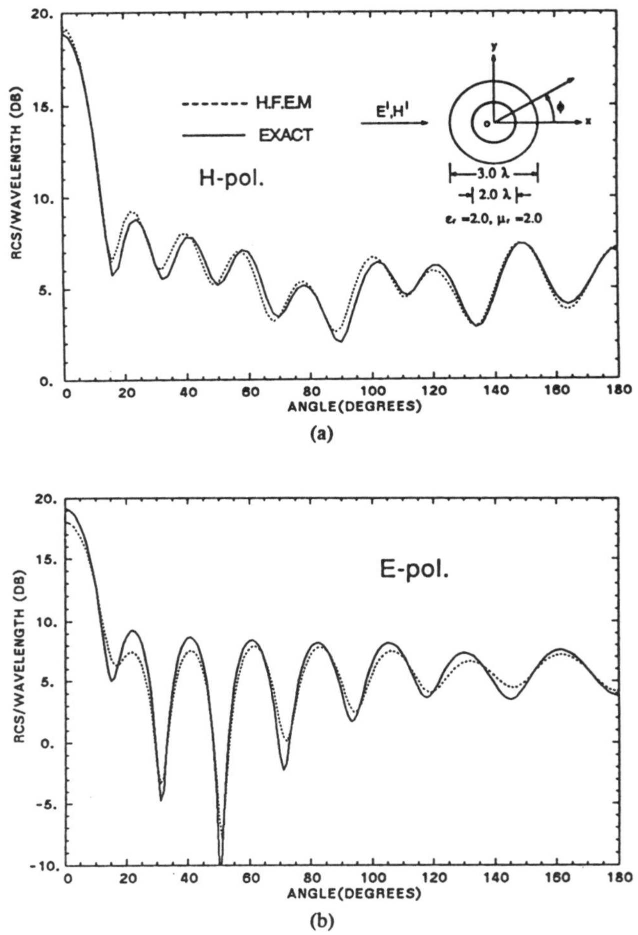

Fig. 2. Bistatic scattering pattern of a coated circular cylinder. (a) *H*-polarization. (b) *E*-polarization.

The second example is an inhomogeneous dielectric circular cylinder having radius 1.2 λ and the permittivity of the dielectric varying in the radial direction according to $\epsilon_r = (2.4\lambda - r)/r$. Since the permittivity tends to infinity at the center, a conducting cylinder of radius 0.2 λ is used to replace the dielectric there. The results for bistatic scattering are shown in Fig. 3. For computation, the cross section of the dielectric cylinder was subdivided into 3948 triangular elements, resulting in $N_A = 94$ and $N_I = 2068$.

The third example is an inhomogeneous rectangular cylinder of size 3.8 λ × 1.2 λ. The permittivity of the cylinder varies along the *y*-direction as $\epsilon_r = 1 + \cos(y\pi/a)$, with $a = 3.8$ λ being the height of the cylinder; the permeability is a constant, having $\mu_r = 1.5$. Both bistatic and backscattering results are given in Fig. 4. In this case, the number of triangular elements is 3648, resulting in $N_A = 200$, $N_I = 1925$.

The last example is an inhomogeneous triangular cylinder with the length of each side 2.77 λ. The permittivity and permeability vary along the *x*-direction as $\epsilon_r = 1.5 + 0.5x/h$ and $\mu_r = 2.0 - 0.5x/h$, respectively, where h (= 2.4 λ) is the height of the triangle. The results are presented in Fig. 5, again for both bistatic and backscattering. In this computation, the number of triangular elements is 3844, resulting in $N_A = 186$, $N_I = 2016$.

All above computations were performed on an Apollo Domain workstation (using Domain series 3000 and 4000 machines). The symmetry property of the first and second examples was not utilized, though it could have been easily adopted to reduce the number of unknowns.

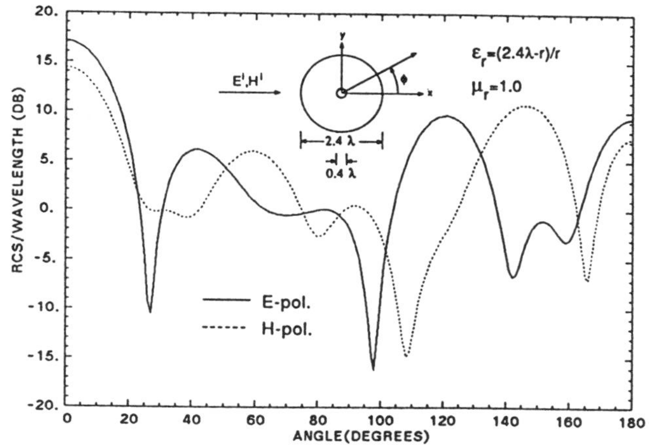

Fig. 3. Bistatic scattering pattern of an inhomogeneous dielectric circular cylinder.

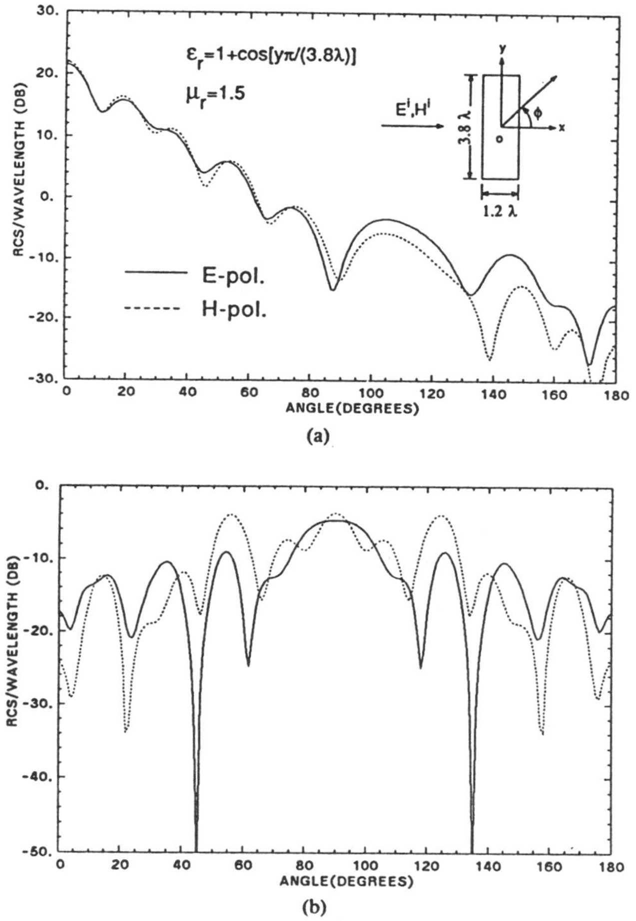

Fig. 4. Plane wave scattering by an inhomogeneous rectangular cylinder. (a) Bistatic scattering. (b) Backscattering.

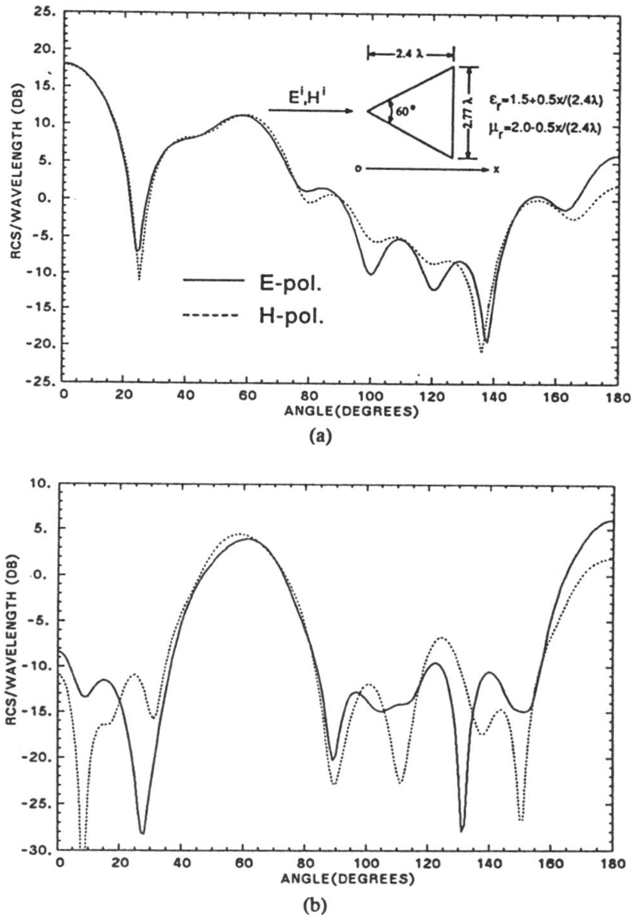

Fig. 5. Plane wave scattering by an inhomogeneous triangular cylinder. (a) Bistatic scattering. (b) Backscattering.

IV. Discussion

In order to give an objective evaluation of the present HFEM, we compare it with two other methods: the unimoment method [12] and the conventional volume integral equation (VIE) method [16], both of which are widely used to solve scattering by inhomogeneous cylinders having both ϵ and μ different from their free-space values.

In Table II we list the number of nonzero matrix elements that need be generated and the size of the matrices to be solved in each of the three methods. In the HFEM, $[K_{II}]$ has about $7N_I$ nonzero elements, since a node is usually connected to six adjacent nodes, and due to the symmetry of the matrix, only $4N_I$ elements need be generated. $2N_A$ is the number of nonzero elements in $[K_{IA}]$, and $2N_A^2$ the number of nonzero elements in $[P_{AA}]$ and $[P_{AI}]$. As previously shown, two matrices need to be solved: one is $[P'_{AA}]$ of size $N_A \times N_A$, and another one is the symmetric and sparse matrix $[K_{II}]$, which in

TABLE II
COMPARISON OF THE METHODS

Method	Modified HFEM	Unimoment method	VIE method
Elements to generate	$4N_I+2N_A+2N_A^2$	$4N+2N_C+4Q^2$	$9N_I^2$
Matrices to solve	$N_I\times B$, $N_A\times N_A$	$N\times B'$, $2Q\times 2Q$	$3N_I\times 3N_I$

banded form is of size $N_I \times B$ with B being the half-bandwidth (including the diagonal) of the matrix. In the unimoment method, as originally developed, an artificial circle is drawn to enclose the entire scatterer. Assume N is the total number of the interior nodes and N_C the total number of boundary nodes. One needs to generate $4N$ nonzero elements for the matrix $[Q]$ and $2N_C$ for the matrix $[T]$ (see [12, eq. (21)]. $4Q^2$ is the number of elements in a matrix of size $2Q \times 2Q$, resulting from the continuity conditions on the circle, where Q denotes the number of harmonics used to expand the field in the exterior region. In this method, also two matrices need to be solved: one is that with the size of $2Q \times 2Q$, and the other is the symmetric and sparse matrix $[Q]$, which in banded form has the size $N \times B'$, where B' is its half-bandwidth. In the VIE method, one needs to generate and solve a matrix of the size $3N_I \times 3N_I$, since at each node there are three unknown equivalent polarized current components.

In order to compare the present HFEM with the unimoment method we need to consider three cases. For a scatterer having its cross section close to a circle, the magnitudes of N_I and N are about the same, and so are the magnitudes of B and B'. Since the number $2Q$ can be smaller than N_A for far-field calculation, the unimoment method in this case is more efficient than the HFEM. However, for a slender scatterer whose cross section substantially deviates from a circle, N is much larger than N_I, and B' is also larger than B. Even though the number $2Q$ is possibly smaller than N_A, the HFEM in this case is expected to be competitive with the unimoment method. Finally, for the case of multiple scatterers, and especially when the scatterers are far apart, the HFEM may prove to be more efficient. This is because in the HFEM the artificial boundary Γ_A can be split into several contours, each enclosing one scatterer, while in the unimoment method a single circle is used to enclose all scatterers. However, a modification to the unimoment method, which uses a boundary conforming to the surface of the scatterer, would enhance its efficiency. One such modification—the hybrid FEM-EBCM was presented by Morgan *et al.* [13] for the case of dielectric bodies of revolution, and if applied to two-dimensional problems it would result in an efficient numerical technique.

Comparing the present HFEM with the conventional VIE method, we find that the HFEM, like the unimoment method and the hybrid FEM-EBCM, is far more efficient than the VIE method. For the problems of size comparable to those considered in Section III, using the VIE method, one would have to generate many more matrix elements, about four hundred times the number needed in the HFEM. The computational difference in solving the matrices is expected to be even larger, though no accurate number can be stated, since the computing time depends on the algorithm chosen for solving the matrix. We note that for the case of dielectric ($\mu_r = 1$) cylinders, the number of unknowns in the VIE method reduces to $2N_I$ for H-polarization and to N_I for E-polarization. We also note that the volume integral equation involving three equivalent polarized current components can be transformed to an integral equation having both volume and surface integrals but only involving one axial field component. Such an equation will result in a technique more efficient than the VIE method; however, it still would not be competitive with either the HFEM or the unimoment method.

V. CONCLUSION

In this paper, a modified formulation of HFEM is presented, which is shown superior to the previous HFEM in that it takes advantage of the important property of the finite element equation, i.e., its coefficient matrix is symmetric, sparse and banded. The modification improves the computational efficiency and enhances the capability of the method, as demonstrated by four numerical examples. The modified HFEM is objectively evaluated by comparing it with other methods. We conclude that, in addition to the unimoment method and the hybrid FEM-EBCM, the modified HFEM provides an efficient method to solve two-dimensional unbounded field problems involving inhomogeneous media.

ACKNOWLEDGMENT

The authors wish to thank the reviewer for his comments, which led to a comparison given in Section IV, and technical writer Mr. Rod Johnson for his help and suggestions.

REFERENCES

[1] P. Silvester and M. S. Hsieh, "Finite-element solution of 2-dimensional exterior field problems," *Proc. Inst. Elec. Eng.*, vol. 118, pp. 1743–1747, Dec. 1971.
[2] B. H. McDonald and A. Wexler, "Finite-element solution of unbounded field problems," *IEEE Trans. Microwave Theory Tech.*, vol. MTT-20, pp. 841–847, Dec. 1972.
[3] S. P. Marin, "Computing scattering amplitudes for arbitrary cylinders under incident plane waves," *IEEE Trans. Antennas Propagat.*, vol. AP-30, pp. 1045–1049, Nov. 1982.
[4] S. K. Jeng and C. H. Chen, "On variational electromagnetics: Theory and application," *IEEE Trans. Antennas Propagat.*, vol. AP-32, pp. 902–907, Sept. 1984.
[5] S. Washisu, I. Fukai, and M. Suzuki, "Extension of finite-element method to unbounded field problems," *Electron. Lett.*, vol. 15, pp. 772–774, 1979.
[6] B. H. McDonald and A. Wexler, *Finite Element in Electrical and Magnetic Field Problem*, M. V. K. Chari and P. P. Silvester, Eds. London and New York, Wiley, 1980, ch. 9.
[7] T. Orikasa, S. Washisu, T. Honma, and I. Fukai, "Finite element method for unbounded field problems and application to two-dimensional taper," *Int. J. Num. Meth. Eng.*, vol. 19, pp. 157–168, 1983.
[8] Y. N. Guo, "Finite-element solution of electromagnetic scattering from inhomogeneous dielectric cylinder with arbitrary cross section," M.S. thesis, Nanjing Univ., 1985.
[9] J. M. Jin and V. V. Liepa, "Application of hybrid finite element method to electromagnetic scattering from coated cylinders," *IEEE Trans. Antennas Propagat.*, vol. 36, pp. 55–70, Jan. 1988.
[10] K. K. Mei, "Unimoment method of solving antenna and scattering problems," *IEEE Trans. Antennas Propagat.*, vol. AP-22, pp. 760–766, Nov. 1974.
[11] M. A. Morgan and K. K. Mei, "Finite-element computation of scattering by inhomogeneous penetrable bodies of revolution," *IEEE Trans. Antennas Propagat.*, vol. AP-27, pp. 202–214, Mar. 1979.
[12] S. K. Chang and K. K. Mei, "Application of the unimoment method to electromagnetic scattering of dielectric cylinders," *IEEE Trans. Antennas Propagat.*, vol. AP-24, pp. 35–42, Jan. 1976.
[13] M. A. Morgan, C. H. Chen, S. C. Hill, and P. W. Barber, "Finite element-boundary integral formulation for electromagnetic scattering," *Wave Motion*, vol. 6, no. 1, pp. 91–103, Jan. 1984.
[14] B. M. Irons, "A frontal solution program for finite element analysis," *Int. J. Num. Meth. Eng.*, vol. 2, pp. 5–32, 1970.
[15] G. Sewell, *Analysis of Finite Element Method: PDE/PROTRAN.* New York: Springer-Verlag, 1985.
[16] R. F. Harrington, *Field Computation by Moment Methods.* New York: Macmillan, 1968.

Coupling of Finite Element and Moment Methods for Electromagnetic Scattering from Inhomogeneous Objects

XINGCHAO YUAN, MEMBER, IEEE, DANIEL R. LYNCH, AND JOHN W. STROHBEHN, SENIOR MEMBER, IEEE

Abstract—A new hybrid formulation is proposed, which combines the method of moments (MM) with the finite element method (FEM) to solve electromagnetic scattering and/or absorption problems involving inhomogeneous media. The basic technique is to apply the equivalence principle and transform the original problem into interior and exterior problems, which are coupled on the exterior dielectric body surface through the continuities of the tangential electric field and magnetic field. The interior problem involving inhomogeneous medium is solved by the FEM and the exterior problem is solved by the MM. The coupling of the interior and exterior problems on their common surface results in a matrix equation for the equivalent current sources for the interior and exterior problems. By combining advantages of both methods, complicated inhomogeneous problems with arbitrary geometry are treated in a straightforward manner. The validity and accuracy of the formulation are checked by two-dimensional numerical results, which are compared with the exact eigenfunction solution, the unimoment solution, and Richmond's pure moment solution.

INTRODUCTION

THE PROBLEM OF ELECTROMAGNETIC scattering and absorption due to inhomogeneous, lossy, and arbitrarily shaped dielectric bodies has been extensively dealt with in the literature because of its presence in many practical situations including microwave hyperthermia for cancer therapy, coupling to missiles with dielectric-filled apertures, and performance of communication antennas in the presence of dielectric and magnetic inhomogeneities. When the size of the dielectric object is neither large nor small compared to the wavelength of the excitation, asymptotic methods cannot be used and a rigorous solution of Maxwell's equations is required.

Many ways have been developed to treat such scattering problems. Among them are the pure method of moments solution using either surface integral [1] or volume integral equations [2]–[5], the finite difference time domain (FD-TD) method [6], [7], the unimoment method [8], [9], and hybrid methods combining the finite method (either finite element [10] or finite difference) with the integral equation method [11]–[16]. It is a well-known fact that integral equation methods such as the moment method treat unbounded problems very effectively but they become computationally intensive when complex inhomogeneities are present. In contrast, inhomogeneities are easily handled by finite methods. The finite element method requires less computer time and storage because it results in a sparse and banded matrix. The FD-TD method uses an iterative approach and is therefore not memory intensive. However, finite methods are most suitable for boundary value problems. Special steps need to be taken if an unbounded region is present. Clearly, any hybrid method that retains the most efficient characteristics of both finite methods and integral equation methods is computationally advantageous. Such hybrid methods developed include the hybrid moment method/FD-TD [11], the hybrid finite element method and extended boundary condition integral method (or FEBI) [12], and the hybrid finite element and the boundary element method (or BEM/FEM) [13]–[15]. A general hybrid formulation referred to as the field feedback formulation (or F^3) can be found in [16]. The basic techniques of these methods are to use a finite method to treat the bounded, inhomogeneous region and to use an integral equation method to treat the unbounded homogeneous region.

In this paper, we present a new hybrid formulation, which combines the method of moments (MM) [17] with the finite element method presented in [15]. This new MM/FEM hybrid has comparable advantages as the above hybrid FEBI and BEM/FEM methods, and in addition facilitates the use of established MM procedures and formulations in terms of equivalent current sources. The particular MM approach utilized here, adapted from [18], [19], is oriented toward scattering of incident waves originating from a distant source. The coupling is achieved explicitly through matching the boundary conditions on the dielectric body surface. Following a detailed formulation, we show numerical examples for two-dimensional problems under transverse magnetic (TM) and transverse electric (TE) plane wave excitations to assess the validity and accuracy of the method.

FORMULATION

An arbitrarily shaped, inhomogeneous, and lossy dielectric body under the illumination of an incident field $(\mathbf{E}^i, \mathbf{H}^i)$ is shown in Fig. 1. The dielectric body bounded by a surface S is characterized by $(\mu(\mathbf{r}), \epsilon(\mathbf{r}))$. The region inside the surface S is called region b and the unbounded region outside S is called region a, which is characterized by (μ_0, ϵ_0) and

Manuscript received August 25, 1988; revised March 15, 1989. This work was supported in part by NIH Grant R01 CA37245; and by NSF Grant CEE-835-2226.

The authors are with the Thayer School of Engineering, Dartmouth College, Hanover, NH 03755.

IEEE Log Number 8933573.

Reprinted from *IEEE Trans. Antennas Propagat.*, vol. 38, no. 3, pp. 386–393, March 1990.

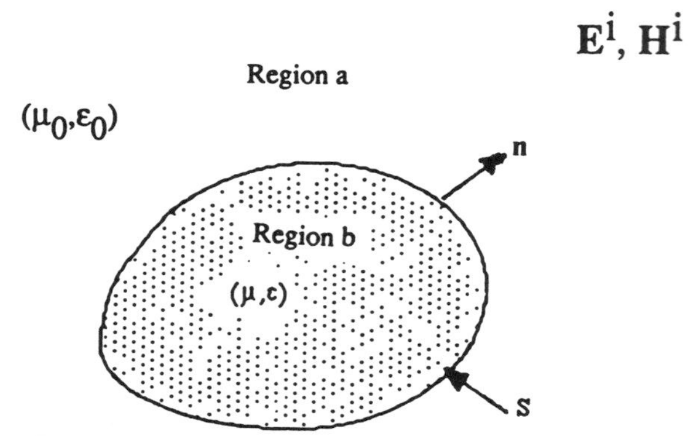

Fig. 1. Original problem: electromagnetic scattering due to an inhomogeneous dielectric body bounded by S.

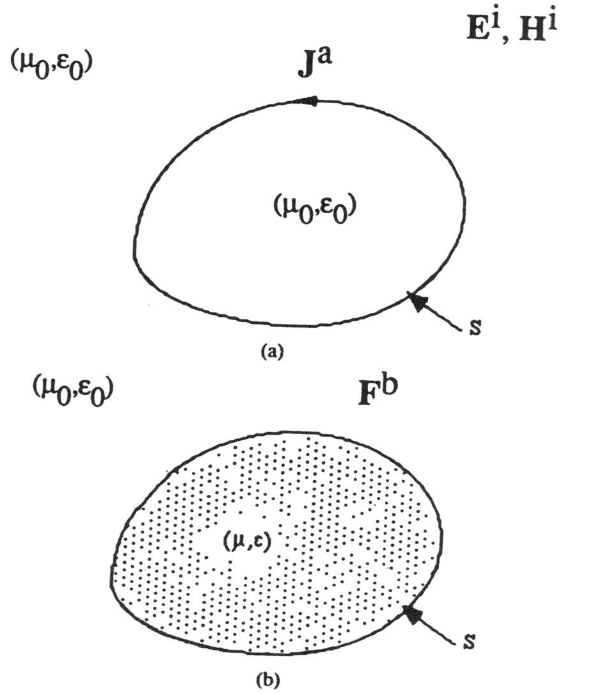

Fig. 2. The equivalence of the original problem. (a) Equivalent problem for region a, where $\mathbf{J}^a$ is the equivalent electric current on S. (b) Boundary value problem for region b, where $\mathbf{F}^b$ is the tangential magnetic field on S.

assumed to be vacuum without loss of generality. The objective is to determine the electromagnetic fields everywhere in space.

The strategy taken here is to use the equivalence principle [20, ch. 3] to transform the original problem into equivalent problems in regions a and b, which are coupled through the continuity conditions on the tangential components of $\mathbf{E}$ and $\mathbf{H}$ on the dielectric body surface S. A similar technique was used in [18] and [19] to solve the problem of electromagnetic scattering due to homogeneous dielectric cylinders partially covered by conductors.

The equivalent problem for region a is shown in Fig. 2(a). The electromagnetic fields ($\mathbf{E}^a$, $\mathbf{H}^a$) in region a can be considered to be the superposition of the incident field ($\mathbf{E}^i$, $\mathbf{H}^i$) and the scattered field ($\mathbf{E}^s$, $\mathbf{H}^s$), where the scattered field is produced by an equivalent surface electric current $\mathbf{J}^a$ on S radiating in the infinite space filled with (μ_0, ϵ_0) everywhere. Specifically,

$$\mathbf{E}^a = \mathbf{E}^i + \mathbf{E}^s(\mathbf{J}^a) \tag{1}$$

$$\mathbf{H}^a = \mathbf{H}^i + \mathbf{H}^s(\mathbf{J}^a). \tag{2}$$

The equivalent problem for region b is shown in Fig. 2(b). The electromagnetic fields in region b are uniquely determined by the unknown tangential magnetic field on S, denoted by $\mathbf{F}^b$. That is,

$$\mathbf{E}^b = \mathbf{E}^b(\mathbf{F}^b) \tag{3}$$

$$\mathbf{H}^b = \mathbf{H}^b(\mathbf{F}^b) \tag{4}$$

where the superscript b on the field quantities indicates those actually present in the inhomogeneous dielectric body.

The equivalent electric current $\mathbf{J}^a$ and the tangential magnetic field $\mathbf{F}^b$ on S can be found by applying the boundary conditions on S. Namely, both the tangential electric field and the tangential magnetic field must be continuous across S, i.e.,

$$\mathbf{n} \times \mathbf{E}^a = \mathbf{n} \times \mathbf{E}^b, \qquad \text{on } S \tag{5}$$

$$\mathbf{n} \times \mathbf{H}^a = \mathbf{n} \times \mathbf{H}^b = \mathbf{F}^b, \qquad \text{on } S \tag{6}$$

where $\mathbf{n}$ is the unit normal on S pointing outward from region b, as indicated in Fig. 1. Substituting (1) and (2) into (5) and (6) and rearranging terms, we obtain

$$-\mathbf{n} \times \mathbf{E}^s(\mathbf{J}^a) + \mathbf{n} \times \mathbf{E}^b(\mathbf{F}^b) = \mathbf{n} \times \mathbf{E}^i, \qquad \text{on } S \tag{7}$$

$$-\mathbf{n} \times \mathbf{H}^s(\mathbf{J}^a) + \mathbf{F}^b = \mathbf{n} \times \mathbf{H}^i, \qquad \text{on } S. \tag{8}$$

Equations (7) and (8), together with the governing Maxwell's equations, are sufficient to determine the equivalent electric current $\mathbf{J}^a$ and the tangential magnetic field $\mathbf{F}^b$. The continuity of the normal components need not be addressed explicitly, since it follows from (7) and (8) and the Maxwell's equations.

Numerical Method

A. Moment Solution

Equations (7) and (8) can be reduced to weak-form matrix equations by the method of moments [17]. To apply the method of moments, we approximate $\mathbf{J}^a$ and $\mathbf{F}^b$ by

$$\mathbf{J}^a = \sum_{j=1}^{N} I_j^a \mathbf{J}_j \tag{9}$$

$$\mathbf{F}^b = \sum_{j=1}^{N} F_j^b \mathbf{J}_j \tag{10}$$

where $\{\mathbf{J}_j\}$ is a set of chosen expansion functions, N is the number of functions chosen, and $\{I_j^a\}$ and $\{F_j^b\}$ are two sets of coefficients to be determined. (Note that we employ the same expansion functions for the equivalent current in region a and the tangential magnetic field in region b. This is convenient but not necessary.) A symmetric product is defined by

$$\langle \mathbf{A}, \mathbf{B} \rangle = \int_S \mathbf{A} \cdot \mathbf{B} \, ds \tag{11}$$

where **A** and **B** are two vector functions defined on S. Assume $\{\mathbf{W}_i, i = 1, 2, \cdots, N\}$ to be a set of weighting functions defined on S. The weak-form discretization of (7) and (8)

$$\begin{bmatrix} [Z^a] & [Z^b] \\ [Y^a] & [Y^b] \end{bmatrix} \begin{bmatrix} I^a \\ F^b \end{bmatrix} = \begin{bmatrix} V \\ U \end{bmatrix} \tag{12}$$

is obtained by substituting (9) and (10) into (7) and (8) and taking the symmetric product (11) of both resulting equations with the weighting functions $\{\mathbf{W}_i\}$. In (12), I^a, F^b, V, and U are $N \times 1$ column vectors. The ith elements of I^a and F^b are simply I_i^a and F_i^b, respectively; and the ith elements of V and U are v_i and u_i given by

$$v_i = \langle \mathbf{W}_i, \mathbf{n} \times \mathbf{E}^i \rangle \tag{13}$$

$$u_i = \langle \mathbf{W}_i, \mathbf{n} \times \mathbf{H}^i \rangle. \tag{14}$$

$[Z^a]$, $[Z^b]$, $[Y^a]$, and $[Y^b]$ are $N \times N$ matrices and their ijth elements are given by

$$Z_{ij}^a = \langle \mathbf{W}_i, -\mathbf{n} \times \mathbf{E}^s(\mathbf{J}_j) \rangle \tag{15}$$

$$Z_{ij}^b = \langle \mathbf{W}_i, \mathbf{n} \times \mathbf{E}^b(\mathbf{J}_j) \rangle \tag{16}$$

$$Y_{ij}^a = \langle \mathbf{W}_i, -\mathbf{n} \times \mathbf{H}^s(\mathbf{J}_j) \rangle \tag{17}$$

$$Y_{ij}^b = \langle \mathbf{W}_i, \mathbf{n} \times \mathbf{H}^b(\mathbf{J}_j) \rangle = \langle \mathbf{W}_i, \mathbf{J}_j \rangle. \tag{18}$$

Equations (13)–(18) have to be evaluated before (12) can be solved to obtain I^a and F^b and in turn to assemble the equivalent sources $\mathbf{J}^a$ and $\mathbf{F}^b$ given in (9) and (10). Once the weighting functions $\{\mathbf{W}_i\}$ and the expansion functions $\{\mathbf{J}_j\}$ are chosen, (13)–(15) and (17) can be evaluated easily since the incident fields $(\mathbf{E}^i, \mathbf{H}^i)$ are given, and the electric field produced by $\mathbf{J}_j$ in the infinite space (μ_0, ϵ_0) is given by [20]

$$\mathbf{E}^s(\mathbf{J}_j) = \frac{-j\omega\mu_0}{4\pi} \int_S \mathbf{J}_j G(|\mathbf{r} - \mathbf{r}'|) \, ds' + \frac{\nabla}{4\pi j\omega\epsilon_0} \int_S \nabla_s' \cdot \mathbf{J}_j G(|\mathbf{r} - \mathbf{r}'|) \, ds' \tag{19}$$

where $G(\cdot)$ is the Green's function given by [20, eq. (5-88)] in two dimensions and by

$$G(|\mathbf{r} - \mathbf{r}'|) = \frac{e^{-jk_0|r-r'|}}{|r - r'|} \tag{20}$$

in three dimensions, where a harmonic time dependence $\exp(j\omega t)$ is assumed and suppressed, and $k_0 = \omega\sqrt{\mu_0\epsilon_0}$ is the wavenumber in region a. The magnetic field $\mathbf{H}^s(\mathbf{J}_j)$ is related to the electric field $\mathbf{E}^s(\mathbf{J}_j)$ by the Maxwell's equations. Specially, if the observation point $\mathbf{r}$ is not on an edge, the tangential magnetic field takes the following form [23, eq. (11.34)]

$$\mathbf{n} \times \mathbf{H}^s(\mathbf{J}_j) = \lim_{r \to S} \frac{\mathbf{n} \times \nabla \times \mathbf{E}^s(\mathbf{J}_j)}{-j\omega\mu_0} = \mathbf{J}_j/2 + \mathbf{n} \times \frac{1}{4\pi} \int_S \mathbf{J}_j \times \nabla' G(|\mathbf{r} - \mathbf{r}'|) \, ds' \tag{21}$$

where $\mathbf{r}$ approaches S from the region a side of S. Equations (19) and (21) may be substituted directly into (15) and (17) to evaluate Z_{ij}^a and Y_{ij}^a.

The evaluation of $\mathbf{n} \times \mathbf{H}^b(\mathbf{J}_j)$ in (18) is trivial since on S $\mathbf{n} \times \mathbf{H}^b(\mathbf{J}_j)$ is simply $\mathbf{J}_j$ itself the tangential magnetic field. Thus, (18) can be easily obtained once $\mathbf{W}_i$ and $\mathbf{J}_j$ are chosen. The evaluation of (16) requires knowledge of $\mathbf{E}^b(\mathbf{J}_j)$. We resort to the finite element method as follows.

B. The Finite Element Solution of $\mathbf{E}^b(\mathbf{J}_j)$

Since $\mathbf{F}^b$ is used to denote the equivalent tangential magnetic field on S and from (3), (4), and (10), $\mathbf{E}^b(\mathbf{J}_j)$ is the electric field due to the equivalent tangential magnetic field $\mathbf{J}_j$ defined everywhere on S. For piecewise approximation, $\mathbf{J}_j$ is usually zero except on a small portion of S (e.g. triangle patches of S). Hence, the problem of finding $\mathbf{E}^b(\mathbf{J}_j)$ in region b is a pure boundary value problem and its general solution is available by the finite element method. As described in [15], we first expand the unknown electric field in region b in terms of conventional real valued, scalar finite element functions ϕ_j defined in the volume enclosed by S:

$$\mathbf{E}^b = \sum_{j=1}^{N^t} \mathbf{E}_j^b \phi_j \tag{22}$$

where N^t is the total number of nodes including all interior nodes and those on the surface S, and $\{\mathbf{E}_j^b\}$ are the unknown electric field values at the nodes. The bases ϕ_j are typically locally supported simple polynomials, continuous with piecewise continuous first derivatives. We then obtain a weak form of the Maxwell's equations by requiring orthogonality with a set of weighting functions. In particular, we employ the Galerkin form [15][1]

$$\left\langle \left(\frac{1}{j\omega\mu} \nabla \times \mathbf{E}^b \right) \times \nabla\phi_i \right\rangle_v + \langle j\omega\epsilon \mathbf{E}^b \phi_i \rangle_v = \int_S \mathbf{n} \times \mathbf{H}^b \phi_i \, ds = \int_S \mathbf{F}^b \phi_i \, ds \tag{23}$$

where $\langle \cdot \rangle_v$ indicates integrating over the entire inhomogeneous volume bounded by S. Note that we do not require the expansion of $\mathbf{H}$ in the interior. Since $\mathbf{n} \times \mathbf{H}$ must be continuous across the internal dielectric interfaces, the boundary integral of (23) mutually cancels along such boundaries and the right-hand side of (23) survives only on the exterior boundary S. Hence, (23) expresses the electric field inside and on S in terms of the tangential magnetic field on S.

Substituting (22) into (23) and requiring the satisfaction of

[1] A harmonic time dependence $\exp(-j\omega t)$ was assumed in [15]. The sign is reversed for internal consistency.

(23) for $i = 1$ through N', we obtain the finite element matrix equation

$$[A][\mathbf{E}^b] = [\mathbf{F}^b] \tag{24}$$

where $[A]$ is a square, sparse, banded, and nonsingular $N' \times N'$ matrix. The detailed form of $[A]$ can be found in [14], [15]. $[\mathbf{E}^b]$ and $[\mathbf{F}^b]$ are $N' \times 1$ column vectors containing the unknown electric field at nodes and the integral of the tangential magnetic field on the exterior boundary S times the weighting function ϕ_i. Therefore, $\mathbf{E}^b(\mathbf{J}_j)$ can be obtained by evaluating the right-hand side of (24) by replacing $\mathbf{F}^b$ in (23) with $\mathbf{J}_j$ and solving the matrix equation (24). N such solutions are required, but the matrix $[A]$ needs to be factored (e.g. LU decomposed) only once.

Finally, the obtained $\mathbf{E}^b(\mathbf{J}_j)$ on S is substituted into (16) to evaluate Z^b_{ij}. Equation (12) can then be solved to obtain I^a and F^b. Hence, the electromagnetic fields at any point in space can be obtained by using (1)–(4), (9), and (10); (19)–(21) for the exterior solution, and (24) for the interior solution. In the later case, retention of the LU decomposition of $[A]$ is efficient.

C. Dual Formulations

In the above formulation, we have used the equivalent sources $\mathbf{J}^a$ and $\mathbf{J}^b$ to produce the correct field in region a and b. Other equivalent sources are also possible. For instance, the combinations of $(\mathbf{M}^a, \mathbf{F}^b)$, $(\mathbf{J}^a, \mathbf{M}^b)$, and $(\mathbf{M}^a, \mathbf{M}^b)$ are all possible and valid equivalent sources, where $\mathbf{M}^a$ and $\mathbf{M}^b$ are equivalent magnetic current and tangential electric field on S, respectively. The determining factor in choosing which set of equivalent sources would be used is the computational simplicity and efficiency. In the two-dimensional TM case, $(\mathbf{J}^a, \mathbf{F}^b)$ is more convenient and $(\mathbf{M}^a, \mathbf{M}^b)$ is more convenient in the two dimensional TE case. For three-dimensional problems, it is not clear there is a preferred choice. In the case of using equivalent tangential electric field source $\mathbf{M}^b$ for the equivalence in region b, the following equation [15]

$$\left\langle \left(\frac{1}{j\omega\epsilon} \nabla \times \mathbf{H}^b\right) \times \nabla\phi_i \right\rangle_v + \langle j\omega\mu \mathbf{H}^b \phi_i\rangle_v = -\int \mathbf{n} \times \mathbf{E}^b \phi_i \, ds \tag{25}$$

should be used to find the magnetic field $\mathbf{H}$ inside S in terms of the tangential electric field $\mathbf{E}$ on S.

Two-Dimensional Scattering

The formulation given above can be applied to both two- and three-dimensional problems. However, only the two-dimensional results are given here. We wish to present the numerical results for three-dimensional problems in a future article.

In the two-dimensional case, the inhomogeneous dielectric body becomes an inhomogeneous infinitely long dielectric cylinder, which can be defined by its cross section on the xy plane assuming the cylinder axis coincides with the z-axis. The incident waves can be separated into transverse magnetic case and transverse electric case. If the incident field is a plane wave impinging upon the cylinder with an angle ϕ^i, then it can be defined by

$$\mathbf{E}^i = \mathbf{u}_z E_z = \mathbf{u}_z e^{jk_0(x \cos\phi^i + y \sin\phi^i)} \tag{26}$$

for the TM case, and

$$\mathbf{H}^i = \mathbf{u}_z H_z = \mathbf{u}_z e^{jk_0(x \cos\phi^i + y \sin\phi^i)} \tag{27}$$

for the TE case, where k_0 is a wavenumber in the free space. Note these two cases are completely dual to each other. That is, one can be obtained from the other by interchanging symbols according to [20, ch. 3]. In either TM or TE case, the equivalent electric or magnetic current sources will also be z-directed. Thus, a simple pulse function (as defined in [19]) can be used as the current expansion function in (9), (10) and also as the weighting function; and the impedance matrix $[Z^a]$ and the admittance matrix $[Y^a]$ may be expressed as in [19]. The admittance matrix $[Y^b]$ is a diagonal matrix since the electric (or magnetic) current expansion function resides only on a single segment of the contour. By a simple matrix manipulation of (12), one needs to solve only an $N \times N$ matrix equation for I^a instead of a $2N \times 2N$ matrix equation for both I^a and F^b simultaneously since the inversion of $[Y^b]$ is trivial. In the FEM solution, the same technique as in [15] is used. The components of $[\mathbf{F}^b]$ for a given $\mathbf{J}_j$ in (24) will, in this case, be nonzero on only two elements. Equation (24) is triangularized only once but the back substitutions are performed N times to obtain $\{\mathbf{E}^b(\mathbf{J}_j), j = 1, 2, \cdots, N\}$ in region b.

To show validity and accuracy of the hybrid MM/FEM method, we have compared our results with the exact eigensolutions and those appearing in the literature. Both far field and near field are examined. The echo width L_e is defined by [20, eq. (7-121)]

$$L_e = \lim_{\rho\to\infty} \left(2\pi\rho \frac{|E^s|^2}{|E^i|^2}\right) \tag{28}$$

for the TM case, and by

$$L_e = \lim_{\rho\to\infty} \left(2\pi\rho \frac{|H^s|^2}{|H^i|^2}\right) \tag{29}$$

for the TE case.

In the following numerical examples involving circular cylinders, the finite element mesh is generated by dividing the radius by N_r divisions and 2π by N_ϕ divisions. The total number of nodes N' is $N_r \times N_\phi + 1$, where the last node resides at the center of the cylinder. The number N in (9) and (10) is equal to N_ϕ. The inner most circle is filled with triangles and the rest of the annular regions are filled with quadrilaterals. Hence, there are N_ϕ triangles and $(N_r - 1) \times N_\phi$ quadrilaterals in the finite element mesh.

Figs. 3 and 4 show the magnitude and phase of the electric field along the x axis (AA') and y axis (BB') inside a circular homogeneous dielectric cylinder, under a TM plane wave incidence ($\phi^i = 180°$). Fig. 5 shows the echo width of the same

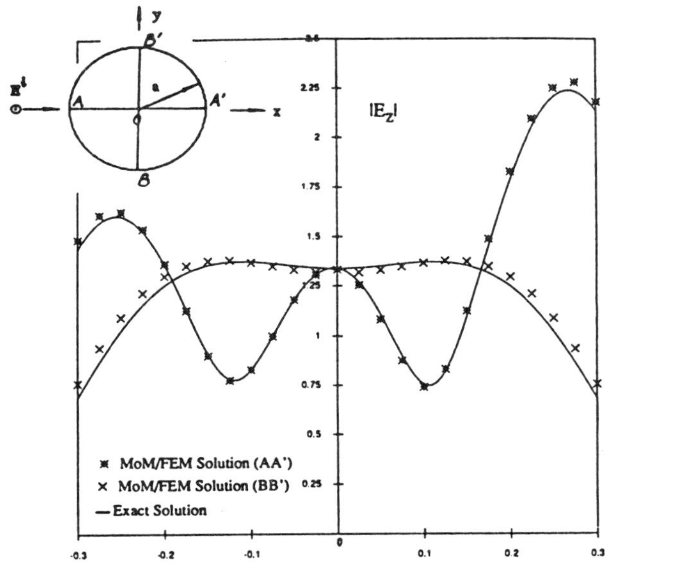

Fig. 3. Amplitude of the electric field along x axis (AA') and y axis (BB') inside a circular homogeneous dielectric cylinder ($a = 0.3\ \lambda$, $\sigma = 0.0$, $\epsilon_r = 4.0$, $f = 300$ MHz, $\phi^i = 180°$, and TM case). (MM/FEM solution compared to exact eigenfunction solution.)

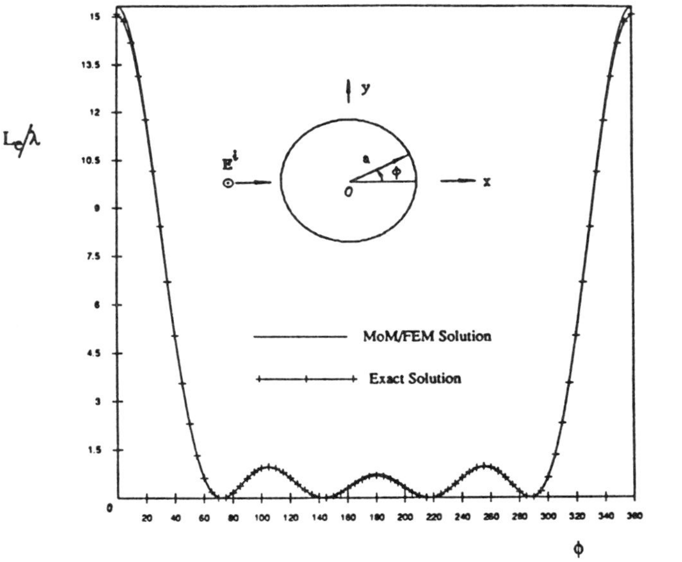

Fig. 5. Echo width of a circular homogeneous dielectric cylinder ($a = 0.3\ \lambda$, $\sigma = 0.0$, $\epsilon_r = 4.0$, $f = 300$ MHz, $\phi^i = 180°$, and TM case). (MM/FEM solution compared to exact eigenfunction solution.)

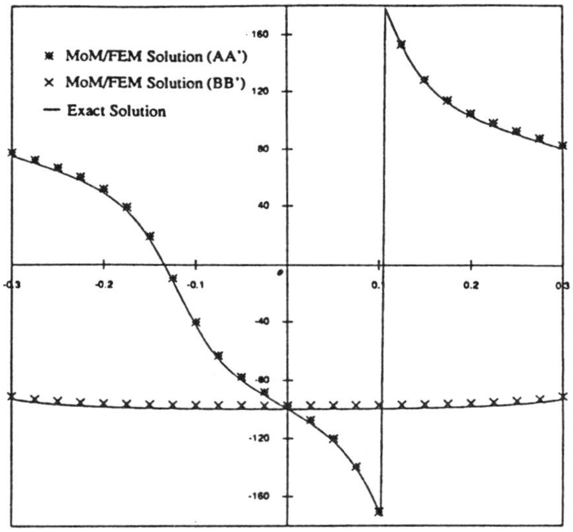

Fig. 4. Phase of the electric field along x axis (AA') and y axis (BB') inside a circular homogeneous dielectric cylinder ($a = 0.3\ \lambda$, $\sigma = 0.0$, $\epsilon_r = 4.0$, $f = 300$ MHz, $\phi^i = 180°$, and TM case). (MM/FEM solution compared to exact eigenfunction solution.)

cylinder under the TM plane wave incidence. In Figs. 3–5, the classical eigenfunction solution is obtained by summing the Fourier series up to 30 terms and the hybrid MM/FEM results are obtained by letting $N_r = 12$ and $N_\phi = 32$. The agreement of those two solutions is good with the maximum relative error within 4%. To show the convergence rate of the method, the numerical results obtained by varying N_r and N_ϕ are given in Table I. As N_ϕ increases from 22 to 42 (fixing N_r at 12), the relative error decreases from 6.74% to 2.068% (at point A). Note there are about 11 subsections per wavelength on the cylinder boundary when $N_\phi = 22$, and the numerical results are already in good agreement with the exact solutions. This is consistent with the general convergence rate of the moment method. When N_r increases from 12 to 20 (fixing N_ϕ at 32), little changes results, which shows the interior finite element solution is satisfactory with more than 20 subsections per wavelength. Finally, one may notice that slightly better results are obtained at the center of the cylinder (point O) than at the boundaries (points A and A'). Further study reveals that the elements span smaller area toward the center. We believe these results give very good evidence of the validity of the method.

To compare the MM/FEM method with other existing techniques, we compute the echo width of an off-centered circular dielectric cylinder under either TM or TE plane wave incidence, which is shown in Fig. 6 and was also obtained by the unimoment method [9]. The agreement is once again excellent. Figs. 7 and 8 show the echo widths, obtained by MM/FEM method and by Richmond [2], [22], of a circular and a semicircular dielectric shell under either TM or TE plane wave illumination. These two solutions generally agree very well except there are some discrepancies in the backscattering region for the semicircular shell under the TE plane wave incidence. The MM/FEM solution is shown to be converged by increasing N' and N. It is not clear what caused the discrepancy. Nevertheless, Figs. 6–8 do demonstrate further the validity and accuracy of the MM/FEM method.

Since the FEM method treats the inhomogeneity in a very natural way, the hybrid MM/FEM method is capable of handling the scattering and/or absorption properties of any dielectric object with arbitrary shape and/or arbitrary inhomogeneity. Once a finite element mesh containing proper material properties is given, (for instance, generated by an automatic mesh generator), the electromagnetic fields at any point in space and other related quantities can be obtained by running the code without other changes required. Figs. 9 and 10 show the echo widths of a four layered circular dielectric cylinder under either TM or TE plane wave incidence. All dielectric layers are assumed to have the same conductivity σ. Also shown are the variations of the echo width with the change of the conductivity.

TABLE I
CONVERGENCE TEST ON CYLINDER IN FIG. 3

	Case 1 $N_r=12,\ N_\phi=22$		Case 2 $N_r=12,\ N_\phi=32$		Case 3 $N_r=12,\ N_\phi=42$		Case 4 $N_r=20,\ N_\phi=32$		
	MM/FEM	Error %	MM/FEM	Error %	MM/FEM	Error %	MM/FEM	Error %	Exact
$\lvert E_z \rvert$ at point A*	1.5276	6.74	1.4793	3.368	1.4607	2.068	1.4773	3.23	1.4311
$\lvert E_z \rvert$ at point O*	1.3160	1.68	1.3313	0.536	1.3369	0.114	1.3287	0.735	1.3385
$\lvert E_z \rvert$ at point A'	2.2295	4.611	2.1798	2.280	2.1614	1.415	2.1768	2.135	2.1312
L_e/λ at $\phi=0$	15.5116	2.842	15.3249	1.604	15.2464	1.06	15.3010	1.446	15.0830

* Points A, O, and A' are defined in Fig. 3.

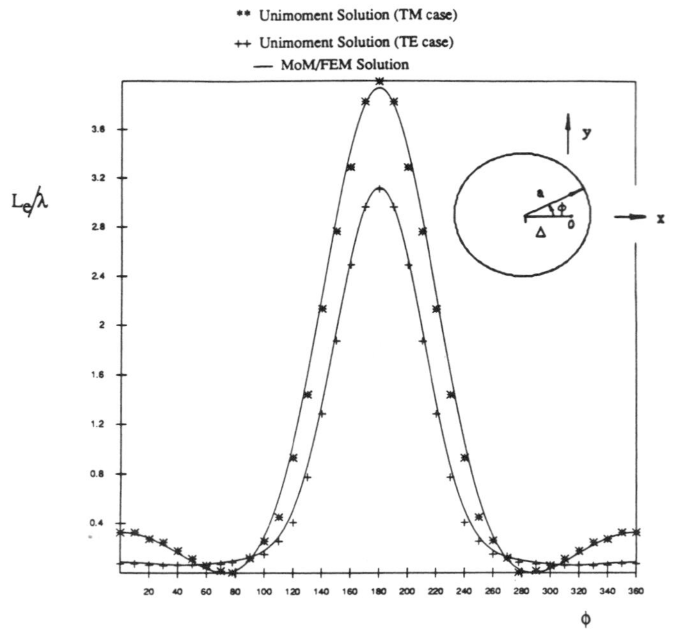

Fig. 6. Echo width of an off-centered circular homogeneous dielectric cylinder $a = 0.3\ \lambda$, $\Delta = 0.2\ \lambda$, $\sigma = 0.0$, $\epsilon_r = 2.0$, $f = 300$ MHz, $\phi^i = 0°$) under either TE or TM plane wave incidence. (MM/FEM solution compared to unimoment solution [9].)

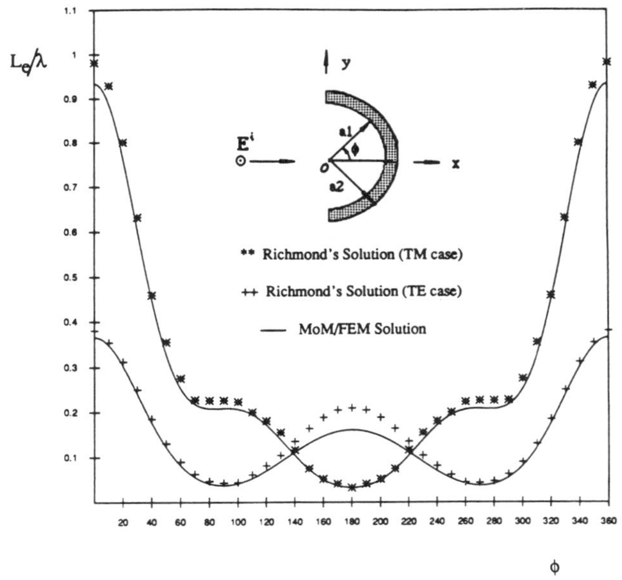

Fig. 8. Echo width of a semicircular dielectric cylindrical shell ($a_1 = 0.25\ \lambda$, $a_2 = 0.3\ \lambda$, $\sigma = 0.0$, $\epsilon_r = 4.0$, $f = 300$ MHz, $\phi^i = 180°$) under either TE or TM plane wave incidence. (MM/FEM solution compared to Richmond's solution [2], [20].)

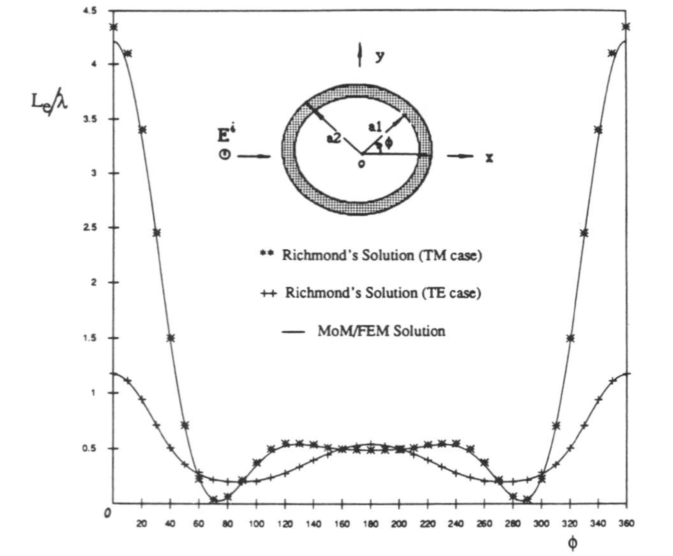

Fig. 7. Echo width of a circular dielectric cylindrical shell ($a_1 = 0.25\ \lambda$, $a_2 = 0.3\ \lambda$, $\sigma = 0.0$, $\epsilon_r = 4.0$, $f = 300$ MHz, $\phi^i = 180°$) under either TE or TM plane wave incidence. (MM/FEM solution compared to Richmond's solution [2], [20].)

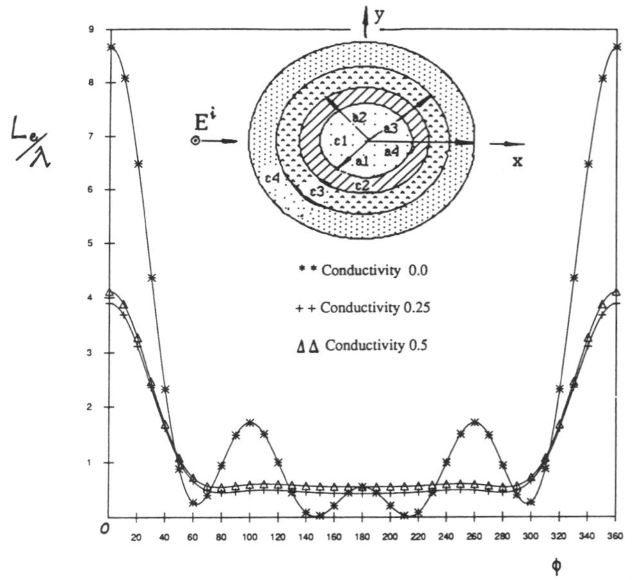

Fig. 9. Echo width of a four layered dielectric cylinder under TM plane wave incidence ($\phi^i = 180°$), where $a_1 = 0.15\ \lambda$, $a_2 = 0.20\ \lambda$, $a_3 = 0.25\ \lambda$, $a_4 = 0.3\ \lambda$, $\epsilon_{r1} = 8.0$, $\epsilon_{r2} = 6.0$, $\epsilon_{r3} = 4.0$, $\epsilon_{r4} = 2.0$, and all layers have the same conductivity σ.

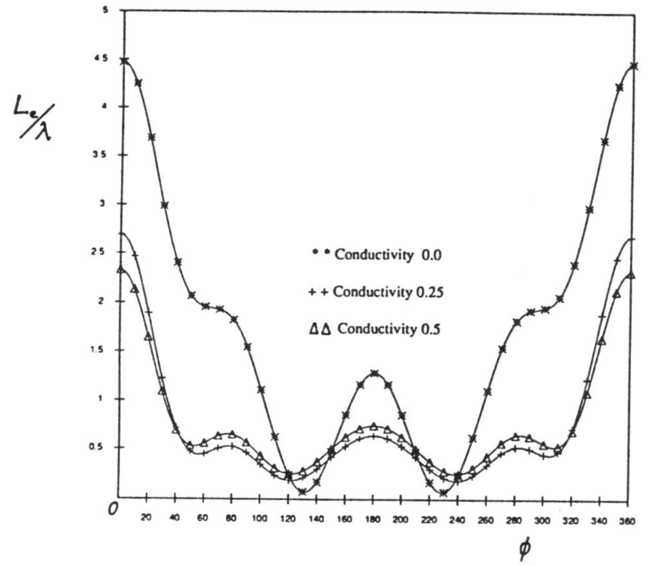

Fig. 10. Echo width of the same layered dielectric cylinder as in Fig. 9 under TE plane wave incidence ($\phi^i = 180°$).

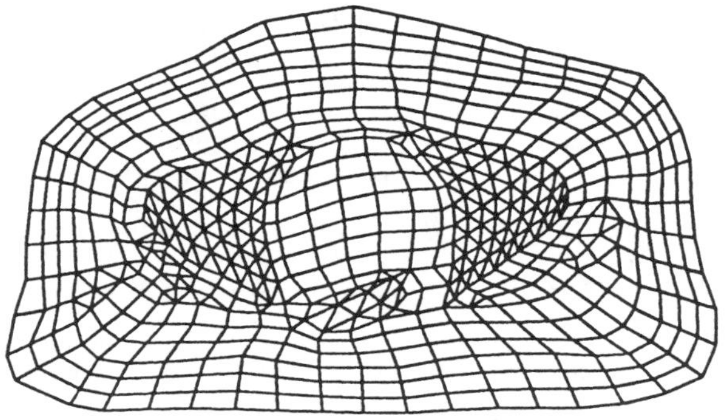

Fig. 11. The finite element mesh of a patient model (pelvic model B of [13]).

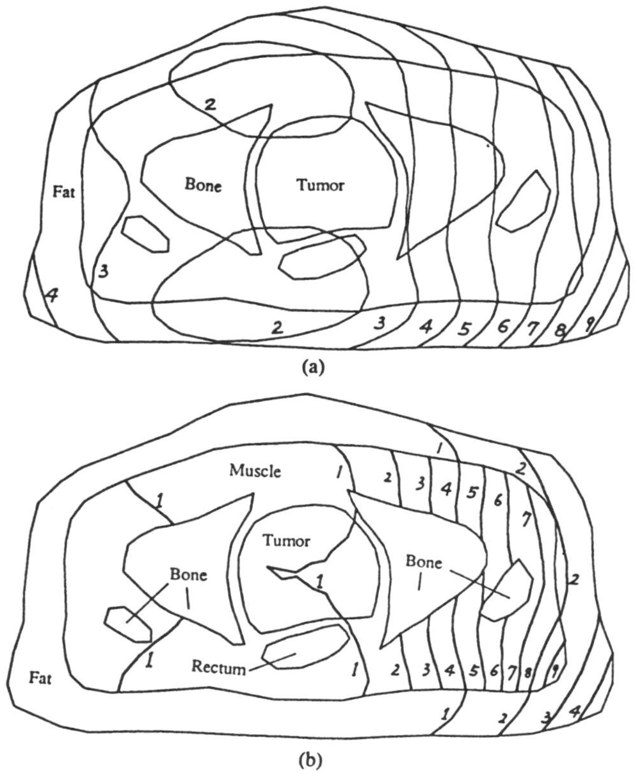

Fig. 12. The magnitude of the electric field and the absorbed power distributions inside the patients model defined in Fig. 12 under TM plane wave incidence ($\phi^i = 0°$). The scale is relative from 1 to 10. (a) Magnitude of the electric field distribution. (b) Absorbed power distribution.

TABLE II
TISSUE PROPERTIES AT 100 MHz

	Bone	Fat	Muscle	Rectum	Tumor
ϵ_r	10	10.5	72	52	72
σ	0.02	0.22	0.89	0.61	0.89
μ_r	1.0	1.0	1.0	1.0	1.0

Finally, Fig. 11 shows the finite element mesh of a two-dimensional ([13, pelvic model B]). The tissue properties of this model at 100 MHz are given in Table II. Fig. 12 shows the distributions of the magnitude of the electric field and the absorbed power under the TM plane wave illumination ($\phi^i = 0°$ and $f = 100$ MHz). The scale is relative from 1 to 10.

CONCLUSION

The method of moments and the finite element method each is a very powerful numerical technique and has found many practical applications. Unfortunately, both have their limitations. The formulation proposed in this paper successfully combines these two methods in a very straightforward manner, while utilizing both methods to their full capabilities. The method is especially suitable for problems involving inhomogeneous materials. Although there is no perfect conductor included here, the formulation can be easily extended to this limiting case, for instance, the dielectric coated with conductor or the conductor coated with dielectric layers. The formulation can even be applied to small aperture coupling problems since the interior solution does not involve the addition of the incident field and the scattered field. We have chosen the finite element method to solve the obtained boundary value problem. However, other numerical techniques, such as the finite difference method, may be used as well. To summarize, the hybrid MM/FEM method has the following advantages:

1) The formulation enables us to solve the interior and exterior problem separately. For the exterior electromagnetic fields, only the integration on the exterior surface is needed. The interior fields have been obtained naturally in the solution without requiring further efforts.

2) Since the interior and exterior problems are coupled only on the exterior surface, the finite element matrix needs to be computed and triangularized only once. Hence, little extra computations are needed if the incident waves change.

3) The versatility is developed in both method of moments and finite element method because on the one hand, the radiation properties are naturally included in the MM method (integral equation approach), and on the other hand the scattering problems involving inhomogeneous medium are easily and systematically handled by the FEM method (differential equation approach). Especially for three-dimensional prob-

lems, the banded and sparse finite element matrix will be computationally very economical.

The validity and accuracy of the solution were confirmed by the exact eigenfunction solution, the unimoment solution, and Richmond's pure method of moments solution. There is virtually no restriction on the shape or inhomogeneity of the scatter or absorber to be solved. Future efforts will be directed to the problem involving perfect conductors and arbitrarily shaped three-dimensional problems.

Finally, we should point out that the above formulation fails at resonant frequencies since at resonant frequencies a pure electric current is not sufficient to produce an arbitrary scattered field. Here, a resonant frequency is a frequency for which S, when covered by a perfect conductor and filled with the external medium, forms a resonant cavity. A magnetic current is also needed in the equivalent problem for region a [21]. Of course, the solution becomes more complex. A simple alternative is to define S as the surface that encloses the entire dielectric body, but does not necessarily coincide with the true dielectric surface. By changing S, we effectively change the resonant frequencies. A good indicator of the solution near resonances is the condition number of the moment matrix in (12).

References

[1] N. Morita, "Surface integral representation for electromagnetic scattering from dielectric cylinders," *IEEE Trans. Antennas Propagat.*, vol. AP-26, no. 2, pp. 261-266, Mar. 1978.

[2] J. H. Richmond, "Scattering by a dielectric cylinder of arbitrary cross section shape," *IEEE Trans. Antennas Propagat.*, vol. AP-13, no. 3, pp. 334-341, May 1965.

[3] D. E. Liversay and K. M. Chen, "Electromagnetic fields induced inside arbitrarily shaped biological bodies," *IEEE Trans. Microwave Theory Tech.*, vol. MTT-22, no. 12, pp. 1273-1280, Dec. 1974.

[4] M. J. Hagmann, O. P. Gandhi, and C. H. Durney, "Numerical calculation of electromagnetic energy deposition for a realistic model of man," *IEEE Trans. Microwave Theory Tech.*, vol. MTT-27, no. 12, pp. 804-809, Sept. 1979.

[5] D. H. Shaubert, D. R. Wilton, and A. W. Glisson, "A tetrahedral modeling method for electromagnetic scattering by arbitrarily shaped inhomogeneous dielectric bodies," *IEEE Trans. Antennas Propagat.*, vol. AP-32, no. 1, pp. 77-85, Jan. 1984.

[6] K. S. Yee, "Numerical solution of initial boundary value problems involving Maxwell's equations in isotropic media," *IEEE Trans. Antennas Propagat.*, vol. AP-14, pp. 302-307, May 1966.

[7] A. Taflove and M. E. Brodwin, "Computation of the electromagnetic fields and induced temperature within a model of the microwave irradiated human eye," *IEEE Trans. Microwave Theory Tech.*, vol. MTT-23, no. 12, pp. 888-896, Nov. 1975.

[8] K. K. Mei, "Unimoment method of solving antenna and scattering problems," *IEEE Trans. Antennas Propagat.*, vol. AP-22, no. 6, pp. 760-766, Nov. 1974.

[9] S. K. Chang and K. K. Mei, "Application of the unimoment method of electromagnetic scattering of dielectric cylinders," *IEEE Trans. Antennas Propagat.*, vol. AP-24, no. 1, pp. 35-42, Jan. 1976.

[10] P. P. Silvester and R. L. Ferrari, *Finite Elements For Electrical Engineers.* Cambridge, Cambridge Univ. Press, 1983.

[11] A. Taflove and K. Umashankar, "A hybrid moment method/finite-difference time-domain approach to electromagnetic coupling and aperture penetration into complex geometries," *IEEE Trans. Antennas Propagat.*, vol. AP-30, no. 4, pp. 617-627, July 1982.

[12] M. A. Morgan, C. H. Chen, S. C. Hill, and P. W. Barber, "Finite element-boundary integral formulation for electromagnetic scattering," *Wave Motion*, vol. 6, pp. 91-103, 1984.

[13] D. R. Lynch, K. D. Paulsen, and J. W. Strohbehn, "Hybrid element method for unbounded electromagnetic problems in hyperthermia," *Int. J. Numerical Methods in Eng.*, vol. 23, pp. 1915-1937, 1986.

[14] K. D. Paulsen, D. R. Lynch, and J. W. Strohbehn, "Three-dimensional finite, boundary, and hybrid elements solutions of the Maxwell equations for lossy dielectric media," *IEEE Trans. Microwave Theory Tech.*, vol. MTT-36, no. 4, pp. 682-693, April 1988.

[15] D. R. Lynch, K. D. Paulsen, and J. W. Strohbehn, "Finite element solution of Maxwell's equation for hyperthermia treatment planning," *J. Comput. Phys.*, vol. 58, no. 2, pp. 246-269, 1985.

[16] M. A. Morgan and B. E. Welch, "The field feedback formulation for electromagnetic scattering computations," *IEEE Trans. Antennas Propagat.*, vol. AP-34, pp. 1377-1382, Dec. 1986.

[17] R. F. Harrington, *Field Computation by Moment Methods.* New York, Macmillan, 1968.

[18] X. C. Yuan, R. F. Harrington, and S. S. Lee, "Electromagnetic scattering by a dielectric cylinder partially covered by conductors," *J. Electromagn. Waves and Appl.*, vol. 2, no. 1, pp. 21-44, Mar. 1987.

[19] X. C. Yuan, "Electromagnetic coupling into slotted TE and TM cylindrical conductors by the pseudo-image method," Ph.D. dissertation, Syracuse Univ., Syracuse, NY, Dec. 1987.

[20] R. F. Harrington, *Time-Harmonic Electromagnetic Fields.* New York, McGraw-Hill, 1961.

[21] J. R. Mautz and R. F. Harrington, "A combined-source solution for radiation and scattering from a perfectly conducting body," *IEEE Trans. Antennas Propagat.*, vol. AP-27, pp. 445-454, July 1979.

[22] J. H. Richmond, "TE-wave scattering by a dielectric cylinder of arbitrary cross section shape," *IEEE Trans. Antennas Propagat.*, vol. AP-14, no. 4, pp. 460-464, July 1966.

[23] J. Van Bladel, *Electromagnetic Fields.* New York: McGraw-Hill, 1964.

Validation of the Hybrid Quasi-Static/Full-Wave Method for Capacitively Loaded Thin-Wire Antennas

ROBERT G. OLSEN, SENIOR MEMBER, IEEE, AND PAUL D. MANNIKKO, STUDENT MEMBER, IEEE

Abstract—Recently, a hybrid method in which quasi-static and full-wave integral equations are combined was developed. The method is useful for electromagnetic scattering problems in which the geometry contains one or more electrically small and geometrically complex subregions which are "capacitive" in nature. Using point collocation, the hybrid method has now been implemented for axially symmetric problems containing thin wires in the full-wave region and conductors and/or dielectrics in the electrically small regions. The method has been successfully validated by comparing predicted input admittances using the hybrid method with measured results for capacitor-loaded antennas. The hybrid method is also shown to be a significant improvement over thin-wire codes in which the electrically small region is replaced with an equivalent lumped load. The improvement is especially apparent in cases for which a detailed knowledge of the field in the electrically small region is needed.

I. Introduction

RECENTLY, a hybrid method in which quasi-static and full-wave integral equations are combined was developed [1]. The method is useful for electromagnetic scattering problems in which the geometry contains one or more electrically small and geometrically complex subregions which are composed of perfect conductors and/or dielectrics and which are "capacitive" in nature. The method is particularly advantageous in problems for which a detailed knowledge of the fields within these regions is necessary. Using quasi-static methods in the electrically small regions, full-wave methods in the remaining regions and coupling the regions as described in [1] results in an improved solution for these problems. These improvements result from 1) improving numerical efficiency via a reduced quasi-static Green's function; 2) reducing the number of unknowns to a single scalar unknown within the electrically small regions rather than up to two vector unknowns if full-wave techniques are used; and 3) reducing the numerical difficulty associated with solving electrically small problems using full-wave techniques [2].

The hybrid method has now been implemented for axially symmetric problems containing thin wires in the full-wave region and conductors and/or dielectrics of arbitrary cross section in the electrically small regions. The purpose of this paper is 1) to present modifications to the theory to include multiple quasi-static regions; 2) to validate the method by comparing the numerical results with measurements; 3) to compare these results to the only competing theory which uses a lumped-load approximation for the electrically small region; and 4) to examine the validity of presently used techniques for calculating field details (or integrals of these fields such as voltages) in the quasi-static region.

Manuscript received November 2, 1988; revised March 30, 1989. This work was supported by the Office of Naval Research under Contract N00015-86-K-0612.

R. G. Olsen is with the Department of Electrical and Computer Engineering, Washington State University, Pullman, WA 99164-2752.

P. D. Mannikko was with the Department of Electrical and Computer Engineering, Washington State University, Pullman, WA. He is now with the Department of Electrical Engineering, Clemson University, Clemson, SC 29634.

IEEE Log Number 9034332.

II. The Hybrid Method

In the full-wave/quasi-static hybrid method presented in [1], the assumption is made that

$$|k_i^2 R \mathbf{J}_s| \ll |\nabla \cdot \mathbf{J}_s| \quad (1)$$

is satisfied on the conductors within the electrically small regions. In this inequality, k_i is the wavenumber of the ith dielectric within the region; R is the distance between source and field points within the region and $\mathbf{J}_s$ is the current density on the conducting surface. Vector quantities are indicated by bold print. Since, by definition, $k_i R \ll 1$ within the electrically small region, the inequality will be satisfied whenever $\mathbf{J}_s$ has "significant" spatial variation within the region. An example of a case where the divergence of $\mathbf{J}_s$ is known *a priori* to be significant is a boundary region on a thin conductor where the magnitude of the current is known to "drop" to zero. Since $\mathbf{J}_s$ is known to have significant spatial variation here, inequality (1) holds. In this case, electrostatic effects dominate; hence, magnetic induction may be ignored. For this reason, regions which satisfy inequality (1) are described as "capacitive."

The portion of Fig. 1 inside the dashed lines is an electrically small region containing both perfect conductors and linear, homogeneous, isotropic dielectrics. A permittivity $\epsilon_n \epsilon_0$ characterizes the nth dielectric, where ϵ_n is the relative permittivity of the dielectric and ϵ_0 is the permittivity of free space. Since electrostatic effects dominate within this region, the integral equation for charge density, valid for points $\mathbf{r}$ on the interface between dielectrics i and j (i.e., s_{ij}), is [1]

$$\frac{\epsilon_i + \epsilon_j}{2\epsilon_0} \rho_s(\mathbf{r}) = -(\epsilon_i - \epsilon_j)\mathbf{n}_i \cdot \mathbf{E}^{\text{inc}}(\mathbf{r}) - (\epsilon_i - \epsilon_j)\mathbf{n}_i/(4\pi\epsilon_0) \cdot \int_{\sum_m \sum_n s_{mn}} \rho_s(\mathbf{r}') \frac{\mathbf{R}}{R^3}\, ds'. \quad (2)$$

Reprinted from *IEEE Trans. Antennas Propagat.*, vol. 38, no. 4, pp. 516–522, April 1990.

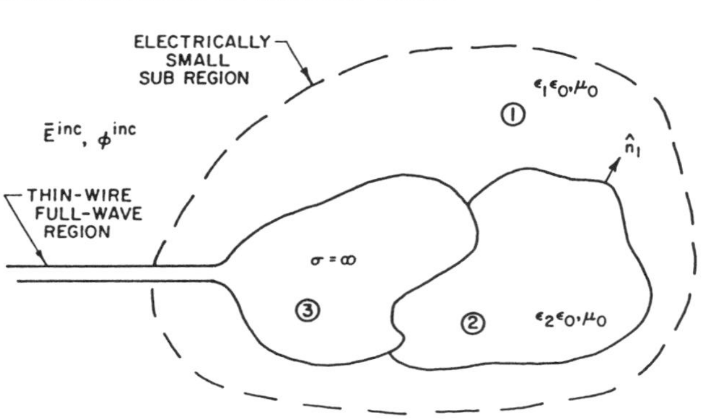

Fig. 1. Typical electrically small capacitive geometry.

Here $\sum_{mn}\sum s_{mn}$ represents the collection of all interfaces (dielectric–dielectric and dielectric–conductor) within the electrically small region. On the former set of interfaces, ρ_s represents the surface polarization charge density, while on the latter set, ρ_s represents free plus polarization charge. $\mathbf{n}_i$ represents the "inward" normal to the ith subregion. The vector $\mathbf{R} = \mathbf{r} - \mathbf{r}'$ is the vector from primed source coordinates to unprimed field coordinates while R is the magnitude of $\mathbf{R}$. The term $\mathbf{E}^{\text{inc}}(\mathbf{r})$ represents the electric field on s_{ij} from all sources in the full-wave region (i.e., from all sources outside the dashed lines in Fig. 1).

The integral equation valid for $\mathbf{r}$ on the surface of a conductor k is [1]

$$V_k = \phi^{\text{inc}}(\mathbf{r}) + \frac{1}{4\pi\epsilon_0}\int_{\sum_m\sum_n s_{mn}} \rho_s(\mathbf{r}')\frac{1}{R}\,ds'. \tag{3}$$

The term ϕ^{inc} represents the incident potential from sources in the full-wave region (i.e., outside the dashed lines). As shown in [1], ϕ^{inc} can be determined only to within a constant; however, this unknown constant may be absorbed into the unknown constant V_k of (3). Note that, in general, the constant associated with ϕ^{inc} may be different for different electrically small regions. Since (3) introduces an additional unknown constant V_k on every conductor within the region, an additional equation is necessary for each conductor. This equation is obtained by enforcing, on the conductor, the conservation of free charge condition [1]

$$\int_{S'} \rho_s(\mathbf{r})\,ds + \epsilon_i \int_{\sum_i s_{ik}} \rho_s(\mathbf{r})\,ds = \kappa. \tag{4}$$

The integral over S' represents all parts of the conductor k that are outside the electrically small region. The constant κ is a known charge on the conductor and is usually assumed to be zero. Note here that conductor k, in general, has portions inside and outside the electrically small region. The charge must be summed over both.

If the full-wave region contains strictly thin wires in an otherwise homogeneous medium, then the full-wave region may be analyzed using well-known thin-wire integral equations augmented with the tangential component of the electric field due to the equivalent charge-density within the electrically small regions. If the thin wires exist in free space and only on the z-axis, then the augmented thin-wire integral equation is [1], [3]

$$-\mathbf{E}_s(z)\cdot\Delta\mathbf{z} = \frac{-j\Delta\mathbf{z}\,\cdot}{4\pi\omega\epsilon_0}\left[k_0^2\int_{\text{wire}}\frac{e^{-jk_0R}}{R}\mathbf{I}(z')\,dz' + \hat{z}\frac{\partial}{\partial z}\int_{\text{wire}}\frac{e^{-jk_0R}}{R}\frac{\partial}{\partial z'}I(z')\,dz'\right] + \mathbf{E}^Q(\rho_s, z)\cdot\Delta\mathbf{z}. \tag{5}$$

In this equation, a suppressed time dependence $\exp\{j\omega t\}$ is assumed. $\Delta\mathbf{z}$ is a vector oriented along the z-axis with a magnitude determined by the segmentation scheme used in the moment method solution. k_0 is the free space wavenumber. $\mathbf{E}_s(z)$ is an impressed electric field used to model the driving source and exists only at the location of the source. The unknown of the integral equation is the total current $\mathbf{I}(z)$ along the thin wire and is related to the surface current density on the wire surface by $\mathbf{I}(z) = 2\pi a J_s(z)$, where a is the radius of the wire. $\mathbf{E}^Q(\rho_s, z)$ represents the electric field due to the charge density within the electrically small region and may be expressed as [1]

$$\mathbf{E}^Q(\rho_s, z) = -\frac{j\omega\mu_0}{4\pi}\int_{\sum_m\sum_n s_{mn}} J_s(\mathbf{r}')\frac{e^{-jk_0R}}{R}\,ds' - \frac{1}{4\pi\epsilon_0}\nabla\int_{\sum_m\sum_n s_{mn}} \rho_s(\mathbf{r}')\frac{e^{-jk_0R}}{R}\,ds'. \tag{6}$$

The surface current density $\mathbf{J}_s$ is determined by integrating the current continuity equation

$$\nabla\cdot\mathbf{J}_s = -j\omega\rho_s. \tag{7a}$$

On the thin wires, this continuity relation can be written as

$$\frac{\partial I(z)}{\partial z} = -j\omega\rho_l(z) \tag{7b}$$

where $\rho_l(z) = 2\pi a\rho_s(z)$ is the line charge density.

A geometry which satisfies the above-mentioned constraints and inequality (1) is the capacitor-loaded antenna of Fig. 2. This geometry consists of an axially symmetric thin-wire antenna loaded with two cylindrical parallel-plate capacitors with an arbitrary dielectric between the plates. Each capacitor represents a separate electrically small region, and each region contains two conducting plates. In the context of this problem, several comments are in order regarding the coupling terms between the full-wave and electrically small regions. First, the only specified unknown in the full-wave region is the wire current. Thus, rewriting (4) in terms of the wire currents in the full-wave region would be useful. Consider first the case for a conductor which has only one part in an electrically small region (e.g. the left-most conductor in Fig. 2 as opposed to the center conductor which contains the source and two electrically small parts). If (4) is multiplied by $j\omega$, then the first term of (4) can be written as

$$I_b = -j\omega\int_0^b \rho_l(z)\,dz \tag{8}$$

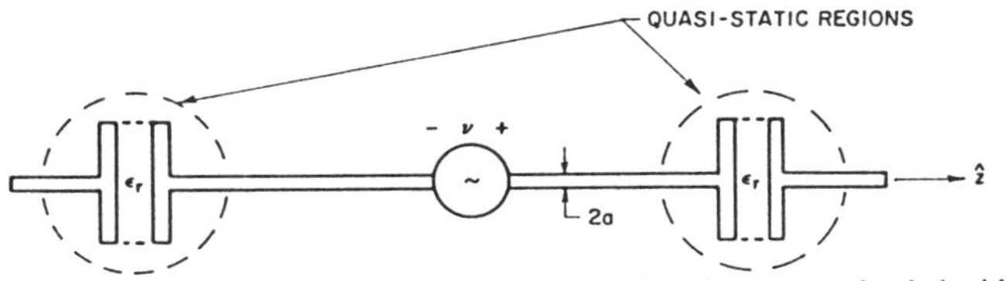

Fig. 2. Cross section of an axially symmetric thin-wire antenna loaded with electrically small capacitors.

which is recognized to be the integration of (7a) from the end of the wire ($z = 0$), where the current is zero, to the boundary between the full-wave region and the electrically small region ($z = b$). The use of (8) in (4) eliminates the need to explicitly define ρ_l in the full-wave region. Alternatively, the use of (8) in (4) can be recognized as the integral form of (7a) where the electrically small portion of the conductor represents a closed surface of integration.

Equations (8) and (4) can also be applied to the right-most conductor. However, consideration must be given to the application of (8) and (4) to the center conductor which contains two electrically small regions. As mentioned previously, since ϕ^{inc} is known only to within an unknown constant which has been absorbed into the unknown constant potential of the conductor V_k, and since this unknown constant may be different for different electrically small regions, two additional equations are necessary for the center conductor of Fig. 2, one for each capacitor plate. However, only one conservation of charge equation is available for this conductor. Equation (8) inserted in (4) can be applied separately to each electrically small region. In this case, the additional equation is interpreted as the integral form of the continuity equation. Applying the continuity equation to each electrically small portion of the center conductor ensures enough equations to determine all V_k.

A second comment can be made about the determination of currents within the electrically small regions for use in (6). The assumption of nonsingular fields on the axis of symmetry implies that $\mathbf{J}_s = 0$ at the intersection of the capacitor plates and the z-axis. Thus, $\mathbf{J}_s$ can be determined in terms of ρ_s by integrating (7a) within the electrically small region subject to the boundary condition at the axis of symmetry. In principle, this can be used to eliminate the need to define $\mathbf{J}_s$ as an explicit unknown within the electrically small region.

III. The Hybrid Code

The hybrid method previously described has been implemented [4] for geometries with axial symmetry. The integral equations are solved using the method of moments with point collocation [5]. The MiniNEC [3] algorithm is used to evaluate interaction within the thin-wire full-wave region, while new code couples the sources within the electrically small region to the thin-wire equations. In the electrically small region, the quasi-electrostatics code, AXIV3 [6], [7], is used to evaluate integral equations (2) and (3) while a newly developed code is used to determine effects of the coupling terms $\mathbf{E}^{\text{inc}}$ and ϕ^{inc} as functions of the currents on the thin wires.

AXIV3 is capable of evaluating axially symmetric conductor and dielectric configurations with arbitrary cross-sectional geometries; however, when coupled with the full-wave problem as is done in the new hybrid program, limitations are placed on the geometry. In particular, any body, whether conducting or dielectric, must contain the axis of symmetry. This requirement allows the use of the boundary condition that the current must be zero on the axis of symmetry and, as mentioned before, is useful when defining currents in the electrically small region for the purpose of finding the incident electric field in the full-wave region. Equations for surfaces which do not have such a boundary condition have not been developed.

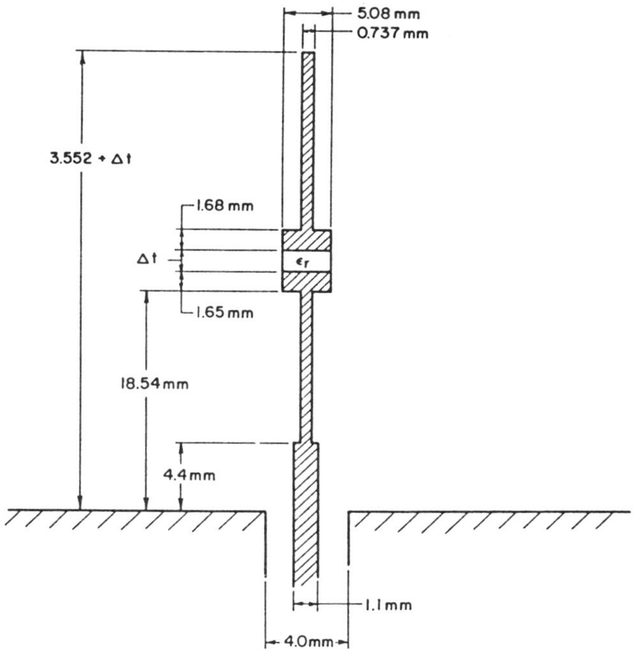

Fig. 3. Capacitor-loaded thin-wire monopole antenna.

IV. Measurements and Models of a Capacitor-Loaded Antenna

A. Experimental Configuration

To validate the hybrid technique, the antenna of Fig. 3 was constructed. Specifically, the measured input admittances of the antenna were compared with the predicted input admittances generated by the hybrid code. Fig. 3 shows the configuration used in the experiment. The antenna was base-driven with a coaxial line through the center of a square aluminum ground sheet with dimensions $6\lambda \times 6\lambda$ at 2 GHz. Meier and Summers [8] have shown that a ground plane of this size will allow reasonable comparison ($\pm 2\ \Omega$) between the measured input impedance of a resonant length antenna and the theory based on an infinite ground plane. Input admittance measurements were made with a Hewlett-Packard 8753A network analyzer fully calibrated for one port measurements. The measurement system is automated through an interface between the network analyzer and an IBM PC AT.

B. Calibration of the Source Model

Whenever experimentally determined admittances are compared with those obtained through an analytical procedure, some consideration should be given to the error in the input susceptance created by the source model and the effect

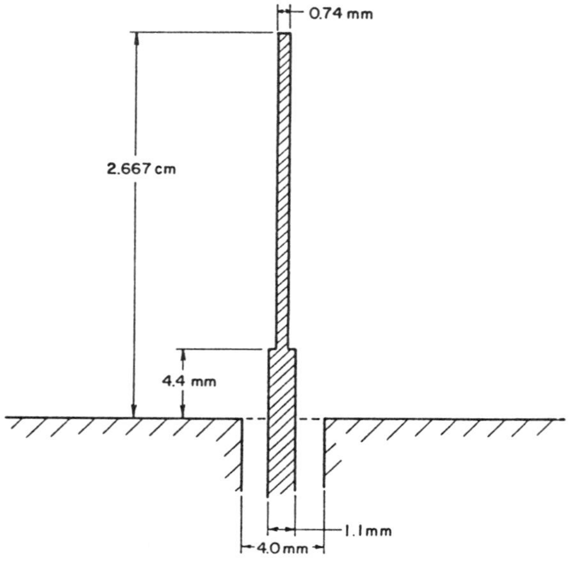

Fig. 4. Thin-wire monopole antenna with radius $a = 0.368$ mm.

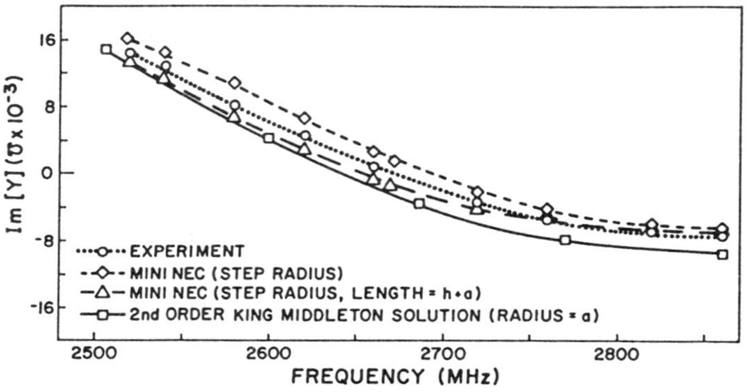

Fig. 5. Input susceptance of the monopole antenna in Fig. 4.

of the thin-wire end model. Thus, the adequacy of the source and wire end models used in the hybrid method was verified by using the full-wave thin-wire algorithm (i.e., MiniNEC) to predict the input admittance of a monopole antenna and by comparing the result with experiment and an alternative theory based on entirely different principles. Use of only the full-wave algorithm to verify the models is sufficient since both the source and the thin-wire ends are assumed to exist in the full-wave region.

The monopole antenna of Fig. 4 was constructed, measured, and modeled numerically. Compared in Fig. 5 are curves of input susceptance as measured, as calculated using the thin-wire algorithm with a source length of 1.71 wire radii and two different wire lengths, and as calculated by King [9, p. 156] using the King–Middleton second-order solution. Convergence of the moment method thin-wire results was established by increasing the number of current "pulses" until the input admittance results stabilized.

There are three reasons that measurements and theory may not coincide. First, the model used in the King–Middleton method uses a delta gap source model, which causes the input susceptance to be infinite [10]. The second-order solution does not exhibit this infinite susceptance, however, because it is an approximate method. This does not necessarily mean that the second-order method will agree with experiment—only that its predicted input susceptance is not infinite. Nevertheless, King points out that there is reasonable agreement between the second-order method and monopole measurements if the radial dimensions of the coaxial feed are electrically small [9, p. 214]. The source model used in MiniNEC is a finite length voltage pulse and is also different from the actual source in the monopole measurements. MiniNEC also results in a finite input susceptance, but again, cannot be assumed *a priori* to give the same result as the monopole measurement [11]. According to Elliot, compensation for the difference between the actual source and the theoretical model source can be made by determining an empirical correction factor from a careful set of measurements [11]. This is the approach used here.

A second reason that measurements and theory may not coincide is that the assumed uniform transmission line parameters of the coaxial feed are not uniform near the transmission line/antenna junction. A correction term has been developed to account for this discrepancy [9, pp. 193–215], [12]. However, these correction terms were small for the parameters used in the experiment reported here.

A third reason for disagreement between measurement and theory is the adequacy of the model for thin-wire ends. In MiniNEC, the ends are modeled as hollow tubes while in the experiment, the wires are actually solid. King [9, p. 70] states that there is an increase in the effective length of the antenna due to the discrepancy in end models; however, the amount of increase "is difficult to determine accurately, but . . . is of the order of magnitude of a [wire radius]."

In light of the above comments, the results of Fig. 5 can now be interpreted. The King–Middleton second-order theory for a monopole with radius $a = 0.368$ mm is in reasonable agreement with experimental results despite the fact that the step change in radius shown in Fig. 4 was not modeled. However, no further conclusions can be drawn from this comparison because of the difference in antenna models. Nevertheless, the reasonable agreement with the King–Middleton theory lends confidence to the experimental results. The same monopole antenna was modeled using the MiniNEC algorithm. In this case, the step radius was modeled. To study the effect of solid wire ends rather than hollow tubes as modeled by MiniNEC, two cases were examined. In the first, the length of the monopole was h while in the second the length of the monopole was $h + a$. Extending the length by one radius demonstrates the increase in effective length explained by King. The two results bracket the experimental results and an appropriate effective length of approximately $h + a/3$ can be inferred from the graph. Comparisons made of the real part of the input admittance suggest the same effective length.

Based on the results discussed above, the decision was made that antennas with a source region as specified for the monopole antenna could be compared to measurements with no correction term added provided that the length of a wire is augmented by $a/3$.

C. Results

The antenna of Fig. 3 was modeled over a band of frequencies which simultaneously preserve the electrically small

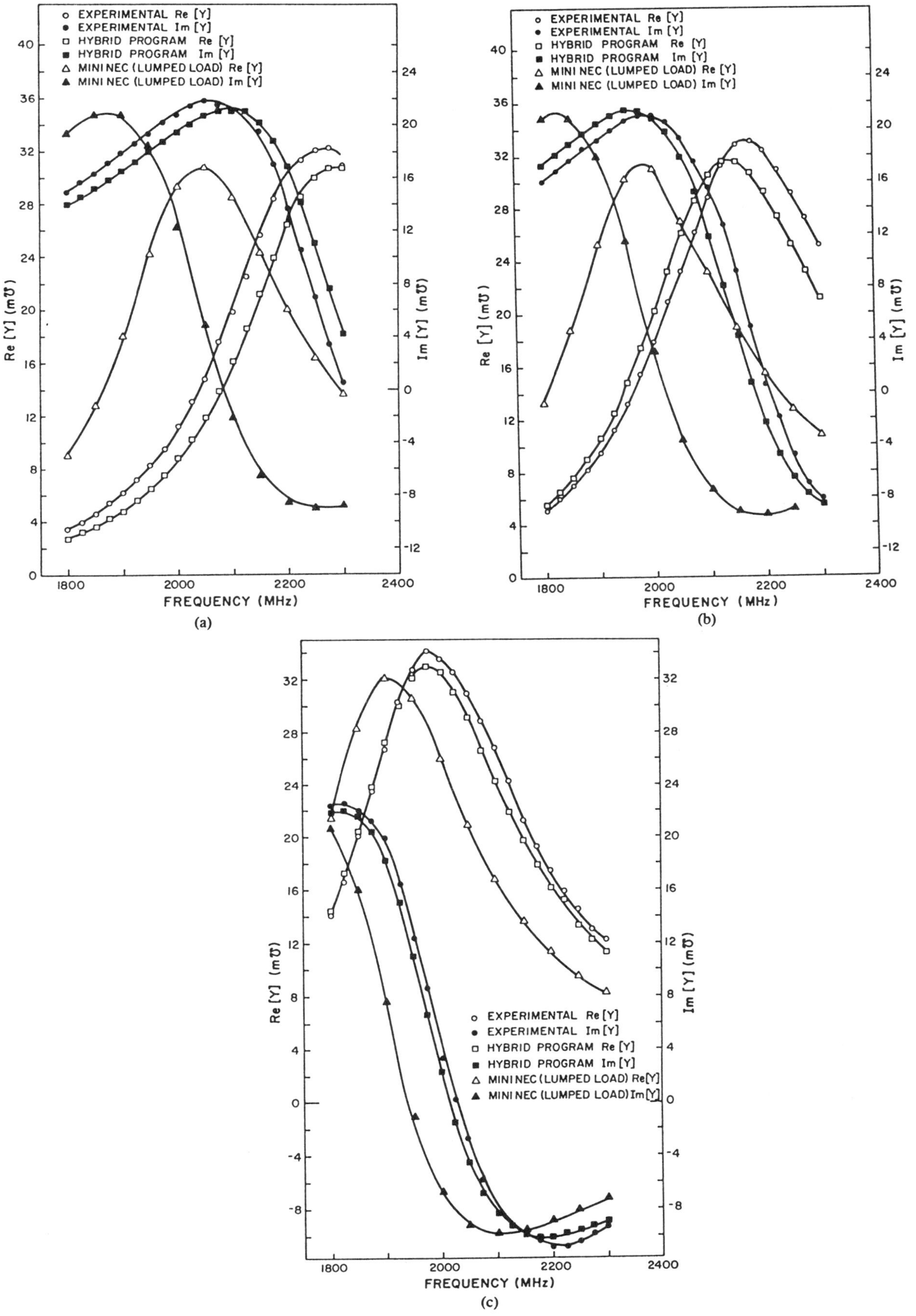

Fig. 6. Input admittance of the capacitor loaded antenna in Fig. 3 for (a) $\epsilon_r = 1$ $\Delta t = 1.31$ mm; lumped load = 0.290 *pF*. (*b*) $\epsilon_r = 2.23$ $\Delta t = 1.47$ mm; lumped load = 0.447 pF. (c) $\epsilon_r = 5.85$ $\Delta t = 1.47$ mm; lumped load = 0.896 pF.

nature of the capacitor region and the adequacy of the finite ground plane. The geometry between $z = 1.774$ cm and $z = 2.409$ cm was modeled quasi-statically. Input admittances, as determined by the hybrid program and as measured are plotted versus frequency in Fig. 6 for relative dielectric constants of 1 (air), 2.23 (duroid), and 5.85 (macor) [13], [14]. Convergence of the hybrid program was ensured by increasing the number of unknowns in the electrically small region until the input admittance stabilized. The number of unknowns used to generate the data reported here was 36. Finally the input admittance, as determined by MiniNEC, of a monopole antenna with a lumped load is shown in the plot. In each case, the capacitance of the lumped-load was determined from the geometry of the entire electrically small region using the electrostatics code AXIV4 [15]. AXIV4 solves the electrostatic integral equations using a Galerkin method (rather than point collocation as used in AXIV3). Calculating the total free charge on all positively charged conductors in the electrically small geometry, and dividing by the electrostatic potential drop across the plates gives the lumped capacitance. It should be noted that the contribution of charge on the wires to this capacitance is negligible but that all sides of the plates make significant contributions. The loaded antenna is modeled in miniNEC by defining a monopole which has a radius equal to the wire radius except in the load region (i.e., length of the capacitor) where it is equal to the radius of the capacitor plates. A lumped capacitive impedance is then placed at the center of the load region. It should be noted that attempts were made to model the load region with a radius equal to the wire radius but the larger radius model proved to be superior in all cases.

It is clear that the new hybrid method follows the actual input admittance much more closely than the lumped-load approximation. The difference between the hybrid method and the experimental admittance, appears as a small frequency shift which can, in part, be explained by an uncertainty in the dielectric constant of the capacitor. This uncertainty results in a change in the effective length of the antenna and, hence, a frequency shift in the input admittance curves. Other uncertainties include the finite ground plane, the finite conductivity of the antenna wire, the end effects of the thin wires, and the effects of the source model.

In some cases, knowledge of the details of the fields within the electrically small region or integrals of these fields is important. For example, a VLF antenna supported by guy wires has insulators on the guy wires in order to reduce induced currents [16]. These insulators are electrically small, capacitive, and geometrically complex and, hence, are a candidate for the hybrid analysis method discussed here. One important parameter in the design of these insulators is the voltage across the insulator. A knowledge of the voltage across the insulator can be used to predict corona or flash-over effects. The voltage is defined as an integral of the field in this electrically small region. Since the only method now available for predicting this voltage is the lumped-load approach, a comparison was made between the lumped-load method and the hybrid method. In Fig. 7, the magnitudes of the potential drop across the capacitor plates as predicted by the hybrid method and by the lumped-load approximation are shown as a function of frequency. The results for small dielectric constant ($\epsilon_r = 1$) suggest that the lumped-load approximation predicts a much lower voltage drop at some frequencies than that predicted by the hybrid method. However, as the dielectric constant increases, the discrepancy is less. This result is expected since the larger the dielectric constant, the more the fields are contained between the parallel plates, and, hence, the larger the capacitance between the plates. If the frequency is high enough, the capacitor plates will effectively become a short circuit. Therefore, in both methods, the voltage drop converges to zero as the dielectric constant increases while a constant frequency is maintained.

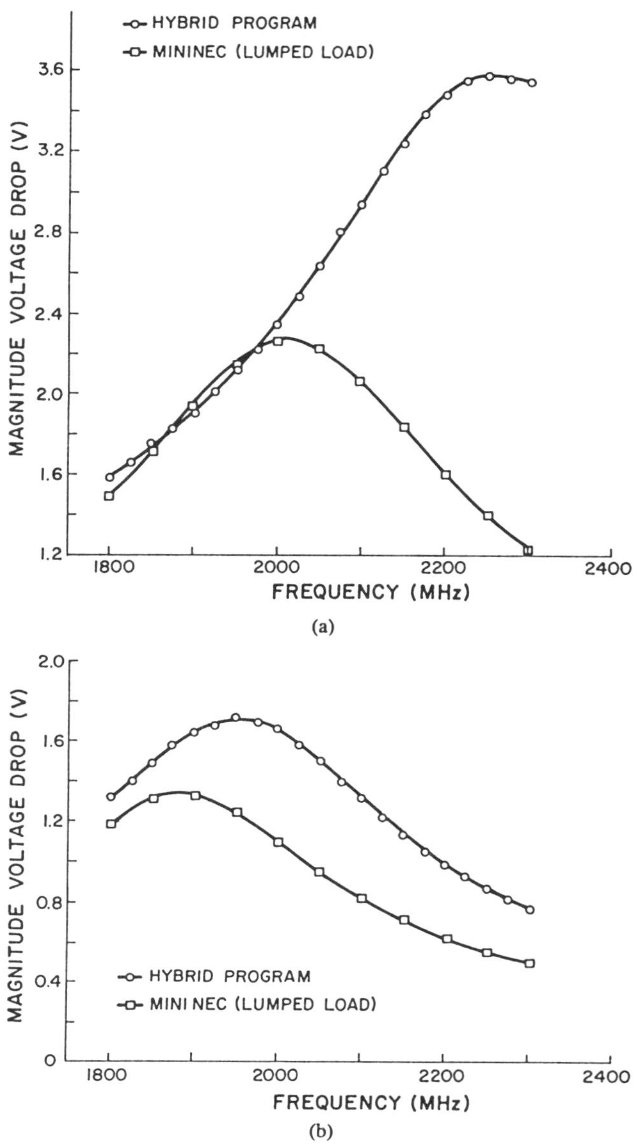

Fig. 7. Voltage drop across capacitor plates in Fig. 3 for (a) $\epsilon_r = 1$; lumped load = 0.290 pF. (b) $\epsilon_r = 5.85$; lumped load = 0.896 pF.

V. Conclusion

The hybrid method for electromagnetic scattering problems where quasi-static methods may be used to evaluate, in detail, the fields of the capacitive electrically small region has been successfully implemented for axially symmetric geometries. Comparing predicted input admittances of the hybrid method with measured results for capacitor-loaded antennas shows excellent agreement. The hybrid method is a significant improvement over lumped-over techniques in these predictions. Comparisons between the hybrid technique and lumped load approximations show considerable disagreement when predicting the voltage drop across the capacitor plates for low dielectric constant capacitors. The above result suggests that the lumped-load approximation may not be sufficient for problems in which a prediction of this voltage drop is consequential.

References

[1] R. G. Olsen, G. L. Hower, and P. D. Mannikko, "A hybrid method for combining quasi-static and full-wave techniques for electromagnetic scattering problems," *IEEE Trans. Antennas Propagat.*, vol. 36, pp. 1180–1184, Aug. 1988.

[2] G. J. Burke and A. J. Poggio, "Numerical Electromagnetics Code (NEC)—Method of moments part III: User's guide," Lawrence Livermore Nat. Lab., Tech. Document UCID-18834, Jan. 1981, pp. 2, 3.

[3] J. C. Logan and J. W. Rockway, "The new MiniNEC (Version 3): A mini-numerical electromagnetics code," Naval Ocean Syst. Center, San Diego, CA, NOSC Tech. Document 938, Sept. 1986.

[4] P. D. Mannikko, "Implementation of a full-wave/quasi-static hybrid method for analysis of axially symmetric thin-wire antennas with capacitive loads," M.S. thesis, Washington State Univ., Pullman, 1988.

[5] A. J. Poggio and E. K. Miller, "Integral equation solutions of three-dimensional scattering problems," in *Computer Techniques for Electromagnetics*, R. Mittra, Ed. Elmsford, NY: Pergamon, 1973, ch. 4.

[6] R. G. Olsen and G. E. Roberts, "Boundary element techniques for quasielectrostatics problems which contain thin conductive coatings on dielectric interfaces," Final Rep. to Bonneville Power Admin. for Contract DE-AC79-838P39831, June 1984.

[7] J. Daffe and R. G. Olsen, "An integral equation technique for solving rotationally symmetric electrostatic problems in conducting and dielectric material," *IEEE Trans. Power Appl. Syst.*, vol. PAS-98, no. 5, pp. 1609–1616, Sept.-Oct. 1979.

[8] A. S. Meier and W. P. Summers, "Measured impedance of vertical antennas over finite ground planes," *Proc. IRE*, pp. 609–616, June 1949.

[9] R. W. P. King, *The Theory of Linear Antennas.* Cambridge, MA: Harvard Univ. Press, 1956, pp. 944.

[10] ——, "The linear antenna—Eighty years of progress," *Proc. IEEE*, vol. 55, pp. 1–16, Jan. 1967.

[11] R. S. Elliot, *Antenna Theory and Design.* Englewood Cliffs, NJ: Prentice-Hall, 1981, pp. 314–320.

[12] R. W. P. King and C. W. Harrison, Jr., *Antennas and Waves: A Modern Approach.* Cambridge, MA: M.I.T. Press, 1969, pp. 225–231.

[13] G. Gonzalez, *Microwave Transistor Amplifiers Analysis and Design.* Englewood Cliffs, NJ: Prentice-Hall, 1984, p. 76.

[14] Macor Corning Machinable Glass Ceramic, Data Bulletin MDS-2, Duramic Products, Inc., Palisades Park, NJ.

[15] R. G. Olsen and O. Einarsson, "Boundary element methods for weakly three dimensional quasi-electrostatic problems," *IEEE Trans. Power Del.*, vol. PWRD-2, no. 4, pp. 1276–84, Oct. 1987.

[16] K. T. Huang and J. C. Squier, "Voltage distribution on guyline insulators in a VLF antenna system," Civil Eng. Lab., Naval Construction Battalion Center, Port Hueneme, CA, TM M-62-79-07, Jan. 1979.

Part 7
Numerical Methods

VARIOUS specialized numerical procedures can be important in making MM models more useful and computationally affordable. Some of the issues discussed in this section include the effects of using various basis and testing functions, performing numerical integration more efficiently and accurately in computing the MM matrix, iteration as a way to solve an MM matrix more efficiently, and the kinds of problems that arise in solving linear systems with increasing numbers of unknowns.

Miller and Burke, in ''Numerical integration methods,'' discuss the kind of integrals needing numerical evaluation in the context of wire modeling, and compare several different procedures, one being adaptive Romberg quadrature, which is presently used in codes such as NEC. The paper ''Analysis of various numerical techniques applied to thin-wire scatterers,'' by Butler and Wilton, describes several basis and weighting functions, their effects on the convergence of thin-wire MM solutions, and the equivalence that arises in the numerical model when interchanging basis and weight functions. Ferguson *et al.* describe one of the first iterative applications to EM MM modeling, using what amounts to banded-matrix preconditioning in ''Efficient solution of large moments problems: Theory and small problem results.'' This paper also explores how the numbering scheme affects the utility of the banded-matrix solution. A survey and analysis of the most common direct and iterative methods for solving the matrix equations that arise in MM solutions, including the conjugate-gradient method, is discussed by Sarkar *et al.* in ''Survey of numerical methods for solution of large systems of linear equations for electromagnetic field problems.'' The paper by Ludwig, ''A comparison of spherical wave boundary value matching versus integral equation scattering solutions for a perfectly conducting body,'' is included as an example of a quite different modeling approach that leads to a matrix (as does an integral-equation model), the computation of which requires no source integration, but rather summing over spherical-wave expansions. The concluding two papers address the common theme of reducing the number of frequency samples needed when using a frequency-domain model to obtain a response over some specified bandwidth. In ''Generation of Wide-Band data from the method of moments by interpolating the impedance matrix,'' Newman uses polynomial interpolation between frequencies at which the MM impedance matrix is explicitly computed to obtain an approximate impedance matrix which is then solved in the usual fashion. Burke *et al.*, in ''Using model-based parameter estimation to increase the efficiency of computing electromagnetic transfer functions,'' describe an alternate method to reduce the number of frequencies at which the impedance and admittance matrices must be explicitly computed. Their approach involves fitting to the coefficients of the inverse, or admittance, matrix a rational function, a physically motivated, analytical model that provides the interpolation function between the MM evaluations.

Numerical Integration Methods

E. K. Miller G. J. Burke
MB Associates, San Ramon, CA 94583

Abstract—**The numerical evaluation of an integral is a frequently encountered problem in electromagnetic theory. This is particularly the case when the current distribution or near field of a scatterer or antenna is required, in which case the far-field approximation cannot be employed. A comparative study of several numerical quadrature techniques is presented. The superiority of a variable interval-width technique based on Romberg's method will be demonstrated for an integral typical of those which arise in scattering and radiation from thin-wire structures.**

Theoretical Approach

Consider the integral whose value is required to find the current induced on a thin-wire ring by an electromagnetic wave incident along the axis of the ring. The integral is

$$F = \int_{-\pi}^{0} f(\phi)\, d\phi \tag{1a}$$

where

$$f(\phi) = \left\{ \frac{\exp(-ikR)}{kR} \left[1 - \sin^2\phi \left(1 + \frac{ikR + 1}{k^2R^2} \right) \right] \right\} \tag{1b}$$

$$R = [s^2 + 2a(s + a)(1 - \cos\phi)]^{1/2}.$$

The ring radius is a, the wire radius s, and k and η denote, respectively, the free-space wavenumber $k = \omega\sqrt{\mu\epsilon}$ and free-space impedance $\eta = \sqrt{\mu/\epsilon}$. The real part of the integrand $f(\phi)$ is seen to be nearly singular at $\phi = 0$ and has in addition a triplet behavior characteristic of the second derivative of a delta function. The behavior of the integrand makes the numerical integration more difficult and, as will be demonstrated below, the efficiency and accuracy of the calculation is very much dependent upon the numerical method used.

Numerical integration (or quadrature) is basically accomplished by approximating an integrand piecewise by polynomials and integrating these exactly. We will restrict our consideration here to quadrature formulas using equally spaced abscissa points within each interval used. Such formulas are known as Newton–Cotes quadrature formulas, special cases of which are the trapezoidal rule and Simpson's (or the parabolic) rule. A particular advantage of using equal abscissa point spacing over each interval is that in any iteration, increase of order or change of interval size, previously calculated integrand values can be used again. This would not otherwise be the case using, for example, Gaussian quadrature.

We may then write a general expression for the integration of $f(x)$ over the integration interval a,b as

$$I = \int_a^b f(x)\, dx \approx \sum_{i=1}^{M} H_i \sum_{j=0}^{N_i} f_{ij} w_j \tag{2}$$

where

$$H_i = [b_i - a_i]/[\sum_{i=0}^{N_i} w_i]$$

$$f_{ij} = f(a_i + jh_i), \quad h_i = (b_i - a_i)/N_i$$

and the w_j are the weight coefficients determined by the quadrature method used. The M intervals a_i,b_i may have varying widths and are ideally chosen to minimize the total number of integrand evaluations required to achieve a desired integration accuracy. The crucial feature of the numerical approach is the development of a meaningful and sensitive test to determine both the optimum interval width and the abscissa point spacing. We will consider two methods which provide built-in tests for this purpose: the parabolic extrapolation technique and the Romberg quadrature.

Parabolic extrapolation (PE) involves applying a three-point parabolic rule formula ($N = 2$, $w_0 = w_2 = 1$, $w_1 = 4$) over the interval a_i,b_i and calculating the quantity

$$f_{i,e} = f_{i,0} - 3f_{i,1} + 3f_{i,2}$$

which is the parabolic extrapolation of f_{ij} to the abscissa $a_i + 3h_i$. The relative difference between $f_{i,e}$ and $f_{i,3}$ is compared to a test ratio T_R and if

$$|(f_{i,e} - f_{i,3})/f_{i,3}| \le T_R \tag{3}$$

is satisfied, the computation of the integral over interval a_i,b_i is considered to be acceptably accurate and the $(i+1)$th interval may be doubled in size relative to the ith interval. On the other hand, if (3) is not satisfied, the ith interval is halved, and the computation is repeated over the new interval. It is apparent that for (3) to be a meaningful and useful test of the integration accuracy, the numerical accuracy actually achieved should be on the order of T_R.

The Romberg integration is based on applying the Richardson extrapolation to successive trapezoidal rule answers over a given interval in which the number of integrand evaluations is doubled for each successive trapezoidal rule answer. The basic result of the extrapolation procedure is to suppress the order-of-the-error term in the trapezoidal rule answers to obtain a more accurate answer without calculating any more integrand values [1], [3]. If we let

$$T_{0,k}^{(i)} = \frac{b_i - a_i}{2^k} [\tfrac{1}{2} f_{i,0} + f_{i,1} + \cdots + f_{i,2^k - 1} + \tfrac{1}{2} f_{i,2^k}] \tag{4}$$

denote the trapezoidal rule answer using 2^k subintervals on a_i,b_i, the Romberg answer $T_{k,0}^{(i)}$ is obtained from an iterative procedure using successive $T_{0,k}^{(i)}$ answers in

$$T_{m,n}^{(i)} = \frac{1}{4^m - 1} [4^m T_{m-1,n+1}^{(i)} - T_{m-1,n}^{(i)}]. \tag{5}$$

The $T_{m,n}^{(i)}$ are conveniently arranged in a triangular matrix of the form

$$\begin{array}{lll} T_{0,0}^{(i)} & & \\ T_{0,1}^{(i)} & T_{1,0}^{(i)} & \\ T_{0,2}^{(i)} & T_{1,1}^{(i)} & T_{2,0}^{(i)} \\ \cdots & \cdots & \cdots \end{array} \tag{6}$$

where it may be seen that the trapezoidal rule answers are contained in the first vertical column and the Romberg answers lie along the diagonal. Convergence to increasingly more accurate answers takes place down the first column and the diagonal as well as towards the right along the rows connecting them. The row convergence generally provides a more realistic indication of error magnitude than two successive trapezoidal rules or Romberg answers, and this indication of error can be used to determine the interval spacing in a fashion similar to that used for parabolic extrapolation. We then determine whether or not the inequality

$$|(T_{k,0}^{(i)} - T_{k-1,1}^{(i)})/T_{k,0}^{(i)}| \le T_R \tag{7}$$

is satisfied. If inequality (7) is satisfied with $k = 1$ or 2, then the $(i+1)$th interval is doubled. If it is not satisfied with $k = 2$, then the interval is halved, and the sequence of calculations is repeated. We allow a maximum of five integrand evaluations per interval (including endpoints) since experience has shown that when a sharp integrand peak is bracketed by a_i,b_i the Romberg answers may

Manuscript received February 17, 1969.

Reprinted from *IEEE Trans. Antennas Propagat.*, vol. AP-17, no. 5, pp. 669–672, September 1969.

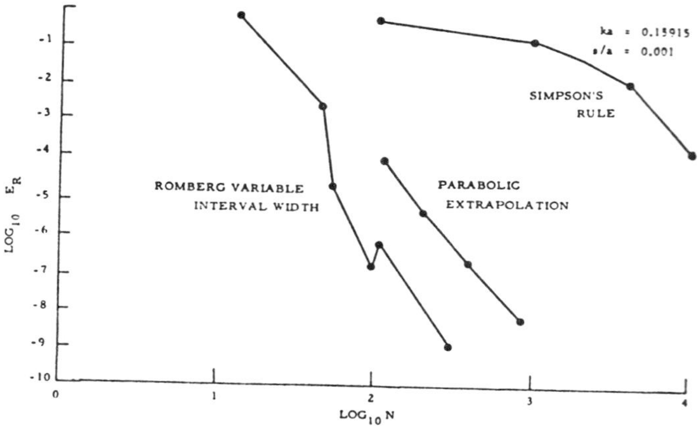

Fig. 1. Comparison of Romberg variable interval width, parabolic extrapolation, and Simpson's rule integration methods.

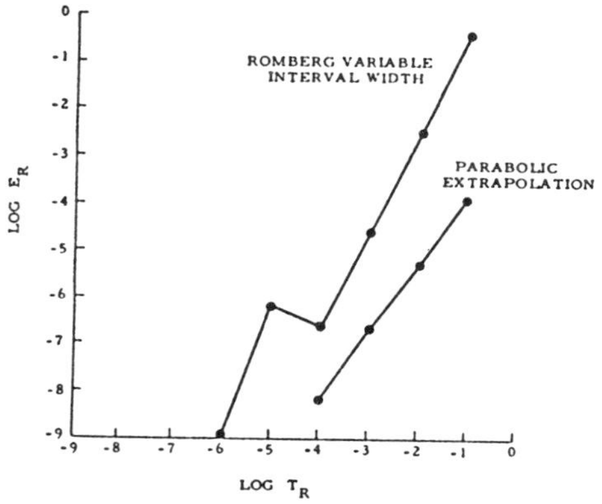

Fig. 2. Relative error as a function of test ratio for Romberg variable interval width and parabolic extrapolation integration.

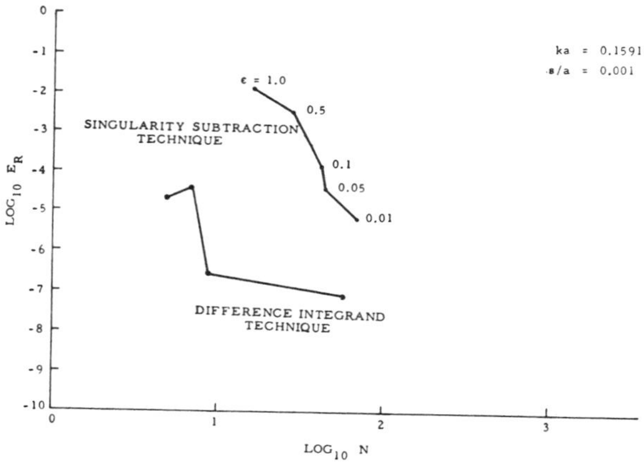

Fig. 3. Relative accuracy as a function of the number of integrand evaluations for Romberg variable interval-width integration.

oscillate and converge more slowly than the trapezoidal rule answers. In addition, an excessive abscissa point density is consequently required outside the peak.

Numerical Results

The two variable interval-width methods described above were applied to the evaluation of the integral in (1) for $ka = 0.15915$ and $s/a = 0.001$. The relative accuracy E_R obtained from applying these two methods to the real part of (1) is shown in Fig. 1 as a function of the number of abscissa evaluations N. (E_R is defined as the absolute value of the difference between the correct answer and the calculated value divided by the correct value. The correct value is the answer obtained with the smallest test ratio.) The Romberg variable interval-width (RVIW) technique can yield the same accuracy as the PE method while using less than half the number of integrand evaluations for $E_R < 10^{-5}$. In order to emphasize the advantage of using a variable interval-width also included on Fig. 1 is the result of evaluating the integral using the parabolic rule and constant abscissa spacing over the interval. The superiority of a variable interval width (and abscissa spacing) over fixed spacing is clearly demonstrated.

A further advantage of the RVIW method over that of the PE method is shown in Fig. 2, where the relative errors for the results of Fig. 1 are plotted as a function of the test ratio T_R. The accuracy actually attained E_R is closer to the specified accuracy T_R for the RVIW technique. A close correlation between T_R and E_R is desirable since E_R is generally proportional to the number of integrand evaluations N. Consequently, a value of E_R less than T_R indicates that N is larger than that required to have E_R just equal T_R. The most efficient quadrature technique, other considerations being equal, is then that method which minimizes the difference between E_R and T_R.

The numerical evaluation of the integral in (1) may be efficiently carried out using a technique which removes the integrand peak by subtracting an analytically integrable function from the actual integrand to cancel the peak at $\phi = 0$ [4]. An appropriate function for this difference integrand technique is

$$f_A(\phi) = -i\left[1 + i\left(\frac{1}{X} - \frac{\phi^2}{X^3}\right)\right] \tag{8}$$

where

$$X = k[s^2 + a(a + s)\phi^2]^{1/2}$$

which is derived by approximating the trigonometric functions by the first term in their Taylor series expansion about $\phi = 0$. The value of (1) may be written then as

$$F = \int_{-\pi}^{0} [f(\phi) - f_A(\phi)]\, d\phi + F_A(\pi) \tag{9}$$

where

$$\begin{aligned} F_A(\theta) &= \int_{-\theta}^{0} f_A(\phi)\, d\phi \\ &= i\left\{\theta + i\left[\frac{1}{c^3}\left(\frac{\theta}{\sqrt{b^2 + \theta^2}}\right)\right.\right. \\ &\quad \left.\left. + \left(\frac{1}{c^2} - 1\right)\frac{1}{c}\ln\left(\frac{b}{\theta + \sqrt{b^2 + \theta^2}}\right)\right]\right\}, \end{aligned}$$

$$b^2 = \frac{(s/a)^2}{1 + (s/a)}, \qquad c = ka\left(1 + \frac{s}{a}\right)^{1/2}. \tag{10}$$

An alternative to (9) is

$$F = \int_{-\pi}^{-\epsilon} f(\phi)\, d\phi + F_A(\epsilon) \tag{11}$$

where we thus avoid numerically integrating over the peak in $f(\phi)$ by analytically integrating the approximation to the actual integrand. We refer to this as the singularity subtraction technique.

The result of using these two techniques with RVIW quadrature to evaluate (1) is shown in Fig. 3. (The PE technique is not shown since the relative difference between it and the Romberg method is similar to that already shown in Fig. 1.) We observe from a comparison of Figs. 1 and 3 that the singularity subtraction technique together with RVIW quadrature provide comparable accuracy to that obtained from PE integration of the actual integrand. The use of RVIW quadrature with the difference integrand method provides

TABLE I

Test Ratio	10 Segments		18 Segments		30 Segments		54 Segments	
	σ/λ^2	E_R	σ/λ^2	E_R	σ/λ^2	E_R	σ/λ^2	E_R
10^{-1}	0.803917108	1×10^{-3}	0.786317177	1×10^{-3}	0.772981040	1×10^{-3}	0.762269368	7×10^{-4}
10^{-2}	0.802959179	7×10^{-5}	0.785241014	7×10^{-5}	0.771974994	6×10^{-5}	0.761750617	1×10^{-5}
10^{-3}	0.802952836	6×10^{-5}	0.785231810	6×10^{-5}	0.771964469	4×10^{-5}	0.761738698	3×10^{-6}
10^{-4}	0.802904694	6×10^{-7}	0.785184329	3×10^{-6}	0.771931160	2×10^{-6}	0.761741102	5×10^{-8}
10^{-5}	0.802904235	2×10^{-8}	0.785182023	7×10^{-8}	0.771929752	9×10^{-8}	0.761741200	6×10^{-8}
10^{-6}	0.802904220	1×10^{-9}	0.785181974	4×10^{-9}	0.771929718	5×10^{-9}	0.761741136	1×10^{-8}
10^{-7}	0.802904219		0.785181972	1×10^{-9}	0.771929714		0.761741139	1×10^{-8}
10^{-8}	0.802904219		0.785181971		0.771929714		0.761741147	

better accuracy for a given N than any other method to about 10^{-7} accuracy. For a higher accuracy than 10^{-7} the RVIW method applied to that actual integrand provides the greatest accuracy principally because of the error which arises from the subtraction of two large numbers using the difference integrand method.

The calculated backscatter radar cross section (RCS) of a half-wave dipole ($kh = \pi/2$, h = half-length) for broadside incidence is presented in Table I as a function of the integration test ratio. The RCS is calculated using the collocation method and sinusoidal interpolation for the current [2]. The numerical integration required is accomplished by RVIW quadrature and the difference integrand technique for the self-field terms. The number of segments into which the scatterer is divided is a parameter varying from 10 to 54. The calculated RCS is numerically convergent and is comparable in accuracy (using the answer obtained from the smallest test ratio as the exact answer) to the value of the test ratio. The answers are more dependent upon the number of collocation segments than upon the test ratio, although the difference between 10 and 54 segments is not very significant in comparison with typical experimental errors. The method used here to calculate the RCS has provided results in good agreement with experimental measurement for thin-wire scatterers up to 15λ long.

References

[1] F. L. Bauer, H. Rutishauer, and E. Stiefel, "New aspects in numerical quadrature," *Proc. 1963 Symp. in Appl. Math.*, Experimental Arithmetic, High Speed Computing and Mathematics, vol. 15. Providence, R. I.: Am. Math. Soc.

[2] Y. S. Yen and K. K. Mei, "Theory of conical equiangular-spiral antennas, pt. I: numerical technique," *IEEE Trans. Antennas and Propagation*, vol. AP-15, pp. 634–639, September 1967.

[3] A. Ralston, *A First Course in Numerical Analysis.* New York: McGraw-Hill, 1965.

[4] S. H. Lin and K. K. Mei, "Numerical solution of dipole radiation in a compressible plasma," *IEEE Trans. Antennas and Propagation*, vol. AP-16, pp. 235–241, March 1968.

Analysis of Various Numerical Techniques Applied to Thin-Wire Scatterers

CHALMERS M. BUTLER AND D. R. WILTON

Abstract—Several numerical schemes for solving Pocklington's and Hallén's equations for thin-wire scatterers are investigated. Convergence rates of solutions obtained from seven methods are given and reasons for different rates are delineated.

Introduction

In moment method solutions of antenna and scattering problems associated with thin-wire structures, both Pocklington (electric field) and Hallén (magnetic vector potential) type integral equations are commonly used. From either, one may obtain solutions for the current on a wire antenna or a scatterer and subsequently calculate all other quantities of interest. Even though the two equations are intimately related [1], each exhibits distinct advantages and disadvantages. The authors present the findings of their investigations of several numerical schemes for solving wire problems. They discuss the relative merits of various techniques, point out pitfalls to be avoided, and summarize their findings.

In the numerical methods, different basis sets for representing the unknown wire current are employed and relative convergences are investigated. So that one may focus attention on the generic feature of the capacity of a given basis set to represent adequately, and converge to, the correct current on the wire, only the case of a scatterer subject to normally incident illumination is considered here. With illumination constant over the length of the scatterer, the problem of inadequate sampling [2] of the integral equation's driving term has no bearing on convergence. Also, in the interest of simplicity and of addressing only the fundamental question of convergence in the sense of how well the actual wire current is represented by the solution, the wire junction problem is not treated nor is any record of computer times given for the various methods.

Thin-Wire Equations

From basic electromagnetic theory, one may readily obtain the following fundamental integro-differential equation:

$$\left(\frac{d^2}{dz^2} + k^2\right) \int_{\zeta=-L/2}^{L/2} i(\zeta) K(z-\zeta)\, d\zeta = -j4\pi\omega\varepsilon E_z^i(z) \quad (1)$$

which relates the unknown total axial current i on a cylinder to the known incident electric field having an axial component E_z^i on the surface of the scatterer. The scatterer is perfectly conducting and resides in a homogeneous space characterized by (μ, ε, $\sigma = 0$), and, as suggested in Fig. 1, it is a tube of length L and radius a. The kernel in (1) is

$$K(\xi) = \frac{1}{2\pi} \int_{\phi'=-\pi}^{\pi} \frac{e^{-jk[\xi^2 + 4a^2 \sin^2(\phi'/2)]^{1/2}}}{[\xi^2 + 4a^2 \sin^2(\phi'/2)]^{1/2}}\, d\phi' \quad (2)$$

where k is 2π/wavelength at the angular frequency ω of the suppressed harmonic time variation $e^{j\omega t}$. For present purposes, the wire radius is looked upon as being very small relative to the

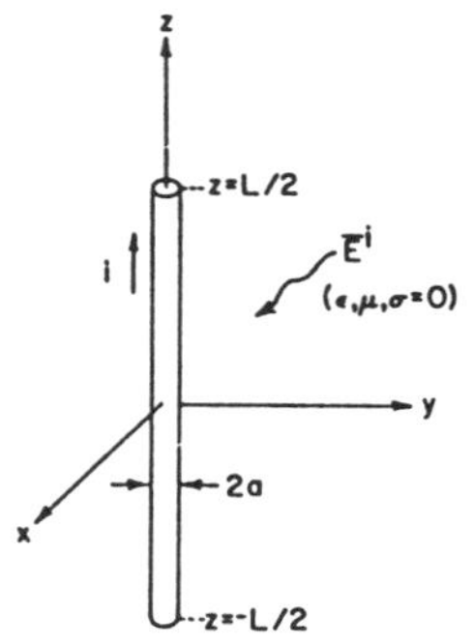

Fig. 1. Straight wire subject to incident illumination.

wavelength λ as well as to the cylinder length. Such restrictions, common in thin-wire analyses, assure one that the current on the cylinder is circumferentially independent and that it can be accounted for by the total axial current i. The thin-wire assumptions also lead to the so-called reduced kernel approximation to $K(\xi)$

$$K(\xi) \doteq \frac{e^{-jk[\xi^2 + a^2]^{1/2}}}{[\xi^2 + a^2]^{1/2}}. \quad (3)$$

From (1), one may readily derive Hallén's equation

$$\int_{\zeta=-L/2}^{L/2} i(\zeta) K(z-\zeta)\, d\zeta = C \cos kz + B \sin kz - j\frac{4\pi}{\eta} \int_{\zeta=0}^{z} E_z^i(\zeta) \sin k(z-\zeta)\, d\zeta \quad (4)$$

where η ($= \sqrt{\mu/\varepsilon}$) is the intrinsic impedance of the medium and where C and B are constants of integration which must be consistent with the boundary condition that the current be zero at the wire ends ($\pm L/2$). In the special case of normally incident plane wave illumination, E_z^i is constant over $z \in (-L/2, L/2)$ allowing one to perform analytically the integration on the right side of (4), and $i(z)$ is an even function of z allowing one to eliminate the constant B. The exact kernel (2) may be used in (4) or, where warranted by thin-wire conditions, the approximation (3) often suffices.

Having accepted the approximation (3), one may interchange at will the integral and differential operators in (1) and, thereby, convert it to any of several equivalent forms usually referred to as Pocklington-type equations. For the present discussion, the form which is most useful is

$$\int_{\zeta=-L/2}^{L/2} \left[\left(\frac{d^2}{d\zeta^2} + k^2\right) i(\zeta)\right] K(z-\zeta)\, d\zeta + \left[i(\zeta)\frac{\partial}{\partial\zeta} K(z-\zeta) - \frac{d}{d\zeta} i(\zeta) K(z-\zeta)\right]_{\zeta=-L/2}^{L/2} = -j4\pi\omega\varepsilon E_z^i(z). \quad (5)$$

Bases

The unknown current $i(z)$ is represented in a numerical solution technique by a linear combination of known elements $i_n(z)$ of a

Manuscript received August 1, 1974; revised December 23, 1974. This work was supported in part by the National Science Foundation under Grant GU3833.

The authors are with the School of Engineering, University of Mississippi, University, Miss. 38677.

Reprinted from *IEEE Trans. Antennas Propagat.*, vol. AP-23, no. 4, pp. 534–540, July 1975.

basis set $\{i_n(z)\}$ selected to approximate the current as

$$i(z) \doteq \sum_n I_n i_n(z). \tag{6}$$

With i in the integral equations replaced by its approximation of (6), Hallén's and Pocklington's equations become, respectively,

$$\sum_n I_n \int_{\zeta=l_n}^{u_n} i_n(\zeta) K(z-\zeta)\, d\zeta = C \cos kz + B \sin kz - j\frac{4\pi}{\eta}\int_{\zeta=0}^{z} E_z^i(\zeta) \sin k(z-\zeta)\, d\zeta \tag{7a}$$

and

$$\sum_n I_n \left\{ \int_{\zeta=l_n}^{u_n} \left[\left(\frac{d^2}{d\zeta^2} + k^2\right) i_n(\zeta)\right] K(z-\zeta)\, d\zeta + \left[i_n(\zeta) \frac{\partial}{\partial \xi} K(z-\zeta) - \frac{d}{d\zeta} i_n(\zeta) K(z-\zeta)\right]_{\zeta=l_n}^{u_n} \right\} = -j4\pi\omega\varepsilon E_z^i(z). \tag{7b}$$

One chooses the set $\{i_n\}$ for its capacity to represent the current well and from the viewpoint of its utility in the numerical procedure. The basis sets considered here are illustrated in Fig. 2 and are defined subsequently where one sees them to be of the subdomain type, since each element i_n differs from zero over only a single subdomain of the total domain of interest $z \in (-L/2, L/2)$.

Piecewise Sinusoidal:

$$i_n(z) = \begin{cases} \dfrac{\sin k(\Delta - |z - z_n|)}{\sin k\Delta}, & z \in (z_{n-1}, z_{n+1}) \\ 0, & z \notin (z_{n-1}, z_{n+1}) \end{cases} \tag{8a}$$

Piecewise Linear:

$$i_n(z) = \begin{cases} \dfrac{\Delta - |z - z_n|}{\Delta}, & z \in (z_{n-1}, z_{n+1}) \\ 0, & z \notin (z_{n-1}, z_{n+1}) \end{cases} \tag{8b}$$

Trigonometric:

$$i_n(z) = \begin{cases} 1 + C_n \cos k(z - z_n^c) + B_n \sin k(z - z_n^c), & z \in (z_n^c - \Delta/2,\ z_n^c + \Delta/2) \\ 0, & z \notin (z_n^c - \Delta/2,\ z_n^c + \Delta/2). \end{cases} \tag{8c}$$

Entire domain basis sets where each element exists over the full domain $(-L/2, L/2)$ are useful in special cases, but, since their application is limited, they are not treated in this paper.

Implicit in the use of (8) in (7) is the division of the scatterer length L into N subintervals of length $\Delta = L/N$. In all cases the boundary conditions $i(\pm L/2) = 0$ are satisfied by (6) *a priori* which, in using (8a) and (8b), means that one sets the I_n associated with the i_n over the intervals $(L/2 - \Delta, L/2 + \Delta)$ and $(-L/2 - \Delta, -L/2 + \Delta)$ equal to zero whereas with (8c) the C_n and B_n of the end-most intervals are adjusted so that $i(\pm L/2) = 0$. For piecewise linear and piecewise sinusoidal basis sets, the interval (l_n, u_n) must be partitioned into (z_{n-1}, z_n) plus (z_n, z_{n+1}) and the terms on the left side of (7b) must be evaluated in each of these two open intervals and summed to obtain the contribution from (l_n, u_n). This partitioning is necessary due to the derivative discontinuity at z_n exhibited by both (8a) and (8b). For the trigonometric set, $l_n = z_n^c - \Delta/2$ and $u_n = z_n^c + \Delta/2$, where z_n^c is the center of the nth subdomain.

As discussed in detail subsequently, the convergence rate of a solution method can be enhanced by modification of (6) through

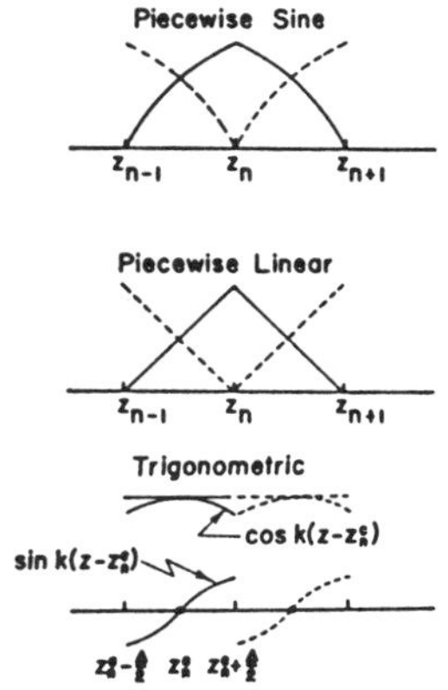

Fig. 2. Bases.

certain constraints applied to $\{i_n\}$. Two such constraints are employed by the authors in methods discussed forthwith. In one case they adjust the C_n and B_n of (8c) to force i of (6) plus its derivative to be continuous at each common boundary point of adjacent subdomains. Subject to these adjustments of the coefficients in (8c), a basis set is obtained which renders the current and its first derivative continuous everywhere in $(-L/2, L/2)$. In another case employing (8c), $i_n(z)$, the current in the nth subdomain, is required to satisfy[1] $i_n(z_{n-1}^c) = i_{n-1}(z_{n-1}^c)$ and $i_n(z_{n+1}^c) = i_{n+1}(z_{n+1}^c)$. One might refer to this as *extrapolated continuity*, and it is evident that the requirements do not force the current or its derivative to be continuous at the boundary of subdomains.

Testing

Substitution of (6) into (4) and (5) yields (7a) and (7b), respectively, in which all integrations and differentiations operate on known functions but into which the additional unknown constants I_n are introduced.

According to the method of moments [4], one can transform (7a) or (7b) into a system of linear algebraic equations with unknowns I_n by testing both sides of either equation with members of a suitable testing set $\{W_m\}$. For example, testing of (7b) yields

$$\sum_n I_n \left\langle \int_{\zeta=l_n}^{u_n} \left[\left(\frac{d^2}{d\zeta^2} + k^2\right) i_n(\zeta)\right] K(z-\zeta)\, d\zeta + \left[i_n(\zeta)\frac{\partial}{\partial\zeta} K(z-\zeta) - \frac{d}{d\zeta} i_n(\zeta) K(z-\zeta)\right]_{\zeta=l_n}^{u_n}, W_m(z) \right\rangle = -j4\pi\omega\varepsilon \langle E_z^i(z), W_m(z)\rangle, \qquad m = 1,2,\cdots,M \tag{9}$$

where

$$\langle f, W \rangle = \int_{z=-L/2}^{L/2} f(z) W(z)\, dz.$$

In the preceding (9) is a system of linear equations which can be solved for the I_n for proper M.

Discussion of Methods

In this section the seven solution methods in Table I are discussed in the order of their listing, and relative convergence data are given in Figs. 3–12. The value of current at the center of the scatterer is the quantity upon which the study of con-

[1] See [3].

vergence is based, since it was ascertained that the convergence rate at points other than the scatterer center was comparable to that of the center current. In all cases, data are normalized with respect to that obtained from Galerkin's method with 31 piecewise sinusoids. Normalized results of each method are plotted against $1/N$ where one recalls N to be the total number of subdomains into which $(-L/2,L/2)$ is divided. In Methods I, II, III, and VI the number of unknown I_n to be determined is $(N - 1)$, while in Method V the number of unknown current coefficients is N. After constraining the C_n and B_n of (8c) in Method IV, one has N unknown I_n to calculate but these constraining equations either augment the system of equations (9) or necessitate that one solve an auxiliary difference equation. Finally, the number of unknown I_n in Method VII is $(N - 1)$ but the two constants C and B are unknown *a priori* so that, effectively, one must solve for $N + 1$ quantities.

Data are given for two radii and for selected wire lengths so that they include special cases of practical interest, e.g., $L = \lambda/2$ and $L = \lambda$. Trends inferred from data given should be representative of typical cases encountered in practice except, of course, for very long wires.

The real and imaginary parts of the center current are investigated separately and each part is normalized with respect to the corresponding part obtained from Galerkin's method as mentioned. The real and imaginary normalization factors, C_R and C_I, are given in each figure. Normalization of each part separately has the advantage that small differences in a part can be independently observed in the data. However, on the other hand, when C_R and C_I differ appreciably, deviations from the norm of data normalized with respect to the smaller may be in large measure due to round-off errors; such a deviation appears extreme to an observer when actually it is of no consequence upon comparison with $\sqrt{C_R^2 + C_I^2}$.

Method I (Pocklington-Piecewise Sinusoid-Collocation)

Since no numerical integration is needed to calculate terms in (9), Pocklington's equation subject to collocation ($W_m(z) = \delta(z - z_m)$) with the piecewise sinusoids (8a) in the current approximation (6) appears, superficially, to be a very attractive method. No numerical integration is needed in (9), since the harmonic differential operator in the integrand applied to (8a) yields zero for $\zeta \in (z_n,z_{n+1})$, leaving one with simply

$$\frac{k}{\sin k\Delta}\sum_n I_n[K(z_m - z_{n-1}) - 2\cos k\Delta\, K(z_m - z_n) + K(z_m - z_{n+1})] = -j4\pi\omega\varepsilon E_z^{\,i}(z_m)$$

where $z_m = m\Delta$, $m = 0, \pm 1, \pm 2,\cdots, \pm(N - 2)/2$, are the match points in $(-L/2,L/2)$ at which (9) is enforced. Unfortunately, the obviation of numerical integration not withstanding, this procedure is unsatisfactory except for large radius and/or resonant length wires. Under close scrutiny the method is not generally satisfactory even for a large radius wire, since the solution converges rapidly for relatively small N, but with further increase in N the system of linear equations (9) which one must solve to determine the coefficients I_n becomes ill-conditioned [5]. These comments are supported by the data of Figs. 3–12 where one sees that for $a = 0.01\lambda$ the method enjoys somewhat better convergence than for $a = 0.001\lambda$ except when $L = 0.5\lambda$; also, P-PS-C exhibits better convergence for $L = 0.5\lambda$ and $a = 0.001\lambda$ than for other lengths and this radius.

TABLE I
NUMERICAL SOLUTION METHODS

Method	Equation	Basis Set	Testing Method	Key
I	Pocklington	Piecewise sinusoidal	Collocation	P-PS-C
II	Pocklington	Piecewise sinusoidal	Collocation/ Galerkin	P-PS-C/G
III	Pocklington	Piecewise sinusoidal	Galerkin	P-PS-G
IV	Pocklington	Trigonometric with continuous current and continuous derivative	Collocation	P-T/C-C
V	Pocklington	Trigonometric with extrapolated continuity	Collocation	P-T/E-C
VI	Pocklington (Difference Equation)	Piecewise linear	Collocation	P-PL-DE
VII	Hallén	Piecewise linear	Collocation	H-PL-C

Method II (Pocklington-Piecewise Sinusoid-Collocation/Galerkin)

Initially the authors attributed the poor convergence of P-PS-C to the fact that the integral equation is enforced at the interior points (match points z_m) but not at the end points ($\pm L/2$) with the consequence that the resulting system (9) may characterize a scatterer of length $L - 2\Delta$ rather than of L, the correct value. As a remedy, they elected to form (9) by testing over $(-L/2, -L/2 + 2\Delta)$ and $(L/2 - 2\Delta, L/2)$ with the piecewise sinusoids of (8a) and point-matching at the interior points (match points z_m). This hybrid testing procedure (Method II) is designated P-PS-C/G. Deceptively, it appears to converge for small N but is little better than P-PS-C for larger N as is seen from the data of Figs. 3–12.

Further investigation of (8) and (9) reveals that the poor convergence rates of P-PS-C and P-PS-C/G are due to the rapidly varying contributions to the scattered electric field arising from the unnatural discontinuities in derivative of the current approximation formed from the piecewise sinusoidal basis set [2]. Notice, however, in Figs. 5 and 6, the acceptable convergence rates of P-PS-C for resonant-length elements, which are due to the fact that resonant current is almost sinusoidal and the discontinuities in derivative almost vanish [1], [2].

A discontinuity in current or its derivative is unphysical and any approximation (6) exhibiting either produces a rapidly varying scattered electric field [2]. In particular, at a discontinuity in current, there is a contribution to scattered field proportional to $(\partial/\partial\zeta)K(z - \zeta)_{z=\zeta}$ as is seen from (5), whereas at a discontinuity of derivative of current there is a contribution proportional to $K(0)$. With either discontinuity, a highly peaked, rapidly varying scattered electric field is produced along the scatterer, which according to (5) must be equal to the constant $-E_z^{\,i}$ (normal incidence). Of course, equating of a constant to a rapidly varying function at discrete points z_m is unlikely to give rise to equality at points other than z_m over the interval $(-L/2, L/2)$.

Method III (Pocklington-Piecewise Sinusoid-Galerkin)

Failure to satisfy (7b) in some sense at all points on the scatterer implies, of course, a poor solution and, in P-PS-C and P-PS-C/G, can be attributed to the unnatural variation of the scattered electric field caused by unphysical discontinuities in the derivative of (6), the representation of current. In an attempt

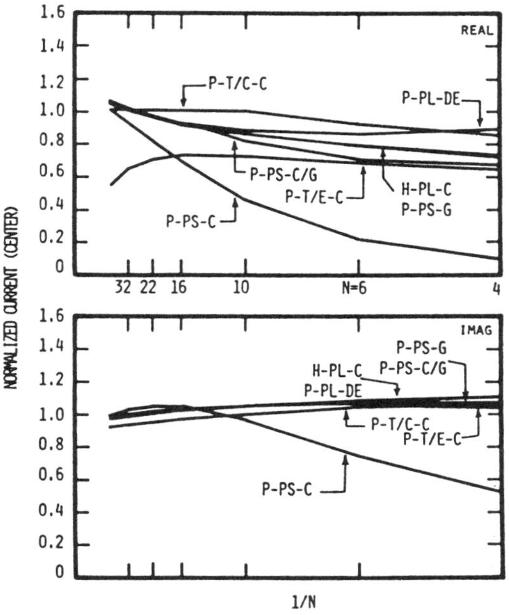

Fig. 3. Convergence data: normalized current at center of scatterer versus 1/number of subdomains ($l = 0.4\lambda$, $a = 0.01\lambda$, $C_R = 3.007$, $C_I = 2.310$).

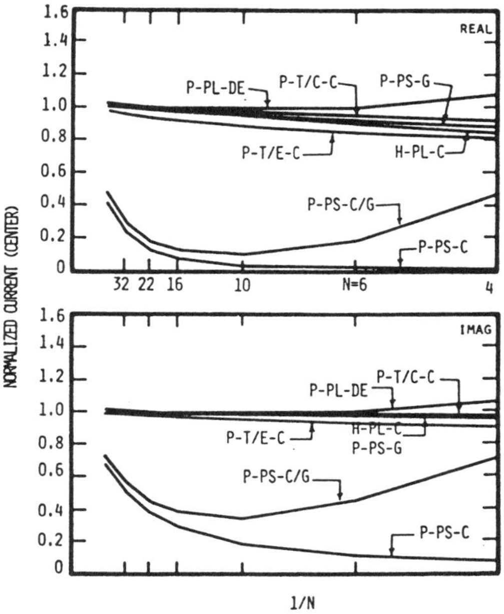

Fig. 4. Convergence data: normalized current at center of scatterer versus 1/number of subdomains ($l = 0.4\lambda$, $a = 0.001\lambda$, $C_R = 0.5498$, $C_I = 1.566$).

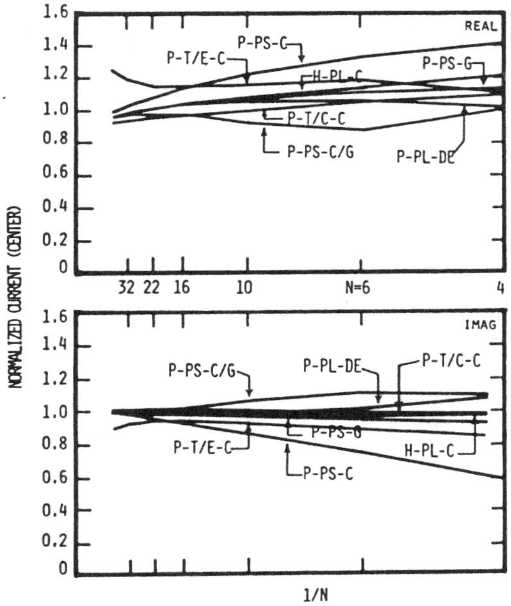

Fig. 5. Convergence data: normalized current at center of scatterer versus 1/number of subdomains ($l = 0.5\lambda$, $a = 0.01\lambda$, $C_R = 2.808$, $C_I = -0.908$).

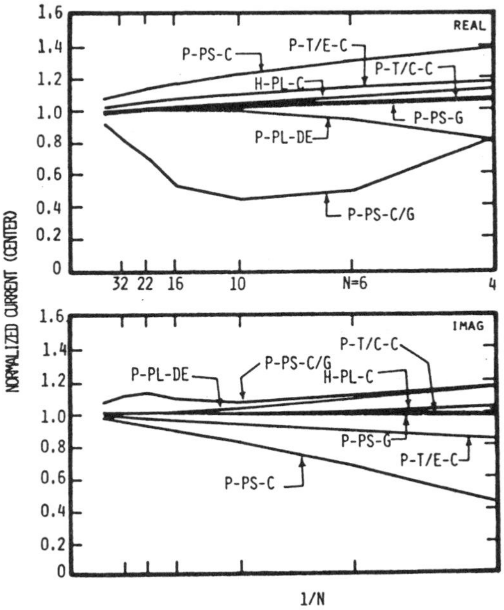

Fig. 6. Convergence data: normalized current at center of scatterer versus 1/number of subdomains ($l = 0.5\lambda$, $a = 0.001\lambda$, $C_R = 2.999$, $C_I = -1.929$).

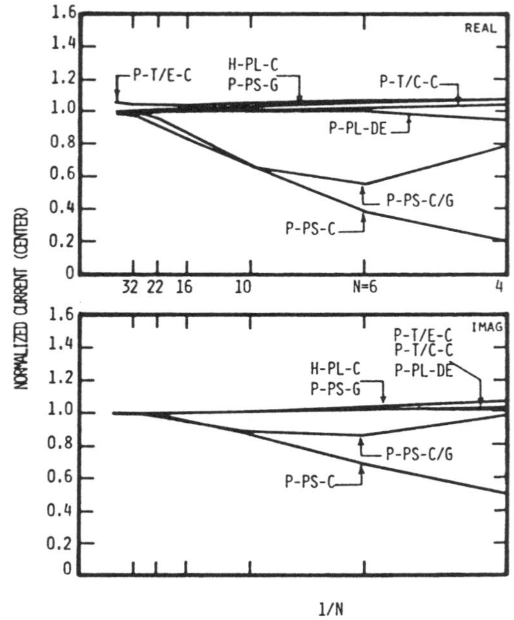

Fig. 7. Convergence data: normalized current at center of scatterer versus 1/number of subdomains ($l = 0.667\lambda$, $a = 0.01\lambda$, $C_R = 1.038$, $C_I = -1.619$).

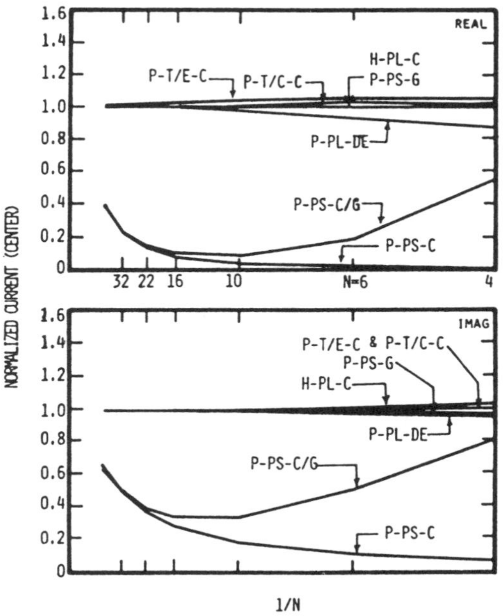

Fig. 8. Convergence data: normalized current at center of scatterer versus 1/number of subdomains ($l = 0.667\lambda$, $a = 0.001\lambda$, $C_R = 0.4658$, $C_I = -1.210$).

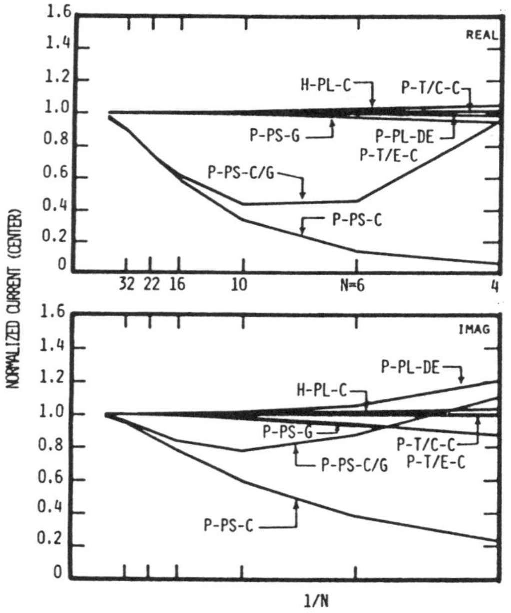

Fig. 9. Convergence data: normalized current at center of scatterer versus 1/number of subdomains ($l = 1.\lambda$, $a = 0.01\lambda$, $C_R = 0.6482$, $C_I = -1.450$).

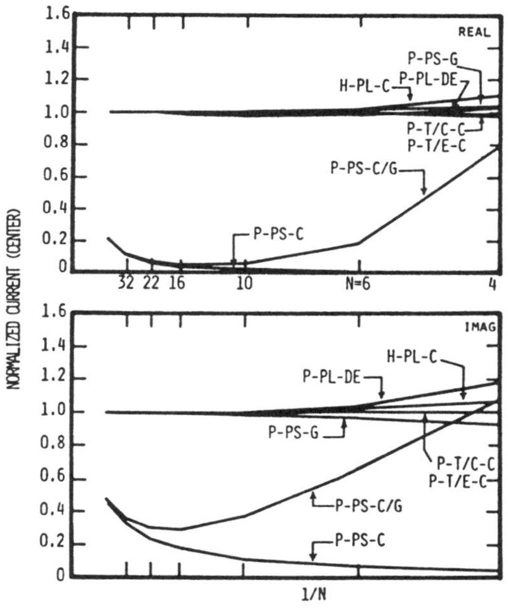

Fig. 10. Convergence data: normalized current at center of scatterer versus 1/number of subdomains ($l = 1.\lambda$, $a = 0.001\lambda$, $C_R = 0.2479$, $C_I = -0.9328$).

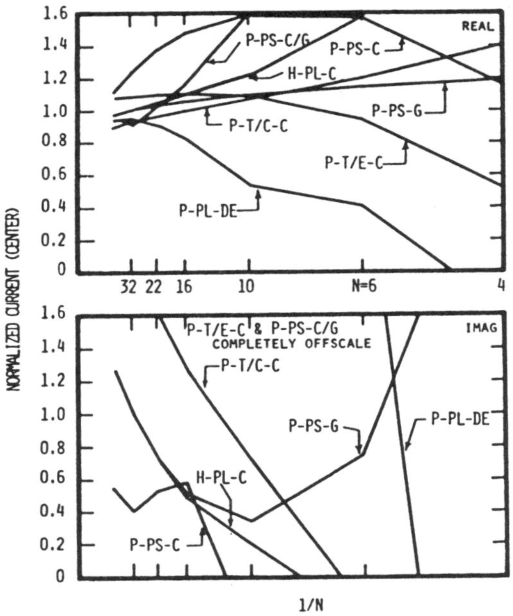

Fig. 11. Convergence data: normalized current at center of scatterer versus 1/number of subdomains ($l = 1.5\lambda$, $a = 0.01\lambda$, $C_R = -1.711$, $C_I = 0.0788$).

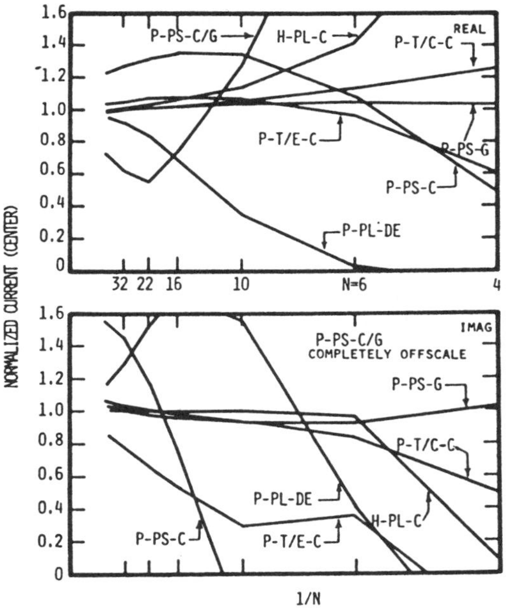

Fig. 12. Convergence data: normalized current at center of scatterer versus 1/number of subdomains ($l = 1.5\lambda$, $a = 0.001\lambda$, $C_R = -2.175$, $C_I = 0.4404$).

to improve solution methods employing basis sets which introduce discontinuities in the approximation to the current or its derivative, one may equate weighted averages of both sides of (7b) in hopes of achieving a better representation of the scattered field over the entire range $(-L/2, L/2)$ compared with its values at discrete z_m. Galerkin's method, which is mentioned under TESTING, effects such averaging. P-PS-G is a technique for solving Pocklington's equation by Galerkin's method with piecewise sinusoids both for representing the current and for testing. Solutions obtained by means of P-PS-G are seen from Figs. 3–12 to converge rapidly. Note that the Galerkin procedure requires two integrations, one over the basis element and another over the testing function. Richmond, who calls this procedure "piecewise sinusoidal reaction matching," has shown, however, that these integrations can be performed analytically for piecewise sinusoidal basis and testing sets [6], [1].

Notice that Methods I, II, and III incorporate (8a) in (6), each with different testing schemes for solving Pocklington's equation. Since, for $\Delta \ll \lambda$, (8b) closely approximates (8a), comments for the piecewise sinusoid also hold for the piecewise linear basis set.

Method IV (Pocklington-Trigonometric/Continuous Current and Derivative-Collocation)

In keeping with the desire to maintain continuity of current and its derivative over the scatterer, the authors investigated the use of the trigonometric set (8c) in Pocklington's equation, but, in addition to point-matching at the center of each subdomain, i and $(di/d\zeta)$ were forced to be equal at the common boundary points of adjacent subdomains.[2] Note that such a current representation may be called a trigonometric spline function. Solutions obtained by this method P-T/C-C are seen to converge rapidly (Figs. 3–12) as one would expect.

Method V (Pocklington-Trigonometric/Extrapolated Continuity-Collocation)

Because the introduction in P-T/C-C of the equations to maintain continuity of (6) and its derivative adds significant complexity to the method, an alternate scheme, P-T/E-C, for suppressing the discontinuities was investigated. Subject to *extrapolated continuity*[3] of P-T/E-C, the discontinuities are not zero but are reduced to levels at which their contributions do not dominate the scattered field. In contrast to the unacceptable complexities encountered in P-T/C-C in constraining the C_n and B_n, P-T/E-C is not difficult to implement, yet one observes from Figs. 3–12 that the convergence rate of solutions obtained by the latter method is near that of those obtained by the former. However, for larger and larger N, P-T/E-C solutions are seen to exhibit trends of diverging. This divergence is understandable, since in P-T/E-C the discontinuities are not zero, and, as $\Delta \to 0$ for larger and larger N, the unnatural contributions due to $K(\Delta/2)$ and $(\partial/\partial\zeta)K(z-\zeta)_{z-\zeta=\Delta/2}$ become more and more significant.

Modifications of P-T/E-C and P-T/C-C where the trigonometric functions of (8c) are replaced by other analytic functions, e.g., three-term power series, also lead to satisfactory solution techniques.

Method VI (Pocklington-Piecewise Linear-Difference Equation)

Methods I–V exhibit convergence rates which strongly depend upon how one handles discontinuities of (6) and its derivative. As an alternative to placing continuity requirements upon (6) to suppress the peaked field variations, one may relax or smooth the derivatives in (1) that operate on

$$\int_{\zeta=-L/2}^{L/2} i(\zeta) K(z-\zeta)\, d\zeta \tag{10}$$

[2] See last paragraph under BASES.

[3] See last paragraph under BASES and [3].

to produce these unnatural contributions to scattered electric field at the discontinuities. In particular in P-PL-DE the harmonic operator of (1) may be replaced by its corresponding difference operator; the difference operator is insensitive to the local variation of (10) due to discontinuities of (6) but does correctly account for desired global derivatives. Specifically, in P-PL-DE, the derivative discontinuities of the piecewise linear representation, (8b) in (6), do not manifest themselves in sharp peaks of the scattered field. Moreover, the difference operator equation approximation to (1) is amenable to use with basis sets which cause discontinuities in (6) as well as in its derivative. The convergence of P-PL-DE is given in Figs. 3–12 [1].

Method VII (*Hallén-Piecewise Linear-Collocation*)

Pocklington's equation relates the electric field to the sources on the scatterer in such a manner that the value of field is very sensitive to both the current and its derivative and, consequently, any efficient method for solving this equation must include special treatment of either the derivatives or of the discontinuities. On the other hand, Hallén's equation is based on the vector potential and the electric field enters the relationship only through integration. In addition, the vector potential is less sensitive to local variation of current than is its second derivative so Hallén's equation yields to solution methods in which the current is approximated (6) by unphysical currents possessing discontinuities. Good convergence is seen in solutions of Hallén's equation when the basis set is piecewise linear (H-PL-C) and in which no attempt is made to lessen the effects of the discontinuous derivatives (Figs. 3–12).

Conclusions

Seven methods for determining the current on a scatterer are presented and the relative convergence rates of the solutions obtained by the methods are investigated. Reasons for differences in rates are delineated. It is shown that, if acceptable convergence rates are to be attained, solution methods applied to Pocklington's equation must incorporate means of suppressing discontinuities in the current approximation and its derivative, or, on the other hand, the deleterious effects of these discontinuities must be circumvented by rendering the equation insensitive to them. The latter can be accomplished by averaging techniques applied to both sides of the equation, e.g., Galerkin's method, or by smoothing the derivatives, e.g., use of difference operator. Hallén's equation governs quantities which are less sensitive to discontinuities and it may be successfully solved numerically with almost any reasonable basis set, even one which causes the approximate current to be discontinuous. For a given basis set, the convergence rate of solutions to Hallén's equation obtained by point-matching is as high as that of solutions to Pocklington's equation with any testing scheme. Solutions by the difference equation method attain a high rate of convergence essentially identical to that of the point-matched Hallén equation. Of further importance, the difference equation procedure is simple, and it is highly amenable to numerical implementation, even when applied to multiple-wire structures.

References

[1] D. R. Wilton and C. M. Butler, "Use of difference equations in conjunction with moment methods," in *USNC/URSI 1974 Spring Meeting*, Atlanta, Ga., June 1974.

[2] L. W. Pearson and C. M. Butler, "Inadequacies of collocation solutions to Pocklington-type models of thin-wire structures," *IEEE Trans. Antennas Propagat.* (Commun.), vol. AP-23, pp. 295–298, Mar. 1975.

[3] Y. S. Yeh and K. K. Mei, "Theory of conical equiangular-spiral antennas, Part I—numerical technique," *IEEE Trans. Antennas Propagat.*, vol. AP-15, pp. 634–639, Sept. 1967.

[4] R. F. Harrington, *Field Computation by Moment Methods.* New York: Macmillan, 1968.

[5] C. D. Taylor and D. R. Wilton, "The extended boundary condition solution of the dipole antenna of revolution," *IEEE Trans. Antennas Propagat.* (Commun.), vol. AP-20, pp. 772–776, Nov. 1972.

[6] J. H. Richmond and N. H. Geary, "Mutual impedance between coplanar-skew dipoles," *IEEE Trans. Antennas Propagat.* (Commun.), vol. AP-18, pp. 414–416, May 1970.

Efficient Solution of Large Moments Problems: Theory and Small Problem Results

THOMAS R. FERGUSON, THEODORE H. LEHMAN, AND ROBERT J. BALESTRI, MEMBER, IEEE

Abstract—**A banded matrix iterative solution method for linear simultaneous equations arising from thin wire moments problems has been applied to a variety of multiple wire configurations. Compared to Gaussian elimination, solution efficiencies around 5 to 10 and up to 23 have been obtained for problems with 100 unknowns for a solution accuracy of 1 percent. Greater efficiencies are obtained for less accurate solutions. Reasonable far-field patterns are found from currents with 30 percent error. The method is intended for large problems, for which good efficiencies are expected. Application of the method requires some expertise at the current stage of development.**

I. INTRODUCTION

Solutions of many problems in electromagnetic theory are most easily obtained using the method of moments [1]. Computer programs are available for solving general problems with the proper input of geometrical and other data. The basic approach is to reduce an integral equation to a set of linear simultaneous equations

$$AX = b \tag{1}$$

where A is a full complex $N \times N$ matrix usually called the impedance matrix, X is the column vector of current coefficients, and b is the excitation column vector. The use of computer programs for solving problems with the method of moments has been restricted in the past to values of N of about 200. Although a large problem requires out-of-core storage manipulations, the main limitation is the execution time required for performing multiplicative operations. The method normally used for solution is Gaussian elimination, for which the execution time increases as the cube of N. It has been proved [2] that no general system of linear algebraic equations can be solved in fewer operations than are required by Gaussian elimination. To solve (1) in fewer operations requires some special feature of the equations. The term "solved" is used here to mean an exact solution. Typical computer solutions are precise to six or even twelve significant figures even though the physical model, mathematical assumptions in the program, and other factors make such precision meaningless. For most problems, an accuracy of one percent would be sufficient. For many problems in antenna design, compatibility, and placement, an accuracy of a few decibels would be adequate. Solutions in this range of accuracy can be obtained by an iterative method with much less computer time than is necessary for Gaussian elimination.

The use of the iterative method requires a limited understanding of the theory and some experience.

II. ITERATIVE METHODS

The basic approach in iterative schemes is to separate the matrix A into the sum of two matrices, one (A_1) containing all of the large elements (and possibly some small elements), and one (A_2) containing only small elements:

$$A = A_1 + A_2. \tag{2}$$

Manuscript received June 8, 1975; revised August 27, 1975. This work was supported by the Rome Air Development Center under Contract F30602-74-C-0182. Portions of this paper were presented at the 1975 International IEEE/AP-S Symposium, Urbana, IL, June 2–5.

The authors are with the BDM Corporation, Albuquerque, NM 87106.

Then (1) can be written as

$$A_1X = b - A_2X, \tag{3}$$

and an iterative scheme is

$$A_1X_{i+1} = b - A_2X_i, \qquad i = 0,1,\cdots. \tag{4}$$

For the scheme to be efficient, this set of equations must be economically solved at each iteration. The list of types of matrices A_1 for which this seems possible is quite short: diagonal, triangular, banded, sparse. The first two yield standard iterative approaches called the Jacobi and Gauss–Seidel methods. The Gauss–Seidel method has been investigated [3] for solving moments equations with the magnetic field integral equation (which produces relatively large diagonal matrix elements). Sparse matrix methods in combination with an iterative method are unlikely to be efficient for moments problems because A_1 can be sparse only if A_2 is dense.

The remaining method is that using banded matrices. Let the matrix A be written as

$$A = L + B + U \tag{5}$$

where B is a banded matrix, L is the triangular matrix below B in A, and U is the triangular matrix above B in A. Several iterative schemes have been proposed [4] using (5). The most simple is to let A_1 be B, so that

$$BX_{i+1} = b - (L + U)X_i. \tag{6}$$

The upper and lower bandwidths of B (numbers of minor diagonals) will be assumed to be equal, with value M. The banded matrix can be decomposed into a product of upper and lower banded triangular matrices, so that only forward and back substitution are required to solve (6) at each iteration.

The cost of decomposition depends on the pivoting strategy used. Full pivoting destroys bandedness, and cannot be used. Partial pivoting doubles either the upper or lower bandwidth, requiring additional storage. Decomposition without pivoting does not increase the size and is the most efficient. It carries the risk of accidental cancellations and large rounding errors. This method was used in all numerical studies reported, and no significant errors occurred.

Theoretically, the banded matrix iterative sequence in (6) will converge if the spectral radius of $-B^{-1}(L + U)$ is less than one. From a practical view, the convergence must be fairly rapid and the solution X_1 must be a reasonable approximation to the exact solution. This requires that most "large" matrix elements be contained in a banded matrix B with a bandwidth M that is considerably less than N. This can be insured for thin wire moments problems of moderate to large size by the proper choice of wire segment numbering.

III. SEGMENT NUMBERING

The numbering scheme is most easily explained for flat objects (all wires in a plane). The basic idea is to superimpose a set of narrow parallel strips on the object of interest. Fig. 1 shows the strips as separated by dashed lines, with the object as an irregular loop made up of short straight segments. Numbering starts at one extreme strip, proceding from left to right (or vice versa). When all segments in the strip are numbered, proceed to the adjacent strip, again numbering from left to right. Continue until all segments are numbered.

Fig. 1 shows the segments as ending at the strip edges, but this need not be the case. For a regular structure, numbering is regular. For an irregular object with unequal segment lengths,

Reprinted from *IEEE Trans. Antennas Propagat.*, vol. AP-24, no. 2, pp. 230–235, March 1976.

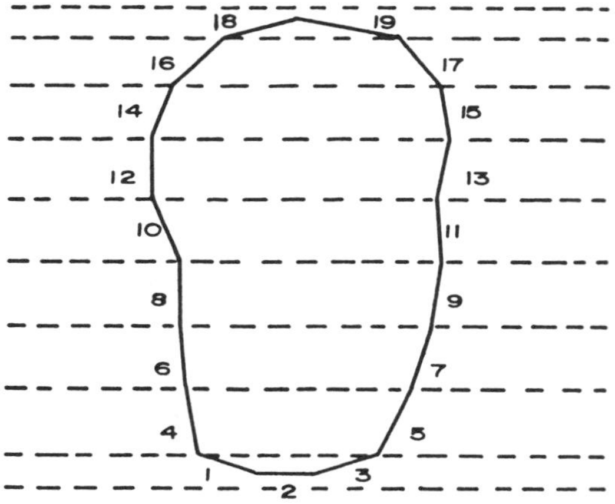

Fig. 1. Numbering scheme for irregular loop.

numbering appears irregular, but follows from the same procedure. The strips should be kept as narrow as possible, so that numbering in each strip proceeds across the object rather than along the object (across the strip). The result is that the difference is small between segment numbers for two close neighboring segments in adjacent strips.

This numbering scheme applies directly to three-dimensional objects. By rolling a wire gridded rectangle into a cylinder about the proper axis, the segment numbering is helix-like. If the cylinder has wire gridded end caps, numbering starts at the center of one end. Numbering is then spiral-helical-spiral. For a wire gridded cube, numbering spirals outward from the center of one face, then helical to the opposite face, and spirals inward. For elongated objects, numbering starts at one extremity.

For some objects, the orientation of the object relative to the strips is critical. For a square loop, placing one edge parallel to the strips will result in numbering sequentially along one side, then alternating between two sides, then along the fourth side. Using equal segment lengths, the matrix bandwidth M would have to be greater than $N/4$ to keep interactions between first neighboring segments at the loop corners in the band. This leads to poor efficiency. A better orientation results from rotating the loop by 45°. Segment numbering then alternates between sides starting at one corner. Crossed wires are handled in a similar manner. For just two wires of equal length intersecting at right angles at their centers, the 45° orientation results in segment numbers alternating between wires.

Other geometries require some imagination and experience with the method. In some cases, more than one numbering scheme may give good solution efficiency. The important thing to bear in mind is that the large interactions are between segments that are close together. To keep these in the band, the difference in segment numbers must be equal to or less than the bandwidth M.

To make the input data simple, most existing computer programs number the wire segments sequentially along each wire, in the order in which wires are entered. For some problems, this numbering is adequate for the iterative method. If it is not, the most obvious way to obtain the desired numbering is to enter the wire segments as separate wires and supply connection data. This process is laborious even for small problems. An alternative is available. Using the numbering supplied by the program, the matrix and excitation column vector are computed. A permutation operation is then performed, yielding a reordered set of equations that can be solved by the iterative method. These equations are identical to the equations that would result from the cumbersome process of entering each segment in the order desired. The permutation operation has been used for several problems [5]. The procedure is straightforward and takes little computer time.[1]

IV. Efficiency

As noted, the primary use of computer central processor (CP) time for numerical calculations is in performing multiplicative operations. Essentially all of the expensive operations in solving (1) or (6) are complex, so comparisons are made using complex multiplicative operations (mo's). Gaussian elimination of an $N \times N$ matrix by decomposition requires [7] about $N^3/3$ mo's. Decomposition of B without pivoting requires about $NM^2 - 2M^3/3$ mo's. Solution of (6) requires N^2 mo's for each iteration. Assuming that adequate convergence of the iterative process occurs in k iterations, the cost of solving (6) is $NM^2 - 2M^3/3 + kN^2$. The efficiency compared to Gaussian elimination is then

$$g \simeq N^3[3(NM^2 - 2M^3/3 + kN^2)]^{-1}. \tag{7}$$

Alternatively, the efficiency can be determined by recorded CP times.

Prediction of efficiencies has not been possible for the problems considered to date. An upper bound for any M is obtained by setting k to zero. A lower bound can sometimes be estimated by substituting a much larger bandwidth than would normally be used. The error in the first solution X_1 is then small and convergence is rapid. The primary cost is then in decomposing B, and k is again set to zero.

At the other extreme, a very narrow bandwidth could be used. The cost of decomposing B is then negligible, but many iterations are required for convergence (or divergence may occur). In this case, g is about $N/3k$.

The best efficiencies are found in the intermediate region. An equal division of time for decomposition and iteration is seldom far from the best efficiency for most of the examples reported here. Although this point cannot be predicted, a few examples provide enough insight to pick a reasonable bandwidth for other problems of the same type.

As used here, the efficiency is defined for the solution process alone. The time required to compute the impedance matrix is larger than the solution time using Gaussian elimination for small problems. The crossover point is generally less than 100 unknowns. Hence, the iterative scheme is most practical for large problems. The early development and application of the method is necessarily limited to small problems for program simplicity.

V. Numerical Procedures

For each example problem, the exact solution X_e of (1) was obtained by Gaussian elimination, and the CP time required for solution was recorded. The iterative method was then used for the same problem. The relative error (RE) for any approximate solution X_n is defined as

$$\mathrm{RE} = [(X_n - X_e)^\dagger (X_n - X_e)/X_e^\dagger X_e]^{1/2} \tag{8}$$

where (†) denotes the complex conjugate transpose. The iteration was always started with X_0 equal to the zero vector, so that the relative error is initially exactly 1, or 100 percent. The solution

[1] Bandwidth minimization schemes have been used in sparse matrix applications [6]. Their use for automating the segment numbering is under consideration. The critical aspect is the effect on overall solution efficiency.

X_1 is then obtained from

$$BX_1 = b. \tag{9}$$

Since B contains all of the large matrix elements, the solution X_1 should be a reasonable solution. The cost of obtaining X_1 is low because B is decomposed, and only forward and back substitution are required.[2] Consequently the procedure yields an inexpensive first solution that is reasonable, without requiring physical intuition to provide a starting condition.

The iteration was terminated[3] in each case at a relative error of one percent, and the CP time required for solution was recorded. For many problems the iterative procedure was repeated using different bandwidths to try to find the bandwidth for optimum solution efficiency. In some instances, the iterative procedure was interrupted at a relative error much larger than one percent to compute radiation patterns using approximate solutions. The CP times required for these calculations were not included in solution times.

All of the reported results were obtained for problems using thin wires. The first results [4] were obtained using impedance matrices generated by the method of Harrington [1], using the potential equation with pulse expansion functions and point matching (collocation). All examples were single straight wire conductors, either antennas or broadside scatterers, with length to wavelength ratios ranging from 2.5 to 10. The results were sufficiently encouraging to justify using a more sophisticated program to generate impedance matrices and to compute far-fields. The program [9] chosen for this purpose uses the Pocklington integral equation, with pulse plus sine plus cosine expansion functions and collocation. With N wire segments in a model, this choice of expansion functions requires $3N$ current coefficients. Continuity conditions at segment midpoints are used to reduce the number of unknowns to N. These N numbers are values of current at the segment midpoints, and are solutions of (1). Numerical results using this program for a variety of multiple wire configurations were reported in detail [5].

All calculations were performed on a CDC 6600 computer. The program was compiled using both the RUN and FTN compilers. On the average, CP times were 50 percent higher for calculations using the RUN compiler. The efficiency may depend on the computer and on the compiler.

VI. Results

During the early studies of single straight wires, it was found that the following hold.

a) The iterative solution converges monotonically at bandwidths from 3 to N. (With 10 segments per wavelength, a bandwidth of 3 means that interactions at distances over 0.3λ are excluded from the band.)

b) Convergence rates depend on the wire radius, as do errors in X_1. Thinner wires result in better efficiency.

c) Relative errors for X_1 are larger for resonant antennas than for nonresonant antennas for the same bandwidth. The difference is greater for smaller wire radii.

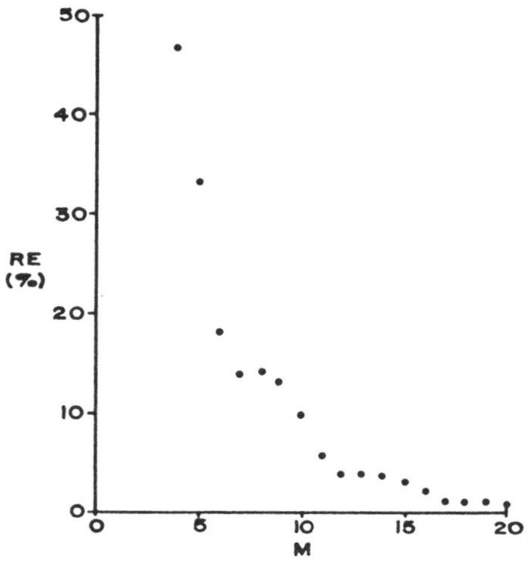

Fig. 2. Relative error (percent) at first iteration versus matrix bandwidth. (3λ centerfed dipole, 31 segments.)

d) The relative error for X_1 does not decrease smoothly and monotonically with increasing bandwidth. Periodic variations occur at changes in bandwidth that correspond to half-wavelength changes in distance along the wires, as exemplified in Fig. 2.

e) The convergence behavior can be approximated with varying accuracy by the formula

$$\text{RE (percent)} \simeq 100e^{-fi}, \tag{12}$$

where i is the number of iterations, and f is a function that is dependent on the bandwidth. The dependence is linear in first approximation, with an oscillation superimposed. However, the slope of the linear part is dependent on the excitation, length, and wire radius.

f) For a length to wavelength ratio of ten, iterative solutions with a relative error of one percent were obtained with an efficiency of about 7. For a relative error of ten percent, an additional reduction in CP time up to a factor of two was obtained.

g) The bandwidth resulting in peak efficiency varies with wire length in a roughly linear fashion. At ten segments per wavelength, the optimum bandwidth is about 4 or 5 for 3λ wires and in the range 10–15 for 10λ wires. With finer segmentation, the bandwidth should be increased proportionally to keep interactions at the same distance within the band.

Using the more sophisticated program for generating the equations, various combinations of thin wire geometries were used to investigate the capabilities and limitations of the banded matrix method. These included

a) one straight wire;
b) two parallel centerfed antennas at varying separations and radii;
c) two collinear centerfed antennas at large and small separations;
d) a linear array of parallel dipoles;
e) square and circular arrays of parallel dipoles;
f) a two-dimensional array of short dipoles;
g) a helix antenna;
h) a vertical half-rhombic antenna over sea water;
i) a wire-gridded rectangular strip;
j) a square loop; and
k) a pair of crossed wires.

In each case, N was restricted to 100 or less for economic reasons.

[2] For small M, the number of mo's for the first iteration is much less than N^2. Then k is replaced by $k-1$ in (7). This is important only for small problems.

[3] A convergence measure that does not require knowledge of the exact solution must eventually be developed. One candidate is the quantity $[(AX_n - b)\dagger(AX_n - b)/b\dagger b]^{1/2}$, called the boundary condition relative error (BCRE). It has been used in descent methods [8]. However, the ratio of the RE to the BCRE is bounded between the condition number of A and its reciprocal [7]. This number can be large for ill-conditioned problems. A small value of the BCRE is then not a valid criterion for convergence in the range of accuracies desired here. Alternative measures of convergence obtained by comparing solutions at each iteration are currently under investigation.

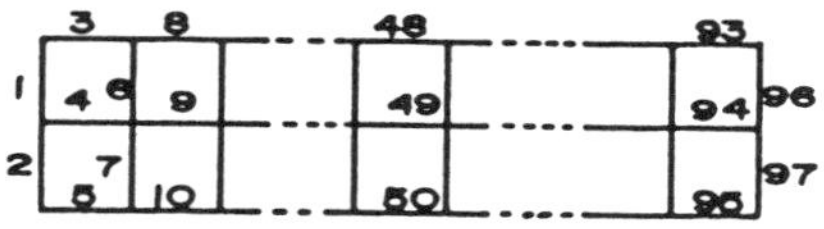

Fig. 3. Wire gridded rectangular strip.

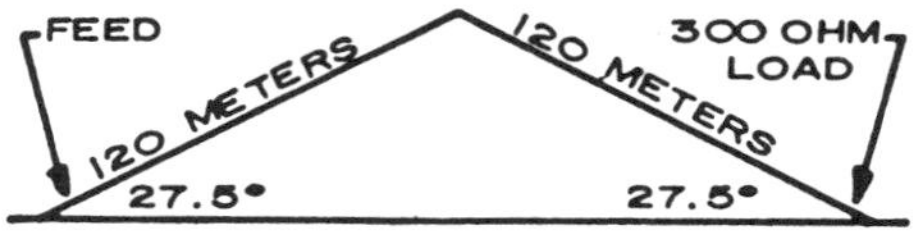

Fig. 4. Vertical half-rhombic antenna over sea water. (ε_r = 80, σ = 5 ℧/m.)

TABLE I
RELATIVE ERROR (PERCENT) AND CP TIME (SECONDS) FOR WIRE GRID CASE 1

ITERATIONS	BANDWIDTH						
	5	10	15	20	25	30	35
1	58.3	29.1	15.8	7.5	6.7	6.8	5.0
2	47.3	11.9	4.9	1.5	0.6	0.4	0.3
3	43.2	4.7	1.1	0.1			
4	39.1	1.4	0.3				
5	35.2	0.5					
15	13.8						
TIME	1.57	0.57	0.58	0.60	0.71	0.89	1.11

TABLE II
RELATIVE ERROR (PERCENT) AND CP TIME (SECONDS) FOR WIRE GRID CASE 2

ITERATIONS	BANDWIDTH						
	5	10	15	20	25	30	35
1	20.8	12.4	7.9	5.8	4.5	3.8	3.2
2	5.3	2.2	1.2	0.6	0.1	0.2	0.2
3	1.7	0.9	0.6				
4	1.0						
TIME	0.38	0.35	0.47	0.50	0.68	0.86	1.09

For many of the examples, the sequence of solutions converged uniformly to the exact solution. Divergence was forced for some examples by using a combination of element numbering, geometry, and bandwidth so that some large matrix elements were not contained in B. Efficiencies around 5 to 10 and up to 23 were obtained for a relative error of 1 percent.

The purpose of the iterative method is to solve large problems. The most important of the example problems may be the wire gridded rectangular strip, because large surfaces can be modeled by grids. The rectangle is 0.2 by 1.9 m, with uniform square mesh 0.1 m on a side. Each square edge was considered as one wire segment. The 97 segments were numbered as shown in Fig. 3. The wire radius was 0.0015. Two cases were investigated at a wavelength of one meter. In case 1, segments 48, 49, and 50 were uniformly excited. The longitudinal currents are then much the same as on a 1.9λ centerfed dipole. In case 2, segments 48 and 50 were excited in opposition. Currents on the central wire are zero, and currents are large only in the vicinity of the excited segments. Tables I and II show the results.

For case 1 at a bandwidth of 5, convergence was slow. Termination was automatic after 15 iterations. For larger bandwidths, convergence was rapid. The exact solution required 4.00 s. The best efficiency based on CP times, about 7, was obtained at bandwidths of 10 to 20.

For case 2, convergence was rapid even at a bandwidth of 5. The exact solution required 3.89 s. The best efficiency was about 11, at a bandwidth of 10. At a bandwidth of 20, the efficiency is down to about 8.

For both case 1 and case 2, a bandwidth of 10 provides peak efficiency. This bandwidth corresponds to a distance of 0.2λ along the grid. From the limited results for single straight wires, the optimum bandwidth for a 1.9λ wire should correspond to a distance of about 0.3λ. This would be equivalent to a bandwidth of about 15 for the wire grid. Interactions at moderate distances may be relatively less important for wire grids than for single wires due to the larger number of neighboring segments at small distances.

Assuming that extrapolation of performance to larger surfaces is valid, efficiencies can be estimated. Consider a square surface of edge length 3λ, modeled with square mesh of size 0.1λ. Numbering is along strips starting at one corner. A bandwidth corresponding to a distance of 0.2λ means that M/N is about 0.2/3. An optimistic upper bound for the efficiency is then 75. If a bandwidth corresponding to a distance of 0.5λ is used, the efficiency is at best 12. Assuming no cost for out-of-core manipulations, the exact solution would require about 7 h. An efficiency of 12 would reduce the time to about 35 min, and an efficiency of 75 would reduce it to about 6 min.

A solution with a relative error much larger than 1 percent was found [5] to be adequate for computing far-fields for many engineering purposes. Using solutions with relative errors as high as 30 percent, patterns have been obtained that have the general appearance of the exact pattern for examples d), f), g), and h) listed above. Solution efficiencies are higher because fewer iterations are required. However, no adequate relationship between pattern accuracy and solution accuracy has been established.

A simple analysis based on traveling waves is available [10] for example h). The geometry is shown in Fig. 4. At a wavelength of 30 m, the total wire length of 240 m is equivalent to 8 wavelengths. With 10 segments per wavelength, N is 80. Segment numbering proceeds from the feed end to the terminal end. Ground effects for sea water are implemented in the computer program used to generate impedance matrices and far-fields [9] by a reflection coefficient method.

The exact solution obtained by Gaussian elimination for example h) required 2.7 s. Results for the iterative solution method are shown in Table III. For the bandwidths shown, the best efficiency based on CP times at an RE of 1 percent was about 10. This was obtained with M equal to 12 and k equal to 2. The time required for banded matrix decomposition is about equal to that required for two iterations for this case. At a bandwidth of 4, the time for decomposition is one-fifth of the time required for a single iteration. Convergence is slow, and the efficiency is less than 3. At a bandwidth of 17, about three-fourths of the time is spent in decomposition, and the efficiency is about 8. For larger bandwidths, only the cost of decomposition is important.

Using the exact solution, the far-field pattern for example h) was computed at 2° increments in the vertical plane containing the antenna. The pattern is shown in Fig. 5. The main lobe is at 88°, or 2° above horizontal. The exact field is listed in the second column of Table IV at angles corresponding to near

TABLE III
RELATIVE ERROR (PERCENT) AND CP TIME FOR VERTICAL HALF-RHOMBIC ANTENNA

ITERATIONS	BANDWIDTH 4	5	6	7	12	17
1	67.67	53.91	40.47	38.91	11.05	6.13
2	46.09	28.89	16.14	15.12	1.36	.59
3	31.62	15.64	6.53	5.94	.18	
4	21.69	8.46	2.64	2.33		
5	14.87	4.57	1.06	.91		
6	10.20	2.47	.43			
7	7.00	1.34				
8	4.80	.72				
9	3.29					
10	2.26					
11	1.55					
12	1.06					
13	.73					
TIME	1.027	0.623	0.476	0.362	0.288	0.338

TABLE IV
FAR-FIELD IN VERTICAL PLANE FOR HALF-RHOMBIC ANTENNA USING EXACT AND APPROXIMATE SOLUTIONS

M: / RE(%):	EXACT	4 / 21.69	5 / 28.89	6 / 16.14	7 / 15.12
Angle (deg)	Far Field (volts/meter)				
8	.046	.045	.044	.044	.044
18	.049	.047	.044	.046	.046
26	.010	.012	.012	.012	.012
34	.218	.216	.205	.204	.200
42	.039	.044	.044	.040	.039
48	.216	.211	.197	.199	.195
56	.022	.022	.124	.022	.022
64	.020	.018	.041	.018	.019
70	.144	.134	.124	.129	.128
76	.043	.042	.041	.041	.041
84	.493	.476	.439	.447	.440
88	.584	.561	.417	.529	.520

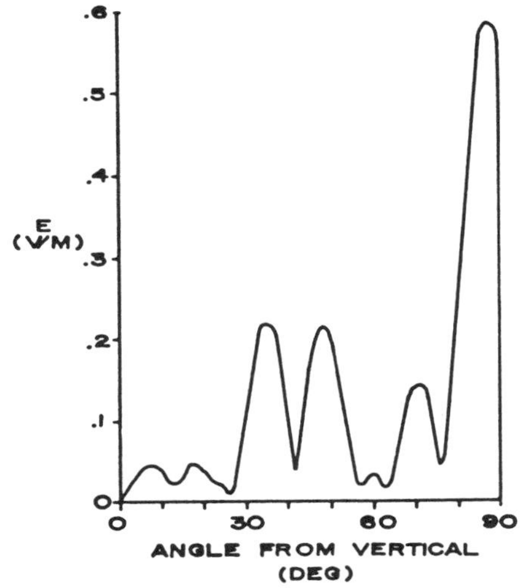

Fig. 5. Far-field pattern for vertical half-rhombic antenna over sea water.

maxima and minima in Fig. 5. Fields calculated from approximate solutions are listed in the remaining columns. (The iterative process was interrupted for field calculations when the relative error was less than 30 percent.) As is evident, the patterns obtained from approximate solutions are quite similar to the exact pattern. Similar results were obtained at bandwidths of 12 and 17. Higher efficiencies are obtained by relaxing the convergence criterion to 30 percent. The efficiency is about 18 to 20 for bandwidths of 5, 6, 7, and 12, and about 9 at M equal to 4. At M equal to 17, only one iteration is saved by the higher error termination. The efficiency is then only about 10, compared to 8 at a convergence criterion of 1 percent.

This same result was found for each example problem. No matter what value is chosen for the convergence criterion, the best efficiency is obtained by using a reasonable balance in times expended for decomposition and iteration. A natural consequence is that the bandwidth for peak efficiency decreases as the convergence criterion is relaxed. Compared to efficiencies for a relative error of one percent, efficiencies for a relative error up to 30 percent were higher by a multiplicative factor of 2 to 5.

VII. CONCLUSIONS

The banded matrix iterative method for solving the linear simultaneous equations arising from thin wire moments problems has provided economical solutions for a variety of small problems at an accuracy of 1 percent. Solution efficiencies relative to Gaussian elimination were generally about 5 to 10 and up to 23. Efficiencies were further increased by multiplicative factors of 2 to 5 by using solutions with errors up to 30 percent. These solutions yielded reasonable far-fields for the problems investigated.

For single straight wires, the solution error decreases exponentially with the number of iterations. The rate of decrease is dependent on the wire excitation, length, and radius, and on the bandwidth. The bandwidth for peak efficiency increases roughly in proportion to the length. The solution error at the first iteration decreases monotonically with increasing bandwidth except for moderate variations at half-wavelength intervals.

For any problem, peak efficiencies are obtained by using a reasonable balance in computer times for banded matrix decomposition and for all iterations. Consequently, the bandwidth for peak efficiency decreases as the convergence criterion is relaxed. Proper segment numbering is required to keep the bandwidth small and obtain high efficiencies. A general approach to segment numbering based on geometrical considerations has been developed. More than one numbering scheme may provide an efficient solution. The choice of bandwidth depends on the geometry of the problem and the choice of numbering schemes, and somewhat on the convergence criterion to be used. Sufficient expertise for reasonable application of the method to many problems can be obtained from documented example problem results.

For wire grids, limited results indicate that the bandwidth for peak efficiency corresponds to a somewhat smaller distance than observed for single straight wires. Projected efficiencies for large wire grids indicate that the banded matrix iterative method may provide solutions for large surfaces at a reasonable cost.

References

[1] R. F. Harrington, *Field Computation by Moment Methods*. New York: Macmillan, 1968.

[2] V. V. Klyuyev and N. I. Kokovkin–Shcherbak, "On the minimization of the number of arithmetic operations for the solution of linear algebraic systems of equations," (transl. by G. J. Tee), Computer Science Department, Stanford University, Stanford, Calif., Tech. Rep. CS24, 1965.

[3] F. K. Oshiro *et al.*, "Calculation of radar cross section, part II, analytical report," Northrup Corporation AFAL-TR-70-21, Part 2, 1970.

[4] T. R. Ferguson, "Iterative techniques in the method of moments," The EMCAP (Electromagnetic Compatability Analysis Program) Third Quarter Tech. Rep., Part 2, The BDM Corporation, Albuquerque, N. Mex. (RADC-TR-75-121).

[5] ——, "Solution of thin wire moments problems by banded matrix iteration," The EMCAP (Electromagnetic Compatability Analysis Program) Fourth Quarter Tech. Rep., Part 2, The BDM Corporation, Albuquerque, N. Mex. RADC-TR-75-189.

[6] D. J. Rose and R. A. Willoughby, Ed., *Sparse Matrices and Their Applications*. New York: Plenum, 1972.

[7] G. E. Forsythe and C. B. Moler, *Computer Solution of Linear Algebraic Systems*. Englewood Cliffs, N.J.: Prentice-Hall, 1967.

[8] J. R. Stewart, "Application of optimization theory to electromagnetic radiation and scattering," Ph.D. dissertation, Syracuse Univ., Syracuse, N.Y., Jan. 1974.

[9] F. J. Deadrick and E. K. Miller, "Program WAMP (Wire Antenna Modeling Program)," Lawrence Livermore Laboratory, Dec. 1973. (UCID-30084). A Modified Version of a Program Created by MB Associates.

[10] M. T. Ma, *Theory and Application of Antenna Arrays*. New York: Wiley, 1974.

Survey of Numerical Methods for Solution of Large Systems of Linear Equations for Electromagnetic Field Problems

TAPAN K. SARKAR, SENIOR MEMBER, IEEE, KENNETH R. SIARKIEWICZ, SENIOR MEMBER, IEEE, AND ROY F. STRATTON

Abstract—**Many of the popular methods for the solution of large matrix equations are surveyed with the hope of finding an efficient method suitable for both electromagnetic scattering and radiation problems and system identification problems.**

I. INTRODUCTION

THE PROBLEM of radiation and scattering from electromagnetic structures may be formulated in terms of the E-field, the H-field, or the combined field integral equations. The integral equations are then reduced to matrix equations by the method of moments. Hence the maximum size of an electromagnetic field problem that can be solved by this technique depends on how efficiently solutions of a set of simultaneous equations are obtained.

In system identification, on the other hand, the problem is formulated in terms of a convolution integral. When any of the standard techniques is utilized to identify the system, one again encounters a set of simultaneous equations. The only difference between the two cases is that in the former, one often encounters a matrix which has large elements on the diagonal whereas in the latter case, the matrix may be nearly singular.

The objective of this paper is to survey many of the popular methods for the solution of large matrix equations with the hope of finding an efficient method suitable for both electromagnetic scattering and radiation problems and system identification problems.

In our discussions of the various methods, only references which are directly relevant are noted. No attempt has been made to cite the earliest sources. In many cases additional references may be found in the papers mentioned.

II. DIRECT METHODS FOR SOLVING MATRIX EQUATIONS [1]–[3]

In this section we present all the direct methods. These include Cramer's rule and two versions of Gaussian elimination [LU decomposition and the compact method]. The Cramer's rule is too well-known to elaborate and too cumbersome for practical use. The other two methods of Gaussian elimination consist in solving $\mathbf{AX} = \mathbf{Y}$ in two parts. This is achieved by decomposing $\mathbf{A}$ into $\mathbf{LU}$ where $\mathbf{L}$ is a lower triangle matrix and $\mathbf{U}$ is an upper triangle matrix. Thus $\mathbf{LUX} = \mathbf{Y}$ represents two triangular systems

$$\mathbf{LG} = \mathbf{Y} \tag{1}$$

$$\mathbf{UX} = \mathbf{G} \tag{2}$$

Manuscript received March 10, 1980; revised January 5, 1981. This work was supported by the RADC post-doctoral program.

T. K. Sarkar is with the Department of Electrical Engineering, Rochester Institute of Technology, Rochester, NY 14623.

K. Siarkiewicz and R. Stratton are with RADC/RBCT, Griffiss Air Force, Rome, NY 13441.

which can very easily be solved. The calculation of $\mathbf{L}$ and $\mathbf{U}$ together with the solution of $\mathbf{LG} = \mathbf{Y}$ is usually called the forward elimination and the solution of $\mathbf{UX} = \mathbf{G}$ is the backward substitution. It is important to note that this method is really optimum in obtaining an exact solution (assuming there is no truncation or roundoff error) if and only if one is interested in handling the elements of the matrices by rows or by columns. Under this condition Klyuyev and Kokovkin-Scherbak [4] have proved that no general system of linear equations can be solved with fewer arithmetic operations than are required by Gaussian elimination. Recently Volker Strassen [5] has shown that it is possible to solve a general system of linear equation with $\theta(N^p)$ arithmetic operations, where in this case $p = \log_2 7 = 2.807$. Here θ is defined as terms of the order of N^p. However it is not known whether this value of p is the minimum exponent.

The advantage of obtaining a solution in a finite number of steps is offset by the building of truncation and round-off errors in direct methods. It can be shown [6] that if $\Delta\mathbf{A}$ are the uncertainties associated with $\mathbf{A}$ and $\Delta\mathbf{Y}$ are the inaccuracies associated with the representation of $\mathbf{Y}$, then the uncertainty $\Delta\mathbf{Y}$ in the solution $\mathbf{X}$ is given by

$$\begin{aligned}\frac{\|\Delta\mathbf{X}\|_2}{\|\mathbf{X}\|_2} &\leqslant \frac{\text{cond}\,[\mathbf{A}]}{1-\sqrt{N}\cdot\text{cond}\,[\mathbf{A}]\cdot 2^{-t}} \cdot \left[\frac{\|\Delta\mathbf{Y}\|_2}{\|\mathbf{Y}\|_2} + \frac{\|\Delta\mathbf{A}\|_2}{\|\mathbf{A}\|_2}\right] \\ &\leqslant \frac{2^{-t}\,[\sqrt{N}+1]\,\text{cond}\,[\mathbf{A}]}{1-\sqrt{N}\cdot\text{cond}\,[\mathbf{A}]\cdot 2^{-t}}\end{aligned} \tag{3}$$

where t is the number of binary digits with which computation is actually carried out in the computer, N is the dimension of $\mathbf{A}$ and cond $[\mathbf{A}]$ represents the condition number of the matrix A, which is the ratio between the maximum and the minimum eigenvalues of $\mathbf{A}$. *Thus there is absolutely no way to recognize an accurate* solution $\mathbf{X}$ given by Gaussian elimination unless [6]

$$\sqrt{N}\cdot 2^{-t}\cdot\text{cond}\,[\mathbf{A}] \ll 1. \tag{4}$$

As an example, consider the solution of the following problem $\mathbf{AX} = \mathbf{Y}$ by Gaussian elimination. Let $\mathbf{A}$ be the ill-conditioned Hilbert matrix and $\mathbf{Y}$ be that vector for which $\mathbf{X} = \{1, 2, 3, \cdots, N\}$. The problem then is to find $\mathbf{X}$ given $\mathbf{A}$ and $\mathbf{Y}$. We solved this problem on the Xerox Sigma-9 computer where computation is carried out using 24 binary digits. For a fourth order Hilbert matrix the condition number is obtained as cond $[\mathbf{A}_4] = 1.55 \times 10^4$ [2]. Thus for a fourth order Hilbert matrix, (3) reduces to

$$\frac{\|\Delta\mathbf{X}\|_2}{\|\mathbf{X}\|_2} \leqslant 0.00278$$

Reprinted from *IEEE Trans. Antennas Propagat.*, vol. AP-29, no. 6, pp. 847–856, November 1981.

and hence a very good accuracy in the results is expected. However for a fifth order Hilbert matrix

$$\frac{\|\Delta \mathbf{X}\|_2}{\|\mathbf{X}\|_2} \leqslant 0.09825$$

since cond $[A_5] = 4.77 \times 10^2$ [2]. This is reflected in the following results:

$$\mathbf{X} \rightarrow 0.996;\ 2.067;\ 2.708;\ 4.442;\ 4.783$$

$$\frac{\|\Delta \mathbf{X}\|_2}{\|\mathbf{X}\|_2} \rightarrow 0.07772.$$

Note that the theortetical error bound is large. For a sixth order Hilbert matrix we have

$$\mathbf{X} \rightarrow 1.007;\ 1.786;\ 4.509;\ -0.0341;\ 9.543;\ 4.182$$

$$\mathbf{X}_{\text{exact}} \rightarrow 1.0;\ 2.0;\ 3.0;\ 4.0;\ 5.0;\ 6.0.$$

In this case

$$\frac{\|\Delta \mathbf{X}\|_2}{\|\mathbf{X}\|_2} \leqslant -2.59!! \qquad \text{(from (3))}$$

since cond $[A_6] = 1.5 \times 10^7$ from [2].

These results prompt us to look for alternate methods in which we could reduce the effects of round-off error in solving a system of equations. The effect of round-off error may become pronounced not only for very ill-conditioned matrices but also for large systems of equations in which a large number of arithmetic operations must be carried out. Iterative methods are good alternatives to rectify this problem of roundoff error. For example, for a seventh order Hilbert matrix for which we know the direct method would not work, we obtained this result by the conjugate gradient method at the end of seven iterations:

$$\mathbf{X} \rightarrow 0.993;\ 2.090;\ 2.768;\ 4.034;\ 5.208;\ 6.121;\ 6.780$$

$$\mathbf{X}_{\text{exact}} \rightarrow 1.0;\ 2.0;\ 3.0;\ 4.0;\ 5.0;\ 6.0;\ 7.0.$$

This is because in iterative methods, the roundoff error is limited only to the most recent stage of iteration.

III. ITERATIVE METHODS FOR SOLVING MATRIX EQUATIONS

The basic philosophy of the iterative methods is discussed in this section. It can be shown [7], [8] that the solution of the set of equations $\mathbf{AX} = \mathbf{Y}$ is equivalent to the maximization/minimization of the functional $F(\mathbf{X}) = 1/2 \langle \mathbf{AX}, \bar{\mathbf{X}}^* \rangle - \langle \mathbf{Y}, \mathbf{X}^* \rangle$ if $\mathbf{A}$ is negative/positive definite, the asterisk denotes the conjugate, and T denotes the transpose conjugate. The contours of constant $F(\mathbf{X})$ are generally N-dimensional ellipsoids. Also, the residuals $\mathbf{R}_n (= \mathbf{AX}_n - \mathbf{Y})$ at the end of each step are normals to the ellipsoid at $\mathbf{X}_n$. The paths $\mathbf{P}$ by which one reaches the center point of the ellipsoid (which is the solution $\mathbf{X}_{\text{exact}}$) are different for the different iterative methods. An iterative process is called linear if the present estimate $\mathbf{X}_n$ is a linear combination of the past estimates $\mathbf{X}_0, \mathbf{X}_1, \mathbf{X}_2, \cdots, \mathbf{X}_{n-1}$. Otherwise the iterative process is nonlinear. A process is called a stationary iterative process if the rule by which $\mathbf{X}_n$ is determined does not change from iteration to iteration. Otherwise the iterative process is called nonstationary. Nonstationary methods are not pursued in this presentation because some ideas are needed about the magnitude of the maximum and the minimum eigenvalues of $\mathbf{A}$ for these methods to be effective. This information is seldomly available.

A. Linear Iterative Methods

In this section we present Gauss's hand relaxation method, Jacobi's cyclical iteration method (simultaneous displacement method), Seidel's method (successive displacement method), back and forth Seidel's method, and the successive overrelaxation method (SOR).

1) *Gauss's Hand Relaxation Method [3]:* This method was developed by Gauss for hand calculation and is mostly of historical significance. In this method the elements of the search direction vector $\mathbf{P}_n$ are chosen corresponding to greatest residual in absolute value. Mathematically, the iterations are obtained as

$$X_{n+1}{}^k = X_n{}^k - \frac{R_n{}^k}{A^{kk}} \cdot P_n{}^k. \tag{5}$$

This method is not very suitable for automatic computation as it is a very laborious process to search for the largest absolute element in the residual.

2) *Jacobi's Cyclical Iteration Method (Simultaneous Displacement Method [9]:* Here the search directions $\mathbf{P}$ now run cyclically through the coordinate directions regardless of the residuals. Thus the ith element of $\mathbf{X}$ at the $n + 1$ iteration is refined by

$$X_{n+1}{}^i = \frac{1}{A^{ii}} \left[Y^i - \sum_{\substack{j=1 \\ j \neq i}}^{N} A^{ij} \cdot X_n{}^j \right] \qquad \text{for } i = 1, 2, \cdots, N. \tag{6}$$

3) *Seidel's Method (Successive Displacement Method) [9] (often incorrectly called the Gauss-Seidel method):* The rate of convergence of Jacobi's method was improved by Seidel in modifying (6) in the following way

$$X_{n+1}{}^k = \frac{1}{A^{kk}} \left\{ Y^k - \sum_{q=1}^{k-1} A^{kq} X_{n+1}{}^q - \sum_{q=k+1}^{N} A^{kq} X_n{}^q \right\} \tag{7}$$

or

$$(\mathbf{D} + \mathbf{L})\mathbf{X}_{n+1} + \mathbf{UX}_n = \mathbf{Y}$$

where $\mathbf{A} = \mathbf{D} + \mathbf{L} + \mathbf{U}$, where $\mathbf{D}, \mathbf{L}, \mathbf{U}$ are the diagonal, lower, and upper triangular matrices of $\mathbf{A}$, respectively. This is achieved by updating the column vector $\mathbf{X}_n$ as soon as one of its components has been calculated and using that updated value in the same iteration rather than waiting for the next iteration as was done by Jacobi. Thus

$$\begin{aligned} \mathbf{X}_{n+1} &= -[\mathbf{D} + \mathbf{L}]^{-1}\mathbf{UX}_n + [\mathbf{D} + \mathbf{L}]^{-1}\mathbf{Y} \\ &\triangleq \mathbf{QX}_n + \mathbf{S}. \end{aligned} \tag{8}$$

TABLE I
VARIOUS COMPONENTS OF X AT THE END OF EACH ITERATION[1]

i	X_0	X_1	X_2	X_3	X_4	X_5	X_6	X_7
1	1.00	−1.60	−0.83	−0.41	−0.18	−0.062	−0.0080	0.01220
2	1.00	0.32	0.00	−0.12	−0.13	−0.107	−0.0761	0.04927
3	1.00	0.80	0.56	0.36	0.21	0.115	0.0577	0.02561
4	1.00	0.21	0.11	0.06	0.03	0.012	0.0034	−0.0007

X_8	X_9	X_{10}
0.01656	0.014576	0.010786
−0.02953	−0.016425	−0.008403
0.00907	0.001421	0.001527
−0.00127	−0.001514	−0.001164

[1] Note that the ratios do not approach any limit. This is shown in Table II.

The necessary and sufficient condition for the convergence of (8) is that the largest eigenvalue of Q be less than unity in magnitude. Other convergence criteria will be discussed later on.

The major drawback of the linear iterative schemes in general and particularly that of Seidel's method is that the convergence is quite irregular if the dominant eigenvalue of Q in (8) is complex. As an example, consider the following problem.

Example 1: Let A be the matrix

$$\begin{bmatrix} 1.0 & 0.7 & 0.7 & 0.2 \\ 0.7 & 1.0 & 0.7 & 0.1 \\ 0.7 & 0.7 & 1.0 & 0.1 \\ 0.2 & 0.1 & 0.1 & 1.0 \end{bmatrix},$$

and we wish to solve the matrix equation

$$AX = 0.$$

Since det $[A] \neq 0$, the only possible solution is $X = 0$. The various iterates are given in Table I. Such erratic behavior is to be expected since the eigenvalues of the matrix Q in (8) are complex. The eigenvalues of Q

$$-Q = [D+L]^{-1}[U] = \begin{bmatrix} 0 & 0 & 0 & 0 \\ 0 & 0.49 & 0.147 & 0.0763 \\ 0 & 0.147 & 0.5341 & 0.0439 \\ 0 & 0.0763 & 0.0439 & 0.0228 \end{bmatrix}$$

are $\lambda = 0, 0.028333, 0.566733 \pm j0.157158$. That is, the dominant eigenvalues of Q are complex conjugates. Thus although the Seidel scheme is convergent in this case, the convergence is erratic. This erratic behavior of the Seidel process is remedied by the "back and forth" Seidel process (see Table II).

4) Back and Forth Seidel Process [9]: The back and forth Seidel process was designed by Aitken and Rosser to overcome the irregular convergence of the Seidel process. This is achieved by making all the eigenvalues of the iterative matrix Q real. It proceeds as follows. Start with a first approximation vector X_0 and then obtain $\bar{X}_1$ by the regular Seidel process as

$$\bar{X}_1 = -[D+L]^{-1}UX_0 + [D+L]^{-1}Y. \tag{9}$$

Then find the next iterate by applying the Siedel process to the equations in reverse order, i.e.,

$$\begin{aligned} X_1 &= -[D+U]^{-1}L\bar{X}_1 + [D+U]^{-1}Y \\ &= [D+U]^{-1}L[D+L]^{-1}UX_0 \\ &\quad - [D+U]^{-1}L[D+L]^{-1}Y + [D+U]^{-1}Y. \end{aligned} \tag{10}$$

Thus we see that for this process, the iteration matrix is

$$S = [D+U]^{-1}L[D+L]^{-1}U. \tag{11}$$

If A is assumed to be symmetric $L = U^T$ and $U = L^T$ (here T denotes transpose) and so the iteration matrix can be rewritten as

$$\begin{aligned} S &= [D+U]^{-1}U^T[D+U^T]^{-1} \\ &= [D+U]^{-1}U^T[\{D+U\}^{-1}]^TU = MM^T \end{aligned} \tag{12}$$

where $M = [D+U]^{-1}U^T$. Thus S is similar to a nonnegative matrix. Since det $[M]$ = det $[U] = 0$, MM^T is semi-definite. So we can conclude that all the eigenvalues of S are on the real half line $x \geqslant 0$ and hence the dominant eigenvalue of S is unique, though possibly it may be a multiple root.

Example 2: We now apply the back and forth Seidel process to Example 1. We again start with the same initial guess. The results are summarized in Table III.

The results from Table IV indicate that these ratios are tending to 0.67, i.e., that the dominant eignevalues of S in this case are 0.65611, 0.36368, 0.02720, and 0. The dominant eigenvalue is real and is equal to

$$\lim_{n\to\infty} \left\{ \frac{x_{n+1}{}^i}{x_n{}^i} \right\}.$$

In this example, the convergence of the back and forth Seidel process, while slower than that of the ordinary Seidel process, is much more regular than the ordinary Seidel process. However the back and forth Seidel process can easily be accelerated. It is not certain, however, which process would give the most accuracy per unit of labor.

Also in most method of moments problems, we encounter a matrix A whose eigenvalues are often complex. The use of the back and forth Seidel process in these problems can be justified; but, as we shall show later, there are faster schemes to treat these problems.

TABLE II
RATIOS OF $X_{n+1}{}^i/X_n{}^i$ FOR SEIDEL'S PROCESS.

i	$X_1{}^i/X_0{}^i$	$X_2{}^i/X_1{}^i$	$X_3{}^i/X_2{}^i$	$X_4{}^i/X_3{}^i$	$X_5{}^i/X_4{}^i$	$X_6{}^i/X_5{}^i$	$X_7{}^i/X_6{}^i$
1	−1.60	0.52	0.49	0.44	0.344	0.1290	−1.525
2	0.32	0.00	−∞	1.08	0.823	0.7112	0.6474
3	0.80	0.70	0.64	0.58	0.548	0.5017	0.4438
4	0.21	0.52	0.55	0.05	0.400	0.2833	−0.0205

i	$X_8{}^i/X_7{}^i$	$X_9{}^i/X_8{}^i$	$X_{10}{}^i/X_9{}^i$
1	1.3574	0.8802	0.739983
2	0.5994	0.5562	0.511597
3	0.3542	0.1567	−1.074603
4	18.1429	1.1114	0.822614

TABLE III
VARIOUS ITERATES OF BACK AND FORTH SEIDEL PROCESS[1]

i	X_0	$\bar{X}_1$	X_1	$\bar{X}_2$	X_2	$\bar{X}_3$	X_3	$\bar{X}_4$
1	1.0	−1.60	−0.99	−0.99	−0.64	−0.64	−0.42	−0.417
2	1.0	0.32	0.48	0.06	0.23	−0.01	0.12	−0.032
3	1.0	0.80	0.88	0.63	0.64	0.43	0.45	0.305
4	1.0	0.21	0.21	0.13	0.13	0.09	0.09	0.056

i	X_4	$\bar{X}_5$	X_5	$\bar{X}_6$	X_6
1	−0.277	−0.27650	−0.18437	−0.18437	−0.12332
2	0.070	−0.02835	0.04305	−0.02144	0.02746
3	0.309	0.20780	0.20966	0.14033	0.14157
4	0.056	0.03736	0.03736	0.02499	0.2499

[1] The ratios X_{k+1}/X_k are given in Table IV.

TABLE IV
RATIO OF X_{n+1}/X_n FOR BACK AND FORTH SEIDEL PROCESS

i	X_1/X_0	X_2/X_1	X_3/X_2	X_4/X_3	X_5/X_4	X_6/X_5
1	−0.99	0.65	0.66	0.660	0.66560	0.66887
2	0.48	0.48	0.52	0.583	0.61500	0.63787
3	0.88	0.73	0.70	0.687	0.67851	0.67524
4	0.21	0.62	0.70	0.622	0.66714	0.66890

5) The Method of Successive Over/Under Relaxation [1], [9], [12]-[16]: In the case of large systems of equations, the Jacobi or Seidel process converges poorly when the maximum absolute eigenvalue (often referred to as the spectral radius) of the iteration matrix (for example Q in (8) lies close to unity. Convergence of the Seidel process could be improved if instead of reaching the minimum point of $F(\mathbf{X})$ we go beyond this point by a certain amount. It seems paradoxical at first to refrain from minimizing the quadratic functional at each iteration step with the goal of achieving better convergence. Let us now modify (8) in the following way:

$$(\mathbf{D}+\mathbf{L})\mathbf{X}_{n+1}+\mathbf{U}\mathbf{X}_n=\mathbf{Y}$$

or

$$\mathbf{D}(\mathbf{X}_{n+1}-\mathbf{X}_n)=\mathbf{Y}-\mathbf{U}\mathbf{X}_n-\mathbf{D}\mathbf{X}_n-\mathbf{L}\mathbf{X}_{n+1}.$$

We now introduce the parameter ω and define the new iteration

$$\omega^{-1}\mathbf{D}(\mathbf{X}_{n+1}-\mathbf{X}_n)=\mathbf{Y}-\mathbf{U}\mathbf{X}_n-\mathbf{D}\mathbf{X}_n-\mathbf{L}\mathbf{X}_{n+1}$$

or

$$\mathbf{X}_{n+1}=[\mathbf{I}-(\omega^{-1}\mathbf{D}+\mathbf{L})^{-1}\mathbf{A}]\mathbf{X}_n+(\omega^{-1}\mathbf{D}+\mathbf{L})^{-1}\mathbf{Y}.$$

In this case the iteration matrix is $[\mathbf{I}-(\omega^{-1}\mathbf{D}+\mathbf{L})^{-1}\mathbf{A}]$ where $\mathbf{I}$ is the identity matrix. We are now interested in determining ω so as to give this matrix a small maximum eigenvalue. It is interesting to note that in symmetric definite systems of equations, the relaxation methods converge to the solution for any fixed value of ω in the range $0<\omega<2$. For $0<\omega<1$, the method is referred to as underrelaxation and for $1<\omega<2$, the method is called overrelaxation. It has been observed by Kahan and Young [11]-[13] that values of $\omega<1$ tends to reduce the rate of convergence whereas $\omega>1$ accelerates the rate of convergence. For $\omega>1$ we overcorrect the solution vectors and hence we speak of overrelaxation methods. Unfortunately for a given problem it is difficult to find the optimum choice of the relaxation parameter ω. For this, additional information about the structure of matrix A is necessary. Nonetheless we can say that the worse the condition of the matrix, the closer the optimum value of ω lies to 2. In such a case, at each iterative step we jump far beyond the minimum point to a new approximation which leaves the quadratic functional $F(\mathbf{X})$ almost as large as it was before. Hence the strategy of making the best improvement in each individual iterative step by going to the minimum point is not the best way of achieving the optimum long-term result.

The successive overrelaxation method has found wide application in the solution of boundary value problems by the finite difference method. In this particular type of problem one often encounters a very sparse matrix. For such matrix equations optimum values of ω have been given by Kahan and Young [11]-[13]. In the case of a full matrix it is difficult to find the optimum ω theoretically unless there is a certain structure to the matrix. Otherwise for each individual problem the optimum value of ω has to be obtained experimentally.

B. Nonlinear Iterative Methods

In nonlinear iterative methods, the refined estimate is no longer a linear function of the past estimates. Newton's method, because of its quadratic convergence $\{\|\mathbf{X}_n - \mathbf{X}_0\| \leq c\|\mathbf{X}_{n-1} - \mathbf{X}_0\|^2\}$, is mathematically the most preferred of the several known nonlinear methods for the solution of systems of equations. Practically however a very important limitation on Newton's method is that it does not generally converge to some solution for an arbitrary starting point. Thus Newton's method may fail to converge if the initial estimate is not sufficiently close to the solution.

The size of the domain of convergence depends upon the system of equations. For real algebraic equations, the size of the domain of convergence is generally inversely related to the degree and the number of equations. Therefore one finds that for two simultaneous second degree equations, almost any initial estimate will lead to one of the solutions; while for eight simultaneous tenth degree equations, the domain becomes much smaller, and it may be very difficult to obtain an initial estimate for which the iteration converges. Kantorovich thus modified Newton's method for optimization problems to become a rapidly converging descent method. This is done by selecting the direction vectors according to Newton's method but moving along them to a point that minimizes $F(\mathbf{X})$ in that direction.

1) Method of Steepest Descent [14]: From a geometric point of view, the steepest descent method involves describing a piecewise linear path with right angled corners in N-dimensional Euclidian space, with the path terminating at the minimum of the quadratic functional $F(\mathbf{X})$. This is achieved by generating

$$\mathbf{X}_{n+1} = \mathbf{X}_n + t_n \mathbf{A}^T \cdot \mathbf{R}_n \tag{13}$$

where

$$\mathbf{R}_n = A X_n - \mathbf{Y} \quad \text{and} \quad t_n = \frac{\langle \mathbf{A}^T\mathbf{R}_n, \ (\mathbf{A}^T\mathbf{R}_n)^*\rangle}{\langle \mathbf{A}\mathbf{A}^T\mathbf{R}_n, \ (\mathbf{A}^T\mathbf{R}_n)^*\rangle}. \tag{14}$$

Unfortunately it turns out that the choice of the best local direction along the largest residual reduction of $F(\mathbf{X})$ in each iteration does not always yield good convergence. This is illustrated by solving the same problem as presented in example 1. The various iterates are in Table V.

2) Conjugate Direction Method [1]: Conjugate direction methods are based on the generation of a set of A-orthogonal vectors and then minimizing successively in the direction of each of them. A set of vectors $\{\mathbf{P}_n\}$, $n = 1, 2, \cdots, N$ is said to be A-conjugate or A-orthogonal if they satisfy

$$\langle \mathbf{A}\mathbf{P}_i, \mathbf{P}_j^*\rangle = 0, \qquad \text{for } i \neq j.$$

Geometrically the method of conjugate directions is equivalent to that of finding the center of an N-dimensional ellipsoid when the starting point is on the surface of the ellipsoid. Thus the minimum point lies on a line parallel to a fixed nonnull vector $\mathbf{P}_k$ which lies on the $(N - 1)$ dimensional hyperplane

$$\langle \mathbf{P}_k^*, \mathbf{A}\mathbf{X} - \mathbf{Y}\rangle = 0 \tag{15}$$

whose normal is $\mathbf{A}\mathbf{P}$. This $(N - 1)$ dimensional plane contains the center point $\mathbf{X}_{\text{exact}} = A^{-1}\mathbf{Y}$ of the ellipsoid in the given space and is said to be conjugate to the vector $\mathbf{P}_k$. Thus the conjugate direction methods are finite step methods. That is, theoretically they all yield the exact solution at the end of a finite number of steps ($\leq N$), assuming no truncation and roundoff error.

The finite number of steps is equivalent to the number of independent eigenvalues of $\mathbf{A}$ provided the dependent eigenvalues do not constitute a Jordan canonical form. Thus if the eigenvalues are equal, $\mathbf{A}$ is proportional to an identity matrix and hence convergence would be obtained in one step.

But the conjugate direction method does not specify how to compute the vector $\mathbf{P}_k$. When the vectors $\mathbf{P}_k$ are obtained by A-orthogonalization of the unit coordinate vectors this particular conjugate direction method yields the popular Gaussian elimination. When the vectors $\mathbf{P}_k$ are obtained by A-orthogonalization of the residual vectors $\mathbf{R}_k$, a conjugate gradient method results. Thus the conjugate gradient method applies more constraints on the iteration process than those imposed by Gaussian elimination. Hence the conjugate gradient method may yield acceptable results under conditions where Gaussian elimination fails.

3) Conjugate Gradient Method [15]-[17]: For the solution of $\mathbf{A}\mathbf{X} = \mathbf{Y}$, the conjugate gradient method starts with an initial guess $\mathbf{X}_0$ and obtains

$$\mathbf{P}_0 = -\mathbf{A}^T\mathbf{R}_0 = -\mathbf{A}^T[\mathbf{A}\mathbf{X}_0 - \mathbf{Y}] \tag{16}$$

and then develops each successive approximation by

$$\mathbf{X}_{n+1} = \mathbf{X}_n + t_n\mathbf{P}_n \tag{17}$$

where

$$t_n = -\frac{\langle \mathbf{A}\mathbf{P}_n, \mathbf{P}_n^*\rangle}{\langle \mathbf{A}\mathbf{P}_n, (\mathbf{A}\mathbf{P}_n)^*\rangle} = \frac{\|\mathbf{A}^T\mathbf{R}_n\|^2}{\|\mathbf{A}\mathbf{P}_n\|^2}. \tag{18}$$

The residuals are generated as

$$\mathbf{R}_{n+1} = \mathbf{R}_n + t_n \cdot \mathbf{A}\mathbf{P}_n. \tag{19}$$

The direction vectors are obtained iteratively as

$$\mathbf{P}_{n+1} = -A^T\mathbf{R}_{n+1} + q_n\mathbf{P}_n \tag{20}$$

where

$$q_n = \frac{\langle \mathbf{A}\mathbf{P}_n, \ (\mathbf{A}\mathbf{A}^T\mathbf{R}_{n+1})^*}{\langle \mathbf{A}\mathbf{P}_n, \ (\mathbf{A}\mathbf{P}_n)^*\rangle} = \frac{\|\mathbf{A}^T\mathbf{R}_{n+1}\|^2}{\|\mathbf{A}^T\mathbf{R}_n\|^2}. \tag{21}$$

TABLE V
RESULTS OF VARIOUS ITERATIONS BY METHOD OF STEEPEST DESCENT[1]

i	X_0	X_1	X_2	X_3	X_4	X_5	X_6
1	1.0	−0.08125	−0.02817	−0.04386	−0.01563	−0.01713	−0.00997
2	1.0	−0.03967	0.04264	0.01892	0.01358	0.00739	0.00690
3	1.0	−0.03967	0.04264	0.01892	0.01358	0.00739	0.00690
4	1.0	0.41779	0.02524	0.01373	−0.00251	−0.00099	0.00202
		X_7	X_8	X_9	X_{10}		
		−0.01002	−0.00584	−0.00587	−0.00342		
		0.00408	0.00404	0.00238	0.00237		
		0.00408	0.00404	0.00238	0.00237		
		0.00122	0.00119	0.00072	0.00069		

[1] Note that the method of steepest descent converges much faster than either of the Seidel methods. Once again the tactic of seeking the most efficient goal by choosing the best local option does not lead to the best overall strategy. The rate of convergence of the method of steepest descent is discussed in Section-III.

TABLE VI
RESULTS OF VARIOUS ITERATIONS GIVEN BY THE CONJUGATE GRADIENT METHOD

i	X_0	X_1	X_2	X_3	X_4
1	1.0	−0.8125	−0.04848	0.6012×10^{-5}	0.7966×10^{-7}
2	1.0	−0.03966	0.02373	0.6217×10^{-5}	0.2905×10^{-6}
3	1.0	−0.03966	0.02373	0.6217×10^{-5}	0.2905×10^{-6}
4	1.0	0.41779	0.00599	0.1281×10^{-5}	-0.9905×10^{-6}

The conjugate gradient method is applied to solve the same problem presented in Example 1. The various iterates for the solution are shown in Table VI.

Observe that there is a sharp increase in the accuracy of the solutions at $\mathbf{X}_3$. One has obtained essentially an exact solution after three iterations. This is because the four eigenvalues of the matrix $\mathbf{A}$ are 2.4372, 0.9725, 0.300, and 0.2903. Note that there are approximately three independent eigenvalues of $\mathbf{A}$. Thus one would expect excellent results at the end of three steps. Hence the conjugate gradient method might converge quite rapidly for a large system of equations if the matrix has quite a few eigenvalues bunched together. This generally happens in matrices which have dominant diagonals (as in the magnetic field integral equation).

Next we derive the various theoretical rates of convergence of the various iterative schemes and show how the conjugate gradient method converges much faster than the others.

IV. ANALYSIS OF CONVERGENCE OF VARIOUS ITERATIVE METHODS [6]

The rates of convergence of the various iterative schemes, both linear and nonlinear, are discussed in this section. We show that for the linear iterative schemes, the rate at which the $\mathbf{X}_n$ approach the exact solution is linear and the $\mathbf{X}_n$ converge geometrically. The nonlinear iterative schemes on the other hand have a geometrical rate of convergence to begin with and possess "superlinear" convergence. The rates of convergence have been given when $\mathbf{A}$ is positive/negative definite.

A. Rate of Convergence for the Linear Iterative Schemes

Let a sequence $\{s_n\}$ converge to α. Then the sequence $\{s_n\}$ is said to have linear convergence if

$$e_{i+1} = (\beta + \sigma_i)e_i \tag{22}$$

where $e_i = \alpha - s_i$ and for a constant β, $|\beta| < 1$ and $\sigma_i \to 0$ as $i \to \infty$. The sequence $\{s_n\}$ is said to have geometric convergence if

$$e_{i+1} = \beta e_i \tag{23}$$

for $|\beta| < 1$. Thus geometric convergence is a special case of convergence in which all $\sigma = 0$. For a large number of iterations the linear iterative schemes converge geometrically as for sufficiently large m and $k > 0$

$$\|\mathbf{X}_{m+k} - \mathbf{X}_0\| < |\lambda_1|^k \|\mathbf{X}_m - \mathbf{X}_0\| \tag{24}$$

where λ_1 is the dominant eigenvalue of the iteration matrix $\mathbf{T}$ in the linear iterative schemes

$$\mathbf{X}_{n+1} = \mathbf{T}\mathbf{X}_n + \mathbf{W}. \tag{25}$$

Thus the smaller the eigenvalue of $\mathbf{T}$, the faster the convergence of the linear iterative schemes. Conversely, when the magnitude of the dominant eigenvalue of $\mathbf{T}$ is close to unity, many iterations are necessary. Hence the number of iterations necessary to reduce the error $\|\mathbf{X}_m - \mathbf{X}_0\|$ by a factor of ten is approximately inversely proportional to $-1/\{\log_{10}\lambda_1\}$. Thus to gain an additional significant decimal place in $\mathbf{X}_m$ demands k iterations from (24).

So the fastest rate of convergence that can be achieved by the linear iterative schemes can at best be geometric and the successive approximations always converge for a definite system of equations. Equivalently the latter condition may also be stated by saying $\|\mathbf{T}\| < 1$. This is also true if the diagonals of the matrix $\mathbf{T}$ are diagonally dominant.

B. Rate of Convergence for Nonlinear Iterative Schemes

1) Method of Steepest Descent: If the matrix $\mathbf{A}$ has N eigenvalues which are ordered as

$$B = \lambda_1 \geq \lambda_2 \geq \lambda_3 \cdots \geq \lambda_N = b > 0, \tag{26}$$

then it has been shown [14]-[18]

$$F(\mathbf{X}_k) - F(\mathbf{X}_{\text{exact}}) \leq \left(\frac{B-b}{B+b}\right)^{2k} \cdot \{F(X_0) - F(\mathbf{X}_{\text{exact}})\} \tag{27}$$

where B and b are the largest and the smallest eigenvalues of **A**, respectively. Also it can be shown that [14]

$$\|\mathbf{X}_k - \mathbf{X}_{exact}\|^2 \leqslant \frac{2}{b}\{F(\mathbf{X}_k) - F(\mathbf{X}_{exact})\}$$
$$\leqslant \frac{F(\mathbf{X}_0) - F(\mathbf{X}_{exact})}{0.5 \times b} \cdot \left(\frac{B-b}{B+b}\right)^{2k} \quad (28)$$

Thus the method of steepest descent converges geometrically to the exact solution. For the case when **A** has N distinct eigenvalues Akaike [19] has shown that this is the best possible estimate.

It has been shown by Daniel [17] and Hayes [9] that whenever **A** is a Legendre operator (i.e., A is a sum of positive definite bounded self-adjoint operator plus a completely continuous operator) the method of steepest descent converges faster than that of a geometric series with ratio greater than $(B - b)/(B + b)^2$. This type of convergence is referred to as "superlinear" convergence. Thus the method of steepest descent converges at worst like a geometric series, and in most cases the convergence is superlinear. For a finite dimensional case **A** is always a Legendre operator.

2) Method of Conjugate Gradient: The method of conjugate gradient generally requires a little more computation than that of steepest descent. However this slight increase in computation required leads to a significant improvement in the rate of convergence over that of steepest descent. It can be shown that [6]

$$\frac{\|\mathbf{X}_k - \mathbf{X}_{exact}\|}{\|\mathbf{X}_0 - \mathbf{X}_{exact}\|} \leqslant \frac{2}{\left(\frac{\sqrt{B}+\sqrt{b}}{\sqrt{B}-\sqrt{b}}\right)^k + \left(\frac{\sqrt{B}-\sqrt{b}}{\sqrt{B}+\sqrt{b}}\right)^k} \quad (29)$$

$$\leqslant 2\left(\frac{\sqrt{B}-\sqrt{b}}{\sqrt{B}+\sqrt{b}}\right)^k. \quad (30)$$

Observe that (29) is a better estimate than (30). As is well known [19], (29) cannot be improved upon.

Also as before, if **A** is a Legendre operator then the method of conjugate gradient converges faster than a geometric series with ratio greather than $(\sqrt{B} - \sqrt{b})/(\sqrt{B} + \sqrt{b})^2$ [19]. In an N-dimensional quadratic problem the error tends to zero in one step with Newton's method and within M steps with the conjugate gradient method, where M is the number of independent eigenvalues of **A**. Hence it is generally stated that the conjugate gradient method yields "quadratic convergence" $[\|\mathbf{X}_{m+1} - \mathbf{X}_0\| \leqslant c\|\mathbf{X}_m - \mathbf{X}_0\|^2]$ in the sense that it converges in M steps for an N-dimensional problem. So it seems appropriate to term this "$1/M$ quadratic convergence" since it requires M steps to achieve the effect of one step of a method with a true quadratic convergence rate.

V. ROUNDOFF ERRORS ASSOCIATED WITH ITERATIVE SCHEMES

In iterative methods the condition number of **A** has very little influence on roundoff error. The roundoff error in iterative methods is confined to the last stage of iteration only.

In a system of linear equations that arises from a practical problem, the elements of matrix **A** and **Y** may not be sharply defined. It is now assumed that all that is known about the typical element A^{ij} of matrix **A** or Y^i for the matrix **Y** is that they are within certain intervals. Here the subscript E denotes exact quantities

$$A_E^{ij} - \Delta A^{ij} \leqslant A^{ij} \leqslant A_E^{ij} + \Delta A^{ij} \quad (31)$$

$$Y_E^i - \Delta Y^i \leqslant Y^i \leqslant Y_E^i + \Delta Y^i. \quad (32)$$

It is also assumed that ΔA^{ij} and ΔY^i are independent distinct quantities due to round-off error and that one does not depend on the other. Thus

$$\mathbf{A}_E\mathbf{X}_E = \mathbf{Y}_E \quad (33)$$

and denote

$$\mathbf{R} = \mathbf{A}_E\mathbf{X} - \mathbf{Y}_E. \quad (34)$$

We check whether

$$\sum_j \Delta A^{ij} \cdot |X^j| + \Delta Y^i \geqslant |R^i| \quad (35)$$

for $i = 1, 2, \cdots, n$. As was shown by Oettli and Prager [21] the inequality (35) is a necessary and sufficient condition for **X** to be a solution of **AX** = **Y** under (31) and (33). So in any iteration method where the residuals are computed routinely, if for a certain residual the inequality (35) is satisfied, we have obtained an excellent solution under the conditions (31) and (32). Note that the introduction of the conditions (31) and (32) does not make the solution of **AX** = **Y** unique. In fact, there are many solutions of **AX** = **Y**. But only those solutions are acceptable which satisfy the inequality (35). Thus the upper and lower bounds for a certain component of the solution X are obtained by solving the linear programming problem [22]

$$\begin{array}{l|l} \min \rightarrow & R^i - \sum_j \Delta A^{ij} \cdot |X^j| - \Delta Y^i \leqslant 0 \\ X^j & \\ \max \rightarrow & -R^i - \sum_j \Delta A^{ij} \cdot X^j - \Delta Y^i \leqslant 0 \end{array}$$

$$\text{for } i = 1, 2, \cdots, N. \quad (36)$$

Iterative methods may be quite advantageous for large systems of matrices or for ill-conditioned matrices as compared to a direct method like Gaussian elimination. This is because *cond* (**A**) *does not arise in round-off error analysis of iterative methods*; and this is the reason we have been able to solve a 7 X 7 system of equations when **A** is a Hilbert matrix by the conjugate gradient method where Gaussian elimination has failed.

VI. CORE STORAGE REQUIRED FOR VARIOUS METHODS [1]

The core storage required for various methods is listed in order of the amount of core storage required, starting with the method requiring the least core storage.

Method	*Core Storage*
Gaussian elimination	$N^2 + 2N$
Seidel's iterative method	$N^2 + 2N$

Gaussian elimination with complete pivoting	$N^2 + 3N$
Jacobi's iterative method	$N^2 + 3N$
Method of Steepest Descent	$N^2 + 4N + 2$
Conjugate gradient method	$N^2 + 6N + 3$.

Note that Gaussian elimination with no pivoting and Seidel's method require the least amount of storage and that the conjugate gradient method requires the largest amount of storage.

VII. OPERATIONS REQUIRED FOR VARIOUS METHODS

The number of divisions, multiplications, and additions/subtractions provide a rough estimate of the efficiency of the algorithm. For each method it is possible to estimate the number of arithmetic operations as a function of N, the order of the matrix. Such functions could be discontinuous if N is large enough that auxiliary storage is required. In the total number of arithmetic operations we have *not* included the timings taken for recording of intermediate results and the time taken in searching for the pivotal element in Gaussian elimination.

Method	Division $\div$	Multiplications $\times$	Additions $+$
Gaussian elimination	N	$\frac{N^3}{3} + N^2 - \frac{N}{3}$	$\frac{N^3}{3} + \frac{N^2}{2} - \frac{5N}{6}$ (total).

(Gaussian elimination with complete pivoting requires $N^3/3 + N^2/2 - 5N/6$ comparisons in addition to the above arithmetic operations)

Jacobi and Seidel	N	N^2	$N^2 - N$ per iteration

(We could do with N divisions only once rather than per iteration at the expense of N more storage spaces.)

Steepest descent for (unsymmetric) matrix **A**	1	$2N^2 + 3N$	$2N^2 + 4N$ per iteration
Conjugate gradient for (unsymmetric) matrix **A**	2	$2N^2 + 6N$	$2N^2 + 6N$ per iteration

For symmetric matrices the work is reduced by half.

For vary large values of N, an iterative method applied to a full symmetric matrix would need to converge in fewer than $N/3$ steps to bring its operations count down to that of a direct method.

VIII. A SPECIAL NOTE ON THE CONJUGATE GRADIENT METHOD

An iterative method called the banded matrix iterative scheme has recently been applied by Ferguson [23] to solve large electromagnetic field problems by the method of moments. The characteristic features of the method applied by Ferguson are as follows.

1) The convergence of the iterative scheme is sensitive to the choice of the numbering scheme used.
2) Because of (1) it requires a person with certain technical background to run the program.
3) The rate of convergence is irregular and sometimes the solution diverges.
4) The banded matrix iterative scheme applied by Ferguson is basically a Jacobi type of iterative scheme and hence it converges slowly [23, p. 16].
5) Theoretically, the method requires an infinite number of steps to converge to the exact result if there is no round-off error.

As we have seen from the previous sections the conjugate gradient method is a nonlinear iterative scheme, in contrast to the linear Jacobi method. The conjugate gradient method also converges at a faster rate than that of a geometric series. Moreover it is highly insensitive to the choice of the initial guess for the solution. Since the conjugate gradient method yields an exact result (assuming no round-off errors) in at most M steps (where M is the number of independent eigenvalues of the $N \times N$ matrix), it has the good points of both an iterative method and a direct method of solution. It has the advantage of an iterative scheme in that roundoff error is limited only to the final step of the solution. It has the advantage of a direct method in that it converges in a finite number of steps.

As a first example consider a wire 3-m in length and 0.01-m in radius. The wire is charged to a potential of $4\pi\epsilon$ volts. The objective is to find the charge distribution on the wire. A method of moments formulation has been employed and the wire is divided into 30 segments. The moment matrix formed by this problem is a typical one which often occurs in the method of moments. The results are presented in Table VII. The first three columns indicate the charge distribution on the wire obtained by the conjugate gradient method. The third column represents the charge distribution corresponding to the segment numbers appearing on column two. The first column states that this result has been obtained at the end of three iterations. The next three columns indicate the charge distribution obtained after eight iterations by the conjugate gradient method. Finally the seventh column gives the result due to Gaussian elimination. As is clear from the data presented in Table VII the conjugate gradient method yields a result better than one percent after three iterations ($M = N/10$).

If for this problem the banded matrix technique is used to yield an accuracy of one percent in one interation, a bandwidth of approximately 15 may be necessary [23, table 10, p. 34]. Hence for the same accuracy, the conjugate gradient method is faster by a factor of 2.5. Also, if the same problem is to be solved by the symmetric Cholesky decomposition, it would have required approximately $N^3/6$ multiplications. The conjugate gradient method required approximately $3N^2$ multiplications as compared to $5N^2$ for Gaussian elimination. Also note that an essentially exact result has been obtained (accuracy better than 10^{-5} in the residuals) after only eight iterations.

As a second example, consider the same problem as above but now the wire is 25-m long. So this time **A** is a 100×100 matrix. Again we obtained an essentially exact result (better than 10^{-5} in the residuals) *after only nine iterations*. This implies that in this type of problem the number of independent eigenvalues is approximately eight or nine. Note that the number of independent distinguishable eigenvalues does not increase as the order of the system is increased considerably. This is an interesting property of diagonally dominant matrices which could easily be exploited by the conjugate gradient method.

As a third example, consider **A** as a 20×20 Hilbert matrix

TABLE VII
COMPARISON OF THE DIFFERENT SOLUTIONS FOR THE CHARGED WIRE

①	②	③	④	⑤	⑥	⑦
3	1	0.1310801E+00	8	1	0.1319899E+00	0.1319883E 00
3	2	0.1079367E+00	8	2	0.1062454E+00	0.1062447E 00
3	3	0.9741676E−01	8	3	0.9698462E−01	0.9698367E−01
3	4	0.9380448E−01	8	4	0.9409392E−01	0.9409255E−01
3	5	0.9250486E−01	8	5	0.9345561E−01	0.9345549E−01
3	6	0.8838725E−01	8	6	0.8868694E−01	0.8868611E−01
3	7	0.8869338E−01	8	7	0.8924234E−01	0.8824180E−01
3	8	0.8770591E−01	8	8	0.8796191E−01	0.8796018E−01
3	9	0.8790541E−01	8	9	0.8807307E−01	0.8807319E−01
3	10	0.8776426E−01	8	10	0.8779728E−01	0.8779657E−01
3	11	0.8714312E−01	8	11	0.8701491E−01	0.8701420E−01
3	12	0.8617383E−01	8	12	0.8591259E−01	0.8592147E−01
3	13	0.8600593E−01	8	13	0.8575714E−01	0.8575720E−01
3	14	0.8614635E−01	8	14	0.8587742E−01	0.8587623E−01
3	15	0.8584630E−01	8	15	0.8543098E−01	0.8542997E−01
3	16	0.8631968E−01	8	16	0.8599222E−01	0.8599192E−01
3	17	0.8616120E−01	8	17	0.8590782E−01	0.8590662E−01
3	18	0.8592218E−01	8	18	0.8567208E−01	0.8567178E−01
3	19	0.8541530E−01	8	19	0.8520442E−01	0.8520377E−01
3	20	0.8547300E−01	8	20	0.8517402E−01	0.8517450E−01
3	21	0.8801979E−01	8	21	0.8832079E−01	0.8832002E−01
3	22	0.8718151E−01	8	22	0.8732998E−01	0.8732986E−01
3	23	0.8758122E−01	8	23	0.8772111E−01	0.8772045E−01
3	24	0.8841532E−01	8	24	0.8893603E−01	0.8893472E−01
3	25	0.8897913E−01	8	25	0.8934522E−01	0.8934456E−01
3	26	0.9207904E−01	8	26	0.9296322E−01	0.9296477E−01
3	27	0.9305668E−01	8	27	0.9314167E−01	0.9314030E−01
3	28	0.9872329E−01	8	28	0.9863120E−01	0.9863114E−01
3	29	0.1081076E+00	8	29	0.1062285E+00	0.1062284E 00
3	30	0.1315722E+00	8	30	0.1324828E+00	0.1324818E 00
	Conjugate gradient method after three iterations			Conjugate gradient method after eight iterations		Gaussian elimination

and let **Y** be chosen in such a way that the solution vector has components 1-20. The problem then is to find **X** given **A** and **Y**. The philosophy behind choosing **A** to be a Hilbert matrix is that nearly singular matrices are often encountered in a system identification problem. So if the conjugate gradient method can efficiently solve such an ill-conditioned problem, then this method may easily be applicable to system-identification problems. The results obtained by two different methods are shown in Table VIII. Calculations were carried out using double precision.

It is clear from Table VIII that the conjugate gradient method yields good results at the end of eight steps. The largest error is only 2.25 percent. The Gaussian elimination method for the same problem completely breaks down. (Note: the Hilbert matrix is extremely ill-conditioned. The condition number of a 20 × 20 Hilbert matrix is of the order of $e^{3.5N} = 2.5 \times 10^{30}$ from [2].)

As a final example, consider the solution of the two components of the current density of a 1λ square metal plate irradiated by a plane wave. When the total number of unknowns for the complex current is 71, we have to solve a 71 × 71 matrix equation. The total time taken for the solution of the complete problem utilizing various techniques is as follows. Gaussian elimination: 27 s (CPU time); conjugate gradient method: 30 s (CPU time) (with 1 percent accuracy in the residual).

Observe that the conjugate gradient method is quite inefficient in this case. However as the dimension of the problem is increased from 71 to 180, the time required by various methods to solve the complete problem is as follows. Gaussian elimination: 500 s (CPU time); conjugate gradient method: for 10^{-2} accuracy in the residual—220 s (CPU time); for 10^{-3} accuracy in the residual—290 s (CPU time); for 10^{-4} accuracy in the residual—390 s (CPU time); for 10^{-5} accuracy in the residual—520 s (CPU time).

So for large systems of equations the conjugate gradient method may prove to be quite useful, especially if one is interested in obtaining an accuracy of 10^{-3} to 10^{-4} in the solutions.

TABLE VIII
COMPARISON OF GAUSSION ELIMINATION AND CONJUGATE GRADIENT METHOD FOR THE SOLUTION VECTOR $X = [A]^{-1}Y$.

Gaussian Elimination	Exact Solution	Conjugate Gradient at the End of Eight Steps
0.9999954	1	1.000289
2.000349	2	1.990388
2.978199	3	3.056398
4.103616	4	3.909776*
4.355928	5	4.981514
55.54727	6	6.056422
−20.17007	7	7.066274
−391.4351	8	8.030005
1050.932	9	8.982147
−397.4495	10	9.947065
212.3952	11	10.93533
1407.415	12	11.94698
−1800.392	13	12.97573
−682.3352	14	14.01228
2770.054	15	15.04653
−1692.660	16	16.06884
637.3160	17	17.07074
559.0285	18	18.04519
78.80465	19	18.98660
97.65299	20	19.8907

* The largest error is about 2.25 percent.

In summary, it is argued that the application of the conjugate gradient method to the analysis of large bodies by method of moments would yield stable, reliable, consistent, and accurate results faster than any methods currently used to obtain a solution. The same is true for problems in system identification. However there may be some buildup of the roundoff error if the residuals are computed iteratively by (19) rather than directly from $\mathbf{AX}_k = \mathbf{Y}$, which would be more time consuming.

NOMENCLATURE

$\mathbf{A}$	$N \times N$ square matrix.
$\mathbf{XY}$	$N \times 1$ matrices.
A^{ij}	element belonging to the ith row and jth column.
X^i	ith element of $\mathbf{X}$.
$\mathbf{X}_n$	matrix obtained after n iterations in interative methods and in direct methods it is just another matrix obtained after processing it n times.
$\|\|X\|\|$	norm of X.
cond [$\mathbf{A}$]	condition number of $\mathbf{A} \triangleq$ (largest eigenvalue of ($\mathbf{A}$))/(minimum eigenvalue of ($\mathbf{A}$)).

REFERENCES

[1] J. Westlake, *A Handbook of Numerical Matrix Inversion and Solution of Linear Equations*. New York: Wiley, 1968.

[2] G. Forsythe and C. B. Moler, *Computer Solution of Linear Algebraic Systems*. Englewood Cliffs, NJ: Prentice–Hall, 1967.

[3] D. K. Fadeev and V. N. Fadeeva, *Computational Methods of Linear Algebra*. San Francisco: W. H. Freeman, 1963, (translated from Russian by R. C. Williams).

[4] V. V. Klyuyev and N. I. Kokovkin—Scherbak, "On the minimization of the number of arithmetic operations for the solution of linear algebraic systems of equations," Comp. Sci. Dept., Stanford University, CA, Tech. Rep.-CS24, 1965, (translated by G. J. Tee).

[5] Volker Strassen, "Gaussian elimination is not optimal," *Numer. Math.*, vol. 13, pp. 354–365, 1969.

[6] T. K. Sarkar *et al.*, "Solution of large systems of linear equations," RADC-TR-81-103, Rochester, NY, Sept. 1979.

[7] M. R. Hestenes, "Applications of the theory of quadratic forms in Hilbert space to the calculus of variations," *Pacific J. Math.*, vol. 1, pp. 525–581, 1951.

[8] R. M. Hayes, "Iterative methods of solving linear problems in Hilbert space," in *Contributions to the solution of systems of linear equations and the determination of eigenvalues*, O. Taussky, ed. Nat. Bur. Standards Appl. Math. Ser., vol. 39, pp. 71–104, 1954.

[9] G. E. Forsythe, "Solving linear equations can be interesting," *Bull. of Am. Math. Society*, pp. 299–329, 1953.

[10] ——, "Theory of selected methods of finite matrix inversion and decomposition," U.S. Dept. of Commerce, INA 52-5.

[11] W. Kahan, "Gauss–Seidel methods for solving large systems of linear equations," Ph.D. dissertation, University of Toronto, 1958.

[12] D. Young, "On the solution of linear systems by iterations," in *Am. Math. Soc. Numerical Analysis Proceedings of Symposia in Applied Mathematics*, J. H. Curtiss, Ed. New York: McGraw—Hill, 1956.

[13] ——, "On the solution of large systems of linear algebraic equations with sparse positive definite matrices," in *Numerical solution of Nonlinear Algebraic equations*, G. D. Byrne, Ed. New York: Academic, 1974.

[14] L. V. Kantorovich, "Functional analysis and applied mathematics," *Uspekhi Matematicheskikh Nauk*, vol. III, no. 6, pp. 89–185, 1948.

[15] J. W. Daniel, "The conjugate gradient method for linear and nonlinear operator equations," *SIAM J. Numer. Anal.*, vol. 4, no. 1, pp. 10–26, 1967.

[16] M. Hestenes and E. Stiefel, "Method of conjugate gradients for solving linear systems," *J. Res. Nat. Bur. Standards*, vol. 49, pp. 409–436, 1952.

[17] J. W. Daniel, "The conjugate gradient method for linear and nonlinear operator equations," Ph.D. dissertation, Stanford University, CA, 1965.

[18] M. Z. Nashed, "Steepest descent for singular linear operator equations," *SIAM J. Numer. Anal.*, vol. 7, no. 3, pp. 358–362, 1970.

[19] H. Akaike, "On a successive transformation of the probability distribution and its application to the analysis of the optimum gradient method," *Ann. Inst. Statistics Math.*, vol. 11, pp. 1–16, Tokyo, 1959.

[20] W. J. Kammerer and M. Z. Nashed, "On the convergence of the conjugate gradient method for singular linear operator equations," *SIAM J. Numer. Anal.*, vol. 9, no. 1, pp. 165–181, 1972.

[21] W. Oettli and W. Prager, "Compatibility of approximate solution of linear equations with given error bounds for coefficients and right hand sides," *Num. Math.*, vol. 6, pp. 405–409, 1964.

[22] W. Oettli, "On the solution set of a linear system with inaccurate coefficients," *J. SIAM Numer. Anal.*, Ser. B., vol. 2, no. 1, pp. 115–119, 1965.

[23] T. R. Ferguson, "The EMCAP iterative techniques in the method of moments," RADC-TR-75-121, May 1975.

[24] M. Marcus, "Basic theorems in matrix theory," U.S. Dept. of Commerce, NBS, *Appl. Math.* ser. 57, 1959.

A Comparison of Spherical Wave Boundary Value Matching Versus Integral Equation Scattering Solutions for a Perfectly Conducting Body

ARTHUR C. LUDWIG

Abstract—A spherical-wave expansion (SPEX) technique for calculating the scattering from a smooth perfectly conducting body is presented. Sample case results are compared with the well-known Lawrence Livermore National Laboratory Numerical Electromagnetics code (NEC), which is based on the integral equation formulation. The internal fields are computed for both results using a third surface current integration program, which is totally independent of both SPEX and NEC. The internal fields, which would be zero for a perfect solution, are much more sensitive to the currents than the scattered fields. The SPEX solution, which uses fewer unknowns and less computer time than NEC, also produces a lower internal field. The SPEX technique also allows a direct check on satisfaction of the boundary condition at any set of points on the surface, independent of the points used to obtain the solution. This provides a valuable built-in test feature for quickly validating results, which is one of the most attractive features of the technique.

I. Introduction

Electromagnetic scattering calculations are commonly performed using an integral equation moment method code such as the Livermore Numerical Electromagnetics Code (NEC) [1]. Although the results obtained are usually satisfactory, this approach can have the following problems.[1]

1) The patch model using the magnetic field integral equation (MFIE) for smooth surfaces breaks down for thin bodies, and it can be difficult to determine when it is accurate.
2) The wire grid model using the thin wire electric-field integral equation (EFIE) is sensitive to the wire diameter selected, and again it is difficult to determine the wire size which will produce the most accurate result.
3) The surface currents are only determined at the patch centers, or wire locations, which for some applications provides insufficient detail of the behavior of the current.
4) It can be difficult to verify that the results are correct.

An alternative spherical-wave expansion (SPEX) method which is less general than the NEC 2 code, but largely overcomes the above problems is presented here. The key advantages are as follows.

1) A direct check on the boundary condition is easily made for an arbitrary set of points on the surface, thereby immediately validating the solution (or invalidating it if the case was run improperly).
2) The surface currents can also be obtained at an arbitrary set of points, allowing study of detail, and also making it relatively easy to accurately calculate the internal fields as a further verification of the results.

The primary disadvantages of the present technique are as follows.

1) Only smooth solid bodies may be treated. Apertures, or sharp edges, invalidate the results (these limitations could possibly be removed in the future).
2) Some trial and error is typically required to obtain an accurate solution (but it is immediately known if the results are inaccurate by using the boundary value check).

In this paper the SPEX approach will be outlined, results will be presented for a sample case of a perfectly conducting cylinder, and a comparison will be made with results using the NEC 2 code. In addition to comparing the surface currents and scattered fields, the fields in the interior—which should be zero—are also computed and compared. The SPEX technique involves spherical wave expansions, which are described in detail elsewhere [2]. This material will not be duplicated here, and readers wishing to try this technique will need to refer to this reference or a similar source for explicit equations, and for detailed information on properties briefly noted here.

II. Approach

The SPEX approach is conceptually very simple: the scattered field is represented by an expansion of outgoing spherical waves, and a matrix is inverted to determine the expansion coefficients such that the boundary condition is satisfied on the surface. This basic approach has been used previously for spherical [3], [4] and cylindrical [5] nominal geometries, but was limited to fat bodies where the ratio of the maximum and minimum dimensions is on the order of two [6]. The problem is that if one attempts to use a single expansion for a long, thin body, high-order terms are needed to match the boundary conditions at the ends of the body. However, these terms blow up near the middle of the body and severe numerical problems occur. The key feature of the technique

Manuscript received September 5, 1985; revised January 27, 1986.
The author is with the General Research Corporation, P.O. Box 6770, Santa Barbara, CA 93160.
IEEE Log Number 8608847.

[1] Many other scattering codes exist and some may avoid one or more of these problems, but the comparison will be restricted to NEC due to its widespread use, and the fact that the author has access to NEC and its documentation.

Reprinted from *IEEE Trans. Antennas Propagat.*, vol. AP-34, no. 7, pp. 857–865, July 1986.

presented here is that *several* spherical wave expansions, with different origins, are used simultaneously to represent the external fields. The concept of using multiple expansions was introduced by Iskander, Lakhtakia, and Durney, where it was applied to expanding the internal fields in a dielectric body [7]. The internal region was subdivided into subvolumes, and the continuity of the internal fields enforced by matching the expansions in an overlap region. Therefore, in each subregion, the fields were basically represented by a single expansion. The external field was represented by a single expansion. The major difference between this previous work and the technique presented in this paper is that here multiple expansions are used in the exterior region, and it is *not* subdivided; all expansions are used together to represent the external field everywhere. This allows more points to be matched on the surface for a given matrix size, since all points are on the surface rather than diverting some to an overlap region.

Another closely related technique is the discrete singularity method (DSM) [8]. Although this method has been applied only for cylindrical geometries using line singularities, the basic idea is the same as the SPEX technique. One major difference in the implementations is that the DSM uses only one unknown coefficient per origin, corresponding to the lowest order cylindrical wave, whereas SPEX typically uses six or more unknown coefficients per origin. A second major difference is that the DSM implementation solves a nonlinear equation to optimize the origin locations, whereas this has not been found necessary using SPEX.

The multiplicity of origins for the outgoing spherical waves can be interpreted as equivalent source locations within the boundary of the conductor. Then, the total scattered field is represented as a sum over the contribution from each origin. Furthermore, the source distribution at each origin generates a field represented as a sum over spherical waves, each of which is generated by a specific source component at that origin. The sources are the multipoles defined in [2, fig. 2.21]. As a result, we are then faced with a distribution of equivalent sources *within* the boundary which create a field *on* the boundary that leads to satisfaction of the boundary condition. Note that, in principle, this is similar to the "thin wire" approximation, wherein finite line sources on the axis of a wire are meant to generate fields equal to those generated by the surface currents on the wire surface at points on and outside the wire.[2]

To illustrate the technique, consider the example given in Fig. 1. Three origins are shown inside a conducting body. The scattered field $\bar{E}_s$ at a point P on (or outside) the surface can be expressed in a shorthand notation as

$$\bar{E}_s(P) = \sum_{n=1}^{N_1} a_n^{(1)} \bar{f}_n(R_1, \theta_1, \phi_1) + \sum_{n=1}^{N_2} a_n^{(2)} \bar{f}_n(R_2, \theta_2, \phi_2) + \sum_{n=1}^{N_3} a_n^{(3)} \bar{f}_n(R_3, \theta_3, \phi_3) \qquad (1)$$

[2] The author is indebted to the reviewer for this description.

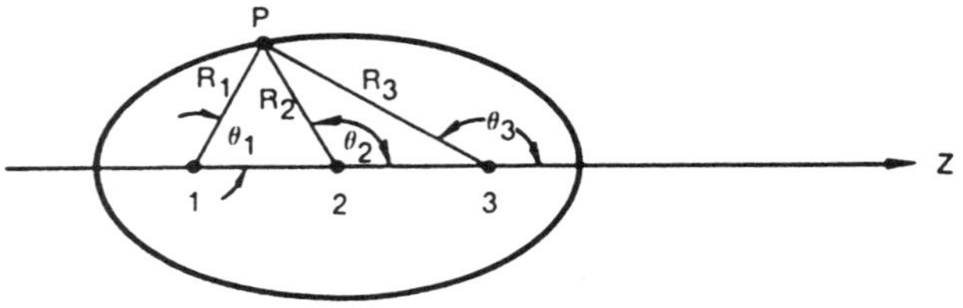

Fig. 1. Use of multiple expansions.

where

$a_n^{(i)}$	the nth outgoing spherical wave coefficient of the expansion at the ith origin
$\bar{f}_n$	the nth outgoing spherical wave
N_i	the truncation point of the ith expansion
R_i, θ_i, ϕ_i	the spherical coordinates of P with respect to the ith origin.

There are $K = N_1 + N_2 + N_3$ complex-valued unknown coefficients in this representation. Let $\hat{i}_t$ and $\hat{i}_\phi$ be two orthogonal unit vectors tangent to the surface at the point P. For a perfect conductor and given incident field $\bar{E}_i$ the boundary condition is

$$[\bar{E}_i(P) + \bar{E}_s(P)] \cdot \hat{i}_t = 0$$
$$[\bar{E}_i(P) + \bar{E}_s(P)] \cdot \hat{i}_\phi = 0. \qquad (2)$$

Therefore, enforcing the boundary condition at each point P results in two equations. By selecting $K/2$ points, one can then develop a matrix equation which may be inverted to obtain the unknown coefficients $a_n^{(i)}$.

Two important properties of outgoing spherical waves as far as this application is concerned are:

1) they are solutions of Maxwell's equations everywhere in space except for $R_i = 0$, so (1) is also a solution everywhere on and outside the body;
2) they satisfy the radiation conditions as $R_i \to \infty$, so (1) will also.

Therefore, *if* a solution of the form of (1) can be obtained which satisfies the boundary condition on the conducting surface, then the uniqueness theorem can be invoked to prove the solution is the one and only correct solution. Note that the uniqueness theorem only guarantees that the solution is the correct solution *outside* the body. Inside the body, (1) will still be a solution of Maxwell's equations everywhere except the origin points $R_i = 0$, but will not give the correct solution inside the body, which in fact is known *a priori* to be zero. The reason for this is the surface currents on the body which cause a discontinuity of the fields as the surface is crossed. Outside of the body, the spherical wave sources (in principal) radiate exactly the same field as the surface currents—i.e., they are equivalent sources for the external fields—but inside the body, the fields are different.

This brief outline of the SPEX technique has intentionally avoided complex derivations and equations, but it is theoretically rigorous as long as we can demonstrate the existence of a solution which satisfies the boundary condition: this issue is considered in the next section and the Appendix.

For a direct point-match solution, it is necessary to select the locations of the points carefully to avoid ill-conditioned matrices. Also, the boundary condition may not be well matched between the points. For these reasons, it has been empirically found advantageous to use two or three times the minimum number of points, and to solve for the coefficients providing the minimum mean square error over the point set—the pseudo-inverse solution. This method was also adopted by Kennaugh [3], Morrison and Cross [4], and Nishimura *et al.* [8]. More will be said about this later.

After solving for the coefficients, the $\bar{E}$-field at an arbitrary point P is easily calculated using (1), and may be used to check the boundary condition over a much denser point set than the one used to obtain the solution. A similar equation provides the $\bar{H}$-field in terms of the same coefficients, so the surface currents are directly computed by adding in the known incident $\bar{H}$-field, using the fact that the interior fields are zero, and computing

$$\bar{J} = \hat{n} \times \bar{H} \tag{3}$$

where $\hat{n}$ is the unit vector normal to the surface. Finally, by evaluating the far-field form of (1), the scattered field is obtained at any desired angle (of course the near-field scattered field could also be obtained if desired).

III. Discussion

This use of spherical waves is very different from the classical applications such as the Mie solution for the sphere [9]. Classical applications make use of the orthogonality and completeness of a single spherical wave expansion. Here we adopt the point of view that the spherical waves are a useful set of basis functions, and are essentially a substitute for the current basis functions used in the integral equation method. *A key advantage of using spherical waves is that it is not necessary to integrate the surface currents to get the field on the surface.* Both the field on the surface and the current are given analytically by the spherical wave equations.

As noted above, the only question regarding the theoretical validity of the SPEX technique is the existence question. For classical applications, this question is answered using the completeness property of the trigonometric and Legendre functions which describe the spherical wave fields on the surface of a sphere. Essentially this guarantees that, as the number of terms increases, the solution will converge to the boundary values as long as the specified boundary values and their first two derivatives are continuous [10]. Here the situation is changed by the fact that the radial variables R_i are not constant over the surface, so the convergence is affected by the spherical wave radial Hankel functions as well. This is precisely the reason that an expansion using a single origin becomes numerically unstable. Referring to the cylindrical case, Mittra and Wilton state that, as the number of terms increases, a single expansion will diverge inside the smallest circle enclosing the surface [11]. Therefore, for noncircular (and similarly nonspherical) boundaries, they question the validity of point-matching techniques using such expansions.[3] Mittra and Wilton use a different analytic continuation technique to avoid this convergence problem. However, the Appendix contains an argument that a sequence of expansions *does* exist which converges to the desired values on the boundary, as long as the surface is smooth (as defined in the Appendix). Therefore, it is claimed here that the technique is theoretically sound for smooth, closed, perfectly conducting bodies.

As a practical matter, numerically evaluated expansions are always finite, and the boundary condition match will be imperfect for any solution. So "success" is defined here in a practical sense as obtaining a finite expansion which matches the boundary condition within some low error. As a further check, the interior fields are computed. It is shown by Waterman that surface currents which produce zero internal fields in any portion (i.e., a small volume) of the interior of a perfectly conducting body are uniquely the correct currents [12]. Again, the internal fields will not be identically zero in practice. It would be nice to be able to bound the error of the scattered fields in terms of the error in the boundary condition match and/or internal fields, but we are not aware of such a relationship. Therefore, strictly speaking, in the absence of such a relationship neither of these checks totally validates the solution. However, correlating changes in these checks with changes in the scattered field strongly indicates that the checks on boundary condition match and internal fields are much more sensitive to errors than the scattered fields. Some results on this are given below. We will now show results for a specific example where the technique has been applied successfully, and also discuss the effect of a lack of smoothness which can cause the method to fail—in the sense of not matching the boundary condition at a discontinuity.

IV. Results for a Sample Case: Half-Wavelength Long Cylinder

Consider the half-wavelength long, hemisphere capped, cylinder shown in Fig. 2. We will calculate the case of a plane wave incident normal to the axis (traveling in the $-x$ direction) polarized parallel to the axis. Results on surface currents, etc., will be shown as a function of the normalized arc length, as illustrated in the figure, for a given angle ϕ. The first two terms of a spherical wave expansion centered at the origin produce a field on the surface as shown in Fig. 3. The full notation of spherical waves is complex; in the shorthand notation of (1) only a single index n is shown. In the full notation, there are two primary indices, m and n. The m index corresponds to sin $m\phi$ and cos $m\phi$ variations, and the n index corresponds to associated Legendre polynomials which are a function of θ, in the usual spherical coordinate system. The case shown in Fig. 3 is for $m = 0$. It is seen that the $n = 1$ mode produces a field that is highly localized. This is generally true for cylinders with a radius small compared to a wavelength. It is clear that if $n = 1$ was the highest order mode used, to match the boundary condition the spherical wave origins should be roughly 0.025 wavelengths apart. A general rule of thumb, for surfaces much less than a wavelength away from the origins, is to space the origins at roughly 1/2 the distance to the surface. Alternatively, one could use a maximum $n = 2$, and then the origin spacing could be roughly twice as much. If the origins are too close together,

[3] Dr. Wilton has brought to my attention a later reference (*IEEE Trans. Antennas Propagat.*, vol. AP-20, pp. 310-317, May 1972) which essentially agrees with the position taken here.

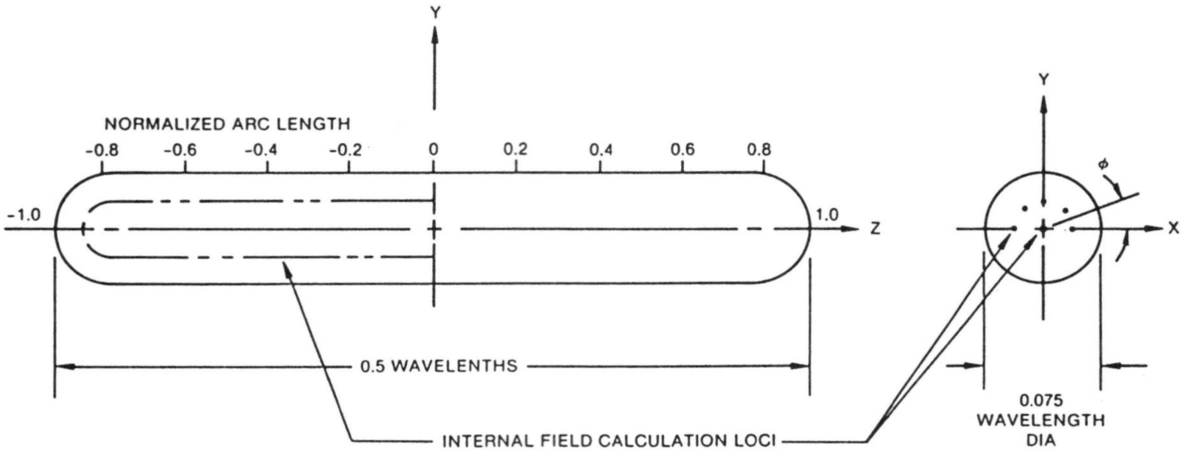

Fig. 2. Sample case scattering body.

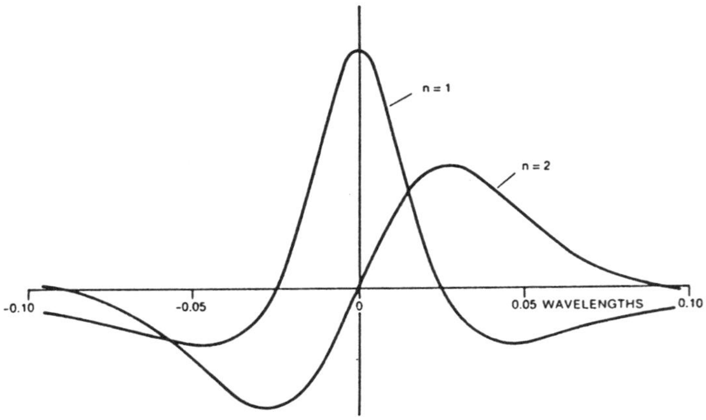

Fig. 3. Fields on the surface produced by two lowest order spherical waves.

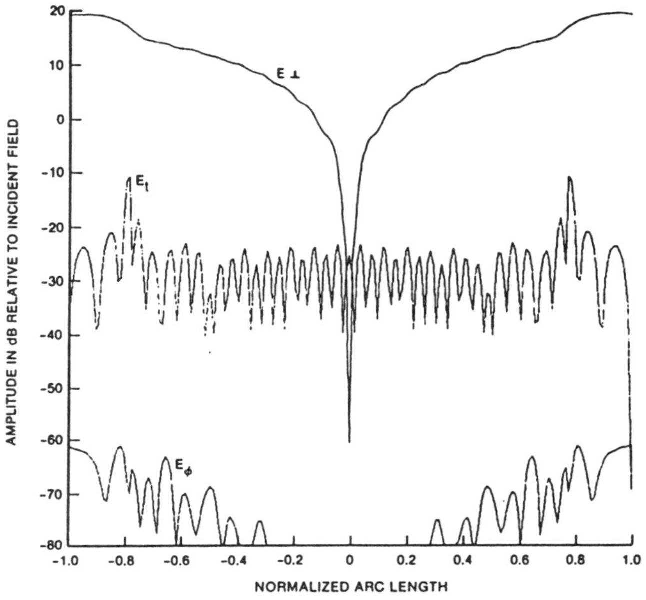

Fig. 4. Boundary condition match, 160 unknowns, 242 surface points.

for a given maximum mode order, the fields from two origins become very similar. Essentially, for a well-conditioned matrix any two modes must be linearly independent, and excessive overlap will cause an ill-conditioned matrix. The maximum value of m cannot exceed the maximum value of n. For small bodies, the $m = 0$ behavior dominates, but it was necessary to go to $m = 2$ to match the ϕ variation for the body shown in Fig. 2 at the endpoints. The adjustment of the origin locations and the maximum m and n indices is the trial and error part of this technique mentioned in the Introduction. Of course, a similar process takes place with, say, developing the wire grid for a moment method code.

A good result was obtained for the sample body with 20 origins equally spaced between the centers of the hemispherical caps. The maximum mode orders were $M = N = 1$ at all origins except two at each end where the maximum was $M = N = 2$. The resulting boundary condition match is shown in Fig. 4 for $\phi = 90°$ (the error for the other ϕ cuts is very similar even though the fields themselves have a large ϕ variation). E_ϕ is the tangential field in the ϕ direction, and E_t is the orthogonal field tangent to the surface. For a perfect match these would both be zero. Also shown is the perpendicular field component $E_\perp$ which is almost 20 dB higher than the incident field at the endpoints. E_ϕ is very low, below -60 dB. E_t is typically down -20 dB with respect to the incident field, except for the spike at ± 0.78 arc length. This is exactly at the joint between the hemispherical caps and the cylinder where the surface second derivative is discontinuous. The surface still satisfies the smoothness condition given in the Appendix, but this apparently makes it difficult to match the field at this point.

A cylinder with flat ends where the first derivative is discontinuous at the sharp corner, does not satisfy the required smoothness condition, and predictably the situation is much worse. An attempt was made to solve for a cylinder with flat ends and the boundary condition match was in fact quite poor. Edge singularities could possibly be handled analytically, but this has not been attempted.

One might expect an even better match for a surface with continuous second derivatives. One case was run for a ellipsoid with the same length and surface area as the cylinder of Fig. 2. The ends had a radius of curvature of 0.009—much smaller than the hemispherical caps—and the perpendicular field component $E_\perp$ was 30 dB above the incident field at the ends. The best boundary condition match obtained had a tangential field component 6 dB below the incident field, but 33 dB below $E_\perp$. The spike in Fig. 4 is 28 dB below $E_\perp$. So the result was actually about the same.

The pseudo-inverse solution for the sample case was obtained using 43 points along the arc, and the 42 minima in Fig. 4 generally coincide with these match points. On the cylinder, five points were equispaced circumferentially, with fewer circumferential points at the ends. A total of 242 match points were used, resulting in 484 equations. There are $2N(N + 2)$ unknowns for each origin with $n \leq N$, so the 16 center origins with $N = 1$ each had six unknowns, and the four outer origins with $N = 2$ each had 16 unknowns, for a total of 160 unknowns. Therefore, there were three times as many equations as unknowns. Even more match points would reduce the peak error shown in Fig. 4, but the matrix fill time would grow proportionally. Fig. 4 also demonstrates the need to check the boundary condition between the match points used to obtain the solution, since the error is 10 to 20 dB higher between the match points.

As mentioned above, the least squares solution is an important part of this technique. For a pure point match, where the number of equations exactly equals the number of unknowns, the matrix can be *very* sensitive to the location of the points. Consider an extreme example: if all of the surface points were located along one line at $\phi = 0$, there is no way to resolve the different trig terms $\sin m\phi$ and $\cos m\phi$, and the matrix will in fact be singular. Obviously, if the maximum m value is M, at least $2M + 1$ ϕ values must be included among the surface points. When different mode orders are used at different origins, it is very tricky to distribute the points over the surface to obtain a well-behaved matrix. For example, for the 160 coefficient case, three solution attempts with 80 surface points (providing 160 equations) resulted in garbage answers. Even a slight overspecification can help. Going from 80 to 87 surface points resulted in the boundary match shown in Fig. 5. The matrix is still not well-conditioned, as evidenced by the asymmetry in Fig. 5, but at least reasonable values were obtained. (In fact, the scattered field at $\phi = 0$ produced by this solution is within 0.02 dB of the other solution, and the peak surface current agrees to within 1.3 percent in magnitude and 1.5° in phase.) However, overspecifying the equations by at least a factor of 1.5, and preferably 2 or 3, greatly relaxes the sensitivity to location and any reasonable point set will provide good results (one must still include at least $2M + 1$ ϕ values).

V. Solution Using NEC 2 Code

The NEC 2 patch code (surface magnetic field integral equation) apparently fails for the selected size of the body shown in Fig. 2. The computed peak current magnitudes were in error by 25 percent or more and the peak current phase was in error by over 100°. The scattered field amplitude was also low by 1 to 9 dB, depending on the patch configuration. Similar results were obtained independently by Stach [13], who also found that the results were unstable as the patch configuration was varied. The patch results will not be considered further.

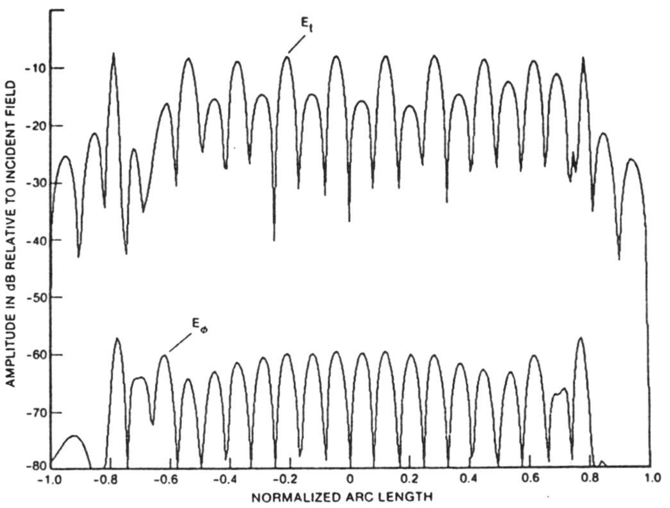

Fig. 5. Boundary condition match, 160 unknowns, 87 match points.

Initial results on the wire grid model (thin wire electric field equation) were better but still somewhat inaccurate. After several trials, good results were obtained with the 294-segment grid configuration shown in Fig. 6. The main improvements were obtained by 1) using more wire segments, and 2) concentrating the segments at the ends. In particular, since the circumferential currents are very low in the center, no circumferential segments were used there, and instead they were bunched up at the ends. It was also found that the results were sensitive to the wire diameter. A rule of thumb is to make the total surface area of the wire equal to the solid surface it represents. However, there seem to be some exceptions to this rule. The longitudinal wires on the cylinder were made 0.01250 wavelengths diameter, which is 1/6 the diameter of the cylinder. On the end caps the diameters were reduced to compensate for the decreasing radius compared to the cylinder. The transverse wires were given a diameter of $1/\pi$ times the separation. When these varying wire diameters were used, the currents shown in Fig. 7 were obtained. Physically, all three curves must be equal at ± 1.0 arc length, so it is clear that there is an error. When the field inside the cylinder was computed (using the NEC code near-field option), and added to the incident plane wave, the field was down only 3 dB from the incident field near the end points.

Using the same diameter of 0.0125 wavelengths for all wires improved the results. The currents shown in Fig. 8 were obtained, which approach equality near ± 1, and the internal field dropped. However, it was still only 6 dB down for the incident field at some points.

A nonzero internal field can arise due to several causes: 1) field coupling to an internal resonance; 2) errors in the surface currents; or 3) poor numerical accuracy in computing the fields from the currents. The diameter of the cylinder is such that it is well below cutoff for any internal resonance, so attention was focused on the latter two possibilities. To improve the numerical accuracy in the computation, a new program was written to interpolate the currents to a much finer

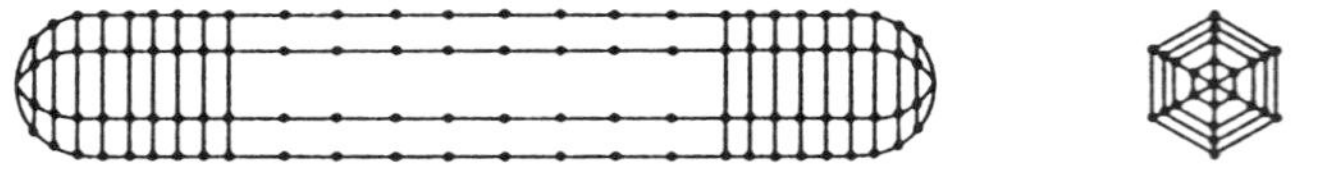

Fig. 6. 294 segment wire grid model of scatterer.

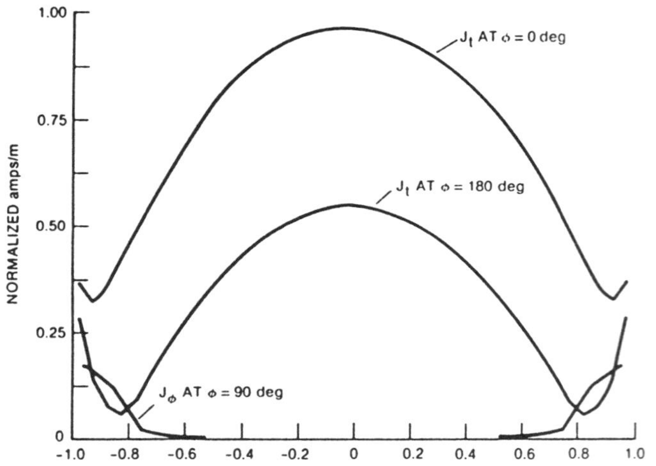

Fig. 7. NEC 2 currents using varying wire diameters.

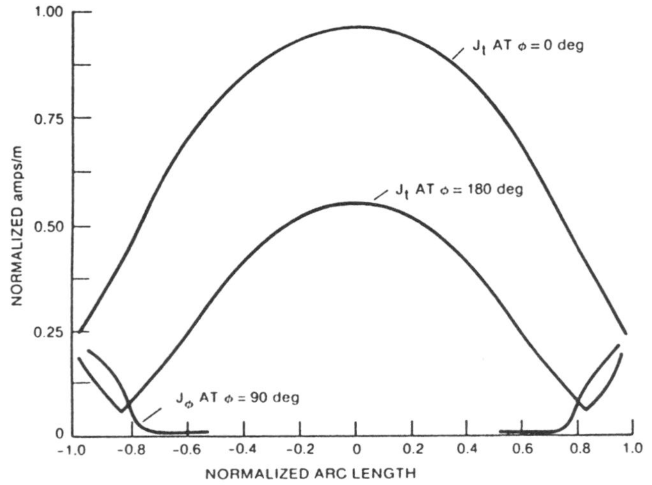

Fig. 8. NEC 2 currents using constant wire diameter.

grid, and then integrate to compute the internal field. Fourier series interpolation was used in the ϕ coordinate, and a taut spline interpolation [14] in the orthogonal coordinate. This lowered one internal field peak from -6.4 dB to -15.7 dB. Another peak near the end was relatively unaffected, indicating that it is probably a real error due to an error in the current. We will now compare the best results obtained using the NEC 2 code to the results using the SPEX technique.

VI. Comparison of Results

The surface currents obtained using the two techniques agree quite well as shown in Fig. 9. As noted in the previous section, the NEC 2 currents have been interpolated. The SPEX currents were directly computed on the same point set. The scale is absolute, and it is seen that the major difference is at the current peaks where the leading edge transverse currents (J_t at $\phi = 0$) differ by 3.5 percent, and the trailing edge

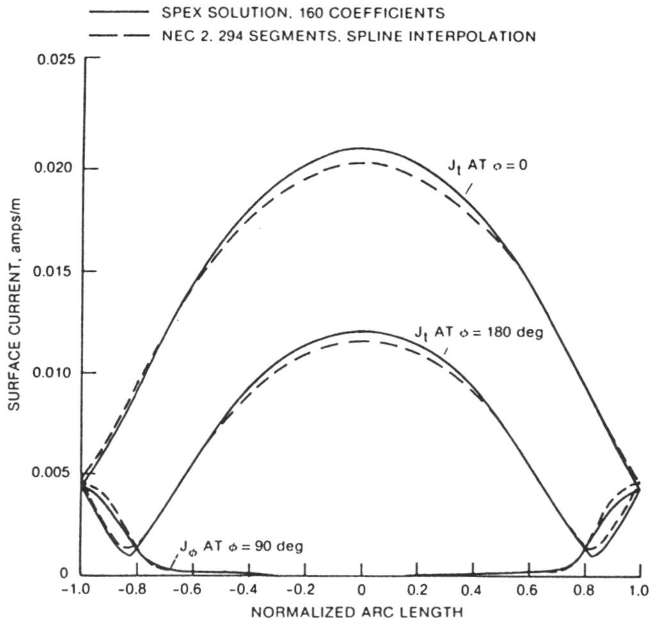

Fig. 9. Surface currents.

currents differ by 4.2 percent. The phase angles of the currents at these peaks agree to within 1.3° and 2.7°, respectively. The azimuthal currents (J_ϕ at $\phi = 90°$) also agree very well (J_ϕ is zero at $\phi = 0°$ and $\phi = 180°$).

The internal field was calculated for the NEC 2 and SPEX results using the same point sets and the same surface current integration program. It would be possible to compute the internal fields using a spherical wave expansion, but integration of the surface currents was used instead for two reasons: 1) to make a fair comparison of results it is preferable to use the same surface integration for both techniques, and 2) since the surface integration program is totally independent of SPEX and NEC, it is a better validation. The field was calculated along six lines as shown in Fig. 2, and an envelope drawn over the six resulting curves, showing the maximum error. The results are shown in Figs. 10 and 11 for the $\bar{E}$-field and $\bar{H}$-field, respectively. Both results are good, the most significant difference being the fairly high peak exhibited by the NEC 2 $\bar{E}$-field at -1.0 arc length. The $\bar{H}$-field is substantially lower at the ends for both techniques. The values for the SPEX technique seem to be generally consistent with the accuracy of the boundary condition match shown in Fig. 4, although the very low internal $\bar{E}$-field values at the center are somewhat surprising.[4]

The scattered field is relatively insensitive to details of the current distribution, and the difference in the results is almost entirely due to the different peak current values. The computed scattered field amplitudes for the SPEX and NEC 2 solutions agreed within 0.2 dB at $\theta = 90°$, for $\phi = 0$ and $\phi = 180°$.

VII. Conclusion

For the selected scatterer, good agreement was obtained between the SPEX technique and the NEC 2 wire model. The

[4] Subsequent runs using SPEX have reduced the internal fields 10–20 dB by using a maximum m value of two for all origins.

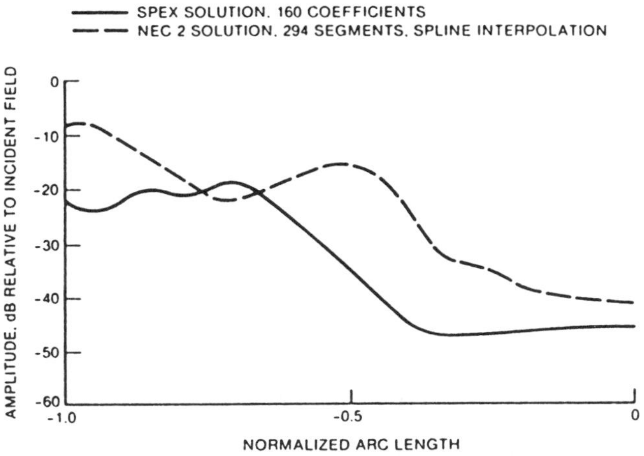

Fig. 10. Internal E-field envelopes.

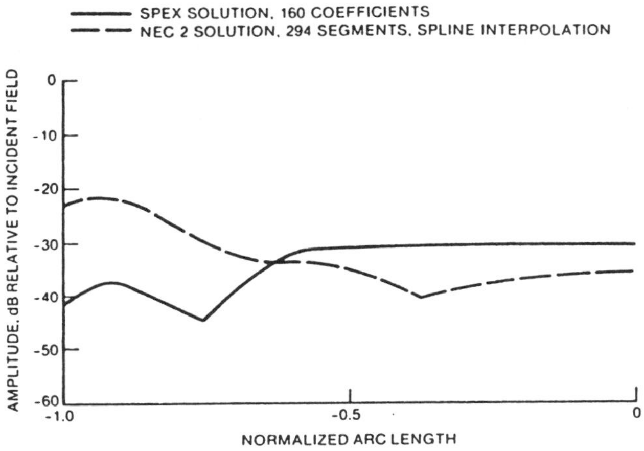

Fig. 11. Internal H-field envelopes.

NEC 2 patch model, based on the surface magnetic field equation, gave poor results. The NEC 2 wire grid model has the disadvantages of a somewhat arbitrary choice for wire diameters, and the necessity to do current interpolation to accurately compute the internal field to check the results. The SPEX solution has the advantages of easily checking the boundary values, and the ability to directly compute the currents at any set of surface points. Although boundary values can also be checked for the integral equation solutions by integrating the surface currents, this is a much trickier numerical calculation. It also should be noted that the peak internal field value was -18.6 dB for the SPEX versus -7.8 dB for the NEC 2 wire grid model, and that the number of unknowns was 160 (242 surface points) versus 294 unknowns and points for the NEC 2 code. Both cases were run on a Digital Equipment Corp. VAX 11/780 computer with a floating-point accelerator. The charged computer time for the two cases as run was 266 s for SPEX and 630 s for NEC 2. Both are single precision, full three-dimensional codes, and no symmetry was utilized in either case, to make an equal comparison. So SPEX not only provided better accuracy, but required less than half of the computer time.

As mentioned in the introduction, the NEC 2 code is significantly more general than the present SPEX code, so the two methods are not directly competitive. At this point, it appears that the verification tools available using the SPEX technique make it an excellent method to produce accurate and reliable solutions, for the cases where it is applicable.

Appendix

Convergence Issues

The convergence of electromagnetic field expansions is similar to the convergence of power series in complex variable theory. Since the mathematical theory of complex variables is so well established, and in some ways easier to deal with, this Appendix will first cite results from this theory. This does not of course prove that the same results hold for electromagnetic fields, but it is hoped it will clarify one main point: the divergence of a series does *not* invalidate the use of a finite series of the same form to approximate a function. This will be established with a very simple example, and then a general theorem will be cited. Having established this, a similar result applicable to electromagnetic fields will be cited.

Consider the function

$$g(z) \equiv \frac{1}{1-z} \tag{4}$$

which is analytic everywhere except for the point $z = 1$. Suppose we want a finite series representation

$$\hat{g}_K(z) \equiv \sum_{k=0}^{K} a_k z^k \tag{5}$$

which closely approximates $g(z)$ everywhere on and within the circular contour C, as shown in Fig. 12. A power series representation of $g(z)$ is

$$g(z) = \sum_{k=0}^{\infty} z^k, \qquad \text{for } |z| < 1. \tag{6}$$

Since this power series diverges everywhere outside the dashed circle of Fig. 12, it might be thought that one could not succeed in obtaining $\hat{g}_K(z)$ as specified above. But, actually, it is easy to construct a sequence of finite series $\hat{g}_K(z)$ which converges absolutely to $g(z)$ everywhere on and within C.

We note that another power series representation for $g(z)$ is

$$g(z) = \frac{1}{2} \sum_{k=0}^{\infty} \frac{(z+1)^k}{2^k}, \qquad \text{for } |z+1| < 2. \tag{7}$$

This series is centered at the center of C, and converges absolutely in a region of radius 2 entirely enclosing C. So we need only define

$$\hat{g}_K(z) \equiv \frac{1}{2} \sum_{k=0}^{K} \frac{(z+1)^k}{2^k} \tag{8}$$

and we have achieved our goal. It is interesting to compare the power series and $\hat{g}_K(z)$ for the first few values of K, as shown in Table I. While the leading terms in $\hat{g}_K(z)$ approach the

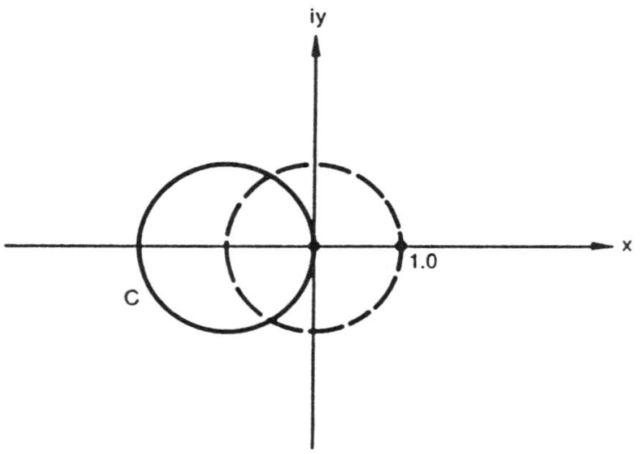

Fig. 12. Complex plane geometry.

TABLE I
POWER SERIES COMPARED TO $\hat{g}_k(z)$

$g(z)$	$= 1 + z + z^2 + z^3 + z^4 + \cdots$
$\hat{g}_0(z)$	$= 1/2$
$\hat{g}_1(z)$	$= 3/4 + (1/4)z$
$\hat{g}_2(z)$	$= 7/8 + (1/2)z + (1/8)z^2$
$\hat{g}_3(z)$	$= 15/16 + (11/16)z + (5/16)z^2 + (1/16)z^3$
$\hat{g}_4(z)$	$= 31/32 + (13/16)z + (1/2)z^2 + (3/16)z^3 + (1/32)z^4$

values of the power series, the higher power terms are quite different. The error between $\hat{g}_K(z)$ and $g(z)$ on and within C is bounded by

$$|g(z) - \hat{g}_K(z)| \le \frac{1}{2^{K+1}}. \tag{9}$$

Now this example was obtained using an analytic continuation analogous to the one employed by Mittra and Wilton [11], and the example was made simple by the fact that a single power series (7) was found that converged everywhere within the desired region. A natural question is: what happens when there are more singularities, so no single power series covers the entire region? There is a theorem that states that there is *still* a sequence of polynomials which achieves the desired result [15].

Theorem: Let S be a closed bounded set that does not separate the plane. Let $g(z)$ be continuous on S and be analytic at interior points of S. Then $g(z)$ may be uniformly approximated on S by polynomials.

So at least for analytic functions it is known that a solution exists. The next question that arises is if a solution can be found by least-squares point-matching on the boundary of C. For a given approximation $\hat{g}(z)$, define the difference function $\epsilon(z) \equiv g(z) - \hat{g}(z)$. Since $g(z)$ and $\hat{g}(z)$ are both analytic inside C, the difference function is also analytic, and by the maximum-modulus theorem, the error must reach its maximum absolute value on the boundary C. Since a sequence of polynomials exists where the error on C becomes arbitrarily small, a sequence of solutions which minimize the mean square error on C should also converge to an arbitrarily small error. (A proof that least squares point-matching achieves this will, however, not be attempted.)

For the electromagnetic field problem considered in this paper, the analog to the above theorem has been proved by Calderon [16], and may be paraphrased as follows.

Theorem: let $\bar{k}$ be a continuous tangential vector field on a surface S, and p be a given point inside S. Then there exists a multipole expansion, satisfying the radiation condition, with origin p, whose tangential component approximates $\bar{k}$ within a vector of magnitude less than ϵ for any given $\epsilon > 0$.

The surface S is the common boundary between two regions D and R, where R is defined as the interior. R must be the union of a finite number of bounded regions; as paraphrased above the theorem applies for a single interior region. S must satisfy a smoothness condition: for each point on S there exist two spheres of fixed radius passing through the point but otherwise contained in R and D, respectively.

If a solution exists in the form of a single expansion, it certainly exists for multiple expansions. Therefore, what remains is the practical question of actually finding an expansion with a viable numerical technique. It would certainly not be claimed that the least squares technique used here will work for any problem, but it is an effective method that works for many useful cases.

ACKNOWLEDGMENT

Dr. Lester Ford of General Research Corporation provided crucial insights which formed the basis for this Appendix. I would also like to thank Dr. E. K. Miller for bringing the discrete singularity method work to my attention, and to express my appreciation for the helpful comments of the reviewer.

REFERENCES

[1] G. J. Burke and A. J. Pogio, "Numerical electromagnetics code (NEC)—Method of moments," NOSC Tech. Document 116, Naval Ocean Syst. Center, San Diego, CA, Jan. 1981.

[2] A. C. Ludwig, "Spherical wave theory," in *The Handbook of Antenna Design*, vol. I, A. W. Rudge *et al.*, Eds., Peter Peregrenus, 1982.

[3] E. M. Kennaugh, "Multipole field expansions and their use in approximate solutions of electromagnetic scattering problems," Ph.D. Dissertation, Ohio State Univ., Dec. 1959. (Also, Report 827-5 of the Antenna Lab., Ohio State Univ., Nov. 1959.)

[4] J. A. Morrison and M. J. Cross, "Scattering of a plane electromagnetic wave by axisymmetric raindrops," *Bell Syst. Tech. J.*, vol. 53, pp. 955–1019, 1974.

[5] C. R. Mullin, R. Sandburg, and C. O. Velline, "A numerical technique for the determination of scattering cross sections of infinite cylinders of arbitrary cross section," *IEEE Trans. Antennas Propagat.*, vol. AP-13, pp. 141–149, Jan. 1965.

[6] G. T. Ruck *et al.*, *Radar Cross Section Handbook*, vol. 1. New York: Plenum, 1970.

[7] M. F. Iskander, A. Lakhtakia, and C. H. Durney, "A new procedure for improving the solution stability and extending the frequency range of the EBCM," *IEEE Trans. Antennas Propagat.*, vol. AP-31, no. 2, pp. 317–324, Mar. 1983.

[8] M. Nishimura, S. Takamatsu, and H. Shigescwa, "A numerical analysis of electromagnetic scattering of a perfect conducting cylinder by means of discrete singularity method improved by optimization process," *Inst. Elec. Comm.*, Japan, vol. 67-B, no. 5, May 1984.

[9] Mie, *Ann. Physik*, vol. 25, p. 377, 1908.

[10] J. A. Stratton, *Electromagnetic Theory*. New York: McGraw-Hill, 1941, sec. 7.3.

[11] R. Mittra and D. R. Wilton, "A numerical approach to the determination of electromagnetic scattering characteristics of perfect conductors," *Proc. IEEE*, pp. 2064–2065, Nov. 1969.
[12] P. C. Waterman, "Matrix formulation of electromagnetic scattering," *Proc. IEEE*, vol. 53, pp. 805–812, 1965.
[13] J. Stach, SRI Int., private communication, June 1985.
[14] C. DeBoor, "A practical guide to splines," in *Applied Math. Sciences*, vol. 27. New York: Springer-Verlag, 1978, p. 310.
[15] P. J. Davis, *Interpolation and Approximation*. New York: Dover, 1963, p. 278.
[16] A. P. Calderon, "The multipole expansion of radiation fields," *J. Rational Mech. Analysis*, vol. 3, no. 5, pp. 523–536, Sept. 1954.

Generation of Wide-Band Data from the Method of Moments by Interpolating the Impedance Matrix

E. H. NEWMAN

Abstract—**In a method of moments (MM) computation, one must compute the impedance matrix at each new frequency. Since computation of the MM impedance matrix at a single frequency can be a very time-consuming process, performing an MM computation over a wide frequency range can require a prohibitive amount of CPU time. Here we describe a method where the impedance matrix is computed at relatively large frequency intervals and then interpolated to approximate its values at intermediate frequencies. Basically, the method trades reduced computer CPU time for increased storage.**

I. Introduction

The method of moments (MM) [1], [2] has proven to be an effective computational tool for the analysis of electromagnetic radiation and scattering problems. The major limitation of MM solutions has always been the computer resources, i.e., computer CPU time and storage, needed to carry out the computation. In many MM solutions the CPU time to compute the impedance matrix $[Z]$ dominates. Here we present a technique for reducing the number of times the impedance matrix must be computed in performing a MM computation over a wide frequency bandwidth. In brief, one computes and stores impedance matrices at three relatively widely spaced frequencies. Quadratic interpolation is then used to find the impedance matrix at intermediate frequencies. This reduces the number of times the impedance matrix must be computed and thus reduces the total CPU time. The method works since in most cases the elements in the impedance matrix are more slowly varying with frequency than are parameters of engineering interest such as input impedance and radiated fields.

The next section describes the impedance matrix interpolation technique and illustrates the method by computing the input impedance of a dipole and the scattering from a perfectly conducting plate. Next, it is shown that an improved interpolation scheme is possible by incorporating known properties of the frequency dependence of the elements in the impedance matrix into the interpolation method.

II. Discussion of the Method

A. Basic [Z] Matrix Interpolation

Often the major computational effort in a method of moments [1] solution is the evaluation of the N^2 elements in the impedance matrix $[Z]$, where N is the number of expansion functions or unknowns in the MM solution. A typical element of $[Z]$ is given by

$$Z_{mn} = -\int_m \mathbf{E_n} \cdot \mathbf{w_m} \, dr, \qquad m, n = 1, 2, \cdots, N \tag{1}$$

where $\mathbf{E}_n$ is the electric field of the current expansion basis function $\mathbf{J}_n$, and $\mathbf{w_m}$ is the mth weighting function, often chosen identical to the expansion functions. The integrals are over the region (line, surface or volume) occupied by the mth weighting function. Here we are concerned with efficiently obtaining the $[Z]$ matrix at many frequencies over a wide frequency range.

Most parameters of interest have a frequency variation consisting of a number of peaks and nulls. These peaks and nulls are typically caused by resonances of the body current, or by the radiated fields from different portions of the body adding or canceling as the frequency is varied. Regardless of the reason for the oscillations, a consequence is that to reproduce accurately the frequency variation, one must make computations at relatively small frequency intervals. Since the MM impedance matrix must be computed at each frequency, the result is that performing a MM computation over a wide bandwidth can be a very time-consuming process.

Considerable saving in CPU time can be made by recognizing that while the currents exhibit resonance behavior and the fields exhibit constructive and destructive interference, the elements in the $[Z]$ matrix normally do not. As a result, the elements of $[Z]$ tend to be more slowly varying with frequency than fields or currents and thus can be interpolated at larger frequency intervals. We will illustrate the method with the examples of the input impedance of a dipole antenna and scattering from a perfectly conducting plate. Both examples will be analyzed with the "electromagnetic surface patch" or ESP code [3]–[5] which uses piecewise sinusoidal subsectional basis and weighting functions. The method has previously been applied by the author to microstrip antennas using entire domain basis and weighting functions [6].

Fig. 1 shows a thin wire dipole of length L which is split into $N + 1$ equal segments of length $d = L/(N + 1)$. Segment n extends from z_n to z_{n+1}, where $z_n = (n - 1)d$. The piecewise sinusoidal expansion function $\mathbf{J_n}$, developed by Richmond [7], is located on the surface of the wire and has surface current density

$$\mathbf{J_n} = \hat{\mathbf{z}} \, \frac{\sin k(d - |z - z_{n+1}|)}{2\pi a \sin kd}, \qquad z_n \le z \le z_{n+2} \tag{2}$$

where a is the wire radius, λ is the free-space wavelength, and $k = 2\pi/\lambda$ is the wavenumber. The weighting functions are comparable to the expansion functions except that they are filaments on the wire centerline.

Consider a thin wire dipole of length $L = 0.5$ m and radius $a = 0.001$ m over the frequency range from $f = 200$ to 1400 MHz. For the purposes of the MM solution the dipole will be split into 20 equal segments, which will result in $N = 19$ piecewise sinusoidal basis functions. Fig. 2 shows the frequency dependence of the magnitude and phase of Z_{11} (i.e., a self-impedance), $Z_{1,10}$, and $Z_{1,19}$. Note that mode 1 is the bottom mode, mode 10 is the center mode, and mode 19 is the top mode. In all cases the magnitude of the impedance element is a slowly varying function of frequency. The phase is an almost linear function of frequency, with the slope increasing as the separation between the modes increases.

The qualitative frequency behavior of the elements in the $[Z]$ matrix can be explained by observing from (1) that Z_{mn} is a weighted average of the electric field of expansion function n over the region of weighting function m. Since for subsectional basis functions the region of the weighting functions are small in terms of a wavelength,

Manuscript received October 13, 1987; revised April 4, 1988. This work was supported in part by the Joint Service Electronics Program under Contract NOOO14-78-C-0049 with The Ohio State University Research Foundation.

The author is with Electroscience Laboratory, Department of Electrical Engineering, The Ohio State University, Columbus, OH 43212.

IEEE Log Number 8823638.

Reprinted from *IEEE Trans. Antennas Propagat.*, vol. 36, no. 12, pp. 1820–1824, December 1988.

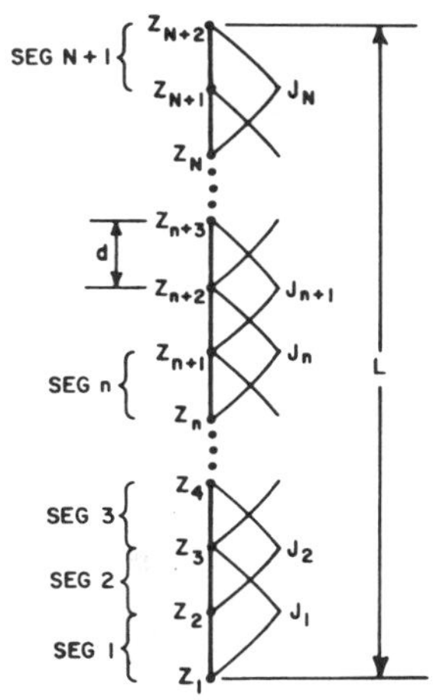

Fig. 1. Geometry for thin wire dipole with N piecewise sinusoidal modes, $N+1$ segments, and $N+2$ points.

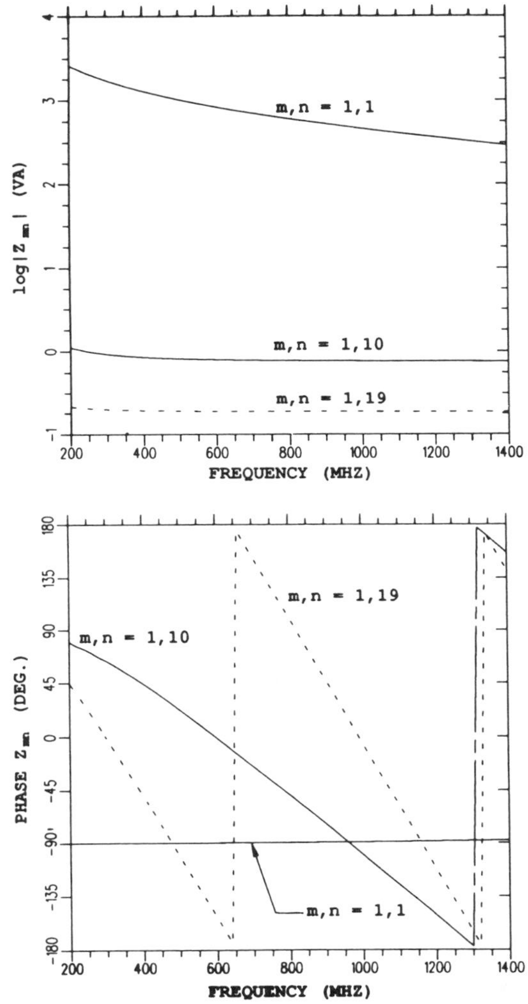

Fig. 2. Frequency dependence of three typical terms in dipole MM impedance matrix.

qualitatively Z_{mn} is simply (minus) the electric field of expansion function n evaluated at the center of weighting function m. Then if modes m and n are not too close, the dominant frequency dependence of Z_{mn} is simply

$$Z_{mn} \propto e^{-jkR_{mn}} \qquad R_{mn} \geq 0.5\lambda, \tag{3}$$

where R_{mn} is the (center-to-center) distance between modes m and n. From (3) it can be seen that as the electrical size of the body increases, the maximum phase variation with frequency of the Z_{mn} also increases.

Consider interpolating the complex exponential function in (3) with a frequency step size Δf. This corresponds to a wavenumber step size of $\Delta k = (2\pi/c)\Delta f$, where c is the speed of light. If we require the interpolation step size to correspond to no more than a phase change of π, then from (3)

$$\Delta k R_{mn} \leq \pi. \tag{4}$$

The largest value of R_{mn} will be denoted L, and in practice is taken as the maximum extent of the body. In this case the maximum interpolation step size is

$$\Delta k_M = \frac{\pi}{L} \text{ or } \Delta f_M = \frac{c}{2L} = \frac{f}{2(L/\lambda)}. \tag{5}$$

Note that (5) provides an estimate of an upper bound on the step size and not a step size which guarantees accurate results. Our experience is that with standard quadratic interpolation, a step size of about one-half the maximum is normally adequate. However, as discussed later, if one can incorporate the dominant frequency dependence of the Z_{mn} into the interpolating scheme, the step size can be increased.

The $[Z]$ matrix interpolation scheme will now be illustrated by computing the input impedance of the $L = 0.5$-m dipole already described from 200 to 1400 MHz, and with $\Delta f = \Delta f_M = 300$ MHz and $\Delta f_M/2 = 150$ MHz. To interpolate the $[Z]$ matrices, it is essential that the MM basis and weighting functions remain unchanged over the frequency range of interest. In practice, this means segmenting the body for acceptable accuracy at the highest frequency of interest. In this case, the dipole is segmented into 20 equal segments, which corresponds to a segment length of about 0.12λ at 1400 MHz, and $N = 19$ basis functions.

For the $\Delta f = 300$ MHz case, the computation is begun by computing and storing the $[Z]$ matrices at 200, 500, and 800 MHz. These frequencies at which the impedance matrix is actually computed and stored are referred to as match points. Quadratic interpolation of the stored impedance matrices at the first three match points is then used to find $[Z]$ and compute the dipole input impedance at intermediate frequencies in the range 200 MHz $\leq f \leq$ 650 MHz. When f exceeds 650 MHz, the $[Z]$ matrix at 1100 MHz (the fourth match point) is computed stored in the locations previously holding the $f = 300$ MHz $[Z]$ matrix. Next the $[Z]$ matrices at 500, 800, and 1100 MHz are interpolated to find $[Z]$ at intermediate frequencies in the range 650 MHz $< f \leq$ 950 MHz. The process is continued until one reaches 1400 MHz, i.e., the end of the frequency range of interest. In this way one is always storing and interpolating the $[Z]$ matrices at the closet three match points. Since it is also necessary to store the $[Z]$ matrix at the frequency at which the MM computation is being made, the $[Z]$ matrix interpolation scheme increases by a factor of four the storage required for the impedance matrix.

Fig. 3 shows the dipole input impedance computed every 10 MHz for $[Z]$ matrix interpolation step sizes of $\Delta f = 300$ MHz and 150 MHz. Comparing Figs. 2 and 3 shows that while the input impedance exhibits resonance behavior, the elements in the impedance matrix do not. The nine heavy dots in Fig. 3 show the match points at which $[Z]$ matrices were actually computed for the $\Delta f = 150$ MHz curve. The five odd-numbered points at 200, 500, 800, 1100, and 1400 MHz are the match points at which the impedance matrices were computed for the $\Delta f = 300$ MHz curve. At a match point the MM computation is exact in the sense that there is no interpolation error. Here the term

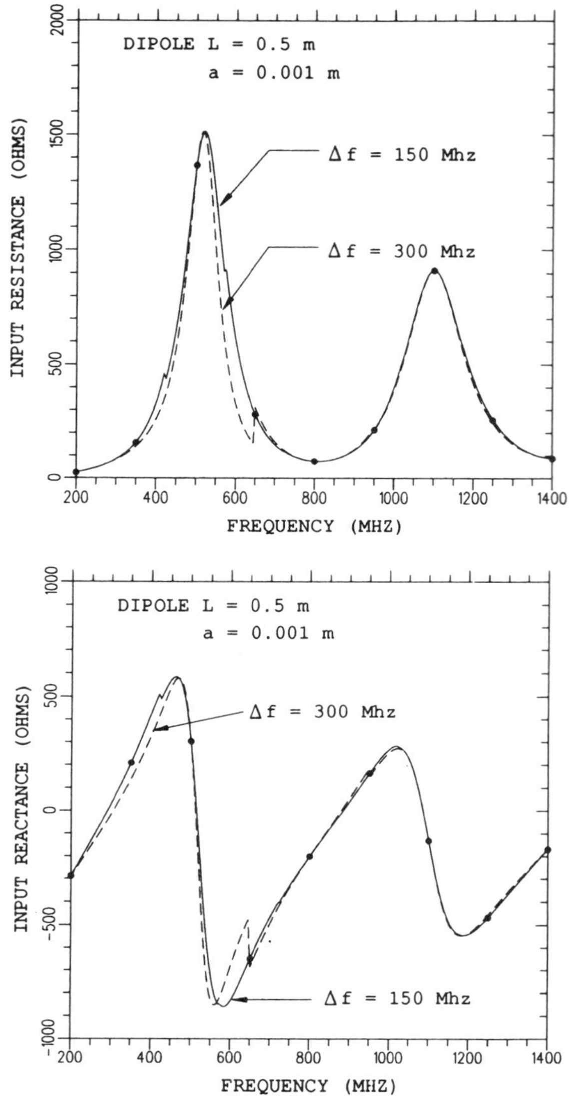

Fig. 3. Dipole input impedance using quadratic interpolation and step sizes of Δf = 300 and 150 MHz.

exact means no interpolation error. Thus the Δf = 150 and Δf = 300 MHz curves agree at the odd numbered match points. The computer CPU time on a VAX 8550 was about 15 and 21 s for the Δf = 300 and 150 MHz curves, respectively. By comparison, about 167 s were required if the $[Z]$ matrix interpolation scheme was not used.

The Δf = 150 MHz curve in Fig. 3 provided very accurate data in the sense that the error in interpolating $[Z]$ was negligible. However, some error is evident in the Δf = 300 MHz curve. In particular, note the discontinuity in the Δf = 300 MHz curve at f = 650 MHz. This discontinuity is caused by the fact that below 650 MHz, we interpolated the $[Z]$ matrices computed at 200, 500, and 800 MHz. Above 650 MHz we interpolated the $[Z]$ matrices computed at 500, 800, and 1100 MHz. The size of the discontinuity is a good measure of the interpolation error at 650 MHz. Note that the Δf = 150 MHz curve also has discontinuities, however, they are much smaller.

As illustrated in the insert in Fig. 4, the next example is the edge on radar cross section (RCS) of a 0.5-m square plate from 200 to 800 MHz. The plate was modeled with 84 piecewise sinusoidal surface patch modes, which corresponds to a segment size of about 0.2λ at 800 MHz [3]–[5]. Curves are shown for Δf = 75, 150, and 300

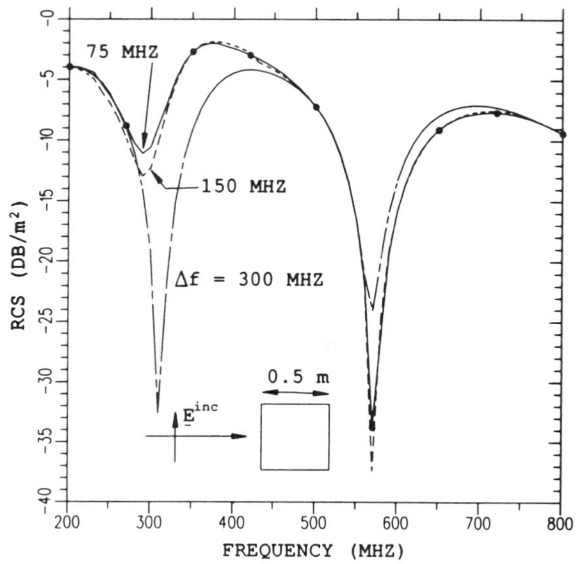

Fig. 4. Edge on RCS of 0.5-m square plate from 200 to 800 MHz.

MHz. Match points are shown every 75 MHz. In each case the RCS was computed every 10 MHz. These data exhibit deep nulls, caused by cancellation of energy diffracted from the front and rear edges of the plate. Although there is a large change in reducing Δf from 300 to 150 MHz, the Δf = 150 and 75 MHz curves are in close agreement. The CPU time for the Δf = 300, 150, and 75 MHz curves was 183, 202, and 242 s, respectively. By comparison, about 840 s would be required to perform this computation without $[Z]$ matrix interpolation.

B. Improved [Z] Matrix Interpolation

In this section the known form of the frequency dependence of the Z_{mn} is used to obtain an improved interpolation scheme. Equation (3) and Fig. 2 both suggest that for subsectional basis functions the major frequency dependence of the Z_{mn} for modes which are not too close electrically is simply $e^{-jkR_{mn}}$. In this case, the function

$$Z'_{mn} = Z_{mn}/e^{-jkR_{mn}} \tag{6}$$

is expected to be more slowly varying with frequency than is Z_{mn}, and thus is easier to interpolate. In the improved interpolation method we use quadratic interpolation of Z'_{mn} if $R_{mn} > \lambda/2$. Once Z'_{mn} is known, (6) is used to obtain Z_{mn}.

For wire antennas it is well known that the imaginary part of self- (or other overlapping) impedances have a logarithmic singularity with frequency [8], [9]. This can be incorporated into the interpolation method by assuming the reactancce of closely spaced elements to have the form

$$X(f) = A + B \ln f + Cf \tag{7}$$

where A, B, and C are constants. These three constants can be determined if we know X at the three evenly spaced frequencies f_1, $f_2 = f_1 + \Delta f$, and $f_3 = f_2 + \Delta f$. Enforcing $X(f)$ in (7) to match at the three frequencies results in three simultaneous linear equations which have the solution

$$B = (X(f_1) - 2X(f_2) + X(f_3))/\ln \frac{f_1 f_3}{f_2^2}$$

$$C = \left[(X(f_2) - X(f_1)) - B \ln \frac{f_2}{f_1} \right] / \Delta f$$

$$A = X(f_1) - B \ln f_1 - Cf_1. \tag{8}$$

TABLE I
EXACT VALUES OF TYPICAL ELEMENTS IN THE DIPOLE IMPEDANCE MATRIX (UNITS ARE VA)

Frequency (MHz)	$m, n = 1, 1$	$m, n = 1, 10$	$m, n = 1, 19$
$f_1 = 200$	$0.2204-j2598$	$0.1921+j1.089$	$0.1520+j0.1499$
$f_2 = 500$	$1.382-j1016$	$0.7541+j0.3193$	$-0.03767-j0.1853$
$f_3 = 800$	$3.590-j607.8$	$0.5094-j0.5893$	$-0.03688+j0.1840$
$f = 640$	$2.278-j779.9$	$0.7816-j0.1358$	$-0.1877-j0.01876$

TABLE II
INTERPOLATED VALUES OF TYPICAL ELEMENTS IN THE DIPOLE IMPEDANCE MATRIX AT 640 MHZ (UNITS ARE VA)

m, n	Quadratic Interpolation	E (percent)	Improved Interpolation	E (percent)
1, 1	$2.282-j679.4$	12.9	$2.282-j753.1$	3.4
1, 10	$0.7403-j0.08750$	8.0	$0.7770-j0.1056$	3.8
1, 19	$-0.06100-j0.1006$	80.1	$-0.1854-j0.02213$	2.2

In applying the improved interpolation method to wire problems, (7) is used to find the imaginary part of the Z_{mn} if $R_{mn} \leq \lambda/2$. The real part of the Z_{mn} for closely spaced elements is determined by standard quadratic interpolation.

The advantages of the improved interpolation method can be illustrated by considering the elements in the impedance matrix for the $\Delta f = 300$ MHz curve in Fig. 3. Table I shows the exact values of the impedance matrix elements $Z_{1,1}$, $Z_{1,10}$, and $Z_{1,19}$ at the three match point frequencies and at 640 MHz. Using standard and improved quadratic interpolation, Table II shows the interpolated values and percent error at 640 MHz. Note that using standard quadratic interpolation, the error in the self-impedance Z_{11} is 12.9 percent and that virtually all of the error is in the imaginary part. By comparison, Table II shows that using the improved interpolation scheme, this error is reduced to 3.4 percent. Also, standard quadratic interpolation fails on $Z_{1,19}$, producing an error of 80.1 percent. However, the improved method is able to interpolate $Z_{1,19}$ with an error of 2.2 percent.

Fig. 5 shows the real and imaginary part of the dipole input impedance from 200 to 1400 MHz, and for $\Delta f = 600$ and 300 MHz with the improved interpolation method. Reducing Δf below 300 MHz produces virtually no change in the input impedance. Note that with the improved interpolation method we are able to get essentially exact results with $\Delta f = \Delta f_M = 300$ MHz, and reasonably accurate results with $\Delta f = 2\Delta f_M = 600$ MHz. It is worth emphasizing that the $\Delta f = 600$ MHz curve interpolates over a 7 to 1 frequency range with only three match points.

III. SUMMARY

In summary, it has been illustrated that in performing an MM computation over a broad frequency range, a saving in CPU time is possible by interpolating the impedance matrix rather than the fields or currents. The basic reason for this is that while currents and fields exhibit resonant and oscillatory behavior with frequency, the elements of $[Z]$ normally do not. The main drawback of the method is that it results in a factor of four increase in the storage required for the impedance matrix.

A logical extension of the $[Z]$ matrix interpolation scheme is to interpolate the $[Y] = [Z]^{-1}$ matrix since this would also reduce matrix solution time. This idea was tried, but it was not found to be useful. The reason is that the elements of $[Y]$ do display resonant and oscillatory behavior with frequency and are thus no easier to interpolate than fields and currents.

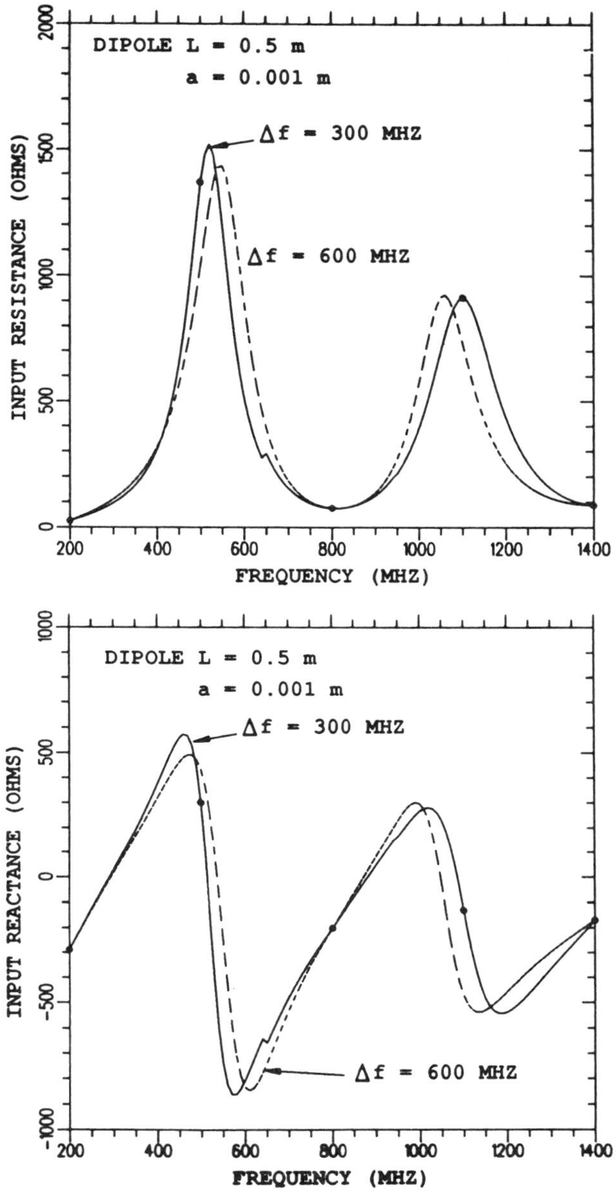

Fig. 5. Dipole input impedance using improved interpolation and step sizes of $\Delta f = 600$ and 300 MHz.

REFERENCES

[1] R. F. Harrington, *Field Computations by Moment Methods.* New York: Macmillan, 1968.
[2] *IEEE Trans. Antennas Propagat., Transactions Cumulative Index*, vol. AP-33, pp. II-169–II-170, 1985.
[3] E. H. Newman and R. L. Dilsavor, "A user's manual for the electromagnetic surface patch code: ESP version III," Ohio State Univ. ElectroSci. Lab. Rep. 716148-19, prepared under Grant NSG 1613 between Ohio State Univ. Res. Foundation and the Na. Aeronautics and Space Admin., Langley Res. Center, Hampton, VA, May 1987.
[4] E. H. Newman, "Electromagnetic modeling of composite wire and surface geometries," *IEEE Trans. Antennas Propagat.*, vol. AP-26, pp. 784–789, Nov. 1978.
[5] E. H. Newman, P. Alexandroupoulos, and E. K. Walton, "Polygonal plate modeling of realistic structures," *IEEE Trans. Antennas Propagat.*, vol. AP-32, pp. 742–747, July 1984.
[6] E. H. Newman and D. Forrai, "Scattering from a microstrip patch,"

IEEE Trans. Antennas Propagat., vol. AP-35, pp. 245–251, Mar. 1987.

[7] Y. T. Lin and J. H. Richmond, "EM modeling of aircraft at low frequencies," *IEEE Trans. Antennas Propagat.*, vol. AP-23, pp. 53–56, Jan. 1975.

[8] W. A. Imbriale and P. G. Ingerson, "On numerical convergence of moment solutions of moderately thick antennas using sinusoidal basis functions," *IEEE Trans. Antennas Propagat.*, vol. AP-28, pp. 42–48, Jan. 1980.

[9] E. H. Newman, "The equivalent separation(s) for the self-impedance of thin strips," *IEEE Trans. Antennas Propagat.*, vol. AP-35, pp. 110–113, 1987.

USING MODEL-BASED PARAMETER ESTIMATION TO INCREASE THE EFFICIENCY OF COMPUTING ELECTROMAGNETIC TRANSFER FUNCTIONS *

G. J. Burke, Lawrence Livermore National Lab., Livermore CA 94550
E. K. Miller, General Research Corporation, Santa Barbara, CA 93111
S. Chakrabarti, K. Demarest, University of Kansas, Lawrence KS 66045

ABSTRACT

Model-based parameter estimation (MBPE) involves fitting physically motivated approximations (the model) to accurately computed or measured electromagnetic quantities from which unknown coefficients (the model parameters) are numerically obtained. The model can then provide a simple, compact representation of the functional behavior of the quantity of interest that can be used for interpolation, extrapolation or other purposes. We examine here approximation of transfer functions from both frequency and frequency-derivative data and applications of the resulting models. Also, a scheme for obtaining frequency derivative information from a Moment-Method solution is outlined.

INTRODUCTION

Model-based parameter estimation (MBPE) offers opportunities for improving the efficiency of Computational Electromagnetics by reducing the computer time needed to perform the basic model computation itself, or by reducing the number of solutions needed to determine a result over some bandwidth. It does this by allowing known properties of the functional behavior of a quantity in space, time or frequency, to be used together with accurate analytic, numerical or measured samples to represent the quantity over a region of the parameter space.

For example, the electromagnetic response of an antenna, propagation path, or a scatterer is most often needed over a spectrum of frequencies rather than at just one or a few isolated frequency points. Often results are computed as discrete frequency samples with linear or low order polynomial interpolation, rather than exploiting the underlying physics of the phenomenon being modeled. The result can be that many more frequency samples are used than should be necessary from sampling requirements based on information theory. By using rational-function models for interpolation, as shown below, the required number of samples can be greatly reduced.

An example of this idea applied in a different context is described in [1] where the spatial variation of Sommerfeld integrals, which occur in the field of a source near an interface, is modeled by a sum of functions suggested by analytic approximations. The function amplitudes are determined by a least-squares fit of their sum to a set of computed field values.

The use of MBPE in the spectral domain with both frequency sampling and frequency-derivative sampling is summarized below, followed by some preliminary numerical results that illustrate application of the latter. Also, a method is outlined to obtain derivatives of the frequency response from a Moment-Method solution. A more through discussion of MBPE applications in electromagnetics is given by Miller in [2].

MBPE IN SPECTRAL AND TIME DOMAINS

The transient behavior of electromagnetic fields, as is true of the response of any linear system in a "waveform domain", is generally well approximated by a function of the form

$$f(x) = f_p(x) + f_{np}(x) = \sum_{\alpha=1}^{W} R_\alpha e^{s_\alpha x} + f_{np}(x). \tag{1}$$

The single-sided Laplace transform of Eq. (1) leads to the generic spectral-domain pole series as a function of the transform variable X as

$$F(X) = F_p(X) + F_{np}(X) = \sum_{\alpha=1}^{W} \frac{R_\alpha}{(X - s_\alpha)} + \sum_{\beta=-Q}^{R} C_\beta X^\beta \tag{2}$$

The function f_{np} represents a "non-pole" component that may be needed to account for part of the driven (as opposed to source free) response of the system and can also account for poles that are not included in the model. In Eq. (2) the non-pole term has been represented as powers of X where Q is usually 0 or 1. The actual number of poles W is generally infinite for electromagnetic response functions, although, as will be seen below, good approximations can be obtained over a limited range of frequency with a model containing a finite number of suitably chosen poles.

A procedure for determining R_α and s_α in Eq. (1) to fit a given function of time was developed by R. Prony in 1795 [3,4]. In Prony's method, samples of the function with equal spacing δ in time are used to generate a matrix equation whose solution yields the coefficients of a polynomial for which the zeros are the terms $\exp(s_\alpha \delta)$. The s_α are then found by solving for the roots of the polynomial, and the solution for R_α follows. A similar approach can yield the parameters in Eq. (1) when derivatives of $f(x)$ with respect to x are known at a single frequency. Alternate methods, that are preferable to Prony's method when the input data is contaminated by noise, are described in [5].

A similar approach can be used to fit the model provided by Eq. (2) to sampled data in the spectral domain with $X = \sigma + j\omega$ a general complex frequency. The procedure, which can be viewed as the frequency domain equivalent of Prony's method, was described by Brittingham, et al. [6]. To determine the parameters s_α and R_α, Eq. (2) is written in least common denominator form as

$$F(X) = \frac{N(X)}{D(X)} = \frac{\sum_{i=0}^{n} N_i X^i}{X^Q \sum_{i=0}^{d} D_i X^i} \tag{3}$$

with polynomials of maximum order $n = W + Q + R$ and $d = W$. Since the numerator and denominator of Eq. (3) can be divided by a constant without changing the equation, any one of the N_i or D_i can be made equal to one, leaving $d + n + 1$ coefficients to be evaluated in the model.

To determine the numerator and denominator coefficients it is convenient to write Eq. (3) as

$$F(X) X^Q \sum_{i=0}^{d} D_i X^i = \sum_{i=0}^{n} N_i X^i. \tag{4}$$

* Work performed under the auspices of the U. S. Department of Energy by the Lawrence Livermore National Laboratory under Contract W-7405-Eng-48.

Reprinted from *IEEE Trans. Magn.*, vol. 25, no. 4, pp. 2807–2809, July 1989.

Then if samples of $F(X)$ are available at $M \geq n+d+1$ frequencies X_j a set of M linear equations can be obtained from Eq. (4) for the N_i and D_i.

The rational approximation of Eq. (3) can also be determined from values of derivatives of $F(X)$ rather than function samples. If t derivative samples are known at a single frequency X_0 the easiest procedure is to transform to a new variable $Y = X - X_0$. Then, using the Padé method [7], the Maclaurin series for $F(Y)$, involving the derivative samples, is substituted into Eq. (4). Equating equal powers of Y results in a rational function having derivatives equal to those of F at $Y = 0$ and yields the set of equations for N_i and D_i.

If derivatives are known at more than one frequency, a rational function matching all of the samples can be obtained by taking derivatives of Eq. (4). The t-th derivative of Eq. (4) with $Q = 0$ is

$$\sum_{k=0}^{t} C_{t,t-k} F^{(t-k)}(X_j) D^{(k)}(X_j) = N^{(t)}(X_j)$$

where $C_{t,t-k} = t!/k!(t-k)!$. Then using T_j derivatives at sample point X_j with D_0 equal to one yields the set of equations

$$\sum_{i=1}^{d} D_i \sum_{k=0}^{\min(i,t_j)} C_{t_j,t_j-k} \frac{i!}{(i-k)!} X_j^{i-k} F^{(t_j-k)}(X_j) - \sum_{i=t_j}^{n} N_i \frac{i!}{(i-t_j)!} X_j^{i-t_j} = -F^{(t_j)}(X_j) \qquad (5)$$

where $j = 1, \ldots, M$; $t_j = 0, \ldots, T_j$ and $\sum_{j=1}^{M}(T_j+1) \geq n+d+1$. The rational function resulting from solving Eq. (5) for the N_i and D_i can provide a continuous interpolation function between several sample points. In addition, the rational function can be fit to a reduced window of two or more sample frequencies, with continuity of value and matched derivatives maintained as the window slides over the full set of samples. In contrast, the error in the single-point Padé result generally increases with distance from the sample frequency.

COMPUTING FREQUENCY DERIVATIVES

One motivation for considering models based on derivatives with respect to frequency is that multiple derivatives of a response at a given frequency can sometimes be computed more easily than the same number of discrete frequency samples. For a moment method solution the procedure can be outlined in terms of the impedance-matrix equation $ZI = V$, where Z is the frequency dependent impedance matrix and V and I are column vectors of the excitation voltages and resulting currents, respectively. Taking successive derivatives of the impedance-matrix equation with respect to frequency, the highest order derivative of I is seen to be multiplied by Z with no derivative. Hence we can conveniently solve for derivatives of the solution I recursively as

$$\begin{aligned} I &= Z^{-1}V \\ I' &= Z^{-1}(V' - Z'I) \\ &\cdots \\ I^{(t)} &= Z^{-1}\left[V^{(t)} - \sum_{r=1}^{t} C_{t,r} Z^{(r)} I^{(t-r)}\right] \end{aligned} \qquad (6)$$

where $C_{t,r}$ is the binomial coefficient.

The advantage of this technique is that each successive derivative involves only the inverse of the zero-derivative matrix. In practice any solution method, such as L-U factorization, can be used in place of the inverse shown. Thus, while the moment-method solution requires a computation time proportional to $AN^2 + BN^3$ to fill and factor the matrix for N unknowns, additional derivatives of the solution with respect to frequency can be obtained in a time proportional to only N^2. The derivative matrices $Z^{(r)}$ can be computed without too much additional computational effort if they are done simultaneously with the Z matrix so that common terms can be shared. The implementation of this method is discussed further in [8].

SOME REPRESENTATIVE RESULTS

A version of the Numerical Electromagnetics Code (NEC-3) for modeling wire antennas [1] has been developed to compute up to four derivatives of the solution for current using the procedure of Eq. (6). Some typical results of rational-function approximation using these derivatives are shown below. In Fig. 1, the input admittance of a dipole antenna has been modeled by fitting a rational function to the value and four derivatives at $L/\lambda = 0.2$ and 2.0. The rational-function curve is mostly hidden by the solid line. This process could be repeated in a sliding window to cover a wide bandwidth efficiently.

The admittance of a dipole over a ground plane, as shown in Fig. 2, has a very narrow resonance requiring an expanded plot scale from L/λ of 0.49 to 0.50. However a rational function with $n = 5$ and $d = 4$ fit to the value and four derivatives at $L/\lambda = 0.25$ and -0.25 shows good agreement. Here we are taking advantage of the fact that the spectral function for negative frequencies must be the complex conjugate of that for positive frequencies, since the time response is real. Hence, when the positive frequency result is known, the negative frequency result can be obtained "for free" by taking the conjugate and reversing the sign of odd derivatives. Including this information in the model improves the result. When only the positive frequency $L/\lambda = 0.25$ is fit with a rational function having $n = d = 2$ the resonance of the rational function moves up to about $L/\lambda = 0.4972$.

The forked monopole, shown in Fig. 3, displays both a normal monopole resonance and a high-Q transmission-line mode resonance. To resolve this resonance behavior accurately with

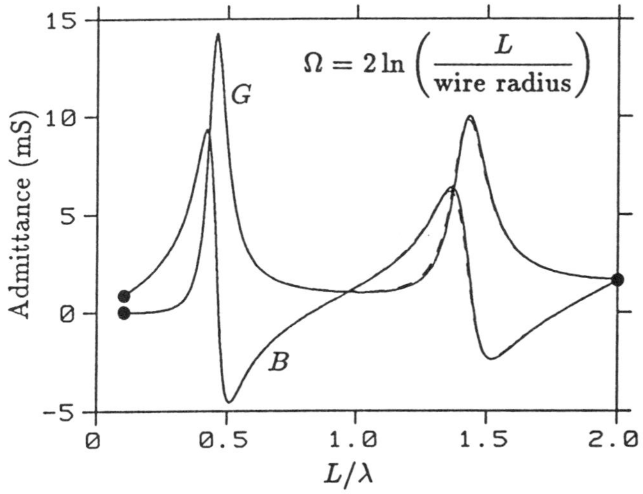

Fig. 1. Input admittance of a dipole antenna with length L and thickness parameter $\Omega = 10$ modeled with 21 segments. Directly computed values (———) are compared with a rational function model (– – – –) with $n = 5$ and $d = 4$, matched to the value and four derivatives at $L/\lambda = 0.2$ and 2.0.

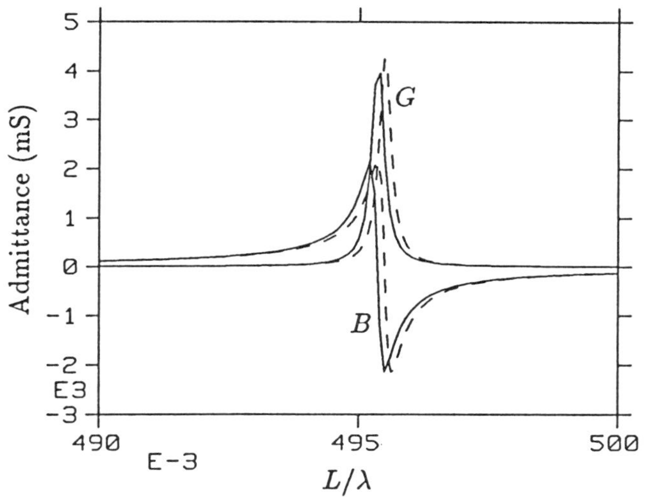

Fig. 2. Input admittance of a dipole antenna with length L and thickness parameter $\Omega = 15$ at height $L/50$ above a perfectly conducting ground. Directly computed values (————) are compared with a rational function model (– – – –) with $n = 5$ and $d = 4$, matched to the value and four derivatives at $L/\lambda = 0.25$ and -0.25.

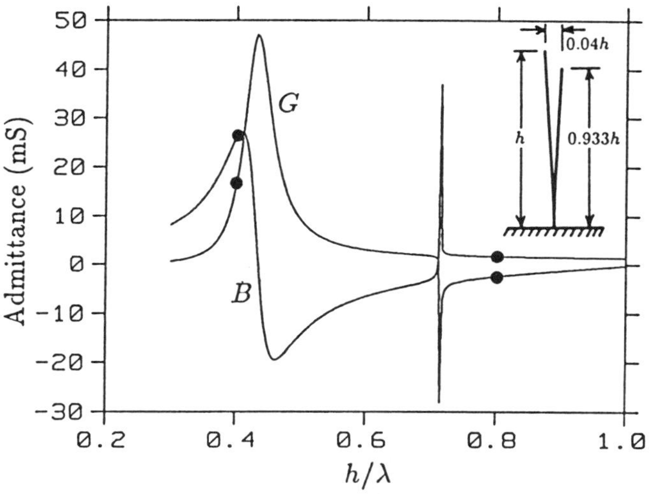

Fig. 3. Input admittance of a forked monopole on a perfectly conducting ground. Directly computed values (————) are compared with a rational function model (– – – –) with $n = 5$ and $d = 4$, matched to the value and four derivatives at $h/\lambda = 0.4$ and 0.8.

discrete frequency samples requires a very large number of evaluations. The narrow resonance can easily be completely missed with course samples in such a case. However a rational function with $n = 5$ and $d = 4$ fit to the value and four derivatives at $L/\lambda = 0.4$ and 0.8 accurately models both resonances. The dashed line is hidden by the solid line, showing only a slight difference if the narrow resonance is plotted on an expanded scale. Similar excellent agreement is obtained with a rational function with $n = d = 4$ fit with no derivatives to nine equally spaced points from $L/\lambda = 0.4$ to 0.8.

CONCLUSIONS

In this paper, we have introduced two main ideas: 1) The use of model-based parameter estimation (MBPE) based on rational function approximations to reduce the number of frequencies at which solutions or samples are required. 2) A sampling approach which employs frequency derivatives of the response using a new analytical technique based on differentiating the moment method impedance equation, and which provides derivative information in a time proportional to N^2 in contrast with the N^3 dependence in solving the original problem. Antenna input admittances were modeled using frequency samples and derivatives. An entire current distribution or some or all of the elements in a moment-method admittance matrix could also be modeled in rational-function form. Fields can be computed from interpolated values of the current. Alternatively, field values and derivatives can be computed directly from the current values and derivatives and then fit to a rational function.

The Rational-function model was shown to offer a large advantage over polynomial interpolation of a frequency response. In addition it offers the opportunity for adaptive sampling in a sliding-window scheme. A measure of error could be developed either by comparing interpolated and extrapolated values from different sample sets or from different model orders with fixed samples.

Application of the frequency-derivative approach has been demonstrated for problems having well defined resonances such as a dipole antenna, and for more challenging problems having narrow resonances. It seems clear that MBPE could develop into an important tool for not only increasing the efficiency of model applications, but in yielding more physically useful representations of those models.

REFERENCES

[1] G. J. Burke and E. K. Miller, "Modeling Antennas Near to and Penetrating a Lossy Interface," *IEEE Trans. Antennas and Prop.*, AP-32, pp. 1040-1049, October 1984.

[2] E. K. Miller, "Model-Based Parameter-Estimation Applications in Electromagnetics," presented at NATO Advanced Study Institute on Modeling and Measurement in Electromagnetic Analysis and Synthesis Problems, Il Ciocco, Italy, to be published by Sijthoff and Noordhoff.

[3] R. Prony, "Essai Experimental et Analytique sur les Lois de la Dilatabilite de Fluides Elastiques et sur Celles del la Force Expansive de la Vapeur de L'alkool a Differentes Temperatures," *J. L'Ecole Polytech. (Paris)*, 1, pp. 24-76, 1795.

[4] F. Hildebrand, *Introduction to Numerical Analysis*, McGraw-Hill, New York, NY, 1956.

[5] D. M. Goodman and D. G. Dudley, "An Output Error Model and Algorithm for Electromagnetic System Identification," *Circuits, Systems and Signal Processing*, Vol. 6, No. 4, pp. 471-505, 1987.

[6] J. N. Brittingham, E. K. Miller and J. L. Willows, "Pole Extraction from Real-Frequency Information," *Proceedings of the IEEE*, 68, pp. 263-273, 1980.

[7] D. S. Jones, *Methods in Electromagnetic Wave Propagation*, Oxford University Press, Engineering Science Series, Oxford, United Kingdom, 1979.

[8] G. J. Burke and E. K. Miller, "Use of Frequency-Derivative Information to Reconstruct an Electromagnetic Transfer Function," *Proceedings of the Fourth Annual ACES Review*, Naval Postgraduate School, Monterey, CA, March 1988.

A Bibliography of Moment Method Papers from 1960 to 1990 *

E. H. Newman, Fellow, IEEE

December 17, 1990

Following is a bibliography of journal articles dealing with the method of moments (MM) from roughly 1960 to 1990, and includes the following journals:

- The IRE and IEEE Transactions on Antennas and Propagation
- The IRE and IEEE Transactions on Microwave Theory and Techniques
- The IEEE Transactions on Electromagnetic Compatability
- The Proceedings of the IEEE
- Radio Science
- IEE Proceedings Part H (England) (1990 not available)
- Electromagnetics.

The survey was done "by hand", rather than by a computerized search.

The articles are listed alphabetically by the last name of the first author, and are grouped by the topics:

1. Theory and General Interest
2. Review Papers
3. Perfectly Conducting Surfaces
4. Material Bodies
5. Apertures
6. Thin Wires
7. Printed Circuit Antennas
8. Waveguides and Transmission Lines
9. Arrays
10. MM/Green's Function Solutions.

---

*This work was sponsored by the Joint Service Electronic Program under Contract N00014-78-C-0049 with The Ohio State University Research Foundation.

The author is with the ElectroScience Lab, The Ohio State University Department of Electrical Engineering.

Theory and General Interest

Al-Badwaihy, K. A. and J. L. Yen, "Extended Boundary Condition Integral for Perfectly Conducting and Dielectric Bodies: Formulation and Uniqueness," *IEEE Trans. Antennas Propagat.*, vol. 23, pp. 546–551, July 1975.

Amitay, N. and V. Galindo, "On Energy Conservation and the Method of Moments in Scattering Problems," *IEEE Trans. Antennas Propagat.*, vol. 17, pp. 747–751, Nov. 1969.

Bates, R. H. T., "Analytic Constraints on Electromagnetic Field Computations," *IEEE Trans. Microwave Th. and Tech.*, vol. 23, pp. 605–623, Aug. 1975.

Bates, R. H. T., "The Point-Matching Method for Interior and Exterior Two-Dimensional Boundary Value Problems," *IEEE Trans. Microwave Th. and Tech.*, vol. 15, pp. 185–187, March 1967.

Chen, H. C. and D. K. Chen, "A Useful Matrix Inversion Formula and Its Application," *Proc. IEEE*, vol. 55, pp. 705–707, May 1967.

Chen, K., "A Mathematical Formulation of the Equivalence Principle," *IEEE Trans. Microwave Th. and Tech.*, vol. 37, pp. 1576–1581, Oct. 1989.

Cheng, D. K. and C. Liang, "Thinning Technique for Moment Method Solutions," *Proc. IEEE*, vol. 71, pp. 265–267, Feb. 1983.

Chuang, C. W., J. H. Richmond, N. Wang, and P. H. Pathak, "New Expressions for Mutual Impedance of Nonplanar-Skew Sinusoidal Monopoles," *IEEE Trans. Antennas Propagat.*, vol. 38, pp. 275–276, Feb. 1990.

Cohen, M. H., "Application of the Reaction Concept to Scattering Problems," *IRE Trans. Antennas Propagat.*, vol. 3, pp. 193–199, Oct. 1955.

Cohoon, D. K., "An Exact Formula for the Accuracy of a Class of Computer Solutions of Integral Equation Formulations of Electromagnetic Scattering Problems," *Electromagnetics*, vol. 7, pp. 153–165, 1987.

Daniel, S. M., "An Optimal Solution to a Scattering Problem," *Proc. IEEE*, vol. 58, pp. 270–272, Feb. 1970.

Djordjevic, A. R. and T. K. Sarkar, "A Theorem on the Moment Methods," *IEEE Trans. Antennas Propagat.*, vol. 35, pp. 353–355, March 1987.

Dudley, D. G., "Error Minimization and Convergence in Numerical Methods," *Electromagnetics*, vol. 5, pp. 89–97, 1985.

Erez, E. and Y. Leviatan, "Analysis of Natural Frequencies of Cavities and Scatterers Using an Imulsive Current Model," *IEEE Trans. Antennas Propagat.*, vol. 38, pp. 534–540, Apr. 1990.

Farden, D. C., "Solution of a Toeplitz Set of Linear Equations," *IEEE Trans. Antennas Propagat.*, vol. 24, pp. 906–907, Nov. 1976.

Ferguson, T. R., T. H. Lehman, and R. J. Balestri, "Efficient Solution of Large Moment Problems: Theory and Small Problem Results," *IEEE Trans. Antennas Propagat.*, vol. 24, pp. 230–235, March 1976.

Garbacz, R. J. and R. H. Turpin, "A Generalized Expansion for Radiated and Scattered Fields," *IEEE Trans. Antennas Propagat.*, vol. 19, pp. 348–357, May 1971.

Gera, A. E., "Simple Expressions for Mutual Impedances," *IEE Proc.*, vol. 135 Pt. H, pp. 395–399, Dec. 1988.

Gething, P. J. D., "Inversion of an Impedance Matrix," *IEEE Trans. Antennas Propagat.*, vol. 13, pp. 830–831, Sep. 1965.

Graglia, R. D., "Static and Dynamic Potential Integrals for Linearly Varying Source Distributions in Two- and Three-Dimensional Problems," *IEEE Trans. Antennas Propagat.*, vol. 35, pp. 662–669, June 1987.

Harrington, R. F., "Small Resonant Scatterers and Their Use for Field Measurements," *IEEE Trans. Microwave Th. and Tech.*, vol. 10, pp. 165–174, May 1962.

Harrington, R. F., "Matrix Methods for Field Problems," *Proc. IEEE*, vol. 55, pp. 136–149, Feb. 1987.

Harrington, R. F. and J. R. Mautz, "Control of Radar Scattering by Reactive Loading," *IEEE Trans. Antennas Propagat.*, vol. 20, pp. 446–454, July 1972.

Herrmann, G. F., "Note on Interpolational Basis Functions in the Method of Moments," *IEEE Trans. Antennas Propagat.*, vol. 38, pp. 134–137, Jan. 1990.

Ikuno, H. and K. Yasuura, "Improved Point-Matching Method with Application to Scattering from a Periodic Surface," *IEEE Trans. Antennas Propagat.*, vol. 21, pp. 657–662, Sep. 1973.

Itoh, T. and R. Mittra, "Relative Convergence Phenomenon Arising in the Solution of Diffraction from Strip Grating on a Dielectric Slab," *Proc. IEEE*, vol. 59, pp. 1363–1365, Sep. 1971.

Jones, D. S., "A Critique of the Variational Method in Scattering Problems," *IRE Trans. Antennas Propagat.*, vol. 4, pp. 297–301, July 1956.

Kastner, R., "On Matrix Partitioning and the "Add-On" Technique," *Electromagnetics*, vol. 9, pp. 331–344, 1989.

Kishi, N. and T. Okoshi, "Proposal for a Boundary-Integral Method without Using Green's Function," *IEEE Trans. Microwave Th. and Tech.*, vol. 35, pp. 887–892, Oct. 1987.

Kleev, A. I. and A. B. Manenkov, "The Convergence of Point-Matching Techniques," *IEEE Trans. Antennas Propagat.*, vol. 37, pp. 50–54, Jan. 1989.

Klein, C. and R. Mittra, "Stability of Matrix Equations Arising in Electromagnetics," *IEEE Trans. Antennas Propagat.*, vol. 21, pp. 902–905, Nov. 1973.

Kress, R., "Numerical Solution of Boundary Integral Equations in Time-Harmonic Electromagnetic Scattering," *Electromagnetics*, vol. 10, pp. 1–20, Jan.-June 1990.

Laxpati, S. R., "Comments on "An Iterative Method of Solving a System of Linear Equations and Its Physical Interpretation from the Point of View of Scattering Theory"," *IEEE Trans. Antennas Propagat.*, vol. 19, p. 566, July 1971.

Leviatan, Y., "Analytic Continuation Considerations when Using Generalized Formulations for Scattering Problems," *IEEE Trans. Antennas Propagat.*, vol. 38, pp. 1259–1263, Aug. 1990.

McDonald, B. H., M. Friedman, and A. Wexler, "Variational Solution of Integral Equations," *IEEE Trans. Microwave Th. and Tech.*, vol. 22, pp. 237–248, March 1974.

Mahadevan, K. and H. A. Auda, "On the Electromagnetic Field of a Rectangular Patch of Uniform and Linear Distributions of Current in the Source Region," *IEEE Trans. Antennas Propagat.*, vol. 38, pp. 1244–1248, Aug. 1990.

Millar, R. F., "On the Legitamacy of an Assumption Underlying the Point-Matching Method," *IEEE Trans. Microwave Th. and Tech.*, vol. 18, pp. 325–326, June 1970.

Miller, E. K., "A Selective Survey of Computational Electromagnetics," *IEEE Trans. Antennas Propagat.*, vol. 36, pp. 1281–1305, Sep. 1988.

Mittra, R., T. Itoh, and T. Li, "Analytical and Numerical Studies of the Relative Convergence Phenomenon Arising in the Solution of an Integral Equation by the Moment Moment," *IEEE Trans. Microwave Th. and Tech.*, vol. 20, pp. 96–104, Feb. 1972.

Nachamkin, J., "Integrating the Dyadic Green's Function Near Sources," *IEEE Trans. Antennas Propagat.*, vol. 38, pp. 919–921, June 1990.

Ney, M. M., "Method of Moments as Applied to Electromagnetic Problems," *IEEE Trans. Microwave Th. and Tech.*, vol. 33, pp. 972–980, Oct. 1985.

Newman, E. H., "Generation of Wide-Band Data from the Method of Moments by Interpolating the Impedance Matrix," *IEEE Trans. Antennas Propagat.*, vol. 36, pp. 1820–1823, Dec. 1988.

Olsen, R. G., G. L. Hower, and P. D. Mannikko, "A Hybrid Method for Combining Quasi-Static and Full-Wave Techniques for Electromagnetic Scattering Problems," *IEEE Trans. Antennas Propagat.*, vol. 36, pp. 1180–1184, Aug. 1988.

Patterson, J. E., T. Cwik, and R. D. Ferraro, "Parallel Computation Applied to Electromagnetic Scattering and Radiation Analysis," *Electromagnetics*, vol. 10, pp. 21–39, Jan.-June 1990.

Pearson, L. W., "A Technique for Organizing Large Moment Calculations for Use with Iterative Solution Methods," *IEEE Trans. Antennas Propagat.*, vol. 33, pp. 1031–1033, Sep. 1985.

Peterson, A. F., "An Analysis of the Spectral Iterative Technique for Electromagnetic Scattering from Individual and Periodic Structures," *Electromagnetics*, vol. 6, pp. 255–276, 1986.

Peterson, A. F., "The "Interior Resonance" Problem Associated with Surface Integral Equations of Electromagnetics: Numerical Consequences and a Survey of Remedies," *Electromagnetics*, vol. 10, pp. 293–312, 1990.

Preis, D. H., "The Toeplitz Matrix: Its Occurrence in Antenna Problems and a Rapid Inversion Algorithm," *IEEE Trans. Antennas Propagat.*, vol. 20, pp. 204–206, March 1972.

Ray, S. L. and A. F. Peterson, "Error and Convergence in Numerical Implementation of Conjugate Gradient Method," *IEEE Trans. Antennas Propagat.*, vol. 36, pp. 1024–1030, Dec. 1988.

Richmond, J. H., "The Numerical Evaluation of Radiation Integrals," *IRE Trans. Antennas Propagat.*, vol. 9, pp. 358–360, July 1961.

Richmond, J. H., "A Reaction Theorem and Its Application to Antenna Impedance Calculations," *IRE Trans. Antennas Propagat.*, vol. 9, pp. 515–520, Nov. 1961.

Rumsey, V. H., "Reaction Concept in Electromagnetic Theory," *Phys. Rev.*, vol. 94, pp. 1483–1491, June 1954.

Sain, M. K., "On "A Useful Matrix Inversion Formula and Its Applications"," *Proc. IEEE*, vol. 55, p. 1753, Oct. 1967.

Sancer, M. I., "Physically Interpretable Alternative to Green's Dyadics, Resulting Representations Theorems, and Integral Equations," *IEEE Trans. Antennas Propagat.*, vol. 38, pp. 564–568, Apr. 1990.

Sancer, M. I., R. L. McClary, and K. J. Glover, "Electromagnetic Computation Using Parametric Geometry," *Electromagnetics*, vol. 10, pp. 85–103, Jan.-June 1990.

Sarkar, T. K., K. R. Siarkiewicz, and R. F. Stratton, "Survey of Numerical Methods for Solution of Large Systems of Linear Equations for Electromagnetic Field Problems," *IEEE Trans. Antennas Propagat.*, vol. 29, pp. 847–856, Nov. 1981.

Sarkar, T. K., "A Note on the Variational Method (Rayleigh-Ritz), Galerkin's Method, and the Method of Least Squares," *Radio Science*, vol. 18, pp. 1207–1224, Nov.-Dec. 1983.

Sarkar, T. K., A. R. Djordjevic, and E. Arvas, "On the Choice of Expansion and Weighting Functions in the Numerical Solution of Operator Equations," *IEEE Trans. Antennas Propagat.*, vol. 33, pp. 988–996, Sep. 1985.

Simpson, T. L., J. C. Logan, and J. W. Rockway, "Equivalent Circuits for Electrically Small Antennas Using LS-Decomposition with the Method of Moments," *IEEE Trans. Antennas Propagat.*, vol. 37, pp. 1632–1635, Dec. 1989.

Singh, S. and R. Singh, "Application of Transforms to Accelerate the Summation of Periodic Free-Space Green's Functions," *IEEE Trans. Microwave Th. and Tech.*, vol. 38, pp. 1746–1748, Nov. 1990.

Tai, C. T., "An Iterative Method of Solving a System of Linear Equations and Its Physical Interpretation from the Point of View of Scattering Theory," *IEEE Trans. Antennas Propagat.*, vol. 18, pp. 713–714, Sep. 1970.

Tew, M. D. and L. L. Tsai, "A Method Toward Improved Convergence of Moment Method Solutions," *Proc. IEEE*, vol. 60, pp. 1436–1440, Nov. 1972.

Thomas, D. T., "Choice of Trial Functions for Matrix Methods," *Proc. IEEE*, vol. 55, pp. 1106–1108, June 1967.

Toyoda, I., M. Matsuhara, and N. Kumagai, "Extended Integral Equation Formulation for Scattering Problems from a Cylindrical Scatterer," *IEEE Trans. Antennas Propagat.*, vol. 36, pp. 1580–1586, Nov. 1988.

Webb, K. J., P. W. Grounds, and R. Mittra, "Convergence in the Spectral Domain Formulation of Waveguide and Scattering Problems," *IEEE Trans. Antennas Propagat.*, vol. 38, pp. 869–877, June 1990.

Whetten, F. L., K. Liu, and C. A. Balanis, "An Efficient Numerical Integral in Three-Dimensional Electromagnetic Field Computations," *IEEE Trans. Antennas Propagat.*, vol. 38, pp. 1512–1514, Sep. 1990.

Wilton, D. R. and C. M. Butler, "Effective Methods for Solving Integral and Integro- Differential Equations," *Electromagnetics*, vol. 1, pp. 289–308, 1981.

Wilton, D. R., K. A. Michalski, and L. W. Pearson, "On the Existence of Branch Points in the Eigenvalues of the Electric Field Integral Equation Operator in the Complex Frequency Plane," *IEEE Trans. Antennas Propagat.*, vol. 31, pp. 86–91, Jan. 1983.

Wilton, D. R., S. M. Rao, A. W. Glisson, D. H. Schaubert, O. M. Al-Bundak, and C. M. Butler, "Potential Integrals for Uniform and Linear Source Distributions on Polygonal and Polyhedral Domains," *IEEE Trans. Antennas Propagat.*, vol. 32, pp. 276–281, March 1984.

Wu, C. and D. K. Cheng, "A Method for Symmetrizing Generalized Impedance Matrices," *IEEE Trans. Electromagnetic Compatibility*, vol. 19, pp. 81–88, May 1977.

Review Papers

Bevensee, R. M., J. N. Brittingham, F. J. Deadrick, T. H. Lehman, E. K. Miller, and A. J. Poggio, "Computer Codes for EMP Interaction and Coupling," *IEEE Trans. Antennas Propagat.*, vol. 26, pp. 156–164, Jan. 1978.

Newman, E. H. and R. J. Marhefka, "Overview of MM and UTM Methods at The Ohio State University," *Proc. IEEE*, vol. 77, pp. 700–708, May 1989.

Perini, J. and D. J. Buchanan, "Assessmenmt of MOM Techniques for Shipboard Applications," *IEEE Trans. Electromagnetic Compatibility*, vol. 24, pp. 32–39, Feb. 1982.

Richmond, J. H., "Digital Computer Solutions of the Rigorous Equations for Scattering Problems," *Proc. IEEE*, vol. 53, pp. 796–804, Aug. 1965.

Ryan, C. E., F. L. Cain, J. J. H. Wang, B. J. Cown, and W. P. Cooke, "Electromagnetic Models for Antenna Performance, EMC, and Biological Effects," *IEEE Trans. Electromagnetic Compatibility*, vol. 22, pp. 244–255, Nov. 1980.

Silvester, P. and Z. J. Csendes, "Numerical Modeling of Passive Microwave Devices," *IEEE Trans. Microwave Th. and Tech.*, vol. 22, pp. 190–201, March 1974.

Strait, B. J. and A. T. Adams, "On Contributions at Syracuse University to the Moment Method," *IEEE Trans. Electromagnetic Compatibility*, vol. 22, pp. 228–237, Nov. 1980.

Wexler, A., "Computation of Electromagnetic Fields," *IEEE Trans. Microwave Th. and Tech.*, vol. 17, pp. 416–439, Aug. 1969.

Perfectly Conducting Surfaces

Albertsen, N. C., J. E. Hansen, and N. E. Jensen, "Computation of Radiation from Wire Antennas on Conducting Bodies," *IEEE Trans. Antennas Propagat.*, vol. 22, pp. 200–206, March 1974.

Andreasen, M. G., "Scattering from Parallel Metallic Cylinders with Arbitrary Cross Section," *IEEE Trans. Antennas Propagat.*, vol. 12, pp. 746–754, Nov. 1964.

Andreasen, M. G., "Scattering from Bodies of Revolution," *IEEE Trans. Antennas Propagat.*, vol. 13, pp. 303–310, March 1965.

Arvas, E., R. F. Harrington, and J. R. Mautz, "Radiation and Scattering from Electrically Small Conducting Bodies of Arbitrary Shape," *IEEE Trans. Antennas Propagat.*, vol. 34, pp. 66–77, Jan. 1986.

Arvas, E., R. F. Harrington, and J. R. Mautz, "Radiation and Scattering from Electrically Small Conducting Bodies of Arbitrary Shape Above an Infinite Ground Plane," *IEEE Trans. Antennas Propagat.*, vol. 35, pp. 378–383, Apr. 1987.

Auckland, D. T. and R. F. Harrington, "Radiation and Scattering from Conducting Cylinders, TM Case," *IEEE Trans. Antennas Propagat.*, vol. 24, p. 544, July 1976.

Axline, R. M. and A. K. Fung, "Numerical Computation of Scattering from a Perfectly Conducting Random Surface," *IEEE Trans. Antennas Propagat.*, vol. 26, pp. 482–487, May 1978.

Azarbar, B. and L. Shafai, "Application of Moment Method to Large Cylindrical Reflector Antennas," *IEEE Trans. Antennas Propagat.*, vol. 26, pp. 500–502, May 1978.

Berthon, A. and R. P. Bills, "Integral Equation Analysis of Radiating Structures of Revolution," *IEEE Trans. Antennas Propagat.*, vol. 37, pp. 159–170, Feb. 1989.

Bhattacharya, S., S. A. Long, and D. R. Wilton, "The Input Impedance of a Monopole Antenna Mounted on a Cubical Conducting Box," *IEEE Trans. Antennas Propagat.*, vol. 35, pp. 756–762, July 1987.

Bornholdt, J. M. and L. N. Medgyesi-Mitschang, "Mixed-Domain Galerkin Expansions in Scattering Problems," *IEEE Trans. Antennas Propagat.*, vol. 36, pp. 216–227, Feb. 1988.

Bostian, C. W. and P. H. Wiley, "Concerning the Moment Solution for the Charge Distribution on a Square Plate," *Proc. IEEE*, vol. 59, p. 1639, Nov. 1971.

Burnside, W. D., C. L. Yu, and R. J. Marhefka, "A Technique to Combine the Geometrical Theory of Diffraction and the Moment Method," *IEEE Trans. Antennas Propagat.*, vol. 23, pp. 393–399, July 1975.

Chang, A. H. and R. Mittra, "Using Half-Plane Solutions in the Context of MM for Analyzing Large Flat Structures with or without Resistive Loading," *IEEE Trans. Antennas Propagat.*, vol. 38, pp. 1001–1009, July 1990.

Cherin, A. H. and J. Goldhirsh, "Impedance and Far Field Characteristics of a Linear Antenna Near a Conducting Cylinder," *IEEE Trans. Electromagnetic Compatibility*, vol. 15, pp. 110–117, Aug. 1973.

Cote, M. G., M. B. Woodworth, and A. D. Yaghjian, "Scattering from the Perfectly Conducting Cube," *IEEE Trans. Antennas Propagat.*, vol. 36, pp. 1321–1329, Sep. 1988.

Davis, W. A. and R. Mittra, "A New Approach to the Thin Scatterer Problem Using the Hybrid Equations," *IEEE Trans. Antennas Propagat.*, vol. 25, pp. 402–406, May 1977.

Djordjevic, A. R., C. K. Allen, T. K. Sarkar, and Z. A. Maricevic, "Inductance of Perfectly Conducting Foils Including Spiral Inductors," *IEEE Trans. Microwave Th. and Tech.*, vol. 38, pp. 1407–1414, Oct. 1990.

Eftimiu, C. and P. L. Huddleston, "Scattering by a Conducting Strip with a Randomly Serrated Edge," *Radio Science*, vol. 20, pp. 1549–1554, Nov.-Dec. 1985.

Elsherbeni, A. Z. and M. Hamid, "Scattering by Parallel Conducting Circular Cylinders," *IEEE Trans. Antennas Propagat.*, vol. 35, pp. 355–358, March 1987.

Fahmy, M. N. I. and A. Z. Botros, "Radiation from Quarter-Wavelength Monopoles on Finite Cylindrical, Conical, and Rocket-Shaped Conducting Bodies," *IEEE Trans. Antennas Propagat.*, vol. 27, pp. 615–623, Sep. 79.

Frenkel, A., "On Entire Domain-Basis Functions with Square-Root Edge Singularity," *IEEE Trans. Antennas Propagat.*, vol. 37, pp. 1211–1214, Sep. 1989.

Glisson, A. W. and D. R. Wilton, "Simple and Efficient Numerical Methods for Problems of Electromagnetic Radiation and Scattering from Surfaces," *IEEE Trans. Antennas Propagat.*, vol. 28, pp. 593–603, Sep. 1980.

Goldhirsh, J., D. L. Knepp, and R. J. Doviak, "Radiation from a Dipole Near a Conducting Cylinder of Finite Length," *IEEE Trans. Electromagnetic Compatibility*, vol. 12, pp. 96–105, Aug. 1970.

Harrington, R. F., "On the Calculation of Scattering by Conducting Cylinders," *IEEE Trans. Antennas Propagat.*, vol. 13, pp. 812–813, Sep. 1965.

Harrington, R. F. and J. R. Mautz, "Theory of Characteristic Modes for Conducting Bodies," *IEEE Trans. Antennas Propagat.*, vol. 19, pp. 622–628, Sep. 1971.

Harrington, R. F. and J. R. Mautz, "Computation of Characteristic Modes for Conducting Bodies," *IEEE Trans. Antennas Propagat.*, vol. 19, pp. 629–639, Sep. 1971.

Harrington, R. F. and J. R. Mautz, "Comments on "Concerning the Moment Solution for the Charge Distribution on a Square Conducting Plate"," *Proc. IEEE*, vol. 60, p. 448, Apr. 1972.

Heedy, D. J. and W. D. Burnside, "An Aperture-Matched Compact Range Feed Horn Design," *IEEE Trans. Antennas Propagat.*, vol. 33, pp. 1249–1255, Nov. 1985.

Helszajn, J., "Contour-Integral Equation Formulation of Complex Gyrator Admittance of Junction Circulators Using Triangular Resonators," *IEE Proc.*, vol. 132 Pt. H, pp. 255–260, July 1985.

Huddleston, P. L., "Scattering from a Conducting Strip with Rough Edges," *Radio Science*, vol. 21, pp. 429–433, May-June 1986.

Janaswamy, R., "A Simplified Expression for the Self/Mutual Impedance Between Coplanar and Parallel Surface Monopoles," *IEEE Trans. Antennas Propagat.*, vol. 35, pp. 1174–1176, Oct. 1987.

Janaswamy, R., "An Accurate Moment Method Model for the Tapered Slot Antenna," *IEEE Trans. Antennas Propagat.*, vol. 37, pp. 1523–1528, Dec. 1989.

Jenn, D. C. and W. V. T. Rusch, "An E-Field Integral Equation Solution for the Radiation from Reflector Antennas with Struts," *IEEE Trans. Antennas Propagat.*, vol. 37, pp. 683–689, June 1989.

Johnson, W. A. and D. R. Wilton, "Modeling Scattering from and Radiation by Arbitrary Shaped Objects with the Electric Field Integral Equation Triangular Surface Patch Code," *Electromagnetics*, vol. 10, pp. 41–63, Jan.-June 1990.

Kanda, M., "Electromagnetic-Field Distortion Due to a Conducting Rectangular Cylinder in a Transverse Electromagnetic Cell," *IEEE Trans. Electromagnetic Compatibility*, vol. 24, pp. 294-301, Aug. 1982.

Kao, C. C., "Electromagnetic Scattering from a Finite Tubular Cylinder: Numerical Solution," *Radio Science*, vol. 5, pp. 617–624, March 1970.

Klein, C. A. and R. Mittra, "An Application of the "Condition Number" Concept to the Solution of Scattering Problems in the Presence of the Interior Resonant Frequencies," *IEEE Trans. Antennas Propagat.*, vol. 23, pp. 431–434, July 1975.

Knepp, D. L. and J. Goldhirsh, "Numerical Analysis of Electromagnetic Radiation Properties of Smooth Conducting Bodies of Arbitrary Shape," *IEEE Trans. Antennas Propagat.*, vol. 20, pp. 383–388, July 1972.

Lee, K. S. H., L. Marin, and J. P. Castillo, "Limitations of Wire Grid Modeling of a Closed Surface," *IEEE Trans. Electromagnetic Compatibility*, vol. 18, pp. 123–129, Aug. 1976.

Lier, E., "Analysis of Soft and Hard Strip-Loaded Horns Using a Circular Cylindrical Model," *IEEE Trans. Antennas Propagat.*, vol. 38, pp. 783–793, June 1990.

Lin, J. L., W. L. Curtis, and M. C. Vincent, "Radar Cross Section of a Rectangular Conducting Plate by Wire Mesh Modeling," *IEEE Trans. Antennas Propagat.*, vol. 22, pp. 718–720, Sep. 1974.

Lin, Y. T. and J. H. Richmond, "EM Modeling of Aircraft at Low Frequencies," *IEEE Trans. Antennas Propagat.*, vol. 23, pp. 53–56, Jan. 1975.

Ludwig, A. C., "A Comparison of Spherical Wave Boundary Value Matching Versus Integral Equation Scattering Solutions for a Perfectly Conducting Body," *IEEE Trans. Antennas Propagat.*, vol. 34, pp. 857–864, July 1986.

Ludwig, A. C., "Wire Grid Modeling of Surfaces," *IEEE Trans. Antennas Propagat.*, vol. 35, pp. 1045–1048, Sep. 1987.

Mautz, J. R. and R. F. Harrington, "Radiation and Scattering from Bodies of Revolution," *Appl. Sci. Res.*, vol. 20, pp. 405–435, June 1969.

Mautz, J. R. and R. F. Harrington, "Generalized Network Parameters, Radiation, and Scattering by Conducting Bodies of Revolution," *IEEE Trans. Antennas Propagat.*, vol. 22, pp. 630–631, July 1974.

Mautz, J. R. and R. F. Harrington, "Radiation and Scattering from Loaded Bodies of Revolution," *IEEE Trans. Antennas Propagat.*, vol. 23, p. 594, July 1975.

Mautz, J. R. and R. F. Harrington, "H-Field, E-Field and Combined Field Solution for Conducting Bodies of Revolution," *Arch. Elek. Ubertragung (AEU Germany)*, vol. 32, pp. 157–164, Apr. 1978.

Mautz, J. R. and R. F. Harrington, "A Combined Source Solution for Radiation and Scattering from a Perfectly Conducting Body," *IEEE Trans. Antennas Propagat.*, vol. 27, pp. 445–454, July 1979.

Mautz, J. R. and R. F. Harrington, "An E-Field Solution for a Conducting Surface Small or Comparable to the Wavelength," *IEEE Trans. Antennas Propagat.*, vol. 32, pp. 330–339, Apr. 1984.

Maystre, D., "Electromagnetic Scattering from Perfectly Conducting Rough Surfaces in the Resonance Region," *IEEE Trans. Antennas Propagat.*, vol. 31, pp. 885–895, Nov. 1983.

Medgyesi-Mitschang, L. N. and D. Wang, "Hybrid Solutions for Scattering from Perfectly Conducting Bodies of Revolution," *IEEE Trans. Antennas Propagat.*, vol. 31, pp. 570–583, July 1983.

Medgyesi-Mitschang, L. N. and J. M. Putnam, "Scattering from Finite Bodies of Translation: Plates, Curved Surfaces, and Noncircular Cylinders," *IEEE Trans. Antennas Propagat.*, vol. 31, pp. 847–852, Nov. 1983.

Medgyesi-Mitschang, L. N. and J. M. Putnam, "Formulation for Wire Radiators on Bodies of Translation With and Without End Caps," *IEEE Trans. Antennas Propagat.*, vol. 31, pp. 853–62, Nov. 1983.

Medgyesi-Mitschang, L. N., "Electromagnetic Scattering from Extended Wires and Two- and Three-Dimensional Surfaces," *IEEE Trans. Antennas Propagat.*, vol. 33, pp. 1090–1100, Oct. 1985.

Medgyesi-Mitschang, L. N. and J. M. Putnam, "Electromagnetic Scattering from Ducts with Irregular Edges: Part I - Circular Case," *IEEE Trans. Antennas Propagat.*, vol. 36, pp. 383–397, March 1988.

Mei, K. K. and J. G. Van Bladel, "Scattering by Perfectly Conducting Rectangular Cylinders," *IEEE Trans. Antennas Propagat.*, vol. 11, pp. 185–192, March 1963.

Michalski, K. A. and D. Zheng, "Electromagnetic Scattering and Radiation by Surfaces of Arbitrary Shape in Layered Media, Part I: Theory," *IEEE Trans. Antennas Propagat.*, vol. 38, pp. 335–344, March 1990.

Michalski, K. A. and D. Zheng, "Electromagnetic Scattering and Radiation by Surfaces of Arbitrary Shape in Layered Media, Part II: Implementation and Results for Contiguous Half-Spaces," *IEEE Trans. Antennas Propagat.*, vol. 38, pp. 345–352, March 1990.

Miron, D. B., "The Singular Integral Problem in Surfaces," *IEEE Trans. Antennas Propagat.*, vol. 31, pp. 507–509, May 1983.

Mittra, R. and D. R. Wilton, "A Numerical Approach to the Determination of Electromagnetic Scattering Characteristics of Perfect Conductors," *Proc. IEEE*, vol. 57, pp. 2064–2065, Nov. 1969.

Mittra, R., Y. Rahmat-Samii, D. V. Jamnejad, and W. A. Davis, "A New Look at the Thin-Plate Scattering Problem," *Radio Science*, vol. 8, pp. 869–875, Oct. 1973.

Mullin, C. R., R. Sandburg, and C. O. Velline, "A Numerical Technique for the Determination of Scattering Cross Sections of Infinite Cylinders of Arbitrary Geometrical Cross Section," *IEEE Trans. Antennas Propagat.*, vol. 13, pp. 141–148, Jan. 1965.

Naishadham, R. and L. W. Pearson, "Numerical Evaluation of Complex Resonances of an Elliptic Cylinder," *IEEE Trans. Antennas Propagat.*, vol. 33, pp. 674–676, June 1985.

Newman, E. H. and D. M. Pozar, "Electromagnetic Modeling of Composite Wire and Surface Geometries," *IEEE Trans. Antennas Propagat.*, vol. 26, pp. 784–788, Nov. 1978.

Newman, E. H. and D. M. Pozar, "Correction to "Electromagnetic Modeling of Composite Wire and Surface Geometries"," *IEEE Trans. Antennas Propagat.*, vol. 27, p. 570, July 1979.

Newman, E. H. and D. M. Pozar, "Considerations for Efficient Wire/Surface Modeling," *IEEE Trans. Antennas Propagat.*, vol. 28, pp. 121–125, Jan. 1980.

Newman, E. H. and P. Tulyathan, "A Surface Patch Model for Polygonal Plates," *IEEE Trans. Antennas Propagat.*, vol. 30, pp. 588–593, July 1982.

Newman, E. H. and M. R. Schrote, "On the Current Distribution for Open Surfaces," *IEEE Trans. Antennas Propagat.*, vol. 31, pp. 515–519, May 1983.

Newman, E. H., P. Alexandropoulos, and E. K. Walton, "Polygonal Plate Modeling of Realistic Structures," *IEEE Trans. Antennas Propagat.*, vol. 32, pp. 742–747, July 1984.

Newman, E. H., "Polygonal Plate Modeling," *Electromagnetics*, vol. 10, pp. 65–83, Jan.-June 1990.

Perala, R. A., T. Rudolph, and F. Eriksen, "Electromagnetic Interaction of Lighting with Aircraft," *IEEE Trans. Electromagnetic Compatibility*, vol. 24, pp. 173–203, May 1982.

Perez-Leal, R. and M. F. Catedra, "Input Impedance of Wire Antennas Attached On Axis to Conducting Bodies of Revolution," *IEEE Trans. Antennas Propagat.*, vol. 36, pp. 1236–1244, Sep. 1988.

Popovic, B. D. and B. M. Kolundzija, "Analysis of a Class of Symmetrical Thin-Plate Triangular Antennas," *IEE Proc.*, vol. 134 Pt. H, pp. 205–210, Apr. 1987.

Pozar, D. M. and E. H. Newman, "Analysis of a Monopole Mounted Near or at the Edge of a Half-Plane," *IEEE Trans. Antennas Propagat.*, vol. 29, pp. 488–495, May 1981.

Pozar, D. M. and E. H. Newman, "Analysis of a Monopole Mounted Near an Edge or a Vertex," *IEEE Trans. Antennas Propagat.*, vol. 30, pp. 401–408, May 1982.

Putnam, J. M. and L. N. Medgyesi-Mitschang, "Electromagnetic Scattering from Ducts with Irregular

Edges: Part II - Noncircular Case," *IEEE Trans. Antennas Propagat.*, vol. 36, pp. 398–404, March 1988.

Rahmat-Samii, Y. and R. Mittra, "Integral Equation Solution and RCS Computation of a Thin Rectangular Plate," *IEEE Trans. Antennas Propagat.*, vol. 22, pp. 608–611, July 1974.

Rahmat-Samii, Y. and R. Mittra, "Correction to "Integral Equation Solution and RCS Computation of a Thin Rectangular Plate," *IEEE Trans. Antennas Propagat.*, vol. 23, p. 302, March 1975.

Rao, S. M., A. W. Glisson, D. R. Wilton, and B. S. Vidula, "A Simple Numerical Solution Procedure for Statics Problems Involving Arbitrary-Shaped Surfaces," *IEEE Trans. Antennas Propagat.*, vol. 27, pp. 604–608, Sep. 1979.

Rao, S. M., D. R. Wilton, and A. W. Glisson, "Electromagnetic Scattering by Surfaces of Arbitrary Shape," *IEEE Trans. Antennas Propagat.*, vol. 30, pp. 409–418, May 1982.

Richmond, J. H., "Scattering by an Arbitrary Array of Parallel Wires," *IEEE Trans. Microwave Th. and Tech.*, vol. 13, pp. 408–412, July 1965.

Richmond, J. H., "A Wire-Grid Model for Scattering by Conducting Bodies," *IEEE Trans. Antennas Propagat.*, vol. 14, pp. 782–786, Nov. 1966.

Richmond, J. H., "TE Radiation and Scattering from Infinitely-Long Noncircular Conducting Cylinders," *IEEE Trans. Antennas Propagat.*, vol. 22, p. 365, Nov. 1974.

Richmond, J. H., D. M. Pozar, and E. H. Newman, "Rigorous Near-Zone Field Expressions for Rectangular Sinusoidal Surface Monopole," *IEEE Trans. Antennas Propagat.*, vol. 26, pp. 509–510, May 1978.

Richmond, J. H., "On the Edge Mode in the Theory of TM Scattering by a Strip or Strip Grating," *IEEE Trans. Antennas Propagat.*, vol. 28, pp. 883–887, Nov. 1980.

Richmond, J. H., "Monopole Antenna on a Circular Disk," *IEEE Trans. Antennas Propagat.*, vol. 32, pp. 1282–1287, Dec. 1984.

Sangster, A. J. and A. H. I. McCormick, "Moment Method Applied to Round-Ended Slots," *IEE Proc.*, vol. 134 Pt. H, pp. 310–314, June 1987.

Sarkar, T. K. and S. R. Rao, "A Simple Technique for Solving E-Field Integral Equations for Conducting Bodies at Internal Resonances," *IEEE Trans. Antennas Propagat.*, vol. 30, pp. 1250–1254, Nov. 1982.

Shaeffer, J. F. and L. N. Medgyesi-Mitschang, "Radiation from Wire Antennas Attached to Bodies of Revolution: The Junction Problem," *IEEE Trans. Antennas Propagat.*, vol. 29, pp. 479–487, May 1981.

Shaeffer, J. F., "EM Scattering from Bodies of Revolution with Attached Wires," *IEEE Trans. Antennas Propagat.*, vol. 30, pp. 426–431, May 1982.

Singh, J. and A. T. Adams, "A Nonrectangular Patch Model for Scattering from Surfaces," *IEEE Trans. Antennas Propagat.*, vol. 27, pp. 531–535, July 1979.

Sun, W., K. Chen, D. P. Nyquist, and E. J. Rothwell, "Determination of the Natural Modes for a Rectangular Plate," *IEEE Trans. Antennas Propagat.*, vol. 38, pp. 643–652, May 1990.

Taflove, A. and K. Umashankar, "Radar Cross Section of General Three-Dimensional Scatterers," *IEEE Trans. Electromagnetic Compatibility*, vol. 25, pp. 443–440, Nov. 1983.

Tanaka, K., "Scattering of Electromagnetic Waves by a Rotating Perfectly Conducting Cylinder with Arbitrary Cross Section: Point-Matching Method," *IEEE Trans. Antennas Propagat.*, vol. 28, pp. 796–803, Nov. 1980.

Tsai, L. L., D. R. Wilton, M. G. Harrison, and E. H. Wright, "A Comparison of Geometrical Theory of Diffraction and Integral Equation Formulation for Analysis of Reflector Antennas," *IEEE Trans. Antennas Propagat.*, vol. 20, pp. 705–711, Nov. 1972.

Tsai, L. L., D. G. Dudley, and D. R. Wilton, "Electromagnetic Scattering by a Three-Dimensional Conducting Rectangular Box," *J. Appl. Phys.*, vol. 45, pp. 4393–4400, Oct. 1974.

Tsai, L. L., "Radar Cross Section of a Simple Target: A Three Dimensional Conducting Rectangular Box," *IEEE Trans. Antennas Propagat.*, vol. 25, pp. 882–884, Nov. 1977.

Wang, J. J. H., "Numerical Analysis of Three-Dimensional Arbitrarily-Shaped Conducting Scatterers by Trilateral Surface Cell Modelling," *Radio Science*, vol. 13, pp. 947–952, Nov.-Dec. 1978.

Wang, J. J. H. and C. J. Drane, "Numerical Analysis of Arbitrarily Shaped Bodies Modeled by Surface Patches," *IEEE Trans. Microwave Th. and Tech.*, vol. 30, pp. 1167–1173, Aug. 1982.

Wang, N. N., J. H. Richmond, and M. C. Gilreath, "Sinusoidal Reaction Formulation for Radiation and Scattering from Conducting Surfaces," *IEEE Trans. Antennas Propagat.*, vol. 23, pp. 376–381, May 1975.

Weiner, M. M., "Monopole Element at the Center of a Circular Ground Plane Whose Radius is Small or Comparable to a Wavelength," *IEEE Trans. Antennas Propagat.*, vol. 35, pp. 488–495, May 1987.

Wilton, D. R. and S. Govind, "Incorporation of Edge Conditions in Moment Method Solutions," *IEEE Trans. Antennas Propagat.*, vol. 25, pp. 845–850, Nov. 1977.

Wu, T., "Fast Convergent Integral Equation Solution of Strip Gratings on Dielectric Substrate," *IEEE Trans. Antennas Propagat.*, vol. 35, pp. 205–207, Feb. 1987.

Yaghjian, A. D. and R. V. McGahan, "Broadside Radar Cross Section of the Perfectly Conducting Cube," *IEEE Trans. Antennas Propagat.*, vol. 33, pp. 321–329, March 1985.

Yee, H. Y., "Application of Point-Matching Method to Low-Frequency Scattering by Conducting Cylinders," *IEEE Trans. Antennas Propagat.*, vol. 13, pp. 818–819, Sep. 1965.

Material Bodies

Al-Badwaihy, K. A. and J. L. Yen, "Extened Boundary Condition Integral Equations for Perfectly Conducting and Dielectric Bodies: Formulation and Uniqueness," *IEEE Trans. Antennas Propagat.*, vol. 23, pp. 546–551, July 1975.

Andreasen, M. G., "Scattering from Cylinders with Arbitrary Surface Impedance," *Proc. IEEE*, vol. 53, pp. 812–817, Aug. 1965.

Antar, Y. M. M., A. A. Kishk, L. Shafai, and L. E. Allan, "Radar Backscattering from Partially Coated Targets with Axial Symmetry," *IEEE Trans. Antennas Propagat.*, vol. 37, pp. 564–575, May 1989.

Appel-Hansen, J. and V. V. Solodukhov, "Echo Width of Foam Supports Used in Scattering Measurements," *IEEE Trans. Antennas Propagat.*, vol. 27, pp. 191–193, March 1979.

Arvas, E., S. M. Rao, and T. K. Sarkar, "E-Field Solution of TM-Scattering from Multiple Perfectly Conducting and Lossy Dielectric Cylinders of Arbitrary Cross-Section," *Proc. IEEE*, vol. 133, pp. 115–121, Apr. 1986.

Arvas, E., M. Ross, and Y. Qian, "TM Scattering from a Conducting Cylinder of Arbitrary Cross-Section Covered by Multiple Layers of Lossy Dielectric," *IEE Proc.*, vol. 135 Pt. H, pp. 226–230, Aug. 1988.

Arvas, E., Y. Qian, T. K. Sarkar, and F. Aslan, "TE Scattering from a Conducting Cylinder of Arbitrary Cross-Section Covered by Multiple Layers of Lossy Dielectrics," *IEE Proc.*, vol. 136 Pt. H, pp. 425–430, Dec. 1989.

Arvas, E. and T. K. Sarkar, "RCS of Two-Dimensional Structures Consisting of both Dielectrics and Conductors of Arbitrary Cross Section," *IEEE Trans. Antennas Propagat.*, vol. 37, pp. 546–554, May 1989.

Arvas, E. and S. Ponnapalli, "Scattering Cross Section of a Small Radome of Arbitrary Shape," *IEEE Trans. Antennas Propagat.*, vol. 37, pp. 655–658, May 1989.

Beker, B. and K. Umashankar, "Analysis of Electromagnetic Scattering by Arbitrarily Shaped Two-Dimensional Anisotropic Objects: Combined Field Surface Integral Equation Formulation," *Electromagnetics*, vol. 9, pp. 215–229, 1989.

Beker, B., K. R. Umashankar, and A. Taflove, "Electromagnetic Scattering by Arbitrarily Shaped Two-Dimensional Perfectly Conducting Objects Coated with Homogeneous Anisotropic Materials," *Electromagnetics*, vol. 10, pp. 387–406, 1990.

Bozzetti, M., F. Corsi, and R. D. Leo, "Electromagnetic Coupling Between Apertures and Biological Structures," *Radio Science*, vol. 16, pp. 1217–1222, Nov.-Dec. 1981.

Beker, B., K. R. Umashankar, and A. Taflove, "Numerical Analysis and Validation of the Combined Field Surface Integral Equations for Electromagnetic Scattering by Arbitrary Shaped Two-Dimentional Anisotropic Objects," *IEEE Trans. Antennas Propagat.*, vol. 37, pp. 1573–1581, Dec. 1989.

Casey, J. P. and R. Bansal, "Square Helical Antenna with a Dielectric Core," *IEEE Trans. Electromagnetic Compatibility*, vol. 30, pp. 429–436, Nov. 1988.

Casey, J. P. and R. Bansal, "Dielectrically Loaded Wire Antennas," *IEE Proc.*, vol. 135 Pt. H, pp. 103–110, Apr. 1988.

Catedra, M. F., "Analysis of Bodies of Revolution Composed of Conductors and Dielectrics Using Only Electric Equivalent Currents: Application to Small Horns with Dielectric Core," *IEEE Trans. Antennas Propagat.*, vol. 36, pp. 1311–1313, Sep. 1988.

Chang, Y. and I. V. Ingvarsson, "A Moment Formulation for Static Field Problems of Dielectric Objects," *Proc. IEEE*, vol. 64, pp. 1732–1733, Dec. 1976.

Chang, Y. and R. F. Harrington, "A Surface Formulation for Characteristic Modes of Material Bodies," *IEEE Trans. Antennas Propagat.*, vol. 25, pp. 789–795, Nov. 1977.

Chatterjee, I., O. P. Gandhi, and M. J. Hagmann, "Numerical and Experimental Results for Near-Field Electromagnetic Absorption in Man," *IEEE Trans. Microwave Th. and Tech.*, vol. 30, pp. 2000–2005, Nov. 1982.

Chen, C. H., "An Integral Equation Formulation of the Direct Scattering Problem for an Inhomogeneous Slab," *IEEE Trans. Antennas Propagat.*, vol. 26, pp. 797–800, Nov. 1978.

Chen, K., D. E. Livesay, and B. S. Guru, "Induced Current in a Scattered Field from a Finite Cylinder with Arbitrary Conductivity and Permittivity,"

IEEE Trans. Antennas Propagat., vol. 24, pp. 330–336, May 1976.

Chen, K. and B. S. Guru, "Internal EM Field and Absorbed Power Density in Human Torsos Induced by 1-500MHz EM Waves," *IEEE Trans. Microwave Th. and Tech.*, vol. 25, pp. 746–756, Sep. 1977.

Cohoon, D. K., "Reduction of the Cost of Solving an Integral Equation Arising in Electromagnetic Scattering Through the Use of Group Theory," *IEEE Trans. Antennas Propagat.*, vol. 28, pp. 104–107, Jan. 1980.

Cooray, M. F. R. and I. R. Ciric, "Electromagnetic Wave Scattering by a System of Two Spheriods of Arbitrary Orientation," *IEEE Trans. Antennas Propagat.*, vol. 37, pp. 608–618, May 1989.

DeFord, J. F., O. P. Gandhi, and M. J. Hagmann, "Moment-Method Solutions and SAR Calculations for Inhomogeneous Models of Man with Large Number of Cells," *IEEE Trans. Microwave Th. and Tech.*, vol. 31, pp. 848–851, Oct. 1983.

Durney, C. H., "Electromagnetic Dosimetry for Models of Humans and Animals: A Review of Theoretical and Numerical Techniques," *Proc. IEEE*, vol. 68, pp. 33–40, Jan. 1980.

Eisler, S. and Y. Leviatan, "Analysis of Electromagnetic Scattering from Metallic and Penetrable Cylinders with Edges Using a Multifilament Current Model," *IEE Proc.*, vol. 136 Pt. H, pp. 431–438, Dec. 1989.

Elsherbeni, A. Z. and H. A. Auda, "Electromagnetic Diffraction by Two Perfectly Conducting Wedges with Dented Edges Loaded with a Dielectric Cylinder," *IEE Proc.*, vol. 136 Pt. H, pp. 225–234, June 1989.

Elsherbeni, A. Z. and M. Hamid, "Scattering by a Perfectly Conducting Strip Loaded with a Dielectric Cylinder (TM Case)," *IEE Proc.*, vol. 136 Pt. H, pp. 185–190, June 1989.

Ghodgaonkar, D. K., O. P. Gandhi, and M. J. Hagmann, "Estimation of Complex Permittivities of Three-Dimensional Inhomogeneous Biological Bodies," *IEEE Trans. Microwave Th. and Tech.*, vol. 31, pp. 442–446, June 1983.

Glisson, A. W., "An Integral Equation for Electromagnetic Scattering from Homogeneous Dielectric Bodies," *IEEE Trans. Antennas Propagat.*, vol. 32, pp. 173–175, Feb. 1984.

Gong, Z. and A. W. Glisson, "A Hybrid Equation Approach for the Solution of Electromagnetic Scattering Problems Involving Two-Dimentional Inhomogeneous Dielectric Cylinders," *IEEE Trans. Antennas Propagat.*, vol. 38, pp. 60–68, Jan. 1990.

Govind, S., D. R. Wilton, and A. W. Glisson, "Scattering from Inhomogeneous Penetrable Bodies of Revolution," *IEEE Trans. Antennas Propagat.*, vol. 32, pp. 1163–1173, Nov. 1984.

Graglia, R. D. and P. L. E. Uslenghi, "Electromagnetic Scattering from Anisotropic Materials, Part II: Computer Code and Numerical Results in Two Dimensions," *IEEE Trans. Antennas Propagat.*, vol. 35, pp. 225–232, Feb. 1987.

Graglia, R. D., P. L. E. Uslenghi, and R. S. Zich, "Moment Method with Isoparametric Elements for Three-Dimentional Anisotropic Scatterers," *Proc. IEEE*, vol. 77, pp. 750–760, May 1989.

Graglia, R. D. and P. L. E. Uslenghi, "Surface Currents on Impedance Bodies of Revolution," *IEEE Trans. Antennas Propagat.*, vol. 36, pp. 1313–1317, Sep. 1988.

Graglia, R. D., "The Use of Parametric Elements in the Moment Method Solution of Static and Dynamic Volume Integral Equations," *IEEE Trans. Antennas Propagat.*, vol. 36, pp. 636–646, May 1988.

Grzesik, J., "Note on Homogeneous and Inhomogeneous Integral Equations in the Theory of Electromagnetic Scattering by Dielectric Obstacles," *Proc. IEEE*, vol. 54, pp. 2028–2029, Dec. 1966.

Guru, B. S. and K. Chen, "Experimental and Theoretical Studies on Electromagnetic Fields Induced Inside Finite Biological Bodies," *IEEE Trans. Microwave Th. and Tech.*, vol. 24, pp. 433–440, July 1976.

Hagmann, M. J. and R. L. Levin, "Accuracy of Block Models for Evaluation of the Deposition of Energy by Electromagnetic Fields," *IEEE Trans. Microwave Th. and Tech.*, vol. 34, pp. 653–659, June 1986.

Hagmann, M. J., "Comments on "A Procedure for Calculating Fields Inside Arbitrarily Shaped, Inhomogeneous Dielectric Bodies Using Linear Basis Functions with the Moment Method"," *IEEE Trans. Microwave Th. and Tech.*, vol. 35, pp. 785–786, Aug. 1987.

Hagmann, M. J. and R. L. Levin, "Convergence of Local and Average Values in Three-Dimensional Moment-Method Solutions," *IEEE Trans. Microwave Th. and Tech.*, vol. 33, pp. 649–654, July 1985.

Hagmann, M. J., "Convergence Tests of Several Moment-Method Solutions," *IEEE Trans. Antennas Propagat.*, vol. 29, pp. 547–550, May 1981.

Hagmann, M. J., O. P. Gandhi, J. A. D'Andrea, and I. Chatterjee, "Head Resonance: Numerical Solutions and Experimental Results," *IEEE Trans. Microwave Th. and Tech.*, vol. 27, pp. 809–813, Sep. 1979.

Hagmann, M. J., O. P. Gandhi, and C. H. Durney, "Improvement of Convergence in Moment-Method Solutions by the Use of Interpolants," *IEEE Trans. Microwave Th. and Tech.*, vol. 26, pp. 904–908, Nov. 1978.

Hagmann, M. J., O. P. Gandhi, and C. H. Durney, "Numerical Calculation of Electromagnetic Energy Deposition for a Realistic Model of Man," *IEEE Trans. Microwave Th. and Tech.*, vol. 27, pp. 804–809, Sep. 1979.

Hagmann, M. J. and O. P. Gandhi, "Numerical Calculation of Electromagnetic Energy Deposition in Models of Man with Grounding and Reflector Effects," *Radio Science*, vol. 14, pp. 23–29, Nov.-Dec. 1979.

Hagmann, M. J. and R. L. Levin, "Procedures for Noninvasive Electromagnetic Property and Dosimetry Measurements," *IEEE Trans. Antennas Propagat.*, vol. 38, pp. 99–106, Jan. 1990.

Hagmann, M. J., O. P. Gandhi, and C. H. Durney, "Upper Bound on Cell Size for Moment-Method Solutions," *IEEE Trans. Microwave Th. and Tech.*, vol. 25, pp. 831–832, Oct. 1977.

Hall, R. C., R. Mittra, and J. R. Mosig, "Analysis of a Parallel Resistive Plate Medium," *IEEE Trans. Antennas Propagat.*, vol. 38, pp. 299–304, March 1990.

Harrington, R. F., J. R. Mautz, and Y. Chang, "Characteristic Modes for Dielectric and Magnetic Bodies," *IEEE Trans. Antennas Propagat.*, vol. 20, pp. 194–201, March 1972.

Harrington, R. F. and J. R. Mautz, "Green's Functions for Surfaces of Revolution," *Radio Science*, vol. 7, pp. 603–611, May 1972.

Harrington, R. F. and J. R. Mautz, "An Impedance Sheet Approximation for Thin Dielectric Shells," *IEEE Trans. Antennas Propagat.*, vol. 23, pp. 531–534, July 1975.

Hill, S. C., C. H. Durney, and D. A. Christensen, "Numerical Calculations of Low-Frequency TE Fields in Arbitrarily Shaped Inhomogeneous Lossy Dielectric Cylinders," *Radio Science*, vol. 18, pp. 328–336, May-June 1983.

Holt, A. R., N. K. Uzunoglu, and B. G. Evans, "An Integral Equation Solution to the Scattering of Electromagnetic Radiation by Dielectric Spheriods and Ellipsoids," *IEEE Trans. Antennas Propagat.*, vol. 26, pp. 706–711, Sep. 1978.

Huddleston, P. L., L. N. Medgyesi-Mitschang, and J. M. Putnam, "Combined Field Integral Equation Formulation for Scattering by Dielectrically Coated Conducting Bodies," *IEEE Trans. Antennas Propagat.*, vol. 34, pp. 510–520, Apr. 1986.

Huddleston, P. L., "Scattering from Conducting Finite Cylinders with Thin Coatings," *IEEE Trans. Antennas Propagat.*, vol. 35, pp. 1128–1136, Oct. 1987.

Hurst, M. P. and L. N. Medgyesi-Mitschang, "Scattering from Partial Bodies of Revolution," *IEEE Trans. Antennas Propagat.*, vol. 38, pp. 69–75, Jan. 1990.

Jin, J. M. and V. V. Liepa, "Simple Moment Method Program for Computing Scattering from Complex Cylindrical Obstacles," *IEE Proc.*, vol. 136 Pt. H, pp. 321–329, Aug. 1989.

Jin, J., J. L. Volakis, and V. V. Liepa, "A Moment Method Solution of a Volume-Surface Integral Equation Using Isoparametric Elements and Point Matching," *IEEE Trans. Microwave Th. and Tech.*, vol. 37, pp. 1641–1645, Oct. 1989.

Jin, J. M. and V. V. Liepa, "A Numerical Technique for Computating TM Scattering by Coated Wedges and Half-Planes," *Electromagnetics*, vol. 9, pp. 201–213, 1989.

Joachimowicz, N. and C. Pichot, "Comparison of Three Integral Formulations for the 2-D TE Scattering Problem," *IEEE Trans. Microwave Th. and Tech.*, vol. 38, pp. 178–185, Feb. 1990.

Joo, K. and M. F. Iskander, "A New Procedure of Point-Matching Method for Calculating the Absorption and Scattering of Lossy Dielectric Objects," *IEEE Trans. Antennas Propagat.*, vol. 38, pp. 1483–1489, Sep. 1990.

Kagami, S. and I. Fukai, "Application of Boundary-Element Method to Electromagnetic Field Problems," *IEEE Trans. Microwave Th. and Tech.*, vol. 32, pp. 455–461, Apr. 1984.

Kajfez, D., A. W. Glisson, and J. James, "Computed Modal Field Distributions for Isolated Dielectric Resonators," *IEEE Trans. Microwave Th. and Tech.*, vol. 32, pp. 1609–1616, Dec. 1984.

Kanda, M., "The Effects of Resistive Loading of "TEM" Horns," *IEEE Trans. Electromagnetic Compatibility*, vol. 24, pp. 245–255, May 1982.

Karimullah, K., K. Chen, and D. P. Nyquist, "Electromagnetic Coupling Between a Thin-Wire Antenna and a Neighboring Biological Body: Theory and Experiment," *IEEE Trans. Microwave Th. and Tech.*, vol. 28, pp. 1218–1225, Nov. 1980.

Kishk, A. A. and L. Shafai, "Different Formulations for Numerical Solution of Single or Multibodies of Revolution with Mixed Boundary Conditions," *IEEE Trans. Antennas Propagat.*, vol. 34, pp. 666–673, May 1986.

Kishk, A. A., Y. M. M. Antar, L. Shafai, and L. E. Allan, "Electromagnetic Scattering from Dielectric

Bodies of Revolution: Theoretical and Experimental Results," *Electromagnetics*, vol. 7, pp. 51–60, 1987.

Kishk, A. A. and L. Shafai, "Improvement of the Numerical Solution of Dielectric Bodies with High Permittivity," *IEEE Trans. Antennas Propagat.*, vol. 37, pp. 1486–1490, Nov. 1989.

Kishk, A. A. and L. Shafai, "Numerical Solution of Scattering from Coated Bodies of Revolution Using Different Integral Equation Formulations," *IEE Proc.*, vol. 133 Pt. H, pp. 227–232, June 1986.

Kishk, A. A. and L. Shafai, "Radiation Characteristics of the Short Dielectric Rod Antenna: A Numerical Solution," *IEEE Trans. Antennas Propagat.*, vol. 35, pp. 139–146, Feb. 1987.

Kleinman, R. E. and G. F. Roach, "New Integral Equations for Scattering by Penetrable Objects, II," *Radio Science*, vol. 19, pp. 1185–1193, Sep.-Oct. 1984.

Kluskens, M. S. and E. H. Newman, "Scattering by a Chiral Cylinder of Arbitrary Cross Section," *IEEE Trans. Antennas Propagat.*, vol. 38, pp. 1448–1455, Sep. 1990.

Knockaert, L. F. and D. D. Zutter, "Integral Equation for the Fields Inside a Dielectric Cylinder Immersed in an Incident E-Wave," *IEEE Trans. Antennas Propagat.*, vol. 34, pp. 1065–1067, Aug. 1986.

Krohn, T. L. and L. N. Medgyesi-Mitschang, "Scattering from Composite Materials: A First-Order Model," *IEEE Trans. Antennas Propagat.*, vol. 37, pp. 219–228, Feb. 1989.

Lakhtakia, A. and M. F. Iskander, "Scattering and Absorption Characteristics of Lossy Dielectric Objects Exposed to the Near Fields of Aperture Sources," *IEEE Trans. Antennas Propagat.*, vol. 31, pp. 111–120, Jan. 1983.

Leviatan, Y., "Analysis of Electromagnetic Scattering from Dielectric Cylinders Using a Multifilament Current Model," *IEEE Trans. Antennas Propagat.*, vol. 35, pp. 1119–1127, Oct. 1987.

Leviatan, Y., A. Boag, and A. Boag, "Analysis of Electromagnetic Scattering from Dielectrically Coated Conducting Cylinders Using a Multifilament Current Model," *IEEE Trans. Antennas Propagat.*, vol. 36, pp. 1602–1607, Nov. 1988.

Leviatan, Y., A. Boag, and A. Boag, "Analysis of TE Scattering from Dielectric Cylinders Using a Multifilament Magnetic Current Model," *IEEE Trans. Antennas Propagat.*, vol. 36, pp. 1026–1031, July 1988.

Leviatan, Y., A. Boag, and A. Boag, "Generalized Formulations for Electromagnetic Scattering form Perfectly Conducting and Homogeneous Materials Bodies–Theory and Numerical Solution," *IEEE Trans. Antennas Propagat.*, vol. 36, pp. 1722–1723, Dec. 1988.

Lin, C., H. Chaung, and K. Chen, "Steady-State and Shock Currents Induced by ELF Electric Fields in a Human Body and a Nearby Vehicle," *IEEE Trans. Electromagnetic Compatibility*, vol. 32, pp. 59–65, Feb. 1990.

Livesay, D. E. and K. Chen, "Electromagnetic Fields Induced Inside Arbitrarily Shaped Biological Bodies," *IEEE Trans. Microwave Th. and Tech.*, vol. 22, pp. 1273–1280, Dec. 1974.

Marx, E., "Integral Equation for Scattering by a Dielectric," *IEEE Trans. Antennas Propagat.*, vol. 32, pp. 166–172, Feb. 1984.

Massoudi, H., C. H. Durney, and M. F. Iskander, "Limitations of the Cubical Block Model of Man in Calculating SAR Distributions," *IEEE Trans. Microwave Th. and Tech.*, vol. 32, pp. 746–752, Aug. 1984.

Mautz, J. R. and R. F. Harrington, "Electromagnetic Coupling to a Conducting Body of Revolution with a Homogeneous Material Region," *Electromagnetics*, vol. 2, pp. 257–308, 1982.

Mautz, J. R., "A Stable Integral Equation for Electromagnetic Scattering from Homogeneous Dielectric Bodies," *IEEE Trans. Antennas Propagat.*, vol. 37, pp. 1070–1071, Aug. 1989.

Medgyesi-Mitschang, L. N. and J. M. Putnam, "Electromagnetic Scattering form Axially Inhomogeneous Bodies of Revolution," *IEEE Trans. Antennas Propagat.*, vol. 32, pp. 797–806, Aug. 1984.

Medgyesi-Mitschang, L. N. and J. M. Putnam, "Electromagnetic Scattering from Electrically Large Coated Flat and Curved Strips: Entire Domain Galerkin Formulation," *IEEE Trans. Antennas Propagat.*, vol. 35, pp. 790–801, July 1987.

Medgyesi-Mitschang, L. N. and J. M. Putnam, "Integral Equation Formulation for Imperfectly Conducting Scatterers," *IEEE Trans. Antennas Propagat.*, vol. 33, pp. 206–214, Feb. 1985.

Medgyesi-Mitschang, L. N. and J. M. Putnam, "Scattering from Composite Laminate Strips," *IEEE Trans. Antennas Propagat.*, vol. 37, pp. 1427–1436, Nov. 1989.

Mei, K. and J. Van Bladel, "Low-Frequency Scattering by Rectangular Cylinders," *IEEE Trans. Antennas Propagat.*, vol. 11, pp. 52–56, Jan. 1963.

Morita, N., "Surface Integral Representations for Electromagnetic Scattering from Dielectric Cylinders," *IEEE Trans. Antennas Propagat.*, vol. 26, pp. 261–266, March 1978.

Morita, N., "Resonant Solutions Involved in the Integral Equation Approach to Scattering from Conducting and Dielectric Cylinders," *IEEE Trans. Antennas Propagat.*, vol. 27, pp. 869–871, Nov. 1979.

Nevels, R. D., C. M. Butler, and W. Yablon, "The Annular Slot Antenna in a Lossy Biological Medium," *IEEE Trans. Microwave Th. and Tech.*, vol. 33, pp. 314–319, Apr. 1985.

Newman, E. H. and M. R. Schrote, "An Open Surface Integral Formulation for Electromagnetic Scattering by Material Plates," *IEEE Trans. Antennas Propagat.*, vol. 32, pp. 672–678, July 1984.

Newman, E. H., "Plane Wave Scattering by a Material Coated Parabolic Cylinder," *IEEE Trans. Antennas Propagat.*, vol. 38, pp. 541–550, Apr. 1990.

Newman, E. H. and P. Tulyathan, "Wire Antennas in the Presence of a Dielectric/Ferrite Inhomogeneity," *IEEE Trans. Antennas Propagat.*, vol. 26, pp. 587–592, July 1978.

Nyquist, D. P., K. Chen, and B. S. Guru, "Coupling Between Small Thin-Wire Antennas and a Biological Body," *IEEE Trans. Antennas Propagat.*, vol. 25, pp. 863–865, Nov. 1977.

Okamoto, N., "Matrix Formulation of Scattering by a Homogeneous Gyrotropic Cylinder," *IEEE Trans. Antennas Propagat.*, vol. 18, pp. 642–649, Sep. 1970.

Papayiannakis, A. G. and E. E. Kriezis, "Scattering from a Dielectric Cylinder of Finite Length," *IEEE Trans. Antennas Propagat.*, vol. 31, pp. 725–731, Sep. 1983.

Peterson, A. F. and P. W. Klock, "An Improved MFIE Formulation for TE-Wave Scattering from Lossy Inhomogeneous Dielectric Cylinders," *IEEE Trans. Antennas Propagat.*, vol. 36, pp. 45–49, Jan. 1988.

Pogorzelski, R. J., "On the Numerical Computation of Scattering from Inhomogeneous Penetrable Objects," *IEEE Trans. Antennas Propagat.*, vol. 26, pp. 616–618, July 1978.

Rao, S. M., E. Arvas, and T. K. Sarkar, "Combined Field Solution for TM Scattering from Multiple Conducting and Dielectric Cylinders of Arbitrary Cross Section," *IEEE Trans. Antennas Propagat.*, vol. 35, pp. 447–451, Apr. 1987.

Richmond, J. H., "Scattering by a Dielectric Cylinder of Arbitrary Cross Section Shape," *IEEE Trans. Antennas Propagat.*, vol. 13, pp. 338–341, May 1965.

Richmond, J. H., "Scattering by Thin Dielectric Strips," *IEEE Trans. Antennas Propagat.*, vol. 33, pp. 64–68, Jan. 1985.

Richmond, J. H., "TE-Wave Scattering by a Dielectric Cylinder of Arbitrary Cross Section Shape," *IEEE Trans. Antennas Propagat.*, vol. 14, pp. 460–464, July 1966.

Richmond, J. H., "Scattering by a Conducting Elliptic Cylinder with Dielectric Coating," *Radio Science*, vol. 23, pp. 1061–1066, Nov.-Dec. 1988.

Ricoy, M. A., S. M. Kilberg, and J. L. Volakis, "Simple Integral Equations for Two-Dimensional Scattering with Further Reductions in Unknowns," *IEE Proc.*, vol. 136 Pt. H, pp. 298–304, Aug. 1989.

Ricoy, M. A. and J. L. Volakis, "Integral Equations with Reduced Unknowns for the Simulation of Two-Dimentional Composite Structures," *IEEE Trans. Antennas Propagat.*, vol. 37, pp. 362–373, March 1989.

Rogers, J. R., "Comments on "Integral Equation Formulations for Imperfectly Conducting Scatterers"," *IEEE Trans. Antennas Propagat.*, vol. 33, pp. 1283–1284, Nov. 1985.

Rojas, R. G., "Scattering by an Inhomogeneous Dielectric/Ferrite Cylinder of Arbitrary Cross-Section Shape – Oblique Incidence Case," *IEEE Trans. Antennas Propagat.*, vol. 36, pp. 238–246, Feb. 1988.

Ruiz, J., M. J. Nunez, A. Navarro, and E. Martin, "Integral Numerical Technique for the Study of Axially Symmetric Resonant Devices," *IEEE Trans. Microwave Th. and Tech.*, vol. 37, pp. 1814–1816, Nov. 1989.

Ruppin, R., "Calculation of Electromagnetic Energy Absorption in Prolate Spheroids by the Point Matching Method," *IEEE Trans. Microwave Th. and Tech.*, vol. 26, pp. 87–90, Feb. 1978.

Ruppin, R., "Electromagnetic Power Deposition in a Dielectric Cylinder in the Presence of a Reflecting Surface," *IEEE Trans. Microwave Th. and Tech.*, vol. 27, pp. 910–914, Nov. 1979.

Rusch, W. V. T. and R. J. Pogorzelski, "A Mixed-Field Solution for Scattering from Composite Bodies," *IEEE Trans. Antennas Propagat.*, vol. 34, pp. 955–958, July 1986.

Sarkar, T. K. and E. Arvas, "Scattering Cross Secion of Composite Conducting and Lossy Dielectric Bodies," *Proc. IEEE*, vol. 77, pp. 788–795, May 1989.

Schaubert, D. H. and P. M. Meaney, "Efficient Computation of Scattering by Inhomogeneous Dielectric Bodies," *IEEE Trans. Antennas Propagat.*, vol. 34, pp. 587–592, Apr. 1986.

Schaubert, D. H., D. R. Wilton, and A. W. Glisson, "A Tetrahedral Modeling Method for Electromagnetic Scattering by Arbitrarily Shaped Inhomogeneous Dielectric Bodies," *IEEE Trans. Antennas Propagat.*, vol. 32, pp. 77–85, Jan. 1984.

Sebak, A. A. and L. Shafai, "Scattering by Imperfectly Conducting and Impedance Spheroids: A Numerical Approach," *Radio Science*, vol. 19, pp. 258–256, Jan.-Feb. 1984.

Sebak, A. and L. Shafai, "Electromagnetic Scattering by Spheroidal Objects with Impedance Boundary Conditions at Axial Incidence," *Radio Science*, vol. 23, pp. 1048–1060, Nov.-Dec. 1988.

Sebak, A. and L. Shafai, "Scattering from Arbitrarily-Shaped Objects with Impedance Boundary Conditions," *IEE Proc.*, vol. 136 Pt. H, pp. 371–376, Oct. 1989.

Senior, T. B. A., "Backscattering from Resistive Strips," *IEEE Trans. Antennas Propagat.*, vol. 27, pp. 808–813, Nov. 1979.

Senior, T. B. A. and J. L. Volakis, "Sheet Simulation of a Thin Dielectric Layer," *Radio Science*, vol. 22, pp. 1261–1272, Dec. 1987.

Smith, G. S., "The Electric-Field Probe Near a Material Interface with Application to the Probing of Fields in Biological Bodies," *IEEE Trans. Microwave Th. and Tech.*, vol. 27, pp. 270–278, March 1979.

Spiegel, R. J., "Numerical Determination of Induced Currents in Humans and Baboons Exposed to 60-Hz Electric Fields," *IEEE Trans. Electromagnetic Compatibility*, vol. 23, pp. 382–390, Nov. 1981.

Spiegel, R. J., "The Thermal Response of a Human in the Near-Zone of a Resonant Thin-Wire Antenna," *IEEE Trans. Microwave Th. and Tech.*, vol. 30, pp. 177–185, Feb. 1982.

Stuchly, M. A., R. J. Spiegel, S. S. Stuchly, and A. Kraszewski, "Exposure of Man in the Near-Field of a Resonant Dipole: Comparison Between Theory and Measurements," *IEEE Trans. Microwave Th. and Tech.*, vol. 34, pp. 26–31, Jan. 1986.

Sultan, M. F. and R. Mittra, "An Iterative Moment Method for Analyzing the Electromagnetic Field Distribution Inside Inhomogeneous Lossy Dielectric Objects," *IEEE Trans. Microwave Th. and Tech.*, vol. 33, pp. 163–168, Feb. 1985.

Tsai, C., H. Massoudi, C. H. Durney, and M. F. Iskander, "A Procedure for Calculating Fields Inside Arbitrarily Shaped Inhomogeneous Dielectic Bodies Using Linear Basis Functions with the Moment Method," *IEEE Trans. Microwave Th. and Tech.*, vol. 34, pp. 1131–1139, Nov. 1986.

Umashankar, K., A. Taflove, and S. M. Rao, "Electromagnetic Scattering by Arbitrary Shaped Three-Dimensional Homogeeous Lossy Dielectric Objects," *IEEE Trans. Antennas Propagat.*, vol. 34, pp. 758–766, June 1986.

Van Doeren, R. E., "An Integral Equation Approach to Scattering by Dielectric Rings," *IEEE Trans. Antennas Propagat.*, vol. 17, pp. 373–374, May 1969.

Wang, J. J. H. and J. R. Dubberley, "Computation of Electromagnetic Fields in Large Biological Bodies by an Iterative Moment Method with a Restart Technique," *IEEE Trans. Microwave Th. and Tech.*, vol. 37, pp. 1918–1923, Dec. 1989.

Wu, T. and L. L. Tsai, "Electromagnetic Fields Induced Inside Arbirary Cylinders of Biological Tissue," *IEEE Trans. Microwave Th. and Tech.*, vol. 25, pp. 61–65, Jan. 1977.

Wu, T. and L. L. Tsai, "Correction to "Electromagnetic Fields Induced Inside Arbitrary Cylinders of Biological Tissue"," *IEEE Trans. Microwave Th. and Tech.*, vol. 25, p. 712, Aug. 1977.

Wu, T., "Electromagnetic Fields and Power Deposition in Body-of-Revolution Models of Man," *IEEE Trans. Microwave Th. and Tech.*, vol. 27, pp. 279–283, March 1979.

Wu, T., "Radar Cross Section of Arbitrarily Shaped Bodies of Revolution," *Proc. IEEE*, vol. 77, pp. 735–740, May 1989.

Wu, T. and L. L. Tsai, "Scattering by Arbitraily Cross-Sectioned Layered Lossy Dielectric Cylinders," *IEEE Trans. Antennas Propagat.*, vol. 25, pp. 518–524, July 1977.

Wu, T. and L. L. Tsai, "Scattering by a Dielectric Wedge: A Numerical Solution," *IEEE Trans. Antennas Propagat.*, vol. 25, pp. 570–571, July 1977.

Yee, H. Y., "Scattering of Electromagnetic Waves by Circular Dielectric-Coated Conducting Cylinders with Arbitrary Cross Sections," *IEEE Trans. Antennas Propagat.*, vol. 13, pp. 822–823, Sep. 1965.

Yuan, X., D. R. Lynch, and J. W. Strohbehn, "Coupling of Finite Element and Moment Methods for Electromagnetic Scattering from Inhomogeneous Objects," *IEEE Trans. Antennas Propagat.*, vol. 38, pp. 386–394, March 1990.

Yuan, X., "Three-Dimensional Electromagnetic Scattering from Inhomogeneous Objects by the Hybrid Moment and Finite Element Method," *IEEE Trans. Microwave Th. and Tech.*, vol. 38, pp. 1053–1058, Aug. 1990.

Zheng, W., "The Null Field Approach to Electromagnetic Scattering from Composite Objects: The Case with Three or More Constituents," *IEEE Trans. Antennas Propagat.*, vol. 36, pp. 1396–1400, Oct. 1988.

Zheng, W. and S. Strom, "The Null Field Approach to Electromagnetic Scattering from Composite Objects: The Case of Concavo-Convex Constituents,"

IEEE Trans. Antennas Propagat., vol. 37, pp. 373–383, March 1989.

Apertures

Arndt, F., K. Wolff, L. Brunjes, R. Heyen, F. Siefken-Herrlich, W. Bothmer, and E. Forgber, "Generalized Moment Method Analysis of Planar Reactively Loaded Rectangular Waveguide Arrays," *IEEE Trans. Antennas Propagat.*, vol. 37, pp. 329–338, March 1989.

Arvas, E. and T. K. Sarkar, "TM Transmission Through Dielectric-Filled Slots in a Conducting Cylindrical Shell of Arbitrary Cross Section," *IEEE Trans. Electromagnetic Compatibility*, vol. 29, pp. 150–156, May 1987.

Arvas, E., "Electromagnetic Diffraction from a Dielectric-Filled Slit-Cylinder Enclosing a Cylinder of Arbitrary Cross Section: TM Case," *IEEE Trans. Electromagnetic Compatibility*, vol. 31, pp. 91–102, Feb. 1989.

Auckland, D. T. and R. F. Harrington, "Electromagnetic Transmission Through a Filled Slit in a Conducting Plane of Finite Thickness, TE Case," *IEEE Trans. Microwave Th. and Tech.*, vol. MTT-26, pp. 499–505, July 1978.

Auckland, D. T. and R. F. Harrington, "A Nonmodal Formulation for Electromagnetic Transmission Through a Filled Slot of Arbitrary Cross Section in a Thick Conducting Screen," *IEEE Trans. Microwave Th. and Tech.*, vol. 28, pp. 548–555, June 1980.

Audone, B. and M. Balma, "Shielding Effectiveness of Apertures in Rectangular Cavities," *IEEE Trans. Electromagnetic Compatibility*, vol. 31, pp. 102–106, Feb. 1989.

Barkeshli, K. and J. L. Volakis, "TE Scattering by a Two-Dimensional Groove in a Ground Plane Using Higher Order Boundary Conditions," *IEEE Trans. Antennas Propagat.*, vol. 38, pp. 1421–1428, Sep. 1990.

Bird, T. S., "Analysis of Mutual Coupling in Finite Arrays of Different-Sized Rectangular Waveguides," *IEEE Trans. Antennas Propagat.*, vol. 38, pp. 166–172, Feb. 1990.

Butler, C. M., "Investigation of a Scatterer Coupled to an Aperture in a Conducting Screen," *IEE Proc.*, vol. 127, Pt. H, pp. 166–169, June 1980.

Butler, C. M., "A Formulation of the Finite-Length Narrow Slot or Strip Equation," *IEEE Trans. Antennas Propagat.*, vol. 30, pp. 1254–1257, Nov. 1982.

Butler, C. M., "General Analysis of a Narrow Slot in a Conducting Screen Between Half-Spaces of Different Electromagnetic Properties," *Radio Science*, vol. 22, pp. 1149–1154, Dec. 1987.

Butler, C. M. and T. L. Keshavamurthy, "Investigation of a Radial, Parallel-Plate Waveguide with an Annular Slot," *Radio Science*, vol. 16, pp. 159–168, March-April 1981.

Butler, C. M. and K. R. Umashankar, "Electromagnetic Penetration Through an Aperture in an Infinite, Planar Screen Separating Two Half Spaces of Different Electromagnetic Properties," *Radio Science*, vol. 11, pp. 611–619, July 1976.

Butler, C. M., Y. Rahmat-Samii, and R. Mittra, "Electromagnetic Penetration Through Apertures in Conducting Surfaces," *IEEE Trans. Antennas Propagat.*, vol. AP-26, pp. 82–93, Jan. 1978.

Butler, C. M. and K. R. Umashankar, "Electromagnetic Excitation of a Wire Through an Aperture-Perforated Conducting Screen," *IEEE Trans. Antennas Propagat.*, vol. AP-24, pp. 456–462, July 1976.

Butler, C. M. and D. R. Wilton, "General Analysis of Narrow Strips and Slots," *IEEE Trans. Antennas Propagat.*, vol. 28, pp. 42–48, Jan. 1980.

Casey, K. F., "Low-Frequency Electromagnetic Penetration of Loaded Apertures," *IEEE Trans. Electromagnetic Compatibility*, vol. 23, pp. 367–377, Nov. 1981.

Catedra, M. F., "A Comparison Between Two Kinds of Equivalent Currents to Analyze Conducting Bodies with Apertures Using Moment Methods: Application to Horns with Symmetry of Revolution," *IEEE Trans. Antennas Propagat.*, vol. 35, pp. 782–789, July 1987.

Chuang, C. W., "Generalized Admittance Matrix for a Slotted Parallel-Plate Waveguide," *IEEE Trans. Antennas Propagat.*, vol. 36, pp. 1227–1231, Sep. 1988.

Chou, T. Y. and A. T. Adams, "The Coupling of Electromagnetic Waves Through Long Slots," *IEEE Trans. Electromagnetic Compatibility*, vol. 19, pp. 65–73, May 1977.

Feng, D. and D. K. Cheng, "Electromagnetic Coupling Characteristics of Two Half Spaces Through a Waveguide with Small Apertures," *Radio Science*, vol. 18, pp. 1243–1254, Nov.-Dec. 1983.

Gajda, G. B. and S. S. Stuchly, "Numerical Analysis of Open-Ended Coaxial Lines," *IEEE Trans. Microwave Th. and Tech.*, vol. 31, pp. 380–384, May 1983.

Goggans, P. M. and T. H. Shumpert, "CFIE MM Solution for TE and TM Incidence on a 2-D Conducting

Body with Dielectric Filled Cavity," *IEEE Trans. Antennas Propagat.*, vol. 38, pp. 1645–1649, Oct. 1990.

Graves, B. D., T. T. Crow, and C. D. Taylor, "On the Electromagnetic Field Penetration Through Apertures," *IEEE Trans. Electromagnetic Compatibility*, vol. 18, pp. 154–162, Nov. 1976.

Hadidi, A. and M. Hamid, "Aperture Field and Circuit Parameters of Cavity-Backed Slot Radiator," *IEE Proc.*, vol. 136 Pt. H, pp. 139–146, Apr. 1989.

Hanyang, W. and W. Wei, "Moment Method Analysis of a Feeding System in a Slotted-Waveguide Antenna," *IEE Proc.*, vol. 135 Pt. H, pp. 313–318, Oct. 1988.

Harrington, R. F., "Resonant Behavior of a Small Aperture Backed by a Conducting Body," *IEEE Trans. Antennas Propagat.*, vol. 30, pp. 205–212, March 1982.

Harrington, R. F. and D. T. Auckland, "Electromagnetic Transmission Through Narrow Slots in Thick Conducting Screens," *IEEE Trans. Antennas Propagat.*, vol. 28, pp. 616–622, Sep. 1980.

Harrington, R. F. and J. R. Mautz, "A Generalized Network Formulation for Aperture Problems," *IEEE Trans. Antennas Propagat.*, vol. 24, pp. 870–873, Nov. 1976.

Harrington, R. F. and J. R. Mautz, "Characteristic Modes for Aperture Problems," *IEEE Trans. Microwave Th. and Tech.*, vol. 33, pp. 500–505, June 1985.

Hejase, H. A. N., A. T. Adams, R. F. Harrington, and T. K. Sarkar, "Shielding Effectivness of "Pigtail" Connections ," *IEEE Trans. Electromagnetic Compatibility*, vol. 31, pp. 63–68, Feb. 1989.

Hongo, K. and G. Ishii, "Diffraction of an Electromagnetic Plane Wave by a Thick Slit," *IEEE Trans. Antennas Propagat.*, vol. 26, pp. 494–499, May 1978.

Hsi, S. W., R. F. Harrington, and J. R. Mautz, "Electromagnetic Coupling to a Conducting Wire Behind an Aperture of Arbitrary Size and Shape," *IEEE Trans. Antennas Propagat.*, vol. 33, pp. 581–587, June 1985.

Jarem, J. M., "The Input Impedance and Antenna Characteristics of a Cavity-Backed Plasma Covered Ground Plane Antenna," *IEEE Trans. Antennas Propagat.*, vol. 34, pp. 262–267, Feb. 1986.

Jarem, J. M. and C. Ma, "The Input Impedance of a Probe-Fed Rectangular Cavity which Transmits Through a Plasma-Covered Cylindrical Body," *IEEE Trans. Antennas Propagat.*, vol. 36, pp. 1157–1161, Aug. 1988.

Jarem, J. M. and F. To, "A K-Space Method of Moments Solution for the Aperture Electromagnetic Fields of a Circular Cylindrical Waveguide Radiating into an Anisotropic Dielectric Half-Space," *IEEE Trans. Antennas Propagat.*, vol. 37, pp. 187–193, Feb. 1989.

Jeng, S., "Scattering from a Cavity-Backed Slit in a Ground Plane - TE Case," *IEEE Trans. Antennas Propagat.*, vol. 38, pp. 1523–1529, Oct. 1990.

Jin, J. and J. L. Volakis, "TE Scattering by an Inhomogeneously Filled Aperture in a Thick Conducting Plane," *IEEE Trans. Antennas Propagat.*, vol. 38, pp. 1280–1286, Aug. 1990.

Jones, J. E. and J. H. Richmond, "Application of an Integral Equation Formulation to the Prediction of Space Shuttle Annular Slot Antenna Radiation Patterns," *IEEE Trans. Antennas Propagat.*, vol. 22, pp. 109–112, Jan. 1974.

Josefsson, L. G., "Analysis of Longitudinal Slots in Rectangular Waveguides," *IEEE Trans. Antennas Propagat.*, vol. 35, pp. 1351–1357, Dec. 1987.

Kabalan, K. Y., R. F. Harrington, H. A. Auda, and J. R. Mautz, "Characteristic Modes for Slots in a Conducting Plane, TE Case," *IEEE Trans. Antennas Propagat.*, vol. 35, pp. 162–168, Feb. 1987.

Kabalan, K. Y., R. F. Harrington, H. A. Auda, and J. R. Mautz, "Characteristic Modes for Slots in a Conducting Plane, TM Case," *IEEE Trans. Antennas Propagat.*, vol. 35, pp. 331–335, March 1987.

Kabalan, K. Y., A. El-Hajj, and R. F. Harrington, "Characteristic Mode Analysis of a Slot in a Conducting Plane Separating Different Media," *IEEE Trans. Antennas Propagat.*, vol. 38, pp. 476–481, Apr. 1990.

Katehi, P. B., "A Space Domain Integral Equation Approach in the Analysis of Dielectric-Covered Slots," *Radio Science*, vol. 24, pp. 253–260, Mar.-Apr. 1989.

Katehi, P. B., "Dielectric-Covered Waveguide Longitudinal Slots with Finite Wall Thickness," *IEEE Trans. Antennas Propagat.*, vol. 38, pp. 1039–1045, July 1990.

Kominami, M. and K. Rokushima, "Analysis of an Antenna Composed of Arbitrarily Located Slots and Wires," *IEEE Trans. Antennas Propagat.*, vol. 32, pp. 154–158, Feb. 1984.

Leong, M. S., P. S. Kool, and Chandra, "Radiation from a Flanged Parallel-Plate Waveguide: Solution by Moment Method with Inclusion of Edge Condition," *IEE Proc.*, vol. 135 Pt. H, pp. 249–256, Aug. 1988.

Leviatan, Y., "Electromagnetic Coupling Between Two Half-Space Regions Separated by Two Slot-Perforated Parallel Conducting Screens," *IEEE Trans. Microwave Th. and Tech.*, vol. 36, pp. 44–46, Jan. 1988.

Leviatan, Y., R. F. Harrington, and J. R. Mautz, "Electromagnetic Transmission Through Apertures in a Cavity in a Thick Conductor," *IEEE Trans. Antennas Propagat.*, vol. 30, pp. 1153–1164, Nov. 1982.

Leong, M., P. S. Kooi, and Chandra, "A New Class of Basis Functions for the Solution of the E-Plane Waveguide Discontinuity Problem," *IEEE Trans. Microwave Th. and Tech.*, vol. 35, pp. 705–709, Aug. 1987.

Leviatan, Y., E. Hudis, and P. D. Einziger, "A Method of Moments Analysis of Electromagnetic Coupling Through Slots Using Gaussian Beam Expansion," *IEEE Trans. Antennas Propagat.*, vol. 37, pp. 1537–1544, Dec. 1989.

Lin, J., W. L. Curtis, and M. C. Vincent, "On the Field Distribution of an Aperture," *IEEE Trans. Antennas Propagat.*, vol. 22, pp. 467–471, May 1974.

Lyon, R. W. and A. J. Sangster, "Efficient Moment Method Analysis of Radiating Slots in a Thick-Walled Rectangular Waveguide," *IEE Proc.*, vol. 128, Pt. H, pp. 197–205, Aug. 1981.

Mahon, J. P., "An Improvement to Stevenson's Derivation of Integral Equations for Waveguide Slot Analysis," *Electromagnetics*, vol. 10, pp. 377–386, 1990.

Mautz, J. R. and R. F. Harrington, "Transmission from a Rectangular Waveguide into Half Space through a Rectangular Aperture," *IEEE Trans. Microwave Th. and Tech.*, vol. 26, pp. 44–45, Jan. 1978.

Naiheng, Y. and R. F. Harrington, "Electromagnetic Coupling to an Infinite Wire Through a Slot in a Conducting Plane," *IEEE Trans. Antennas Propagat.*, vol. 31, pp. 310–316, March 1983.

Nevels, R. D. and C. M. Butler, "Electromagnetic Penetration Through a Slot in a Screen Covered by a Dielectric Slab," *Electromagnetics*, vol. 2, pp. 147–159, 1982.

Nevels, R. D. and C. M. Butler, "Electromagnetic Diffraction by a Slot in a Ground Screen Covered by a Dielectric Slab," *IEEE Trans. Antennas Propagat.*, vol. 30, pp. 390–395, May 1982.

Nevels, R. D., C. M. Butler, and W. Yablon, "The Annular Slot Antenna in a Lossy Biological Medium," *IEEE Trans. Microwave Th. and Tech.*, vol. 33, pp. 314–319, Apr. 1985.

Newman, E. H., J. R. Birchmeier, and K. A. Shubert, "Remote Detection of a Thin Slit in a Thick Ground Plane," *IEEE Trans. Antennas Propagat.*, vol. 35, pp. 116–120, Jan. 1987.

Okon, E. E. and R. F. Harrington, "The Polarizabilities of Electrically Small Apertures of Arbitrary Shape," *IEEE Trans. Electromagnetic Compatibility*, vol. 23, pp. 359–366, Nov. 1981.

Parkinson, J. R. and M. J. Mehler, "Uniqueness of Numerical Solutions to Electric-, Magnetic- and Combined-Field Integral Equations for Open-Ended Circular Waveguides," *IEE Proc.*, vol. 136 Pt. H, pp. 269–275, June 1989.

Peterson, A. F. and R. Mittra, "Mutual Admittance Between Slots in Cylinders of Arbitrary Shape," *IEEE Trans. Antennas Propagat.*, vol. 37, pp. 858–864, July 1989.

Pozar, D. M., "Analysis of an Infinite Phased Array of Aperture Coupled Microstrip Patches," *IEEE Trans. Antennas Propagat.*, vol. 37, pp. 418–425, Apr. 1989.

Rahmat-Samii, Y., "Electromagnetic Pulse Coupling Through an Aperture into a Two-Parallel-Plate Region," *IEEE Trans. Electromagnetic Compatibility*, vol. 20, pp. 436–442, Aug. 1978.

Rengarajan, S. R., "Compound Radiating Slots in a Broad Wall of a Waveguide," *IEEE Trans. Antennas Propagat.*, vol. 37, pp. 1116–1123, Sep. 1989.

Richmond, J. H. and M. C. Gilreath, "Flush-Mounted Dielectric-Loaded Axial Slot on Circular Cylinder," *IEEE Trans. Antennas Propagat.*, vol. 23, pp. 348–351, May 1975.

Richmond, J. H., "Axial Slot Antenna on a Dielectric-Coated Elliptic Cylinder," *IEEE Trans. Antennas Propagat.*, vol. 37, pp. 1235–1241, Oct. 1989.

Sadigh, A. and E. Arvas, "Transmission Through Dielectric-Filled Slots in a Conducting Cylindrical Shell of Arbitrary Cross Section: TE Case," *IEEE Trans. Electromagnetic Compatibility*, vol. 32, pp. 240–245, Aug. 1990.

Safavi-Naini, S., S. Lee, and R. Mittra, "Transmission of an EM Wave Through the Aperture of a Cylindrical Cavity," *IEEE Trans. Electromagnetic Compatibility*, vol. 19, pp. 74–81, May 1977.

Scharstein, R. W. and A. T. Adams, "Galerkin Solution for the Thin Circular Iris in a TE11-Mode Circular Waveguide," *IEEE Trans. Microwave Th. and Tech.*, vol. 36, pp. 106–113, Jan. 1988.

Scharstein, R. W., "Two Numerical Solutions for the Parallel Plate-Fed Slot Antenna," *IEEE Trans. Antennas Propagat.*, vol. 37, pp. 1415–1426, Nov. 1989.

Scharstein, R. W., "Mutual Coupling in a Slotted Phased Array, Infinite in E-Plane and Finite in H-Plane," *IEEE Trans. Antennas Propagat.*, vol. 38, pp. 1186–1192, Aug. 1990.

Schuman, H. K. and D. E. Warren, "Aperture Coupling in Bodies of Revolution," *IEEE Trans. Antennas Propagat.*, vol. 26, pp. 778–783, Nov. 1978.

Seidel, D. B., "Aperture Excitation of a Wire in a Rectangular Cavity," *IEEE Trans. Microwave Th. and Tech.*, vol. 26, pp. 908–914, Nov. 1978.

Senior, T. B. A., "Electromagnetic Field Penetration into a Cylindrical Cavity," *IEEE Trans. Electromagnetic Compatibility*, vol. 18, pp. 71–73, May 1976.

Senior, T. B. A., K. Sarabandi, and J. R. Natzke, "Scattering by a Narrow Gap," *IEEE Trans. Antennas Propagat.*, vol. 38, pp. 1102–1110, July 1990.

Silvestro, J. W., "Mutual Coupling in a Finite Array with Interelement Holes Present," *IEEE Trans. Antennas Propagat.*, vol. 37, pp. 791–794, June 1989.

Sinha, S. N., "A Generalised Network Formulation for a Class of Waveguide Coupling Problems," *IEE Proc.*, vol. 134, Pt. H, pp. 502–508, Dec. 1987.

Sinha, S. N., D. K. Mehra, and R. P. Agarwal, "Radiation from a Waveguide-Backed Aperture in an Infinite Ground Plane in the Presence of a Thin Conducting Plate," *IEEE Trans. Antennas Propagat.*, vol. 34, pp. 539–545, Apr. 1986.

Taflove, A. and K. Umashankar, "A Hybrid Moment Method/Finite-Difference Time-Domain Approach to Electromagnetic Coupling and Aperture Penetration into Complex Geometries," *IEEE Trans. Antennas Propagat.*, vol. 30, pp. 617–627, July 1982.

Umashankar, K. R. and J. R. Wait, "Electromagnetic Coupling to an Infinite Cable Placed Behind a Slot-Perforated Screen," *IEEE Trans. Electromagnetic Compatibility*, vol. 20, pp. 406–411, Aug. 1978.

Van Bladel, J., "The Matrix Formulation of Scattering Problems," *IEEE Trans. Microwave Th. and Tech.*, vol. 14, pp. 130–135, March 1966.

Wallenberg, R. F. and R. F. Harrington, "Radiation from Apertures in Conducting Cylinders of Arbitrary Cross Section," *IEEE Trans. Antennas Propagat.*, vol. 17, pp. 56–62, Jan. 1969.

Wang, T., R. F. Harrington, and J. R. Mautz, "Quasi-Static Analysis of a Microstrip Via Through a Hole in a Ground Plane," *IEEE Trans. Microwave Th. and Tech.*, vol. 36, pp. 1008–1013, June 1988.

Wang, T., A. Cuevas, and H. Ling, "RCS of a Partially Open Box in the Resonant Region," *IEEE Trans. Antennas Propagat.*, vol. 38, pp. 1498–1504, Sep. 1990.

Warne, L. K. and K. C. Chen, "Equivalent Antenna Radius for Narrow Slot Apertures Having Depth," *IEEE Trans. Antennas Propagat.*, vol. 37, pp. 824–834, July 1989.

Wheeler, J. E. and R. D. Nevels, "Reflection and Transmission Properties of Annular Slot Covered by Dielectric Hemisphere," *IEE Proc.*, vol. 136 Pt. H, pp. 59–63, Feb. 1989.

Wright, D. B., R. Lee, and D. D. Dudley, "Transient Current on a Wire Penetrating a Cavity-Backed Circular Aperture in an Infinite Screen," *IEEE Trans. Electromagnetic Compatibility*, vol. 32, pp. 197–204, Aug. 1990.

Wu, C. P., "Integral Equation Solutions for the Radiation from a Waveguide Through a Dielectric Slab," *IEEE Trans. Antennas Propagat.*, vol. 17, pp. 733–739, Nov. 1969.

Wu, T. and L. L. Tsai, "Low-Frequency Shielding Properties of Conducting Cylindrical Shells of Arbitrary Cross Section," *IEEE Trans. Electromagnetic Compatibility*, vol. 20, pp. 349–351, May 1978.

Ziokowski, R. W., D. P. Marsland, and L. F. Libelo, "Scattering from an Open Spherical Shell Having a Circular Aperture and Enclosing a Concentric Dielectric Sphere," *IEEE Trans. Antennas Propagat.*, vol. 36, pp. 985–999, July 1988.

Thin Wires

Abul-Kassem, A. S. and D. C. Chang, "On Two Parallel Loop Antennas," *IEEE Trans. Antennas Propagat.*, vol. 28, pp. 491–496, July 1980.

Adams, A. T., T. E. Baldwin, and D. E. Warren, "Near Fields of Thin-Wire Antennas–Computation and Experiment," *IEEE Trans. Electromagnetic Compatibility*, vol. 20, pp. 259–266, Feb. 1978.

Adams, A. T., J. Perini, M. Miyabayashi, D. H. Shau, and K. Heidary, "Electromagnetic Field-to-Wire Coupling in the SHF Frequency Range and Beyond," *IEEE Trans. Electromagnetic Compatibility*, vol. 29, pp. 126–131, May 1987.

Adams, A. T., B. J. Strait, D. E. Warren, D. Kuo, and T. E. Baldwin, "Near Fields of Wire Antennas by Matrix Methods," *IEEE Trans. Antennas Propagat.*, vol. 21, pp. 602–609, Sep. 1973.

Agrawal, P. K., "A Hybrid Technique for Wire Antennas in a Cavity," *IEEE Trans. Antennas Propagat.*, vol. 26, pp. 434–438, May 1978.

Aronson, E. A., C. D. Taylor, and R. F. Harrington, "Matrix Methods for Solving Antenna Problems," *IEEE Trans. Antennas Propagat.*, vol. 15, pp. 696–697, Sep. 1967.

Atia, A. E. and K. K. Mei, "Analysis of Multiple-Arm Conical Long-Spiral Antennas," *IEEE Trans. Antennas Propagat.*, vol. 19, pp. 320–331, May 1971.

Baghdasarian, A. and D. J. Angelakos, "Scattering from Conducting Loops and Solution of Circular Loop Antennas by Numerical Methods," *Proc. IEEE*, vol. 53, pp. 818–822, Aug. 1965.

Baker, H. C. and A. H. LaGrone, "Digital Computation of the Mutual Impedance Between Thin Dipoles," *IRE Trans. Antennas Propagat.*, vol. 10, pp. 172–178, March 1962.

Bhattacharyya, T. and O. P. Rustogi, "Theoretical Formulation for Current Distribution on Crossed Coplanar Skew Dipoles Illuminated by Plane Wave," *IEEE Trans. Antennas Propagat.*, vol. 20, pp. 94–96, Jan. 1972.

Bhojwani, H. R. and L. W. Zelby, "Spiral Top-Loaded Antenna: Characteristics and Design," *IEEE Trans. Antennas Propagat.*, vol. 21, pp. 293–297, May 1973.

Butler, C. M., "Currents Induced on a Pair of Skew Crossed Wires," *IEEE Trans. Antennas Propagat.*, vol. 20, pp. 731–735, Nov. 1972.

Butler, C. M., "Evaluation of Potential Integral at Singularity of Exact Kernel in Thin-Wire Calculations," *IEEE Trans. Antennas Propagat.*, vol. 23, pp. 293–295, March 1975.

Butler, C. M. and L. L. Tsai, "An Alternate Frill Field Formulation," *IEEE Trans. Antennas Propagat.*, vol. 21, pp. 115–116, Jan. 1973.

Butler, C. M. and D. R. Wilton, "Analysis of Various Numerical Techniques Applied to Thin-Wire Scatterers," *IEEE Trans. Antennas Propagat.*, vol. 23, pp. 534–540, July 1975.

Butler, C. M. and D. R. Wilton, "A Useful Transformation of the Free-Space Green's Function with Applications to Analyses of Wire Structures," *IEEE Trans. Antennas Propagat.*, vol. 28, pp. 108–111, Jan. 1980.

Cambrell, G. K. and C. T. Carson, "On Mei's Integral Equation of Thin Wire Antennas," *IEEE Trans. Antennas Propagat.*, vol. 19, pp. 781–782, Nov. 1971.

Casey, J. P. and R. Bansal, "Finite Length Helical Sheath Antenna in a General Homogeneous Medium," *Radio Science*, vol. 23, pp. 1141–1151, Nov.-Dec. 1988.

Cassedy, E. S. and J. Fainberg, "Back Scattering Cross Sections of Cylindrical Wires of Finite Conductivity," *IRE Trans. Antennas Propagat.*, vol. 8, pp. 1–6, Jan. 1960.

Chang, D. C., "On the Electrically Thick Cylindrical Antenna," *Radio Science*, vol. 2, pp. 1043–1060, Sep. 1967.

Chang, D. C., "On the Electrically Thick Monopole Part I-Theoretical Solution," *IEEE Trans. Antennas Propagat.*, vol. 16, pp. 58–63, Jan. 1968.

Chang, D. C., C. W. Harrison, and E. A. Aronson, "Tubular Monopole of Arbitrary Dimensions: The Radiation Field," *IEEE Trans. Antennas Propagat.*, vol. 17, pp. 534–540, Sep. 1969.

Chao, H. H. and B. J. Strait, "Radiation and Scattering by Configurations of Bent Wires with Junctions," *IEEE Trans. Antennas Propagat.*, vol. 19, p. 701, Sep. 1971.

Choe, W. and J. K. Lee, "Analysis of Higher Order Regular Polygonal Loop Antennas," *IEEE Trans. Antennas Propagat.*, vol. 38, pp. 1114–1117, July 1990.

Collin, R. E., "Equivalent Line Current for Cylindrical Dipole Antennas and its Asymptotic Behavior," *IEEE Trans. Antennas Propagat.*, vol. 32, pp. 200–204, Feb. 1984.

Demarest, K. R. and R. J. Garbacz, "Anomalous Behavior of Near Fields Calculated by the Method of Moments," *IEEE Trans. Antennas Propagat.*, vol. 27, pp. 609–614, Sep. 1979.

Demarest, K. R. and J. H. Richmond, "The Analysis of the RF Response of a Solid Wire Excited by a Focused Laser Beam," *IEEE Trans. Antennas Propagat.*, vol. 30, pp. 177–182, March 1982.

Fante, R. L., K. K. Hazard, and J. Dolan, "RCS of Bent Wires," *IEEE Trans. Antennas Propagat.*, vol. 16, pp. 130–132, Jan. 1968.

Fukuzawa, K., M. Tada, T. Yoshikawa, K. Ouchi, and R. Sato, "A New Method of Calculating 3-Meter Site Attenuation," *IEEE Trans. Electromagnetic Compatibility*, vol. 24, pp. 389–397, Nov. 82.

Garbacz, R. J. and E. H. Newman, "Characteristic Modes of a Symmetric Wire Cross," *IEEE Trans. Antennas Propagat.*, vol. 28, pp. 712–715, Sep. 1980.

Hansen, R. C., "Formulation of Echelon Dipole Mutual Impedance for Computer," *IEEE Trans. Antennas Propagat.*, vol. 20, pp. 780–781, Nov. 1972.

Harrington, R. F. and J. R. Mautz, "Straight Wires with Arbitrary Excitation and Loading," *IEEE Trans. Antennas Propagat.*, vol. 15, pp. 502–515, July 1967.

Harrison, C. W. and E. A. Aronson, "On the Evaluation of Potential Integrals Occuring in Antenna Theory Using Digital Computers," *IEEE Trans. Antennas Propagat.*, vol. 15, p. 576, July 1967.

Hilbert, M., M. A. Tilston, and K. G. Balmain, "Resonance Phenomena of Log-Periodic Antennas: Characteristic-Mode Analysis," *IEEE Trans. Antennas Propagat.*, vol. 37, pp. 1224–1235, Oct. 1989.

Hirasawa, K., "Bounds of Uncertain Interference Between Closely Located Antennas," *IEEE Trans. Electromagnetic Compatibility*, vol. 26, pp. 129–133, Aug. 84.

Imbriale, W. A. and P. G. Ingerson, "On Numerical Convergence of Moment Solutions of Moderately Thick Wire Antennas Using Sinusoidal Basis Functions," *IEEE Trans. Antennas Propagat.*, vol. 21, pp. 363–366, May 1973.

Inagaki, N. and T. Sekiguchi, "A Note on the Antenna Integral Equation," *IEEE Trans. Antennas Propagat.*, vol. 17, pp. 223–224, March 1969.

Jaggard, D. L., "On Bounding the Equivalent Radius," *IEEE Trans. Antennas Propagat.*, vol. 28, pp. 384–388, May 1980.

Janaswamy, R. and S. Lee, "Scattering from Dipoles Loaded with Diodes," *IEEE Trans. Antennas Propagat.*, vol. 26, pp. 1649–1651, Nov. 1988.

Judasz, T. J., W. L. Ecklund, and B. B. Balsley, "The Coaxial Collinear Antenna: Current Distribution from the Cylindrical Antenna Equation," *IEEE Trans. Antennas Propagat.*, vol. 35, pp. 327–331, March 1987.

Kalafus, R. M., "Broad-Band Dipole Design Using the Method of Moments," *IEEE Trans. Antennas Propagat.*, vol. 19, pp. 771–773, Nov. 1971.

Kawana, T., S. Horiguchi, and Y. Yamanaka, "Evaluation of 3-m Site Attenuation by the Moment Method," *IEEE Trans. Electromagnetic Compatibility*, vol. 28, pp. 117–124, Aug. 1986.

Klein, C. A. and R. Mittra, "The Effect of Different Testing Functions in the Moment Method Solution of Thin-Wire Antenna Problems," *IEEE Trans. Antennas Propagat.*, vol. 23, pp. 258–261, March 1975.

Kominami, M. and K. Rokushima, "On the Integral Equation of Piecewise Linear Antennas," *IEEE Trans. Antennas Propagat.*, vol. 29, pp. 787–791, Sep. 1981.

Kraft, U. R. and G. Monich, "Main-Beam Polarization Properties of Modified Helical Antennas," *IEEE Trans. Antennas Propagat.*, vol. 38, pp. 589–597, May 1990.

Kuo, D. C., H. H. Chao, J. R. Mautz, B. J. Strait, and R. F. Harrington, "Analysis of Radiation and Scattering by Arbitrary Configurations of Thin Wires," *IEEE Trans. Antennas Propagat.*, vol. 20, pp. 814–815, Nov. 1972.

Lee, S. H. and K. K. Mei, "Analysis of Zigzag Antennas," *IEEE Trans. Antennas Propagat.*, vol. 18, pp. 760–764, Nov. 1970.

Lin, J., "The Imperfectly Conducting Circular-Loop Antenna," *Radio Science*, vol. 8, pp. 251–257, March 1973.

Lin, J. and D. J. Matson, "On the Backscattering from Two Arbitrarily Located Identical Parallel Wires," *IEEE Trans. Antennas Propagat.*, vol. 19, pp. 697–700, Sep. 1971.

Mao, Z. and D. K. Cheng, "Scattering and Pattern Perturbation by a Conducting Tubular Cylinder of Finite Length," *IEEE Trans. Electromagnetic Compatibility*, vol. 25, pp. 441–447, Nov. 1983.

Medgyesi-Mitschang, L. N. and C. Eftimiu, "Scattering from Wires and Open Circular Cylinders of Finite Length Using Entire Domain Galerkin Expansions," *IEEE Trans. Antennas Propagat.*, vol. 30, pp. 628–636, July 1982.

Mei, K. K., "On the Integral Equations of Thin Wire Antennas," *IEEE Trans. Antennas Propagat.*, vol. 13, pp. 374–378, May 1965.

Merewether, D. E., "The Arbitrarily Driven Long Cylindrical Antenna," *IEEE Trans. Antennas Propagat.*, vol. 16, pp. 769–771, Nov. 1968.

Miller, E. K. and G. J. Burke, "Numerical Integration Methods," *IEEE Trans. Antennas Propagat.*, vol. 17, pp. 669–672, Sep. 1969.

Miller, E. K., G. J. Burke, B. J. Maxum, G. M. Pjerrou, and A. R. Neureuther, "Radar Cross Section of a Long Wire," *IEEE Trans. Antennas Propagat.*, vol. 17, pp. 381–384, May 1969.

Miller, E. K., G. J. Burke, and E. S. Selden, "Accuracy-Modeling Guideline for Integral-Equation Evaluation of Thin-Wire Scattering Structures," *IEEE Trans. Antennas Propagat.*, vol. 19, pp. 534–536, July 1971.

Miller, E. K., F. J. Deadrick, and W. O. Henry, "Computer Evaluation of Large Low-Frequency Antennas," *IEEE Trans. Antennas Propagat.*, vol. 21, pp. 386–389, May 1973.

Miller, E. K. and J. B. Morton, "The RCS of a Metal Plate with a Resonant Slot," *IEEE Trans. Antennas Propagat.*, vol. 18, pp. 290–292, March 1970.

Morgan, M. A., R. C. Hurley, and F. K. Schwering, "Computation of Monopole Antenna Currents Using Cylindrical Harmonics," *IEEE Trans. Antennas Propagat.*, vol. 38, pp. 1130–1133, July 1990.

Nauwelaers, B. K. J. C. and A. R. VanDeCapelle, "Integrals for the Mutual Coupling Between Dipoles or Between Slots: With or Without Complex Conjugate?," *IEEE Trans. Antennas Propagat.*, vol. 36, pp. 1375–1381, Oct. 1988.

Newman, E. H., "Small Antenna Location Synthesis Using Characteristic Modes," *IEEE Trans. Antennas Propagat.*, vol. 27, pp. 530–531, July 1979.

Newman, E. H., "Simple Examples of the Method of Moments in Electromagnetics," *IEEE Transactions on Education*, vol. 31, pp. 193–200, Aug. 1988.

Newman, E. H., "The Equivalent Separation(s) for the Self-Impedance of Thin Strips," *IEEE Trans. Antennas Propagat.*, vol. 35, pp. 110–113, Jan. 1987.

Newman, E. H., M. R. Schrote, A. R. Djordjevic, B. D. Popovic, and M. B. Dragovic, "Some Effects of the Circumferential Polarization of Current on Thin-Wire Antennas," *IEEE Trans. Antennas Propagat.*, vol. 29, pp. 815–817, Sep. 1981.

Olsen, R. G. and P. D. Mannikko, "Validation of the Hybrid Quasi-Static/Full-Wave Method for Capacitively Loaded Thin-Wire Antennas," *IEEE Trans. Antennas Propagat.*, vol. 38, pp. 516–522, Apr. 1990.

Otto, D. V. and J. H. Richmond, "Rigorous Field Expression for Piecewise-Sinusoidal Line Sources," *IEEE Trans. Antennas Propagat.*, vol. 17, p. 98, Jan. 1969.

Pearson, L. W., "A Separation of the Logarithmic Singularity in the Exact Kernel of the Cylindrical Antenna Integral Equation," *IEEE Trans. Antennas Propagat.*, vol. 23, pp. 256–258, March 1975.

Pearson, L. W. and C. M. Butler, "Inadequacies of Collocation Solutions to Pocklington-Type Models of Thin-Wire Structures," *IEEE Trans. Antennas Propagat.*, vol. 23, pp. 295–298, March 1975.

Popovic, B. D. and A. Nesic, "Generalisation of the Concept of Equivalent Radius of Thin Cylindrical Antennas," *IEE Proc.*, vol. 131 Pt. H, pp. 153–158, June 1984.

Preis, D. H., "A Comparison of Methods to Evaluate Potential Integrals," *IEEE Trans. Antennas Propagat.*, vol. 24, pp. 223–229, March 1976.

Richmond, J. H., "Scattering by Imperfectly Conducting Wires," *IEEE Trans. Antennas Propagat.*, vol. 15, pp. 802–805, Nov. 1967.

Richmond, J. H., "Coupled Linear Antennas with Skew Orientation," *IEEE Trans. Antennas Propagat.*, vol. 18, pp. 694–696, Sep. 1970.

Richmond, J. H. and N. H. Geary, "Mutual Impedance Between Coplanar-Skew Dipoles," *IEEE Trans. Antennas Propagat.*, vol. 18, pp. 414–416, May 1970.

Richmond, J. H., L. M. Schwab, and R. G. Wickliff, "Tumble-Average Radar Backscatter of Some Thin-Wire Chaff Elements," *IEEE Trans. Antennas Propagat.*, vol. 24, pp. 124–127, Jan. 1974.

Richmond, J. H., "Radiation and Scattering by Thin-Wire Structures in a Homogeneous Conducting Medium," *IEEE Trans. Antennas Propagat.*, vol. 22, p. 365, Nov. 1974.

Richmond, J. H. and N. H. Geary, "Mutual Impedance of Nonplanar-Skew Dipoles," *IEEE Trans. Antennas Propagat.*, vol. 23, pp. 412–414, May 1975.

Richmond, J. H. and E. H. Newman, "Dielectric Coated Wire Antennas," *Radio Science*, vol. 11, pp. 13–20, Jan. 1976.

Richmond, J. H., "On the Edge Mode in the Theory of Thick Cylindrical Monopole Antennas," *IEEE Trans. Antennas Propagat.*, vol. 28, pp. 916–921, Nov. 1980.

Rogers, P. G. and M. W. Gunn, "An Entire-Domain Galerkin Analysis of the Moderately Thick Dipole," *IEEE Trans. Antennas Propagat.*, vol. 28, pp. 117–121, Jan. 1980.

Sarkar, T. K., "A Study of the Various Methods for Computing Electromagnetic Field Utilizing Thin Wire Integral Equations," *Radio Science*, vol. 18, pp. 29–38, Jan. - Feb. 1983.

Sato, G., H. Kawakami, and F. Funatsu, "Theoretical Analysis and Some Experiments on Vierergruppe Antenne (Two-Dipole Antennas)," *IEEE Trans. Antennas Propagat.*, vol. 31, pp. 413–418, May 1983.

Seshardi, S. R. and T. T. Wu, "An Integral Equation for the Current in an Asymmetrically Driven Cylindrical Antenna," *Proc. IEEE*, vol. 55, p. 1097, June 1967.

Shumpert, T. H. and D. J. Galloway, "Capacitance Bounds and Equivalent Radius," *IEEE Trans. Antennas Propagat.*, vol. 25, pp. 284–286, March 1977.

Siarkiewicz, K. R. and A. T. Adams, "Near- and Far-Field Thin-Wire Coupling–Theory and Experiement," *IEEE Trans. Electromagnetic Compatibility*, vol. 19, pp. 394–401, Nov. 1977.

Simpson, T. L., "The Theory of Top-Loaded Antennas: Integral Equations for the Currents," *IEEE Trans. Antennas Propagat.*, vol. 19, pp. 186–190, March 1971.

Strait, B. J. and A. T. Adams, "Analysis and Design of Wire Antennas with Applications to EMC," *IEEE Trans. Electromagnetic Compatibility*, vol. 12, pp. 45–54, May 1970.

Strait, B. J. and K. Hirasawa, "On Long Wire Antennas with Multiple Excitations and Loadings," *IEEE Trans. Antennas Propagat.*, vol. 18, pp. 699–700, Sep. 1970.

Taylor, C. D., "Electromagnetic Scattering From Arbitrary Configurations of Wires," *IEEE Trans. Antennas Propagat.*, vol. 17, pp. 662–663, Sep. 1969.

Taylor, C. D. and C. W. Harrison, "On Thin-Wire Multiturn Loop Antennas," *IEEE Trans. Antennas Propagat.*, vol. 22, pp. 407–413, May 1974.

Taylor, C. D., S. Lin, and H. V. McAdams, "Scattering from Crossed Wires," *IEEE Trans. Antennas Propagat.*, vol. 18, pp. 133–136, Jan. 1970.

Taylor, C. D. and D. R. Wilton, "The Extended Boundary Condition Solution of the Dipole Antenna of Revolution," *IEEE Trans. Antennas Propagat.*, vol. 22, pp. 772–776, Nov. 1972.

Tesche, F. M., "The Effect of the Thin-Wire Approximation and the Source Gap Model on the High-Frequency Integral Equation Solution of Radiating Antennas," *IEEE Trans. Antennas Propagat.*, vol. 20, pp. 210–211, March 1972.

Tesche, F. M., "Evaluation of the Surface Integral Occurring in the E-Field Integral Equations for Wire Antennas," *IEEE Trans. Electromagnetic Compatibility*, vol. 16, pp. 209–210, Nov. 1974.

Tew, M. D., "Correction to WIRES Program," *IEEE Trans. Antennas Propagat.*, vol. 23, pp. 450–451, May 1975.

Thiele, G. A., "Calculation of the Current Distribution on a Thin Linear Antenna," *IRE Trans. Antennas Propagat.*, vol. 14, pp. 648–650, Sep. 1966.

Thiele, G. A., "The Maximum Echo Area of Imperfectly Conducting Dipoles," *IEEE Trans. Antennas Propagat.*, vol. 17, pp. 379–381, May 1969.

Thiele, G. A., "Radar Cross Section of a Small Low-Pitch Angle Helix," *IEEE Trans. Antennas Propagat.*, vol. 18, pp. 707–709, Sep. 1970.

Tilston, M. A. and K. G. Balmain, "On the Suppression of Asymmetric Artifacts Arising in an Implementation of the Thin-Wire Method of Moments," *IEEE Trans. Antennas Propagat.*, vol. 38, pp. 281–286, Feb. 1990.

Tilston, M. A. and K. G. Balmain, "A Multiradius, Reciprocal Implementation of the Thin-Wire Moment Method," *IEEE Trans. Antennas Propagat.*, vol. 38, pp. 1636–1644, Oct. 1990.

Tulyathan, P. and E. H. Newman, "The Circumferential Variation of the Axial Component of Current in Closely Spaced Thin-Wire Antennas," *IEEE Trans. Antennas Propagat.*, vol. 27, pp. 46–50, Jan. 1979.

Turpin, R. H., "A Basis Transformation Technique with Application to Scattering by Wires," *IEEE Trans. Antennas Propagat.*, vol. 20, pp. 80–82, Jan. 1972.

Vollmer, E. and J. H. Hinken, "Synthesis Method for Broad-Band Tapered Wire Antennas and Its Experimental Verification," *IEEE Trans. Antennas Propagat.*, vol. 37, pp. 959–965, Aug. 1989.

Warren, D. E. and T. E. Baldwin, "Near Electric and Magnetic Fields of Wire Antennas," *IEEE Trans. Antennas Propagat.*, vol. 22, p. 364, March 1974.

Williams, O. C. and C. E. Hickman, "Computer Determination of Current Distribution on Arbitrarily Located Parallel Center-Fed Dipoles with Terminals in a Common Plane," *IEEE Trans. Antennas Propagat.*, vol. 20, pp. 540–541, July 1972.

Wilton, D. R. and C. M. Butler, "Efficient Numerical Techniques for Solving Pocklington's Equation and Their Relationships to Other Methods," *IEEE Trans. Antennas Propagat.*, vol. 24, pp. 83–86, Jan. 1976.

Yeh, Y. S. and K. K. Mei, "Theory of Conical Equiangular-Spiral Antennas Part I–Numerical Technique," *IEEE Trans. Antennas Propagat.*, vol. 15, pp. 634–638, Sep. 1967.

Printed Circuit Antennas

Aberle, J. T. and D. M. Pozar, "Analysis of Infinite Arrays of Probe-Fed Rectangular Microstrip Patches Using a Rigorous Feed Model," *IEE Proc.*, vol. 136 Pt. H, pp. 110–119, Apr. 1989.

Agrawal, P. K. and M. C. Bailey, "An Analysis Technique for Microstrip Antennas," *IEEE Trans. Antennas Propagat.*, vol. 25, pp. 756–759, Nov. 1977.

Alexopoulos, N. G. and I. E. Rana, "Mutual Impedance Computation Between Printed Dipoles," *IEEE Trans. Antennas Propagat.*, vol. 29, pp. 106–111, Jan. 1981.

Alexopoulos, N. G., P. L. E. Uslenghi, and N. K. Uzunoglu, "Microstrip Dipoles on Cylindrical Structures," *Electromagnetics*, vol. 3, pp. 311–326, 1983.

Alexopoulos, N. G., P. B. Katehi, and D. B. Rutledge, "Substrate Optimization for Integrated Circuit Antennas," *IEEE Trans. Microwave Th. and Tech.*, vol. 31, pp. 550–557, July 1983.

Alexopoulos, N. G., "Integrated-Circuit Structures on Anisotropic Substrates," *IEEE Trans. Microwave Th. and Tech.*, vol. 33, pp. 847–881, Oct. 1985.

Bailey, M. C. and M. D. Deshpande, "Integral Equation Formulation of Microstrip Antennas," *IEEE Trans. Antennas Propagat.*, vol. 30, pp. 651–656, July 1982.

Bailey, M. C. and M. D. Deshpande, "Analysis of Elliptical and Circular Microstrip Antennas Using Moment Method," *IEEE Trans. Antennas Propagat.*, vol. 33, pp. 954–959, Sep. 1985.

Barkeshli, S., P. H. Pathak, and M. Marin, "An Asymptotic Closed-Form Microstrip Surface Green's Function for the Efficient Moment Method Analysis of Mutual Coupling in Microstrip Antennas," *IEEE Trans. Antennas Propagat.*, vol. 38, pp. 1374–1383, Sep. 1990.

Barlatey, L., J. R. Mosig, and T. Sphicopoulos, "Analysis of Stacked Microstrip Patches with a Mixed Potential Integral Equation," *IEEE Trans. Antennas Propagat.*, vol. 38, pp. 608–615, May 1990.

Chew, W. C. and Q. Liu, "Resonance Frequency of a Rectangular Microstrip Patch," *IEEE Trans. Antennas Propagat.*, vol. 36, pp. 1045–1056, Aug. 1988.

Chi, C. and N. G. Alexopoulos, "An Efficient Numerical Approach for Modeling Microstrip-Type Antennas," *IEEE Trans. Antennas Propagat.*, vol. 38, pp. 1399–1404, Sep. 1990.

Chi, C. L. and N. G. Alexopoulos, "An Image Extraction Approach to Modeling Printed Circuit Antennas," *Electromagnetics*, vol. 6, pp. 161–170, 1986.

Compton, R. C., R. C. McPhedran, Z. Popovic, G. M. Rebeiz, P. P. Tong, and D. B. Rutledge, "Bow-Tie Antennas on a Dielectric Half-Space: Theory and Experiment," *IEEE Trans. Antennas Propagat.*, vol. 35, pp. 622–631, June 1987.

Damiano, J. P., "Comuptation of Input Impedance in Microstrip Antennas. Graphic Representation and Numerical Integration of Oscillating Functions," *IEE Proc.*, vol. 134 Pt. H, pp. 456–466, Oct. 1987.

Das, N. K. and D. M. Pozar, "Analysis and Design of Series-Fed Arrays of Printed-Dipoles Proximity-Coupled to a Perpendicular Microstripline," *IEEE Trans. Antennas Propagat.*, vol. 37, pp. 435–444, Apr. 1989.

Davidovitz, M. and Y. T. Lo, "Rigorous Analysis of a Circular Patch Antenna Excited by a Microstrip Transmission Line," *IEEE Trans. Antennas Propagat.*, vol. 37, pp. 949–958, Aug. 1989.

Deshpande, M. D. and M. C. Bailey, "Input Impedance of Microstrip Antennas," *IEEE Trans. Antennas Propagat.*, vol. 30, pp. 645–650, July 1982.

Deshpande, M. D. and Y. R. Rao, "Analysis of Reactively Loaded Microstrip Disk Antenna," *IEE Proc.*, vol. 136 Pt. H, pp. 417–419, Oct. 1989.

Gang, X., "Engineering Model for Arbitrarily Shaped Microstrip Antennas Based on the Collocation Method," *IEE Proc.*, vol. 134 Pt. H, pp. 83–86, Feb. 1987.

Habashy, T. M. and J. A. Kong, "Coupling Between Two Circular Microstrip Disk Resonators," *Electromagnetics*, vol. 3, pp. 347–370, 1983.

Habashy, T. M., S. M. Ali, and J. A. Kong, "Input Impedance and Radiation Pattern of Cylindrical-Rectangular and Wraparound Microstrip Antennas," *IEEE Trans. Antennas Propagat.*, vol. 38, pp. 722–731, May 1990.

Habashy, T. M., J. A. Kong, and W. C. Chew, "Resonance and Radiation of the Elliptic Disk Microstrip Structure Part I: Formulation," *IEEE Trans. Antennas Propagat.*, vol. 35, pp. 877–886, Aug. 1987.

Hall, R. C. and J. R. Mosig, "The Analysis of Coaxially Fed Microstrip Antennas with Electrically Thick Substrates," *Electromagnetics*, vol. 9, pp. 367–384, 1989.

Hall, R. C. and J. R. Mosig, "Vertical Monopoles Embedded in a Dielectric Substrate," *IEE Proc.*, vol. 136 Pt. H, pp. 462–468, Dec. 1989.

Hansen, V., "Finite Array of Printed Dipoles with a Dielectric Cover," *IEE Proc.*, vol. 134 Pt. H, pp. 261–269, June 1987.

Herault, J., R. Moini, A. Reineix, and B. Jecko, "A New Approach to Microstrip Antennas Using a Mixed Analysis: Transient-Frequency," *IEEE Trans. Antennas Propagat.*, vol. 38, pp. 1166–1175, Aug. 1990.

Itoh, T., "Analysis of Microstrip Resonators," *IEEE Trans. Microwave Th. and Tech.*, vol. 22, pp. 946–952, Nov. 1974.

Itoh, T. and R. Mittra, "A New Method for Calculating the Capacitance of a Circular Disk for Microwave Integrated Circuits," *IEEE Trans. Microwave Th. and Tech.*, vol. 21, pp. 431–432, June 1973.

Jackson, D. R. and N. G. Alexopoulos, "Analysis of Planar Strip Geometries in a Substrate-Superstrate Configuration," *IEEE Trans. Antennas Propagat.*, vol. 34, pp. 1430–1438, Dec. 1986.

Jackson, D. R., "The RCS of a Rectangular Microstrip Patch in a Substrate-Superstrate Geometry," *IEEE Trans. Antennas Propagat.*, vol. 38, pp. 2–8, Jan. 1990.

Jackson, D. R., W. F. Richards, and A. Ali-Khan, "Series Expansions for the Mutual Coupling in Microstrip Patch Arrays," *IEEE Trans. Antennas Propagat.*, vol. 37, pp. 269–274, March 1989.

Jackson, R. W. and D. M. Pozar, "Full-Wave Analysis of Microstrip Open-End and Gap Discontinuities," *IEEE Trans. Microwave Th. and Tech.*, vol. 33, pp. 1036–1042, Oct. 1985.

Janaswamy, R. and D. H. Schaubert, "Characteristic Impedance of a Wide Slotline on Low-Permittivity Substrates," *IEEE Trans. Microwave Th. and Tech.*, vol. 34, pp. 900–902, Aug. 1986.

Katehi, P. B. and N. G. Alexopoulos, "Frequency-Dependent Characteristics of Microstrip Discontinuities in Millimeter-Wave Integrated Circuits," *IEEE Trans. Microwave Th. and Tech.*, vol. 33, pp. 1029–1035, Oct. 1985.

Katehi, P. B., "A Generalized Method for the Evaluation of Mutual Coupling in Microstrip Arrays," *IEEE Trans. Antennas Propagat.*, vol. 35, pp. 125–133, Feb. 1987.

Katehi, P. B. and N. G. Alexopoulos, "On the Modeling of Electromagnetically Coupled Microstrip Antennas - The Printed Strip Dipole," *IEEE Trans. Antennas Propagat.*, vol. 32, pp. 1179–1186, Nov. 1984.

Katehi, P. B., "Mutual Coupling Between Microstrip Dipoles in Multielement Arrays," *IEEE Trans. Antennas Propagat.*, vol. 37, pp. 275–280, March 1989.

Kishk, A. A. and L. Shafai, "The Effect of Various Parameters of Circular Microstrip Antennas on Their Radiation Efficiency and the Mode Excitation," *IEEE Trans. Antennas Propagat.*, vol. 34, pp. 969–976, Aug. 1986.

Krowne, C. M., "Determination of the Green's Function in the Spectral Domain Using a Matrix Method: Application to Radiators or Resonators Immersed in a Complex Anisotropic Layered Medium," *IEEE Trans. Antennas Propagat.*, vol. 34, pp. 247–253, Feb. 1986.

Lepeltier, P. M., J. Citerne, and J. M. Floch, "On The EMC Dipole Feed-Line Parasitic Radiation," *IEEE Trans. Antennas Propagat.*, vol. 38, pp. 878–882, June 1990.

Lin, Y. and L. Shafai, "Moment-Method Solution of the Near-Field Distribution and Far-Field Patterns of Microstrip Antennas," *IEE Proc.*, vol. 132 Pt. H, pp. 369–374, Oct. 1985.

Luk, K. M., W. Y. Tam, and C. L. Yip, "Analysis of Circular Microstrip Antennas with Superstrate," *IEE Proc.*, vol. 136 Pt. H, pp. 261–262, June 1989.

Mosig, J. R. and F. E. Gardiol, "Analytical and Numerical Techniques in the Green's Function Treatment of Microstrip Antennas and Scatterers," *IEE Proc.*, vol. 130 Pt. H, pp. 175–182, March 1983.

Mosig, J. R. and F. E. Gardiol, "General Integral Equation Formulation for Microstrip Antennas and Scatterers," *IEE Proc.*, vol. 132 Pt. H, pp. 424–432, Dec. 1985.

Mosig, J. R. and T. K. Sarkar, "Comparison of Quasi-Static and Exact Electromagnetic Fields from a Horizontal Electric Dipole Above a Lossy Dielectric Backed by an Imperfect Ground Plane," *IEEE Trans. Microwave Th. and Tech.*, vol. 34, pp. 379–387, Apr. 1986.

Mosig, J. R., "Arbitrarily Shaped Microstrip Structures and Their Analysis with a Mixed Potential Integral Equation," *IEEE Trans. Microwave Th. and Tech.*, vol. 36, pp. 314–323, Feb. 1988.

Mosig, J. R. and F. E. Gardiol, "Integral Equation Techniques for the Dynamic Analysis of Microstrip Discontinuities," *Alta Frequenza*, vol. 57, pp. 171–181, June 1988.

Nakano, H., S. R. Kerner, and N. G. Alexopoulos, "The Moment Method Solution for Printed Wire Antennas of Arbitrary Configuration," *IEEE Trans. Antennas Propagat.*, vol. 36, pp. 1667–1673, Dec. 1988.

Nakano, H., K. Hirose, T. Suzuki, S. R. Kerner, and N. G. Alexopoulos, "Numerical Analyses of Printed Line Antennas," *IEE Proc.*, vol. 136 Pt. H, pp. 98–104, Apr. 1989.

Nakatani, A., N. G. Alexopoulos, N. K. Uzunoglu, and P. L. E. Uslenghi, "Accurate Green's Function Computation for Printed Circuit Antennas on Cylindrical Substrates," *Electromagnetics*, vol. 6, pp. 243–254, 1986.

Nakatani, A. and N. G. Alexopoulos, "Microstrip Elements on Cylindrical Substrates - General Algorithm and Numerical Results," *Electromagnetics*, vol. 9, pp. 405–426, 1989.

Nelson, R. M., D. A. Rogers, and A. G. D'Assuncao, "Resonant Frequency of a Rectangular Microstrip Patch on Several Uniaxial Substrates," *IEEE Trans. Antennas Propagat.*, vol. 38, pp. 973–981, July 1990.

Newman, E. H., "Strip Antennas in a Dielectric Slab," *IEEE Trans. Antennas Propagat.*, vol. 26, pp. 647–653, Sep. 1978.

Newman, E. H. and P. Tulyathan, "Analysis of Microstrip Antennas Using Moment Methods," *IEEE Trans. Antennas Propagat.*, vol. 29, pp. 47–53, Jan. 1981.

Newman, E. H., J. H. Richmond, and B. W. Kwan, "Mutual Impedance Computation Between Microstrip Antennas," *IEEE Trans. Microwave Th. and Tech.*, vol. 31, pp. 941–944, Nov. 1983.

Newman, E. H. and D. Forrai, "Scattering from a Microstrip Patch," *IEEE Trans. Antennas Propagat.*, vol. 35, pp. 245–251, March 1987.

Nie, Z., W. C. Chew, and Y. T. Lo, "Analysis of the Annular-Ring-Loaded Circular-Disk Microstrip Antenna," *IEEE Trans. Antennas Propagat.*, vol. 38, pp. 806–813, June 1990.

Pinhas, S., S. Shtrikman, and D. Treves, "Moment-Method Solution of the Center-Fed Microstrip Disk Antenna Invoking Feed and Edge Current Singularities," *IEEE Trans. Antennas Propagat.*, vol. 37, pp. 1516–1522, Dec. 1989.

Pozar, D. M., "Input Impedance and Mutual Coupling of Rectangular Microstrip Antennas," *IEEE Trans. Antennas Propagat.*, vol. 30, pp. 1191–1196, Nov. 1982.

Pozar, D. M., "Improved Computational Efficiency for the Moment Method Solution of Printed Dipoles and Patches," *Electromagnetics*, vol. 3, pp. 299–309, 1983.

Pozar, D. M., "Considerations for Millimeter Wave Printed Antennas," *IEEE Trans. Antennas Propagat.*, vol. 31, pp. 740–747, Sep. 1983.

Pozar, D. M., "A Reciprocity Method of Analysis for Printed Slot and Slot-Coupled Microstip Antennas," *IEEE Trans. Antennas Propagat.*, vol. 34, pp. 1439–1446, Dec. 1986.

Pozar, D. M., "Radiation and Scattering from a Microstrip Patch on a Uniaxial Substrate," *IEEE Trans. Antennas Propagat.*, vol. 35, pp. 613–621, June 1987.

Pozar, D. M. and S. M. Voda, "A Rigorous Analysis of a Microstripline Fed Patch Antenna," *IEEE Trans. Antennas Propagat.*, vol. 35, pp. 1343–1350, Dec. 1987.

Ragheb, H. A. and L. Shafai, "Analysis of Arbitrary Shape Printed Line Microstrip Antennas," *IEEE Trans. Antennas Propagat.*, vol. 38, pp. 269–274, Feb. 1990.

Rana, I. E. and N. G. Alexopoulos, "Current Distribution and Input Impedance of Printed Dipoles," *IEEE Trans. Antennas Propagat.*, vol. 29, pp. 99–105, Jan. 1981.

Rana, I. E., N. G. Alexopoulos, and P. L. Katehi, "Theory of Microstrip Yagi-Uda Arrays," *Radio Science*, vol. 16, pp. 1077–1079, Nov.-Dec. 1981.

Rautio, J. C. and R. F. Harrington, "An Electromagnetic Time-Harmonic Analysis of Shielded Microstrip Circuits," *IEEE Trans. Microwave Th. and Tech.*, vol. 35, pp. 726–729, Aug. 1987.

Ruiz, J., M. J. Nunez, M. C. Sanchez, A. Navarro, and J. M. Zamarro, "Study of Resonant Frequency of Shielded Dielectric Resonators Coupled to a Microstrip Line," *IEE Proc.*, vol. 136 Pt. H, pp. 305–309, Aug. 1989.

Sarkar, T. K., S. M. Rao, and A. R. Djordjevic, "Electromagnetic Scattering and Radiation from Finite Microstrip Structures," *IEEE Trans. Microwave Th. and Tech.*, vol. 38, pp. 1568–1575, Nov. 1990.

Sarkar, T. K. and E. Arvas, "An Integral Equation Approach to the Analysis of Finite Microstrip Antennas: Volume/Surface Formulation," *IEEE Trans. Antennas Propagat.*, vol. 38, pp. 305–312, March 1990.

Shafai, L. and A. A. Sebak, "Radiation Characteristics and Polarisation of Undulated Microstrip Line Antennas," *IEE Proc.*, vol. 132 Pt. H, pp. 433–439, Dec. 1985.

Soares, A. J. M., S. B. A. Fonesca, and A. J. Giarola, "The Effect of a Dielectric Cover on the Current Distribution and Input Impedance of Printed Dipoles," *IEEE Trans. Antennas Propagat.*, vol. 32, pp. 1149–1153, Nov. 1984.

Splitt, G. and M. Davidovitz, "Guidelines for Design of Electromagnetically Coupled Microstrip Patch Antennas on Two-Layer Substrates," *IEEE Trans. Antennas Propagat.*, vol. 38, pp. 1136–1140, July 1990.

Sullivan, P. L. and D. H. Schaubert, "Analysis of an Aperture Coupled Microstrip Antenna," *IEEE Trans. Antennas Propagat.*, vol. 34, pp. 977–984, Aug. 1986.

Wood, C., "Analysis of Microstrip Circular Patch Antennas," *IEE Proc.*, vol. 128 Pt. H, pp. 69–76, Apr. 1981.

Yang, H., N. G. Alexopoulos, P. M. Lepeltier, and G. J. Stern, "Design of Transversely Fed EMC Microstrip Dipole Arrays Including Mutual Coupling," *IEEE Trans. Antennas Propagat.*, vol. 38, pp. 145–151, Feb. 1990.

Yang, H. Y. and J. A. Castaneda, "Printed Dipole Characteristics in a Two-Layer Geometry with Uniaxial Anisotropy," *Electromagnetics*, vol. 9, pp. 439–450, 1989.

Yang, H., A. Nakatani, and J. A. Castaneda, "Efficient Evaluation of Spectral Integrals in the Moment Method Solution of Microstrip Antennas and Circuits," *IEEE Trans. Antennas Propagat.*, vol. 38, pp. 1127–1130, July 1990.

Waveguides and Transmission Lines

Alexopoulos, N. G. and C. M. Krowne, "Characteristics of Single and Coupled Microstrips on Anisotropic Substrates," *IEEE Trans. Microwave Th. and Tech.*, vol. 26, pp. 387–393, June 1978.

Alexopoulos, N. G. and A. Nakatani, "Cylindrical Substrate Microstrip Line Characterization," *IEEE Trans. Microwave Th. and Tech.*, vol. 35, pp. 843–849, Sep. 1987.

Alphones, A. and G. S. Sanyal, "Propagation Characteristics of Eccentric Core Fibers Using Point-Matching Method," *Proc. IEEE*, vol. 74, pp. 1456–1458, Oct. 1986.

Auda, H. and R. F. Harrington, "A Moment Solution for Waveguide Junction Problems," *IEEE Trans. Microwave Th. and Tech.*, vol. 31, pp. 515–520, July 1983.

Ayasli, Y., "Analysis of Wide-Band Stripline Circulators by Integral Equation Technique," *IEEE Trans.*

Microwave Th. and Tech., vol. 28, pp. 200–209, March 1980.

Bagby, J. S., D. P. Nyquist, and B. C. Drachman, "Integral Formulation for Analysis of Integrated Dielectric Waveguides," *IEEE Trans. Microwave Th. and Tech.*, vol. 33, pp. 906–915, Oct. 1985.

Bates, R. H. T., "The Theory of the Point-Matching Method for Perfectly Conducting Waveguides and Transmission Lines," *IEEE Trans. Microwave Th. and Tech.*, vol. 17, pp. 294–301, June 1969.

Brooke, R. T. and J. E. Cruz, "Current Distribution and Impedance of Lossless Conductor Systems," *IEEE Trans. Microwave Th. and Tech.*, vol. 15, pp. 358–364, June 1967.

Burke, J. J. and R. W. Jackson, "Surface-to-Surface Transition via Electromagnetic Couping of Microstrip and Coplanar Waveguide," *IEEE Trans. Microwave Th. and Tech.*, vol. 37, pp. 519–525, March 1989.

Chan, C. H. and R. Mittra, "The Propagation Characteristics of Signal Lines Embedded in a Multilayered Structure in the Presence of a Periodically Perforated Ground Plane," *IEEE Trans. Microwave Th. and Tech.*, vol. 36, pp. 968–975, June 1988.

Chan, C. H., K. T. Ng, and A. B. Kouki, "A Mixed Spectral-Domain Approach for Dispersion Analysis of Suspended Planar Transmission Lines with Pedestals," *IEEE Trans. Microwave Th. and Tech.*, vol. 37, pp. 1716–1723, Nov. 1989.

Chow, Y. L. and S. Wu, "A Moment Method with Mixed Basis Functions for Scatterings by Waveguide Junctions," *IEEE Trans. Microwave Th. and Tech.*, vol. 21, pp. 333–340, May 1973.

Clements, J. C., C. R. Paul, and A. T. Adams, "Computation of the Capacitance Matrix for Systems of Dielectric-Coated Clyindrical Conductors," *IEEE Trans. Electromagnetic Compatibility*, vol. 17, pp. 238–248, Nov. 1975.

Das, N. K. and D. M. Pozar, "A Generalized Spectral-Domain Green's Function for Multilayer Dielectric Substractes with Application to Multilayer Transmission Lines," *IEEE Trans. Microwave Th. and Tech.*, vol. 35, pp. 326–335, March 1987.

Deklava, J. and V. Roje, "Accurate Numerical Solution of Coupled Integral Equations for Microstrip Transmission Line," *IEE Proc.*, vol. 134, Pt. H., pp. 163–168, Apr. 1987.

Delbare, W. and D. DeZutter, "Space-Domain Green's Function Approach to the Capacitance Calculation of Multiconductor Lines in Multilayered Dielectrics with Improved Surface Charge Modeling," *IEEE Trans. Microwave Th. and Tech.*, vol. 37, pp. 1562–1568, Oct. 1989.

Djordjevic, A. R., T. K. Sarkar, and S. M. Rao, "Analysis of Finite Conductivity Cylindrical Conductors Excited by Axially-Independent TM Electromagnetic Field," *IEEE Trans. Microwave Th. and Tech.*, vol. 33, pp. 960–966, Oct. 1985.

Dunleavy, L. P. and P. B. Katehi, "A Generalized Method for Analyzing Shielded Thin Microstrip Discountinuities," *IEEE Trans. Microwave Th. and Tech.*, vol. 36, pp. 1758–1766, Dec. 1988.

Elmoazzen, Y. E. and L. Shafai, "Numerical Solution of Coupling Between Two Collinear Parallel-Plate Waveguides," *IEEE Trans. Microwave Th. and Tech.*, vol. 23, pp. 871–876, Nov. 1975.

El-Sharawy, E. and R. W. Jackson, "Coplanar Waveguide and Slot Line on Magnetic Substrates: Analysis and Experiment," *IEEE Trans. Microwave Th. and Tech.*, vol. 36, pp. 1071–1079, June 1988.

El-Sharawy, E. and R. W. Jackson, "Full-Wave Analysis of an Infinitely Long Magnetic Surface Wave Transducer," *IEEE Trans. Microwave Th. and Tech.*, vol. 38, pp. 730–738, June 1990.

Fache, N. and D. D. Zutter, "Rigorous Full-Wave Space-Domain Solution for Dispersive Microstrip Lines," *IEEE Trans. Microwave Th. and Tech.*, vol. 36, pp. 731–737, Apr. 1988.

Fache, N. and D. D. Zutter, "Full-Wave Analysis of a Perfectly Conducting Wire Transmission Line in a Double-Layered Conductor-Backed Medium," *IEEE Trans. Microwave Th. and Tech.*, vol. 37, pp. 512–518, March 1989.

Faraji-Dana, R. and Y. L. Chow, "The Current Distribution and AC Resistance of a Microstrip Structure," *IEEE Trans. Microwave Th. and Tech.*, vol. 38, pp. 1268–1277, Sep. 1990.

Farrar, A. and A. T. Adams, "Characteristic Impedance of Microstrip by the Method of Moments," *IEEE Trans. Microwave Th. and Tech.*, vol. 18, pp. 65–66, Jan. 1970.

Farrar, A. and A. T. Adams, "Matrix Methods for Microstrip Three-Dimensional Problems," *IEEE Trans. Microwave Th. and Tech.*, vol. 20, pp. 497–504, Aug. 1972.

Finch, K. L. and N. G. Alexopoulos, "Shunt Posts in Microstrip Transmission Lines," *IEEE Trans. Microwave Th. and Tech.*, vol. 38, pp. 1585–1594, Nov. 1990.

Fuller, J. A. and N. F. Audeh, "The Point-Matching Solution of Uniform Nonsymmetric Waveguides," *IEEE Trans. Microwave Th. and Tech.*, vol. 17, pp. 114–115, Feb. 1969.

Geshiro, M. and T. Itoh, "Analysis of Double-Layered Finlines Containing a Magnetized Ferrite," *IEEE Trans. Microwave Th. and Tech.*, vol. 35, pp. 1377–1381, Dec. 1987.

Giri, D. V., S. Chang, and F. M. Tesche, "A Coupling Model for a Pair of Skewed Transmission Lines," *IEEE Trans. Electromagnetic Compatibility*, vol. 22, pp. 20–28, Feb. 1980.

Giri, D. V., F. M. Tesche, and S. Chang, "The Transverse Distribution of Surface Charge Densities on Multiconductor Transmission Lines," *IEEE Trans. Electromagnetic Compatibility*, vol. 21, pp. 220–227, Aug. 1979.

Gillespie, E. S. and J. J. Gustincic, "The Scattering of an Axial Cylindrical Surface Wave by a Perfectly Conducting Plane Annulus," *IEEE Trans. Microwave Th. and Tech.*, vol. 16, pp. 334–341, June 1968.

Gillespie, E. S. and F. J. Kilburg, "The Impedance and Scattering Properties of a Perfectly Conducting Strip Above a Plane Surface-Wave System," *IEEE Trans. Microwave Th. and Tech.*, vol. 21, pp. 413–419, June 1973.

Glandorf, F. J. and I. Wolff, "A Spectral-Domain Analysis of Periodically Nonuniform Microstrip Lines," *IEEE Trans. Microwave Th. and Tech.*, vol. 35, pp. 336–343, March 1987.

Harokopus, W. P. and P. B. Katehi, "Characterization of Microstrip Discontinuities on Multilayer Dielectric Substrates Including Radiation Losses," *IEEE Trans. Microwave Th. and Tech.*, vol. 37, pp. 2058–2066, Dec. 1989.

Harrington, R. F. and C. Wei, "Losses on Multiconductor Transmission Lines in Multilayered Dielectric Media," *IEEE Trans. Microwave Th. and Tech.*, vol. 32, pp. 705–710, July 1984.

Hashimoto, M. and K. Fujisawa, "Considerations on Matrix Methods and Estimation of Thier Errors," *IEEE Trans. Microwave Th. and Tech.*, vol. 18, pp. 352–359, July 1970.

Homentcovschi, D., "An Analytical Solution to the Microstrip Line Problem," *IEEE Trans. Microwave Th. and Tech.*, vol. 38, pp. 766–769, June 1990.

Howard, A. Q. and D. B. Seidel, "Singularity Extraction in Kernel Functions in Closed Region Problems," *Radio Science*, vol. 13, pp. 425–429, May-June 1978.

Huting, W. A. and K. J. Webb, "Numerical Solution of the Continuous Waveguide Transition Problem," *IEEE Trans. Microwave Th. and Tech.*, vol. 37, pp. 1802–1808, Nov. 1989.

Iskander, M. F. and T. S. Lind, "Electromagnetic Coupling of Coplanar Waveguides and Microstrip Lines to Highly Lossy Dielectric Media," *IEEE Trans. Microwave Th. and Tech.*, vol. 37, pp. 1910–1917, Dec. 1989.

Itoh, T. and R. Mittra, "Special-Domain Approach for Calculating the Dispersion Characteristics of Microstrip Lines," *IEEE Trans. Microwave Th. and Tech.*, vol. 21, pp. 496–499, July 1973.

Itoh, T., R. Mittra, and R. D. Ward, "A Method for Computing Edge Capacitance of Finite and Semi-Infinite Microstrip Lines," *IEEE Trans. Microwave Th. and Tech.*, vol. 20, pp. 847–849, Dec. 1972.

Jackson, R. W., "Full-Wave, Finite Element Analysis of Irregular Microstrip Discountinuities," *IEEE Trans. Microwave Th. and Tech.*, vol. 37, pp. 81–89, Jan. 1989.

Jackson, R. W. and D. M. Pozar, "Full-Wave Analysis of Microstrip Open-End and Gap Discontinuities," *IEEE Trans. Microwave Th. and Tech.*, vol. 33, pp. 1036–1042, Oct. 1985.

Janaswamy, R., "Even-Mode Characteristics of the Bilateral Slotline," *IEEE Trans. Microwave Th. and Tech.*, vol. 38, pp. 760–765, June 1990.

Janaswamy, R. and D. H. Schaubert, "Characteristic Impedance of a Wide Slotline on Low-Permittivity Substrates," *IEEE Trans. Microwave Th. and Tech.*, vol. 34, pp. 900–902, Aug. 1986.

Jansen, R. H., D. Ing, and C. Eng, "Hybrid Mode Analysis of End Effects of Planar Microwave and Millimetrewave Transmission Lines," *IEE Proc.*, vol. 128 Pt. H., pp. 77–86, Apr. 1981.

Katehi, P. B., "Radiation Losses in MM-Wave Open Microstrip Filters," *Electromagnetics*, vol. 7, pp. 137–152, 1987.

Katehi, P. B. and N. G. Alexopoulos, "Frequency-Dependent Characteristics of Microstrip Discontinuities in Millimeter-Wave Integated Circuits," *IEEE Trans. Microwave Th. and Tech.*, vol. 33, pp. 1029–1036, Oct. 1985.

Kawano, K., "Hybrid-Mode Analysis of a Broadside-Coupled Microstrip Line," *IEE Proc.*, vol. 131 Pt. H., pp. 21–24, Feb. 1984.

Khilla, A. and I. Wolff, "The Point-Matching Solution for Magnetically Tunable Cylindrical Cavities and Ferrite Planar Resonators," *IEEE Trans. Microwave Th. and Tech.*, vol. 27, pp. 592–598, June 1979.

Kiang, J., "On Resonance and Shielding of Printed Traces on a Circuit Board," *Proc. IEEE*, vol. 32, pp. 269–276, Nov. 1990.

Kiang, J., S. M. Ali, and J. A. Kong, "Integral Equation Solution to the Guidance and Leakage Properties of Coupled Dielectric Strip Waveguides," *IEEE Trans. Microwave Th. and Tech.*, vol. 38, pp. 193–203, Feb. 1990.

Kobayashi, M. and F. Ando, "Dispersion Characteristics of Open Microstrip Lines," *IEEE Trans. Microwave Th. and Tech.*, vol. 35, pp. 101–105, Feb. 1987.

Kobayashi, M. and T. Iijima, "Frequency-Dependent Characteristics of Current Distributions on Microstrip Lines," *IEEE Trans. Microwave Th. and Tech.*, vol. 37, pp. 799–801, Apr. 1989.

Kolk, E. W., N. H. G. Baken, and H. Blok, "Domain Integral Equation Analysis of Integrated Optical Channel and Ridge Waveguides in Stratified Media," *IEEE Trans. Microwave Th. and Tech.*, vol. 38, pp. 78–85, Jan. 1990.

Kosslowski, S., F. Bogelsack, and I. Wolff, "The Application of the Point Matching Method to the Analysis of Microstrip Lines with Finite Metallization Thickness," *IEEE Trans. Microwave Th. and Tech.*, vol. 36, pp. 1265–1271, Aug. 1988.

Kuo, J. and C. Tzuang, "Complex Modes in Shielded Suspended Coupled Microstrip Lines," *IEEE Trans. Microwave Th. and Tech.*, vol. 38, pp. 1278–1286, Sep. 1990.

Laura, P. A., "Application of the Point-Matching Method in Waveguide Problems," *IEEE Trans. Microwave Th. and Tech.*, vol. 14, p. 251, May 1966.

Lee, S. W., W. R. Jones, and J. J. Campbell, "Convergence of Numerical Solutions of Iris-Type Discontinuity Problems," *IEEE Trans. Microwave Th. and Tech.*, vol. 19, pp. 528–536, June 1971.

Leung, T. and C. A. Balanis, "Pulse Dispersion Distortion in Open and Shielded Microstrips Using the Spectral-Domain Method," *IEEE Trans. Microwave Th. and Tech.*, vol. 36, pp. 1223–1226, July 1988.

Lewin, L., "On the Resolution of a Class of Waveguide Discontinuity Problems by the Use of Singular Integral Equations," *IEEE Trans. Microwave Th. and Tech.*, vol. 9, pp. 321–332, July 1961.

Maia, M. R. D., A. G. D'Assuncao, and A. J. Giarola, "Dynamic Analysis of Microstrip Lines and Finlines on Uniaxial Anisotropic Substrates," *IEEE Trans. Microwave Th. and Tech.*, vol. 35, pp. 881–886, Oct. 1987.

Marin, M., S. Barkeshli, and P. H. Pathak, "Efficient Analysis of Planar Microstrip Geometries Using a Closed-Form Asymptotic Representation of the Grounded Dielectric Slab Green's Function," *IEEE Trans. Microwave Th. and Tech.*, vol. 37, pp. 669–679, Apr. 1989.

Matsumoto, M., M. Tsutsumi, and N. Kumagai, "Radiation Characteristics of a Dielectric Slab Waveguide Periodically Loaded with Thick Metal Strips," *IEEE Trans. Microwave Th. and Tech.*, vol. 35, pp. 89–95, Feb. 1987.

Mautz, J. R., R. F. Harrington, and C. G. Hsu, "The Inductance Matrix of a Multiconductor Transmission Line in Multiple Magnetic Media," *IEEE Trans. Microwave Th. and Tech.*, vol. 36, pp. 1293–1295, Aug. 1988.

Medina, F. and M. Horno, "Capacitance and Inductance Matrices for Multistrip Structures in Multilayered Anisotropic Dielectrics," *IEEE Trans. Microwave Th. and Tech.*, vol. 35, pp. 1002–1008, Nov. 1987.

Medina, F., M. Horno, and H. Baudrand, "Generalized Spectral Analysis of Planar Lines on Layered Media Including Uniaxial and Biaxial Dielectric Substrates," *IEEE Trans. Microwave Th. and Tech.*, vol. 37, pp. 504–511, March 1989.

Michalski, K. A. and D. Zheng, "Rigorous Analysis of Open Microstrip Lines of Arbitrary Cross Section in Bound and Leaky Regimes," *IEEE Trans. Microwave Th. and Tech.*, vol. 37, pp. 2005–2010, Dec. 1989.

Mittra, R. and T. Itoh, "A New Technique for the Analysis of the Dispersion Characteristics of Microstrip Lines," *IEEE Trans. Microwave Th. and Tech.*, vol. 19, p. 47, Jan. 1971.

Mostafa, A. A., C. M. Krowne, and K. A. Zaki, "Numerical Spectral Matrix Method for Propagation in General Layered Media: Application to Isotropic and Anisotopic Substrates," *IEEE Trans. Microwave Th. and Tech.*, vol. 35, pp. 1399–1407, Dec. 1987.

Mrozowski, M. and M. Okoniewski, "Comments on "Computation of Cutoff Wavenumbers of TE and TM Modes in Waveguides of Arbitrary Cross Sections Using a Surface Integral Formulation"," *IEEE Trans. Microwave Th. and Tech.*, vol. 38, pp. 1761–1762, Nov. 1990.

Nakatani, A. and N. G. Alexopoulos, "Coupled Microstrip Lines on a Cylindrical Substrate," *IEEE Trans. Microwave Th. and Tech.*, vol. 35, pp. 1392–1398, Dec. 1987.

Ng, F. L., "Tabulation of Methods for the Numerical Solution of the Hollow Waveguide Problem," *IEEE Trans. Microwave Th. and Tech.*, vol. 22, pp. 322–329, March 1974.

Ng, K. T. and C. H. Chan, "Unified Solution of Various Dielectric-Loaded Ridge Waveguides with a Mixed

Spectral-Domain Method," *IEEE Trans. Microwave Th. and Tech.*, vol. 37, pp. 2080–2085, Dec. 1989.

Papatheodorou, S., R. F. Harrington, and J. R. Mautz, "The Equivalent Circuit of a Microstrip Crossover in a Dielectic Substrate," *IEEE Trans. Microwave Th. and Tech.*, vol. 38, pp. 135–140, Feb. 1990.

Papatheodorou, S., R. F. Harrington, and J. R. Mautz, "Full-Wave Analysis of a Strip Crossover," *IEEE Trans. Microwave Th. and Tech.*, vol. 38, pp. 1439–1448, Oct. 1990.

Paul, C. R., "Reference Potential Terms in Static Capacitance Calculations via the Method of Moments," *IEEE Trans. Electromagnetic Compatibility*, vol. 20, pp. 267–269, Feb. 1978.

Paul, C. R. and A. E. Feather, "Computation of the Transmission Line Inductance and Capacitance Matrices from the Generalized Capacitance Matrix," *IEEE Trans. Electromagnetic Compatibility*, vol. 18, pp. 175–183, Nov. 1976.

Plumb, R. G. and R. F. Harrington, "An Electromagnetic Model for Multiconductor Connectors," *IEEE Trans. Electromagnetic Compatibility*, vol. 32, pp. 38–52, Feb. 1990.

Rahmat-Samii, Y., T. Itoh, and R. Mittra, "A Spectral Domain Analysis for Solving Microstrip Discontinuity Problems," *IEEE Trans. Microwave Th. and Tech.*, vol. 22, pp. 372–378, Apr. 1974.

Rengarajan, S. R., "Analysis of a Centered-Inclined Waveguide Slot Coupler," *IEEE Trans. Microwave Th. and Tech.*, vol. 37, pp. 884–889, May 1989.

Rubin, B. J., "Electromagnetic Modeling of Waveguides Involving Finite-Size Dielectric Regions," *IEEE Trans. Microwave Th. and Tech.*, vol. 38, pp. 807–812, June 1990.

Rubin, B. J. and H. L. Bertoni, "Waves Guided by Conductive Strips Above a Periodically Perforated Ground Plane," *IEEE Trans. Microwave Th. and Tech.*, vol. 31, pp. 541–549, July 1983.

Saad, S. M., "Review of Numerical Methods for the Analysis of Arbitrarily-Shaped Microwave and Optical Dielectric Waveguides," *IEEE Trans. Microwave Th. and Tech.*, vol. 33, pp. 894–899, Oct. 1985.

Shafai, L. and E. E. M. Hassan, "Field Solution and Electical Characteristics of Slotted Waveguides," *IEE Proc.*, vol. 128 Pt. H., pp. 87–94, Apr. 1981.

Shalaby, A. T. K. and A. Kumar, "Dispersion in Unilateral Finlines on Anisotropic Substrates," *IEEE Trans. Microwave Th. and Tech.*, vol. 35, pp. 448–450, Apr. 1987.

Shockley, T. D., C. R. Haden, and C. E. Lewis, "Application of the Point-Matching Method in Determining the Reflection and Transmission Coeffieients in Linearly Tapered Waveguides," *IEEE Trans. Microwave Th. and Tech.*, vol. 16, pp. 562–564, Aug. 1968.

Swaminathan, M., E. Arvas, T. K. Sarkar, and A. R. Djordjevic, "Computation of Cutoff Wavenumbers of TE and TM Modes in Waveguides of Arbitrary Cross Sections Using a Surface Integral Formulation," *IEEE Trans. Microwave Th. and Tech.*, vol. 38, pp. 154–159, Feb. 1990.

Tanaka, K. and M. Kojima, "Volume Integral Equations for Analysis of Dielectic Branching Waveguides," *IEEE Trans. Microwave Th. and Tech.*, vol. 36, pp. 1239–1245, Aug. 1988.

Taylor, C. D. and J. P. Castillo, "On Electromagnetic-Field Excitation of Unshielded Multiconductor Cables," *IEEE Trans. Electromagnetic Compatibility*, vol. 20, pp. 495–500, Nov. 1978.

Thong, V. K., "Solutions for Some Waveguide Discontinuities by the Method of Moments," *IEEE Trans. Microwave Th. and Tech.*, vol. 20, pp. 416–418, June 1972.

Tsalamengas, J. L., N. K. Uzunoglu, and N. G. Alexopoulos, "Propagation Characteristics of a Microstrip Line Printed on a General Anisotropic Substrate," *IEEE Trans. Microwave Th. and Tech.*, vol. 33, pp. 941–945, Oct. 1985.

Uchida, K., T. Noda, and T. Matsunaga, "New Type of Spectral-Domain Analysis of a Microstrip Line," *IEEE Trans. Microwave Th. and Tech.*, vol. 37, pp. 947–953, June 1989.

Uwano, T., "Accurate Characterization of Microstrip Resonator Open End with New Current Expression in Spectral-Domain Approach," *IEEE Trans. Microwave Th. and Tech.*, vol. 37, pp. 630–633, March 1989.

Van Deventer, T. E., P. B. Katehi, and A. C. Cangellaris, "An Integral Equation Method for the Evaluation of Conductor and Dielectric Losses in High-Frequency Interconnects," *IEEE Trans. Microwave Th. and Tech.*, vol. 37, pp. 1964–1972, Dec. 1989.

Venkataraman, J., S. M. Rao, A. R. Djordjevic, T. K. Sarkar, and Y. Naiheng, "Analysis of Arbitrarily Oriented Microstrip Transmission Lines in Arbitrarily Shaped Dielectric Media Over a Finite Ground Plane," *IEEE Trans. Microwave Th. and Tech.*, vol. 33, pp. 952–959, Oct. 1985.

Wei, C., R. F. Harrington, J. R. Mautz, and T. K. Sarkar, "Multiconductor Transmission Lines in Multilayered Dielectric Media ," *IEEE Trans. Microwave Th. and Tech.*, vol. 32, pp. 439–449, Apr. 1984.

Wu, S. and Y. L. Chow, "An Application of the Moment Method to Waveguide Scattering Problems," *IEEE Trans. Microwave Th. and Tech.*, vol. 20, pp. 744–749, Nov. 1972.

Xu, Q., K. J. Webb, and R. Mittra, "Study of Modal Solution Procedures for Microstrip Step Discontinuities," *IEEE Trans. Microwave Th. and Tech.*, vol. 37, pp. 381–387, Feb. 1989.

Yamashita, E. and K. Atsuki, "Analysis of Microstrip-Like Transmission Lines by Nonuniform Discretization of Integral Equations," *IEEE Trans. Microwave Th. and Tech.*, vol. 24, pp. 195–200, Apr. 1976.

Yang, H. and N. G. Alexopoulos, "Characterization of the Finline Step Discontinuity on Anisotropic Substrates," *IEEE Trans. Microwave Th. and Tech.*, vol. 35, pp. 956–963, Nov. 1987.

Yang, H. and N. G. Alexopoulos, "A Dynamic Model for Microstrip-Slotline Transition and Related Structures," *IEEE Trans. Microwave Th. and Tech.*, vol. 36, pp. 286–293, Feb. 1988.

Yang, H., N. G. Alexopoulos, and D. R. Jackson, "Microstrip Open-End and Gap Discontinuities in a Substrate–Superstrate Structure," *IEEE Trans. Microwave Th. and Tech.*, vol. 37, pp. 1542–1546, Oct. 1989.

Yee, H. Y. and N. F. Audeh, "Uniform Waveguides with Arbitrary Cross-Section Considered by the Point-Matching Method," *IEEE Trans. Microwave Th. and Tech.*, vol. 13, pp. 847–851, Nov. 1965.

Yuan, Y. and D. P. Nyquist, "Full-Wave Perturbation Theory Based Upon Electric Field Integral Equations for Coupled Microstrip Transmission Lines," *IEEE Trans. Microwave Th. and Tech.*, vol. 38, pp. 1576–1584, Nov. 1990.

Arrays

Aberle, J. T. and D. M. Pozar, "Analysis of Infinite Arrays of One- and Two-Probe-Fed Circular Patches," *IEEE Trans. Antennas Propagat.*, vol. 38, pp. 421–432, Apr. 1990.

Adams, A. T., P. C. Hsi, and A. Farrar, "Random Effects in Planar Arrays of Thin-Wire Dipoles," *IEEE Trans. Electromagnetic Compatibility*, vol. 20, pp. 223–232, Feb. 1978.

Agrawal, P. K., G. A. Richards, G. A. Thiele, and J. H. Richmond, "Analysis and Design of TEM-Line Antennas," *IEEE Trans. Antennas Propagat.*, vol. 20, pp. 561–568, Sep. 1972.

Allam, A. M. M. A. and E. A. Parker, "Application of Pocklington's Equation to Analysis of Dipole Frequency-Selective Surfaces of Finite Size," *IEE Proc.*, vol. 134 Pt. H, pp. 521–526, Dec. 1987.

Boag, A., Y. Leviatan, and A. Boag, "Analysis of Two-Dimensional Electromagnetic Scattering from Nonplanar Periodic Surfaces Using a Strip Current Model," *IEEE Trans. Antennas Propagat.*, vol. 37, pp. 1437–1446, Nov. 1989.

Borgiotti, G. V., "Modal Analysis of Periodic Planar Phased Arrays of Apertures," *Proc. IEEE*, vol. 56, pp. 1881–1892, Nov. 1968.

Cha, A. G. and J. K. Hsiao, "A Matrix Formulation for Large Scale Numerical Computation of the Finite Planar Waveguide Array Problem," *IEEE Trans. Antennas Propagat.*, vol. 22, pp. 106–108, Jan. 1974.

Chan, C. H. and R. Mittra, "On the Analysis of Frequency-Selective Surfaces Using Subdomain Basis Functions," *IEEE Trans. Antennas Propagat.*, vol. 38, pp. 40–50, Jan. 1990.

Chang, V. W. H., "Infinite Phased Dipole Array," *Proc. IEEE*, vol. 56, pp. 1892–1900, Nov. 1968.

Chen, C., "Scattering by a Two-Dimensional Periodic Array of Conducting Plates," *IEEE Trans. Antennas Propagat.*, vol. 18, pp. 660–665, Sep. 1970.

Chen, C., "Transmission Through a Conducting Screen Perforated Periodically with Apertures," *IEEE Trans. Microwave Th. and Tech.*, vol. 18, pp. 627–632, Sep. 1970.

Chen, C., "Diffraction of Electromagnetic Waves by a Conducting Screen Perforated Periodically with Circular Holes," *IEEE Trans. Microwave Th. and Tech.*, vol. 19, pp. 475–481, May 1971.

Cheng, D. K. and F. I. Tseng, "Gain Optimization for Arbitrary Antenna Arrays," *IEEE Trans. Antennas Propagat.*, vol. 13, pp. 973–974, Nov. 1965.

Cheng, D. K., "Optimization Techniques for Antenna Arrays," *Proc. IEEE*, vol. 59, pp. 1664–1674, Dec. 1971.

Cheng, D. K. and C. A. Chen, "Optimum Element Spacings for Yagi-Uda Arrays," *IEEE Trans. Antennas Propagat.*, vol. 21, pp. 615–622, Sep. 1973.

Clarricoats, P. J., S. M. Tun, and C. G. Parini, "Effects of Mutual Coupling in Conical Horn Arrays," *IEE Proc.*, vol. 131 Pt. H, pp. 165–171, June 1984.

Deshpande, M. D. and M. C. Bailey, "Analysis of Finite Phased Arrays of Circular Microstrip Patches," *IEEE Trans. Antennas Propagat.*, vol. 37, pp. 1355–1360, Nov. 1989.

Eftimiu, C. and P. L. Huddleston, "Scattering of Electromagnetic Waves by Conducting Periodic Surfaces: A Comparison of Exact Integral Equation Methods," *Radio Science*, vol. 22, pp. 815–824, Sep.-Oct. 1987.

Fan, D., "A New Approach to Diffraction Analysis of Conductor Grids, Part I – Parallel-Polarized Incident Plane Waves," *IEEE Trans. Antennas Propagat.*, vol. 37, pp. 84–88, Jan. 1989.

Fan, D., "A New Approach to Diffraction Analysis of Conductor Grids, Part II – Perpendicular-Polarized Incident Plane Waves," *IEEE Trans. Antennas Propagat.*, vol. 37, pp. 89–93, Jan. 1989.

Fenn, A. J., G. A. Thiele, and B. A. Munk, "Moment Method Analysis of Finite Rectangular Waveguide Phased Arrays," *IEEE Trans. Antennas Propagat.*, vol. 30, pp. 554–564, July 1982.

Fenn, A. J., "Theoretical and Experimental Study of Monopole Phased Array Antennas," *IEEE Trans. Antennas Propagat.*, vol. 33, pp. 1118–1126, Oct. 1985.

Fenn, A. J., "Element Gain Pattern Prediction for Finite Arrays of V-Dipole Antennas Over Ground Plane," *IEEE Trans. Antennas Propagat.*, vol. 36, pp. 1629–1633, Nov. 1988.

Gallagher, J. G. and D. J. Brammer, "Scattering from an Infinite Array of Periodic Broken Wires Buried in a Dielectric Sheet," *Radio Science*, vol. 20, pp. 50–62, Jan.-Feb. 1985.

Gong, Z. and K. G. Balmain, "Reduction of the Anomalous Resonances of Symmetric Log-Periodic Dipole Antennas," *IEEE Trans. Antennas Propagat.*, vol. 34, pp. 1404–1410, Dec. 1986.

Gulick, J. J. and R. S. Elliott, "The Design of Linear and Planar Arrays of Wageguide-Fed Longitudinal Slots," *Electromagnetics*, vol. 10, pp. 327–347, 1990.

Hall, R. C., R. Mittra, and K. M. Mitzner, "Scattering from Finite Thickness Resistive Strip Gratings," *IEEE Trans. Antennas Propagat.*, vol. 36, pp. 504–510, Apr. 1988.

Harrington, R. F., "Antenna Excitation for Maximum Gain," *IEEE Trans. Antennas Propagat.*, vol. 13, pp. 896–903, Nov. 1965.

Harrington, R. F. and J. R. Mautz, "Pattern Synthesis for Loaded N-Port Scatterers," *IEEE Trans. Antennas Propagat.*, vol. 22, pp. 184–190, March 1974.

Harrington, R. F. and J. R. Mautz, "Optimization of Radar Cross Section of N-Port Loaded Scatterers," *IEEE Trans. Antennas Propagat.*, vol. 22, pp. 697–700, Sep. 1974.

Hirasawa, K. and B. J. Strait, "On a Method for Array Design by Matrix Inversion," *IEEE Trans. Antennas Propagat.*, vol. 19, pp. 446–447, May 1971.

Hirasawa, K. and D. H. Sinnott, "On the Relationship Between Classical and Matrix Design Methods for Arrays of Wire Antennas," *IEEE Trans. Antennas Propagat.*, vol. 20, pp. 661–663, Sep. 1972.

Jackson, D. R., A. E. Dinbergs, and S. A. Long, "A Moment-Method Design Procedure for an Array of EMC Dipoles," *IEEE Trans. Antennas Propagat.*, vol. 38, pp. 766–770, May 1990.

Jensen, N. E., "Plane-Wave Scattering from Half-Wave Dipole Arrays," *IEEE Trans. Antennas Propagat.*, vol. 18, pp. 829–831, Nov. 1970.

Johansson, F. S., "Convergence Phenomenon in the Solution of Dichroic Scattering Problems by Galerkin's Method," *IEE Proc.*, vol. 134 Pt. H, pp. 87–92, Feb. 1987.

Jorgenson, R. E. and R. Mittra, "Oblique Scattering from Lossy Strip Structures with One-Dimensional Periodicity," *IEEE Trans. Antennas Propagat.*, vol. 38, pp. 212–219, Feb. 1990.

Kalhor, H. A. and M. Ilyas, "Scattering of Plane Electromagnetic Waves by a Grating of Conducting Cylinders Embedded in a Dielectric Slab in Free Space," *IEE Proc.*, vol. 128 Pt. H, pp. 155–158, June 1981.

Kalhor, H. A. and M. Ilyas, "Scattering of Plane Electromagnetic Waves by a Grating of Conducting Cylinders Embedded in a Dielectric Slab over a Ground Plane," *IEEE Trans. Antennas Propagat.*, vol. 30, pp. 576–579, July 1982.

Kastner, R., "Analysis of Microstrip Antenna Structures Using the "Add-On" Technique," *IEEE Trans. Antennas Propagat.*, vol. 38, pp. 114–117, Jan. 1990.

Kishk, A. A. and L. Shafai, "Gain Enhancement of Antennas Over Finite Ground Plane Covered by a Dielectric Sheet," *IEE Proc.*, vol. 134 Pt. H, pp. 60–64, Feb. 1987.

Kominami, M., D. M. Pozar, and D. H. Schaubert, "Dipole and Slot Elements and Arrays on Semi-Infinite Substrates," *IEEE Trans. Antennas Propagat.*, vol. 33, pp. 600–607, June 1985.

Lee, J. J., "Effects of Metal Fences on the Scan Performance of an Infinite Dipole Array," *IEEE Trans. Antennas Propagat.*, vol. 38, pp. 683–692, May 1990.

Lee, K. M. and R. Chu, "Analysis of Mutual Coupling Between a Finite Phased Array of Dipoles and Its Feed Network," *IEEE Trans. Antennas Propagat.*, vol. 36, pp. 1681–1699, Dec. 1988.

Lee, S., "Scattering by Dielectric-Loaded Screen," *IEEE Trans. Antennas Propagat.*, vol. 19, pp. 656–665, Sep. 1971.

Liu, D., R. J. Garbacz, and D. M. Pozar, "Antenna Synthesis and Optimization Using Generalized Characteristic Modes," *IEEE Trans. Antennas Propagat.*, vol. 38, pp. 862–866, June 1990.

Luzwick, J. L., E. C. Ngai, and A. T. Adams, "Analysis of a Large Linear Antenna Array of Uniformly Spaced Thin-Wire Dipoles Parallel to a Perfectly Conducting Plane," *IEEE Trans. Antennas Propagat.*, vol. 30, pp. 230–234, March 1982.

Luzwick, J. L. and R. F. Harrington, "Mutual Coupling Analysis in a Finite Planar Rectangular Waveguide Antenna Array," *Electromagnetics*, vol. 2, pp. 25–42, 1982.

Mailloux, R. J., "On the Use of Metallized Cavities in Printed Slot Arrays with Dielectric Substrates," *IEEE Trans. Antennas Propagat.*, vol. 35, pp. 477–487, May 1987.

Mautz, J. R. and R. F. Harrington, "Modal Analysis of Loaded N-Port Scatterers," *IEEE Trans. Antennas Propagat.*, vol. 21, pp. 188–198, March 1973.

Mayhan, J. T. and L. L. Tsai, "Reflection and Transmission Characteristics of Thin Periodic Interfaces," *IEEE Trans. Antennas Propagat.*, vol. 24, pp. 449–455, July 1976.

Mittra, R., C. H. Chan, and T. Cwik, "Techniques for Analyzing Frequency Selective Surfaces–A Review," *Proc. IEEE*, vol. 76, pp. 1593–1615, Dec. 1988.

Montgomery, J. P., "Scattering by an Infinite Periodic Array of Thin Conductors on a Dielectric Sheet," *IEEE Trans. Antennas Propagat.*, vol. 23, pp. 70–75, Jan. 1975.

Munk, B. A., R. G. Kouyoumjian, and L. Peters, "Reflection Properties of Periodic Surfaces of Loaded Dipoles," *IEEE Trans. Antennas Propagat.*, vol. 19, pp. 612–617, Sep. 1971.

Munk, B. A. and R. J. Luebbers, "Reflection Properties of Two-Layer Dipole Arrays," *IEEE Trans. Antennas Propagat.*, vol. 22, pp. 766–773, Nov. 1974.

Munk, B. A., R. J. Leubbers, and R. D. Fulton, "Transmission Through a Two-Layer Array of Loaded Slots," *IEEE Trans. Antennas Propagat.*, vol. 22, pp. 804–809, Nov. 1974.

Munk, B. A. and G. A. Burrell, "Plane-Wave Expansion for Arrays of Arbitrarily Oriented Piecewise Linear Elements and Its Application in Determining the Impedance of a Single Linear Antenna in a Lossy Half-Space," *IEEE Trans. Antennas Propagat.*, vol. 27, pp. 331–343, May 1979.

Munk, B. A., T. W. Kornbau, and R. D. Fulton, "Scan Independent Phased Arrays," *Radio Science*, vol. 14, pp. 979–990, Nov.-Dec. 1979.

Ormsby, J. F. A., "Antenna, Load, and Field Effects on the Bistatic Scattering Patterns from a Linear Dipole Array," *IEEE Trans. Antennas Propagat.*, vol. 27, pp. 116–122, Jan. 1979.

Orta, R., R. Tascone, and R. Zich, "Three-Dimensional Periodic Arrays of Thin Conductors," *Electromagnetics*, vol. 7, pp. 185–203, 1987.

Orta, R., P. Savi, and R. Tascone, "The Effect of Finite Conductivity on Frequency Selective Surface Behavior," *Electromagnetics*, vol. 10, pp. 213–227, 1990.

Ott, R. H., R. G. Kouyoumjian, and L. Peters, "Scattering by a Two-Dimensional Periodic Array of Narrow Plates," *Radio Science*, vol. 2, pp. 1347–1359, Nov. 1967.

Pelton, E. L. and B. A. Munk, "Scattering from Periodic Arrays of Crossed Dipoles," *IEEE Trans. Antennas Propagat.*, vol. 27, pp. 323–330, May 1979.

Poey, P. and P. Guigue, "Determination of the Current Distribution on a Bidimensional Infinite Periodic Structure of Thin Metallic Wires Using the Method of the Singular Integral Equation," *IEEE Trans. Antennas Propagat.*, vol. 35, pp. 221–224, Feb. 1987.

Pozar, D. M., "Antenna Synthesis and Optimization Using Weighted Inagaki Modes," *IEEE Trans. Antennas Propagat.*, vol. 32, pp. 159–165, Feb. 1984.

Pozar, D. M. and D. H. Schaubert, "Scan Blindness in Infinite Phased Arrays of Printed Dipoles," *IEEE Trans. Antennas Propagat.*, vol. 32, pp. 602–610, June 1984.

Pozar, D. M. and D. H. Schaubert, "Analysis of an Infinite Array of Rectangular Microstrip Patches with Idealized Probe Feeds," *IEEE Trans. Antennas Propagat.*, vol. 32, pp. 1101–1107, Oct. 1984.

Pozar, D. M., "Analysis of Finite Phased Arrays of Printed Dipoles," *IEEE Trans. Antennas Propagat.*, vol. 33, pp. 1045–1053, Oct. 1985.

Pozar, D. M., "Finite Phased Arrays of Rectangular Microstrip Patches," *IEEE Trans. Antennas Propagat.*, vol. 34, pp. 658–665, May 1986.

Richmond, J. H., "Admittance Matrix of Coupled V Antennas," *IEEE Trans. Antennas Propagat.*, vol. 18, pp. 820–821, Nov. 1970.

Richmond, J. H. and R. J. Garbacz, "Surface Waves on Periodic Array of Imperfectly Conducting Vertical Dipoles over the Flat Earth," *IEEE Trans. Antennas Propagat.*, vol. 27, pp. 783–787, Nov. 1979.

Riggs, L. S. and R. G. Smith, "Efficient Current Expansion Modes for the Triarm Frequency-Selective Surface," *IEEE Trans. Antennas Propagat.*, vol. 36, pp. 1172–1177, Aug. 1988.

Rubin, B. J. and H. L. Bertoni, "Reflection from a Periodically Perforated Plane Using a Subsectional Current Approximation," *IEEE Trans. Antennas Propagat.*, vol. 31, pp. 829–836, Nov. 1983.

Rubin, B. J. and H. L. Bertoni, "Scattering from a Periodic Array of Conducting Bars of Finite Surface Resistance," *Radio Science*, vol. 20, pp. 827–832, Jul.-Aug. 1985.

Sahalos, J., "Synthesis and Optimization for Arrays of Nonparallel Wire Antennas by the Orthogonal Method," *IEEE Trans. Antennas Propagat.*, vol. 26, pp. 886–991, Nov. 1978.

Sangster, A. J. and A. H. I. McCormick, "Theoretical Design/Synthesis of Slotted Waveguide Arrays," *IEE Proc.*, vol. 136 Pt. H, pp. 39–46, Feb. 1989.

Sanzgiri, S. M. and J. A. Cummins, "E-Plane Synthesis of Dipole Array Antennas," *IEEE Trans. Antennas Propagat.*, vol. 21, pp. 380–382, May 1973.

Saoudy, S. A. S. and M. Hamid, "Optimal Design of Multiply Fed Dipole Antennas," *IEEE Trans. Antennas Propagat.*, vol. 35, pp. 1001–1009, Sep. 1987.

Sarkar, T. K., "An Optimization Program for Linear Arrays of Parallel Wires," *IEEE Trans. Antennas Propagat.*, vol. 22, pp. 631–632, July 1974.

Sarkar, T. K. and B. J. Strait, "Optimization Methods for Arbitrarily Oriented Arrays of Antennas in Any Environment," *Radio Science*, vol. 11, pp. 959–967, Dec. 1976.

Sarkar, T. K., M. F. Costa, C. I, and R. F. Harrington, "Electromagnetic Transmission Through Mesh Covered Apertures and Arrays of Apertures in a Conducting Screen," *IEEE Trans. Antennas Propagat.*, vol. 32, pp. 908–913, Sep. 1984.

Savov, S. V., "Cavity-Backed Slot Array Analysis," *IEE Proc.*, vol. 134 Pt. H, pp. 280–284, June 1987.

Scherer, J. P. and A. R. Neruerther, "Mutual Coupling in Linear Dipole Arrays," *IEEE Trans. Antennas Propagat.*, vol. 20, pp. 651–653, Sep. 1972.

Schuman, H. K., D. R. Pflug, and L. D. Thompson, "Infinite Planar Arrays of Arbitrarily Bent Thin Wire Radiators," *IEEE Trans. Antennas Propagat.*, vol. 32, pp. 364–377, Apr. 1984.

Siakavara, K. and J. N. Sahalos, "A Simplification of the Synthesis of Parallel Wire Antenna Arrays," *IEEE Trans. Antennas Propagat.*, vol. 37, pp. 936–940, July 1989.

Sinnott, D. H., "Matrix Analysis of Linear Antenna Arrays of Equally Spaced Elements," *IEEE Trans. Antennas Propagat.*, vol. 21, pp. 385–386, May 1973.

Sinnott, D. H. and R. F. Harrington, "Analysis and Design of Circular Antenna Arrays by Matrix Methods," *IEEE Trans. Antennas Propagat.*, vol. 21, pp. 610–614, Sep. 1973.

Sinnott, D. H., "Multiple-Frequency Computer Analysis of the Log-Periodic Dipole Antenna," *IEEE Trans. Antennas Propagat.*, vol. 22, pp. 592–594, July 1974.

Strait, B. J. and K. Hirasawa, "Array Design for a Specified Pattern by Matrix Methods," *IEEE Trans. Antennas Propagat.*, vol. 17, pp. 237–239, March 1969.

Strait, B. J. and K. Hirasawa, "Constrained Optimization of the Gain of an Array of Thin Wire Antennas," *IEEE Trans. Antennas Propagat.*, vol. 20, pp. 665–666, Sep. 1972.

Taylor, C. D., E. A. Aronson, and C. W. Harrison, "Theory of Coupled Monopoles," *IEEE Trans. Antennas Propagat.*, vol. 18, pp. 360–366, May 1970.

Thiele, G. A., "Analysis of Yagi–Uda-Type Antennas," *IEEE Trans. Antennas Propagat.*, vol. 17, pp. 24–30, Jan. 1969.

Walker, W. A. and C. M. Butler, "A Method for Computing Scattering by Large Arrays of Narrow Strips," *IEEE Trans. Antennas Propagat.*, vol. 32, pp. 1327–1334, Dec. 1984.

Wickliff, R. G. and R. J. Garbacz, "The Average Backscattering Cross Section of Clouds of Randomized Resonant Dipoles," *IEEE Trans. Antennas Propagat.*, vol. 22, pp. 503–505, May 1974.

Wolter, J., "Solution of Maxwell's Equation for Log-Periodic Dipole Antennas," *IEEE Trans. Antennas Propagat.*, vol. 18, pp. 734–740, Nov. 1970.

Wu, C. P., "Analysis of Finite Parallel-Plate Waveguide Arrays," *IEEE Trans. Antennas Propagat.*, vol. 18, pp. 328–344, May 1970.

Wu, K. and V. Dzougaiev, "Complete Theoretical and Experimental Analysis on Properties of Planar Periodic Waveguides," *IEE Proc.*, vol. 135 Pt. H, pp. 27–33, Feb. 1988.

Yang, H., J. A. Castaneda, and N. G. Alexopoulos, "An Integral Equation Analysis of an Infinite Array of Rectangular Dielectric Waveguides," *IEEE Trans. Microwave Th. and Tech.*, vol. 38, pp. 873–880, July 1990.

MM/Green's Function Solutions

Auda, H. and R. F. Harrington, "Inductive Posts and Diaphragms of Arbitrary Shape and Number in a Rectangular Waveguide," *IEEE Trans. Microwave Th. and Tech.*, vol. 32, pp. 606–613, June 1984.

Awadalla, K. H. and T. S. MacLean, "Input Impedance of a Monopole Antenna at the Center of a Finite Ground Plane," *IEEE Trans. Antennas Propagat.*, vol. 26, pp. 244–247, March 1978.

Awadalla, K. H. and T. S. MacLean, "Monopole Antenna at Center of Circular Ground Plane: Input

Impedance and Radiation Pattern," *IEEE Trans. Antennas Propagat.*, vol. 27, pp. 151–153, March 1979.

Burke, G. J., W. A. Johnson, and E. K. Miller, "Modeling of Simple Antennas Near to and Penetrating an Interface," *Proc. IEEE*, vol. 71, pp. 174–175, Jan. 1983.

Burke, G. J. and E. K. Miller, "Modeling Antennas Near to and Penetrating a Lossy Interface," *IEEE Trans. Antennas Propagat.*, vol. 32, pp. 1040–1049, Oct. 1984.

Burnside, W. D., C. L. Yu, and R. J. Marhefka, "A Technique to Combine the Geometrical Theory of Diffraction and the Moment Method," *IEEE Trans. Antennas Propagat.*, vol. 23, pp. 551–558, July 1975.

Butler, C. M. and T. L. Keshavamurthy, "Analysis of a Wire Antenna in the Presence of Sphere," *IEEE Trans. Electromagnetic Compatibility*, vol. 22, pp. 113–118, May 1980.

Butler, C. M. and T. L. Keshavamurthy, "Corrections to "Analysis of a Wire Antenna in the Presence of a Sphere," *IEEE Trans. Electromagnetic Compatibility*, vol. 22, p. 329, Nov. 1980.

Butler, C. M., "Current Induced on a Conducting Strip Which Resides on the Planar Interface Between Two Semi-Infinite Half-Spaces," *IEEE Trans. Antennas Propagat.*, vol. 32, pp. 226–231, March 1984.

Butler, C. M., X. Xu, and A. W. Glisson, "Current Induced on a Conducting Cylinder Located Near the Planar Interface Between Two Semi-Infinite Half-Spaces," *IEEE Trans. Antennas Propagat.*, vol. 33, pp. 616–624, June 1985.

Chang, D. C. and J. R. Wait, "Theory of a Vertical Tubular Antenna Located Above a Conducting Half-Space," *IEEE Trans. Antennas Propagat.*, vol. 18, pp. 182–188, March 1970.

Chi, C. and N. G. Alexopoulos, "Radiation by a Probe Through a Substrate," *IEEE Trans. Antennas Propagat.*, vol. 34, pp. 1080–1091, Sep. 1986.

Ekelman, E. P. and G. A. Thiele, "A Hybrid Technique for Combining the Moment Method Treatment of Wire Antennas with the GTD for Curved Surfaces," *IEEE Trans. Antennas Propagat.*, vol. 28, pp. 831–839, Nov. 1980.

Glisson, A. W. and C. M. Butler, "Analysis of a Wire Antenna in the Presence of a Body of Revolution," *IEEE Trans. Antennas Propagat.*, vol. 28, pp. 604–609, Sep. 1980.

Green, H. E., "Impedance of a Monopole on the Base of a Large Cone," *IEEE Trans. Antennas Propagat.*, vol. 17, pp. 703–706, Nov. 1969.

Henderson, L. W. and G. A. Thiele, "A Hybrid MM-GTD Technique for the Treatment of Wire Antennas Near a Curved Surface," *Radio Science*, vol. 16, pp. 1125–1130, Nov.-Dec. 1981.

Henderson, L. W. and G. A. Thiele, "A Hybrid MM-Geometrical Optics Technique for the Treatment of Wire Antennas Mounted on a Curved Surface," *IEEE Trans. Antennas Propagat.*, vol. 30, pp. 1257–1261, Nov. 1982.

Hongo, K. and A. Hamamura, "Asymptotic Solutions for the Scattered Field of Plane Wave by a Cylindrical Obstacle Buried in a Dielectric Half-Space," *IEEE Trans. Antennas Propagat.*, vol. 34, pp. 1306–1312, Nov. 1986.

Hsu, C. G. and H. A. Auda, "Multiple Dielectric Posts in a Rectangular Waveguide," *IEEE Trans. Microwave Th. and Tech.*, vol. 34, pp. 883–891, Aug. 1986.

Jarem, J. M., "A Multifiliment Method-of-Moments Solution for the Input Impedance of a Probe-Excited Semi-Infinite Waveguide," *IEEE Trans. Microwave Th. and Tech.*, vol. 35, pp. 14–19, Jan. 1987.

Joshi, J. S. and J. A. Cornick, "Analysis of Waveguide Post Configurations: Part I - Gap Immittance Matrices," *IEEE Trans. Microwave Th. and Tech.*, vol. 25, pp. 169–173, March 1977.

Joshi, J. S. and J. A. Cornick, "Analysis of Waveguide Post Configurations: Part II - Dual-Gap Cases," *IEEE Trans. Microwave Th. and Tech.*, vol. 25, pp. 173–181, March 1977.

Johnson, W. A., "Analysis of a Vertical, Tubular Cylinder Which Penetrates an Air-Dielectric Interface and Which is Excited by an Azimuthally Symmetric Source," *Radio Science*, vol. 18, pp. 1273–1281, Nov.-Dec. 1983.

Karunaratne, M. D. G., K. A. Michalski, and C. M. Butler, "TM Scattering from a Conducting Strip Loaded by a Dielectric Cylinder," *IEE Proc.*, vol. 132 Pt. H, pp. 115–122, Apr. 1985.

Karunaratne, M. D. G., K. A. Michalski, and C. M. Butler, "TE Scattering from a Conducting Strip Loaded by a Dielectric Cylinder," *IEE Proc.*, vol. 132 Pt. H, pp. 375–383, Oct. 1985.

Karwowski, A., "Low-Frequency Approach to the Problem of a Horizontal Wire Antenna Above an Imperfect Ground," *IEE Proc.*, vol. 131 Pt. H, pp. 214–216, June 1984.

Karwowski, A., "Low-Frequency Approach to the Problem of a Vertical Wire Antenna Above an Imperfect Ground," *IEE Proc.*, vol. 132 Pt. H, pp. 123–126, Apr. 1985.

Karwowski, A. and K. A. Michalski, "A Comparative Numerical Study of Several Techniques for Modeling a Horizontal Wire Antenna Over a Lossy Half-Space," *Radio Science*, vol. 22, pp. 922–928, Nov. 1987.

Kuo, W. C. and K. K. Mei, "Numerical Approximations of the Sommerfeld Integral for Fast Convergence," *Radio Science*, vol. 13, pp. 407–415, May-June 1978.

Leviatan, Y., P. G. Li, A. T. Adams, and J. Perini, "Single-Post Inductive Obstacle in Rectangular Waveguide," *IEEE Trans. Microwave Th. and Tech.*, vol. 31, pp. 806–811, Oct. 1983.

Li, P. G., A. T. Adams, Y. Leviatan, and J. Perini, "Multiple-Post Inductive Obstacles in Rectangular Waveguide," *Proc. IEEE*, vol. 32, pp. 365–373, Apr. 1984.

Lin, C. C. and K. K. Mei, "Radiation and Scattering From Partially Buried Vertical Wires," *Electromagnetics*, vol. 2, pp. 309–334, 1982.

Marin, M. and M. F. Catedra, "A Study of a Monopole Arbitrarily Located on a Disk Using Hybrid MM/GTD Techniques," *IEEE Trans. Antennas Propagat.*, vol. 35, pp. 287–292, March 1987.

Medgyesi-Mitschang, L. N. and D. Wang, "Hybrid Methods for Analysis of Complex Scatterers," *Proc. IEEE*, vol. 77, pp. 770–779, May 1989.

Michalski, K. A. and C. M. Butler, "Determination of Current Induced on a Conducting Strip Embedded in a Dielectric Slab," *Radio Science*, vol. 18, pp. 1195–1206, Nov.-Dec. 1983.

Michalski, K. A., C. E. Smith, and C. M. Butler, "Analysis of a Horizontal Two-Element Antenna Array Above a Dielectric Halfspace," *IEE Proc.*, vol. 132 Pt. H, pp. 335–338, Aug. 1985.

Michalski, K. A., R. D. Nevels, and D. Zheng, "Comparison of Lorentz and Coulomb Gauge Formulations for Transverse-Electric Wave Scattering by Two-Dimensional Surfaces of Arbitrary Shape in the Presence of a Circular Cylinder," *IEEE Trans. Antennas Propagat.*, vol. 38, pp. 732–739, May 1990.

Newman, E. H., "TM Scattering by a Dielectric Cylinder in the Presence of a Half-Plane," *IEEE Trans. Antennas Propagat.*, vol. 33, pp. 773–782, July 1985.

Newman, E. H., "TM and TE Scattering by a Dielectric/Ferrite Cylinder in the Presence of a Half-Plane," *IEEE Trans. Antennas Propagat.*, vol. 34, pp. 804–813, June 1986.

Newman, E. H., "An Overview of the Hybrid MM/Green's Function Method in Electromagnetics," *Proc. IEEE*, vol. 76, pp. 270–282, March 1988.

Newman, E. H. and J. L. Blanchard, "TM Scattering by an Impedance Sheet Extension of a Parabolic Cylinder," *IEEE Trans. Antennas Propagat.*, vol. 36, pp. 527–534, Apr. 1988.

Nicol, J. L., "The Input Impedance of Horizontal Antennas Above an Imperfect Plane Earth," *Radio Science*, vol. 15, pp. 471–477, May-June 1980.

Omar, A. S. and K. Schunemann, "Scattering by Material and Conducting Bodies Inside Waveguides, Part I: Theoretical Formulations," *IEEE Trans. Microwave Th. and Tech.*, vol. 34, pp. 266–272, Feb. 1986.

Parhami, P. and R. Mittra, "Wire Antennas Over a Lossy Half-Space," *IEEE Trans. Antennas Propagat.*, vol. 28, pp. 397–404, May 1980.

Rao, R., "A Two-Element Yagi-Type Array in a Parallel-Plate Waveguide – Theoretical and Experimental Studies," *IEEE Trans. Antennas Propagat.*, vol. 13, pp. 675–682, Sep. 1965.

Richmond, J. H. and E. H. Newman, "Mutual Impedance Between Vertical Dipoles Over a Flat Earth," *Radio Science*, vol. 14, pp. 957–959, Nov.-Dec. 1979.

Richmond, J. H., "Monopole Antenna on Circular Disk Over Flat Earth," *IEEE Trans. Antennas Propagat.*, vol. 33, pp. 633–637, June 1985.

Sahalos, J. N. and G. A. Thiele, "On the Application of the GTD-MM Technique and its Limitations," *IEEE Trans. Antennas Propagat.*, vol. 29, pp. 780–786, Sep. 1981.

Sarkar, T. K., "Analysis of Radiation by Arrays of Parallel Vertical Wire Antennas Over Imperfect Ground," *IEEE Trans. Antennas Propagat.*, vol. 23, p. 749, July 1975.

Sarkar, T. K., "Analysis of Radiation by Arrays of Parallel Vertical Wire Antennas Over Plane Imperfect Ground (Sommerfeld Formulation)," *IEEE Trans. Antennas Propagat.*, vol. 24, pp. 544–545, July 1976.

Sarkar, T. K., "Analysis of Radiation by Arrays of Parallel Horizontal Wire Antennas Over Plane Imperfect Ground (Sommerfeld Formulation)," *IEEE Trans. Antennas Propagat.*, vol. 24, pp. 545–546, July 1976.

Sarkar, T. K., "Analysis of Radiation by Arrays of Arbitrarily Oriented Wire Antennas Over Plane Imperfect Ground," *IEEE Trans. Antennas Propagat.*, vol. 24, p. 546, July 1976.

Sarkar, T. K., "Analysis of Radiation by Linear Arrays of Parallel Horizontal Wire Antennas Over Imperfect Ground (Reflection Coefficient Method)," *IEEE Trans. Antennas Propagat.*, vol. 24, pp. 907–908, Nov. 1976.

Sarkar, T. K., D. D. Weiner, and R. F. Harrington, "Analysis of Nonlinearly Loaded Multiport Antenna Structures Over an Imperfect Ground Plane Using the Volterra-Series Method," *IEEE Trans. Electromagnetic Compatibility*, vol. 20, pp. 278–287, May 1978.

Sarkar, T. K. and R. F. Harrington, "Radar Cross Sections of Conducting Bodies Over a Lossy Half Space," *Radio Science*, vol. 15, pp. 581–585, May-June 1980.

Siakavara, K. and J. N. Sahalos, "A Hybrid MM-GTD Technique for the Optimization of the Power Gain of Arrays of Wire Antennas Near an Elliptic Cylinder," *Proc. IEEE*, vol. 73, pp. 1426–1428, Sep. 1985.

Sinha, S., "Analysis of Multiple-Strip Discontinuity in a Rectangular Waveguide," *IEEE Trans. Microwave Th. and Tech.*, vol. 34, pp. 696–700, June 1986.

Taylor, C. D., "Thin Wire Receiving Antenna in a Parallel Plate Waveguide," *IEEE Trans. Antennas Propagat.*, vol. 15, pp. 572–574, July 1967.

Tesche, F. M. and A. R. Neureuther, "Radiation Patterns for Two Monopoles on a Perfectly Conducting Sphere," *IEEE Trans. Antennas Propagat.*, vol. 18, pp. 692–694, Sep. 1970.

Tesche, F. M., "On the Behavior of Thin-Wire Antennas and Scatterers Arbitrarily Located Within a Parallel Plate Region," *IEEE Trans. Antennas Propagat.*, vol. 20, pp. 482–487, July 1972.

Tesche, F. M. and A. R. Neureuther, "The Analysis of Monopole Antennas Located on a Spherical Vehicle: Part 1, Theory," *IEEE Trans. Electromagnetic Compatibility*, vol. 18, pp. 2–8, Feb. 1976.

Tesche, F. M., A. R. Neureuther, and R. E. Stovall, "The Analysis of Monopole Antenas Located on a Spherical Vehicle: Part 2, Numerical and Experimental Results," *IEEE Trans. Electromagnetic Compatibility*, vol. 18, pp. 8–15, Feb. 1976.

Thiele, G. A. and T. H. Newhouse, "A Hybrid Technique for Combining Moment Methods with the Geometrical Theory of Diffraction," *IEEE Trans. Antennas Propagat.*, vol. 23, pp. 62–69, Jan. 1975.

Thiele, G. A. and G. K. Chan, "Application of the Hybrid Technique to Time Domain Problems," *IEEE Trans. Antennas Propagat.*, vol. 26, pp. 151–155, Jan. 1978.

Tiberio, R., G. Manara, and G. Pelosi, "A Hybrid Technique for Analyzing Wire Antennas in the Presence of a Plane Interface," *IEEE Trans. Antennas Propagat.*, vol. 33, pp. 881–885, Aug. 1985.

Wang, B., "Mutual Impedance Between Probes in a Circular Waveguide," *IEEE Trans. Microwave Th. and Tech.*, vol. 37, pp. 1006–1011, June 1989.

Wang, J. H., "Analysis of a Three-Dimensional Arbitrarily Shaped Dielectric or Biological Body Inside a Rectangular Waveguide," *IEEE Trans. Microwave Th. and Tech.*, vol. 26, pp. 457–462, July 1978.

Williamson, A. G., "Variable-Length Cylindrical Post in a Rectangular Waveguide," *IEE Proc.*, vol. 133 Pt. H, pp. 1–9, Feb. 1986.

Xu, X. and C. M. Butler, "Current Induced by TE Excitation on a Conducting Cylinder Located Near the Planar Interface Between Two Semi-Infinite Half-Spaces," *IEEE Trans. Antennas Propagat.*, vol. 34, pp. 880–890, July 1986.

Xu, X. and C. M. Butler, "Scattering of TM Excitation by Coupled and Partially Buried Cylinders at the Interface Between Two Media," *IEEE Trans. Antennas Propagat.*, vol. 35, pp. 529–538, May 1987.

Zhang, Q. and T. Itoh, "Spectral-Domain Analysis of Scattering from E-Plane Circuit Elements," *IEEE Trans. Microwave Th. and Tech.*, vol. 35, pp. 138–150, Feb. 1987.

Author Index

Subject Index

L

M

N

O

P

Q

R

S

Editors' Biographies

Edmund K. Miller (S'60–M'66–SM'70–SM'81–F'84) earned the B.S.E.E. degree in 1957 from Michigan Technological University, Houghton, and the M.S. degree in nuclear engineering in 1958, the M.S.E.E. degree in 1961, and the Ph.D.E.E. degree in 1965, all from the University of Michigan.

He was employed from 1959–1965 as an Assistant Research Engineer at the Radiation Laboratory during his Ph.D. candidacy at the University of Michigan, and as an Associate Research Engineer at the University of Michigan High Altitude Engineering Laboratory from 1966–1968. He then spent three years as Senior Scientist at MBAssociates, San Ramon, CA from 1968–1971, and nearly 15 years at Lawrence Livermore National Laboratory, from 1971–1985, as Group Leader, Associate Division Leader of the Nuclear Energy Systems Division. Subsequent positions include Regents-Distinguished Professor of Electrical and Computer Engineering at the University of Kansas, 1985–1987, Manager of Electromagnetics, Rockwell Science Center, Thousand Oaks, CA, 1987–1988, Director of Electromagnetics Research Operation, General Research Corporation, Santa Barbara, CA, 1988–1989, and his present assignment as Leader of the Instrumentation Group at Los Alamos National Laboratory, Los Alamos, NM. His current research interests include computational electromagnetics, signal-processing applications in electromagnetics, and education applications of computer graphics.

Dr. Miller has published numerous journal articles and book chapters in the areas of computational electromagnetics, signal processing, and computer graphics, and he edited the book, "Time-Domain Measurements in Electromagnetics" (1986). He is a member of several United States Commissions of the International Scientific Radio Union. He served as first President of the Applied Computational Electromagnetics Society, and as Chairman of Commission A, Metrology, of United States URSI. In 1985 he received (with G. J. Burke) a Certificate of Achievement from IEEE Electromagnetic Compatibility Society for Contributions to Development of NEC (Numerical Electromagnetics Code), and the 1989 Best Paper Award from IEEE Education Society for 1988 Transactions Paper "Computer Movies for Education" (with R. Cole and R. Merrill). He has served as IEEE Antennas and Propagation Society Distinguished Lecturer (1982–1985), and has chaired and organized numerous special sessions at various AP-S and URSI meetings. He has also chaired the AP-S Education Committee and EM Modeling Software Committee. He is currently serving on the IEEE PRESS Editorial Board.

Louis N. Medgyesi-Mitschang (S'60–M'63–SM'86) received the B.S.E.E. and M.S.E.E. degrees in 1961 and 1963, respectively, and the D.Sc. degree in electrical engineering, minoring in mathematics and physics, in 1967, all from Washington University, St. Louis, MO.

His graduate work was done at the Bio-Medical Computer Laboratory of the Washington University School of Medicine. His work dealt with the application of statistical communication theory for analysis and reconstruction of biophysical phenomena. In 1967 he joined the McDonnell Douglas Research Laboratories (MDRL). He was principal investigator on a series of research programs for the U.S. Army, Air Force, and Navy, dealing with airborne antenna systems and scattering problems. His special interest is the mathematical modeling of nonspecular effects in absorbers and the development of hybrid methods for radar cross section analysis and reduction.

Dr. Medgyesi-Mitschang is a member of the IEEE Aerospace and Electronic Systems Society, AIAA, Eta Kappa Nu, Tau Beta Pi, and Sigma Xi. He is former PAC Chairman of the St. Louis IEEE Section, a member of its executive committee, and past chairman of the combined Antennas and Propagation, Electron Devices, and Microwave Theory and Techniques Group of the St. Louis Section.

Edward H. Newman (S'67–M'74–SM'86–F'89) was born in Cleveland, OH on July 9, 1946. He received the B.S.E.E., M.S., and Ph.D. degrees in electrical engineering from The Ohio State University in 1969, 1970, and 1974, respectively.

Since 1974, he has been a member of The Ohio State University, Department of Electrical Engineering, where he is currently an Associate Professor. His primary research interest is in the development of method of moments techniques for the analysis of general antenna or scattering problems, and he is the primary author of the "Electromagnetic Surface Patch Code" (ESP). Other research interests include electrically small antennas, arrays, printed circuit antennas, antennas in inhomogeneous media, and scattering from material coated bodies. He has published over 40 journal articles in these areas.

Dr. Newman is a member of Commission B of URSI and a member of the Electromagnetics Institute. He is a recipient of the 1986 College of Engineering Research Award, and is a past chairman of the Columbus sections of the IEEE Antennas and Propagation and Microwave Theory and Techniques Societies.